Fundamentals of Multiphase Heat Transfer and Flow

Amir Faghri · Yuwen Zhang

Fundamentals of Multiphase Heat Transfer and Flow

Springer

Amir Faghri
Department of Mechanical Engineering
University of Connecticut
Storrs, CT, USA

Yuwen Zhang
Department of Mechanical
and Aerospace Engineering
University of Missouri
Columbia, MO, USA

ISBN 978-3-030-22136-2 ISBN 978-3-030-22137-9 (eBook)
https://doi.org/10.1007/978-3-030-22137-9

The first edition of this text was published by Academic Press in 2006.

This Springer imprint is published by the registered company Springer Nature Switzerland AG
The registered company address is: Gewerbestrasse 11, 6330 Cham, Switzerland

To Our Families

Pouran, Tanaz, and Ali Faghri

Jennifer, Angela, and Joanna Zhang

Whose Love and Support

Make All Things Possible

Preface to the Second Edition

We have received numerous requests over the last several years to publish a new edition of our book, *Transport Phenomena in Multiphase Systems*. Two excellent reviews of this book were also published in the *International Journal of Heat and Mass Transfer* and *Heat Transfer Engineering*. We have changed the title of the second edition to *Fundamentals of Multiphase Heat Transfer and Flow* to more appropriately reflect the contents of the new edition. The new edition was updated extensively as we put more emphasis on the fundamentals, eliminated some applications, and added two new chapters. The new edition is not limited to *two-phase flow and heat transfer* but also spans a much broader range of "multiphase" systems than other textbooks. Thus, student, instructor, and practitioner can benefit from our discussion of multiphase system processes such as interfacial phenomena, melting/solidification, and sublimation/vapor deposition, in addition to boiling, evaporation, condensation, and two-phase flow and heat transfer.

Several books over the last 20 years have summarized the state of the art in liquid $\leftrightarrows$ vapor systems. No serious attempts have been made to bring all three forms of phase change, i.e., liquid $\leftrightarrows$ vapor, solid $\leftrightarrows$ liquid, and solid $\leftrightarrows$ vapor, into one volume and to describe them from one perspective. Please note that in this text, pairs of arrows, $\leftrightarrows$, are used to portray energy and mass exchange associated with the multiphase transfer between the phases listed. Furthermore, most of the existing books were developed as monographs for research purposes rather than textbooks. This textbook provides fundamental principles related to multiphase heat transfer and flow with phase changes.

This textbook presents modern heat and mass transfer in the context of all phase changes among solid, liquid, and vapor, including interfacial phenomena. *Fundamentals of Multiphase Heat Transfer and Flow* is developed as a textbook for senior undergraduate and graduate students in a wide variety of engineering disciplines including mechanical engineering, chemical engineering, material science and engineering, nuclear engineering, biomedical engineering, petroleum engineering, and civil and environmental engineering. This book can also be used as a textbook to teach modern and new applications of heat and mass transfer—a topic which was previously restricted to classical texts, a status that remained unchanged for five decades. Materials presented are very useful for graduate students and practicing engineers working in technical areas related to both macro- and microscale systems with an emphasis on multiphase, multicomponent, nonconventional geometries with coupled heat and mass transfer and phase changes, with the possibility of full numerical simulation.

Chapters 1 and 3 present fundamental materials that are prerequisites to understanding the rest of the book. These two chapters provide the foundation for other chapters. However, the other chapters can be selected and chosen by the instructor based on the needs of students and curriculum requirements. Chapter 1 begins with the fundamentals of multiphase flow regimes as well as the classification of multiphase systems with phase change. The processes of phase change between solid, liquid, and vapor are discussed. A review of heat and mass transfer with a detailed emphasis in multicomponent systems and scaling are presented in Chap. 1 as well as multiphase notations and concepts.

Chapter 2 presents the thermodynamics of multiphase systems, which begins with the fundamentals of equilibrium and stability. This is followed by a discussion of multicomponent multiphase systems and the metastable equilibrium that exists in the multiphase system. Chapter 2 concludes with a discussion of thermodynamics at the interface, surface tension, disjoining pressure, and superheat.

Chapter 3 presents the generalized macroscopic (integral) and microscopic (differential) conservation equations for multiphase systems for both local-instance and averaged formulations. The instantaneous formulation requires a differential balance for each phase, combined with an appropriate jump and boundary conditions to match the solution of these governing equations at the interfaces. The averaged formulations are obtained by averaging the governing conservation equations within a small time interval (time average) or a small control volume (spatial average). The governing conservation equations for the multidimensional, multifluid, homogeneous, mixture, and separated models are also discussed as well as area-averaged governing conservation equations for one-dimensional flows.

Vector and tensor notations are used in the development of generalized governing equations in Chap. 3. The neatness, generality, and compactness of vector and tensor notations are considered sufficient to overcome the criticism of those who may consider the subject too sophisticated. Examples in Chap. 3 with applications of these in nonvectorial one-, two-, or three-dimensional forms for various geometries complement and support related information in the following chapters thereby providing an adequate learning experience. In many examples, equations for simple one-dimensional processes are also developed based on actual physical mass, momentum, and energy balances, so that students may appreciate the physical significance of various terms.

Chapter 4 introduces the interfacial concepts of surface tension, wetting phenomena, and contact angle, which are followed by a discussion on the motion induced by capillarity. Additional detailed descriptions are presented for interfacial balances and boundary conditions for mass, momentum, energy, and species for multicomponent and multiphase systems. Chapter 4 also delineates heat and mass transfer through the thin-film region during evaporation and condensation, including the effect of interfacial resistance and disjoining pressure. The dynamics of interfaces, including stability and wave effects, are presented in Chap. 4. Finally, a review is given on the numerical simulation of interfaces.

Solid–liquid phase changes, including melting and solidification, are treated in Chap. 5, starting with the classification of solid–liquid phase changes and generalized boundary conditions at the interface. Different approaches to the solution of melting and solidification problems are introduced, including exact, integral, and numerical solutions. Finally, solidification in contact melting is presented. Solid–vapor phase change, including sublimation and vapor deposition, is introduced in Chap. 6. The discussion begins with a brief overview of solid–vapor phase change and proceeds to detailed analyses of sublimation with and without chemical reaction as well as chemical vapor deposition.

Chapter 7 begins with a discussion of the two main modes of liquid droplet embryo formation in condensation: homogeneous and heterogeneous, followed by a detailed examination of dropwise and filmwise condensation with different approximations and methods. Chapter 8 presents criteria and classification of evaporation, which include evaporation from an adiabatic wall, evaporation from a heated wall, and direct-contact evaporation. Chapter 9 introduces the pool boiling curve and characterizes the various boiling regimes (free convection, nucleate, transition, and film boiling), followed by detailed discussions and an analysis of each pool boiling regime, critical heat flux, minimum heat flux, and direct numerical simulation of film boiling. Chapter 9 also discusses the Leidenfrost phenomenon.

Chapter 10 starts with definitions of various parameters for two-phase flow and flow patterns in vertical and horizontal tubes. This is followed by two-phase flow models as well as a prediction of pressure drops and void fractions. Finally, the two-phase flow regimes and heat transfer characteristics for forced convective condensation and boiling at both macro- and microscale levels are presented. Chapter 11 starts with a discussion about size distribution of particles and the interaction of dry particles, followed by a discussion of fluid-particle interactions. This chapter also covers the fundamentals and applications of various liquid-particle and gas-particle systems. Chapter 12 considers single- and multiphase heat and mass transfer in porous media as well as boiling, condensation, melting, and solidification in porous media.

The International System of Units (SI) is used throughout the book, and the conversion factors for different unit systems are provided in Appendix A. The complete thermophysical properties of various substances for all phases are provided in Appendix B. Appendix C provides a brief review of vector and tensor operations. Appendix D provides convective heat transfer correlations for various heat and mass transfer modes and geometries.

The only prerequisite courses necessary for the material are undergraduate thermodynamics and heat transfer or transport phenomena. No graduate course in convection, conduction, or transport phenomena is required. In fact, convection, conduction, and/or transport phenomena are special cases of the general material presented here, if taught properly. A solution manual and Microsoft PowerPoint presentation package are provided only to those instructors who adopt the book for the course.

The authors would like to express their deep appreciation to the following distinguished members of the multiphase heat transfer and flow community who shared their expertise and time in reviewing the second edition of the book: Rafid Al-khoury, C. Thomas Avedisian, Mohamed Awad, Theodore L. Bergman, Yiding Cao, Louis Chow, Asghar Esmaeeli, Timothy S. Fisher, James Guo, Qing Hao, Gisuk Hwang, Yogendra Joshi, Xianglin Li, Charlie Lin, Issam Mudawar, Vinod Narayanan, Ugur Pasaogullari, Laurent Pilon, Joel Plawsky, Rui Qiao, Muhammad Rahmann, Hamidreza Shabgard, Ying Sun, Hongwei Sun, Shimin Wang, and Yi Zheng. We are grateful to these dedicated professionals for their support, sage advice, improvements, and additions, which resulted in a superior and more comprehensive textbook than we envisioned.

This textbook provides an opportunity to cover fundamental principles of multiphase heat transfer and flow with all forms of phase change from one perspective. It is our hope that this textbook will influence some engineering colleges to treat multiphase heat transfer and flow as a core requirement of the graduate curriculum in mechanical, chemical, environmental, nuclear, biomedical, and the materials science disciplines. Your recommendations, comments, and criticisms are truly appreciated.

Storrs, CT, USA — Amir Faghri
Columbia, MO, USA — Yuwen Zhang
2019

Preface to the First Edition

Transport phenomena in multiphase systems with phase change is of great interest to scientists and engineers working in the power, nuclear, chemical processes, environmental, microelectronics, biotechnology, nano-technology, polymer science, food processing, cryogenics, space, and many other industries, from the established to emerging multidisciplinary technologies. For example, almost two-thirds of industrial heat exchangers undergo phase change; therefore, physical understanding and development of the first principal models are not only of interest in fundamental research, they also are greatly needed for a more accurate and reliable design of multiphase thermal systems.

The subject of transport phenomena in a multiphase system with phase change is important, because a unified physical/mathematical treatment is essential for engineering practitioners in the 21st century, who must cope with issues such as high heat flux and micro- or nanoscale systems for various applications.

Our motive in preparing this new textbook was to address the challenges and opportunities facing graduate education and teaching in thermal sciences within the mechanical engineering discipline and/or advanced transport phenomena in chemical engineering, which have remained basically unchanged for five decades. For example, the convection and/or conduction courses offered by most mechanical engineering departments as core courses in thermal sciences focus almost exclusively on single-phase, single-component, simple geometry such as channel flows or flat plates with the goal of an analytical solution with the continuum approach. Similarly, advanced transport phenomena in chemical engineering are based mostly on the excellent classical book by Bird *et al.*, which was originally published in 1960.

In contrast with their educational training, practicing engineers working in the thermal sciences or scientists in academia and the private sector have in recent years focused mostly on multiphase, multicomponent, non-conventional geometries, with coupled heat and mass transfer and phase change, with the goal of developing a numerical simulation using a continuum or non-continuum approach. We therefore developed this new textbook with the intention of helping instructors to bridge classroom learning and engineering practice by offering them advanced fundamental and general course materials that can replace conventional, limited, approaches for teaching advanced heat and mass transfer or transport phenomena.

The purpose of this textbook is to accurately present the basic principles for analyzing transport phenomena in multiphase systems and to demonstrate their wide variety of possible applications. Since it would take many book volumes to do justice to all aspects of multiphase systems, the scope of this book is limited to thermodynamics and momentum, heat and mass transfer fundamentals, with emphases on melting, solidification, sublimation, vapor deposition, condensation, evaporation, boiling, and two-phase flow.

Several books over the last 20 years have summarized the state of the art in liquid $\leftrightarrows$ vapor systems. No serious attempts were made to bring all three forms of phase change, i.e., liquid $\leftrightarrows$ vapor, solid $\leftrightarrows$ liquid, and solid $\leftrightarrows$ vapor, into one volume and to describe them from one perspective (in this text, pairs of arrows, $\leftrightarrows$, are used to portray energy and mass exchange associated with multiphase transfer between the phases listed). Furthermore, most of the existing texts were developed as monographs rather than textbooks. In writing this textbook, our goal was to provide basic engineering fundamentals related to transport phenomena in multiphase systems with phase change, including microscale and porosity effects. In most cases, the basic physical phenomena are presented with different mathematical models.

Historically, the field of transport phenomena has developed successful textbooks for momentum, heat and mass transfer in single-phase systems because these are straightforward and well developed concepts, in terms of physical and mathematical modeling. The same is not true for multiphase systems, which involve some components of the semi-empirical approach, are much more complex, and are thus less well understood. However, because of significant developments in transport phenomena in multiphase systems with phase change during the last two decades, we have much better physical, analytical, and numerical tools to model these types of problems: this is the purpose of our textbook.

Furthermore, traditionally three approaches were used to present transport phenomena: microscopic (differential), macroscopic (integral) and molecular level. Most heat transfer textbooks place the emphasis on microscopic and/or macroscopic. With the importance of microscale heat transfer or transport phenomena in applications of nanotechnology and biotechnology, as well as molecular dynamic simulations, it is important to discuss the molecular approach and the connection between the molecular and microscopic approaches. In this textbook, an attempt is made to better describe this relationship. For example, the generalized conservation equations in Chapter 3 have been developed not only microscopically and macroscopically using the continuum approach, but also using the Boltzmann equation.

There are three types of information available in the area of transport phenomena in multiphase systems that can be covered in a textbook of this nature:

1. Significant existing experimental work and correlations
2. Analytical and physical models
3. Numerical simulation modeling due to recent significant advances in digital computers and computational methodologies

We have not presented much in the way of item 1 except well established semi-empirical correlations that have been accepted in practice. The emphasis in this book is on the last two items. With respect to the final item, note that this is not a numerical method book; however, we have set up the framework so that students who wish to pursue this approach are equipped with the basic background material necessary to use existing commercial computer codes. Numerical methodologies and approaches are presented if they are specific to multiphase systems with phase change. Analytical and numerical physical models of transport phenomena in multiphase systems are the main focus in this textbook.

Chapters 1 through 4 present materials that are fundamental to the entire text. These chapters should be considered before proceeding to other chapters.

Chapter 1 begins with a review of the concept of phases of matter and a discussion of the role of phases in systems that include, simultaneously, more than one phase. This is followed by a review of transport phenomena with detailed emphasis in multicomponent systems, microscale heat transfer, dimensional analysis, and scaling. The processes of phase change between solid, liquid, and vapor are also reviewed, and the classification of multiphase systems is presented. Finally, some typical practical applications are described, which require students to understand the operational principles of these multiphase devices for further understanding and application in homework and examples in future chapters.

The thermodynamics of multiphase systems is presented in Chapter 2, which begins with a review of single-phase thermodynamics, including thermodynamic laws and relations, and proceeds to the concepts of equilibrium and stability. This is followed by discussion of thermodynamic surfaces and phase diagrams for single- and multicomponent systems. Also discussed are equilibrium criteria for single and multicomponent multiphase systems and the metastable equilibrium that exists in a multiphase system. Chapter 2 concludes with a discussion of thermodynamics at the interface and the effects of surface tension and disjoining pressure, including the superheat effect.

Chapter 3 presents the generalized macroscopic (integral) and microscopic (differential) governing equations for multiphase systems in local-instance formulations. The instantaneous formulation requires a differential balance for each phase, combined with appropriate jump and boundary conditions to match the solution of these differential equations at the interfaces. Also discussed in Chapter 3 are a rarefied vapor self-diffusion model and the application of the differential formulations to combustion. The generalized governing equations for multiphase systems in averaged formulations are presented in Chapter 4. The averaged formulations are obtained by averaging the governing equations within a small time interval (time average) or a small control volume (spatial average). The governing equations for the multidimensional multi-fluid and homogeneous models, as well as area-averaged governing equations for one-dimensional flows, are also discussed. Chapter 4 also covers single- and multiphase transport phenomena in porous media, including multi-fluid and mixture models. Finally, Boltzmann statistical averaging, including a detailed discussion of the Boltzmann equation and the Lattice Boltzmann method for modeling both single and multiphase systems, is presented.

Vector and tensor notations have been used in the development of generalized governing equations in Chapters 3 and 4. The neatness, generality, and compactness of vector and tensor notations are considered sufficient to overcome the criticism of those who may consider the subject too sophisticated. Examples in Chapters 3 and 4 and applications of these in non-vectorial one-, two-, or three-dimensional forms for various geometries in following chapters will provide adequate experience. In many examples, equations for simple one-dimensional processes are also developed based on actual physical mass, momentum, and energy balance, so that students appreciate the physical significance of various terms.

Chapter 5 introduces the concepts of surface tension, wetting phenomena, and contact angle, which are followed by a discussion on motion induced by capillarity. Additional detailed descriptions are presented for interfacial balances and boundary conditions for mass, momentum, energy, and species for multicomponent and multiphase interface. Also considered in Chapter 5 are heat and mass transfer through the thin film region during evaporation and condensation, including the effect of interfacial resistance and disjoining pressure. The dynamics of interfaces, including stability and wave effects, are presented. Finally, numerical simulations of interfaces and free surfaces using both continuum and non-continuum approaches are provided.

Solid–liquid phase change, including melting and solidification, is treated in Chapter 6, starting with the classification of solid–liquid phase changes and generalized boundary conditions at the interface. Different approaches to the solution of melting and solidification problems, including exact, integral approximate, and numerical solutions, are introduced. Solidification in binary solution systems, contact melting, melting and solidification in porous media, applications of solid–liquid phase change, and microscale solid–liquid phase change are also presented. Solid–vapor phase change, including sublimation and vapor deposition, is introduced in Chapter 7. The discussion begins with a brief overview of solid–vapor phase change and proceeds to detailed analyses on sublimation without and with chemical reaction, as well as physical and chemical vapor deposition.

Chapter 8 begins with a discussion of two main modes of liquid droplet embryo formation in condensation: homogeneous and heterogeneous, followed by a detailed examination of dropwise and filmwise condensation at both macro- and microscale levels. Applications of condensation in microgravity and condensation in porous media are also discussed. Chapter 9 presents criteria and classification of evaporation, evaporation from an adiabatic wall, evaporation from a heated wall, evaporation in porous media, evaporation in micro/miniature channels, as well as direct-contact evaporation.

Chapter 10 introduces the pool boiling curve and characterizes the various boiling regimes (free convection, nucleate, transition, and film boiling), followed by detailed discussions of each of the four pool boiling regimes, critical heat flux, minimum heat flux, and direct numerical simulation. Also discussed in Chapter 10 are the Leidenfrost phenomena as well as physical phenomena of boiling in porous media. Chapter 11 starts with definitions of various parameters for two-phase flow and flow patterns in vertical and horizontal tubes. This is followed by two-phase flow models as well as prediction of pressure drops and void fractions. Finally, the two-phase flow regimes and heat transfer characteristics for forced convective condensation and boiling at both macro- and microscale levels are presented.

The International System of Units (SI) is used throughout the book, and the conversion factors for different unit systems are provided in Appendix A. The complete thermophysical properties for all phases of various substances, along with empirical correlations of thermal properties as functions of temperature, are provided in Appendix B. Appendix C provides a brief review of vector and tensor operations.

We have used consistent symbols throughout the book. However, we have used some symbols for more than one purpose in a number of cases. We believe the context, as well as the nomenclature section, will clarify the meaning of the symbols used in these cases.

This textbook is designed for use as an advanced-level undergraduate or graduate textbook in mechanical engineering, chemical engineering, material science and engineering, nuclear engineering, biomedical engineering, or environmental engineering. It offers examples and homework problems as well as references from engineering and research applications related to multiphase systems. The only prerequisite courses necessary for the material are undergraduate thermodynamics, and heat transfer or transport phenomena. No graduate course in convection, conduction, or transport phenomena is required. In fact, convection, conduction, and/or transport phenomena are special cases of the general material presented here, if taught properly. We recognize a new trend at a number of universities to offer a single course in transport phenomena of multiphase system for all disciplines, and therefore we have tried to cover the materials that various departments might wish to have included in such a course. The materials included in this text may require more than one semester of instruction depending on the desired level of completeness. Therefore, it is recommended that the instructor choose the materials to be covered based on the background and needs of the students.

This text is not intended as a reference tool or handbook summarizing the state-of-the-art, nor does it to detail the history of multiphase systems with phase change. Part of the text was developed originally from lecture notes prepared by one of the authors (AF) who was teaching a graduate-level course at the University of Connecticut. Materials have been considerably rewritten by both authors and used as lecture notes for senior elective and/or graduate-level courses taught by the authors at the University of Connecticut, New Mexico State University, and the University of Missouri-Columbia. This textbook is suitable for students from a wide variety of backgrounds. The examples and homework problems were added to provide students a better physical understanding of theoretical concepts and uses for various applications. While the examples are designed to confer a better physical understanding, including mathematical modeling and a feeling for the order of magnitude of variables, end-of-chapter homework problems will help students appreciate fundamental concepts. There are three types of problems we have developed for this textbook: (1) simple numerical manipulation, (2) detailed physical and analytical models, and (3) open-ended problems. It is important that students gain experience in solving all three types of problems. A copyrighted solution manual and Microsoft PowerPoint presentation package are provided only to those instructors who adopt the book for the course.

The authors would like to express their deep thanks to a number of distinguished members of the heat transfer community who shared their expertise and time in reviewing this book: Thomas Avedisian, Christopher Beckermann, Arthur Bergles, F.B. Cheung, John Howell, Raymond Viskanta, and Ralph Webb. In addition, we wish to thank the following individuals who generously reviewed individual chapters or part of the book: Yutaka Asako, Theodore Bergman, Yiding Cao, Baki Cetegen, Wilson Chiu, Emily Green, Hongbin Ma, Robert McGurgan, Dmitry Khrustalev, Roop Mahajan, Gregory Jewett, Ugur Pasaogullari, Ranga Pitchumani, Jeremy Rice, Scott Thomas, and Kambiz Vafai. We are grateful to these dedicated professionals for their support, sage advice, improvements, and additions, which resulted in a superior and more comprehensive text than we envisioned. It is important to acknowledge the contributions of students over the last several years who were taught from the manuscripts out of which this book evolved. Our special thanks to Nan Cooper and Emily Jerome for their expert editing of the manuscripts.

This textbook provides an opportunity to cover fundamentals of transport phenomena in multiphase systems with all forms of phase change from one perspective. It is our hope that this textbook will influence some engineering colleges to treat transport phenomena in multiphase systems as a core requirement of the graduate curriculum in mechanical, chemical, environmental, nuclear, biomedical, and materials science disciplines. Your recommendations, comments, and criticisms are appreciated.

Amir Faghri
Yuwen Zhang

Contents

Nomenclature

A	area (m^2)
A_ℓ	Cross-sectional area of liquid flow passage (m^2)
A_v	Cross-sectional area of vapor flow passage (m^2)
A'	Dispersion constant (J)
Bi	Biot number, hL/k (k is thermal conductivity of solid)
Bo	Bond number $[(\rho_\ell - \rho_v)gL^2/\sigma]$
C	Specific heat (J/(kg-K)); wave velocity (m/s); speed of sound (m/s)
$\mathbf{C}$	Particle velocity (m/s)
c_i	Molar concentration of the ith species (kmol/m^3)
c_p	Specific heat at constant pressure (J/kg-K)
c_v	Specific heat at constant volume (J/kg-K)
C	Heat capacity (J/K); parameter in Chisholm correlation
C_D	Drag coefficient
C_f	Friction coefficient
Co	Convective number $\left((\rho_v/\rho_\ell)^{0.5}[(1-x)/x]^{0.5}\right)$
d	Particle diameter (m)
D	Diameter (m); self-diffusion coefficient (m^2/s)
$\mathbf{D}$	Rate of strain tensor (1/s)
D_h	Hydraulic diameter (m)
D_{ij}	Binary diffusivity (m^2/s)
D/Dt	Substantial derivative
e	Specific internal energy (J/kg)
E	Internal energy or surface free energy (J); emissive power (W/m^2)
$\hat{E}$	Total energy (J)
f	Degree of freedom; solid fraction; wave frequency (1/s); molecular velocity distribution function
F	Force (N); Helmholtz free energy (J/kg-K)
$\mathbf{F}$	Force vector (N)
$\tilde{f}_\mathrm{n}$	Discrete number frequency
f_n	Frequency function
$F_n(d_i)$	Accumulative distribution
Fo	Fourier number ($\alpha t/L^2$)
Fr	Froude number ($U/\sqrt{gL}$ or U^2/gL)
G	Gravitational acceleration (m/s^2); specific Gibbs free energy (J/kg)
G	Gibbs free energy (J); coupling factor (W/m^3-K)

Gr	Grashof number ($g\beta\Delta TL^3/\nu^2$)
$g(r)$	Radial distribution function (RDF) ($dN(r)/(4\pi r^2 dr\rho)$)
h	Heat transfer coefficient (W/(m^2-K)); specific enthalpy (J/kg)
$\overline{h}$	Average heat transfer coefficient (W/m^2-K)
$\tilde{h}$	Average enthalpy of the multiphase mixture (J/kg)
$h_{\ell v}$	Latent heat of vaporization (J/kg)
$h'_{\ell v}$	Modified latent heat of vaporization (J/kg)
h_G	Heat transfer coefficient in noncondensable gas section (W/m^2-K)
h_m	Convective mass transfer coefficient (m/s)
$h_{m,G}$	Mass transfer coefficient in noncondensable gas section (m/s)
$h_{s\ell}$	Latent heat of fusion (J/kg)
h_{sv}	Latent heat of sublimation (J/kg)
h_x	Local heat transfer coefficient (W/m^2-K)
H	Enthalpy (J); height (m); Henry's constant
Ha	Hamaker constant (J)
He_b	Hedstrom number ($D^2\rho_m\tau_b/\mu_b^2$)
$\boldsymbol{I}$	Identity tensor
I_i	Inertial moment of particle i ($0.4m_iR_i^2$)
J	Volume flux (m/s)
j	Superficial velocity (m/s)
$\mathbf{J}_i$	Mass flux of the *ith* species relative to mass-averaged velocity (kg/m^2-s)
$\mathbf{J}_i^*$	Molar flux of the *ith* species relative to molar-averaged velocity (kmol/m^2-s)
Ja	Jakob number ($c_p\Delta T/h_{\ell v}$)
k	Thermal conductivity (W/(m-K))
k'	Reaction rate constant
k_b	Boltzmann constant (1.38×10^{-23}J/K)
K	Interface curvature (1/m); permeability (m^2)
K_0'	Arrhenius constant
K_{jk}	Momentum exchange coefficient between phases j and k (kg/(m^3-s))
Ka	Kapitza number ($\mu_\ell^4 g/[(\rho_\ell - \rho_v)\sigma^3]$)
L	(Characteristic) Length (m)
Le	Lewis number (α/D)
L_b	Bubble or capillary scale ($\sqrt{\sigma/[g(\rho_\ell - \rho_v)]}$, m)
m	Mass (kg)
$\dot{m}$	Mass flow rate (kg/s)
$\dot{m}''$	Absolute mass flux relative to stationary coordinate system (kg/m^2-s)
$\dot{\boldsymbol{m}}''$	Mass flux vector (kg/m^2-s)
$\dot{m}'''$	Mass source per unit volume (kg/m^3-s)
M	Molecular mass (kg/kmol)
Ma	Mach number (U/c); Marangoni number (($(dT/dy)(d\sigma/dT)\delta^2/(\alpha_\ell\mu_\ell)$))
n	Number of moles; number of horizontal tubes in an array
$\boldsymbol{n}$	Unit normal vector

n''_b	Number of vapor bubbles released per unit area and release cycle (1/m^2)
n''_D	Liquid droplet size distribution, (1/m^3)
n_i	Number of moles for *i*th component in a multicomponent system
$\dot{n}''_i$	Absolute molar flux of *i*th component relative to stationary coordinate system (kmol/s)
N	Number of components, number of molecules
$\mathfrak{N}$	Number density of the molecules
N_A	Avogadro's number (6.022×10^{23} mol^{-1})
N''_a	Number density of nucleation sites
Nu	Nusselt number (hL/k)
$\overline{Nu}$	Average Nusselt number ($\bar{h}L/k$)
Nu_x	Local Nusselt number ($h_x x/k$)
p	Pressure (Pa)
$\mathrm{p_d}$	Disjoining pressure (Pa)
P	Perimeter (m); laser power (W)
Pe	Peclet number (UL/α)
Pr	Prandtl number (ν/α)
Pr_t	Turbulent Prandtl number
q	Heat rate (W)
q'	Heat rate per unit length (W/m)
q''	Heat flux (W/m^2)
$\mathbf{q}''$	Heat flux vector (W/m^2)
$q''_{\max}$	Maximum (critical) heat flux in boiling (W/m^2)
$q''_{\min}$	Minimum heat flux in boiling (W/m^2)
q'''	Internal heat generation per unit volume (W/m^3)
Q	Total heat transfer (J); volume flow rate (m^3/s)
r	Radial coordinate (m)
r_eff	Effective pore radius (m)
R	Radius (m); radius of curvature (m); dimensionless radius (r/r_i); resistance (K/W)
R_b	Bubble radius (m)
R_g	Gas constant (R_u/M, kJ/kg-K)
$\mathfrak{R}_j$	Net reaction rate of *j*th chemical reaction
R_men	Radius of curvature of the meniscus (m)
R_u	Universal gas constant (8.3144 kJ/kmol-K)
R_v	Vapor space radius (m)
R_δ	Interfacial thermal resistance (m^2-K/W)
Ra	Rayleigh number ($g\beta\Delta TL^3/(\nu\alpha)$)
Re	Reynolds number (UL/ν or $\rho UL/\mu$; $4\Gamma/\mu_\ell$) (for film condensation or evaporation)
s	Specific entropy (J/(kg-K)); interface location (m)
S	Entropy (J/K); nondimensional interface location (s/L); source intensity in microscale melting; slip ratio (w_v/w_ℓ); solubility (kmol/Pa-m^3)
Sc	Schmidt number (ν/D); subcooling parameter ($c_{ps}(T_m - T_i)/h_{s\ell}$)
Sh	Sherwood number ($h_m L/D_{12}$)

Sk	Stokes number (t_v/t_F)
St	Stanton number ($h/(\rho c_p U)$)
Ste	Stefan number ($c_p\|T_w - T_m\|/h_{s\ell}$)
t	Time (s)
$\mathbf{t}$	Unit tangential vector
t_{pp}	Average time between particle–particle collisions (s)
t_T	Thermal response time
t_v	Momentum response time
T	Temperature (K)
$\mathbf{T}$	Torque vector (N-m)
T_m	Melting point (K)
T_{sat}	Saturation temperature (K)
T_w	Wall temperature (K)
T_∞	Temperature of environment (K)
U	Velocity in the x-direction (m/s)
U	Velocity (m/s)
U_t	Terminal velocity (m/s)
$\bar{U}$	Mean velocity (m/s)
u_c	Critical Helmholtz velocity (m/s)
u_f	Frictional velocity ($(\tau_w/\rho_\ell)^{1/2}$)
$\bar{u}$	Mean velocity (m/s)
v	Specific volume (m^3/kg); velocity in the y-direction (m/s); radial velocity (m/s); or vapor velocity along the η-coordinate (m/s)
V	Volume (m^3)
$\mathbf{V}$	Velocity vector (m/s)
$\tilde{\mathbf{V}}$	Mass-averaged velocity vector ($\left(\sum_{k=1}^{\Pi} \varepsilon_k \langle\rho_k\rangle^k \langle\mathbf{V}_k\rangle^k\right)/\langle\rho\rangle$, m/s)
w	Velocity in the z-direction or axial velocity (m/s)
w_ℓ	Liquid phase axial velocity (m/s)
w_v	Vapor phase axial velocity (m/s)
w_w	Wave velocity (m/s)
W	Work (J); width of the cavity (m); width of a capillary groove (m)
We	Weber number ($\rho U^2 L/\sigma$)
x	Cartesian coordinate (m); vapor quality
x_i	Molar fraction of ith species
X	Material coordinate (m); dimensionless coordinate (x/L)
$\mathbf{X}_{k,i}$	Body force per unit mass acting on ith species in kth phase (m/s^2)
y	Cartesian coordinate (m)
Y	Material coordinate (m); Young's modulus (N/m^2)
z	Cartesian coordinate, axial coordinate (m)
Z	Material coordinate (m); compressibility factor (pv/RT)

Greek Symbols

α	Thermal diffusivity (m^2/s); accommodation coefficient; wave number ($2\pi/\lambda$); void fraction; volume fraction

α_a	Absorptivity
β	Coefficient of thermal expansion (1/K); contact angle measured in degrees; volumetric flow fraction $(j_v/(j_v+j_\ell))$
β_m	Composition coefficient of volume expansion
γ	Damping coefficient (s)
Γ	Liquid mass flow rate per unit width (kg/m-s)
δ	Liquid or vapor film thickness (m); thermal penetration depth (m); laser irradiation penetration depth (m); thickness of the deposited film (m)
δ^+	Nondimensional film thickness $(\delta u_f/v_\ell)$
δ^*	Nondimensional film thickness $(\delta/L_F$, or $\delta(v_\ell^2/g)^{-1/3})$
Δ	Dimensionless thermal penetration depth $(\delta/(\alpha_1\rho_1 h_{s\ell}/q_0''))$; dimensionless liquid layer thickness in contact melting (δ/W)
Δt	Time interval for time average; time step in numerical solution (s)
ΔT	Half of the width of phase-change temperature range (K); temperature difference (K)
ΔV	Volume element for volume average (m^3)
ε	Porosity; volume fraction; emissivity; eddy diffusivity (m^2/s)
ζ	Transformed coordinate $((2\pi/\lambda)(x-ct))$
η	Dimensionless variable for binary solidification $(X/(2\sqrt{\tau}))$; dimensionless coordinate (y/δ); coordinate normal to the solid–liquid interface (m)
θ	Inclination angle (rad); contact or wetting angle (rad); dimensionless temperature
θ_f	Contact angle obtained from the smooth-surface model
θ_{men}	Meniscus contact angle
θ_R	Radiation temperature (K)
κ	Surface roughness (m)
κ_T	Isothermal compressibility $(-(\partial V/\partial p)_T/V)$
λ	Wavelength (m); constant in solid–liquid phase change $(S/(2\tau^{1/2}))$; friction factor
λ_c	Critical wavelength (m)
λ_D	Most dangerous wavelength (m)
μ	Dynamic viscosity (kg/(m-s)); chemical potential (J/mol); friction coefficient
μ_b	Limiting viscosity of Bingham plastic fluid (kg/(m-s))
μ_g	Geometric mean diameter or number median diameter of particle (m)
μ_n	Mean number diameter of the particles (m)
μ_s	Solid loading
v	Kinematic viscosity (m^2/s)
ξ_t	Tangential displacement (m)
Π	Number of phases
π_s	Surface pressure (N/m^2)
ζ	Degree of advancement of the chemical reaction
ρ	Density (kg/m^3)
ρ_i	Mass concentration of species i (kg/m^3)
σ	Surface tension force (N/m); collision diameter (Å); Poisson's ratio

σ_n	Standard deviation of particle size (m)
σ_n^2	Number variance of particle size (m^2)
σ_{SB}	Stefan–Boltzmann constant (5.67×10^{-8} W/m^2 - K^4)
τ	Shear stress (N/m^2); thermal relaxation time (s); dimensionless time ($\alpha t/L^2$)
$\boldsymbol{\tau}'$	Stress tensor (N/m^2)
$\boldsymbol{\tau}$	Viscous stress tensor (N/m^2)
τ_b	Yield stress of Bingham plastic fluid (N/m^2)
τ_I, τ_δ	Interfacial shear stress (N/m^2)
τ_w	Shear stress at wall (N/m^2)
ϕ	Lennard-Jones potential (J); specific value of general property (Φ); Wave amplitude; pressure drop multiplier in two-phase flow
φ	Similarity variable for sublimation or evaporation; dependent variable in combustion of spherical droplet; volume fraction
Φ	General property
ψ	Stream function (m^2/s), particle sphericity
Ψ	Availability (kJ)
$\boldsymbol{\omega}_i$	Angular velocity of particle i
ω_i	Mass fraction of species i
$\boldsymbol{\Omega}$	General vector quantity; surface tension parameter
∇	Laplace operator vector

Subscripts

0	Reference variables; initial condition; reservoir conditions
a	Adiabatic, acceleration
c	Critical, condenser
cap	Capillary
CV	Control volume
DPC	Dilute pneumatic conveying
DSU	Dense suspension up-flow
e	Equilibrium; evaporator
f	Final; fuel; thin film
g	Gas (vapor); gravity
H	Homogeneous
i	ith component; initial; inner
I	Interface
k	kth phase in a multiphase system
ℓ	Liquid
ℓo	Single liquid phase
L	Left; characteristic length; laminar
M	Melting point or mushy zone
max	Maximum
mb	Minimum bubbling
men	Meniscus
mf	Minimum fluidization
min	Minimum

ms	Minimum slugging
o	Outer
opt	Optimum
p	Product; particle; pulse
r	Reduced; reactant; radiation
R	Right
ref	Reference frame
rel	Relative
s	Shaft; solid; surface; settling
sat	Saturation
t or T	Turbulent
TP	Two-phase
tr	Transition; transport
v	Vapor phase
vo	Single vapor phase
w	Wall
δ	Liquid–vapor interface
∞	Ambient or bulk property

Superscripts

e	Electron
K	Phase index
ℓ	Lattice
n	Normal component
t	Tangential component
$'$	Fluctuation
*	Dimensionless
+	Dimensionless

Others

~	Same order of magnitude
$< >$	Volume averaged
$< >^k$	Phase average
$\bar{}$	Time averaged, mean
[illegible]	Mass-averaged

Introduction 1

1.1 Introduction

Multiphase heat transfer and fluid flow must be considered in the design and optimization of many engineering systems, such as heat exchangers, heat pipes, electronics cooling devices, biotechnology, nanotechnology, food processing equipment, and fuel cells. While some of these examples make intentional use of phase change to transfer large quantities of heat over small temperature differences, others involve phase change as an inevitable consequence rather than an intended design feature of the process. In each of these cases, and in many others cited in this text, the presence of multiple phases and of phase change has a profound impact on system performance, and it must be accounted for in order to achieve the system design objectives in the most efficient manner. The presence of multiple phases within a single system may significantly alter the systems' performance characteristics in terms of: (1) the pressure drop that determines a flowing system's power requirements, (2) the heat transfer rate that controls its capacity, and (3) the flow stability that in turn affects its operational characteristics. The presence of multiple phases inevitably makes systems more complicated and affects their reliability.

In making the progression to the more realistic, i.e., complex, design space of multiphase systems, the students or practicing engineers confront some characteristic features that, usually in some combination, distinguish multiphase systems from the single-phase, single-component world of typical undergraduate heat transfer courses and texts:

1. Thermodynamic equilibrium between phases.
2. Multiple sets of thermophysical properties and field variables.
3. Interfaces between phases including mass transfer where jump conditions prevail.
4. A prominent role for latent heat transfer.
5. Simultaneous coupled heat and mass transfer.
6. Complex dynamics such as bubble growth and collapse.
7. A wide range of physical scales within a single design problem.

In spite of these complexities, the theory of multiphase systems relies on the familiar laws of thermodynamics, fluid mechanics, and heat transfer, such as the first and second laws of thermodynamics, Newton's laws, Fourier's law, etc. Moreover, the definition of reliable predictive algorithms for multiphase systems requires the use of many familiar analytical tools such as control volume or differential analysis, Lagrangian or Eulerian reference systems, and dimensional and scale

A. Faghri and Y. Zhang, *Fundamentals of Multiphase Heat Transfer and Flow*, https://doi.org/10.1007/978-3-030-22137-9_1

analysis. Typically predicted events, as given or implied by the design problems posed, include heat transfer rates, temperature histories, and steady-state temperature profiles, as well as rates of mass transfer—all issues that are relevant regardless of the number of phases present in the system.

However, while most of the physical principles, analytical tools, and predicted outputs are common to single- and multiphase systems, the latter require more complex expressions of the basic laws, more elaborate application of the analytical tools, and some unique terms in order to accommodate distinguishing factors such as those listed above. Depending on the designer's chosen modeling approach, this more complex expression may take the form of discrete sets of continuity, momentum, or energy equations, with one set for each phase in the system. Alternatively, the increased complexity may appear as added or modified terms—latent heat or homogeneous density, for example—within the same number of equations as a single-phase system would require. When applying analytical tools, special care must be taken to perform control volume analyses in the multiphase context within the limits of their applicability, i.e., within regions where suitable averaging schemes are feasible and in which the presence of interfaces between phases has been taken into account. The analyst must also develop interfacial equations that express jump conditions between phases. New concepts that will be encountered while deriving multiphase system models include terms such as void fraction, which describes the relationship between the volumes, and slip ratio, which describes velocities of the phases comprising the system. As is clear from these examples, readers may expect to encounter in multiphase heat transfer, much that is familiar from more basic subject matter, as well as new concepts that are challenging in nature.

It is the aim of this book to develop, within a single volume, these complex analytical expressions from the familiar conservation laws, and then to demonstrate their use for *all types* of multiphase flows, including phase changes (solid $\leftrightarrows$ liquid $\leftrightarrows$ vapor $\leftrightarrows$ solid). In the process, the text aims to provide a thorough understanding of: (a) the physical principles governing the experimental and analytical bases of multiphase heat transfer and flow; (b) the generalized governing equations in multiphase systems for physical and analytical modeling of multiphase heat and mass transfer; and (c) the analysis of phase-change heat and mass transfer for various multiphase systems.

This chapter begins with an introduction to multiphase flow regimes and classification of multiphase systems. This is followed by a brief review of transport phenomena, with the detailed emphasis in multicomponent systems and scaling. Finally, multiphase flow notations and concepts are presented.

1.2 Multiphase Flows and Multiphase Systems

1.2.1 Types of Multiphase Flow Regimes

Multiphase flow is the flow of a mixture of multiple phases of matter, where the three possible phases are solid, liquid, and gas. In a multiphase flow, a phase can also be considered an identifiable material in the flow that has a particular inertial response to an interaction with the flow in which it is present. For example, different sized solid particles of the same material can be treated as a different phase because each size particle will interact differently within the flow. Some multiphase flow regimes/system changes are due to phase change and/or mass transfer. Several types of multiphase flow regimes/systems are detailed below (Michaelides et al. 2016; ANSYS Fluent Theory Guide 2017):

1. **Dispersed Phase and Separated Flows**: Dispersed-phase flows are flows in which one of the phases present is in the form of individual, distinct elements. One type of dispersed-phase flows is a slurry, in which solid particles are carried by the liquid. A dispersed-phase flow can be observed

in all types of two-phase flows—gas–liquid, gas–solid, liquid–liquid, and solid–liquid—(Serizawa 2010). A separated flow is a flow in which the two phases are separated by a plane of contact. One type of separated flow is the annular flow, in which a layer of liquid is present on the wall of the pipe through which the multiphase flow is traveling, and the center of the flow is a gas.

2. **Gas–Liquid or Liquid–Liquid Flows**: Gas–liquid flows can be in the form of a continuous fluid or gas flow with gas bubbles in a liquid, or liquid droplets in a gas (droplet flow). liquid–liquid flows can be a flow of fluid bubbles in a continuous fluid. A slug flow is the flow of large bubbles in a continuous fluid. In this category, stratified/free-surface flow is the flow of immiscible substances, which do not form a homogeneous mixture when combined and are clearly separated in the flow. Examples of gas–liquid flow used for different purposes follow:
   - Bubble columns are a common example of gas–liquid flows used in industry.
   - Atomization used to generate small droplets for combustion is important in power generation systems.
   - Controlling the process of droplet formation and their impact when forming spray is important in material processing.
   - The gas–liquid steam-water flow is important in power systems.
3. **Gas–Solid Flows**: Gas–solid flow is when solid particles are suspended in a gas (particle-laden flow). Granular flow is the motion of particles down an inclined plane where the interactions between particles with each other and with the walls are more important than the interstitial gas forces. Pneumatic transport is when the flow pattern depends on solid loading, Reynolds numbers, and/or particle properties. Fluidized beds consist of suspended solid particles in a vertical cylinder into which a gas is added through a distributor, resulting in the particles being suspended in the gas as the gas rises. Examples of gas–solid flows include:
   - Pollution control devices remove particulates from exhaust gases, which can be achieved by a cyclone separator or an electrostatic precipitator.
   - Coal combustion in power plants are dependent on the dispersion and burning of coal particles.
   - Micron-size particles in solid-propellant rocket exhaust affect the rocket's performance.
4. **Liquid-Solid Flows**: Flows in which the liquid phase carries solid particles are called slurry flows. In this case, the behavior of the flow changes with the relation of solid particles properties to liquid-particle properties. The Stokes number is normally less than 1. When the Stokes number is larger than 1, the flow is a solid–liquid fluidization. Hydrotransport is when the distribution of solid particles in the continuous liquid is dense. Sedimentation is when a uniformly dispersed mixture of particles is present in a tall column. When reaching the bottom of the column, particles will slow down, forming a sludge layer, while a clear interface appears at the top and the middle becomes a constant settling zone. Liquid-solid flows can also be classified as dispersed-phase flows, which are explained later. One example is the:
   - Transport of coals and ores through mud.
5. **Three-Phase Flows**: Three-phase flows are combinations of other flow regimes under certain conditions/circumstances. One example is when:
   - Three phases result from bubbles in a slurry flow.

1.2.2 Classifications of the Multiphase Systems

A *multiphase system* is one characterized by the simultaneous presence of several phases, the two-phase system being the simplest case. The term "two-component" is sometimes used to describe flows in which the phases consist of different chemical substances. For example, steam-water flows are two-phase, while air–water flows are two-component. Some two-component flows (mostly liquid–liquid) technically consist of a single phase but are identified as two-phase flows in which the term "phase" is applied to each of the components. Since the same mathematics describes two-phase and two-component flows, the two expressions will be treated as synonymous. This book deals with a variety of multiphase systems, in which the phases passing through the system may be solid, liquid or gas, or a combination of these three.

The analysis of multiphase systems can include consideration of *multiphase flow* and *multiphase heat and mass transfer*. When all of the phases in a multiphase system exist at the same temperature, multiphase flow is the only concern. However, when the temperatures of the individual phases are different, interphase heat transfer also occurs. If different phases of the same pure substance are present in a multiphase system, interphase heat transfer will result in a change of phase, which is always accompanied by interphase mass transfer. The combination of heat transfer with mass transfer during phase change makes multiphase systems distinctly more challenging than simpler systems.

Based on the phases that are involved in the system, phase-change problems can be classified as: (1) solid–liquid phase change (melting and solidification), (2) solid–vapor phase change (sublimation and deposition), and (3) liquid–vapor phase change (boiling/evaporation and condensation). Melting and sublimation are also referred to as fluidification because both liquid and vapor are regarded as fluids.

Phase-change problems can also be classified on the basis of the system's geometric configurations and the structures of the interfaces separating different phases. From the geometric configuration of the system, one can classify multiphase problems as (1) external phase-change problems in which one phase extends to infinity, and (2) internal phase-change problems in which the different phases are confined to a limited space. Examples belonging to the former class include melting and solidification in semi-infinite regions, pool boiling, and film condensation. Some examples belonging to the latter class are melting and solidification in the finite slabs, forced convective boiling, and condensation in channels.

Another method for classifying multiphase systems considers the structure of the interfaces. Multiphase systems can be classified as (1) separated phase, (2) mixed phase, and (3) dispersed phase, as summarized in Table 1.1. The *separated-phase* case has two immiscible phases separated by a clearly defined geometrically simple interface (Cases A through E). Such systems can be further classified according to whether phase change occurs on a plane surface or inside a channel. Phase change occurring on a plane surface can include combinations of different phases, as indicated in Table 1.1. Liquid-gas jet flow may involve a liquid jet in a gas phase or a gas jet in a liquid phase, while phase change in a channel includes liquid–vapor annular flow as well as melting and solidification occurring at a single temperature. At the other extreme of interfacial complexity are the *dispersed phases* (cases L through N), including bubbly flow—discrete gaseous bubbles in a continuous fluid; droplet flow—discrete fluid droplet in a continuous liquid–vapor (gas) system; and solid–particle flow—discrete particles in a liquid or gas carrier. Change in an interfacial structure from separated phase to dispersed phase can occur gradually; as a result, there are *mixed phases* (Cases F through K) in which both separated and dispersed phases coexist. For a liquid–vapor annular flow with a vapor core surrounded by a liquid film, thin film evaporation occurs when heat is applied to the external surface of the tube—Case C of the separated-phase type. If the wall temperature is increased to a sufficient level, vapor bubbles can be generated in the liquid layer, so the system transforms to case G of the mixed-phase type: bubbly annular flow. If the wall temperature is further increased, the flow changes to a liquid–vapor droplet form—Case M of the dispersed-flow type.

Characteristic features may be associated with the behavior of each of the three possible phases comprising the multiphase systems of Table 1.1. The solid phase can be regarded as incompressible because the density of the solid phase can be treated as constant for most cases. In cases where no fluidification or solidification occurs, the solid phase has a nondeformable interface with the fluid

Table 1.1 Classification of multiphase systems

Type	Case	Typical regimes	Geometry	Configuration	Examples
Separated phases	A	Phase change on plane surface	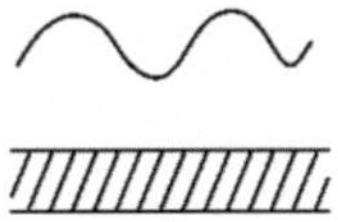	(a) Liquid layer in vapor (b) Vapor layer in liquid (c) Solid layer in liquid (d) Liquid layer in solid (e) Solid layer in vapor	(a) Film condensation (b) Film boiling (c) Solidification (d) Melting (e) Sublimation and deposition
	B	Liquid-gas jet flow	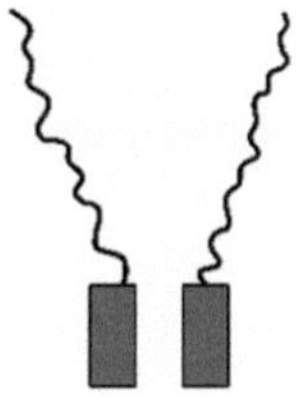	(a) Liquid jet in gas (b) Gas jet in liquid	(a) Atomization (b) Jet condenser
	C	Liquid–vapor annular flow	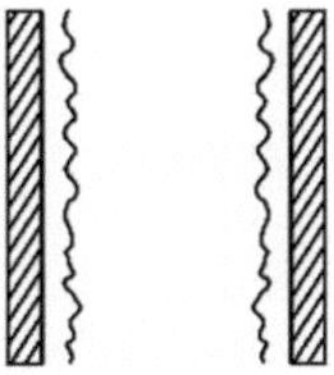	(a) Liquid core and vapor film (b) Vapor core and liquid film	(a) Film boiling (b) Film condensation or evaporation
	D	Melting at a single melting point	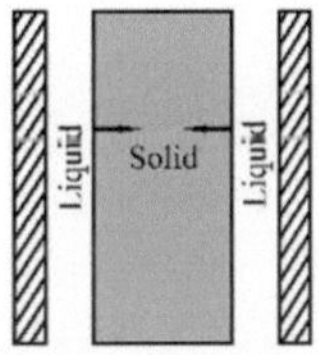	Solid core and liquid annular layer	Melting of ice in a duct
	E	Solidification at a single melting point	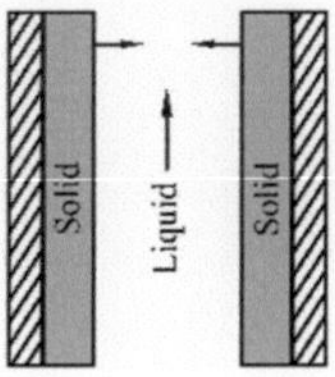	Liquid core and solid annular layer	Freezing water in a duct
Mixed phases	F	Slug or plug flow	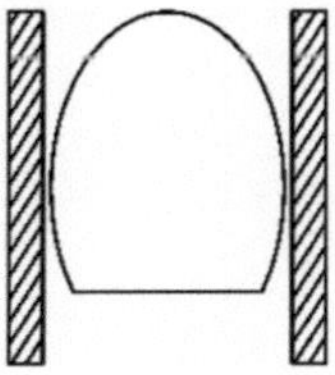	Large vapor bubbles in a continuous liquid	Pulsating heat pipes

(continued)

Table 1.1 (continued)

Type	Case	Typical regimes	Geometry	Configuration	Examples
	G	Bubbly annular flow	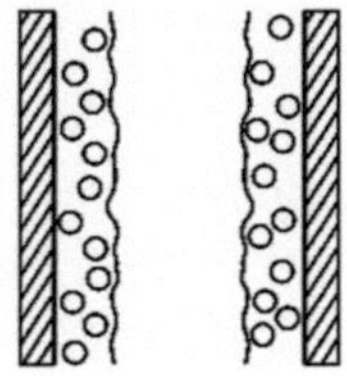	Vapor bubbles in liquid film with vapor core	Film evaporation with wall nucleation
	H	Droplet annular flow	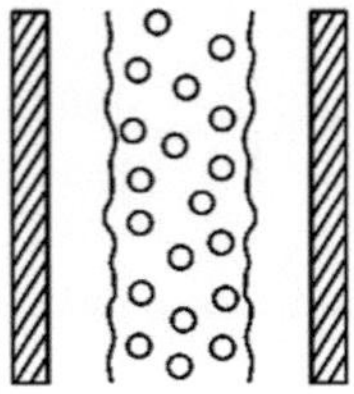	Vapor core with liquid droplets and annular liquid film	Steam generator in boiler
	I	Bubbly droplet annular flow	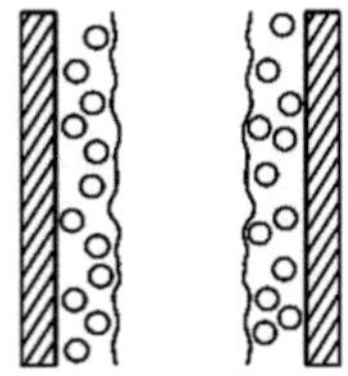	Vapor bubbles in liquid film with vapor core	Boiling nuclear reactor channel
	J	Melting over a temperature range	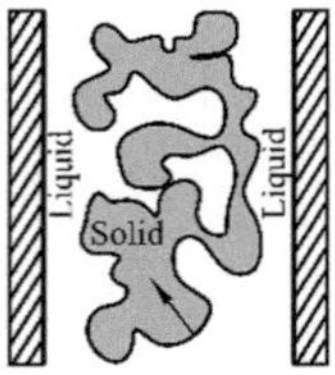	Solid and mushy zone in liquid	Melting of binary solid
	K	Solidification over a temperature range	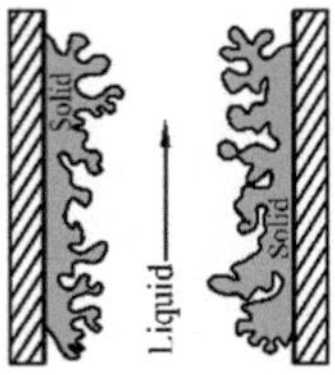	Liquid core with layer of solid and mushy zone	Freezing of binary solution
	L	Liquid–vapor (gas) bubbly flow	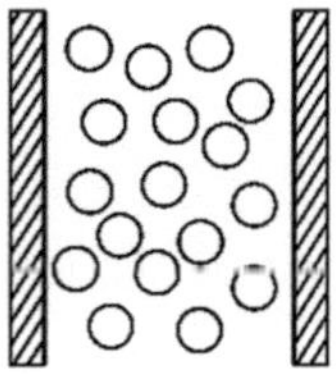	Discrete vapor bubbles in a liquid	Chemical reactors Absorbers Evaporators Separating devices

(continued)

Table 1.1 (continued)

Type	Case	Typical regimes	Geometry	Configuration	Examples
Dispersed phases	M	Liquid–vapor (gas) droplet flow	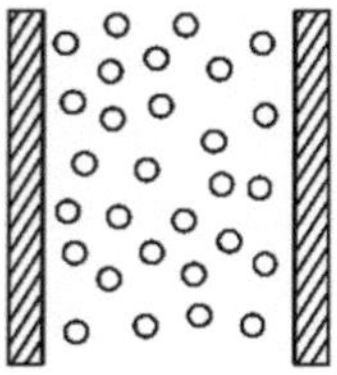	Discrete liquid droplets in a vapor	Spray cooling Atomizers Combustors
	N	Particulate flow	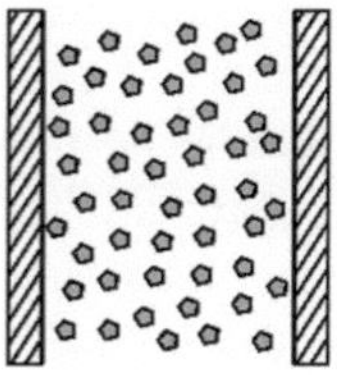	(a) Solid particles in liquid (slurry flow) (b) Discrete solid particles in gas (c) Fluidized beds	(a) Melting, solidification of PCM suspension in liquid (b) Combustion of solid fuels (c) Fluidized bed reactors

phase, or phases, flowing over it. The flow characteristics depend strongly on the size of the individual solid elements and on the motion of the associated fluids. When melting and solidification are involved, the volume and shape of the solid can change with time. For melting and solidification occurring at a single melting point, the liquid phase is continuous, while the solid phase is discontinuous in a mushy zone formed by melting or solidification of a binary substance. The solid phase is also discontinuous in cases of particulate flow because the solid particles are dispersed in either liquid or gas phases.

In multiphase systems containing a liquid phase, the liquid can be the continuous phase, containing dispersed elements of solids (particles), gases (bubbles), or other liquids (drops). The liquid can also be discontinuous, for example, in the form of drops suspended in a gas or in another liquid such as in liquid–vapor droplet flow. A liquid also differs from a solid because its interface with other fluids (gases or other liquids) is readily deformable. The vapor (gas) phase in a multiphase system can be continuous, as in film evaporation or condensation, or as in liquid–vapor annular flow. It can also be discontinuous, as in liquid–vapor bubbly flow. Compared with the liquid phase, a vapor (gas) is highly compressible because its density is a strong function of the temperature and pressure. Notwithstanding this behavior, many multiphase flows containing vapor (gases) can be treated as essentially incompressible, especially if the pressure is reasonably high and the Mach number for the gas phase is low, say less than 0.3.

A multiphase system with separated phases can be considered as a field that is divided into single-phase regions with interfaces between the phases. The governing equations for a multiphase system with separated phases can be written using the standard local instantaneous differential balance for each single-phase region, with appropriate jump conditions to match the solution of these differential equations at the interfaces. This method, which is referred to as the interface tracking method, involves solving the single-phase equations in each separate phase. By contrast, explicit tracking of the interfaces in mixed-phase and dispersed-phases is more complex and sometimes even impossible. In this case, spatial averaging of the governing equations is performed over each phase or simultaneously over the phases within a multiphase control volume.

1.3 Physical Concepts

1.3.1 Sensible Heat

Multiphase heat and mass transfer is concerned primarily with the study of interactions between energy and matter. In light of this fact, it is instructive to briefly review the historical background of the concepts of sensible and latent heat. As will be explained in the subsequent development of this text, energy is a property possessed by particles of matter, and heat is the transfer of this energy between particles.

It is common sense that heat flows from an object at a higher temperature to one at a lower temperature. Before the nineteenth century, it was believed that heat was a fluid substance named *caloric*. The temperature of an object was thought to increase when caloric flowed into an object and to decrease when caloric flowed out of the object. Combustion was believed to be a process during which a large amount of caloric was released. Because heat flow never produced a detectable change in mass, and because caloric could not be detected by any other means, it was logical to assume that caloric was massless, odorless, tasteless, and transparent. Although the caloric theory explained many observations, such as heat flow from an object with a high temperature to one at a lower temperature, it was unable to account for other phenomena, such as heat generated by friction. For example, one can rub together two pieces of metal for a long time and generate heat indefinitely, a process that is inconsistent with a characterization of heat as a substance of finite quantity contained within an object.

In 1800s, an English brewer, James Prescott Joule, established the correct concept of heat through a number of experiments. One of his experiments is shown in Fig. 1.1. The paddle wheel turns when the weight lowers, and friction between the paddle wheel and the water causes the water temperature to rise. The same temperature rise can also be obtained by heating the water on a stove. From this and many other experiments, Joule found that one joule (J) of work always equals 4.18 calories (cal) of heat, which is well known today as the mechanical equivalent of heat. Therefore, heat, like work, is a transfer of thermal energy rather than the flow of a substance. In a process where heat is transferred

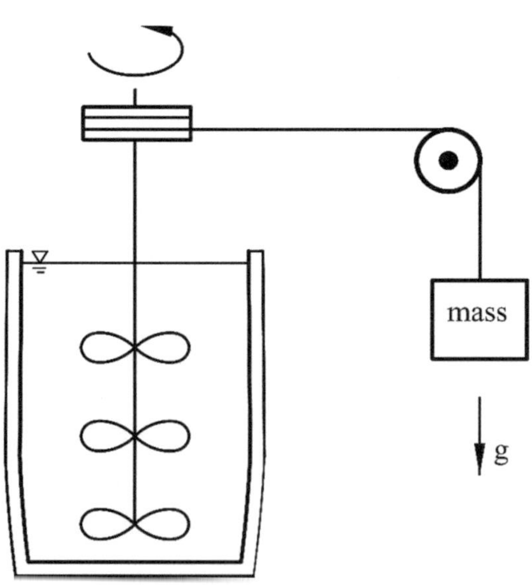

Fig. 1.1 Schematic of Joule's experiment demonstrating the mechanical equivalent of heat

from a high-temperature object to a low-temperature object, thermal energy, not a substance, is transferred from the former to the latter. In fact, the unit for heat in the SI system is the joule, which is also the unit for work.

The amount of sensible heat Q required to raise the temperature of a system from T_1 to T_2 is proportional to the mass of the system and the temperature rise, i.e.,

$$Q = mc(T_2 - T_1) \tag{1.1}$$

where the proportionality constant c is called the specific heat and is a property of the material. Specific heat is defined as the amount of heat required to raise the temperature of a unit mass of the substance by one degree. For example, $c = 1.00\,\text{kcal/kg}\,°\text{C} = 4.18\,\text{kJ/kg}\,°\text{C}$ for water at 15 °C, which means that it takes 1 kcal of heat to raise the temperature of 1 kg of water from 15 to 16 °C.

The definition of the specific heat for a gas differs from that of a liquid and solid because the value of the gas specific heat depends on how the process is carried out. The specific heat values for two particular processes are of special interest to scientists and engineers: constant volume and constant pressure. The values of the specific heats at constant volume, c_v, and at constant pressure, c_p, for gases are quite different. The relationship between these two specific heats for an ideal gas is given by

$$c_p - c_v = R_g \tag{1.2}$$

where R_g, the gas constant, is related to the universal gas constant, R_u, by $R_g = R_u/M$ with M being the molecular mass of the ideal gas. The molecular masses and specific heats for some selected substances are listed in Table 1.2.

Clearly, the values of the specific heats for a given gas at constant pressure and constant volume are quite different. For liquids and solids, the specific heat can be assumed to be process-independent because these phases are nearly incompressible. Therefore, the specific heats of liquids and solids at constant pressure are assumed to apply to all real processes.

Table 1.2 Specific heats of different substances at 20 °C

Substance	M (kg/kmol)	c_p (kJ/kg °C)	c_v (kJ/kg °C)
Air	28.97	1.005	0.718
Aluminum	26.9815	0.90	
Carbon dioxide	44.01	0.84	0.65
Copper	63.546	0.39	
Glass		0.84	
Hydrogen	2.016	14.27	10.15
Ice (−5 °C)	18.015	2.10	
Iron	55.847	0.45	
Lead	207.2	0.13	
Marble		0.86	
Nitrogen	28.013	1.04	0.74
Oxygen	31.999	0.92	0.66
Steam (100 °C)	18.015	2.02	1.47
Silver	107.868	0.23	
Mercury	200.59	0.14	
Water	18.015	4.18	

1.3.2 Latent Heat and Phase Change

Although phase-change phenomena such as the solidification of lava, the melting of ice, the evaporation of water, and the fall of rain have been observed by mankind for centuries, the scientific methods used to study phase change were not developed until the seventeenth century because a flawed understanding of temperature, energy, and heat prevailed. It was incorrectly believed that the addition or removal of heat could always be measured by the change of temperature. Based on the misconception that temperature change always accompanies heat addition, a solid heated to its melting point was thought to require only a very small amount of additional heat to completely melt. Likewise, it was thought that only a small amount of extra cooling was required to freeze a liquid at its melting point. In both of these examples, the heat transferred during phase change was believed to be very small because the temperature of the substance undergoing phase change did not change by a significant amount.

Between 1758 and 1762, an English professor of medicine, Dr. Joseph Black, conducted a series of experiments measuring the heat transferred during phase-change processes. He found that the quantity of heat transferred during phase change was in fact very large, a phenomenon that could not be explained in terms of sensible heat. He demonstrated that the conventional wisdom about the amount of heat transferred during phase change was wrong, and he used the term "latent heat" to define heat transferred during phase change. Latent heat is a *hidden* heat, and it is not evident until a substance undergoes a phase change. Perhaps the most significant application of Dr. Black's latent heat theory was James Watt's 500% improvement of steam engine thermal efficiency. James Watt was an engineer and Dr. Black's assistant for a time.

The concept of latent heat can be demonstrated by tracing the phase change of water from subcooled ice below 0 °C, to superheated vapor above 100 °C. Let us consider a 1-kg mass of ice with an initial temperature of −20 °C. When heat is added to the ice, its temperature gradually increases to 0 °C, at which point the temperature stops increasing even when heat is continuously added. During the ensuing interval of constant temperature, the change of phase from ice to liquid water can be observed. After the entire mass of ice is molten, further heating produces an increase in the now-liquid temperature up to 100 °C. Continued heating of the liquid water at 100 °C does not yield any increase of temperature; instead, the liquid water is vaporized. After the last drop of the water is vaporized, continued heating of the vapor will result in the increase of its temperature. The phase-change process from subcooled ice to superheated vapor is shown in Fig. 1.2. It is seen that a substantial amount of heat is required during a change of phase, an observation consistent with Dr. Black's latent heat theory.

The heat required to melt a solid substance of unit mass is defined as the latent heat of fusion, and it is represented by $h_{s\ell}$. The latent heat of fusion for water is about 335 kJ/kg, which is very high

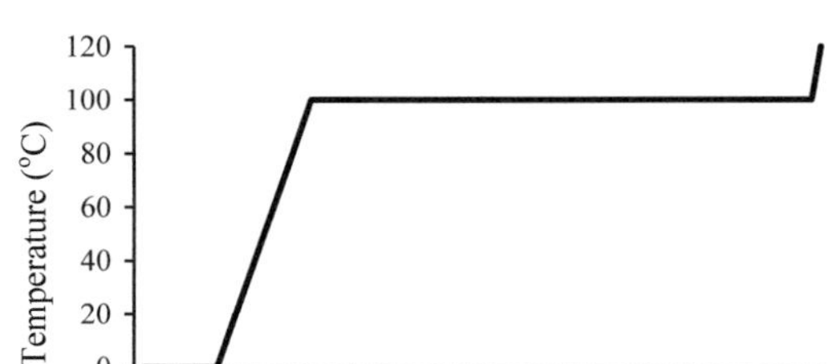

Fig. 1.2 Temperature profile for phase change from subcooled ice to superheated steam

Table 1.3 Latent heat of fusion and vaporization for selected materials at 1 atm

Substance	Melting point (°C)	$h_{s\ell}$ (kJ/kg)	Boiling point (°C)	$h_{\ell v}$ (kJ/kg)
Oxygen	−218.18	14	−183	220
Ethyl alcohol	−114	105	78	870
Water	0	335	100	2251
Lead	327	25	1750	900
Silver	961	88	2193	2300
Tungsten	3410	184	5900	4800

compared with the most substances. The heat required to vaporize a liquid substance of unit mass is defined as the latent heat of vaporization, and it is represented by $h_{\ell v}$. The latent heat of vaporization for water is about 2251 kJ/kg. The latent heats of other materials, shown in Table 1.3, demonstrate that the latent heats of vaporization for all materials are much larger than their latent heats of fusion because the molecular spacing for vapor is much larger than that for solid or liquid. The latent heat for deposition/sublimation, h_{sv}, for water is about 2847 kJ/kg.

When a process involves phase change, the heat absorbed or released can be expressed as the product of the mass quantity and the latent heat of the phase-change material. For liquid–vapor phase change, such as evaporation, boiling, or condensation, the heat transferred can be expressed as

$$q = \dot{m}h_{\ell \mathrm{v}} \tag{1.3}$$

where $\dot{m}$ is the mass of material changing phase per unit time. The latent heat of vaporization, $h_{\ell \mathrm{v}}$, is the difference between the enthalpy of vapor and of liquid, i.e.,

$$h_{\ell \mathrm{v}} = h_{\mathrm{v}} - h_{\ell} \tag{1.4}$$

During melting and solidification processes, heat transfer can be expressed as

$$q - \dot{m}h_{s\ell} \tag{1.5}$$

where $h_{s\ell}$ is the latent heat of fusion.

1.4 Review of Fundamentals of Transport Phenomena

1.4.1 Continuum Flow Limitations

The transport phenomena are usually modeled in continuum states for most applications—the materials are assumed to be continuous and the fact that matter is made of atoms is ignored. Recent development in fabrication and utilization of nanotechnology, microdevices, and microelectromechanical systems (MEMS) requires noncontinuum modeling of transport phenomena in nano- and microchannels. When the dimension of the nano- or microchannel, D, is small compared to the molecular mean free path λ, which is defined as average distance between collisions for a molecule, the traditional Navier–Stokes equation and the energy equation based on the continuum assumption have failed to provide accurate results.

The continuum assumption may also not be valid in conventional systems—for example, the early stages of high-temperature heat pipe start-up from a frozen state and microscale heat pipes (Faghri

2016). During the early stage of start-up of high-temperature heat pipes, the vapor density in the heat pipe core is very low and partly loses its continuum characteristics. The vapor flow in this condition is usually referred to as rarefied vapor flow. Because of the low density, the vapor in the rarefied state is somewhat different from the conventional continuum state. Also, the vapor density gradient is very large along the axial direction of the heat pipe. The vapor flow along the axial direction is caused mainly by the density gradient via vapor molecular diffusion.

The validity of the continuum assumption can also be violated in micro heat pipes. As the size of the heat pipe decreases, the vapor in the heat pipe may lose its continuum characteristics. The heat transport capability of a heat pipe operating under noncontinuum vapor flow conditions is very limited, and a large temperature gradient exists along the heat pipe length. This is especially true for a miniature or microheat pipe, whose dimensions may be extremely small.

The continuum criterion is usually expressed in terms of the Knudsen number

$$Kn = \frac{\lambda}{D} \tag{1.6}$$

Based on the degree of rarefaction of gas, the flow regimes in microchannel can be classified into four regimes:

1. *Continuum regime* ($Kn < 0.01$). The Navier–Stokes equation is valid.
2. *Slip flow regime* ($0.01 < Kn < 0.1$). The Navier–Stokes equation can be used with the application of slip condition, i.e., allowing nonzero axial fluid velocity along near the wall of the microchannel.
3. *Transition regime* ($0.01 < Kn < 0.1$). The Navier–Stokes equation is not valid, and the flow must be solved using the Boltzmann equation or Direct Monte Carlo Simulation (DMCS).
4. *Free molecular flow regime* ($Kn > 3$). The collision between molecules can be neglected and a collisionless Boltzmann equation can be used.

The mean free path for dilute gases based on the kinetic theory in terms of temperature and pressure is:

$$\lambda = \frac{1.051 k_b T}{\sqrt{2}\pi\sigma^2 p} \tag{1.7}$$

The transition density under which the continuum assumption is invalid can be obtained by combining Eqs. (1.6)–(1.7) and $Kn = 0.01$, i.e.,

$$\rho_{tr} = \frac{1.051 k_b}{\sqrt{2}\pi\sigma^2 R_g D Kn} \tag{1.8}$$

where the ideal gas equation of state, $p = \rho R_g T$ was used.

Assuming that the vapor is in the saturation state, the transition vapor temperature T_{tr} corresponding to the transition density can be obtained by using the Clausius–Clapeyron equation (see Chap. 2) combined with the equation of state:

$$T_{tr} = \frac{p_{sat}}{\rho R_g} \exp\left[-\frac{h_{\ell v}}{R_g}\left(\frac{1}{T_{tr}} - \frac{1}{T_{sat}}\right)\right] \tag{1.9}$$

where p_{sat} and T_{sat} are the saturation pressure and temperature, $h_{\ell v}$ is the latent heat of vaporization, and the vapor density ρ is given by Eq. (1.8). Equation (1.9) can be rewritten as

$$\ln\left(\frac{T_{tr}\rho R_g}{p_{sat}}\right)+\frac{h_{\ell v}}{R_g}\left(\frac{1}{T_{tr}}-\frac{1}{T_{sat}}\right)=0 \tag{1.10}$$

and solved iteratively for T_{tr} using the Newton–Raphson/secant method. The transition vapor temperature is the boundary between continuum regime and noncontinuum regime.

1.4.2 Transport Phenomena

Transport phenomena include momentum transfer, heat transfer, and mass transfer, all of which are fundamental to an understanding of both single and multiphase systems (Bird et al. 2006; Faghri et al. 2010; Kulacki 2018). It is assumed that the reader has a good undergraduate-level understanding of transport phenomena as applied to single-phase systems, as well as the associated thermodynamics, fluid mechanics, and heat transfer. Therefore, this subsection provides only an overview of these transport phenomena in single-phase system.

1.4.2.1 Fluid Mechanics

A fluid at rest can resist a normal force but not a shear force, while fluid in motion can also resist a shear force. The fluid continuously deforms under the action of shear force. A fluid's resistance to shear or angular deformation is measured by viscosity, which can be thought of as the internal "stickiness" of the fluid. The force and the rate of strain (i.e., rate of deformation) produced by the force are related by a constitutive equation. For a Newtonian fluid, the shear stress in the fluid is proportional to the time rate of deformation of a fluid element or particle. Figure 1.3 shows a Couette flow where the plate at the bottom is stationary and the fluid is driven by the upper moving plate. This flow is one-dimensional since velocity does not vary in the x- and z-directions are zero. The constitutive relation for Couette flow can be expressed as

$$\tau_{yx}=-\mu\frac{du}{dy} \tag{1.11}$$

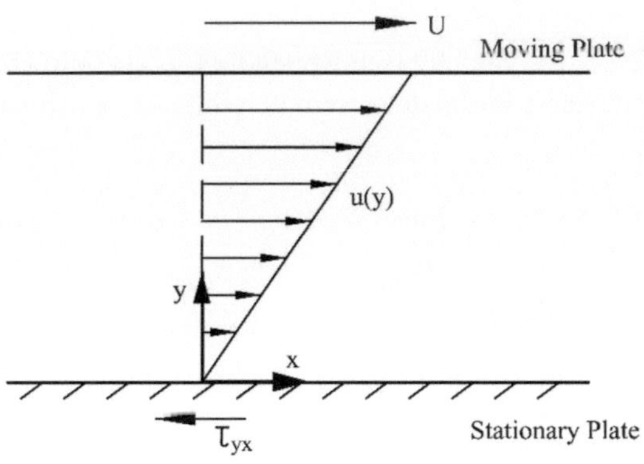

Fig. 1.3 Couette flow

where τ_{yx} is the shear stress (N/m^2), μ is the dynamic viscosity (N s/m^2), which is a fluid property, and du/dy is the velocity gradient in the y-direction, also known as the rate of deformation. If the shear stress τ_{yx} and the rate of deformation du/dy have a linear relationship, as shown in Eq. (1.11), the fluid is referred to as Newtonian and Eq. (1.11) is called Newton's law of viscosity. It is found that the resistance to flow for all gases and liquids with molecular mass less than 5000 is well presented by Eq. (1.11). For non-Newtonian fluids, such as polymeric liquids, slurries, or other complex fluids, for example in biological applications, such as blood or operation of joints, drag-reducing slimes on marine animals, and digesting foodstuffs, the shear stress and the rate of deformation no longer have a linear relationship. Bird et al. (2006) provided detailed information for treatment for non-Newtonian fluids.

The viscous force comes into play whenever there is velocity gradient in a fluid. For a three-dimensional fluid flow problem, the stress $\boldsymbol{\tau}'$ is a tensor of rank two with nine components. It can be expressed as summation of an isotropic thermodynamic stress, $-p\mathbf{I}$, and a viscous stress, $\boldsymbol{\tau}$:

$$\boldsymbol{\tau}' = -p\mathbf{I} + \boldsymbol{\tau} \tag{1.12}$$

where p is thermodynamic pressure, and $\mathbf{I}$ is the unit tensor defined as

$$\mathbf{I} = \begin{bmatrix} 1 & 0 & 0 \\ 0 & 1 & 0 \\ 0 & 0 & 1 \end{bmatrix} \tag{1.13}$$

In a Cartesian coordinate system, the viscous stresses are

$$\boldsymbol{\tau} = \begin{bmatrix} \tau_{xx} & \tau_{xy} & \tau_{xz} \\ \tau_{yx} & \tau_{yy} & \tau_{yz} \\ \tau_{zx} & \tau_{zy} & \tau_{zz} \end{bmatrix} \tag{1.14}$$

where the first subscript represents the axis normal to the face on which the stress acts, and the second subscript represents the direction of the stress (see Fig. 1.4).

The components of the shear stresses are symmetric ($\tau_{xy} = \tau_{yx}$, $\tau_{xz} = \tau_{zx}$, and $\tau_{yz} = \tau_{zy}$).

The viscous stress tensor can be expressed by Newton's law of viscosity:

$$\boldsymbol{\tau} = 2\mu\mathbf{D} - \frac{2}{3}\mu(\nabla \cdot \mathbf{V})\mathbf{I} \tag{1.15}$$

where $\nabla \cdot \mathbf{V}$ is the divergence of the velocity (see Appendix C). The rate of deformation, or strain rate $\mathbf{D}$, presented below for a Cartesian coordinate system in a three-dimensional flow is another tensor of rank two (see Appendix C).

$$\mathbf{D} = \frac{1}{2}\left[\nabla\mathbf{V} + (\nabla\mathbf{V})^{\mathrm{T}}\right] = \begin{bmatrix} \dfrac{\partial u}{\partial x} & \dfrac{1}{2}\left(\dfrac{\partial u}{\partial y} + \dfrac{\partial v}{\partial x}\right) & \dfrac{1}{2}\left(\dfrac{\partial u}{\partial z} + \dfrac{\partial w}{\partial x}\right) \\ \dfrac{1}{2}\left(\dfrac{\partial v}{\partial x} + \dfrac{\partial u}{\partial y}\right) & \dfrac{\partial v}{\partial y} & \dfrac{1}{2}\left(\dfrac{\partial v}{\partial z} + \dfrac{\partial w}{\partial y}\right) \\ \dfrac{1}{2}\left(\dfrac{\partial w}{\partial x} + \dfrac{\partial u}{\partial z}\right) & \dfrac{1}{2}\left(\dfrac{\partial w}{\partial y} + \dfrac{\partial v}{\partial z}\right) & \dfrac{\partial w}{\partial z} \end{bmatrix} \tag{1.16}$$

where $(\nabla\mathbf{V})^{\mathrm{T}}$ is the transverse tensor of $\nabla\mathbf{V}$ as defined in Appendix C.

For example, in a Cartesian coordinate system, the normal and shear viscous stresses can be expressed as

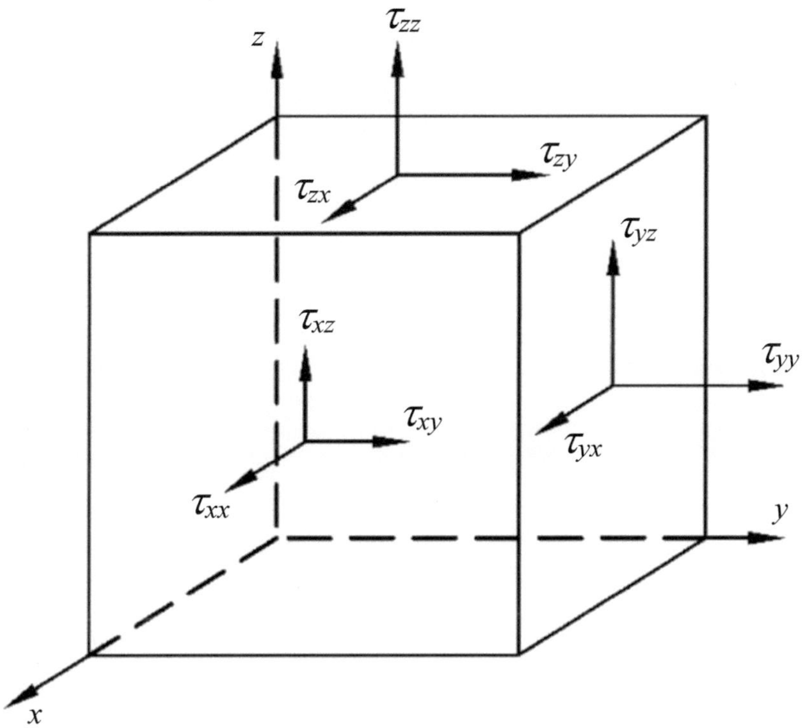

Fig. 1.4 Components of the stress tensor in a fluid

$$\tau_{xx} = 2\mu\frac{\partial u}{\partial x} - \frac{2}{3}\mu\left(\frac{\partial u}{\partial x} + \frac{\partial v}{\partial y} + \frac{\partial w}{\partial z}\right) \tag{1.17}$$

$$\tau_{xy} = \mu\left(\frac{\partial u}{\partial y} + \frac{\partial v}{\partial x}\right) \tag{1.18}$$

For one-dimensional flow in Fig. 1.3, Eq. (1.18) is reduced to Eq. (1.11).

The stress–strain rate relationships in Eqs. (1.11) and (1.15) are valid for laminar flow only. For turbulent flow, Eqs. (1.11) or (1.15) can still be used provided that the time-averaged velocity is used, and the turbulent effects are included in the viscosity (White 2005; Kays et al. 2004; Faghri et al. 2010).

For a multicomponent system, the viscosity of the mixture is related to the viscosity of the individual component by

$$\mu = \sum_{i=1}^{N} \frac{x_i \mu_i}{\sum_{j=1}^{N} x_i \phi_{ij}} \tag{1.19}$$

where x_i is molar fraction of component i and

$$\phi_{ij} = \frac{1}{\sqrt{8}}\left(1 + \frac{M_i}{M_j}\right)^{-1/2}\left[1 + \left(\frac{\mu_i}{\mu_j}\right)^{1/2}\left(\frac{M_j}{M_i}\right)^{1/4}\right]^2 \tag{1.20}$$

where M_i is molecular mass of the ith species. Equation (1.19) can reproduce the viscosity for the mixtures with an averaged deviation of 2%. Additional correlations to estimate the viscosities of various gases and gas mixtures as well as liquids can be found from the standard reference by Poling et al. (2000).

1.4.2.2 Heat Transfer

Heat transfer is a process whereby thermal energy is transferred in response to a temperature difference. There are three modes of heat transfer: conduction, convection, and radiation (Faghri et al. 2010). Conduction is heat transfer across a *stationary* medium, either solid or fluid. For an electrically nonconducting solid, conduction is attributed to atomic activity in the form of lattice vibration, while the mechanism of conduction in an electrically conducting solid is a combination of lattice vibration and translational motion of electrons. Heat conduction in a liquid or gas is due to the random motion and interaction of the molecules. For most engineering problems, it is impractical and unnecessary to track the motion of individual molecules and electrons, which may instead be described using the macroscopic averaged temperature. The heat transfer rate is related to the temperature gradient by Fourier's law. For the one-dimensional heat conduction problem shown in Fig. 1.5, in which temperature varies along the y-direction only, the heat transfer rate is obtained by Fourier's law.

$$q''_y = -k\frac{\mathrm{d}T}{\mathrm{d}y} \tag{1.21}$$

where q''_y is the heat flux along the y-direction, i.e., the heat transfer rate in the y-direction per unit area (W/m^2), and $\mathrm{d}T/\mathrm{d}y$ (K/m) is the temperature gradient. The proportionality constant k is thermal conductivity (W/m K), a property of the medium.

For heat conduction in a multidimensional isotropic system, Eq. (1.21) can be rewritten in the following generalized form:

$$\mathbf{q}'' = -k\nabla T \tag{1.22}$$

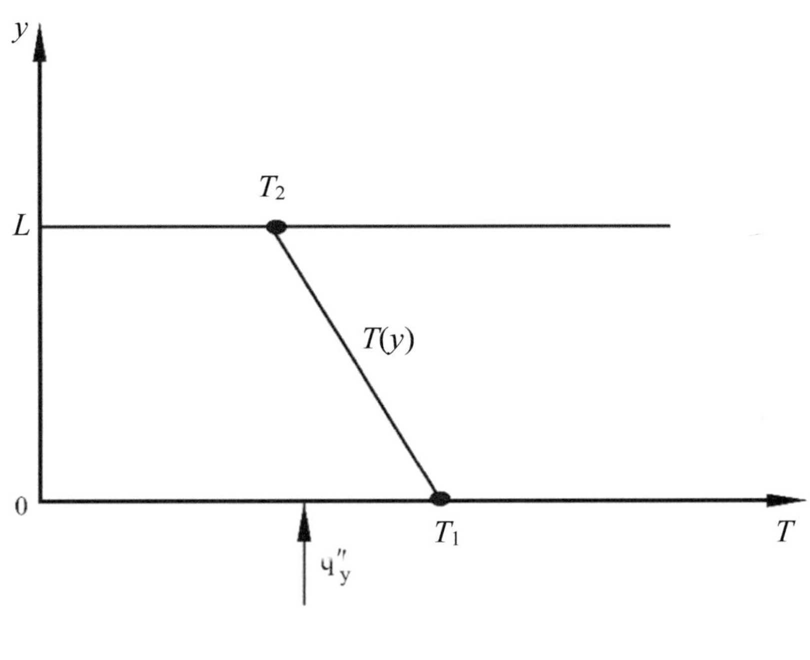

Fig. 1.5 One-dimensional conduction

where both the heat flux and the temperature gradient are vectors, i.e.,

$$\mathbf{q}'' = \mathbf{i}q''_x + \mathbf{j}q''_y + \mathbf{k}q''_z \tag{1.23}$$

While the thermal conductivity for *isotropic* materials does not depend on the direction, it is dependent on direction for *anisotropic* materials. Unlike the isotropic materials whose thermal conductivity is a scalar, the thermal conductivity of the anisotropic material is a tensor of the second order:

$$\mathbf{k} = \begin{bmatrix} k_{xx} & k_{xy} & k_{xz} \\ k_{yx} & k_{yy} & k_{yz} \\ k_{zx} & k_{zy} & k_{zz} \end{bmatrix} \tag{1.24}$$

and Eq. (1.22) will become

$$\mathbf{q}'' = -\mathbf{k} \cdot \nabla T \tag{1.25}$$

The second mode of heat transfer is convection, which occurs between a wall at one temperature, T_{w}, and a moving fluid at another temperature, T_∞; this is exemplified by forced convective heat transfer over a flat plate, as shown in Fig. 1.6. The mechanism of convection heat transfer is a combination of random molecular motion (conduction) and bulk motion (advection) of the fluid. Newton's law of cooling is used to describe the heat transfer rate:

$$q'' = h(T_{\mathrm{w}} - T_\infty) \tag{1.26}$$

where h is the convective heat transfer coefficient (W/m^2 K), which depends on many factors including fluid properties, flow velocity, geometric configuration, and any fluid phase change that may occur as a result of heat transfer. Unlike thermal conductivity, the convective heat transfer coefficient is not a property of the fluid. Typical values of mean convective heat transfer coefficients for various heat transfer modes are listed in Table 1.4. Convective heat transfer is often measured using the Nusselt number defined by

$$Nu = \frac{hL}{k} \tag{1.27}$$

where L and k are characteristic length and thermal conductivity of the fluid, respectively.

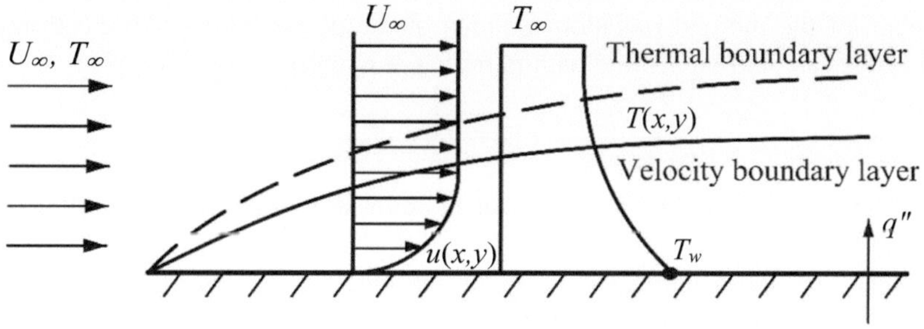

Fig. 1.6 Forced convective heat transfer

Table 1.4 Typical values of mean convective heat transfer coefficients

Mode	Geometry	$\bar{h}$ (W/m^2 K)
Forced convection	Air flows at 2 m/s over a 0.2 m square plate	12
	Air at 2 atm flowing in a 2.5-cm-diameter tube with a velocity of 10 m/s	65
	Water flowing in a 2.5-cm-diameter tube with a mass flow rate of 0.5 kg/s	3500
	Airflow across 5-cm-diameter cylinder with velocity of 50 m/s	180
Free convection ($\Delta T = 20\,°\text{C}$)	Vertical plate 0.3 m high in air	4.5
	Horizontal cylinder with a diameter of 2 cm in water	890
Evaporation	Falling film on a heated wall	6000–27,000
Condensation of water at 1 atm	Vertical surface	4000–11,300
	Outside horizontal tube	9500–25,000
Boiling of water at 1 atm	Pool	2500–3500
	Forced convection	5000–10,000
Natural convection-controlled melting and solidification	Melting in a rectangular enclosure	500–1500
	Solidification around a horizontal tube in a superheated liquid phase-change material	1000–1500

The third mode of heat transfer is radiation. The transmission of thermal radiation does not require the presence of a material medium because radiation heat transfer can occur in a vacuum. Thermal radiation is a form of energy emitted by matter at a nonzero temperature and its wavelength is primarily in the range between 0.1 to 10 μm. The emission can be from a solid surface as well as from a liquid or gas. Thermal radiation may be considered as the propagation of electromagnetic waves or alternatively of a collection of particles, such as photons or quanta of photons. When matter is heated, some of its molecules or atoms are excited to a higher energy level. Thermal radiation occurs when these excited molecules or atoms return to lower energy states. Although thermal radiation can result from changes of the energy states of electrons, as well as vibrational and rotational energy of molecules or atoms, all of these radiant energies travel at the speed of light. The wavelength λ is related to the frequency, v, by

$$\lambda v = c \tag{1.28}$$

where c is the speed of light with a value of 2.998×10^8 m/s in a vacuum. A quantitative description of the mechanism of the thermal radiation requires quantum mechanics. An electromagnetic wave with frequency of v can also be viewed as a particle—a photon—with energy of

$$e = hv \tag{1.29}$$

where $h = 6.626068 \times 10^{-34}$ m^2 kg/s is the Planck's constant. A photon is massless, and its charge is zero.

For a blackbody, defined as an ideal surface that emits the maximum energy that can be emitted by any surface at the same temperature, the spectral emissive power, $E_{b,\lambda}$ (W/m^3), can be obtained by Planck's law

$$E_{b,\lambda} = \frac{c_1}{\lambda^5 (e^{c_2/(\lambda T)} - 1)} \tag{1.30}$$

where $c_1 = 3.742 \times 10^{-16}$ W m^2 and $c_2 = 1.4388 \times 10^{-2}$ m K are radiation constants. The unit of the surface temperature T in Eq. (1.30) is K.

The emissive power for a blackbody, E_b (W/m^2), is

$$E_b = \int_0^{\infty} E_{b,\lambda} d\lambda \tag{1.31}$$

Substituting Eq. (1.30) into Eq. (1.31), Stefan–Boltzmann's law is obtained

$$E_b = \sigma_{SB} T^4 \tag{1.32}$$

where σ_{SB} is the Stefan–Boltzmann constant, 5.67×10^{-18} W/m^2 K^4.

For a real surface, the emissive power is obtained by

$$E = \varepsilon E_b \tag{1.33}$$

where ε is the emissivity, defined as the ratio of emissive power of the real surface to that of a blackbody at the same temperature. Since a blackbody is the best emitter, the emissivity of any surface must be less than or equal to 1. A simple but important case of radiation heat transfer is the radiation heat exchange between a small surface with area A, emissivity ε, and temperature T_w, and a much larger surface surrounding the small surface (see Fig. 1.7). If the temperature of the surroundings is T_{sur}, the heat transfer rate per unit area from the small object is obtained by

$$q'' = \varepsilon \sigma_{SB} (T_w^4 - T_{sur}^4) \tag{1.34}$$

For a detailed treatment of radiation heat transfer, including radiation of nongray surfaces and participating media, the readers should consult Howell et al. (2015).

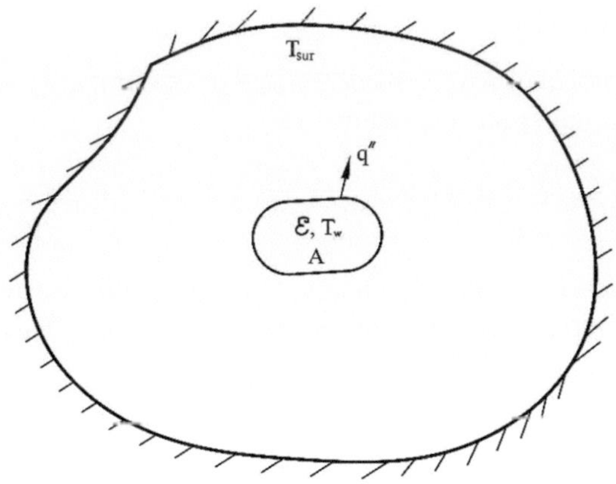

Fig. 1.7 Radiation heat transfer between a small surface and its surroundings

1.4.2.3 Mass Transfer

When there is a species concentration difference in a multicomponent mixture, mass transfer occurs. There are two modes of mass transfer: diffusion and convection. Diffusion results from random molecular motion at the microscopic level, and it can occur in a solid, liquid, or gas. Similar to convective heat transfer, convective mass transfer is due to a combination of random molecular motion at the microscopic level and bulk motion at the macroscopic level, and it can occur only in a liquid or gas.

The species concentration in a mixture can be measured by concentration ρ_i, which is defined as the mass of species i per unit volume of the mixture (kg/m^3). The density of the mixture equals the sum of the concentrations of all N species, i.e.,

$$\rho = \sum_{i=1}^{N} \rho_i \tag{1.35}$$

The concentration of the ith species can also be represented by the mass fraction of the ith species, defined as

$$\omega_i = \frac{\rho_i}{\rho} \tag{1.36}$$

It follows from Eq. (1.35) that

$$\sum_{i=1}^{N} \omega_i = 1 \tag{1.37}$$

The concentration of the ith species can also be represented by molar concentration, defined as the number of moles of the ith species per unit volume, c_i (kmol/m^3), which is related to the mass concentration by

$$c_i = \frac{\rho_i}{M_i} \tag{1.38}$$

The molar concentration of the mixture equals the sum of the molar concentrations of all N species, i.e.,

$$c = \sum_{i=1}^{N} c_i \tag{1.39}$$

The molar fraction of the ith species is defined as

$$x_i = \frac{c_i}{c} \tag{1.40}$$

which is identical to the *molecular number fraction*—the fraction of the number of molecules of the ith species to the number of molecules of all species in a given volume. This concept is essential when kinetic theory is used to describe the mass transfer process. Equation (1.40) leads to

$$\sum_{i=1}^{N} x_i = 1 \tag{1.41}$$

The mean molecular mass of the mixture can be expressed as

$$\overline{M} = \frac{\rho}{c} = \sum_{i=1}^{N} x_i M_i \tag{1.42}$$

The mass fraction is related to the molar fraction by

$$\omega_i = \frac{x_i M_i}{\sum_{j=1}^{N} x_j M_j} = \frac{x_i M_i}{\overline{M}} \tag{1.43}$$

where $\overline{M}$ and is the molar-averaged molecular mass of the mixture.

The molar fraction is related to the mass fraction by

$$x_i = \frac{\omega_i / M_i}{\sum_{j=1}^{N} \omega_j / M_j} \tag{1.44}$$

Mass diffusion of the ith component in the mixture will result in a velocity, $\mathbf{V}_i$, of the ith component relative to the stationary coordinate axes. The local mass-averaged velocity of all species, $\tilde{\mathbf{V}}$, is defined as

$$\tilde{\mathbf{V}} = \frac{\sum_{i=1}^{N} \rho_i \mathbf{V}_i}{\sum_{i=1}^{N} \rho_i} = \frac{\sum_{i=1}^{N} \rho_i V_i}{\rho} = \sum_{i=1}^{N} \omega_i \mathbf{V}_i \tag{1.45}$$

which demonstrates that the local mass flux due to diffusion, $\rho\tilde{\mathbf{V}}$, is equal to the summation of mass flux for each species, $\sum_{i=1}^{N} \rho_i \mathbf{V}_i$.

The molar-averaged velocity can be defined in a similar manner:

$$\tilde{\mathbf{V}}^* = \frac{\sum_{i=1}^{N} c_i \mathbf{V}_i}{c} = \sum_{i-1}^{N} x_i \mathbf{V}_i \tag{1.46}$$

The velocity of the ith species relative to the mass or molar-averaged velocity, $\mathbf{V}_i - \tilde{\mathbf{V}}$ or $\mathbf{V}_i - \tilde{\mathbf{V}}^*$ is defined as *diffusion velocity*. The mass flux and molar flux relative to stationary coordinate axes are defined as

$$\dot{\mathbf{m}}''_i = \rho_i \mathbf{V}_i \tag{1.47}$$

$$\dot{\mathbf{n}}''_i = c_i \mathbf{V}_i \tag{1.48}$$

The fluxes defined in Eqs. (1.47) and (1.48) are related by

$$\dot{\mathbf{n}}''_i = \frac{\dot{\mathbf{m}}''_i}{M_i} \tag{1.49}$$

Applying Eqs. (1.47) and (1.48) into Eqs. (1.45) and (1.46), the total mass flux and molar flux are obtained.

$$\dot{\mathbf{m}}'' = \sum_{i=1}^{N} \dot{\mathbf{m}}_i'' = \rho \tilde{\mathbf{V}} \tag{1.50}$$

$$\dot{\mathbf{n}}'' = \sum_{i=1}^{N} \dot{\mathbf{n}}''_i = c \tilde{\mathbf{V}}^* \tag{1.51}$$

The mass flux relative to the mass-averaged velocity is

$$\mathbf{J}_i = \rho_i (\mathbf{V}_i - \tilde{\mathbf{V}}) \tag{1.52}$$

and the molar flux relative to the molar-averaged velocity is

$$\mathbf{J}_i^* = c_i (\mathbf{V}_i - \tilde{\mathbf{V}}^*) \tag{1.53}$$

According to Eqs. (1.45) and (1.46), we have

$$\sum_{i=1}^{N} \mathbf{J}_i = \sum_{i=1}^{N} \mathbf{J}_i^* = 0 \tag{1.54}$$

Although any one of the four fluxes defined in Eqs. (1.47)–(1.48) and (1.52)–(1.53) are adequate to describe mass diffusion under all circumstances, there is usually a preferred definition of flux that can lead to less algebraic complexity. When mass diffusion is coupled with advection, Eq. (1.52) is preferred because the mass-averaged velocity, $\tilde{\mathbf{V}}$, is the velocity used in the momentum and energy equations. On the other hand, Eq. (1.53) is preferred for the multicomponent system with constant molar density c, resulting from constant pressure and temperature.

According to Eqs. (1.47)–(1.53), the following relationships between different fluxes are valid:

$$\dot{\mathbf{m}}_i'' = \rho_i \tilde{\mathbf{V}} + \mathbf{J}_i = \omega_i \sum_{j=1}^{N} \dot{\mathbf{m}}_j'' + \mathbf{J}_i \tag{1.55}$$

$$\dot{\mathbf{n}}''_i = c_i \tilde{\mathbf{V}}^* + \mathbf{J}_i^* = x_i \sum_{i=1}^{N} \dot{\mathbf{n}}''_i + \mathbf{J}_i^* \tag{1.56}$$

For a binary system that is uniform with all aspects except concentration, i.e., no temperature or pressure gradient, the diffusive mass flux can be obtained by Fick's law:

$$\mathbf{J}_1 = -\rho D_{12} \nabla \omega_1 \tag{1.57}$$

or alternatively

$$\mathbf{J}_1^* = -c D_{12} \nabla x_1 \tag{1.58}$$

where D_{12} is binary diffusivity (m^2/s).

The mass and molar flux relative to a stationary coordinate axes are

$$\dot{\mathbf{m}}_1'' = \omega_1(\dot{\mathbf{m}}_1'' + \dot{\mathbf{m}}_2'') - \rho D_{12}\nabla\omega_1 \tag{1.59}$$

$$\dot{\mathbf{n}}''_1 = x_1(\dot{\mathbf{n}}''_1 + \dot{\mathbf{n}}''_2) - cD_{12}\nabla x_1 \tag{1.60}$$

Equation (1.59) is widely applied in systems with constant density, while Eq. (1.60) is more appropriate for systems with constant molar concentration. It should be pointed out that from Eqs. (1.59) and (1.60), the absolute fluxes of species ($\dot{\mathbf{m}}''_1$ or $\dot{\mathbf{n}}''_1$) for a binary system can always be presented as summation of two parts: one part due to convection [the first terms in Eqs. (1.59) and (1.60)], and another part due to diffusion [the second terms in Eqs. (1.59) and (1.60)]. For an isothermal and isobaric steady-state one-dimensional binary system shown in Fig. 1.8, in which surface ($y = 0$) is impermeable to species 2, the mass flux of the species 1 at the surface ($y = 0$) is

$$\dot{m}''_{1y} = -\frac{\rho D_{12}}{1-\omega_1}\frac{\partial\omega_1}{\partial y} \tag{1.61}$$

For a system with more than two components, Fick's law is no longer appropriate and one must find other approaches to relate mass flux and concentration gradient. For a multicomponent low-density gaseous mixture, the following Maxwell–Stefan relation can be used to relate the molar fraction gradient of the ith component and the molar flux by:

$$\begin{aligned}\nabla x_i &= -\sum_{j=1(j\neq i)}^{N}\frac{x_i x_j}{D_{ij}}(\mathbf{V}_i - \mathbf{V}_j)\\ &= -\sum_{j=1(j\neq i)}^{N}\frac{1}{cD_{ij}}(x_j\dot{\mathbf{n}}''_i - x_i\dot{\mathbf{n}}''_j), \quad i = 1, 2, \ldots, N-1\end{aligned} \tag{1.62}$$

where D_{ij} is binary diffusivity from species i to species j. Equation (1.62) was originally suggested by Maxwell for a binary mixture based on kinetic theory and extended to diffusion of gaseous mixture of N species by Stefan. For an N-component system, $N(N-1)/2$ diffusivities are required. The diffusion in a multicomponent system is different from diffusion in a binary system because the movement of the ith species is no longer proportional to the negative concentration gradient of the ith species. It is possible that (1) a species moves against its own concentration gradient, referred to as reverse diffusion; (2) a species can diffuse even when its concentration gradient is zero, referred to as osmotic diffusion; or (3) a species does not diffuse although its concentration gradient is favorable to such diffusion, referred to as diffusion barrier (Bird et al. 2006).

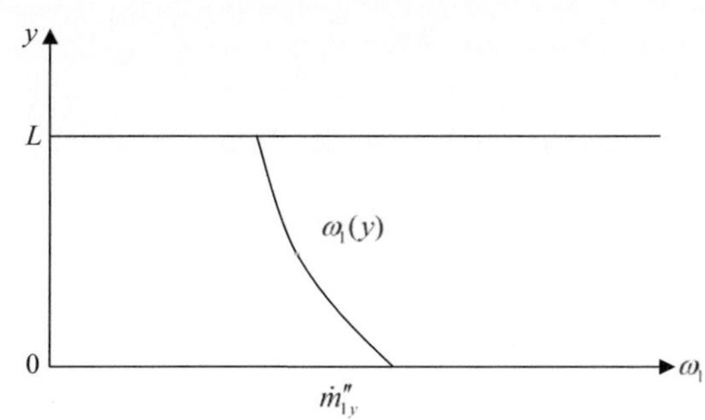

Fig. 1.8 One-dimensional mass diffusion

The Maxwell–Stefan relation can also be rewritten in terms of mass fraction and mass flux

$$\nabla \omega_i + \omega_i \nabla(\ln \overline{M}) = \frac{\overline{M}}{\rho} \sum_{j=1(j\neq i)}^{N} \frac{\omega_i \mathbf{J}_j - \omega_j \mathbf{J}_i}{D_{ij} M_j}, \quad i = 1, 2, \ldots, N-1 \tag{1.63}$$

The diffusive mass fluxes for a system that contains N components can be obtained by solving a set of $N - 1$ Eqs. (1.63) and Eq. (1.54) (Kleijn et al. 1989)

$$\begin{aligned} \mathbf{J}_i = & -\rho D_{im} \nabla \omega_i - \rho \omega_i D_{im} \nabla(\ln \overline{M}) \\ & + \overline{M} \omega_i D_{im} \sum_{j=1(j\neq i)}^{N} \frac{\mathbf{J}_j}{M_j D_{ij}}, \quad i = 1, 2, \ldots, N \end{aligned} \tag{1.64}$$

where the effective mass diffusivity from species i to the mixture D_{im} is

$$D_{\mathrm{im}} = \left(\sum_{j=1(j\neq i)}^{N} \frac{\overline{M} \omega_j}{M_j D_{ij}} \right)^{-1}, \quad i = 1, 2, \ldots, N \tag{1.65}$$

It is noted that Eq. (1.62) is valid for a system that is uniform in all regards except for mass concentration. For a multicomponent system with concentration, temperature and pressure gradients, the mass flux can be expressed as a linear combination of concentration gradients—referred to as generalized Fick's law—or a linear combination of mass fluxes—referred to as Maxwell's expression.

The second mode of mass transfer, convective mass transfer, may be expressed in a manner analogous to Eq. (1.26):

$$\dot{m}'' = h_{\mathrm{m}}(\rho_{1,\mathrm{w}} - \rho_{1,\infty}) = \rho h_m(\omega_{1,\mathrm{w}} - \omega_{1,\infty}) \tag{1.66}$$

where the species mass flux, $\dot{m}''$, is again exemplified by transport from a flat surface to a vapor stream flowing over that surface. The term h_{m} (m/s) in Eq. (1.66) is the convection mass transfer coefficient, $\rho_{1,\mathrm{w}}$ is the species mass concentration at the surface, $\rho_{1,\infty}$ is the species mass concentration in the free stream, and $\omega_{1,\mathrm{w}}$ and $\omega_{1,\infty}$ are the species mass fractions at the surface and in the free stream, respectively.

As is the case for the convective heat transfer coefficient, the convective mass transfer coefficient is a function of fluid properties, the flow field characteristics, and the geometric configuration. The results are frequently expressed in a dimensionless form that also reflects the analogy between heat and mass transfer, the Sherwood number:

$$Sh = \frac{h_{\mathrm{m}} L}{D_{12}} \tag{1.67}$$

Equations (1.11) and (1.21) can be rewritten in the forms analogous to Eq. (1.61), i.e.,

$$\tau_{yx} = -\nu \frac{\partial(\rho u)}{\partial y} \tag{1.68}$$

Table 1.5 Summary of fundamental laws in transport phenomena

Flux	Equations	Requirements	Comments/assumptions
Momentum	Newton's law of viscosity, Eqs. (1.15) and (1.12) $\boldsymbol{\tau}' = -p\mathbf{I} + \mu\left[\nabla\mathbf{V} + (\nabla\mathbf{V})^{\mathrm{T}}\right] - \frac{2}{3}\mu(\nabla\cdot\mathbf{V})\mathbf{I}$	Newtonian fluid Laminar	Single or multicomponent
	Simplified Newton's law of viscosity $\boldsymbol{\tau}' = -p\mathbf{I} + \mu\left[\nabla\mathbf{V} + (\nabla\mathbf{V})^{\mathrm{T}}\right]$	Newtonian fluid Laminar	Single or multicomponent Incompressible
Energy	Fourier's law, Eq. (1.22) $\mathbf{q}'' = -k\nabla T$	Isotropic materials	Conductivity is a scalar
	Fourier's law, Eq. (1.25) $\mathbf{q}'' = -\mathbf{k}\cdot\nabla T$	Anisotropic materials	Conductivity is a tensor of the second order
Mass	Fick's law, Eqs. (1.57) and (1.58) $\mathbf{J}_1 = -\rho D_{12}\nabla\omega_1$ $\mathbf{J}_1^* = -cD_{12}\nabla x_1$	Binary only	No temperature and pressure gradients. Same body force for both components
	Maxwell–Stefan equation (1.62) $\nabla x_i = -\sum_{j=1(j\neq i)}^{N}\frac{x_i x_j}{D_{ij}}(\mathbf{V}_i - \mathbf{V}_j)$ $= -\sum_{j=1(j\neq i)}^{N}\frac{1}{cD_{ij}}(x_j\dot{\mathbf{n}}''_i - x_i\dot{\mathbf{n}}''_j)$	Multi-component gaseous mixture	No temperature and pressure gradients. Same body forces for all N components
	Maxwell–Stefan equation (1.63) $\nabla\omega_i + \omega_i\nabla(\ln\overline{M})$ $= \frac{\overline{M}}{\rho}\sum_{j=1(j\neq i)}^{N}\frac{\omega_i\mathbf{J}_j - \omega_j\mathbf{J}_i}{D_{ij}M_j}$	Multi-component gaseous mixture	No temperature and pressure gradients. Same body forces for all N components

$$q''_y = -\alpha\frac{\partial(\rho c_{\mathrm{p}}T)}{\partial y} \tag{1.69}$$

While Eq. (1.61) states that mass transfer occurs due to a concentration (ρ_1) gradient, momentum, and heat transfer are due to momentum (ρu) and energy $(\rho c_{\mathrm{p}}T)$ gradients, as indicated in Eqs. (1.68) and (1.69). Therefore, momentum, heat, and mass transfer are analogous to each other. This analogous relationship can be used to predict one transport phenomenon on the basis of knowledge of another transport phenomenon. For example, the empirical correlation of turbulent heat transfer can be obtained by applying correlations of friction through the Reynolds analogy (Bergman and Lavine 2017).

Various fluxes in terms of the transport properties in multicomponent systems are summarized in Table 1.5, which will be used as constitutive equations to obtain generalized governing equations in Chap. 3. The transport properties are presented in Appendix B. Various empirical equations for transport properties of gases and liquids can also be found in Poling et al. (2000).

1.4.3 Scaling

Multiphase heat transfer is, like all engineering disciplines, quantitative in nature. We seek to define, for example, the amount of energy that can be transferred by a given heat pipe design in order to determine its suitability for a particular application. Or we seek to restrict the dimensions of a certain flat-plate heat exchanger in order to maintain a stable flow in a system. The student or practicing engineer is well advised to pursue such "bottom-line" answers deliberately and systematically so that all of the physics relevant to a given problem can be identified and prioritized early in the problem-solving process. If attention has not been first paid to the "big picture," plunging hastily into "crunching the numbers" is counterproductive. Two highly recommended analytical tools can accomplish the critical task of organizing and filtering information so the design process can be efficiently and effectively executed. These important tools are dimensional analysis and scale analysis, and they are applicable to both single- and multiphase systems.

Dimensional analysis is based on the Buckingham's Π theorem (Buckingham 1914), which stated that the number of dimensionless variables required to describe the problem equals the number of dimensional variables, less the number of primary dimensions required to describe the problem. Dimensional analysis is described in detail by Welty et al. (2014). Table 1.6 provides a summary of the definitions, physical interpretations, and areas of significance of the important dimensionless numbers for transport phenomena in multiphase systems. At reduced-length scale, the effects of gravitational and inertial forces become less important while the surface tension plays a dominant role (Eijkel and van den Berg 2005).

Table 1.6 Summary of dimensionless numbers for transport phenomena in multiphase systems

Name	Symbol	Definition	Physical interpretation	Area of significance
Bond number	Bo	$(\rho_\ell - \rho_v)gL^2/\sigma$	Buoyancy force/surface tension	Boiling and condensation
Brinkman number	Br	$\mu U^2/(k\Delta T)$	Viscous dissipation/enthalpy change	High-speed flow
Capillary number	Ca	$\mu U/\sigma$	We/Re	Two-phase flow
Eckert number	Ec	$U^2/(c_p\Delta T)$	Kinetic energy/enthalpy change	High-speed flow
Fourier number	Fo	$\alpha t/L^2$	Dimensionless time	Transient problems
Froude number	Fr	$U^2/(gL)$	Interfacial force/gravitational force	Flow with a free surface
Grashof number	Gr	$g\beta\Delta TL^3/\nu^2$	Buoyancy force/viscous force	Natural convection
Jakob number	Ja_ℓ	$c_{p\ell}\Delta T/h_{\ell v}$	Sensible heat/latent heat	Film condensation and boiling
	Ja_v	$c_{pv}\Delta T/h_{\ell v}$		
Kapitza number	Ka	$\mu_\ell^4 g/[(\rho_\ell - \rho_v)\sigma^3]$	Surface tension/viscous force	Wave on liquid film
Knudsen number	Kn	λ/L	Mean free path/characteristic length	Noncontinuum flow
Mach number	Ma	U/c	Velocity/speed of sound	Compressible flow
Nusselt number	Nu	hL/k	Thermal resistance of conduction/thermal resistance of convection	Single and multiphase convective heat transfer
Peclet number	Pe	UL/α	$RePr$	Forced convection
Prandtl number	Pr	ν/α	Rate of diffusion of viscous effect/rate of diffusion of heat	Single and multiphase convection
Rayleigh number	Ra	$g\beta\Delta TL^3/(\nu\alpha)$	$GrPr$	Natural convection
Reynolds number	Re	UL/ν	Inertial force/viscous force	Forced convection
Schmidt number	Sc	ν/D	Rate of diffusion of viscous effect/rate of diffusion of mass	Convective mass transfer

(continued)

Table 1.6 (continued)

Name	Symbol	Definition	Physical interpretation	Area of significance
Sherwood number	*Sh*	$h_m L/D_{12}$	Resistance of diffusion/resistance of convection	Convective mass transfer
Stanton number	*St*	$h/(\rho c_p U)$	*Nu/(Re Pr)*	Forced convection
Stanton number (mass transfer)	St_m	h_m/U	*Sh/(Re Sc)*	Mass transfer
Stefan number	*Ste*	$c_p \Delta T/h_{s\ell}$	Sensible heat/latent heat	Melting and solidification
Stokes number	*Sk*	t_v/t_F	Characteristic time of a particle to a characteristic time of the flow	Particle flow
Weber number	*We*	$\rho U^2 L/\sigma$	Inertial force/surface tension force	Liquid–vapor phase change

The convective heat transfer coefficient can be obtained analytically, numerically, or experimentally, and the results are often expressed in terms of the Nusselt number. The heat transfer coefficient depends on surface geometry, the driving force of the fluid motion, and thermal properties of the fluid, as well as flow properties. The existing heat transfer correlations in literature for various heat transfer modes for both single-phase and two-phase systems in conventional geometric configurations are summarized in Appendix D.

Scaling, or scale analysis, is a process that uses the basic principles of heat transfer (or other engineering disciplines) to provide order of magnitude estimates for quantities of interest. For example, scale analysis of a boundary-layer type flow can provide the order of magnitude of the boundary-layer thickness. In addition, scale analysis can provide the order of magnitude of the heat transfer coefficient or Nusselt number, as well as the form of the functions that describe these quantities. Scale analysis confers remarkable capability because its result is within percentage points of the results produced by the exact solution (Bejan 2013).

Scaling will be demonstrated by analyzing a heat conduction problem and a contact melting problem. The first example is a scale analysis of a thermal penetration depth for conduction in a semi-infinite solid as shown in Fig. 1.9. The initial temperature of the semi-infinite body is T_i. At $t = 0$, the surface temperature is suddenly increased to T_0. At a given time t, the thermal penetration depth is δ, beyond which the temperature of the solid is not affected by the surface temperature, i.e., the temperature satisfies the following two conditions at the thermal penetration depth.

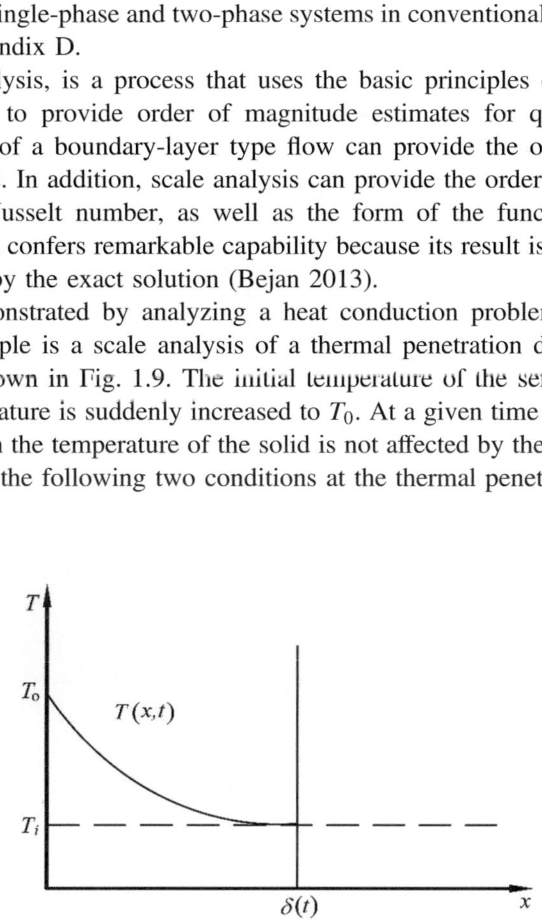

Fig. 1.9 Thermal penetration depth for conduction in a semi-infinite solid

$$T(\delta, t) = T_{\mathrm{i}} \tag{1.70}$$

$$\left.\frac{\partial T}{\partial x}\right|_{x=\delta} = 0 \tag{1.71}$$

The energy equation for this problem and the corresponding initial and boundary conditions are:

$$\frac{\partial^2 T}{\partial x^2} = \frac{1}{\alpha}\frac{\partial T}{\partial t} \quad x > 0,\ t > 0 \tag{1.72}$$

$$T(x, t) = T_0 \quad x = 0,\ t > 0$$

$$T(x, t) = T_i \quad x > 0,\ t = 0 \tag{1.74}$$

Since temperature difference occurs only within $0 \le x < \infty$, the order of magnitude of x is the same as δ, i.e.,

$$x \sim \delta \tag{1.75}$$

The order of magnitude of the term on the left-hand side of Eq. (1.72) is

$$\frac{\partial^2 T}{\partial x^2} = \frac{\partial}{\partial x}\left(\frac{\partial T}{\partial x}\right) \sim \frac{1}{\delta}\frac{\Delta T}{\delta} = \frac{\Delta T}{\delta^2} \tag{1.76}$$

where $\Delta T = T_0 - T_{\mathrm{i}}$.

The order of magnitude of the right-hand side of Eq. (1.72) is

$$\frac{1}{\alpha}\frac{\partial T}{\partial t} \sim \frac{1}{\alpha}\frac{\Delta T}{t} \tag{1.77}$$

Equation (1.72) requires that the two orders of magnitude represented by Eqs. (1.76) and (1.77) equal each other, i.e.,

$$\frac{\Delta T}{\delta^2} \sim \frac{1}{\alpha}\frac{\Delta T}{t} \tag{1.78}$$

The order of magnitude of the thermal penetration depth is then

$$\delta \sim \sqrt{\alpha t} \tag{1.79}$$

Equation (1.79) indicates that the thermal penetration depth is proportional to the square root of time, which agrees with the results obtained by the integral approximate solution (see Chap. 5). This simple example demonstrates the additional capability of scale analysis over dimensional analysis, which could not provide the form of the functions.

Scale analysis can also be used to analyze heat transfer problems with phase change. Figure 1.10 shows a schematic of a contact melting problem in which a solid phase-change material (PCM) at its melting point, T_{m}, sits on top of a heating surface at temperature $T_{\mathrm{w}}(T_{\mathrm{w}} > T_{\mathrm{m}})$. The width of the PCM is L. Melting occurs at the contact area between the PCM and heating surface. The liquid PCM produced by melting is in the form of a thin layer since gravitational force acts on the solid PCM. The solid PCM melts as it contacts the heating surface, and thus, the entire solid moves downward at a velocity of V_{s}.

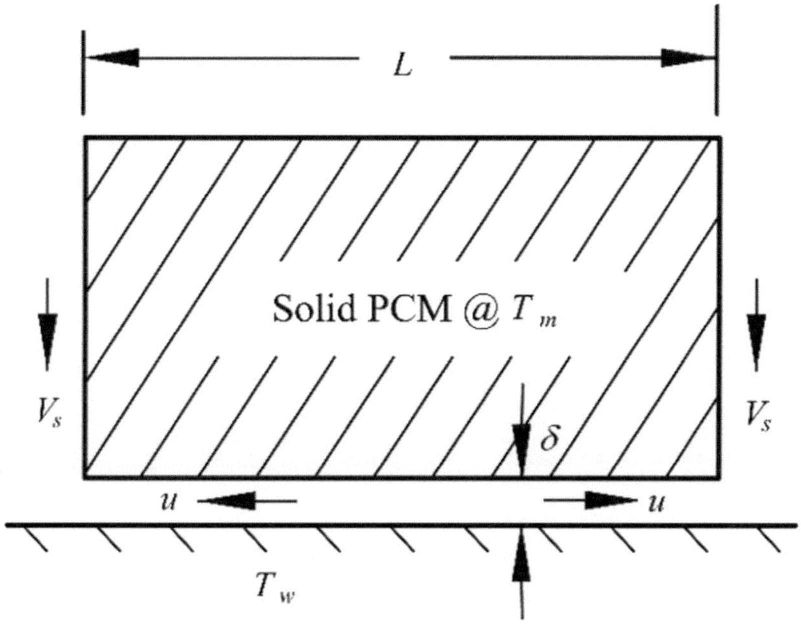

Fig. 1.10 Contact melting

The continuity and momentum equations of the liquid phase are

$$\frac{\partial u}{\partial x} + \frac{\partial v}{\partial y} = 0 \quad (1.80)$$

$$0 = -\frac{\partial p}{\partial x} + \mu \frac{\partial^2 u}{\partial y^2} \quad (1.81)$$

where the inertia terms in Eq. (1.81) have been neglected because the liquid velocity is very low.

The orders of magnitude of the two terms in Eq. (1.80) must be the same, i.e.,

$$\frac{u}{L} \sim \frac{V_s}{\delta} \quad (1.82)$$

where V_s and δ are solid downward velocity and the thickness of the liquid film underneath the solid, respectively.

Similarly, the following relationship can be obtained by equating orders of magnitude of two terms in Eq. (1.81):

$$\frac{p}{L} \sim \mu \frac{u}{\delta^2} \quad (1.83)$$

Supposing that the heat transfer in the thin liquid film is due only to conduction, then the energy balance at the solid–liquid interface gives us

$$-k_\ell \frac{\partial T}{\partial y} = \rho_s h_{s\ell} V_s \tag{1.84}$$

The scale analysis of Eq. (1.84) yields

$$k_\ell \frac{\Delta T}{\delta} \sim \rho_s h_{s\ell} V_s \tag{1.85}$$

where $\Delta T = T_w - T_m$.

Combining Eqs. (1.82) and (1.83) yields

$$\delta \sim \left(\frac{\mu V_s L^2}{\Delta p} \right)^{1/3} \tag{1.86}$$

Substituting Eq. (1.86) into Eq. (1.85), the solid PCM velocity is obtained:

$$V_s \sim \left(\frac{k_\ell \Delta T}{\rho_s h_{s\ell}} \right)^{3/4} \left(\frac{p}{\mu L^2} \right)^{1/4} \tag{1.87}$$

Equation (1.87) can be nondimensionalized as

$$\frac{V_s L}{\alpha} \sim \left(\frac{\rho}{\rho_s} Ste \right)^{3/4} \left(\frac{pL^2}{\mu \alpha} \right)^{1/4} \tag{1.88}$$

where the left-hand side is the Peclet number based on L.

The order of magnitude of the pressure in the liquid layer is defined as the net weight of the solid PCM over the horizontal projected area of the solid PCM. The horizontal projected length of the PCM is L in this example. The order of magnitude of the downward velocity obtained from scaling, Eq. (1.87), has the same form as the results obtained by analytical solution as will be shown in Chap. 5. In a similar analysis for contact melting of a PCM encapsulated in a circular tube, the horizontal projected length of the PCM has the same order of magnitude as the diameter of the circular tube.

1.5 Multiphase Notations and Concepts

1.5.1 Continuous and Dispersed Phases

The dispersed phase is the particles, droplets, and/or bubbles that are present in multiphase flow. The continuous phase is the phase that the particles, droplets, or bubbles are dispersed in.

1.5.2 Densities and Volume Fractions

The volume fraction of a phase is the volume of that particular phase present in the entire volume. Unlike density in a continuum flow, volume fraction cannot be defined at a point. Volume fraction is defined as:

$$\varepsilon_k = \lim_{V \to V^o} (V_k/V) \tag{1.89}$$

where ε is the volume fraction, k is the phase index, V is the volume, V_k is the volume of phase k present in volume V, and V^o is the limiting volume that ensures a stationary average.

The volume fraction of the continuous phase is sometimes referred to as the void fraction, and the volume fraction of the dispersed phase is sometimes referred to as holdup. The sum of the volume fractions must be equal to 1. Therefore,

$$\sum_{k=1}^{\Pi} \varepsilon_k = 1 \tag{1.90}$$

where Π is the number of phases present. If the phases are only the dispersed phase and the continuous phase represented with subscripts d and c, respectively; then,

$$\varepsilon_{\mathrm{d}} + \varepsilon_{\mathrm{c}} = 1 \tag{1.91}$$

The apparent density (or bulk density) of a phase is the mass of that particular phase per unit volume of the multiphase mixture. Apparent density is defined as:

$$\overline{\rho_k} = \lim_{V \to V^o} (m_k/V) = \varepsilon_k \rho_k \tag{1.92}$$

where $\overline{\rho_k}$ is the apparent density of phase k in volume V, m_k is the mass of phase k present in volume V, and ρ is material density.

The apparent density of a phase is also equal to the volume fraction of the phase multiplied by the material density. The sum of the apparent densities is equal to the mixture density. Therefore,

$$\sum_{k=1}^{\Pi} \varepsilon_k \rho_k = \rho_{\mathrm{m}} \tag{1.93}$$

where ρ_{m} is the mixture density. In the discussion of dispersed and continuous phases,

$$\overline{\rho}_{\mathrm{d}} + \overline{\rho}_{\mathrm{c}} = \rho_{\mathrm{m}} \tag{1.94}$$

1.5.3 Superficial and Phase Velocities

Superficial velocity is the velocity of a phase if that phase fills the entire flow area. For multiphase flow in a pipe, the superficial velocity of each phase is the mass flow rate of that phase divided by the cross-sectional area of the pipe and material density of that phase:

$$\mathbf{j}_k = \dot{\mathbf{m}}_k/(\rho_k A) \tag{1.95}$$

where $\mathbf{j}_k$ is the superficial velocity of phase k, and A is the cross-sectional area of the pipe. The phase velocity is the actual velocity of the phase; thus,

$$\mathbf{v}_k = \mathbf{j}_k / \varepsilon_k \tag{1.96}$$

where $\mathbf{v}_k$ is the phase velocity of phase k.

1.5.4 Quality and Concentration

The concentration of a multiphase flow is the apparent density ratio of the dispersed phase to the continuous phase in the multiphase mixture. The dispersed-phase mass concentration is:

$$C = \overline{\rho}_{\mathrm{d}} / \overline{\rho}_{\mathrm{c}} \tag{1.97}$$

The quality of a liquid–vapor mixture (where the liquid is the dispersed phase) is the ratio of apparent density of the dispersed phase divided by the mixture density:

$$x = \overline{\rho}_{\mathrm{d}} / \rho_{\mathrm{m}} \tag{1.98}$$

1.5.5 Response Times

The response time of a particle or droplet is the time it takes to respond to changes in flow velocity or temperature (Michaelides et al. 2016). The momentum response time is the time it takes for a particle or droplet to respond to a change in velocity. More specifically, the momentum response time is the time required for a particle, after being released from rest, to reach 63% of the free-stream velocity. For low Reynolds numbers, momentum response time is defined as:

$$t_{\mathrm{v}} = \left(\rho_{\mathrm{d}} D_{\mathrm{d}}^2\right) / 18\mu_{\mathrm{c}} \tag{1.99}$$

where t_{v} is the momentum response time (particle relaxation time), D is the pipe diameter, and μ is viscosity. Therefore, the change in velocity over change in time is

$$\frac{\mathrm{d}u_{\mathrm{p}}}{\mathrm{d}t} = (1/t_{\mathrm{v}})\left(u_{\mathrm{g}} - u_{\mathrm{p}}\right) \tag{1.100}$$

where u_{p} is the particle velocity, and u_{g} is the gas velocity.

Thermal response time is the time it takes for a particle or droplet to respond to a temperature change of the carrier fluid. More specifically, thermal response time is the time required for the particle to reach 63% of the initial temperature difference between the dispersed and carrier phase (Michaelides et al. 2016). Assuming the temperature throughout a particle is uniform and radiative effects are negligible, thermal response time for low Reynolds number flows is:

$$t_{\mathrm{T}} = \left(\rho_{\mathrm{d}} c_{\mathrm{d}} D^2\right) / 12k_{\mathrm{c}} \tag{1.101}$$

where c_{d} is the specific heat of the dispersed phase, and k_{c} is the thermal conductivity of the continuous phase. Therefore, the change in temperature of the dispersed phase is

$$\frac{dT_d}{dt} = \frac{1}{t_T}(T_c - T_d) \tag{1.102}$$

1.5.6 Stokes Number

The Stokes number is a dimensionless number related to the particle velocity, representing the time that a particle has to respond to changes in flow velocity, and is defined as:

$$Sk = t_v / t_F \tag{1.103}$$

where t_F is the characteristic time of the flow field, $t_F = L_s/V_s$, and L_s and V_s are the characteristic length and characteristic velocity, respectively.

If $Sk \ll 1$, the response time of the particles in the flow is much less than the characteristic time. Therefore, the particles will have enough time to respond to changes in flow velocity and the velocities of the particles and fluid will be nearly equal. If $Sk \gg 1$, the particles will not have enough time to respond to the velocity change of the fluid; furthermore, the particle velocity will not be noticeably affected by the change in the fluid velocity.

1.5.7 Dispersed Versus Dense Flows

A dispersed flow is a flow in which the motion of the particles is controlled by fluid forces (drag and lift). A dense flow is one where the motion of the particles is controlled by collisions. The ratio of the momentum response time and average time between particle–particle collisions can be used to determine whether the flow is mainly dense or dispersed, or whether the particle motion is dominated by fluid forces or collisions. The flow is dispersed; therefore, particle motion is dominated by fluid forces if

$$\left(t_v / t_{pp}\right) < 1 \tag{1.104}$$

where t_{pp} is the average time between particle–particle collisions (Michaelides et al. 2016).

This means that the particles have enough time to respond to fluid forces before the next collision. However, if

$$\left(t_v / t_{pp}\right) > 1 \tag{1.105}$$

the flow is dense, and the particles do not have enough time to respond to fluid forces before the next collision.

Problems

1.1. Why can severe skin burns be caused by hot steam?

1.2. When 0.01 kg of ice at 0 °C is mixed with 0.1 kg of steam at 100 °C and 1 atm, what is the phase of the final mixture? What is the final temperature of the final mixture?

1.3. An ice skater moving at 8 m/s glides to a stop. The ice in immediate contact with the skates absorbs the heat generated by friction and melts. If the temperature of the ice is 0 °C and the weight of the ice skater is 60 kg, how much ice melts?

1.4. A lead bullet with 30-g mass traveling at 600 m/s hits a thin iron wall and emerges at a speed of 300 m/s. Suppose 50% of the heat generated is absorbed by the bullet, and the initial temperature of the bullet is 20 °C. Find the final phase and temperature of the lead bullet after the impact. The specific heat of the liquid lead can be assumed to be the same as that of the solid lead.

1.5. A rigid tank filled with a mixture of liquid and vapor in equilibrium (Fig. P1.5) is heated at a rate of q (kW) and evaporation takes place in the rigid tank. To maintain a constant pressure during evaporation, a relief valve at the top of the tank is opened and the mass flow rate of vapor through the valve is $\dot{m}$ (kg/s). Show that the heating rate q is related to the mass flow rate $\dot{m}$ and properties of liquid and vapor as follows:

$$q = \dot{m}\left[h_{\ell v} + \left(h_\ell - \frac{v_v e_\ell - v_\ell e_v}{v_v - v_\ell}\right)\right] \tag{1.106}$$

where v_v, v_ℓ, e_v and e_ℓ are specific volumes and internal energies of vapor and liquid, respectively.

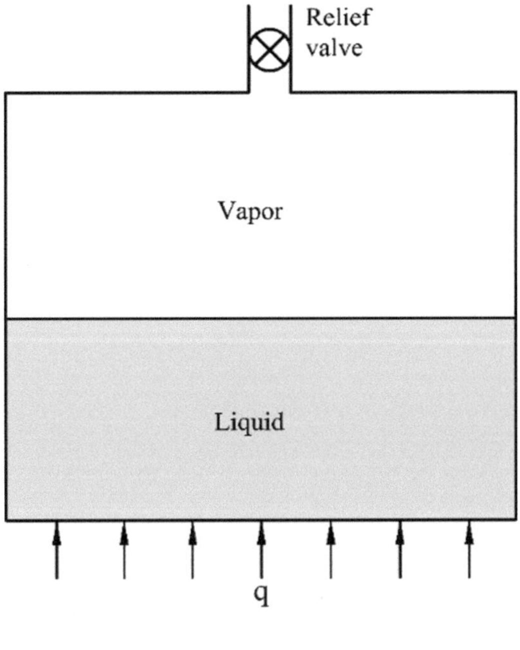

Fig. P 1.5

1.6. Assume an ideal two-component gas mixture A and B, and develop a relation for mole and mass fraction for A and B in terms of partial pressure and total pressure for the mixture.

1.7. Verify that the following equation is valid for a binary system:

$$\frac{\mathbf{J}_1}{\rho\omega_1\omega_2} = \frac{\mathbf{J}_1^*}{cx_1x_2}$$

1.8. Prove that the sum of mass fluxes by diffusion for a multicomponent mixture is zero.

$$\sum_{i=1}^{N} \mathbf{J}_i = 0$$

Show that the sum of mass flux relative to stationary coordinate axes for multicomponent mixtures is different from zero.

1.9. Show that the Maxwell–Stefan equation (1.62) for multicomponent system can be reduced to Fick's law, Eq. (1.58), for a binary system.

1.10. Develop the mass flux relationship for component A, leaving from a solid or liquid wall made of component A only to a binary (Fig. P1.10) mixture of gases A and B in terms of (a) mass fractions and mass density of component A, and (b) molar fractions and molar concentration of component A. Describe the assumptions in arriving at your final result for each case.

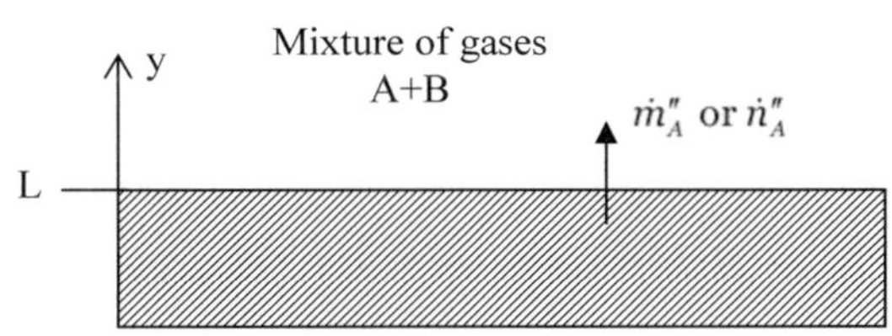

Fig. P1.10

1.11. Air with a velocity of 10 m/s, temperature 22 °C at 50% relative humidity, flows over a swimming pool (40 m by 20 m) along its length direction. The water temperature in the pool is 32 °C. Calculate the average heat loss due to (a) sensible heat using Appendix D and (b) latent heat of vaporization.

1.12. A full water container with a 1 m^2 surface area is maintained at 60 °C with surrounding air at 30 °C and 20% relative humidity. Calculate the convection and evaporation heat losses.

1.13. Determine the thickness of an ice layer on a large rectangular swimming pool of 7 × 10 m^2 in winter when air blows along the length direction over the pool at 16 km/h and −23 °C. Assume the pool temperature is 5 °C and the inner ice surface temperature is at 0 °C. The convection heat transfer coefficient at the bottom of the ice is 50 W/m^2 K. Use Appendix D to calculate the convective heat transfer coefficient on the top of the ice layer. What is the surface temperature on the outer layer of ice?

1.14. A water droplet is placed in dry stagnant air of uniform temperature, T_∞. Show that

(a) $\frac{k_{\text{air}}}{h_{\ell v}\rho_{\text{air}}} Nu(T_\infty - T_{\text{sat}}) = D_{12} Sh \frac{\omega_{1\delta} - \omega_{1\infty}}{1 - \omega_{1\delta}}$

(b) $c_{\text{p,air}} \frac{(T_\infty - T_{\text{sat}})}{h_{\ell v}} = Le \frac{\omega_{1\delta} - \omega_{1\infty}}{1 - \omega_{1\delta}}$

What are the assumptions made to produce the results found in (a) and (b)?

1.15. Compare the orders of magnitude of the Prandtl, Schmidt, and Lewis numbers for various gases and liquids using data from Appendix B. Can any conclusions be made?

1.16. The convective heat transfer coefficient is a function of the thermal properties of the fluid, the geometric configuration, flow velocities, and driving forces. As a simple example, let us consider forced convection in a circular tube with a length L and a diameter D. The flow is assumed to be incompressible and natural convection is negligible compared with forced convection. The heat transfer coefficient can be expressed as

$$h = h(k, \mu, c_\text{p}, \rho, U, \Delta T, D, L) \quad (1.107)$$

where k is the thermal conductivity of the fluid, μ is viscosity, c_p is specific heat, ρ is density, U is velocity, and ΔT is the temperature difference between the fluid and tube wall. Equation (1.107) can also be rewritten as

$$F(h, k, \mu, c_\text{p}, \rho, U, \Delta T, D, L) = 0 \quad (1.108)$$

Using Buckingham's Π theorem show that the number of nondimensional variables is 4, as opposed to the 9 dimensional variables in Eq. (1.108). Furthermore, use dimensional analysis to show that

$$Nu = f(Re, Pr, L/D) \quad (1.109)$$

1.17. In Problem 1.16, if the flow is assumed to be compressible and natural convection is not negligible compared with forced convection, the heat transfer coefficient can be expressed as

$$h = h(k, \mu, c_\text{p}, \rho, U, g, \beta, \Delta T, D, L, c) \quad (1.110)$$

where β is thermal expansion coefficient, and c is local speed of sound. Equation (1.110) can also be rewritten as

$$F(h, k, \mu, c_\text{p}, \rho, U, g, \beta, \Delta T, D, L, c) = 0 \quad (1.111)$$

Using Buckingham's Π theorem and dimensional analysis to show that

$$Nu = f(Re, Gr, Pr, L/D, Ma, Fr) \quad (1.112)$$

where Gr, Ma, and Fr are the Grashof number, Mach number, and Froude number, respectively, i.e.,

$$Gr = \frac{\rho^2 g \beta \Delta T D^3}{\mu^2}, \quad Ma = \frac{U}{c}, \quad Fr = \frac{U^2}{gL} \quad (1.113)$$

1.18. Estimate the order of magnitude of the thickness of the momentum boundary layer for forced convection flow over a flat plate as shown in Fig. P1.18 using scale analysis.

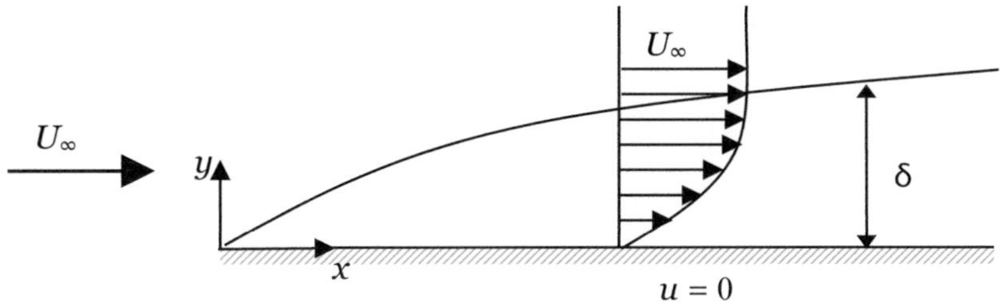

Fig. P1.18

1.19. The skin friction coefficient of Problem 1.18 defined as $C_f = \tau/(\rho U_\infty^2/2)$, where τ is the shear stress at the surface of the flat plate. Estimate the order of magnitude of the friction coefficient using scale analysis.

1.20. A solid PCM with a uniform initial temperature at the melting point, T_m, is in a half space, $x > 0$. At time $t = 0$, the temperature at the boundary $x = 0$ is suddenly increased to a temperature, T_0, which is above the melting point of the PCM. Perform a scale analysis to obtain the order of magnitude of the location of the melting front.

1.21. Perform an Internet search with the phrases "multiphase heat transfer" as well as "multiphase flow" and write a brief summary of what you find and the differences.

1.22. Plasma is regarded as a state of matter in addition to gas, liquid, and solid. Use the Internet to find information about plasmas. How do you compare the properties of plasma with those of a gas, liquid and solid?

1.23. The temperature and dew point of a city in summer are 80 and 40 °F, respectively. What is the relative humidity?

1.24. A two-phase heat sink is a very effective device for electronic cooling. What are the advantages and disadvantages of a two-phase heat sink in comparison with a single-phase heat sink?

1.25. Give two examples of two-phase flow from everyday life and industrial applications, and briefly discuss the physical phenomena involved in your examples.

1.26. Laser welding is a process that uses a laser beam to join two metal workpieces together. Analyze the phase changes involved in the laser welding process. You can use the internet to find related information.

1.27. The working fluid for a high-temperature heat pipe is liquid metal, which is solid at room temperature. Qualitatively analyze the phase-change phenomena involved during the start-up and shutdown of the high-temperature heat pipes.

References

ANSYS Fluent Theory Guide. (2017). ANSYS, Inc.

Bejan, A. (2013). *Convection heat transfer* (4th ed.). New York: Wiley.

Bergman, T. L., & Lavine, A. S. (2017). *Fundamentals of heat and mass transfer* (8th ed.). New York: Wiley.

Bird, R. B., Stewart, W. E., & Lightfoot, E. N. (2006). *Transport phenomena* (Revised 2nd ed.). New York: Wiley.

Buckingham, E. (1914). On physically similar systems: Illustrations of the use of dimensional equations. *Physical Review, 4,* 345–376.

Eijkel, J. C. T., & van den Berg, A. (2005). Nanofluidics: What is it and what can we expect from it? *Microfluidics and Nanofluidics, 1*(3), 249–267.

Faghri, A. (2016). *Heat pipe science and technology* (2nd ed.). Columbia, MO: Global Digital Press.

Faghri, A., Zhang, Y., & Howell, J. R. (2010). *Advanced heat and mass transfer*. Columbia, MO: Global Digital Press.

Howell, J., Menguc, M. P., & Siegel, R. (2015). *Thermal radiation heat transfer* (6th ed.). Boca Raton, FL: CRC Press.

Kays, W. M., Crawford, M. E., & Weigand, B. (2004). *Convective heat transfer* (4th ed.). New York, NY: McGraw-Hill.

Kleijn, C. R., van Der Meer, H., & Hoogendoorn, C. J. (1989). A mathematical model for LPCVD in a single wafer reactor. *Journal of the Electrochemical Society, 136,* 3423–3432.

Kulacki, F. A. (2018). *Handbook of thermal science and engineering*. New York, NY: Springer International Publishing.

Michaelides, E., Crowe, C. T., & Schwarzkopf, J. D. (2016). *Multiphase flow handbook* (2nd ed.). Boca Raton: CRC Press.

Poling, B. E., Prausnitz, J. M., & O'Connell, J. P. (2000). *The properties of gases and liquids* (5th ed.). New York, NY: McGraw-Hill.

Serizawa, A. (2010). Dispersed flow. In *Thermopedia, Begell House*, http://www.thermopedia.com/content/5/. Accessed May 15, 2018.

Welty, J. R., Rorrer, G. L., & Foster, D. G. (2014). *Fundamentals of momentum, heat and mass transfer* (6th ed.). New York, NY: Wiley.

White, F. M. (2005). *Viscous fluid flow* (3rd ed.). New York, NY: McGraw-Hill.

Thermodynamics of Multiphase Systems

2

2.1 Introduction

The primary objective of this chapter is to define the concept of thermodynamic equilibrium and to describe the conditions under which equilibrium exists in thermodynamic systems. The system considered may consist of either a single phase or multiple phases and may include one or more components. In its simplest form, a system is said to be in equilibrium when its measurable properties do not change over time. It must be clearly pointed out that steady-state open systems that exchange mass, heat, or work with the surroundings do not meet this criterion and therefore are not in equilibrium. There are no unbalanced driving potentials within a system in the thermodynamic equilibrium state, and the system in equilibrium experiences no change when it is isolated from its surroundings. The system is said to be at equilibrium if conditions for all types of thermodynamic equilibrium are satisfied. The equilibriums that can be encountered in a multiphase system include (1) thermal equilibrium (no change occurs in temperature), (2) mechanical equilibrium (no change occurs in pressure), (3) chemical equilibrium (no change occurs in chemical composition), and (4) phase equilibrium (no phase change occurs).

For a thermodynamic system to be in equilibrium, a mathematical treatment of the combination of the first two laws of thermodynamics must ascertain that certain intensive properties are uniform throughout the system. These intensive properties include the temperature, pressure, and chemical potential. In other words, this uniformity implies no heat, mass, or mechanical work transfer between the system and surroundings. To describe a system that is in equilibrium, a certain number of independent intensive thermodynamic variables must be specified. Another related, but different, concept is *stability*, which characterizes the result of small perturbations to a system in thermodynamic equilibrium. The system in thermodynamic equilibrium is said to be thermodynamically stable if equilibrium can be maintained after a small perturbation. In practice, the equilibrium properties are used in solving engineering problems where mass and energy are continuously exchanged; in these cases, *local* thermodynamic equilibrium is assumed.

Section 2.2 starts with a discussion of the general criteria for equilibrium in a closed system with a single-phase substance. The discussion of equilibrium criteria is then extended to a closed system with a multicomponent single-phase substance; this is followed by a discussion of the thermal, mechanical, and chemical potential stability for a single-phase system.

Section 2.3 extends the discussion of the equilibrium criteria to a system consisting of two phases and a single component, and then develops the Clapeyron equation, which relates the temperature and pressure of a thermodynamic system in the saturation region. In addition, equilibrium criteria for a

A. Faghri and Y. Zhang, *Fundamentals of Multiphase Heat Transfer and Flow*, https://doi.org/10.1007/978-3-030-22137-9_2

closed system with a multicomponent, multiphase substance are presented. Section 2.3 also addresses the question of how far from equilibrium a two-phase system can deviate before the system becomes unstable. This small region, termed the metastable region, is important to engineers because it affects safety. Section 2.4 presents a discussion on the thermodynamic equilibrium conditions at an interface, followed by thermodynamic definitions of surface tension, a discussion of the surface tension effects at microscale vapor bubble/liquid droplets, disjoining pressure, and superheat.

2.2 Fundamentals of Equilibrium and Stability

For a single-component closed system (fixed mass), the first law of thermodynamics gives us (Borgnakke and Sonntag 2017)

$$d\hat{E} = \delta Q - \delta W \tag{2.1}$$

where $\hat{E}$ is the total energy of the closed system, δQ is heat transferred to a system, and δW is the work done by the system to the surroundings. The contribution to the total energy is due to internal (E), kinetic, potential, electromagnetic, surface tension or other forms of energies. If change of all other forms of energies can be neglected, then $\hat{E} = E$. Heat transfer to a system is positive (system receives heat), whereas heat transfer from the system is negative (system loses heat). In contrast, work done by a system is positive (system loses work), and the work done to the system is negative (system receives work). The mechanical work for a closed system is usually expressed as $\delta W = pdV$, where p is the pressure and V is the volume of the system—both are thermodynamic properties of the system.

The second law of thermodynamics for the single-component closed system can be described by the Clausius inequality, i.e.,

$$dS \geq \frac{\delta Q}{T} \tag{2.2}$$

where the equal sign designates a reversible process, which is defined as an ideal process that after taking place can be reversed without leaving any change to either system or surroundings. The greater-than sign denotes an irreversible process. dS is the change of entropy of the closed system.

Combining these general forms of the first two laws of thermodynamics results in an expression that is very useful for determining the conditions for equilibrium and stability of systems, namely the fundamental relation of thermodynamics:

$$dE \leq TdS - \delta W \tag{2.3}$$

where the inequality is used for irreversible processes and the equality for reversible processes. For a finite change in a system, the fundamental thermodynamic relationship becomes

$$\Delta E \leq T\Delta S - W \tag{2.4}$$

where $W = \int_{V_1}^{V_2} pdV$ is the work done by the system to the surroundings.

The equilibrium criteria for a single-phase closed system under the following three sets of constraints are discussed in this section: (1) constant-volume isolated system with no heat or work transfer between the system and its surroundings, (2) constant-volume, constant-temperature system, and (3) constant-pressure, constant-temperature system.

2.2.1 Equilibrium Criteria for Pure Substances

2.2.1.1 Constant-Volume Isolated Systems

For a constant-volume isolated system that exchanges neither heat nor works with its surroundings,

$$Q = 0 \tag{2.5}$$

$$\Delta V = 0 \tag{2.6}$$

$$W = 0 \tag{2.7}$$

It follows from the first law of thermodynamics that a system that has no heat or work interaction with the surroundings also has no change in internal energy. Thus

$$\Delta E = 0 \tag{2.8}$$

Applying Eqs. (2.5)–(2.8) with the Clausius inequality, Eq. (2.4) gives

$$\Delta S_{E,V} \geq 0 \tag{2.9}$$

Equation (2.9) asserts that system entropy always increases for a spontaneous and irreversible finite process occurring in a system with constant internal energy, E, and constant volume V. These spontaneous processes continuously move the system toward an equilibrium state where the entropy will reach a maximum value. When the system reaches an equilibrium state, any infinitesimal change in the system will result in a zero change of entropy, i.e.,

$$\mathrm{d}S_{E,V} = 0 \tag{2.10}$$

2.2.1.2 Constant-Temperature and Volume Systems

Since the temperature and volume of the closed system are constants, we have

$$\Delta T = 0 \tag{2.11}$$

$$\Delta V = 0 \tag{2.12}$$

Assuming that the only work present in this closed system is of type pV, the work exchange between the system and the surroundings must be zero, i.e., $W = 0$. The fundamental thermodynamic relationship, Eq. (2.4), simplifies as

$$\Delta E - T\Delta S \leq 0 \tag{2.13}$$

Recalling the well-known Helmholtz free energy function, F,

$$F = E - TS \tag{2.14}$$

and expanding Eq. (2.14) to define a finite change in the system yields

$$\Delta F = \Delta E - T\Delta S - S\Delta T \tag{2.15}$$

Substituting Eqs. (2.15) and (2.11) into Eq. (2.13), a second equilibrium criterion is obtained:

$$\Delta F_{T,V} \leq 0 \tag{2.16}$$

Therefore, for a closed system at constant temperature and volume, the Helmholtz free energy must decrease with any spontaneous system change and be minimal at equilibrium. At equilibrium conditions, any infinitesimal change from constant-temperature, constant-volume equilibrium must result in zero change in the Helmholtz free energy.

$$\mathrm{d}F_{T,V} = 0 \tag{2.17}$$

2.2.1.3 Constant-Temperature and Pressure Systems

For a closed system with constant temperature and constant pressure,

$$\Delta T = 0 \tag{2.18}$$

$$\Delta p = 0 \tag{2.19}$$

Once again, the goal is to determine the equilibrium criteria for such a system, with the assumption that the only work is of the pV type. The work exchange between the system and its surroundings is $W = p\Delta V$. The fundamental thermodynamic relationship, Eq. (2.4), can be written as

$$\Delta E - T\Delta S + p\Delta V \leq 0 \tag{2.20}$$

To determine a useful equilibrium criterion for such a system, another common thermodynamic property, the Gibbs free energy, is recalled:

$$G = E - TS + pV \tag{2.21}$$

Expanding Eq. (2.21) for a system undergoing a finite system change results in

$$\Delta G = \Delta E - T\Delta S - S\Delta T + p\Delta V + V\Delta p \tag{2.22}$$

Substituting Eqs. (2.22) and (2.18)–(2.19) into Eq. (2.20) results in the well-known criterion of equilibrium at constant temperature and pressure:

$$\Delta G_{T,p} \leq 0 \tag{2.23}$$

Thus, for a closed system at constant temperature and pressure, the Gibbs free energy of the system must decrease with any spontaneous finite system change and will be at its minimum value at equilibrium. Finally, if a system of constant temperature and pressure is at equilibrium, any infinitesimal system change will result in zero change in the system's Gibbs free energy.

$$\mathrm{d}G_{T,p} \leq 0 \tag{2.24}$$

2.2.1.4 Summary of the Equilibrium Criteria

Other sets of inequality constraints exist for system equilibrium and can be found in a manner similar to the one detailed above but have limited applications. In summary, the equilibrium constraints for a system undergoing a finite change as determined above are as follows:

$$\Delta S_{E,V} \geq 0 \tag{2.25}$$

$$\Delta F_{T,V} \leq 0 \tag{2.26}$$

$$\Delta G_{T,p} \leq 0 \tag{2.27}$$

$$\Delta E_{S,V} \leq 0 \tag{2.28}$$

$$\Delta S_{H,p} \geq 0 \tag{2.29}$$

$$\Delta H_{S,p} \leq 0 \tag{2.30}$$

where $H = E + pV$ is the enthalpy of the system.

During an infinitesimal change in the system, this change can be assumed to be reversible ($\mathrm{d}S = 0$). Therefore, the criteria of equilibrium for a system undergoing an infinitesimal change from equilibrium conditions become

$$\mathrm{d}S_{E,V} = 0 \tag{2.31}$$

$$\mathrm{d}F_{T,V} = 0 \tag{2.32}$$

$$\mathrm{d}G_{T,p} = 0 \tag{2.33}$$

$$\mathrm{d}U_{S,V} = 0 \tag{2.34}$$

$$\mathrm{d}S_{H,p} = 0 \tag{2.35}$$

$$\mathrm{d}H_{S,p} = 0 \tag{2.36}$$

Example 2.1

Show that the equilibrium criterion for a system with constant entropy and volume is $\Delta E_{S,V} \leq 0$.

Solution For a system with constant entropy and volume, we have $\Delta S = 0$ and $\Delta V = 0$. The work exchange between the system and its surrounding is $W = 0$. The fundamental relationship, Eq. (2.4), becomes

$$\Delta E \leq T\Delta S \tag{2.37}$$

Since entropy is constant, i.e., $\Delta S = 0$, it follows that

$$\Delta E_{S,V} \leq 0 \tag{2.38}$$

2.2.2 Maxwell Relations

The fundamental thermodynamic relation for a reversible process in a single-component system, where the only work term considered is $p\mathrm{d}V$, is obtained from Eq. (2.3), i.e.,

$$\mathrm{d}E = T\mathrm{d}S - p\mathrm{d}V \tag{2.39}$$

which can also be rewritten in terms of enthalpy ($H = E + pV$), Helmholtz free energy ($F = E - TS$), and Gibbs free energy ($G = H - TS$) as

$$\mathrm{d}H = T\mathrm{d}S + V\mathrm{d}p \tag{2.40}$$

$$\mathrm{d}F = -S\mathrm{d}T - p\mathrm{d}V \tag{2.41}$$

$$\mathrm{d}G = -S\mathrm{d}T + V\mathrm{d}p \tag{2.42}$$

which all have the form of

$$\mathrm{d}z = M\mathrm{d}x + N\mathrm{d}y \tag{2.43}$$

where

$$M = \left(\frac{\partial z}{\partial x}\right)_y \tag{2.44}$$

$$N = \left(\frac{\partial z}{\partial y}\right)_x \tag{2.45}$$

and dz is an exact differential, as thermodynamic properties like E, H, F, and G are path-independent functions.

Since Eq. (2.43) is the total differential of function z, M and N are related by

$$\left(\frac{\partial M}{\partial y}\right)_x = \left(\frac{\partial N}{\partial x}\right)_y = \frac{\partial^2 z}{\partial x \partial y} \tag{2.46}$$

Applying Eq. (2.46) to Eqs. (2.39)–(2.42), the following relationships are obtained:

$$\left(\frac{\partial T}{\partial V}\right)_S = -\left(\frac{\partial p}{\partial S}\right)_V \tag{2.47}$$

$$\left(\frac{\partial T}{\partial p}\right)_S = \left(\frac{\partial V}{\partial S}\right)_p \tag{2.48}$$

$$\left(\frac{\partial S}{\partial V}\right)_T = \left(\frac{\partial p}{\partial T}\right)_V \tag{2.49}$$

$$\left(\frac{\partial S}{\partial p}\right)_T = -\left(\frac{\partial V}{\partial T}\right)_p \tag{2.50}$$

which are referred to as Maxwell relations. The goal of Maxwell relations is to find equivalent partial derivatives containing p, T, and V that can be physically measured and therefore provide a means of determining the change of entropy, which cannot be measured directly.

2.2.3 Closed Systems with Compositional Change

The internal energy in Eq. (2.39) is a function of only two independent variables, $E = E(S, V)$, when dealing with a single-phase, single-component system. When a compositional change is possible, i.e., for multicomponent systems, internal energy must also be a function of the number of moles of each of the N components:

$$E = E(S, V, n_1, n_2, \ldots n_N) \tag{2.51}$$

Expanding Eq. (2.51) in terms of each independent variable, while holding all other properties constant, produces the following:

$$\mathrm{d}E = \left(\frac{\partial E}{\partial S}\right)_{V,n_i} \mathrm{d}S + \left(\frac{\partial E}{\partial V}\right)_{S,n_i} \mathrm{d}V + \sum_{i=1}^{N} \left(\frac{\partial E}{\partial n_i}\right)_{S,V,n_{j \neq i}} \mathrm{d}n_i \tag{2.52}$$

where $j \neq i$. The first two terms on the right side of Eq. (2.52) refer to conditions of constant composition, as represented by Eq. (2.39). Comparing Eqs. (2.52) and (2.39), the coefficients of the first two terms in Eq. (2.52) are

$$\left(\frac{\partial E}{\partial S}\right)_{V,n_i} = T \tag{2.53}$$

$$\left(\frac{\partial E}{\partial V}\right)_{S,n_i} = -p \tag{2.54}$$

The third term on the right-hand side of Eq. (2.52) corresponds to the effects of the presence of multiple components. The chemical potential can be defined as

$$\mu_i = \left(\frac{\partial E}{\partial n_i}\right)_{S,V,n_{j \neq i}} \tag{2.55}$$

Therefore, the above expanded fundamental equation for a multicomponent system, as seen in Eq. (2.52), can be rewritten as

$$\mathrm{d}E = T\mathrm{d}S - p\mathrm{d}V + \sum_{i=1}^{N} \mu_i \mathrm{d}n_i \tag{2.56}$$

which is known as the internal energy representation of the fundamental thermodynamic equation of multicomponent systems. Other representations can be directly obtained from Eq. (2.56) by using the definitions of enthalpy ($H = E + PV$), Helmholtz free energy ($F = E - TS$), and Gibbs free energy ($G = E - TS + pV$), i.e.,

$$\mathrm{d}H = V\mathrm{d}p + T\mathrm{d}S + \sum_{i=1}^{N} \mu_i \mathrm{d}n_i \tag{2.57}$$

$$\mathrm{d}F = -S\mathrm{d}T - p\mathrm{d}V + \sum_{i=1}^{N} \mu_i \mathrm{d}n_i \tag{2.58}$$

$$\mathrm{d}G = V\mathrm{d}p - S\mathrm{d}T + \sum_{i=1}^{N} \mu_i \mathrm{d}n_i \tag{2.59}$$

It is therefore readily determined from Eqs. (2.57)–(2.59) that other expressions of chemical equilibrium exist; these are

$$\mu_i = \left(\frac{\partial H}{\partial n_i}\right)_{p,S,n_{j\neq i}} = \left(\frac{\partial F}{\partial n_i}\right)_{T,V,n_{j\neq i}} = \left(\frac{\partial G}{\partial n_i}\right)_{T,p,n_{j\neq i}} \tag{2.60}$$

In addition, the following expressions for the fundamental thermodynamic properties are valid:

$$T = \left(\frac{\partial E}{\partial S}\right)_{V,n_i} = \left(\frac{\partial H}{\partial S}\right)_{p,n_i} \tag{2.61}$$

$$-p = \left(\frac{\partial E}{\partial V}\right)_{S,n_i} = \left(\frac{\partial F}{\partial V}\right)_{T,n_i} \tag{2.62}$$

$$V = \left(\frac{\partial H}{\partial p}\right)_{S,n_i} = \left(\frac{\partial G}{\partial p}\right)_{T,n_i} \tag{2.63}$$

which will be very useful in stability analysis in the next subsection.

2.2.4 Stability Criteria

Equation (2.10) demonstrates that for a constant-volume isolated system, the equilibrium condition requires that the entropy of the system must be stationary, i.e., $dS = 0$. Equation (2.10) is also valid if the entropy of the system is at either maximum or minimum. To ensure that the system is at a stable equilibrium, i.e., equilibrium in the system can be maintained after a small perturbation, it is necessary for the system to satisfy Eq. (2.9) as well. All other equilibrium criteria listed in Eqs. (2.32)–(2.36) are similarly inadequate to ensure stable equilibrium, and the inequalities stated in Eqs. (2.26)–(2.30) are necessary additionally. We will now focus on the use of the energy minimum principle (in E, F, G, and H representation) to address the stability of a simple system in equilibrium. For a simple system to be in equilibrium, as analyzed in detail above, the system must be stable. If the system is not stable, it will spontaneously change state to become stable. Stability for a single-phase system can be broken down into three distinct types: (1) thermal, (2) mechanical, and (3) chemical.

2.2.4.1 Thermal Stability

A simple system (homogeneous, single phase) confined by an adiabatic, rigid, and impermeable boundary is shown in Fig. 2.1. Under these constraints, the energy and volume of the system remain constant during any process. In its initial state, the system is divided by an adiabatic partition into two halves of equal volume, with the left side at a slightly higher temperature than the right. This temperature difference is accounted for in the representation of the internal energy by $(E + \Delta E)$ for the left half and $(E - \Delta E)$ for the right half. At some arbitrary time, heat is allowed to be exchanged between the two halves and sufficient time elapses for the system to reach equilibrium. Since no energy has been added to the system, the total energy in the initial state matches the total energy in the final state, i.e.,

$$E_i = (E + \Delta E) + (E - \Delta E) = 2E = E_f \tag{2.64}$$

If the entropies of the left and right halves in the final state are S_L and S_R, respectively, they must be identical, because the two halves possess the same internal energy and volume at the final state, i.e.,

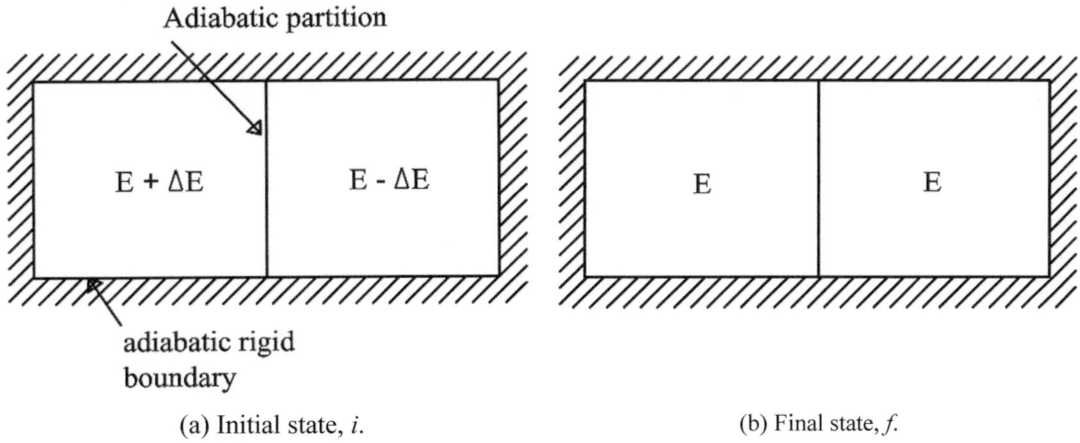

Fig. 2.1 Isolated system illustrating thermal stability

$$S_{\mathrm{L}} = S_{\mathrm{R}} = S(E, V) \tag{2.65}$$

Since the volume of the left half is not changed between the initial and final states, the entropy of the left half in the initial state can be found by Taylor series expansion as follows:

$$S_{\mathrm{L},i} = S(E + \Delta E, V) = S + \left(\frac{\partial S}{\partial E}\right)_V (\Delta E) + \frac{1}{2}\left(\frac{\partial^2 S}{\partial E^2}\right)_V (\Delta E)^2 + \cdots \tag{2.66}$$

Similarly, the entropy in the right half at the initial state can be represented by

$$S_{\mathrm{R},i} = S(E - \Delta E, V) = S - \left(\frac{\partial S}{\partial E}\right)_V (\Delta E) + \frac{1}{2}\left(\frac{\partial^2 S}{\partial E^2}\right)_V (\Delta E)^2 - \cdots \tag{2.67}$$

According to Eq. (2.9), the final system entropy must be equal to or greater than the initial total system entropy. In other words, the entropy generated by this process must not be negative.

$$S_{\mathrm{gen}} = (S_{\mathrm{L}} + S_{\mathrm{R}}) - (S_{\mathrm{L},i} + S_{\mathrm{R},i}) \geq 0 \tag{2.68}$$

Combining Eqs. (2.65)–(2.68) results in the following expression:

$$S_{\mathrm{gen}} = -\left(\frac{\partial^2 S}{\partial E^2}\right)_V (\Delta E)^2 \geq 0 \tag{2.69}$$

Since $(\Delta E)^2$ will always be finite positive, approaching zero as the two halves steadily approach uniform total internal energy, the second-order partial derivative in Eq. (2.69) must be negative and finite.

$$\left(\frac{\partial^2 S}{\partial E^2}\right)_V < 0 \tag{2.70}$$

Equation (2.70) can be rewritten as follows with the use of Eq. (2.53):

$$\frac{\partial}{\partial E}\left(\frac{1}{T}\right)_V = -\frac{1}{T^2}\left(\frac{\partial T}{\partial E}\right)_V < 0 \tag{2.71}$$

i.e.,

$$\frac{1}{T^2}\left(\frac{\partial T}{\partial E}\right)_V > 0 \tag{2.72}$$

so thermal stability requires that internal energy of the system must be an increasing function of temperature.

Considering the definition of the specific heat at constant volume,

$$c_V = \frac{1}{m}\left(\frac{\partial E}{\partial T}\right)_V \tag{2.73}$$

Equation (2.72) can be written as

$$\frac{1}{T^2 m c_V} > 0 \tag{2.74}$$

Since both T^2 and m in Eq. (2.74) are positive, Eq. (2.74) implies that the specific heat of the system at constant volume must always be positive in order for the system to move from its initial state—as defined above—to its final state, rather than in the opposite direction.

$$c_V > 0 \tag{2.75}$$

In other words, positive c_V ensures that the system cannot spontaneously segregate itself into two thermally dissimilar regions.

2.2.4.2 Mechanical Stability

The mechanical stability criteria can be determined by considering a simple system that initially has an internal pressure discontinuity. Figure 2.2 shows a simple system with constant total volume and temperature. The constant temperature is maintained by contact with a thermal reservoir at a constant temperature, T. Initially, the system is divided into two parts of slightly different volume by an

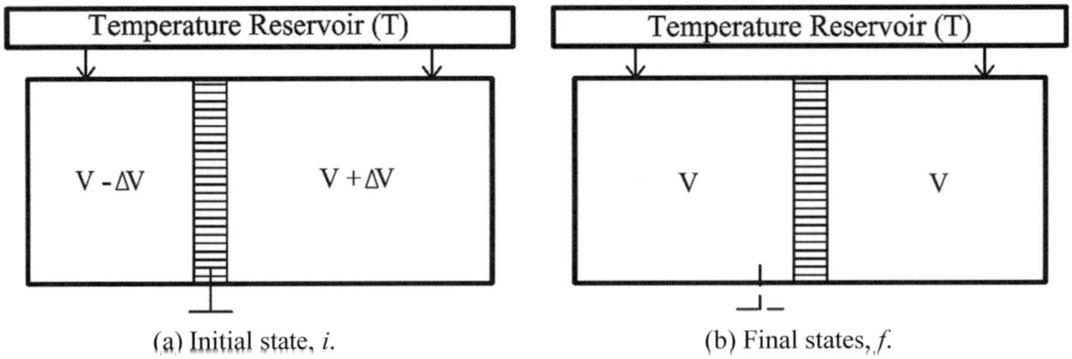

Fig. 2.2 Constant-temperature and constant-volume system illustrating mechanical stability

off-center partition held in place by a locking mechanism. The left side is at a slightly higher pressure and lower volume. When the locking mechanism is disengaged, the partition will gradually float to the midpoint of the system, at which point the pressures and volumes of the two halves are equalized. In a manner similar to the analysis of thermal stability, the total volume of the system in the initial and final states can be represented as

$$V_i = (V - \Delta V) + (V + \Delta V) = V + V = V_f \tag{2.76}$$

Since the temperature and total volume of this system remain constant, the minimum Helmholtz free energy principle, Eq. (2.16), can be applied to this problem. In the final state, the volumes and temperatures of the two halves are the same; therefore, the Helmholtz free energies of the left and right halves in the final state must be the same, i.e.,

$$F_{\mathrm{L}} = F_{\mathrm{R}} = F(V, T) \tag{2.77}$$

where F is the final Helmholtz free energy of either half of the system. As was the case with entropy in the preceding thermal stability analysis, the Helmholtz free energies of the initial two parts can be related to the final values by noting the constant-temperature volume changes experienced by each of the two parts:

$$F_{\mathrm{L},i} = F(V - \Delta V, T) = F - \left(\frac{\partial F}{\partial V}\right)_T (\Delta V) + \frac{1}{2}\left(\frac{\partial^2 F}{\partial V^2}\right)_T (\Delta V)^2 - \cdots \tag{2.78}$$

$$F_{\mathrm{R},i} = F(V + \Delta V, T) = F + \left(\frac{\partial F}{\partial V}\right)_T (\Delta V) + \frac{1}{2}\left(\frac{\partial^2 F}{\partial V^2}\right)_T (\Delta V)^2 + \cdots \tag{2.79}$$

From the equilibrium analysis above, it can be concluded that the Helmholtz free energy of the final system must be less than or equal to the initial total system Helmholtz free energy:

$$(F_{\mathrm{L}} + F_{\mathrm{R}}) - (F_{\mathrm{L},i} + F_{\mathrm{R},i}) \le 0 \tag{2.80}$$

Combining Eqs. (2.77)–(2.80) results in the following expression:

$$\left(\frac{\partial^2 F}{\partial V^2}\right)_T (\Delta V)^2 \ge 0 \tag{2.81}$$

Since $(\Delta V)^2$ will always be finite positive, approaching zero as the Helmholtz free energies of the two portions steadily approach uniformity, the second-order partial derivative of Eq. (2.81) must be positive and finite:

$$\left(\frac{\partial^2 F}{\partial V^2}\right)_T > 0 \tag{2.82}$$

Equation (2.82) can be rewritten in the following form by considering Eq. (2.62):

$$-\left(\frac{\partial p}{\partial V}\right)_T = \frac{1}{\kappa_T V} > 0 \tag{2.83}$$

where κ_T is isothermal compressibility:

$$\kappa_T = -\frac{1}{V}\left(\frac{\partial V}{\partial p}\right)_T \tag{2.84}$$

From Eq. (2.83), the criterion for mechanical stability for the system shown in Fig. 2.2 is found to be

$$\kappa_T > 0 \tag{2.85}$$

Therefore, a simple system at equilibrium is mechanically stable if the isothermal compressibility factor is positive, i.e., the volume of the system shrinks with increasing pressure.

2.2.4.3 Chemical Stability

For a simple system in equilibrium to be stable, the system must also have chemical stability. In other words, certain conditions will prevent the system from spontaneously separating into two or more subsystems of varying chemical composition. As an aid to analyzing the criteria for chemical stability, the system shown in Fig. 2.3 is presented. The simple system depicted is in contact with a constant-temperature reservoir and its boundary is impermeable to all species present. Two frictionless pistons ensure that the pressure of the system is always a constant value. Internally, the system in its initial state consists of a semipermeable membrane partition that prevents the movements of only species i between the two portions of the system. Initially, the left portion of the system contains more moles of species i than the right portion. At some arbitrary time, the membrane is made permeable to permit the flow of species i between the two compartments, and the system reaches an equilibrium condition with respect to matter flow.

Conservation of the number of moles of species i during the process shown in Fig. 2.3 can be written as

$$n_{i,i} = (n_i + \Delta n_i) + (n_i - \Delta n_i) = 2n_i = n_{i,f} \tag{2.86}$$

Since this system maintains constant temperature and pressure, the energy minimum principle that governs this system is the Gibbs free energy principle shown in Eq. (2.23). Therefore, denoting the Gibbs free energy of the left and right halves at the equilibrium final state as G_L and G_R, respectively, the following can be stated:

$$G_L = G_R = G(n_i, T, p) \tag{2.87}$$

where G is the final Gibbs free energy of either half of the system. As demonstrated above with entropy in the thermal stability analysis, the Gibbs free energies of the initial two parts can be related

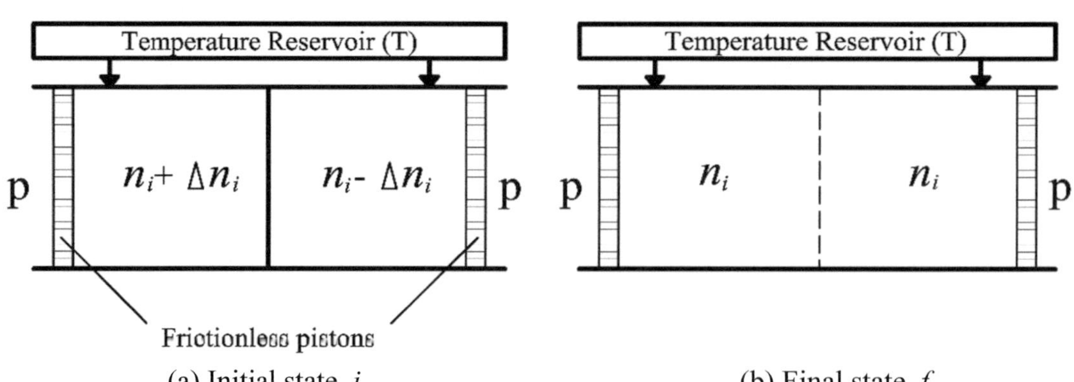

Fig. 2.3 Constant-temperature and constant-pressure system with composition change illustrating chemical stability

to the final values by noting the change of the Gibbs free energy with the change in the mole number i of the left- and right-hand sides when temperature, pressure, and all other mole numbers are held constant:

$$G_{\mathrm{L},i} = G(n_i + \Delta n_i, T, p) = G + \left(\frac{\partial G}{\partial n_i}\right)_{T,p,n_j}(\Delta n_i) + \frac{1}{2}\left(\frac{\partial^2 G}{\partial n_i^2}\right)_{T,p,n_j}(\Delta n_i)^2 + \cdots \tag{2.88}$$

$$G_{\mathrm{R},i} = G(n_i + \Delta n_i, T, p) = G - \left(\frac{\partial G}{\partial n_i}\right)_{T,p,n_j}(\Delta n_i) + \frac{1}{2}\left(\frac{\partial^2 G}{\partial n_i^2}\right)_{T,p,n_j}(\Delta n_i)^2 - \cdots \tag{2.89}$$

In the equilibrium analysis followed above, it was established that the Gibbs free energy of the final system must be less than or equal to the initial total system Gibbs free energy, i.e.,

$$(G_{\mathrm{L}} + G_{\mathrm{R}}) - \left(G_{\mathrm{L},i} + G_{\mathrm{R},i}\right) \leq 0 \tag{2.90}$$

Combining Eqs. (2.86)–(2.90) results in the following expression:

$$\left(\frac{\partial^2 G}{\partial n_i^2}\right)_{T,p,n_j}(\Delta n_i)^2 \geq 0 \tag{2.91}$$

Since $(\Delta n_i)^2$ will always be finite positive, approaching zero as the Gibbs free energy of the two portions steadily approaches uniformity, the second-order partial derivative in Eq. (2.91) must be positive and finite:

$$\left(\frac{\partial^2 G}{\partial n_i^2}\right)_{T,p,n_j} > 0 \tag{2.92}$$

Substituting Eq. (2.60) into Eq. (2.92) results in the chemical stability criterion for the simple system presented in Fig. 2.3:

$$\left(\frac{\partial \mu_i}{\partial n_i}\right)_{T,p,n_j} > 0 \tag{2.93}$$

Therefore, a simple system in equilibrium, such as the final state of the system shown in Fig. 2.3, is chemically stable if the chemical potential of the ith species increases with an increase in mole number of the ith species.

2.2.5 Systems with Chemical Reactions

2.2.5.1 Chemical Reaction and Combustion

The change in the chemical composition in a system discussed in the preceding subsection can result from a chemical reaction. During a chemical reaction, some of the chemical bonds binding the atoms into molecules are broken, and new ones are formed. Since the chemical energies associated with the chemical bonds in reactants and products are generally different, the resulting change in chemical energy and its effect on the overall energy balance of the system must be accounted for. For a single-phase reacting system, the change of energy can be due to the change of sensible internal energy associated with temperature and pressure change. It may also reflect the change of chemical energy associated with chemical reactions. The reacting system starts with the mixture of reactants, and the chemical reaction in the system produces new components that will coexist in the mixture. Therefore, a system undergoing chemical reaction can be considered as a mixture that contains both reactants and products. In practical applications, one particular kind of chemical reaction, namely combustion, is particularly important. Combustion is a chemical reaction during which a fuel is oxidized and a large amount of thermal energy is released. Most fuels (such as coal, gasoline, and diesel fuel) consist of hydrogen and carbon, and are called hydrocarbon fuels. The oxidant gas for most combustion processes is air, which can be treated as a mixture of 21% oxygen and 79% nitrogen (each kmol of oxygen in the air is accompanied by 0.79/0.21 = 3.76 kmol of nitrogen).

For a given reaction, the chemical equation establishes the relationship between the mole numbers of the reactants consumed and the mole numbers of the products generated. For example, combustion of 1 kmol of methane with air that contains 2 kmol of oxygen can be represented by the following chemical equation:

$$CH_4 + 2O_2 + 7.52N_2 \rightarrow CO_2 + 2H_2O + 7.52N_2 \tag{2.94}$$

where nitrogen is present on both sides of the equation and is a nonreacting species that is carried to the products. At high combustion temperature, a small amount of nitrogen N_2 is oxidized to nitrogen oxides (NO, NO_2). However, this amount is negligibly small as far as the overall combustion reaction is concerned. Combustion is complete if all of the carbon in the fuel burns to carbon dioxide and all of the hydrogen burns to water. The combustion in Eq. (2.94) is a complete combustion. On the other hand, the combustion is incomplete if the product contains any unburned fuel or C, H_2, or CO. Common causes of incomplete combustion include an insufficient supply of oxygen and inadequate mixing between fuel and oxygen.

The first law of thermodynamics is a general law that applies to any process, including combustion. The first law of thermodynamics for a closed system is expressed as

$$\delta Q = \mathrm{d}E + p\mathrm{d}V \tag{2.95}$$

or, for a finite process,

$$Q = (E_2 - E_1) + \int_{V_1}^{V_2} p\mathrm{d}V \tag{2.96}$$

where Q is the heat transfer between the system and its surroundings, $\int_{V_1}^{V_2} p\mathrm{d}V$ is the work done by the system on its surroundings, and E_1 and E_2 are total internal energies, including sensible and latent internal energy and chemical energy, before and after the chemical reaction. Equation (2.96) is a general expression that is applicable to any combustion process. Combustion occurring under two specific conditions is of the greatest interest for practical applications: constant volume and constant pressure.

For combustion at constant volume, the heat transfer is

$$Q = E_2 - E_1 \tag{2.97}$$

When combustion occurs at constant pressure, Eq. (2.96) becomes

$$Q = (E_2 - E_1) + p(V_2 - V_1) = H_2 - H_1 \tag{2.98}$$

Equation (2.98) is also valid for combustion in an open system provided that H_1 and H_2 represent enthalpy at the inlet and exit, respectively, of the system.

2.2.5.2 Chemical Equilibrium

Before the chemical equilibrium theory was established, it was believed that all chemical reactions would proceed until all reactants were completely converted into products. In fact, chemical reactions proceed only until they reach an equilibrium state, referred to as *chemical equilibrium*. At the equilibrium state, the chemical reaction then proceeds incrementally in both directions, so that there is no net change in composition. Under these specific conditions, and if the chemical equilibrium is stable, the equilibrium will not change with time.

Chemical equilibrium is another cause of incomplete combustion and cannot be prevented. At chemical equilibrium, the combustion of methane can be represented by

$$CH_4 + 2O_2 + 7.52N_2 \rightleftarrows CO_2 + 2H_2O + 7.52N_2 \tag{2.99}$$

where the two arrows in Eq. (2.99) indicate that the reaction takes place in both directions at the equilibrium state. Therefore, chemical reaction does not cease at chemical equilibrium, but the reaction rate is the same in both directions. Consequently, there is no notable change of composition for a reacting system at chemical equilibrium.

For a system containing N_r reactants and N_p products, the generalized chemical equation can be expressed as

$$a_{r1}A_{r1} + a_{r2}A_{r2} + \cdots + a_{rN_r}A_{rN_r} \rightleftarrows a_{p1}A_{p1} + a_{p2}A_{p2} + \cdots + a_{pN_p}A_{pN_p} \tag{2.100}$$

where $A_{ri}(i = 1, 2, \ldots N_r)$ are chemical symbols for the reactants and $A_{pi}\,(i = 1, 2, \ldots N_p)$ are chemical symbols for the products, with their corresponding stoichiometric coefficients $a_{r,i}$ and $a_{p,i}$. In the chemical reaction shown in Eq. (2.99), the total numbers of reactants and products are $N_r = 3$ and $N_p = 3$, respectively. The chemical symbols of reactants and products are $A_{r1} = CH_4$, $A_{r2} = O_2$, $A_{r3} = N_2$, $A_{p1} = CO_2$, $A_{p2} = H_2O$, and $A_{p3} = N_2$. The coefficients preceding the chemical symbols in Eq. (2.100) are referred to as stoichiometric coefficients. These coefficients describe the proportion of the mole numbers of reactants disappearing and mole numbers of products appearing during the reaction process. The stoichiometric coefficients for reactants CH_4, O_2, and N_2 are, respectively 1, 2, and 7.52.

Equation (2.100) can also be written as a more compact form (Bejan 2016):

$$0 \rightleftarrows \sum_{i=1}^{N_r+N_p} a_i A_i \tag{2.101}$$

where

$$a_i = \begin{cases} -a_{ri} & i = 1, 2, \cdots N_r \\ a_{p(i-N_r)} & i = N_r + 1, N_r + 2, \ldots N_r + N_p \end{cases} \tag{2.102}$$

and

$$A_i = \begin{cases} A_{ri} & i = 1, 2, \cdots N_r \\ A_{p(i-N_r)} & i = N_r + 1, N_r + 2, \ldots N_r + N_p \end{cases} \tag{2.103}$$

During a chemical reaction process, the decreasing mole number of reactants and increasing mole number of products must be proportional to the corresponding stoichiometric coefficients. If a chemically reacting system initially contains N_r reactants and the mole number of ith reactant is n_i^0, when the chemical reaction reaches chemical equilibrium, the mole number of ith reactants becomes

$n_{ri} = n_{ri}^0 - a_{ri}\zeta\,(i = 1, 2, \ldots, N_r)$ and the mole number of the ith product is $n_{pi} = a_{pi}\zeta\,(i = 1, 2, \ldots, N_p)$. In the above notation, ζ indicates the degree of advancement of the chemical reaction (i.e., $\zeta = 0$ means no reaction and a very large ζ represents a large mole number of products). The maximum value of ζ_{max} is reached when at least one of the reactants is exhausted.

When the system is at chemical equilibrium with a degree of advancement ζ, and the mole number of each of the reactants and products is represented by $n_i\,(i = 1, 2, \cdots N_r, N_r + 1, \ldots N_r + N_p)$, a slight advancement of the chemical reaction will bring the system to a new equilibrium state represented by $\zeta + d\zeta$, in which state the new mole numbers of each of the components become $n_i + dn_i\,(i = 1, 2, \cdots N_r,\ N_r + 1, \ldots N_r + N_p)$. The change of mole number of each component is then

$$dn_i = a_i d\zeta \quad i = 1, 2, \ldots N_r, N_r + 1, \ldots N_r + N_p \tag{2.104}$$

The change in internal energy for the chemically reacting system can be obtained by substituting Eq. (2.104) into Eq. (2.56), i.e.,

$$dE = TdS - pdV + \left(\sum_{i=1}^{N_r+N_p} \mu_i a_i\right) d\zeta \tag{2.105}$$

Introducing De Donder's affinity function (Bejan 2016),

$$Y = -\sum_{i=1}^{N_r+N_p} \mu_i a_i \tag{2.106}$$

Equation (2.105) becomes

$$dE = TdS - pdV - Yd\zeta \tag{2.107}$$

The affinity function is a linear combination of the chemical potentials of reactants and products; therefore, the affinity function itself is a property of the chemically reacting system. Equation (2.107) suggests that the internal energy of a chemically reactive system is a function of entropy, volume, and degree of affinity, i.e.,

$$E = E(S, V, \zeta) \tag{2.108}$$

Expanding Eq. (2.108) in terms of each independent variable while holding all other properties constant produces the following:

$$dE = \left(\frac{\partial E}{\partial S}\right)_{V,\zeta} dS + \left(\frac{\partial E}{\partial V}\right)_{S,\zeta} dV + \left(\frac{\partial E}{\partial \zeta}\right)_{S,V} d\zeta \tag{2.109}$$

Comparing the third terms of Eqs. (2.107) and (2.109) yields

$$Y = -\left(\frac{\partial E}{\partial \zeta}\right)_{S,V} \tag{2.110}$$

Other representations of the fundamental relation for chemically reactive systems can be directly obtained from Eq. (2.107) by using the definitions of enthalpy, Helmholtz free energy, and Gibbs free energy, i.e.,

$$dH = Vdp + TdS - Yd\zeta \tag{2.111}$$

$$dF = -SdT - pdV - Yd\zeta \tag{2.112}$$

$$dG = Vdp - SdT - Yd\zeta \tag{2.113}$$

It can be readily determined from Eqs. (2.111)–(2.113) that other expressions of chemical equilibrium are

$$Y = -\left(\frac{\partial H}{\partial \zeta}\right)_{S,p} = -\left(\frac{\partial F}{\partial \zeta}\right)_{T,V} = -\left(\frac{\partial G}{\partial \zeta}\right)_{T,p} \tag{2.114}$$

For a typical process wherein pressure and temperature are constant, the equilibrium condition requires that [see Eq. (2.33)]

$$dG_{T,p} = 0 \tag{2.115}$$

Comparison of Eqs. (2.115) and (2.113) reveals that for a chemical reaction occurring at constant pressure and temperature, the degree of affinity at equilibrium is zero.

$$Y = -\sum_{i=1}^{N_r + N_p} \mu_i a_i = 0 \tag{2.116}$$

In order for the chemical equilibrium of the reacting system with constant pressure and temperature to be stable, it is necessary for the Gibbs free energy to shown in Fig. 2.4. Mathematically, the condition for stability can be expressed as satisfy Eq. (2.27) as well, i.e., the Gibbs free energy must be at its minimum as:

$$\left(\frac{\partial^2 G}{\partial \zeta^2}\right)_{T,p} > 0 \tag{2.117}$$

Substituting Eq. (2.114) into Eq. (2.117), the condition for stability becomes

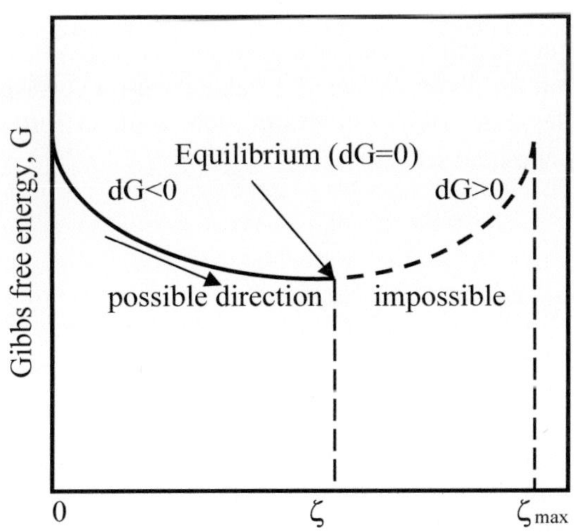

Fig. 2.4 Chemical equilibrium at constant temperature and pressure

$$\left(\frac{\partial Y}{\partial \zeta}\right)_{T,p} < 0 \tag{2.118}$$

Example 2.2

At chemical equilibrium, the combustion of methane can be represented by Eq. (2.99). What is the relationship among the chemical potential of the reactants and products?

Solution The total number of reactants is $N_r = 3$ and the total number of products is $N_p = 3$. The stoichiometric coefficients for the reactants are $a_{CH_4} = -1$, $a_{O_2} = -2$, and $a_{N_2} = -7.52$, respectively. The stoichiometric coefficients for the products are $a_{CO_2} = 1$, $a_{H_2O} = 2$, and $a_{N_2} = 7.52$, respectively. At equilibrium, the degree of affinity must be zero, i.e.,

$$\begin{aligned} Y &= -\sum_{i=1}^{3+3} \mu_i a_i \\ &= -(\mu_{CH_4} a_{CH_4} + \mu_{O_2} a_{O_2} + \mu_{N_2} a_{N_2} + \mu_{CO_2} a_{CO_2} + \mu_{H_2O} a_{H_2O} + \mu_{N_2} a_{N_2}) \\ &= \mu_{CH_4} + 2\mu_{O_2} + 7.52\mu_{N_2} - \mu_{CO_2} - 2\mu_{H_2O} - 7.52\mu_{N_2} = 0 \end{aligned}$$

Thus, the chemical potentials of the reactants and products at equilibrium satisfy

$$\mu_{CH_4} + 2\mu_{O_2} + 7.52\mu_{N_2} = \mu_{CO_2} + 2\mu_{H_2O} + 7.52\mu_{N_2}$$

2.3 Equilibrium and Stability of Multiphase Systems

Fundamentals of thermodynamic equilibrium criteria with emphasis on single-phase systems were discussed in Sect. 2.2. This section focuses on thermodynamic equilibrium and stability criteria for multiphase systems.

2.3.1 Two-Phase Single-Component Systems

The criteria for the equilibrium of two phases of a pure substance can be developed from any of the criteria for equilibrium equalities given in Sect. 2.2, along with a corresponding fundamental relationship. An infinitesimal departure from equilibrium will result in zero entropy change when the internal energy and volume are held constant, i.e.,

$$\mathrm{d}S_{E,V} = 0 \tag{2.119}$$

For a closed system containing two phases, Eq. (2.119) can be written as

$$\left(\sum_{k=1}^{2} \mathrm{d}S_k\right)_{E,V} = 0 \tag{2.120}$$

where the subscript k identifies the individual phases. This generic subscript is chosen so that Eq. (2.120) can represent any type of phase-change combination, including liquid–vapor, solid–liquid, or solid–vapor.

The fundamental relation, Eq. (2.56), in terms of internal energy, where $N_i = 1$ (single component) for both phases under consideration, is as follows:

$$dE_k = T_k dS_k - p_k dV_k + \mu_k dn_k \quad (k = 1, 2) \tag{2.121}$$

Since E and V are held constant in this analysis, the change in internal energy of the two phases must sum to zero, i.e.,

$$dE = \sum_{k=1}^{2} dE_k = 0 \tag{2.122}$$

$$dV = \sum_{k=1}^{2} dV_k = 0 \tag{2.123}$$

Also, because the system is closed by definition, the change in the number of moles resulting from the two-phase changes must sum to zero:

$$dn = \sum_{k=1}^{2} dn_k = 0 \tag{2.124}$$

Combining Eqs. (2.119)–(2.124), the following expression is obtained:

$$dS = 0 = \left(\frac{1}{T_1} - \frac{1}{T_2}\right) dE_1 + \left(\frac{p_1}{T_1} - \frac{p_2}{T_2}\right) dV_1 - \left(\frac{\mu_1}{T_1} - \frac{\mu_2}{T_2}\right) dn_1 \tag{2.125}$$

It follows directly from Eq. (2.125) that for equilibrium to exist between two phases of a single component,

$$T_1 = T_2 \tag{2.126}$$

$$p_1 = p_2 \tag{2.127}$$

$$\mu_1 = \mu_2 \tag{2.128}$$

In other words, the pressure, temperature, and chemical potential of the two phases must be identical in order for equilibrium to exist. Although the equilibrium conditions specified in Eqs. (2.126)–(2.128) are derived by applying the fundamental relation, Eq. (2.56), to a two-phase system with constant internal energy and volume, they are valid phase equilibrium conditions for any two-phase systems.

2.3.2 Van der Waals Equation

The equation of state for ideal gas is

$$pV = nR_u T \quad \text{or} \quad pv = R_g T \tag{2.129}$$

where $R_u = 8.3143$ kJ/kmol K is the universal gas constant, n is mole number, and $R_g = R_u$ (M is molecular mass in kg/kmol) is the particular gas constant. The equation of state for an ideal gas is only applicable to a situation where the pressure of the gas is very low. At higher pressures, the behavior of a gas or vapor deviates substantially from that of the ideal gas. In addition, the ideal gas law is also invalid in the two-phase region represented by the "hump" on a p–v diagram, where liquid and vapor phases coexist. As is shown in the p–v diagram, the properties experience a discontinuous change when liquid–vapor phase change takes place. It was known that if the critical isotherm—which passed through the critical point on the p–v diagram—was followed, the change from vapor to liquid would be continuous. It is noted that continuous transition isotherms below the critical isotherm must exist in order to bridge the gap between ideal gas and incompressible liquid. This idea was realized in the theoretical van der Waals equation constructed by Johannes Diderik van der Waals in 1873. The theoretical equation of state developed by van der Waals is a cubic polynomial:

$$p = \frac{R_g T}{v - b} - \frac{a}{v^2} \tag{2.130}$$

where the constant b represents the minimum volume occupied by the pure substance in the limit as $p \to \infty$, i.e., the volume occupied by the molecules of the substance. The a/v^2 term is the additional pressure term that accounts for mutual attraction between the molecules and is proportional to the density squared. The van der Waals equation bridges the gap between ideal gas behavior and incompressible liquid behavior, which can be shown in two limiting cases. The first is the limit as the specific volume of the substance approaches infinity, i.e., the very dilute gas limit, where the van der Waals equation reduces to the ideal gas equation of state, i.e.,

$$pv = R_g T \quad v \to \infty \tag{2.131}$$

The second is the limit as the pressure of the system approaches infinity, and the volume of the system approaches the minimum volume that the molecules of the substance can occupy, i.e.,

$$v = b \quad p \to \infty \tag{2.132}$$

The van der Waals equation is a cubic polynomial that provides three roots for v at any given pressure. This can be seen in Fig. 2.5, which shows the shape of the van der Waals isotherms below the critical temperature of the substance. While an isotherm passing the two-phase hump on a p–v diagram makes a straight line, the van der Waals isotherm has minimum and maximum points below and above the straight line, respectively. It will be shown later in this section that this wavy line is inherently unstable, and therefore, the van der Waals equation of state is not a good approximation in the two-phase region.

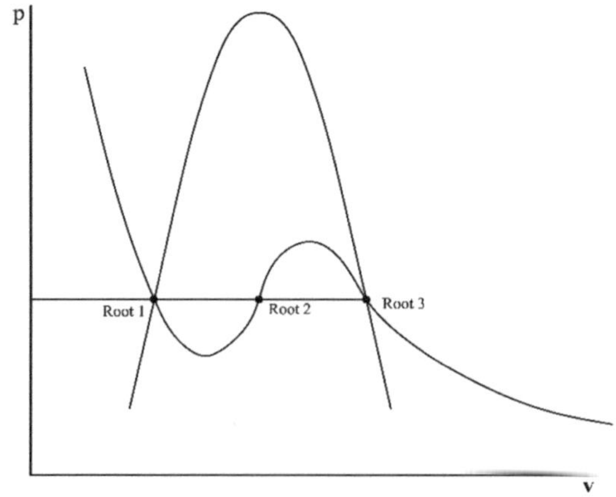

Fig. 2.5 Van der Waals polynomial showing three roots for v at a given p

Table 2.1 Critical point properties for selected fluids

Substance	Formula	Molecular mass	Critical temperature (K)	Critical pressure (MPa)
Air	–	28.97	133.2	3.77
Ammonia	NH_3	17.031	405.5	11.35
Carbon dioxide	CO_2	44.01	304.1	7.38
Carbon monoxide	CO	28.01	132.9	3.5
Nitrogen	N_2	28.013	126.2	3.39
Oxygen	O_2	31.999	154.6	5.04
Water	H_2O	18.015	647.3	22.12
Propane	C_3H_8	44.094	369.8	4.25
R-12	CCl_2F_2	120.914	385.0	4.14
R-134a	CF_3CH_2F	102.03	374.2	4.06

However, it is a very helpful equation because on either side of the two-phase "hump," it very accurately represents the ideal gas and incompressible liquid behavior of a given substance.

It is necessary to know the constants a and b for different substances in order to use the van der Waals equation to evaluate the p–v–T relation. Because the critical isotherm passes through a point of inflection at the critical point, and the slope is zero at this point, the van der Waals equation can be differentiated with respect to v at constant temperature:

$$\left(\frac{\partial p}{\partial v}\right)_T = -\frac{R_g T}{(v-b)^2} + \frac{2a}{v^3} = 0 \tag{2.133}$$

$$\left(\frac{\partial^2 p}{\partial v^2}\right)_T = \frac{2R_g T}{(v-b)^3} - \frac{6a}{v^4} = 0 \tag{2.134}$$

These two derivatives, along with the van der Waals equation at the critical conditions,

$$p_c = \frac{R_g T_c}{v_c - b} - \frac{a}{v_c^2} \tag{2.135}$$

can be solved to find the two constants and the critical specific volume, which are not as amenable to measurement as are the critical temperature and pressure. These values are as follows:

$$v_c = 3b \tag{2.136}$$

$$a = \frac{27}{64}\frac{R_g^2 T_c^2}{p_c} \tag{2.137}$$

$$b = \frac{R_g T_c}{8 p_c} \tag{2.138}$$

The critical properties of selected substances are shown in Table 2.1.

Example 2.3

A 1-m^3 vessel is filled with 80 kg of propane at a temperature of 120 °C. The gas constant for propane is 0.188 kJ/kg K. Find the pressure of the propane by using the van der Waals equation. What will the pressure of the propane be if the ideal gas law is used?

Solution The critical pressure and temperature for propane can be found from Table 2.1; they are $p_c = 4.25$ MPa and $T_c = 369.8$ K. The constants a and b can be determined using Eqs. (2.137) and (2.138):

$$a = \frac{27}{64}\frac{R_g^2 T_c^2}{p_c} = \frac{27}{64}\frac{(0.188 \times 10^3)^2 369.8^2}{4.25 \times 10^6} = 479.78 \text{ Pa m}^6/\text{kg}^2$$

$$b = \frac{R_g T_c}{8 p_c} = \frac{0.188 \times 10^3 \times 369.8}{8 \times 4.25 \times 10^6} = 2.04 \times 10^{-3} \text{ m}^3/\text{kg}$$

The specific volume of the propane is

$$v = \frac{V}{m} = \frac{1}{80} = 0.0125 \text{ m}^3/\text{kg}$$

The pressure of the propane can be found by using the van der Waals equation, Eq. (2.130), i.e.,

$$p = \frac{R_g T}{v - b} - \frac{a}{v^2} = \frac{0.188 \times 10^3 \times (120 + 273.15)}{0.0125 - 0.00204} - \frac{479.78}{0.0125^2} = 4.00 \text{ MPa}$$

If the ideal gas law is used, the pressure of propane will be

$$p = \frac{R_g T}{v} = \frac{0.188 \times 10^3 \times (120 + 273.15)}{0.0125} = 5.91 \text{ MPa}$$

It can be seen that the ideal gas law significantly overpredicts the pressure in this case.

2.3.3 Clapeyron Equation

If the temperature of a two-phase system in equilibrium is slightly changed, the pressure of the system will be affected; this relationship is described by the Clapeyron equation. This simple relation between pressure and temperature for two phases in equilibrium is derived in this subsection, and common forms are presented in this section.

As shown in detail in the preceding subsection, the equilibrium conditions for two phases of a pure substance are represented by Eqs. (2.126)–(2.128). The two statements represented by Eqs. (2.126) and (2.127) are the conditions for thermal and mechanical equilibrium. Equation (2.128) is automatically satisfied when Eqs. (2.126) and (2.127) are satisfied, because the intensive chemical potential property, μ, at equilibrium can be expressed as a function of two other intensive properties, T and p. Furthermore, the temperature, T, and pressure, p, are not independent of each other in a system that contains two phases in equilibrium. An explicit expression for the slope of these equilibrium lines can be found in terms of easily measurable variables, including temperature and pressure. Suppose a two-phase equilibrium system at temperature T and pressure p experiences an infinitesimal change of temperature to $T + \mathrm{d}T$, so that the corresponding pressure changes to $p + \mathrm{d}p$. Since the two-phase system is at equilibrium at the new temperature and pressure, the new chemical potentials of the two phases must also be equal:

$$\mu_1 + \mathrm{d}\mu_1 = \mu_2 + \mathrm{d}\mu_2 \tag{2.139}$$

Substituting Eq. (2.128) into Eq. (2.139), one obtains

$$\mathrm{d}\mu_1 = \mathrm{d}\mu_2 \tag{2.140}$$

The fundamental relations, Eq. (2.59), in terms of Gibbs free energy for both phases under consideration, are

$$\mathrm{d}G_1 = V_1 \mathrm{d}p - S_1 \mathrm{d}T + \mu_1 \mathrm{d}n_1 \tag{2.141}$$

$$\mathrm{d}G_2 = V_2 \mathrm{d}p - S_2 \mathrm{d}T + \mu_2 \mathrm{d}n_2 \tag{2.142}$$

These equations can be considered in light of the following relationships between the extensive and intensive properties:

$$V = n\bar{v}, S = n\bar{s}, \ G = n\bar{g} \tag{2.143}$$

where $\bar{v}, \bar{s},$ and $\bar{g}$ are specific molar volume (m^3/mol), specific molar entropy (J/mol K), and specific molar Gibbs free energy (J/mol), respectively.

Equations (2.141) and (2.142) can be rewritten as

$$\mathrm{d}\bar{g}_1 = \bar{v}_1 \mathrm{d}p - \bar{s}_1 \mathrm{d}T + \frac{\mu_1 - \bar{g}_1}{n_1} \mathrm{d}n_1 \tag{2.144}$$

$$\mathrm{d}\bar{g}_2 = \bar{v}_2 \mathrm{d}p - \bar{s}_2 \mathrm{d}T + \frac{\mu_2 - \bar{g}_2}{n_2} \mathrm{d}n_2 \tag{2.145}$$

For a single-component system, Eqs. (2.60) and (2.143) indicate that

$$\mu = \left(\frac{\partial G}{\partial n}\right)_{T,p} = \bar{g} \tag{2.146}$$

Thus, Eqs. (2.144) and (2.145) can be rewritten as

$$\mathrm{d}\mu_1 = \bar{v}_1 \mathrm{d}p - \bar{s}_1 \mathrm{d}T \tag{2.147}$$

$$\mathrm{d}\mu_2 = \bar{v}_2 \mathrm{d}p - \bar{s}_2 \mathrm{d}T \tag{2.148}$$

Substituting Eqs. (2.147) and (2.148) into Eq. (2.140) yields

$$\bar{v}_1 \mathrm{d}p - \bar{s}_1 \mathrm{d}T = \bar{v}_2 \mathrm{d}p - \bar{s}_2 \mathrm{d}T \tag{2.149}$$

which can be rearranged as

$$\frac{\mathrm{d}p}{\mathrm{d}T} = \frac{\bar{s}_1 - \bar{s}_2}{\bar{v}_1 - \bar{v}_2} \tag{2.150}$$

Equation (2.150) can be rewritten in terms of specific entropy and specific volume per unit mass, i.e.,

$$\frac{\mathrm{d}p}{\mathrm{d}T} = \frac{s_1 - s_2}{v_1 - v_2} \tag{2.151}$$

The specific entropy may be advantageously replaced with the more usable term of specific enthalpy. The equilibrium condition in terms of the specific Gibbs free energy is

$$\bar{g}_1 = \bar{g}_2 \quad \text{or} \quad g_1 = g_2 \tag{2.152}$$

i.e.,

$$h_1 - Ts_1 = h_2 - Ts_2 \tag{2.153}$$

Equation (2.153) can be rearranged to yield

$$s_1 - s_2 = \frac{h_1 - h_2}{T} \tag{2.154}$$

Substituting Eq. (2.154) into Eq. (2.150), one obtains

$$\frac{dp}{dT} = \frac{h_1 - h_2}{(v_1 - v_2)T} = \frac{h_{21}}{v_{21}T} \tag{2.155}$$

where h_{21} is the latent heat of phase change and v_{21} is the change of specific volume during phase change. Equation (2.155), which is referred to as the Clapeyron equation, describes a general relationship among the pressure, temperature, volume change, and enthalpy change for a single-component, two-phase system at equilibrium. All of the properties in Eq. (2.155) are experimentally measurable; the equation itself has been repeatedly tested and found to be valid (Kyle 1999). The Clapeyron equation applies to any two phases in equilibrium, such as solid/liquid, solid/vapor, and liquid/vapor, which are signified by the general superscripts 1 and 2. For a liquid–vapor system, the Clapeyron equation can be written as

$$\frac{\mathrm{d}p}{\mathrm{d}T} = \frac{h_{\ell v}}{(v_v - v_\ell)T_{sat}} \tag{2.156}$$

Example 2.4

Water boils on the top of a mountain at 95 °C. Estimate the height of the mountain.

Solution At sea level, where the pressure is equal to 1 atm, the boiling point of the water is 100 °C. The latent heat of vaporization and the change of specific volume for water at 1 atm are $h_{\ell v} = 2257.03$ kJ/kg and $v_{\ell v} = 1.67185\ \mathrm{m^3/kg}$. The Clapeyron equation (2.156) for a liquid–vapor system can be approximated as

$$\frac{\Delta p}{\Delta T} = \frac{h_{\ell v}}{v_{\ell v}T_{sat}}$$

The height of the mountain, H, is related to the pressure difference by

$$\Delta p = \rho_{\rm air} g H$$

where the density of the air can be approximated as the density of the air at sea level at 25 °C, i.e., $\rho_{\rm air} = 1.169\,{\rm kg/m^3}$. Combining the above two equations and substituting the given values yield

$$H = \frac{h_{\ell v}\Delta T}{v_{\ell v} T_{\rm sat}\rho_{\rm air} g} = \frac{2257.03 \times 10^3 \times (100 - 95)}{1.67185 \times 373.15 \times 1.169 \times 9.8} = 1573\,{\rm m}$$

For liquid–vapor equilibrium at *low pressure*, the specific volume of the liquid, v_ℓ, is negligible in comparison with the specific volume of the vapor, $v_{\rm v}$. It is further assumed that the vapor behaves like an ideal gas at low pressure, and therefore, the specific volume of vapor can be obtained using the ideal gas law $v_{\rm v} = R_{\rm g}T/p$. In this case, the Clapeyron equation (2.156) reduces to

$$\frac{{\rm d}p}{{\rm d}T} = \frac{h_{\ell v} p}{R_{\rm g} T^2} \tag{2.157}$$

which is referred to as the Clausius–Clapeyron equation.

If the saturation temperature corresponding to any reference pressure, p_0, is T_0, the relationship between the saturation temperature and pressure at the vicinity of a point (p_0, T_0) can be obtained by integrating Eq. (2.157), i.e.,

$$\ln\frac{p}{p_0} = -\frac{h_{\ell v}}{R_{\rm g}}\left(\frac{1}{T} - \frac{1}{T_0}\right) \tag{2.158}$$

Rearranging Eq. (2.158) yields the saturation pressure at temperature T:

$$p = p_0 \exp\left[-\frac{h_{\ell v}}{R_{\rm g}}\left(\frac{1}{T} - \frac{1}{T_0}\right)\right] \tag{2.159}$$

Equation (2.159) is also applicable to a mixture of vapor and gas, provided that the pressure is the partial pressure of the vapor in the mixture; this is related to the total pressure, p, by

$$p_{\rm v} = x_{\rm v} p \tag{2.160}$$

and the molar fraction of the saturated vapor in the mixture, x, is defined as

$$x_{\rm v} = \frac{\rho_{\rm v}}{M_{\rm v}}\bigg/\left(\frac{\rho_{\rm v}}{M_{\rm v}} + \frac{\rho_{\rm g}}{M_{\rm g}}\right) \tag{2.161}$$

where $\rho_{\rm v}$ and $\rho_{\rm g}$ are concentrations of vapor and gas in the mixture, respectively.

If the total pressure of the vapor–gas mixture remains constant (as is the case with fog in air) when the temperature of the mixture is changed from T_0 to T, the vapor molar fraction consistent with the new saturated vapor state becomes

$$x_{\rm v} = x_{\rm v0} \exp\left[-\frac{h_{\ell v}}{R_{\rm g}}\left(\frac{1}{T} - \frac{1}{T_0}\right)\right] \tag{2.162}$$

where x_{v0} is the molar fraction of vapor at the reference temperature T_0.

Equation (2.162) can be rearranged using the ideal gas law:

$$x_v = \frac{R_g T}{p} \rho_v \tag{2.163}$$

After substituting Eq. (2.163) into Eq. (2.161) and assuming the total pressure is constant, the concentration of vapor in the mixture becomes

$$\rho_v = \rho_{v0} \frac{T_0}{T} \exp\left[-\frac{h_{\ell v}}{R_g} \left(\frac{1}{T} - \frac{1}{T_0} \right) \right] \tag{2.164}$$

where ρ_{v0} is the density of vapor at the reference temperature T_0.

2.3.4 Multiphase Multicomponent Systems

The requirements for equilibrium can be expressed in terms of different thermodynamic variables in a number of ways. One of the more common thermodynamic situations is a system with constant temperature and pressure.

The equilibrium criterion for a closed system with compositional changes in terms of the Gibbs free energy is expressed by Eq. (2.60). For a simplified case of a two-component, two-phase system, Eq. (2.60) can be written as follows:

$$\mathrm{d}G_k = -S_k \mathrm{d}T + V_k \mathrm{d}p + \mu_{k,A} \mathrm{d}n_{k,A} + \mu_{k,B} \mathrm{d}n_{k,B} \quad (k = 1, 2) \tag{2.165}$$

where the subscript k denotes the kth phase, and the subscripts A and B denote the components A and B. As stated above, the temperature and pressure of the system are assumed to be constant, which simplifies Eq. (2.165). The system is defined further by allowing a very small amount of components A and B to be transferred from phase 1 to phase 2.

$$\mathrm{d}n_{2,A} = -\mathrm{d}n_{1,A} \tag{2.166}$$

$$\mathrm{d}n_{2,B} = -\mathrm{d}n_{1,B} \tag{2.167}$$

Since the system is assumed to be in equilibrium, $\mathrm{d}G_{T,p} = 0$, and therefore

$$\mathrm{d}G = \sum_{k=1}^{2} \mathrm{d}G_k = 0 \tag{2.168}$$

Equations (2.165) and (2.168) can be combined to create the following expression:

$$\begin{aligned} \mathrm{d}G &= \sum_{k=1}^{2} \mu_{k,A} \mathrm{d}n_{k,A} + \sum_{k=1}^{2} \mu_{k,B} \mathrm{d}n_{k,B} \\ &= \mathrm{d}n_{1,A} \left(\mu_{1,A} - \mu_{2,A} \right) + \mathrm{d}n_{1,B} \left(\mu_{1,B} - \mu_{2,B} \right) = 0 \end{aligned} \tag{2.169}$$

which takes into account that T and p are constant. Since $\mathrm{d}n_{1,A}$ and $\mathrm{d}n_{1,B}$ are independent and are not necessarily equal to zero, it follows that at equilibrium,

$$\mu_{1,A} = \mu_{2,A} \tag{2.170}$$

$$\mu_{1,B} = \mu_{2,B} \tag{2.171}$$

Thus, the equilibrium requires that the chemical potential of each component be the same in all phases. As stated above, there are many ways of arriving at the most general form of equilibrium criteria.

This principle may be easily extended to a system that includes multiple phases and components, and where all components may be transferred from one phase to another. Therefore, Eqs. (2.170) and (2.171) can be extended to state that the chemical potential of each component must be identical in all phases for systems to be in equilibrium at constant temperature and pressure, i.e.,

$$\begin{aligned} \mu_{1,A} &= \mu_{2,A} = \mu_{3,A} \\ \mu_{1,B} &= \mu_{2,B} = \mu_{3,B} \\ \mu_{1,C} &= \mu_{2,C} = \mu_{3,C} \end{aligned} \tag{2.172}$$

which can be repeated up to N components. This means that the chemical potential for a particular component must be equal in all phases at equilibrium.

The number of independent intensive thermodynamic variables and the number of phases for a system are related by the Gibbs phase rule:

$$\Pi + f = N + 2 \tag{2.173}$$

where Π is the number of phases present and N is the number of components present. The degrees of freedom, f, designate the number of intensive independent properties that must be specified to fix the state of a system for each phase.

The Gibbs phase rule can be proved based on our discussion of the thermodynamic equilibrium of a multiphase system. Consider a system that has N components and Π phases in equilibrium at a given temperature and pressure, and assume that each component can exist in each phase. The system in each phase could be completely specified if the concentration of each component in each phase, the temperature, and the pressure were specified, i.e., the number of degrees of freedom is

$$f = N\Pi + 2 \tag{2.174}$$

However, we know that at equilibrium, the chemical potential of each component is the same in all the phases, so we reduce the degrees of freedom by $N(\Pi - 1)$.

Finally, we also recognize that since the sum of the mole fractions equals unity in each of the Π phases, we may also reduce the degrees of freedom by Π additional intensive properties. Subtracting these two corrections from the original degrees of freedom in Eq. (2.174) gives

$$f = N\Pi + 2 - \Pi - N(\Pi - 1) = N + 2 - \Pi \tag{2.175}$$

which can be rearranged to give the Gibbs phase rule, Eq. (2.173).

2.3.5 Metastable Equilibrium and Nucleation

Equilibriums can be classified as (a) stable equilibrium, (b) metastable equilibrium, and (c) unstable equilibrium; these equilibriums can be illustrated using analogy examples in the mechanical equilibrium of a ball shown in Fig. 2.6. The ball in Fig. 2.6a is in stable equilibrium because it can always return to equilibrium after displacement. The ball in Fig. 2.6b is in metastable equilibrium because it can return to equilibrium after *small* displacement. If the displacement is large, the ball will move to a new equilibrium position. The ball in Fig. 2.6c is in unstable equilibrium because equilibrium cannot be maintained after any displacement.

The nature of the metastable equilibrium is defined as stable equilibrium restricted to small systematic and environmental changes. If the changes of the systematic or environmental variables exceed the restricted range, the metastable system becomes unstable. When imbalances in the intensive variables are large enough, a spontaneous change must occur in the system to bring the system to a new equilibrium state. However, many situations arise in which the changes proceed slowly enough that departures from stable equilibrium are small. Consequently, the unstable intermediate states may closely approximate a stable equilibrium path, and time is no longer an important factor.

In thermodynamics, metastable regions play an important role in determining equilibrium states. Figure 2.7 shows a p–v diagram for a pure substance—an isothermal slice through a surface on the p–v–T diagram. The liquid–vapor phase change occurs along an isotherm ($1 \rightarrow 2 \rightarrow 4 \rightarrow 5$) that consists of three states: liquid, two-phase mixture, and vapor. Under stable conditions, the liquid phase at point 1 may expand along the isotherm $1 \rightarrow 2$. At point 2, the fluid reaches the saturated liquid state, and continued expansion under stable conditions results in vaporization, represented by the path $2 \rightarrow 4$. On the other hand, the superheated vapor phase at point 5 may be compressed along the same isotherm $5 \rightarrow 4$. At point 4, the fluid becomes saturated vapor and further compression results in condensation, represented by path $4 \rightarrow 2$. These single-phase paths ($1 \rightarrow 2$ and $5 \rightarrow 4$), as well as the phase-change path $2 \leftrightarrow 4$, are completely reversible under stable conditions.

However, the volume of the liquid can be increased along line $2 \rightarrow 2'$ instead of going through process $2 \rightarrow 3$. Therefore, it is possible in the absence of vapor bubble nucleation to superheat the liquid above the saturation temperature. The volume of the vapor can also be decreased along line $4 \rightarrow 4'$, which means that, in the absence of liquid droplet nucleation, the vapor can be subcooled

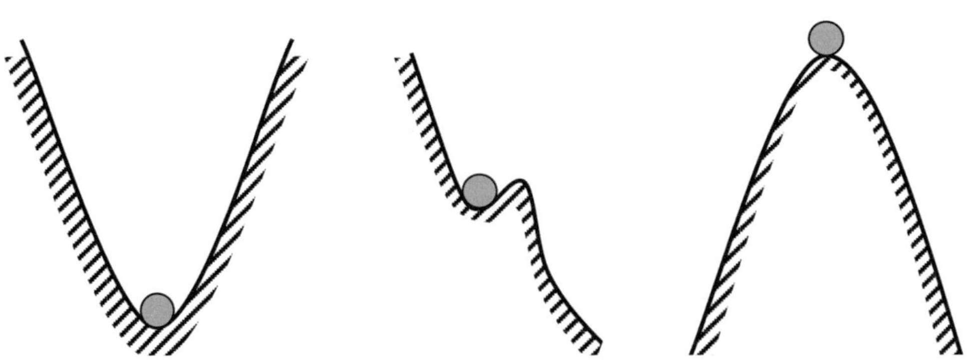

(a) stable equilibrium. (b) metastable equilibrium. (c) unstable equilibrium.

Fig. 2.6 Schematic of stability

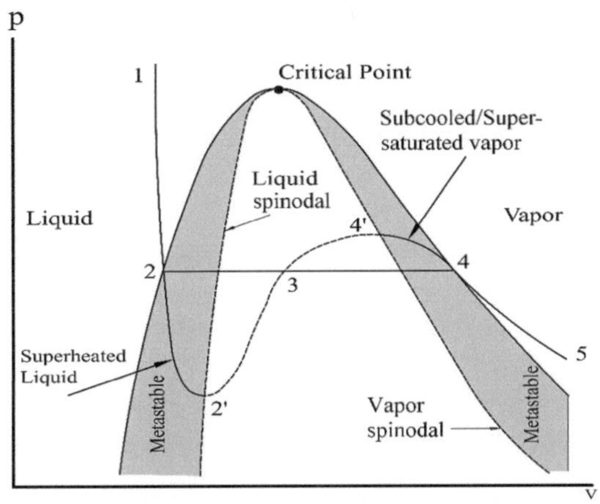

Fig. 2.7 p-v diagram for a pure substance illustrating metastable equilibrium

below its saturation temperature. The superheated liquid and subcooled vapor are both in metastable equilibrium because the criterion for mechanical stability represented by Eq. (2.83) is satisfied. However, the states along the path $2' \rightarrow 3 \rightarrow 4'$ are completely unstable because when moving along this path, $(\partial p/\partial v)_T > 0$, which violates Eq. (2.83). Therefore, the path $2' \leftrightarrow 4'$ is not accessible for boiling and condensation. The loci of the limiting points $2'$ and $4'$, where $(\partial p/\partial v)_T = 0$, are called liquid and vapor spinodals, respectively.

Since the states along the path $2' \rightarrow 3 \rightarrow 4'$ are not in an equilibrium condition, the equations of state are not valid to describe them. However, this path is very similar to the isotherm obtained by using the van der Waals equation. If it is assumed that the van der Waals equation (2.130) is valid to describe this path, one can estimate the parameters on the spinodal by using

$$\left(\frac{\partial p}{\partial v}\right)_T = -\frac{R_g T}{(v-b)^2} + \frac{2a}{v^3} = 0 \tag{2.176}$$

The thermodynamic parameters on the spinodal can be determined by using Eqs. (2.130) and (2.176).

Example 2.5

A 1-m^3 rigid vessel is filled with 80 kg of propane. The constants a and b in the van der Waals equation are $a = 479.78\,\text{Pa m}^6/\text{kg}^2$ and $b = 2.04 \times 10^{-3}\,\text{m}^3$. The vessel is cooled in order to condense the propane. Determine the temperature at which condensation will occur. What is the corresponding pressure?

Solution Since the vessel is rigid, the specific volume of the propane will remain constant in the cooling process. The specific volume of the propane is

$$v = \frac{V}{m} = \frac{1}{80} = 0.0125\,\text{m}^3/\text{kg}$$

When the vessel is cooled isochorically, the propane gas becomes supercooled and enters a metastable state. Continued cooling below the temperature corresponding to the temperature at the vapor spinodal will make the system unstable and result in condensation. This temperature can be found using Eq. (2.176), i.e.,

$$T = \frac{2a(v-b)^2}{R_g v^3} = \frac{2 \times 479.78 \times (0.0125 - 0.00204)^2}{0.188 \times 10^3 \times 0.0125^3} = 285.92\,\mathrm{K} = 12.77\,^\circ\mathrm{C}$$

The corresponding pressure of the propane can be found using the van der Waals equation, Eq. (2.130), i.e.,

$$p = \frac{R_g T}{v-b} - \frac{a}{v^2} = \frac{0.188 \times 10^3 \times 285.92}{0.0125 - 0.00204} - \frac{479.78}{0.0125^2} = 2.07\,\mathrm{MPa}$$

For liquid heated at constant pressure above its corresponding saturation temperature, the liquid spinodal (point 2′) represents a maximum upper limit of superheat based on thermodynamic consideration; it is referred to as the thermodynamic limit of superheat. Similarly, the spinodal limit for supercooled vapor (point 4′) is the maximum thermodynamic limit for supercooling of vapor. While the spinodal limits provide maximum limits on the superheat or supercooling, nucleation of new phases occurs in temperature range defined by the saturation temperature and the spinodal limits. Nucleation of vapor that occurs completely in liquid, or nucleation of liquid that occurs completely in vapor, is referred to as homogeneous nucleation. On the other hand, if nucleation occurs at an interface between the metastable phase (liquid or vapor) and solid, it is called heterogeneous nucleation. The conditions for nucleation of the liquid phase in vapor (condensation) and nucleation of the vapor phase in liquid (boiling) will be discussed in detail in Chaps. 7 and 9, respectively.

2.4 Thermodynamics of the Interfaces

2.4.1 Equilibrium at the Interface

Two bulk fluids of large extent, separated by an interfacial region, constitute a system in equilibrium. This very general description can be used to consider, for example, the cases of two immiscible liquids in contact with each other; a single substance in two phases; or a mixture of gases in contact with a solid, with a chemical reaction occurring on the surface of the solid. When analyzing such systems, the interface is a unique region that requires special attention. When mass transport occurs between the two bulk substances, the interface problem becomes significantly more complicated. Any mass exchange between the two bulk substances also requires consideration of momentum and energy exchange. A single substance undergoing a phase change is the simplest case of mass transport across an interfacial surface. To develop an understanding of the unique and significant effects of interfacial surfaces on the interaction of two bulk systems, a thermodynamic analysis will be performed here.

Surfaces 1 and 2 constitute a demarcation of the region that possesses all of the properties of the bulk fluids. The dividing surface I is at an arbitrary location within the region between surfaces 1 and 2 (see Fig. 2.8). For a single-component system, the fundamental thermodynamic relation represented by Eq. (2.56) can be simplified as

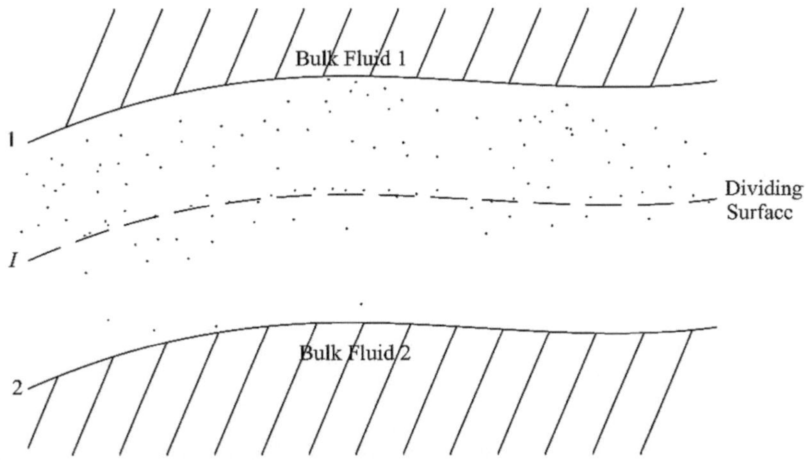

Fig. 2.8 Interfacial region between two fluids

$$dE = TdS - pdV + \mu dn \tag{2.177}$$

Suppose the configuration of the volume and surfaces 1, 2, and I is fixed. In this case, the internal energy is only a function of S and n, and

$$dE = TdS + \mu dn \tag{2.178}$$

for each 1, 2, and I within the interfacial region.

Now consider a process wherein some mass and energy exchange occurs between the bulk fluid 2 and the interfacial surface I, with bulk fluid 1 remaining unchanged. The total energy of the entire (larger) system, comprised of bulk fluids 1 and 2 and the interface, remains constant. Thus, an energy balance for the system requires of the new equilibrium that

$$E_{\text{total}} = E_I + E_1 + E_2 = \text{constant} \tag{2.179}$$

The energy in the region between 2 and I, with the energy at 1 held constant, is then related by

$$dE_{\text{total}} = dE_I + dE_2 = 0 \tag{2.180}$$

i.e.,

$$0 = T_I dS_I + \mu_I dn_I + T_2 dS_2 + \mu_2 dn_2 \tag{2.181}$$

When mass and energy are exchanged between bulk region 2 and surface I, with fluid 1 remaining unchanged, the total mole number for bulk region 2 and surface I satisfies

$$n_{\text{total}} = n_I + n_2 = \text{constant} \tag{2.182}$$

i.e.,

$$\mathrm{d}n_{\mathrm{total}} = 0 = \mathrm{d}n_I + \mathrm{d}n_2 \tag{2.183}$$

$$\mathrm{d}n_I = -\mathrm{d}n_2 \tag{2.184}$$

The entropy balance for the system requires that

$$S_{\mathrm{total}} = S_I + S_1 + S_2 = \text{constant} \tag{2.185}$$

$$\mathrm{d}S_{\mathrm{total}} = 0 = \mathrm{d}S_I + \mathrm{d}S_2 \tag{2.186}$$

$$\mathrm{d}S_I = -\mathrm{d}S_2 \tag{2.187}$$

Substituting these into the energy-accounting Eq. (2.181) gives the conditions for equilibrium of the process described above as:

$$(T_2 - T_I)\mathrm{d}S_2 + (\mu_2 - \mu_I)\mathrm{d}n_2 = 0 \tag{2.188}$$

Since Eq. (2.188) must be valid for any $\mathrm{d}S_2$ and $\mathrm{d}n_2$, the following conditions must be satisfied:

$$T_2 = T_I \tag{2.189}$$

$$\mu_2 = \mu_I \tag{2.190}$$

The same procedure can be used to show that

$$T_1 = T_I \tag{2.191}$$

$$\mu_1 = \mu_I \tag{2.192}$$

Thus, at equilibrium, the temperatures and chemical potentials for each of the three regions must be equal, i.e.,

$$T_1 = T_I = T_2 \tag{2.193}$$

$$\mu_1 = \mu_I = \mu_2 \tag{2.194}$$

2.4.2 Surface Tension

The liquid–vapor (gas) interface is often treated as a sharp discontinuity in macroscale thermodynamics and heat transfer. However, the change of properties between different phases actually occurs over a very thin but finite region, as shown in Fig. 2.8. Since the density of the liquid is higher than that of the vapor, the molecules in the liquid are closer to each other and the intermolecular forces are attractive in nature.

By formulating the problem of the phase interface in terms of the *surface excess* quantities (Carey 2016), classical thermodynamics can then be used to determine the relationship between surface tension and other macroscopic variables. Equilibrium conditions establish the equality of T and μ for the three regions, i.e., the bulk fluids and the interface. The total mole number of the two-phase system shown in Fig. 2.8 can be written as

$$n_{\text{total}} = n_I + n_1 + n_2 \tag{2.195}$$

The mole number of the interface is then

$$n_I = n_{\text{total}} - n_1 - n_2 \tag{2.196}$$

Likewise, for internal energy,

$$E_I = E_{\text{total}} - E_1 - E_2 \tag{2.197}$$

The surface I possesses all the information required to analyze this region. The exact location of this surface—referred to as the dividing surface—does not need to be known at this time. An imaginary surface defined to possess these properties is sufficient.

As stated by Eq. (2.178), the internal energy of the surface is

$$E_I = E_I(S_I, n_I) \tag{2.198}$$

which is valid when the shape and location of the interface are fixed and flat. If the interface is deformable, both the shape and the area of the interface will affect the internal energy of the interface. The internal energy of the deformable interface becomes

$$E_I = E_I(S_I, n_I, A, K) \tag{2.199}$$

where A and K are the surface area and the curvature of the interface, respectively.

The change in E is then

$$\mathrm{d}E = \left(\frac{\partial E}{\partial S}\right)_{n,A,K} \mathrm{d}S + \left(\frac{\partial E}{\partial n}\right)_{S,A,K} \mathrm{d}n + \left(\frac{\partial E}{\partial A}\right)_{S,n,K} \mathrm{d}A + \left(\frac{\partial E}{\partial K}\right)_{S,n,A} \mathrm{d}K \tag{2.200}$$

where the subscript I is dropped for ease of notation. Considering Eq. (2.178), Eq. (2.200) can be rewritten in the following form:

$$\mathrm{d}E = T\mathrm{d}S + \mu \mathrm{d}n + \left(\frac{\partial E}{\partial A}\right)_{S,n,K} \mathrm{d}A + \left(\frac{\partial E}{\partial K}\right)_{S,n,A} \mathrm{d}K \tag{2.201}$$

If the variation of the curvature effect is negligible, Eq. (2.201) simplifies to

$$\mathrm{d}E = T\mathrm{d}S + \mu \mathrm{d}n + \sigma \mathrm{d}A \tag{2.202}$$

where σ is the surface tension, defined as

$$\sigma = \left(\frac{\partial E}{\partial A}\right)_{S,n} \tag{2.203}$$

Another representation of the surface tension uses the Helmholtz free energy for the surface region, which is defined as

$$F_I = E_I - TS_I \tag{2.204}$$

Differentiating the above equation gives

$$\mathrm{d}F_I = \mathrm{d}E_I - T\mathrm{d}S_I - S_I\mathrm{d}T \tag{2.205}$$

Substituting Eq. (2.202) into Eq. (2.205), one obtains

$$\mathrm{d}F = -S\mathrm{d}T + \mu\mathrm{d}n + \sigma\mathrm{d}A \tag{2.206}$$

If the differential of the Helmholtz function is as given above, then the function can be considered as

$$F = F(T, n, A) \tag{2.207}$$

which implies that

$$\sigma = \left(\frac{\partial F}{\partial A}\right)_{T,n} \tag{2.208}$$

As the force acting on the interface, surface tension tends to resist an increase in the interfacial area. The work done on the system to increase the area of the interface is given by

$$\text{work} = \sigma\mathrm{d}A \tag{2.209}$$

which supplies a second definition of surface tension: the work per unit area required to produce a new surface. When the interface area is increased by dA, the perimeter of the interface is increased by dP. Therefore, the work done on the system, Eq. (2.209), can also be expressed as a product of the force per unit length of the perimeter and increase of the perimeter, dP. It follows that there are two equivalent interpretations of surface tension, σ: (1) energy per unit area of the surface and (2) force per unit perimeter of the surface.

Equation (2.209) relates σ to the work required to increase the area of a surface. Thermodynamics establishes that work is a path-dependent function. It is convenient to shift our emphasis from work done *on* the system to work done *by* the system when considering change of area. If work done by the system when its area is changed is defined as δW_σ, Eq. (2.209) becomes

$$\delta W_\sigma = -\sigma\mathrm{d}A \tag{2.210}$$

According to Eq. (2.210), a decrease in area (dA negative) corresponds to work done *by* the system, whereas an increase in area requires work to be done *on* the system (dA positive and δW_σ negative). The quantity δW_σ can also be related to other thermodynamic variables. According to the first law, the change in the internal energy E of the system equals

$$\mathrm{d}E = \delta Q - \delta W \tag{2.211}$$

in which δW is the work done by the system and δQ is the thermal energy absorbed by the system. The quantity δW is conveniently divided into a pressure–volume term and a nonpressure-volume term:

$$\delta W = \delta W_{pV} + \delta W_{\text{non}-pV} = p\text{d}V + \delta W_{\text{non}-pV} \tag{2.212}$$

The nonpressure-volume types of work include electrical and chemical, as well as other non-pV types of mechanical work. The work defined by Eq. (2.210) may also be classified as nonpressure-volume work.

The second law of thermodynamics indicates that for reversible processes,

$$\delta Q_{\text{rev}} = T\text{d}S \tag{2.213}$$

Substituting Eqs. (2.212) and (2.213) into Eq. (2.211), with the stipulation of reversibility as required by Eq. (2.213), one obtains

$$\text{d}E = T\text{d}S - p\text{d}V - \delta W_{\text{non}-pV} \tag{2.214}$$

Recalling the definition of the Gibbs free energy G,

$$G = H - TS = E + pV - TS \tag{2.215}$$

Differentiating Eq. (2.215) yields

$$\text{d}G = \text{d}E + p\text{d}V + V\text{d}p - T\text{d}S - S\text{d}T \tag{2.216}$$

Substituting Eq. (2.214) into Eq. (2.216) gives

$$\text{d}G = V\text{d}p - S\text{d}T - \delta W_{\text{non}-pV} \tag{2.217}$$

which shows that for a constant temperature, constant pressure, reversible process,

$$\text{d}G = -\delta W_{\text{non}-pV} \tag{2.218}$$

that is, dG equals the maximum nonpressure-volume work derivable from such a process, since maximum work is associated with reversible processes. We have already seen through Eq. (2.210) that changes in surface area entail nonpressure-volume work. Therefore, we can identify δW_σ from Eq. (2.210) with $\delta W_{\text{non}-pV}$ in Eq. (2.218) and write

$$\text{d}G = \sigma\text{d}A \tag{2.219}$$

Considering the stipulations made in going from Eq. (2.216) to Eq. (2.218), we obtain

$$\sigma = \left(\frac{\partial G}{\partial A}\right)_{T,p} \tag{2.220}$$

which identifies the surface tension as the increment in Gibbs free energy per unit increment in area. The path-dependent variable δW_σ is replaced by a state variable as a result of this analysis. The three definitions of surface tension given by Eqs. (2.203), (2.208), and (2.220) are equivalent to each other, and different definitions can be applied to different systems.

It is worthwhile to discuss surface tension from the molecular perspective in order to understand the mechanism of surface tension for different substances. Surface tension can be considered as the summation of two parts: One part is due to dispersion force, and the other part is due to specific

Table 2.2 Surface tensions for different liquids at liquid–vapor interface

Types of liquid	Liquid	Temperature (°C)	Surface tension (mN/m)
Nonpolar liquid	Helium	−271	0.26
	Nitrogen	−153	0.20
Hydrogen-bonded liquid (polar)	Ammonia	−40	35.4
	Water	20	72.9
Metallic liquid	Mercury	20	484
	Silver	1100	878

forces, like metallic or hydrogen bonding (Fowkes 1965). Surface tension force in a nonpolar liquid is due only to the dispersion force; therefore, the surface tension for a nonpolar fluid is very low. For a hydrogen-bonded liquid, surface tension is slightly higher because the surface tension is due to both dispersion forces and hydrogen bonding. The surface tension for a liquid metal is highest because the surface tension is due to a combination of dispersion forces and metallic bonding, and metallic bonding is much stronger than the hydrogen bonding. The surface tensions for different liquids are quantitatively demonstrated in Table 2.2.

2.4.3 Microscale Vapor Bubbles and Liquid Droplets

Surface tension effects on system equilibrium between phases are very important. However, analysis of these systems usually requires individual attention; it is difficult to arrive at a simple form of expression that covers all cases. In this subsection, the effect of interfacial surface tension between a liquid and its vapor is considered.

For the development of this problem, we will consider phase change in an isolated rigid system which is occupied by a mixture of liquid and vapor (see Fig. 2.9). The vapor at temperature T_v and pressure p_v is contained in a microscale spherical vapor bubble of radius R_b; liquid at a constant temperature T_ℓ and pressure p_ℓ surrounds the vapor bubble. At phase equilibrium, the temperatures of

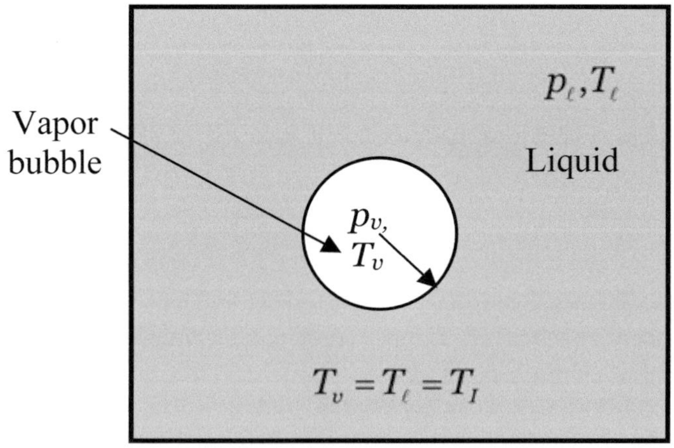

Fig. 2.9 Vapor bubble suspending in a liquid phase in a rigid vessel

the two phases are the same, i.e., $T_\ell = T_v = T$. According to Eq. (2.128), the chemical potentials of the liquid and vapor phases are also the same at equilibrium, i.e., $\mu_\ell = \mu_v = \mu$. The condition for equilibrium is chosen to be that the total Helmholtz free energy function will be at its minimum. By applying the fundamental thermodynamic relation of Helmholtz free energy function, Eq. (2.58), in the liquid and vapor phases, one obtains

$$\mathrm{d}F_\ell = -S_\ell \mathrm{d}T - p_\ell \mathrm{d}V_\ell + \mu \mathrm{d}n_\ell \tag{2.221}$$

$$\mathrm{d}F_v = -S_v \mathrm{d}T - p_v \mathrm{d}V_v + \mu \mathrm{d}n_v \tag{2.222}$$

The fundamental relation for Helmholtz free energy at the interface, as indicated by Eq. (2.206), is

$$\mathrm{d}F_I = -S_I \mathrm{d}T + \sigma \mathrm{d}A + \mu_I \mathrm{d}n_I \tag{2.223}$$

For a reversible phase-change process under constant volume and temperature, Eq. (2.32) must be satisfied, i.e.,

$$\mathrm{d}F = \mathrm{d}F_\ell + \mathrm{d}F_v + \mathrm{d}F_I = 0 \tag{2.224}$$

Substituting Eqs. (2.221)–(2.223) into Eq. (2.224), the equilibrium condition becomes

$$-(S_\ell + S_v)\mathrm{d}T - (p_\ell \mathrm{d}V_\ell + p_v \mathrm{d}V_v) + \sigma \mathrm{d}A + \mu(\mathrm{d}n_\ell + \mathrm{d}n_v + \mathrm{d}n_I) = 0 \tag{2.225}$$

The total volume does not change because the system is rigid, i.e.,

$$\mathrm{d}V = \mathrm{d}V_\ell + \mathrm{d}V_v = 0 \tag{2.226}$$

Conservation of mass requires that

$$\mathrm{d}n = \mathrm{d}n_\ell + \mathrm{d}n_v + \mathrm{d}n_I = 0 \tag{2.227}$$

Since the phase change occurs at constant temperature, we have

$$\mathrm{d}T = 0 \tag{2.228}$$

Substituting Eqs. (2.226)–(2.228) into Eq. (2.225) yields a relationship between the pressures in the liquid and vapor phases at equilibrium:

$$p_v - p_\ell = \sigma \frac{\mathrm{d}A}{\mathrm{d}V_v} \tag{2.229}$$

Since the surface area and volume of the vapor bubbles are, respectively, $A = 4\pi R_b^2$ and $V_v = 4\pi R_b^3/3$, Eq. (2.229) becomes

$$p_v - p_\ell = \frac{2\sigma}{R_b} \tag{2.230}$$

which is the Laplace–Young equation. Although Eq. (2.230) was obtained by analyzing a vapor bubble suspended in a liquid within a rigid system, it can be demonstrated that it is also valid for any other system (see Chap. 4).

At phase equilibrium, the chemical potentials of both phases must be equal. As indicated by Eq. (2.146), the chemical potential for the two-phase system is the specific Gibbs energy. Therefore, the specific Gibbs free energy functions must be equal for the liquid and vapor phases, i.e.,

$$g_\ell(p_\ell, T) = g_v(p_v, T) \tag{2.231}$$

By differentiating Eq. (2.231) and considering the fact that the temperature is constant in the phase-change process, one can obtain

$$\left(\frac{\partial g_\ell}{\partial p_\ell}\right)_T \mathrm{d}p_\ell = \left(\frac{\partial g_v}{\partial p_v}\right)_T \mathrm{d}p_v \tag{2.232}$$

Considering Eq. (2.63), Eq. (2.232) can be rewritten as

$$v_\ell \mathrm{d}p_\ell = v_v \mathrm{d}p_v \tag{2.233}$$

The relationship between changes in the liquid and vapor pressures, due to an infinitesimally small change of the vapor bubble radius, can be obtained by differentiating the Laplace–Young equation:

$$\mathrm{d}p_v - \mathrm{d}p_\ell = -\frac{2\sigma}{R_b^2}\mathrm{d}R_b \tag{2.234}$$

Substituting Eq. (2.234) into Eq. (2.233), the pressure in the vapor phase can be eliminated:

$$v_\ell\left(\mathrm{d}p_v + \frac{2\sigma}{R_b^2}\mathrm{d}R_b\right) = v_v \mathrm{d}p_v \tag{2.235}$$

If the vapor behaves like an ideal gas ($v_v = R_g T/p_v$), Eq. (2.235) can be rewritten as

$$R_g T\frac{\mathrm{d}p_v}{p_v} - v_\ell \mathrm{d}p_v = \frac{2\sigma}{R_b^2} v_\ell \mathrm{d}R_b \tag{2.236}$$

where R_g is the gas constant of the vapor. If the radius of the vapor bubble goes to infinity ($1/R_b \to 0$), the vapor pressure equals the saturation pressure corresponding to the temperature, $p_{sat}(T)$. Integrating Eq. (2.236) from an equilibrium state for a flat surface, one obtains

$$R_g T \ln\left[\frac{p_v}{p_{sat}(T)}\right] - v_\ell[p_v - p_{sat}(T)] = -\frac{2\sigma}{R_b} v_\ell \tag{2.237}$$

i.e.,

$$p_v = p_{sat}(T)\exp\left\{\frac{v_\ell[p_v - p_{sat}(T) - 2\sigma/R_b]}{R_g T}\right\} \tag{2.238}$$

which indicates that the bubble is in equilibrium only if the pressure of the vapor phase exceeds the saturation pressure $p_{sat}(T)$. In another words, the vapor phase must be superheated. If the pressure inside the vapor bubble is below the pressure required by Eq. (2.238), the vapor bubble will shrink by condensation. On the other hand, the vapor bubble will grow by evaporation if the pressure inside the bubble is higher than that required by Eq. (2.238).

Equation (2.238) can be inverted to obtain the equilibrium bubble radius:

$$R_b = \frac{2\sigma}{R_g T \ln[p_{sat}(T)/p_v]/v_\ell + p_v - p_{sat}(T)} \tag{2.239}$$

Only vapor bubbles with radii equal to that given by Eq. (2.239) will be in equilibrium with the surrounding superheated liquid at T_ℓ and p_ℓ.

For most cases, $p_v - p_{sat}(T) \ll 2\sigma/R_b$, Eqs. (2.238) and (2.239) can be simplified as

$$p_v = p_{sat}(T)\exp\left\{-\frac{2\sigma v_\ell}{R_b R_g T}\right\} = p_{sat}(T)\exp\left\{-\frac{2\sigma/R_b}{p_v}\frac{v_\ell}{v_v}\right\} \tag{2.240}$$

$$R_b = \frac{2\sigma v_\ell}{R_g T \ln[p_{sat}(T)/p_v]} \tag{2.241}$$

Since $2\sigma/R_b \ll p_v$ and $v_\ell \ll v_v$, Eq. (2.240) can be simplified as

$$p_v \simeq p_{sat}(T) \tag{2.242}$$

The pressure in the liquid phase can be readily obtained by using the Laplace–Young equation, Eq. (2.230), i.e.,

$$p_\ell = p_{sat}(T) - \frac{2\sigma}{R_b} \tag{2.243}$$

which means that the liquid pressure must be less than the saturation pressure corresponding to the system temperature T, i.e., the liquid phase must be superheated in order to maintain phase equilibrium. The superheated liquid is in metastable equilibrium, as represented by the region between 2 and 2′ in Fig. 2.7.

Example 2.6

Nucleate boiling is characterized as generation, growth, and departure of vapor bubbles. A 0.5-mm-diameter vapor bubble is observed in superheated liquid water at a temperature of 102 °C. Find the pressures in the vapor bubble and in the liquid pool.

Solution According to Eq. (2.242), the pressure in the vapor bubble is equal to the saturation pressure corresponding to the vapor temperature. Therefore,

$$p_v = p_{sat}(T) = 1.1102 \times 10^5 \text{ Pa}$$

The surface tension of water at 102 °C is $\sigma = 58.52 \times 10^{-3}$ N/m. The pressure in the liquid pool can be obtained by using Eq. (2.243):

$$p_\ell = p_{sat}(T) - \frac{2\sigma}{R_b} = 1.1102 \times 10^5 - \frac{2 \times 58.52 \times 10^{-3}}{0.5 \times 10^{-3}} = 1.1079 \times 10^5 \text{ Pa}$$

The ratio of the vapor pressure over the liquid pressure is

$$\frac{p_{\mathrm{v}}}{p_{\ell}} = \frac{1.1102 \times 10^5}{1.1079 \times 10^5} = 1.002$$

For a liquid droplet suspended in the vapor phase (Fig. 2.10), the pressure inside the liquid droplet is related to the vapor pressure by the Laplace–Young equation

$$p_{\ell} - p_{\mathrm{v}} = \frac{2\sigma}{R_{\mathrm{d}}} \tag{2.244}$$

which can be differentiated to yield

$$\mathrm{d}p_{\ell} - \mathrm{d}p_{\mathrm{v}} = -\frac{2\sigma}{R_{\mathrm{d}}^2}\mathrm{d}R_{\mathrm{d}} \tag{2.245}$$

At phase equilibrium, the specific Gibbs free energy functions must be equal for the liquid and vapor phases [see Eq. (2.231)]. By following a procedure similar to that for the case of the vapor bubble, it can be shown that Eq. (2.233) is also valid for the case of liquid droplet in vapor phase.

Substituting Eq. (2.245) into Eq. (2.233), the pressure in the vapor phase can be eliminated:

$$v_{\ell}\left(\mathrm{d}p_{\mathrm{v}} - \frac{2\sigma}{R_{\mathrm{d}}^2}\mathrm{d}R_{\mathrm{d}}\right) = v_{\mathrm{v}}\mathrm{d}p_{\mathrm{v}} \tag{2.246}$$

Applying ideal gas laws to the vapor phase, Eq. (2.246) can be rewritten as

$$R_{\mathrm{g}}T\frac{\mathrm{d}p_{\mathrm{v}}}{p_{\mathrm{v}}} - v_{\ell}\mathrm{d}p_{\mathrm{v}} = -\frac{2\sigma}{R_{\mathrm{d}}^2}v_{\ell}\mathrm{d}R_{\mathrm{d}} \tag{2.247}$$

If the radius of the liquid droplet equals infinity ($1/R_{\mathrm{d}} \to 0$), the vapor pressure equals the saturation pressure corresponding to the temperature, $p_{\mathrm{sat}}(T)$. Integrating Eq. (2.247) from an equilibrium state for a planar surface, the pressure in the vapor phase is obtained:

$$p_{\mathrm{v}} = p_{\mathrm{sat}}(T)\exp\left\{\frac{v_{\ell}[p_{\mathrm{v}} - p_{\mathrm{sat}}(T) + 2\sigma/R_{\mathrm{d}}]}{R_{\mathrm{g}}T}\right\} \tag{2.248}$$

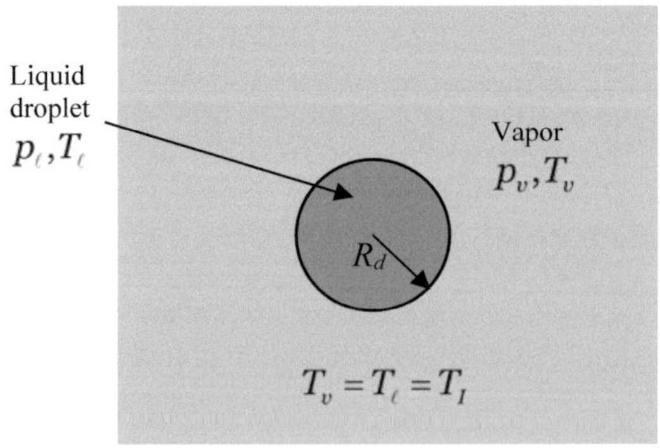

Fig. 2.10 Liquid droplet suspending in vapor phase

If the vapor pressure is below the pressure required by Eq. (2.248), the liquid droplet will shrink by evaporation. The liquid droplet will grow by condensation if the vapor pressure is higher than that required by Eq. (2.248).

For most cases, $p_v - p_{sat}(T) \ll 2\sigma/R_d$, Eq. (2.248) can be simplified as

$$p_v = p_{sat}(T)\exp\left(\frac{2\sigma v_\ell}{R_d R_g T}\right) \tag{2.249}$$

The equilibrium droplet radius can be obtained by reverting Eq. (2.249), i.e.,

$$R_d = \frac{2\sigma v_\ell}{R_g T \ln[p_v/p_{sat}(T)]} \tag{2.250}$$

which is similar to Eq. (2.241) except the pressure ratio in the denominator is different.

It can be seen that equilibrium of a liquid droplet requires that the vapor phase be supersaturated. The degree of supersaturation, as measured by $p_v/p_{sat}(T)$, is dependent from the size of the vapor bubble or liquid droplet as indicated by Eqs. (2.240) and (2.249).

Example 2.7

A liquid water droplet with radius, R_d, is suspended in steam at 1 atm. Quantitatively demonstrate the dependence of the degree of supersaturation on the radius of the droplet and the number of molecules in each droplet.

Solution At 1 atm, the properties of the water can be found from Table B.48 as $v_\ell = 1/958.77 = 1.043 \times 10^{-3}\,\text{m}^3/\text{kg}$, $\sigma = 58.91 \times 10^{-3}\,\text{N/m}$. The gas constant of the water vapor is $R_g = R_u/M = 8.3143 \times 10^3/18.0\ 461.9\,\text{J/kg K}$. The degree of supersaturation is therefore expressed as:

$$\frac{p_v}{p_{sat}} = \exp\left(\frac{2\sigma v_\ell}{R_d R_g T}\right)$$
$$= \exp\left(\frac{2 \times 58.91 \times 10^{-3} \times 1.043 \times 10^{-3}}{R_d \times 461.9 \times 373}\right) = \exp\left(\frac{7.13 \times 10^{-10}}{R_d}\right)$$

The number of molecules in the droplet is estimated bywhere $N_A = 6.022 \times 10^{23}$ is the Avogadro's number, which is the number of molecules per mole; m is the mass of the liquid droplet; and $M = 18.0\,\text{kg/kmol}$ is the molecular mass of the water. Thus

$$N = \frac{1000 \times 6.022 \times 10^{23}}{18.0} \times 958.77 \times \frac{4}{3}\pi R_d^3 = 1.34 \times 10^{29} R_d^3$$

The dependence of the degree of supersaturation on the size of the droplet and the number of molecules per droplet can be shown in Table 2.3. The results in Table 2.3 can be interpreted as follows. If the degree of supersaturation is very negligible (e.g., $p_v/p_{sat} = 1.000007$), 1.34×10^{17} water molecules must come together spontaneously for the liquid phase pressure to be nucleated. As the degree of the supersaturation increases, the number of molecules needed to come together to nucleate the liquid phase will be significantly reduced. In reality, foreign nuclei often act as the seed for the liquid droplet.

Table 2.3 Dependence of supersaturation pressure ratio of water at 1 atm on the size of the droplet

Droplet radii (m)	Number of molecules per droplet	Supersaturation pressure ratio (p_v/p_{sat})
1.0×10^{-9}	1.34×10^{2}	2.040612
1.0×10^{-8}	1.34×10^{5}	1.07393
1.0×10^{-7}	1.34×10^{8}	1.007158
1.0×10^{-6}	1.34×10^{11}	1.000714
1.0×10^{-5}	1.34×10^{14}	1.000071
1.0×10^{-4}	1.34×10^{17}	1.000007

2.4.4 Disjoining Pressure

Disjoining pressure is a phenomenon that occurs in thin liquid films (Israelachvilli 1992). When ultra-thin liquid films are in contact with a solid surface, there is attraction between the liquid molecules and the solid molecules. The pressure in the liquid must balance the ambient pressure and the attractive forces between the liquid and solid. When a film is very thin, the liquid-solid attractive forces act to pull the liquid away, and the balancing pressure that counteracts this force is the disjoining pressure.

Disjoining pressure theory has been used in ultra-thin films on solid surfaces to model the molecular force interactions between the liquid-solid interfaces. This has been used most extensively in modeling thin film transport in microheat pipes, axially grooved evaporators, and microheat pump loops (Faghri 2016). The idea of disjoining pressure is well known as an explanation of the effect of wall–fluid force interaction in thin films.

Carey and Wemhoff (2005) analyzed the effects of disjoining pressure in ultra-thin layers and films, as discussed below. An example of a situation that involves disjoining pressure in micropassages is shown in Fig. 2.11. The deeper Region A carries most of the liquid while Region B carries a thin-film flow with a thickness δ. Most of the interface (Region A) is flat and separates the deeper liquid flow from the vapor at equilibrium

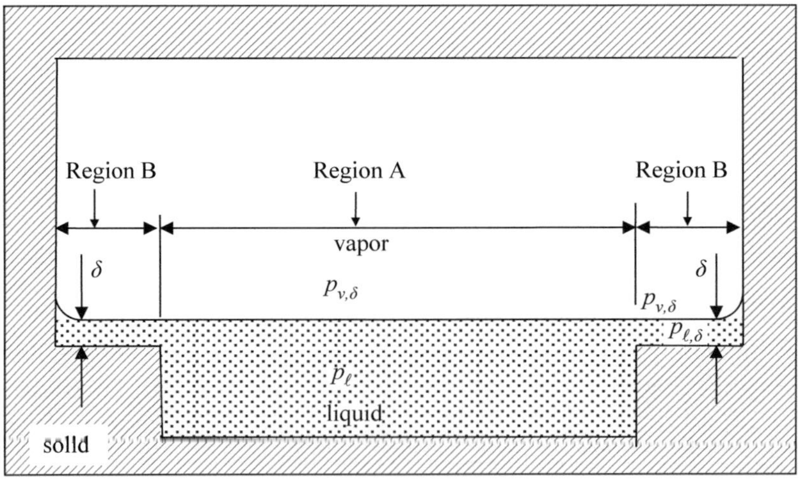

Fig. 2.11 Analysis of disjoining pressure in a cross section of a micropassage containing thin liquid films for a stratified configuration

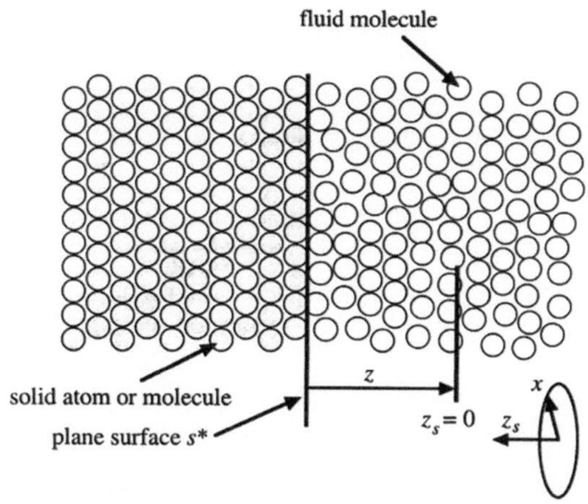

Fig. 2.12 Schematic used for derivation of disjoining pressure (Carey and Wemhoff 2005)

$$p_{v,\delta} = p_\ell \tag{2.252}$$

where p_ℓ is the liquid pressure at the interface in the absence of attractive wall forces and $p_{v,\delta}$ is the vapor pressure at the interface. The pressure in the thin film at Region B is changed by force interactions between the liquid and solid. The general disjoining pressure is derived using the potential energy from the forces that the wall exerts on the fluid. To find the pressure variation in the film, the interface between solid and fluid as shown in Fig. 2.12 is considered.

The interactions between fluid and metallic solid molecules are modeled using the Lennard–Jones interaction potential:

$$\phi_{fs}(r) = -\frac{C_{\phi,fs}}{r^6}\left[1 - \frac{D_m^6}{r^6}\right] \tag{2.253}$$

where ϕ_{fs} is the solid–fluid intermolecular potential and r is the distance between molecules. The long-range attraction between a pair of two fluid molecules is assumed to have a similar form:

$$\phi_{ff}(r) = -\frac{C_{\phi,ff}}{r^6}\left[1 - \frac{D_f^6}{r^6}\right] \tag{2.254}$$

where ϕ_{ff} is the fluid-fluid intermolecular potential. The constants D_m and D_f are the closest approach distance of fluid to solid molecules and two fluid molecules, respectively. To find the total effect of all the solid molecules on a free fluid molecule, the product of density and molecular potential is integrated. The mean-field potential energy, Φ_{fmf}, felt by the free fluid molecule due to interactions with all the solid molecules, as seen in Fig. 2.12, is

$$\Phi_{fmf} = \int_{z_s=z}^{\infty}\int_{x=0}^{\infty} \mathfrak{N}_s \phi_{fs}(2\pi x)\mathrm{d}x\mathrm{d}z_s \tag{2.255}$$

where $\mathfrak{N}_s$ is the solid wall molecular number density. Substituting Eq. (2.253) and $x^2 + z^2$ for r^2 reduces Eq. (2.255) to

$$\Phi_{\mathrm{fmf}} = -\frac{\pi \mathfrak{N}_{\mathrm{s}} C_{\phi,\mathrm{fs}}}{6z^3} + \frac{\pi \mathfrak{N}_{\mathrm{s}} C_{\phi,\mathrm{fs}} D_{\mathrm{m}}^6}{45z^9} \tag{2.256}$$

The above relation is reorganized in terms of a modified Hamaker constant A_{H}

$$A_{\mathrm{H}} = \pi^2 \mathfrak{N}_{\mathrm{f}} \mathfrak{N}_{\mathrm{s}} C_{\phi,\mathrm{fs}} \tag{2.257}$$

where $\mathfrak{N}_f$ is the fluid molecular number density. Considering Eq. (2.257), Eq. (2.256) becomes

$$\Phi_{\mathrm{fmf}} = \frac{A_{\mathrm{H}}}{6\pi \mathfrak{N}_{\mathrm{f}} D_{\mathrm{m}}^3} \left[\frac{2}{15} \left(\frac{D_{\mathrm{m}}}{z} \right)^9 - \left(\frac{D_{\mathrm{m}}}{z} \right)^3 \right] \tag{2.258}$$

which is equivalent to a body force similar to the hydrostatic variation of pressure caused by gravity. A force balance is used to obtain the pressure gradient

$$\frac{\mathrm{d}p}{\mathrm{d}z} \mathbf{n}_z = -\frac{\mathfrak{N}_{\mathrm{f}} M}{N_A} \mathbf{f}_{\mathrm{fs}} \tag{2.259}$$

where $\mathbf{n}_z$ is the unit vector in the z-direction and $\mathbf{f}_{\mathrm{fs}}$ is the force per unit mass on the fluid system. The force exerted on a single molecule by the entire wall is given by

$$\mathbf{F}_{\mathrm{fs}} = -\frac{\mathrm{d}\Phi_{\mathrm{fmf}}}{\mathrm{d}z} = \frac{A_{\mathrm{H}}}{2\pi \mathfrak{N}_{\mathrm{f}} D_{\mathrm{m}}^4} \left[\frac{2}{5} \left(\frac{D_{\mathrm{m}}}{z} \right)^{10} - \left(\frac{D_{\mathrm{m}}}{z} \right)^4 \right] \mathbf{n}_z \tag{2.260}$$

The corresponding force per unit mass is

$$\mathbf{f}_{fs} = \frac{N_A \mathbf{F}_{\mathrm{fs}}}{M} = \frac{N_{\mathrm{A}} A_{\mathrm{H}}}{2\pi M \mathfrak{N}_{\mathrm{f}} D_{\mathrm{m}}^4} \left[\frac{2}{5} \left(\frac{D_{\mathrm{m}}}{z} \right)^{10} - \left(\frac{D_{\mathrm{m}}}{z} \right)^4 \right] \mathbf{n}_z \tag{2.261}$$

where N_{A} is Avogadro's number. Substituting Eq. (2.261) into Eq. (2.259) and considering that the force only acts in the z-direction,

$$\frac{\mathrm{d}p}{\mathrm{d}z} = -\frac{A_{\mathrm{H}}}{2\pi D_{\mathrm{m}}^4} \left[\frac{2}{5} \left(\frac{D_{\mathrm{m}}}{z} \right)^{10} - \left(\frac{D_{\mathrm{m}}}{z} \right)^4 \right] \tag{2.262}$$

Integrating both sides of Eq. (2.262) from z to ∞, and considering the pressure at ∞ is p_ℓ, the pressure profile close to the wall is

$$p(z) = p_\ell + \frac{A_{\mathrm{H}}}{6\pi D_{\mathrm{m}}^3} \left[\frac{2}{15} \left(\frac{D_{\mathrm{m}}}{z} \right)^9 - \left(\frac{D_{\mathrm{m}}}{z} \right)^3 \right] \tag{2.263}$$

The z^{-9} term in Eq. (2.263) may be neglected because D_{m} is on the order of a molecular diameter. Equation (2.263) simplifies to

$$p(z) = p_\ell - \frac{A_H}{6\pi z^3} \tag{2.264}$$

The pressure in the thin liquid film at Region B in Fig. 2.11 is expected to vary with distance from the lower wall. At the interface $(z = \delta)$, the pressure in the liquid $p_{\ell,\delta}$ must be

$$p_{\ell,\delta} = p(\delta) = p_\ell - \frac{A_H}{6\pi\delta^3} \tag{2.265}$$

Combining Eqs. (2.252) and (2.265) and solving for $p_{v,\delta}–p_{\ell,\delta}$ yields

$$p_{v,\delta} - p_{\ell,\delta} = \frac{A_H}{6\pi\delta^3} \tag{2.266}$$

The disjoining pressure p_d is the amount that $p_{v,\delta}$ differs from $p_{\ell,\delta}$

$$p_d = -\frac{A_H}{6\pi\delta^3} \tag{2.267}$$

The pressure difference across the interface in Fig. 2.11 is equal to the disjoining pressure, which quickly increases in magnitude as the film thins. The disjoining pressure has been found to alter thermodynamic equilibrium conditions at the liquid–vapor interface of thin films. The change in vapor pressure versus temperature relation must be considered when modeling thin film evaporation and condensation in micropassages of microheat pipes and microcapillary-pumped loops.

Disjoining pressure can also be developed using the classical thermodynamic analysis by integrating the Gibbs–Duhem equation at a constant temperature from saturation conditions to an arbitrary state in the liquid and vapor phases.

$$d\bar{\mu} = -\bar{s}dT + \bar{v}dp \tag{2.268}$$

where $\bar{s}$ is the molar entropy and $\bar{v}$ is the molar specific volume. The liquid is assumed incompressible, and the vapor is an ideal gas. This yields the following for liquid and vapor molar chemical potentials, $\bar{\mu}_\ell$ and $\bar{\mu}_v$ respectively, without wall attraction effects:

$$\bar{\mu}_v = \bar{\mu}_{v,sat} + R_u T \ln(p_{v,\delta}/p_{sat}) \tag{2.269}$$

$$\bar{\mu}_\ell = \bar{\mu}_{\ell,sat} + \bar{v}_\ell \ln(p_\ell - p_{sat}) \tag{2.270}$$

where $\bar{\mu}_{\ell,sat}$ and $\bar{\mu}_{v,sat}$ are the molar chemical potential of bulk liquid and vapor at saturation, respectively. The chemical potential is equal to the specific Gibbs function for a pure liquid:

$$\bar{\mu}_\ell = \bar{g}_\ell \tag{2.271}$$

When considering wall attraction effects in the liquid film, the potential energy associated with the interaction of fluid and surface molecules is added to the Gibbs function. The liquid chemical potential becomes

$$\bar{\mu}_\ell = \bar{\mu}_{\ell,\text{sat}} + \bar{v}_\ell \ln(p_\ell - p_{\text{sat}}) + N_A \Phi_{\text{fmf}} \tag{2.272}$$

where Φ_{fmf} is the potential energy per fluid molecule due to interactions with the wall, and $\bar{v}_\ell$ is the liquid molar specific volume. For equilibrium conditions at the interface, the liquid pressure must be equal to the vapor pressure

$$p_\ell = p_{v,\delta} \tag{2.273}$$

Setting p_ℓ equal to $p_{v,\delta}$ and equating the right sides of Eqs. (2.269) and (2.272)

$$R_u T \ln\left(p_{v,\delta}/p_{\text{sat}}\right) = \bar{v}_\ell\left(p_{v,\delta} - p_{\text{sat}}\right) + N_A \Phi_{\text{fmf}} \tag{2.274}$$

where p_{sat} is the normal saturation pressure corresponding to vapor bulk of the system. Rearranging Eq. (2.274), one obtains

$$p_{v,\delta}/p_{\text{sat}} = \exp\left[\frac{p_{\text{sat}}\bar{v}_\ell}{R_u T}\left(\frac{p_{v,\delta}}{p_{\text{sat}}} - 1\right) + \frac{N_A \Phi_{\text{fmf}}}{R_u T}\right] \tag{2.275}$$

Neglecting the first term in the brackets at the right hand of Eq. (2.275) gives

$$p_{v,\delta}/p_{\text{sat}} = \exp\left(\frac{N_A \Phi_{\text{fmf}}}{k_b T}\right) \tag{2.276}$$

Substituting Eq. (2.258) into (2.276), neglecting z^{-9} and setting z equal to δ at the interface,

$$p_{v,\delta}/p_{\text{sat}} = \exp\left(-\frac{A_H}{6\pi\rho_f \delta^3 k_b T}\right) \tag{2.277}$$

Example 2.8

Show that the disjoining pressure obtained by thermodynamic analysis is consistent with Eq. (2.267).

Solution The liquid and vapor molar chemical potentials, $\bar{\mu}_v$ and $\bar{\mu}_\ell$, with attraction effects are

$$\bar{\mu}_v = \bar{\mu}_{v,\text{sat}} + R_u T \ln\left(p_{v,\delta}/p_{\text{sat}}\right) \tag{2.278}$$

$$\bar{\mu}_\ell = \bar{\mu}_{\ell,\text{sat}} + \bar{v}_\ell \ln\left(p_{\ell,\delta} - p_{\text{sat}}\right) \tag{2.279}$$

The vapor and liquid pressures are related by

$$p_{v,\delta} - p_{\ell,\delta} = p_{\text{cap}} - p_d \tag{2.280}$$

where p_{cap} is the capillary pressure. If the liquid film is flat, the capillary pressure can be neglected and Eq. (2.280) is simplified as

$$p_{v,\delta} - p_{\ell,\delta} = -p_d \tag{2.281}$$

Substituting Eq. (2.281) into Eq. (2.279) and equating the right sides of Eqs. (2.278) and (2.279) yield

$$R_u T \ln\left(p_{v,\delta}/p_{sat}\right) = \bar{v}_\ell\left(p_{v,\delta} + p_d - p_{sat}\right) \tag{2.282}$$

which can be rearranged to obtain

$$p_{v,\delta}/p_{sat} = \exp\left[\frac{p_{sat}\bar{v}_\ell}{R_u T}\left(\frac{p_{v,\delta}}{p_{sat}} - 1\right) + \frac{p_d \bar{v}_\ell}{R_u T}\right] \tag{2.283}$$

Neglecting the first term in the brackets at the right hand of Eq. (2.283) gives

$$p_{v,\delta}/p_{sat} = \exp\left(\frac{p_d \bar{v}_\ell}{R_u T}\right) \tag{2.284}$$

Comparing Eqs. (2.277) and (2.284), we have

$$p_d = -\frac{A_H}{6\pi\delta^3}\frac{N_A}{\Re_f \bar{v}_\ell} \tag{2.285}$$

where $N_A/\Re_f\bar{v}_\ell = 1$, and Eq. (2.285) will become identical to Eq. (2.267).

2.4.5 Superheat-Thermodynamic and Kinetic Limit Definitions

From the classical thermodynamic point of view, phase transformation occurs at the equilibrium normal saturation condition as a quasi-equilibrium process. However, real phase transformation usually occurs as a nonequilibrium process. For example, in vaporization processes, a superheated liquid may exist in part of the system. Similarly, in the condensation process, part of the vapor usually has been supercooled below its equilibrium normal saturation temperature.

In Fig. 2.7, it was shown that metastable conditions correspond to situations where vapor is supercooled below its normal saturation temperature or liquid is superheated above its normal equilibrium temperature. As shown in Sect. 2.2.4, mechanical stability requires that

$$\left.\frac{\partial p}{\partial v}\right|_T \leq 0 \tag{2.286}$$

Based on the above, the liquid or vapor in metastable domain is not mechanically unstable, even though it is not in thermodynamic equilibrium. In Fig. 2.7 between 2′ and 4′, $(\partial p/\partial v)_T > 0$ and this domain corresponds to a region between the liquid and vapor spinodal where it does violate mechanical stability. The superheat limit is the maximum temperature that a liquid can be heated before it homogeneously nucleates. This superheat limit can be determined thermodynamically or kinetically using kinetic and nucleation theory. The degree of superheat $(T_t - T_{sat})$ ranges from less than one to a few hundred degrees and depends on factors such as the type and amount of liquid, surface conditions, and the type and rate of heating.

A superheated liquid is one that does not follow its normal or saturated equilibrium phase boundary. Normal refers to a special case of equilibrium across a flat-plate boundary, $R \to \infty$, where R is the radius of curvature of the phase boundary.

Initial bubbles in superheated liquids are in mechanical equilibrium.

$$p_{\mathrm{v}} - p_{\ell} = \frac{2\sigma}{R} > 0 \tag{2.287}$$

Vapor pressure, p_{v}, is not the same as equilibrium vapor pressure, p_{sat} at temperature T. This is due to equilibrium across flat and curved boundaries (see Sect. 2.4.3):

$$p_{\mathrm{v}} = p_{\mathrm{sat}}(T)\exp\left\{\frac{v_{\ell}}{R_{\mathrm{g}}T}\left[p_{\mathrm{v}} - p_{\mathrm{sat}}(T) - \frac{2\sigma}{R}\right]\right\} \tag{2.288}$$

The external liquid pressure, p_{ℓ}, may be either compressive or extensive. However, for this problem, it will be a compressive pressure on the bubble wall. For a vapor bubble in which $p_{\mathrm{v}} > p_{\ell}$ and $R < \infty$, $T(p_{\ell}) > T_{\mathrm{sat}}(p_{\ell})$.

2.4.5.1 Thermodynamic Limit of Superheat

There is a limit to the extent of isobaric heating that the liquid can undergo. At this limit, the liquid is unstable and any perturbation will cause a phase transition.

The limit of stability, also called the thermodynamic limit of superheat, occurs when the entropy of an isolated system is at its maximum point in a stable equilibrium state with respect to small variations of its natural variables. The Helmholtz function, F, assumes a minimum stable equilibrium state for an open system with respect to variations. For variations from a stable state

$$\Delta F > 0 \tag{2.289}$$

The thermodynamic limit of superheat is the limit of mechanical stability since it is already thermally stable, [see Eq. (2.84)]

$$\left.\frac{\partial p}{\partial v}\right|_{T} \leq 0 \tag{2.290}$$

The calculated value of superheat depends upon the equation of state used for the calculation of stability. For example, consider van der Waals equation of state for a pure substance [see Eq. (2.130)]

$$\left(p + \frac{a}{v^2}\right)(v - b) = R_{\mathrm{g}}T \tag{2.291}$$

where a and b are constants. This equation inaccurately represents the saturation state of most substances. The spinodal curve for the van der Waals equation of state is

$$p = \frac{a}{v^2} - \frac{2ab}{v^3} \tag{2.292}$$

If a pressure is given, then v may be eliminated from Eqs. (2.291) and (2.292) to give the thermodynamic limit of superheat, $T \rightarrow T_{\mathrm{t}}$. This generally requires an iterative solution except when $p \rightarrow 0$, in which case $T_{\mathrm{t}} = 27T_{\mathrm{c}}/32$. A simple correlation for T_{t} is given by Lienhard (1976) to eliminate iteration needed, as

$$T_{\mathrm{t}} \simeq T_{\mathrm{c}}\left[\frac{27}{32} + \frac{5}{32}\left(\frac{T_{\mathrm{sat}}}{T_{\mathrm{c}}}\right)^{5.16}\right] \tag{2.293}$$

Table 2.4 Thermodynamic limit of superheat of some pure liquids at atmosphere pressure[a] (Avedisian 1986)

Substance	T_{sat}	T_{t1}	T_{t2}	T_m	T_c	$J\ (T_{t2})$
n-pentane	309	405	431	426	470	8×10^{24}
n-heptane	372	468	499	494	540	8×10^{26}
n-octane	399	494	525	514	569	2×10^{26}
Methanol	338	442	477	466	513	10^{29}
Ethanol	352	447	482	472	516	10^{30}
Water	373	552	596	575	647	9×10^{28}

[a]T_{sat}—Normal boiling point (K) at 0.101 MPa
T_{t1}—Calculated thermodynamic limit of superheat (K) at 0.101 MPa using van der Waals equation of state
T_{t2}—Calculated thermodynamic limit of superheat (K) at 0.101 MPa using Peng–Robinson equation of state
T_m—Highest measured liquid phase temperature (K) at 0.101 MPa
J—Nucleation rate (nuclei/cm^3 s) at T_{t2} and 0.101 MPa

Table 2.4 lists the thermodynamic limit of superheat for some substances at 0.10 MPa (Avedisian 1986). These values are significantly higher than their respective boiling temperatures, which prove that they can undergo substantial superheating. This can be proved experimentally; however, the best experiments can only be expected to yield maximum temperatures of

$$T_t(p_\ell) > T_m(p_\ell) \tag{2.294}$$

The van der Waals limit is not valid for calculations because calculated superheat limits would fall into unstable regions, violating the second law. Using the Peng–Robinson equation of state (Peng and Robinson 1976), a different thermodynamic limit of superheat is obtained.

$$p = \frac{R_g T}{(v-b)} - \frac{a}{v^2 + 2bv + b^2} \tag{2.295}$$

where a and b are constants. The spinodal curve of Eq. (2.295) is

$$\frac{R_g T}{(v-b^2)} - \frac{2a(v+b)}{(v^2+2bv-b^2)^2} = 0 \tag{2.296}$$

The superheat temperature derived by the above equation, T_{t2}, in Table 2.4 is appropriate since it is higher than measured data. Use of another equation of state would yield another thermodynamic limit of superheat. This shows the challenges involved in trying to solve the thermodynamic limit of superheat. The thermodynamic superheat limit is the upper limit of stability of superheated liquid.

2.4.5.2 Kinetic Limit of Superheat

Superheated liquids are not quiescent at the microscopic level. Random molecular motion creates local variations in density. The fluctuation in density creates "holes" or "nuclei" within which the molecules may resemble gas in terms of spacing and potential energy. These nuclei grow and decay until a certain size nucleus is created that is in unstable equilibrium with its surroundings. These bubbles are the initial condition for bubble growth within a liquid, and the critical size nuclei is known.

Homogeneous nucleation theory can predict the rate of formation of a critical size nucleus at a given pressure, temperature, and composition. The nucleation rate is the mean rate at which nuclei are formed and grow to macroscopic size.

Kinetic theory mechanically formulates the critical nucleus. The theory states that the steady-state nucleation rate is proportional to the exponential of the formation energy

$$J = \Gamma k_{f(N^*)} \mathfrak{N}_\ell \exp\left[-\frac{\Delta\Phi^*}{k_b T_\ell}\right] \tag{2.297}$$

where $k_{f(N^*)}$ is the molecular condensation rate in a critical size nucleus (with N^* molecules), $\mathfrak{N}_\ell$ is the total number density of molecules. In terms of T_ℓ

$$T_\ell \equiv T_k = \frac{\Delta\Phi^*}{k_b}\left[\ln\left(\frac{\Gamma k_{f(N^*)} \mathfrak{N}_\ell}{J}\right)\right]^{-1} \tag{2.298}$$

T_k is the kinetic limit of superheat and $\Delta\Phi^*$ is the minimum energy necessary for formation of a critical size nucleus and is given as

$$\Delta\Phi^* = \frac{16\pi\sigma^3}{3\left(p_{sat(T_\ell)} - p_\ell\right)^2} \tag{2.299}$$

Factor Γ takes into account the detailed mechanism by which critical nuclei are formed within the molecular network of the liquid. To determine Γ, the following issues must be solved.

1. The energy of a nucleus is a function of the number of molecules in it.
2. The exponential dependence of J on $\Delta\Phi^*$ must be determined.
3. The mechanism by which critical size nuclei form must be described.

The theory of homogeneous nucleation, including the determination of the superheat and appropriate experiments, is investigated extensively (Blander and Katz 1975; Skripov 1974; Avedisian 1986; Debenedetti 1996; Iida et al. 1997). In nucleation theory, the net rate of embryo growth from the size N(N molecules) to size $N + 1$, per unit volume per unit time is (Skripov 1974)

$$J = \frac{p_v/(2\pi m k_b T_\ell)^{1/2}}{\int_0^\infty (AN_s)^{-1} dN} \tag{2.300}$$

where m is mass of a single molecule, k_b is the Boltzmann constant, A is the interfacial area of the embryo, and N_s is the number density of embryos containing N molecules. The numerator in Eq. (2.300) is the vaporization rate for the surface of the embryo (see Sect. 4.5.2). Debenedetti (1996) simplified the above relation in terms of easily measured properties

$$J = N_\ell\left(\frac{2\sigma}{\pi m B}\right)^{1/2} \exp\left[-\frac{16\pi}{3k_b T}\frac{\sigma^3}{\delta^2\left(p_{sat(T_\ell)} - p_\ell\right)^2}\right] \tag{2.301}$$

Table 2.5 Limit of superheat and nucleation rate of water at atmospheric pressure[a] (Avedisian 1986)

T	p_v	p_{sat}	$R \times 10^7$	J	Waiting time/cm^3 ($\sim 1/J$)
500	25.8	25.2	25.2	$<10^{-99}$	$>10^{91}$ years
550	59.1	61.0	6.76	$<10^{-99}$	$<10^{91}$ years
560	68.3	71.0	5.2	1.7×10^{-76}	1.2×10^{68} years
570	78.5	82.0	3.9	8.5×10^{-20}	3.7×10^{11} years
575	83.9	88.0	3.4	5.7×10^{-3}	1.8×10^{2} s
580	89.6	94.4	2.9	4.3×10^{9}	2.3×10^{-10} s
590	101.6	108.9	2.1	4.3×10^{23}	2.3×10^{-24} s

[a] T temperature (K), p_v pressure in vapor nucleus (atm), p_{sat} equilibrium vapor pressure (atm), R radius of critical nucleus (cm), J nucleation rate (nuclei/cm^3 s)

$$\delta = 1 - \frac{p_v v_\ell}{k_b T_\ell} \tag{2.302}$$

$$B = \frac{2}{3} + \frac{1}{3}\frac{p_\ell}{p_{sat(T_\ell)}} \tag{2.303}$$

The effect of temperature is significant in rate of nucleation considering the exponential form of Eq. (2.301). Avedisian (1986) showed the variation of J with change in temperature for superheated water at 1 atm in Table 2.5. The waiting time (the reciprocal of J assuming volume of liquid 1 cm3) is also shown in the same table. The data for water at 1 atm pressure indicates that the bubble nucleation is rare at temperatures less than 570 K. At higher temperatures, J increases while waiting time decreases with increasing temperature. This is a range in which homogeneous nucleation does not occur and above that temperature it occurs spontaneously. In practice to use Eq. (2.301), a suitable threshold is chosen. It is found that a threshold of $J = 10^{30}$ nuclei/cm^3 s is appropriate for most working fluids and operating conditions. The detailed discussion of superheat limit can be found in Avedisian (1986).

Problems

2.1. Show that the equilibrium criterion for a system with constant entropy and pressure is $\Delta H_{S,p} \leq 0$.

2.2. A simple system is confined by an adiabatic, rigid, and impermeable boundary. In its initial state, the system is divided by a diathermal partition (a partition that allows heat to penetrate) into two equal halves. The internal energies of the left and right halves satisfy

$$E_L = \frac{3}{2} n_L R_u T_L \quad E_R = \frac{3}{2} n_R R_u T_R$$

where the mole numbers of the left and right sides are $n_L = 2$ kmol and $n_R = 3$ kmol, respectively. The initial temperatures of the two halves are, respectively, $T_L = 250$ K and $T_R = 350$ K. After thermal equilibrium has been established, find (a) the values of E_L and E_R, and (b) the equilibrium temperature.

2.3. Two cylinders are fitted with two pistons coupled together so that the linear displacement of the two pistons must be the same (see Fig. P2.3). The ratio of the cross-sectional area of the two cylinders is $A_L/A_R = 0.5$. A bar with high thermal conductivity connects the two cylinders so that the final temperatures of the two cylinders are the same. Determine the ratio of the pressures in the two cylinders, p_L/p_R, at equilibrium.

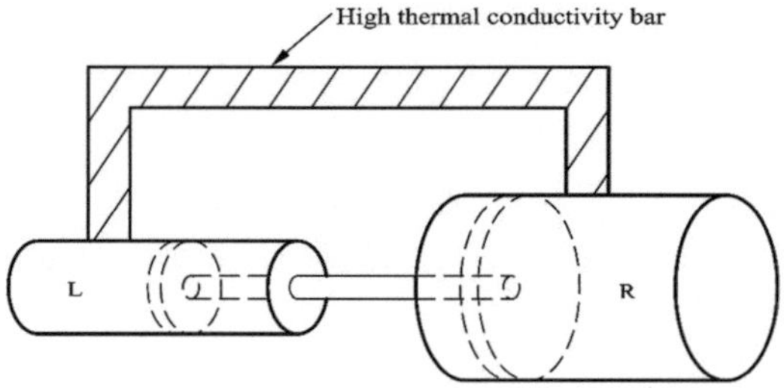

Fig. P2.3

2.4. A closed system is initially divided into two halves. The pressures of the two halves are identically equal to p. The initial entropies of the left and right sides are both equal to S, and the total entropy of the system is $2S$. After an infinitesimal change, the entropy of the left and right halves becomes $S_L = S + \Delta S$ and $S_R = S - \Delta S$. The pressures of the two halves remain unchanged throughout the process. Using the enthalpy minimum principle, show that the system is stable if the specific heat at constant pressure is positive, $c_p > 0$.

2.5. Show that the specific heat at constant pressure is always greater than that at constant volume for any substance, i.e., $c_p > c_v$.

2.6. The following two equations are valid for an ideal gas:

$$pV = nR_uT \quad E = cnR_uT$$

where R_u is the universal gas constant and c is a constant which depends on the molecular structure of the ideal gas. What is the entropy of the ideal gas?

2.7. A 1-m^3 vessel is filled with propane at room temperature of 20 °C and pressure of 100 kPa. Find the mass of the propane by using (a) the ideal gas law and (b) the van der Waals equation.

2.8. Reduced pressure, temperature, and specific volume can be defined as

$$p_r = \frac{p}{p_c} \quad T_r = \frac{T}{T_c} \quad v_r = \frac{v}{v_c}$$

where p_c, T_c, and v_c are the pressure, temperature, and specific volume at the critical point. Show that the van der Waals equation in terms of reduced pressure, temperature, and specific volume is

$$p_r = \frac{8T_r}{3v_r - 1} - \frac{3}{v_r^2}$$

2.9. For a fluid that does not satisfy the ideal gas law, a compressibility factor Z is introduced: $Z = pv/RT$; this can also be written in terms of reduced pressure, temperature, and specific volume, i.e., $Z = Z_c p_r v_r / T_r$, where Z_c is the compressibility factor at the critical point. Determine the compressibility of the fluid that satisfies the van der Waals equation of state.

2.10. For a gas that satisfies the van der Waals equation, show that its internal energy can be expressed as

$$e = e_0 + c_v(T - T_0) + a\left(\frac{1}{v_0} - \frac{1}{v}\right)$$

if the specific heat at constant volume, c_v, is constant. The subscript 0 in the equation denotes a reference state.

2.11. Redlich–Kwong equation is generally considered to be the best among the two-constant equations of state:

$$p = \frac{R_g T}{v - b} - \frac{a}{v(v + b)T^{1/2}}$$

Obtain the two constants a and b in the Redlich–Kwong equation by applying the critical point conditions: $(\partial p/\partial v)_T = 0$ and $(\partial^2 p/\partial v^2)_T = 0$.

2.12. Using the Clapeyron equation (2.156), estimate the value of the latent heat of vaporization of water at 100 °C, and compare it with the value from Table B.48.

2.13. Determine the boiling point of water in a city where the elevation is 1500 m. The atmospheric pressure at this elevation is 84.56 kPa.

2.14. The temperature of ice in an ice rink is −5 °C. An ice skater is standing on one foot, and the contact area between the skate and ice is 280 × 2 mm^2. The weight of the ice skater is 60 kg. What is the melting point of the ice underneath the skate?

2.15. The saturation temperature of water at 1 atm is 100 °C. Use the Clausius–Clapeyron equation to find the saturation pressure if the temperature is increased to 110 °C and compare your result with that obtained by using a steam table

2.16. A system with constant temperature and pressure contains two phases of the same substance. Show that the two phases in the system are in equilibrium if Eqs. (2.126)–(2.128) are satisfied.

2.17. The chemical potential of a single-component system is its Gibbs free energy, $\bar{g}$ [see Eq. (2.146)]. According to Eq. (2.128), the equilibrium chemical potential of the liquid and vapor at the liquid–vapor interface must be equal: $g_v = g_\ell$. Start from this relation, and show that the latent heat of vaporization is $h_{\ell v} = T(s_v - s_\ell)$ where s_v and s_ℓ are the entropies of saturated vapor and liquid, respectively.

2.18. A mixture of saturated liquid and vapor is in a piston–cylinder system as shown in Fig. P2.18. The piston is frictionless, so the piston–cylinder system is a system with constant pressure and temperature. If the masses of liquid and vapor are, respectively, m_ℓ and m_v, the Gibbs free energy of the mixture is $G = m_\ell g_\ell + m_v g_v$. Show that the condition for the liquid and vapor phases at phase equilibrium is $g_\ell = g_v$.

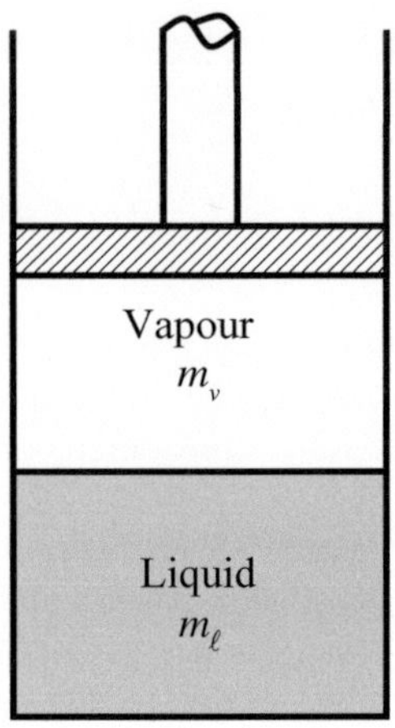

Fig. P2.18

2.19. A rigid tank filled with saturated nitrogen vapor at $T = 100\,\text{K}$ is cooled to condense the vapor. It is assumed that the van der Waals equation is valid and the two constants are $a = 167.15\,\text{Pa m}^6/\text{kg}^2$ and $b = 1.35 \times 10^{-3}\,\text{m}^3$. The system is in a metastable state before condensation starts. Determine the temperature and the corresponding pressure at which condensation will occur.

2.20. A mixture of liquid and vapor water fills a chamber with the wall temperature maintained at a constant level, T_w (see Fig. P2.20). A valve near the bottom of the chamber is opened, and 1 kg of water is drained from the chamber. Find an expression, in terms of the saturation properties of liquid and vapor, for the increase in the volume of the vapor phase.

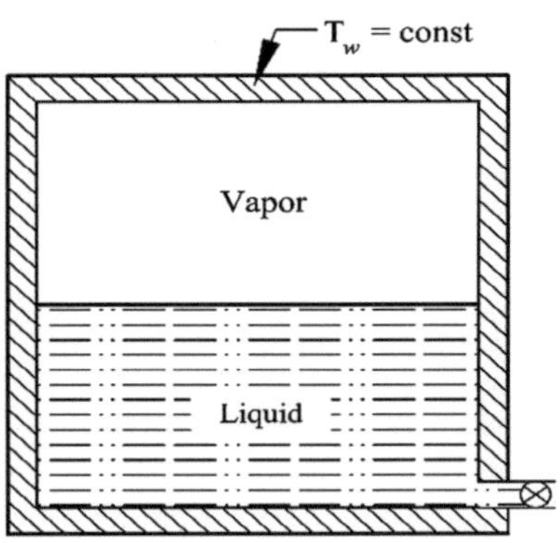

Fig. P2.20

2.21. Surface tension has been represented in terms of internal energy [Eq. (2.203)], Helmholtz free energy [Eq. (2.208)], and Gibbs free energy [Eq. (2.220)]. Find the representation of the surface tension using enthalpy.

2.22. The degree of supersaturation was expressed in terms of the ratio of the vapor pressure and saturation pressure in Example 2.7. Express the degree of supersaturation in Example 2.7 in terms of temperature using the Clapeyron–Clausius equation.

References

Avedisian, C. T. (1986). Bubble growth in superheated liquid droplets. In *Encyclopedia of fluid mechanics*. Houston, TX: Gulf Publishing Company.

Bejan, A. (2016). *Advanced engineering thermodynamics* (4th ed.). New York, NY: Wiley.

Blander, M., & Katz, J. P. (1975). Bubble nucleation in liquids. *AIChE Journal, 21*(5), 833–848.

Borgnakke, C., & Sonntag, R. E. (2017). *Fundamentals of thermodynamics*. New York, NY: Wiley.

Carey, V. P. (2016). *Liquid-vapor phase-change phenomena: An introduction to the thermophysics of vaporization and condensation processes in heat transfer equipment* (3rd ed.). New York, NY: Taylor & Francis.

Carey, V. P., & Wemhoff, A. P. (2005). Disjoining Pressure effects in ultra-thin liquid films in micropassages-comparison of thermodynamic theory with predictions of molecular dynamics simulations. In *Proceedings of 2005 ASME International Mechanical Engineering Congress and Exposition*, Orlando, FL.

Debenedetti, P. G. (1996). *Metastable liquids—Concepts and principles*. Princeton, NJ: Princeton University Press.

Faghri, A. (2016). *Heat pipe science and technology* (2nd ed.). Columbia, MO: Global Digital Press.

Fowkes, F. M. (1965). Attractive forces at interfaces. In *Chemistry and physics of interfaces*. Washington, DC: American Chemical Society.

Iida, Y., Okuyama, K., & Nishizawa, T. (1997). Heat transfer during boiling initiated by fluctuation nucleation on a platinum film rapidly heated to the limit of liquid superheat. *Journal of the Japan Society of Mechanical Engineers, B, 63*(613), 3048–3054.

Israelachvilli, J. (1992). *Intermolecular & surface forces* (2nd ed.). London: Academic Press.

Kyle, B. G. (1999). *Chemical and process thermodynamics* (3rd ed.). Englewood Cliffs, New Jersey: Prentice-Hall Inc.

Lienhard, J. H. (1976). Correlation for the limiting liquid superheat. *Chemical Engineering Science, 31,* 847–849.

Peng, D. Y., & Robinson, D. B. (1976). A new two constant equation of state. *Industrial Engineering Fundamentals, 15,* 59–64.

Skripov, V. P. (1974). *Metastable liquids*. New York: Wiley.

Modeling Multiphase Flow and Heat Transfer 3

3.1 Introduction

A multiphase system, which is distinguished from a single-phase system by the presence of one or more interfaces separating the phases, can be considered as a field that is divided into single-phase regions by those interfaces, or moving boundaries, between phases. A straightforward method of describing a multiphase system is to use the standard *local instance* formulation for each single-phase subregion, with appropriate jump conditions to match the solution of these differential equations at the interfaces. This method requires tracking the location of interfaces, and for this reason, it is referred to as the *interface tracking method*. Since the locations of the interfaces are unknown a priori, the governing equations are solved separately for each phase; then, iterations are performed at the anticipated location of the phase interface until convergence of certain interfacial balances is obtained. These interfacial balances must fulfill the requirement of mass, momentum, and energy conservation across the interface. With this approach, the position of the phase interface may be clearly identified and used to separate the solutions of the governing -equations on both sides of the interface. This method is most appropriate for dealing with separated-phase cases (see Table 1.1), such as melting, solidification, film evaporation, film condensation, and single bubble (or droplet) growth or collapse. For the multiphase systems that have extremely complicated and deformable interfaces, it is not possible to solve the local instance fluid flow. Various averaging techniques can be applied to obtain the mean values of properties and eliminate the need to explicitly track interfaces and/or the local instance fluctuations of properties.

Two common techniques are used for describing fluid flow: *Lagrangian* and *Eulerian*. The Lagrangian approach requires that the properties of a particular element of the fluid particles be tracked as it traverses the flow. This approach is similar to those used in particle and rigid-body dynamics. The location of a fluid element is described by its coordinates (x, y, z), which are functions of time. The fluid element can be identified by tracking it from its initial location (x_0, y_0, z_0) at time $t = 0$, and the velocity of this element at an arbitrary time t is expressed as $\mathbf{V} = \mathbf{V}(x_0, y_0, z_0, t)$. In order to describe a fluid flow using the Lagrangian approach, the sensors that monitor fluid properties would have to move at the same velocity as the fluid element; this is an impractical and often impossible requirement to meet, especially for such complex cases as three-dimensional transient flow. Therefore, the Lagrangian approach is rarely used in description of fluid flow.

The Eulerian approach, on the other hand, observes the flow properties from a fixed location relative to a reference frame, which can be stationary or more generally moves at its own velocity. The Eulerian approach gives the values of the fluid variable at a given point (x, y, z) at a given time

A. Faghri and Y. Zhang, *Fundamentals of Multiphase Heat Transfer and Flow*, https://doi.org/10.1007/978-3-030-22137-9_3

t. For example, the velocity can be expressed as $\mathbf{V} = \mathbf{V}(x, y, z, t)$, where x, y, and z are independent of t. The Eulerian approach requires that the fluid properties be measured at spatial locations that are fixed relative to the reference frame in the fluid field, so the sensors are not required to move with the individual particles. Since the Eulerian approach is consistent with conventional experimental observation techniques, it is widely used in fluid mechanics and will also be adopted in this textbook.

The conservation laws for flow and heat transfer can be expressed for a fixed mass or for a control volume. While the former is a fixed collection of particles with constant mass (fixed mass), the latter is a defined region relative to the reference frame in space (fixed volume). From a thermodynamic point of view, the fixed mass and the control volume can be considered as closed and open systems, respectively. Writing the governing equations for a fixed mass requires tracking the motion of the particles, i.e., the Lagrangian approach. As we have already mentioned, although the Lagrangian description is applicable to some fluid mechanics problems, it is not a very practical way to describe multiphase systems, with a few exceptions such as liquid droplet tracking in a thermal plasma. The governing equations written for a control volume, on the other hand, express the relationship between the change of properties inside the control volume and the property of the flow into or out of the control volume. The governing equations obtained by writing the conservation laws for a control volume are consistent with the Eulerian approach. However, all fundamental laws of mechanics, including conservation of mass, momentum, and energy, are formulated for a collection of particles with fixed identity; that is to say, they are Lagrangian in nature. Therefore, it is necessary to apply the fundamental laws to the fixed mass first, and then to convert them into expressions for the control volume.

Section 3.2 presents local instance macroscopic (integral) formulations of the governing equations of the heat and mass transfer. The macroscopic (integral) formulation is obtained by performing mass, momentum, energy, and species mass balances over a control volume that includes different phases as well as interfaces. The local instance microscopic (differential) formulation is presented in Sect. 3.3. The local instance microscopic (differential) formulations are obtained by simplifying the integral formulations for control volumes with only one phase. The local instance differential equations must be supplemented by jump conditions at the interface. The classification of PDEs and boundary conditions, as well as a jump conditions at the interface, is also discussed. Section 3.3 is closed by discussion of a rarefied vapor self-diffusion model and application of the differential formulations to combustion. Section 3.4 presents a thorough overview of various averaging methods used to describe multiphase flow and heat and mass transfer problems; this is followed by the governing conservation equations for the multifluid, homogeneous, mixture, and separated models. Also discussed in Sect. 3.4 are area-averaged governing equations for one-dimensional liquid-gas (vapor), and guidance and comparison of multiphase flow models.

3.2 Macroscopic (Integral) Local Instance Formulations

A fixed mass system describes an amount of matter that can move, flow, and interact with the surroundings, but the control volume approach depicts a region or volume of interest in a flow field that is not unique and depends on the user. So conservation laws for a fixed mass system need to be transformed to apply to a control volume system. A relation that allows one to mathematically and physically link the conservation laws for a control volume with that of a fixed mass system will be derived.

Figure 3.1 shows the flow field under consideration. At time t, the control volume shown by the solid line coincides with a single-phase fixed mass system depicted by the dashed line. At time $t + dt$, a portion of the fixed mass system moves outside of the boundaries of the control volume. It can be

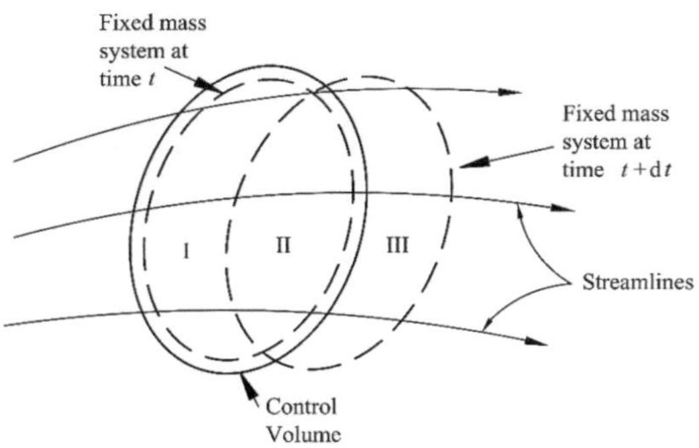

Fig. 3.1 Relation between a fixed mass system and a control volume

seen from Fig. 3.1 that region I is occupied by the system at time t only, region II is common to the system at both t and $t + dt$, and region III is occupied by the system at $t + dt$ only. For a system with a fixed mass, the change of the general property Φ, which has a specific value (per unit mass) ϕ, can be written as (Welty et al. 2014):

$$\left.\frac{d\Phi}{dt}\right|_{\text{system}} = \frac{\Phi|_{t+dt} - \Phi|_t}{dt} \tag{3.1}$$

Considering that the mass occupies regions I and II at time t, one obtains $\Phi|_t = \Phi|_{\text{I}} + \Phi|_{\text{II}}$. At time $t + dt$, the fixed mass system occupies both regions II and III, i.e., $\Phi|_{t+dt} = \Phi|_{\text{II}} + \Phi|_{\text{III}}$. Therefore, Eq. (3.1) can be rewritten as

$$\left.\frac{d\Phi}{dt}\right|_{\text{system}} = \frac{\Phi_{\text{II}}|_{t+dt} - \Phi_{\text{II}}|_t}{dt} + \frac{\Phi_{\text{III}}|_{t+dt}}{dt} - \frac{\Phi_{\text{I}}|_t}{dt} \tag{3.2}$$

The first term on the right-hand side of Eq. (3.2) may be written as

$$\frac{\Phi_{\text{II}}|_{t+dt} - \Phi_{\text{II}}|_t}{dt} = \frac{d\,\Phi_{\text{II}}}{dt}$$

and represents the rate of change of property Φ within the control volume, because region II becomes coincident with the control volume as $dt \to 0$. For the general case of variable Φ within a control volume, it relates to its specific value ϕ by $\Phi = \int_V \rho\phi\, dV$ and it is appropriate to write the time derivative of Φ as

$$\frac{d\Phi|_{\text{II}}}{dt} = \left(\frac{d\Phi}{dt}\right)_{CV} = \frac{\partial}{\partial t}\int_V \rho\phi\, dV$$

where V is the volume of the control volume.

The second and third terms on the right-hand side of Eq. (3.2), respectively, represent the property Φ leaving and entering the control volume due to mass flow across its boundary. If the absolute velocity is $\mathbf{V}$, and the reference frame moves with a constant velocity $\mathbf{V}_{\text{ref}}$, the relative velocity is $\mathbf{V}_{\text{rel}} = \mathbf{V} - \mathbf{V}_{\text{ref}}$. For the control volume's entire surface area, A, the rate of movement of property Φ due to mass flow may be written as (see Fig. 3.2)

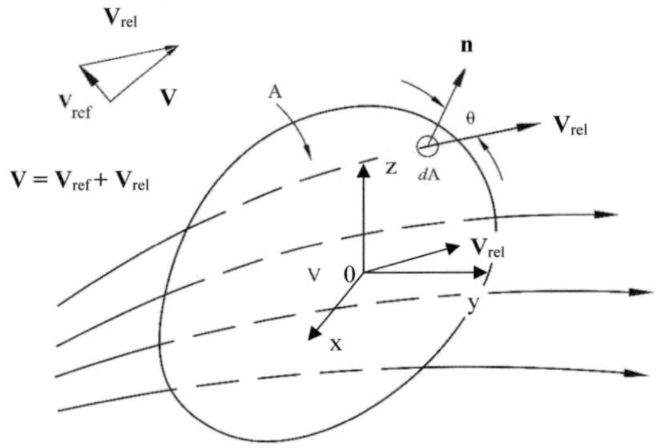

Fig. 3.2 Control volume in a flow field

$$\frac{\Phi_{\mathrm{III}}|_{t+\Delta t}}{\mathrm{d}t} - \frac{\Phi_{\mathrm{I}}|_{t}}{\mathrm{d}t} = \int_A \rho(\mathbf{V}_{\mathrm{rel}} \cdot \mathbf{n})\phi \, \mathrm{d}A$$

where $\mathbf{n}$ is the normal direction of the control volume.

Therefore, Eq. (3.2) can be rewritten as

$$\left.\frac{\mathrm{d}\Phi}{\mathrm{d}t}\right|_{\mathrm{system}} = \frac{\partial}{\partial t}\int_V \rho\phi \, \mathrm{d}V + \int_A \rho(\mathbf{V}_{\mathrm{rel}} \cdot \mathbf{n})\phi \, \mathrm{d}A \tag{3.3}$$

which is the final form of the transformation formula that relates the change of property for a fixed mass system to that of the control volume (Welty et al. 2014; Faghri et al. 2010). It states that the rate of change of a property Φ for a fixed mass system is equal to the rate of change of Φ in the control volume (the first term on the right-hand side) plus the net rate of efflux of Φ by mass flow into or out of the control volume (the second term on the right-hand side).

It should be pointed out that the control volume moves with the reference frame, which moves with a constant velocity, $\mathbf{V}_{\mathrm{ref}}$. The coordinate system is attached to and moves with the reference frame. In other words, the coordinate system is stationary relative to the reference frame. The reference frame is inertial (with zero acceleration), so Newton's second law is valid in the coordinate system that moves with the reference frame. Equation (3.3) will be used to obtain the macroscopic (integral) formulation of the basic laws for a control volume.

In the development of Eq. (3.3), it was assumed that the control volume is located in a region occupied only by one phase. For multiphase flow and heat transfer problems, the control volume may include multiple phases separated by several interfaces. Figure 3.3 shows a fixed control volume V with surface area A that contains three phases (solid, liquid, and gas): (1) solid with volume $V_{\mathrm{s}}(t)$ bounded by $A_{\mathrm{s}}(t)$, $A_{\mathrm{I}}^{\mathrm{sg}}(t)$ (interfacial area between solid and gas), and $A_{\mathrm{I}}^{\mathrm{s}\ell}(t)$ (interfacial area between solid and liquid); (2) liquid with volume $V_{\ell}(t)$ bounded by $A_{\ell}(t)$, $A_{\mathrm{I}}^{\ell\mathrm{g}}(t)$ (interfacial area between liquid and gas), and $A_{\mathrm{I}}^{\mathrm{s}\ell}(t)$; and (3) gas with volume $V_{\mathrm{g}}(t)$ bounded by $A_{\mathrm{g}}(t)$, $A_{\mathrm{I}}^{\mathrm{sg}}(t)$, and $A_{\mathrm{I}}^{\ell\mathrm{g}}(t)$. Since the interfaces that separate different phases move, the volume and surface area of each phase are functions of time. The volume and area of each phase satisfy $V_{\mathrm{s}}(t) + V_{\ell}(t) + V_{\mathrm{g}}(t) = V$ and $A_{\mathrm{s}}(t) + A_{\ell}(t) + A_{\mathrm{g}}(t) = A$, because the control volume is fixed in space. For a control volume that contains three phases, Eq. (3.3) becomes

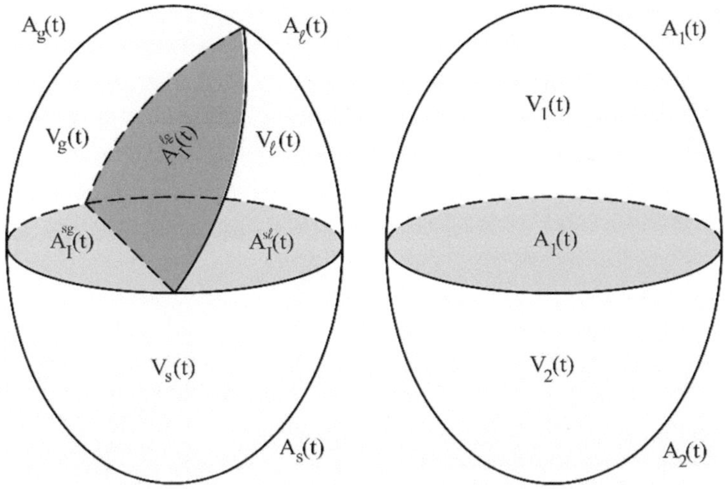

Fig. 3.3 Control volumes that enclose multiple phases separated by interfaces (the subscripts 1 and 2 can be s, ℓ, or g, depending on the phases in the control volume)

$$\begin{aligned}\left.\frac{d\Phi}{dt}\right|_{\text{system}} &= \frac{\partial}{\partial t}\int_{V_s(t)} \rho_s\phi_s dV + \frac{\partial}{\partial t}\int_{V_\ell(t)} \rho_\ell\phi_\ell dV + \frac{\partial}{\partial t}\int_{V_g(t)} \rho_g\phi_g dV \\ &+ \int_{A_s(t)} \rho_s(\mathbf{V}_{s,\text{rel}}\cdot\mathbf{n}_s)\phi_s dA + \int_{A_\ell(t)} \rho_\ell(\mathbf{V}_{\ell,\text{rel}}\cdot s\mathbf{n}_\ell)\phi_\ell dA \\ &+ \int_{A_g(t)} \rho_g(\mathbf{V}_{g,\text{rel}}\cdot\mathbf{n}_g)\phi_g dA\end{aligned} \tag{3.4}$$

If the control volume contains only two phases, as shown in Fig. 3.3(b), Eq. (3.4) may be reduced to

$$\begin{aligned}\left.\frac{d\Phi}{dt}\right|_{\text{system}} &= \frac{\partial}{\partial t}\int_{V_1(t)} \rho_1\phi_1 dV + \frac{\partial}{\partial t}\int_{V_2(t)} \rho_2\phi_2 dV \\ &+ \int_{A_1(t)} \rho_1(\mathbf{V}_{1,\text{rel}}\cdot\mathbf{n}_1)\phi_1 dA + \int_{A_2(t)} \rho_2(\mathbf{V}_{2,\text{rel}}\cdot\mathbf{n}_2)\phi_2 dA\end{aligned} \tag{3.5}$$

Therefore, for control volumes containing only one phase, the local instance integral formulation can be obtained by using Eq. (3.3). For the control volume that contains more than one phase separated by interfaces, the local instance integral formulation can be obtained by using Eq. (3.4) or (3.5). In Sects. 3.2.1–3.2.4, the integral form of the governing equations for a control volume that contains only one phase will be derived first, followed by an extension of the integral form of the governing equations to a control volume that contains multiple phases.

3.2.1 Conservation of Mass

The law of the conservation of mass dictates that mass may be neither created nor destroyed. For a control volume that contains only one phase, conservation of mass can be obtained by setting the general and specific property forms to $\Phi = m$ and $\phi = 1$ in Eq. (3.3), i.e.,

$$\left.\frac{\mathrm{d}m}{\mathrm{d}t}\right|_{\text{system}} = \frac{\partial}{\partial t}\int_V \rho \mathrm{d}V + \int_A \rho(\mathbf{V}_{\text{rel}} \cdot \mathbf{n})\mathrm{d}A \tag{3.6}$$

where the first term on the right-hand side represents the time rate of change of the mass of the contents of the control volume, and the second term on the right-hand side represents the net rate of the mass flow through the control surface. The term $(\mathbf{V}_{\text{rel}} \cdot \mathbf{n})\mathrm{d}A$ in the mass flow integral represents the product of the velocity component perpendicular to the control surface and differential area. This term is the volume flow rate through $\mathrm{d}A$ and becomes the mass flow rate when multiplied by density, ρ.

Since the mass of a fixed mass system is constant by definition, and the fixed mass system contains only one phase, the resulting formulation of conservation of mass is

$$\frac{\partial}{\partial t}\int_V \rho \mathrm{d}V + \int_A \rho(\mathbf{V}_{\text{rel}} \cdot \mathbf{n})\mathrm{d}A = 0 \tag{3.7}$$

which shows that the time rate of change of the mass of the control volume plus the net rate of mass flow through the control surface must be equal to zero. In other words, the sum of the mass flow rate into and out of the control volume must be equal to the accumulation and depletion of the mass within the control volume.

For a control volume containing multiple phases separated by interfaces, the conservation of mass can be similarly obtained by using Eq. (3.4) or (3.5), i.e.,

$$\sum_{k=1}^{\Pi}\left[\frac{\partial}{\partial t}\int_{V_k(t)} \rho_k \mathrm{d}V + \int_{A_k(t)} \rho_k(\mathbf{V}_{k,\text{rel}} \cdot \mathbf{n}_k)\mathrm{d}A\right] = 0 \tag{3.8}$$

where k denotes the kth phase in the multiphase system, and Π is the number of phases.

3.2.2 Momentum Equation

Newton's second law states that, in an inertial reference frame, the time rate of momentum change of a fixed mass system is equal to the net force acting on the system, and it takes place in the direction of the net force. Mathematically, Newton's second law of motion for fixed mass in a reference frame that moves at a constant velocity $\mathbf{V}_{\text{ref}}$ (see Fig. 3.2) is written as

$$\sum \mathbf{F} = \frac{\mathrm{d}(m\mathbf{V})_{\text{rel}}}{\mathrm{d}t} \tag{3.9}$$

where the left-hand side is the net force vector acting on the fixed mass system, and the right-hand side is the rate of momentum change.

For control volumes that contain only one phase, the integral form of Newton's second law can be obtained by using Eq. (3.3). With the applicable value of Φ and ϕ in Eq. (3.3) defined as $\Phi = m\mathbf{V}_{\mathrm{rel}}$ and $\phi = \mathbf{V}_{\mathrm{rel}}$, one obtains

$$\sum \mathbf{F} = \frac{\partial}{\partial t}\int_V \rho \mathbf{V}_{\mathrm{rel}} \mathrm{d}V + \int_A \rho(\mathbf{V}_{\mathrm{rel}} \cdot \mathbf{n})\mathbf{V}_{\mathrm{rel}} \mathrm{d}A \tag{3.10}$$

which is in vector form and is valid in all three directions.

Forces acting on the control volume include body forces and contact forces that act on its surface. For example, for a multicomponent system that contains N components, if the body force per unit mass acting on the ith species is $\mathbf{X}_i$, the total body force acting on the control volume is $\int_{\mathrm{V}} \left[\sum_{i=1}^{N} \rho_i \mathbf{X}_i\right] \mathrm{d}V$, where ρ_i is the mass concentration (kg/m^3) of the ith species. If the body force per unit mass is the same for different species (as is the case with gravity), the body force term is reduced to $\int_V \rho \mathbf{X} \mathrm{d}V$, where ρ is the density of the mixture. The stress tensor acts on the surface of a fluid control volume and includes both normal and shear stresses. The net force may be written as

$$\sum \mathbf{F} = \int_V \left[\sum_{i=1}^{N} \rho_i \mathbf{X}_i\right] \mathrm{d}V + \int_A \boldsymbol{\tau}' \cdot \mathbf{n} \mathrm{d}A \tag{3.11}$$

where $\boldsymbol{\tau}'$ is the total stress tensor and $\mathbf{n}$ is the local normal unit vector on surface A. The dot product of a tensor of rank two, $\boldsymbol{\tau}'$, and a vector, $\mathbf{n}$, is a vector that represents the force acting on the surface of the control volume per unit area.

Combining Eqs. (3.10) and (3.11), we obtain the momentum equation for the control volume of a single-phase system:

$$\begin{aligned} &\int_V \left[\sum_{i=1}^{N} \rho_i \mathbf{X}_i\right] \mathrm{d}V + \int_A \boldsymbol{\tau}' \cdot \mathbf{n} \mathrm{d}A \\ &= \frac{\partial}{\partial t}\int_V \rho \mathbf{V}_{\mathrm{rel}} \mathrm{d}V + \int_A \rho(\mathbf{V}_{\mathrm{rel}} \cdot \mathbf{n})\mathbf{V}_{\mathrm{rel}} \mathrm{d}A \end{aligned} \tag{3.12}$$

where the two terms on the left-hand side represent, respectively, the body force and stress on the control volume, and the two terms on the right-hand side represent, respectively, the rate of momentum change in the control volume and the rate of the momentum flow into or out of the control volume. $\mathbf{V}_{\mathrm{rel}}$ is the bulk relative velocity of the mixture that contains N components.

When the control volume includes multiple phases and components, integrations must be performed for each subvolume. In that case, the momentum equation becomes:

$$\begin{aligned} &\sum_{k=1}^{\Pi} \left[\int_{V_k(t)} \left[\sum_{i=1}^{N} \rho_{k,i} \mathbf{X}_{k,i}\right] \mathrm{d}V + \int_{A_k(t)} \boldsymbol{\tau}'_k \cdot \mathbf{n}_k \mathrm{d}A\right] \\ &= \sum_{k=1}^{\Pi} \left[\frac{\partial}{\partial t}\int_{V_k(t)} \rho_k \mathbf{V}_{k,\mathrm{rel}} \mathrm{d}V + \int_{A_k(t)} \rho_k(\mathbf{V}_{k,\mathrm{rel}} \cdot \mathbf{n}_k)\mathbf{V}_{k,\mathrm{rel}} \mathrm{d}A\right] \end{aligned} \tag{3.13}$$

Equations (3.12) and (3.13) are momentum equations in a coordinate system that is attached to and moves with an inertial reference frame. For a fixed coordinate system that does not move with the reference frame (while the control volume still moves with the reference frame at velocity $\mathbf{V}_{\text{ref}}$), one can substitute the general variables $\Phi = m\mathbf{V}$ and $\phi = \mathbf{V}$ into Eq. (3.3) to obtain the momentum equation.

3.2.3 Energy Equation

The first law of thermodynamics for a fixed mass system (closed system) states that

$$\left.\frac{\mathrm{d}\hat{E}}{\mathrm{d}t}\right|_{\text{system}} = \frac{\delta Q}{\mathrm{d}t} - \frac{\delta W}{\mathrm{d}t} \tag{3.14}$$

where Q is positive if heat is transferred into the system, and W is positive if the work is done by the system to the surrounding. The mass of system can store energy internally in a number of different forms. Therefore, the total energy $\hat{E}$ is due to internal energy, E, kinetic, potential, electromagnetic, surface tension, and other forms. For the development of equations in this chapter, the contributions of internal and kinetic energies are considered. Other contributions can also be added as a source term or boundary condition based on the physical model.

For a control volume including only one phase, the energy equation for the control volume can be obtained by setting the general property as $\Phi = E + m\mathbf{V}_{\text{rel}}^2/2$ and $\phi = e + \mathbf{V}_{\text{rel}}^2/2$ in Eq. (3.3), i.e.,

$$\left.\frac{\mathrm{d}\hat{E}}{\mathrm{d}t}\right|_{\text{system}} = \frac{\partial}{\partial t}\int_V \rho\left(e + \frac{\mathbf{V}_{\text{rel}}^2}{2}\right)\mathrm{d}V + \int_A \rho(\mathbf{V}_{\text{rel}}\cdot\mathbf{n})\left(e + \frac{\mathbf{V}_{\text{rel}}^2}{2}\right)\mathrm{d}A \tag{3.15}$$

where $\hat{E}$ and e are, respectively, total internal energy and specific internal energy, and $\mathbf{V}_{\text{rel}}^2 = \mathbf{V}_{\text{rel}}\cdot\mathbf{V}_{\text{rel}}$.

Substituting Eq. (3.15) into Eq. (3.14), an integral form of the energy equation may be written as

$$\frac{\delta Q}{\mathrm{d}t} - \frac{\delta W}{\mathrm{d}t} = \frac{\partial}{\partial t}\int_V \rho\left(e + \frac{\mathbf{V}_{\text{rel}}^2}{2}\right)\mathrm{d}V + \int_A \rho(\mathbf{V}_{\text{rel}}\cdot\mathbf{n})\left(e + \frac{\mathbf{V}_{\text{rel}}^2}{2}\right)\mathrm{d}A \tag{3.16}$$

The heat flow, $\delta Q/\mathrm{d}t$, is attributed to conduction (and radiation, for the case of participating medium) across the boundary and/or to an internal heat generation, i.e.,

$$\frac{\delta Q}{\mathrm{d}t} = \int_A -\mathbf{q}''\cdot\mathbf{n}\mathrm{d}A + \int_V q'''\mathrm{d}V$$

where $\mathbf{q}''$ is the heat flux vector at the control volume surface. The dot product of the heat flux vector, $\mathbf{q}''$, with the unit normal vector, $\mathbf{n}$, gives heat conducted out of the control volume. The term q''' is the internal heat generation per unit volume, and it can be caused by chemical reaction, electrical heating, etc.

There are four types of work included in the work rate term. The first is shaft work, $\delta W_{\text{sh}}/\mathrm{d}t$, which is done by the control volume to its surroundings that could cause a shaft to rotate or raise a weight through a distance. The second type is the work due to body force, $\delta W_{\text{b}}/\mathrm{d}t$. The third type is associated with pressure, $\delta W_\text{p}/\text{d}t$, and the remaining part of work is necessary to overcome viscous force due to normal and shear stresses, $\delta W_\text{s}/\text{d}t$:

$$\frac{\delta W}{\text{d}t} = \frac{\delta W_\text{sh}}{\text{d}t} + \frac{\delta W_\text{b}}{\text{d}t} + \frac{\delta W_\text{p}}{\text{d}t} + \frac{\delta W_\text{s}}{\text{d}t} \tag{3.17}$$

The work done by the body force is

$$\frac{\delta W_\text{b}}{\text{d}t} = -\int_V \left[\sum_{i=1}^{N} \rho_i \mathbf{X}_i\right] \cdot \mathbf{V}_\text{rel}\text{d}V \tag{3.18}$$

The work associated with pressure can be written as

$$\frac{\delta W_\text{p}}{\text{d}t} = -\int_A p(\mathbf{V}_\text{rel} \cdot \mathbf{n}) \cdot \text{d}A \tag{3.19}$$

The viscous work is evaluated by taking the scalar product of the force acting on the surface of the control volume per unit area, $\mathbf{n}_k \cdot \boldsymbol{\tau}_k$, and the velocity, $\mathbf{V}_\text{rel}$, over the entire surface area of the control volume:

$$\frac{\delta W_\text{s}}{\text{d}t} = -\int_A (\mathbf{n} \cdot \boldsymbol{\tau}) \cdot \mathbf{V}_\text{rel}\text{d}A \tag{3.20}$$

Substituting Eq. (3.17) into Eq. (3.16), one obtains

$$\begin{aligned} &\frac{\delta Q}{\text{d}t} - \frac{\delta W_\text{sh}}{\text{d}t} - \frac{\delta W_\text{b}}{\text{d}t} - \frac{\delta W_\text{p}}{\text{d}t} - \frac{\delta W_\text{s}}{\text{d}t} \\ &= \frac{\partial}{\partial t}\int_V \rho\left(e + \frac{\mathbf{V}_\text{rel}^2}{2}\right)\text{d}V + \int_A \rho(\mathbf{V}_\text{rel} \cdot \mathbf{n})\left(e + \frac{\mathbf{V}_\text{rel}^2}{2} + \frac{p}{\rho}\right)\text{d}A \end{aligned} \tag{3.21}$$

For a single-phase system, neglecting the shaft work and including pressure work as part of $\boldsymbol{\tau}'$, the integral form of the energy Eq. (3.16) becomes

$$\begin{aligned} &\frac{\partial}{\partial t}\int_V \rho\left(e + \frac{\mathbf{V}_\text{rel}^2}{2}\right)\text{d}V + \int_A \rho(\mathbf{V}_\text{rel} \cdot \mathbf{n})\left(e + \frac{\mathbf{V}_\text{rel}^2}{2}\right)\text{d}A \\ &= -\int_A \mathbf{q}'' \cdot \mathbf{n}\text{d}A + \int_V q'''\text{d}V + \int_A (\mathbf{n} \cdot \boldsymbol{\tau}_\text{rel}') \cdot \mathbf{V}_\text{rel}\text{d}A + \int_V \left[\sum_{i=1}^{N} \rho_i \mathbf{X}_i\right] \cdot \mathbf{V}_\text{rel}\text{d}V \end{aligned} \tag{3.22}$$

where the left-hand side represents the rate of total energy change in the control volume and the rate of total energy flow into or out of the control volume. The four terms on the right-hand side represent heat transfer across the boundary of the control volume, internal heat generation, work done by stress (including normal, shear stresses, and pressure at the boundary of control volume), and the work done by body force.

For control volumes including multiple phases, the energy equation becomes:

$$\sum_{k=1}^{\Pi}\left[\frac{\partial}{\partial t}\int\limits_{V_k(t)}\rho_k\left(e_k+\frac{\mathbf{V}_{k,\mathrm{rel}}^2}{2}\right)\mathrm{d}V+\int\limits_{A_k(t)}\rho_k(\mathbf{V}_{k,\mathrm{rel}}\cdot\mathbf{n}_k)\left(e_k+\frac{\mathbf{V}_{k,\mathrm{rel}}^2}{2}\right)\mathrm{d}A\right]$$
$$=\sum_{k=1}^{\Pi}\left[-\int\limits_{A_k(t)}\mathbf{q}''_k\cdot\mathbf{n}_k\mathrm{d}A+\int\limits_{V_k(t)}q_k'''\mathrm{d}V+\int\limits_{A_k(t)}(\mathbf{n}_k\cdot\boldsymbol{\tau}'_{k,\mathrm{rel}})\cdot\mathbf{V}_{k,\mathrm{rel}}\mathrm{d}A\right.$$
$$\left.+\int\limits_{V_k(t)}\left[\sum_{i=1}^{N}\rho_{k,i}\mathbf{X}_{k,i}\right]\cdot\mathbf{V}_{k,\mathrm{rel}}\mathrm{d}V\right] \tag{3.23}$$

where the second integral in the bracket on the left-hand side is advection of energy due to mass flow.

Example 3.1

An incompressible fluid flows in a circular tube of cross-sectional area A_1, which dumps into a large tube of cross-sectional area A_2 as shown in Fig. 3.4. Find the change in internal energy between (1) and (2) for steady condition. Neglect shear stress at the wall.

Solution The control volume is chosen as Sect. 3.2 with the length away from the inlet. For a steady-state flow problem, the continuity Eq. (3.7) becomes

$$\int\limits_{A}\rho(\mathbf{V}_{\mathrm{rel}}\cdot\mathbf{n})\mathrm{d}A=0 \tag{3.24}$$

i.e.,

$$-\rho u_1A_1+\rho u_2A_2=0 \tag{3.25}$$

Therefore, the velocity in Sect. 3.2 is

$$u_2=\frac{A_1}{A_2}u_1 \tag{3.26}$$

Neglecting the body force and considering steady-state condition, the momentum Eq. (3.12) becomes

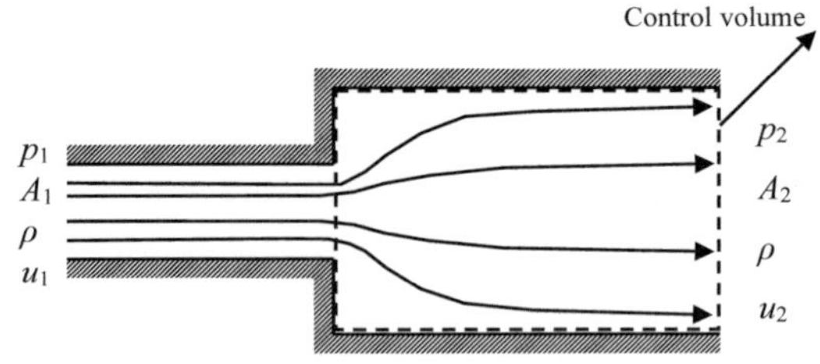

Fig. 3.4 Flow through a sudden enlargement

$$\int_A \boldsymbol{\tau}'_{\mathrm{rel}} \cdot \mathbf{n} dA = \int_A \rho(\mathbf{V}_{\mathrm{rel}} \cdot \mathbf{n})\mathbf{V}_{\mathrm{rel}} dA \tag{3.27}$$

For control volume shown in Fig. 3.4, we have,

$$p_1 A_2 - p_2 A_2 = \rho u_2^2 A_2 - \rho u_1^2 A_1 \tag{3.28}$$

which can be rearranged to

$$\frac{p_1 - p_2}{\rho} = u_2^2 - u_1^2 \frac{A_1}{A_2} \tag{3.29}$$

Since there is no heat transfer ($\delta Q/\mathrm{d}t = 0$) and no shaft, body force, and viscous work, $\delta W_{\mathrm{sh}}/\mathrm{d}t = \delta W_{\mathrm{b}}/\mathrm{d}t = \delta W_{\mathrm{s}}/\mathrm{d}t = 0$, and problem is steady state, $\partial/\partial t = 0$, the energy Eq. (3.21) becomes

$$\int_A \rho(\mathbf{V}_{\mathrm{rel}} \cdot \mathbf{n})\left(e + \frac{\mathbf{V}_{\mathrm{rel}}^2}{2} + \frac{p}{\rho}\right) dA = 0 \tag{3.30}$$

i.e.,

$$\rho u_2 A_2 \left(e_2 + \frac{u_2^2}{2} + \frac{p_2}{\rho}\right) = \rho u_1 A_1 \left(e_1 + \frac{u_1^2}{2} + \frac{p_1}{\rho}\right) \tag{3.31}$$

Substituting Eq. (3.25) into Eq. (3.31), one obtains

$$e_2 - e_1 = \frac{p_1 - p_2}{\rho} + \frac{u_1^2 - u_2^2}{2} \tag{3.32}$$

Considering Eq. (3.29), Eq. (3.32) becomes

$$e_2 - e_1 = \frac{u_1^2}{2}\left(1 - \frac{A_1}{A_2}\right)^2 \tag{3.33}$$

3.2.4 Species

The continuity equation states that the total mass for a closed system is constant. For a system containing one phase but more than one component, the total mass of the system is composed of different species. If the concentrations of each of these species are not uniform, mass transfer occurs in a way that makes the concentrations more uniform. Therefore, it is necessary to track the individual components by applying the principle of conservation of species mass. For a system of multiple components, each component can have its own mass density and velocity. The conservation of mass for the ith species in a single-phase system for a fixed mass system is obtained by applying Eq. (3.3) for the ith species with $\Phi = m_i$ and $\phi = \rho_i/\rho$ (where ρ_i is the concentration of the ith component), i.e.,

$$\left.\frac{\mathrm{d}m_i}{\mathrm{d}t}\right|_{\text{system}} = \frac{\partial}{\partial t}\int_V \rho_i \mathrm{d}V + \int_A \rho_i(\mathbf{V}_{i,\text{rel}} \cdot \mathbf{n})\mathrm{d}A \tag{3.34}$$

If there is no chemical reaction, the total mass of the ith species for a closed system remains constant. Chemical reactions, on the other hand, will result in the production or consumption of the ith species, which can be modeled as a mass source or sink for the ith species, i.e.,

$$\left.\frac{\mathrm{d}m_i}{\mathrm{d}t}\right|_{\text{system}} = \int_V \dot{m}_i''' \mathrm{d}V \tag{3.35}$$

where $\dot{m}_i'''$ is the mass production rate for the ith species, which can be either positive (mass production) or negative (mass consumption). For the cases without chemical reaction, $\dot{m}_i''' = 0$. The conservation of species mass for a control volume is therefore

$$\frac{\partial}{\partial t}\int_V \rho_i \mathrm{d}V + \int_A \rho_i(\mathbf{V}_{i,\text{rel}} \cdot \mathbf{n})\mathrm{d}A = \int_V \dot{m}_i''' \mathrm{d}V \tag{3.36}$$

which is valid for each component in the control volume. If the total number of species is N, summation of the conservation of mass of all species results in

$$\frac{\partial}{\partial t}\int_V \left[\sum_{i=1}^N \rho_i\right] \mathrm{d}V + \int_A \left[\sum_{i=1}^N (\rho_i \mathbf{V}_{i,\text{rel}}) \cdot \mathbf{n}\right] \mathrm{d}A = \int_V \left[\sum_{i=1}^N \dot{m}_i'''\right] \mathrm{d}V \tag{3.37}$$

The summation of the densities of the individual species is equal to the bulk density of the multicomponent substance:

$$\rho = \sum_{i=1}^N \rho_i \tag{3.38}$$

The bulk velocity of the multicomponent substance is the mass-averaged velocity of each individual species:

$$\rho \mathbf{V}_{\text{rel}} = \sum_{i=1}^N (\rho_i \mathbf{V}_{i,\text{rel}}) \tag{3.39}$$

The right-hand side of Eq. (3.37) must be zero, because the total mass of species produced is equal to the total mass of species consumed, i.e.,

$$\sum_{i=1}^N \dot{m}_i''' = 0 \tag{3.40}$$

Substituting Eqs. (3.38)–(3.40) into Eq. (3.37), the continuity Eq. (3.7) is obtained. This means that we can write conservation of species mass for N species, but only $N - 1$ of these equations are independent. In practice, one can use these $N - 1$ equations for conservation of species mass in combination with the continuity equation, Eq. (3.7), to describe multicomponent systems.

In each equation of the conservation of species mass, there are two new unknowns: mass density of the ith species ρ_i, and the species mass velocity $\mathbf{V}_{i,\mathrm{rel}}$. This yields more unknowns than the number of equations. So in order to have a properly posed equation set, it is necessary to reduce the number of unknowns in each equation from two to one. The second term on the left-hand side of Eq. (3.36) represents the species i mass flow across the surface of the control volume, which results from convection by bulk flow and diffusion relative to the bulk convection.

$$\int_A \rho_i(\mathbf{V}_{i,\mathrm{rel}} \cdot \mathbf{n})\mathrm{d}A = \int_A \rho_i(\mathbf{V}_{\mathrm{rel}} \cdot \mathbf{n})\mathrm{d}A + \int_A \mathbf{J}_i \cdot \mathbf{n}\mathrm{d}A \tag{3.41}$$

where $\mathbf{J}_i$ is the diffusive mass flux vector of species i (see Table 1.5).

Substituting Eq. (3.41) into Eq. (3.36), one obtains an expression for the conservation of species mass that contains only one new additional variable, ρ_i:

$$\frac{\partial}{\partial t}\int_V \rho_i \mathrm{d}V + \int_A \rho_i(\mathbf{V}_{\mathrm{rel}} \cdot \mathbf{n})\mathrm{d}A = -\int_A \mathbf{J}_i \cdot \mathbf{n}\mathrm{d}A + \int_V \dot{m}_i''' \mathrm{d}V \tag{3.42}$$

The above analysis is based on the assumption that the control volume contains only one phase. For a control volume containing Π phases, the conservation of species mass is:

$$\begin{aligned}&\sum_{k=1}^{\Pi}\left[\frac{\partial}{\partial t}\int_{V_k(t)} \rho_{k,i}\mathrm{d}V + \int_{A_k(t)} \rho_{k,i}(\mathbf{V}_{k,\mathrm{rel}} \cdot \mathbf{n}_k)\mathrm{d}A\right]\\ &= \sum_{k=1}^{\Pi}\left[-\int_{A_k(t)} (\mathbf{J}_{k,i} \cdot \mathbf{n}_k)\mathrm{d}A + \int_{V_k(t)} \dot{m}_{k,i}''' \mathrm{d}V\right]\end{aligned} \tag{3.43}$$

3.3 Microscopic (Differential) Local Instance Formulations

The microscopic (differential) formulations presented in this section include conservation equations and jump conditions. The former apply within a particular phase, and the latter are valid at the interface that separates two phases. The phase equations for a particular phase should be the same as those for a single-phase system. Most textbooks (e.g., White 2005; Bergman and Lavine 2017; Bejan 2013; Kays et al. 2004) obtain the differential governing equations for a single-phase system by performing mass, momentum, and energy balances for a microscopic element. We will obtain the differential conservation equations by using the integral equations for a finite control volume and applying the divergence theorem (Faghri et al. 2010). Jump conditions at the interface will be obtained by applying the conservation laws at the interfaces.

In order to obtain the differential formulations, it is necessary to apply the divergence theorem from vector calculus. For a general vector quantity, $\mathbf{\Omega}$, which is continuously differentiable, and for a control volume V, enclosed by a piecewise smooth control surface A, the divergence theorem states that

$$\int_A \boldsymbol{\Omega}\cdot \mathbf{n}\mathrm{d}A = \int_V \nabla\cdot \boldsymbol{\Omega}\mathrm{d}V \tag{3.44}$$

The desired differential equations may be obtained by applying this relationship to the integral form of the basic laws.

Furthermore, since the control volume shape and size are fixed in time, Leibniz's rule for the specific general quantity, ϕ, is also valid:

$$\frac{\mathrm{d}}{\mathrm{d}t}\int_V \rho\phi\mathrm{d}V = \int_V \frac{\partial(\rho\phi)}{\partial t}\mathrm{d}V \tag{3.45}$$

3.3.1 Conservation of Mass

The surface integral in Eq. (3.7) may be converted to a volume integral by applying Eq. (3.44) as follows:

$$\int_A \rho(\mathbf{V}_{\mathrm{rel}}\cdot \mathbf{n})\mathrm{d}A = \int_V \nabla\cdot \rho\mathbf{V}_{\mathrm{rel}}\mathrm{d}V \tag{3.46}$$

Substituting Eq. (3.46) into Eq. (3.7) and considering Eq. (3.45), the entire left-hand side of Eq. (3.7) is included in a single volume integral, i.e.,

$$\int_V \left(\frac{\partial\rho}{\partial t} + \nabla\cdot \rho\mathbf{V}_{\mathrm{rel}}\right)\mathrm{d}V = 0 \tag{3.47}$$

The only condition that ensures that Eq. (3.47) is true for any arbitrary shape and size of the control volume is that the integrand must be equal to zero, i.e.,

$$\frac{\partial\rho}{\partial t} + \nabla\cdot \rho\mathbf{V}_{\mathrm{rel}} = 0 \tag{3.48}$$

which is the differential form of the law of the conservation of mass.

Equation (3.48) can also be rewritten as

$$\frac{D\rho}{Dt} + \rho\nabla\cdot \mathbf{V}_{\mathrm{rel}} = 0 \tag{3.49}$$

where D/Dt is the *substantial derivative* (or *material derivative*) defined by

$$\frac{D}{Dt} \equiv \frac{\partial}{\partial t} + \mathbf{V}_{\mathrm{rel}}\cdot \nabla \tag{3.50}$$

For a stationary reference frame, $\mathbf{V}_{\mathrm{rel}} = \mathbf{V}$, and Eq. (3.49) becomes

$$\frac{D\rho}{Dt} + \rho \nabla \cdot \mathbf{V} = 0 \tag{3.51}$$

For incompressible flow, in which the density of the fluid is constant ($\rho \equiv$ const.), Eq. (3.51) simplifies as

$$\nabla \cdot \mathbf{V} = 0 \tag{3.52}$$

For steady-state *compressible* flow, the continuity equation can be obtained by simplifying Eq. (3.48), i.e.,

$$\nabla \cdot \rho \mathbf{V}_{\text{rel}} = 0 \tag{3.53}$$

Example 3.2

For a stationary reference frame, present the continuity equation in the Cartesian coordinate system for both compressible and incompressible fluid flow.

Solution The ∇ operator in the Cartesian coordinate system is [see Eq. (C.6) in Appendix C]

$$\nabla = \mathbf{i}\frac{\partial}{\partial x} + \mathbf{j}\frac{\partial}{\partial y} + \mathbf{k}\frac{\partial}{\partial z} \tag{3.54}$$

where **i**, **j**, and **k** are the unit vectors in the x-, y-, and z-directions, respectively. The velocity vector in a Cartesian system is

$$\mathbf{V} = \mathbf{i}u + \mathbf{j}v + \mathbf{k}w \tag{3.55}$$

where u, v, and w are the velocity components in the x-, y-, and z-directions, respectively. The continuity equation for compressible fluid flow in a stationary reference frame can be obtained by substituting Eqs. (3.54) and (3.55) into Eq. (3.51), i.e.,

$$\frac{D\rho}{Dt} + \rho\left(\frac{\partial u}{\partial x} + \frac{\partial v}{\partial y} + \frac{\partial w}{\partial z}\right) = 0 \tag{3.56}$$

where

$$\frac{D\rho}{Dt} = \frac{\partial \rho}{\partial t} + u\frac{\partial \rho}{\partial x} + v\frac{\partial \rho}{\partial y} + w\frac{\partial \rho}{\partial z} \tag{3.57}$$

Equation (3.56) can also be rearranged to:

$$\frac{\partial \rho}{\partial t} + \frac{\partial(\rho u)}{\partial x} + \frac{\partial(\rho v)}{\partial y} + \frac{\partial(\rho w)}{\partial z} = 0 \tag{3.58}$$

For incompressible flow, Eqs. (3.56) and (3.58) simplify as

$$\frac{\partial u}{\partial x} + \frac{\partial v}{\partial y} + \frac{\partial w}{\partial z} = 0 \tag{3.59}$$

3.3.2 Momentum Equation

The integral form of Newton's second law for a control volume that includes only one phase is expressed by Eq. (3.12). The surface integral terms in Eq. (3.12) can be rewritten using the divergence theorem:

$$\int_A \boldsymbol{\tau}' \cdot \mathbf{n} dA = \int_V \nabla \cdot \boldsymbol{\tau}' dV \tag{3.60}$$

$$\int_A \rho(\mathbf{V}_{\text{rel}} \cdot \mathbf{n})\mathbf{V}_{\text{rel}} dA = \int_V \nabla \cdot \rho \mathbf{V}_{\text{rel}} \mathbf{V}_{\text{rel}} dV \tag{3.61}$$

Substituting Eq. (3.60) and (3.61) into Eq. (3.12), and considering Eq. (3.45), the entire equation can be rewritten as a volume integral:

$$\int_V \left[\nabla \cdot \boldsymbol{\tau}' + \sum_{i=1}^{N} \rho_i \mathbf{X}_i - \frac{\partial}{\partial t}(\rho_k \mathbf{V}_{\text{rel}}) - \nabla \cdot \rho \mathbf{V}_{\text{rel}} \mathbf{V}_{\text{rel}} \right] dV = 0 \tag{3.62}$$

As was the case for the continuity equation, the integrand must equal zero to assure the general validity of Eq. (3.62); so, one obtains the desired differential form of the momentum equation:

$$\frac{\partial}{\partial t}(\rho \mathbf{V}_{\text{rel}}) + \nabla \cdot \rho \mathbf{V}_{\text{rel}} \mathbf{V}_{\text{rel}} = \nabla \cdot \boldsymbol{\tau}' + \sum_{i=1}^{N} \rho_i \mathbf{X}_i \tag{3.63}$$

The derivatives on the left-hand side of Eq. (3.63) may be expanded to yield

$$\mathbf{V}_{\text{rel}} \left[\frac{\partial \rho}{\partial t} + \nabla \cdot \rho \mathbf{V}_{\text{rel}} \right] + \rho \left[\frac{\partial \mathbf{V}_{\text{rel}}}{\partial t} + \mathbf{V}_{\text{rel}} \cdot \nabla \mathbf{V}_{\text{rel}} \right] = \nabla \cdot \boldsymbol{\tau}' + \sum_{i=1}^{N} \rho_i \mathbf{X}_i \tag{3.64}$$

The first bracketed term on the left vanishes, as required by the continuity Eq. (3.48). The second term may be written more simply in substantial derivative form, and the entire equation becomes

$$\rho \frac{D\mathbf{V}_{\text{rel}}}{Dt} = \nabla \cdot \boldsymbol{\tau}' + \sum_{i=1}^{N} \rho_i \mathbf{X}_i \tag{3.65}$$

The stress tensor, $\boldsymbol{\tau}'_{\text{rel}}$, is the sum of an isotropic thermodynamic stress, $-p\mathbf{I}$, and the viscous stress tensor, $\boldsymbol{\tau}$ [defined in Eq. (1.12)], i.e.,

$$\boldsymbol{\tau}' = -p\mathbf{I} + \boldsymbol{\tau} \tag{3.66}$$

Substituting Eq. (3.66) into Eq. (3.65), the momentum equation becomes

$$\rho \frac{D\mathbf{V}_{\text{rel}}}{Dt} = -\nabla p + \nabla \cdot \boldsymbol{\tau} + \sum_{i=1}^{N} \rho_i \mathbf{X}_i \tag{3.67}$$

The viscous stress tensor measured in the reference frame, $\boldsymbol{\tau}$, can be determined by using Newton's law of viscosity [see Eq. (1.15)]:

$$\boldsymbol{\tau} = 2\mu\mathbf{D} - \frac{2}{3}\mu(\nabla \cdot \mathbf{V}_{\text{rel}})\mathbf{I} \tag{3.68}$$

where **D** is the rate of strain tensor, i.e.,

$$\mathbf{D} = \frac{1}{2}\left[\nabla\mathbf{V}_{\text{rel}} + (\nabla\mathbf{V}_{\text{rel}})^T\right] \tag{3.69}$$

and **I** in Eq. (3.68) is the unit tensor that satisfies $\mathbf{a}\cdot\mathbf{I} = \mathbf{I} \cdot \mathbf{a} = \mathbf{a}$ for any tensor **a**. The diagonal components of **I** are equal to one, and all other components are zero:

$$I_{ij} = \begin{cases} 1 & i = j \\ 0 & i \neq j \end{cases} \quad (i,j = 1, 2, 3) \tag{3.70}$$

If the fluid is incompressible ($\rho =$ const.), the second term on the right-hand side of Eq. (3.68) will be zero according to Eq. (3.52). The momentum Eq. (3.65) then becomes

$$\rho \frac{D\mathbf{V}_{\text{rel}}}{Dt} = \sum_{i=1}^{N} \rho_i \mathbf{X}_i - \nabla p + \nabla \cdot (\mu\nabla\mathbf{V}_{\text{rel}}) \tag{3.71}$$

where the left-hand side is the inertial term (mass per unit volume ρ times acceleration, $D\mathbf{V}_{\text{rel}}/Dt$). The three terms on the right-hand side represent body force per unit volume, pressure force per unit volume, and viscous force per unit volume, respectively.

For $D\mathbf{V}_{\text{rel}}/Dt = 0$, we have Stokes' flow or creep flow, and Eq. (3.71) becomes elliptic and is similar to the steady-state conduction equation.

In a Cartesian coordinate system, the vector form of the momentum equation, Eq. (3.71), for incompressible and Newtonian fluid with constant viscosity can be written as three equations in the *x*-, *y*-, and *z*-directions:

$$\rho \frac{Du_{\text{rel}}}{Dt} = \sum_{i=1}^{N} \rho_i X_i - \frac{\partial p}{\partial x} + \mu\left(\frac{\partial^2 u_{\text{rel}}}{\partial x^2} + \frac{\partial^2 u_{\text{rel}}}{\partial y^2} + \frac{\partial^2 u_{\text{rel}}}{\partial z^2}\right) \tag{3.72}$$

$$\rho \frac{Dv_{\text{rel}}}{Dt} = \sum_{i=1}^{N} \rho_i Y_i - \frac{\partial p}{\partial y} + \mu\left(\frac{\partial^2 v_{\text{rel}}}{\partial x^2} + \frac{\partial^2 v_{\text{rel}}}{\partial y^2} + \frac{\partial^2 v_{\text{rel}}}{\partial z^2}\right) \tag{3.73}$$

$$\rho\frac{Dw_{\mathrm{rel}}}{Dt}=\sum_{i=1}^{N}\rho_i Z_i-\frac{\partial p}{\partial z}+\mu\left(\frac{\partial^2 w_{\mathrm{rel}}}{\partial x^2}+\frac{\partial^2 w_{\mathrm{rel}}}{\partial y^2}+\frac{\partial^2 w_{\mathrm{rel}}}{\partial z^2}\right) \tag{3.74}$$

where X_i, Y_i and Z_i are the components of body force per unit volume acting on the ith species in the x-, y-, and z- directions, respectively.

For the case that the only body force is gravity, $\mathbf{X}_i=\mathbf{g}$, Eq. (3.71) becomes

$$\rho\frac{D\mathbf{V}_{\mathrm{rel}}}{Dt}=\rho\mathbf{g}-\nabla p+\nabla\cdot(\mu\nabla\mathbf{V}_{\mathrm{rel}}) \tag{3.75}$$

For natural convection problem, it is often assumed that the fluid is incompressible except in the first term on the right-hand side of Eq. (3.75); this is referred to as the *Boussinesq assumption*. The density of a mixture is a function of temperature and mass fractions of species. It can be expanded using a Taylor's series near the vicinity of a reference point $(\bar{T},\bar{\omega}_1,\bar{\omega}_2,\ldots\bar{\omega}_N)$:

$$\begin{aligned}\rho&=\bar{\rho}+\frac{\partial\rho}{\partial T}(T-\bar{T})+\sum_{i=1}^{N}\frac{\partial\rho}{\partial\omega_i}(\omega_i-\bar{\omega}_i)+\cdots\\&\approx\bar{\rho}-\bar{\rho}\beta(T-\bar{T})-\bar{\rho}\sum_{i=1}^{N}\beta_{\mathrm{m}}(\omega_i-\bar{\omega}_i)\end{aligned} \tag{3.76}$$

where $\bar{\rho}$ is density at the reference point, $\beta=-(\partial\rho/\partial T)/\bar{\rho}$ is the coefficient of thermal expansion, and $\beta_{\mathrm{m}}=-(\partial\rho/\partial\omega_i)/\bar{\rho}$ is the composition coefficient of volume expansion. Substituting Eq. (3.76) into Eq. (3.75), the momentum equation for natural convection is obtained

$$\rho\frac{D\mathbf{V}_{\mathrm{rel}}}{Dt}=(-\nabla p+\bar{\rho}\mathbf{g})-\bar{\rho}\mathbf{g}\beta(T-\bar{T})-\bar{\rho}\mathbf{g}\sum_{i=1}^{N}\beta_{\mathrm{m}}(\omega_i-\bar{\omega}_i)+\nabla\cdot(\mu\nabla\mathbf{V}_{\mathrm{rel}}) \tag{3.77}$$

where the second and third terms on the right-hand side of Eq. (3.77) describe the effect of buoyancy force due to temperature and composition variation within the system, respectively.

3.3.3 Energy Equation

The surface integrals on the right-hand side of Eq. (3.22) can be rewritten as volume integrals using Eq. (3.44), i.e.,

$$\int_A\rho(\mathbf{V}_{\mathrm{rel}}\cdot\mathbf{n})\left(e+\frac{\mathbf{V}_{\mathrm{rel}}^2}{2}\right)\mathrm{d}A=\int_V\nabla\cdot\left[\rho\mathbf{V}_{\mathrm{rel}}\left(e+\frac{\mathbf{V}_{\mathrm{rel}}^2}{2}\right)\right]\mathrm{d}V \tag{3.78}$$

$$\int_A-\mathbf{q}''\cdot\mathbf{n}\mathrm{d}A=\int_V-\nabla\cdot\mathbf{q}''\mathrm{d}V \tag{3.79}$$

$$\int_A \mathbf{n} \cdot \boldsymbol{\tau}' \cdot \mathbf{V}_{\text{rel}} \mathrm{d}A = \int_V \nabla \cdot (\boldsymbol{\tau}' \cdot \mathbf{V}_{\text{rel}}) \mathrm{d}V \tag{3.80}$$

Substituting Eqs. (3.78)–(3.80) into Eq. (3.22) and considering Eq. (3.45) yields

$$\begin{aligned} \int_V \Bigg\{ & \frac{\partial}{\partial t}\left[\rho\left(e + \frac{\mathbf{V}_{\text{rel}}^2}{2}\right)\right] + \nabla \cdot \left[\rho \mathbf{V}_{\text{rel}}\left(e + \frac{\mathbf{V}_{\text{rel}}^2}{2}\right)\right] \\ & + \nabla \cdot \mathbf{q}'' - q''' - \nabla \cdot (\boldsymbol{\tau}' \cdot \mathbf{V}_{\text{rel}}) - \left(\sum_{i=1}^{N} \rho_i \mathbf{X}_i\right) \cdot \mathbf{V}_{\text{rel}} \Bigg\} \mathrm{d}V = 0 \end{aligned} \tag{3.81}$$

For Eq. (3.81) to be true for any arbitrary control volume, the integrand must be zero. The result is the general differential form of the energy equation:

$$\begin{aligned} & \frac{\partial}{\partial t}\left[\rho\left(e + \frac{\mathbf{V}_{\text{rel}}^2}{2}\right)\right] + \nabla \cdot \left[\rho \mathbf{V}_{\text{rel}}\left(e + \frac{\mathbf{V}_{\text{rel}}^2}{2}\right)\right] \\ & = -\nabla \cdot \mathbf{q}'' + q''' + \nabla \cdot (\boldsymbol{\tau}' \cdot \mathbf{V}_{\text{rel}}) + \left(\sum_{i=1}^{N} \rho_i \mathbf{X}_i\right) \cdot \mathbf{V}_{\text{rel}} \end{aligned} \tag{3.82}$$

The left-hand side of Eq. (3.82) can be rewritten to yield:

$$\begin{aligned} & \left(e + \frac{V_{\text{rel}}^2}{2}\right)\left[\frac{\partial \rho}{\partial t} + \nabla \cdot (\rho \mathbf{V}_{\text{rel}})\right] + \rho\left[\frac{\partial}{\partial t}\left(e + \frac{\mathbf{V}_{\text{rel}}^2}{2}\right) + \mathbf{V}_{\text{rel}} \cdot \nabla\left(e + \frac{\mathbf{V}_{\text{rel}}^2}{2}\right)\right] \\ & = -\nabla \cdot \mathbf{q}'' + q''' + \nabla \cdot (\boldsymbol{\tau}' \cdot \mathbf{V}_{\text{rel}}) + \left(\sum_{i=1}^{N} \rho_i \mathbf{X}_i\right) \cdot \mathbf{V}_{\text{rel}} \end{aligned} \tag{3.83}$$

According to the continuity Eq. (3.48), the first bracketed term on the left-hand side of Eq. (3.83) is zero. The second term on the left-hand side may be written more simply in substantial derivative form, i.e.,

$$\rho \frac{D}{Dt}\left(e + \frac{\mathbf{V}_{\text{rel}}^2}{2}\right) = -\nabla \cdot \mathbf{q}'' + q''' + \nabla \cdot (\boldsymbol{\tau}' \cdot \mathbf{V}_{\text{rel}}) + \left(\sum_{i-1}^{N} \rho_i \mathbf{X}_i\right) \cdot \mathbf{V}_{\text{rel}} \tag{3.84}$$

which is the total energy (including thermal and mechanical) balance equation.

It would be more convenient to remove the mechanical energy terms from Eq. (3.84). The mechanical energy balance equation can be obtained by forming a dot (scalar) product of the momentum Eq. (3.65) with the velocity vector $\boldsymbol{V}_{\text{rel}}$, i.e.,

$$\rho \frac{D\mathbf{V}_{\text{rel}}}{Dt} \cdot \mathbf{V}_{\text{rel}} = (\nabla \cdot \boldsymbol{\tau}') \cdot \mathbf{V}_{\text{rel}} + \left(\sum_{i=1}^{N} \rho_i \mathbf{X}_i\right) \cdot \mathbf{V}_{\text{rel}} \tag{3.85}$$

which can be rearranged to obtain

$$\rho \frac{D}{Dt}\left(\frac{\mathbf{V}_{\text{rel}}^2}{2}\right) = \nabla \cdot (\boldsymbol{\tau}' \cdot \mathbf{V}_{\text{rel}}) - \nabla \mathbf{V} : \boldsymbol{\tau} + \left(\sum_{i=1}^{N} \rho_i \mathbf{X}_i\right) \cdot \mathbf{V}_{\text{rel}} \tag{3.86}$$

Subtracting Eq. (3.86) from Eq. (3.84) yields the following thermal energy equation:

$$\rho \frac{De}{Dt} = -\nabla \cdot \mathbf{q}'' + q''' - p\nabla \cdot \mathbf{V}_{\text{rel}} + \nabla \mathbf{V}_{\text{rel}} : \boldsymbol{\tau} \tag{3.87}$$

where the left-hand side represents the rate of gain of energy per unit volume. The terms in the right-hand side are the rate of energy input by heat transfer per unit volume, the internal heat generation per unit volume, the reversible rate of energy increase per unit volume by compression, and the irreversible rate of energy increase per unit volume by viscous dissipation, respectively; the viscous dissipation is the contraction [see Eq. (C.35)] of two tensors of rank two: $\nabla \mathbf{V}_{\text{rel}}$ and $\boldsymbol{\tau}$, i.e.,

$$\begin{aligned} \nabla \mathbf{V}_{\text{rel}} : \boldsymbol{\tau} = & \frac{\partial u_{\text{rel}}}{\partial x}\tau_{xx} + \frac{\partial u_{\text{rel}}}{\partial y}\tau_{xy} + \frac{\partial u_{\text{rel}}}{\partial z}\tau_{xz} + \frac{\partial v_{\text{rel}}}{\partial x}\tau_{yx} + \frac{\partial v_{\text{rel}}}{\partial y}\tau_{yy} \\ & + \frac{\partial v_{\text{rel}}}{\partial z}\tau_{yz} + \frac{\partial w_{\text{rel}}}{\partial x}\tau_{zx} + \frac{\partial w_{\text{rel}}}{\partial y}\tau_{zy} + \frac{\partial w_{\text{rel}}}{\partial z}\tau_{zz} \end{aligned} \tag{3.88}$$

Equation (3.87) is the energy equation expressed in terms of internal energy. In order to obtain an equation that contains enthalpy, the definition of enthalpy is employed:

$$h = e + \frac{p}{\rho} \tag{3.89}$$

Substituting Eq. (3.89) into Eq. (3.87), and considering the continuity Eq. (3.48), the energy equation in terms of enthalpy is obtained:

$$\rho \frac{Dh}{Dt} = -\nabla \cdot \mathbf{q}'' + \frac{Dp}{Dt} + q''' + \nabla \mathbf{V}_{\text{rel}} : \boldsymbol{\tau} \tag{3.90}$$

For a multicomponent system, the enthalpy can be expressed as

$$h = \sum_{i=1}^{N} \omega_i h_i \tag{3.91}$$

where ω_i and h_i are the mass fraction and specific enthalpy of the ith component. Substituting Eq. (3.91) into Eq. (3.90), the energy equation for a multicomponent system becomes

$$\sum_{i=1}^{N}\left[\rho \frac{D}{Dt}(\omega_i h_i)\right] = -\nabla \cdot \mathbf{q}'' + \frac{Dp}{Dt} + q''' + \nabla \mathbf{V}_{\text{rel}} : \boldsymbol{\tau} \tag{3.92}$$

The heat flux vector $\mathbf{q}''$ can be obtained from Eq. (1.22)

$$\mathbf{q}'' = -k\nabla T \tag{3.93}$$

Substituting Eq. (3.93) into Eq. (3.92), the energy equation becomes

$$\sum_{i=1}^{N}\left[\rho\frac{D}{Dt}(\omega_i h_i)\right] = \nabla\cdot(k\nabla T) + \frac{Dp}{Dt} + q''' + \nabla\mathbf{V}_{\text{rel}} : \boldsymbol{\tau} \tag{3.94}$$

For a pure substance, Eq. (3.94) is reduced to:

$$\rho\frac{Dh}{Dt} = \nabla\cdot(k\nabla T) + \frac{Dp}{Dt} + q''' + \nabla\mathbf{V}_{\text{rel}} : \boldsymbol{\tau} \tag{3.95}$$

The enthalpy of a pure substance (single component) can be expressed as a function of temperature and pressure, $h = f(T, p)$, i.e.,

$$\frac{Dh}{Dt} = \left(\frac{\partial h}{\partial T}\right)_{\text{p}}\frac{DT}{Dt} + \left(\frac{\partial h}{\partial p}\right)_{T}\frac{Dp}{Dt} \tag{3.96}$$

Thermodynamic relations give us

$$\left(\frac{\partial h}{\partial T}\right)_{\text{p}} = c_{\text{p}}\,,\quad \left(\frac{\partial h}{\partial p}\right)_{T} = \frac{1-\beta T}{\rho} \tag{3.97}$$

where

$$\beta = \frac{1}{\rho}\left(\frac{\partial \rho}{\partial T}\right)_{\text{p}} \tag{3.98}$$

is the coefficient of thermal expansion.

Therefore, Eq. (3.96) becomes

$$\frac{Dh}{Dt} = c_{\text{p}}\frac{DT}{Dt} + \frac{1-\beta T}{\rho}\frac{Dp}{Dt} \tag{3.99}$$

Substituting Eq. (3.99) into Eq. (3.95), the energy equation becomes

$$\rho c_{\text{p}}\frac{DT}{Dt} = \nabla\cdot(k\nabla T) + T\beta\frac{Dp}{Dt} + q''' + \nabla\mathbf{V}_{\text{rel}} : \boldsymbol{\tau} \tag{3.100}$$

which can be simplified for different substances as demonstrated below.

For ideal gases, substituting the equation of state for ideal gas, $\rho = p/(R_{\text{a}}T)$, into Eq. (3.98) yields $\beta = 1/T$, and Eq. (3.100) becomes

$$\rho c_{\text{p}}\frac{DT}{Dt} = \nabla\cdot(k\nabla T) + \frac{Dp}{Dt} + q''' + \nabla\mathbf{V}_{\text{rel}} : \boldsymbol{\tau} \tag{3.101}$$

For incompressible fluids, the constant density yields $\beta = 0$ and Eq. (3.100) becomes

$$\rho c_{\text{p}}\frac{DT}{Dt} = \nabla\cdot(k\nabla T) + q''' + \nabla\mathbf{V}_{\text{rel}} : \boldsymbol{\tau} \tag{3.102}$$

For solids, the density is constant and the velocity is zero, and the energy equation becomes

$$\rho_s c_{ps} \frac{\partial T_s}{\partial t} = \nabla \cdot (k_s \nabla T_s) + q_s''' \tag{3.103}$$

which is the heat conduction equation.

Example 3.3

Determine the viscous dissipation for incompressible flow in the Cartesian coordinate system. The reference velocity can be assumed to be zero.

Solution Since the reference velocity is zero, the viscous dissipation is $\nabla \mathbf{V} : \boldsymbol{\tau}$. According to Eq. (C.33),

$$\nabla \mathbf{V} = \begin{bmatrix} \frac{\partial u}{\partial x} & \frac{\partial u}{\partial y} & \frac{\partial u}{\partial z} \\ \frac{\partial v}{\partial x} & \frac{\partial v}{\partial y} & \frac{\partial v}{\partial z} \\ \frac{\partial w}{\partial x} & \frac{\partial w}{\partial y} & \frac{\partial w}{\partial z} \end{bmatrix}$$

The shear stress tensor for incompressible flow is

$$\boldsymbol{\tau} = 2\mu \mathbf{D} = \begin{bmatrix} 2\mu \frac{\partial u}{\partial x} & \mu\left(\frac{\partial u}{\partial y} + \frac{\partial v}{\partial x}\right) & \mu\left(\frac{\partial u}{\partial z} + \frac{\partial w}{\partial x}\right) \\ \mu\left(\frac{\partial v}{\partial x} + \frac{\partial u}{\partial y}\right) & 2\mu \frac{\partial v}{\partial y} & \mu\left(\frac{\partial v}{\partial z} + \frac{\partial w}{\partial y}\right) \\ \mu\left(\frac{\partial w}{\partial x} + \frac{\partial u}{\partial z}\right) & \mu\left(\frac{\partial w}{\partial y} + \frac{\partial v}{\partial z}\right) & 2\mu \frac{\partial w}{\partial z} \end{bmatrix} \tag{3.104}$$

Therefore, the viscous dissipation is

$$\nabla \mathbf{V} : \boldsymbol{\tau} = \begin{bmatrix} \frac{\partial u}{\partial x} & \frac{\partial u}{\partial y} & \frac{\partial u}{\partial z} \\ \frac{\partial v}{\partial x} & \frac{\partial v}{\partial y} & \frac{\partial v}{\partial z} \\ \frac{\partial w}{\partial x} & \frac{\partial w}{\partial y} & \frac{\partial w}{\partial z} \end{bmatrix} : \begin{bmatrix} 2\mu \frac{\partial u}{\partial x} & \mu\left(\frac{\partial u}{\partial y} + \frac{\partial v}{\partial x}\right) & \mu\left(\frac{\partial u}{\partial z} + \frac{\partial w}{\partial x}\right) \\ \mu\left(\frac{\partial v}{\partial x} + \frac{\partial u}{\partial y}\right) & 2\mu \frac{\partial v}{\partial y} & \mu\left(\frac{\partial v}{\partial z} + \frac{\partial w}{\partial y}\right) \\ \mu\left(\frac{\partial w}{\partial x} + \frac{\partial u}{\partial z}\right) & \mu\left(\frac{\partial w}{\partial y} + \frac{\partial v}{\partial z}\right) & 2\mu \frac{\partial w}{\partial z} \end{bmatrix}$$

The contraction of two tensors of rank two yields

$$\begin{aligned} \nabla \mathbf{V} : \boldsymbol{\tau} = {} & 2\mu\left(\frac{\partial u}{\partial x}\right)^2 + \mu\left[\left(\frac{\partial u}{\partial y}\right)^2 + \frac{\partial u}{\partial y}\frac{\partial v}{\partial x}\right] + \mu\left[\left(\frac{\partial u}{\partial z}\right)^2 + \frac{\partial u}{\partial z}\frac{\partial w}{\partial x}\right] \\ & + \mu\left[\left(\frac{\partial v}{\partial x}\right)^2 + \frac{\partial v}{\partial x}\frac{\partial u}{\partial y}\right] + 2\mu\left(\frac{\partial v}{\partial y}\right)^2 + \mu\left[\left(\frac{\partial v}{\partial z}\right)^2 + \frac{\partial v}{\partial z}\frac{\partial w}{\partial y}\right] \\ & + \mu\left[\left(\frac{\partial w}{\partial x}\right)^2 + \frac{\partial w}{\partial x}\frac{\partial u}{\partial z}\right] + \mu\left[\left(\frac{\partial w}{\partial y}\right)^2 + \frac{\partial w}{\partial y}\frac{\partial v}{\partial z}\right] + 2\mu\left(\frac{\partial w}{\partial z}\right)^2 \end{aligned} \tag{3.105}$$

The final form of the viscous dissipation for incompressible flow is obtained by rearranging Eq. (3.105), i.e.,

$$\begin{aligned}\nabla\mathbf{V}:\boldsymbol{\tau}=\mu\Bigg[&2\left(\frac{\partial u}{\partial x}\right)^2+2\left(\frac{\partial v}{\partial y}\right)^2+2\left(\frac{\partial w}{\partial z}\right)^2\\&+\left(\frac{\partial v}{\partial x}+\frac{\partial u}{\partial y}\right)^2+\left(\frac{\partial w}{\partial y}+\frac{\partial v}{\partial z}\right)^2+\left(\frac{\partial u}{\partial z}+\frac{\partial w}{\partial x}\right)^2\Bigg]\end{aligned} \tag{3.106}$$

The viscous dissipation in cylindrical and spherical coordinate systems can be found by similar means using information in Appendix C.

3.3.4 Species

The surface integrals in Eq. (3.42) may be converted to volume integrals by applying Eq. (3.44) as follows:

$$\int_A \rho_i(\mathbf{V}_{\text{rel}}\cdot\mathbf{n})\text{d}A=\int_V \nabla\cdot\rho_i\mathbf{V}_{\text{rel}}\text{d}V \tag{3.107}$$

$$\int_A \mathbf{J}_i\cdot\mathbf{n}\text{d}A=\int_V \nabla\cdot\mathbf{J}_i\text{d}V \tag{3.108}$$

Substituting Eqs. (3.107) and (3.108) into Eq. (3.42) and considering Eq. (3.45), the entire left-hand side and right-hand side of Eq. (3.42) are included in a single volume integral, i.e.,

$$\int_V\left(\frac{\partial\rho_i}{\partial t}+\nabla\cdot\rho_i\mathbf{V}_{\text{rel}}+\nabla\cdot\mathbf{J}_i-\dot{m}_i'''\right)\text{d}V=0 \tag{3.109}$$

The only condition that makes Eq. (3.109) true regardless of the shape and size of the control volume is that the integrand must equal zero, i.e.,

$$\frac{\partial\rho_i}{\partial t}+\nabla\cdot\rho_i\mathbf{V}_{\text{rel}}=-\nabla\cdot\mathbf{J}_i+\dot{m}_i''',\quad i=1,2,\ldots,N \tag{3.110}$$

The first term on the left-hand side is the rate of increase of mass of the species i per unit volume, and the second term is net rate of additions of mass of the ith species per unit volume by convection. The terms on the right-hand side are the net rate of mass of ith species per unit volume by diffusion and rate of production of species i by chemical reaction.

Equation (3.110) is the equation of conservation of mass for species. If the total number of species is N, a total of $N-1$ independent equations for conservation of species mass can be obtained.

After defining the mass fraction of species i as

$$\omega_i = \frac{\rho_i}{\rho} \tag{3.111}$$

Equation (3.110) can be rewritten as

$$\left[\frac{\partial \rho}{\partial t} + \nabla \cdot \rho \mathbf{V}_{\text{rel}}\right]\omega_i + \rho\left[\frac{\partial \omega_i}{\partial t} + \nabla \cdot \omega_i \mathbf{V}_{\text{rel}}\right] = -\nabla \cdot \mathbf{J}_i + \dot{m}_i''' \tag{3.112}$$

According to Eq. (3.48), the first bracket on the left-hand side of Eq. (3.112) is zero. The second bracket on the left-hand side is the substantial derivative of the mass fraction. Therefore, the conservation of species mass in terms of mass fraction becomes

$$\rho \frac{D\omega_i}{Dt} = -\nabla \cdot \mathbf{J}_i + \dot{m}_i''' \tag{3.113}$$

Assuming binary system of A and B, one can use Fick's law in Eq. (3.113) to yield

$$\rho \frac{D\omega_A}{Dt} = -\rho \nabla \cdot (D_{AB} \nabla \omega_A) + \dot{m}_A''' \tag{3.114}$$

which is useful in determining the diffusion in dilute liquid solution at constant temperature and pressure. Equation (3.114), with $\dot{m}_A''' = 0$, is similar to energy Eq. (3.102) with no internal heat source and viscous dissipation, and therefore, it is used for analogy between heat and mass transfer analysis.

In the preceding discussion to develop Eq. (3.110), the mass fraction and mass flux were used. The species equation can also be developed in terms of molar concentration (or molar fraction) and molar flux. By following a similar procedure, the species equation is

$$\frac{\partial c_i}{\partial t} = -\nabla \cdot \dot{\mathbf{n}}''_i + \dot{n}_i''' \tag{3.115}$$

where the molar flux relative to the stationary coordinate axes can be obtained from Eq. (1.56), i.e.,

$$\dot{\mathbf{n}}_i'' = c_i \tilde{\mathbf{V}}^* + \mathbf{J}_i^* \tag{3.116}$$

Substituting Eq. (3.116) into Eq. (3.115), we have

$$\frac{\partial c_i}{\partial t} + \nabla \cdot (c_i \tilde{\mathbf{V}}^*) = -\nabla \cdot \mathbf{J}_i^* + \dot{n}_i''' \tag{3.117}$$

where the first term on the left-hand side is rate of increase of mole of the species i per unit volume, and the second term is net rate of additions of mole of the ith species per unit volume by convection. The terms on the right-hand side are net rate of mole of ith species per unit volume by diffusion and the molar rate of production of species i by chemical reaction.

For a binary system of components A and B with constant pressure, Eq. (3.117) reduces to

$$\frac{\partial c_A}{\partial t} + \nabla \cdot (c_A \tilde{\mathbf{V}}^*) = c \nabla \cdot (D_{AB} \nabla x_A) + \dot{n}_A''' \tag{3.118}$$

where c is the mixture molar concentration and $\tilde{\mathbf{V}}^*$ is molar-averaged velocity defined in Chap. 1. Equation (3.118) can be applied to low-density gases with constant temperature and pressure.

The production rate of the *i*th species, $\dot{m}_i'''$(or $\dot{n}_i'''$), is still unknown at this point, but it can be obtained by analyzing the chemical reaction. If the number of chemical reactions taking place in the system is N_c, the mass production rate is (Kleijn 1991; Mahajan 1996)

$$\dot{m}_i''' = \sum_{j=1}^{N_c} a_{ij} M_i \Re_j \tag{3.119}$$

where a_{ij} is the stoichiometric coefficient that describes the proportions of the mole numbers of reactants disappearing and mole numbers of products appearing as a result of the reaction process (see Sect. 2.3.3), and M_i is molar mass of the *i*th component. The net reaction rate of the *j*th chemical reaction $\Re_j$ is the difference between the forward and backward reactions, i.e.,

$$\Re_j = \Re_{j^+} - \Re_{j^-} \tag{3.120}$$

If chemical reactions take place in gas mixture, the forward and backward reaction rates are

$$\Re_{j^+} = r_{j^+}(p, T) \prod_{i=1}^{N_r} \left(\frac{p}{R_u T} x_i \right)^{|a_{ij}|} \tag{3.121}$$

$$\Re_{j^-} = r_{j^-}(p, T) \prod_{i=1}^{N_p} \left(\frac{p}{R_u T} x_i \right)^{|a_{ij}|} \tag{3.122}$$

where N_r and N_p are number of reactants and products, respectively. The two reaction rate constants, r_{j^+} and r_{j^-}, depend on the specific chemical reaction under consideration. These constants are related by

$$\frac{r_{j^-}(p, T)}{r_{j^+}(p, T)} = \frac{1}{K_j(T)} \left(\frac{R_u T}{p^0} \right)^{\sum_{i=1}^{N} a_{ij}} \tag{3.123}$$

where p^0 is the standard pressure (101.3 kPa) and $K_j(T)$ is the thermodynamic equilibrium constant of the *j*th chemical reaction:

$$K_j(T) = \exp\left[\frac{-\Delta \bar{G}_j^0(T)}{R_u T} \right] \tag{3.124}$$

where $\Delta \bar{G}_j^0(T)$ is the standard Gibbs energy for the *j*th chemical reaction; its value depends on the specific chemical reaction considered.

For the case in which only one chemical reaction ($N_c = 1$) leads to production of the *i*th component, the molar production rate can be obtained by a simplified form

$$\dot{n}_i''' = k_n''' c_i^n \tag{3.125}$$

where the index *n* represents the order of the reaction and k_n''' is a rate constant ($1/s^n$) that depends on the temperature.

Equation (3.113) is valid for the case that chemical reaction occurs in the entire volume—referred to as *homogeneous* reacting system. For the case where the chemical reaction takes place on a surface

—referred to as *heterogeneous* reaction—the source term in Eq. (3.113) will not appear, and the rate of production will be accounted as a boundary condition. More discussion about modeling of the heterogeneous reacting system can be found in discussion of Chemical Vapor Deposition (CVD) in Sect. 6.3.2.

Example 3.4

Couette flow is found in many engineering applications, such as lubricant flow in a journal bearing. Consider the laminar viscous Couette flow where two plates are of infinite extent, steady-state conditions apply (see Fig. 3.5). The fluid is Newtonian and incompressible. Starting from the basic differential equations and using mathematical, physical arguments to make suitable simplifying assumptions, derive the differential equations governing the fluid velocity and temperature, as well as boundary conditions. Obtain the velocity and temperature distributions. The fluid properties can be assumed to be constant, but viscous dissipation should be accounted for.

Solution To solve the problem, the following assumptions are made:

1. The flow and heat transfer are steady state, $\partial/\partial t = 0$
2. Thermophysical properties are constants
3. The flow is laminar
4. No pressure gradient in the x-direction, $\partial p/\partial x = 0$
5. No edge effect in the z-direction, $\partial/\partial z = 0$
6. No end effect in the x-direction, $\partial/\partial x = 0$
7. Gravity is the only body force, $X = Z = 0, Y = -g$

If the reference frame is stationary, the continuity, momentum, and energy equations can be simplified based on the above assumptions

$$\underbrace{\frac{\partial u}{\partial x}}_{\substack{0\\ \text{Assumptions (6)}}} + \frac{\partial v}{\partial y} + \underbrace{\frac{\partial w}{\partial z}}_{\substack{0\\ (5)}} = 0 \tag{3.126}$$

$$\rho\left(\underbrace{\frac{\partial u}{\partial t}}_{\substack{0\\ \text{Assumptions (1)}}} + \underbrace{u\frac{\partial u}{\partial x}}_{\substack{0\\ (6)}} + v\frac{\partial u}{\partial y} + \underbrace{w\frac{\partial u}{\partial z}}_{\substack{0\\ (5)}}\right) = \underbrace{\rho X}_{\substack{0\\ (7)}} - \underbrace{\frac{\partial p}{\partial x}}_{\substack{0\\ (4)}} + \mu\left(\underbrace{\frac{\partial^2 u}{\partial x^2}}_{\substack{0\\ (6)}} + \frac{\partial^2 u}{\partial y^2} + \underbrace{\frac{\partial^2 u}{\partial z^2}}_{\substack{0\\ (5)}}\right) \tag{3.127}$$

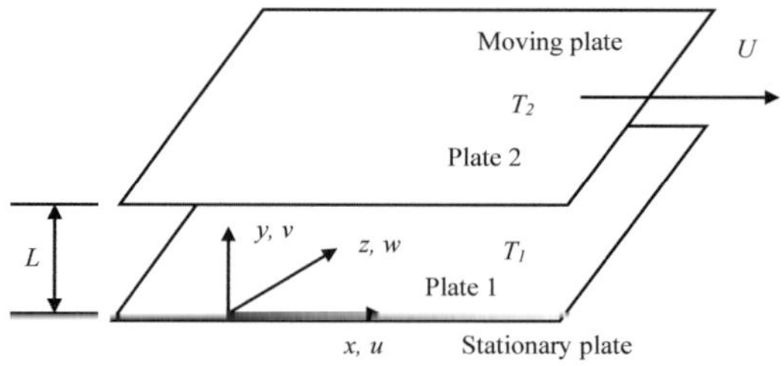

Fig. 3.5 Couette flow

$$\rho\left(\cancel{\frac{\partial v}{\partial t}}+\cancel{u\frac{\partial v}{\partial x}}+v\frac{\partial v}{\partial y}+\cancel{w\frac{\partial v}{\partial z}}\right)=\cancel{\rho Y}-\frac{\partial p}{\partial y}+\mu\left(\cancel{\frac{\partial^2 v}{\partial x^2}}+\frac{\partial^2 v}{\partial y^2}+\cancel{\frac{\partial^2 v}{\partial z^2}}\right) \tag{3.128}$$

0 0 0 $-\rho g$ 0 0

Assumptions (1) (6) (5) (7) (6) (5)

$$\rho\left(\cancel{\frac{\partial w}{\partial t}}+\cancel{u\frac{\partial w}{\partial x}}+v\frac{\partial w}{\partial y}+\cancel{w\frac{\partial w}{\partial z}}\right)=\cancel{\rho Z}-\cancel{\frac{\partial p}{\partial z}}+\mu\left(\cancel{\frac{\partial^2 w}{\partial x^2}}+\frac{\partial^2 w}{\partial y^2}+\cancel{\frac{\partial^2 w}{\partial z^2}}\right) \tag{3.129}$$

0 0 0 0 0 0 0

Assumptions (1) (6) (5) (7) (5) (6) (5)

$$\rho c_{\mathrm{p}}\left(\cancel{\frac{\partial T}{\partial t}}+\cancel{u\frac{\partial T}{\partial x}}+v\frac{\partial T}{\partial y}+\cancel{w\frac{\partial T}{\partial z}}\right)=k\left(\cancel{\frac{\partial^2 T}{\partial x^2}}+\frac{\partial^2 T}{\partial y^2}+\cancel{\frac{\partial^2 T}{\partial z^2}}\right)+\mu\left[\cancel{2\left(\frac{\partial u}{\partial x}\right)^2}\right.$$

0 0 0 0 0 0

Assumptions (1) (6) (5) (6) (5) (6)

$$\left.+2\left(\frac{\partial v}{\partial y}\right)^2+\cancel{2\left(\frac{\partial w}{\partial z}\right)^2}+\left(\cancel{\frac{\partial v}{\partial x}}+\frac{\partial u}{\partial y}\right)^2+\left(\cancel{\frac{\partial w}{\partial y}}+\frac{\partial v}{\partial z}\right)^2+\left(\cancel{\frac{\partial u}{\partial z}}+\cancel{\frac{\partial w}{\partial x}}\right)^2\right] \tag{3.130}$$

0 0 0 0 0

Assumptions (5) (6) (5) (5) (6)

Therefore, Eqs. (3.126)–(3.130) can become

$$\frac{\partial v}{\partial y}=0 \tag{3.131}$$

$$\rho v\frac{\partial u}{\partial y}=\mu\frac{\partial^2 u}{\partial y^2} \tag{3.132}$$

$$\rho v\frac{\partial v}{\partial y}=-\rho g-\frac{\partial p}{\partial y}+\mu\frac{\partial^2 v}{\partial y^2} \tag{3.133}$$

$$\rho v\frac{\partial w}{\partial y}=\mu\frac{\partial^2 w}{\partial y^2} \tag{3.134}$$

$$\rho c_{\mathrm{p}} v\frac{\partial T}{\partial y}=k\frac{\partial^2 T}{\partial y^2}+\mu\left[2\left(\frac{\partial v}{\partial y}\right)^2+\left(\frac{\partial u}{\partial y}\right)^2+\left(\frac{\partial w}{\partial y}\right)^2\right] \tag{3.135}$$

The boundary conditions for Eqs. (3.131)–(3.135) are

$$u=w=0,\quad y=0\quad \text{(no-slip)} \tag{3.136}$$

$$v=0,\quad y=0\quad \text{(impermeable wall)} \tag{3.137}$$

$$T = T_1, \quad y = 0 \quad \text{(no temperature jump)} \tag{3.138}$$

$$u = U, \quad w = 0, \quad y = L \quad \text{(no-slip)} \tag{3.139}$$

$$v = 0, \quad y = L \quad \text{(impermeable wall)} \tag{3.140}$$

$$T = T_2, \quad y = L \quad \text{(no temperature jump)} \tag{3.141}$$

The solution of continuity Eq. (3.131) with boundary conditions specified by Eq. (3.137) and (3.140) is

$$v = 0 \tag{3.142}$$

Substituting Eq. (3.142) into Eq. (3.134), one obtains

$$\frac{\partial^2 w}{\partial y^2} = 0 \tag{3.143}$$

The solution of Eq. (3.143) with boundary conditions specified by Eqs. (3.136) and (3.139) is

$$w = 0 \tag{3.144}$$

Substituting Eqs. (3.142) and (3.144) into Eqs. (3.132), (3.133), and (3.135), one obtains

$$\frac{\partial^2 u}{\partial y^2} = 0 \tag{3.145}$$

$$\frac{\partial p}{\partial y} = -\rho g \tag{3.146}$$

$$k\frac{\partial^2 T}{\partial y^2} = -\mu\left(\frac{\partial u}{\partial y}\right)^2 \tag{3.147}$$

The solutions of Eqs. (3.145) with boundary conditions specified by Eqs. (3.136) and (3.139) is

$$u = \frac{y}{L}U \tag{3.148}$$

The solution of Eq. (3.146) is

$$p = p_0 - \rho g y \tag{3.149}$$

where p_0 is pressure at $y = 0$.

Substituting Eq. (3.148) into Eq. (3.147) and considering Eqs. (3.138) and (3.141), the solution for temperature distribution is obtained

$$\frac{T - T_1}{T_2 - T_1} = \frac{y}{L} + \frac{\mathrm{Ec}Pr}{2}\frac{y}{L}\left(1 - \frac{y}{L}\right) \tag{3.150}$$

where $\mathrm{Ec} = U^2/[c_\mathrm{p}(T_2 - T_1)]$ is Eckert number.

Example 3.5

In a multicomponent system, mass transfer can also contribute to heat transfer via interdiffusional convection and the heat flux can be expressed as:

$$\mathbf{q}'' = k\nabla T - \sum_{i=1}^{N} \mathbf{J}_i h_i \tag{3.151}$$

where $\mathbf{J}_\mathrm{i}$ is the mass flux relative to the mass-averaged velocity. Assuming that the viscous dissipation is negligible, simplify the energy equation for a multicomponent system, Eq. (3.92), for the case that the interdiffusional convection needs to be considered. The reference frame is stationary ($\mathbf{V}_\mathrm{ref} = 0$), and there is no internal heat generation.

Solution Substituting Eq. (3.151) into Eq. (3.92) and neglecting the viscous dissipation and internal heat generation, the energy equation for a multicomponent becomes:

$$\sum_{i=1}^{N}\left[\rho\frac{D}{Dt}(\omega_i h_i)\right] = \nabla\cdot(k\nabla T) - \sum_{i=1}^{N}\nabla\cdot(\mathbf{J}_i h_i) + \frac{Dp}{Dt} \tag{3.152}$$

or

$$\sum_{i=1}^{N}\left[\frac{\partial}{\partial t}(\omega_i\rho h_i) + \nabla\cdot(\mathbf{V}\omega_i\rho h_i)\right] = \nabla\cdot(k\nabla T) - \sum_{i=1}^{N}\nabla\cdot(\mathbf{J}_i h_i) + \frac{Dp}{Dt} \tag{3.153}$$

The diffusive mass flux can be rewritten in terms of the velocity of one component, $\mathbf{V}_i$, compared to the mass-averaged velocity, $\mathbf{V}$ [see Eq. (1.52)]

$$\mathbf{J}_i = \omega_i\rho(\mathbf{V}_i - \mathbf{V}) \tag{3.154}$$

The advective and diffusive term can be combined as

$$\omega_i\rho\mathbf{V} + \mathbf{J}_i = \omega_i\rho\mathbf{V}_i \tag{3.155}$$

The mass-averaged velocity, $\mathbf{V}$, is a function of the component velocities

$$\mathbf{V} = \sum_{i=1}^{N}\omega_i\mathbf{V}_i \tag{3.156}$$

Using the component velocity, $\mathbf{V}_i$, defined in Eq. (3.155), in the energy equation, Eq. (3.153), yields:

$$\sum_{i=1}^{N}\left[\frac{\partial}{\partial t}(\omega_i\rho h_i)+\nabla\cdot(\omega_i\rho\mathbf{V}_i h_i)\right]=\nabla\cdot(k\nabla T)+\frac{Dp}{Dt} \tag{3.157}$$

For a stationary reference frame, the species conservation equation, (3.110), becomes

$$\frac{\partial}{\partial t}(\omega_i\rho)+\nabla\cdot(\omega_i\rho\mathbf{V})=-\nabla\cdot\mathbf{J}_i+\dot{m}_i''' \tag{3.158}$$

Substituting Eq. (3.155) into Eq. (3.158), the conservation of species equation can be written in terms of the velocity of the ith component

$$\frac{\partial}{\partial t}(\omega_i\rho)+\nabla\cdot(\omega_i\rho\mathbf{V}_i)=\dot{m}_i''' \tag{3.159}$$

For convenience, let's define a substantial derivative in terms of the velocity of the ith component

$$\frac{D_i}{Dt}()=\frac{\partial}{\partial t}()+\mathbf{V}_i\cdot\nabla() \tag{3.160}$$

Using the species equation, Eq. (3.159), the energy Eq. (3.157) can be rewritten as

$$\sum_{i=1}^{N}\left[\omega_i\rho\frac{D_i h_i}{Dt}\right]=\nabla\cdot(k\nabla T)+\frac{Dp}{Dt}-\sum_{i=1}^{N}\left[\dot{m}_i''' h_i\right] \tag{3.161}$$

The specific enthalpy of the ith component is a function of temperature and pressure, $h_i = h_i(T,p)$. The substantial derivative of enthalpy is [see Eq. (3.99)]

$$\frac{D_i h_i}{Dt}=c_{\mathrm{p},i}\frac{D_i T}{Dt}+\frac{1-\beta_i T}{\rho_i}\frac{D_i p}{Dt} \tag{3.162}$$

where

$$\beta_i=-\frac{1}{\rho_i}\left(\frac{\partial\rho_i}{\partial T}\right)_{\mathrm{p}} \tag{3.163}$$

is the coefficient of thermal expansion for the ith component in the kth phase. Substituting Eq. (3.162) into Eq. (3.161) yields:

$$\sum_{i=1}^{N}\left[\omega_i\rho c_{\mathrm{p},i}\frac{D_i T}{Dt}+(1-\beta_i T)\frac{D_i p}{Dt}\right]=\nabla\cdot(k\nabla T)+\frac{Dp}{Dt}-\sum_{i=1}^{N}\left[\dot{m}_i''' h_i\right] \tag{3.164}$$

Since the mass-averaged velocity is a function of the component velocity [see Eq. (3.156)], a simplification can be made on one of the pressure terms.

$$\frac{Dp}{Dt}-\sum_{i=1}^{N}\frac{D_i p}{Dt}=\mathbf{V}\cdot\nabla p-\sum_{i=1}^{N}\mathbf{V}_i\cdot\nabla p=-\sum_{i=1}^{N}(1-\omega_i)\mathbf{V}_i\cdot\nabla p \tag{3.165}$$

Substituting this relation into the energy equation, Eq. (3.164), yields

$$\sum_{i=1}^{N}\left[\omega_i\rho c_{p,i}\frac{D_iT}{Dt}\right] - \sum_{i=1}^{N}(1-\omega_i)\mathbf{V}_i\cdot\nabla p + \sum_{i=1}^{N}\left[\beta_i T\frac{D_i p}{Dt}\right] - \sum_{i=1}^{N}\left[\dot{m}_i'''h_i\right] \tag{3.166}$$

where the second and third terms on the right-hand side of the energy equation above are due to the difference in pressure work from the mass average velocity, and the pressure work from each component. The last term is due to species generation or consumption.

For a single-component system, the energy equation reduces to

$$\rho c_p\frac{DT}{Dt} = \nabla\cdot(k\nabla T) + \beta T\frac{Dp}{Dt} \tag{3.167}$$

3.3.5 Classification of Multiphase Governing Equations

A general transport equation, whether mass, momentum, energy, or species, can be written as:

$$\frac{\partial}{\partial t}(\rho\Phi) + \nabla\cdot(\rho\mathbf{V}\Phi) = \nabla\cdot(\Gamma\nabla\Phi) + F(\mathbf{x}, t, \Phi, \ldots) \tag{3.168}$$

where the dependent variable, Φ, can be one for mass, any component of velocity (u, v, w), enthalpy, h, or the species mass fraction, ω_i. The gradient operator is the partial derivative of the dependent variable with respect to all the spatial directions $\mathbf{x}$ for a given coordinate system.

A partial differential equation (PDE) is an equation of a function and its partial derivatives. In general, a PDE is classified by its linearity or nonlinearity, and by its order. Its order is considered by its highest derivative. A general second-order PDE for two independent variables, η and ζ, is

$$A\frac{\partial^2\Phi}{\partial\eta^2} + 2B\frac{\partial^2\Phi}{\partial\eta\partial\zeta} + C\frac{\partial^2\Phi}{\partial\zeta^2} + D\frac{\partial\Phi}{\partial\eta} + E\frac{\partial\Phi}{\partial\zeta} + F\Phi + G = 0 \tag{3.169}$$

This equation is linear if all the coefficients (A, B, C, D, E, F, and G) are a function of η and ζ, a constant, or zero. If they are a function of Φ or any of its derivatives, then the PDE is nonlinear. A partial differential equation is called quasi-linear if it is linear in the highest derivatives. In Eq. (3.169), this means that A, B, and C are a function of η and ζ, a constant, or zero. If a differential equation is quasi-linear, it can be classified as an elliptic $(AC - B^2 > 0)$, parabolic $(AC - B^2 = 0)$, or hyperbolic $(AC - B^2 < 0)$ equation.

The classification of a PDE describes how disturbances or changes propagate through a domain. If you are interested in a single point on the domain, you need to know two things: What region in the domain affects that point, and, if you do something at that point, what region in the domain it will affect. What region affects a point and what region that point affects are called the zone of dependence and the zone of influence, respectively. A representation of the zone of dependence and the zone of influence of elliptic, parabolic, and hyperbolic equations is displayed in Fig. 3.6. A multidimensional PDE that varies with time may be classified by looking at only the relation of two independent variables at a time. For instance, look at how the PDE would be classified only in the x- and

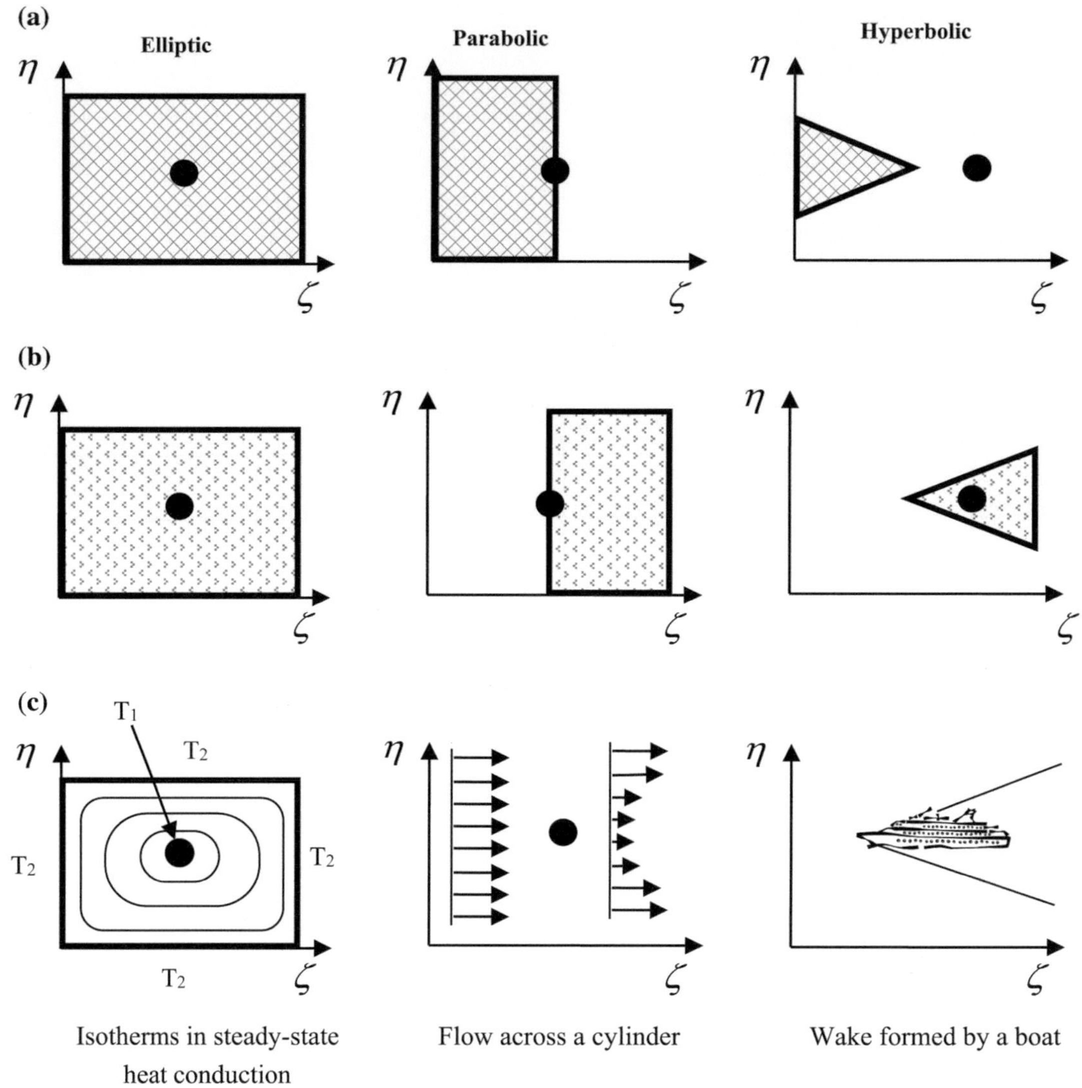

Fig. 3.6 Propagation of disturbances for different types of PDEs displayed as the **a** zone of dependence, **b** zone of influence, and **c** a physical example

y-directions. Then, examine the PDE in x and t. Finally, examine the PDE in y and t. The zone of dependence is where the zone of dependence for the three cases intersects. The zone of influence is where the zone of influence for the three cases intersects.

An example of an elliptic equation is the Laplace equation. This equation governs physical problems, such as two-dimensional steady-state heat conduction, electrical potential, the free stream characteristics in boundary layer equations, and the pressure field in a porous medium with Darcy's assumption. The independent variables, η and ζ, are spatial coordinates in these problems, and Φ is the potential field, representing temperature, voltage, the stream function, or pressure, respectively, for the given cases.

Some examples of parabolic problems are unsteady heat conduction, boundary layer equations for momentum, energy, and species as well as small vibrations in an elastic beam. For the unsteady heat equation, η is a spatial coordinate, and ζ is time. For the boundary layer equations, η is a spatial coordinate normal to the free stream velocity, and ζ is a spatial coordinate parallel to the free stream.

A hyperbolic function is often called the wave equation. The most common examples of the wave equation are the propagation of shock waves and sound waves. The interface between phases is also hyperbolic. Another less intuitive example of the use of a hyperbolic function is heat conduction in micro- and nanoscale. It becomes relevant in these situations because heat actually conducts at a finite rate, and the limits of these rates are relevant at such small scales.

For any problem to be well defined, there are boundary/initial conditions that must be applied. There are three basic boundary conditions for second-order PDEs. These boundary conditions are the Dirichlet [$\Phi = f(\eta, \zeta)$], the Neumann [$\partial\Phi/\partial\eta = f(\zeta)$, or $\partial\Phi/\partial\zeta = f(\eta)$], and the mixed [$a\,(\partial\Phi/\partial\eta) + b\Phi = f(\zeta)$, or $a(\partial\Phi/\partial\zeta) + b\Phi = f(\eta)$] type.

The number of boundary conditions that must be applied for each independent variable is equal to the highest order of that variable. For example, if the PDE is second order in both η and ζ, then two boundary conditions are needed for each independent variable for a total of four boundary conditions. If ζ is time and first order, then an initial Dirichlet-type boundary condition is needed, and if it is second order, then both a Dirichlet boundary condition and Neumann boundary condition are needed.

There are implications to experimental measurements and numerical analysis based on the classification of the governing PDE. For example, experimental measurements in an incompressible flow field with moderate to low Reynolds numbers are very difficult, because the governing equations are highly elliptic. The elliptic nature means that disturbances downstream greatly affect the upstream flow field. Therefore, any measurements that require a device in the flow field may change the nature of that flow field and are inherently inaccurate. Numerical simulations are very reliable for these cases when the flow is in the laminar regime. In the laminar flow regime, the full Navier–Stokes equations can be directly solved as an elliptic problem with no approximations.

In the compressible flow regime with high Reynolds numbers, the characteristic of the flow is parabolic or hyperbolic, depending on the Mach number. In these cases, disturbances produced downstream do not affect the upstream flow field; therefore, measurement devices in the flow will give an accurate depiction of the flow field without the device. In numerical simulations, it is computationally efficient to solve a parabolic flow field, because disturbances propagate in one direction. However, a flow field that is truly parabolic usually has a high Reynolds number and therefore is turbulent, which means it is three-dimensional and transient. Turbulence modeling either involves a very fine mesh, which is very computationally expensive, or an averaging technique, which requires a closure problem. The classification of a particular problem will help reduce the time it takes to get reliable results.

3.3.6 Jump Conditions at Interfaces

The phase equations introduced above can be applied within each phase and up to an interface. However, they are not valid across the interface, where sharp changes in various properties occur. Appropriate boundary conditions at the interface must be specified in order to solve the governing equations for heat, mass, and momentum transfer in the two adjoining phases. The interface conditions will serve as boundary conditions for the transport equations in the adjacent phases. These will be formulated based on the integral formulation for a control volume that includes two phases separated by an interface, as shown in Fig. 3.3b. Jump conditions at the interface can also be obtained by applying the basic laws (conservation of mass, momentum, and energy) at the interface. These interfacial basic laws are very important and are discussed further in detail in Sect. 4.5, including the effects of surface tension and disjoining pressure.

The mass balance at the interface can be obtained by considering conservation of mass for a control volume that includes two phases as given by Eq. (3.8) with $\Pi = 2$, and the result can be expressed as (see Problem 3.23):

$$\dot{m}_1'' + \dot{m}_2'' = 0 \tag{3.170}$$

where the mass fluxes of the two phases at the interface are

$$\dot{m}_1'' = \rho_1(\mathbf{V}_1 - \mathbf{V}_\mathrm{I}) \cdot \mathbf{n}_1 \tag{3.171}$$

$$\dot{m}_2'' = \rho_2(\mathbf{V}_2 - \mathbf{V}_\mathrm{I}) \cdot \mathbf{n}_2 \tag{3.172}$$

Since the thickness of the interface is considered to be zero, unit normals at an interfacial point in two adjacent phases are related by (see Fig. 3.7)

$$\mathbf{n}_1 = -\mathbf{n}_2 \tag{3.173}$$

Substituting Eq. (3.171)–(3.177) into Eq. (3.170) yields

$$\rho_1(\mathbf{V}_1 - \mathbf{V}_\mathrm{I}) \cdot \mathbf{n}_1 = \rho_2(\mathbf{V}_2 - \mathbf{V}_\mathrm{I}) \cdot \mathbf{n}_1 \tag{3.174}$$

If there is no phase change at the interface (a water-air two-phase system, for example), the mass flow rate at the interface should be zero, i.e.,

$$\dot{m}_1'' = -\dot{m}_2'' = 0$$

and the mass balance equation at the interface simply reduces to

$$\mathbf{V}_1 = \mathbf{V}_2 = \mathbf{V}_\mathrm{I} \tag{3.175}$$

The momentum balance at the interface can be obtained by considering the integral momentum equation for a control volume containing two phases separated by an interface as given by Eq. (3.13) with $\Pi = 2$, and the result is (see Problem 3.25)

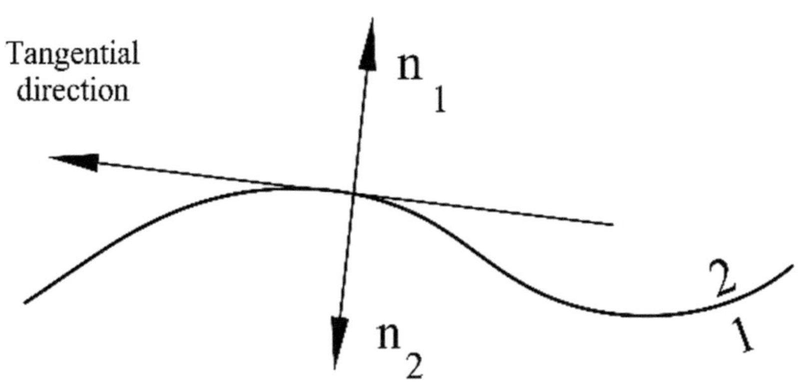

Fig. 3.7 Unit normal vectors at interface

$$(\boldsymbol{\tau}'_1 - \boldsymbol{\tau}'_2) \cdot \mathbf{n}_1 = \dot{m}''_1(\mathbf{V}_1 - \mathbf{V}_2) \quad (3.176)$$

Similarly, the energy balance at the interface can be obtained by considering the integral energy equation for a control volume that includes two phases as expressed by Eq. (3.23) with $\Pi = 2$(see Problem 3.26):

$$\sum_{k=1}^{2}\left[\dot{m}''_k\left(e_k + \frac{\mathbf{V}_{k,\mathrm{rel}}^2}{2}\right) + \mathbf{q}''_k \cdot \mathbf{n}_k - (\mathbf{n}_k \cdot \boldsymbol{\tau}'_k) \cdot \mathbf{V}_{k,\mathrm{rel}}\right] = 0 \quad (3.177)$$

where the three terms in the bracket represent the contributions of advection, heat transfer, and work done by normal and shear stresses.

Example 3.6

A container with an open top is exposed to superheated water vapor at temperature T_v, with the bottom and the side walls insulated. At time $t = 0$, the bottom surface of the vessel comes into contact with the air at temperature T_∞, which is below the vapor temperature T_v. The sides and top of the vessel remain insulated for $t > 0$ (see Fig. 3.8a, b). Formulate the governing equations for the condensation problem. If the vapor is saturated ($T_v = T_{sat}$), find the instantaneous thickness of the condensate.

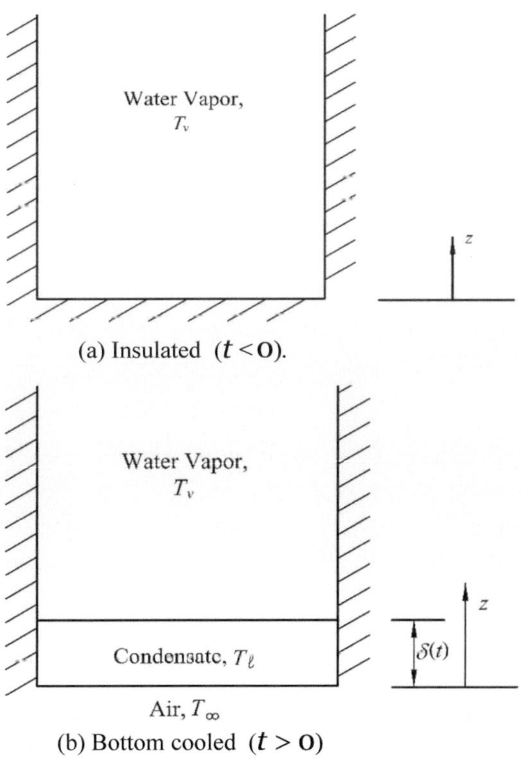

Fig. 3.8 Thin-walled vessel containing two-phase mixture

Solution This is a one-dimensional condensation problem because the sides of the vessel are insulated. The thickness of the bottom wall is very small, and therefore, the temperature drop across the bottom wall can be neglected. In addition, it is assumed that the thermal properties of both liquid and vapor are constants. The interfacial velocity is related to the instantaneous liquid thickness by

$$w_{\mathrm{I}} = \frac{\mathrm{d}\delta}{\mathrm{d}t} \tag{3.178}$$

The continuity equation at the interface can be obtained by using Eq. (3.178) in the z-direction, i.e.,

$$\rho_\ell (w_\ell - w_{\mathrm{I}}) = \rho_{\mathrm{v}} (w_{\mathrm{v}} - w_{\mathrm{I}})$$

The condensate is stationary, so $w_\ell = 0$. Therefore, the vapor phase velocity is

$$w_{\mathrm{v}} = \left(\frac{\rho_\ell}{\rho_{\mathrm{v}}} - 1\right)\frac{\mathrm{d}\delta}{\mathrm{d}t} \tag{3.179}$$

For the liquid phase, the governing energy equation and the corresponding initial and boundary conditions are

$$\frac{\partial T_\ell}{\partial t} = \alpha_\ell \frac{\partial^2 T_\ell}{\partial z^2} \tag{3.180}$$

$$T_\ell(z, 0) = T_{\mathrm{sat}} \tag{3.181}$$

$$k_\ell \frac{\partial T_\ell}{\partial z} = h[T_\ell(0, t) - T_\infty] \tag{3.182}$$

$$\frac{\partial T_{\mathrm{v}}}{\partial t} + \left(\frac{\rho_\ell}{\rho_{\mathrm{v}}} - 1\right)\frac{\mathrm{d}\delta}{\mathrm{d}t}\frac{\partial T_{\mathrm{v}}}{\partial z} = \alpha_{\mathrm{v}} \frac{\partial^2 T_{\mathrm{v}}}{\partial z^2} \tag{3.183}$$

$$T_{\mathrm{v}}(z, 0) = T_{\mathrm{v}} \tag{3.184}$$

$$T_{\mathrm{v}}(\infty, t) = T_{\mathrm{v}} \tag{3.185}$$

The liquid and vapor temperatures at the interface must be the same, i.e.,

$$T_{\mathrm{v}}(\delta, t) = T_\ell(\delta, t) = T_{\mathrm{sat}} \tag{3.186}$$

The energy balance at the interface can be obtained by using Eq. (3.177). Since the velocity in the liquid is zero and the velocity in the vapor phase is not very high, the kinetic energy and the viscous dissipation in Eq. (3.177) can be neglected. Applying Fourier's law for heat flux, the resulting energy balance equation at the interface is

$$-k_{\mathrm{v}} \frac{\partial T_{\mathrm{v}}(\delta, t)}{\partial z} + k_\ell \frac{\partial T_\ell(\delta, t)}{\partial z} = \rho_\ell h_{\ell \mathrm{v}} \frac{\mathrm{d}\delta}{\mathrm{d}t} \tag{3.187}$$

where $h_{\ell\mathrm{v}}$ is latent heat of liquid–vapor phase change. The analytical solution of this problem will be very difficult. If the vapor is saturated, the governing equations for the vapor phase will not be necessary because the vapor temperature will be uniformly equal to the saturation temperature.

This is a one-domain problem because only the liquid domain needs to be investigated. The problem can be further simplified by introducing the quasi-steady-state assumption. This step eliminates the transient term in the energy equation for the liquid phase because condensation is a very slow process and the change of internal energy in the liquid can be neglected. The energy equation in the liquid phase is thus reduced to

$$0 = \alpha_\ell \frac{\partial^2 T_\ell}{\partial z^2} \tag{3.188}$$

which is subject to the boundary conditions specified by Eqs. (3.182) and (3.187). The energy balance at the interface is reduced to

$$k_\ell \frac{\partial T_\ell(\delta, t)}{\partial z} = \rho_\ell h_{\ell v} \frac{\mathrm{d}\delta}{\mathrm{d}t} \tag{3.189}$$

The temperature in the liquid film can be obtained by integrating Eq. (3.188) twice:

$$T(z) = c_1 z + c_2 \tag{3.190}$$

where c_1 and c_2 are integral constants. They can be determined using Eqs. (3.181) and (3.182), i.e.,

$$k_\ell c_1 = h(c_2 - T_\infty) \tag{3.191}$$

$$T_{\text{sat}} = c_1 \delta + c_2 \tag{3.192}$$

which can be solved to yield

$$c_1 = \frac{h(T_{\text{sat}} - T_\infty)}{k_\ell + h\delta(t)} \qquad c_2 = T_{\text{sat}} - \frac{h(T_{\text{sat}} - T_\infty)}{k_\ell + h\delta(t)} \delta(t) \tag{3.193}$$

Substituting Eq. (3.193) into Eq. (3.190) yields the temperature distribution in the liquid film

$$T_\ell(z, t) = T_{\text{sat}} + \frac{h(T_{\text{sat}} - T_\infty)}{k_\ell + h\delta(t)} [z - \delta(t)] \tag{3.194}$$

The temperature distribution in the vessel can be illustrated by Fig. 3.9. At this point, Eq. (3.194) can be substituted into Eq. (3.189). An ordinary differential equation for the liquid thickness is obtained as a result:

$$\rho_\ell h_{\ell v} \frac{\mathrm{d}\delta}{\mathrm{d}t} = \frac{k_\ell h(T_{\text{sat}} - T_\infty)}{k_\ell + h\delta(t)} \tag{3.195}$$

which is subject to the following initial condition:

$$\delta(0) = 0 \tag{3.196}$$

Rearranging Eq. (3.195) so that δ appears only on the left-hand side, and t appears only on the right-hand side, gives

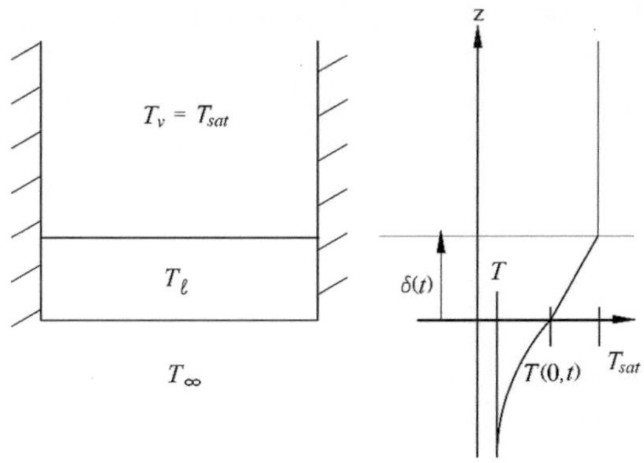

Fig. 3.9 Temperature distribution for a two-phase mixture in a thin-walled vessel

$$\left[\frac{h}{k_\ell}\delta + 1\right]\mathrm{d}\delta = \frac{h(T_{\mathrm{sat}} - T_\infty)}{\rho_\ell h_{\ell\mathrm{v}}}\mathrm{d}t \tag{3.197}$$

Integrating Eq. (3.197) with the boundary condition, Eq. (3.196) yields

$$\frac{h}{2k_\ell}\delta^2 + \delta - \frac{h(T_{\mathrm{sat}} - T_\infty)}{\rho_\ell h_{\ell\mathrm{v}}}t = 0 \tag{3.198}$$

which is a quadratic equation in the form of $ax^2 + bx + c = 0$. This allows for one positive solution and one negative solution of δ, and the latter does not physically make sense. Therefore, the liquid film thickness is

$$\frac{h\delta(t)}{k_\ell} = -1 + \left[1 + \frac{2h^2(T_{\mathrm{sat}} - T_\infty)t}{k_\ell \rho_\ell h_{\ell\mathrm{v}}}\right]^{\frac{1}{2}} \tag{3.199}$$

3.3.7 Rarefied Vapor Self-diffusion Model

The discussions so far are limited to the case where the density of the fluid is sufficiently high to permit the continuum assumption. There are cases in which the assumption of continuum is not valid ($Kn \geq 0.001$, see Sect. 1.4.1). For example, in the early stage of heat pipe start-up from the frozen state, the vapor pressure and density are very small in the heat pipe core. Because of the low density, the vapor in the rarefied state is somewhat different from the conventional continuum state. Also, the vapor density gradient is very large along the axial direction of the heat pipe. The vapor flow along the axial direction is mainly caused by the density gradient via vapor molecular diffusion. The low-density vapor state that has partly lost its continuum characteristics is referred to as rarefied vapor.

Neglecting the presence of noncondensable gases, the rarefied vapor flow can be simulated by a self-diffusion model. The term self-diffusion here means the interdiffusion of particles of the same mass due to a gradient in density. The governing equations for start-up of a heat pipe from the frozen state were derived by applying the principles of the conservation of mass and energy in a differential cylindrical control volume in conjunction with the definition of mass flux (Faghri 2016).

The mass self-diffusion equation is

$$\frac{\partial \rho}{\partial t} - \frac{\partial}{\partial z}\left(D_{\mathrm{v}}\frac{\partial \rho}{\partial z}\right) - \frac{1}{r}\frac{\partial}{\partial z}\left(rD_{\mathrm{v}}\frac{\partial \rho}{\partial r}\right) = 0 \tag{3.200}$$

and the energy equation is

$$\frac{\partial(\rho c_{\mathrm{v}} T)}{\partial t} + \frac{1}{r}\frac{\partial}{\partial r}(\dot{m}_r'' c_{\mathrm{p}} rT) + \frac{\partial}{\partial z}(\dot{m}_z'' c_{\mathrm{p}} T) = \frac{1}{r}\frac{\partial}{\partial z}\left(rk_{\mathrm{v}}\frac{\partial T}{\partial r}\right) + \frac{\partial}{\partial z}\left(D_{\mathrm{v}}\frac{\partial T}{\partial z}\right) \tag{3.201}$$

where the mass fluxes $\dot{m}_r''$ and $\dot{m}_z''$ are

$$\dot{m}_r'' = \rho v = -D_{\mathrm{v}}\frac{\partial \rho}{\partial r} \tag{3.202}$$

$$\dot{m}_z'' = \rho w = -D_{\mathrm{v}}\frac{\partial \rho}{\partial z} \tag{3.203}$$

where D_{v} is the self-diffusion coefficient, and k_{v} is the vapor molecular conductivity. The evaluation of low-density properties such as D_{v} is carried out using the kinetic theory of gases. The coefficient of self-diffusion is obtained from the relation based on the Chapman–Enskog kinetic theory (Hirschfelder et al. 1966):

$$D_{\mathrm{v}} = 2.628 \times 10^{-7}\frac{\sqrt{T^3/MW}}{p\sigma^2\Omega_D(k_{\mathrm{b}}T/\varepsilon)} \tag{3.204}$$

where p is the pressure in atmospheres, σ is the collision diameter in Å, ε is the maximum energy of attraction between a pair of molecules, k_{b} is the Boltzmann constant, and Ω_D is the collision integral for mass diffusion. For sodium, both σ and $\varepsilon/k_{\mathrm{b}}$ can be found from the table for constants of the Lennard-Jones potential model (Edwards et al. 1979), which gives σ = 3.567 Å, and $\varepsilon/k_{\mathrm{b}} = 1375\,\mathrm{K}$. The value of the collision integral Ω_D can also be found from the same reference, which is listed as a function of $\varepsilon/k_{\mathrm{b}}$.

3.3.8 An Extension: Combustion

Combustion is an exothermic chemical reaction process between fuel and oxidant. If combustion involves a liquid fuel, the liquid fuel does not actually burn as a liquid; it is vaporized first and diffuses away from the liquid–vapor surface. Meanwhile, the gaseous oxidant diffuses toward the liquid–vapor interface. Under the right conditions, the mass fluxes of vapor fuel and gaseous oxidant meet and the chemical reaction occurs at a certain location known as the flame (Lock 1994; Avedisian 1997, 2000). The flame is usually a very thin region with a color dictated by the temperature of the combustion.

The temperature and mass concentration distributions during a combustion process can be represented as shown in Fig. 3.10. The initial temperature of a liquid fuel is T_0, and the temperature of the liquid–vapor interface, T_{I}, is at the dew point temperature of the fuel. The temperature reaches a maximum at the location of the flame and decreases with increasing x until it reaches T_∞. The mass

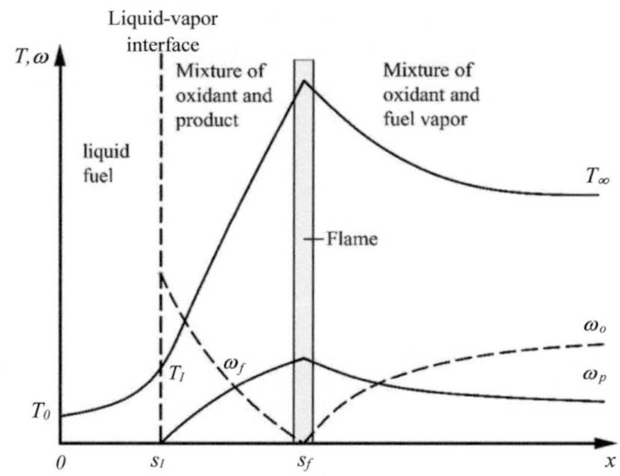

Fig. 3.10 Combustion near a planar surface

fraction of fuel, ω_{f}, is maximal at the liquid–vapor interface and decreases as location of the flame is approached. The mass fraction of the oxidant, ω_{o}, on the other hand, is maximal at infinity and decreases as location of the flame is approached. The mass fractions of both vapor fuel and oxidant reach their minimum values at the location of the flame. The mass fractions of the products of combustion, ω_{p}, are at their maximum at the location of the flame and decrease as x either increases or decreases.

The transport phenomena involved in combustion include heat and mass transfer in both the liquid fuel and the gaseous mixture. The governing equations for the gaseous phase will be discussed below. Firstly, the gaseous mixture must satisfy the continuity equation, i.e.,

$$\frac{D\rho}{Dt} + \rho\nabla\cdot\mathbf{V} = 0 \tag{3.205}$$

and the gas flow is governed by the momentum equation,

$$\rho\frac{D\mathbf{V}}{Dt} = \nabla\cdot\boldsymbol{\tau} + \rho\mathbf{g} \tag{3.206}$$

where the shear stress can be obtained from Eq. (3.68) and gravity is the only body force considered.

The energy equation for combustion is

$$\rho\frac{Dh}{Dt} = \nabla\cdot(k\nabla T) \tag{3.207}$$

which is a simplified version of Eq. (3.90), with internal heat generation and viscous dissipation neglected. The effect of the substantial derivative of pressure on the energy is also neglected. The specific enthalpy for the mixture is related to the specific enthalpy of each component in the mixture by

$$h = \sum_{i=1}^{N}\omega_i h_i \tag{3.208}$$

It is a common practice in thermodynamic and heat transfer analyses to consider the change in enthalpy during a chemical reaction process rather than the absolute values of enthalpy. For a process that does not involve chemical reaction, we can choose any reference state for an individual substance and define the enthalpy at that reference state as zero. For example, the reference state for water is often chosen at the triple point, while the reference state for an ideal gas is often chosen as zero *K*. However, when chemical reaction is involved in a process, as is the case with combustion, the composition of the system changes during the chemical reaction; therefore, the reference states for all reactants and products must be the same.

One convenient option in such a situation is to segregate the enthalpy of any component into two parts: (1) the enthalpy due to its chemical composition at the *standard reference state* (at 25 °C and 1 atm) and (2) the sensible enthalpy due to any temperature deviation from the standard reference state. Therefore, the enthalpy for the *i*th component in the mixture can be expressed as

$$h_i = h_i^\circ + h_i^T \tag{3.209}$$

where h_i° is the enthalpy of formation for the *i*th component, i.e., the enthalpy due to its chemical composition at the standard reference state. The enthalpy of formation for selected substances is shown in Table 3.1.

The sensible enthalpy, h_i^T, is related to temperature by

$$h_i^T = \int_{T^\circ}^{T} c_{pi} \mathrm{d}T \tag{3.210}$$

Table 3.1 Enthalpy of formation h° for selected substances at 25 °C and 1 atm

Substance	Formula (phase)	Enthalpy of formation (kJ/kmol)
Acetylene	C_2H_2 (g)	226,730
Benzene	C_6H_6 (g)	82,930
Carbon	C (s)	0
Carbon monoxide	CO (g)	−110,530
Carbon dioxide	CO_2 (g)	−393,520
Ethyl alcohol	C_2H_5OH (g)	−235,310
Ethyl alcohol	C_2H_5OH (ℓ)	−277,690
Ethylene	C_2H_4 (g)	52,280
Ethane	C_2H_6 (g)	−84,680
Hydrogen	H_2 (g)	0
Methane	CH_4 (g)	−74,850
Nitrogen	N_2 (g)	0
Nitrogen	N (g)	472,650
n-Dodecane	$C_{12}H_{26}$ (ℓ)	−291,010
n-Octane	C_8H_{18} (g)	−208,450
n-Octane	C_8H_{18} (ℓ)	−249,950
Oxygen	O_2 (g)	0
Propane	C_3H_8 (g)	−103,850
Water	H_2O (ℓ)	−241,820
Water vapor	H_2O (g)	−285,830

Substituting Eqs. (3.209) and (3.210) into Eq. (3.208), the enthalpy of the mixture becomes

$$h = \sum_{i=1}^{N} \omega_i h_i^\circ + \int_{T^\circ}^{T} c_p \mathrm{d}T \tag{3.211}$$

where c_p is the average specific heat of the mixture, defined as

$$c_p = \sum_{i=1}^{N} \omega_i c_{pi} \tag{3.212}$$

Substituting Eq. (3.211) into Eq. (3.207), the energy equation for combustion becomes

$$\rho \frac{D(c_p T)}{Dt} = \nabla \cdot k\nabla T - \rho \sum_{i=1}^{N} h_i^\circ \frac{D\omega_i}{Dt} \tag{3.213}$$

If the fuel is consumed at a rate of $\dot{m}_f'''$ per unit volume, h_c—the heat of combustion—is defined by

$$h_c = -\frac{\rho}{\dot{m}_f'''} \sum_{i=1}^{N} h_i \frac{D\omega_i}{Dt} \tag{3.214}$$

Substituting Eq. (3.230) into Eq. (3.235) gives

$$h_c = -\frac{\rho}{\dot{m}_f'''} \sum_{i=1}^{N} h_i^\circ \frac{D\omega_i}{Dt} - \frac{\rho}{\dot{m}_f'''} \sum_{i=1}^{N} h_i^T \frac{D\omega_i}{Dt} \tag{3.215}$$

The contribution of the second term on the right-hand side of Eq. (3.215) is negligible, since $h_c \gg h_i^T$. Dropping the second term from the right-hand side of Eq. (3.215) and substituting the result into Eq. (3.208), the energy equation becomes

$$\rho c_p \frac{DT}{Dt} = \nabla \cdot k\nabla T + \dot{m}_f''' h_c \tag{3.216}$$

The mass fraction of each component (fuel, oxidant, and product) is dominated by

$$\rho \frac{D\omega_i}{Dt} = -\nabla \cdot \mathbf{J}_i + \dot{m}_i''' \tag{3.217}$$

where the subscript i can be f (fuel), o (oxidant), or p (product). The ratio of the rates of oxygen and fuel consumption is defined as the oxygen/fuel ratio:

$$\gamma = \frac{\dot{m}_o'''}{\dot{m}_f'''} \tag{3.218}$$

The above analysis applies to combustion occurring on a planar surface. For many applications, combustion of the liquid fuel is usually preceded by breaking up a fuel jet into liquid droplets so that combustion occurs around a spherical liquid droplet. Combustion of a falling liquid droplet, as shown in Fig. 3.11, is analyzed here. The liquid fuel droplet vaporizes at the dew point, T_d, which is the saturation temperature corresponding to the partial pressure of the fuel vapor in the mixture near the

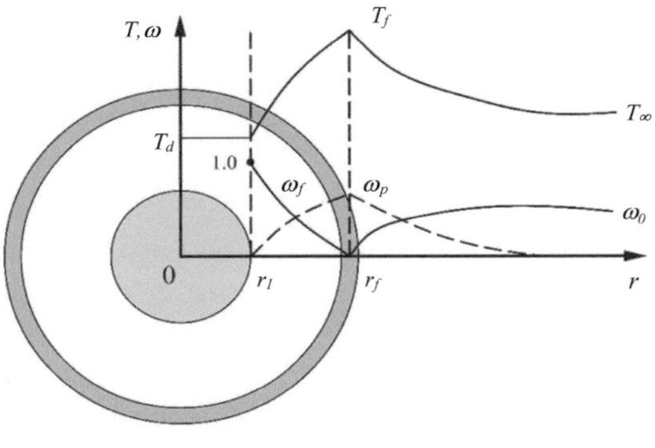

Fig. 3.11 Combustion near a liquid fuel droplet

liquid–vapor interface. To simplify the analysis, it is assumed that the temperature in the liquid fuel droplet is uniformly equal to the saturation temperature of the fuel at the total system pressure, i.e., the mass fraction of the fuel at the liquid–vapor interface equals one. It is further assumed that the shapes of both the liquid fuel droplet and the flame are spherical, which allows for application of a one-dimensional symmetric model (Lock 1994), which is presented here.

If the combustion process is assumed to be in a quasi-steady state (neglecting the transient term in the governing equation), the energy equation for combustion becomes

$$\frac{1}{r^2}\frac{\mathrm{d}}{\mathrm{d}r}(r^2\rho c_{\mathrm{p}}uT) = \frac{1}{r^2}\frac{\mathrm{d}}{\mathrm{d}r}\left(r^2k\frac{\mathrm{d}T}{\mathrm{d}r}\right) + \dot{m}_{\mathrm{f}}'''h_{\mathrm{c}} \tag{3.219}$$

where u is the velocity of the gaseous mixture in the radial direction.

The mass fraction of the oxidant can be obtained by simplifying Eq. (3.217), i.e.,

$$\frac{1}{r^2}\frac{\mathrm{d}}{\mathrm{d}r}(r^2\rho u\omega_{\mathrm{o}}) = \frac{1}{r^2}\frac{\mathrm{d}}{\mathrm{d}r}\left[r^2\rho D_{\mathrm{o}}\frac{\mathrm{d}\omega_{\mathrm{o}}}{\mathrm{d}r}\right] - \dot{m}_{\mathrm{o}}''' \tag{3.220}$$

where D_{o} is the mass diffusivity of the oxidant in the gaseous mixture. Multiplying Eq. (3.220) by h_{c}/γ and combining the resulting equation with Eq. (3.219) yields

$$\frac{1}{r^2}\frac{\mathrm{d}}{\mathrm{d}r}\left[r^2\rho c_{\mathrm{p}}u\left(T+\omega_{\mathrm{o}}\frac{h_{\mathrm{c}}}{\gamma c_{\mathrm{p}}}\right)\right] = \frac{1}{r^2}\frac{\mathrm{d}}{\mathrm{d}r}\left[r^2k\frac{\mathrm{d}T}{\mathrm{d}r} + r^2\rho D_{\mathrm{o}}\frac{\mathrm{d}}{\mathrm{d}r}\left(\omega_{\mathrm{o}}\frac{h_{\mathrm{c}}}{\gamma}\right)\right] \tag{3.221}$$

where Eq. (3.218) was used to eliminate the rates of fuel and oxygen consumption.

Assuming that the Lewis number $[Le = k/(\rho c_{\mathrm{p}}D_{\mathrm{o}})]$ equals one and the specific heat is constant, Eq. (3.221) can be simplified as follows:

$$\frac{1}{r^2}\frac{\mathrm{d}}{\mathrm{d}r}(r^2\rho c_{\mathrm{p}}uT^*) = \frac{1}{r^2}\frac{\mathrm{d}}{\mathrm{d}r}\left(r^2k\frac{\mathrm{d}T^*}{\mathrm{d}r}\right) \tag{3.222}$$

where

$$T^* = T + \omega_{\mathrm{o}}\frac{h_{\mathrm{c}}}{\gamma c_{\mathrm{p}}} \tag{3.223}$$

is a modified temperature in the combustion process.

Equation (3.222) can be rearranged to the following form:

$$\frac{\mathrm{d}}{\mathrm{d}r}\left[r^2\left(\rho c_\mathrm{p} u T^* - k\frac{\mathrm{d}T^*}{\mathrm{d}r}\right)\right] = 0 \tag{3.224}$$

The continuity equation requires that

$$4\pi r^2 \rho u = 4\pi r_\mathrm{I}^2 (\rho u)_\mathrm{I} = \dot{m} \tag{3.225}$$

Since it has been assumed that the mass fraction of the fuel in the mixture at the liquid–vapor interface equals one, the mass flow rate $\dot{m}$ reflects the mass flow rate of the fuel. The mass flux is often used in combustion analysis, and it is defined as

$$\dot{m}'' = \frac{\dot{m}}{4\pi r^2} \tag{3.226}$$

Equations (3.224) and (3.225) can be rewritten in terms of mass flux, i.e.,

$$\frac{\mathrm{d}}{\mathrm{d}r}\left[r^2\left(\dot{m}'' c_\mathrm{p} T^* - k\frac{\mathrm{d}T^*}{\mathrm{d}r}\right)\right] = 0 \tag{3.227}$$

$$4\pi r^2 \dot{m}'' = 4\pi r_\mathrm{I}^2 \dot{m}_\mathrm{I}'' = \dot{m} \tag{3.228}$$

Integrating Eq. (3.227) from the liquid fuel droplet surface ($r = r_\mathrm{I}$) to an arbitrary radius ($r > r_\mathrm{I}$) yields

$$r^2\left(\dot{m}'' c_\mathrm{p} T^* - k\frac{\mathrm{d}T^*}{\mathrm{d}r}\right) = r_\mathrm{I}^2\left(\dot{m}'' c_\mathrm{p} T^* - k\frac{\mathrm{d}T^*}{\mathrm{d}r}\right)_\mathrm{I} \tag{3.229}$$

Substituting Eq. (3.228) into Eq. (3.229), one obtains

$$r^2 k\frac{\mathrm{d}T^*}{\mathrm{d}r} = r_\mathrm{I}^2 \dot{m}_\mathrm{I}'' c_\mathrm{p}(T^* - T_\mathrm{I}^*) + r_\mathrm{I}^2\left(k\frac{\mathrm{d}T^*}{\mathrm{d}r}\right)_{r=r_\mathrm{I}} \tag{3.230}$$

The mass flow rate at the surface of the liquid fuel droplet, $\dot{m}_\mathrm{I}''$, is the same as the fuel burning rate, $\dot{m}_\mathrm{f}''$, because the mass fraction at the surface of the droplet is one. Introducing the excess modified temperature,

$$\theta = T^* - T_\mathrm{I}^* \tag{3.231}$$

Equation (3.230) becomes

$$r^2 k\frac{\mathrm{d}\theta}{\mathrm{d}r} = r_\mathrm{I}^2 \dot{m}_\mathrm{f}'' c_\mathrm{p}\left[\theta + \frac{1}{\dot{m}_\mathrm{f}'' c_\mathrm{p}}\left(k\frac{\mathrm{d}\theta}{\mathrm{d}r}\right)_{r=r_\mathrm{I}}\right] \tag{3.232}$$

Introducing a new dependent variable,

$$\varphi = \theta + \frac{1}{\dot{m}_{\mathrm{f}}'' c_{\mathrm{p}}} \left(k \frac{\mathrm{d}\theta}{\mathrm{d}r} \right)_{r=r_{\mathrm{I}}} \tag{3.233}$$

Equation (3.232) becomes

$$\frac{1}{\varphi} d\varphi = \frac{r_{\mathrm{I}}^2 \dot{m}_{\mathrm{f}}'' c_{\mathrm{p}}}{k} \frac{1}{r^2} \mathrm{d}r \tag{3.234}$$

Integrating Eq. (3.234) over the interval of (r, ∞), one obtains

$$\ln \frac{\varphi_{\infty}}{\varphi} = \frac{r_{\mathrm{I}}^2 \dot{m}_{\mathrm{f}}'' c_{\mathrm{p}}}{k} \frac{1}{r} \tag{3.235}$$

The fuel burning rate, $\dot{m}_{\mathrm{f}}''$, in Eq. (3.235) is related to the heat transfer at the liquid droplet surface as follows

$$\dot{m}_{\mathrm{f}}'' h_{\ell \mathrm{v}} = q_{\mathrm{I}}'' = k \left(\frac{\mathrm{d}\theta}{\mathrm{d}r} \right)_{r=r_{\mathrm{I}}} \tag{3.236}$$

Substituting Eq. (3.236) into Eq. (3.233), the new dependent variable φ becomes

$$\varphi = \theta + \frac{h_{\ell \mathrm{v}}}{c_{\mathrm{p}}} \tag{3.237}$$

Substituting Eq. (3.237) into Eq. (3.235) and letting $r = r_{\mathrm{I}}$, an equation for the fuel burning rate is obtained:

$$\dot{m}_{\mathrm{f}}'' = \frac{k}{r_{\mathrm{I}} c_{\mathrm{p}}} \ln \left(1 + \frac{c_{\mathrm{p}} \theta_{\infty}}{h_{\ell \mathrm{v}}} \right) \tag{3.238}$$

Equation (3.238) can also be used to determine the transient liquid fuel droplet size, because the fuel burning rate is related to the size of the droplet by

$$\dot{m}_{\mathrm{f}}'' = -\frac{\mathrm{d}(\rho_{\mathrm{f}} 4\pi r_{\mathrm{I}}^3/3)/\mathrm{d}t}{4\pi r_{\mathrm{I}}^2} = -\rho_{\mathrm{f}} \frac{\mathrm{d}r_{\mathrm{I}}}{\mathrm{d}t} \tag{3.239}$$

where ρ_{f} is the density of the liquid fuel.

Substituting Eq. (3.239) into Eq. (3.238), one obtains

$$r_{\mathrm{I}} \frac{\mathrm{d}r_{\mathrm{I}}}{\mathrm{d}t} = -\frac{k}{\rho_{\mathrm{f}} c_{\mathrm{p}}} \ln \left(1 + \frac{c_{\mathrm{p}} \theta_{\infty}}{h_{\ell \mathrm{v}}} \right) \tag{3.240}$$

Integrating Eq. (3.240) and considering the initial condition $r_{\mathrm{I}} = r_i$ at $t = 0$, the liquid droplet radius becomes

$$r_{\mathrm{I}}^2 = r_i^2 - \frac{2kt}{\rho_{\mathrm{f}} c_{\mathrm{p}}} \ln \left(1 + \frac{c_{\mathrm{p}} \theta_{\infty}}{h_{\ell \mathrm{v}}} \right) \tag{3.241}$$

Equation (3.241) can be used to estimate the time needed to completely burn the liquid fuel droplet:

$$t_{\mathrm{f}} = \frac{\rho_{\mathrm{f}} c_{\mathrm{p}} r_i^2}{2k \ln\left(1 + \frac{c_{\mathrm{p}}\theta_\infty}{h_{\ell \mathrm{v}}}\right)} \tag{3.242}$$

Example 3.7

Liquid hydrocarbon fuel is suspended in air in the form of small droplet, and the oxygen/fuel ratio is $\gamma = 3.5$. Before the liquid fuel is sprayed into the air, the mass fraction of oxygen is $\omega_{\mathrm{o},\infty} = 0.21$, and the temperature of the air is $T_\infty = 25\,°\mathrm{C}$. The latent heat of evaporation for the fuel is $h_{\ell\mathrm{v}} = 360$ kJ/kg, and the enthalpy of combustion is $h_{\mathrm{c}} = 4.4 \times 10^4$ kJ/kg. The specific heat and thermal conductivity of the gaseous mixture are $c_{\mathrm{p}} = 1.15$ kJ/kg K and $k = 0.07$ W/m K, respectively. The initial temperature of the liquid droplet can be assumed to be equal to the saturation temperature of the fuel, 187 °C. The density of the liquid fuel is $\rho_{\mathrm{f}} = 700$ kg/m^3. Estimate the time needed to completely burn a liquid fuel droplet with a diameter of 40 μm.

Solution The excess modified temperature at the locations far away from the fuel droplet can be obtained by substituting Eq. (3.223) into Eq. (3.231):

$$\begin{aligned}
\theta_\infty &= T_\infty^* - T_{\mathrm{I}}^* = (T_\infty - T_{\mathrm{I}}) + (\omega_{\mathrm{o},\infty} - \omega_{\mathrm{o,I}})\frac{h_{\mathrm{c}}}{\gamma c_{\mathrm{p}}} \\
&= (25 - 187) + (0.21 - 0) \times \frac{4.4 \times 10^4}{3.5 \times 1.15} = 2133.65\,°\mathrm{C}
\end{aligned}$$

The time required to completely burn the liquid fuel droplet is then obtained by using Eq. (3.242), i.e.,

$$t_{\mathrm{f}} = \frac{\rho_{\mathrm{f}} c_{\mathrm{p}} r_i^2}{2k \ln\left(1 + \frac{c_{\mathrm{p}}\theta_\infty}{h_{\ell\mathrm{v}}}\right)} = \frac{700 \times 1.15 \times 10^3 \times (20 \times 10^{-6})^2}{2 \times 0.07 \times \ln\left(1 + \frac{1.15 \times 2133.65}{360}\right)} = 0.00112\mathrm{s}$$

3.4 Multiphase Flow Modeling

3.4.1 Overview of Averaging Approaches

The objectives of the various averaging methods are twofold: (1) to define the average properties for the multiphase system and correlate the experimental data and (2) to obtain solvable governing conservation equations that can be used to predict the macroscopic properties of the multiphase system. This chapter will address the application of averaging methods to the governing equations.

Based on the physical concepts used to formulate multiphase transport phenomena, the averaging methods can be classified into three major groups: (1) *Eulerian averaging*, (2) *Lagrangian averaging*, and (3) *molecular statistical averaging*. These averaging techniques are briefly reviewed below.

3.4.1.1 Eulerian Averaging

Eulerian averaging is the most important and widely used method of averaging, because it is consistent with the control volume analysis that we used to develop the governing equations in the preceding section. It is also applicable to the most common techniques of experimental observations. Eulerian averaging is based on time-space description of physical phenomena. In the Eulerian description, changes in the various dependent variables, such as velocity, temperature, and pressure, are expressed as functions of time and space coordinates, which are considered to be independent variables. One can average these independent variables over both space and time. The integral operations associated with these averages smooth out the local spatial or instant variations of the properties within the domain of integration.

For a generalized function $\Phi = \Phi(x, y, z, t)$, the most widely used Eulerian averaging includes *time averaging* and *volumetric averaging*. The Eulerian time average is obtained by averaging the flow properties over a certain period of time, Δt, at a fixed point in the reference frame, i.e.,

$$\bar{\Phi} = \frac{1}{\Delta t} \int_{\Delta t} \Phi(x, y, z, t) \mathrm{d}t \tag{3.243}$$

For this equation, the time period Δt is chosen so that it is larger than the largest timescale of the local properties' fluctuation, yet small enough in comparison with the process macroscopic timescale. During this time period, different phases can flow through the fixed point. Eulerian time averaging is particularly useful for a turbulent multiphase flow as well as for the dispersed-phase systems.

Eulerian volumetric averaging is usually performed over a volume element, ΔV, around a point (x, y, z) in the flow. For a multiphase system that includes Π different phases, the total volume equals the summation of the individual phase volumes, i.e.,

$$\Delta V = \sum_{k=1}^{\Pi} \Delta V_k \tag{3.244}$$

The volume fraction of the kth phase, ε_k, is defined as the ratio of the elemental volume of the kth phase to the total elemental volume for all phases, i.e.,

$$\varepsilon_k = \frac{\Delta V_k}{\Delta V} \tag{3.245}$$

The volume fraction of all phases must sum to unity:

$$\sum_{k=1}^{\Pi} \varepsilon_k = 1 \tag{3.246}$$

Eulerian volume averaging is expressed as

$$\langle \Phi \rangle = \frac{1}{\Delta V} \sum_{k=1}^{\Pi} \int_{\Delta V_k} \Phi_k(x, y, z, t) \mathrm{d}V \tag{3.247}$$

where the volume element ΔV must be much smaller than the total volume of the multiphase system so that the average can provide a local value of Φ in the flow field. The volume element ΔV must also be large enough to yield a stationary average. Since the volume element includes different phases, information about the spatial variation of Φ for each individual phase is lost and $\langle \Phi \rangle$ represents the average for all phases.

For any variable or property that is associated with a particular phase, Φ_k, the phase-average value of any variable or property for that phase is obtained with the following equations:

Intrinsic phase average:

$$\langle \Phi_k \rangle^k = \frac{1}{\Delta V_k} \int_{\Delta V_k} \Phi_k dV \tag{3.248}$$

Extrinsic phase average:

$$\langle \Phi_k \rangle = \frac{1}{\Delta V} \int_{\Delta V_k} \Phi_k dV \tag{3.249}$$

Intrinsic means that it forms to the inherent part of a phase and is independent of other phases in the volume element. In contrast, extrinsic means it is a property that depends on the phase's relationship with other phases in the volume element.

While the intrinsic phase average is taken over only the volume of the kth phase in Eq. (3.248), the extrinsic phase average for a particular phase is taken over an entire elemental volume in Eq. (3.249). These two phase averages are related by

$$\langle \Phi_k \rangle = \varepsilon_k \langle \Phi_k \rangle^k \tag{3.250}$$

The intrinsic and extrinsic phase averages defined in Eqs. (3.248) and (3.249) are related to the volume average defined in Eq. (3.247) by

$$\langle \Phi \rangle = \sum_{k=1}^{\Pi} \langle \Phi_k \rangle = \sum_{k=1}^{\Pi} \varepsilon_k \langle \Phi_k \rangle^k \tag{3.251}$$

The deviation from a respective intrinsic phase-average value is

$$\hat{\Phi}_k \equiv \Phi_k - \langle \Phi_k \rangle^k \tag{3.252}$$

When the products of two variables are phase-averaged, the following relations are needed:

$$\langle \Phi_k \Psi_k \rangle^k = \langle \Phi_k \rangle^k \langle \Psi_k \rangle^k + \left\langle \hat{\Phi}_k \hat{\Psi}_k \right\rangle^k \tag{3.253}$$

$$\langle \Phi_k \Psi_k \rangle = \varepsilon_k \langle \Phi_k \rangle^k \langle \Psi_k \rangle^k + \left\langle \hat{\Phi}_k \hat{\Psi}_k \right\rangle \tag{3.254}$$

In order to obtain the volume-averaged governing equations, the volume average of the partial derivative with respect to time and gradient must be obtained. For a control volume ΔV shown in

Fig. 3.12, the volume averaging of the partial derivative with respect to time is obtained by the following general transport theorem:

$$\left\langle \frac{\partial \Omega_k}{\partial t} \right\rangle = \frac{\partial \langle \Omega_k \rangle}{\partial t} - \frac{1}{\Delta V} \int_{A_k} \Omega_k \mathbf{V}_\mathrm{I} \cdot \mathbf{n}_k dA_k \tag{3.255}$$

where A_k is the interfacial area surrounding the kth phase within control volume ΔV, ΔV_k is the volume occupied by the kth phase in the control volume and ΔV, $\mathbf{V}_\mathrm{I}$ is the interfacial velocity, and $\mathbf{n}_k$ is the unit normal vector at the interface directed outward from phase k (see Fig. 3.12).

The volume average of the gradient is

$$\langle \nabla \Omega_k \rangle = \nabla \langle \Omega_k \rangle + \frac{1}{\Delta V} \int_{A_k} \Omega_k \mathbf{n}_k dA_k \tag{3.256}$$

and the volume average of a divergence is

$$\langle \nabla \cdot \boldsymbol{\Omega}_k \rangle = \nabla \cdot \langle \boldsymbol{\Omega}_k \rangle + \frac{1}{\Delta V} \int_{A_k} \boldsymbol{\Omega}_k \cdot \mathbf{n}_k dA_k \tag{3.257}$$

The general quantity $\boldsymbol{\Omega}_k$ in Eqs. (3.255) and (3.256) can be a scalar, vector, or tensor of the second order. It can be a vector or tensor of the second order in Eq. (3.257).

The formulation of macroscopic equations for multiphase systems can be classified into two groups: (1) the *multifluid model* (Sect. 3.4.2) and (2) the *homogeneous model* (Sect. 3.4.3), also known as the mixture or diffuse model. If the averaging is performed for each individual phase within a multiphase control volume, as shown in Eqs. (3.248) and (3.249), one obtains the multifluid model, in which Π sets of averaged conservation equations—each set includes continuity, momentum, and

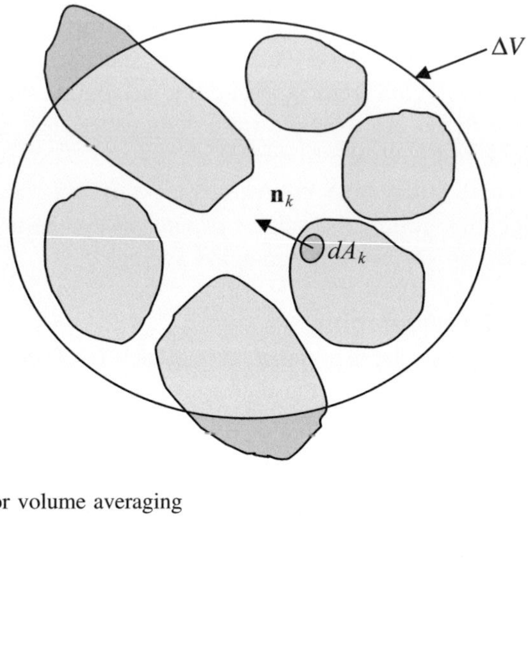

Fig. 3.12 Control volume for volume averaging

energy equations—describe the flow of a Π-phase system. The equations will also include source terms that account for the transfer of momentum, energy, and mass between phases. If only two phases are present, the multifluid model is referred to as the *two-fluid model*. However, if spatial averaging is performed over both phases simultaneously within a multiphase control volume, as indicated in Eq. (3.247), the homogeneous model is obtained; in this case, the mixture of a two-phase fluid would be considered a whole. The governing equations for the homogeneous model comprise a single set of equations including continuity, momentum, and energy equations, with one additional diffusion equation to account for the concentration change due to interphase mass transfer by phase change. Continuity, momentum, and energy equations for the mixture model can be obtained by adding together the governing equations for the multifluid models; a diffusion model must be developed to account for mass transfer between phases. In this section, it is assumed for the sake of simplicity that the reference frame is stationary.

3.4.1.2 Lagrangian Averaging

Lagrangian averaging is directly related to the Lagrangian description of a system, which requires tracking the motion of each individual fluid particle. Therefore, Lagrangian averaging is a very useful tool when the dynamics of individual particles are of interest. To obtain Lagrangian time averaging, it is necessary to follow a specific particle and observe its behavior for a certain time interval. Then, the behavior of this particle is averaged over the time interval.

For a generalized function $\Phi = \Phi(X, Y, Z, t)$, X, Y, and Z are material coordinates moving with the particle, and X, Y, Z are functions of the spatial coordinates x, y, z, and time t, i.e.,

$$X = X(x, y, z, t), \quad Y = Y(x, y, z, t), \quad Z = Z(x, y, z, t)$$

The most widely used Lagrangian averaging is *time averaging*, where the time average of the function Φ in time interval of Δt is

$$\bar{\Phi} = \frac{1}{\Delta t}\int_{\Delta t} \Phi(X, Y, Z, t)\mathrm{d}t \tag{3.258}$$

Lagrangian time averaging is performed for a distinct particle moving in the field; therefore, X, Y, and Z in the time interval Δt are not fixed in space. This focus on specific particles moving in space and time distinguishes Lagrangian averaging from Eulerian time averaging, which treats a fixed point in space relative to the reference frame. An example from daily experience will serve to illustrate this difference. In order to monitor traffic on the highway, the speed of all cars passing a point can be measured and averaged over a certain time interval—a case of Eulerian averaging. To catch an individual speeder, the police must follow the vehicle of interest to measure its speed as it moves in space over a certain time interval—a case of Lagrangian averaging.

3.4.1.3 Molecular Statistical Averaging

When the collective mechanics of a large number of particles is of interest, molecular statistical averaging may be employed. This relies on the concept of *particle number density*, which is the number of particles per unit volume. For a system with a large number of particles, the behavior of each individual particle is random because random collisions occur. To describe the behavior of each particle, it is necessary to track the motion resulting from each collision—an impractical and often unnecessary task. Although the behavior of each particle is random, the collection of particles may demonstrate some statistical behaviors that are different from those of the individual particles. When the number of molecules involved in the averaging process is large enough, the statistical average value becomes independent of the number of molecules involved. The statistical average value of the microscopic properties for a large number of molecules is related to the macroscopic properties of the system. For example, temperature is a statistical measure of the kinetic energy of the individual molecules, and the pressure of a gas in a container is the result of many molecules' collisions with the wall. For some engineering problems, the macroscopic properties of the fluid as well as the microscopic properties are required for design or analysis.

Most numerical codes are based on the Navier–Stokes equations, which treats a fluid as a continuous field. It is well known that a fluid is made of a discrete number of particles or molecules. Since the number of molecules is extremely large (Avogadro's number = 6.022×10^{23} atoms/mole) for almost all practically sized systems, it may never be computationally viable to track each particle and its interactions with other particles. The number of molecules in a given region and the molecular interaction are described through the fluid's density and transport coefficients (i.e., viscosity) in the continuous model. Modeling the individual molecules for a small system over a small period of time has been achieved by molecular dynamic simulations (MDS). The computational requirements needed in these simulations can be greatly reduced if the degrees of freedom of the system are reduced. Also, instead of considering individual molecules, groups of molecules can be considered. The degrees of freedom can be reduced by restricting the movement of the molecules to a lattice. A lattice is simply a predefined direction in which a molecule can move.

From this standpoint, the independent variables are space, velocity, and time, while the dependent variable is a molecular distribution function for species i, $f_i(\mathbf{x}, \mathbf{c}, t)$. The Boltzmann equation relates the distribution function at $(\mathbf{x}, \mathbf{c}, t)$ to the distribution function at $(\mathbf{x} + \Delta\mathbf{x},\ \mathbf{c} + \Delta\mathbf{c},\ t + \Delta t)$. The location in space is $\mathbf{x}$, and the particle velocity is $\mathbf{c}$. It is important to note that the particle velocity is directly related to the mass average velocity, $\mathbf{V}$, that is used throughout this book. This distribution function can be related to the Navier–Stokes equations as well as other transport equations; these relationships give insight to the origin of transport coefficients such as viscosity.

3.4.2 Multifluid Model

If spatial averaging is performed for each individual phase within a multiphase control volume, the multifluid model is obtained. Additional source terms are needed in these equations to account for the interaction between phases.

3.4.2.1 Continuity Equation

The volume average of the continuity equation for the kth phase is obtained by taking extrinsic phase averaging from Eq. (3.48)

$$\left\langle \frac{\partial \rho_k}{\partial t} \right\rangle + \langle \nabla \cdot \rho_k \mathbf{V}_k \rangle = 0 \tag{3.259}$$

where the two terms on the left-hand side can be obtained by using Eqs. (3.255) and (3.257), i.e.,

$$\left\langle \frac{\partial \rho_k}{\partial t} \right\rangle = \frac{\partial \langle \rho_k \rangle}{\partial t} - \frac{1}{\Delta V} \int_{A_k} \rho_k \mathbf{V}_\mathrm{I} \cdot \mathbf{n}_k \mathrm{d}A_k$$

$$\langle \nabla \cdot \rho_k \mathbf{V}_k \rangle = \nabla \cdot \langle \rho_k \mathbf{V}_k \rangle + \frac{1}{\Delta V} \int_{A_k} \rho_k \mathbf{V}_k \cdot \mathbf{n}_k \mathrm{d}A_k$$

Substituting the above expressions into Eq. (3.259), the volume-averaged continuity equation becomes

$$\frac{\partial \langle \rho_k \rangle}{\partial t} + \nabla \cdot \langle \rho_k \mathbf{V}_k \rangle = -\frac{1}{\Delta V} \int_{A_k} \rho_k (\mathbf{V}_k - \mathbf{V}_\mathrm{I}) \cdot \mathbf{n}_k \mathrm{d}A_k \tag{3.260}$$

The right-hand side of Eq. (3.260) represents the *net* mass transfer per unit volume between all other phases to the kth phase; it can be rewritten as

$$-\frac{1}{\Delta V} \int_{A_k} \rho_k (\mathbf{V}_k - \mathbf{V}_\mathrm{I}) \cdot \mathbf{n}_k \mathrm{d}A_k = \sum_{j=1(j \neq k)}^{\Pi} (\dot{m}'''_{jk} - \dot{m}'''_{kj}) \tag{3.261}$$

where $\dot{m}'''_{jk}$ represents mass transfer per unit volume from the jth to the kth phase, and $\dot{m}'''_{kj}$ represents mass transfer per unit volume from the kth to the jth phase. The difference between $\dot{m}'''_{jk}$ and $\dot{m}'''_{kj}$ represents the *net* mass transfer per unit volume from the jth phase to the kth phase. The mass transfer between phases can be due to phase change (more common), dissolution, or outgassing.

The extrinsic phase-averaged density, $\langle \rho_k \rangle$, is related to the intrinsic phase-averaged density, $\langle \rho_k \rangle^k$, by

$$\langle \rho_k \rangle = \varepsilon_k \langle \rho_k \rangle^k \tag{3.262}$$

Furthermore, the intrinsic phase-averaged density is equal to the density ρ_k.

Substituting Eqs. (3.261) and (3.262) into Eq. (3.260), and considering Eq. (3.254), the continuity equation for the kth phase becomes

$$\frac{\partial}{\partial t} \left(\varepsilon_k \langle \rho_k \rangle^k \right) + \nabla \cdot \left(\varepsilon_k \langle \rho_k \rangle^k \langle \mathbf{V}_k \rangle^k + \langle \hat{\rho}_k \hat{\mathbf{V}}_k \rangle \right) = \sum_{j=1(j \neq k)}^{\Pi} (\dot{m}'''_{jk} - \dot{m}'''_{kj}) \tag{3.263}$$

The dispersive term in Eq. (3.263), $\langle \hat{\rho}_k \hat{\mathbf{V}}_k \rangle$, is generally small compared with $\varepsilon_k \langle \rho_k \rangle^k \langle \mathbf{V}_k \rangle^k$; it is assumed that it can be neglected. The continuity equation for the kth phase becomes

$$\frac{\partial}{\partial t} \left(\varepsilon_k \langle \rho_k \rangle^k \right) + \nabla \cdot \left(\varepsilon_k \langle \rho_k \rangle^k \langle \mathbf{V}_k \rangle^k \right) = \sum_{j=1(j \neq k)}^{\Pi} (\dot{m}'''_{jk} - \dot{m}'''_{kj}) \tag{3.264}$$

3.4.2.2 Momentum Equation

The extrinsic phase-averaged momentum equation for the kth phase can be obtained by performing extrinsic phase averaging on the momentum Eq. (3.63):

$$\left\langle \frac{\partial(\rho_k \mathbf{V}_k)}{\partial t} \right\rangle + \langle \nabla \cdot (\rho_k \mathbf{V}_k \mathbf{V}_k) \rangle = \langle \nabla \cdot \boldsymbol{\tau}'_k \rangle + \langle \rho_k \mathbf{X}_k \rangle \tag{3.265}$$

where the body force per unit mass is assumed to be the same for different species for the sake of simplicity. Evaluating each term in Eq. (3.265) using Eqs. (3.255)–(3.257), the momentum equation becomes:

$$\begin{aligned} \frac{\partial \langle \rho_k \mathbf{V}_k \rangle}{\partial t} + \nabla \cdot \langle \rho_k \mathbf{V}_k \mathbf{V}_k \rangle = \nabla \cdot \langle \boldsymbol{\tau}'_k \rangle - \frac{1}{\Delta V} \int_{A_k} \rho_k \mathbf{V}_k (\mathbf{V}_k - \mathbf{V}_\mathrm{I}) \cdot \mathbf{n}_k \mathrm{d}A_k \\ + \frac{1}{\Delta V} \int_{A_k} \boldsymbol{\tau}'_k \cdot \mathbf{n}_k \mathrm{d}A_k + \langle \rho_k \rangle \mathbf{X}_k \end{aligned} \tag{3.266}$$

Considering Eq. (3.254) and ignoring the product of the deviations, Eq. (3.266) becomes

$$\begin{aligned} \frac{\partial}{\partial t} \left(\varepsilon_k \langle \rho_k \rangle^k \langle \mathbf{V}_k \rangle^k \right) + \nabla \cdot \left(\varepsilon_k \langle \rho_k \rangle^k \langle \mathbf{V}_k \mathbf{V}_k \rangle^k \right) = \nabla \cdot \left(\varepsilon_k \langle \boldsymbol{\tau}'_k \rangle^k \right) \\ - \frac{1}{\Delta V} \int_{A_k} \rho_k \mathbf{V}_k (\mathbf{V}_k - \mathbf{V}_\mathrm{I}) \cdot \mathbf{n}_k \mathrm{d}A_k + \frac{1}{\Delta V} \int_{A_k} \boldsymbol{\tau}'_k \cdot \mathbf{n}_k \mathrm{d}A_k + \varepsilon_k \langle \rho_k \rangle^k \mathbf{X}_k \end{aligned} \tag{3.267}$$

The second and third terms on the right-hand side of Eq. (3.267) represent the momentum exchanges and interactive forces between all other phases and the kth phase, described as:

$$-\frac{1}{\Delta V} \int_{A_k} \rho_k \mathbf{V}_k (\mathbf{V}_k - \mathbf{V}_\mathrm{I}) \cdot \mathbf{n}_k \mathrm{d}A_k = \sum_{j=1 (j \neq k)}^{\Pi} \left(\dot{m}'''_{jk} \langle \mathbf{V}_{k,\mathrm{I}} \rangle^k - \dot{m}'''_{kj} \langle \mathbf{V}_{j,\mathrm{I}} \rangle^k \right) \tag{3.268}$$

$$\frac{1}{\Delta V} \int_{A_k} \boldsymbol{\tau}'_k \cdot \mathbf{n}_k \mathrm{d}A_k = \sum_{j=1 (j \neq k)}^{\Pi} \langle \mathbf{F}_{jk} \rangle \tag{3.269}$$

where $\langle \mathbf{V}_{k,\mathrm{I}} \rangle^k$ and $\langle \mathbf{V}_{j,\mathrm{I}} \rangle^k$ are the intrinsic phase-averaged velocity of the kth and jth phases at the interface. The difference between two adjacent phases results solely from the density difference between the two phases. $\langle \mathbf{F}_{jk} \rangle$ is an interactive force between the jth and the kth phases and depends on the friction, pressure, and cohesion between different phases. Newton's third law requires that the interactive forces satisfy

$$\langle \mathbf{F}_{jk} \rangle = -\langle \mathbf{F}_{kj} \rangle \tag{3.270}$$

The interactive force can be determined by

$$\langle \mathbf{F}_{jk} \rangle = K_{jk} \left(\langle \mathbf{V}_j \rangle^j - \langle \mathbf{V}_k \rangle^k \right) \tag{3.271}$$

where K_{jk} is the momentum exchange coefficient between phases j and k. Determining the momentum exchange coefficient is a very challenging task because interphase momentum exchange depends on the structure of the interfaces. If a secondary phase j is dispersed in the primary phase k, as is the case with the dispersed-phase system summarized in Table 1.1, one can assume that the secondary phase is spherical in shape and an appropriate empirical correlation can be used to obtain the momentum exchange coefficient.

Substituting Eqs. (3.268) and (3.269) into Eq. (3.267), the multifluid volume-averaged momentum equation becomes:

$$\begin{aligned}\frac{\partial}{\partial t}\left(\varepsilon_k\langle\rho_k\rangle^k\langle\mathbf{V}_k\rangle^k\right)+\nabla\cdot\left(\varepsilon_k\langle\rho_k\rangle^k\langle\mathbf{V}_k\mathbf{V}_k\rangle^k\right)\\ =\nabla\cdot\left(\varepsilon_k\left\langle\boldsymbol{\tau}_k'\right\rangle^k\right)+\varepsilon_k\langle\rho_k\rangle^k\mathbf{X}_k\\ +\sum_{j=1(j\neq k)}^{\Pi}\left(\left\langle\mathbf{F}_{jk}\right\rangle+\dot{m}_{jk}'''\left\langle\mathbf{V}_{k,\mathrm{I}}\right\rangle^k-\dot{m}_{jk}'''\left\langle\mathbf{V}_{j,\mathrm{I}}\right\rangle^k\right)\end{aligned} \tag{3.272}$$

Since liquid–vapor flow is widely used in various applications, we will use liquid–vapor flow as an example to explain the determination of the momentum exchange coefficient. If liquid is the primary phase and vapor is the secondary phase, the vapor phase is dispersed in the liquid as vapor bubbles. If vapor is the primary phase and liquid is the secondary phase, the liquid phase is dispersed in the vapor as liquid droplets. Boysan (1990) suggested that the momentum exchange coefficient could be estimated by

$$K_{jk}=\frac{3}{4}C_D\frac{\varepsilon_j\langle\rho_k\rangle^k}{d_j}\left|\left\langle\mathbf{V}_j\right\rangle^j-\langle\mathbf{V}_k\rangle^k\right| \tag{3.273}$$

where phase k is the primary phase, and phase j is the secondary phase, and d_j is the diameter of vapor bubbles or liquid droplets of the secondary phase j. C_D is the drag coefficient based on the relative Reynolds number, which is obtained by the following empirical correlations:

$$C_\mathrm{D}=\begin{cases}\dfrac{24}{Re}\left(1+0.15Re^{0.687}\right) & Re\leq 1000\\ 0.44 & Re>1000\end{cases} \tag{3.274}$$

where

$$Re=\frac{\langle\rho_k\rangle\left|\left\langle\mathbf{V}_j\right\rangle^j-\langle\mathbf{V}_k\rangle^k\right|d_j}{\mu_k} \tag{3.275}$$

Example 3.8

A mixture of water and its vapor at 1 atm flows in a 0.1 m ID tube with a mass flow rate of 0.1 kg/s. The liquid water is dispersed in the vapor phase in the form of 0.1-mm-diameter droplets and the quality of the mixture is $x = 0.8$. While the vapor is saturated, the liquid droplets are subcooled, both at 95 °C. The volume fraction of the liquid phase is $\varepsilon_\ell = 0.01$. Find the interactive force between the liquid and vapor phases.

Solution Since the liquid droplets are dispersed in the vapor phase, vapor (v) is the primary phase and liquid (ℓ) is the secondary phase. At 1 atm, the properties of the saturated vapor are

$$\mu_v = 16.698 \times 10^{-6}\,\text{N s/m}^2 \quad \rho_v = 0.596\,\text{kg/m}^3$$

The density of the subcooled water at 95 °C is $\rho_\ell = 965.35\,\text{kg/m}^3$
The mass flow rates of the liquid and vapor phases are, respectively,

$$\dot{m}_v = x\dot{m} = 0.8 \times 0.1 = 0.08\,\text{kg/s}$$

$$\dot{m}_\ell = (1-x)\dot{m} = (1-0.8) \times 0.1 = 0.02\,\text{kg/s}$$

The total cross-sectional area of the tube is

$$A = \frac{1}{4}\pi D^2 = \frac{1}{4} \times \pi \times 0.1^2 = 7.85 \times 10^{-3}\,\text{m}^2$$

The mass flow rates of the liquid and vapor phases are, respectively,

$$\dot{m}_\ell = \rho_\ell \langle w_\ell \rangle^\ell A_\ell = \rho_\ell \langle w_\ell \rangle^\ell \varepsilon_\ell A$$

$$\dot{m}_v = \rho_v \langle w_v \rangle^v A_v = \rho_v \langle w_v \rangle^v (1-\varepsilon_\ell) A$$

The average velocities of the liquid and vapor phases can be obtained by

$$\langle w_\ell \rangle^\ell = \frac{\dot{m}_\ell}{\rho_\ell \varepsilon_\ell A} = \frac{0.02}{965.35 \times 0.01 \times 7.85 \times 10^{-3}} = 0.264\,\text{m/s}$$

$$\langle w_v \rangle^v = \frac{\dot{m}_v}{\rho_v (1-\varepsilon_\ell) A} = \frac{0.08}{0.596 \times (1-0.01) \times 7.85 \times 10^{-3}} = 17.27\,\text{m/s}$$

The relative Reynolds number of the liquid droplets is

$$\begin{aligned} Re &= \frac{\rho_v \left| \langle w_\ell \rangle^\ell - \langle w_v \rangle^v \right| d_\ell}{\mu_v} \\ &= \frac{0.596 \times |0.264 - 17.27| \times 0.1 \times 10^{-3}}{16.698 \times 10^{-6}} = 60.70 \end{aligned}$$

The drag coefficient is then

$$C_D = \frac{24}{Re}(1 + 0.15 Re^{0.687}) = \frac{24}{60.70} \times (1 + 0.15 \times 60.70^{0.687}) = 1.39$$

The momentum exchange coefficient can be obtained from Eq. (3.273), i.e.,

$$\begin{aligned} K_{\ell v} &= \frac{3}{4} C_D \frac{\varepsilon_\ell \langle \rho_v \rangle^v}{d_\ell} \left| \langle w_\ell \rangle^\ell - \langle w_v \rangle^v \right| \\ &= \frac{3}{4} \times 1.39 \times \frac{0.01 \times 0.596}{0.1 \times 10^{-3}} \times |0.264 - 17.27| = 1056.63\,\text{kg/(m}^3\,\text{s)} \end{aligned}$$

The interactive force between liquid droplets and vapor is

$$\begin{aligned}\langle F_{\ell \mathrm{v}}\rangle &= K_{\ell \mathrm{v}}\left(\langle w_\ell\rangle^{\ell} - \langle w_{\mathrm{v}}\rangle^{\mathrm{v}}\right)\\ &= 1056.63 \times (0.264 - 17.27) = -1.797 \times 10^4\,\mathrm{N/m^3}\end{aligned}$$

which is negative because the liquid droplet moves slower than the vapor phase.

3.4.2.3 Energy Equation

The extrinsic phase average of the energy Eq. (3.90) is

$$\begin{aligned}&\left\langle \frac{\partial(\rho_k h_k)}{\partial t}\right\rangle + \langle \nabla \cdot \rho_k \mathbf{V}_k h_k\rangle\\ &\quad = -\langle \nabla \cdot \mathbf{q}_k''\rangle + \langle q_k'''\rangle + \left\langle \frac{\partial p_k}{\partial t}\right\rangle + \langle \mathbf{V}_k \cdot \nabla p_k\rangle + \langle \nabla \mathbf{V}_k : \boldsymbol{\tau}_k\rangle\end{aligned} \tag{3.276}$$

Evaluating each term in Eq. (3.276) using Eqs. (3.255)–(3.257), the energy equation becomes:

$$\begin{aligned}&\frac{\partial\langle \rho_k h_k\rangle}{\partial t} + \nabla \cdot \langle \rho_k \mathbf{V}_k h_k\rangle\\ &\quad = -\nabla \cdot \langle \mathbf{q}_k''\rangle + \langle q_k'''\rangle + \frac{\partial \langle p_k\rangle}{\partial t} + \langle \mathbf{V}_k\rangle \cdot \nabla\langle p_k\rangle + \nabla\langle \mathbf{V}_k\rangle : \langle \boldsymbol{\tau}_k\rangle\\ &\quad\quad - \frac{1}{\Delta V}\int_{A_k} \rho_k h_k(\mathbf{V}_k - \mathbf{V}_{\mathrm{I}}) \cdot \mathbf{n}_k \mathrm{d}A_k - \frac{1}{\Delta V}\int_{A_k} \mathbf{q}_k'' \cdot \mathbf{n}_k \mathrm{d}A_k\\ &\quad\quad + \frac{1}{\Delta V}\left[-\int_{A_k} p_k \mathbf{V}_{\mathrm{I}} \cdot \mathbf{n}_k \mathrm{d}A_k + \langle \mathbf{V}_k\rangle \cdot \int_{A_k} p_k \mathbf{n}_k \mathrm{d}A_k + \left(\int_{A_k} \mathbf{V}_k \mathbf{n}_k \mathrm{d}A_k\right) : \langle \boldsymbol{\tau}_k\rangle\right]\end{aligned} \tag{3.277}$$

where the terms in the square bracket on the right-hand side represent the work per unit volume done by the pressure and shear stress at the interface. These terms reflect conversion of mechanical energy to the thermal energy at the interface and are usually negligible compared to the other terms.

Considering Eq. (3.254) and ignoring the product of the deviations, the volume-averaged energy Eq. (3.277) becomes

$$\begin{aligned}&\frac{\partial}{\partial t}\left(\varepsilon_k \langle \rho_k\rangle^k \langle h_k\rangle^k\right) + \nabla \cdot \left(\varepsilon_k \langle \rho_k\rangle^k \langle \mathbf{V}_k h_k\rangle^k\right)\\ &\quad = -\nabla \cdot \langle \mathbf{q}_k''\rangle + \langle q_k'''\rangle + \varepsilon_k \frac{D\langle p_k\rangle^k}{Dt} + \nabla\langle \mathbf{V}_k\rangle : \langle \boldsymbol{\tau}_k\rangle\\ &\quad\quad - \frac{1}{\Delta V}\int_{A_k} \rho_k h_k(\mathbf{V}_k - \mathbf{V}_{\mathrm{I}}) \cdot \mathbf{n}_k \mathrm{d}A_k - \frac{1}{\Delta V}\int_{A_k} \mathbf{q}_k'' \cdot \mathbf{n}_k \mathrm{d}A_k\end{aligned} \tag{3.278}$$

The fifth term on the right-hand side of Eq. (3.278) is interphase enthalpy exchange between all other phases and the kth phase, i.e.,

$$-\frac{1}{\Delta V}\int_{A_k}\rho_k h_k(\mathbf{V}_k-\mathbf{V}_\mathrm{I})\cdot\mathbf{n}_k dA_k=\sum_{j=1(j\neq k)}^{\Pi}\left(\dot{m}_{jk}'''\langle h_{k,\mathrm{I}}\rangle^k-\dot{m}_{kj}'''\langle h_{j,\mathrm{I}}\rangle^k\right)\tag{3.279}$$

where $\langle h_{k,\mathrm{I}}\rangle^k$ and $\langle h_{j,\mathrm{I}}\rangle^k$ are the intrinsic phase-averaged enthalpy of the kth and jth phases at the interface. The sixth term on the right-hand side of Eq. (3.278) is heat transfer from all other phases and the kth phase,

$$-\frac{1}{\Delta V}\int_{A_k}\mathbf{q}_k''\cdot\mathbf{n}_k dA_k=\sum_{j=1(j\neq k)}^{\Pi}\left\langle q_{jk}'''\right\rangle\tag{3.280}$$

where $\left\langle q_{jk}'''\right\rangle$ is the intensity of heat exchange between phases j and k. It can be obtained by using Newton's law of cooling:

$$\left\langle q_{jk}'''\right\rangle=\frac{h_c\Delta A_j\left(\langle T_j\rangle^j-\langle T_k\rangle^k\right)}{\Delta V_j}\tag{3.281}$$

where h_c is the convective heat transfer coefficient, ΔA_j is the area of the interface between phases j and k, and ΔV_j is the volume of the secondary phase in the volume element. Like the momentum exchange coefficient, the interphase heat transfer also depends on the structure of the interfaces. If a secondary phase j is dispersed in the primary phase k—as in the dispersed phase summarized in Table 1.1—the following empirical correlation recommended is widely used:

$$Nu=2+(0.4Re^{1/2}+0.06Re^{2/3})Pr_k^{0.4}\left(\frac{\mu_k}{\mu_{k,s}}\right)^{1/4}\tag{3.282}$$

where the Reynolds number, Re, is obtained by Eq. (3.275), the Nusselt number is defined as

$$Nu=\frac{h_c d_j}{k_k}\tag{3.283}$$

and all thermal properties of the primary phase are evaluated at T_k except $\mu_{k,s}$, which is evaluated at T_j. Equation (3.282) is valid for $3.5<Re<7.6\times10^4$ and $0.71<Pr_k<380$, which covers a wide variety of problems.

If the secondary phase is liquid and the primary phase is vapor (gas), Eq. (3.282) can be simplified to

$$Nu=2+0.6Re^{1/2}Pr_k^{1/3}\tag{3.284}$$

Substituting Eqs. (3.279) and (3.280) into Eq. (3.278), the volume-averaged energy equation becomes

$$\frac{\partial}{\partial t}\left(\varepsilon_k\langle\rho_k\rangle^k\langle h_k\rangle^k\right)+\nabla\cdot\left(\varepsilon_k\langle\rho_k\rangle^k\langle\mathbf{V}_k h_k\rangle^k\right)=-\nabla\cdot\langle\mathbf{q}_k''\rangle+\langle q_k'''\rangle$$
$$+\varepsilon_k\frac{D\langle p_k\rangle^k}{Dt}+\nabla\langle\mathbf{V}_k\rangle:\langle\boldsymbol{\tau}_k\rangle+\sum_{j=1(j\neq k)}^{\Pi}\left[\left\langle q_{jk}'''\right\rangle+\dot{m}_{jk}'''\langle h_{k,\mathrm{I}}\rangle^k-\dot{m}_{kj}'''\langle h_{j,\mathrm{I}}\rangle^k\right] \tag{3.285}$$

Example 3.9

Estimate the interphase heat transfer rate between the liquid and vapor phases in Example 3.8.

Solution The thermal conductivity and the Prandtl number of the vapor phase are $k_{\mathrm{v}}=0.0248\,\mathrm{W/m\,K}$ and $Pr_{\mathrm{v}}=0.984$, respectively. Since the primary phase is vapor (v) and the secondary phase is liquid (ℓ), Eq. (3.284) can be used to estimate the heat transfer coefficient, i.e.,

$$Nu=2+0.6Re^{1/2}Pr_{\mathrm{v}}^{1/3}=2+0.6\times 60.70^{1/2}\times 0.984^{1/3}=6.65$$

The heat transfer coefficient is then

$$h_{\mathrm{c}}=\frac{Nu\cdot k_{\mathrm{v}}}{d_\ell}=\frac{6.65\times 0.0248}{0.1\times 10^{-3}}=1649.2\,\mathrm{W/m^2\ K}$$

Therefore, the interphase heat transfer between the liquid and vapor phases can be obtained by Eq. (3.281):

$$\begin{aligned}\left\langle q_{\ell\mathrm{v}}'''\right\rangle&=\frac{h_{\mathrm{c}}A_\ell\left(\langle T_\ell\rangle^\ell-\langle T_{\mathrm{v}}\rangle^{\mathrm{v}}\right)}{V_\ell}=\frac{h_{\mathrm{c}}\pi d_\ell^2\left(\langle T_\ell\rangle^\ell-\langle T_{\mathrm{v}}\rangle^{\mathrm{v}}\right)}{\pi d_\ell^3/6}\\&=\frac{6h_{\mathrm{c}}\left(\langle T_\ell\rangle^\ell-\langle T_{\mathrm{v}}\rangle^{\mathrm{v}}\right)}{d_\ell}=\frac{6\times 1649.2\times(95-100)}{0.1\times 10^{-3}}=-4.95\times 10^8\,\mathrm{W/m^3}\end{aligned}$$

3.4.2.4 Species

If the fluid undergoing phase change involves multiple components, it is also necessary to write the equation for conservation of the species mass in the kth phase. The extrinsic phase average of conservation of species mass can be obtained by

$$\left\langle\frac{\partial\rho_{k,i}}{\partial t}\right\rangle+\langle\nabla\cdot\rho_{k,i}\mathbf{V}_k\rangle=-\langle\nabla\cdot\mathbf{J}_{k,i}\rangle+\left\langle\dot{m}_{k,i}'''\right\rangle \tag{3.286}$$

where each term can be obtained by

$$\left\langle\frac{\partial\rho_{k,i}}{\partial t}\right\rangle=\frac{\partial\langle\rho_{k,i}\rangle}{\partial t}-\frac{1}{\Delta V}\int_{A_k}\rho_{k,i}\mathbf{V}_{\mathrm{I}}\cdot\mathbf{n}_k dA_k$$

$$\langle \nabla \cdot \rho_{k,i} \mathbf{V}_k \rangle = \nabla \cdot \langle \rho_{k,i} \mathbf{V}_k \rangle + \frac{1}{\Delta V} \int_{A_k} \rho_{k,i} \mathbf{V}_k \cdot \mathbf{n}_k dA_k$$

$$\langle \nabla \cdot \mathbf{J}_{k,i} \rangle = \nabla \cdot \langle \mathbf{J}_{k,i} \rangle + \frac{1}{\Delta V} \int_{A_k} \mathbf{J}_{k,i} \cdot \mathbf{n}_k dA_k$$

Substituting the above expression into Eq. (3.286), one obtains

$$\begin{aligned} \frac{\partial \langle \rho_{k,i} \rangle}{\partial t} &+ \nabla \cdot \langle \rho_{k,i} \mathbf{V}_k \rangle = -\nabla \cdot \langle \mathbf{J}_{k,i} \rangle + \langle \dot{m}'''_{k,i} \rangle \\ &- \frac{1}{\Delta V} \int_{A_k} \rho_{k,i} (\mathbf{V}_k - \mathbf{V}_{\mathrm{I}}) \cdot \mathbf{n}_k dA_k + \frac{1}{\Delta V} \int_{A_k} \mathbf{J}_{k,i} \cdot \mathbf{n}_k dA_k \end{aligned} \tag{3.287}$$

where the third and fourth terms in the right-hand side of Eq. (3.287) represent mass source (or sink) of the ith component in the kth phase due to phase change from other phases to the kth phase, as well as mass transfer at the interface due to diffusion, respectively, i.e.,

$$-\frac{1}{\Delta V} \int_{A_k} \rho_{k,i} (\mathbf{V}_k - \mathbf{V}_{\mathrm{I}}) \cdot \mathbf{n}_k dA_k + \frac{1}{\Delta V} \int_{A_k} \mathbf{J}_{k,i} \cdot \mathbf{n}_k dA_k = \sum_{j=1 (j \neq k)}^{\Pi} (\dot{m}'''_{jk,i} - \dot{m}'''_{kj,i}) \tag{3.288}$$

where $\dot{m}'''_{jk,i}$ and $\dot{m}'''_{kj,i}$ represent the mass source (or sink) of the ith component in phase k due to phase change between phases j and k, as well as diffusive mass transfer at the interface between phases j and k.

Substituting Eq. (3.288) into Eq. (3.287) and considering Eq. (3.254), the volume-averaged species equation becomes

$$\frac{\partial}{\partial t} \left(\varepsilon_k \langle \rho_{k,i} \rangle^k \right) + \nabla \cdot \left(\varepsilon_k \langle \rho_{k,i} \rangle^k \langle \mathbf{V}_k \rangle^k \right) = -\nabla \cdot \langle \mathbf{J}_{k,i} \rangle + \langle \dot{m}'''_{k,i} \rangle + \sum_{j=1 (j \neq k)}^{\Pi} (\dot{m}'''_{jk,i} - \dot{m}'''_{kj,i}) \tag{3.289}$$

where the three terms on the right-hand side represent the effects of mass diffusion, species source/sink due to chemical reaction, and phase change at the interfaces.

Example 3.10

A cooling tower is a multiphase system. Warm water is sprayed at the top, in the form of water droplets. Gravity drives these droplets downward. The water droplets are cooled by evaporated effects as they flow through the air. Air enters the cooling towers from the bottom on the sides. The system being investigated has a planar geometry. The schematic for this system is presented in Fig. 3.13. The temperature of the water collected at the bottom of the tank is desired, as well as the amount of water lost due to evaporation. Develop the governing equations by the volume-average method, as well as the boundary conditions.

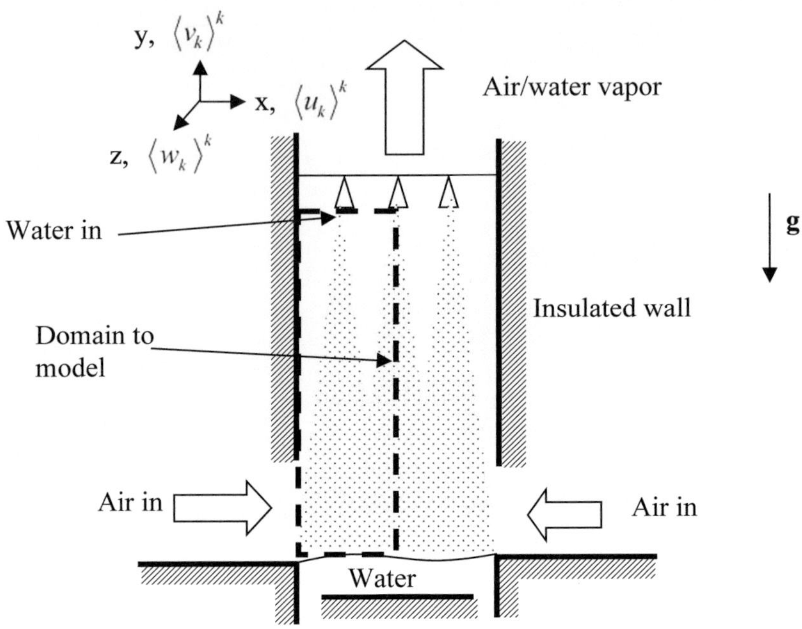

Fig. 3.13 Schematic of the cooling of droplets

Solution This problem can be solved using a volume-averaged multifluid formulation. The basic assumptions are that the flow field is steady. The planar geometry is long enough that the changes in the *z*-direction can be neglected; therefore, a two-dimensional geometry is utilized. Also, since this is the case, a plane of symmetry exists half the distance between the two adiabatic walls. The pressure effects on energy are considered negligible. The final assumption is that intrinsic phase average of the products of deviation is zero, $\langle \hat{\mathbf{V}}_k \hat{\mathbf{V}}_k \rangle^k = \langle \hat{\mathbf{V}}_k \hat{h}_k \rangle^k = \left\langle \hat{\mathbf{V}}_k \hat{\Omega}_k \right\rangle^k = 0.$

The volume-averaged continuity equation, Eq. (3.264), for two phases is:

$$\frac{\partial}{\partial x}\left(\varepsilon_\ell \langle \rho_\ell \rangle^\ell \langle u_\ell \rangle^\ell\right) + \frac{\partial}{\partial y}\left(\varepsilon_\ell \langle \rho_\ell \rangle^\ell \langle v_\ell \rangle^\ell\right) = \dot{m}'''_{\mathrm{v}\ell} \tag{3.290}$$

$$\frac{\partial}{\partial x}\left(\varepsilon_\mathrm{v} \langle \rho_\mathrm{v} \rangle^\mathrm{v} \langle u_\mathrm{v} \rangle^\mathrm{v}\right) + \frac{\partial}{\partial y}\left(\varepsilon_\mathrm{v} \langle \rho_\mathrm{v} \rangle^\mathrm{v} \langle v_\mathrm{v} \rangle^\mathrm{v}\right) = \dot{m}'''_{\ell\mathrm{v}} = -\dot{m}'''_{\mathrm{v}\ell} \tag{3.291}$$

The volume-averaged momentum equations for both phases in the *x*- and *y*-directions are obtained from Eq. (3.272):

Liquid x-momentum

$$\begin{aligned} &\varepsilon_\ell \langle \rho_\ell \rangle^\ell \langle u_\ell \rangle^\ell \frac{\partial \langle u_\ell \rangle^\ell}{\partial x} + \varepsilon_\ell \langle \rho_\ell \rangle^\ell \langle v_\ell \rangle^\ell \frac{\partial \langle u_\ell \rangle^\ell}{\partial y} \\ &= -\frac{\partial \varepsilon_\ell \langle p_\ell \rangle^\ell}{\partial x} + \frac{\partial}{\partial x}\left(\varepsilon_\ell \mu_\ell \frac{\partial \langle u_\ell \rangle^\ell}{\partial x}\right) + \frac{\partial}{\partial y}\left(\varepsilon_\ell \mu_\ell \frac{\partial \langle u_\ell \rangle^\ell}{\partial y}\right) \\ &+ \langle F_{\mathrm{v}\ell,x} \rangle + \langle \dot{m}'''_{\mathrm{v}\ell} \rangle \langle u_{\ell,\mathrm{I}} \rangle^\ell \end{aligned} \tag{3.292}$$

Liquid y-momentum

$$\begin{aligned}&\varepsilon_\ell\langle\rho_\ell\rangle^\ell\langle u_\ell\rangle^\ell\frac{\partial\langle v_\ell\rangle^\ell}{\partial x}+\varepsilon_\ell\langle\rho_\ell\rangle^\ell\langle v_\ell\rangle^\ell\frac{\partial\langle v_\ell\rangle^\ell}{\partial y}\\&=-\frac{\partial\varepsilon_\ell\langle p_\ell\rangle^\ell}{\partial y}+\frac{\partial}{\partial x}\left(\varepsilon_\ell\mu_\ell\frac{\partial\langle v_\ell\rangle^\ell}{\partial x}\right)+\frac{\partial}{\partial y}\left(\varepsilon_\ell\mu_\ell\frac{\partial\langle v_\ell\rangle^\ell}{\partial y}\right)\\&+\langle F_{\mathrm{v}\ell,y}\rangle+\langle\dot m'''_{\mathrm{v}\ell}\rangle\langle v_{\ell,\mathrm{I}}\rangle^\ell-\varepsilon_\ell\langle\rho_\ell\rangle^\ell g\end{aligned}\tag{3.293}$$

Vapor x-momentum

$$\begin{aligned}&\varepsilon_\mathrm{v}\langle\rho_\mathrm{v}\rangle_\mathrm{v}\langle u_\mathrm{v}\rangle^\mathrm{v}\frac{\partial\langle u_\mathrm{v}\rangle^\mathrm{v}}{\partial x}+\varepsilon_\mathrm{v}\langle\rho_\mathrm{v}\rangle^\mathrm{v}\langle v_\mathrm{v}\rangle\frac{\partial\langle u_\mathrm{v}\rangle^\mathrm{v}}{\partial y}=-\frac{\partial\varepsilon_\mathrm{v}\langle p_\mathrm{v}\rangle^\mathrm{v}}{\partial x}\\&+\frac{\partial}{\partial x}\left(\varepsilon_\mathrm{v}\mu_\mathrm{v}\frac{\partial\langle u_\mathrm{v}\rangle^\mathrm{v}}{\partial x}\right)+\frac{\partial}{\partial y}\left(\varepsilon_\mathrm{v}\mu_\mathrm{v}\frac{\partial\langle u_\mathrm{v}\rangle^\mathrm{v}}{\partial y}\right)+\langle F_{\ell\mathrm{v},x}\rangle+\langle\dot m'''_{\ell\mathrm{v}}\rangle\langle u_{\mathrm{v,I}}\rangle^\mathrm{v}\end{aligned}\tag{3.294}$$

Vapor y-momentum

$$\begin{aligned}&\varepsilon_\mathrm{v}\langle\rho_\mathrm{v}\rangle^\mathrm{v}\langle u_\mathrm{v}\rangle^\mathrm{v}\frac{\partial\langle v_\mathrm{v}\rangle^\mathrm{v}}{\partial x}+\varepsilon_\mathrm{v}\langle\rho_\mathrm{v}\rangle^\mathrm{v}\langle v_\mathrm{v}\rangle^\mathrm{v}\frac{\partial\langle v_\mathrm{v}\rangle^\mathrm{v}}{\partial y}\\&=-\frac{\partial\varepsilon_\mathrm{v}\langle p_\mathrm{v}\rangle^\mathrm{v}}{\partial y}+\frac{\partial}{\partial x}\left(\varepsilon_\mathrm{v}\mu_\mathrm{v}\frac{\partial\langle v_\mathrm{v}\rangle^\mathrm{v}}{\partial x}\right)+\frac{\partial}{\partial y}\left(\varepsilon_\mathrm{v}\mu_\mathrm{v}\frac{\partial\langle v_\mathrm{v}\rangle^\mathrm{v}}{\partial y}\right)\\&+\langle F_{\ell\mathrm{v},y}\rangle+\langle\dot m'''_{\ell\mathrm{v}}\rangle\langle v_{\mathrm{v,I}}\rangle^\mathrm{v}-\varepsilon_\mathrm{v}\langle\rho_\mathrm{v}\rangle^\mathrm{v}g\end{aligned}\tag{3.295}$$

The volume-averaged energy equations for the liquid and vapor phases are obtained from Eq. (3.285):
Liquid energy

$$\begin{aligned}&\varepsilon_\ell\langle\rho_\ell\rangle^\ell\langle u_\ell\rangle\frac{\partial\langle h_\ell\rangle^\ell}{\partial x}+\varepsilon_\ell\langle\rho_\ell\rangle^\ell\langle v_\ell\rangle^\ell\frac{\partial\langle h_\ell\rangle^\ell}{\partial y}\\&=\frac{\partial}{\partial x}\left(\varepsilon_\ell k_\ell\frac{\partial\langle T_\ell\rangle^\ell}{\partial x}\right)+\frac{\partial}{\partial y}\left(\varepsilon_\ell k_\ell\frac{\partial\langle T_\ell\rangle^\ell}{\partial y}\right)+\langle q'''_{\mathrm{v}\ell}\rangle+\langle\dot m'''_{\mathrm{v}\ell}\rangle h_{\ell\mathrm{v}}\end{aligned}\tag{3.296}$$

Vapor energy

$$\begin{aligned}&\varepsilon_\mathrm{v}\langle\rho_\mathrm{v}\rangle\langle u_\mathrm{v}\rangle^\mathrm{v}\frac{\partial\langle h_\mathrm{v}\rangle^\mathrm{v}}{\partial x}+\varepsilon_\mathrm{v}\langle\rho_\mathrm{v}\rangle^\mathrm{v}\langle v_\mathrm{v}\rangle^\mathrm{v}\frac{\partial\langle h_\mathrm{v}\rangle^\mathrm{v}}{\partial y}\\&=\frac{\partial}{\partial x}\left(\varepsilon_\mathrm{v}k_\mathrm{v}\frac{\partial\langle T_\mathrm{v}\rangle^\mathrm{v}}{\partial x}\right)+\frac{\partial}{\partial y}\left(\varepsilon_\mathrm{v}k_\mathrm{v}\frac{\partial\langle T_\mathrm{v}\rangle^\mathrm{v}}{\partial y}\right)+\langle q'''_{\ell\mathrm{v}}\rangle+\langle\dot m'''_{\ell\mathrm{v}}\rangle h_{\ell\mathrm{v}}\end{aligned}\tag{3.297}$$

where the enthalpy is defined as:

$$\langle h_\ell\rangle^\ell=\int c_{\mathrm{p}\ell}\mathrm{d}T\tag{3.298}$$

$$\langle h_{\mathrm{v}}\rangle^{\mathrm{v}}=\int c_{p\mathrm{v}}\mathrm{d}T \tag{3.299}$$

There are multiple components in the vapor phase; therefore, the specific heat is a function of the mass fraction of those components.

$$c_{p\mathrm{v}}=\sum_{i=1}^{N}\omega_i c_{pi} \tag{3.300}$$

Air is considered to be one component, and water vapor is considered to be the other component. Therefore, only one species equation is needed. The species equation for water vapor, Eq. (3.289), is:

$$\begin{aligned}&\varepsilon_{\mathrm{v}}\langle\rho_{\mathrm{v}}\rangle^{\mathrm{v}}\langle u_{\mathrm{v}}\rangle^{\mathrm{v}}\frac{\partial\langle\omega_{\mathrm{v}}\rangle^{\mathrm{v}}}{\partial x}+\varepsilon_{\mathrm{v}}\langle\rho_{\mathrm{v}}\rangle^{\mathrm{v}}\langle v_{\mathrm{v}}\rangle^{\mathrm{v}}\frac{\partial\langle\omega_{\mathrm{v}}\rangle^{\mathrm{v}}}{\partial y}\\&=\frac{\partial}{\partial x}\left(\varepsilon_{\mathrm{v}}D_{\mathrm{v,air}}\frac{\partial\langle\omega_{\mathrm{v}}\rangle^{\mathrm{v}}}{\partial x}\right)+\frac{\partial}{\partial y}\left(\varepsilon_{\mathrm{v}}D_{\mathrm{v,air}}\frac{\partial\langle\omega_{\mathrm{v}}\rangle^{\mathrm{v}}}{\partial y}\right)+\langle\dot{m}'''_{\ell\mathrm{v}}\rangle\end{aligned} \tag{3.301}$$

It is also known that

$$\varepsilon_\ell+\varepsilon_{\mathrm{v}}=1 \tag{3.302}$$

$$\dot{m}'''_{\mathrm{v}\ell}=\frac{h_{\mathrm{m}}A_{\ell\mathrm{v}}\left(\omega_{\mathrm{sat}}-\omega_{\mathrm{v},i}\right)}{\Delta V(1-\omega_{\mathrm{sat}})} \tag{3.303}$$

$$\omega_{\mathrm{sat}}=\frac{p_0}{R_{\mathrm{g}}\langle T_{\mathrm{v}}\rangle^{\mathrm{v}}\langle\rho_{\mathrm{v}}\rangle^{\mathrm{v}}}\exp\left(-\frac{h_{\ell\mathrm{v}}}{R_{\mathrm{g}}}\left(\frac{1}{\langle T_{\mathrm{v}}\rangle^{\mathrm{v}}}-\frac{1}{T_0}\right)\right) \tag{3.304}$$

$$\dot{q}'''_{\mathrm{v}\ell}=\frac{h_{\mathrm{c}}A_{\ell\mathrm{v}}(T_{\mathrm{v}}-T_\ell)}{\Delta V} \tag{3.305}$$

$$\langle\mathbf{F}_{\mathrm{v}\ell}\rangle=-\langle\mathbf{F}_{\ell\mathrm{v}}\rangle=K_{\mathrm{v}\ell}\left(\langle\mathbf{V}_{\mathrm{v}}\rangle^{\mathrm{v}}-\langle\mathbf{V}_\ell\rangle^\ell\right) \tag{3.306}$$

$$\langle\rho_{\mathrm{v}}\rangle^{\mathrm{v}}=\frac{P}{R_{\mathrm{g}}\langle T_{\mathrm{v}}\rangle^{\mathrm{v}}} \tag{3.307}$$

The surface-to-volume ratio can be calculated by assuming that the liquid is in the form of spherical droplets. Therefore, the surface-to-volume ratio is defined in terms of the liquid droplet diameter, d_ℓ, and the volume fraction of the liquid.

$$\frac{A_{\ell\mathrm{v}}}{\Delta V}=\varepsilon_\ell\frac{6}{d_\ell} \tag{3.308}$$

There are 15 equations and 16 unknowns (9 equations and 6 constitutive relations). To solve this problem, it is assumed that the liquid is in droplets, and there is no pressure gradient within each droplet. Therefore, the liquid pressure gradient is assumed to be equal to the vapor pressure gradient.

$$\frac{\partial\langle p_\ell\rangle^\ell}{\partial x}=\frac{\partial\langle p_{\mathrm{v}}\rangle^{\mathrm{v}}}{\partial x},\quad\frac{\partial\langle p_\ell\rangle^\ell}{\partial y}=\frac{\partial\langle p_{\mathrm{v}}\rangle^{\mathrm{v}}}{\partial y} \tag{3.309}$$

Now, the liquid pressure is eliminated; therefore, we have 13 equations and 13 unknowns, which are well posed.

The boundary conditions are:

At walls

$$\langle u_v\rangle^v = \langle v_v\rangle^v = \langle u_\ell\rangle^\ell = \langle v_\ell\rangle^\ell = 0 \tag{3.310}$$

$$\frac{\partial\langle T_\ell\rangle^\ell}{\partial x} = \frac{\partial\langle T_v\rangle^v}{\partial x} = \frac{\partial\langle \omega_v\rangle^v}{\partial x} = 0 \tag{3.311}$$

At air inlet

$$\langle p_v\rangle^v = \langle \rho_v\rangle^v gh \tag{3.312}$$

$$\langle T_v\rangle^v = T_{in} \tag{3.313}$$

$$\varepsilon_\ell = 0 \tag{3.314}$$

At top

$$\langle T_\ell\rangle^\ell = T_{top} \tag{3.315}$$

$$\varepsilon_\ell = \varepsilon_{\ell,in} \tag{3.316}$$

$$\langle p_v\rangle^v = 0 \tag{3.317}$$

At symmetry

$$\frac{\partial\phi}{\partial x} = 0, \quad \phi = \langle v_k\rangle^k, \langle T_k\rangle^k, \varepsilon_\ell, \langle p_k\rangle^k, \langle \omega_v\rangle^v \tag{3.318}$$

$$\langle u_k\rangle^k = 0, \quad k = \ell, v \tag{3.319}$$

At bottom surface

$$\langle \omega_v\rangle^v = \omega_{v,sat} \tag{3.320}$$

$$-\frac{\varepsilon_v\langle \rho_v\rangle^v D_{v,i,air}}{1-\langle \omega_v\rangle^v}\frac{\partial\langle \omega_v\rangle^v}{\partial y} = \varepsilon_v\langle \rho_v\rangle^v\langle v_v\rangle^v \tag{3.321}$$

$$-\varepsilon_v k_v\frac{\partial\langle T_v\rangle^v}{\partial y} = \varepsilon_v\langle \rho_v\rangle^v\langle v_v\rangle^v h_{\ell v} \tag{3.322}$$

$$\langle u_v\rangle^v = \langle u_\ell\rangle^\ell \approx 0 \tag{3.323}$$

$$\langle T_v\rangle^v = \langle T_\ell\rangle^\ell \tag{3.324}$$

3.4.3 Homogeneous Model

The multifluid model presented above is obtained by performing phase averaging as defined in Eqs. (3.248) and (3.249). If spatial averaging is performed for all phases within a multiphase control volume, the homogeneous model can be obtained. The relationship between volume averaging and phase averaging is given in Eq. (3.251), which indicates that the homogeneous model can be obtained by summing the individual phase equations of the multifluid model.

3.4.3.1 Continuity Equation

The continuity equation for phase k in the multifluid model is expressed by Eq. (3.264). Summing the continuity equations for all Π phases together, one obtains

$$\frac{\partial}{\partial t}\left(\sum_{k=1}^{\Pi}\varepsilon_k\langle\rho_k\rangle^k\right)+\nabla\cdot\sum_{k=1}^{\Pi}\varepsilon_k\langle\rho_k\rangle^k\langle\mathbf{V}_k\rangle^k=\sum_{k=1}^{\Pi}\sum_{j=1(j\neq k)}^{\Pi}(\dot{m}_{jk}'''-\dot{m}_{jk}''') \tag{3.325}$$

The right-hand side of Eq. (3.325) must be zero because the total mass of all phases produced by phase change must equal the total mass of all phases consumed by phase change. Considering this fact and Eq. (3.262), the continuity equation becomes

$$\frac{\partial\langle\rho\rangle}{\partial t}+\nabla\cdot\sum_{k=1}^{\Pi}\varepsilon_k\langle\rho_k\rangle^k\langle\mathbf{V}_k\rangle^k=0 \tag{3.326}$$

The bulk velocity of the multiphase mixture is the mass-averaged velocity of all the individual phases:

$$\tilde{\mathbf{V}}=\frac{1}{\langle\rho\rangle}\sum_{k=1}^{\Pi}\varepsilon_k\langle\rho_k\rangle^k\langle\mathbf{V}_k\rangle^k \tag{3.327}$$

Substituting Eq. (3.327) into Eq. (3.326), the final form of the continuity equation for a multiphase mixture is

$$\frac{\partial\langle\rho\rangle}{\partial t}+\nabla\cdot\left(\langle\rho\rangle\tilde{\mathbf{V}}\right)=0 \tag{3.328}$$

It can be seen that Eq. (3.328) has the same form as the local continuity Eq. (3.48), except that the volume-averaged density and velocity are used in Eq. (3.328), where $\langle\rho\rangle=\sum_{k=1}^{\Pi}\varepsilon_k\langle\rho_k\rangle^k$.

3.4.3.2 Momentum Equation

The momentum equation for phase k in the multifluid model is expressed in Eq. (3.272). By adding together the momentum equations for all Π phases, one obtains

$$\begin{aligned}\frac{\partial}{\partial t}\left(\sum_{k=1}^{\Pi}\varepsilon_k\langle\rho_k\rangle^k\langle\mathbf{V}_k\rangle^k\right)+\nabla\cdot\left(\sum_{k=1}^{\Pi}\varepsilon_k\langle\rho_k\rangle^k\langle\mathbf{V}_k\mathbf{V}_k\rangle^k\right)=\nabla\cdot\sum_{k=1}^{\Pi}\varepsilon_k\langle\boldsymbol{\tau}_k'\rangle^k\\+\sum_{k=1}^{\Pi}\left(\varepsilon_k\langle\rho_k\rangle^k\right)\mathbf{X}+\sum_{k=1}^{\Pi}\sum_{j=1(j\neq k)}^{\Pi}\left(\langle\mathbf{F}_{jk}\rangle+\dot{m}_{jk}'''\langle\mathbf{V}_{k,\mathrm{I}}\rangle^k-\dot{m}_{kj}'''\langle\mathbf{V}_{j,\mathrm{I}}\rangle^{k}\right)\end{aligned} \tag{3.329}$$

The stress tensor of the multiphase mixture is

$$\langle \boldsymbol{\tau}' \rangle = \sum_{k=1}^{\Pi} \varepsilon_k \langle \boldsymbol{\tau}_k \rangle^k = -\langle p \rangle \mathbf{I} + \mu \left[\nabla \tilde{\mathbf{V}} + \nabla \tilde{\mathbf{V}}^T \right] - \frac{2}{3} \mu (\nabla \cdot \tilde{\mathbf{V}}) \mathbf{I} \tag{3.330}$$

The summation of all interphase forces must be zero since $\langle \mathbf{F}_{jk} \rangle = -\langle \mathbf{F}_{kj} \rangle$, i.e.,

$$\sum_{k=1}^{\Pi} \sum_{j=1 (j \neq k)}^{\Pi} \langle \mathbf{F}_{jk} \rangle = 0 \tag{3.331}$$

Considering Eqs. (3.327), (3.330), and (3.331), the momentum equation becomes

$$\frac{\partial}{\partial t} (\langle \rho \rangle \tilde{\mathbf{V}}) + \nabla \cdot \left(\sum_{k=1}^{\Pi} \varepsilon_k \langle \rho_k \rangle^k \langle \mathbf{V}_k \mathbf{V}_k \rangle^k \right) = \nabla \cdot \langle \boldsymbol{\tau}' \rangle + \langle \rho \rangle \mathbf{X} + \dot{\mathbf{M}}_{\mathrm{I}}''' \tag{3.332}$$

where

$$\dot{\mathbf{M}}_{\mathrm{I}}''' = \sum_{k=1}^{\Pi} \sum_{j=1 (j \neq k)}^{\Pi} \left(\dot{m}_{jk}''' \langle \mathbf{V}_{k,\mathrm{I}} \rangle^k - \dot{m}_{kj}''' \langle \mathbf{V}_{j,\mathrm{I}} \rangle^k \right) \tag{3.333}$$

Equation (3.333) represents the momentum production rate due to interaction between different phases along their separating interfaces. It must be specified according to the combination of phases in the multiphase system that is under consideration.

3.4.3.3 Energy Equation

By summing the energy equations for all Π phases in the multifluid model, Eq. (3.285), one obtains

$$\begin{aligned} &\frac{\partial}{\partial t} \left(\sum_{k=1}^{\Pi} \varepsilon_k \langle \rho_k \rangle^k \langle h_k \rangle^k \right) + \nabla \cdot \left(\sum_{k=1}^{\Pi} \varepsilon_k \langle \rho_k \rangle^k \langle \mathbf{V}_k h_k \rangle^k \right) = -\nabla \cdot \left(\sum_{k=1}^{\Pi} \langle \mathbf{q}_k'' \rangle \right) + \sum_{k=1}^{\Pi} \langle q_k''' \rangle \\ &+ \sum_{k=1}^{\Pi} \varepsilon_k \frac{D \langle p_k \rangle^k}{Dt} + \sum_{k=1}^{\Pi} \nabla \langle \mathbf{V}_k \rangle^k : \varepsilon_k \langle \boldsymbol{\tau}_k \rangle^k + \sum_{k=1}^{\Pi} \sum_{j=1 (j \neq k)}^{\Pi} \langle q_{jk}''' \rangle \\ &+ \sum_{k=1}^{\Pi} \sum_{j=1 (j \neq k)}^{\Pi} \left(\dot{m}_{jk}''' \langle h_{k,\mathrm{I}} \rangle^k - \dot{m}_{kj}''' \langle h_{j,\mathrm{I}} \rangle^k \right) \end{aligned} \tag{3.334}$$

The mass average enthalpy of the multiphase mixture is

$$\tilde{h} = \frac{1}{\langle \rho \rangle} \sum_{k=1}^{\Pi} \varepsilon_k \langle \rho_k \rangle^k \langle h_k \rangle^k \tag{3.335}$$

The fifth term on the right-hand side of Eq. (3.334) is for summation of all interphase heat transfer, and it must be zero. The last term on the right-hand side of Eq. (3.334) accounts for contribution of interphase phase-change energy flux due to phase change; it can be defined as

$$\sum_{k=1}^{\Pi}\sum_{j=1(j\neq k)}^{\Pi}\left(\dot{m}_{jk}'''\left\langle h_{k,\mathrm{I}}\right\rangle^{k}-\dot{m}_{kj}'''\left\langle h_{j,\mathrm{I}}\right\rangle^{k}\right)=q_{\mathrm{I}}''' \tag{3.336}$$

It is usually not zero although $\sum_{k=1}^{\Pi}\sum_{j=1(j\neq k)}^{\Pi}(\dot{m}_{jk}'''-\dot{m}_{kj}''')=0$.

Considering Eqs. (3.335) and (3.336), the energy Eq. (3.334) becomes

$$\frac{\partial}{\partial t}(\langle\rho\rangle\tilde{h})+\nabla\cdot\left(\sum_{k=1}^{\Pi}\varepsilon_k\langle\rho_k\rangle^k\langle\mathbf{V}_k h_k\rangle^k\right)=-\nabla\cdot\langle\mathbf{q}''\rangle+\frac{D\langle p\rangle}{Dt}+\langle q'''\rangle+\nabla\tilde{\mathbf{V}}:\langle\boldsymbol{\tau}\rangle+q_{\mathrm{I}}''' \tag{3.337}$$

3.4.3.4 Species

Summing the equations for conservation of species mass, Eq. (3.289), for all phases yields

$$\begin{aligned}&\frac{\partial}{\partial t}\left(\sum_{k=1}^{\Pi}\varepsilon_k\left\langle\rho_{k,i}\right\rangle^k\right)+\nabla\cdot\left(\sum_{k=1}^{\Pi}\varepsilon_k\left\langle\rho_{k,i}\right\rangle^k\langle\mathbf{V}_k\rangle^k\right)\\&=-\nabla\cdot\left(\sum_{k=1}^{\Pi}\left\langle\mathbf{J}_{k,i}\right\rangle\right)+\sum_{k=1}^{\Pi}\left\langle\dot{m}_{k,i}'''\right\rangle+\sum_{k=1}^{\Pi}\sum_{j=1(j\neq k)}^{\Pi}(\dot{m}_{jk,i}'''-\dot{m}_{kj,i}''')\end{aligned} \tag{3.338}$$

By applying Eq. (3.262) to the mass density of the ith component, one obtains

$$\langle\rho_i\rangle=\sum_{k=1}^{\Pi}\varepsilon_k\left\langle\rho_{k,i}\right\rangle^k \tag{3.339}$$

In accordance with the conservation of mass, the mass source (or sink) of the ith component due to phase change in all phases must add up to zero, i.e.,

$$\sum_{k=1}^{\Pi}\sum_{j=1(j\neq k)}^{\Pi}(\dot{m}_{jk,i}'''-\dot{m}_{kj,i}''')=0 \tag{3.340}$$

Substituting Eqs. (3.339) and (3.340) into Eq. (3.338), and using the mass-averaged velocity defined in Eq. (3.327), the conservation of species mass becomes

$$\frac{\partial\langle\rho_i\rangle}{\partial t}+\nabla\cdot\langle\rho_i\rangle\tilde{\mathbf{V}}=-\nabla\cdot\mathbf{J}_i+\dot{m}_i''' \tag{3.341}$$

Example 3.11

Solve problem in Example 3.10 using the homogeneous model and the same assumptions made in Example 3.10. The water droplets are small; therefore, make further reasonable assumptions to find a relationship between the liquid velocity and vapor velocity so that the homogeneous model can be applied. The flow variables to be solved are the mass-averaged velocities, $\tilde{u}$ and $\tilde{v}$, as well as the average temperature, $\langle T\rangle$.

Solution Use the same assumptions listed in Example 3.10. To make this problem a homogenous problem, more assumptions need to be made. The inertial terms in the liquid momentum equation can be assumed to be negligible because the droplets are very small. It can also be assumed that the evaporation rates and pressure gradients do not affect liquid momentum equation. Also, assume that the temperature of the liquid and vapor is in local equilibrium, $\langle T_\ell \rangle^\ell = \langle T_v \rangle^v$.
Based on the assumptions above, the liquid momentum equation in the *x-direction* reduces to a statement that the velocities of the liquid and vapor are equivalent.

$$\langle u_\ell \rangle^\ell = \langle u_v \rangle^v = \tilde{u} \tag{3.342}$$

Similarly, the liquid momentum equation in the *y-direction* reduces to a statement that drag force between the liquid and vapor is equivalent to the gravitational force on the liquid droplets.

$$\langle v_\ell \rangle^\ell = \langle v_v \rangle^v - \frac{\varepsilon_\ell \langle \rho_\ell \rangle^\ell g}{K_{v\ell}} \tag{3.343}$$

Therefore, the *y-component* of the mass-averaged velocity is

$$\tilde{v} = \frac{\left(\varepsilon_v \langle \rho_v \rangle^v + \varepsilon_\ell \langle \rho_\ell \rangle^\ell\right) \langle v_v \rangle^v - \left(\varepsilon_\ell^2 \langle \rho_\ell \rangle^\ell\right)^2 g K_{v\ell}^{-1}}{\left(\varepsilon_v \langle \rho_v \rangle^v + \varepsilon_\ell \langle \rho_\ell \rangle^\ell\right)} = \langle v_v \rangle^v - \frac{\varepsilon_\ell^2 \left(\langle \rho_\ell \rangle^\ell\right)^2 g}{\langle \rho \rangle K_{v\ell}} \tag{3.344}$$

Therefore, the continuity equations, Eq. (3.328), for the liquid phase and the mixture are:
Liquid

$$\frac{\partial}{\partial x}(\varepsilon_\ell \rho_\ell \tilde{u}) + \frac{\partial}{\partial y}\left(\varepsilon_\ell \rho_\ell \tilde{v} + \frac{\varepsilon_\ell^2 \langle \rho_\ell \rangle^{\ell 2} g}{K_{v\ell}} \left(\frac{\varepsilon_\ell \langle \rho_\ell \rangle^\ell}{\langle \rho \rangle} - 1\right)\right) = \dot{m}_{v\ell}''' \tag{3.345}$$

Mixture

$$\frac{\partial}{\partial x}(\langle \rho \rangle \tilde{u}) + \frac{\partial}{\partial y}(\langle \rho \rangle \tilde{v}) = 0 \tag{3.346}$$

The homogeneous momentum equations, Eq. (3.332), are:

$$\begin{aligned} &\varepsilon_v \langle \rho_v \rangle^v \tilde{u} \frac{\partial \tilde{u}}{\partial x} + \varepsilon_v \langle \rho_v \rangle^v \left(\tilde{v} + \frac{\varepsilon_\ell^2 \left(\langle \rho_\ell \rangle^\ell\right)^2 g}{\langle \rho \rangle K_{v\ell}}\right) \frac{\partial \tilde{u}}{\partial y} \\ &= -\frac{\partial \langle p \rangle}{\partial x} + \frac{\partial}{\partial x}\left(\varepsilon_v \mu_v \frac{\partial \tilde{u}}{\partial x}\right) + \frac{\partial}{\partial y}\left(\varepsilon_v \mu_v \frac{\partial \tilde{u}}{\partial y}\right) \end{aligned} \tag{3.347}$$

$$\varepsilon_{\mathrm{v}}\langle\rho_{\mathrm{v}}\rangle^{\mathrm{v}}\tilde{u}\frac{\partial}{\partial x}\left(\tilde{v}+\frac{\varepsilon_{\ell}^{2}\left(\langle\rho_{\ell}\rangle^{\ell}\right)^{2}g}{K_{\mathrm{v}\ell}}\right)+\varepsilon_{\mathrm{v}}\langle\rho_{\mathrm{v}}\rangle^{\mathrm{v}}\left(\tilde{v}+\frac{\varepsilon_{\ell}^{2}\left(\langle\rho_{\ell}\rangle^{\ell}\right)^{2}g}{\langle\rho\rangle K_{\mathrm{v}\ell}}\right)$$
$$\cdot\frac{\partial}{\partial y}\left(\tilde{v}+\frac{\varepsilon_{\ell}^{2}\left(\langle\rho_{\ell}\rangle^{\ell}\right)^{2}g}{\langle\rho\rangle K_{\mathrm{v}\ell}}\right)=-\frac{\partial\langle p\rangle}{\partial y}+\frac{\partial}{\partial x}\left(\varepsilon_{\mathrm{v}}\mu_{\mathrm{v}}\frac{\partial}{\partial x}\left(\tilde{v}+\frac{\varepsilon_{l}^{2}\left(\langle\rho_{\ell}\rangle^{\ell}\right)^{2}g}{\langle\rho\rangle K_{\mathrm{v}\ell}}\right)\right) \tag{3.348}$$
$$+\frac{\partial}{\partial y}\left(\varepsilon_{\mathrm{v}}\mu_{\mathrm{v}}\frac{\partial}{\partial y}\left(\tilde{v}+\frac{\varepsilon_{l}^{2}\left(\langle\rho_{\ell}\rangle^{\ell}\right)^{2}g}{\langle\rho\rangle K_{\mathrm{v}\ell}}\right)\right)-\varepsilon_{\mathrm{v}}\rho_{\mathrm{v}}g$$

The homogenous energy equation, Eq. (3.337), is:

$$\tilde{u}\left(\varepsilon_{\mathrm{v}}\langle\rho_{\mathrm{v}}\rangle^{\mathrm{v}}\frac{\partial\langle h_{\mathrm{v}}\rangle^{\mathrm{v}}}{\partial x}+\varepsilon_{\ell}\langle\rho_{\ell}\rangle^{\ell}\frac{\partial\langle h_{\ell}\rangle^{\ell}}{\partial x}\right)+\varepsilon_{\mathrm{v}}\langle\rho_{\mathrm{v}}\rangle^{\mathrm{v}}\left(\tilde{v}+\frac{\varepsilon_{\ell}^{2}\left(\langle\rho_{\ell}\rangle^{\ell}\right)^{2}g}{\langle\rho\rangle K_{\mathrm{v}\ell}}\right)\frac{\partial\langle h_{\mathrm{v}}\rangle^{\mathrm{v}}}{\partial y}$$
$$+\varepsilon_{\ell}\langle\rho_{\ell}\rangle^{\ell}\left(\tilde{v}+\frac{\varepsilon_{\ell}\langle\rho_{\ell}\rangle^{\ell}g}{K_{\mathrm{v}\ell}}\left(\frac{\varepsilon_{\ell}\langle\rho_{\ell}\rangle^{\ell}}{\langle\rho\rangle}-1\right)\right)\frac{\partial h_{\ell}}{\partial y}=\frac{\partial}{\partial x}\left((\varepsilon_{\mathrm{v}}k_{\mathrm{v}}+\varepsilon_{\ell}k_{\ell})\frac{\partial\langle T\rangle}{\partial x}\right) \tag{3.349}$$
$$+\frac{\partial}{\partial y}\left((\varepsilon_{\mathrm{v}}k_{\mathrm{v}}+\varepsilon_{\ell}k_{\ell})\frac{\partial\langle T\rangle}{\partial y}\right)+\left\langle\dot{m}_{\mathrm{v}\ell}'''h_{\ell\mathrm{v}}\right\rangle$$

The enthalpies of the liquid and the vapor are:

$$\langle h_{\ell}\rangle^{\ell}=\int c_{\mathrm{p}\ell}\mathrm{d}T \tag{3.350}$$

$$\langle h_{\mathrm{v}}\rangle^{\mathrm{v}}=\int c_{\mathrm{pv}}\mathrm{d}T \tag{3.351}$$

There are multiple components in the vapor phase; therefore, the specific heat is a function of the mass fraction of those components.

$$c_{\mathrm{pv}}=\sum_{i=1}^{N}\omega_{i}c_{\mathrm{p}i} \tag{3.352}$$

The homogenous species equation, Eq. (3.341), is:

$$\varepsilon_{\mathrm{v}}\langle\rho_{\mathrm{v}}\rangle^{\mathrm{v}}\tilde{u}\frac{\partial\langle\omega_{\mathrm{v}}\rangle^{\mathrm{v}}}{\partial x}+\varepsilon_{\mathrm{v}}\langle\rho_{\mathrm{v}}\rangle^{\mathrm{v}}\left(\tilde{v}+\frac{\varepsilon_{\ell}^{2}\left(\langle\rho_{\ell}\rangle^{\ell}\right)^{2}g}{\langle\rho\rangle K_{\mathrm{v}\ell}}\right)\frac{\partial\langle\omega_{\mathrm{v}}\rangle^{\mathrm{v}}}{\partial y}$$
$$=\frac{\partial}{\partial x}\left(\varepsilon_{\mathrm{v}}D_{\mathrm{v,air}}\frac{\partial\langle\omega_{\mathrm{v}}\rangle^{\mathrm{v}}}{\partial x}\right)+\frac{\partial}{\partial y}\left(\varepsilon_{\mathrm{v}}D_{\mathrm{v,air}}\frac{\partial\langle\omega_{\mathrm{v}}\rangle^{\mathrm{v}}}{\partial y}\right)+\left\langle\dot{m}_{\ell\mathrm{v}}'''\right\rangle \tag{3.353}$$

Also,

$$\varepsilon_\ell + \varepsilon_v = 1 \quad (3.354)$$

$$\dot{m}'''_{v\ell} = \frac{h_m A_{\ell v}(\omega_{sat} - \langle\omega_v\rangle^v)}{\Delta V(1 - \omega_{sat})} \quad (3.355)$$

$$\omega_{sat} = \frac{p_0}{R_g \langle T_v\rangle^v \langle\rho_v\rangle^v} \exp\left(-\frac{h_{\ell v}}{R_g}\left(\frac{1}{\langle T_v\rangle^v} - \frac{1}{T_0}\right)\right) \quad (3.356)$$

$$\langle\rho_v\rangle^v = \frac{P}{R_g \langle T_v\rangle^v} \quad (3.357)$$

The surface-to-volume ratio can be calculated by assuming that the liquid is in the form of spherical droplets. Therefore, the surface-to-volume ratio is defined in terms of the liquid droplet diameter, d_ℓ, and the volume fraction of the liquid.

$$\frac{A_{\ell v}}{\Delta V} = \varepsilon_\ell \frac{6}{d_\ell} \quad (3.358)$$

The problem was reduced to six equations with three constitutive relations from nine equations. In total, the present formulation has nine equations and nine unknowns and is well posed.

At walls

$$\tilde{u} = \tilde{v} = 0 \quad (3.359)$$

$$\frac{\partial\langle T\rangle}{\partial x} = \frac{\partial\langle\omega_v\rangle^v}{\partial x} = 0 \quad (3.360)$$

At top

$$\langle T\rangle = T_{top} \quad (3.361)$$

$$\langle p\rangle = 0 \quad (3.362)$$

$$\varepsilon_\ell = \varepsilon_{\ell,in} \quad (3.363)$$

At air inlet

$$\langle p\rangle = \langle\rho_v\rangle^v gh \quad (3.364)$$

$$\langle T\rangle = T_{in} \quad (3.365)$$

$$\varepsilon_\ell = 0 \quad (3.366)$$

At symmetry

$$\frac{\partial\phi}{\partial x} = 0,\ \phi = \tilde{v}, \langle T\rangle, \varepsilon_\ell, \langle\omega_v\rangle^v \quad (3.367)$$

$$\tilde{u} = 0 \quad (3.368)$$

At bottom surface

$$\langle \omega_{\mathrm{v}} \rangle^{\mathrm{v}} = \omega_{\mathrm{v,sat}} \tag{3.369}$$

$$\langle \rho \rangle \tilde{v} = -\left(\varepsilon_{\mathrm{v}} \langle \rho_{\mathrm{v}} \rangle^{\mathrm{v}} + \varepsilon_{\ell} \langle \rho_{\ell} \rangle^{\ell} \right) \frac{D_{\mathrm{v,air}}}{1 - \langle \omega_{\mathrm{v}} \rangle^{\mathrm{v}}} \frac{\partial \langle \omega_{\mathrm{v}} \rangle^{\mathrm{v}}}{\partial y} - \varepsilon_{\ell} \langle \rho_{\ell} \rangle^{\ell} \frac{\varepsilon_{\ell} \langle \rho_{\ell} \rangle^{\ell} g}{K_{\mathrm{v}\ell}} \tag{3.370}$$

$$(\varepsilon_{\mathrm{v}} k_{\mathrm{v}} + \varepsilon_{\ell} k_{\ell}) \frac{\partial \langle T \rangle}{\partial y} = \varepsilon_{\mathrm{v}} \langle \rho_{\mathrm{v}} \rangle^{\mathrm{v}} \left(\tilde{v} + \frac{\varepsilon_{\ell}^2 \left(\langle \rho_{\ell} \rangle^{\ell} \right)^2 g}{\langle \rho \rangle K_{\mathrm{v}\ell}} \right) h_{\ell \mathrm{v}} \tag{3.371}$$

From the above problem formulation, the homogenous model can still be used for counterflowing phases. However, more assumptions were made to solve the problem as a homogeneous model. So the trade-off for reducing the number of equations is the accuracy of the solution.

3.4.4 Mixture Model

The homogeneous model can be used to model flows with very strong coupling and phases moving at the same velocity. The mixture model is a generalization of the homogeneous model, which includes the effect of relative velocities on the dispersed phases. Therefore, the mixture model can be used to model multiphase flows where the phases move at different velocities. However, local equilibrium is assumed over short distances. The phases in the mixture model are treated as interpenetrating continua, as in the multifluid model.

The mixture model uses relative velocities to describe the dispersed flows in the mixture momentum equation. Continuity, momentum, and energy equations for the mixture, volume fraction equations for the secondary phases, and algebraic expressions for the relative velocities (if the phases are moving at different velocities) are used to model n fluid or particulate phases.

3.4.4.1 Relative (Slip) Velocity and the Drift Velocity

The relative (slip) velocity is the velocity of the secondary phase (j) relative to the velocity of the primary phase (k):

$$\mathbf{V}_{jk} = \mathbf{V}_j - \mathbf{V}_k \tag{3.372}$$

where $\mathbf{V}$ is the velocity vector, and $\mathbf{V}_{jk}$ is the relative velocity of phase j with respect to phase k.

The drift velocity is the relative velocity minus the sum of the mass fractions multiplied by the relative velocities of each phase:

$$\mathbf{V}_{dr,j} = \mathbf{V}_{jk} - \sum_{k=1}^{\Pi} \bar{\omega}_k \mathbf{V}_{jk} \tag{3.373}$$

where $\mathbf{V}_{dr,j}$ is the drift velocity of phase j, and ω_k is the mass fraction of phase k; thus, $\omega_k = (\varepsilon_k \rho_k)/\rho_{\mathrm{m}}$.

An algebraic slip formulation can be used in the mixture model, which assumes that in order to prescribe an algebraic relation for the relative velocity, a local equilibrium must exist between phases over a short distance (Manninen et al. 1996). The relative velocity is then given by:

$$\mathbf{V}_{jk} = (t_{\mathrm{v}}/F_{drag})[(\rho_j - \rho_{\mathrm{m}})/\rho_j]\mathbf{a} \tag{3.374}$$

where t_{v} is the particle relaxation time (momentum response time, $t_{\mathrm{v}} = \left(\rho_j D_j^2\right)/18\mu_k$), and F_{drag} is the drag force (Schiller and Naumann 1935):

$$F_{\mathrm{drag}} = \begin{cases} 1 + 0.15Re^{0.687} & Re \leq 1000 \\ 0.0183Re & Re > 1000 \end{cases} \tag{3.375}$$

where Re is the Reynolds number, and $\boldsymbol{a}$ is the acceleration vector for the secondary phase,

$$\mathbf{a} = \mathbf{g} - (\tilde{\mathbf{V}} \cdot \nabla)\tilde{\mathbf{V}} - \frac{\partial \tilde{\mathbf{V}}}{\partial t} \tag{3.376}$$

where $\tilde{\mathbf{V}}$ is the mass-averaged velocity,

$$\tilde{\mathbf{V}} = \sum_{k=1}^{\Pi} (\varepsilon_k \rho_k \mathbf{V}_k) \Big/ \sum_{k=1}^{\Pi} (\varepsilon_k \rho_k) \tag{3.377}$$

and $\mathbf{g}$ is the gravitational acceleration, and ∇ is the Laplace operator vector.

The drift flux model is the simplest algebraic slip formulation. In this model, the acceleration of the particle is given by gravity and/or a centrifugal force. The particle relaxation time should be modified to take the presence of other particles into account.

3.4.4.2 Continuity Equation

The continuity equation for the mixture is:

$$\frac{\partial \langle \rho \rangle}{\partial t} + \nabla \cdot \left(\langle \rho \rangle \tilde{\mathbf{V}}\right) = 0 \tag{3.378}$$

The volume-averaged density and velocity are used in the continuity equation, where $\langle \rho \rangle = \sum_{k=1}^{\Pi} \varepsilon_k \langle \rho_k \rangle^k$.

3.4.4.3 Momentum Equation

The momentum equation for the mixture model as shown below is much like the homogeneous model except the last term is added to account for the effect of drift velocity (ANSYS Fluent Theory Guide 2017):

$$\frac{\partial}{\partial t}\left(\langle \rho \rangle \tilde{\mathbf{V}}\right) + \nabla \cdot \left(\sum_{k=1}^{\Pi} \varepsilon_k \langle \rho_k \rangle^k \langle \mathbf{V}_k \mathbf{V}_k \rangle^k\right) = \nabla \cdot \langle \boldsymbol{\tau}' \rangle + \langle \rho \rangle \mathbf{X} + \nabla \cdot \sum_{k=1}^{\Pi} \varepsilon_k \rho_k \mathbf{V}_{dr,k} \mathbf{V}_{dr,k} \tag{3.379}$$

3.4.4.4 Energy Equation

The energy equation for the mixture is:

$$\frac{\partial}{\partial t}\left(\langle\rho\rangle\tilde{h}\right)+\nabla\cdot\left(\sum_{k=1}^{\Pi}\varepsilon_k\langle\rho_k\rangle^k\langle\mathbf{V}_k h_k\rangle^k\right)=-\nabla\cdot\langle\mathbf{q}''\rangle+\frac{D\langle p\rangle}{Dt}+\langle q'''\rangle+\nabla\tilde{\mathbf{V}}:\langle\boldsymbol{\tau}\rangle+q_{\mathrm{I}}''' \quad (3.380)$$

3.4.4.5 Volume Fraction Equation for the Secondary Phases

The tracking of the interface(s) between the phases is accomplished by the solution of a continuity equation for the volume fraction of one (or more) of the phases. For the kth phase, this equation has the following form:

$$\frac{\partial}{\partial t}\left(\varepsilon_j\rho_j\right)+\nabla\cdot\left(\varepsilon_j\rho_j\tilde{\mathbf{V}}\right)=-\nabla\cdot\left(\varepsilon_j\rho_j\mathbf{V}_{dr,j}\right)+\sum_{k=1(j\neq k)}^{\Pi}\left(\dot{m}_{kj}'''-\dot{m}_{jk}'''\right) \quad (3.381)$$

where $\dot{m}_{kj}'''$ is mass transfer per unit volume from phase k to phase j.

Because the mixture model is a simpler model, it can be used in place of the full multifluid model in some cases. When there is a wide distribution of the particulate phase or when the interphase laws are unknown or unreliable, a full multifluid model may not be practical. The mixture model can perform as well as a full multifluid model with a smaller number of variables for some applications as stated below.

Mixture model applications include

- Particle-laden flows with low loading
- Bubbly flows with low gas volume fractions
- Sedimentation
- Cyclone separators.

3.4.5 Separated Model

The separated flow model is a surface tracking technique which tracks the volume fraction of each fluid in the domain by solving a single set of momentum equations and is used when the interface between two or more immiscible fluids is of interest.

3.4.5.1 Continuity Equation

The separated flow model solves the continuity equation for the volume fraction of each phase to track the interface between phases. For the kth phase, this equation is formed as:

$$\frac{1}{\langle\rho_k\rangle^k}\left[\frac{\partial}{\partial t}\left(\varepsilon_k\langle\rho_k\rangle^k\right)+\nabla\cdot\left(\varepsilon_k\langle\rho_k\rangle^k\langle\mathbf{V}_k\rangle^k\right)=\sum_{j=1(j\neq k)}^{\Pi}\left(\dot{m}_{jk}'''-\dot{m}_{kj}'''\right)\right] \quad (3.382)$$

Since the sum of the volume fraction of all the fluid phases present in each computational cell must sum to 1, the volume fraction is solved for each phase except the primary phase. The volume fraction of the primary phase is determined by solving for the volume fraction such that the sum of all volume fractions is 1:

$$\sum_{k=1}^{\Pi} \varepsilon_k = 1 \tag{3.383}$$

3.4.5.2 Momentum Equation

A single momentum equation is solved throughout the domain, which is dependent on the volume fractions of all the phases. The velocity field is shared among the phases. Therefore, this model may not be accurate in calculating velocities at the interface if there are large velocity differences between the phases. The properties in the momentum equation are calculated based on the phases in each control volume. For example, in a two-phase system, the mixture density is:

$$\rho_{\mathrm{m}} = \varepsilon_2 \rho_2 + (1 - \varepsilon_2)\rho_1 \tag{3.384}$$

where ρ_1 and ρ_2 represent the two phases present in the system, and the subscript m in ρ_m denotes the mixture property. The mixture density equation applies to all other properties, including viscosity.

The momentum equation has the form:

$$\begin{aligned} &\frac{\partial}{\partial t}\left(\langle\rho\rangle\tilde{\mathbf{V}}\right) + \nabla\cdot\left(\sum_{k=1}^{\Pi} \varepsilon_k \langle\rho_k\rangle^k \langle\mathbf{V}_k\mathbf{V}_k\rangle^k\right) \\ &= \nabla\cdot\boldsymbol{\tau}' + \langle\rho\rangle\mathbf{X} + \sum_{k=1}^{\Pi}\sum_{j=1(j\neq k)}^{\Pi}\left(\dot{m}'''_{jk}\langle\mathbf{V}_{k,\mathrm{I}}\rangle^k - \dot{m}'''_{kj}\langle\mathbf{V}_{j,\mathrm{I}}\rangle^k\right) \end{aligned} \tag{3.385}$$

3.4.5.3 Energy Equation

The energy equation, like the momentum equation, is shared among the phases. The energy equation has the form:

$$\begin{aligned} &\frac{\partial}{\partial t}\left(\langle\rho\rangle\tilde{h}\right) + \nabla\cdot\left(\sum_{k=1}^{\Pi} \varepsilon_k \langle\rho_k\rangle^k \langle\mathbf{V}_k h_k\rangle^k\right) \\ &= -\nabla\cdot\langle\mathbf{q}''_k\rangle + \langle q'''_k\rangle + \frac{D\langle p\rangle}{Dt} + \nabla\langle\ddot{\mathbf{V}}\rangle : \langle\boldsymbol{\tau}\rangle + q'''_{\mathrm{I}} \end{aligned} \tag{3.386}$$

The mixture model is different from the separated flow model because

- The mixture model allows the phases to be interpenetrating, and the volume fractions for the phases in a control volume can be any value between 0 and 1.
- The mixture model allows the phases to move at different velocities by using the concept of slip velocities. The phases can also be assumed to move at the same velocity, in which case the mixture model simplifies to the homogeneous model.

Separated model applications include

- Stratified flow
- Free-surface flow
- Filling
- Sloshing
- Motion of large bubbles in a liquid
- Motion of liquid after a dam break
- Prediction of jet breakup (surface tension)
- Steady or transient tracking of any liquid-gas interface.

The mixture model uses a single-fluid approach, as shown in the separated flow model. The separated model is also called the volume of fluid (VOF) model (ANSYS Fluent Theory Guide 2017).

3.4.6 Area-Averaged Model

The volume-averaged governing equations are discussed in detail in the preceding subsections. For channel flow, area-averaged governing equations are more useful. Area-averaged governing equations are obtained by performing area averaging of the governing equations over the *entire* cross section of the channel; this is distinct from volume averaging, wherein the average is performed over a small-volume ΔV. The resulting area-averaged governing equations will be one-dimensional in nature, rather than three-dimensional like volume-averaged equations. Therefore, information about the distribution of properties in the cross section is lost in the area-averaged governing equations. The jump conditions at the interfaces between two phases, as well as the exchange of momentum and energy between the channel wall and the fluid, must be expressed in the form of empirical correlations.

For a multiphase system that includes Π different phases flowing in a channel, the total cross-sectional area of the channel equals the summation of the areas occupied by individual phase, i.e.,

$$A_c = \sum_{k=1}^{\Pi} A_k \tag{3.387}$$

The volume fraction of the *k*th phase, ε_k, for channel flow is defined as the ratio of the cross-sectional area of the *k*th phase to the total area, i.e.,

$$\varepsilon_k = \frac{A_k}{A_c} \tag{3.388}$$

It follows from Eq. (3.387) that

$$\sum_{k=1}^{\Pi} \varepsilon_k = 1 \tag{3.389}$$

Eulerian area averaging is performed over the cross section of a channel, and the governing equations are reduced to one-dimensional

$$\langle \Phi \rangle = \frac{1}{A_c} \sum_{k=1}^{\Pi} \int_{A_k} \Phi_k \mathrm{d}A \tag{3.390}$$

If the average is performed for the kth phase only, the phase averages are obtained:

Intrinsic phase average

$$\langle \Phi_k \rangle^k = \frac{1}{A_k} \int_{A_k} \Phi_k \mathrm{d}A \tag{3.391}$$

Extrinsic phase average

$$\langle \Phi_k \rangle = \frac{1}{A_c} \int_{A_k} \Phi_k \mathrm{d}A \tag{3.392}$$

The area averaging and phase averaging defined in Eq. (3.390) and Eqs. (3.391) and (3.392) are related by

$$\langle \Phi \rangle = \sum_{k=1}^{\Pi} \langle \Phi_k \rangle = \sum_{k=1}^{\Pi} \varepsilon_k \langle \Phi_k \rangle^k \tag{3.393}$$

The area-averaged homogeneous model treats the flow as a homogeneous mixture with all phases flowing at the same area-averaged velocity. Since averaging is not performed in the axial direction, the governing equations are still local instantaneous in the axial direction. The basic equations for one-dimensional *homogeneous* equilibrium flow in a channel will be introduced here.

3.4.6.1 Homogeneous Model

The area-averaged continuity equation is

$$\frac{\partial \langle \rho \rangle}{\partial t} + \frac{\partial}{\partial z}(\langle \rho \rangle \langle w \rangle) = 0 \tag{3.394}$$

where $\langle \rho \rangle$ and $\langle w \rangle$ are, respectively, the area-averaged density and axial velocities of the multiphase mixture. The product of the deviations of density and velocity is neglected. For a mixture of liquid and vapor, it can be expressed as

$$\langle \rho \rangle = \alpha \langle \rho_v \rangle^v + (1-\alpha) \langle \rho_\ell \rangle^\ell \tag{3.395}$$

where α is the vapor or holdup void fraction, $\alpha = \varepsilon_v$, which represents the time-averaged volumetric fraction of vapor in a two-phase mixture.

The area-averaged momentum equation is

$$\langle \rho \rangle \left(\frac{\partial \langle w \rangle}{\partial t} + \langle w \rangle \frac{\partial \langle w \rangle}{\partial z} \right) = -\frac{\partial \langle p \rangle}{\partial z} - \langle \rho \rangle g \cos \theta - \frac{P}{A_c} \tau_w \tag{3.396}$$

where gravity is the only body force considered, τ_w is the average shear stress at the wall of the channel, P is the perimeter of the duct, θ is the inclination of the axial direction of the duct to the vertical, and A_c is the cross-sectional area of the channel.

The area-averaged energy equation is

$$\frac{\partial}{\partial t}\left[\langle\rho\rangle\left(\langle e\rangle+\frac{\langle w\rangle^2}{2}\right)\right]+\frac{\partial}{\partial z}\left[\langle\rho\rangle\langle w\rangle\left(\langle h\rangle+\frac{\langle w\rangle^2}{2}\right)\right] \\ =\frac{P}{A_c}q''_w-\langle\rho\rangle\langle w\rangle g\cos\theta \tag{3.397}$$

where $\langle e\rangle$ and $\langle h\rangle$ are the area-averaged internal energy and enthalpy of the multiphase flow, and q''_w is the heat flux at the wall of the channel. Equation (3.397) is the total energy (including internal energy, kinetic energy, and potential energy) balance for the channel. The mechanical energy balance can be obtained by multiplying Eq. (3.396) by $\langle w\rangle$, i.e.,

$$\langle\rho\rangle\left[\frac{\partial}{\partial t}\left(\frac{\langle w\rangle^2}{2}\right)+\langle w\rangle\frac{\partial}{\partial z}\left(\frac{\langle w\rangle^2}{2}\right)\right] \\ =-\frac{\partial\langle p\rangle}{\partial z}\langle w\rangle-\langle\rho\rangle g\langle w\rangle\cos\theta-\frac{P}{A_c}\tau_w\langle w\rangle \tag{3.398}$$

Subtracting Eq. (3.398) from Eq. (3.397) and replacing internal energy with enthalpy $(h=e+p/\rho)$, one obtains the final form of the energy equation:

$$\frac{\partial\langle h\rangle}{\partial t}+\langle w\rangle\frac{\partial\langle h\rangle}{\partial z}=\frac{1}{\langle\rho\rangle}\left(\frac{\partial\langle p\rangle}{\partial t}+\langle w\rangle\frac{\partial\langle p\rangle}{\partial z}\right)+\langle w\rangle\frac{P}{A_c}\frac{\tau_w}{\langle\rho\rangle}+\frac{Pq''_w}{A_c\langle\rho\rangle} \tag{3.399}$$

where the three terms on the right-hand side represent the effects of compressibility, viscous dissipation, and heat transfer from the wall.

The above governing equations are valid for a channel with constant cross-sectional area. The following example shows the derivation of the governing equations for a channel with variable cross-sectional area.

Example 3.12

Derive the continuity, momentum, and energy equations for the channel with an inclination angle of θ as shown in Fig. 3.14, using the area-averaged homogeneous model by making mass, momentum, and energy balances on a channel element. The cross-sectional area A_c is changing along the length.

Solution The governing equation for a channel with variable cross-sectional area can be obtained by applying various conservation laws to the channel element dz (control volume). For a channel element without mass generation, the mass balance in the channel can be written as

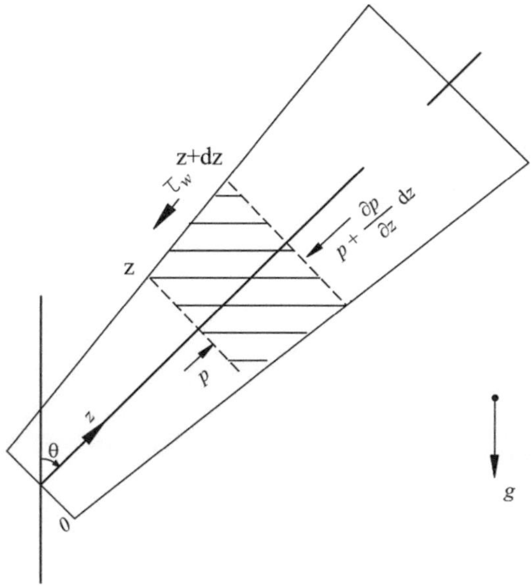

Fig. 3.14 Homogeneous model for a channel with variable cross section

$$\begin{bmatrix}\text{mass}\\\text{inflow}\\\text{rate}\end{bmatrix}-\begin{bmatrix}\text{mass}\\\text{outflow}\\\text{rate}\end{bmatrix}=\begin{bmatrix}\text{mass}\\\text{storage}\\\text{rate}\end{bmatrix}\tag{3.400}$$

The mass inflow rate is the mass flow rate at z, and it is $\langle\rho\rangle\langle w\rangle A_c$. The mass outflow rate is the mass flow rate at z + dz, written as $\langle\rho\rangle\langle w\rangle A_c+\frac{\partial}{\partial z}(\langle\rho\rangle\langle w\rangle A_c)\mathrm{d}z$. The mass storage rate is $A_c\mathrm{d}z\frac{\partial\langle\rho\rangle}{\partial t}$. Therefore, Eq. (3.400) becomes

$$\langle\rho\rangle\langle w\rangle A_c-\langle\rho\rangle\langle w\rangle A_c-\frac{\partial}{\partial z}(\langle\rho\rangle\langle w\rangle A_c)\mathrm{d}z-A_c\mathrm{d}z\frac{\partial\langle\rho\rangle}{\partial t}\tag{3.401}$$

which can be rearranged to obtain the continuity equation:

$$A_c\frac{\partial\langle\rho\rangle}{\partial t}+\frac{\partial}{\partial z}(\langle\rho\rangle\langle w\rangle A_c)=0\tag{3.402}$$

The momentum balance for the channel element dz can be written as

$$\begin{bmatrix}\text{momentum}\\\text{storage rate}\end{bmatrix}+\begin{bmatrix}\text{momentum}\\\text{outflow rate}\end{bmatrix}-\begin{bmatrix}\text{momentum}\\\text{inflow rate}\end{bmatrix}=\begin{bmatrix}\text{sum of forces acting}\\\text{on control volume}\end{bmatrix}\tag{3.403}$$

The momentum storage rate is $\partial(\langle\rho\rangle A_c\mathrm{d}z\langle w\rangle)/\partial t$. The momentum inflow rate is the momentum flow rate at z, and it is $\langle\rho\rangle\langle w\rangle A_c\langle w\rangle$. The momentum outflow rate is the momentum flow rate at z + dz, and it can be obtained by $\langle\rho\rangle\langle w\rangle A_c\langle w\rangle+\frac{\partial}{\partial z}(\langle\rho\rangle\langle w\rangle A_c\langle w\rangle)\mathrm{d}z$. The forces acting on the control volume include the pressure acting on surfaces z and z + dz, gravity, and the shear force acting on the wall of the channel, i.e.,

$$\begin{bmatrix}\text{sum of forces acting}\\ \text{on control volume}\end{bmatrix}$$
$$= \langle p\rangle A_c - \left[\langle p\rangle A_c + \frac{\partial(\langle p\rangle A_c)}{\partial z}\mathrm{d}z\right] - \langle\rho\rangle g\cos\theta A_c\mathrm{d}z - \tau_w\mathrm{d}zP$$

Therefore, Eq. (3.403) becomes

$$\begin{aligned}&\frac{\partial(\langle\rho\rangle A_c\mathrm{d}z\langle w\rangle)}{\partial t} + \left[\langle\rho\rangle\langle w\rangle A_c\langle w\rangle + \frac{\partial}{\partial z}(\langle\rho\rangle\langle w\rangle A_c\langle w\rangle)\mathrm{d}z\right] - \langle\rho\rangle\langle w\rangle A_c\langle w\rangle\\ &= \langle p\rangle A_c - \left[\langle p\rangle A_c + \frac{\partial(\langle p\rangle A_c)}{\partial z}\mathrm{d}z\right] - \langle\rho\rangle g\cos\theta A_c\mathrm{d}z - \tau_w\mathrm{d}zP\end{aligned} \tag{3.404}$$

The momentum equation is obtained by rearranging Eq. (3.404), i.e.,

$$A_c\frac{\partial(\langle\rho\rangle\langle w\rangle)}{\partial t} + \frac{\partial}{\partial z}(\langle\rho\rangle\langle w\rangle A_c\langle w\rangle) = -\frac{\partial(\langle p\rangle A_c)}{\partial z} - \langle\rho\rangle g\cos\theta A_c - \tau_w P \tag{3.405}$$

Rewriting the left-hand side of Eq. (3.405) yields

$$\begin{aligned}&\langle w\rangle\left[A_c\frac{\partial\langle\rho\rangle}{\partial t} + \frac{\partial}{\partial z}(\langle\rho\rangle A_c\langle w\rangle)\right] + A_c\langle\rho\rangle\left[\frac{\partial\langle w\rangle}{\partial t} + \langle w\rangle\frac{\partial\langle w\rangle}{\partial z}\right]\\ &= -\frac{\partial(\langle p\rangle A_c)}{\partial z} - \langle\rho\rangle g\cos\theta A_c - \tau_w P\end{aligned} \tag{3.406}$$

According to Eq. (3.402), the first bracket at the left-hand side of Eq. (3.406) is zero. Therefore, the momentum equation is

$$\langle\rho\rangle\left[\frac{\partial\langle w\rangle}{\partial t} + \langle w\rangle\frac{\partial\langle w\rangle}{\partial z}\right] = -\frac{1}{A_c}\frac{\partial(\langle p\rangle A_c)}{\partial z} - \langle\rho\rangle g\cos\theta - \tau_w\frac{P}{A_c} \tag{3.407}$$

The energy equation can be obtained by applying the conservation of energy to the control volume, i.e.,

$$\begin{bmatrix}\text{energy}\\ \text{inflow}\\ \text{rate}\end{bmatrix} - \begin{bmatrix}\text{energy}\\ \text{outflow}\\ \text{rate}\end{bmatrix} = \begin{bmatrix}\text{energy}\\ \text{storage}\\ \text{rate}\end{bmatrix} \tag{3.408}$$

The energy inflow rate includes the energy flow into the control volume at surface z, $\langle\rho\rangle A_c\langle w\rangle\left(\langle h\rangle + \langle w\rangle^2/2 + gz\cos\theta\right)$, and the heat transferred into the control volume through the wall, $q''_w P\mathrm{d}z$. The energy outflow rate is the energy flow out from the control volume at surface z + dz, and it is $\langle\rho\rangle A_c\langle w\rangle\left(\langle h\rangle + \langle w\rangle^2/2 + gz\cos\theta\right) + \frac{\partial}{\partial z}\left[\langle\rho\rangle A_c\langle w\rangle\left(\langle h\rangle + \langle w\rangle^2/2 + gz\cos\theta\right)\right]\mathrm{d}z$. The energy storage rate is $\frac{\partial}{\partial t}\left[\langle\rho\rangle A_c dz\left(\langle e\rangle + \langle w\rangle^2/2\right)\right]$. Thus, the energy balance for the channel element becomes

$$\begin{aligned}&\langle\rho\rangle A_c\langle w\rangle\left(\langle h\rangle + \frac{\langle w\rangle^2}{2} + gz\cos\theta\right) + q_w'' P dz\\&-\left\{\langle\rho\rangle A_c\langle w\rangle\left(\langle h\rangle + \frac{\langle w\rangle^2}{2} + gz\cos\theta\right)\right.\\&\left.+\frac{\partial}{\partial z}\left[\langle\rho\rangle A_c\langle w\rangle\left(\langle h\rangle + \frac{\langle w\rangle^2}{2} + gz\cos\theta\right)\right]dz\right\} = \frac{\partial}{\partial t}\left[\langle\rho\rangle A_c dz\left(\langle e\rangle + \frac{\langle w\rangle^2}{2}\right)\right]\end{aligned} \tag{3.409}$$

Rearranging Eq. (3.409) yields the energy equation:

$$\begin{aligned}&\frac{\partial}{\partial t}\left[\langle\rho\rangle\left(\langle e\rangle + \frac{\langle w\rangle^2}{2}\right)\right] + \frac{1}{A_c}\frac{\partial}{\partial z}\left[\langle\rho\rangle A_c\langle w\rangle\left(\langle h\rangle + \frac{\langle w\rangle^2}{2}\right)\right]\\&= \frac{P}{A_c}q_w'' - \langle\rho\rangle\langle w\rangle g\cos\theta\end{aligned} \tag{3.410}$$

When the channel's cross-sectional area is constant, Eqs. (3.402), (3.407), and (3.410) reduce to Eqs. (3.394), (3.396), and (3.397).

3.4.6.2 Separated Flow Model

In the area-averaged homogeneous model, all flow parameters—both velocities and fluid properties—are assumed to be the same across all phases. A slightly more complex approach, the separated flow model (Hewitt 1998) is the intermediate between the homogeneous and multifluid models. It accommodates selected parameters that differ between phases, but maintains the single set of conservation equations that is characteristic of the homogeneous model.

Recall that the two-fluid model, the simplest case of the multifluid model, recognizes that the two phases can have different properties and velocities. Eulerian phase averaging is performed for each individual phase, yielding separate equations of continuity, momentum, energy, entropy, and species mass for each phase. These equations are solved simultaneously and in conjunction with rate equations that describe how the phases interact with each other and the system boundaries.

A simple version of the two-fluid model is the separated flow model, in which only one parameter, such as velocity, is allowed to differ between the two phases. That parameter is obtained for each phase via phase averaging, as defined in Eq. (3.391). A single set of conservation equations is written for the combined flow, but it uses both velocities. Therefore, the separated flow model is neither multifluid nor homogeneous, but has features of both. The multifluid feature is the recognition of different velocities between phases. The homogeneous feature is the definition of only one governing equation for each of the quantities—mass, momentum, energy, entropy, and species mass—which applies to all phases and across the entire flow cross section. In addition, the pressure across any given cross section of a channel carrying a multiphase flow is assumed to be the same for both phases. The following example shows the separated flow model derivation of the governing equations for steady flow in a channel with variable cross-sectional area by using mass, momentum, and energy balance for an element.

Example 3.13

Derive the steady-state continuity, momentum, and energy equations for liquid-gas two-phase flow in the channel shown in Fig. 3.15 by using the area-averaged separated flow model as discussed above. The channel has a cross-sectional area of $A_c(z)$ and an inclination angle of θ.

Solution For a channel element without mass generation, the mass balance in the channel satisfies Eq. (3.400).

The mass inflow rate is the mass flow rate at z, and it is $\langle\rho_\ell\rangle^\ell\langle w_\ell\rangle^\ell A_\ell+\langle\rho_{\rm v}\rangle^{\rm v}\langle w_{\rm v}\rangle^{\rm v}A_{\rm v}$. The mass outflow rate is the mass flow rate at z + dz, and it can be obtained by $\langle\rho_\ell\rangle^\ell\langle w_\ell\rangle^\ell A_\ell+\langle\rho_{\rm v}\rangle^{\rm v}\langle w_{\rm v}\rangle^{\rm v}A_{\rm v}+\frac{\rm d}{{\rm d}z}(\langle\rho_\ell\rangle^\ell\langle w_\ell\rangle^\ell A_\ell+\langle\rho_{\rm v}\rangle^{\rm v}\langle w_{\rm v}\rangle^{\rm v}A_{\rm v}){\rm d}z$. The mass storage rate for steady-state flow is zero. Therefore, Eq. (3.400) becomes

$$\begin{aligned}\langle\rho_\ell\rangle^\ell\langle w_\ell\rangle^\ell A_\ell+\langle\rho_{\rm v}\rangle^{\rm v}\langle w_{\rm v}\rangle^{\rm v}A_{\rm v}-\Big[&\langle\rho_\ell\rangle^\ell\langle w_\ell\rangle^\ell A_\ell+\langle\rho_{\rm v}\rangle^{\rm v}\langle w_{\rm v}\rangle^{\rm v}A_{\rm v}\\ &+\frac{\rm d}{{\rm d}z}\Big(\langle\rho_\ell\rangle^\ell\langle w_\ell\rangle^\ell A_\ell+\langle\rho_{\rm v}\rangle^{\rm v}\langle w_{\rm v}\rangle^{\rm v}A_{\rm v}\Big){\rm d}z\Big]=0\end{aligned} \tag{3.411}$$

The above equation can be rearranged to obtain the continuity equation:

$$\frac{\rm d}{{\rm d}z}\Big(\langle\rho_\ell\rangle^\ell\langle w_\ell\rangle^\ell A_\ell+\langle\rho_{\rm v}\rangle^{\rm v}\langle w_{\rm v}\rangle^{\rm v}A_{\rm v}\Big)=0 \tag{3.412}$$

The momentum balance for the channel element dz satisfies Eq. (3.403), except that the momentum storage rate is zero for a steady-state flow. The momentum inflow rate is the momentum flow rate at z, and it is $\langle\rho_\ell\rangle^\ell\langle w_\ell\rangle^\ell A_\ell\langle w_\ell\rangle^\ell+\langle\rho_{\rm v}\rangle^{\rm v}\langle w_{\rm v}\rangle^{\rm v}A_{\rm v}\langle w_{\rm v}\rangle^{\rm v}$. The momentum

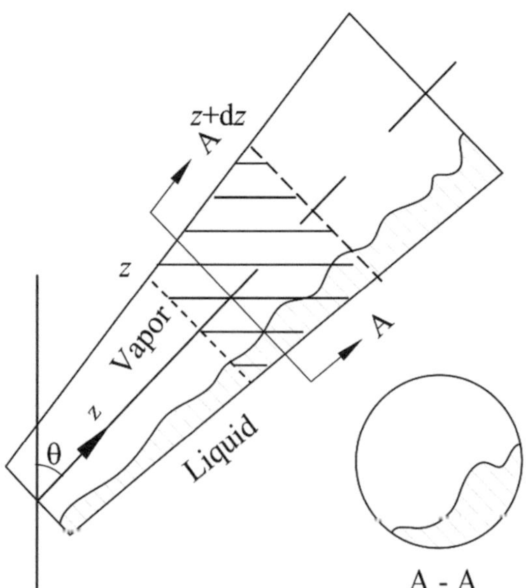

Fig. 3.15 Separated flow model for a channel with variable cross section

outflow rate is the momentum flow rate at z + dz, and it can be obtained by $\langle\rho_\ell\rangle^\ell\langle w_\ell\rangle^\ell A_\ell\langle w_\ell\rangle^\ell + \langle\rho_v\rangle^v\langle w_v\rangle^v A_v\langle w_v\rangle^v + \frac{d}{dz}(\langle\rho_\ell\rangle^\ell\langle w_\ell\rangle^\ell A_\ell\langle w_\ell\rangle^\ell + \langle\rho_v\rangle^v\langle w_v\rangle^v A_v\langle w_v\rangle^v)dz$. The forces acting on the control volume include pressures acting on surfaces z and z + dz, gravity, and the shear force on the wall of the channel, i.e.,

$$\begin{bmatrix}\text{sum of forces acting}\\ \text{on control volume}\end{bmatrix} = \langle p\rangle A_c - \left[\langle p\rangle A_c + \frac{d(\langle p\rangle A_c)}{dz}dz\right] - [\alpha\langle\rho_\ell\rangle^\ell + (1-\alpha)\langle\rho_v\rangle^v]g\cos\theta A_c dz - \tau_w dz P$$

Therefore, Eq. (3.403) becomes

$$\begin{aligned}\frac{d}{dz}\left(\langle\rho_\ell\rangle^\ell\langle w_\ell\rangle^\ell A_\ell\langle w_\ell\rangle^\ell + \langle\rho_v\rangle^v\langle w_v\rangle^v A_v\langle w_v\rangle^v\right)dz = \langle p\rangle A_c\\ - \left[\langle p\rangle A_c + \frac{d(\langle p\rangle A_c)}{dz}dz\right] - \left[\alpha\langle\rho_\ell\rangle^\ell + (1-\alpha)\langle\rho_v\rangle^v\right]g\cos\theta A_c dz - \tau_w dz P\end{aligned} \tag{3.413}$$

The momentum equation can be obtained by rearranging Eq. (3.413),

$$\begin{aligned}\frac{1}{A_c}\frac{d}{dz}\left(\langle\rho_\ell\rangle^\ell\langle w_\ell\rangle^\ell A_\ell\langle w_\ell\rangle^\ell + \langle\rho_v\rangle^v\langle w_v\rangle^v A_v\langle w_v\rangle^v\right)\\ = -\frac{1}{A_c}\frac{d(\langle p\rangle A_c)}{dz} - \frac{P}{A_c}\tau_w - \left[\alpha\langle\rho_\ell\rangle^\ell + (1-\alpha)\langle\rho_v\rangle^v\right]g\cos\theta\end{aligned} \tag{3.414}$$

The energy equation balance in the control volume satisfies Eq. (3.408), except that the energy storage rate is zero. The energy inflow rate includes energy flow into the control volume at surface z, $\langle\rho_\ell\rangle^\ell A_\ell\langle w_\ell\rangle^\ell\cdot(\langle h_\ell\rangle^\ell + \langle w_\ell\rangle^{\ell 2}/2 + gz\cos\theta) + \langle\rho_v\rangle^v A_v\langle w_v\rangle^v(\langle h_v\rangle^v + \langle w_v\rangle^{v2}/2 + gz\cos\theta)$, and heat transferred into the control volume through the wall, $q''_w P dz$. The energy outflow rate is the energy flow out from the control volume at surface z + dz, and it is $\langle\rho_\ell\rangle^\ell A_\ell\langle w_\ell\rangle^\ell(\langle h_\ell\rangle^\ell + \langle w_\ell\rangle^{\ell 2}/2 + gz\cos\theta) + \langle\rho_v\rangle^v A_v\langle w_v\rangle^v(\langle h_v\rangle^v + \langle w_v\rangle^{v2}/2 + gz\cos\theta) + \frac{d}{dz}[\langle\rho_\ell\rangle^\ell A_\ell\langle w_\ell\rangle^\ell\cdot(\langle h_\ell\rangle^\ell + \langle w_\ell\rangle^{\ell 2}/2 + gz\cos\theta) + \langle\rho_v\rangle^v A_v\langle w_v\rangle^v(\langle h_v\rangle^v + \langle w_v\rangle^{v2}/2 + gz\cos\theta)]dz$. Thus, the energy balance for the channel element becomes

$$\begin{aligned}q''_w P - \frac{d}{dz}\left[\langle\rho_\ell\rangle^\ell A_\ell\langle w_\ell\rangle^\ell\left(\langle h_\ell\rangle^\ell + \frac{\langle w_\ell\rangle^{\ell 2}}{2} + gz\cos\theta\right)\right.\\ \left. + \langle\rho_v\rangle^v A_v\langle w_v\rangle^v\left(\langle h_v\rangle^v + \frac{\langle w_v\rangle^{v2}}{2} + gz\cos\theta\right)\right] = 0\end{aligned} \tag{3.415}$$

Rearranging Eq. (3.415) yields the following energy equation:

$$\begin{aligned}\frac{d}{dz}\left[\langle\rho_\ell\rangle^\ell A_\ell\langle w_\ell\rangle^\ell\left(\langle h_\ell\rangle^\ell + \frac{\langle w_\ell\rangle^{\ell 2}}{2}\right) + \langle\rho_v\rangle^v A_v\langle w_v\rangle^v\left(\langle h_v\rangle^v + \frac{\langle w_v\rangle^{v2}}{2}\right)\right]\\ = q''_w P - \left(\langle\rho_\ell\rangle^\ell A_\ell\langle w_\ell\rangle^\ell + \langle\rho_v\rangle^v A_v\langle w_v\rangle^v\right)g\cos\theta\end{aligned} \tag{3.416}$$

3.4.7 Guidelines and Comparison of Multiphase Models

3.4.7.1 Guidelines

The multiphase flow modeling approaches discussed thus far are based on the Euler–Euler approach, which models all phases in the multiphase system based on Eulerian approach. Another type of approach is the Euler–Lagrange approach, which describes the continuous phase with the Euler approach, while the particles are described using the Lagrange approach. This is referred to as a discrete phase model. Under this approach, the discrete particles are tracked based on Newton's laws of motion (see Chap. 11).

In some cases, the choice of a multiphase model is not straightforward. In these cases, the particulate loading, β , and the Stokes number, Sk, are used to help identify the appropriate multiphase model to use (ANSYS Fluent Theory Guide 2017). Particulate loading is the mass density ratio of the dispersed phase to the carrier phase, and it has a major impact on phase interactions. Particulate loading is algebraically expressed as:

$$\beta = (\varepsilon_{\mathrm{d}}\rho_{\mathrm{d}})/(\varepsilon_{\mathrm{c}}\rho_{\mathrm{c}}) \tag{3.417}$$

The material density ratio is:

$$\gamma = \rho_{\mathrm{d}}/\rho_{\mathrm{c}} \tag{3.418}$$

Furthermore, it is greater than 1000 for gas–solid flows, approximately 1 for liquid-solid flows, and less than 0.001 for gas–liquid flows. These parameters are used to estimate the average distance between the individual particles of the dispersed phase, which is given by:

$$(L/d_{\mathrm{d}}) = ((\pi(1+\kappa))/6\kappa)^{1/3} \tag{3.419}$$

where $\kappa = \beta/\gamma$.

This information is important when determining how to treat the dispersed phase. The degree of interaction between phases can be divided into three categories:

1. **Very low loading**: Here, the continuous phase influences the dispersed phase via drag and turbulence, but the dispersed-phase particles have no effect on the continuous phase fluid (the coupling is one-way). The discrete phase, mixture, and multifluid models can all handle this type of problem. However, the discrete phase model or mixture model is recommended because they are less expensive than the multifluid model.
2. **Intermediate loading**: The continuous phase influences the dispersed phase via drag and turbulence, and the dispersed-phase particles influence the carrier fluid by reducing the mean momentum and turbulence (two-way coupling). The discrete phase, mixture, and multifluid models can all handle this type of problem, but other factors, such as the Stokes number, should be taken into account when choosing which model to use.
3. **High loading**: Two-way coupling plus particle pressure and viscous stresses from the particles when combined are referred to as four-way coupling. Only, the multifluid model can be used for this type of problem.

For systems with intermediate particulate loading, estimating the Stokes number can help in selecting which multiphase model to use. The Stokes number is:

$$Sk = t_v / t_F \tag{3.420}$$

Thus, the following Stokes number relationships should be considered:

- If the $Sk \ll 1.0$, the particles will follow the flow closely and any of the discrete, multifluid, or mixture models are appropriate. In most cases, the mixture model will be the most practical choice and should be used unless other factors need to be considered
- If the $Sk > 1.0$, the particles will move independently of the flow, and the discrete or multifluid models are applicable
- If the $Sk \approx 1.0$, any of the discrete, multifluid, or mixture models are appropriate; moreover, the least expensive should be chosen, unless other factors must be considered.

3.4.7.2 Multiphase Flow Model Comparison

The following are guidelines and recommendations for choosing appropriate multiphase models (ANSYS Fluent Theory Guide 2017):

- In general, when dealing with complex multiphase flows involving multiple flow regimes, choose a model based on the aspect of the flow that is of the most interest. The separated flow model is appropriate for stratified or free-surface flows, and the mixture and multifluid models are appropriate for flows in which the phases mix or separate and/or the dispersed-phase volume fractions exceed 10%.
- If accuracy is more important than computational effort, the multifluid model should be used. However, the complexity of the multifluid model can make it less stable than the mixture model. To choose between the mixture model and the multifluid model, consider the following guidelines:
- If there is a wide distribution of the dispersed phase (the particles vary in size and the largest particles do not separate from the primary flow field), the mixture model is preferable.
- If the dispersed phases are concentrated in portions of the domain, the multifluid model should be used.
- If interphase drag laws are applicable, the multifluid model is usually more accurate than the mixture model. If the interphase drag laws are unknown or their applicability to the system is questionable, the mixture model should be used.
- For simpler problems requiring less computational effort, the mixture model is a better option because it solves using fewer equations than the multifluid model.
- For bubbly, droplet, or particle laden flows with phase mix and/or dispersed-phase volume fractions exceeding 10%, the mixture model or multifluid model can be used.
- For slug flows, use the separated flow model.
- For stratified/free-surface flows, use the separated flow model.
- For pneumatic transport, use the mixture model for homogeneous flow or the multifluid model for granular flow.
- For fluidized beds, use the multifluid model.
- For slurry flows and hydrotransport, use the mixture or multifluid model.
- For sedimentation, use the multifluid model.

Problems

3.1. The local instance continuity equation can also be obtained by performing a mass balance for the differential control volume shown in Fig. P3.1. Derive the continuity equation and compare your result with Eq. (3.51).

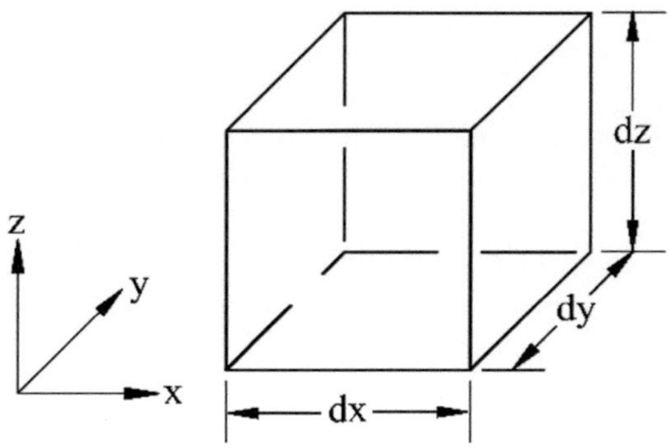

Fig. P3.1

3.2. Derive the local instance continuity equation in a cylindrical coordinate system using the control volume shown in Fig. P3.2.

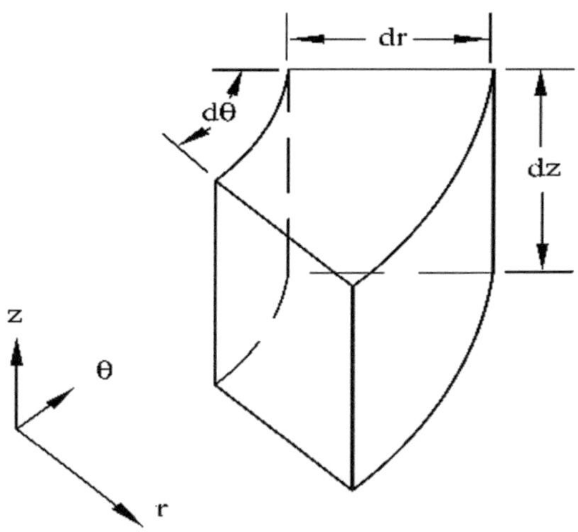

Fig. P3.2

3.3. The continuity equation for incompressible flow is Eq. (3.52). Is it valid for transient flow? Why or why not?

3.4. In order to take water without stopping the train, a narrow trough of several thousand feet long can be placed in the midway between the rails of a railroad track. As a locomotive passes over the narrow trough, the water can be forced up into the tender through a scoop by the speed of the locomotive (see Fig. P3.4). Assuming the cross-sectional area of the scoop is A, find the force acting on the train by the water using momentum balance in (a) a coordinate system that is attached to and moves with the locomotive, and (b) a fixed coordinate system.

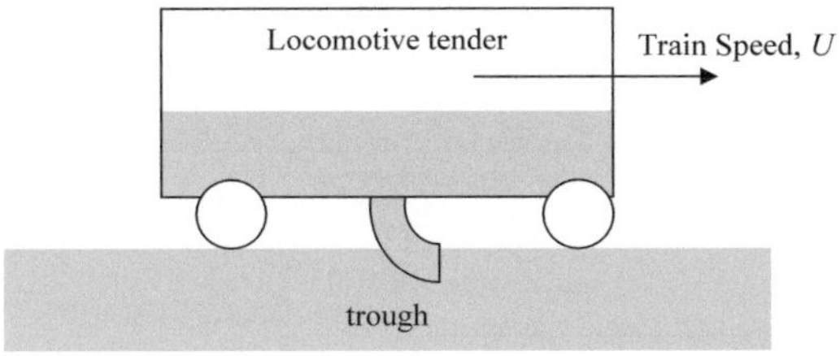

Fig. P3.4

3.5. Use the integral form of the energy equation to calculate the steady mass flow rate of water in a pump that delivers 4 horsepower to the fluid, as given in Fig. P3.5, for the control volume shown. The pump inlet and outlet diameters are 10. and 4 in., respectively.

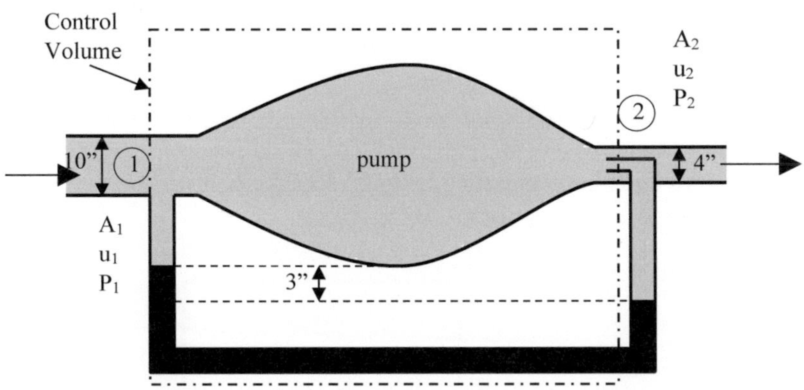

Fig. P3.5

3.6. Consider a liquid heated in an upward flow configuration in a vertical 10-m-long tube of constant diameter. The inlet average velocity and pressure are 2 m/s and 350 kPa, respectively. Determine the heat needed to be added to the fluid so that the losses in internal energy equal 180 kJ/kg. Assume the density is 1000 kg/m^3 and outlet pressure is 1 atm.

3.7. Water flows through a 3-in.-diameter pipe at a rate of 45 gal/min. Frictional forces produce a pressure drop of 8 psi. Determine the temperature change of the water if the heat transfer to the pipe can be neglected.

3.8. Determine the change in water temperature before (1) and (3) in terms of the inlet parameters A_1, A_2, u_1, u_2, c_v, and θ for the configuration shown in Fig. P3.8. Assume $T_1 = T_2$ and $p_1 = p_2$. The change in internal energy can be approximated as $de = c_v dT$.

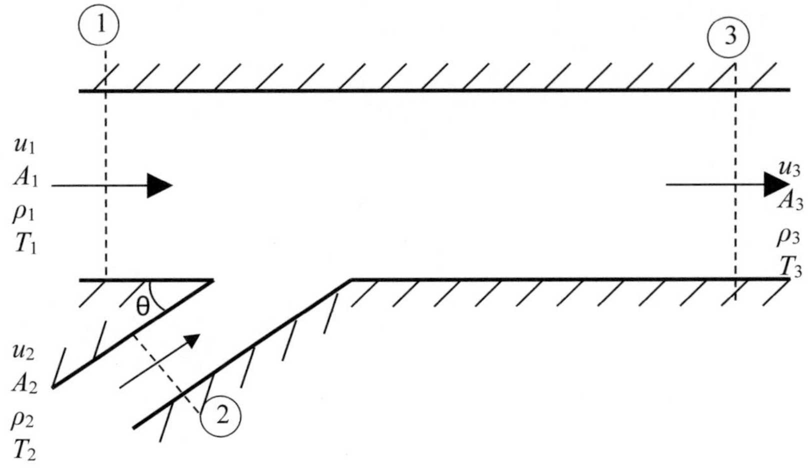

Fig. P3.8

3.9. Gas enters an insulated tank with total volume V through a pipe with diameter D with an inlet velocity and temperature of u_i and T_i, respectively, as shown in Fig. P3.9. Using the integral form of the energy equation, determine a first-order equation for the rate of change in temperature in the tank as a function of (V, D, u_i, T_i).

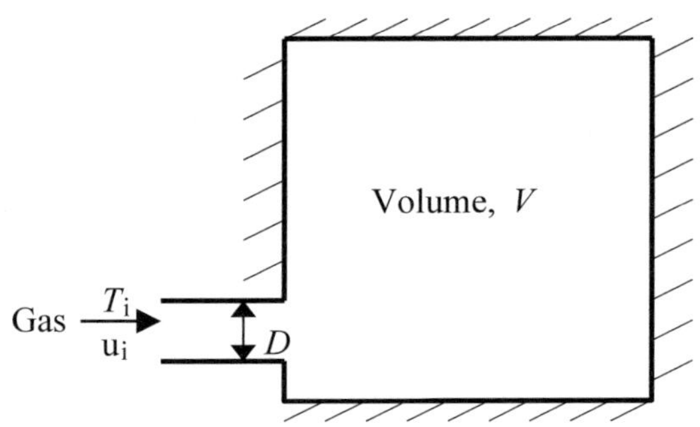

Fig. P3.9

3.10. Determine viscous dissipation for *compressible* flow in the Cartesian coordinate system.

3.11. Write down the governing equations for two-dimensional, compressible flow with the effect of viscosity being accounted for in a Cartesian coordinate system. Assume that the constant properties assumptions apply and that the only body force is gravity.

3.12. Fourier's law of conduction [see Eq. (1.22)] assumes that heat flux responds immediately to a temperature gradient. However, for cases with a very high temperature gradient or extremely short duration, heat is propagated at a finite speed, which can be accounted for by a modified heat flux model:

$$\mathbf{q}'' + \tau \frac{\partial \mathbf{q}''}{\partial t} = -k\nabla T$$

where τ is the thermal relaxation time. Derive the energy equation for a pure substance using the above model and Eq. (3.90). Internal heat generation and viscous dissipation can be neglected.

3.13. The enthalpy for the ith component is h_i is function of temperature, T, and pressure, p. Show that the substantial derivative in terms of the velocity of the ith component, V_i, can be obtained from Eq. (3.162).

3.14. Formulate the hydrodynamics for developing flow in a duct formed by two infinite flat plates. Assume the flow to be laminar, incompressible, 2D, and steady. It can also be assumed that the inlet velocity at the entrance of the duct is uniform, as shown in Fig. P3.14. Using the integral method, find out the hydrodynamic entrance and the skin friction coefficient in terms of Reynolds number. Is the analysis valid for a circular pipe?

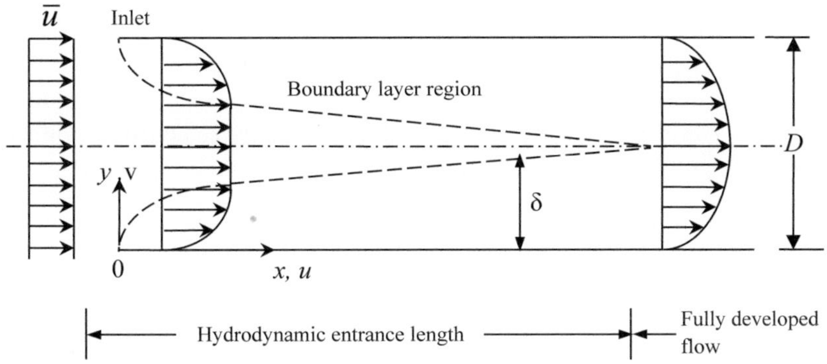

Fig. P3.14

3.15. Formulate the hydrodynamics when flow is fully developed in a duct formed by two infinite flat plates and a round pipe. Assume the flow to be laminar, incompressible, two-dimensional, and steady. Because flow is fully developed, it can be assumed that the velocity in the duct does not change in the x-direction. Determine the skin friction coefficient. Clearly, describe the methods used to solve the momentum and continuity equations.

3.16. Obtain the Nusselt number for constant wall temperature and constant wall heat flux cases for fully developed flow in terms of temperature and velocity profiles in a circular pipe. Assume laminar, incompressible, two-dimensional, and steady state.

3.17. Develop analytical expressions for the Nusselt number for constant wall temperature and constant heat flux cases in a circular pipe with a fully developed velocity profile but a developing temperature profile. Assume laminar, incompressible, two-dimensional, and steady state with uniform inlet velocity.

3.18. Consider steady-state, two-dimensional, fully developed thermal and hydrodynamic laminar flow with constant properties in a circular pipe. Let there be heat transfer to or from the fluid at a constant rate per unit of tube length. Determine the Nusselt number if the effect of frictional heating due to viscosity (viscous dissipation) is included in the analysis. How does frictional heating affect the Nusselt number? Identify the pertinent nondimensional parameters appearing in this problem.

3.19. The Couette flow was discussed in Example 3.4. When both surfaces are stationary and flow is caused by an applied pressure gradient in the x-direction, $\mathrm{d}p/\mathrm{d}x$, it is called Poiseuille flow (see Fig. P3.19). The fluid is Newtonian and incompressible. Temperature in the upper and lower plates is kept constant at T_1 and T_2, respectively. Determine the fluid temperature distribution and heat flux with the following assumptions (Poiseuille flow): (1) steady state, (2) constant properties, (3) laminar flow, (4) constant $\partial p/\partial x$, (5) no end effects, (6) no edge effects, and (7) gravity is the only body force.

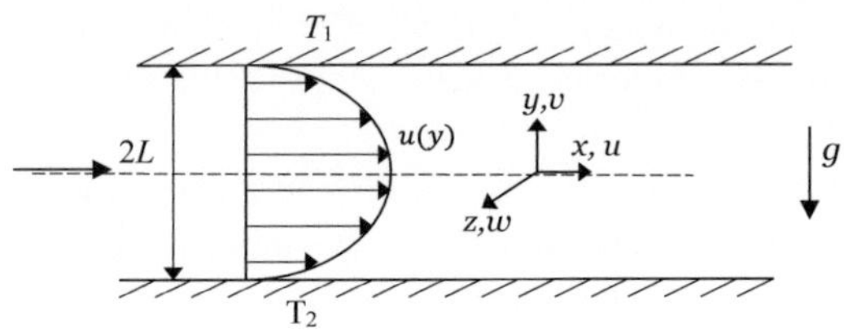

Fig. P3.19

3.20. Consider the steady flow of an incompressible fluid making an angle θ with the vertical direction with constant liquid film thickness δ along an inclined flat surface with length L and width W as shown in Fig. P3.20. Determine the average velocity, volumetric flow rate, film thickness δ, and the force of the fluid on the surface. Neglect the end, exit, and edge effects and assume constant properties. The air is stagnant next to the liquid free surface.

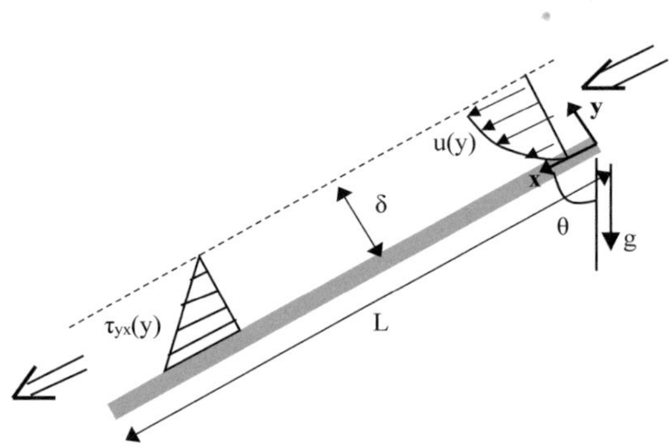

Fig. P3.20

3.21. Develop the energy equation and boundary conditions, for Problem 3.20 for the case of constant wall temperature T_w and uniform inlet temperature T_0. Determine the temperature distribution and heat transfer rate for the wall to film for start-up period (short contact time) that the fluid temperature changes only in the vicinity of the wall.

3.22. Develop formulation for the case of heating with no evaporation for the flow over a rotating disk (Fig. P3.22) Assume that the flow is laminar, incompressible, steady, axisymmetric, and the disk has a constant wall temperature. The flow is introduced from a central collar that directs the liquid radially outward with velocity u_0 over a gap height of h_0. δ is the liquid film thickness and r_0 is the initial radius of the disk at the outlet of the collar, h_0 is the collar height and δT is the thickness of the thermal boundary layer. Obtain Nusselt number $Nu = hr_0/k$, and film thickness δ/h_0 as a function of the Reynolds number and Rossby numbers $[Re = u_0 r_0/v$ and $Ro = u_0^2/(\omega^2 r_0^2)]$ using the integral method.

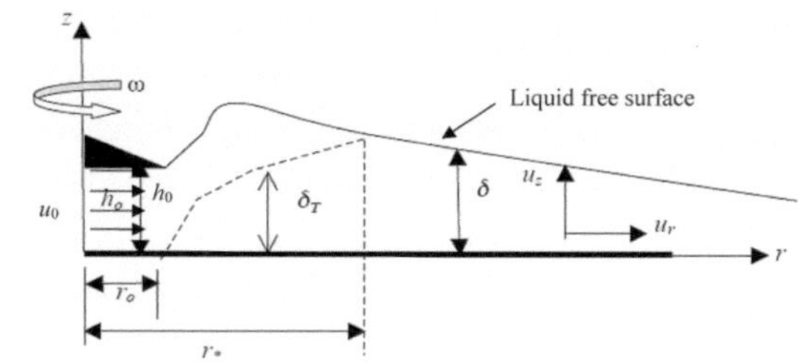

Fig. P3.22

3.23. Derive the continuity equation at a liquid–vapor interface, Eq. (3.174), by considering conservation of mass for a control volume that includes two phases as given by Eq. (3.8) with $\Pi = 2$.

3.24. Eq. (3.174) is valid for an interface between liquid and vapor phases. Derive the conservation of mass for solid–liquid interface that is applicable to the melting problem.

3.25. Derive the momentum equation at an interface, Eq. (3.176), from the integral momentum equation for a control volume containing two phases separated by an interface as given by Eq. (3.13) with $\Pi = 2$.

3.26. Derive the energy equation at an interface, Eq. (3.177), from the integral energy equation for a control volume containing two phases separated by an interface as given by Eq. (3.23) with $\Pi = 2$.

3.27. The momentum balance equation at a three-dimensional interface was discussed in Sect. 3.3.6. Derive the momentum balance equation for a two-dimensional interface between phases 1 and 2. The effect of surface tension can be neglected.

3.28. A tube with radius R_o and zero thickness passes through a liquid phase-change material at its melting point, T_m (see Fig. P3.28). A cold fluid with a temperature of T_c, which is below T_m, flows through the inside of the tube. Solidification will occur on the outer surface of the tube. Formulate the problem by giving the governing equations and corresponding boundary conditions in the cylindrical coordinate system. If the problem can be assumed to be quasi-steady state, solve for the instantaneous location of the solid–liquid interface.

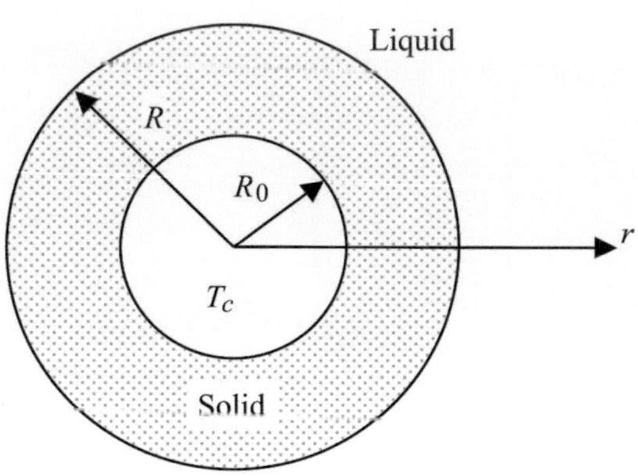

Fig. P3.28

3.29. The time it takes to completely burn a fuel droplet of specified size as estimated by Eq. (3.242) is valid for conduction-controlled combustion processes. For a liquid droplet that has sufficient inertia, the heat transfer between vapor and liquid is dominated by convection, i.e., Equation (3.236) is replaced by $\dot{m}_{\mathrm{f}}'' h_{\ell \mathrm{v}} = q_{\mathrm{I}}'' = h\theta_\infty$, where h is the heat transfer coefficient between the gaseous mixture and liquid droplet, and can be obtained by $Nu_D = hD/k = CRe_D^{1/2}$. What is the time it takes to completely burn the liquid fuel droplet? Compare your result with Eq. (3.242).

3.30. Redo Example 3.7 using the result from Problem 3.29 and compare your result with the result in Example 3.7. The relative velocity of the fuel droplet is $V_{\mathrm{p}} = 5\,\mathrm{m/s}$, and the heat transfer correlation is $Nu_D = 0.6Re_D^{1/2}$. The viscosity of the gaseous mixture is $4 \times 10^{-5}\,\mathrm{m^2/s}$.

3.31. A fuel droplet with an initial diameter of D_{p} moves at a velocity $\mathbf{v}_{\mathrm{p}}$ in an oxidant gas. The gravitational acceleration vector is $\mathbf{g}$, and the local oxidant gas velocity vector is $\mathbf{v}_{\mathrm{g}}$. The fuel droplet evaporates as it moves in the combustor. If the drag coefficient of the fuel droplet is C_{D}, write the momentum equation for the fuel droplet.

3.32. If the temperature of the fuel droplet in Problem 3.31 can be assumed to be uniform at any time, and the latent heat of evaporation of the fuel is $h_{s\ell}$, derive the energy equation for the fuel droplet.

3.33. A saturated mixture of water and its vapor at 180 °C flows in a 0.1 m ID tube with a mass flow rate of 0.5 kg/s. The liquid water is dispersed in the vapor phase in the form of 0.1-mm-diameter droplets, and the quality of the mixture is $x = 0.75$. The average velocity of the vapor phase is 30 m/s. Find (a) the average velocity of the liquid droplets and (b) the interactive force between the liquid and vapor phase.

3.34. If the liquid droplet is subcooled at a temperature of 175 °C, estimate the interphase heat transfer rate between liquid and vapor phases in the above problem.

3.35. The mass flux in the channel is defined as the mass flow rate per unit area. What is the total mass flow rate for the multiphase flow discussed in Example 3.10?

3.36. Consider flow entering from different sides of a *tee-junction* in a duct. The fluid entering is a mixture of air and liquid water, with known velocities and volume fraction at each inlet. The outlet is held at a constant pressure. A schematic of this problem is presented in Fig. P3.36. Model the problem using a two-dimensional Cartesian homogeneous model to find the mass-averaged velocity. Assume that the flow is in the laminar regime and that gravitational effects are important. Also, assume that the flow field is isothermal. The volume fraction of air is small; therefore, the air exists in the form of bubbles. The diameter of these bubbles is known from experimental observation.

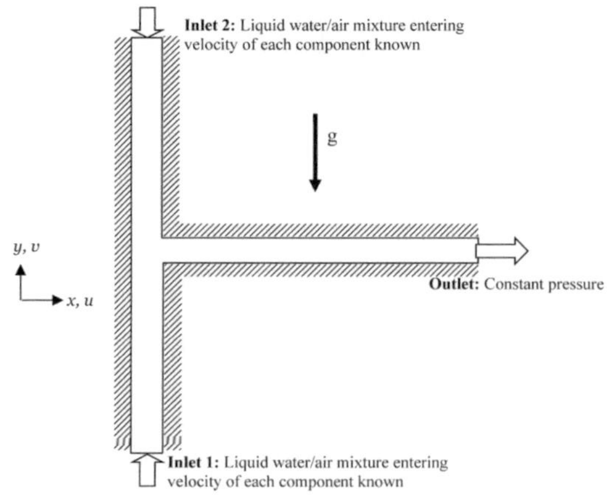

Fig. P3.36

3.37. A parametric study on the effect of the inlet velocities and gas volume fraction from Problem 3.36 is needed. However, before performing this study using the homogeneous model, the inaccuracies of the model must first be quantified. The inaccuracies should be quantified by comparing the homogeneous results to a volume-averaged multifluid model. Therefore, set up the governing equation using a volume-averaged multifluid model with the corresponding boundary conditions with the same assumptions used in Problem 3.36.

3.38. Consider a case of annular flow in which the liquid flows as a film on the wall of a pipe while the gas flows down the central cylindrical core. For a pipe of diameter D, let the interfacial and wall shear stresses be τ_δ and τ_W as given inputs, respectively. Assuming vertical cocurrent flow configuration as shown in Fig. P3.38, develop the formulation for one-dimensional, unsteady, laminar flow with multifluid modeling.

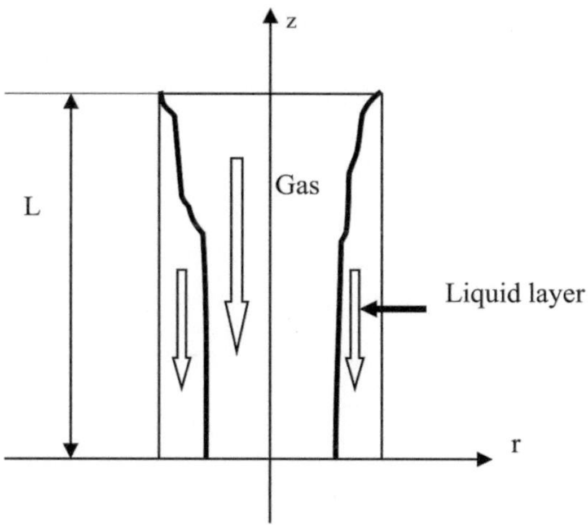

Fig. P3.38

3.39. Deduce the general one-dimensional governing equations for mass and momentum for stratified turbulent flow of two phases (1) and (2) with evaporation over a flat plate as shown in Fig. P3.39.

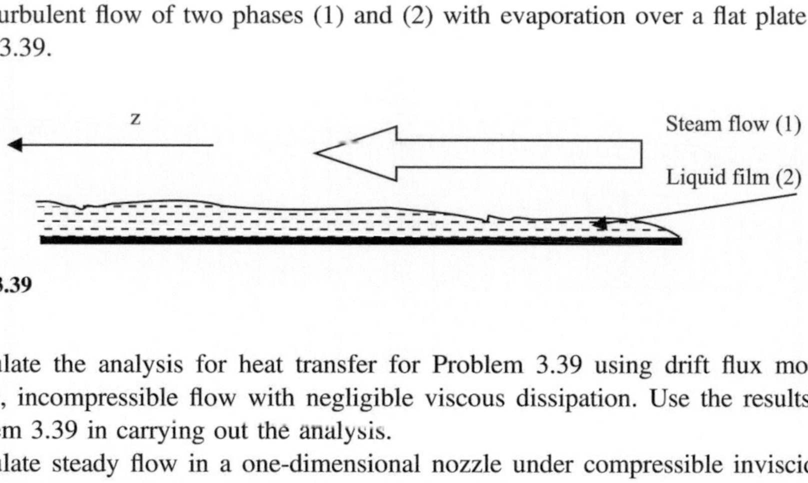

Fig. P3.39

3.40. Formulate the analysis for heat transfer for Problem 3.39 using drift flux model. Assume steady, incompressible flow with negligible viscous dissipation. Use the results obtained in Problem 3.39 in carrying out the analysis.

3.41. Formulate steady flow in a one-dimensional nozzle under compressible inviscid conditions. The problem should be modeled as a homogeneous two-component flow with no phase change. Neglect the effects of surface tension. Obtain velocity and volume fraction distribution along the nozzle.

3.42. Derive the equations for mass, momentum, and energy for the flow of compressible gas through a channel containing dispersed particles. The flow occurs at high speed. Neglect the following effects: exchange of momentum, heat and mass with the channel walls, phase change, chemical reactions, viscous drag and viscous dissipation, and gravity. Assume the flow is one-dimensional, steady, and constant particle phase density. Do the analysis using the area-averaged multifluid models for channel flows.

3.43. For a coal classifier with $\boldsymbol{L_s} = \mathbf{1}\,\mathrm{m}$ and $\boldsymbol{V_s} = \mathbf{10}\,\mathrm{m/s}$, **Sk** $= 0.04$ for particles with D = 30 microns, what is the **Sk** for particles with D = 300 microns? Is the mixture model appropriate for the larger diameter particles?

3.44. For a mineral processing system, $L_\mathrm{s} = 0.2\,\mathrm{m}$ and $V_\mathrm{S} = 2\,\mathrm{m/s}$, $\mathrm{Sk} = 0.005$ for particles with D = 300 microns. Which multiphase model(s) is/are appropriate?

3.45. Saturated liquid water at 1 atm flows upward with a (uniform) velocity of 1 m/s ($U_\ell = 1\,\mathrm{m/s}$) between two parallel plates with a large distance in between. A heated vertical plate is inserted into the liquid, and film boiling occurs on both surfaces of the plate (see Fig. P3.45). For simplicity, we can assume that the liquid flow is not affected by the vapor motion and no slip condition is valid at the liquid–vapor interface (i.e., $u_\mathrm{v}(\delta) = U_\ell$). Assuming that the convective terms in energy and momentum equations are negligible, the governing equations in the vapor film become:

$$\mu_\mathrm{v}\frac{\partial^2 u_\mathrm{v}}{\partial y^2} + g(\rho_\mathrm{v} - \rho_\ell) = 0, \quad \frac{\partial^2 T_\mathrm{v}}{\partial y^2} = 0$$

(a) Determine the momentum and energy boundary conditions at the plate surface (y = 0), as well as the interfacial conditions at the liquid–vapor interface ($y = \delta(x)$).

(b) Determine the mass flow rate of vapor per unit width, $\Gamma(\delta)$ as a function of film thickness, δ.

(c) Calculate the film thickness at 0.25 m from the bottom edge of the plate, if the heat flux applied by the flat plate is 1 kW/m^2 on each side.

(d) How would the film thickness change with increasing liquid velocity, U_ℓ?

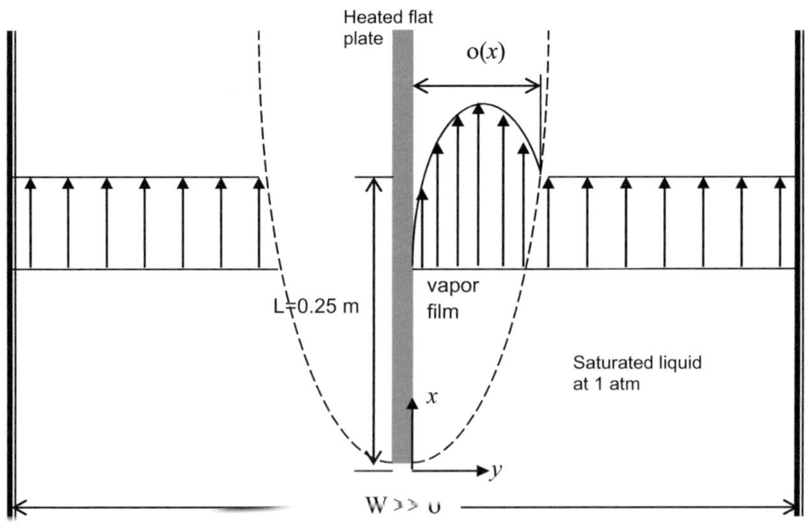

Fig. P3.45

3.46. Saturated liquid water-gas mixture at 80 °C (353 K) flows through the manifold channels of a polymer electrolyte fuel cell (PEFC). The pressure of the manifold can be assumed constant at 100 kPa. The gas mixture contains water vapor, and the remainder is 80% nitrogen (N_2, MW_{N2} = 32 gm/mol) and 20% oxygen (O_2, MW_{O2} = 32 gm/mol) by weight.

a. Assuming ideal gas mixture, calculate the (intrinsic) density of the gas mixture. *Saturation pressure of water at* 80 °C *is* 47 kPa. MW_{H2O} = 18 gm/mol.
b. Intrinsic average velocity of gas mixture is 2 m/s. If the area fraction of liquid is $\varepsilon_l = 0.05$ and the quality of mixture is $x = 0.75$, calculate the intrinsic average velocity of the liquid phase.
c. If the manifold is of 1-mm square cross section, calculate the thickness of the film.
d. Assuming Boysan's correlation for momentum exchange coefficient, K_{jk} is applicable in the modified form below, calculate the interactive force $\langle \boldsymbol{F}_{jk} \rangle$ between the liquid and vapor.

$$K_{jk} = \frac{3}{4} f \frac{\varepsilon_j \langle \rho_k \rangle^k}{\delta_j} \left| \langle \mathbf{V}_j \rangle^j - \langle \mathbf{V}_k \rangle^k \right|, \quad f = \frac{24}{Re}$$

$$Re = \frac{\langle \rho_k \rangle^k \left| \langle \mathbf{V}_j \rangle^j - \langle \mathbf{V}_k \rangle^k \right| \delta_j}{\mu_k}, \quad \langle \mathbf{F}_{jk} \rangle = K_{jk} \left(\langle \mathbf{V}_j \rangle^j - \langle \mathbf{V}_k \rangle^k \right)$$

$$\mu_\ell = 3.55 \times 10^{-6} \text{ Pa s}; \quad \mu_g = 2.15 \times 10^{-5} \text{ Pa s}$$

3.47. Electroosmosis is flow induced by an external electric field. When surfaces acquire a finite charge density and are in contact with a polar solution, they induce a distribution of electrical charges within the electrically charged solution. Electroosmosis is used in biochip technology as a pumping mechanism in MEMS devices for chemical and biological analysis and diagnoses. Two-fluid electroosmotic flow is used in many microfluids. The working fluid needs to be a conducting liquid with significant electrical conductivity to function properly. As shown in Fig. P3.47, a high electroosmotic mobility liquid is at the bottom of the channel, and a low electroosmotic mobility liquid is in the upper section of the microchannel. When an electric field is applied across the channel, pressure and electroosmosis effects drive the liquids. The flow depends on the viscosity ratio of the two fluids, the external electric field, electroosmosis characteristics of the high mobility fluid, and the interfacial curvature between the two fluids. Develop the steady formulation for velocity and the electric potential due to an electrically applied field caused by charges on the wall.

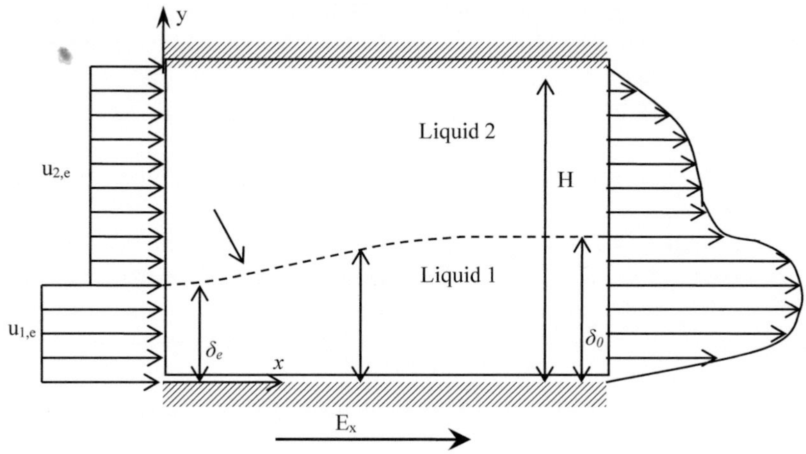

Fig. P3.47

References

ANSYS fluent theory guide. (2017). ANSYS, Inc.

Avedisian, C. T. (1997). Soot formation in spherically symmetric droplet combustion. In I. Irvin Glassman, F. L. Dryer, & R. F. Sawyer (eds) *Physical and chemical aspects of combustion* (pp. 135–160). Gordon and Breach Publishers.

Avedisian, C. T. (2000). Recent advances in soot formation from spherical droplet flames at atmospheric pressure. *Journal of Propulsion and Power, 16,* 628–656.

Bejan, A. (2013). *Convection heat transfer* (4th ed.). New York: Wiley.

Bergman, T. L., & Lavine, A. S. (2017). *Fundamentals of heat and mass transfer* (8th ed.). New York: Wiley.

Boysan, F. (1990). *A two-fluid model for fluent*. Sheffield, England: Flow Simulation Consultants Ltd.

Edwards, D. K., Denny, V. E., & Mills, A. F. (1979). *Transfer process*. New York: Hemisphere.

Faghri, A. (2016). *Heat pipe science and technology* (2nd ed.). Columbia, MO: Global Digital Press.

Faghri, A., Zhang, Y., & Howell, J. R. (2010). *Advanced heat and mass transfer*. Columbia, MO: Global Digital Press.

Hewitt, G. F. (1998). *"Multiphase fluid flow and pressure drop", heat exchanger design handbook* (Vol. 2). New York, NY: Begell House.

Hirschfelder, J. O., Curtiss, C. F., & Bird, R. B. (1966). *Molecular theory of gases and liquids*. New York: Wiley.

Kays, W. M., Crawford, M. E., & Weigand, B. (2004). *Convective heat transfer* (4th ed.). New York, NY: McGraw-Hill.

Kleijn, C. R. (1991). A mathematical model of the hydrodynamics and gas phase reaction in silicon LPCVD in a single wafer reactor. *Journal of the Electrochemical Society, 138,* 2190–2200.

Lock, G. S. H. (1994). *Latent heat transfer*. Oxford University, Oxford, UK: Oxford Science Publications.

Mahajan, R. L. (1996). Transport phenomena in chemical vapor-deposition systems. In *Advances in heat transfer*. Academic Press, San Diego, CA.

Manninen, M., Taivassalo, V., & Kallio, S. (1996). On the mixture model for multiphase flow. VTT Publications 288, Technical Research Centre of Finland.

Schiller, L., & Naumann, A. (1935). A drag coefficient correlation. *Zeitschrift des Vereins Deutscher Ingenieure, 77,* 318–320.

Welty, J. R., Rorrer, G. L., & Foster, D. G. (2014). *Fundamentals of momentum, heat and mass transfer* (6th ed.). New York, NY: Wiley.

White, F. M. (2005). *Viscous fluid flow* (3rd ed.). New York: McGraw-Hill.

Interfacial Phenomena

4

4.1 Introduction

The interfacial region between two homogeneous phases contains matter in a distinct physical state; that is to say, matter in the *interfacial state* exhibits properties different from those matters in the gaseous, liquid, or solid states. As a result, as soon as interfaces are considered explicitly, new variables—for example, interfacial surface tension—enter into the classical thermodynamic description of equilibrium systems. Interfaces in equilibrium systems need not be considered explicitly unless the surface-to-volume ratio is large because the contribution of interfacial free energy to the total free energy is usually small. However, interfacial effects on the dynamic behavior of flow systems can be profound, even when the proportion of matter in interfacial regions is extremely small. Furthermore, motion may originate in an interface in systems that are not in thermal or compositional equilibrium.

When two adjacent fluids are at rest, their interface ordinarily behaves as if it is in a state of uniform tension. It is both possible and convenient to mathematically represent the interface as a geometric surface in tension. This representation is also appropriate for many flows with free boundaries; indeed, it is the basis of the treatment of capillarity in classical hydrodynamics. Considerations of equilibrium surface tension lead to the conclusion that the normal component of fluid stress, or pressure, is discontinuous at a curved interface, while the shear stress is continuous. Classical hydrodynamics also recognizes—in connection with the calming action of oil on water waves—that extension and contraction of a surface film produces longitudinal variations in surface tension. This in turn gives rise to discontinuities in the tangential components of fluid stress at the interface.

Certain phenomena can be observed in the interfacial region between two distinct material regions. These can be demonstrated by considering the idealized problems shown in Fig. 4.1. The first two problems, (a) and (b), use classical methods of fluid mechanics and thermodynamics that result in closed-form solutions with appropriate assumptions. Study of the combustion problem presented in (c) has produced meaningful solutions, but usually in the form of more complex versions of the idealized problem stated above. The feature common to these problems is that all of them require input information at the boundary between the two regions. Problems (b–d) all involve mass transfer across the boundary. The distinguishing feature of these problems, which is also the primary source of complication in the final formulation, is the introduction of additional terms through the boundary conditions. These additional terms account for the flux of mass, momentum, and energy from one region to another; hence, the influence of interfacial phenomena on these fluxes at the boundary becomes part of the problem. Many devices utilizing two-phase heat transfer are designed so that

A. Faghri and Y. Zhang, *Fundamentals of Multiphase Heat Transfer and Flow*, https://doi.org/10.1007/978-3-030-22137-9_4

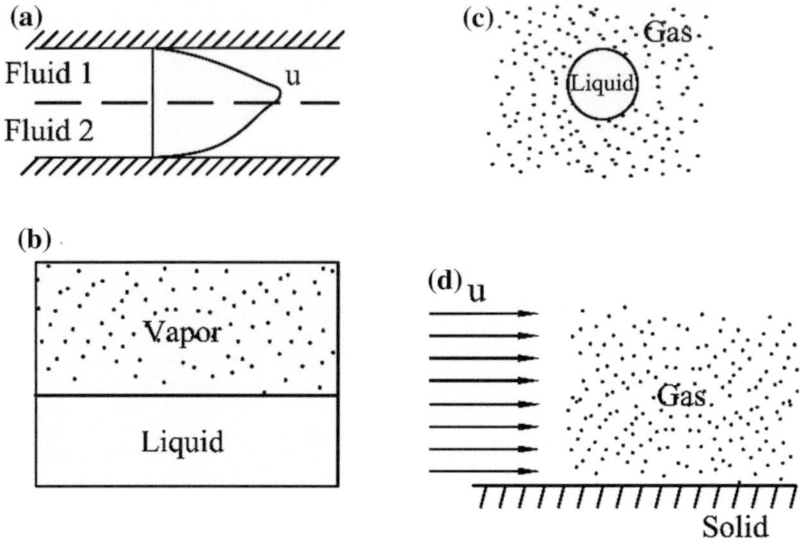

Fig. 4.1 Examples of interfacial phenomena: **a** flow of two immiscible fluids between two parallel plates; **b** pure substance in two phases at equilibrium; **c** combustion of a liquid fuel droplet in gas; **d** solid surface reacting with gas in a surrounding atmosphere

these terms are the most significant ones; therefore, they create the driving potential as well as the limiting conditions for performance.

A typical approach to treating interfacial phenomena effects in two-phase heat transfer is to apply the kind of knowledge used in formulating and solving Problems (a) and (b).

A design engineer must have some understanding of these and other phenomena to design effective devices. The field of interfacial phenomena has been the realm of researchers primarily in chemical and mechanical engineering, physical chemistry, and material science. Much of the analytical basis for their work comes from a sub continuum view of the physical world. Models of molecular interaction and the use of statistical mechanics are typical in the literature. The earliest practical work on interfacial phenomena used equilibrium thermodynamics. Today, the ad hoc use of thermodynamics and simple molecular interaction models constitute the most useful treatment of two-phase heat transfer problems. However, a significant effort has been made to develop a unified approach via the extension of continuum mechanics using the conservation equations and some form of the flux laws. Direct solution methods for interfacial phenomena do not always follow from this type of investigation, but one frequently gains valuable insight into the behavior at the interface. The assumption is that one is not interested in the precise details of the interfacial region but rather in how it affects the bulk regions.

Section 4.2 discusses capillary pressure and interface shape at equilibrium. Also considered in Sect. 4.2 are the effects of interfacial tension gradient on the fluid flow and the temperature-dependence of surface tension. Wetting phenomena and contact angles in solid-liquid-vapor systems are taken up in Sect. 4.3. Section 4.4 addresses phase equilibrium in a microscale interfacial system; this includes the effects of disjoining pressure on ultra-thin liquid films and the change in saturated vapor pressure over a curved interface due to the combined effects of surface tension and disjoining pressure. Section 4.5 introduces interfacial mass, momentum, and energy balance as well as thermal resistance across ultra-thin films. The instability of liquid–vapor interface and waves on liquid film are discussed in Sect. 4.6. Finally, Sect. 4.7 discusses numerical simulation of interfaces and free surfaces.

4.2 Surface Tension

4.2.1 Capillary Pressure

Since the distance between the molecules in the vapor phase is much greater than that in the liquid phase, the intermolecular force between the molecules in the vapor phase is very weak. The intermolecular attractive force in the liquid phase holds the molecules in the liquid close to each other. For the molecules within the liquid phase, the intermolecular forces from all directions are balanced. Although the forces acting on the molecules at the liquid–vapor interface are balanced along the tangential direction, the attractive force from the molecules in the liquid phase, F_{i} (in normal direction), tends to pull the molecules at the liquid–vapor interface toward the liquid phase because the attractive force from the vapor phase, F_{o}, is much weaker. The net inward force $F_{\mathrm{i}} - F_{\mathrm{o}}$ causes movement of the liquid molecules until the maximum number of molecules is in the interior, which leads to an interface of minimum area (Fig. 4.2).

It is generally necessary to specify two radii of curvature to describe an arbitrarily curved surface, R_{I} and R_{II}, as shown in Fig. 4.3. The surface section is taken to be small enough that R_{I} and R_{II} are approximately constant. If the surface is now displaced outward by a small distance, the change in area is

$$\Delta A = (x + \mathrm{d}x)(y + \mathrm{d}y) - xy \tag{4.1}$$

If $\mathrm{d}x\mathrm{d}y \approx 0$, then

$$\Delta A = y\,\mathrm{d}x + x\,\mathrm{d}y \tag{4.2}$$

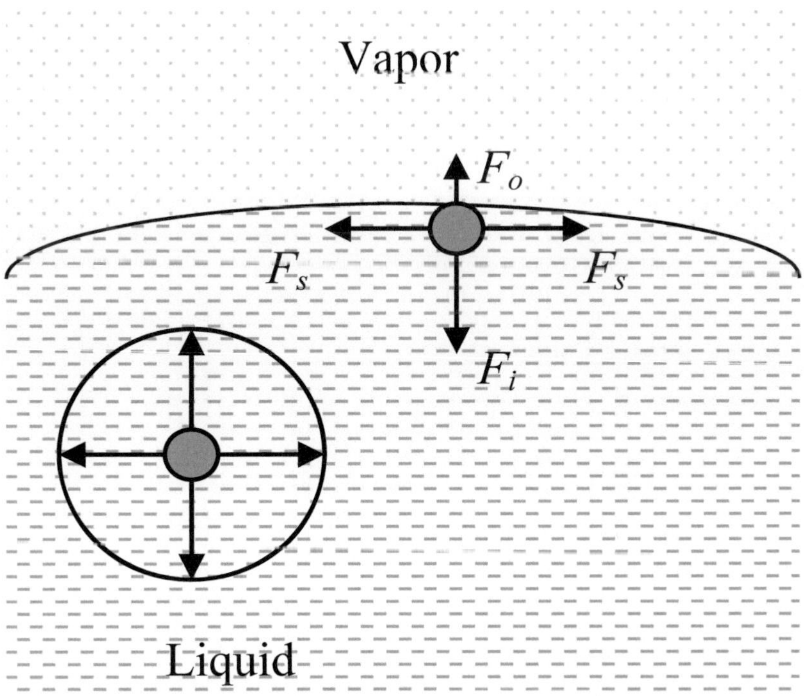

Fig. 4.2 Origin of surface tension at liquid–vapor interface

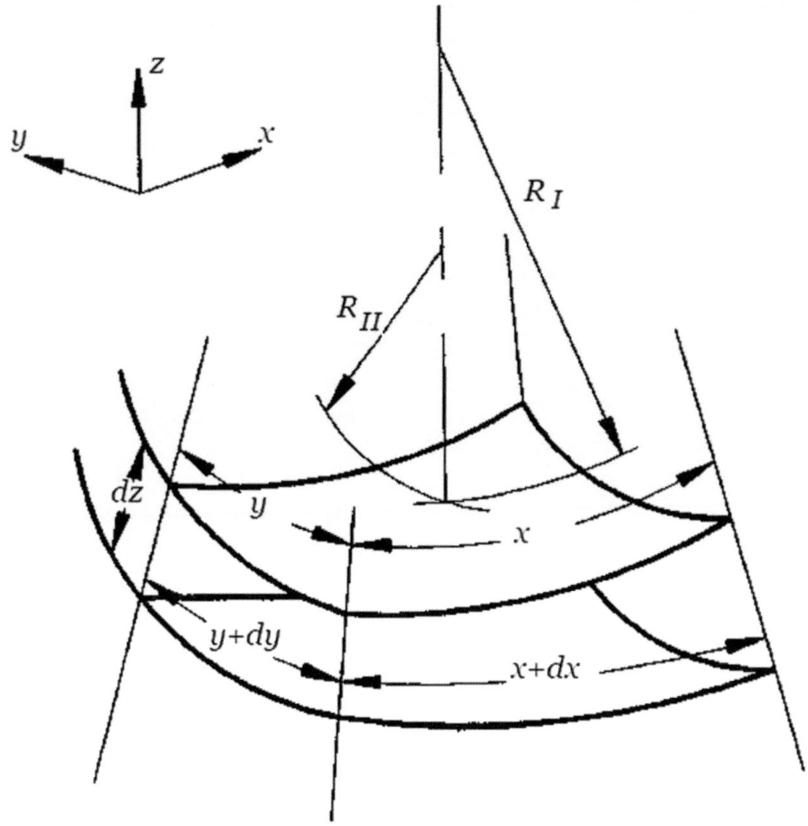

Fig. 4.3 Arbitrarily curved surface with two radii of curvature R_{I} and R_{II}

The work required to displace the surface is obtained from Eq. (2.209), i.e.,

$$\delta W = \sigma(x\,\mathrm{d}y + y\,\mathrm{d}x) \tag{4.3}$$

Displacement acting on the area xy over the distance dz also creates a pressure difference Δp across the surface—capillary pressure (p_{cap}). The work attributed to generating this pressure difference is

$$\delta W = \Delta p\,xy\,\mathrm{d}z = p_{\mathrm{cap}}xy\,\mathrm{d}z \tag{4.4}$$

From the geometry of Fig. 4.3, it follows that

$$\frac{x + \mathrm{d}x}{R_{\mathrm{I}} + \mathrm{d}z} = \frac{x}{R_{\mathrm{I}}} \tag{4.5}$$

or

$$\mathrm{d}x = \frac{x\,\mathrm{d}z}{R_{\mathrm{I}}} \tag{4.6}$$

Similarly,

$$dy = \frac{y\,dz}{R_{II}} \tag{4.7}$$

For the surface to be in equilibrium across this differential change, the two expressions for the work must be equal:

$$\sigma(x\,dy + y\,dx) = \Delta p\,xy\,dz \tag{4.8}$$

i.e.,

$$\sigma\left(\frac{xy\,dz}{R_I} + \frac{xy\,dz}{R_{II}}\right) = \Delta p\,xy\,dz \tag{4.9}$$

The pressure difference between two phases becomes

$$p_{cap} = \Delta p = \sigma\left(\frac{1}{R_I} + \frac{1}{R_{II}}\right) = \sigma(K_1 + K_2) \tag{4.10}$$

where K_1 and K_2 are curvatures of the surface.

This expression is called the Young-Laplace equation, and it is the fundamental equation for capillary pressure. It can be seen that when the two curvature radii are equal, in which case the curved surface is spherical, Eq. (4.10) can be reduced to Eq. (2.230).

4.2.2 Interface Shapes at Equilibrium

Surface tension effects on the shape of liquid–vapor interfaces can be demonstrated by considering a free liquid surface of a wetting liquid meeting a planar vertical wall (see Fig. 4.4). Since the interface is two-dimensional, the second principal radius of curvature is infinite. For the first principal radius of curvature R_I, it follows from analytical geometry that

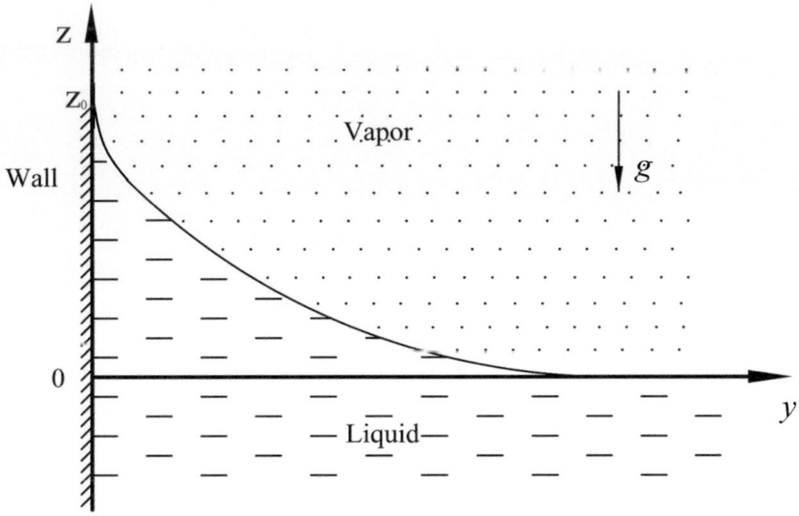

Fig. 4.4 Shape of the liquid–vapor interface near a vertical wall

$$\frac{1}{R_{\mathrm{I}}}=\frac{\mathrm{d}^2 z/\mathrm{d}y^2}{\left[1+(\mathrm{d}z/\mathrm{d}y)^2\right]^{3/2}} \tag{4.11}$$

The Young-Laplace equation, Eq. (4.10), becomes

$$p_{\mathrm{v}}-p_{\ell}=\sigma/R_{\mathrm{I}} \tag{4.12}$$

At a point on the interface that is far away from the wall ($y \to \infty$, $z = 0$), the liquid and vapor pressures are the same, i.e.,

$$p_{\mathrm{v}}(\infty,0)=p_{\ell}(\infty,0) \tag{4.13}$$

The liquid and vapor pressures near the wall ($z > 0$) are

$$p_{\mathrm{v}}(z)=p_{\mathrm{v}}(\infty,0)-\rho_{\mathrm{v}} g z \tag{4.14}$$

$$p_{\ell}(z)=p_{\ell}(\infty,0)-\rho_{\ell} g z \tag{4.15}$$

The relationship between the pressures in two phases can be obtained by combining Eqs. (4.13)–(4.15), i.e.,

$$p_{\mathrm{v}}-p_{\ell}=g(\rho_{\ell}-\rho_{\mathrm{v}})z \tag{4.16}$$

Combining Eqs. (4.12) and (4.16), the equation for the interface shape becomes

$$\frac{g(\rho_{\ell}-\rho_{\mathrm{v}})z}{\sigma}-\left[1+\left(\frac{\mathrm{d}z}{\mathrm{d}y}\right)^2\right]^{-3/2}\frac{\mathrm{d}^2 z}{\mathrm{d}y^2}=0 \tag{4.17}$$

Multiplying this equation by dz/dy and integrating gives

$$\frac{g(\rho_{\ell}-\rho_{\mathrm{v}})z^2}{\sigma}+\left[1+\left(\frac{\mathrm{d}z}{\mathrm{d}y}\right)^2\right]^{-1/2}=C_1 \tag{4.18}$$

Since at $y \to \infty$ both z and dz/dy equal zero, $C_1 = 1$. The boundary condition for Eq. (4.18) is

$$(\mathrm{d}z/\mathrm{d}y)_{y=0}\to\infty \tag{4.19}$$

Equations (4.18) and (4.19) can then be solved for z at $y = 0$:

$$z_0=z(0)=\left[\frac{2\sigma}{(\rho_{\ell}-\rho_{\mathrm{v}})g}\right]^{1/2} \tag{4.20}$$

Using Eq. (4.20) as a boundary condition, integrating Eq. (4.18) yields the following relation to the shape of the interface:

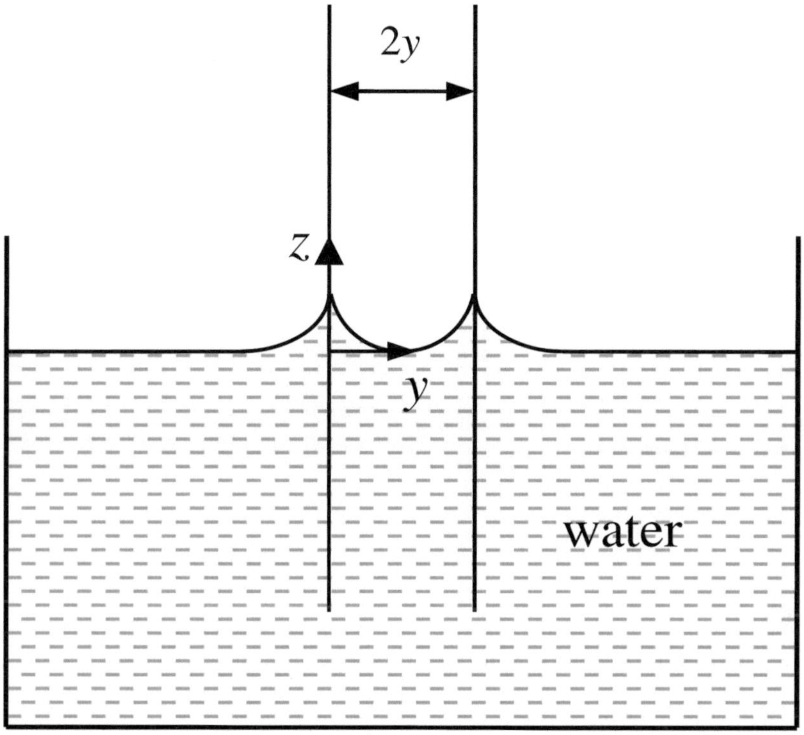

Fig. 4.5 Two parallel plates in bulk water

$$\frac{y}{L_c} = \left[\cosh\left(\frac{2L_c}{z_0}\right)\right]^{-1} - \left[\cosh\left(\frac{2L_c}{z}\right)\right]^{-1} + \left(4 + \frac{z_0^2}{L_c^2}\right)^{1/2} - \left(4 + \frac{z^2}{L_c^2}\right)^{1/2} \tag{4.21}$$

where

$$L_c = \left[\frac{\sigma}{(\rho_\ell - \rho_v)g}\right]^{1/2} \tag{4.22}$$

is a characteristic length for the capillary scale. For small tubes with $R \ll L_c$, the interfacial radius of curvature is approximately constant along the interface and equal to $r/\cos\theta$, where θ is the contact angle.

Example 4.1

Two parallel plates are put into bulk water at the bottom end (see Fig. 4.5). Estimate the minimal distance between the plates at which the central point of the liquid-air interface between the plates is not elevated from the bulk water level at equilibrium condition. Assume that water completely wets the material of the plates. The system temperature is 20 °C, $\sigma = 0.07288$ N/m, and $\rho_\ell = 999$ kg/m.

Solution Since the density of vapor is much less than that of the liquid—$\rho_v \ll \rho_\ell$—Eq. (4.22) can be simplified to obtain

$$L_c = \left[\frac{\sigma}{(\rho_\ell - \rho_v)g}\right]^{1/2} \doteq \left[\frac{\sigma}{\rho_\ell g}\right]^{1/2} = \left[\frac{0.07288}{999 \times 9.8}\right]^{1/2} = 2.728 \times 10^{-3}\ \text{m}$$

The rise of the liquid surface at $y = 0$ is obtained from Eq. (4.20):

$$z_0 = z(0) = \left[\frac{2\sigma}{(\rho_\ell - \rho_v)g}\right]^{1/2} = \left[\frac{2\sigma}{\rho_\ell g}\right]^{1/2} = \left[\frac{2 \times 0.07288}{999 \times 9.8}\right]^{1/2} = 3.859 \times 10^{-3}\ \text{m}$$

The value of y for $z = 0$ can be found by using Eq. (4.21), i.e.,

$$\begin{aligned}\frac{y}{L_c} &= \left[\cosh\left(\frac{2L_c}{z_0}\right)\right]^{-1} - \left[\cosh\left(\frac{2L_c}{z}\right)\right]^{-1} + \left(4 + \frac{z_0^2}{L_c^2}\right)^{1/2} - \left(4 + \frac{z^2}{L_c^2}\right)^{1/2} \\ &= \left[\cosh\left(\frac{2 \times 2.728 \times 10^{-3}}{3.859 \times 10^{-3}}\right)\right]^{-1} - \left[\cosh\left(\frac{2 \times 2.728 \times 10^{-3}}{0}\right)\right]^{-1} \\ &\quad + \left(4 + \left(\frac{3.859 \times 10^{-3}}{2.728 \times 10^{-3}}\right)^2\right)^{1/2} - \left(4 + \left(\frac{0}{2.728 \times 10^{-3}}\right)^2\right)^{1/2} \\ &= 0.909\end{aligned}$$

i.e.,

$$y = 0.909L_c = 0.909 \times 2.728 \times 10^{-3} = 2.48 \times 10^{-3}\ \text{m} = 2.48\ \text{mm}$$

This value of y is one-half the minimal distance between the plates. Therefore, the minimal distance is about 5.96 mm.

4.2.3 Effects of Interfacial Tension Gradients

Since surface tension depends on temperature, a permanent nonuniformity of temperature or concentration (for a multicomponent system) at a liquid–vapor interface causes a surface tension gradient. The interfacial area with small surface tension expands at the expense of an area with greater surface tension, which in turn establishes a steady flow pattern in the liquid; this flow caused by the surface tension gradient along the liquid–vapor (gas) interface is referred to as the *Marangoni effect*. The surface tension of a multicomponent liquid that is in equilibrium with the vapor is a function of temperature and composition of the mixture, i.e.,

$$\sigma = \sigma(T, x_1, x_2, \ldots, x_{N-1}) \tag{4.23}$$

where x_i is the molar fraction of the ith component in the liquid phase and N is the total number of components in the liquid phase. The change of surface tension can be caused by either change of temperature or composition, i.e.,

$$\mathrm{d}\sigma = \left(\frac{\partial \sigma}{\partial T}\right)_{x_i} \mathrm{d}T + \sum_{i=1}^{N-1} \left(\frac{\partial \sigma}{\partial x_i}\right)_{T, x_{j \neq i}} \mathrm{d}x_i \tag{4.24}$$

Recall from thermodynamics that as the critical temperature for a given fluid is approached, the properties of the liquid and vapor phases of the fluid become identical, i.e., σ vanishes. It therefore follows that the surface tension decreases with temperature, i.e., $\left[(\partial \sigma / \partial T)_{x_i} < 0\right]$. For a pure substance, surface tension is a function of temperature only. The curve-fit equations for surface tension are almost linear for most fluids and can have the form

$$\sigma = C_0 - C_1 T \tag{4.25}$$

where C_0 and C_1 are empirical constants that vary between substances. For water, $C_0 = 75.83 \times 10^{-3}$ N/m and $C_1 = 0.1477 \times 10^{-3}$ N/m °C. The unit of the temperature T in Eq. (4.25) is °C.

Since surface tension varies with temperature, the surface tension will not be uniform if the temperature, along the liquid–vapor interface, is nonuniform. Consequently, the liquid in the lower surface tension region near the interface will be pulled toward the region with higher surface tension. Furthermore, because the surface tension usually decreases with increasing temperature, the flow driven by surface tension moves away from the interface with high temperature and towards the interface with low temperature. As noted before, this motion of a liquid caused by a surface tension gradient at the interface is referred to as the *Marangoni effect*. The most well-known example of surface-tension-driven flow is *Bernard cellular flow*, which occurs in a thin horizontal liquid layer heated from below. Figure 4.6 illustrates steady cellular flow driven by the Marangoni effect. Once a

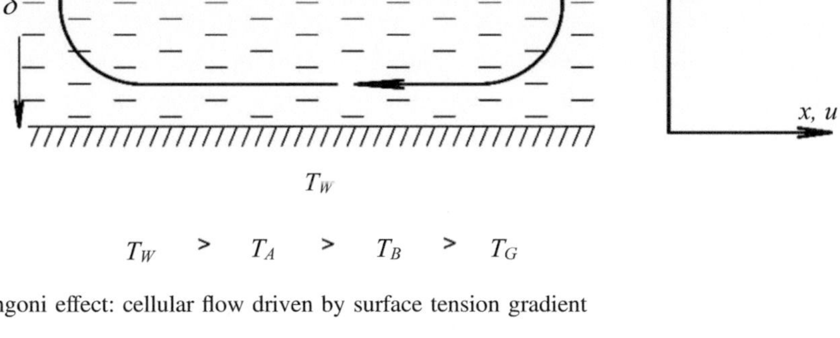

Fig. 4.6 Marangoni effect: cellular flow driven by surface tension gradient

steady Marangoni flow is established, the liquid velocity is upward at point A and downward at point B. Since the liquid at surface point A comes directly from the hot surface, the temperature at point A is higher than that at point B. Consequently, the surface tension at point A is smaller than that at point B, and the fluid at point A is pulled toward point B. Although a temperature gradient exists in the vertical direction, the actual driving force is the surface tension gradient in the horizontal direction. The liquid surface loses heat to the gas phase at temperature, T_G, through convection with a heat transfer coefficient of h_δ.

For mixtures, the surface tension is a function of both the temperature and concentration. Therefore, in heat pipes with mixture working fluids, one needs to consider both the thermal and concentration contributions to the Marangoni effect. These effects can oppose or favor the liquid motion toward the evaporator (hotter end). In general, for all pure working fluids used in conventional heat pipes and thermosyphons, the surface tension is a decreasing function of temperature. Therefore, one expects the Marangoni effect for pure fluids in heat pipes/thermosyphons to be unfavorable for the return of liquid to the evaporator because the temperature gradient is toward the condenser (cold end). This phenomenon was observed to be true by experiments performed by Kuramae and Suzuki (1993) by using water-ethanol solutions as the working fluid in a large-scale drop tower. Binary mixtures as working fluids are suggested for conventional thermosyphons for improved performance. Savino et al. (2007) performed numerical and experimental investigation to better understand the role of the Marangoni effect in heat pipes with aqueous solutions of long-chain alcohols.

An important boundary condition at the interface explains the liquid flow field caused by surface tension variation along the interface:

$$\tau_{yx}\big|_{y=\delta} = \mu_\ell \left(\frac{\partial u}{\partial y}\right)_{y=\delta} = \left(\frac{\partial \sigma}{\partial T}\right)\left(\frac{\partial T}{\partial x}\right)_{y=\delta} \tag{4.26}$$

Bernard cellular flow resulting from the Marangoni effect can occur only if certain conditions are met. The conditions for the onset of cellular motion can be predicted using linear stability analysis as presented by Carey (2016). The local temperature is assumed to equal the sum of the basic temperature $\overline{T}$—given by the initial linear profile—and a small sinusoidal fluctuation T' that represents a Fourier component of random disturbances, i.e.,

$$T = \overline{T} + T' = (T_w - \zeta y) + T' \tag{4.27}$$

where $\zeta = (T_w - T_\delta)/\delta$.

Since all base flow velocities are zero, the velocities are

$$v = \overline{v} + v' = v' \tag{4.28}$$

$$u = \overline{u} + u' = u' \tag{4.29}$$

Substituting Eqs. (4.27)–(4.29) into the continuity, momentum, and energy equations, and subtracting the corresponding base flow equations, the resulting equations are solved assuming the following forms for T' and v':

$$T' = \theta(y)e^{i\alpha x + \beta t} \tag{4.30}$$

$$v' = V(y)e^{i\alpha x + \beta t} \tag{4.31}$$

where $\alpha = 2\pi/\lambda$ is the wave number and β can be a complex number with its real part representing the amplification factor, and its imaginary part representing the temporal frequency. The resulting stationary wave solutions (for $\beta = 0$), which are not amplified or dampened with time, are therefore interpreted as stable perturbations. The conditions for which these solutions are obtained are assumed to correspond to the onset of instability, which leads to cellular convection. The case of marginal stability corresponds to

$$Ma = \frac{8\hat{\alpha}(\hat{\alpha}\cosh\hat{\alpha} + Bi\sinh\hat{\alpha})(\hat{\alpha} - \sinh\hat{\alpha}\cosh\hat{\alpha})}{\hat{\alpha}^3\cosh\hat{\alpha} - \sinh^3\hat{\alpha} - (8Cr\,\hat{\alpha}^5\cosh\hat{\alpha})/(Bo + \hat{\alpha}^2)} \tag{4.32}$$

where Ma is the Marangoni number:

$$Ma = \frac{\zeta(\mathrm{d}\sigma/\mathrm{d}T)\delta^2}{\alpha_\ell \mu_\ell} \tag{4.33}$$

and α_ℓ is the thermal diffusivity of the liquid.

The other dimensionless numbers in Eq. (4.32) are the wave number $\hat{\alpha}$, the Biot number Bi, the Bond number Bo, and the Crispation number Cr. Their definitions are

$$\hat{\alpha} = \alpha\delta \tag{4.34}$$

$$Bi = \frac{h_\delta \delta}{k_\ell} \tag{4.35}$$

$$Bo = \frac{(\rho_\ell - \rho_g)g\delta^2}{\sigma} \tag{4.36}$$

$$Cr = \frac{\mu_\ell \alpha_\ell}{\sigma\delta} \tag{4.37}$$

The liquid film is stable if the Marangoni number is below that predicted by Eq. (4.32). However, the liquid film is not stable if its Marangoni number is above that obtained by Eq. (4.32). The marginal stability predicted by Eq. (4.32) for some typical combinations of parameters is illustrated in Fig. 4.7. Since the Fourier components of all wavelengths can be contained in a random disturbance, the system becomes unstable when it is unstable at any wavelength. For a system with $Cr < 10^{-4}$ and $Bi \to 0$, $Bo \to 0$, Fig. 4.7 indicates that the critical Marangoni number is about 80 and the associated dimensionless wave number $\hat{\alpha} = 2$.

The Marangoni effect can have an important influence on heat and mass transfer processes, including the evaporation of a falling film, as well as causing vapor bubbles in a liquid with a temperature gradient to move toward the high temperature region during boiling.

Example 4.2

A 0.3-mm-thick water film sits on a surface held at a temperature of T_w = 80 °C. The top of the liquid film is exposed to air at a bulk temperature of T_G = 20 °C, and the convective heat transfer coefficient between the liquid film and the air is h_δ = 10 W/m^2 K. Determine whether the water film is stable.

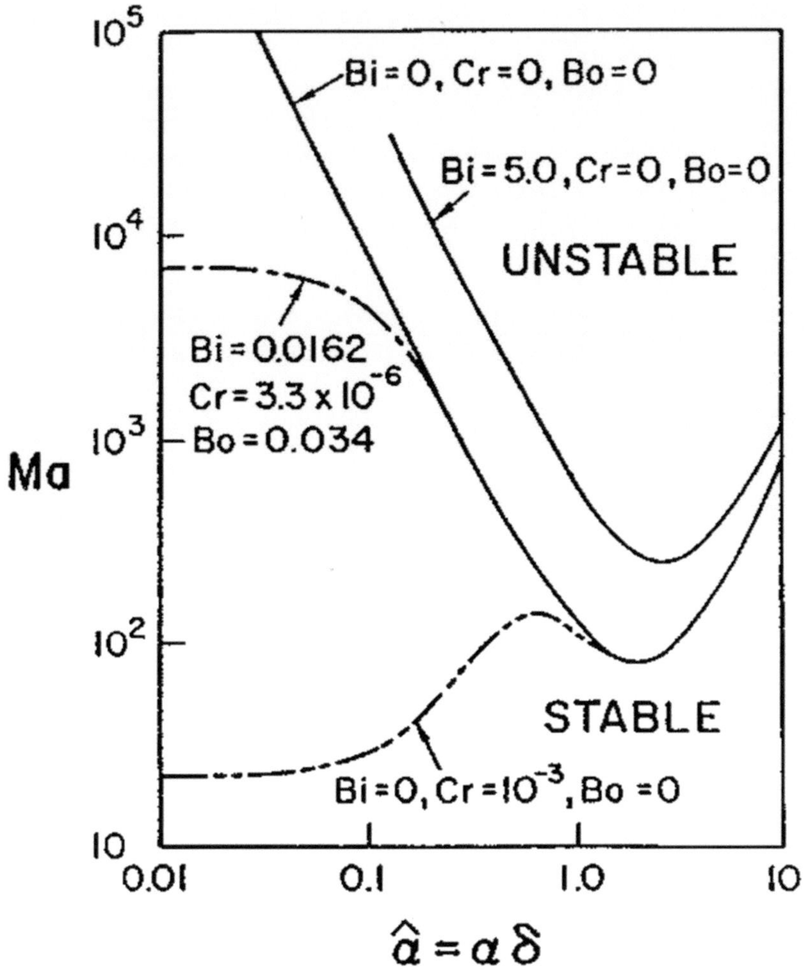

Fig. 4.7 Stability plane for the onset of cellular motion (Carey 2016; Reproduced by permission of Routledge/Taylor & Francis Group, LLC)

Solution Since the liquid film is very thin, the temperature drop across the liquid film will be very small. The properties of the liquid film can be determined at the wall temperature of 80 °C, i.e., $k_\ell = 0.67$ W/m K, $\rho_\ell = 971.9$ kg/m^3, $\alpha_\ell = 1.64 \times 10^{-7}$ m^2/s, $\mu_\ell = 351.1 \times 10^{-6}$ N s/m^2, $\sigma = 0.0626$ N/m, and $d\sigma/dT = -1.7 \times 10^{-4}$ N/m °C.

The Biot number is

$$Bi = \frac{h_\delta \delta}{k_\ell} = \frac{10 \times 0.3 \times 10^{-3}}{0.67} = 0.00425$$

At steady state, the surface temperature of the interface, T_δ, satisfies

$$k_\ell \frac{T_w - T_\delta}{\delta} = h_\delta (T_\delta - T_G)$$

i.e.,

$$\begin{aligned} T_\delta &= \left(T_w + \frac{h_\delta \delta}{k_\ell} T_G\right) \bigg/ \left(1 + \frac{h_\delta \delta}{k_\ell}\right) = \frac{T_w + Bi\, T_G}{1 + Bi} \\ &= \frac{80 + 4.25 \times 10^{-3} \times 20}{1 + 4.25 \times 10^{-3}} = 79.75\ ^\circ\mathrm{C} \end{aligned}$$

which demonstrates that the temperature drop across the liquid film is minimal. The Bond number is obtained by using Eq. (4.37), i.e.,

$$Bo = \frac{(\rho_\ell - \rho_g) g \delta^2}{\sigma} \doteq \frac{\rho_\ell g \delta^2}{\sigma} = \frac{971.9 \times 9.8 \times 0.0003^2}{0.0626} = 0.014$$

The Crispation number Cr is

$$Cr = \frac{\mu_\ell \alpha_\ell}{\sigma \delta} = \frac{351.1 \times 10^{-6} \times 1.64 \times 10^{-7}}{0.0626 \times 0.0003} = 3.07 \times 10^{-6}$$

which is less than 10^{-4}. The critical Marangoni number, Ma_c, below which the liquid film is stable, can be obtained from Fig. 4.7; its value is 80 at $\hat{\alpha} = 2$.

The Marangoni number of the system can be obtained from Eq. (4.33), i.e.,

$$\begin{aligned} Ma &= \frac{\zeta (\mathrm{d}\sigma/\mathrm{d}T) \delta^2}{\alpha_\ell \mu_\ell} = \frac{T_\delta - T_w}{\delta} \frac{(\mathrm{d}\sigma/\mathrm{d}T) \delta^2}{\alpha_\ell \mu_\ell} \\ &= \frac{79.75 - 80}{0.0003} \times \frac{(-1.7 \times 10^{-4}) \times 0.0003^2}{1.64 \times 10^{-7} \times 351.1 \times 10^{-6}} \\ &= 221.4 > Ma_c \end{aligned}$$

Therefore, the system is unstable and Marangoni convection will occur.

4.3 Wetting Phenomena and Contact Angles

4.3.1 Apparent Contact Angles

In addition to the surface tension at a liquid–vapor (ℓv) interface discussed above, surface tensions can also exist at interfaces between solid–liquid interface ($s\ell$) and solid–vapor interface (sv); this can be demonstrated using a liquid-vapor-solid system in Fig. 4.8. The contact line is the locus of points where the three phases intersect. The contact angle, θ, is the angle through the liquid between the tangent to the liquid–vapor interface and the tangent to the solid surface. The contact angle is defined for the equilibrium condition. In 1805, Young published the basic equation for the contact angle on a smooth, insoluble, and homogeneous solid:

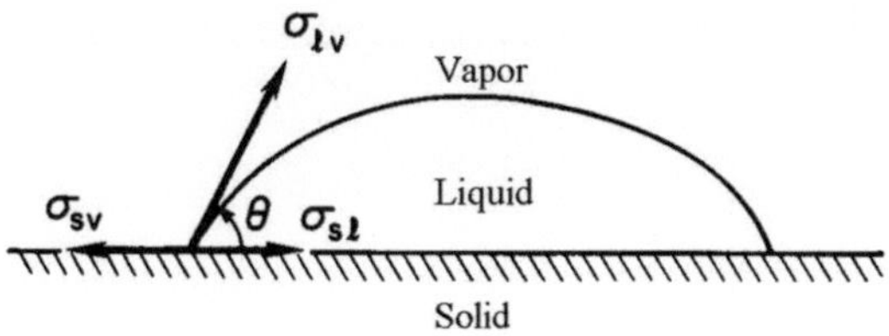

Fig. 4.8 Drop of liquid on a planar surface

$$\cos\theta = \frac{\sigma_{sv} - \sigma_{s\ell}}{\sigma_{\ell v}} \tag{4.38}$$

which follows from a balance of the horizontal force components (Faghri 2016), as shown in Fig. 4.8. When there is relative motion of a liquid drop over a solid surface, a different contact angle can be expected. When the relative motion stops, an angle different from the apparent (equilibrium) contact angle is seen; it depends upon the direction of the previous motion, i.e., whether it was a receding or advancing surface, as shown in Fig. 4.9. The minimum wetting contact angles for different solid–liquid combinations obtained by Stepanov et al. (1977) are reproduced in Table 4.1. All contact angle approaches are based on the following assumptions (Kwok and Neumann 1999; Yang et al. 2003): (1) validity of Young's equation (4.38), (2) pure liquid, (3) constant values of σ_{sv}, $\sigma_{s\ell}$, and $\sigma_{\ell v}$, (4) the value of liquid surface tension which should be higher than the anticipated solid surface tension, and (5) a value of σ_{sv} independent of the liquid used.

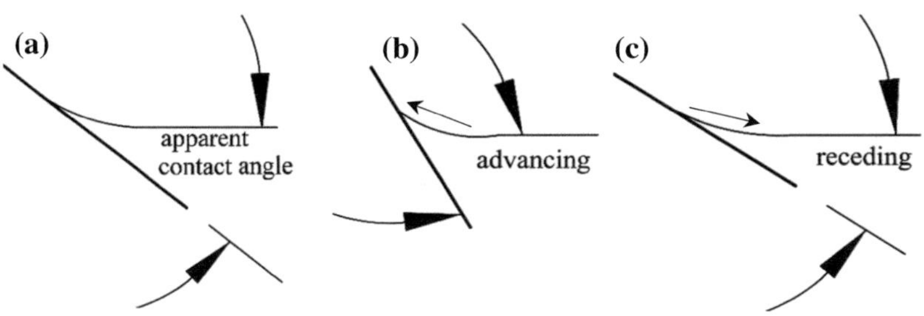

Fig. 4.9 Schematic of apparent contact angle θ: **a** stationary liquid; **b** liquid flows upward; **c** liquid flows downward

Table 4.1 Minimum wetting contact angle in arc degree (the upper and lower values are for advancing and receding liquid front, respectively; Stepanov et al. 1977)

	Acetone	Water	Ethanol	R-113
Aluminum		73/34		
Beryllium	25/11	63/7	0/0	
Brass		82/35	18/8	
Copper		84/33	15/7	
Nickel	16/7	79/34	16/7	
Silver		63/38	14/7	
Steel	14/6	72/40	19/8	16/5
Titanium		73/40	18/8	

For a rough surface, the contact angle θ_{rough} is related to the contact angle on a smooth surface θ by

$$\cos\theta_{\text{rough}} = \gamma\cos\theta \tag{4.39}$$

where γ is the ratio of the rough surface area to the smooth surface area. Since γ is always greater than 1, $\cos\theta_{\text{rough}}$ is greater than $\cos\theta$. Therefore, the contact angle θ_{rough} is less than the contact angle on a smooth surface, θ. Equilibrium contact angles can vary depending on the motion history of the contact line, particularly for rough surfaces. Equilibrium contact angles may be used in calculations for various heat transfer devices.

4.3.2 Wettability and Adsorption

Depending on the contact angle, liquids can be classified as *nonwetting* ($90° < \theta < 180°$), *partially wetting* ($0° < \theta < 90°$), or *completely wetting* ($\theta = 0°$). When a small amount of liquid is brought into contact with an initially dry solid surface, the liquid behaves in one of two ways: (1) if the liquid does not wet the solid, it may break up into small droplets, or (2) if the liquid wets the solid, it may spread over the solid surface and form a thin liquid film. Wettability can be attributed to a strong intermolecular attractive force near the interface between the solid and liquid. The thermodynamic definition of surface tension, Eq. (2.203), establishes that there is a significant decrease in the surface free energy per unit area in a wetting liquid. Spreading of a liquid on a solid surface can be described by the spreading coefficient Sp, defined as

$$\text{Sp}_{\ell s} = \sigma_{sv} - \sigma_{\ell v} - \sigma_{s\ell} \tag{4.40}$$

which is a measure of the ability of a liquid to spread over a solid surface. The spreading coefficient defined by Eq. (4.40) is difficult to evaluate, because data for σ_{sv} and $\sigma_{s\ell}$ are not available for most substances. Substituting Eq. (4.38) into Eq. (4.40), one obtains

$$\text{Sp}_{\ell s} = -\sigma_{\ell v}(1 - \cos\theta) \tag{4.41}$$

For a partially wetting liquid ($0° \le \theta \le 90°$), $\cos\theta \le 1$ and the spreading coefficient $\text{Sp}_{\ell s}$ are always negative, which means that the liquid will partially wet the solid surface and an equilibrium contact angle can be established.

The discussion thus far has treated the three phases—solid, liquid, and vapor—as though their boundaries were sharply delineated lines or surfaces. This idealization, which serves as a useful analytical device at the macroscopic level, does not hold at the microscopic level. At that level, the interfaces between phases appear as regions over which properties vary continuously, rather than as lines or surfaces with discontinuous property changes. Intermolecular forces of both repulsion and attraction influence how material in the various phases is distributed throughout these interfacial regions. Adsorption—which is one of the consequences of this intermolecular action—occurs when a liquid or solid phase adjacent to a second phase (solid, liquid, or gas) retains molecules, atoms, or ions of the second phase at the shared interface. Adsorption affects the wetting process because it alters the interfacial tension of the solid–liquid interface. Introducing the *surface pressure* of the adsorbed material on the solid surface, p_s:

$$p_s = \sigma_{sv} - \sigma_{sv,a} \tag{4.42}$$

where $\sigma_{sv,a}$ is the interfacial tension with the absorbed substance present. Young's equation, Eq. (4.38), can be rewritten as

$$\sigma_{\ell v} \cos\theta = (\sigma_{sv} - p_s) - \sigma_{s\ell} \tag{4.43}$$

which indicates that adsorption changes equilibrium contact angles and surface tension.

Surface tension, equilibrium contact angles, and capillary pressure determine the behavior of liquids in small-diameter tubes, slots, and porous media. The best example is the capillary rise of a wetting liquid in a small tube (see Fig. 4.10), in which case the capillary force is balanced by gravitational force. The pressure difference across the liquid–vapor interface in such tube is given by Eq. (4.10). For tubes with a very small radius, the two radii of curvature R_I and R_{II} are the same, i.e.,

$$R_I = R_{II} = \frac{r}{\cos\theta} \tag{4.44}$$

Substituting Eq. (4.44) into Eq. (4.10), one obtains the pressure difference between liquid and vapor at point C

$$(p_v - p_\ell)_C = p_{cap,C} = \frac{2\sigma}{r}\cos\theta \tag{4.45}$$

The pressure at the flat surface A (see Fig. 4.10a) is related to the vapor pressure by

$$p_A = p_v + \rho_v g H \tag{4.46}$$

where H is the height of the meniscus above the flat surface A. The pressure at point B inside the tube must be equal to that at point A because they are on the same horizontal surface, i.e.,

$$p_B = p_v + \rho_\ell g H - p_{cap,C} = p_A \tag{4.47}$$

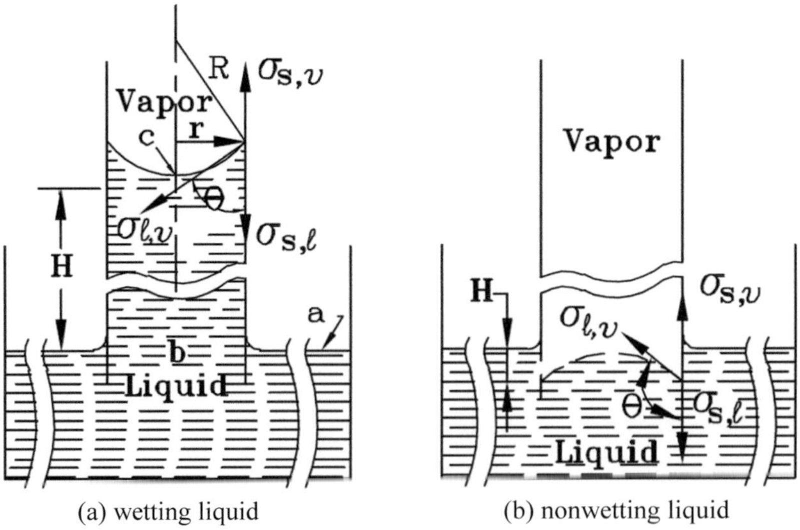

Fig. 4.10 Capillary phenomenon in an open tube (Faghri 2016)

Combining Eqs. (4.45)–(4.47), one obtains

$$\frac{2\sigma\cos\theta}{r} = (\rho_\ell - \rho_\mathrm{v})gH \tag{4.48}$$

Capillary rise phenomena can be observed when the liquid wets the tube wall ($\theta < 90°$). If the liquid cannot wet the tube wall ($\theta < 90°$), the capillary rise H obtained by Eq. (4.48) is negative, which indicates that there is a capillary depression, as shown in Fig. 4.10b.

In general, solid materials have only two types of behavior when they interact with water. They are either hydrophobic or hydrophilic. Hydrophobic materials have little or no tendency to absorb water, while hydrophilic materials have an affinity for water and readily absorb it. The criteria for hydrophobic or hydrophilic properties of a material is based on its contact angle.

$$\theta_{\mathrm{Hydrophilic}} < \frac{\pi}{2}, \theta_{\mathrm{Hydrophobic}} \geq \frac{\pi}{2}$$

Hydrophobic materials can be observed as beading of water on a surface, such as a freshly waxed car surface. Hydrophilic materials allow water to wet its surface forming a film or coating. Hydrophilic materials are usually charged or have polar side groups to their structure that attract water.

There are many cases of hydrophobic surfaces in nature, including some plant leaves, butterfly wings, duck feathers, and some insects' exoskeletons. There are many synthetic hydrophobic materials available including waxes, alkanes, oils, Teflon, and Gortex. There are numerous applications for using these materials such as the protection of stone, wood, and cement from the effects of rain, waterproofing fabrics and the removal of water from glass surfaces, such as a windshield, to increase transparency. Hydrophobic materials are also used for cleaning up oil spills, removal of oil from water and for chemical separation processes to remove nonpolar from polar compounds.

A hydrophilic material's ability to absorb and transport water gives it numerous applications in cleaners, housings, cables, tubes and hoses, waterproofing, catheters, surgical garments, etc. A hydrophilic coating on a tube or hose eliminates the need for other lubricants, which is useful to prevent cross-contamination. Hydrophilic coatings on plugs and o-rings increase their ability to stop leaks; this is the basis for water-stop and sealants.

Another use for hydrophobic and hydrophilic materials is the storage and distribution of water and methanol in miniature direct methanol fuel cells (DMFCs). For the distribution of methanol, a material is chosen that wets methanol but is hydrophobic to water. This type of a material is a preferential wicking material. This allows neat methanol to be stored and distributed in a DMFC without water diffusing into the methanol storage layers. The water storage layer at the anode of the fuel cell is a hydrophilic material. It is not preferential to either water or methanol and provides a layer in which they can mix.

Example 4.3

A 0.25-mm-diameter tube is placed vertically in a pool of water as shown in Fig. 4.8. The density of water is 1000 kg/m^3 and its surface tension is 0.06 N/m. It is assumed that the water can completely wet the tube. Find the capillary rise.

Solution Since the water can completely wet the tube, the contact angle is $\theta = 0$. Considering that the density of the vapor is much less than that of the liquid, Eq. (4.48) can be simplified as

$$\frac{2\sigma}{r} \approx \rho_\ell g H$$

Therefore, the capillary rise is

$$H = \frac{2\sigma}{\rho_\ell g r} = \frac{2 \times 0.06}{1000 \times 9.8 \times 0.125 \times 10^{-3}}$$
$$= 0.098 \text{ m} = 98 \text{ mm}$$

4.4 Phase Equilibrium in Microscale Interfacial Systems

4.4.1 Ultra-Thin Liquid Films

When a liquid film on a solid surface becomes very thin, the intermolecular attractive force between the molecules in the liquid and those in the solid surface turns to pull the liquid back to the liquid film. For a flat liquid film in which capillary pressure is absent, the pressure in the liquid film, p_ℓ, is changed by an amount, p_d ($p_d < 0$), referred to as *disjoining pressure* which causes the liquid pressure to be less than the vapor pressure ($p_\ell < p_v$). At equilibrium, the liquid pressure becomes $p_\ell = p_v + p_d$; in this case, the disjoining pressure is negative when the solid draws the liquid into the film. The disjoining pressure is a product of long-range intermolecular forces composed mainly of molecular (dispersion) and electrostatic interactions. They can be effective as long-range forces when the intermolecular spacing in the liquid film is about 0.2–10 nm. These forces are present even in nonpolar liquids because they are quantum-mechanical in origin. A pressure gradient is generated within the thin layer of liquid that covers the solid section in contact with the vapor. The disjoining pressure becomes more important for ultra-thin liquid films, because the liquid–vapor interface is closer to the solid surface. As a result, the attractive forces from the solid to liquid will have a greater effect on the liquid–vapor interface. As an example, Fig. 4.11 illustrates the variation of disjoining pressure with the liquid film thickness for CCl_4 on glass at 77 °C.

The dependence of disjoining pressure, p_d, on liquid film thickness, δ, can be obtained by analyzing the molecular forces and the reversible work required to establish a liquid microlayer:

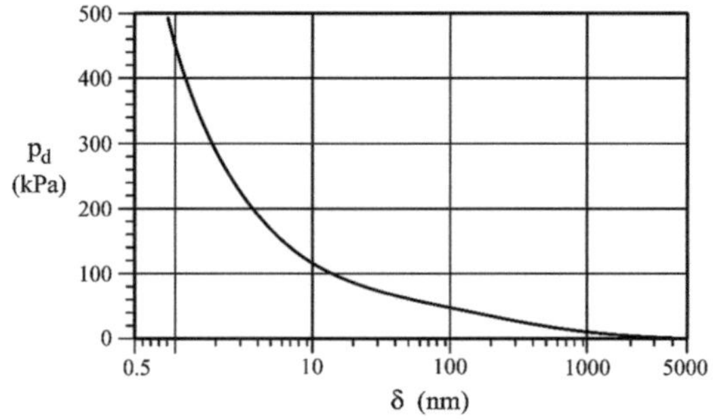

Fig. 4.11 Disjoining pressure of CCl_4 on glass at 77 °C (Potash and Wayner 1972)

$$p_d(\delta) = -\frac{2A_H}{\pi(n-2)(n-3)}\delta^{3-n} \tag{4.49}$$

where A_H is a constant, and n is the exponent in the long-range interaction potential $\phi(r) = -C_\phi/r^n$ that characterizes the force interaction between the liquid and solid molecules. This long-range interaction potential can be represented by the second term of the Lennard-Jones 6–12 potential. This results in the exponent in the long-range potential, n, equaling 6. Thus Eq. (4.49) becomes

$$p_d(\delta) = -\frac{A_H}{6\pi}\delta^{-3} \tag{4.50}$$

A more generalized relationship between the disjoining pressure and the film thickness can be expressed as (Potash and Wayner 1972)

$$p_d = -A'\delta^{-B} \tag{4.51}$$

where A' and B are constants that characterize the molecular and electrostatic interactions. Several components of the disjoining pressure are distinguished: dispersion (molecular), adsorption, electrostatic, and structural (Derjaguin 1955, 1989). Values of dispersion constants strongly depend on liquid types. For ammonia, for example, $A' = 1 \times 10^{-21}$ J. The nature of this phenomenon produces increasing (absolute value) pressure with decreasing film thickness. The pressure is considered to be positive for repulsion and negative for the attraction of the surface film. Due to the extremely high values of p_d in ultra-thin films, the liquid transport can be significant; thus the role of disjoining pressure in evaporation can be essential, particularly for low-temperature fluids. Disjoining pressure is one of the fundamental phenomena that affect the formation of thin evaporating films and the magnitude of the contact angles.

When a wetting liquid is in contact with a solid, the liquid forms a curved liquid/vapor interface, as shown in Fig. 4.12. When the liquid has a high wetting capability, the liquid spreads on the solid wall to form an extended meniscus. The extended meniscus can be divided into three regions: the equilibrium thin film, microfilm region, and intrinsic region. In the equilibrium thin film, the liquid molecules adhere to the solid, and the interfacial temperature is equal to the wall temperature. No evaporation occurs in this region because the liquid/vapor interfacial equilibrium temperature is elevated to the wall temperature due to the disjoining pressure effects. Virtually all of the evaporation occurs in the microfilm region. In this region, the disjoining and capillary pressure significantly affect its shape. When the wall temperature is greater than the saturation temperature for a given vapor pressure, the interfacial temperature lies between these two values, i.e., $T_{sat}(p_v) < T_\delta < T_w$. In the intrinsic region, the effect of disjoining pressure is negligible and surface tension effect is dominating. The liquid flow that continually feeds the evaporating thin film is driven by a pressure gradient. The pressure gradient near the equilibrium region is primarily caused by changes in disjoining pressure as the film thickens. While the film thickens, the curvature of the surface increases to a maximum. Once the surface curvature is at its maximum, it starts to decrease, and both the disjoining pressure gradient and the change in curvature drive the flow. As the film thickens further, the disjoining pressure effects become negligible and curvature change alone drives the flow. As the film thickness increases further, the evaporation rate drops to zero, and the curvature stays at a constant value. This region of the interface can be referred to as the meniscus, and is the fourth region labeled in Fig. 4.12. The ultra-thin film phenomena and the effects of disjoining pressure are very important in two-phase micro/miniature devices such as micro heat pipes (Khrustalev and Faghri 1994), rotating miniature heat pipes (Lin and Faghri 1999), and pulsating heat pipes (Zhang and Faghri 2008).

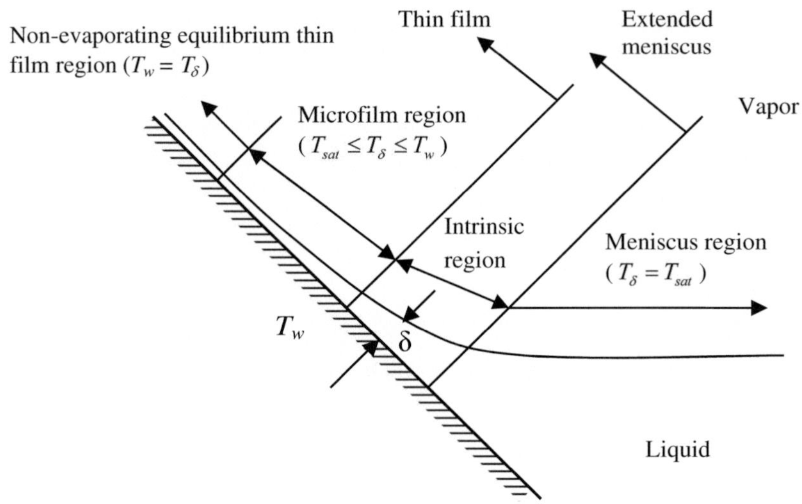

Fig. 4.12 Thin liquid film on a solid surface

The pressure difference between vapor and liquid at the liquid–vapor interface is due to the capillary and disjoining pressures, and is expressed by using the augmented Young-Laplace equation:

$$p_{\mathrm{v}} - p_\ell = p_{\mathrm{cap}} + p_{\mathrm{d}} \tag{4.52}$$

The heat and mass transfer that occurs at small scales are affected by interfacial and molecular forces. Surface roughness creates additional intermolecular interactions due to the increased surface area. Ojha et al. (2010) performed a detailed experimental studies to tie roughness to the thermal film thickness. They developed a direct relationship between the wetting properties of the liquid and surface properties. The modified capillary and disjoining pressure forces due to roughness influenced the evaporative heat transfer. There is also a large amount of ongoing work in the area of engineered roughness at the micro- and nanoscales for improved heat transfer (Plawsky et al. 2014).

4.4.2 Change in Saturated Vapor Pressure

In very thin films, as noted above, the attractive force from the solid surface to the liquid produces a pressure difference (disjoining pressure) across the liquid–vapor interface, in addition to the capillary effect. These two effects reduce the saturated vapor pressure over a thin film with curvature in comparison with the normal saturated condition. Consider a thin liquid film with liquid thickness δ over a substrate with liquid interface temperature T_δ and normal saturation vapor pressure corresponding to T_δ of $p_{\mathrm{sat}}(T_\delta)$. At equilibrium, the chemical potential in the two phases must be equal (see Chap. 2):

$$\mu_\ell = \mu_{\mathrm{v}} \tag{4.53}$$

Integrating the Gibbs-Duhem equation,

$$\mathrm{d}\mu = -s\mathrm{d}T + v\mathrm{d}p \tag{4.54}$$

at constant temperature from the normal saturated pressure $p_{\mathrm{sat}}(T_\delta)$ to an arbitrary pressure gives

$$\mu - \mu_{\text{sat}} = \int_{p_{\text{sat}}(T\delta)}^{p} \upsilon dp \tag{4.55}$$

Using the ideal gas law $\left(\upsilon_\upsilon = R_{\text{g}}T_\delta/P_{\text{v}}\right)$ for the vapor phase, and assuming the liquid phase is incompressible $(\upsilon = \upsilon_\ell)$, one obtains the following relations upon integration of Eq. (4.55) for the vapor and liquid chemical potentials, respectively:

$$\mu_{\text{v},\delta} = \mu_{\text{sat,v}} + R_{\text{g}}T_\delta \ln \frac{p_{\text{v},\delta}}{p_{\text{sat}}(T_\delta)} \tag{4.56}$$

$$\mu_{\ell,\delta} = \mu_{\text{sat},\ell} + \upsilon_\ell[p_\ell - p_{\text{sat}}(T_\delta)] \tag{4.57}$$

Since $\mu_{\text{sat},\ell} = \mu_{\text{sat,v}}$, substituting Eqs. (4.56) and (4.57) into Eq. (4.54) yields

$$p_{\text{v},\delta} = p_{\text{sat}}(T_\delta) \exp\left\{\frac{\upsilon_\ell[p_\ell - p_{\text{sat}}(T_\delta)]}{R_{\text{g}}T_\delta}\right\} \tag{4.58}$$

The pressure difference in the vapor phase $p_{\text{v},\delta}$ and the liquid phase p_ℓ due to capillary and disjoining effects are related as follows:

$$p_{\text{v},\delta} - p_\ell = p_{\text{cap}} - p_{\text{d}} \tag{4.59}$$

where p_{cap} is capillary pressure. Equation (4.59) can be used to eliminate p_ℓ in Eq. (4.58).

$$p_{\text{v},\delta} = p_{\text{sat}}(T_\delta) \exp\left\{\frac{[p_{\text{v},\delta} - p_{\text{sat}}(T_\delta) - p_{\text{cap}} + p_{\text{d}}]}{\rho_\ell R_{\text{g}} T_\delta}\right\} \tag{4.60}$$

When the interface is flat and $p_{\text{d}} = 0$, $p_{\text{v},\delta} = p_{\text{sat}}(T_\delta)$. For a curved interface and $p_{\text{d}} = 0$, Eq. (4.60) is reduced to the following Kelvin equation:

$$p_{\text{v},\delta} = p_{\text{sat}}(T_\delta) \exp\left\{\frac{[p_{\text{v},\delta} - p_{\text{sat}}(T_\delta) - p_{\text{cap}}]}{\rho_\ell R_{\text{g}} T_\delta}\right\} \tag{4.61}$$

4.5 Interfacial Heat and Mass Transfer

4.5.1 Interfacial Mass, Momentum, Energy, and Species Balances

The conservation laws for transport phenomena can be reduced to local partial differential equations if they are considered at a point that does not belong to a surface of discontinuity, such as an interface. When considering a discontinuous point, appropriate jump conditions relating to the values of the fundamental quantities on both sides of the interface should be considered. Jump conditions at an interface were discussed in Sect. 3.3.6, but the effects of surface tension and disjoining pressure associated with a non-flat liquid–vapor interface were not taken into account. It is the objective of this subsection to specify mass, momentum, and energy balance at a non-flat liquid–vapor interface, as well as species balance in solid-liquid-vapor interfaces.

4.5.1.1 Mass Balance

At a liquid–vapor interface, the mass balance is

$$\dot{m}_\delta'' = \rho_\ell(\mathbf{V}_\ell - \mathbf{V}_\mathrm{I}) \cdot \mathbf{n} = \rho_\mathrm{v}(\mathbf{V}_\mathrm{v} - \mathbf{V}_\mathrm{I}) \cdot \mathbf{n} \tag{4.62}$$

where $\dot{m}_\delta''$ is mass flux at the interface due to phase change and $\mathbf{V}_\mathrm{I}$ is the velocity of the interface. For a three-dimensional interface, there are three components of velocity: the normal direction, and two tangential directions, denoted by $\mathbf{n}$, $\mathbf{t}_1$, and $\mathbf{t}_2$, respectively. Therefore, the velocity components should be defined according to these directions, as follows:

$$\mathbf{V} \cdot \mathbf{n} = V_\mathbf{n} \tag{4.63}$$

$$\mathbf{V} \cdot \mathbf{t}_1 = V_{\mathbf{t}_1} \tag{4.64}$$

$$\mathbf{V} \cdot \mathbf{t}_2 = V_{\mathbf{t}_2} \tag{4.65}$$

The interfacial mass balance can be rewritten in these terms by:

$$\dot{m}_\delta'' = \rho_\ell\left(V_{\ell,\mathbf{n}} - V_{\mathrm{I},\mathbf{n}}\right) = \rho_\mathrm{v}\left(V_{\mathrm{v},\mathbf{n}} - V_{\mathrm{I},\mathbf{n}}\right) \tag{4.66}$$

4.5.1.2 Momentum Balance

For a situation where the effects of surface tension and disjoining pressure are negligible, the momentum balance at the liquid–vapor interface can be described by Eq. (3.176), i.e.,

$$\left(\boldsymbol{\tau}_\ell' - \boldsymbol{\tau}_\upsilon'\right) \cdot \mathbf{n} = \dot{m}_\delta''(\mathbf{V}_\ell - \mathbf{V}_\mathrm{v}) \tag{4.67}$$

where $\mathbf{n}$ is the normal direction of the interface and points toward the vapor. The left-hand side of Eq. (4.67) represents the force per unit area acting on the interface, while the right-hand side represents the change of momentum across the interface. For applications involving thin film evaporation and condensation, the effects of surface tension and disjoining pressure will create additional forces on the interface; in these cases, the left-hand side of Eq. (4.67) should be modified. The force per unit area created by the surface tension as indicated by Eq. (4.10) is $\sigma(T)(1/R_\mathrm{I} + 1/R_\mathrm{II})\mathbf{n}$. The force per unit area created by the disjoining pressure is $-p_\mathrm{d}\mathbf{n}$. For the situations where the interfacial temperature is not a constant, the contribution of the Marangoni effect, $(\mathrm{d}\sigma/\mathrm{d}T)\nabla T_\delta$, should also be included. Therefore, the momentum balance at the interface becomes

$$\begin{aligned} &\left(\boldsymbol{\tau}_\ell' - \boldsymbol{\tau}_\upsilon'\right) \cdot \mathbf{n} + \sigma(T)\left(\frac{1}{R_\mathrm{I}} + \frac{1}{R_\mathrm{II}}\right)\mathbf{n} - p_\mathrm{d}\mathbf{n} \\ &\quad - \left(\frac{\mathrm{d}\sigma}{\mathrm{d}T}\right)(\nabla T_\delta \cdot \mathbf{t})\mathbf{t} = \dot{m}_\delta''(\mathbf{V}_\ell - \mathbf{V}_\mathrm{v}) \end{aligned} \tag{4.68}$$

On the left-hand side, the first term is the stress tensor, the second term is the capillary pressure, the third term is the disjoining pressure, and the fourth term is the Marangoni stress. The right-hand side is the momentum transfer due to inertia. In this equation, the tangential direction, $\mathbf{t}$, can either be $\mathbf{t}_1$ or $\mathbf{t}_2$. The stress tensor is:

$$\boldsymbol{\tau}' = -p\mathbf{I} + 2\mu\mathbf{D} - \frac{2}{3}\mu(\nabla \cdot \mathbf{V})\mathbf{I} \tag{4.69}$$

The deformation tensor can be written for a reference frame that is adjusted to the interface:

$$\mathbf{D} = \frac{1}{2}\left[\nabla\mathbf{V} + (\nabla\mathbf{V})^T\right] = \begin{bmatrix} \frac{\partial V_{\mathbf{n}}}{\partial x_{\mathbf{n}}} & \frac{1}{2}\left(\frac{\partial V_{\mathbf{n}}}{\partial x_{\mathbf{t}_1}} + \frac{\partial V_{\mathbf{t}_1}}{\partial x_{\mathbf{n}}}\right) & \frac{1}{2}\left(\frac{\partial V_{\mathbf{n}}}{\partial x_{\mathbf{t}_2}} + \frac{\partial V_{\mathbf{t}_2}}{\partial x_{\mathbf{n}}}\right) \\ \frac{1}{2}\left(\frac{\partial V_{\mathbf{n}}}{\partial x_{\mathbf{t}_1}} + \frac{\partial V_{\mathbf{t}_1}}{\partial x_{\mathbf{n}}}\right) & \frac{\partial V_{\mathbf{t}_1}}{\partial x_{\mathbf{t}_1}} & \frac{1}{2}\left(\frac{\partial V_{\mathbf{t}_1}}{\partial x_{\mathbf{t}_2}} + \frac{\partial V_{\mathbf{t}_2}}{\partial x_{\mathbf{t}_1}}\right) \\ \frac{1}{2}\left(\frac{\partial V_{\mathbf{n}}}{\partial x_{\mathbf{t}_2}} + \frac{\partial V_{\mathbf{t}_2}}{\partial x_{\mathbf{n}}}\right) & \frac{1}{2}\left(\frac{\partial V_{\mathbf{t}_1}}{\partial x_{\mathbf{t}_2}} + \frac{\partial V_{\mathbf{t}_2}}{\partial x_{\mathbf{t}_1}}\right) & \frac{\partial V_{\mathbf{t}_2}}{\partial x_{\mathbf{t}_2}} \end{bmatrix} \tag{4.70}$$

The normal direction of the interface is [1 0 0], the first tangential direction is [0 1 0] and the second tangential direction is [0 0 1]. Therefore,

$$\begin{aligned} \boldsymbol{\tau}' \cdot \mathbf{n} = {} & [-p \quad 0 \quad 0] \\ & + 2\mu\left[\frac{\partial V_{\mathbf{n}}}{\partial x_{\mathbf{n}}} \quad \frac{1}{2}\left(\frac{\partial V_{\mathbf{n}}}{\partial x_{\mathbf{t}_1}} + \frac{\partial V_{\mathbf{t}_1}}{\partial x_{\mathbf{n}}}\right) \quad \frac{1}{2}\left(\frac{\partial V_{\mathbf{n}}}{\partial x_{\mathbf{t}_2}} + \frac{\partial V_{\mathbf{t}_2}}{\partial x_{\mathbf{n}}}\right)\right] \\ & - \frac{2}{3}\mu[\nabla \cdot \mathbf{V} \quad 0 \quad 0] \end{aligned} \tag{4.71}$$

This can be reduced to the three components to obtain:

$$\begin{aligned} \boldsymbol{\tau}' \cdot \mathbf{n} \cdot \mathbf{n} &= -p + 2\mu\frac{\partial V_{\mathbf{n}}}{\partial x_{\mathbf{n}}} - \frac{2}{3}\mu\nabla \cdot \mathbf{V} \\ &= -p + \frac{4}{3}\mu\frac{\partial V_{\mathbf{n}}}{\partial x_{\mathbf{n}}} - \frac{2}{3}\mu\left(\frac{\partial V_{\mathbf{t}_1}}{\partial x_{\mathbf{t}_1}} + \frac{\partial V_{\mathbf{t}_2}}{\partial x_{\mathbf{t}_2}}\right) \end{aligned} \tag{4.72}$$

$$\boldsymbol{\tau}' \cdot \mathbf{n} \cdot \mathbf{t}_1 = \mu\left(\frac{\partial V_{\mathbf{n}}}{\partial x_{\mathbf{t}_1}} + \frac{\partial V_{\mathbf{t}_1}}{\partial x_{\mathbf{n}}}\right) \tag{4.73}$$

$$\boldsymbol{\tau}' \cdot \mathbf{n} \cdot \mathbf{t}_2 = \mu\left(\frac{\partial V_{\mathbf{n}}}{\partial x_{\mathbf{t}_2}} + \frac{\partial V_{\mathbf{t}_2}}{\partial x_{\mathbf{n}}}\right) \tag{4.74}$$

The momentum equation balance at the interface is then broken into its three components, as follows:

Normal Direction

$$\begin{aligned} & -p_\ell + p_\mathrm{v} + \frac{4}{3}\left(\mu_\ell\frac{\partial V_{\ell,\mathbf{n}}}{\partial x_{\ell,\mathbf{n}}} - \mu_\mathrm{v}\frac{\partial V_{\mathrm{v},\mathbf{n}}}{\partial x_{\mathrm{v},\mathbf{n}}}\right) \\ & - \frac{2}{3}\left[\mu_\ell\left(\frac{\partial V_{\ell,\mathbf{t}_1}}{\partial x_{\mathbf{t}_1}} + \frac{\partial V_{\ell,\mathbf{t}_2}}{\partial x_{\mathbf{t}_2}}\right) - \mu_\mathrm{v}\left(\frac{\partial V_{\mathrm{v},\mathbf{t}_1}}{\partial x_{\mathbf{t}_1}} + \frac{\partial V_{\mathrm{v},\mathbf{t}_2}}{\partial x_{\mathbf{t}_2}}\right)\right] \\ & + \sigma\left(\frac{1}{R_\mathrm{I}} + \frac{1}{R_\mathrm{II}}\right) - p_\mathrm{d} = \dot{m}''_\delta\left(V_{\ell,\mathbf{n}} - V_{\mathrm{v},\mathbf{n}}\right) \end{aligned} \tag{4.75}$$

Tangential 1

$$\mu_\ell\left(\frac{\partial V_{\ell,\mathbf{n}}}{\partial x_{\mathbf{t}_1}}+\frac{\partial V_{\ell,\mathbf{t}_1}}{\partial x_{\mathbf{n}}}\right)-\mu_{\mathrm{v}}\left(\frac{\partial V_{\mathrm{v},\mathbf{n}}}{\partial x_{\mathbf{t}_1}}+\frac{\partial V_{\mathrm{v},\mathbf{t}_1}}{\partial x_{\mathbf{n}}}\right)-\left(\frac{\mathrm{d}\sigma}{\mathrm{d}T}\right)\left(\frac{\partial T_\delta}{\partial x_{\mathbf{t}_1}}\right)=\dot{m}''_\delta\left(V_{\ell,\mathbf{t}_1}-V_{\mathrm{v},\mathbf{t}_1}\right) \tag{4.76}$$

Tangential 2

$$\mu_\ell\left(\frac{\partial V_{\ell,\mathbf{n}}}{\partial x_{\mathbf{t}_2}}+\frac{\partial V_{\ell,\mathbf{t}_2}}{\partial x_{\mathbf{n}}}\right)-\mu_{\mathrm{v}}\left(\frac{\partial V_{\mathrm{v},\mathbf{n}}}{\partial x_{\mathbf{t}_2}}+\frac{\partial V_{\mathrm{v},\mathbf{t}_2}}{\partial x_{\mathbf{n}}}\right)-\left(\frac{\mathrm{d}\sigma}{\mathrm{d}T}\right)\left(\frac{\partial T_\delta}{\partial x_{\mathbf{t}_2}}\right)=\dot{m}''_\delta\left(V_{\ell,\mathbf{t}_2}-V_{\mathrm{v},\mathbf{t}_2}\right) \tag{4.77}$$

The non-slip condition at the liquid–vapor interface requires that $V_{\ell,\mathbf{t}_1}=V_{\mathrm{v},\mathbf{t}_1}$ and $V_{\ell,\mathbf{t}_2}=V_{\mathrm{v},\mathbf{t}_2}$. The momentum balance at the tangential directions becomes

$$\mu_\ell\left(\frac{\partial V_{\ell,\mathbf{n}}}{\partial x_{\mathbf{t}_1}}+\frac{\partial V_{\ell,\mathbf{t}_1}}{\partial x_{\mathbf{n}}}\right)=\mu_{\mathrm{v}}\left(\frac{\partial V_{\mathrm{v},\mathbf{n}}}{\partial x_{\mathbf{t}_1}}+\frac{\partial V_{\mathrm{v},\mathbf{t}_1}}{\partial x_{\mathbf{n}}}\right)+\left(\frac{\mathrm{d}\sigma}{\mathrm{d}T}\right)\left(\frac{\partial T_\delta}{\partial x_{\mathbf{t}_1}}\right) \tag{4.78}$$

$$\mu_\ell\left(\frac{\partial V_{\ell,\mathbf{n}}}{\partial x_{\mathbf{t}_2}}+\frac{\partial V_{\ell,\mathbf{t}_2}}{\partial x_{\mathbf{n}}}\right)=\mu_{\mathrm{v}}\left(\frac{\partial V_{\mathrm{v},\mathbf{n}}}{\partial x_{\mathbf{t}_2}}+\frac{\partial V_{\mathrm{v},\mathbf{t}_2}}{\partial x_{\mathbf{n}}}\right)+\left(\frac{\mathrm{d}\sigma}{\mathrm{d}T}\right)\left(\frac{\partial T_\delta}{\partial x_{\mathbf{t}_2}}\right) \tag{4.79}$$

For most applications, the evaporation or condensation rate—$\dot{m}''_\delta$—is not very high; therefore, it can be assumed that $\dot{m}''_\delta(\mathbf{V}_\ell-\mathbf{V}_{\mathrm{v}})\doteq 0$. If the liquid and vapor phases are further assumed to be inviscid ($\boldsymbol{\tau}_\ell=\boldsymbol{\tau}_{\mathrm{v}}\doteq 0$), the momentum equation at the interface can be reduced to

$$p_{\mathrm{v}}-p_\ell=\sigma(T)\left(\frac{1}{R_{\mathrm{I}}}+\frac{1}{R_{\mathrm{II}}}\right)-p_{\mathrm{d}} \tag{4.80}$$

4.5.1.3 Energy

The energy balance at the interface can be obtained from Eq. (3.177), i.e.,

$$\left(\mathbf{q}''_\ell-\mathbf{q}''_{\mathrm{v}}\right)\cdot\mathbf{n}-\left(\mathbf{n}\cdot\boldsymbol{\tau}'_\ell\right)\cdot\mathbf{V}_{\ell,\mathrm{rel}}+\left(\mathbf{n}\cdot\boldsymbol{\tau}'_{\mathrm{v}}\right)\cdot\mathbf{V}_{\mathrm{v,rel}}=\dot{m}''_\delta\left[\left(e_{\mathrm{v}}+\frac{\mathbf{V}^2_{\mathrm{v,rel}}}{2}\right)-\left(e_\ell+\frac{\mathbf{V}^2_{\ell,\mathrm{rel}}}{2}\right)\right] \tag{4.81}$$

where the work done by surface tension and disjoining pressure is neglected. If the velocity of the reference frame is taken as the interfacial velocity $\mathbf{V}_I$, Eq. (4.81) can be rewritten as

$$
\begin{aligned}
&(k_v \nabla T_v - k_\ell \nabla T_\ell) \cdot \mathbf{n} - (\mathbf{n} \cdot \boldsymbol{\tau}'_\ell) \cdot (\mathbf{V}_\ell - \mathbf{V}_I) + (\mathbf{n} \cdot \boldsymbol{\tau}'_v) \cdot (\mathbf{V}_v - \mathbf{V}_I) \\
&= \dot{m}''_\delta \left\{ \left[e_v + \frac{(\mathbf{V}_v - \mathbf{V}_I)^2}{2} \right] - \left[e_\ell + \frac{(\mathbf{V}_\ell - \mathbf{V}_I)^2}{2} \right] \right\}
\end{aligned}
\tag{4.82}
$$

where the heat fluxes in the liquid and vapor phases have been determined by Fourier's law of conduction.

Equation (4.82) can be rewritten in terms of enthalpy ($h = e + p/\rho$), i.e.,

$$
\begin{aligned}
&(k_v \nabla T_v - k_\ell \nabla T_\ell) \cdot \mathbf{n} - (\mathbf{n} \cdot \boldsymbol{\tau}'_\ell) \cdot (\mathbf{V}_\ell - \mathbf{V}_I) + (\mathbf{n} \cdot \boldsymbol{\tau}'_\upsilon) \cdot (\mathbf{V}_v - \mathbf{V}_I) \\
&= \dot{m}''_\delta \left[h_{\ell v} - \left(\frac{p_v}{\rho_v} - \frac{p_\ell}{\rho_\ell} \right) + \frac{1}{2}\mathbf{V}_v^2 - \mathbf{V}_v \cdot \mathbf{V}_I - \frac{1}{2}\mathbf{V}_\ell^2 + \mathbf{V}_\ell \cdot \mathbf{V}_I \right]
\end{aligned}
\tag{4.83}
$$

where $h_{\ell v}$ is the difference between the enthalpy of vapor and liquid at saturation, i.e., the latent heat of vaporization. The stress tensor in Eq. (4.83) can be expressed as

$$
\boldsymbol{\tau}' = -p\mathbf{I} + 2\mu\mathbf{D} - \frac{2}{3}\mu(\nabla \cdot \mathbf{V})\mathbf{I} = -p\mathbf{I} + \boldsymbol{\tau}
\tag{4.84}
$$

Since the relative velocities at the interface satisfy $(\mathbf{V}_\ell - \mathbf{V}_I) \cdot \mathbf{n} = \dot{m}''_\delta / \rho_\ell$ and $(\mathbf{V}_v - \mathbf{V}_I) \cdot \mathbf{n} = \dot{m}''_\delta / \rho_v$, we have

$$
p_\ell (\mathbf{V}_\ell - \mathbf{V}_I) \cdot \mathbf{n} - p_v (\mathbf{V}_v - \mathbf{V}_I) \cdot \mathbf{n} = \dot{m}''_\delta \left[-\left(\frac{p_v}{\rho_v} - \frac{p_\ell}{\rho_\ell} \right) \right]
\tag{4.85}
$$

Substituting Eqs. (4.84) and (4.85) into Eq. (4.83), the energy balance can be written as

$$
\begin{aligned}
&(k_v \nabla T_v - k_\ell \nabla T_\ell) \cdot \mathbf{n} - (\mathbf{n} \cdot \boldsymbol{\tau}_\ell)(\mathbf{V}_\ell - \mathbf{V}_I) + (\mathbf{n} \cdot \boldsymbol{\tau}_v)(\mathbf{V}_v - \mathbf{V}_I) \\
&= \dot{m}''_\delta \left[h_{\ell v} + \frac{1}{2}\mathbf{V}_v^2 - \mathbf{V}_v \cdot \mathbf{V}_I - \frac{1}{2}\mathbf{V}_\ell^2 + \mathbf{V}_\ell \cdot \mathbf{V}_I \right]
\end{aligned}
\tag{4.86}
$$

To simplify the energy equation, the kinetic energy terms are considered negligible and no-slip conditions are assumed at the interface,

$$
V_{\ell,\mathbf{t}} = V_{v,\mathbf{t}} = V_{I,\mathbf{t}}
\tag{4.87}
$$

Therefore, the energy equation can be rewritten as

$$
\begin{aligned}
\left(k_v \frac{\partial T_v}{\partial x_\mathbf{n}} - k_\ell \frac{\partial T_\ell}{\partial x_\mathbf{n}} \right) = \dot{m}''_\delta \bigg[& h_{\ell v} + \frac{4}{3}\nu_\ell \frac{\partial V_{\ell,\mathbf{n}}}{\partial x_\mathbf{n}} - \frac{2}{3}\nu_\ell \left(\frac{\partial V_{\ell,\mathbf{t}_1}}{\partial x_{\mathbf{t}_1}} + \frac{\partial V_{\ell,\mathbf{t}_2}}{\partial x_{\mathbf{t}_2}} \right) \\
& - \frac{4}{3}\nu_v \frac{\partial V_{v,\mathbf{n}}}{\partial x_\mathbf{n}} + \frac{2}{3}\nu_v \left(\frac{\partial V_{v,\mathbf{t}_1}}{\partial x_{\mathbf{t}_1}} + \frac{\partial V_{v,\mathbf{t}_2}}{\partial x_{\mathbf{t}_2}} \right) \bigg]
\end{aligned}
\tag{4.88}
$$

where ν_ℓ and ν_v are kinematic viscosities of liquid and vapor phases, respectively.

The energy balance at the interface can be simplified by assuming that the change in the kinetic energy across the interface is negligible, i.e.,

$$(k_{\rm v}\nabla T_{\rm v} - k_\ell \nabla T_\ell)\cdot \mathbf{n} = \dot{m}''_\delta h_{\ell \rm v} \tag{4.89}$$

Equations (4.62), (4.80), and (4.89) are widely used in the analysis of evaporation and condensation.

4.5.1.4 Species Balance

In general, to obtain a detailed solution, one needs to solve the continuity, species, momentum, and energy equations in each phase (presented in Chap. 3) and use the above interfacial balances to couple the conditions in each phase. This provides information for dependent variables, such as pressure, density, mass fraction, velocity, and temperature, as functions of time and space. Such an approach requires detailed numerical simulation in all phases. One may often be interested in a restricted solution for making an order of magnitude analysis or doing an analytical solution for a limiting case, in contrast to a detailed numerical solution. In some applications, dependent variables are also coupled, preventing the solution of one dependent variable independent of the others. In practice, one often must make assumptions to simplify the interface conditions or complexity of the multiphase problem. In other cases one can neglect, based on physical significance, the resistance to mass transfer and/or heat transfer in one of the phases; this permits simplification of the solution techniques and the solution of conservation equations in only one phase.

Some dependent variables, such as temperature, are continuous across the phases; other variables, such as concentration, are discontinuous.

For a general interface between phases k and j in a multi-component system, a local balance in mass flux of species i must be upheld. The total species mass flux, $\dot{m}''_i$, at an interface is:

$$\dot{m}''_i = \rho_{k,i}(\mathbf{V}_{k,i} - \mathbf{V}_{\rm I})\cdot \mathbf{n} = \rho_{j,i}(\mathbf{V}_{j,i} - \mathbf{V}_{\rm I})\cdot \mathbf{n} \tag{4.90}$$

The velocity of species i in phase k and phase j is $\mathbf{V}_{k,i}$ and $\mathbf{V}_{j,i}$, respectively. These velocities are defined as:

$$\rho_{k,i}\mathbf{V}_{k,i} = \mathbf{J}_{k,i} + \omega_{k,i}\rho_k \mathbf{V}_k \tag{4.91}$$

$$\rho_{j,i}\mathbf{V}_{j,i} = \mathbf{J}_{j,i} + \omega_{j,i}\rho_j \mathbf{V}_j \tag{4.92}$$

Using the interfacial species mass balance, and substituting the definition of the species velocity, the interfacial species mass flux is:

$$\begin{aligned}\dot{m}''_i &= \mathbf{J}_{k,i}\cdot \mathbf{n} + \omega_{k,i}\rho_k(\mathbf{V}_k - \mathbf{V}_{\rm I}) \\ &= \mathbf{J}_{j,i}\cdot \mathbf{n} + \omega_{j,i}\rho_j(\mathbf{V}_{jk} - \mathbf{V}_{\rm I})\end{aligned} \tag{4.93}$$

Remembering the overall mass conservation at the interface,

$$\dot{m}'' = \rho_k(\mathbf{V}_k - \mathbf{V}_{\rm I})\cdot \mathbf{n} = \rho_j(\mathbf{V}_j - \mathbf{V}_{\rm I})\cdot \mathbf{n} \tag{4.94}$$

The interfacial species mass flux is:

$$\dot{m}''_i = \mathbf{J}_{k,i}\cdot \mathbf{n} + \omega_{k,i}\dot{m}'' = \mathbf{J}_{j,i}\cdot \mathbf{n} + \omega_{j,i}\dot{m}'' \tag{4.95}$$

In some problems, the species mass flux will be specified, and the total mass flux is simply a sum of all the species mass fluxes.

$$\dot{m}'' = \sum_i \dot{m}_i'' \tag{4.96}$$

If the species mass flux is not specified, the total mass flux at an interface can be calculated from the interfacial species mass flux equation, Eq. (4.95):

$$\dot{m}'' = \frac{\left(\mathbf{J}_{j,i} - \mathbf{J}_{k,i}\right) \cdot \mathbf{n}}{\omega_{k,i} - \omega_{j,i}} \tag{4.97}$$

Since the species equation is second order in space, and there are two phases, there needs to be a secondary condition relating the species mass fraction in phases k and j. This can be done by expressing the mass fraction of species i in phase k as a function of the species mass fraction i in phase j, or vice versa.

$$\omega_{k,i} = \omega_{k,i}\left(\omega_{j,i}\right) \tag{4.98}$$

$$\omega_{j,i} = \omega_{j,i}\left(\omega_{k,i}\right) \tag{4.99}$$

Note that, in general, the species mass fraction is not a continuous function, and there is almost always a jump condition at the interface when two or more species are present. To make this point clear, the simplest case of a binary mixture is considered. For a binary mixture, the species diffusion flux, **J**, can be calculated by Fick's law.

$$\mathbf{J}_{k,1} \cdot \mathbf{n} = -\rho_k D_{k,12} \nabla \omega_{k,1} \cdot \mathbf{n} \tag{4.100}$$

$$\mathbf{J}_{j,1} \cdot \mathbf{n} = -\rho_j D_{j,12} \nabla \omega_{j,1} \cdot \mathbf{n} \tag{4.101}$$

There are several examples of different phenomena of a binary mixture in the presence of a liquid/vapor, solid/vapor, or liquid/solid interface. In other examples, the species flux in one of the phases is unimportant; therefore, only one phase must be considered.

A classical example of species flux is the condition of zero mass flux due to an impermeable surface where species A does not diffuse to the stationary media ($\nabla x_A \cdot \mathbf{n} = 0$) or ($\partial x_A / \partial y = 0$). The case of constant surface concentration is more typical; however, all types of mass transfer boundary conditions are demonstrated in three categories below.

a. The mass transfer from solid or liquid to a gas stream

Evaporation and sublimation are typical examples of mass transfer from liquid or solid to a gas mixture, as shown in Fig. 4.13, where $x_{A,\ell}$, $x_{A,s}$, $x_{A,g}$ are molar fractions in liquid, solid, and gas, respectively, and $\dot{m}_A''$ is mass flux for species A. There are two cases shown in Fig. 4.13. In case I, a pure liquid/solid is evaporating/sublimating into a gaseous mixture and the evaporation/sublimation rate is controlled purely by the species gradient in the gas phase. In case II, the liquid/solid is not pure, therefore concentration gradients also exist in the liquid/solid phases. The mass transfer rate is limited by both phases. Also note that the concentration gradient in the solid (Fig. 4.13b) is steeper and does not penetrate as deeply when compared to the liquid concentration gradient (Fig. 4.13a). Also, the liquid concentration gradient is steeper and does not penetrate as far compared to the gas. These trends are due to the ratio of the mass diffusivities in each of the phases.

A simplified relationship for the mass concentration condition at the interface can be obtained assuming the gas mixture is approximated by the ideal gas and the solid or liquid has a high

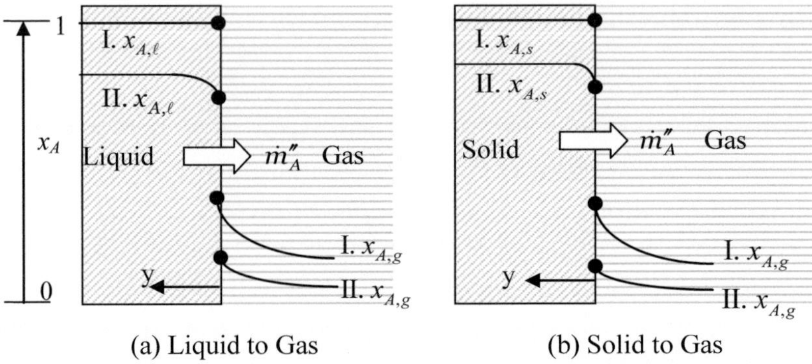

Fig. 4.13 Species concentration and mass transfer from solid and or liquid to a gas mixture

concentration of A. The condition within the gas stream is of interest because the main resistance to mass transfer is within that region. With the preceding assumptions, the partial vapor pressure of A in the gas mixture at the interface can be approximated from Raoult's Law:

$$p_A\big|_{y=0^-} = x_{A,\ell}\big|_{y=0^+}\left(p_{A,\mathrm{sat}}\right) \quad \text{Liquid to gas} \tag{4.102}$$

$$p_A\big|_{y=0^-} = x_{A,s}\big|_{y=0^+}\left(p_{A,\mathrm{sat}}\right) \quad \text{Solid to gas} \tag{4.103}$$

where p_A is the partial vapor pressure of A in the gas mixture, $x_{A,\ell}\big|_{y=0^+}$ is the mole fraction of species A in liquid, $x_{A,s}\big|_{y=0^+}$ is the mole fraction of species in solid, and $p_{A,\mathrm{sat}}$ is the normal saturation pressure of species A at the surface interface. Clearly, if the solid or liquid is made of pure species A, then $x_{A,\ell}\big|_{y=0} = x_{A,s}\big|_{y=0} = 1$ and Eqs. (4.102) and (4.103) reduce to

$$p_A\big|_{y=0^-} = p_{A,\mathrm{sat}} \tag{4.104}$$

This means that the partial vapor pressure of species A in the gas mixture at the interface is equal to the normal saturation pressure for species A, which is a function of interfacial temperature and can be obtained from thermodynamic tables in Appendix B. At the interface, species A and B in the gas phase must be in equilibrium with species A and B in the liquid or solid phase, except in extreme circumstances. Knowing $p_A\big|_{y=0^-}$ and total pressure p, one can easily calculate the mole fraction $x_{A,g}$ and mass fraction $\omega_{A,g}$ of species A at the interface on the gas side by the following relation

$$x_{A,g} = \frac{p_A\big|_{y=0^-}}{p} \tag{4.105}$$

$$\omega_{A,g} = \frac{x_{A,g}M_A}{x_{A,g}M_A + x_{B,g}M_B} \tag{4.106}$$

where M_A and M_B are molecular weights of species A and B. A conventional example for the case in Fig. 4.13a is the evaporation of water to a water-air mixture; obviously, one can imagine that the water absorbed a small amount of air. Sublimation of iodine in air is an application for the case of Fig. 4.13b.

Mass transfer from a solid to a gaseous state sometimes requires the specification of diffusion molar flux, rather than concentration, at the solid surface.

$$J_A^* = -cD_{AB}\frac{\partial x_A}{\partial y} \tag{4.107}$$

where J_A^* can be a function of concentration and not necessarily constant. One application is related to catalytic surface reaction. For this case, $\dot{m}'' = 0$ because the mass flux of reactants equals the mass flux of products. Therefore, from Eq. (4.95), $\dot{m}_i'' = J_{k,i} \cdot \mathbf{n}$. Catalytic surfaces are used to promote heterogeneous reactions (see Chap. 3), which occur at the surface; the appropriate boundary condition is

$$J_A^* = -k_1'\left(c_A\big|_{y=0}\right)^n \tag{4.108}$$

where k_1' is the reaction rate constant and n is the order of reaction.

b. **Mass transfer from gases to liquids or solids**

There are two major forms of mass transfer from gas to liquid, shown in Fig. 4.14. Case I is for condensable species in the liquid phase, or for species that can deposit vapor into the solid phase. In this case, a species of low concentration in the gas phase can condense at a higher concentration in the liquid phase. This phenomenon is important in distillation processes. Case II is for species that are weakly soluble in a liquid or a solid. In the liquid ($x_{A,\ell}$ is small), Henry's law will relate the mole fraction of species A in the liquid to the partial vapor pressure of A in the gas mixture at the interface by following relation (Fig. 4.14a).

$$x_{A,\ell}\big|_{y=0^+} = \frac{p_A\big|_{y=0^-}}{H} \tag{4.109}$$

Henry's constant, H, is dependent primarily on the temperature of the aqueous solution and the absorption of two species. The pressure dependence is usually small and in fact, negligible up to pressure equal to +5 bar. Table B.55 provides H for selected aqueous solutions. For binary soluble gases, Henry's law is not appropriate, and solubility data are usually presented in term of gas-phase partial pressure to liquid-phase mole fraction. Such data are given in Appendix B for NH_3-water and SO_2-water systems.

In chemical engineering applications, there are many cases in which gas is absorbed into a liquid; two examples are the absorption of hydrogen sulfide from an H_2S-air mixture into liquid water and the absorption of O_2 or chlorine into liquid water.

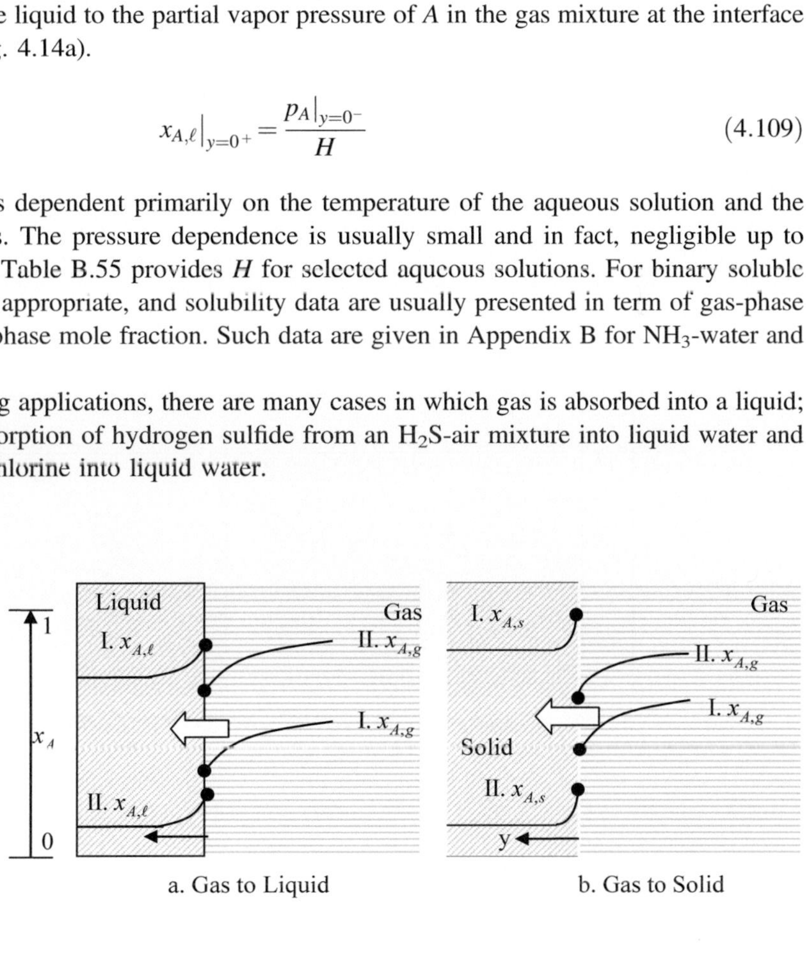

Fig. 4.14 Species concentration and mass transfer from gas mixture to liquid or solid

The concentration of gas in a solid at the interface is usually obtained by the use of a property known as the solubility, S, defined below.

$$c_{A,s}\big|_{y=0^+} = S p_{A,g}\big|_{y=0^-} \qquad (4.110)$$

where $c_{A,s}$ is the mole concentration of species A in the solid at the interface and $p_{A,g}\big|_{y=0^-}$ is the vapor partial pressure of species A in the gas at the interface. The values of S (kmol/Pa m^3) for vapor gas–solid combinations are given in chemistry handbooks; some are in Appendix B (Tables B.54, B.56, B.57). Special care should be exercised when using Eq. (4.110) for the variation in form and units—two different versions of S are presented in Appendix B. An example of this case is the diffusion of helium in a glass. The dissolution of gas in metals is much more complex and depends on the type of metal used. The dissolution of gas in some metals can be reversed, such as the case of hydrogen into titanium. In contrast to hydrogen, oxygen dissolution in titanium is irreversible but is complicated because it forms a layer of scale TiO_2 on the surface. This topic is beyond the scope of this book, and interested readers should refer to a metallurgy or materials handbook to consult appropriate phase diagrams.

c. **Mass transit from solid to liquid or liquid to solid**

Two cases are also presented for the case of mass transfer between a solid and a liquid (Fig. 4.15). Case I represents a pure solid dissolving into a liquid, or a pure liquid diffusing into a solid. Case II represents a solid mixture melting into a liquid mixture, or a liquid mixture solidifying into a solid mixture. For the first case, the mass transfer is related to the solubility, which can be found in chemistry handbooks (Lide 2004) and some are reproduced in Appendix B. A conventional example for this case is the dissolution of salt into water.

For a liquid/solid interface during melting and solidification, the ratio of the species concentration of the solid in liquid phases is called the partition ratio, K_p.

$$K_p = \frac{c_{A,s}}{c_{A,\ell}} \qquad (4.111)$$

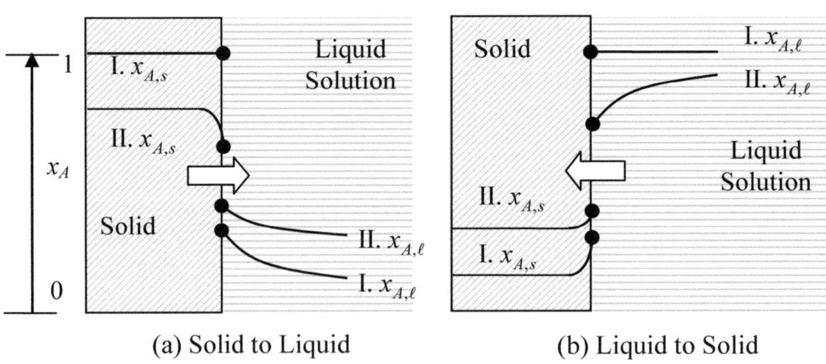

Fig. 4.15 Species concentration and mass transfer from solid to liquid and liquid to solid

The partition coefficient is a function of temperature. Furthermore, when the liquid and solid lines are nearly straight, it is a constant. This information can be obtained from phase diagrams for different solid/liquid mixtures.

Example 4.4

Write the continuity, momentum, energy, and species equations and the necessary boundary conditions for the horizontal evaporating capillary tube in Fig. 4.16, which is open at one end and closed at the other. The capillary tube is on the order of 100 μm. The evaporation is driven by the concentration gradient of vapor in the air. The evaporation cools the interface while the wall heats the fluid, causing a temperature gradient along the interface. Since the surface tension is a function of temperature, Maragoni stresses are created due to the temperature gradient along the interface. The wall is at a constant temperature, which is the ambient air temperature. The meniscus of a volatile liquid in ambient conditions with no forced heating recedes into the tube because of evaporation. The fluid density and viscosity are constant. Gravitational effects are negligible because the tube lies horizontally and the pore diameter is small enough that free convection can be considered negligible. The disjoining pressure effects are put into the contact angle because the thin film region of the capillary diameter has minimum effects on the total heat transfer and evaporation. The gas in the vapor region is assumed to be an ideal gas. Assume quasi-steady state and that the specific heats of the vapor and air are the same.

Solution Figure 4.17 shows the control volume, which was chosen due to symmetry, and the boundaries identified with numerical indices of the problem. The incompressible Navier-Stokes equations are solved, and the fluid is considered to be Newtonian. The continuity and momentum equations are

$$\nabla \cdot \mathbf{V} = 0 \tag{4.112}$$

$$\rho \frac{D\mathbf{V}}{Dt} = -\nabla p + \nabla \cdot (\mu \nabla \mathbf{V}) \tag{4.113}$$

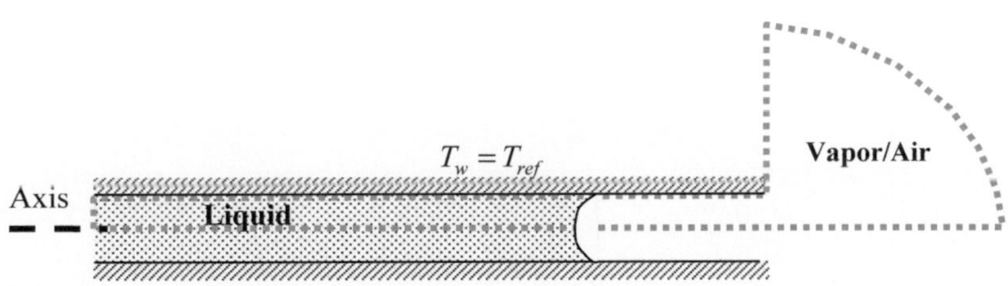

Fig. 4.16 Physical model of a horizontal evaporating capillary tube

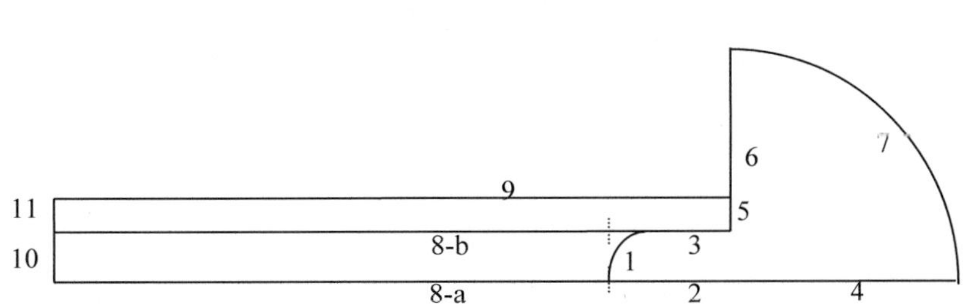

Fig. 4.17 Control volumes and boundaries for the horizontal evaporating capillary tube

The energy equation in the liquid region is

$$\frac{\partial \rho h}{\partial t} + \mathbf{V} \cdot \nabla(\rho h) = \nabla \cdot (k \nabla T) \tag{4.114}$$

If the solid wall is being modeled, only the conduction in the solid is modeled, since only the steady-state solution is of interest.

$$\nabla^2 T = 0 \tag{4.115}$$

The species equation for vapor phase is

$$\frac{\partial \rho \omega}{\partial t} + \mathbf{V} \cdot \nabla(\rho \omega) = \nabla \cdot (\rho D \nabla \omega) \tag{4.116}$$

In the energy equation, the enthalpy, h, is

$$h = c_\mathrm{p}(T - T_\mathrm{ref}) \tag{4.117}$$

This condition is used for both the liquid and the gaseous regions because the vapor has a specific heat similar to air. The density in the liquid is constant, and the density of the gas is determined from the ideal gas law.

$$\rho = \frac{p_\mathrm{ref}}{\left(\dfrac{\omega}{M_v} + \dfrac{1-\omega}{M_\mathrm{air}}\right) R_u T} \tag{4.118}$$

The diffusion coefficient, D, in the species equation is

$$D = B \frac{T^{3/2}}{p} \tag{4.119}$$

The viscosity and thermal conductivity are considered to be constant in the liquid region, while the mass weighted average of these properties is used in the gaseous region.

$$\phi = \omega \phi_\mathrm{v} + (1-\omega)\phi_\mathrm{air} \tag{4.120}$$

At the liquid/vapor interface (boundary 1 in Fig. 4.17), the interfacial boundary conditions are

Conservation of Mass

$$\begin{aligned} \dot{m}'' &= \rho D \frac{\nabla \omega \cdot \mathbf{n}}{1-\omega} = \rho_\ell (\mathbf{V}_\ell - \mathbf{V}_I) \cdot \mathbf{n} \\ &= \rho_\mathrm{g} (\mathbf{V}_\mathrm{g} - \mathbf{V}_I) \cdot \mathbf{n} \end{aligned} \tag{4.121}$$

The subscript g refers to the mixture of vapor and air.

Conservation of Normal Momentum

$$\dot{m}''_I(\mathbf{V}_\ell - \mathbf{V}_g) \cdot \mathbf{n} + p_\ell - p_g = \sigma(K_1 + K_2) \tag{4.122}$$

Conservation of Tangential Momentum

$$\left(-\mu_\ell \nabla(\mathbf{V}_\ell \cdot \mathbf{t}) + \mu_g \nabla(\mathbf{V}_g \cdot \mathbf{t})\right) \cdot \mathbf{n} = \gamma \nabla T \cdot \mathbf{t} \tag{4.123}$$

where γ is the surface temperature change with temperature $\mathrm{d}\sigma/\mathrm{d}T$.

Conservation of Energy

$$-k_\ell \nabla T \cdot \mathbf{n} + k_g \nabla T \cdot \mathbf{n} = h_{\ell v} \dot{m}'' \tag{4.124}$$

Saturation Mass Fraction

$$\omega = \omega_{\mathrm{sat}} \tag{4.125}$$

No-Slip Velocity in the *Tangential Direction*

$$\mathbf{V}_\ell = \mathbf{V}_g \tag{4.126}$$

Continuity of Temperature

$$T_\ell = T_g \tag{4.127}$$

The surface tension is a function of temperature and is approximated by the following equation:

$$\sigma_T = \sigma_{T_{\mathrm{ref}}} + \gamma \cdot (T - T_{\mathrm{ref}}) \tag{4.128}$$

The mass fraction at the liquid/vapor interface was found using the molar concentration of an ideal gas

$$x_g = \frac{p_{\mathrm{sat}}}{p} \tag{4.129}$$

$$\omega_{\mathrm{sat}} = \frac{x_g M_g}{x_g M_g + (1 - x_g) M_{\mathrm{air}}} \tag{4.130}$$

The partial pressure of the vapor is a very crucial parameter in the evaporation process. The function of pressure related to temperature from Yaws (1992) is used.

$$\log \frac{p}{132.953} = a - \frac{b}{(T + c)} \tag{4.131}$$

Since the free surface is always a function of the radial location, f, the two curvatures, K_1 and K_2, are:

$$K_1 = -\frac{|f''|}{(f'^2 + 1)^{3/2}} \quad (4.132)$$

$$K_2 = -\frac{f'}{r\sqrt{f'^2 + 1}} \quad (4.133)$$

The first and second derivates of f should be taken by differentiating the face center locations of the interface. The pressure at the face closest to the wall is calculated with a central difference scheme, but the point of intersection between the wall and the liquid/vapor interface is calculated from the prescribed contact angle, θ.

$$x_w = \frac{\frac{1}{\tan\theta}\left(a - \frac{a^2}{b}\right) + x_a - \frac{a^2}{b^2}x_b}{1 - \frac{a^2}{b^2}} \quad (4.134)$$

where a is the radial distance from the wall to its adjacent face and b is the radial distance from the wall to the second face from the wall. If the interface is close to the tube mouth, and x_w exceeds x_m, then the prescribed location of intersection between the interface and the wall to calculate the pressure is x_m. The conditions at the inner wall (the boundaries labeled 3, 5, and 8b in Fig. 4.17) are

$$T = T_{ref} \quad (4.135)$$

$$\mathbf{V} = 0 \quad (4.136)$$

$$\nabla\omega \cdot \mathbf{n} = 0 \quad (4.137)$$

Far from the tube mouth, ambient air conditions of pressure, temperature, and mass fraction are constants. These conditions apply to a quarter circle with a radius of $6r_0$, as shown in Fig. 4.17 (boundaries 6 and 7):

$$p = p_{ref} \quad (4.138)$$

$$\text{If } \mathbf{V} \cdot \mathbf{n} \le 0 \rightarrow T = T_{ref} \text{ else } \nabla T \cdot \mathbf{n} = 0 \quad (4.139)$$

$$\omega = 0 \quad (4.140)$$

At the tube liquid entrance (boundary 10), the boundary conditions are:

$$\mathbf{V} \cdot \mathbf{n} = \frac{\sum_j \dot{m}''_{I,j}|A_j|}{\rho\pi r_0^2} \quad (4.141)$$

$$\nabla T \cdot \mathbf{n} = 0 \quad (4.142)$$

At the axis of axi-symmetric geometry (boundaries 2, 4, and 8a), the boundary conditions are

$$\nabla(\mathbf{V} \cdot \mathbf{t}) \cdot \mathbf{n} = 0 \quad (4.143)$$

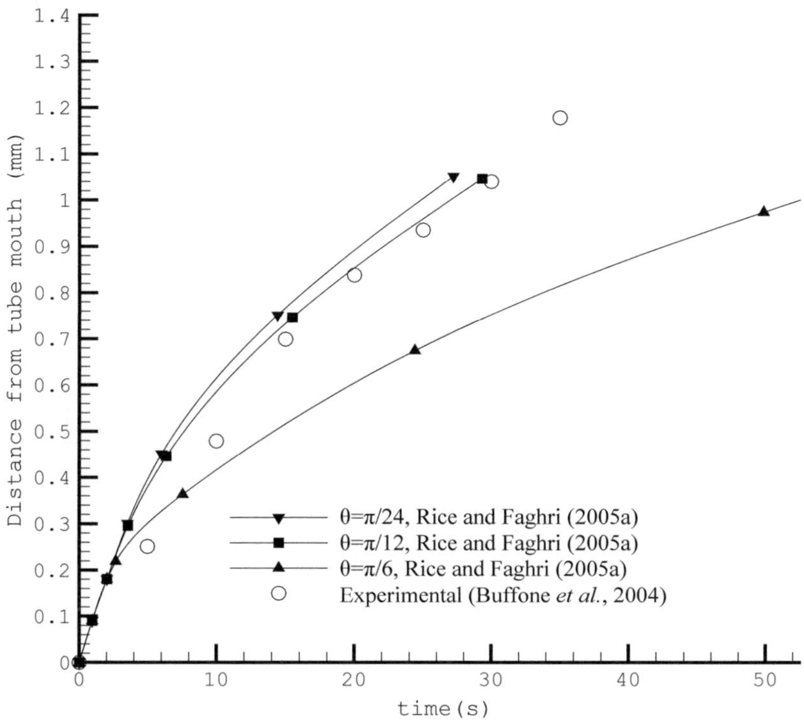

Fig. 4.18 Distance of meniscus center distance from mouth versus time for acetone, for a tube with a diameter of 600 μm (Rice and Faghri 2005a)

$$\mathbf{V} \cdot \mathbf{n} = 0 \tag{4.144}$$

$$\nabla T \cdot \mathbf{n} = 0 \tag{4.145}$$

Rice and Faghri (2005a) developed a computational liquid/vapor interface tracking technique to model an interface between a liquid and a vapor, including mass transfer, for this problem. The effect of contact angle on the distance the meniscus recedes inside the capillary over time is presented in Fig. 4.18. The rate at which the meniscus recedes inside the tube increases with decreasing contact angle. The distance the meniscus recedes over time asymptotically approaches the solution with a contact angle of 0°. The time it takes the center of the meniscus to be approximately 1.1 mm away from the tube mouth is more than 100% different for the contact angles of $\pi/6.0$ and $\pi/12.0$. However, the difference reduces to about 6% between contact angles of $\pi/12.0$ and $\pi/24.0$. This result verifies the assumption of including all of the solid/liquid/vapor interaction into the contact angle. The evaporation rate increases with decreasing contact angle, because the diffusion length scale decreases with decreasing contact angle. With a contact angle of zero, if the meniscus has a spherical shape, the meniscus spans r_0 in the axial direction. The experimental results of Buffone et al. (2004) are also plotted in Fig. 4.18 for comparison. With the smaller contact angles, the numerical work has a slightly higher evaporation rate than experimentally observed. The increased evaporation rate can be attributed to the constant temperature boundary condition, whereas the experimental condition is natural convection with the ambient air. Therefore, the numerical work has a slightly higher temperature, and therefore a slightly higher vapor pressure at the interface, leading to a higher evaporation rate.

4.5.2 Interfacial Resistances in Vaporization and Condensation

High-heat transfer coefficients, typically associated with evaporation and condensation processes in heat transfer devices are restricted by interfacial resistance. When condensation occurs at an interface, the flux of vapor molecules into the liquid must exceed the flux of liquid molecules escaping to the vapor phase. When evaporation occurs, on the other hand, the flux of liquid molecules escaping to the vapor phase must exceed the flux of vapor molecules into the liquid. Schrage (1953) used the kinetic theory of gases to describe condensation and evaporation processes and considered separately the fluxes of condensing and vaporizing molecules in each direction.

From kinetic theory, the mass flow rate (of molecules) passing in either direction (right or left) through an imagined plane is given by

$$|\dot{m}_\delta| = \left(\frac{M_v}{2\pi R_u}\right)^{1/2} \frac{p}{T^{1/2}} \tag{4.146}$$

where $\dot{m}''_\delta$ is the flux of molecules, R_u is the universal gas constant, and M_v is the molecular mass of the vapor. The net molecular flux through an interface is

$$\dot{m}''_\delta = \dot{m}''_{\delta+} - \dot{m}''_{\delta-} \tag{4.147}$$

Actually, the Boltzmann transport equation should be solved with appropriate boundary conditions of thermal equilibrium at several mean free path distances from the interface. However, a reasonable approximation can be obtained by means of correction factors if it is assumed that the interaction between molecules leaving the interface and those approaching was at equilibrium. This obtains the following relation for the net mass flux at the interface:

$$\dot{m}''_\delta = \frac{q''_\delta}{h_{\ell v}} = \alpha\sqrt{\frac{M_v}{2\pi R_u}}\left(\frac{\Gamma p_v}{\sqrt{T_v}} - \frac{p_\ell}{\sqrt{T_\ell}}\right) \tag{4.148}$$

where α is the accommodation coefficient ($\alpha \leq 1$), and the function Γ is given by Schrage (1953):

$$\Gamma(a) = \exp(a^2) + a\sqrt{\pi}[1 + \mathrm{erf}(a)] \tag{4.149}$$

$$\Gamma(-a) = \exp(a^2) - a\sqrt{\pi}[1 - \mathrm{erf}(a)] \tag{4.150}$$

where

$$a = \frac{q''_\delta}{\rho_v h_{\ell v}}\sqrt{\frac{M_v}{2R_u T_v}} \tag{4.151}$$

and

$$\mathrm{erf}(a) = \frac{2}{\pi}\int_0^a e^{-x^2}\,\mathrm{d}x \tag{4.152}$$

is the Gaussian error function.

The heat flux to the interface is equal to the net mass flux multiplied by the latent heat $\left(q''_\delta = \dot{m}''_\delta h_{\ell v}\right)$. Since Γ is a function of q''_δ, Eq. (4.148) does not provide an explicit relation for the interfacial heat flux. Assuming that p_ℓ and p_v are the saturation pressures corresponding to T_ℓ and T_v, Eq. (4.148) can be represented in the following form:

$$q''_\delta = \alpha h_{\ell v}\sqrt{\frac{M_v}{2\pi R_u}}\left[\frac{\Gamma p_{sat}(T_v)}{\sqrt{T_v}} - \frac{p_{sat}(T_\ell)}{\sqrt{T_\ell}}\right] \tag{4.153}$$

For evaporation and condensation of working fluids at moderate and high temperatures, a is usually very small according to its definition [see Eq. (4.151)]. In such a case, Eq. (4.147) can be approximated by

$$\Gamma = 1 + a\sqrt{\pi} \tag{4.154}$$

An explicit relation for q''_δ and $\dot{m}''_\delta$ was obtained by Silver and Simpson (1961) by substituting Eq. (4.154) into Eq. (4.148) and using $\rho_v = p_v M_v / R_u T_v$:

$$\dot{m}''_\delta = \frac{q''_\delta}{h_{\ell v}} = \left(\frac{2\alpha}{2-\alpha}\right)\sqrt{\frac{M_v}{2\pi R_u}}\left(\frac{p_v}{\sqrt{T_v}} - \frac{p_\ell}{\sqrt{T_\ell}}\right) \tag{4.155}$$

which is referred to as the Kucherov-Rikenglaz equation (Kucherov and Rikenglaz 1960) in the Soviet literature.

One can develop an alternative form of Eq. (4.153) for small a by assuming that $(p_v - p_\ell)/p_v \ll 1$, $(T_v - T_\ell)/T_v \ll 1$, and by using the Clausius-Clapeyron relation.

$$q''_\delta = \left(\frac{2\alpha}{2-\alpha}\right)\left(\frac{h_{\ell v}^2}{T_v v_{\ell v}}\right)\sqrt{\frac{M_v}{2\pi R_u T_v}}\left(1 - \frac{p_v v_{\ell v}}{2h_{\ell v}}\right)(T_v - T_\ell) \tag{4.156}$$

Using the above relation, the heat transfer coefficient at the interface h_δ is obtained from the following equation:

$$h_\delta = \frac{q''_\delta}{(T_v - T_\ell)} = \left(\frac{2\alpha}{2-\alpha}\right)\left(\frac{h_{\ell v}^2}{T_v v_{\ell v}}\right)\sqrt{\frac{M_v}{2\pi R_u T_v}}\left(1 - \frac{p_v v_{\ell v}}{2h_{\ell v}}\right) \tag{4.157}$$

If h_δ is of the same order of magnitude as the other h values, the effects of interfacial resistances should be accounted for. It should also be noted that the above equations relating q''_δ, h_δ, and $(T_v - T_\ell)$ apply equally well to evaporation and to condensation, with the convention that q''_δ is positive for condensation and negative for evaporation.

It is clear that predicting the interfacial resistances using any of the above equations depends on the value of the accommodation coefficient a, which varies widely in the literature. Paul (1962) compiled the accommodation coefficients for evaporation for a large number of working fluids. Mills (1965) recommended that α should be less than unity when the working fluids or the interface is contaminated.

The local heat flux through the liquid film due to heat conduction is:

$$q'' = k_\ell \frac{T_{\mathrm{w}} - T_\delta}{\delta} \tag{4.158}$$

where δ and T_δ are the local thickness of the liquid layer, and the temperature of the free liquid film surface, respectively.

The interfacial temperature T_δ, which is affected by the disjoining and capillary pressure also depends on the value of the interfacial resistance. For the case of a comparatively small heat flux, interfacial resistance is defined by the following relation [see Eq. (4.155)]:

$$q''_\delta = -\left(\frac{2\alpha}{2-\alpha}\right)\frac{h_{\ell \mathrm{v}}}{\sqrt{2\pi R_{\mathrm{g}}}}\left[\frac{p_{\mathrm{v}}}{\sqrt{T_{\mathrm{v}}}} - \frac{(p_{\mathrm{sat}})_\delta}{\sqrt{T_\delta}}\right] \tag{4.159}$$

where p_v and $(p_{\mathrm{sat}})_\delta$ are the saturation pressures corresponding to T_v in the bulk vapor and at the thin liquid film interface, respectively.

The relation between the vapor pressure over the thin evaporating film, $(p_{\mathrm{sat}})_\delta$—which is affected by the disjoining pressure—and the saturation pressure corresponding to T_δ, $p_{\mathrm{sat}}(T_\delta)$, is given by Eq. (4.60):

$$(p_{\mathrm{sat}})_\delta = p_{\mathrm{sat}}(T_\delta)\exp\left[\frac{(p_{\mathrm{sat}})_\delta - p_{\mathrm{sat}}(T_\delta) + p_{\mathrm{d}} - \sigma K}{\rho_\ell R_{\mathrm{g}} T_\delta}\right] \tag{4.160}$$

which is referred to as extended Kelvin equation and it reflects the fact that under the influence of disjoining and capillary pressures, the liquid-free surface saturation pressure $(p_{\mathrm{sat}})_\delta$ differs from the normal saturation pressure $p_{\mathrm{sat}}(T_\delta)$.

Under steady-state conditions, the heat flux obtained by Eqs. (4.158) and (4.159) is the same. Combining these two expressions yields

$$T_\delta = T_{\mathrm{w}} + \frac{\delta}{k_\ell}\left(\frac{2\alpha}{2-\alpha}\right)\frac{h_{\ell \mathrm{v}}}{\sqrt{2\pi R_{\mathrm{g}}}}\left[\frac{p_{\mathrm{v}}}{\sqrt{T_{\mathrm{v}}}} - \frac{(p_{\mathrm{sat}})_\delta}{\sqrt{T_\delta}}\right] \tag{4.161}$$

Equations (4.160) and (4.161) determine the interfacial temperature T_δ and pressure $p_{\mathrm{sat}}(T_\delta)$.

As the liquid film thins, the disjoining pressure p_{d} and the interfacial temperature T_δ increase. A non-evaporating film thickness is present under specific conditions, as demonstrated in Sect. 4.4.1, that gives the equality of the liquid–vapor interface and solid surface temperatures:

$$T_\delta = T_{\mathrm{w}} \tag{4.162}$$

Substituting Eq. (4.162) into Eq. (4.161), one obtains

$$(p_{\mathrm{sat}})_\delta = p_{\mathrm{v}}\sqrt{\frac{T_{\mathrm{w}}}{T_{\mathrm{v}}}} \tag{4.163}$$

Substituting Eq. (4.163) into Eq. (4.160), one obtains the disjoining pressure for the non-evaporating film thickness as

$$p_{\mathrm{d}} = -p_{\mathrm{v}}\sqrt{\frac{T_{\mathrm{w}}}{T_v}} + p_{\mathrm{sat}}(T_{\mathrm{w}}) + \rho_\ell R_{\mathrm{g}} T_{\mathrm{w}} \ln\left(\frac{p_{\mathrm{v}}}{p_{\mathrm{sat}}(T_{\mathrm{w}})}\sqrt{\frac{T_{\mathrm{w}}}{T_{\mathrm{v}}}}\right) + \sigma K \tag{4.164}$$

At the non-evaporating film thickness, the disjoining pressure can also be obtained by Eq. (4.51), i.e.,

$$p_{\mathrm{d}} = -A'\delta_0^{-B} \tag{4.165}$$

Combining Eqs. (4.164) and (4.165) yields the non-evaporating film thickness:

$$\delta_0 = \left\{\frac{1}{A'}\left[p_{\mathrm{v}}\sqrt{\frac{T_{\mathrm{w}}}{T_{\mathrm{v}}}} - p_{\mathrm{sat}}(T_{\mathrm{w}}) - \rho_\ell R_{\mathrm{g}} T_{\mathrm{w}} \ln\left(\frac{p_{\mathrm{v}}}{p_{\mathrm{sat}}(T_{\mathrm{w}})}\sqrt{\frac{T_{\mathrm{w}}}{T_{\mathrm{v}}}}\right) - \sigma K\right]\right\}^{-1/B} \tag{4.166}$$

which is applicable to nonpolar liquids.

For water, however, the following logarithmic dependence of disjoining pressure on the liquid film thickness is preferable (Holm and Goplen 1979):

$$p_{\mathrm{d}} = \rho_\ell R_{\mathrm{g}} T_\delta \ln\left[a\left(\frac{\delta}{3.3}\right)^b\right] \tag{4.167}$$

where a = 1.5336 and b = 0.0243.

4.6 Dynamic Behaviors of Interfaces

The instability of a horizontal or vertical interface between a liquid and vapor at different velocities will be discussed in Sect. 4.6.1, followed by a discussion of waves at liquid–vapor interface in Sect. 4.6.2.

4.6.1 Rayleigh-Taylor and Kelvin-Helmholtz Instabilities

A given physical state is said to be *stable* if it can withstand a disturbance and still return to its original state. Otherwise, the particular state is unstable. The objective of stability analysis is to analyze the effect of a particular disturbance on the physical state. If ϕ (it can be velocity, pressure, or temperature) represents a basic solution, a disturbance ϕ' is added to this basic solution and $\phi + \phi'$ will be substituted into the governing equations. The governing equations with ϕ as a dependent variable are then subtracted from the governing equation with $\phi + \phi'$ as dependent variables to yield disturbance equation for ϕ'. If the disturbance ϕ' damps out, ϕ is stable; otherwise, if the disturbance ϕ' grows with increasing time, ϕ is unstable.

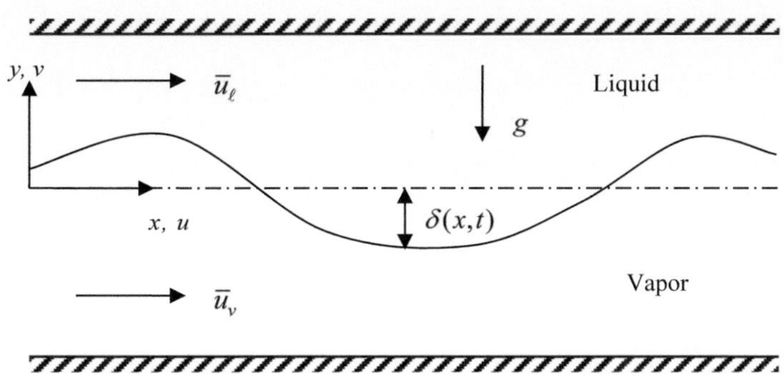

Fig. 4.19 Rayleigh-Taylor instability for dense liquid overlay less dense vapor

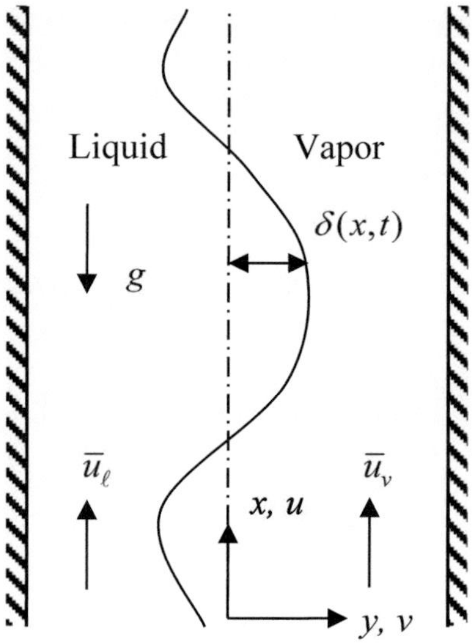

Fig. 4.20 Kelvin-Helmholtz instability in vertical liquid–vapor interface for concurrent flow

Interface morphology is important for heat and mass transfer. In a horizontal co-current flow system where a dense liquid phase overlays a less dense vapor phase (see Fig. 4.19), both phases are incompressible, inviscid, and immiscible; the interface may become unstable if there is a disturbance $\delta(x,t)$. This instability is referred to as *Rayleigh-Taylor* instability. On the other hand, if the gravity is parallel to the directions of the liquid–vapor co-current flow (see Fig. 4.20), the instability is referred to as the *Kelvin-Helmholtz* instability. The conditions for which the interfaces are stable with respect to an arbitrary perturbation $\delta(x,t)$ will be presented in this section (Carey 2016).

Assuming that the liquid and vapor flows are two-dimensional, the governing equations for configuration in Fig. 4.19 are

$$\frac{\partial u}{\partial x}+\frac{\partial \upsilon}{\partial y}=0 \tag{4.168}$$

$$\rho\left[\frac{\partial u}{\partial t}+u\frac{\partial u}{\partial x}+\upsilon\frac{\partial u}{\partial y}\right]=-\frac{\partial p}{\partial x} \tag{4.169}$$

$$\rho\left[\frac{\partial v}{\partial t}+u\frac{\partial v}{\partial x}+\upsilon\frac{\partial v}{\partial y}\right]=-\frac{\partial p}{\partial y}-\rho g \tag{4.170}$$

The velocities and the pressure are decomposed as follows into base flow and perturbed components:

$$u=\overline{u}+u', \quad \upsilon=\overline{\upsilon}+\upsilon' \quad p=\overline{p}+p' \tag{4.171}$$

Substituting Eq. (4.171) into Eqs. (4.168)–(4.170) and considering the fact that the base flow should also satisfy Eqs. (4.168)–(4.170), the equations for the flow simplify to

$$\frac{\partial u'}{\partial x}+\frac{\partial \upsilon'}{\partial y}=0 \tag{4.172}$$

$$\rho\left[\frac{\partial u'}{\partial t}+\overline{u}\left(\frac{\partial u'}{\partial x}\right)\right]=-\frac{\partial p'}{\partial x} \tag{4.173}$$

$$\rho\left[\frac{\partial \upsilon'}{\partial t}+\overline{u}\left(\frac{\partial \upsilon'}{\partial x}\right)\right]=-\frac{\partial p'}{\partial y} \tag{4.174}$$

In arriving at Eqs. (4.172)–(4.174), the products of perturbation (primed) terms are neglected, and we recognize that $\partial\overline{u}/\partial x=\partial\overline{u}/\partial y=\overline{\upsilon}=0$. Differentiating Eqs. (4.173) and (4.174) with respect to x and y, respectively, then summing them and substituting the continuity equation, yields the Laplace equation for the pressure perturbation field:

$$\frac{\partial^2 p'}{\partial x^2}+\frac{\partial^2 p'}{\partial y^2}=0 \tag{4.175}$$

The shape of the interface at time t can be described by

$$\delta(x,t)=A\mathrm{e}^{i\alpha z+\beta t} \tag{4.176}$$

and the perturbation quantities υ' and p' can be postulated to have the following forms:

$$\upsilon'(x,y,t)=\hat{\upsilon}\mathrm{e}^{i\alpha x+\beta t} \tag{4.177}$$

$$p'(x,y,t)=\hat{p}\mathrm{e}^{i\alpha x+\beta t} \tag{4.178}$$

where $\hat{\upsilon}$ and $\hat{p}$ are the magnitudes of the perturbation.

Employing the Young-Laplace equation and the equation for the curvature of the liquid film, $p_\mathrm{v}-p_\ell=\sigma/R$, Carey (2016) used the perturbation analysis to obtain the following condition for an unstable interface

$$|\overline{u}_\ell-\overline{u}_\mathrm{v}|^2>\frac{[\sigma\alpha+(\rho_\ell-\rho_\mathrm{v})g/\alpha](\rho_\ell+\rho_\mathrm{v})}{\rho_\ell\rho_\mathrm{v}} \tag{4.179}$$

where $\alpha=2\pi/\lambda$ is the wave number.

Surface tension and gravity tend to stabilize the interface (for this configuration only). The right side of this inequality has a minimum when the wave number is equal to a critical wave number α_crit:

$$\alpha_\mathrm{crit}=\left[\frac{(\rho_\ell-\rho_\mathrm{v})g}{\sigma}\right]^{1/2} \tag{4.180}$$

It follows from Eq. (4.179) that

$$|\overline{u}_\ell-\overline{u}_\mathrm{v}|_\mathrm{crit}=\left[\frac{2(\rho_\ell-\rho_\mathrm{v})}{\rho_\ell}\right]^{1/2}\left[\frac{\sigma(\rho_\ell-\rho_\mathrm{v})g}{\rho_\mathrm{v}^2}\right]^{1/4} \tag{4.181}$$

For motionless liquid over motionless vapor ($\overline{u}_\ell=\overline{u}_\mathrm{v}=0$), we obtain from Eq. (4.181)

$$\alpha>\alpha_\mathrm{crit}=\left[\frac{(\rho_\ell-\rho_\mathrm{v})g}{\sigma}\right]^{1/2} \tag{4.182}$$

The critical wavelength corresponding to the critical wave number is

$$\lambda_c = 2\pi\left[\frac{\sigma}{(\rho_\ell - \rho_v)g}\right]^{1/2} \tag{4.183}$$

A perturbation with a wavelength greater than λ_c will grow and result in instability. If the length of the interface in the x-direction is less than λ_c, the interface is stable because a perturbation of wavelength greater than λ_c cannot arise.

A specific value of α exists where β in Eqs. (4.176)–(4.178) is at its maximum. Its value is

$$\alpha_{max} = \left[\frac{(\rho_\ell - \rho_v)g}{3\sigma}\right]^{1/2} \tag{4.184}$$

The disturbance wavelength corresponding to α_{max} is referred to as the most dangerous wavelength, λ_D—

$$\lambda_D = 2\pi\left[\frac{3\sigma}{(\rho_\ell - \rho_v)g}\right]^{1/2} = \sqrt{3}\lambda_c \tag{4.185}$$

which has many applications, including derivation of the critical heat flux for pool boiling.

When the direction of the gravity is parallel to the direction of the liquid–vapor co-current flow as shown in Fig. 4.19, the gravity will not have a significant effect on the pressures in the liquid and vapor phases. Equation (4.179) becomes

$$|\bar{u}_\ell - \bar{u}_v|^2 > \frac{\sigma\alpha(\rho_\ell + \rho_v)}{\rho_\ell \rho_v} \tag{4.186}$$

which is the condition for an unstable vertical interface, and this condition is termed the *Kelvin-Helmholtz* instability.

Example 4.5

In a stratified horizontal two-phase flow system, saturated water vapor flows above saturated liquid water. The pressure of the two-phase system is at one atmosphere. Determine the critical velocity, the corresponding critical wavelength λ_c, and the most dangerous wavelength λ_D.

Solution The properties of saturated vapor and liquid water at one atmosphere are $\rho_v = 0.5974$ kg/m^3, $\rho_\ell = 958.77$ kg/m^3, and $\sigma = 0.05891$ N/m.

The critical velocity is obtained by Eq. (4.181), i.e.,

$$\begin{aligned}|\bar{u}_\ell - \bar{u}_v|_{crit} &= \left[\frac{2(\rho_\ell - \rho_v)}{\rho_\ell}\right]^{1/2}\left[\frac{\sigma(\rho_\ell - \rho_v)g}{\rho_v^2}\right]^{1/4} \\ &= \left[\frac{2 \times (958.77 - 0.5974)}{958.77}\right]^{1/2}\left[\frac{0.05891 \times (958.77 - 0.5974) \times 9.8}{0.5974^2}\right]^{1/4} \\ &= 8.87 \text{ m/s}\end{aligned}$$

The critical wavelength is obtained by Eq. (4.183), i.e.,

$$\lambda_c = 2\pi\left[\frac{\sigma}{(\rho_\ell - \rho_v)g}\right]^{1/2}$$
$$= 2\pi\left[\frac{0.05891}{(958.77 - 0.5974) \times 9.8}\right]^{1/2} = 0.0157\ \text{m} = 15.7\ \text{mm}$$

The most dangerous wavelength can be obtained by Eq. (4.185), i.e.,

$$\lambda_D = \sqrt{3}\lambda_c = \sqrt{3} \times 0.0157 = 0.0273\ \text{m} = 27.3\ \text{mm}$$

4.6.2 Surface Waves on Liquid Film Flow

As will become evident in later chapters, condensation or evaporation at the interface of a thin liquid film flowing over a solid surface is often encountered in engineering applications. This type of process is commonly found in chemical and mechanical engineering equipment such as condensers, long-tube evaporators, wetted wall columns, and cooling towers. The fact that current literature shows continued interest in this general area affirms its continued importance. The problem of the transport from a liquid layer running down a vertical wall has been considered in literature with particular attention to heating, evaporation, condensation, and gas absorption for various applications. The flow in such a liquid layer may be either (a) laminar with a constant thickness, or (b) laminar with variable thickness due to waves, or turbulent with waves moving down the interface.

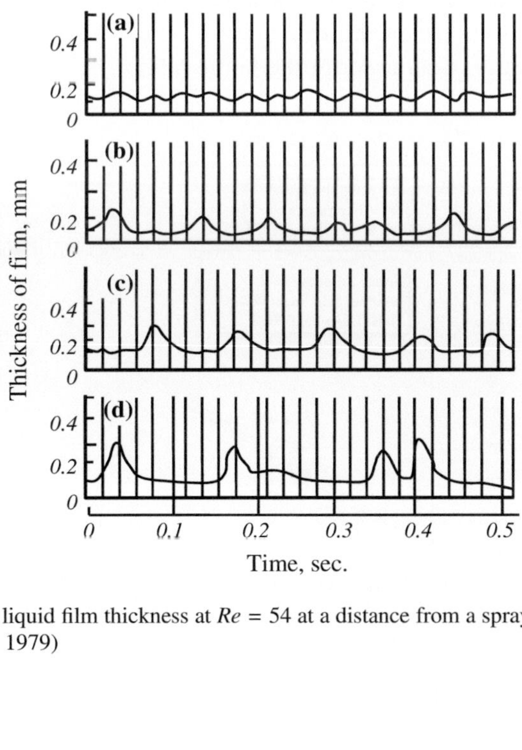

Fig. 4.21 Change in the local liquid film thickness at Re = 54 at a distance from a sprayer of **a** 130, **b** 310, **c** 445, and **d** 670 mm (Faghri and Payvar 1979)

Kapitza (1964) predicted theoretically that the mean film thickness is reduced by approximately 8% due to ripples on the interface. Many experiments have focused on determining the values of the wave properties in both laminar and turbulent flows. Experimental data on the mean film thickness of wavy laminar flow falls between the predictions of Kapitza and Nusselt (smooth interface) theories. Although the wave effect does not have any significant effect on the mean film thickness in wavy laminar regions, it plays a major role in heat and mass transfer processes. The structure of the liquid film running down a plane usually shows a randomly distributed wave on the surface, except at low flow rate at the entry region, where a periodic motion exists. Figure 4.21, which is taken originally from Rogovan et al. (1969), shows such a variation as obtained at a Reynolds number of 54 for water.

Comparisons of the experimental and theoretical values of the transport in falling film should be supported by a specification of the nature of the waves, for it is through the interaction of the normal velocities produced by the waves and the distribution of the transported quantity that the additional transport arises. But this is complicated by the existence of an initial wave-free length, depending upon the way in which the film was formed by the liquid supply, and by a region of wave development that ostensibly culminates in a steady regime at a distance far enough down the height of the film. Brauer (1956), for example, made measurements on water, and mixtures of water and diethylene glycol, at a distance of 1.3 m from the point of film initiation, and found sinusoidal waves to begin at a Reynolds number, $Re = 4\Gamma/\mu = 1.2/Ka^{1/10}$, where $Ka = \nu^4\rho^3 g/\sigma^3$ is the Kapitza number. The ratio of the crest height of the waves to the mean film thickness increased with the Reynolds number to about twice for this Reynolds number than for higher Reynolds number. This ratio remained the same up to a Reynolds number of $140/Ka^{1/10}$. Between these Reynolds numbers, the waves become distorted and no longer sinusoidal; above the higher Reynolds number, there were secondary capillary waves on the surface. Throughout, up to $4\Gamma/\mu = 1600$, the average film-thickness remains essentially that given by the Nusselt analysis. But in the range where the ratio of the crest height to the mean film thickness is almost constant, the wave frequency increases with the Reynolds number to indicate a basis for a proportional increase in heat and mass transfer.

Koizumi et al. (2005) studied the behavior of a liquid film flowing down the inner surface of a pipe with countercurrent gas flow. In the experiment, the gas used was air, and silicone oils of 500, 1000, and 3000 cSt as well as water, were used for the liquid phase.

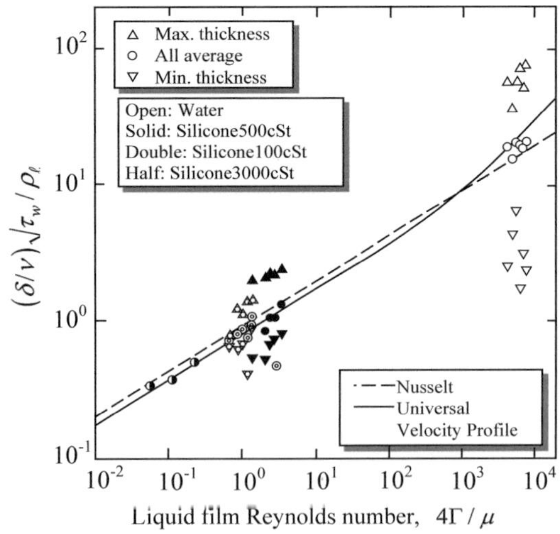

Fig. 4.22 Dimensionless liquid film thickness as a function of Reynolds number for falling liquid film (adopted from Koizumi et al. 2005)

Three primary parameters were correlated: the film thickness, wave velocity, and wavelength. The film thickness was given three different values: minimum, maximum, and mean film thickness. The film height was determined by taking photographs of the film and measuring its thickness. The wave velocity was determined by seeing how far a wave peak traveled during the time interval between picture frames. The wavelength was determined in a similar manner; that is, photographs were taken, and the distance between peaks of the wave was measured.

Figure 4.22 presents experimental results for water and silicone oil with no airflow for dimensionless δ^* versus the Reynolds number, where $\delta^* = (y/v)\sqrt{\tau_w/\rho_\ell}$, δ is the film thickness, τ_w is the wall shear stress and ρ_ℓ is the liquid density. When there is no airflow, the maximum film thickness of the 500 cSt and the 1000 cSt silicone and water films is much greater.

To correlate the wave properties using a non-dimensional analysis, the non-dimensional wave velocity, $N_w = w_w/v_g^{1/3}$, and maximum film thickness, $\delta^+_{max} = \delta_{max}(g/v^2)^{1/3}$, are found according to the Buckingham pi theorem. They are functions of the Reynolds number ($Re = 4\Gamma/\mu$), the Morton number ($K_F = \rho^3 v^4 g/\sigma^3$) and the non-dimensional wavelength ($N_\lambda = \lambda(g/v^2)^{1/3}$), where w_w is the wave velocity and λ is the wavelength.

Nosoko et al. (1996) developed the following correlations for wave velocity and maximum film thickness:

$$N_w = 0.68 K_F^{0.02} N_\lambda^{0.31} Re^{0.37} \tag{4.187}$$

$$\delta^+_{max} = 0.26 K_F^{0.044} N_\lambda^{0.39} Re^{0.46} \tag{4.188}$$

Equations (4.187) and (4.188) require information about the wavelength; therefore, the correlations are not closed. The wave velocity and maximum film thickness should be evaluated only from the film flow rate and physical dimensions, if possible. The wavelength correlation was incorporated into the Nosoko correlations by Koizumi et al. (2005), and the constants and exponents were modified for better accuracy. The final forms are

$$N_\lambda = 14.9 Fr^{1.29} Re^{-0.133} \tag{4.189}$$

$$N_w = 0.88 K_F^{-0.008} Fr^{0.977} Re^{0.214} \tag{4.190}$$

$$\delta^+_{max} = 1.09 K_F^{0.021} Fr^{0.316} Re^{0.424} \tag{4.191}$$

where $Fr = w_w/\sqrt{g\bar{\delta}}$ and $\bar{\delta}$ is the mean film thickness.

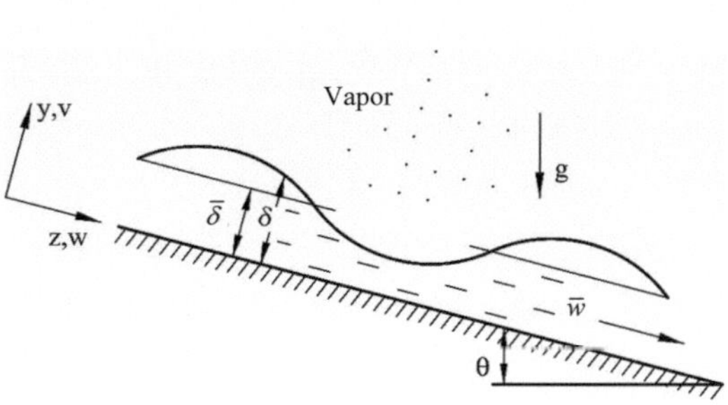

Fig. 4.23 Liquid film on an inclined surface

The thickness of these liquid films is small enough that boundary layer assumptions are valid. However, these films are not thin enough for disjoining pressure to play any significant role. Figure 4.23 shows a slow liquid flow on an inclined surface with negligible shear stress at the free surface. The gravitational and liquid viscous forces, as well as surface tension effects, are dominant. As the mass flow rate of the liquid increases, waves on the liquid film interface can be observed, as indicated by Fig. 4.23.

The flow regimes of the liquid film include smooth laminar, wavy laminar, and turbulent, which correspond to different film Reynolds numbers. The Reynolds number is defined as

$$Re_\delta = \frac{4\Gamma}{\mu_\ell} = \frac{4\overline{w}\delta}{\mu_\ell} \tag{4.192}$$

where Γ is the mass flow rate per unit width and $\overline{w}$ is the mean velocity of the liquid film. The liquid film is laminar if the film Reynolds number is below 30. A wavy flow regime can be observed when the film Reynolds number is between 30 and 1600. The liquid flow becomes turbulent when the film Reynolds number is above 1600.

For a fully developed steady-state downward laminar flow, the momentum equation reads

$$\frac{(\rho_\ell - \rho_v)g\sin\theta}{\rho_\ell} + v_\ell \frac{d^2w}{dy^2} = 0 \tag{4.193}$$

Integrating Eq. (4.193) twice and considering the boundary conditions $w = 0$ at $y = 0$ and $dw/dy = 0$ at $y = \delta$, the velocity distribution in the liquid film is obtained.

$$w = \frac{(\rho_\ell - \rho_v)(2\delta y - y^2)g\sin\theta}{2\mu_\ell} \tag{4.194}$$

The mean velocity in the liquid film is then

$$\overline{w} = \frac{1}{\delta}\int_0^\delta w(y)dy = \frac{(\rho_\ell - \rho_v)g\sin\theta\delta^2}{3\mu_\ell} \tag{4.195}$$

which is obtained under assumption that the liquid flow is steady state. Under these assumptions (Nusselt analysis) and assuming $\rho_\ell \gg \rho_v$, the dimensionless film thickness, and film velocity at the surface δ, w_δ and the mean velocity $\overline{w}$ are given for $\theta = 90°$ below:

$$\delta^+ = \delta\left(\frac{g}{v^2}\right)^{1/3} = 0.908Re^{1/3} \tag{4.196}$$

$$w_\delta = \frac{g}{2v}\delta^2 \tag{4.197}$$

$$\overline{w} = \frac{2}{3}w_\delta \tag{4.198}$$

When waves are present, the liquid film flow will be unsteady. However, it is assumed that Eq. (4.195) is also valid for unsteady film flow.

Consideration of hydrostatics and the Young-Laplace equation at the interface yields:

$$\frac{\partial p_\ell}{\partial z} = \frac{\partial p_\mathrm{v}}{\partial z} + (\rho_\ell - \rho_\mathrm{v})g\cos\theta\frac{\partial \delta}{\partial z} - \sigma\frac{\partial^3 \delta}{\partial z^3} \tag{4.199}$$

Since there is no motion in the vapor phase, the pressure gradient in the vapor phase satisfies

$$\frac{\partial p_\upsilon}{\partial z} = \rho_\mathrm{v} g\sin\theta \tag{4.200}$$

The momentum equation for an unsteady-state flow in the liquid film is

$$\rho_\ell\left(\frac{\partial w}{\partial t} + \upsilon\frac{\partial w}{\partial y} + w\frac{\partial w}{\partial z}\right) = -\frac{\partial p_\ell}{\partial z} + \rho_\ell g\sin\theta + \mu_\ell\frac{\partial^2 w}{\partial y^2} \tag{4.201}$$

Substituting Eqs. (4.199)–(4.200) into Eq. (4.201), the momentum equation becomes

$$\begin{aligned}\frac{\partial w}{\partial t} + \upsilon\frac{\partial w}{\partial y} + w\frac{\partial w}{\partial z} &= \frac{(\rho_\ell - \rho_\mathrm{v})g}{\rho_\ell}\left(\sin\theta - \cos\theta\frac{\partial \delta}{\partial z}\right)\\ &\quad + \frac{\sigma}{\rho_\ell}\frac{\partial^3\delta}{\partial z^3} + \upsilon_\ell\frac{\partial^2 w}{\partial y^2}\end{aligned} \tag{4.202}$$

Integrating Eq. (4.202) and considering the continuity equation, two governing equations for $\delta(z,t)$ can be obtained:

$$\frac{3\overline{w}^2}{\delta}\frac{\partial\delta}{\partial z} = \frac{1}{\rho_\ell}(\rho_\ell - \rho_\mathrm{v})g\left(\cos\theta\frac{\partial\delta}{\partial z} - \sin\theta\right) - \frac{\sigma}{\rho_\ell}\frac{\partial^3\delta}{\partial z^3} = \upsilon_\ell\frac{3\overline{w}}{\delta^2} \tag{4.203}$$

$$\frac{\partial\delta}{\partial t} = 3\overline{w}\frac{\partial\delta}{\partial z} \tag{4.204}$$

These equations can be solved along with the energy equation. Our purpose, however, is to find thc conditions of stability of thc film. Assuming that $\delta - \delta_0 + \delta'$ and $\overline{w} - \overline{w}_0$, and noticing from Eq. (4.195) that

$$\frac{1}{\rho_\ell}(\rho_\ell - \rho_v)g\sin\theta = \frac{3\overline{w}_0\nu_\ell}{\delta_0^2} \tag{4.205}$$

the following is obtained:

$$\frac{3\overline{w}_0^2}{\delta_0}\frac{\partial\delta'}{\partial z} = \frac{1}{\rho_\ell}(\rho_\ell - \rho_\mathrm{v})g\cos\theta\frac{\partial\delta'}{\partial z} - \frac{\sigma}{\rho_\ell}\frac{\partial^3\delta'}{\partial z^3} \tag{4.206}$$

The fluctuating component δ' is assumed to be a sinusoidal wave of the form

$$\delta' = \delta^* \mathrm{e}^{i\alpha(z-ct)} \tag{4.207}$$

where α is the wave number and c is the wave velocity. For $\alpha > 0$, which is the condition under which waves exist, it follows from Eqs. (4.206) and (4.207) that

$$\bar{\delta} > \left[\frac{3\mu_\ell^2\cos\theta}{(\rho_\ell - \rho_v)\rho_\ell g\sin^2\theta}\right]^{1/3} \tag{4.208}$$

which can also be written in terms of film Reynolds number, $Re_\delta = 4\rho_\ell \overline{w}_0 \bar{\delta}/\mu_\ell$, i.e.,

$$Re_\delta > 4\cot\theta \tag{4.209}$$

Equation (4.209) indicates that the critical Reynolds number at which waves appear equals $\cot\theta$. For a vertical surface where $\theta = 90°$, Eq. (4.209) becomes $Re_\delta > 0$, which means that waves can be present no matter how small the film Reynolds number is. This result contradicts experimental observation, because waves are present only when the film Reynolds number reaches a certain value. This contradiction can be solved by considering the amplification rate of wave amplitude. If amplification over the time interval required to travel 100 times, the liquid film thickness is considered, the amplification factor is (Benjamin 1957; Carey 2016)

$$\frac{|\delta|}{|\delta|_{t=0}} = \exp\left\{\left[0.31\nu_\ell^{4/3}g^{1/3}\left(\frac{\sigma}{\rho_\ell}\right)^{-1}\right]Re_\delta^{8/3}\right\} \tag{4.210}$$

For saturated water at one atmosphere, $\nu_\ell = 2.91\times10^{-7}\ \mathrm{m^2/s}$,, $\sigma = 0.0589$ N/m, and $\rho_\ell = 958.7\ \mathrm{kg/m^2}$, and Eq. (4.210) becomes

$$\frac{|\delta|}{|\delta|_{t=0}} = \exp\left(2.08\times10^{-5}Re_\delta^{8/3}\right) \tag{4.211}$$

The dependence of disturbance amplification versus film Reynolds number is shown in the following table:

Re_δ	10	20	40	80	120	160
$\|\delta\|/\|\delta\|_{t=0}$	1.010	1.063	1.476	11.85	1466.2	6.588×10^6

It can be seen from the above table that the wave becomes significant after Re_δ is greater than 40. Waves increase the heat transfer coefficient during evaporation (also condensation) of wavy films as compared to a smooth surface due to an increase in interfacial surface area and mixing action (Faghri and Seban 1985). Heat transfer during condensation and vaporization over a wavy liquid film will be discussed in detail in Chaps. 7 and 8, respectively.

4.7 Numerical Simulation of Interfaces and Free Surfaces

4.7.1 Interface Tracking Techniques

Many engineering applications involve interfacial phenomena, because of the high heat transfer rates that can be achieved. There are many complexities that need to be addressed when modeling an interface, since an interface is generally irregular, involves mass transfer, is three-dimensional, and is not at a fixed location. The interface can be assumed to be a planar surface, and therefore a continuum approach is valid. In these problems, it is necessary to solve for both the phases as well as the interfacial location. These problems can be solved numerically on an Eulerian or Lagrangian mesh.

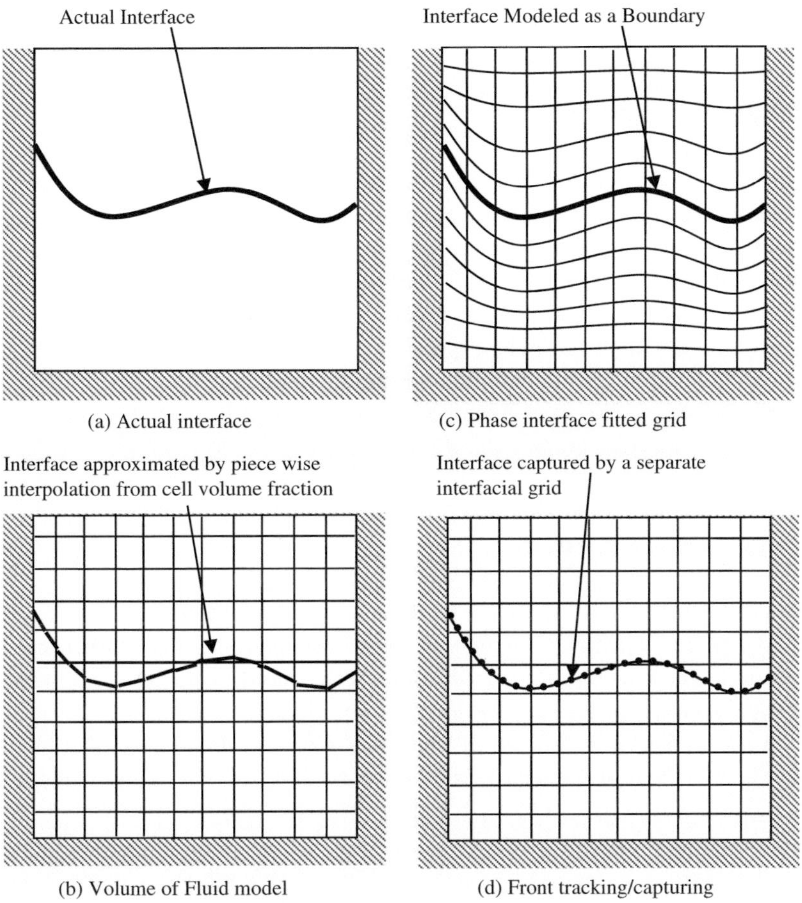

Fig. 4.24 Interfacial representations for different interface tracking techniques

A Eulerian mesh is stationary and defined prior to the start of a solution. When using a Eulerian mesh, the interface is tracked by solving an additional scalar equation. In the Lagrangian approach, a boundary of the mesh is aligned with the interface, and this boundary moves with the interface.

When thinking of a multiphase system from a continuum approach, in the bulk region, a phase is continuous and is discontinuous at an interface between different phases. In general, the interface is free to deform based on the nature of the flow; therefore, it is difficult to efficiently capture an interface between phases with just one model. Consequently, there have been strong efforts to track an interface based on several different techniques, each with its own pros and cons. An actual interface, represented by the different numerical techniques, is presented in Fig. 4.24. The techniques are as follows:

- **Lagrangian Approach**: grid and fluid move together, interface is directly captured.
- **Stationary Grid Approach**: standard CFD modeling in cells that contain only a single phase, special consideration taken in cells in the vicinity of an interface.
- **Phase Interface Fitted Grid Approach**: interface directly tracked as boundary, rest of grid moves as a function of interfacial movement, "semi-Lagrangian".
- **Front Tracking Approach**: stationary grid is used and modified near the interface so that the grid is aligned with the interface; combination of stationary grid approach and phase interface fitted grid approach.

Each of the above methods will be briefly described below.

4.7.2 Lagrangian Approach

In the Lagrangian approach, fluid particles are tracked directly. Therefore, a particle on the interface will also be tracked directly, so the interface will automatically be resolved. The Lagrangian approach is used by Shopov et al. (1990) to model a droplet interacting with a wall. Even though a Lagrangian approach directly captures an interface, an Eulerian approach is often preferable to a Lagrangian approach to solve the bulk flow of a fluid. An Eulerian approach is generally preferable because fluid flow is usually complex, and fluid particle pathlines often intersect and/or disperse, making fluid interaction very difficult to manage with a Lagrangian method. Stationary grid approach, phase interface fitted grid approach, and front tracking approach use an Eulerian point of view to solve the governing equations in the bulk fluids with special consideration to track the interface.

4.7.3 Stationary Grid Approach

The second class of numerical techniques for multiphase fluids is based on a stationary grid where the fluid interface is captured directly. The first of such approaches is the marker-and-cell approach originated by Harlow and Welch (1965). In this approach, massless particles are introduced into the flow field, and the locations are projected from their interpolated velocities. Cells with a particle are considered to have one phase in them, and cells without a particle do not have that phase in them. An interface is considered where cells with particles are neighbored with cells without particles. For efficiency, this method was extended to only track particles on the surface (Nichols and Hirt 1975). Further development of this class of models leads to the Volume of Fluid method (VOF) by Hirt and Nichols (1981); they used a donor-acceptor method to effectively eliminate numerical diffusion at an interface.

The VOF method is one of the most widely used routines to solve a free surface problem on a Eulerian mesh. In this method, there is one velocity, pressure, and temperature field, and it is shared by all of the phases modeled. The interface between the phases is tracked by the volume fraction of phase k, ε_k. The volume fraction equation is the continuity equation of phase k divided by the density of that phase.

$$\frac{1}{\rho_k}\left[\frac{\partial}{\partial t}(\varepsilon_k\rho_k)+\nabla\cdot(\varepsilon_k\rho_k\mathbf{V})=\sum_{j=1(j\neq k)}^{\Pi}\dot{m}_{jk}'''\right] \tag{4.212}$$

When the volume fraction is between 0 and 1 in a computational cell, that cell is considered to be an interfacial cell. For a two-phase system, the phases are k and j. When the volume fraction is 1, that cell is occupied by only phase k, and when it is 0, that cell is occupied by only phase j. The sum of the volume fractions is unity.

$$\sum_{k=1}^{\Pi}\varepsilon_k=1 \tag{4.213}$$

Therefore, $(k-1)$ volume fraction equations need to be solved. The fluid properties, such as density, viscosity, and thermal conductivity, are calculated by their volume weighted average.

$$\Phi_{\text{eff}} = \sum_{k=1}^{\Pi} \varepsilon_k \Phi_k \tag{4.214}$$

The overall continuity equation in conjunction with the VOF method is:

$$\frac{\partial}{\partial t}(\rho_{\text{eff}}) + \nabla \cdot (\rho_{\text{eff}} \mathbf{V}) = 0 \tag{4.215}$$

The continuity equation is the same as Eq. (3.48), except it uses the effective density and the velocity field is shared by both phases; therefore, subscript k is dropped. The momentum equation is

$$\frac{\partial}{\partial t}(\rho_{\text{eff}} \mathbf{V}) + \nabla \cdot (\rho_{\text{eff}} \mathbf{V}\mathbf{V}) = \nabla \cdot \boldsymbol{\tau}'_{\text{eff}} + \rho_{\text{eff}} \mathbf{X} + \sum_{k=1}^{\Pi} \sum_{j=1 (j \neq k)}^{\Pi} \langle \dot{m}''' \mathbf{V}_{jk} \rangle \tag{4.216}$$

The momentum equation is also the same as Eq. (3.63), with the same exceptions as the continuity equation and the momentum transfer due to phase change. In the energy equation, the enthalpy is mass averaged instead of volume averaged.

$$h_{\text{eff}} = \frac{1}{\rho_{\text{eff}}} \sum_{k=1}^{\Pi} \varepsilon_k \rho_k h_k \tag{4.217}$$

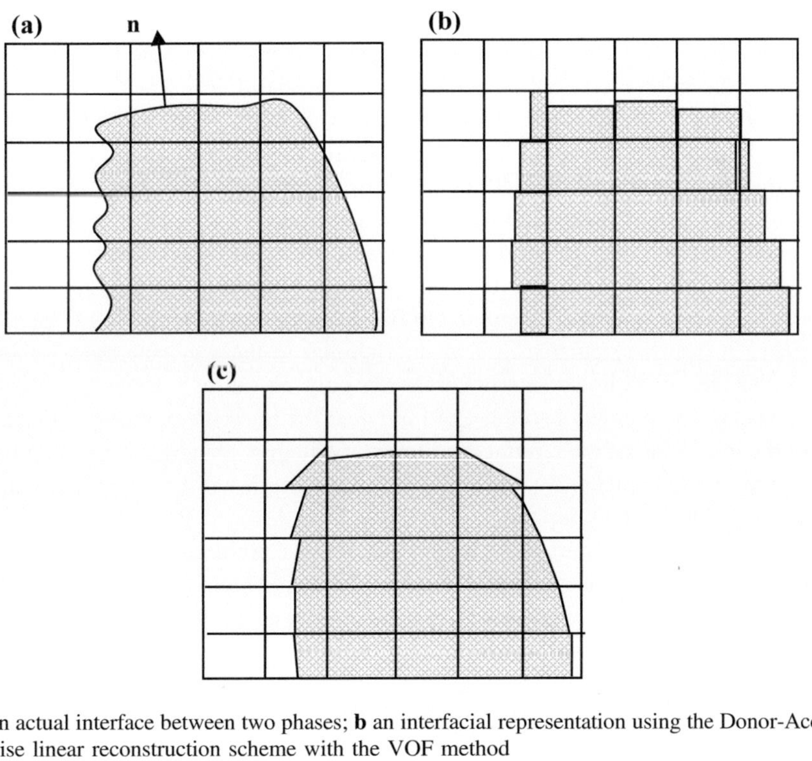

Fig. 4.25 **a** An actual interface between two phases; **b** an interfacial representation using the Donor-Acceptor scheme, and **c** a piecewise linear reconstruction scheme with the VOF method

Therefore, the energy equation, neglecting pressure effects and viscous dissipation, is:

$$\frac{\partial}{\partial t}(\rho_{\text{eff}} h_{\text{eff}}) + \nabla \cdot (\rho_{\text{eff}} \mathbf{V} h_{\text{eff}}) = -\nabla \cdot \mathbf{q}''_{\text{eff}} + \sum_{k=1}^{\Pi} \sum_{j=1(j \neq k)}^{\Pi} \dot{m}'''_{jk} h_k + \dot{q}'''_{\text{eff}} \tag{4.218}$$

The energy equation has the same form as Eq. (3.90), except effective properties are incorporated and a latent heat term due to phase change is added.

The continuity, momentum, and energy equations can be solved by standard solution procedures, such as the SIMPLE class of algorithms in Patankar (1980). However, if the volume fraction equation is solved using a standard implicit or explicit scheme, the interface will quickly lose resolution due to numerical diffusion. The numerical diffusion will lead to inaccurate solutions and/or a solution that will not converge. Therefore, special consideration must be taken on interfacial cells to construct the interface so that the advection transport of fluid is representative of the physical problem. An actual interface between two phases, as well as the corresponding interface represented by two special interpolation schemes, is presented in Fig. 4.25.

As noted before, the first of such methods is the donor-acceptor scheme proposed by Hirt and Nichols (1981). If a cell is an interfacial cell, $0 < \varepsilon_k < 1$, the fluid will be rearranged in the cell to be on one face, as shown in Fig. 4.25b. The face on which the fluid will be rearranged will depend on the normal direction of the interface. The normal direction of the interface with respect to phase k can be calculated by the gradient of the volume fraction:

$$\mathbf{n}_k = -\frac{\nabla \varepsilon_k}{|\nabla \varepsilon_k|} \tag{4.219}$$

The component in which the normal is the greatest will occur where the fluid is perpendicular to the interface while the interface is either horizontal or vertical. Once this is done, one cell is designated as a donor, and its neighbor is designated as an acceptor. The amount of fluid leaving the donor cell is exactly equal to the amount of fluid entering the acceptor through each computational face. Also, the maximum amount of fluid that can leave a cell is equal to the amount of fluid in that cell or the amount that would make another cell fill with fluid.

A more refined interface interpolation scheme was developed by Youngs (1982), in which the interface was approximated as piecewise linear in each cell. The normal of the reconstructed interface in each cell is the same as the normal calculated in Eq. (4.219), as shown in Fig. 4.25c. The advection of fluid through each face is calculated in a manner similar to the donor-acceptor scheme. Both the donor-acceptor and piecewise linear interpolation can only be run in transient simulations, and the time step must be kept small enough so that the fluid near the interface will only advance by one cell at a time. Also, the resolution of the interface is limited to the grid spacing of the computational mesh. Any surface waves that are smaller than the spacing of the mesh will be smoothed out, as shown in Fig. 4.25 on the left side of the interface.

Surface tension effects are important in many free surface problems. The surface tension effects can be applied as a body force in the momentum equations. This method is called the continuum surface force (CSF) model and was proposed by Brackbill et al. (1992). The curvature is defined as the divergence of the surface normal vector.

$$K_k = \nabla \cdot \mathbf{n}_k \tag{4.220}$$

The volume force due to surface tension, F_σ, is

$$F_\sigma = \sum_{k=1}^{\Pi}\sum_{j=1}^{k-1} -\sigma_{jk}\frac{\varepsilon_j\rho_j K_k \nabla\varepsilon_j + \varepsilon_k\rho_k K_j \nabla\varepsilon_k}{\frac{1}{2}\left(\rho_j+\rho_k\right)} \tag{4.221}$$

If only two phases are present, the body force reduces to

$$F_\sigma = \sigma_{jk}\frac{\rho_{\text{eff}} K_k \nabla\varepsilon_j}{\frac{1}{2}\left(\rho_k+\rho_j\right)} \tag{4.222}$$

It can be seen that the body force is proportional to the cell density. It is important to note that when surface tension forces are large compared to other flow characteristics, numerical inaccuracies of the surface tension forces in the CSF model create artificial currents called parasitic currents. Much work has been done to eliminate parasitic currents, such as the second gradient method proposed by Jamet et al. (2002).

Despite the adverse effects of parasitic currents, the VOF method has been widely used and can give reasonably accurate results for a wide range of applications. It is robust in its handling of free surface problems with large interface distortion and can easily handle problems in which the free surface breaks apart, such as droplet formation. One other drawback of the VOF method is that the interface resolution is limited to the grid spacing. Therefore, a refined mesh is needed anywhere the interface is going to travel. This refinement can lead to many computational cells in regions of the mesh resided in by the interface for a short period of time, which will increase the total computational time of the solution. Therefore, advanced remeshing algorithms are needed for problems of this type.

4.7.4 Phase Interface Fitted Grid Approach

The third method, in which a phase interface fitted grid is employed, can be very useful for multiphase problems in which the interface does not deform greatly or break apart (although it can be handled with complex algorithms). The benefit of these models is that they directly capture an interface; therefore, a phenomenon occurring at a phase interface is not subject to interpolation error, such as the interface location itself.

The third class of numerical method used to capture a free surface is the semi-Lagrangian approach. Since a Lagrangian approach is not feasible for most fluid mechanics problems, only the

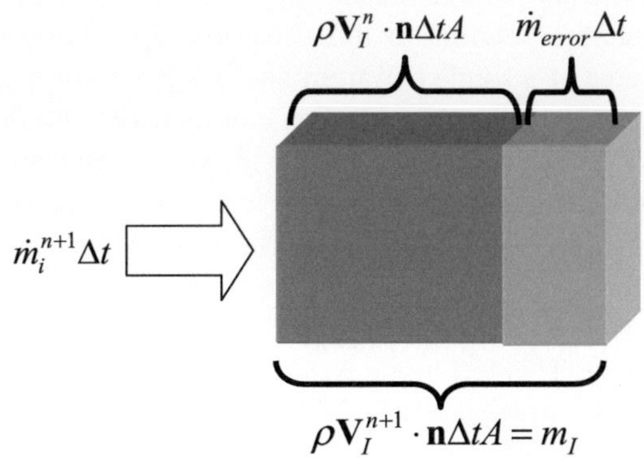

Fig. 4.26 Error in interfacial location due to the lag in time

interface is tracked in a Lagrangian manner, while the rest of the mesh can be manipulated to maintain good mesh qualities. Therefore, this type of technique can be deemed "semi-Lagrangian," and is a front tracking technique. Rice and Faghri (2005a) developed a semi-Lagrangian approach that works effectively on the interface between phases with phase change. They used a transient method in which the present interfacial velocity, $\mathbf{V}_{\mathrm{I}}$, is calculated based on the error in mass flux through each computational face from the previous time step. Therefore, the interfacial velocity is lagging by one time-step. The conservation of mass between phase k and phase j is

$$\dot{m}'' = \rho_k(\mathbf{V}_k - \mathbf{V}_{\mathrm{I}}) \cdot \mathbf{n} = \rho_j(\mathbf{V}_j - \mathbf{V}_{\mathrm{I}}) \cdot \mathbf{n} = \dot{m}''_{\mathrm{actual}} \tag{4.223}$$

The interfacial continuity equation represents the actual mass flux at the interface. Instead of implicitly calculating the interfacial velocity of the next time step ($n + 1$), the interfacial velocity from the previous time step (n) is used. By lagging the interfacial velocity, there is an inherent error in mass flux at the interface.

$$\dot{m}''^{n+1}_{\mathrm{error}} = \rho_k\left(\mathbf{V}_k^{n+1} - \mathbf{V}_{\mathrm{I}}^n\right) \cdot \mathbf{n} - \dot{m}''^{n+1}_{\mathrm{actual}} = \rho_k\left(\mathbf{V}_{\mathrm{I}}^{n+1} - \mathbf{V}_{\mathrm{I}}^n\right) \cdot \mathbf{n} \tag{4.224}$$

The normal velocity of phase j at the interface is fixed by continuity. However, the normal velocity of phase k at the interface is not fixed, and fluid is allowed to pass through this interface without regard to the interfacial mass balance. Figure 4.26 demonstrates how the interfaces move and show that there is an error in the mass of the system due to the interfacial movement. In order to compensate for the error in mass through each computational face, the new velocity can be found by rearranging Eq. (4.224) and integrating over time and area.

$$\Delta t \int_{A_{\mathrm{I}}} \rho_k \mathbf{V}_{\mathrm{I}}^{n+1} \cdot \mathbf{n}\, \mathrm{d}A = \Delta t \int_{A_{\mathrm{I}}} \left(\dot{m}''^{n+1}_{\mathrm{error}} + \rho_k \mathbf{V}_{\mathrm{I}}^n \cdot \mathbf{n}\right) \mathrm{d}A \tag{4.225}$$

In discretized form, Eq. (4.225) becomes

$$\left[\Delta t\left(\rho_k \mathbf{V}_{\mathrm{I}}^{n+1} \cdot \mathbf{n}\right)A\right]_i = \left[\Delta t\left(\dot{m}''^{n+1}_{\mathrm{error}} + \rho_k \mathbf{V}_{\mathrm{I}}^n \cdot \mathbf{n}\right)A\right]_i = m_i \tag{4.226}$$

The subscript i represents a single face on the interface. Rice and Faghri (2005a) suggest moving the interface at the beginning of each time step based on the error in mass flux and interfacial velocity of the previous time step, so that the mass the interface gains by movement is equal to m_i. They specified a direction in which the interface location (δ_{I}) will always be a function of $\delta_{\mathrm{I}} = f(\eta)$, and moved the nodes on the interface normal to this direction. Specifying the coordinate system can prevent the interfacial face area of a single cell from becoming zero or negative. Some knowledge of the problem is necessary in order to specify the direction in which the interface can move.

The pressure of phase k at the interface should be defined from the interfacial momentum equation. When a second phase also influences the interface, this interface will move with the interface of phase k. However, the mass flux that passes through this interface should be identically equal to $\dot{m}''_{\mathrm{actual}}$. This side of the interface affects the interfacial movement through the momentum equations. When surface tension effects are important, they are directly applied as an interfacial boundary condition in the interfacial momentum equation. This technique was shown by Rice and Faghri (2005b) to eliminate the parasitic currents caused by surface tension effects that are experienced in the VOF method. Also, this technique directly tracks the interface; therefore, the interface is always directly resolved, no matter what the mesh spacing is. For problems with large fluid distortion, it is very difficult to define a coordinate system of which the interface will always be a function; therefore, this coordinate system will need to be able to move over time. If the interface is largely distorted, advanced remeshing techniques will be needed for the computational domain away from the interface to maintain a good quality of computational cells.

Rice and Faghri's (2005a, b) technique used a transient solver that allowed a small amount of liquid mass to pass through the interface. During each new time-step, the liquid that passed through the interface during the previous time-step was recaptured in a manner in which mass is directly conserved. Even though this technique is generally not suitable for interfaces undergoing large deformations, such as jet breakup, it has the advantage of directly capturing the interface. Therefore, this technique offers a great advantage in problems involving multiple components, whose interfacial values are not continuous.

4.7.5 Front Tracking Approach

The fourth class is a combination of the second and third class of multiphase and interface modeling techniques and is called the front tracking approach. One form was developed by Glimm et al. (2001). This method uses a fixed grid that is modified only near the interface. The interface is modeled as a separate moving grid. The interface can move and it separates the stationary grid cells into two cells near the interface, representing each fluid. Therefore, each fluid is treated separately. Another front tracking technique developed by Tryggvason et al. (2001) also uses a separate grid to track the interface; however, the phases are considered together, and a single set of governing equations is solved for the whole field. Therefore, the stationary cells that lie below the interface have a shared density and velocity field. Also, since the interfacial area is free to expand or contract, the subgrid that tracks the interface must have the ability to add or subtract nodes as needed. Further consideration of this method is discussed in the following section.

The main concept of the front tracking method is to identify a fluid with a Heaviside function, H. This function is one at a location where a particular fluid exists and zero elsewhere. The interface is located where the H function changes from zero to one, which is the location where the delta function, δ, is nonzero. H can be expressed as a function of the δ function.

$$H(\mathbf{x}, t) = \int_{\Delta V} \delta(\mathbf{x} - \mathbf{x}_\mathrm{I}) \mathrm{d}V \tag{4.227}$$

The integration is performed over an entire volume, V, with δ being nonzero only on the interface between the phases, A_I, at location x_I. The calculation of the fluid properties, Φ, is straightforward with the Heaviside function.

$$\Phi(\mathbf{x}, t) = \Phi_k H(\mathbf{x}, t) + \Phi_j(1 - H(\mathbf{x}, t)) \tag{4.228}$$

Note that when these values are integrated over a finite volume, the properties used in the calculation are volume averaged. The calculation of the interface is done in a Lagrangian fashion.

$$\frac{\mathrm{d}\mathbf{x}_\mathrm{I}}{\mathrm{d}t} \cdot \mathbf{n} = \mathbf{V}_\mathrm{I} \cdot \mathbf{n} \tag{4.229}$$

To write the equivalent expression using a kinematic equation, the gradient of H must be defined

$$\nabla H = \int_{A_{\mathrm{I}}} \mathbf{n}\delta(\mathbf{x} - \mathbf{x}_{\mathrm{I}})\mathrm{d}s \tag{4.230}$$

where s is a surface that is integrated over A_{I}. Therefore, the kinematic equation using H to represent the interface is:

$$\frac{\partial H}{\partial t} = -\mathbf{V}_{\mathrm{I}} \cdot \nabla H = -\int_{A_{\mathrm{I}}} \mathbf{V}_I \cdot \mathbf{n}\delta(\mathbf{x} - \mathbf{x}_I)\mathrm{d}s \tag{4.231}$$

To calculate the interfacial velocity, the overall continuity equation can be considered.

$$\frac{\partial}{\partial t}(\rho) = -\nabla \cdot (\rho \mathbf{V}) \tag{4.232}$$

The evaluation of fluid properties and flow values uses the H function as shown in Eq. (4.228). The overall continuity equation can be written as:

$$\frac{\partial}{\partial t}(\rho_k H + \rho_j(1 - H)) + \nabla \cdot (\rho_k \mathbf{V}_k H + \rho_j \mathbf{V}_j(1 - H)) = 0 \tag{4.233}$$

Using the chain rule, the continuity equation is broken into three regions: the bulk regions for phase k and phase j, and the interface. The interfacial continuity equation is:

$$(\rho_k - \rho_j)\frac{\partial}{\partial t}(H) = -(\rho_k \mathbf{V}_k + \rho_j \mathbf{V}_j) \cdot \nabla(H) \tag{4.234}$$

Applying Eqs. (4.230) and (4.231) to Eq. (4.234) yields:

$$\int_{A_{\mathrm{I}}} (\rho_k - \rho_j)\mathbf{V}_{\mathrm{I}} \cdot \mathbf{n}\delta(\mathbf{x} - \mathbf{x}_{\mathrm{I}})\mathrm{d}s = \int_{A_{\mathrm{I}}} (\rho_k \mathbf{V}_k + \rho_j \mathbf{V}_j) \cdot \mathbf{n}\delta(\mathbf{x} - \mathbf{x}_{\mathrm{I}})\mathrm{d}s \tag{4.235}$$

The interface velocity can be calculated using Eq. (4.229). The momentum equation is:

$$\frac{\partial \rho \mathbf{V}}{\partial t} + \nabla \cdot (\rho \mathbf{V}\mathbf{V}) = \nabla \cdot \boldsymbol{\tau} + \rho \mathbf{g} + \int_{A_{\mathrm{I}}} \sigma \kappa \mathbf{n}\delta(\mathbf{x} - \mathbf{x}_s)\mathrm{d}s \tag{4.236}$$

The last term represents the surface tension effects where the twice the mean curvature is κ. Also, the energy equation can be written as:

$$\frac{\partial}{\partial t}(\rho c_{\mathrm{p}} T) + \nabla \cdot (\rho \mathbf{V} c_{\mathrm{p}} T) = -\nabla \cdot \mathbf{q}'' - \int_{A_{\mathrm{I}}} \dot{m}'' \left[h_{kj} + (c_{\mathrm{p},k} - c_{\mathrm{p},j})\delta(\mathbf{x} - \mathbf{x}_{\mathrm{s}})\right]\mathrm{d}s \tag{4.237}$$

Note that pressure effects and viscous dissipation are neglected. The last term represents the latent heat release in a phase change process.

The interface is made of a separate mesh that goes through the stationary mesh. It generally does not lie on top of the points in which the conservation laws are applied. Therefore, the interfacial values are calculated by grid interpolation, and mass and momentum of the advection front are not necessarily conserved. The error in mass and momentum conservation at the interface can be reduced to an acceptable level by refining the mesh.

Problems

4.1. Liquid droplets that condense to form fog in the atmosphere can be as small as 2 μm in diameter when they first form. Determine the pressure in a droplet of this size at 20 °C (101.3 kPa). Note that for water in contact with air, the surface tension σ is 0.0728 N/m. Describe any assumption you need to make in order to get the final answer.

4.2. A 0.2-mm-thick water film sits on a surface held at a temperature of 60 °C. The liquid film is exposed to air at a bulk temperature of $T_G = 20$ °C, and the convective heat transfer coefficient between the liquid film and the air is $h_\delta = 15$ W/m^2 K. Suppose the wave number is $\hat{\alpha} = 2$. What is the critical Marangoni number, Ma_c ? Is the liquid film stable?

4.3. A 0.2-mm-diameter tube is vertically placed in a pool of water at 20 °C. Find the capillary rise in tubes of the following materials: (a) aluminum, (b) brass, (c) copper, and (d) steel.

4.4. Consider the smooth flat plates shown in the sketch. The angle between plates is 2γ ($\gamma < 90°$). The contact angle of the liquid with the surface is θ ($\theta < 90°$). For the cases showing in Fig. P4.4, sketch the shape of the liquid–vapor interface between the plates relative the surface of the liquid in the reservoir. *Assume that the radius of the curvature along the interface is constant.*

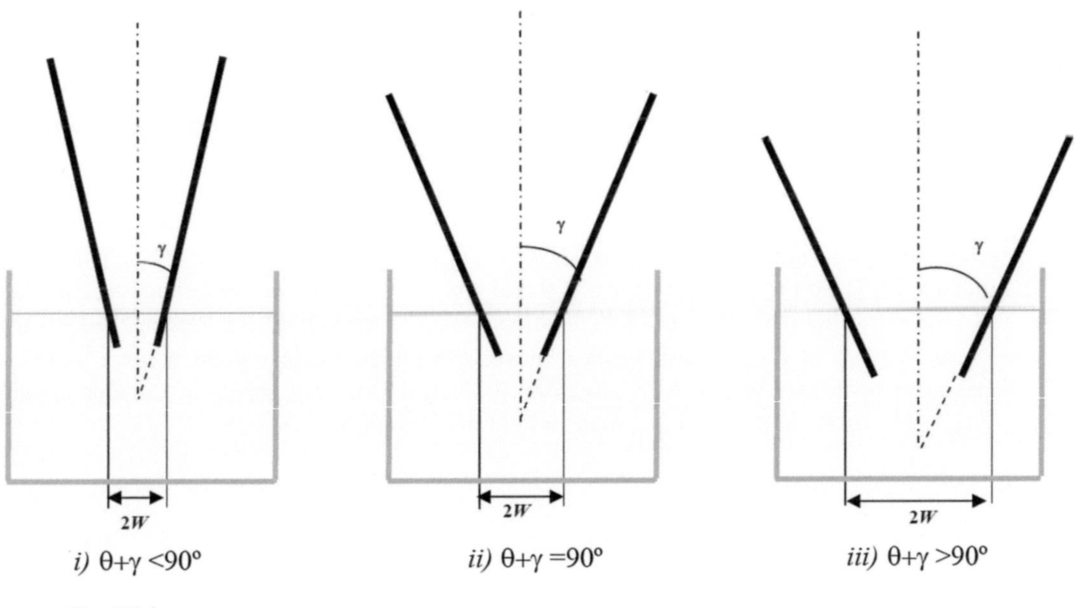

Fig. P4.4

4.5. Consider water slug in the copper tube with 1-mm inner diameter (see Fig. P4.5). The receding and advancing contact angles of copper at 20 °C are given as 30° and 60°, respectively (θ_r = 30°, θ_a = 60°). Calculate the maximum stable length of the slug (i.e., before liquid slug starts to move) between the plates. ρ_l = 1000 kg/m^3; ρ_g = 1 kg/m^3, σ_{lv} = 0.0725 N/m.

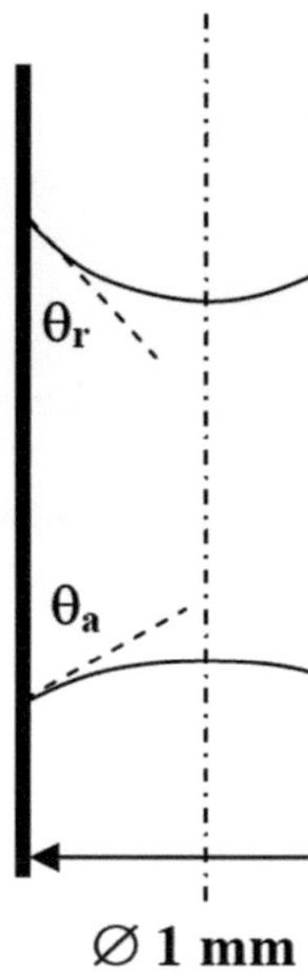

Fig. P4.5

4.6. Two nearly vertical but nonparallel plates touch a pool of water with their parallel bottom edges. The angle between the plates is 2γ so that these plates would intersect somewhere under the pool surface. The distance between the plates at the pool surface level, 2 W, is small. The apparent contact angle is denoted by θ. The liquid–vapor meniscus between the plates is elevated from the pool surface level due to capillary pressure. Assume that the meniscus curvature does not change along the liquid–vapor interface. Derive an algebraic equation for the height of the capillary rise H for this situation (for $\theta \neq 0$ and $\gamma \neq 0$). Using the derived equation, estimate H for the following data: $W = 0.5$ mm, $\theta = 0, \gamma = 0, \rho_l = 1000$ kg/m^3, $\sigma = 0.06$ N/m.

4.7. A small amount of liquid metal resides between two high melting point powder particles separated by a distance of ℓ ($\ell > 2r_0$) (see Fig. P4.7). If the effect of gravity on the liquid metal is negligible, estimate the force required to hold the two particles together.

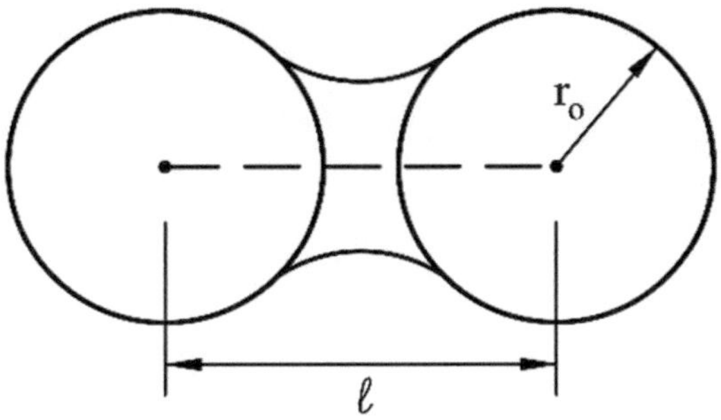

Fig. P4.7

4.8. Write the continuity, momentum, energy, and species equations and the necessary boundary conditions for a very long capillary tube that is open at both ends, as shown in Fig. P4.8. The evaporation is driven by the concentration gradient of vapor in the air. The evaporation cools the interface while the wall heats the fluid, causing a temperature gradient along the interface. The assumptions are the same as Example 4.4.

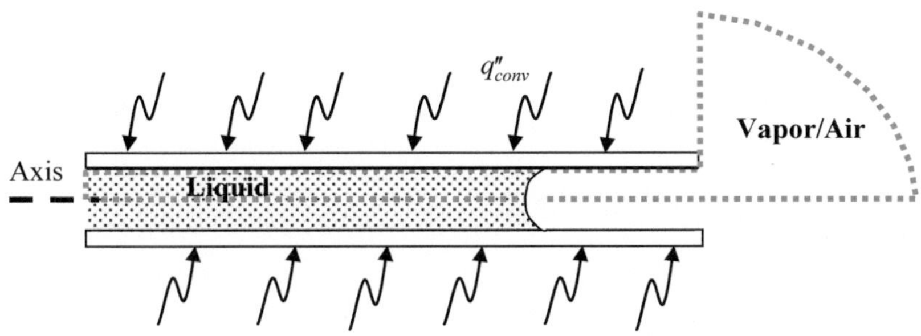

Fig. P4.8

4.9. The generalized mass balance on a liquid–vapor interface is given in Eq. (4.62). Rewrite it in a three-dimensional Cartesian coordinate system.

4.10. The jump conditions at an interface can also be obtained by applying the various conservation laws to a control volume that includes the interface. Apply the conservation of mass principle to the control volume shown in Fig. P4.10, and show that when the thickness of the control volume Δx goes to zero, the conservation of mass is $\rho_1(u_1 - u_{\mathrm{I}}) = \rho_2(u_2 - u_{\mathrm{I}})$.

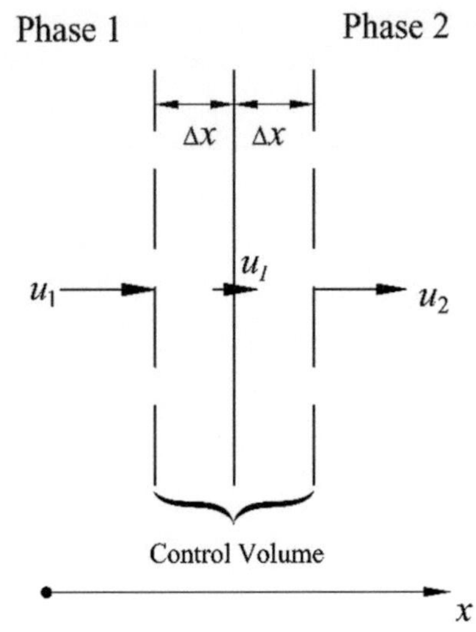

Fig. P4.10

4.11. For the control volume shown in Fig. P4.10, show that the momentum balance at the interface is $p_1 - p_2 = \rho_2(u_2 - u_\mathrm{I})u_2 - \rho_1(u_1 - u_\mathrm{I})u_1$ when the thickness of the control volume Δx goes to zero.

4.12. In a counter-current condenser shown in Fig. P4.12, the liquid flows downward with a mass flow rate of $\dot{m}_\ell$, while the vapor flows upward with a mass flow rate of $\dot{m}_\mathrm{v}$. The heat flux at the external wall is q''_w. Derive the jump conditions using mass and energy balances at the liquid–vapor interface.

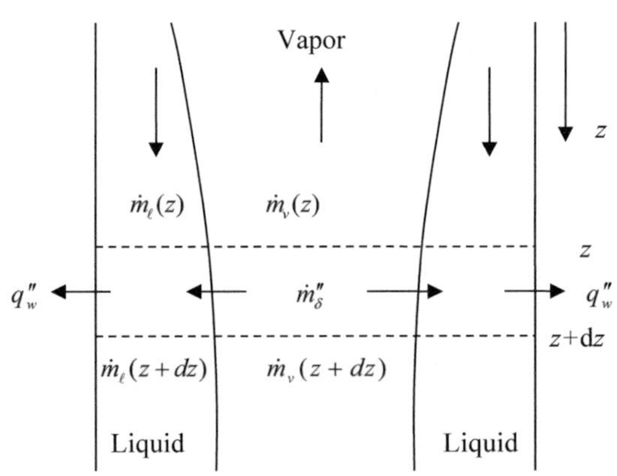

Fig. P4.12

4.13. Redo the Problem 4.12 for the case of co-current condensation, i.e., both liquid and vapor flow downward.

4.14. Redo Problem 4.12 for the case of counter-current evaporation and discuss the effect of the heat transfer direction on the energy balance at the interface.

4.15. Consider an ultrathin evaporating liquid film on a hot solid surface of temperature T_w in the presence of a saturated vapor of temperature T_v. The film thickness δ is small and the disjoining pressure p_d is significant. The film is flat. Does the disjoining pressure increase or decrease the evaporation rate in comparison with the hypothetical case where $p_d = 0$ (for the same film thickness and temperatures)? Assuming that $T_w - T_v$ is extremely small, explain the effect of p_d on the evaporation rate by referring to the extended Kelvin equation and interfacial resistance.

4.16. A liquid film of ammonia with thickness $\delta = 0.375 \times 10^{-8}$ m is evaporating from a smooth solid surface into bulk vapor. The solid surface temperature is T_w = 250.12 K. The bulk vapor temperature is T_v = 250 K. The constants in the correlation for disjoining pressure $p_d = -A'\delta^{-B}$ are $A' = 10^{-21}$ and $B = 3$. The accommodation coefficient is equal to unity. In this particular case, the perfect gas law can be used instead of the saturation tables, for convenience. Estimate the heat flux at the liquid interface.

4.17. A slug of water 4.5 mm in length sits in a vertical copper tube with an inner radius of 2 mm at room temperature. The temperature of the whole system is slowly increasing. At what temperature will the water slug start to move?

4.18. Nusselt-type condensation takes place on a flat vertical plate made from aluminum. The thickness of the plate is 5 mm and there are 1-mm ID microchannels inside it containing cooling liquid. You must increase the rate of condensation, which is restricted primarily by the conduction through the condensate film. What would you do?

4.19. The effective heat transfer during evaporation of liquid is more extensive from a solid surface with small grooves than from a plain surface. How would you explain this?

4.20. A vertical capillary tube is connected to bulk liquid at the bottom end and to a heat source at the upper end. The liquid evaporates from the liquid–vapor meniscus, so a quasi-steady state exists. What is the heat load that would initiate dry-out of the upper end and lead to unbounded increase of its temperature? Write corresponding general formulas or equations.

4.21. The temperature at the meniscus of Problem 4.20 equals the saturation temperature, and the bulk liquid is at a lower temperature T_b. Repeat Problem 4.20 accounting for the conduction heat loss from the meniscus to the bulk liquid.

4.22. A capillary tube of inner radius r is vertically dipped into a liquid. The depth from the free surface to the lower end of the tube is a. When the liquid rises in the capillary tube due to capillary force, it can be viewed as a variable mass system with capillary force, gravity, and viscous force acting upon it. What is the governing equation for the liquid slug in the capillary tube? The shape of the meniscus in the tube can be assumed to be the same at all times.

4.23. A Newtonian liquid is in an open cylindrical container with a radius of R and height of H (see Fig P4.23). A stationary continuous axisymmetric laser beam with an intensity of $I = I_0 \exp(-r^2/r_0^2)$ irradiates the top of the liquid surface. The side of the cylindrical container is maintained at a constant temperature, T_0, and the bottom of the container is adiabatic. Supposing the surface tension is a linear function of temperature $\sigma = C_0 - C_1 T$, write the governing equations and the corresponding boundary conditions of the problem. The effect of gravity is negligible, and the liquid flow induced by the laser beam can be assumed to be axisymmetric.

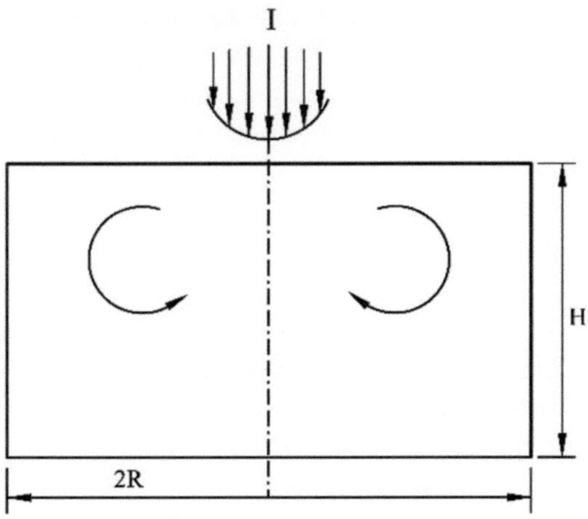

Fig. P4.23

4.24. Introduce appropriate non-dimensional variables and non-dimensionalize the governing equations and boundary conditions in Problem 4.25. Identify the dominant dimensionless variables in the problem.

4.25. Saturated vapor enters a miniature channel with a radius of R and a length of L (see Fig. P4.25). Condensation takes place on the wall of the channel, since the wall temperature T_w is below the saturation temperature of the working fluid T_{sat}. The condensate fluid flows in the positive x-direction due to the effects of shear force and surface tension. The problem is condensation on the inner surface of the miniature channel with cocurrent vapor flow. The average vapor velocity is not constant along the x-direction because condensation occurring on the wall reduces the amount of vapor flow in the core of the tube. Capillary blocking occurs when capillary force causes the liquid to block the tube cross-section at some distance, L_δ, from the condenser entrance. Supposing the effect of gravity on the two-phase flow in the miniature tube is negligible, write the governing equations and boundary conditions for the liquid and vapor flows.

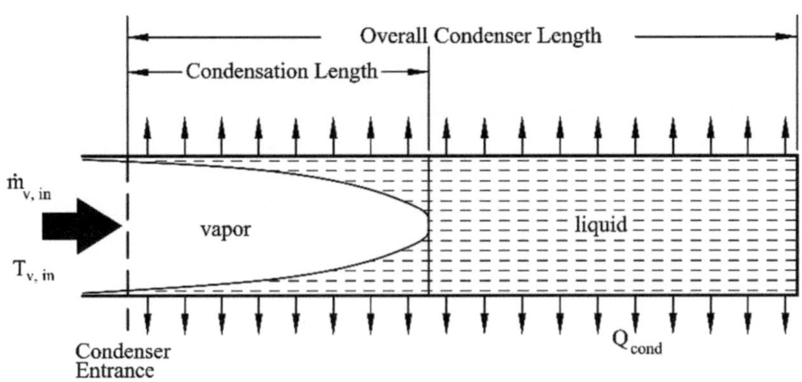

Fig. P4.25

4.26. Heat is applied to the outer surface of a microchannel with inner and outer radii of r_i and r_o, respectively, as shown in Fig. P4.26. Specify the governing equations that describe evaporation in the microchannel. The effects of disjoining pressure and interfacial thermal resistance must be taken into account.

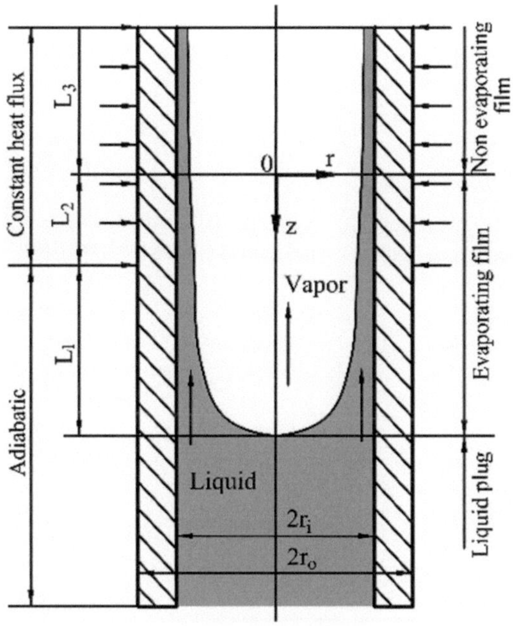

Fig. P4.26

4.27. The physical model of the condenser section of the pulsating heat pipe (PHP) with an open end is shown in Fig. P4.27. It is assumed that the liquid film in the condenser section is divided into two regions: a thin film region and a meniscus region. These two regions are separated by a transition point where the film thickness is $\delta = \delta_{tr}$. The curvature of the liquid film in the meniscus region is assumed to be constant. Assuming the effect of disjoining pressure is negligible, specify the governing equation for the liquid film thickness in $0 < x < L_2$.

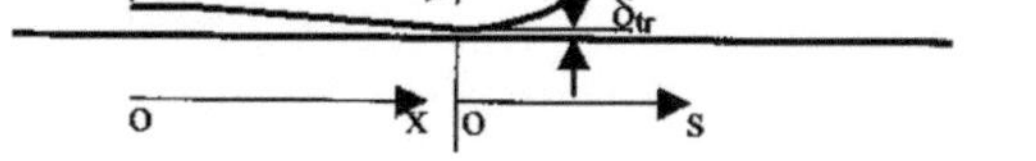

Fig. P4.27

4.28. The fluid mechanics of capillary tubes and microchannels is of great interest for modeling transport phenomena in microdevices. Develop the physical formulation, including continuity and momentum equations with appropriate boundary and initial conditions, for a typical miniature channel with a free surface as depicted in Fig. P4.28 for both submerged and blocked end configuration. Neglect mass transfer and Marangoni effects.

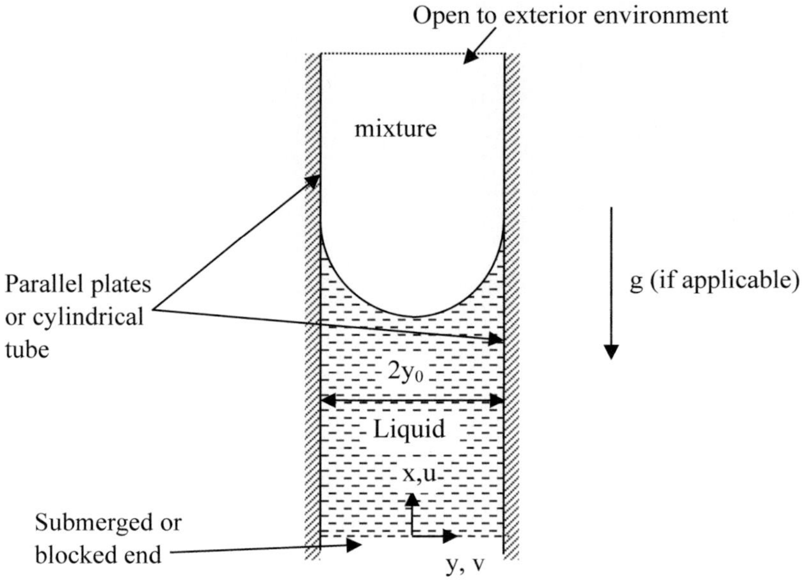

Fig. P4.28

4.29. Describe the physical/mathematical formulation to model the problem related to instability growth of a free-flowing cylindrical column of fluid as shown in Fig. P4.29.

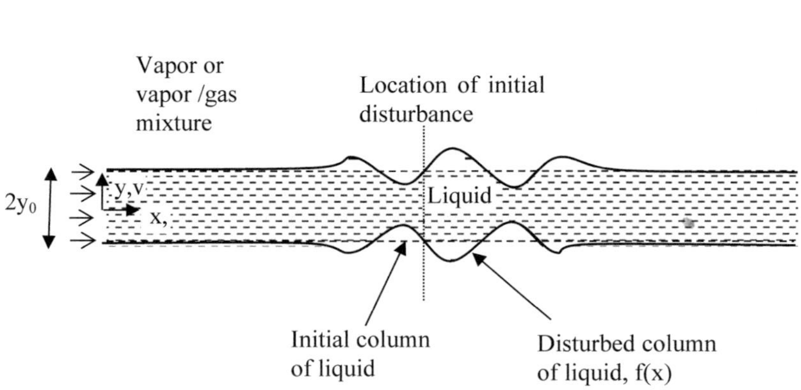

Fig. P4.29

4.30. Obtain an analytical solution for the shape of the meniscus in a capillary tube in 1 g environment for mechanical equilibrium, as shown in Fig. P4.30 for both parallel plates and circular tubes. Assume the effect of the gaseous region is negligible with vapor pressure constant. You should get a second-order ordinary differential equation for $f(y)$ which can be solved by the Runge-Kutta method. Obtain numerical results for the working fluid as acetone with $y_0 = 0.001$ m and contact angle $\theta = 0.0$. Obtain a closed analytical solution for the situation with no gravity.

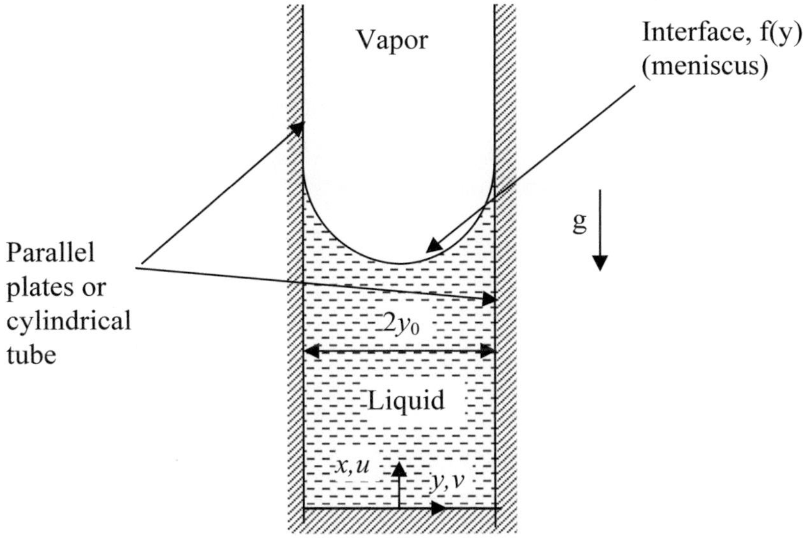

Fig. P4.30

4.31. Obtain an analytical solution for the rise of a meniscus in a capillary tube with both ends open, as shown in Fig. P4.31, for both parallel plates and a circular tube. If $\bar{u}$ is the bulk motion or the average velocity, obtain your solution in terms of $\bar{u} = \mathrm{d}H/\mathrm{d}t$ where H is the mean height of the meniscus.

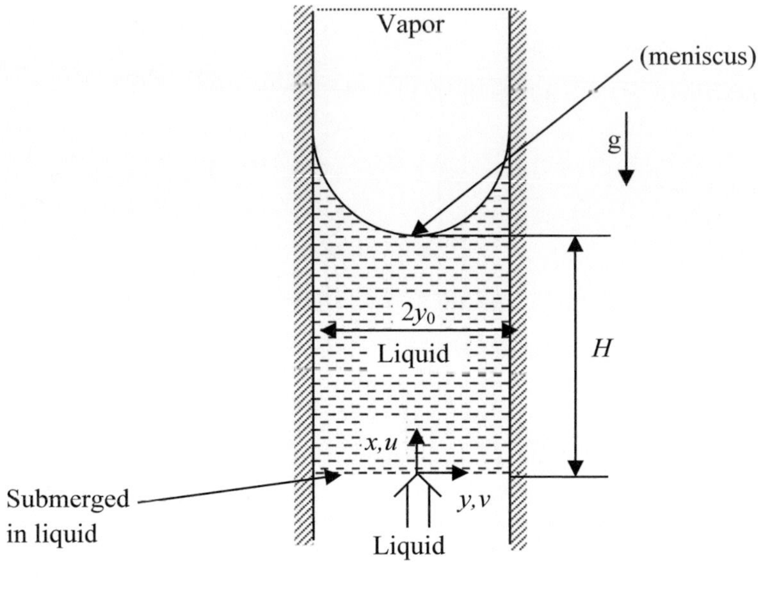

Fig. P4.31

4.32. Consider the steady, laminar flow of a liquid film down an inclined flat plate (see Fig. P4.32). Assume the flow is fully developed. However, the presence of a small temperature gradient in the axial direction dT/dz produces a constant surface tension gradient $d\sigma/dz = C$. Assume the surface tension effect is only at the interface, which produces a nonzero shear stress and no other effect on physical properties of the system. Solve the momentum equation to determine the shear stress variation and velocity distribution as a function of distance perpendicular to the plate.

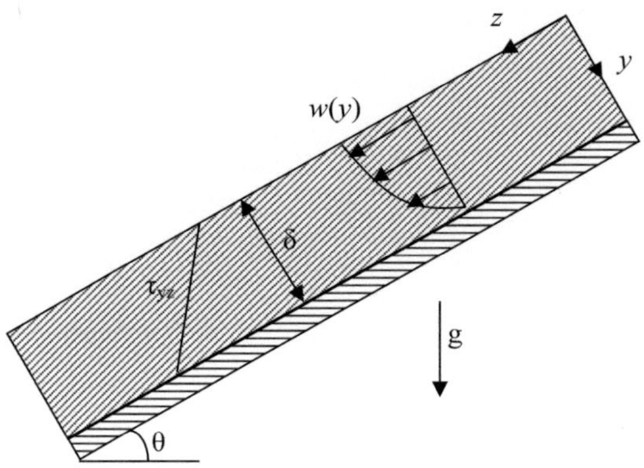

Fig. P4.32

4.33. Consider the absorption of gas A (O_2) to a laminar, steady, fully developed, isothermal falling film of liquid B (water) (see Fig. P4.33). Since O_2 is slightly soluble in water, one can assume the properties of water are not affected. Since the diffusion of O_2 is taking place slowly and thermal diffusion is in the immediate vicinity of liquid, O_2 does not penetrate very far into the liquid. Obtain the total mole flow rate of gas A to liquid B.

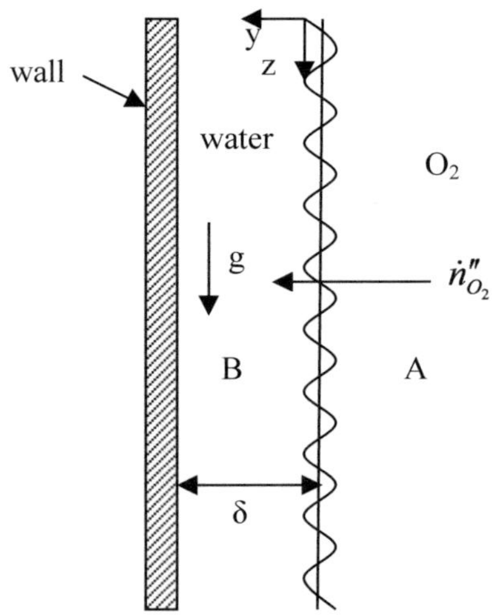

Fig. P4.33

4.34. Consider the solid dissolution of a component A for a solid wall to a steady, laminar, isothermal falling liquid film B (see Fig. P4.34). Assume that species A is slightly soluble in B. Obtain the physical formulation of the simplified model.

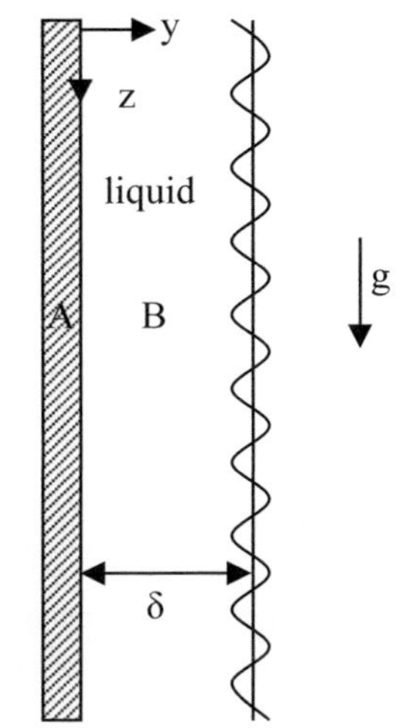

Fig. P4.34

4.35. Saturated vapor at 100 °C flows inside a horizontal tube cooled at the external surface. A pool of the liquid water exists at the bottom of the tube and flows with a velocity of 0.5 m/s. Determine the critical velocity of the vapor when waves appear on the liquid–vapor interface.

4.36. In a stratified, horizontal, two-phase flow system, the saturated water vapor flows above the saturated liquid water. The pressure of the two-phase system is at 5 bar. What is the most dangerous wavelength, λ_D?

References

Benjamin, T. B. (1957). Wave formation in a laminar flow down an inclined plane. *Journal of Fluid Mechanics, 16,* 554–574.

Brackbill, J. U., Kothe, D. B., & Zemach, C. (1992). A continuum method for modeling surface tension. *Journal of Computational Physics, 100,* 335–354.

Brauer, H. (1956). Stromung and Warmenbergang bei Rieselfilmen. In *VDI Forschunsheft, Dusseldorf,* 457.

Buffone, C., Sefiane, K., & Christy, J. R. E. (2004). Experimental investigation of the hydrodynamics and stability of an evaporating wetting film placed in a temperature gradient. *Applied Thermal Engineering, 24,* 1157–1170.

Carey, V. P. (2016). *Liquid-vapor phase-change phenomena: An introduction to the thermophysics of vaporization and condensation processes in heat transfer equipment* (3rd ed.). New York, NY: Taylor & Francis.

Derjaguin, B. V. (1955). Definition of the concept of and magnitude of the disjoining pressure and its role in the statics and kinetics of thin layers of liquid. *Kolloidny Zhurnal, 17,* 191–197.

Derjaguin, B. V. (1989). *Theory of stability of colloids and thin films*. New York, NY: Plenum.

Faghri, A. (2016). *Heat pipe science and technology* (2nd ed.). Columbia, MO: Global Digital Press.

Faghri, A., & Payvar, P. (1979). Transport to thin falling liquid films. *Reg Journal of Energy Heat and Mass Transfer, 1,* 153–173.

Faghri, A., & Seban, R. A. (1985). Heat transfer in wavy liquid films. *International Journal of Heat and Mass Transfer, 28,* 506–508.

Glimm, J., Li, X. L., Liu, Y., & Zhao, N. (2001). Conservative front tracking and level set algorithms. In *Proceedings of National Academic of Science: Applied Mathematics* (Vol. 98, pp. 14198–14201).

Harlow, F. H., & Welch, J. E. (1965). Numerical calculation of time-dependent viscous incompressible flow. *Physics of Fluids, 8,* 2182–2189.

Hirt, C. W., & Nichols, B. D. (1981). Volume of fluid (VOF) method for the dynamics of free boundaries. *Journal of Computational Physics, 39,* 201–225.

Holm, F. W., & Goplen, S. P. (1979). Heat transfer in the meniscus thin-film transition region. *Journal of Heat Transfer, 101*(3), 543–547.

Jamet, D., Torres, D., & Brackbill, J. U. (2002). On the theory and computation of surface tension: The elimination of parasitic currents through energy conservation in the second-gradient method. *Journal of Computational Physics, 182,* 262–276.

Kapitza, P. L. (1964). *Wave flow of thin layers of a viscous fluid*. Kapitza, MacMillan, NY: Collected Papers of P.L.

Khrustalev, D., & Faghri, A. (1994). Thermal analysis of a micro heat pipe. *Journal of Heat Transfer, 116,* 189–198.

Koizumi, Y., Enari, R., & Ohtake, H. (2005). Correlations of characteristics of waves on a film falling down on a vertical wall. In *Proceeding of* the *2005 ASME International Mechanical Engineering Congress and Exposition*. Orlando, FL, November 5–11, 2005.

Kucherov, R. Y., & Rikenglaz, L. E. (1960). The problem of measuring the condensation coefficient. *Doklady Akademii Nauk SSSR, 133,* 1130–1131.

Kuramae, M., & Suzuki, M. (1993). Two-component heat pipes utilizing the Marangoni effect. *Chemical Engineering of Japan, 26,* 230–231.

Kwok, D. Y., & Neumann, A. W. (1999). Contact angle measurement and contact angle interpretation. *Advances in Colloid and Interface Science, 81,* 167–249.

Lide, D. R. (Ed.). (2004). *CRC handbook of chemistry and physics* (85th ed.). Boca Raton, FL: CRC Press.

Lin, L., & Faghri, A. (1999). Heat transfer in the micro region of a rotating miniature heat pipe. *International Journal of Heat and Mass Transfer, 42,* 1363–1369.

Mills, A. F. (1965). *The condensation of steam at low pressures* (Report No. NSF GP-2520, Series No. 6, Issue No. 39). Berkeley: Space Sciences Laboratory, University of California, Berkeley.

Nichols, B. D., & Hirt, C. W. (1975). Methods for calculating multidimensional, transient free surface flows past bodies. In *Proceedings of the First International Conference on Numerical Ship Hydrodynamics*. Gaithersburg, MD, October 20–23.

Nosoko, T., Yoshimura, P. N., Nagata, T., & Oyakawa, K. (1996). Characteristics of two-dimensional waves on a falling film. *Chemical Engineering Science, 51,* 725–732.

Ojha, M., Chatterjee, A., Dalakos, G., Wayner, P. C., & Plawsky, J. L. (2010). Role of solid surface structure on evaporative phase change from a completely wetting corner meniscus. *Physics of Fluids, 22,* 052101.

Patankar, S. V. (1980). *Numerical heat transfer and fluid flow*. Washington, DC: Hemisphere.

Paul, B. (1962). Compilation of evaporation coefficients. *ARSJ, 32,* 1321–1328.

Plawsky, J. L., Fedorov, A. G., Garimella, S. V., Ma, H. B., Maroo, S. C., Chen, L., et al. (2014). Nano and microstructures for thin-films evaporation: A review. *Nanoscale and Microscale Thermophysical Engineering, 18* (3), 251–269.

Potash, M., & Wayner, P. C. (1972). Evaporation from a two-dimensional extended meniscus. *International Journal of Heat and Mass Transfer, 15,* 1851–1863.

Rice, J., & Faghri, A. (2005a). A new computational method to track a liquid/vapor interface with mass transfer, demonstrated on the concentration driven evaporation from a capillary tube, and the Marangoni effect. In *Proceeding of the 2005 ASME International Mechanical Engineering Congress and Exposition*. Orlando, FL, November 5–11.

Rice, J., & Faghri, A. (2005b). A new computational method for free surface problems. In *Proceeding of the 2005 ASME Summer Heat Transfer Conference*. San Francisco, CA, July 17–22.

Rogovan, I. A., Olevskii, V. M., & Runova, N. G. (1969). Measurement of the parameters of film type wavy flow on a vertical plate. *Theoretical Foundations of Chemical Engineering, 3,* 164–171.

Savino, R., Francescantonioa, N., Fortezzab, R., & Abe, Y. (2007). Heat pipes with binary mixtures and inverse Marangoni effects for microgravity applications. *Acta Astronautica, 61,* 16–26.

Schrage, R. W. (1953). *A thermal study of interface mass transfer*. New York: Columbia University Press.

Shopov, P. J., Minev, P. D., Bazhekov, I. B., & Zapryanov, Z. D. (1990). Interaction of a deformable bubble with a rigid wall at moderate Reynolds numbers. *Journal of Fluid Mechanics, 219,* 241–271.

Silver, R. S., & Simpson, H. C. (1961). The condensation of superheated steam. In *Proceedings of the Conference on National Engineering Laboratory*. Glasgow, Scotland.

Stepanov, V. G., Volyak, L. D., & Tarlakov, Y. V. (1977). Wetting contact angles for some systems. *Journal of Engineering Physics and Thermophysics, 32,* 1000–1003.

Tryggvason, G., Bunner, B., Esmaeeli, A., Juric, D., Al-Rawahi, N., Tauber, W., et al. (2001). A front-tracking method for the computations of multiphase flow. *Journal of Computational Physics, 169,* 707–759.

Yang, J., Han, J., Isaacson, K., & Kwok, D. Y. (2003). Effects of surface defects, polycrystallinity, and nanostructure of self-assembled monolayers for octadecanethiol adsorbed on to Au on wetting and its surface energetic interpretation. *Langmuir, 19,* 9231–9238.

Youngs, D. L. (1982). Time-dependent multimaterial flow with large fluid distortion. In K. W. Morton & M. J. Baines (Eds.), *Numerical methods for fluid dynamics* (pp. 273–285). Cambridge: Academic Press.

Yaws, C. L. (1992). *Thermodynamic and physical property data*. Houston, Texas: Gulf Publication Co.

Zhang, Y., & Faghri, A. (2008). Advances and unresolved issues in pulsating heat pipes. *Heat Transfer Engineering, 29* (1), 20–44.

Melting and Solidification 5

5.1 Introduction

Melting and solidification find applications in the geophysical sciences; industrial forming operations such as casting and laser drilling; latent heat energy storage systems; and food and pharmaceutical processing. Any manmade metal products must undergo liquid forms at some point during manufacturing processes and solidify to form intermediate or final products. Melting and solidification processes can be classified as one of three types: one-region, two-region, and multiple-region. The classification depends on the properties of the phase-change material (PCM) involved and the initial conditions. For a single-component PCM, melting or solidification occurs at a single temperature. Pure water, for example, melts at a uniform temperature of 0 °C, while pure n-Octadecane ($C_{18}H_{38}$) melts at 28 °C. For the solid–liquid phase-change process of a PCM with a single melting point, the solid–liquid interface appears as a clearly observable sharp border. Initial conditions for the solid–liquid phase-change process of a single-component PCM determine whether the problem will be classified as a one- or two-region problem.

For the melting (or solidification) process, if the initial temperature of the PCM, T_i, equals the melting point, T_m, the temperature in the solid (liquid) phase remains uniformly equal to the melting point throughout the process. In this case, only the temperature distribution in the liquid (solid) phase needs to be determined. Thus, the temperature of only one phase is unknown, and the problem is called a *one-region problem*. Figure 5.1 shows the temperature distribution of one-dimensional, one-region melting, and solidification problems. The surface temperature, T_0, is greater than T_m for melting and is below T_m for solidification.

For the melting (solidification) process, if the initial temperature of the PCM, T_i, is below (above) the melting point of the PCM, T_m, the temperature distribution of both the liquid and solid phases must be determined; this is called a *two-region problem*. Figure 5.2 shows the temperature distribution of one-dimensional two-region melting and solidification problems.

For a multicomponent PCM, the solid–liquid phase-change process occurs over a range of temperatures (T_{m1}, T_{m2}), instead of a single temperature. The PCM is liquid if its temperature is above T_{m2} and solid when its temperature is below T_{m1}. Between the solid and liquid phases, there is a *mushy zone* where the temperature falls within the phase-change temperature range (T_{m1}, T_{m2}). Successful solution of phase-change problems involving these substances requires determination of the temperature distribution in the liquid, solid, and mushy zones; therefore, these are referred to as *multiregion problems*. The temperature distribution of one-dimensional solidification in a

A. Faghri and Y. Zhang, *Fundamentals of Multiphase Heat Transfer and Flow*, https://doi.org/10.1007/978-3-030-22137-9_5

Fig. 5.1 One-region melting and solidification

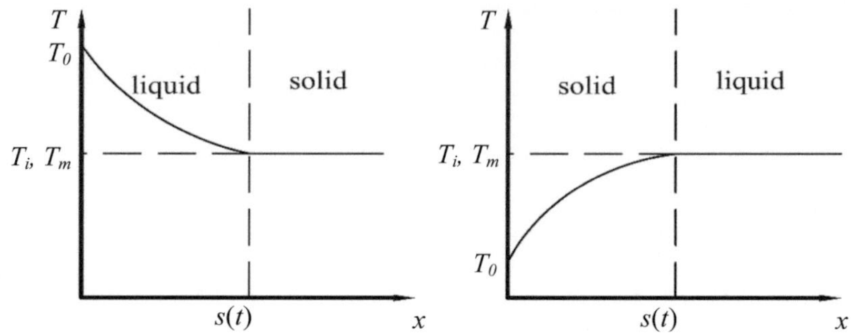

(a) Melting ($T_0 > T_m$) (b) Solidification ($T_0 < T_m$)

Fig. 5.2 Two-region melting and solidification

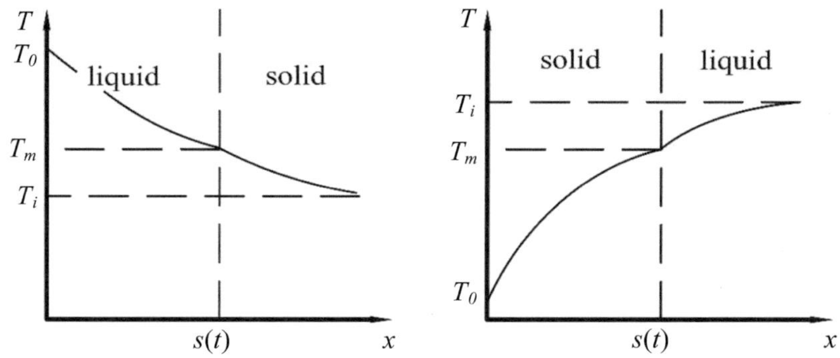

(a) Melting ($T_0 > T_m > T_i$) (b) Solidification ($T_0 < T_m < T_i$)

multicomponent PCM is shown in Fig. 5.3, where it can be seen that the solution requires the tracking of two interfaces.

In solid–liquid phase-change problems, the location of the solid–liquid interface is unknown before the final solution is obtained and this presents a special difficulty. Since the interface also moves during melting or solidification, such problems are referred to as *moving boundary problems* and always have time as an independent variable.

Natural convection induced by the temperature gradient in the liquid and mushy zones can have a significant effect on melting and solidification processes. The importance of natural convection in solid–liquid phase-change problems was not recognized until the late 1970s. Figure 5.4 shows the effect of natural convection on a melting process in a rectangular cavity. The left wall temperature is at $T_h (T_h > T_m)$ and right wall temperature is at the initial temperature, $T_i (T_i \leq T_m)$. When conduction controls the melting process, the melting front moves parallel to the heated vertical wall, while natural convection in the liquid causes the front to accelerate in the upper portion of the cavity; this occurs because the warmer liquid near the heated wall rises while the colder fluid near the interface descends. As a result, the shape of the melting front is inclined, and the rate of melting increases due to the improved heat transfer in the liquid. Convection has a very strong influence on the growth of microstructure during solidification of pure metal or alloys.

This chapter presents a number of prototypical problems of melting and solidification; its goal is to establish the physics of this form of phase change and to demonstrate the variety of tools available for its solution. Section 5.2 discusses boundary conditions at the solid–liquid interface, including continuity of temperature and conservation of energy at the interface. Section 5.3 presents the exact solutions of one-dimensional solid–liquid phase problems, including both one-region and two-region

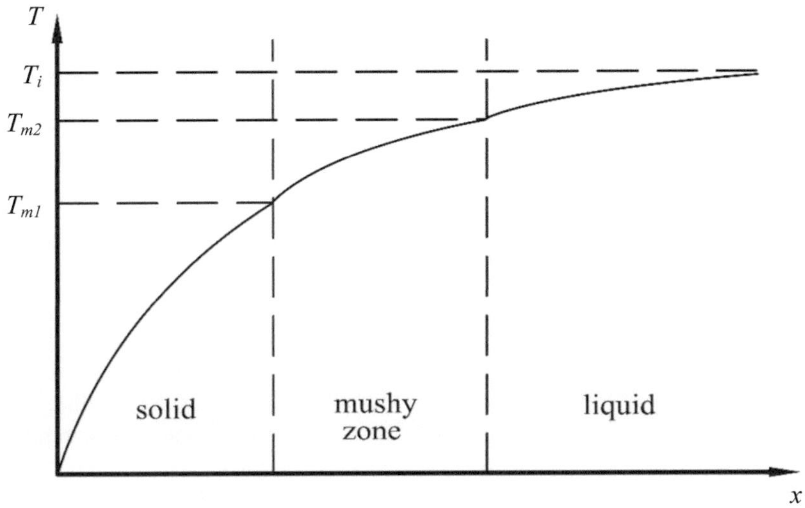

Fig. 5.3 Solidification of a multicomponent PCM

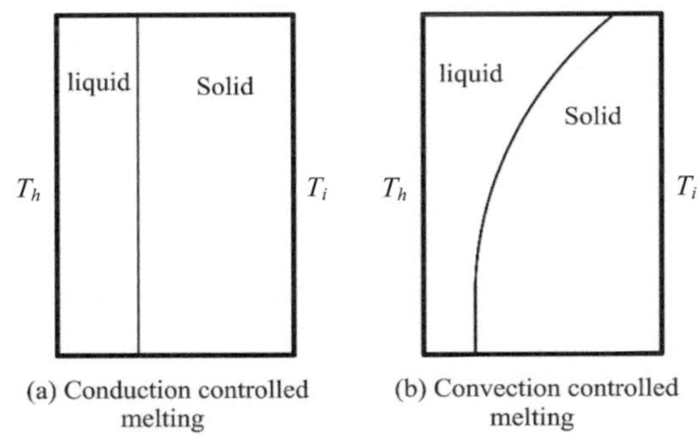

Fig. 5.4 Conduction- and convection-controlled melting

problems. Integral solutions of various solid–liquid phase-change problems are presented in Sect. 5.4. Integral solution is applied to one-region and two-region problems under different boundary conditions and coordinate systems, as well as to solidification of binary solutions. Section 5.5 presents an analysis of contact melting in a cavity, which is a topic of importance in latent heat thermal energy storage applications. Since the known analytical solutions only cover a limited number of situations, numerical simulation methods are also addressed in Sect. 5.6 through three approaches: the enthalpy method, the equivalent heat capacity method, and the temperature-transforming model. Melting and solidification in porous media are discussed in Chap. 12.

5.2 Boundary Conditions at the Solid–Liquid Interface

For a solid–liquid phase change of a PCM with a single melting temperature, the solid–liquid interface clearly delineates the liquid and solid phases. The boundary conditions at this interface must be specified in order to solve the problem. The boundary conditions to be specified in this section are special cases of those discussed in Chap. 3.

As shown for the one-dimensional melting problem illustrated in Fig. 5.2a, the solid–liquid interface separates the liquid and solid phases. The temperatures of the liquid and solid phases near the interface must equal to the temperature of the interface, which is at the melting point, $T_m T_m$. Therefore, the boundary conditions at the interface can be expressed as

$$T_\ell(x,t) = T_s(x,t) = T_m, \qquad x = s(t) \tag{5.1}$$

where $T_\ell(x,t)$ and $T_s(x,t)$ are the temperatures of the liquid and solid phases, respectively.

The density of the PCM usually differs between the liquid and solid phases; therefore, density change always accompanies the phase-change process. The solid PCM is usually denser than the liquid PCM, except for a few substances such as water and gallium. For example, the volume of paraffin, a very useful PCM for energy storage systems, expands about 10% when it melts. Therefore, the density of the liquid paraffin, ρ_ℓ, is less than the density of the solid paraffin, ρ_s. When water freezes, however, its volume increases, so the density of ice is less than that of liquid water.

The density change that occurs during solid–liquid phase change will produce an extra increment of motion in the solid–liquid interface. For the melting problem in Fig. 5.2a, where the liquid phase velocity at $x = 0$ is zero, if the density of the solid is larger than the density of the liquid (i.e., $\rho_s > \rho_\ell$) the resulting extra motion of the interface is along the positive direction of the x-axis. Assume that the velocity of the solid–liquid interface is $u_p = ds/dt$, while the extra velocity of the solid–liquid interface due to density change is u_ρ. The density change must satisfy the conservation of mass at the interface, i.e.,

$$\rho_s(u_p - u_\rho) = \rho_\ell u_p \tag{5.2}$$

which follows from Eq. (3.174). The extra velocity induced by the density change can be obtained by rearranging Eq. (5.2) as

$$u_\rho = \frac{\rho_s - \rho_\ell}{\rho_s} u_p \tag{5.3}$$

which is also valid for case $\rho_s > \rho_\ell$, except the extra velocity becomes negative.

Another necessary boundary condition is the energy balance at the solid–liquid interface, which is a special case of Eq. (3.177). If the enthalpy of the liquid and solid phases at the melting point are h_ℓ and h_s, the energy balance at the solid–liquid interface can be expressed as:

$$q''_\ell - q''_s = \rho_\ell u_p h_\ell - \rho_s(u_p - u_\rho)h_s, \qquad x = s(t) \tag{5.4}$$

where q''_ℓ and q''_s are the heat fluxes in the x-direction in the liquid and solid phases, respectively.

Substituting Eq. (5.3) into Eq. (5.4), one obtains

$$q''_\ell - q''_s = \rho_\ell h_{s\ell} \frac{ds}{dt}, \qquad x = s(t) \tag{5.5}$$

where $h_{s\ell} = h_\ell - h_s$ is the latent heat of melting. If convection in the liquid phase can be neglected and heat conduction is the only heat transfer mechanism in both the liquid and solid phases, the heat flux in both phases can be determined by Fourier's law of conduction:

$$q''_\ell = -k_\ell \frac{\partial T_\ell(x,t)}{\partial x}\bigg|_{x=s(t)} \tag{5.6}$$

$$q''_s = -k_s \frac{\partial T_s(x,t)}{\partial x}\bigg|_{x=s(t)} \tag{5.7}$$

The energy balance at the solid–liquid interface for a melting problem can be obtained by substituting Eqs. (5.6) and (5.7) into Eq. (5.5), i.e.,

$$k_s \frac{\partial T_s(x,t)}{\partial x} - k_\ell \frac{\partial T_\ell(x,t)}{\partial x} = \rho_\ell h_{s\ell} \frac{ds(t)}{dt}, \qquad x = s(t) \tag{5.8}$$

For a solidification process, the energy balance equation at the interface can be obtained by a similar procedure:

$$k_s \frac{\partial T_s(x,t)}{\partial x} - k_\ell \frac{\partial T_\ell(x,t)}{\partial x} = \rho_s h_{s\ell} \frac{ds(t)}{dt}, \qquad x = s(t) \tag{5.9}$$

The only difference between Eqs. (5.8) and (5.9) is the density on the right-hand side of the equation. If the temperature distributions in the liquid and solid phases are known, the location of the solid–liquid interface can be obtained by solving Eqs. (5.8) or (5.9). It should be noted that density change causes advection in the liquid phase, which further complicates the problem.

For the melting problem, if the heat transfer mechanism in the liquid phase is convection, the heat flux in the liquid phase can be obtained by Newton's law of cooling:

$$q''_\ell = h(T_\infty - T_m) \tag{5.10}$$

where h and T_∞ in Eq. (5.10) are the convective heat transfer coefficient and the bulk temperature of the liquid phase, respectively. The energy balance at the solid–liquid interface is

$$k_s \frac{\partial T_s(x,t)}{\partial x} + h(T_\infty - T_m) = \rho_\ell h_{s\ell} \frac{ds(t)}{dt}, \qquad x = s(t) \tag{5.11}$$

For the solidification process, the heat flux in the liquid phase is

$$q''_\ell = h(T_m - T_\infty) \tag{5.12}$$

The energy balance equation at the solid–liquid interface becomes

$$k_s \frac{\partial T_s(x,t)}{\partial x} - h(T_\infty - T_m) = \rho_s h_{s\ell} \frac{ds(t)}{dt}, \qquad x = s(t) \tag{5.13}$$

For the natural-convection controlled melting in a rectangular cavity, the solid–liquid interface is inclined along the height of the rectangular cavity as shown in Fig. 5.4b. For solidification in a rectangular cavity, on the other hand, the interface will also be inclined, because natural convection accelerates solidification at the lower portion but decelerates solidification in the upper portion. Assuming **n** is a unit vector along the normal direction of the solid–liquid interface, the boundary conditions at the interface can be expressed as

$$T_\ell(x, y, t) = T_s(x, y, t) = T_m, \qquad x = s(y, t) \tag{5.14}$$

$$k_s \frac{\partial T_s(x, y, t)}{\partial n} - k_\ell \frac{\partial T_\ell(x, y, t)}{\partial n} = \rho_\ell h_{s\ell} v_n, \qquad x = s(y, t) \tag{5.15}$$

where v_n is the solid–liquid interface velocity along the **n**-direction.

It is apparent that Eq. (5.15) is not convenient for numerical solution because it contains temperature derivatives along the **n**-direction. Suppose the shape of the solid–liquid interface can be expressed as

$$x = s(y, t) \tag{5.16}$$

Equation (5.16) can then become the following form (see Problem 5.6; Ozisik 1993):

$$\left[1 + \left(\frac{\partial s}{\partial y}\right)^2\right]\left[k_s \frac{\partial T_s}{\partial x} - k_\ell \frac{\partial T_\ell}{\partial x}\right] = \rho_\ell h_{s\ell} \frac{\partial s}{\partial t}, \qquad x = s(y, t) \tag{5.17}$$

Similarly, for a three-dimensional melting problem with an interface described by

$$z = s(x, y, t) \tag{5.18}$$

The energy balance at the interface is

$$\left[1 + \left(\frac{\partial s}{\partial x}\right)^2 + \left(\frac{\partial s}{\partial y}\right)^2\right]\left[k_s \frac{\partial T_s}{\partial z} - k_\ell \frac{\partial T_\ell}{\partial z}\right] = \rho_\ell h_{s\ell} \frac{\partial s}{\partial t}, \qquad z = s(x, y, t) \tag{5.19}$$

For solidification problems, it is necessary to replace the liquid density ρ_ℓ in Eqs. (5.17) and (5.19) with the solid-phase density ρ_s. The density change in solid–liquid phase change is often neglected in the literature in order to eliminate the additional interface motion discussed earlier.

5.3 Exact Solution

5.3.1 Governing Equations of the Solidification Problem

The physical model of the solidification problem to be investigated in this subsection is shown in Fig. 5.2b, where a liquid PCM with a uniform initial temperature T_i, which exceeds the melting point T_m, is in a half-space $x > 0$. At time $t = 0$, the temperature at the boundary $x = 0$ is suddenly decreased to a temperature T_0, which is below the melting point of the PCM. Solidification occurs from the time $t = 0$. This is a two-region solidification problem as the temperatures of both the liquid and solid phases are unknown and must be determined. It is assumed that the densities of the PCM for both phases are the same. Natural convection in the liquid phase is neglected, and therefore, the heat transfer mechanism in both phases is pure conduction (Faghri et al. 2010).

The temperature in the solid phase must satisfy

$$\frac{\partial^2 T_1}{\partial x^2} = \frac{1}{\alpha_1} \frac{\partial T_1}{\partial t}, \qquad 0 < x < s(t), \quad t > 0 \tag{5.20}$$

$$T_1(x,t) = T_0, \qquad x = 0, \quad t > 0 \tag{5.21}$$

For the liquid phase, the governing equations are

$$\frac{\partial^2 T_2}{\partial x^2} = \frac{1}{\alpha_2}\frac{\partial T_2}{\partial t}, \qquad s(t) < x < \infty, \quad t > 0 \tag{5.22}$$

$$T_2(x,t) \to T_{\rm i}, \qquad x \to \infty, \quad t > 0 \tag{5.23}$$

$$T_2(x,t) = T_{\rm i}, \qquad x > 0, \quad t = 0 \tag{5.24}$$

The boundary conditions at the interface are

$$T_1(x,t) = T_2(x,t) = T_{\rm m}, \qquad x = s(t), \quad t > 0 \tag{5.25}$$

$$k_1\frac{\partial T_1}{\partial x} - k_2\frac{\partial T_2}{\partial x} = \rho h_{s\ell}\frac{{\rm d}s}{{\rm d}t}, \qquad x = s(t), \quad t > 0 \tag{5.26}$$

Before obtaining the solution of the above problem, a scale analysis of the energy balance Eq. (5.26) is performed. At the solid–liquid interface, the scales of derivatives of solid and liquid temperature are

$$\frac{\partial T_1}{\partial x} \sim \frac{T_{\rm m} - T_0}{s} \tag{5.27}$$

$$\frac{\partial T_2}{\partial x} \sim \frac{T_i - T_{\rm m}}{s} \tag{5.28}$$

The scale of the interface velocity is

$$\frac{{\rm d}s}{{\rm d}t} \sim \frac{s}{t} \tag{5.29}$$

Substituting Eqs. (5.27)–(5.29) into Eq. (5.26), one obtains

$$k_1\frac{T_{\rm m} - T_0}{s} - k_2\frac{T_i - T_{\rm m}}{s} \sim \rho h_{s\ell}\frac{s}{t} \tag{5.30}$$

Rearranging Eq. (5.30), the scale of the location of the solid–liquid interface is obtained as follows:

$$\frac{s^2}{\alpha_1 t} \sim Ste\left(1 - \frac{k_2}{k_1}\frac{T_{\rm i} - T_{\rm m}}{T_{\rm m} - T_0}\right) \tag{5.31}$$

where

$$Ste = \frac{c_{p1}(T_{\rm m} - T_0)}{h_{s\ell}} \tag{5.32}$$

is the Stefan number. Named after J. Stefan, a pioneer in discovery of the solid–liquid phase-change phenomena, the Stefan number is a very important dimensionless variable in solid–liquid phase-change phenomena. The Stefan number represents the ratio of sensible heat, $c_{p_1}(T_{\rm m} - T_0)$, to latent heat, $h_{s\ell}$. For a latent heat thermal energy storage system, the Stefan number is usually very small because the temperature difference in such a system is small, while the latent heat $h_{s\ell}$ is very high. Therefore, the effect of the sensible heat transfer on the motion of the solid–liquid interface is very weak, and various approximate solutions to the phase-change problem can be introduced without incurring significant error.

It can be seen from Eq. (5.31) that the effect of heat conduction in the liquid phase can be neglected if $T_{\rm i} - T_{\rm m} \ll T_{\rm m} - T_0$ or $k_2 \ll k_1$. In that case, Eq. (5.31) can be simplified as

$$\frac{s^2}{\alpha_1 t} \sim Ste \tag{5.33}$$

which means that the interfacial velocity increases with increasing $\Delta T = |T_{\rm m} - T_0|$ or decreasing $h_{s\ell}$.

5.3.2 Dimensionless Form of the Governing Equations

The governing Eqs. (5.20)–(5.26) can be nondimensionalized by introducing the following dimensionless variables (Faghri et al. 2010):

$$\left.\begin{array}{l} \theta = \dfrac{T_{\rm m} - T}{T_{\rm m} - T_0} \quad \theta_{\rm i} = \dfrac{T_{\rm m} - T_{\rm i}}{T_{\rm m} - T_0} \quad X = \dfrac{x}{L} \quad S = \dfrac{s}{L} \quad \tau = \dfrac{\alpha_1 t}{L^2} \\ N_\alpha = \dfrac{\alpha_2}{\alpha_1} \quad N_k = \dfrac{k_2}{k_1} \quad Ste = \dfrac{c_{p_1}(T_{\rm m} - T_0)}{h_{s\ell}} \end{array}\right\} \tag{5.34}$$

where L in Eq. (5.34) is a characteristic length of the problem and can be determined by the nature of the problem or requirement of the solution procedure. The dimensionless governing equations are as follows:

$$\frac{\partial^2 \theta_1}{\partial X^2} = \frac{\partial \theta_1}{\partial \tau}, \qquad 0 < X < S(\tau), \quad \tau > 0 \tag{5.35}$$

$$\theta_1(X,\tau) = 1, \qquad X = 0, \quad \tau > 0 \tag{5.36}$$

$$\frac{\partial^2 \theta_2}{\partial X^2} = \frac{1}{N_\alpha}\frac{\partial \theta_2}{\partial \tau}, \qquad S(\tau) < X < \infty, \quad \tau > 0 \tag{5.37}$$

$$\theta_2(X,\tau) \rightarrow \theta_{\rm i}, \qquad X \rightarrow \infty, \quad \tau > 0 \tag{5.38}$$

$$\theta_2(X,\tau) = \theta_{\rm i}, \qquad X > 0, \quad \tau = 0 \tag{5.39}$$

$$\theta_1(X,\tau) = \theta_2(X,\tau) = 0, \qquad X = S(\tau), \quad \tau > 0 \tag{5.40}$$

$$-\frac{\partial \theta_1}{\partial X} + N_k \frac{\partial \theta_2}{\partial X} = \frac{1}{Ste}\frac{dS}{d\tau}, \qquad X = S(\tau), \quad \tau > 0 \tag{5.41}$$

Dimensionless temperature distribution in a PCM can be qualitatively illustrated in Fig. 5.5. It can be seen that the dimensionless temperature distribution is similar to that of a melting problem. Equations (5.35)–(5.41) are also valid for melting problems, provided that the subscripts "1" and "2" represent liquid and solid, respectively. The following solutions of one-region and two-region problems will be valid for both melting and solidification problems.

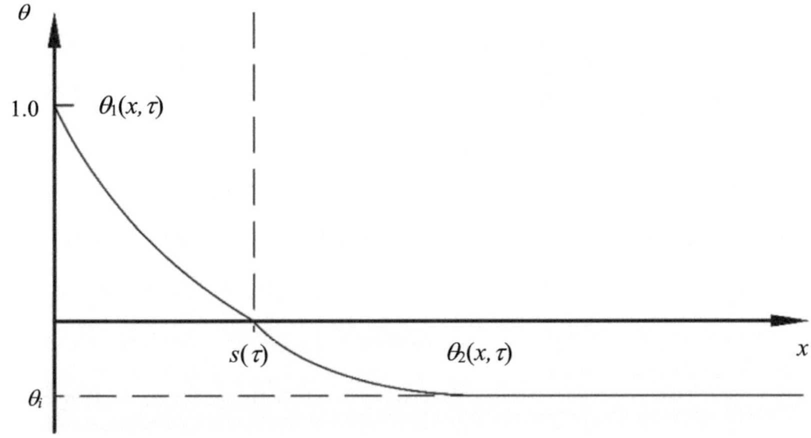

Fig. 5.5 Dimensionless temperature distribution in the PCM

5.3.3 Exact Solution of the One-Region Problem

If the initial temperature of a PCM equals its melting point of the PCM ($T_{\rm i} = T_{\rm m}$), the melting or solidification problem is a one-region problem referred to as a Stefan problem. The temperature distribution in one phase is unknown while the temperature in the other phase remains at the melting point. Therefore, the temperature in only one phase needs to be solved. The mathematical description of a Stefan problem can be obtained by simplifying Eqs. (5.35)–(5.41), i.e.,

$$\frac{\partial^2 \theta}{\partial X^2} = \frac{\partial \theta}{\partial \tau}, \qquad 0 < X < S(\tau), \quad \tau > 0 \tag{5.42}$$

$$\theta(X, \tau) = 1, \qquad X = 0, \quad \tau > 0 \tag{5.43}$$

$$\theta(X, \tau) = 0, \qquad X - S(\tau), \quad \tau > 0 \tag{5.44}$$

$$-\frac{\partial \theta}{\partial X} = \frac{1}{Ste}\frac{{\rm d}S}{{\rm d}\tau}, \qquad X = S(\tau), \quad \tau > 0 \tag{5.45}$$

where the subscript "1" for the liquid phase has been dropped for ease of notation.

The temperature distribution of this problem can be constructed on the basis of the heat conduction solution in a semi-infinite body (Ozisik 1993), i.e.,

$$\theta(X, \tau) = 1 + B{\rm erf}(X/2\tau^{1/2}) \tag{5.46}$$

where B is an unspecified constant. The error function in Eq. (5.46) is defined as:

$${\rm erf}(z) = \frac{2}{\sqrt{\pi}} \int_0^z {\rm e}^{-z^2} {\rm d}z \tag{5.47}$$

It can be verified that Eq. (5.46) satisfies Eqs. (5.42) and (5.43). The constant B in Eq. (5.46) can be determined by substituting Eq. (5.46) into Eq. (5.44), i.e.,

$$0 = 1 + B\text{erf}(S/2\tau^{1/2}) \tag{5.48}$$

Since B is a constant, $S/2\tau^{1/2}$ must also be a constant in order for Eq. (5.48) to be satisfied. This constant can be represented by λ, so

$$\lambda = S/2\tau^{1/2} \tag{5.49}$$

Thus,

$$B = -\frac{1}{\text{erf}(\lambda)} \tag{5.50}$$

By substituting Eq. (5.50) into Eq. (5.46), the dimensionless temperature distribution in phase 1 is obtained:

$$\theta(X, \tau) = 1 - \frac{\text{erf}(X/2\tau^{1/2})}{\text{erf}(\lambda)} \tag{5.51}$$

Substituting Eq. (5.51) and Eq. (5.49) into Eq. (5.45), the following equation is obtained for the constant λ:

$$\lambda e^{\lambda^2}\text{erf}(\lambda) = \frac{Ste}{\sqrt{\pi}} \tag{5.52}$$

The constant λ is a function of the Stefan number only; this constitutes the appeal of solving the problem by using the dimensionless form of governing equations. Once the constant λ is obtained, the temperature distribution $\theta(X, \tau)$ and the location of the solid–liquid interface $S(\tau)$ can be obtained by Eqs. (5.51) and (5.49), respectively. The exact solution of the one-region phase-change problem can also be obtained by using a similarity solution (see Problem 5.8).

Example 5.1

A solid PCM with a uniform initial temperature equal to the melting point, T_m, is in a half-space, $x > 0$. At time $t = 0$,, a constant heat flux q''_o is suddenly applied to the surface of the semi-infinite body. Assume that the densities of the PCM for both phases are the same and that natural convection in the liquid phase is negligible. Find the transient location of the solid–liquid interface.

Solution The melting problem under consideration is shown in Fig. 5.6. Since the temperature of the solid phase is equal to the melting point and only the temperature in the liquid phase is unknown, this is a one-region melting problem. The governing equation and the corresponding initial and boundary conditions of the problem are as follows

$$\frac{\partial^2 T}{\partial x^2} = \frac{1}{\alpha}\frac{\partial T}{\partial t}, \qquad 0 < x < s(t), \quad t > 0 \tag{5.53}$$

$$-k\frac{\partial T(x,t)}{\partial x} = q''_0, \qquad x = 0, \quad t > 0 \tag{5.54}$$

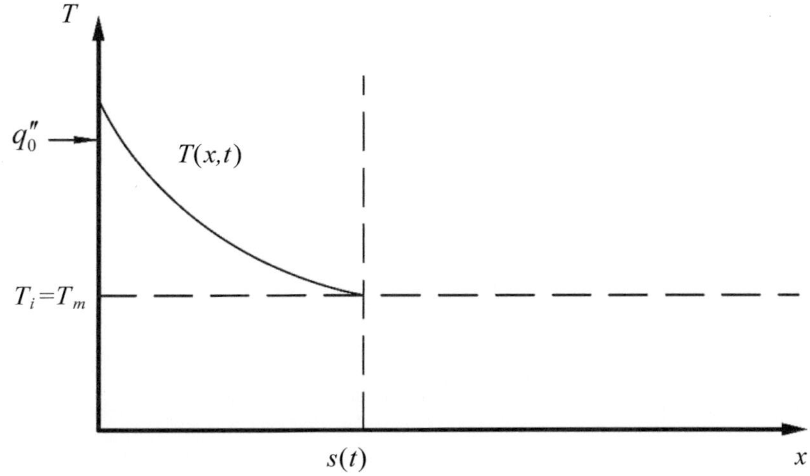

Fig. 5.6 Melting in a semi-infinite region under constant heat flux

$$T(x,t) = T_{\mathrm{m}}, \qquad x = s(t), \quad t > 0 \tag{5.55}$$

$$-k\frac{\partial T}{\partial x} = \rho h_{s\ell}\frac{\mathrm{d}s}{\mathrm{d}t}, \qquad x = s(t), \quad t > 0 \tag{5.56}$$

where the subscript 1 for liquid phase has been dropped to simplify the notation. Introducing the following dimensionless variables:

$$\theta = \frac{T - T_{\mathrm{m}}}{q_0''L/k}, X = \frac{x}{L}, S = \frac{s}{L}, \quad \tau = \frac{\alpha t}{L^2} \quad Ste = \frac{c_{\mathrm{p}}(q_0''L/k)}{h_{s\ell}} \tag{5.57}$$

where L is a characteristic length of the problem, Eqs. (5.53)–(5.56) are nondimensionalized as

$$\frac{\partial^2\theta}{\partial X^2} = \frac{\partial\theta}{\partial\tau}, \qquad 0 < X < S(\tau), \quad \tau > 0 \tag{5.58}$$

$$\frac{\partial\theta(X,\tau)}{\partial X} = -1, \qquad X = 0, \quad \tau > 0 \tag{5.59}$$

$$\theta(X,\tau) = 0, \qquad X = S(\tau), \quad \tau > 0 \tag{5.60}$$

$$-\frac{\partial\theta}{\partial X} = \frac{1}{Ste}\frac{\mathrm{d}S}{\mathrm{d}\tau}, \qquad X = S(\tau), \quad \tau > 0 \tag{5.61}$$

It is assumed that the temperature distribution in the liquid phase region can be constructed in a fashion similar to the exact solution of conduction in a semi-infinite body with boundary conditions of the second kind.

$$\theta(X,\tau) = A + 2\sqrt{\tau}\,\mathrm{ierfc}\left(\frac{X}{2\sqrt{\tau}}\right) \tag{5.62}$$

where ierfc is a differential complementary error function:

$$\mathrm{ierfc}(Z) = \frac{1}{\sqrt{\pi}} \mathrm{e}^{-z^2} - Z\mathrm{erfc}(Z) \tag{5.63}$$

The unknown variable A in Eq. (5.62) can be obtained by substituting Eq. (5.62) into Eq. (5.60), i.e.,

$$A = -2\sqrt{\tau}\mathrm{ierfc}\left(\frac{S}{2\sqrt{\tau}}\right) \tag{5.64}$$

Thus, the temperature distribution in the liquid phase is

$$\theta(X,\tau) = 2\sqrt{\tau}\left[\mathrm{ierfc}\left(\frac{X}{2\sqrt{\tau}}\right) - \mathrm{ierfc}\left(\frac{S}{2\sqrt{\tau}}\right)\right] \tag{5.65}$$

Substituting Eq. (5.65) into Eq. (5.61) yields an ordinary differential equation about the location of the solid–liquid interface, i.e.,

$$\frac{\mathrm{d}S}{\mathrm{d}\tau} = Ste \cdot \mathrm{erfc}\left(\frac{S}{2\sqrt{\tau}}\right) \tag{5.66}$$

which is subject to the following initial condition:

$$S(\tau) = 0, \qquad \tau = 0 \tag{5.67}$$

A closed-form expression of $S(\tau)$ cannot be obtained from Eq. (5.66) because the separation of variables in Eq. (5.66) is impossible. Equation (5.66) can be solved numerically to determine the transient location of the solid–liquid interface.

The above exact solution in dimensional form was first obtained by El-Genk and Cronenberg (1979). Cho and Sunderland (1981) pointed out that there is a contradiction in this solution, since Eq. (5.62) does not satisfy the partial differential Eq. (5.58) unless A in Eq. (5.62) is a constant. However, A in Eq. (5.62) is not a constant because it is a function of dimensionless time τ [see Eq. (5.64)]. The interested reader can find the detailed solution of El-Genk and Cronenberg's (1979) solution in Cho and Sunderland (1981). Nevertheless, this is a first attempt to obtain the exact solution of melting/solidification in a semi-infinite body with boundary conditions other than those of the first kind.

In addition to the melting/solidification under boundary conditions of the first and second kinds discussed above, the boundary condition of the third kind (convective heating or cooling at surface) is a more generalized boundary condition for many applications, such as freezing of biological materials. Melting and solidification under the boundary condition of the third kind can be solved by a procedure similar to that in Example 5.1 (see Problem 5.9). A very simplified solution, which will be given in Example 5.2, can provide the first-order estimation to the phase-change problem.

Example 5.2

A slab of biological material with a thickness of $2L$ is cooled by a fluid with temperature of T_∞, and the convective heat transfer coefficient is h. It is assumed that the biological material can be treated as single-component PCM with well-defined melting point, T_m, and its initial temperature is at T_m. The frozen process is so slow that heat transfer in a frozen layer can be regarded as a pseudo-steady-state process. Estimate the time it takes for the entire slab to be frozen.

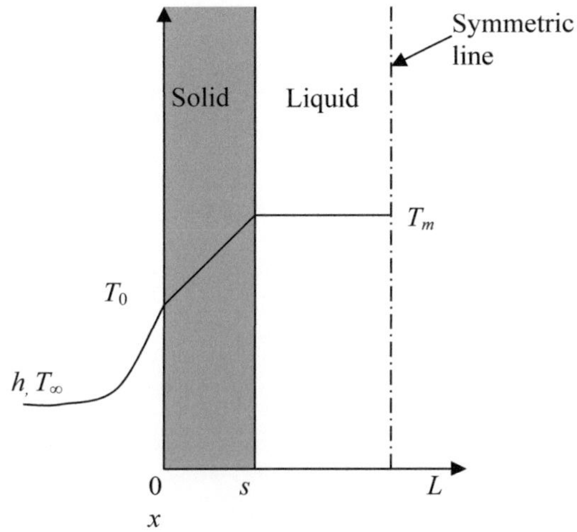

Fig. 5.7 Freezing of a slab under convective cooling

Solution Since both sides of the slab are cooled, only half of the problem needs to be considered (see Fig. 5.7). The latent heat released by the biological material due to freezing is

$$q'' = \rho h_{s\ell} \frac{ds}{dt} \tag{5.68}$$

Assuming steady-state conduction in the frozen layer, the latent heat released by freezing must be conducted through the frozen layer, i.e.,

$$q'' = (T_m - T_\infty)\left(\frac{s}{k} + \frac{1}{h}\right)^{-1} \tag{5.69}$$

Combining Eqs. (5.68) and (5.69), we have

$$\frac{T_m - T_\infty}{\rho h_{s\ell}} dt = \left(\frac{s}{k} + \frac{1}{h}\right) ds \tag{5.70}$$

The time it takes to freeze the entire slab, t_f, can be obtained by integrating Eq. (5.70)

$$\int_0^{t_f} \frac{T_m - T_\infty}{\rho h_{s\ell}} dt = \int_0^L \left(\frac{s}{k} + \frac{1}{h}\right) ds \tag{5.71}$$

i.e.,

$$t_f = \frac{\rho h_{s\ell}}{T_m - T_\infty}\left(\frac{L^2}{2k} + \frac{L}{h}\right) \tag{5.72}$$

which can be nondimensionalized as

$$\tau_{\mathrm{f}} = \frac{1}{Ste}\left(\frac{1}{2} + \frac{1}{Bi}\right) \tag{5.73}$$

where $\tau_{\mathrm{f}} = \alpha t_{\mathrm{f}}/L^2$, $Ste = c_{\mathrm{p}}(T_{\mathrm{m}} - T_{\infty})/h_{s\ell}$, $Bi = hL/k$ are dimensionless time, Stefan number and Biot number. When $Bi \to \infty$, the problem becomes the boundary condition of the first kind $[T(0,t) = T_{\infty}]$ and Eq. (5.73) becomes

$$\tau_{\mathrm{f}} = \frac{1}{2Ste} \tag{5.74}$$

5.3.4 Exact Solution of the Two-Region Problem

If the initial temperature of the PCM is not equal to its melting point of the PCM ($T_{\mathrm{i}} \neq T_{\mathrm{m}}$), the melting or solidification problem becomes a two-region problem, referred to as a *Neumann problem* in the literature. Temperature distributions in both the liquid and solid phases are unknown and must be solved. Equations (5.35)–(5.41) provide the complete mathematical description of a Neumann problem.

Based on the heat conduction solution of a semi-infinite body, the temperature distribution in the PCM can be constructed as follows:

$$\theta_1(X,\tau) = 1 + A\mathrm{erf}(X/2\tau^{1/2}) \tag{5.75}$$

$$\theta_2(X,\tau) = \theta_i + B\mathrm{erfc}[X/2(N_{\alpha}\tau)^{1/2}] \tag{5.76}$$

where A and B in Eqs. (5.75) and (5.76) are unspecified constants, and erfc is the complementary error function, defined as

$$\mathrm{erfc}(z) = 1 - \mathrm{erf}(z) \tag{5.77}$$

It should be noted that Eqs. (5.75) and (5.76) satisfy Eqs. (5.35)–(5.39). The constants A and B can be determined by using boundary condition (5.40), i.e.,

$$1 + A\mathrm{erf}(S/2\tau^{1/2}) = 0 \tag{5.78}$$

$$\theta_{\mathrm{i}} + B\mathrm{erfc}[S/2(N_{\alpha}\tau)^{1/2}] = 0 \tag{5.79}$$

Since A and B are constants, $S/2\tau^{1/2}$ must also be a constant in order for Eqs. (5.78) and (5.79) to be satisfied. This constant can be represented by λ, so

$$\lambda = S/2\tau^{1/2} \tag{5.80}$$

Thus, the constants A and B can be determined as

$$A = -\frac{1}{\mathrm{erf}(\lambda)} \tag{5.81}$$

$$B = -\frac{\theta_i}{\mathrm{erfc}(\lambda/N_\alpha^{1/2})} \tag{5.82}$$

Substituting Eqs. (5.81) and (5.82) into Eqs. (5.75) and (5.76), the temperature distributions in both phases are determined as follows:

$$\theta_1(X,\tau) = 1 - \frac{\mathrm{erf}[X/2\tau^{1/2}]}{\mathrm{erf}(\lambda)} \tag{5.83}$$

$$\theta_2(X,\tau) = \theta_\mathrm{i}\left[1 - \frac{\mathrm{erfc}[X/2(N_\alpha\tau)^{1/2}]}{\mathrm{erfc}(\lambda/N_\alpha^{1/2})}\right] \tag{5.84}$$

Substituting Eqs. (5.83), (5.84), and (5.80) into Eq. (5.41), the following equation is obtained for the constant λ:

$$\frac{\mathrm{e}^{-\lambda^2}}{\mathrm{erf}(\lambda)} + \frac{N_\mathrm{k}\theta_\mathrm{i}}{N_\alpha^{1/2}}\frac{\mathrm{e}^{-\lambda^2/N_\alpha}}{\mathrm{erfc}(\lambda/N_\alpha^{1/2})} = \frac{\lambda\sqrt{\pi}}{Ste} \tag{5.85}$$

Equation (5.85) can be solved by using an iterative method with under-relaxation because its left-hand side is a very complicated function of λ. Once λ is obtained, the temperature distributions —$\theta_1(X,\tau)$ and $\theta_2(X,\tau)$—and the location of the solid–liquid interface $S(\tau)$ can be obtained from Eqs. (5.83), (5.84), and (5.80), respectively.

5.4 Integral Solution

The melting and solidification problems that can be solved by the exact solution are very limited, so it is necessary to introduce some approximate solution techniques. The integral approximate method proposed by Goodman (1958) is one of the most attractive techniques, because it is very simple and its physical concept is very clear with good prediction. After the integral solution technique is introduced by solving heat conduction in a semi-infinite body, and its applications to various melting/solidification problems will be discussed.

5.4.1 Heat Conduction in a Semi-infinite Body

The early applications of integral solutions to heat transfer problems included integral solutions of boundary-layer momentum and energy equations. The method can also be used to solve linear or nonlinear transient conduction problems. The integral solution will be employed to solve the heat conduction problem in a semi-infinite body with a boundary condition of the first kind. The physical model of the problem is illustrated in Fig. 5.8, and the governing equation of the heat conduction problem and the corresponding initial and boundary conditions are:

$$\frac{\partial^2 T(x,t)}{\partial x^2} = \frac{1}{\alpha}\frac{\partial T(x,t)}{\partial t}, \qquad x > 0, \quad t > 0 \tag{5.86}$$

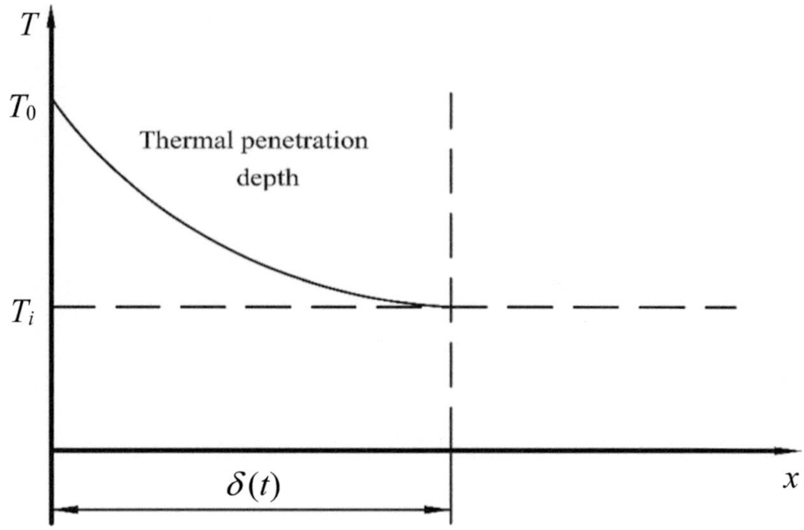

Fig. 5.8 Heat conduction in a semi-infinite body with constant wall temperature

$$T(x,t) = T_0 \qquad x = 0, \quad t > 0 \tag{5.87}$$

$$T(x,t) = T_{\mathrm{i}}, \qquad x > 0, \quad t = 0 \tag{5.88}$$

For transient heat conduction in a semi-infinite body, the initial temperature is uniformly at T_{i}. At time $t = 0$, the surface temperature of the semi-infinite body is suddenly increased to a temperature T_0. The temperature near the surface of the semi-infinite body will increase as a result of the surface temperature change, while the temperature far from the surface of the semi-infinite body is not affected and remains at the initial temperature T_{i}. It is useful here to introduce a concept similar to the thermal boundary layer for convective heat transfer, the *thermal penetration depth*. Assuming the thickness of the thermal penetration depth at time t is δ, the temperature of the semi-infinite body at $x < \delta$ will be affected, but the temperature at $x > \delta$ will remain unchanged (see Fig. 5.8). It should be pointed out that the thermal penetration depth, δ, increases with increasing time.

According to the definition of the thermal penetration depth, the temperature at the thermal penetration depth should satisfy

$$\frac{\partial T(x,t)}{\partial x} = 0, \qquad x = \delta(t) \tag{5.89}$$

$$T(x,t) = T_{\mathrm{i}}, \qquad x = \delta(t) \tag{5.90}$$

Integrating Eq. (5.86) in the interval $(0, \delta)$, one obtains

$$\left.\frac{\partial T}{\partial x}\right|_{x=\delta(t)} - \left.\frac{\partial T}{\partial x}\right|_{x=0} = \frac{1}{\alpha}\int_0^{\delta(t)} \frac{\partial T(x,t)}{\partial t}\mathrm{d}x \tag{5.91}$$

The right-hand side of Eq. (5.91) can be rewritten using Leibnitz's rule, i.e.,

$$\left.\frac{\partial T}{\partial x}\right|_{x=\delta(t)} - \left.\frac{\partial T}{\partial x}\right|_{x=0} = \frac{1}{\alpha}\left[\frac{\mathrm{d}}{\mathrm{d}t}\left(\int_0^{\delta} T\mathrm{d}x\right) - T|_{x=\delta}\frac{\mathrm{d}\delta}{\mathrm{d}t}\right] \tag{5.92}$$

which represents the energy balance within the thermal penetration depth.

Substituting Eqs. (5.89) and (5.90) into Eq. (5.92) yields

$$-\alpha\left.\frac{\partial T}{\partial x}\right|_{x=0} = \frac{\mathrm{d}}{\mathrm{d}t}(\Theta - T_{\mathrm{i}}\delta) \tag{5.93}$$

where

$$\Theta(t) = \int_0^{\delta(t)} T(x,t)\mathrm{d}x \tag{5.94}$$

Equation (5.93) is the *integral energy equation* of the conduction problem, and this equation pertains for the *entire* thermal penetration depth. It follows that a temperature distribution that satisfies Eq. (5.93) does not necessarily satisfy differential Eq. (5.86), which describes the energy balance at any and all points in the domain of the problem.

In order to use the integral Eq. (5.93) to solve the conduction problem, the temperature profile in the thermal penetration depth is needed. The next step of the integral solution is to assume a temperature distribution in the thermal penetration depth. The assumed temperature distribution can be any arbitrary function provided that the boundary conditions at $x = 0$ and $x = \delta$ are satisfied. Let us assume that the temperature distribution in the thermal penetration depth is a third-order polynomial function of x, i.e.,

$$T(x,t) = A_0 + A_1x + A_2x^2 + A_3x^3 \tag{5.95}$$

where A_0, A_1, A_2, and A_3 are four constants to be determined using the boundary conditions. Since there are only three boundary conditions available—Eqs. (5.87), (5.89) and (5.90)—one more condition must be identified so that all four constants in Eq. (5.95) can be determined. The surface temperature of the semi-infinite body, T_0, is not a function of time t, so

$$\frac{\partial T(x,t)}{\partial t} = 0 \qquad x = 0 \tag{5.96}$$

Substituting Eq. (5.86) into Eq. (5.96) yields

$$\frac{\partial^2 T(x,t)}{\partial x^2} = 0 \qquad x = 0 \tag{5.97}$$

Substituting Eq. (5.95) into Eqs. (5.87), (5.89), (5.90) and (5.97) yields four equations for the constants in Eq. (5.95). Solving for the four constants and substituting the results into Eq. (5.95), the temperature distribution in the thermal penetration depth becomes

$$\frac{T(x,t) - T_i}{T_0 - T_i} = 1 - \frac{3}{2}\left(\frac{x}{\delta}\right) + \frac{1}{2}\left(\frac{x}{\delta}\right)^3 \tag{5.98}$$

where the thermal penetration depth, δ, is still unknown.

Substituting Eq. (5.98) into Eq. (5.93), an ordinary differential equation for δ is obtained:

$$4\alpha = \delta \frac{d\delta}{dt} \qquad t > 0 \tag{5.99}$$

Since the thermal penetration depth equals zero at the beginning of the heat conduction, Eq. (5.99) is subject to the following initial condition:

$$\delta = 0 \qquad t = 0 \tag{5.100}$$

The solution of Eqs. (5.99) and (5.100) is

$$\delta = \sqrt{8\alpha t} \tag{5.101}$$

which is consistent with the result of scale analysis, $\delta \sim \sqrt{\alpha t}$ (see Eq. 1.88).

The temperature distribution in the thermal penetration depth can be obtained by Eq. (5.98), and the temperature in the semi-infinite body beyond the thermal penetration depth equals the initial temperature, T_i. The temperature distribution in the thermal penetration depth obtained here, as well as the thermal penetration depth thickness, depends on the assumed temperature distribution. The degree of the polynomial function for the temperature distribution in the thermal penetration depth should not be higher than four, or the obtained temperature will oscillate around the actual temperature profile, giving erroneous results.

From the above analysis, we can summarize the procedure of the integral solution as follows:

1. Obtain the integral equation of the problem by integrating the partial differential equation over the thermal penetration depth.
2. Assume an appropriate temperature distribution—usually a polynomial function—in the thermal penetration depth, and determine the unknown constants in the polynomial by using the boundary conditions at $x = 0$ and $x = \delta$. Additional boundary conditions, if needed, can be obtained by further analysis of the boundary conditions and the conduction equation.
3. Obtain an ordinary differential equation of thermal penetration depth by substituting the temperature distribution into the integral equation. The thermal penetration depth thickness can be obtained by solving this ordinary differential equation.
4. The temperature distribution in the thermal penetration depth can be obtained by combining the thermal penetration depth thickness from step 3 into the temperature distribution from step 2.

5.4.2 One-Region Problem

The one-region phase-change problem to be studied here is the same as the one-region problem in Sect. 5.3.1. The dimensionless governing equations for melting and solidification problems have the same form if the dimensionless variables defined in the preceding section are used. The procedure for solving one-region phase-change problems will be demonstrated by solving a one-region melting problem. The dimensionless governing equations of the melting problem are Eqs. (5.42)–(5.45).

To solve the problem by integral solution, the thermal penetration depth must be specified. For a one-region melting problem, only the temperature distribution in the liquid phase needs to be solved, because the temperature in the solid phase remains uniformly equal to the melting point of the PCM. Therefore, the thickness of the liquid phase is identical to the thermal penetration depth.

Integrating Eq. (5.42) with respect to X in the interval of $(0, S)$, one obtains

$$\left.\frac{\partial \theta(X,\tau)}{\partial X}\right|_{X=S(\tau)} - \left.\frac{\partial \theta(X,\tau)}{\partial X}\right|_{X=0} = \frac{\mathrm{d}\Theta(\tau)}{\mathrm{d}\tau} \tag{5.102}$$

where

$$\Theta(\tau) = \int_0^{S(\tau)} \theta(X,\tau)\mathrm{d}X \tag{5.103}$$

Substituting Eq. (5.45) into Eq. (5.102) yields the integral equation of the one-region problem:

$$-\frac{1}{Ste}\frac{\mathrm{d}S}{\mathrm{d}\tau} - \left.\frac{\partial \theta(X,\tau)}{\partial X}\right|_{X=0} = \frac{d\Theta(\tau)}{\mathrm{d}\tau} \tag{5.104}$$

Assume now that the temperature distribution in the liquid phase is the following second-degree polynomial:

$$\theta(X,\tau) = A_0 + A_1\left(\frac{X-S}{S}\right) + A_2\left(\frac{X-S}{S}\right)^2 \tag{5.105}$$

where A_0, A_1, and A_2 are three unknown constants to be determined. Equations (5.43) and (5.44) can be used to determine the constants in Eq. (5.105). However, Eq. (5.45) is not suitable for determining the constant in Eq. (5.105) because $\mathrm{d}S/\mathrm{d}\tau$. in Eq. (5.45) is unknown. Another appropriate boundary condition at the solid–liquid interface is needed.

Differentiating Eq. (5.44), one obtains

$$\mathrm{d}\theta = \frac{\partial \theta}{\partial X}\mathrm{d}X + \frac{\partial \theta}{\partial \tau}\mathrm{d}\tau = 0, \qquad X = S(\tau) \tag{5.106}$$

i.e.,

$$\frac{\partial \theta}{\partial X}\frac{\mathrm{d}S}{\mathrm{d}\tau} + \frac{\partial \theta}{\partial \tau} = 0, \qquad X = S(\tau) \tag{5.107}$$

Substituting Eqs. (5.42) and (5.45) into Eq. (5.107) yields

$$Ste\left(\frac{\partial \theta}{\partial X}\right)^2 = \frac{\partial^2 \theta}{\partial X^2} \tag{5.108}$$

Equation (5.108) is an additional boundary condition at the solid–liquid interface; it can be used to determine the coefficients in Eq. (5.105). Thus, the constants in Eq. (5.105) can be determined by Eqs. (5.43), (5.44) and (5.108). After the constants in Eq. (5.105) are determined, the temperature distribution in the liquid phase becomes

$$\theta(X,\tau) = \frac{1-\sqrt{1+2Ste}}{Ste}\left(\frac{X-S}{S}\right) + \left(\frac{1-\sqrt{1+2Ste}}{Ste}+1\right)\left(\frac{X-S}{S}\right)^2 \tag{5.109}$$

Substituting Eq. (5.109) into the integral Eq. (5.104) leads to an ordinary differential equation for $S(\tau)$:

$$S\frac{dS}{d\tau} = 6\frac{1-\sqrt{1+2Ste}+2Ste}{5+\sqrt{1+2Ste}+2Ste} \tag{5.110}$$

The initial condition of Eq. (5.110) is

$$S(\tau) = 0, \qquad \tau = 0 \tag{5.111}$$

The right-hand side of Eq. (5.110) is a constant and its solution is

$$S = 2\lambda\tau^{\frac{1}{2}} \tag{5.112}$$

where

$$\lambda = \left[3\frac{1-\sqrt{1+2Ste}+2Ste}{5+\sqrt{1+2Ste}+2Ste}\right]^{\frac{1}{2}} \tag{5.113}$$

Figure 5.9 shows the comparison of λ obtained by the exact solution and the integral solution. The integral solution agrees very well for a small Stefan number, but the difference increases along

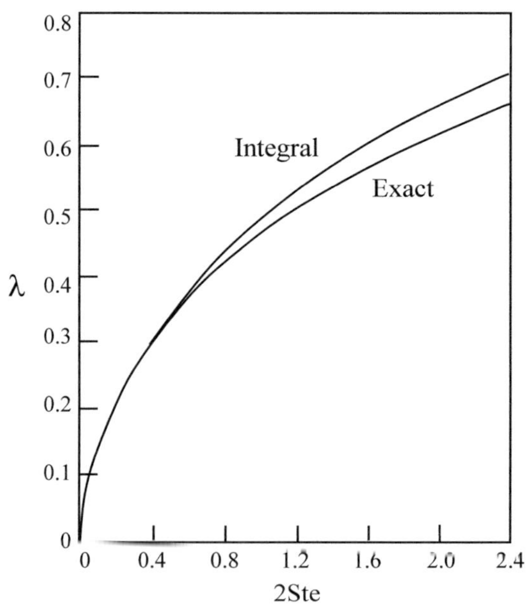

Fig. 5.9 Comparison between integral and exact solutions of the one-region conduction-controlled melting problem

with increasing Stefan number. For a latent heat thermal energy storage system, the Stefan number is usually less than 0.2, so the integral solution can provide sufficiently accurate results for that case.

Example 5.3

A solid PCM with a uniform initial temperature at its melting point, T_m, is in a half-space, $x > 0$. At time $t = 0$, a variable heat flux, $q''_0(t)$, is suddenly applied to the surface of the semi-infinite body. Assume that the densities of the PCM for both phases are the same and that natural convection in the liquid phase is negligible. Find the transient location of the solid–liquid interface.

Solution This problem is the same as Example 5.1 except for the heat flux, which is

$$q''(t) = f(t)q''_0 \tag{5.114}$$

where q''_0 is a reference heat flux and $f(t)$ is a given function. The dimensionless governing equation and the corresponding initial and boundary conditions of the problem are

$$\frac{\partial^2\theta}{\partial X^2} = \frac{\partial\theta}{\partial\tau}, \qquad 0 < X < S(\tau), \quad \tau > 0 \tag{5.115}$$

$$\frac{\partial\theta}{\partial X} = -f(\tau), \qquad X = 0, \quad \tau > 0 \tag{5.116}$$

$$\theta(X,\tau) = 0, \qquad X = S(\tau), \quad \tau > 0 \tag{5.117}$$

$$-\frac{\partial\theta}{\partial X} = \frac{1}{Ste}\frac{dS}{d\tau}, \qquad X = S(\tau), \quad \tau > 0 \tag{5.118}$$

where the nondimensional variables in the above Eqs. (5.115)–(5.118) are defined by Eq. (5.57). Since Eq. (5.116) is a nonhomogeneous boundary condition, the exact solution in Example 5.1 cannot be directly applied here. The integral solution, on the other hand, has the capability of dealing with nonlinear and nonhomogeneous boundary conditions.

Integrating Eq. (5.115) with respect to X in the interval (0, S), and considering the boundary conditions at the surface, the integral equation of the problem is obtained:

$$-\frac{1}{Ste}\frac{dS(\tau)}{d\tau} + f(\tau) - \frac{d\Theta(\tau)}{d\tau} \tag{5.119}$$

where

$$\Theta(\tau) = \int_0^{S(\tau)} \theta(X,\tau)dX \tag{5.120}$$

Integrating Eq. (5.119) with respect to τ in the interval $(0, \tau)$ yields

$$\frac{S(\tau)}{Ste} + \Theta(\tau) = \int_0^{\tau} f(\tau)d\tau \tag{5.121}$$

Assuming that the temperature distribution in the liquid phase is

$$\theta(X,\tau) = A_0 + A_1\left(\frac{X-S}{S}\right) + A_2\left(\frac{X-S}{S}\right)^2 \tag{5.122}$$

where the constants, A_0, A_1, and A_2 are three unspecified constants that can be determined from Eqs. (5.116), (5.117) and (5.108). After all unknown constants are determined, the temperature distribution in the liquid phase becomes

$$\theta(X,\tau) = \frac{1}{2}\frac{\left[1-(1+4\mu)^{1/2}\right]}{Ste}\left(\frac{X-S}{S}\right) + \frac{1}{8}\frac{\left[1-(1+4\mu)^{1/2}\right]^2}{Ste}\left(\frac{X-S}{S}\right)^2 \tag{5.123}$$

where

$$\mu = f(\tau)S(\tau)Ste \tag{5.124}$$

Substituting Eq. (5.123) into Eq. (5.121), one obtains

$$Ste^2 f(\tau)\int_0^\tau f(\tau)d\tau = \frac{\mu}{6}\left[\mu + 5 + (1+4\mu)^{1/2}\right] \tag{5.125}$$

For the case of constant heat flux, i.e., $f(\tau) = 1$, Eq. (5.125) can be simplified to

$$S\left[SteS + 5 + (1+4SteS)^{1/2}\right] = 6Ste\tau \tag{5.126}$$

5.4.3 Two-Region Problem

The physical model of a melting problem is shown in Fig. 5.10: a solid PCM with a uniform initial temperature T_i, which is below its melting point T_m, is in a half-space, $x > 0$. At time $t = 0$, a constant heat flux, q_0'', is suddenly applied to the surface of the semi-infinite body. Because the initial temperature of the PCM is below its melting point, melting does not begin until after the wall

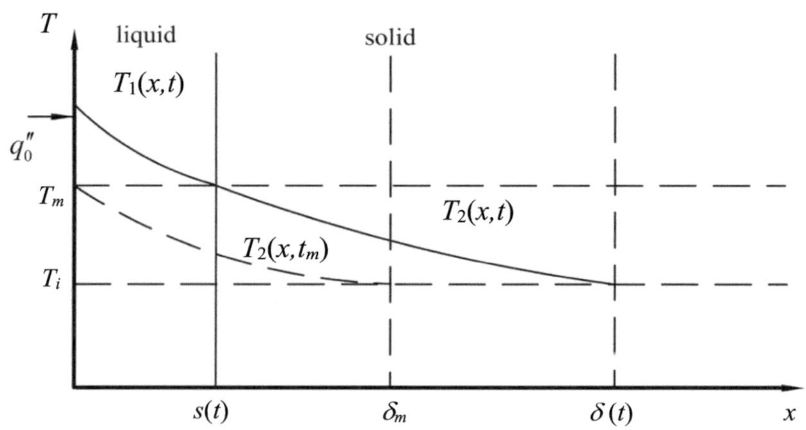

Fig. 5.10 Melting in a subcooled semi-infinite body under constant heat flux

temperature reaches the melting point. Therefore, the problem can be divided into two subproblems: (1) heat conduction over the duration of preheating and (2) the actual melting process. An integral approximate method will be employed to solve both subproblems (Zhang et al. 1993).

5.4.3.1 Duration of Preheating

At the beginning of heating, no melting occurs, and the problem is a pure conduction problem with a boundary condition of the second kind. Its mathematical description is as follows:

$$\frac{\partial^2 T_2}{\partial x^2} = \frac{1}{\alpha_2}\frac{\partial T_2}{\partial t}, \quad 0<x<\infty, \quad 0<t<t_\mathrm{m} \tag{5.127}$$

$$\frac{\partial T_2}{\partial x} = -\frac{1}{k_2}q_0'', \quad x=0, \quad 0<t<t_\mathrm{m} \tag{5.128}$$

$$T_2(x,t) \to T_\mathrm{i}, \quad x\to\infty, \quad 0<t<t_\mathrm{m} \tag{5.129}$$

$$T_2(x,t) = T_\mathrm{i}, \quad 0<x<\infty, \quad t=0 \tag{5.130}$$

where t_m is the duration of preheating. This problem is solved by integral approximate method. Assuming the temperature profile is a second-degree polynomial, one obtains the temperature profile:

$$T_2(x,t) = T_i + \frac{q_0''\delta}{2k_2}\left(1-\frac{x}{\delta}\right)^2 \tag{5.131}$$

where δ is the thermal penetration depth. It can be obtained by substituting Eq. (5.131) into the integral equation, which can in turn be obtained by integrating Eq. (5.127) in the interval of (0, δ). The result is

$$\delta = \sqrt{6\alpha_2 t} \tag{5.132}$$

The highest temperature of the semi-infinite body occurs at its surface ($x = 0$) and can be expressed as

$$T_\mathrm{s}(t) = T_\mathrm{i} + \frac{q_0''\delta}{2k_2} \tag{5.133}$$

Melting occurs when the surface temperature reaches the melting point, T_m, and the corresponding thermal penetration depth is

$$\delta_\mathrm{m} = \frac{2k_2(T_\mathrm{m}-T_\mathrm{i})}{q_0''} \tag{5.134}$$

Then the duration of preheating can be obtained by combining Eqs. (5.132) and (5.134),

$$t_\mathrm{m} = \frac{2k_2^2(T_\mathrm{m}-T_\mathrm{i})^2}{3\alpha_2 q_0''^2} \tag{5.135}$$

The temperature distribution at time t_m is

$$T_2(x, t_\mathrm{m}) = T_\mathrm{i} + (T_\mathrm{m} - T_\mathrm{i})\left(1 - \frac{x}{\delta_m}\right)^2 \tag{5.136}$$

5.4.3.2 Governing Equations for the Melting Stage

After melting starts, the governing equations in the different phases must be specified separately. The temperature in the liquid phase satisfies

$$\frac{\partial^2 T_1}{\partial x^2} = \frac{1}{\alpha_1}\frac{\partial T_1}{\partial t}, \qquad 0<x<s(t), \quad t > t_\mathrm{m} \tag{5.137}$$

$$\frac{\partial T_1(x,t)}{\partial x} = -\frac{1}{k_1}q_0'', \qquad x = 0, \quad t > t_\mathrm{m} \tag{5.138}$$

The governing equation and corresponding boundary and initial conditions for the solid phase are

$$\frac{\partial^2 T_2}{\partial x^2} = \frac{1}{\alpha_2}\frac{\partial T_2}{\partial t}, \qquad s(t)<x<\infty, \quad t > t_\mathrm{m} \tag{5.139}$$

$$T_2(x,t) \to T_i, \qquad x \to \infty, \quad t > t_m \tag{5.140}$$

$$T_2(x,t) = T_\mathrm{i} + (T_\mathrm{m} - T_\mathrm{i})\left(1 - \frac{x}{\delta_\mathrm{m}}\right)^2, \qquad x > 0, \quad t = t_\mathrm{m} \tag{5.141}$$

At the solid–liquid interface, the following boundary conditions are necessary to link solutions in the liquid and solid phases:

$$T_1(x,t) = T_2(x,t) = T_\mathrm{m}, \qquad x = s(t), \quad t > t_\mathrm{m} \tag{5.142}$$

$$k_2\frac{\partial T_2}{\partial x} - k_1\frac{\partial T_1}{\partial x} = \rho h_{s\ell}\frac{\mathrm{d}s}{\mathrm{d}t}, \qquad x = s(t), \quad t > t_\mathrm{m} \tag{5.143}$$

By defining the dimensionless variables as follows:

$$\left.\begin{array}{lll}
\theta_1 = \dfrac{c_{p1}(T_1 - T_\mathrm{m})}{h_{s\ell}} & \theta_2 = \dfrac{c_{p2}(T_2 - T_\mathrm{m})}{h_{s\ell}} & Sc = \dfrac{c_{p2}(T_\mathrm{m} - T_\mathrm{i})}{h_{s\ell}} \\
X = \dfrac{x}{\alpha_1\rho_1 h_{s\ell}/q_0''} & S = \dfrac{s}{\alpha_1\rho_1 h_{s\ell}/q_0''} & \Delta = \dfrac{\delta}{\alpha_1\rho_1 h_{s\ell}/q_0''} \\
\Delta_\mathrm{m} = \dfrac{\delta_\mathrm{m}}{\alpha_1\rho_1 h_{s\ell}/q_0''} & \tau = \dfrac{\alpha_1 t}{(\alpha_1\rho_1 h_{s\ell}/q_0'')^2} & N_\alpha = \dfrac{\alpha_2}{\alpha_1}
\end{array}\right\} \tag{5.144}$$

where Sc is subcooling parameter that signifies the ratio of the sensible preheat need to raise the temperature of the body before melting takes place, Eqs. (5.137)–(5.143) become

$$\frac{\partial^2 \theta_1}{\partial X^2} = \frac{\partial \theta_1}{\partial \tau}, \qquad 0<X<S(\tau), \quad \tau > \tau_m \tag{5.145}$$

$$\frac{\partial \theta_1(X,\tau)}{\partial X} = -1, \qquad X = 0, \quad \tau > \tau_m \tag{5.146}$$

$$\frac{\partial^2 \theta_2}{\partial X^2} = \frac{1}{N_\alpha}\frac{\partial \theta_2}{\partial \tau}, \qquad S(\tau) < X < \infty, \quad \tau > \tau_\mathrm{m} \tag{5.147}$$

$$\theta_2(X,\tau) \to -Sc, \qquad X \to \infty, \quad \tau > \tau_\mathrm{m} \tag{5.148}$$

$$\theta_2(X,\tau) = Sc\left[\left(1 - \frac{X}{\Delta_\mathrm{m}}\right)^2 - 1\right], \qquad X > 0, \quad \tau = \tau_\mathrm{m} \tag{5.149}$$

$$\theta_1(X,\tau) = \theta_2(X,\tau) = 0, \qquad X = S(\tau), \quad \tau > \tau_\mathrm{m} \tag{5.150}$$

$$-\frac{\partial \theta_1}{\partial X} + \frac{1}{N_\alpha}\frac{\partial \theta_2}{\partial X} = \frac{\mathrm{d}S}{\mathrm{d}\tau}, \qquad X = S(\tau), \quad \tau > \tau_\mathrm{m} \tag{5.151}$$

where Δ_m and τ_m can be obtained by substituting Eq. (5.144) into Eqs. (5.134) and (5.135), i.e.,

$$\Delta_\mathrm{m} = 2N_\alpha Sc \tag{5.152}$$

$$\tau_\mathrm{m} = \frac{2}{3}N_\alpha Sc^2 \tag{5.153}$$

5.4.3.3 Integral Solution

Figure 5.11 shows the physical model represented by the above dimensionless governing equations. After the melting process starts, the solid–liquid interface moves along the positive *x*-axis, and the thermal penetration depth continuously increases. The solid phase temperature over the range of $S(\tau) < X < \Delta(\tau)$ is affected by the boundary condition at the surface. The temperature in the region beyond the thermal penetration depth, Δ, is not affected and remains at—*Sc*.

Integrating Eq. (5.147) over the interval (S, Δ), then applying the definition of the thermal penetration depth and Eq. (5.150), yields the integral equation of the solid phase:

$$-\left.\frac{\partial \theta_2}{\partial X}\right|_{X=S} = \frac{1}{N_\alpha}\frac{d}{\mathrm{d}\tau}(\Theta_2 + Sc\Delta) \tag{5.154}$$

Fig. 5.11 Dimensionless temperature distribution for melting in a semi-infinite body under constant heat flux

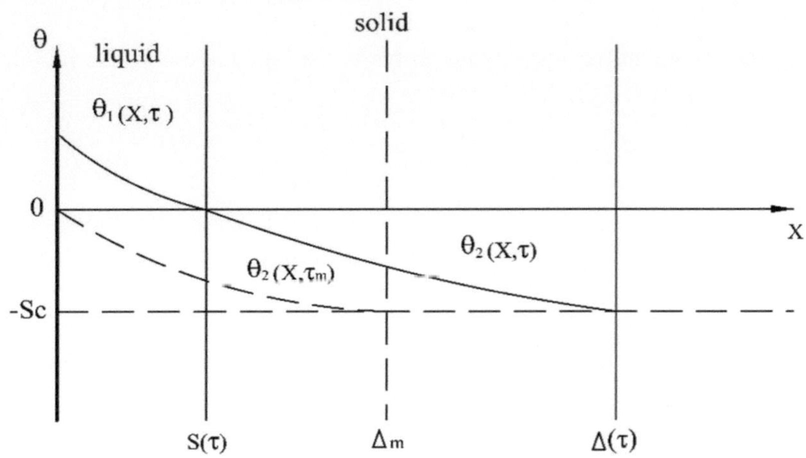

where

$$\Theta_2 = \int_S^\Delta \theta_2 \mathrm{d}X \tag{5.155}$$

The temperature distribution in the solid phase is assumed to be a second-degree polynomial, i.e.,

$$\theta_2 = B_0 + B_1(X - S) + B_2(X - S)^2 \tag{5.156}$$

The constants in Eq. (5.156) can be obtained from the boundary condition, Eq. (5.150) and the definition of the thermal penetration depth ($\theta_2|_{X=\Delta} = -Sc$ and $\partial\theta_2/\partial X|_{X=\Delta} = 0$). The final form of the temperature distribution in the solid phase is

$$\theta_2(X, \tau) = Sc\left[\left(\frac{\Delta - X}{\Delta - S}\right)^2 - 1\right] \tag{5.157}$$

Substituting Eq. (5.157) into Eq. (5.154) yields a relationship between the location of the solid–liquid interface and the thermal penetration depth:

$$\frac{6N_\alpha}{\Delta - S} = 2\frac{\mathrm{d}S}{\mathrm{d}\tau} + \frac{\mathrm{d}\Delta}{\mathrm{d}\tau} \tag{5.158}$$

Integrating the differential equation of liquid phase Eq. (5.145) over the interval of (*0, S*) and applying the boundary conditions Eqs. (5.146) and (5.150) yields the integral equation of the liquid phase:

$$\left.\frac{\partial\theta_1}{\partial X}\right|_{X=S} + 1 = \frac{\mathrm{d}\Theta_1}{\mathrm{d}\tau} \tag{5.159}$$

where

$$\Theta_1 = \int_0^S \theta_1 \mathrm{d}X \tag{5.160}$$

Assuming that the temperature profile in the liquid phase is a second-degree polynomial function,

$$\theta_1 = A_0 + A_1\left(\frac{X - S}{S}\right) + A_2\left(\frac{X - S}{S}\right)^2 \tag{5.161}$$

where A_0, A_1, and A_2 need to be determined using appropriate boundary conditions. Equations (5.146) and (5.150) provided two conditions; and the third condition needs to be obtained by differentiating Eq. (5.150), i.e.,

$$\mathrm{d}\theta_1 - \frac{\partial\theta_1}{\partial X}\mathrm{d}X + \frac{\partial\theta_1}{\partial\tau}\mathrm{d}\tau = 0, \qquad X = S(\tau) \tag{5.162}$$

i.e.,

$$\frac{\partial \theta_1}{\partial X}\frac{\mathrm{d}S}{\mathrm{d}\tau}+\frac{\partial \theta_1}{\partial \tau}=0, \qquad X=S(\tau) \tag{5.163}$$

Substituting Eqs. (5.145) and (5.151) into Eq. (5.163) yields

$$\left(\frac{\partial \theta_1}{\partial X}\right)^2-\frac{1}{N_\alpha}\frac{\partial \theta_1}{\partial X}\frac{\partial \theta_2}{\partial X}=\frac{\partial^2 \theta_1}{\partial X^2} \tag{5.164}$$

which is an additional boundary condition at the solid–liquid interface and can be used to determine the coefficients in Eq. (5.161). After the constants in Eq. (5.161) are determined, the temperature distribution in the liquid phase becomes

$$\theta_1(X,\tau)=\frac{S}{2}\left(\frac{X-S}{S}\right)^2-\frac{1}{2}p\frac{X^2-S^2}{S^2} \tag{5.165}$$

where

$$p=\left(\frac{N_\alpha ScS}{\Delta-S}-\frac{1}{2}\right)+\sqrt{\left(\frac{N_\alpha ScS}{\Delta-S}-\frac{1}{2}\right)^2+S} \tag{5.166}$$

Substituting the integral Eqs. (5.159) and (5.154) into Eq. (5.151) yields

$$\frac{\mathrm{d}S}{\mathrm{d}\tau}+\frac{d}{\mathrm{d}\tau}(\Theta_2+Sc\Delta)+\frac{d\Theta_1}{\mathrm{d}\tau}=1 \tag{5.167}$$

Integrating both sides of Eq. (5.167) with respect to τ within (τ_m,τ),, one obtains

$$S+[\Theta_2(\tau)-\Theta_2(\tau_\mathrm{m})]+Sc(\Delta-\Delta_\mathrm{m})+[\Theta_1(\tau)-\Theta_1(\tau_\mathrm{m})]=(\tau-\tau_\mathrm{m}) \tag{5.168}$$

Substituting Eqs. (5.155), (5.157), (5.160), and (5.165) into Eq. (5.168), the resulting interface equation is

$$\frac{1}{2}S^2+(p+3+2Sc)S+Sc(\Delta-\Delta_\mathrm{m})=3(\tau-\tau_\mathrm{m}) \tag{5.169}$$

Equation (5.158) is a first-order ordinary differential equation, while Eq. (5.169) is an algebraic equation. They can be solved simultaneously with an implicit numerical method to determine the location of the solid–liquid interface and the thermal penetration depth. If there is no subcooling, $Sc = 0$, the solid–liquid interface location can be obtained by the following expression:

$$S\left(S+5+\sqrt{1+4S}\right)=6\tau \tag{5.170}$$

5.4.4 Ablation

Ablation is defined as the removal of material from the surface of an object by vaporization, chipping, or other erosive process. One application is when tire manufacturers design tires for automotive race events, such as NASCAR races. They have to balance the expected tire wear on the track versus the amount of heat that will be removed via ablation. If they underestimate the wear, the tire overheats and fails as too little heat is removed. If they overestimate the wear, the tire temperature never reaches its optimal operating condition to maximize adhesion. Ablation is also an effective means of protecting the surfaces of missiles and space shuttles from high-rate aerodynamic heating during atmospheric reentry. It is a sacrificial cooling method, because the protective layer is partially destroyed. The advantage of the ablative cooling process is its self-regulation: the rate of ablation is automatically adjusted in response to the heating rate. The most commonly used materials for ablative cooling are PCMs with higher melting points (such as glass, carbon, or polymer fiber) in combination with an organic binder.

From the heat transfer point of view, ablation is a special case of melting in which the liquid phase is completely removed immediately upon its production. Therefore, ablation can be considered as a one-region melting problem where heating occurs directly on a solid–liquid interface. If the spacecraft maintains a constant velocity during reentry, one can assume that the entire process occurs under constant heat flux heating. Similar to the melting of a subcooled solid under constant heat flux, ablation does not start simultaneously with heating, because the initial temperature of the ablating material, T_i, is usually below its melting point, T_m. The melting will start only after a period of preheating that allows the surface temperature to reach the melting point of the ablating material, T_m. The preheating problem is a pure conduction problem and can be solved analytically. Since the surface of the ablating material is constantly moving and the liquid phase does not accumulate, ablation can be described by using a coordinate system, x, that is attached to and moves with the ablating velocity, U_a (see Fig. 5.12). The ablation material moves with a velocity—U_a in the moving coordinate system. The energy equation in a moving coordinate is (Eckert and Drake 1987):

$$\frac{\partial}{\partial x}\left(k\frac{\partial T}{\partial x}\right) = -U_a\rho c_p\frac{\partial T}{\partial x} + \rho c_p\frac{\partial T}{\partial t} \tag{5.171}$$

Since the ablating materials usually have very low thermal conductivity, it is reasonable to assume that the ablation occurs in a semi-infinite body. Therefore, the initial and boundary conditions of Eq. (5.171) are

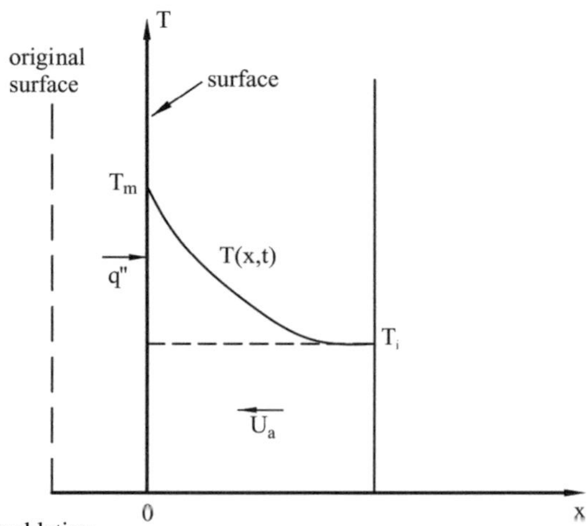

Fig. 5.12 Physical model for ablation

$$T = T_{\mathrm{i}}, \quad t = 0 \tag{5.172}$$

$$T = T_{\mathrm{m}}, \quad x = 0 \tag{5.173}$$

$$T = T_{\mathrm{i}}, \quad \frac{\partial T}{\partial x} = 0, \quad x \to \infty \tag{5.174}$$

The energy balance at the surface is

$$q'' - \rho U_{\mathrm{a}} h_{s\ell} = -k \frac{\partial T}{\partial x}\bigg|_{x=0}, \quad x = 0 \tag{5.175}$$

Shortly after ablation begins, the process enters steady-state and the ablation velocity becomes a constant. Equation (5.171) can be simplified as

$$\frac{\mathrm{d}^2 T}{\mathrm{d}x^2} = -\frac{U_{\mathrm{a}}}{\alpha} \frac{\mathrm{d}t}{\mathrm{d}x} \tag{5.176}$$

where the thermal properties are assumed to be independent of temperature.

Equation (5.176) can be treated as a first-order differential equation of d*t*/d*x*, and its solution is

$$\frac{\mathrm{d}t}{\mathrm{d}x} = C_1 \mathrm{e}^{-\frac{U_{\mathrm{a}}}{\alpha} x} \tag{5.177}$$

where C_1 is an integrating constant. Equation (5.177) can be further integrated to yield

$$T = -\frac{C_1 \alpha}{U_{\mathrm{a}}} \mathrm{e}^{-\frac{U_{\mathrm{a}}}{\alpha} x} + C_2 \tag{5.178}$$

The constants C_1 and C_2 can be determined from Eqs. (5.173)–(5.174), and the temperature distribution in the ablating material becomes

$$T = T_{\mathrm{i}} + (T_{\mathrm{m}} - T_{\mathrm{i}}) \mathrm{e}^{-\frac{U_{\mathrm{a}}}{\alpha} x} \tag{5.179}$$

Substituting Eq. (5.179) into Eq. (5.175) yields

$$U_{\mathrm{a}} h_{s\ell} = \rho c_{\mathrm{p}} U_{\mathrm{a}} (T_{\mathrm{m}} - T_{\mathrm{i}}) \tag{5.178}$$

which can be rearranged to obtain the ablation velocity:

$$U_{\mathrm{a}} = \frac{q''}{\rho h_{s\ell}(1 + Sc)} \tag{5.181}$$

where *Sc* is the subcooling parameter defined as

$$Sc = \frac{c_{\mathrm{p}}(T_{\mathrm{m}} - T_{\mathrm{i}})}{h_{s\ell}} \tag{5.182}$$

The fraction of heat removed by ablation over the total heat flux is

$$\frac{q''_{\mathrm{a}}}{q''} = \frac{\rho U_{\mathrm{a}} h_{s\ell}}{q''} \tag{5.183}$$

Substituting the ablation velocity from Eq. (5.181) into Eq. (5.183), one obtains

$$\frac{q''_{\mathrm{a}}}{q''} = \frac{1}{1 + Sc} \tag{5.184}$$

If the subcooling parameter is $Sc = 0.2$, 83% of the total heat flux will be removed by ablation.

5.4.5 Solidification in Cylindrical Coordinate Systems

Application of the integral solution to one-dimensional solid–liquid phase-change problems—including ablation—in a Cartesian coordinate system has been discussed in the preceding sections. Since tubes are widely used in shell-and-tube thermal energy storage devices, it is necessary to study melting and solidification in cylindrical coordinate systems as well. The polynomial temperature distribution is a very good approximation of the one-dimensional problem in the Cartesian coordinate system, but it can result in a very significant error if it is used to solve for the phase-change heat transfer in a cylindrical coordinate system. This is because heat transfer area for a cylindrical coordinate system varies with the coordinate r instead of remaining constant. Therefore, the temperature distribution in the coordinate has to be modified by taking into account the variation of the heat transfer area.

Solidification around a cylinder with radius of r_{i}, as shown in Fig. 5.13, will be investigated in order to demonstrate application of the integral solution in the cylindrical coordinate system. An infinite liquid PCM has a uniform initial temperature equal to the melting point of the PCM, T_{m}. At time $t = 0$, the inner surface of the cylinder suddenly decreases to a temperature T_0, which is below the melting point of the PCM. Since the temperature of the liquid PCM equals the melting point of the PCM and the temperature in the solid phase is unknown, this is a one-region solidification problem.

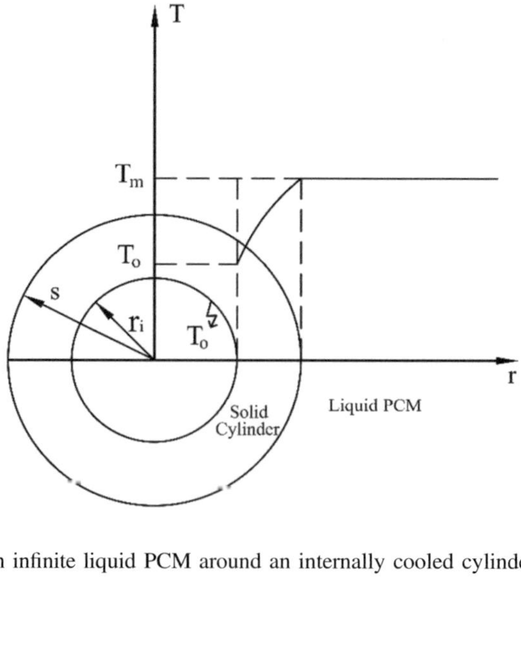

Fig. 5.13 Solidification of an infinite liquid PCM around an internally cooled cylinder

Conduction controls the solidification process, because the temperature in the liquid phase is uniformly equal to the melting point of the PCM. The governing equation and the initial and boundary conditions of this problem are

$$\frac{1}{r}\frac{\partial}{\partial r}\left(r\frac{\partial T}{\partial r}\right)=\frac{1}{\alpha}\frac{\partial T}{\partial t},\quad r_i<r<s(t),\quad t>0 \tag{5.185}$$

$$T(r,t)=T_0,\quad r=r_i,\quad t>0 \tag{5.186}$$

$$T(r,t)=T_m,\quad r=s,\quad t>0 \tag{5.187}$$

$$k\frac{\partial T}{\partial r}=\rho h_{s\ell}\frac{ds}{dt},\quad r=s,\quad t>0 \tag{5.188}$$

where the subscript 1 for solid phase has been dropped for ease of notation.

Defining dimensionless variables,

$$\theta=\frac{T_m-T}{T_m-T_0}\quad R=\frac{r}{r_i}\quad S=\frac{s}{r_i}\quad \tau=\frac{\alpha t}{r_i^2}\quad Ste=\frac{c_p(T_m-T_0)}{h_{s\ell}} \tag{5.189}$$

Equations (5.185)–(5.188) become

$$\frac{1}{R}\frac{\partial}{\partial R}\left(R\frac{\partial \theta}{\partial R}\right)=\frac{\partial \theta}{\partial \tau},\quad 1<R<S(\tau),\quad \tau>0 \tag{5.190}$$

$$\theta(R,\tau)=1,\quad R=1,\quad \tau>0 \tag{5.191}$$

$$\theta(R,\tau)=0,\quad R=S(\tau),\quad \tau>0 \tag{5.192}$$

$$-\frac{\partial \theta}{\partial R}=\frac{1}{Ste}\frac{dS}{d\tau},\quad R=S,\quad \tau>0 \tag{5.193}$$

The above Eqs. (5.190)–(5.193) are also valid for melting around a hollow cylinder when used with the appropriate dimensionless variables.

From elementary heat transfer, it is known that the logarithmic function is the exact solution of the steady-state heat conduction problem in a cylindrical wall. Therefore, we can assume that the temperature distribution has a second-order logarithmic function of the form (Zhang and Faghri 1996a):

$$\theta=1+\varphi\left(\frac{\ln R}{\ln S}\right)-(1+\varphi)\left(\frac{\ln R}{\ln S}\right)^2 \tag{5.194}$$

where φ is an unknown variable. Equation (5.194) automatically satisfies Eqs. (5.191) and (5.192) and φ can be obtained by differentiating Eq. (5.192)

$$d\theta=\frac{\partial \theta}{\partial R}dR+\frac{\partial \theta}{\partial \tau}d\tau=0,\ R=S$$

which can be rearranged to get

$$\frac{\partial \theta}{\partial R}\frac{dS}{d\tau} + \frac{\partial \theta}{\partial \tau} = 0 \tag{5.195}$$

Substituting Eqs. (5.190) and (5.192) into Eq. (5.195) yields the following expression:

$$-Ste\left(\frac{\partial \theta}{\partial R}\right)^2 + \frac{1}{R}\frac{\partial}{\partial R}\left(R\frac{\partial \theta}{\partial R}\right) = 0, \quad R = S, \quad \tau > 0 \tag{5.196}$$

Substituting Eq. (5.194) into Eq. (5.196) gives the following expression for φ:

$$2 + \varphi = \frac{\sqrt{1 + 2Ste} - 1}{Ste} \tag{5.197}$$

Substituting Eqs. (5.194) and (5.197) into Eq. (5.193) leads to an ordinary differential equation for the locations of the solid–liquid interface

$$\frac{dS}{d\tau} = \frac{\sqrt{1 + 2Ste} - 1}{S \ln S} \tag{5.198}$$

which is subjected to the following initial condition:

$$S(\tau) = 1, \qquad \tau = 0 \tag{5.199}$$

Integrating Eq. (5.198) over the time interval $(0, \tau)$ results in the following equation for the location of the solid–liquid interface:

$$\frac{1}{2}S^2 \ln S - \frac{1}{4}(S^2 - 1) = \left(\sqrt{1 + 2Ste} - 1\right)\tau \tag{5.200}$$

which is valid for solidification around a cylinder with a constant inner temperature. However, in practical applications, cooling inside the tube is usually achieved by the flow of cooling fluid through the tube. Therefore, the boundary condition at the inner surface of the tube should be a convective boundary condition instead of an isothermal boundary condition. The following example demonstrates an application of the integral approximate method to the solution of solidification around a cylindrical tube convectively cooled from inside.

Example 5.4

An infinite liquid PCM has a uniform initial temperature equal to the melting point of the PCM, T_m. At time $t = 0$, a cooling fluid with temperature T_i flows inside the tube. The heat transfer coefficient between the cooling fluid and the inner surface of the tube is h_i. Find the transient location of the solid–liquid interface.

Solution The governing equations of the problem can also be represented by Eqs. (5.185)–(5.188) except that Eq. (5.186) needs to be replaced by the following expression:

$$k\frac{\partial T}{\partial r} = h_i(T - T_i), \quad r = r_i, \quad t > 0 \tag{5.201}$$

The dimensionless governing equations of the problem are the same as Eqs. (5.190)–(5.193), but Eq. (5.191) needs to be replaced by the dimensionless form of Eq. (5.201), i.e.,

$$\frac{\partial \theta}{\partial R} = Bi(\theta - 1), \quad R = 1, \quad \tau = 0 \tag{5.202}$$

where Bi in Eq. (5.202) is the Biot number defined as

$$Bi = \frac{hr_{\mathrm{i}}}{k} \tag{5.203}$$

and θ is defined as

$$\theta = \frac{T_{\mathrm{m}} - T}{T_{\mathrm{m}} - T_{\mathrm{i}}} \tag{5.204}$$

It is also assumed that the temperature distribution has a second-order logarithmic function of the form

$$\theta = A + B\left(\frac{\ln R}{\ln S}\right) - (A + B)\left(\frac{\ln R}{\ln S}\right)^2 \tag{5.205}$$

The two unknown variables A and B in Eq. (5.205) can be determined by substituting Eq. (5.205) into Eqs. (5.202) and (5.196), with the result that

$$B = -Bi(1 - A)\ln S \tag{5.206}$$

$$2A + B = \frac{\sqrt{1 + 2ASte} - 1}{Ste} \tag{5.207}$$

Substituting Eq. (5.206) into Eq. (5.207), an equation for A is obtained as follows:

$$2A - Bi(1 - A)\ln S = \frac{\sqrt{1 + 2ASte} - 1}{Ste} \tag{5.208}$$

Substituting Eq. (5.205) into Eq. (5.193), the ordinary differential equation for the location of the solid–liquid interface is obtained:

$$\frac{\mathrm{d}S}{\mathrm{d}\tau} = \frac{\sqrt{1 + 2ASte} - 1}{S\ln S} \tag{5.209}$$

which is subject to the initial condition specified by Eq. (5.199).

The temperature of the inner surface of the tube is

$$\theta(1, \tau) = A \tag{5.210}$$

The solid–liquid interface location can be obtained by numerical solution of Eq. (5.209) and (5.210). It should be noted that A becomes 1 if the Biot number becomes infinite [see Eq. (5.208)]. In that case, the temperature of the inner surface becomes 1 and Eq. (5.209) reduces to Eq. (5.198).

Phase-change thermal energy storage systems can store thermal energy while being subjected to heat input, and then release it to the environment over a long period of time. Therefore, they are especially suitable for space applications involving pulsed power loads, such as a large amount of heat rejection from a power cycle in a short period of time. Heat transfer during solidification around a horizontal tube with internal convective cooling (Zhang et al. 1997) will be discussed.

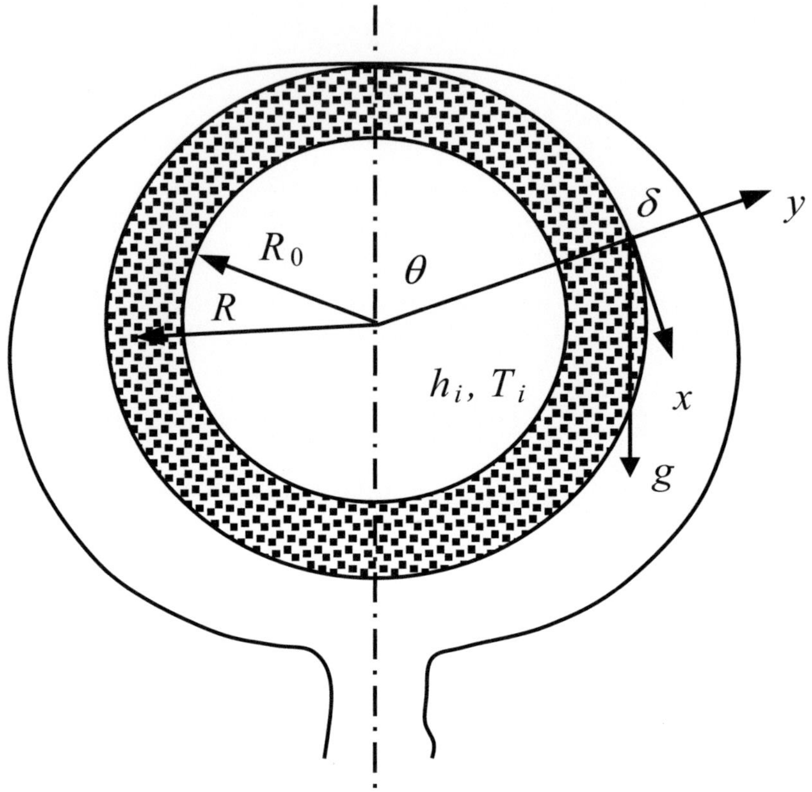

Fig. 5.14 Solidification around a horizontal tube

The theoretical model employed in this study is shown in Fig. 5.14. At the very beginning of the process ($t = 0$), the tube, which has a radius of R_0, is surrounded by a liquid phase-change material with uniform temperature $T_f > T_m$. The temperature of the working fluid inside the tube is T_i, and the convective heat transfer coefficient between the working fluid and the internal tube wall is h_i. Both h_i and T_i are kept constant throughout the process. The thickness of the tube is assumed to be very thin, so the thermal resistance of the tube wall can be neglected. The phase-change material can be treated as if it were directly in contact with the coolant inside the tube. Liquid adjacent to the cooled surface will be solidified, and the temperature difference between the solid–liquid interface and the otherwise quiescent liquid will drive natural convection in the liquid region. The liquid region is assumed to be sufficiently large so that the convection can be treated as natural convection between the surface and an extensive fluid medium. Conduction is the only heat transfer mode in the solidified region. The analysis is based on several additional assumptions:

1. The liquid is Newtonian and Boussinesq as well as incompressible. The Prandtl number of the liquid phase-change material is greater than unity.
2. The solid is homogeneous and isotropic.
3. The liquid motion induced by volumetric variation during solidification is neglected, i.e., the density of the solid is equal to the density of the liquid. In addition, the phase-change material's properties are constant in the liquid and solid regions.
4. The solid–liquid interface is assumed to be a smooth cylinder concentric with the cooled tube.

5. The effect of natural convection is restricted within the boundary layer and the bulk liquid has a uniform temperature, T_f.

Compared to the thickness of the solidified layer, the thickness of the natural convective boundary layer on the solid–liquid interface is very thin, except at the onset of freezing (the boundary-layer thickness in Fig. 5.14 is significantly exaggerated so as to be clearly visible). However, since the onset of freezing is a very short period compared to the whole solidification process, it is reasonable to neglect the curvature effect in the equations of the boundary layer. The boundary-layer equations of the problem can be written as follows:

$$\frac{\partial u}{\partial x} + \frac{\partial v}{\partial y} = 0 \tag{5.211}$$

$$\nu \frac{\partial^2 u}{\partial y^2} + g\beta(T - T_m)\sin\theta = 0 \tag{5.212}$$

$$\frac{\partial T}{\partial t} + \frac{\partial(uT)}{\partial x} + \frac{\partial(vT)}{\partial y} = \alpha_f \frac{\partial^2 T}{\partial y^2} \tag{5.213}$$

Since the Prandtl number of the liquid is much larger than unity, the inertia terms in the momentum equation have been neglected. These equations were solved by an integral method. Assuming a polynomial profile, the temperature and velocity profiles inside the boundary layer of thickness δ are expressed as

$$T = T_f - (T_f - T_m)\left(1 - \frac{y}{\delta}\right)^2 \tag{5.214}$$

$$u = U\frac{y}{\delta}\left(1 - \frac{y}{\delta}\right)^2 \tag{5.215}$$

where U is a characteristic velocity, δ is the boundary-layer thickness, and both of them are functions of time t and angle θ. Integrating both sides of Eq. (5.212) with respect to y within the interval $(0, \delta)$, the expression for U in Eq. (5.215) should be

$$U = \frac{g\beta\sin\theta}{3\nu}(T_f - T_m)\delta^2 \tag{5.216}$$

Integrating Eq. (5.213) in the same manner, one obtains

$$-\frac{1}{3}\frac{\partial\delta}{\partial t} + \frac{dR}{dt} - \frac{1}{90}(T_f - T_m)\frac{g\beta\sin\theta}{\nu R}\frac{\partial^3\delta}{\partial\theta^3} + \frac{2\alpha_f}{\delta} = 0 \tag{5.217}$$

Heat transfer in the solidified layer is dominated by conduction. The governing equation of the solid layer and corresponding boundary conditions are as follows:

$$\frac{1}{r}\frac{\partial}{\partial r}\left(r\frac{\partial T}{\partial r}\right) - \frac{1}{\alpha_s}\frac{\partial T}{\partial t}, \quad R_0 < r < R, \quad t > 0 \tag{5.218}$$

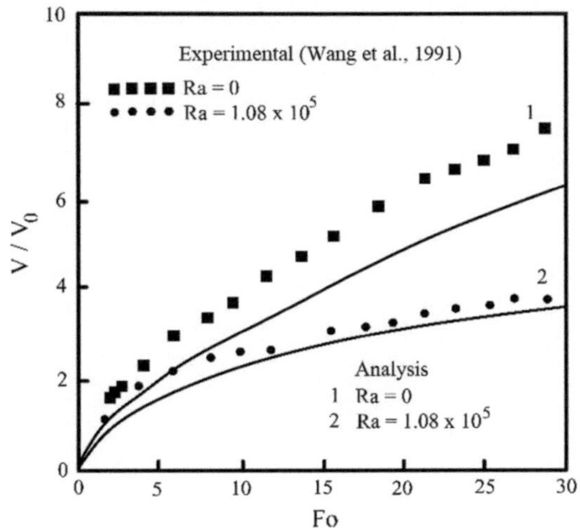

Fig. 5.15 Comparison of predicted solidification rate with experiments (Zhang et al. 1997)

$$k_s \frac{\partial T}{\partial r} = h_i (T - T_i), \quad r = R_0, \quad t > 0 \tag{5.219}$$

$$T_s(r, t) = T_m, \quad r = R, \quad t > 0 \tag{5.220}$$

$$k_s \left. \frac{\partial T_s}{\partial r} \right|_{r=R} - k_f \left. \frac{\partial T}{\partial y} \right|_{y=0} = \rho h_{s\ell} \frac{dR}{dt}, \quad r = R, \quad t > 0 \tag{5.221}$$

Equations (5.217)–(5.221) can be nondimensionalized and solved using integral approximate method. When $Bi \to \infty$, it corresponds to boundary conditions of the first kind, i.e., the tube wall temperature is T_i and is kept steady throughout the process. Wang et al. (1991) experimentally investigated the solidification process around a horizontal cooled tube. A comparison of the predicted solidification rate, $V/V_0 = (R/R_0)^2$ and the experimental results is shown in Fig. 5.15. When $Ra = 0$, i.e., no superheat exists in the liquid region or the solidification process is dominated by conduction, the predicted value is 18% lower than the experimental data. During the conduction-dominated freezing process, the front of the freezing layer is a dendritic layer. Therefore, it is believed that the solid–liquid interface is extended by the dentric layer. As the Rayleigh number increases, natural convection occurs in the liquid region, and the solid–liquid interface becomes smooth because of the natural motion of the liquid. When $Ra = 1.8 \times 10^5$, the predicted value is only 8% lower than the experimental data, so the agreement is satisfactory.

5.4.6 Binary Solidification

Solid–liquid phase diagrams of binary alloys are extremely useful for material scientists and mechanical engineers. Phase diagrams for multicomponent substances differ considerably from single-component phase diagrams. A mixture of two metals is called a binary alloy and constitutes a two-component system, since each metallic element in an alloy is considered a separate component. The phase diagrams of binary alloy systems are usually presented with the temperature of the system as the ordinate and the chemical composition of the system as the abscissa.

In a eutectic binary alloy, the two components have limited solid solubility in each other. Although some experimental results regarding solid–liquid phase change of binary metallic alloys appear in the

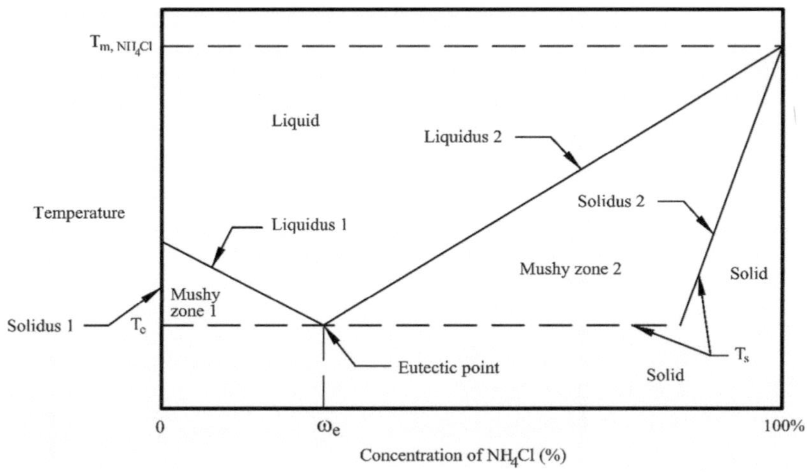

Fig. 5.16 Phase diagram for eutectic binary solution (solid–liquid), aqueous ammonium chloride ($NH_4Cl–H_2O$) at constant pressure

literature, many researchers conduct experiments using transparent phase-change materials (PCMs) such as $NH_4Cl–H_2O$ solution, because their solidification is quite similar to that of alloys and, moreover, it is easy to observe (Beckerman and Viskanta 1988; Braga and Viskanta 1990). The equilibrium phase diagram for aqueous ammonium chloride is shown in Fig. 5.16. At the eutectic point, the temperature and composition ($NH_4Cl–H_2O$ mass fraction) are $T_e = -15.4$ °C and $\omega_e = 19.7\%$, respectively. The eutectic point is also the point of intersection of the two liquidus lines, above which the binary solution is in the liquid phase. When the temperature is below the eutectic temperature, or the temperature corresponding to the second solidus line, the solid phase is present. There are two mushy zones where solid and liquid coexist. Mushy zone 1 is for a *subeutectic* concentration of $NH_4Cl–H_2O$ in water ($\omega < \omega_e$) and is bounded by the solidus 1, the liquidus 1, and the eutectic line. The solid phase in the mushy zone 1 is pure ice. Mushy zone 2 is for the *supereutectic* concentration ($\omega > \omega_e$) and is bounded by liquidus 2, solidus 2, and the eutectic line. The solid phase contained in the mushy zone 2 is solid $NH_4Cl–H_2O$. The phase diagram can be used, in conjunction with knowledge of the mixture concentration and temperature, to relate the phase concentrations to the mass fraction of the phase on the basis of local thermodynamic equilibrium.

Solidification of multicomponent PCMs has been investigated due to its importance in the fields of metallurgy (Viskanta 1988), crystal growth (Bardsley et al. 1979), and oceanography (Hobbs 1974). It is characterized by the presence of a variety of microscopically complex interfacial structures; as a result, the macroscopic solid–liquid interface is not smooth, as is the case in the solidification of single-component PCMs. The most popular microstructure is known as the dendrite, which can exist in either cocolumnar or equiaxed forms (Beckermann and Wang 1995). From the heat transfer point of view, the microscopic region where solid and liquid coexist is considered a mushy zone that exists between pure solid and pure liquid phase. The temperature at the interface between the mushy zone and the liquid region is the liquidus line temperature, while the interfacial temperature between the solid and the mushy zone is the solidus temperature (see Fig. 5.16). Therefore, phase change of multicomponent PCMs can be considered to occur over a range of temperatures. The presence of a mushy zone in solid–liquid phase-change system not only characterizes the phase change of multicomponent mixtures, but also poses the primary challenge in analyzing and modeling of binary and multicomponent solid–liquid phase-change systems.

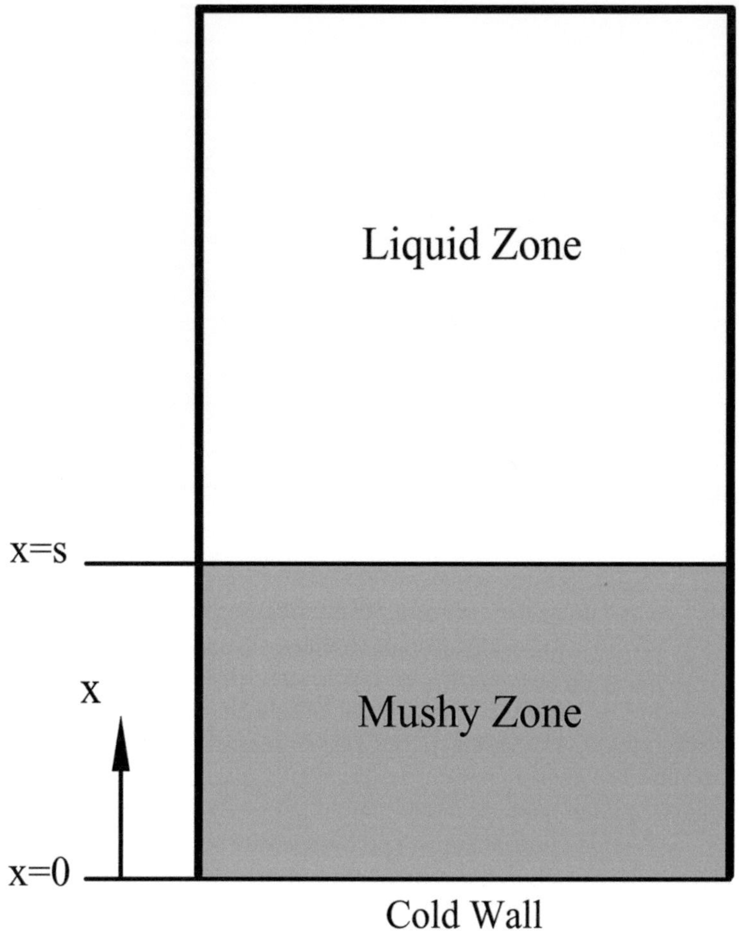

Fig. 5.17 Solidification of a binary solution (Zhang and Faghri 1998)

Many researchers conduct experiments using a transparent PCM, such as $NH_4Cl–H_2O$ solution, because its solidification is quite similar to the solidification of alloys and is easy to observe. An integral solution of solidification of a binary solution on a cold isothermal surface will be discussed.

The mushy zone formation in binary solutions on a horizontal cold surface will be studied using an integral approximate method. The physical model of the solidification problem is shown in Fig. 5.17 (Zhang and Faghri 1998). An ammonium chloride water solution with initial temperature T_i and mass fraction ω_i fills a half-space $x > 0$. At time $t = 0$, the wall temperature at $x = 0$ is suddenly reduced to a temperature T_w, which is higher than the eutectic temperature of the ammonium chloride water solution. The solidification process starts from the cold wall and the mushy zone grows upward. There is no solid phase because the cold wall temperature, T_w, is above the eutectic temperature. Therefore, this is a two-region problem with temperatures in the mushy and liquid zones, as well as the location of interface between these two zones, as unknowns. During the solidification process, a denser and colder solution appears in the mushy zone because the solution near the ice surface rejects the solute (Braga and Viskanta 1990). Thus, the liquid phase is hydrodynamically stable and no natural convection occurs in the liquid phase. Furthermore, the following assumptions are made in order to simplify the analysis:

1. Mass diffusion in the mushy zone is neglected because the thermal diffusivity of the salt solution is 100 times greater than the mass diffusivity (Fang et al. 1984).
2. The properties of the solid and liquid are constant within the liquid and solid, but different from each other, and the properties of the mushy zone are weighted according to the local solid fraction.
3. The densities of the solid and liquid phases are the same, i.e., the density change during the solidification process is neglected.

Based on the above assumptions, the governing energy equations of the solidification problem can be given as follows:

Mushy zone:

$$\rho_{\rm mu}\frac{\partial}{\partial t}(c_{\rm mu}T_{\rm mu}) = \frac{\partial}{\partial x}\left(k_{\rm mu}\frac{\partial T_{\rm mu}}{\partial x}\right) + \rho_{\rm mu}h_{s\ell}\frac{\partial f}{\partial t}, \qquad 0 \le x \le s \tag{5.222}$$

$$T = T_{\rm w}, \qquad x = 0 \tag{5.223}$$

Liquid zone:

$$\rho_\ell c_{p\ell}\frac{\partial T_\ell}{\partial t} = k_\ell\frac{\partial^2 T_\ell}{\partial x^2}, \qquad x \ge s \tag{5.224}$$

$$T_\ell = T_{\rm i}, \qquad x \to \infty \tag{5.225}$$

$$T_\ell = T_{\rm i}, \qquad t = 0 \tag{5.226}$$

Mushy zone and liquid zone interface:

$$T_{\rm mu} = T_\ell = T_{\rm s}, \qquad x = s \tag{5.227}$$

$$k_{\rm mu}\frac{\partial T_{\rm mu}}{\partial x} = k_\ell\frac{\partial T_\ell}{\partial x}, \qquad x = s \tag{5.228}$$

$$s = 0, \qquad t = 0 \tag{5.229}$$

where the last term in Eq. (5.222) represents the latent heat released during solidification. Since phase change occurs within the entire mushy zone, the latent heat appears as a source term in the energy Eq. (5.222), instead of appearing in a boundary condition at interface, as was the case for solid–liquid phase change of single-component PCMs. f is the local solid mass fraction in the mushy zone, which is the same as the volume fraction of solid if the density is not changed during the phase-change process. It can be determined from the phase diagram by

$$f = \frac{\omega(T_{\rm mu}) - \omega}{\omega(T_{\rm mu})} \tag{5.230}$$

where $\omega(T_{\rm mu})$ is the mass fraction obtained by liquidus 1 in the phase diagram, Fig. 5.16. $T_{\rm s}$ in Eq. (5.227) is a liquidus line temperature distinguishing the mushy zone and the liquid zone. The liquidus line temperature can be determined from the liquidus line equation of the phase diagram of ammonium chloride water solution based on the initial concentration (see Problem 5.27). The properties in the mushy zone are mass weighted according to the local solid fraction as follows:

$$\rho_{\text{mu}} = \rho_\ell = \rho_\text{s} \tag{5.231}$$

$$c_{\text{pmu}} = (c_{\text{ps}} - c_{\text{p}\ell})f + c_{\text{p}\ell} \tag{5.232}$$

$$k_{\text{mu}} = (k_\text{s} - k_\ell)f + k_\ell \tag{5.233}$$

By defining the following dimensionless variables,

$$\left.\begin{array}{lllll} \theta_{\text{mu}} = \dfrac{T_{\text{mu}} - T_\text{s}}{T_\text{s} - T_\text{w}} & \theta_\ell = \dfrac{T_\ell - T_\text{s}}{T_\text{s} - T_\text{w}} & \theta_\text{i} = \dfrac{T_\text{i} - T_\text{s}}{T_\text{s} - T_\text{w}} & X = \dfrac{x}{L} & \\ S = \dfrac{s}{L} & \tau = \dfrac{\alpha_\ell t}{L^2} & Ste = \dfrac{c_{p\ell}(T_\text{s} - T_w)}{h_{\text{s}\ell}} & R_\text{c} = \dfrac{c_{\text{ps}}}{c_{\text{p}\ell}} & R_\text{k} = \dfrac{k_\text{s}}{k_\ell} \end{array}\right\} \tag{5.234}$$

the governing equations and boundary conditions become

$$\begin{aligned} &\frac{\partial}{\partial \tau}[((R_\text{c} - 1)f + 1)\theta_{\text{mu}}] \\ &= \frac{\partial}{\partial X}\left[((R_\text{k} - 1)f + 1)\frac{\partial \theta_{\text{mu}}}{\partial X}\right] + \frac{1}{Ste}\frac{\partial f}{\partial \tau}, 0 \le X \le S \end{aligned} \tag{5.235}$$

$$\theta_{\text{mu}} = -1, \qquad X = 0 \tag{5.236}$$

$$\frac{\partial \theta_\ell}{\partial \tau} = \frac{\partial^2 \theta_\ell}{\partial X^2} \quad X > S \tag{5.237}$$

$$\theta_\ell = \theta_\text{i} \qquad X \to \infty \tag{5.238}$$

$$\theta_\ell = \theta_\text{i} \qquad \tau = 0 \tag{5.239}$$

$$\theta_{\text{mu}} = \theta_\ell = 0 \qquad X = S \tag{5.240}$$

$$\frac{\partial \theta_{\text{mu}}}{\partial X} = \frac{\partial \theta_\ell}{\partial X} \qquad X = S \tag{5.241}$$

$$S = 0 \quad \tau = 0 \tag{5.242}$$

The exact analytical solution of Eq. (5.237) can be obtained by using the exact solution of heat conduction in a semi-infinite body, i.e.,

$$\theta_\ell = \theta_\text{i}\left[1 - \frac{\text{erfc}(X/\sqrt{4\tau})}{\text{erfc}(S/\sqrt{4\tau})}\right] \tag{5.243}$$

which exactly satisfies Eqs. (5.237)–(5.240).

For solidification on a cold isothermal surface, the thickness of the mushy zone, S, can be expressed as (Braga and Viskanta 1990; Ozisik 1993; Tien and Geiger 1967)

$$\text{S} = 2\lambda\sqrt{\tau} \tag{5.244}$$

where λ is a constant.

Integrating Eq. (5.235) over the interval of (0, S) yields the integral equation of the mushy zone:

$$\begin{aligned}&\frac{\mathrm{d}}{\mathrm{d}\tau}\int_0^S[((R_\mathrm{c}-1)f+1)\theta_\mathrm{mu}]\mathrm{d}X\\&=\left.\frac{\partial\theta_\mathrm{mu}}{\partial X}\right|_{X=S}-[(R_\mathrm{k}-1)f+1]\left.\frac{\partial\theta_\mathrm{mu}}{\partial X}\right|_{X=0}+\frac{1}{Ste}\frac{d}{\mathrm{d}\tau}\int_0^S f\mathrm{d}X\end{aligned} \tag{5.245}$$

where $(\partial\theta_\mathrm{mu}/\partial X)|_{X=S}$ can be found from Eq. (5.242) via Eq. (5.238).

Assuming that the temperature distribution in the mushy zone is a second-order polynomial function, and determining the coefficients, one obtains the temperature distribution in the mushy zone:

$$\theta_\mathrm{mu}=\left[\frac{2\lambda\mathrm{e}^{-\lambda^2}\theta_\mathrm{i}}{\mathrm{erfc}(\lambda)}\right]\left(\frac{X-S}{S}\right)+\left[\frac{2\lambda\mathrm{e}^{-\lambda^2}\theta_\mathrm{i}}{\mathrm{erfc}(\lambda)}-1\right]\left(\frac{X-S}{S}\right)^2 \tag{5.246}$$

To solve the solidification problem using the integral approximate method, it is necessary to know the distribution of the solid mass fraction f in the mushy zone. The expression of the solid fraction as a single value of temperature in the mushy zone—as in Braga and Viskanta (1990)—is impossible to use here because the integral term in Eq. (5.246) will be very difficult to obtain. In their integral solution of the solidification in a semi-infinite region without liquid superheat, Tien and Geiger (1967) assume a linear distribution of the solid fraction in the mushy zone. In that instance, the solid fraction distribution is found to have no significant effect. Cao and Poulikakos (1991) assumed that the solid fraction, and its derivative with respect to temperature are constants when solving freezing problems of a binary alloy saturating a packed bed of spheres. For the sake of simplicity, it is also assumed here that the solid fraction is a linear function in the mushy zone, i.e.,

$$f=f_\mathrm{w}\left(1-\frac{X}{S}\right) \tag{5.247}$$

where f_w in Eq. (5.247) is the solid mass fraction at the cold isothermal surface, which depends on the cold wall temperature T_w and the initial concentration of the ammonium chloride water solution ω_i. It can be determined from the phase diagram by the lever rule (see Example 2.4):

$$f_\mathrm{w}=1-\frac{\omega_\mathrm{i}}{\omega(T_\mathrm{w})} \tag{5.248}$$

where $\omega(T_\mathrm{w})$ can be determined by the liquidus line equation:

$$\omega(T_\mathrm{w})=1.678\times10^{-3}-1.602\times10^{-2}T_\mathrm{w}-2.857\times10^{-4}T_\mathrm{w}^2-4.491\times10^{-6}T_\mathrm{w}^3 \tag{5.249}$$

Substituting Eqs. (5.246) and (5.247) into Eq. (5.245) and considering Eq. (5.244), an algebraic equation of λ is obtained:

Table 5.1 Comparison of the mushy zone thickness obtained by different methods (Zhang and Faghri 1998)

$\omega_i(\%)$	$T_i(°C)$	$T_w(°C)$	t(min)	s (mm)		
				Braga and Viskanta (1990)		Equation (5.244)
				Analytical	Experimental	
5	15.3	−13.3	500	51.0	48.0	50.7
10	20.2	−14.6	540	36.6	32.0	33.6
15	19.5	−14.7	540	16.3	13.0	14.8

$$\lambda = \left\{ \frac{2[(R_k - 1)f_w + 1] - \frac{2\lambda e^{-\lambda^2}\theta_0}{\operatorname{erfc}(\lambda)}[(R_k - 1)f_w + 2]}{\frac{2\lambda e^{-\lambda^2}\theta_0}{\operatorname{erfc}(\lambda)}\frac{f_w + 2}{12} + \frac{3(R_c - 1)f_w + 4}{12} + \frac{f_w}{2Ste}} \right\}^{\frac{1}{2}} \tag{5.250}$$

which can be solved iteratively. After the value of λ is obtained, the dimensionless thickness of the mushy zone can be obtained from Eq. (5.244), and the temperature distribution in the liquid and mushy zones can be obtained by Eqs. (5.243) and (5.246). In order to compare these results based on the integral method with Braga and Viskanta's (1990) experimental results, the following dimensionless variable should be introduced:

$$\eta = \frac{X}{2\sqrt{\tau}} \tag{5.251}$$

The temperature distributions in the liquid and mushy zones then become

$$\theta_\ell = \theta_i \left[1 - \frac{\operatorname{erfc}(\eta)}{\operatorname{erfc}(\lambda)}\right] \tag{5.252}$$

$$\theta_m = \left[\frac{2\lambda e^{-\lambda^2}\theta_i}{\operatorname{erfc}(\lambda)}\right]\left(\frac{\eta - \lambda}{\lambda}\right) + \left[\frac{2\lambda e^{-\lambda^2}\theta_i}{\operatorname{erfc}(\lambda)} - 1\right]\left(\frac{\eta - \lambda}{\lambda}\right)^2 \tag{5.253}$$

The thermal properties of the ammonium chloride water solution at specified concentrations of 5, 10, and 15% can be found in Zeng and Faghri (1994) and Cao and Poulikakos (1991) and will not be repeated here. In Braga and Viskanta's (1990) experimental investigation, the solidification of $NH_4Cl–H_2O$ was performed for three different cases. Table 5.1 compares the mushy zone thickness obtained using the integral method by Zhang and Faghri (1998) and by Braga and Viskanta (1990). It can be seen that the result obtained by the integral solution agreed well with Braga and Viskanta's similarity solution and the experimental results.

5.5 Contact Melting

5.5.1 Fixed Melting and Contact Melting

Two types of melting patterns may be observed during the experimental investigation of melting in a two-dimensional rectangular cavity: fixed melting (Fig. 5.18) and contact melting (Fig. 5.19). The solid core is fixed during the fixed melting process, and heat transfer is controlled mainly by natural

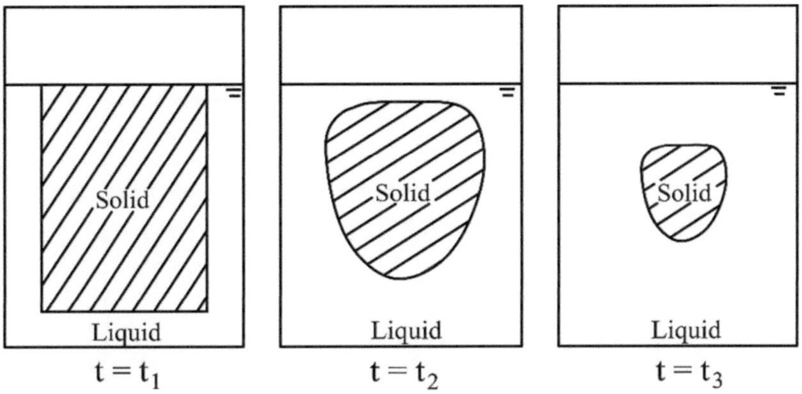

Fig. 5.18 Fixed melting in a rectangular cavity heated from two sides and bottom ($t_1 < t_2 < t_3$)

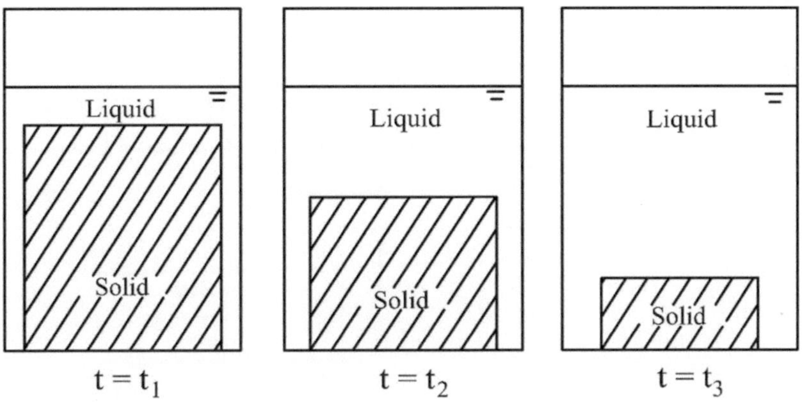

Fig. 5.19 Contact melting in a rectangular cavity heated from two sides and bottom ($t_1 < t_2 < t_3$)

convection. During the contact melting process, on the other hand, gravity causes the solid core to fall so that it maintains constant contact with the bottom surface of the cavity. Heat transfer in the contact melting process is controlled primarily by conduction between the heated bottom surface and the solid PCM. Since the unmelted solid core always maintains contact with the bottom surface of the cavity, the melting proceeds more quickly in contact melting than it does in fixed melting. A complete review of contact melting in various geometries and their applications can be found in Bejan (1994, 2013).

5.5.2 Contact Melting in a Rectangular Cavity

The contact melting in a rectangular cavity shown in Fig. 5.20 was modeled by Hirata et al. (1991). The initial temperature of the solid is assumed to be at the melting point of the PCM, T_m At time $t = 0$, the temperatures of the sidewall and bottom wall suddenly increase to T_w, which is higher than T_m. Melting occurs at the sidewalls and bottom wall simultaneously, and the solid falls due to gravity, always keeping in good contact with the bottom surface of the cavity. As a result, the liquid layer between the solid and the bottom surface of the cavity is very thin. The liquid layer thickness in Fig. 5.20 is exaggerated for clarity of presentation. The following assumptions are needed in order to solve the problem:

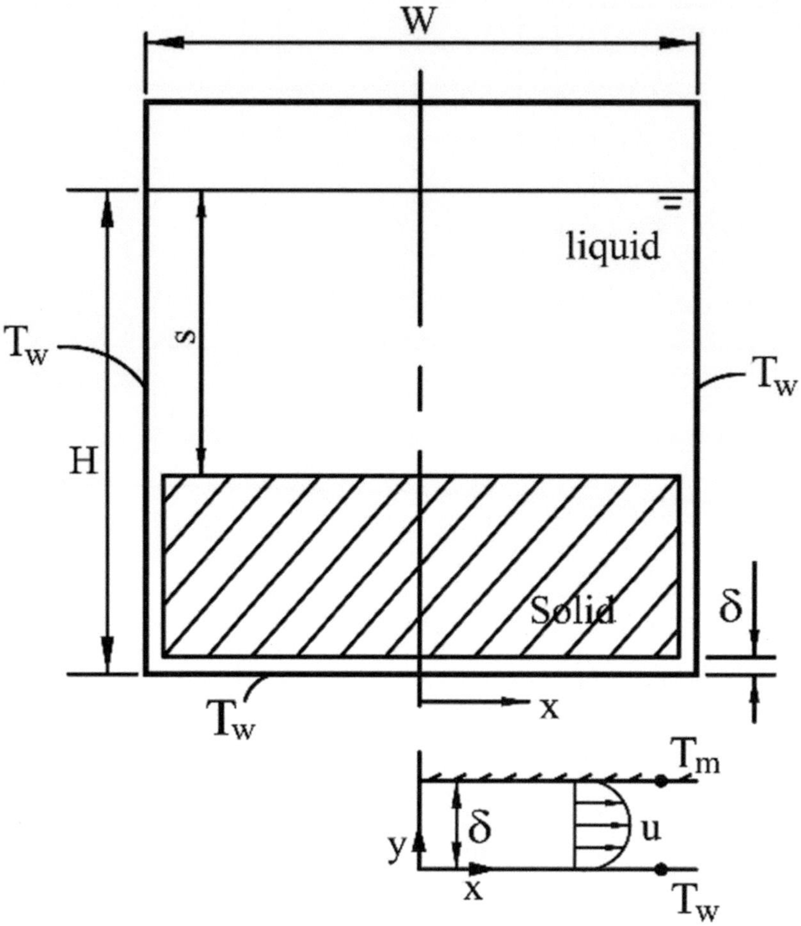

Fig. 5.20 Physical model of contact melting

1. Compared with melting at the bottom wall, melting at the two sides and above the solid can be neglected.
2. The liquid is Newtonian.
3. Compared with the viscous term, the inertia term of the momentum equation is negligible.
4. Compared with heat conduction, heat transfer by convection is negligible.
5. The thermophysical properties of the liquid are constant.
6. The flow of the liquid layer is symmetric at $x = 0$.

Nusselt liquid film theory is applicable to this melting process. The momentum and energy equations for the liquid film are

$$\mu_\ell \frac{\partial^2 u}{\partial y^2} = \frac{\mathrm{d}p}{\mathrm{d}x} \tag{5.254}$$

$$\frac{\partial^2 T}{\partial y^2} = 0 \tag{5.255}$$

The corresponding boundary conditions are

$$y = 0: \quad u = 0, \quad T = T_w \tag{5.256}$$

$$y = \delta: \quad u = 0, \quad T = T_m \tag{5.257}$$

Integrating Eqs. (5.254) and (5.255) and considering boundary conditions (5.256) and (5.257), the velocity and temperature distribution in the liquid layer are

$$u = \frac{1}{2\mu_\ell}\frac{dp}{dx}(y^2 - \delta y) \tag{5.258}$$

$$T = T_w + (T_m - T_w)\frac{y}{\delta} \tag{5.259}$$

The heat balance equation over the range (0, x) in the liquid layer can be given by

$$-\int_0^x k_\ell \left.\frac{\partial T}{\partial y}\right|_{y=0} dx = \rho h_{s\ell}\int_0^\delta u dy \tag{5.260}$$

The transport of sensible heat has been neglected because the Stefan number is assumed to be much less than unity.

Substituting Eqs. (5.258) and (5.259) into Eq. (5.260) yields

$$-k_\ell(T_m - T_w)\int_0^x \frac{dx}{\delta} = -\frac{\rho_\ell h_{s\ell}}{12\mu_\ell}\frac{dp}{dx}\delta^3 \tag{5.261}$$

Experimental observation (Dong et al. 1991) indicates that the variation of the liquid layer thickness with respect to x is negligible, and therefore, δ can be treated as a constant across x.

Neglecting the effect of sensible heat in the liquid layer, the heat balance equation for the range of (0, x) can be given as

$$-k_\ell(T_m - T_w)\int_0^x \frac{dx}{\delta} = \int_0^x h_{s\ell}\dot{m}'' dx \tag{5.262}$$

where $\dot{m}''$ is the molten mass per unit area and unit time, which is independent of x. Integrating Eq. (5.262) with respect to x, one obtains the liquid layer thickness as

$$\delta = \frac{k_\ell(T_w - T_m)}{\dot{m}'' h_{s\ell}} \tag{5.263}$$

Substituting Eq. (5.263) into Eq. (5.261) and integrating it with boundary condition $p = 0$ at $x = W/2$ yields the pressure distribution in the liquid layer:

$$p = -6\dot{m}''^4 \nu_\ell \left[\frac{h_{s\ell}}{k_\ell(T_w - T_m)}\right]^3 \left[x^2 - \left(\frac{W}{2}\right)^2\right] \tag{5.264}$$

The force balance of the buoyancy and pressure for the solid is

$$2\int_{0}^{W/2} p\mathrm{d}x = g(\rho_s - \rho_\ell)W(H - s) \tag{5.265}$$

Substituting Eq. (5.264) into Eq. (5.265) yields

$$\dot{m}''^4 \nu_\ell \left[\frac{h_{s\ell}W}{k_\ell(T_w - T_m)}\right]^3 = g(\rho_s - \rho_\ell)W(H - s) \tag{5.266}$$

The relationship between $\dot{m}''$ and s is (see Fig. 5.20)

$$\dot{m}'' = \rho_s \frac{\mathrm{d}s}{\mathrm{d}t} = \rho_s V_s \tag{5.267}$$

where V_s is the downward velocity of the solid.

Substituting Eq. (5.267) into (5.266), the following ordinary differential equation for s is obtained:

$$\left(\rho_s \frac{\mathrm{d}s}{\mathrm{d}t}\right)^4 \nu_\ell \left[\frac{h_{s\ell}W}{k_\ell(T_w - T_m)}\right]^3 = g(\rho_s - \rho_\ell)W(H - s) \tag{5.268}$$

The downward velocity of the solid is then

$$V_s = \left[\frac{k_\ell(T_w - T_m)}{\rho_s h_{s\ell}}\right]^{\frac{3}{4}} \left[\frac{g(\rho_s - \rho_\ell)(H - s)}{\rho_s W^2 \nu_\ell}\right]^{\frac{1}{4}} \tag{5.269}$$

The order of magnitude of the downward velocity of the solid was obtained by scale analysis in Chap. 1 and the result was [see Eq. (1.87)]

$$V_s \sim \left(\frac{k_\ell \Delta T}{\rho_s h_{s\ell}}\right)^{\frac{3}{4}} \left(\frac{p}{\mu L^2}\right)^{\frac{1}{4}} \tag{5.270}$$

We will show below that the result of analytical solution, Eq. (5.269), is consistent with the results of the scale analysis, Eq. (5.270). It follows from the force balance of the solid, Eq. (5.265), that the order of magnitude of the pressure in the liquid layer is

$$p \sim g(\rho_s - \rho_\ell)(H - s) \tag{5.271}$$

The order of magnitude of the second bracket on the right-hand side of Eq. (5.269) is therefore

$$\left[\frac{g(\rho_s - \rho_\ell)(H - s)}{\rho_s W^2 \nu_\ell}\right]^{\frac{1}{4}} \sim \left[\frac{p}{\rho_s W^2 \nu_\ell}\right]^{\frac{1}{4}} \tag{5.272}$$

Since $\rho_s \sim \rho_\ell$ and $W \sim L$, Eq. (5.272) becomes

$$\left[\frac{g(\rho_s - \rho_\ell)(H - s)}{\rho_s W^2 \nu_\ell}\right]^{\frac{1}{4}} \sim \left[\frac{p}{\mu_\ell L^2}\right]^{\frac{1}{4}} \tag{5.273}$$

Substituting Eq. (5.273) into Eq. (5.269) and considering $\Delta T = T_\mathrm{w} - T_\mathrm{m}$, Eq. (5.270) can be obtained.

In general, it is important to find the time it takes to completely melt the solid. Introducing the following dimensionless variables

$$\left.\begin{array}{l} S = \dfrac{s}{W} \quad B = \dfrac{H}{W} \quad Ar = \dfrac{\rho_\mathrm{s} - \rho_\ell}{\rho_\ell}\dfrac{gW^3}{v_\ell^2} \quad R_\rho = \dfrac{\rho_\ell}{\rho_\mathrm{s}} \quad \Delta = \dfrac{\delta}{W} \\ Ste = \dfrac{c_{p\ell}(T_w - T_\mathrm{m})}{h_{s\ell}} \quad \tau = Ste\dfrac{\alpha_\ell t}{W^2} \quad A(\tau) = \dfrac{s}{H} = \dfrac{S}{B} \end{array}\right\} \tag{5.274}$$

where Ar is the Archimedes number and $A(\tau)$ is the melting rate, Eq. (5.268) becomes

$$\frac{\mathrm{d}\tau}{\mathrm{d}S} = \left(\frac{Ste}{ArPrB}\right)^{\frac{1}{4}}\left(1 - \frac{S}{B}\right)^{-\frac{1}{4}} \tag{5.275}$$

Substituting Eq. (5.267) into Eq. (5.263) yields

$$\delta = \frac{k_\ell(T_\mathrm{w} - T_\mathrm{m})}{\rho_s h_{s\ell}}\frac{\mathrm{d}t}{\mathrm{d}s} \tag{5.276}$$

Substituting Eq. (5.274) into Eq. (5.276), the dimensionless liquid film thickness becomes

$$\Delta = \frac{\delta}{W} = \frac{\mathrm{d}\tau}{\mathrm{d}S} \tag{5.277}$$

Thus, the dimensionless liquid layer thickness can be obtained by substituting Eq. (5.275) into Eq. (5.277):

$$\Delta = \left[\frac{Ste}{ArPrB[1 - A(\tau)]}\right]^{\frac{1}{4}} \tag{5.278}$$

Equation (5.275) can be rewritten as

$$d\tau = \left(\frac{Ste}{ArPrB}\right)^{\frac{1}{4}}[1 - A(\tau)]^{-\frac{1}{4}}BdA(\tau) \tag{5.279}$$

Integrating Eq. (5.279) with respect to τ, and considering the initial condition of $A(0) = 0$, gives

$$\tau = \frac{4}{3}\left(\frac{Ste}{ArPrB}\right)^{\frac{1}{4}}B\left\{1 - [1 - A(\tau)]^{\frac{3}{4}}\right\} \tag{5.280}$$

Thus, the melting rate at any time can be obtained by solving Eq. (5.280), and then the dimensionless liquid layer thickness at any time can be obtained by using Eq. (5.278). When all of the solid has melted, the melting rate $A(\tau) = s/H = 1$. The corresponding dimensionless time is

$$\tau_f = \frac{4}{3}\left(\frac{Ste}{ArPr}\right)^{\frac{1}{4}}B^{\frac{3}{4}} \tag{5.281}$$

which gives the time it takes to completely melt the solid; it can be used to estimate the working time of the latent heat thermal energy storage system with contact melting.

5.6 Numerical Solutions

5.6.1 Overview

Due to the strong nonlinearity of solid–liquid phase-change phenomena, and the moving boundary, the problems that can be solved via the analytical method are very limited. Exact solutions and some integral approximate solutions for melting and solidification have been introduced in preceding sections. The limitation of these methods is that they apply to conduction-controlled one-dimensional problems. Although some investigators have attempted to solve two-dimensional phase-change problems by using analytical methods (Poots 1962; Rathjen and Jiji 1971; Budhia and Kreith 1973), the cases investigated represented very simple and special geometries, such as a semi-infinite corner or a semi-infinite wedge. Generally, the analytical method will not work for two- or three-dimensional problems; this is particularly true for convection-controlled phase-change processes. Most real applications of phase change occur in complex two- or three-dimensional geometries. In addition, natural convection in the liquid phase often plays a significant role in the phase-change processes. For such complex solid–liquid phase-change processes, analytical solutions will not work. Therefore, it is necessary to employ numerical methods.

A large number of effective numerical methods have been developed, and they can be divided into two groups (Voller 1997). The first group is called *strong numerical solution* or *classical solution*. For this group, transformed coordinate systems or deformed grids are employed to deal with the location of the solid–liquid interface. In this methodology, complex geometric regions of solid or liquid phases are transformed into fixed, simple geometric regions through the coordinate transformation technique. At the same time, the governing equations and the boundary conditions become more complicated. The governing equations of the problem can be solved using the finite difference method. Such methods also deal successfully with multidimensional one-region or two-region problems with or without natural convection in the liquid phase. The disadvantage of these methods is that they are often difficult to program and thus require a significant amount of computer time.

The second group is called the *weak solution* or *the fixed-grid solution*. In these methods, the governing equations are written for the entire phase-change region, including liquid and solid phases. The location of the solid–liquid interface is determined after the converged temperature distribution is obtained. The three principal important methods in this group are the enthalpy method (Shamsunder and Sparrow 1975; Crank 1984; Voller 1997; Voller et al. 1987; Sultana et al. 2018), the equivalent heat capacity method (Bonacina et al. 1973), and the temperature-transforming model (Cao and Faghri 1990). The temperature-transforming model has the advantages of both the enthalpy method and the equivalent heat capacity method and can also account for the effect of natural convection in the liquid phase. The most significant advantage of the weak solution, as compared with the strong numerical solution, is its simplicity. In this section, various fixed-grid numerical solutions of the solid–liquid phase problem will be introduced.

5.6.2 Enthalpy Method

In this methodology, the governing energy equation is written for the entire region of the PCM, including solid and liquid phases and the interface. The enthalpy method is introduced by analyzing a conduction-controlled, two-region melting problem in a finite slab, as shown in Fig. 5.21. It is assumed that the densities of the liquid and solid phase are identical ($\rho_s = \rho_\ell$). The energy equation can be written as

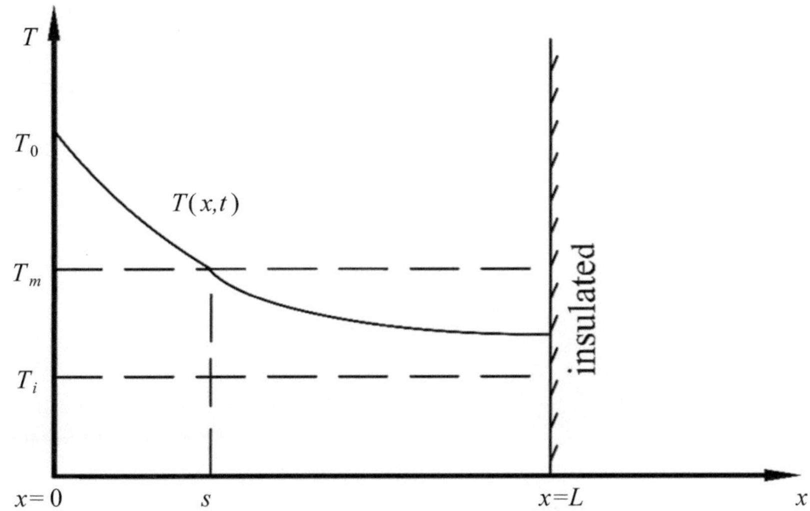

Fig. 5.21 Conduction-controlled melting in a finite slab

$$\rho \frac{\partial h}{\partial t} = \frac{\partial}{\partial x}\left(k \frac{\partial T}{\partial x}\right) \tag{5.282}$$

The enthalpy, h, is a function of temperature, T:

$$h(T) = \begin{cases} c_{p_s}(T - T_m), & T < T_m \\ c_{p_\ell}(T - T_m) + h_{s\ell}, & T > T_m \end{cases} \tag{5.283}$$

The variation of enthalpy h with temperature T can be plotted as shown in Fig. 5.22. At the melting point of the PCM, the enthalpies of solid and liquid phases at the melting point are 0 and $h_{s\ell}$, respectively; there is a jump of enthalpy equal to the latent heat of the PCM. The thermal conductivity of the PCM, k, can be expressed as

$$k(T) = \begin{cases} k_s, & T < T_m \\ k_\ell, & T > T_m \end{cases} \tag{5.284}$$

It can be verified that the energy equation for liquid and solid phases can be obtained simply by substituting Eqs. (5.283) and (5.284) into Eq. (5.282). The energy balance at the solid–liquid interface can be obtained by analyzing the energy balance for a control volume, which includes the solid–liquid interface.

Solving for temperature, T, from Eq. (5.283) yields

$$T = \begin{cases} T_m + h/c_{ps}, & h \le 0 \\ T_m, & 0 < h < h_{s\ell} \\ T_m + (h - h_{s\ell})/c_{p\ell}, & h \ge h_{s\ell} \end{cases} \tag{5.285}$$

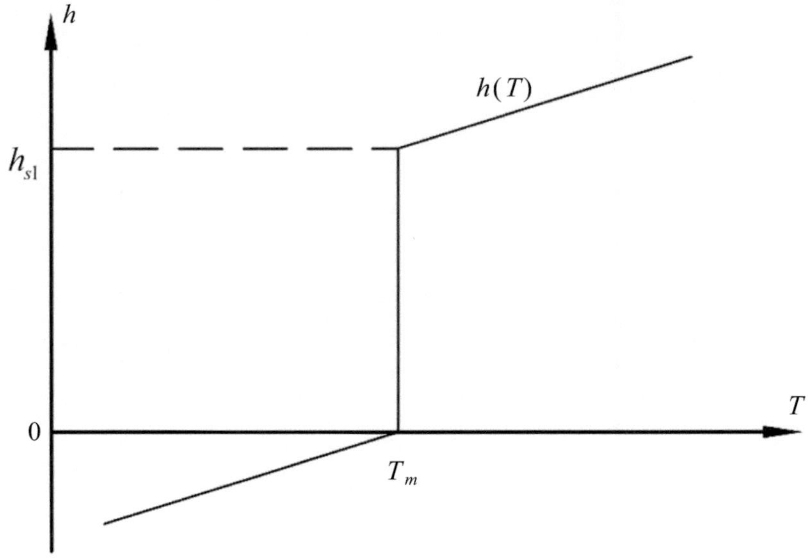

Fig. 5.22 Enthalpy of the PCM as a function of temperature

The initial and the boundary conditions of the melting problem are

$$T(x,t) = T_\mathrm{i} < T_\mathrm{m}, \qquad t = 0 \tag{5.286}$$

$$T(x,t) = T_0 > T_\mathrm{m}, \qquad x = 0 \tag{5.287}$$

$$\frac{\partial T(x,t)}{\partial x} = 0, \qquad x = L \tag{5.288}$$

The discretization of space and time is shown in Fig. 5.23, in which location and time are represented by j and n. The temperature at $x = (j-1)\Delta x$ and time $t = n\Delta t$ can be represented by the symbol T_j^n. Equation (5.282) is discretized by using an explicit scheme:

$$\rho \frac{h_j^{n+1} - h_j^n}{\Delta t} = \frac{k_{j+\frac{1}{2}}(T_{j+1}^n - T_j^n) - k_{j-\frac{1}{2}}(T_j^n - T_{j-1}^n)}{(\Delta x)^2} \qquad j = 2, \ldots, N-1 \tag{5.289}$$

which uses forward difference in time and central difference in space.

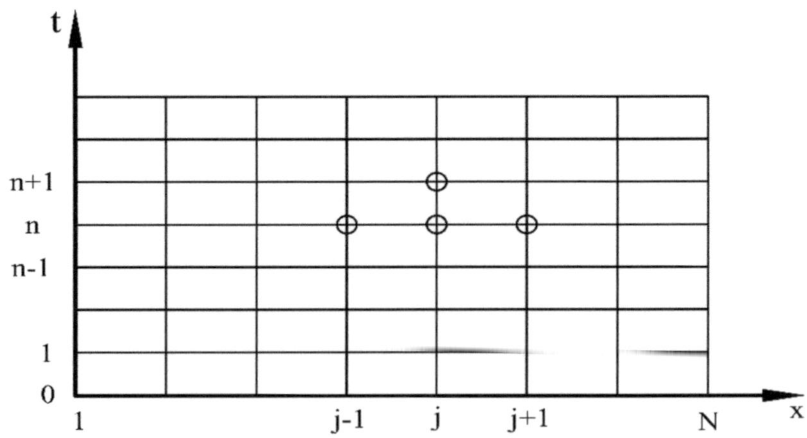

Fig. 5.23 Discretization of space and time for the solution of the two-region melting problem

The thermal conductivities at the half-grid, $k_{j+\frac{1}{2}}$ and $k_{j-\frac{1}{2}}$, are calculated using the harmonic mean method:

$$k_{j+\frac{1}{2}} = \frac{2k_j k_{j+1}}{k_j + k_{j+1}} \tag{5.290}$$

$$k_{j-\frac{1}{2}} = \frac{2k_{j-1} k_j}{k_{j-1} + k_j} \tag{5.291}$$

Equation (5.289) can be rewritten as

$$\begin{aligned} h_j^{n+1} &= \frac{\Delta t}{\rho\,(\Delta x)^2}\left(k_{j+\frac{1}{2}}T_{j+1}^n + k_{j-\frac{1}{2}}T_{j-1}^n\right) \\ &+ \left(h_j^n - \frac{\Delta t\left(k_{j+\frac{1}{2}} + k_{j-\frac{1}{2}}\right)}{\rho\,(\Delta x)^2}T_j^n\right), \quad j = 2,\ldots,N-1 \end{aligned} \tag{5.292}$$

After the enthalpy distribution of the $(n+1)$th time step is obtained, the temperature distribution of the $(n+1)$th time zone can be obtained from Eq. (5.285), i.e.,

$$T_j^{n+1} = \begin{cases} T_{\mathrm{m}} + h_j^{n+1}/c_{\mathrm{ps}}, & h_j^{n+1} \le 0 \quad \text{(solid)} \\ T_m, & 0 < h_j^{n+1} < h_{s\ell} \quad \text{(interface)}\, j = 2,\ldots,N-1 \\ T_{\mathrm{m}} + (h_j^{n+1} - h_{\mathrm{s}\ell})/c_{\mathrm{p}\ell}, & h_j^{n+1} \ge h_{\mathrm{s}\ell} \quad \text{(liquid)} \end{cases} \tag{5.293}$$

The initial and boundary conditions are discretized as

$$T_j^0 = T_i, \quad j = 1, 2, \ldots, N \tag{5.294}$$

$$T_1^{n+1} = T_0 \tag{5.295}$$

$$T_N^{n+1} = T_{N-1}^{n+1} \tag{5.296}$$

The location of the solid–liquid interface at the $(n+1)$th time zone can easily be determined according to the temperature distribution. For example, if the enthalpy at a certain point $j = M$ satisfies $0 < h_{\mathrm{M}}^{n+1} < h_{s\ell}$, the solid–liquid interface will be

$$s^{n+1} = (M-1)\Delta x - \frac{\Delta x}{2} + \frac{h_M^{n+1}}{h_{s\ell}}\Delta x \tag{5.297}$$

The solution procedure can be summarized as follows:

1. Determine the initial enthalpy at every node h_j^0 by using Eq. (5.283).
2. Calculate the enthalpy after the first time step at nodes $j = 2,\ldots,N-1$ by using Eq. (5.292).
3. Determine the temperature after the first time step at node $j = 1,\ldots,N$ by using Eqs. (5.295), (5.293), and (5.296).
4. Find a control volume in which the enthalpy falls between 0 and $h_{s\ell}$ and determine the location of the solid–liquid interface by using Eq. (5.297).
5. Solve the phase-change problem at the next time step with the same procedure.

The solution procedure using the explicit scheme is attractive in its simplicity. However, there is a major drawback of this scheme—the limitation of stability. The limitation of stability should be familiar from prior experience in undergraduate heat transfer dealing with the numerical solution of unsteady state heat conduction. To satisfy the stability condition, the following Neumann stability criterion has to be satisfied:

$$\frac{\max(\alpha_s, \alpha_\ell)\Delta t}{(\Delta x)^2} \leq \frac{1}{2} \tag{5.298}$$

where α_s and α_ℓ are thermal diffusivity of solid and liquid, respectively. Equation (5.298) is valid for one-dimensional problems only.

For a multidimensional problem, the energy Eq. (5.282) becomes:

$$\rho\frac{\partial h}{\partial t} = \nabla \cdot (k\nabla T) \tag{5.299}$$

and Eqs. (5.283)–(5.285) are still valid. Equation (5.299) can be discretized as:

$$\rho a_{\mathrm{P}}^0 h_{\mathrm{P}} = \rho a_{\mathrm{P}}^0 h_{\mathrm{P}}^0 + \sum_{\mathrm{nb}} a_{\mathrm{nb}} T_{\mathrm{nb}}^0 - a_{\mathrm{P}} T_{\mathrm{P}}^0 \tag{5.300}$$

for enthalpy and temperature at node *P* and *a*'s are coefficients. The subscript *nb* represents the nodes that are neighbors of node *P* and the superscript 0 represents the previous time step that are known. Equation (5.300) is similar to Eq. (5.292) that the enthalpy at the current time step, h_P.

In addition to the above explicit scheme, one can use an implicit scheme—which is unconditionally stable—to overcome the drawback of potential instability of the explicit scheme:

$$\rho a_{\mathrm{P}}^0 h_{\mathrm{P}} = \rho a_{\mathrm{P}}^0 h_{\mathrm{P}}^0 + \sum_{\mathrm{nb}} a_{\mathrm{nb}} T_{\mathrm{nb}} - a_{\mathrm{P}} T_{\mathrm{P}} \tag{5.301}$$

which is different from Eq. (5.300) that the temperatures at the current time step appear on the right-hand side. Equation (5.301) is unconditionally stable under any time step, but its solution process for an implicit scheme will be more complex than the solution for an explicit scheme because two unknown variables (enthalpy and temperature) are involved. Equation (5.301) can be solved by employing an enthalpy linearization method that assumes the phase change occurs over an arbitrarily small range of temperature (Voller 1997). Sultana et al. (2018) reviewed fixed-grid enthalpy method based on the solution techniques of conduction and convection-related phase-change problems.

The above enthalpy model treated enthalpy as a dependent variable in addition to the temperature and discretized the energy equation into a set of equations which contain both *h* and *T*. For the implicit schemes, they actually treated all of the terms containing $T = T(h)$ as a constant heat source term in the energy equation during iterations at each time step. This may cause some problems with convergence when $T = T(h)$ is complicated and physical properties change significantly, as is the case for frozen heat pipe start-up, or when the boundary conditions are severe. Furthermore, when the energy equation contains a convective term, the previous methods have difficulties in handling the relationship between the convective and diffusive terms because of the two-dependent variable nature of the equation.

5.6.3 Equivalent Heat Capacity Method

During the solid–liquid phase-change process, the PCM can absorb or release heat at a constant temperature, T_m. This means that the temperature of the PCM does not change while it absorbs or releases heat, implying that the heat capacity of the PCM at temperature T_m is infinite. In the equivalent heat capacity method, it is assumed that the melting or solidification processes occur over a temperature range $(T_m - \Delta T, T_m + \Delta T)$ instead of at a single temperature T_m. For a multicomponent system, ΔT can be chosen based on the range of phase-change temperature. For a single-component with well-defined melting point, ΔT should be as small as possible. Also, the latent heat is converted to an equivalent heat capacity of the PCM in the assumed temperature range. Thus, the specific heat of the PCM can be expressed as

$$c_p(T) = \begin{cases} c_{ps}, & T < T_m - \Delta T \\ \dfrac{h_{s\ell}}{2\Delta T} + \dfrac{c_{ps} + c_{p\ell}}{2}, & T_m - \Delta T < T < T_m + \Delta T \\ c_{p\ell}, & T > T_m + \Delta T \end{cases} \tag{5.302}$$

which assumes that the temperature of the PCM is changed from $T_m - \Delta T$ to $T_m + \Delta T$ when latent heat, $h_{s\ell}$, is absorbed by the PCM during melting. During the solidification process, the PCM releases the latent heat and its temperature decreases from $T_m + \Delta T$ to $T_m - \Delta T$. The equivalent specific heat in the mushy zone ($T_m - \Delta T < T < T_m + \Delta T$) includes the effect of both latent heat (the first term) and sensible heat (the second term).

The relationship between specific heat and temperature in the equivalent heat capacity method is plotted in Fig. 5.24. For a three-dimensional conduction-controlled melting/solidification problem in the Cartesian coordinate system, the energy equation for the entire region of the PCM can be expressed as

$$\rho c_p \left(\frac{\partial T}{\partial t} + u\frac{\partial T}{\partial x} + v\frac{\partial T}{\partial y} + w\frac{\partial T}{\partial z} \right) = \nabla \cdot (k \nabla T) \tag{5.303}$$

where the thermal conductivity k is a function of temperature, T. Assuming that the thermal conductivity of the PCM in the two-phase region is a linear function of temperature, one obtains

Fig. 5.24 Dependence of specific heat to temperature for equivalent heat capacity model

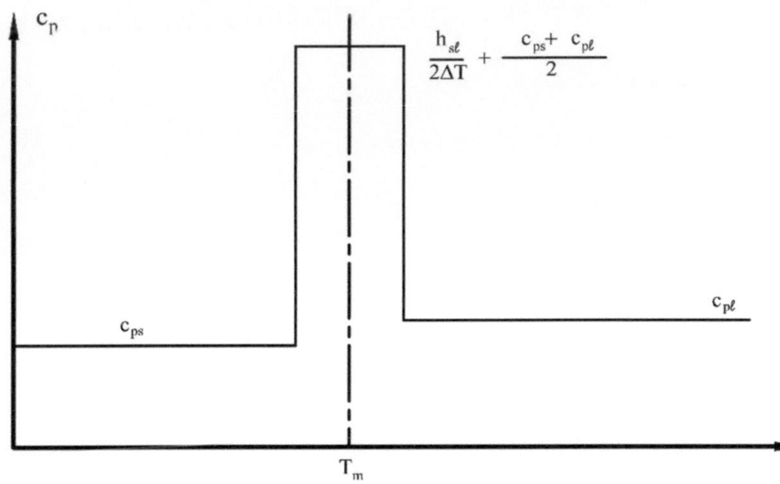

$$k(T) = \begin{cases} k_s, & T < T_m - \Delta T \\ k_s + \frac{k_\ell - k_s}{2\Delta T}(T - T_m + \Delta T), & T_m - \Delta T < T < T_m + \Delta T \\ k_\ell, & T > T_m + \Delta T \end{cases} \tag{5.304}$$

The advantage of the equivalent heat capacity model is its simplicity. Equation (5.303) is simply the nonlinear heat conduction equation, and it appears that a conventional computational methodology for conduction problems is adequate for solving a solid–liquid phase-change problem. However, many studies have revealed difficulties in the selection of time step, Δt, grid size, (Δx, Δy, Δz) and the phase-change temperature range, ΔT. If these variables cannot be properly selected, the predicted location of the solid–liquid interface and the temperature may include some unrealistic oscillation. Therefore, although the equivalent heat capacity model leads to simple code development, it is not used as widely as the enthalpy model.

5.6.4 Temperature-Transforming Model

The temperature-transforming model proposed by Cao and Faghri (1990) combines the advantages of the enthalpy and equivalent heat capacity models. For a three-dimensional conduction-controlled phase-change problem, the governing equation in enthalpy form is

$$\frac{\partial \rho h}{\partial t} + \frac{\partial(\rho u h)}{\partial x} + \frac{\partial(\rho v h)}{\partial y} + \frac{\partial \rho w h}{\partial z} = \nabla \cdot (k \nabla T) \tag{5.305}$$

For a phase change occurring over a temperature range ($T_m - \Delta T$, $T_m + \Delta T$) with the specific heats assumed to be constant for each phase, the relationship between enthalpy and temperature can be plotted as in Fig. 5.25. This relationship can be analytically expressed as (see problem 5.33)

$$h(T) = \begin{cases} c_{ps}(T - T_m) + c_{ps}\Delta T, & T < T_m - \Delta T \\ \left(\frac{c_{ps} + c_{p\ell}}{2} + \frac{h_{s\ell}}{2\Delta T}\right)(T - T_m) + \frac{c_{ps} + c_{p\ell}}{2}\Delta T + \frac{h_{s\ell}}{2}, & T_m - \Delta T < T < T_m + \Delta T \\ c_{p\ell}(T - T_m) + c_{ps}\Delta T + h_{s\ell}, & T > T_m + \Delta T \end{cases} \tag{5.306}$$

By defining specific heat in the mushy zone as

$$c_m = \frac{c_{ps} + c_{p\ell}}{2} \tag{5.307}$$

Equation (5.306) becomes

$$h(T) = \begin{cases} c_{ps}(T - T_m) + c_{ps}\Delta T, & T < T_m - \Delta T \\ \left(c_m + \dfrac{h_{s\ell}}{2\Delta T}\right)(T - T_m) + c_m\Delta T + \dfrac{h_{s\ell}}{2}, & T_m - \Delta T < T < T_m + \Delta T \\ c_{p\ell}(T - T_m) + c_{ps}\Delta T + h_{s\ell}, & T > T_m + \Delta T \end{cases} \tag{5.308}$$

which can be rewritten as

$$h(T) = c_p(T)(T - T_m) + b(T) \tag{5.309}$$

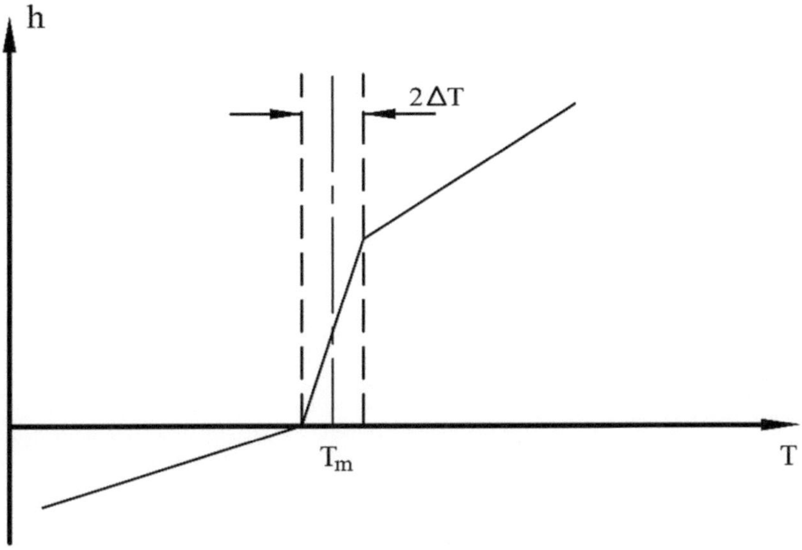

Fig. 5.25 Dependence of enthalpy on temperature for phase change occurring over a range of temperature

where $c_p(T)$ and $S(T)$ can be determined from Eq. (5.308):

$$c_p(T) = \begin{cases} c_{ps}, & T < T_m - \Delta T \\ c_m + \dfrac{h_{s\ell}}{2\Delta T}, & T_m - \Delta T < T < T_m + \Delta T \\ c_{p\ell}, & T > T_m + \Delta T \end{cases} \tag{5.310}$$

$$b(T) = \begin{cases} c_{ps}\Delta T, & T < T_m - \Delta T \\ c_m \Delta T + \dfrac{h_{s\ell}}{2}, & T_m - \Delta T < T < T_m + \Delta T \\ c_{ps}\Delta T + h_{s\ell}, & T > T_m + \Delta T \end{cases} \tag{5.311}$$

Substituting Eq. (5.309) into Eq. (5.305) yields

$$\frac{\partial(\rho c_p T)}{\partial t} + \frac{\partial(\rho u c_p T)}{\partial x} + \frac{\partial(\rho v c_p T)}{\partial y} + \frac{\partial(\rho w c_p T)}{\partial z} = \nabla \cdot (k \nabla T) + B \tag{5.312}$$

where

$$B = -\left[\frac{\partial(\rho b)}{\partial t} + \frac{\partial(\rho u b)}{\partial x} + \frac{\partial(\rho v b)}{\partial y} + \frac{\partial(\rho w b)}{\partial z}\right] \tag{5.313}$$

The thermal conductivity, k, is a function of temperature and can be obtained by Eq. (5.304). The energy equation has been transformed into a nonlinear equation with a single dependent variable T. The nonlinearity of the phase-change problem results from the large amount of heat released or absorbed at the solid–liquid interface. Also, in either the liquid or solid phases at some distance away from the interface, Eq. (5.312) reduces to a linear equation.

The temperature-transforming model and the equivalent heat capacity model differ in significant ways. Comparing Eqs. (5.312) and (5.303), one can conclude that the equivalent heat capacity model is a special case of Eq. (5.312), with $B = 0$ and c_p independent of the spatial variables x, y, z, and

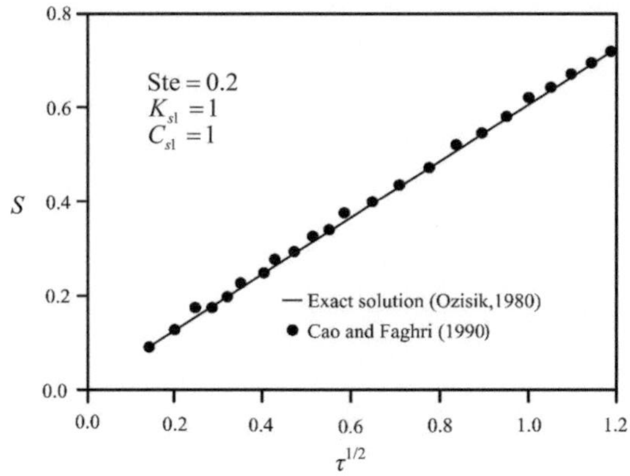

Fig. 5.26 Comparison of interface location for the Stefan problem obtained by exact and numerical solutions using temperature-transforming model

time. This is the underlying reason why many studies using the equivalent heat capacity model have encountered difficulty in selecting the grid size and time step and have often produced physically unrealistic oscillatory results. In order to satisfy $\partial b/\partial t = 0$ and $\partial(c_{\mathrm{p}}T)/\partial t = c_{\mathrm{p}}\partial T/\partial t$ the time step has to be small enough to assure that c_{p} and b are independent of both time and space variables—a difficult criterion to satisfy.

Equation (5.312) can be solved by various numerical methods, including the finite volume approach by Patankar (1980). To verify the accuracy of the temperature-transforming model, a numerical solution of the Stefan problem is presented here. The physical model of the Stefan problem has been given in the previous section. Figure 5.26 shows a comparison between the exact solution and the numerical solution using the temperature-transforming model for the dimensionless location of the melting front as a function of the square root of dimensionless time. The definition of dimensionless melting front location and dimensionless time is Eq. (5.34).

The temperature-transforming model eliminates the time step and grid-size limitations and is insensitive to the phase-change temperature range. Therefore, the model can properly handle phase change occurring over a temperature range (as in the case of multicomponent PCM phase change) or at a single temperature (as in the case of single-component PCM phase change). The temperature-transforming model has been used to solve many different solid–liquid phase-change problems—including melting in a latent heat thermal energy storage system (Zhang and Faghri 1996b), convection controlled problems (Ma and Zhang 2006), selective laser sintering (Xiao and Zhang 2007), and solidification in porous media (Damronglerd et al. 2012).

Problems

5.1. A horizontal cylinder is embedded in infinite solid paraffin. The initial temperature of the paraffin is equal to its melting temperature and the surface temperature of the cylinder is greater than the melting point of paraffin. The melting occurs around the cylinder. The three possible shapes of solid–liquid interfaces are given in Fig. P5.1. Discuss which one should form and for what reason. If the PCM becomes ice (the surface temperature of the cylinder is less than 4 °C), what is your answer?

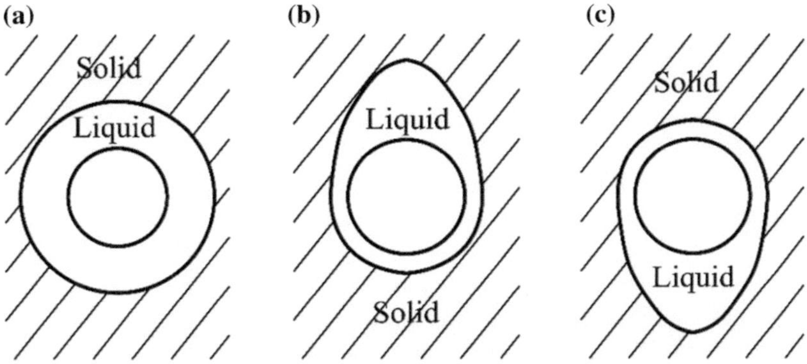

Fig. P5.1

5.2. A horizontal cylinder is embedded in liquid paraffin with a temperature above its melting point. From time $t = 0$, the surface temperature of the cylinder is suddenly decreased to a temperature below the melting point, and then the solidification process occurs around the cylinder. Discuss the effect of natural convection on the solid–liquid interface shape.

5.3. In the thermal energy storage system that is shown in Fig. P5.3, the PCM is n-Octadecane (the thermal conductivity of solid is 0.358 W/m-K) and the transfer fluid is water. The liquid with a temperature below the melting point flows through the tube. The solidification process occurs in the annulus. To enhance the heat transfer in the system, one engineer suggests installing longitudinal fins on the PCM side. Others suggest installing longitudinal fins on the fluid side. In your opinion, who is right?

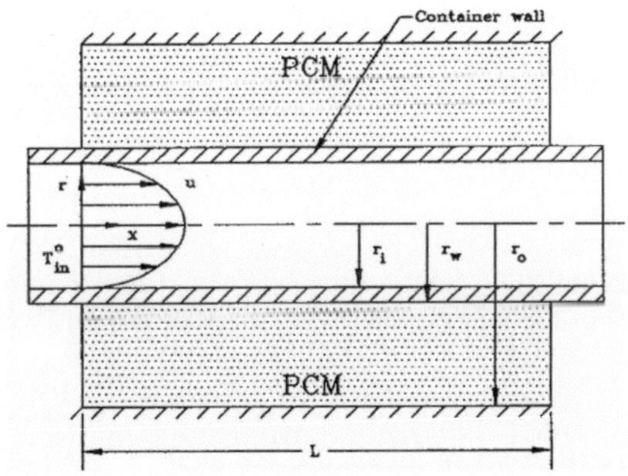

Fig. P5.3

5.4. The energy balance at the solid–liquid interface for a melting problem given by Eq. (5.8) is obtained by assuming that Fourier's conduction model is valid. When the temperature gradient is very high or the duration of heating is extremely short, Fourier's conduction model is no longer valid. Instead, the heat flux and temperature gradient are related by

$$q''(x,t) + \tau \frac{\partial q''(x,t)}{\partial t} = -k \frac{\partial T(x,t)}{\partial x}$$

where τ is the thermal relaxation time. Apply the above equation for heat flux in both solid and liquid phases and derive the energy balance at the interface. The thermal relaxation times for both phases can be assumed to be the same.

5.5. When heat flux and temperature gradient are related by the equation of the previous problem, there may be a discontinuity of temperature at the solid–liquid interface (see Fig. P5.5). Modify the energy balance that you obtained in Problem 5.4 to account for the contribution of sensible heat to the interfacial energy balance.

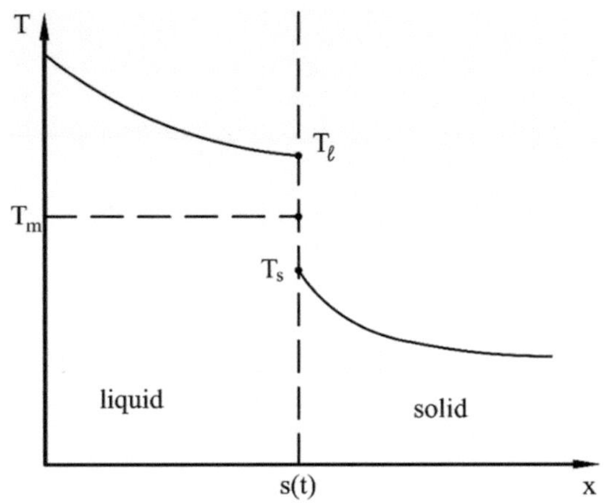

Fig. P5.5

5.6. For a two-dimensional melting/solidification problem, the energy balance equation at the interface is

$$k_s \frac{\partial T_s(x,y,t)}{\partial n} - k_\ell \frac{\partial T_\ell(x,y,t)}{\partial n} = \rho h_{s\ell} v_n \qquad x = s(y,t)$$

The above equation is not convenient for the analytical solution or numerical solution because it contains derivatives of the temperature along the **n**-direction. Prove that the above equation can be rewritten as

$$\left[1 + \left(\frac{\partial s}{\partial y}\right)^2\right]\left[k_s \frac{\partial T_s}{\partial x} - k_\ell \frac{\partial T_\ell}{\partial x}\right] = \rho h_{s\ell} \frac{\partial s}{\partial t} \qquad x = s(y,t)$$

5.7. A solid PCM with a uniform initial temperature T_i below the melting point T_m is in a half-space $x > 0$. At time $t = 0$, the temperature at the boundary $x = 0$ is suddenly increased to a temperature T_0 above the melting point of the PCM. The melting process occurs from the time $t = 0$. The densities of the PCM for both phases are assumed to be the same and the natural convection in the liquid phase can be neglected. Show that Eqs. (5.35)–(5.41), which are nondimensional governing equations for solidification in a semi-infinite region, are also valid for this problem provided that subscripts 1 and 2 represent liquid and solid, respectively.

5.8. The governing equation of the one-region phase-change problem discussed in Sect. 5.3.3, Eq. (5.42) can also be reduced to an ordinary differential equation by introducing a similarity variable: $\eta = X/(2\sqrt{\tau})$, which is referred to as the similarity solution. Obtain the exact solution of the one-region problem using the similarity solution.

5.9. A solid PCM with a uniform initial temperature equal to the melting point T_m is in a half-space $x > 0$. At time $t = 0$, the surface of the semi-infinite body is exposed to a fluid at temperature T_∞. The convective heat transfer coefficient between the PCM and the fluid is h. Assume that the densities of the PCM for both phases are the same and that natural convection in the liquid phase is negligible. The temperature distribution in the liquid phase can be constructed based on the solution of heat conduction in a semi-infinite region subject to the boundary condition of the third kind. Find the transient location of the solid–liquid interface.

5.10. To solve heat conduction in a semi-infinite body with a convective boundary condition by using the integral solution, the temperature distribution in the thermal penetration depth must be determined. If it can be assumed to be a second-order polynomial function, determine the unknown constants in the polynomial function.

5.11. Find thermal penetration depth for heat conduction in a semi-infinite body with convective boundary condition. The temperature distribution in the thermal penetration depth can be assumed to be a second-order polynomial function.

5.12. Solve for melting in a semi-infinite body with convective heating. The temperature distribution in the liquid phase can be assumed to be a second-order polynomial function.

5.13. Solve for melting in a subcooled semi-infinite solid with constant wall temperature by using the integral approximate method. The temperature distribution in the liquid phase can be assumed to be a linear distribution. The temperature distribution in the solid phase can be assumed to be a second-order polynomial function.

5.14. The dimensionless governing equations of melting in a subcooled semi-infinite solid with constant heat flux heating were given in Eqs. (5.145)–(5.151). Solve the problem using a semi-exact solution: exact solution in the liquid (see Example 5.1) and integral solution in the solid (see Sect. 5.4.3).

5.15. You are given a one-dimensional heat sink to analyze. It works by melting a block of encased wax. The block of wax (PCM) is one centimeter thick. The heat sink operates by keeping one wall at 10 °C above the wax melt temperature. Assume that the solid PCM is uniformly at the melting point at time $t = 0$. The heat sink is used up when all of the PCM is liquid. The wax has the following properties: $T_m = 0°$ C, $c_p = 1$ J/kg-K, $h_{s\ell} = 100$ J/kg, In calculating thermal capacity, what error would result if only the latent heat—and not also the sensible heat rise—is accounted for?

5.16. Since the surface temperature of an ablation material is very high, the surface is subjected to convective and radiation cooling. Modify the energy balance at the surface of the ablating material—Eq. (5.175)—and obtain the ablation velocity with the convective and radiation cooling effects considered.

5.17. The surface heat flux in the ablation problem discussed in Sect. 5.4.4 was assumed as constant, and only steady-state solution was discussed. The heat flux during ablation is usually a function of time. A finite slab with thickness L and an initial temperature of T_i ($T_i < T_m$) is subject to a variable heat flux $q_0''(t) = Ae^t$ at the surface $x = 0$. The surface $x = L$ can be assumed to be adiabatic. Since the initial temperature of the slab is below the melting point, it must be preheated before ablation begins. Analyze the preheating and ablation problem using integral approximate method. It is assumed that the thermal penetration depth in the ablating material never reaches to $x = L$ so the ablating material can be treated as a semi-infinite solid. The temperature distribution in the solid can be assumed as the second-degree polynomial function.

5.18. Derive the quasi-steady formulation for the temperature of the ablative material and substrate when a constant heat flux q''_w is applied to the surface of the ablative material, as shown in Fig. P5.18. Develop the governing equations, including the boundary conditions and initial conditions needed. The two materials in this problem are assumed to be homogenous, isotropic, and in good contact with each other. The inner surface is assumed to be adiabatic. The temperature distribution is initially uniform and then exposed to a constant external heat flux on the outer surface of the ablative material. A phase change occurs at the outer surface, and the new material is immediately removed upon formation. Find the ablation rate for the limiting case of infinite thickness of the ablation material at steady state.

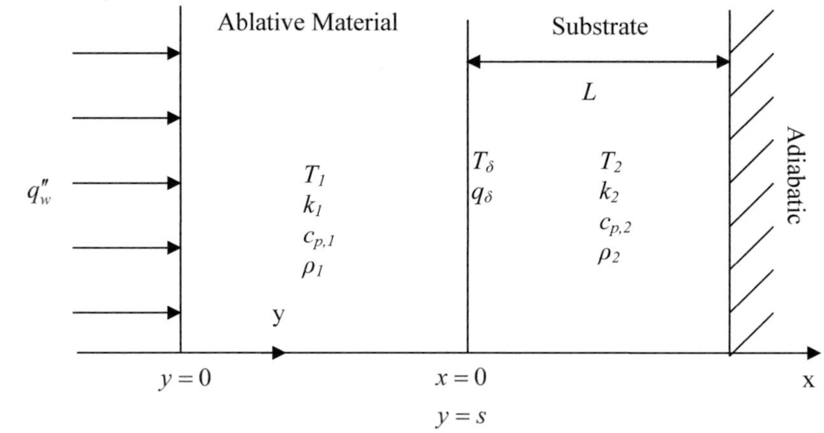

Fig. P5.18

5.19. A horizontal cylinder is embedded in solid paraffin with a temperature equal to the melting point of paraffin. From time $t = 0$, a constant heat flux is applied to the inner surface of the tube so that melting occurs around the tube. Assume that the thickness of the tube is negligible. Solve the melting problem using the integral approximate method.

5.20. The physical model of melting in an eccentric annulus is shown in Fig. P5.20. The PCM fills the eccentric annulus with outer radius r_o, inner radius r_i, and eccentricity e. The initial temperature of the PCM is assumed to be at its freezing temperature, T_m. At the beginning, $t = 0$, the temperatures of the outer and inner walls of the eccentric annulus are suddenly reduced to a temperature T_c, below the freezing temperature. The freezing process will occur simultaneously at the outer and inner walls of the eccentric annulus. It is assumed that the freezing at the inside of the outer wall and at the outside of the inner wall can be solved independently. Before the freezing fronts meet, the freezing volume is the sum of the volumes at the inside of the outer wall and the outside of the inner wall. After the freezing fronts meet, the freezing rate will decrease since the freezing process stopped at the locations where the freezing fronts met. Therefore, the freezing rate needs to be calculated for two different stages: before and after the freezing fronts meet. The radius of the inner and outer freezing fronts can be determined by an integral solution. Obtain the rate of freezing as a function of time.

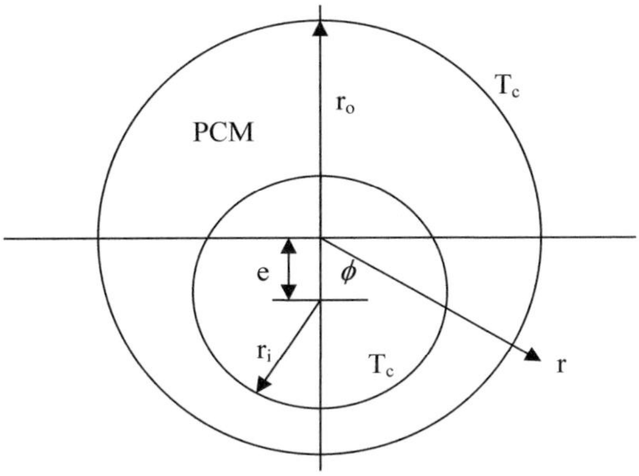

Fig. P5.20

5.21. In a shell-and-tube latent heat thermal energy storage system, as shown in Fig. P5.3, the solid PCM at melting point fills in the shell side, while the transfer fluid flows inside the tube. It is assumed that (1) the inlet velocity of the transfer fluid is fully developed, but the heat transfer on the tube side is in the thermal entry region, (2) the quasi-steady assumption is applicable to convective heat transfer inside the tube, (3) the axial conductions in both the tube side and shell side are assumed to be negligible, and (4) the tube wall is assumed to be very thin so that its heat capacity and axial conduction can be neglected. Find the transient location of the melting front before it reaches the adiabatic wall of the thermal energy storage device by analyzing forced convective heat transfer inside the tube and melting on the shell side.

5.22. A line heat source with intensity of q' (W/m) is located at $r = 0$ in an infinite solid as shown in Fig. P5.22. The initial temperature of the solid, T_i, is below the melting point, T_m, so that this is a two-region melting problem. Develop the governing equations of the problem and obtain the transient location of the melting front. The temperature distributions in the liquid and solid phases can be constructed using an exponential integral function in the form of $\mathrm{Ei}[-r^2/(4\alpha t)]$, where $Ei(z) = \int_z^{\infty} \frac{e^{-z}}{z} dz$.

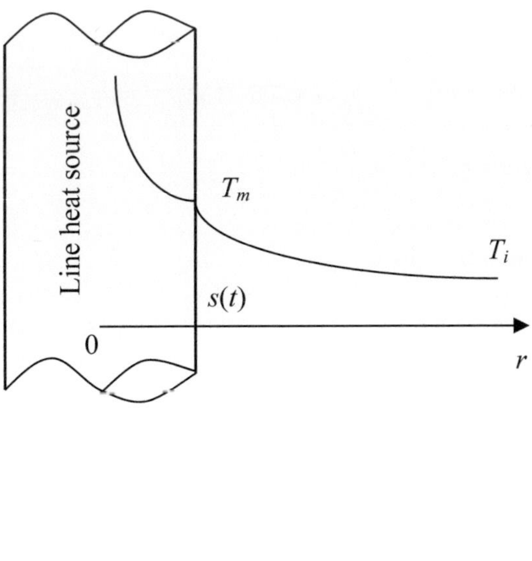

Fig. P5.22

5.23. One-region melting in a semi-infinite body with constant surface temperature can be described by Eqs. (5.42)–(5.45). In a quasi-stationary approximate solution, the derivative of the temperature in the liquid phase with respect to time is assumed to be zero ($\partial\theta/\partial\tau = 0$). Consequently, the energy Eq. (5.42) is simplified as

$$\frac{d^2\theta}{dX^2} = 0 \qquad 0 < X < S(\tau), \quad \tau > 0$$

Solve the melting problem using quasi-stationary approximate solution and compare your results with integral solution for Ste = 0.1, 0.2, and 0.4.

5.24. A very long cylindrical biological material with a radius of r_0 is placed in a freezer at temperature T_0, which is assumed to be the surface temperature of the cylinder during freezing. Assume that the biological material can be treated as single-component PCM with a well-defined melting point, T_m, and that its initial temperature is at T_m. The freezing process is so slow that heat transfer in a frozen layer can be regarded as a quasi-steady-state process. Estimate the time it takes to freeze the entire cylinder.

5.25. Solve melting around a hollow sphere with boundary condition of the first kind by using the quasi-stationary method. The initial temperature of the PCM is assumed to be equal to the melting point.

5.26. A liquid PCM with initial temperature T_i ($T_i > T_m$) enters a circular tube with diameter D and wall temperature T_w ($T_w < T_m$, see Fig. P5.26). The liquid PCM is cooled in the tube and solidification begins at $x = L_s$. While the diameter of the liquid channel shrinks as a result of solidification, the Nusselt number is assumed to be a constant before and after solidification begins. Assuming that axial conduction in the liquid and solid PCM are negligible and that the solidification inside the tube is steady-state, obtain the length at which solidification begins and the thickness of the solid layer. The properties of the liquid and solid PCM can be assumed to be the same.

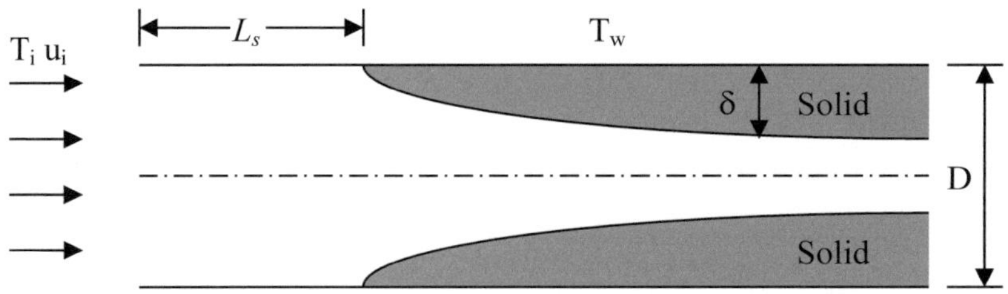

Fig. P5.26

5.27. The liquidus 1 in the phase diagram, Fig. 5.16, can be approximated by a $\omega(T) = 1.678 \times 10^{-3} - 1.602 \times 10^{-2}T - 2.857 \times 10^{-4}T^2 - 4.491 \times 10^{-6}T^3$, where the unit of temperature is °C. The local temperature and concentration of NH_4Cl at a point in the mushy zone formed by solidification are T_m = −10 °C and $\omega\, m$ = 10%, respectively. What is the local solid fraction,f, (i.e., mass fraction of the solid) in the mushy zone?

5.28. An NH_4Cl–H_2O solution is used as a phase-change material (PCM) in a cold storage system designed to be operated at −10 °C. In order to maintain the mobility of the mushy PCM, the maximum allowable solid fraction is 0.5. What is the appropriate concentration of NH_4Cl?

5.29. Show that the energy equation for the enthalpy model, Eq. (5.282), satisfies the energy equations for both the liquid and solid phases, Eqs. (5.20) and (5.22), as well as, energy balance, Eq. (5.26).

5.30. In the enthalpy method, we obtained an enthalpy equation for the numerical solution:

$$h_j^{n+1} = \frac{\Delta t}{\rho\,(\Delta x)^2}\left(k_{j+\frac{1}{2}}T_{j+1}^n + k_{j-\frac{1}{2}}T_{j-1}^n\right) + \left(h_j^n - \frac{\Delta t\left(k_{j+\frac{1}{2}} + k_{j-\frac{1}{2}}\right)}{\rho\,(\Delta x)^2}T_j^n\right), \quad j = 2,,N-1$$

Prove that the above equation is stable only if the following equation is satisfied:

$$\frac{max(\alpha_s, \alpha_\ell)\Delta t}{(\Delta x)^2} \le \frac{1}{2}$$

5.31. Phase change of a PCM occurs over a range of temperatures ($T_m - \Delta T$, $T_m + \Delta T$). A finite slab of this PCM with thickness L has an initial temperature T_i, which is below $T_m - \Delta T$. At time $t = 0$, the temperature of the left surface suddenly increases to $T_0 > T_m + \Delta T$, while the right surface remains insulated. Give the governing equations using enthalpy method and design the solution procedure.

5.32. You are given a one-dimensional melting problem that you need to solve with an explicit numerical scheme. The phase-change material is n-Octadecane.

(a) Given a time step of 0.1 s, find the minimum Δx necessary for stable numerical solution.
(b) Given a Δx of 1 mm, find the minimum time step necessary for stable numerical solution.
(c) What is the requirement for numerical simulation using the implicit method?

5.33. Show that the enthalpy–temperature relation in Fig. 5.25 can be represented analytically by Eq. (5.306).

5.34. Most materials that are used as solid-to-liquid PCMs have poor thermal conductivity. Describe at least three ways of increasing thermal conductivity to melt the material located "far" from the heat source.

5.35. In a convection-controlled melting or solidification process, the heat transfer coefficient is a function of the thermal properties of the PCMs, a characteristic length L, the temperature difference ΔT, the buoyancy force $g\beta\Delta T$, and the latent heat of fusion $h_{s\ell}$:

$$h = h(k, \mu, c_p, \rho, L, \Delta T, g\beta\Delta T, h_{s\ell}, t)$$

where t is time. Using Buckingham's Π theorem shows that the number of nondimensional variables is 5, as opposed to the 10 dimensional variables in the above equation. Furthermore, use dimensional analysis to show that

$$Nu = f(Gr, \Pr, Ste, Fo)$$

where Gr and Fo are Grashof and Fourier numbers.

References

Bardsley, W., Hurle, D. T. J., & Mullin, T. B. (1979). *Crystal growth: A tutorial approach*. Amsterdam: North-Holland.

Beckerman, C. & Viskanta, R. (1988). An experimental study of solidification of binary mixture with double diffusion in the liquid. In *Proceedings of the 1988 National Heat Transfer Conference*, vol. 3, pp. 67–79.

Beckermann, C., & Wang, C. Y. (1995). multi-phase/-scale modeling of transport phenomena in alloy solidification. *Annual Review of Heat Transfer, 6,* 115–198.

Bejan, A. (1994). Contact melting heat transfer and lubrication. *Advances in Heat Transfer, 24,* 1–38.

Bejan, A., 2013, *Convection heat transfer* (4th ed.). New York: Wiley Inc.

Bonacina, C., Comini, G., Fasano, A., & Primicerio, A. (1973). Numerical solution of phase change problems. *International Journal of Heat and Mass Transfer, 16,* 1285–1832.

Braga, S. L., & Viskanta, R. (1990). Solidification of a binary solution on a cold isothermal surface. *International Journal of Heat and Mass Transfer, 33,* 745–754.

Budhia, H., & Kreith, F. (1973). Heat transfer with melting or freezing in a wedge. *International Journal of Heat and Mass Transfer, 16,* 195–211.

Cao, Y., & Faghri, A. (1990). A numerical analysis of phase change problem including natural convection. *Journal of Heat Transfer, 112,* 812–815.

Cao, W. Z., & Poulikakos, D. (1991). Freezing of a binary alloy saturating a packed bed of spheres. *Journal of Thermophysics and Heat Transfer, 5,* 46–53.

Cho, S. H., & Sunderland, J. E. (1981). Approximate temperature distribution for phase change of semi-infinite body. *Journal of Heat Transfer, 103,* 401–403.

Crank, J. (1984). *Free and moving boundary problems*. Oxford: Clarendon Press.

Damronglerd, P., Zhang, Y., & Yang, M. (2012). Numerical simulation of solidification of liquid copper saturated in porous structures fabricated by sintered steel particles. *International Journal of Numerical Methods for Heat and Fluid Flow, 22*(1), 94–111.

Dong, Z. F., Chen, Z. Q., Wang, Q. J., & Ebadian, M. A. (1991). Experimental and analytical study of contact melting in a rectangular cavity. *Journal of Thermophysics and Heat Transfer, 5,* 347–354.

Eckert, E. R. G. & Drake, R. M. (1987). *Analysis of heat and mass transfer.* Washington, DC: Hemisphere.

El-Genk, M. S., & Cronenberg, A. W. (1979). Solidification in a semi-infinite region with boundary condition of the second kind: an exact solution. *Letters in Heat and Mass Transfer, 6,* 321–327.

Faghri, A., Zhang, Y., & Howell, J. R. (2010). *Advanced heat and mass transfer*. Columbia, MO: Global Digital Press.

Fang, L. J., Cheung, F. B., Linehan, J. H., & Pedersen, D. R. (1984). Selective freezing of a dilute salt solution on a cold ice surface. *Journal of Heat Transfer, 106,* 385–393.

Goodman, T. R. (1958). The heat-balance integral and its application to problems involving a change of phase. *Transactions of ASME, 80,* 335–342.

Hirata, T., Makino, Y., & Kaneko, Y. (1991). Analysis of close-contact melting for octadecane and ice inside isothermally heated horizontal rectangular capsule. *International Journal of Heat and Mass Transfer, 34,* 3097–3106.

Hobbs, P. V. (1974). *Ice physics*. Oxford: Clarendon Press.

Ma, Z. H., & Zhang, Y. (2006). Solid velocity correction schemes for a temperature transforming model for convection phase change. *International Journal of Numerical Methods for Heat and Fluid Flow, 16*(2), 204–225.

Ozisik, M. N. (1993). *Heat conduction* (2nd ed.). New York: Wiley-Interscience.

Patankar, S. V. (1980). *Numerical heat transfer and fluid flow*. New York: McGraw-Hill.

Poots, G. (1962). An approximate treatment of heat conduction problem involving a two-dimensional solidification front. *International Journal of Heat and Mass Transfer, 5,* 339–348.

Rathjen, K. A., & Jiji, L. M. (1971). Heat conduction with melting or freezing in a corner. *Journal of Heat Transfer, 93,* 101–109.

Shamsunder, N., & Sparrow, E. M. (1975). Analysis of multidimensional conduction phase change via the enthalpy model. *Journal of Heat Transfer, 97,* 333–340.

Sultana, K. R., Dehghani, S. R., Pope, K., & Muzychka, Y. S. (2018). Numerical techniques for solving solidification and melting phase change problems. *Numer. Heat Transfer, B, 73*(3), 129–145.

Tien, R. H., & Geiger, G. E. (1967). A heat-transfer analysis of the solidification of a binary eutectic system. *Journal of Heat Transfer, 91,* 230–234.

Viskanta, R. (1988). Heat transfer during melting and solidification of metals. *Journal of Heat Transfer, 110,* 1205–1219.

Voller, V. R. (1997). An overview of numerical methods for solving phase change problems. In W. J. Minkowycz, & E. M. Sparrow (Eds.), *Advances in numerical heat transfer* (vol. 1). Basingstoke: Taylor & Francis.

Voller, V. R., Cross, M., & Markatos, N. C. (1987). An enthalpy method for convection/diffusion phase change. *International Journal for Numerical Methods in Engineering, 24,* 271–284.

Wang, Q. J., Wang, C., & Chen, Z. Q. (1991). Heat transfer during superheated solidification around bare tube and finned tubes. In *Experimental heat transfer, fluid mechanics and thermodynamics 1991* (pp. 1171–1176). Elsevier Science Publishing Co., Inc.

Xiao, B., & Zhang, Y. (2007). Laser sintering of metal powders on top of sintered layers under multiple-line laser scanning. *Journal of Physics. D. Applied Physics, 40*(21), 6725–6734.

Zeng, X., & Faghri, A. (1994). Temperature-transforming model for binary solid-liquid phase-change problems part I: Physical and numerical scheme. *Numerical Heat Transfer Part B, 25,* 467–480.

Zhang, Y., Chen, Z. Q., & Faghri, A. (1997). Heat transfer during solidification around a horizontal tube with internal convective cooling. *Journal of Solar Energy Engineering, 119,* 44–47.

Zhang, Y., Chen, Z., & Wang, Q. (1993). Analytical solution of melting in a subcooled semi-infinite sold with boundary conditions of the second kind. *Journal of Thermal Science, 2,* 111–115.

Zhang, Y., & Faghri, A. (1996a). Semi-analytical solution of thermal energy storage system with conjugate laminar forced convection. *International Journal of Heat and Mass Transfer, 39,* 717–724.

Zhang, Y., & Faghri, A. (1996b). Heat transfer enhancement in latent heat thermal energy storage system by using an external radial finned tube. *Journal of Enhanced Heat Transfer, 3,* 119–127.

Zhang, Y., & Faghri, A. (1998). A thermal model for mushy zone formation in binary solutions. *Journal of Solar Energy Engineering, 120,* 144–147.

6 Sublimation and Vapor Deposition

6.1 Introduction

When the pressure and temperature of ice are above the triple point pressure and temperature of the water is heated, melting occurs as discussed in Chap. 5. However, when the ice is exposed to moist air with a partial pressure of water below its triple point pressure, heating of the ice will result in a phase change from ice directly to vapor without first going through the liquid phase. Spacecrafts and space suits can reject heat by sublimating ice into the vacuum of space. Another application for sublimation of ice is the preparation of specimens using freeze-drying for a scanning electron microscope (SEM) or a transmission electron microscope (TEM). This type of phase change is referred to as sublimation. The opposite process is deposition, which describes the process of vapor changing directly to solid without going through the condensation and freezing. The phase-change processes related to solids can be illustrated by a phase diagram in Fig. 6.1. Sublimation and deposition will be the subjects of this chapter.

When a subcooled solid is exposed to its superheated vapor, as shown in Fig. 6.2a, the vapor phase temperature is above the temperature of the solid–vapor interface and the temperature in the solid is below the interfacial temperature. The boundary condition at the solid–vapor interface is

$$k_{\mathrm{s}} \frac{\partial T_{\mathrm{s}}}{\partial x} - h_{\delta}(T_{\infty} - T_{\delta}) = \rho_{\mathrm{s}} h_{\mathrm{sv}} \frac{\mathrm{d}\delta}{\mathrm{d}t} \tag{6.1}$$

where h_δ is the convective heat transfer coefficient at the solid–vapor interface, h_{sv} is the latent heat of sublimation, and δ is the thickness of the sublimable or deposited material. The interfacial velocity $\mathrm{d}\delta/\mathrm{d}t$ in Eq. (6.1) can be either positive or negative, depending on the direction of the overall heat flux at the interface. While a negative interfacial velocity signifies sublimation, a positive interfacial velocity signifies deposition. When the vapor phase is superheated, as shown in Fig. 6.2a, the solid–vapor interface is usually smooth and stable.

In another possible scenario, as shown in Fig. 6.2b, the solid temperature is above the interfacial temperature, and the vapor phase is supercooled. The interfacial energy balance for this case can still be described by Eq. (6.1). Depending on the degrees of superheat in the solid phase and supercooling in the vapor phase [the relative magnitude of the first and second terms in Eq. (6.1)], both sublimation and deposition are possible. During sublimation, a smooth and stable interface can be obtained. During deposition, on the other hand, the interface is dendritic and not stable, because supercooled vapor is not stable. The solid formed by deposition of supercooled vapor has a porous structure.

A. Faghri and Y. Zhang, *Fundamentals of Multiphase Heat Transfer and Flow*, https://doi.org/10.1007/978-3-030-22137-9_6

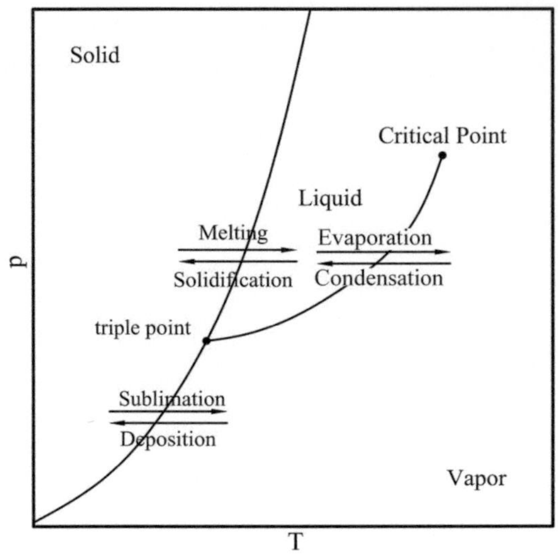

Fig. 6.1 Phase diagram for solid–liquid and solid–vapor phase change

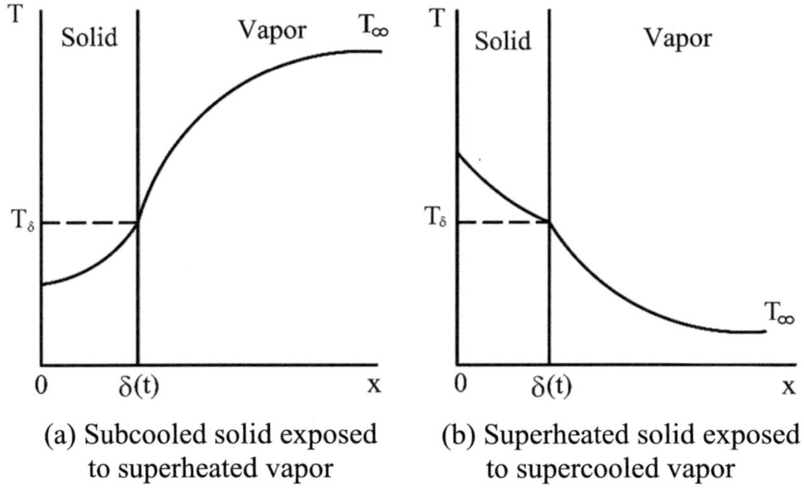

Fig. 6.2 Temperature distribution in sublimation and deposition

During sublimation or deposition, the latent heat of sublimation can be supplied from or absorbed by either the solid phase or the vapor phase, depending on the temperature distributions in both phases.

Naphthalene sublimation is also a technique whereby a heat transfer coefficient can be obtained through the measurement of a mass transfer coefficient and the analogy between heat and mass transfer (Eckert and Goldstein 1976). The significant advantages of this method include its high accuracy and the simplicity of the experimental apparatus. In addition, the local heat transfer coefficient can be obtained by measuring the local sublimed depth of the specimen.

Vapor deposition, which finds applications in coating and thermal manufacturing processes, is classified into two broad categories: physical vapor deposition (PVD) and chemical vapor deposition (CVD). PVD operates at a very low pressure and transports the species generated by one of two means:

(1) evaporation or (2) bombarding the target materials to the substrate through free molecular flow or transition flow. CVD, on the other hand, is a process in which material is formed on a substrate by chemical reaction of gaseous precursors using activation energy. The deposited film thickness can range from a few nanometers, as applied to optical coating, to tens of microns, as applied to wear-resistance coating (Jenson et al. 1991). Conventional CVD has been extensively investigated by many researchers, and a detailed literature review is given by Mahajan (1996). In a pyrolytic CVD process, the entire substrate is heated and vapor deposition occurs over the whole substrate. When a laser beam is used to heat the substrate, only a very small spot on the substrate is heated by the laser beam and vapor deposition occurs only on the heated spot. In this case, the activation energy is provided by the laser beam, and it is therefore referred to as laser chemical vapor deposition (LCVD). LCVD can also be based on chemical reactions initiated photolytically, which involves tuning the laser to an electrical or vibrational level of the gas (Bauerle 1996). The irradiated material decomposes, and the products deposit on the cooler substrate to form the solid film (Mazumder and Kar 1995).

Section 6.2 presents analytical solutions of sublimation over a flat plate in parallel flow and inside a tube; these problems are treated as a conjugated heat and mass transfer problem. Section 6.2 also includes a detailed analysis of a sublimation process with chemical reaction. Section 6.3 presents an in-depth discussion of CVD, including various CVD configurations, governing equations, transport properties, and LCVD.

6.2 Sublimation

6.2.1 Sublimation Over a Flat Plate

Sublimation over a flat plate can find its application in analogy between heat and mass transfer (Kurosaki 1973; Zhang et al. 1996). Figure 6.3 shows the physical model of the sublimation problem considered by Zhang et al. (1996). A flat plate is coated with a layer of sublimable material and is subject to constant heat flux heating underneath. A gas with the ambient temperature T_∞ and mass fraction of sublimable material ω_∞ flows over the flat plate at a velocity of u_∞. The heat flux applied from the bottom of the flat plate will be divided into two parts: one part is used to supply the latent heat of sublimation, and another part is transferred to the gas through convection. The sublimated vapor is injected into the boundary layer and is removed by the gas flow.

The following assumptions are made in order to solve the problem:

1. The flat plate is very thin, and so the thermal resistance of the flat plate can be neglected.
2. The gas is incompressible, with no internal heat source in the gas.
3. The sublimation problem is two-dimensional steady state.

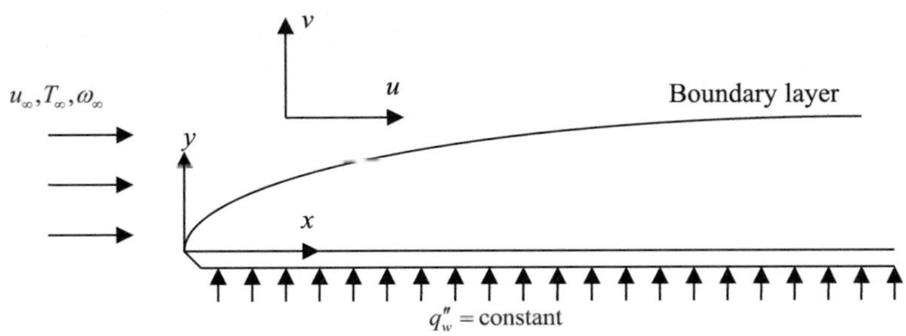

Fig. 6.3 Sublimation on a flat plate with constant heat flux

The governing conservation equations for mass, momentum, energy, and species of the problem are (Faghri et al. 2010):

$$\frac{\partial u}{\partial x}+\frac{\partial v}{\partial y}=0 \tag{6.2}$$

$$u\frac{\partial u}{\partial x}+v\frac{\partial u}{\partial y}=\nu\frac{\partial^2 u}{\partial y^2} \tag{6.3}$$

$$u\frac{\partial T}{\partial x}+v\frac{\partial T}{\partial y}=\alpha\frac{\partial^2 T}{\partial y^2} \tag{6.4}$$

$$u\frac{\partial \omega}{\partial x}+v\frac{\partial \omega}{\partial y}=D\frac{\partial^2 \omega}{\partial y^2} \tag{6.5}$$

Nonslip condition at the surface of the flat plate requires that

$$u=0,\quad y=0 \tag{6.6}$$

For a binary mixture that contains the vapor sublimable substance and gas, the molar flux of the sublimable substance at the surface of the flat plate is (see Eq. 1.61)

$$\dot{m}''=-\frac{\rho D}{1-\omega}\frac{\partial\omega}{\partial y},\quad y=0 \tag{6.7}$$

Since the mass fraction of the sublimable substance in the mixture is very low, i.e., $\omega \ll 1$, the mass flux at the wall can be simplified to

$$\dot{m}''=-\rho D\frac{\partial\omega}{\partial y},\quad y=0 \tag{6.8}$$

Sublimation at the surface causes a normal blowing velocity, $v_w=\dot{m}''/\rho$, at the surface. The normal velocity at the surface of the flat plate is therefore

$$v=v_{\mathrm{w}}=-D\left.\frac{\partial\omega}{\partial y}\right|_{y=0},\quad y=0 \tag{6.9}$$

The energy balance at the surface of the flat plate is

$$-k\frac{\partial T}{\partial y}-\rho h_{\mathrm{sv}}D\frac{\partial\omega}{\partial y}=q''_{\mathrm{w}},\quad y=0 \tag{6.10}$$

Another reasonable, practical, and representable boundary condition at the surface of the flat plate emerges by setting the mass fraction at the wall as the saturation mass fraction at the wall temperature. The mass fraction and the temperature at the surface of the flat plate have the following relationship (Kurosaki 1973, 1974):

$$\omega=aT+b,\quad y=0 \tag{6.11}$$

where a and b are constants that depend on the sublimable material and its temperature.

As $y \to \infty$, the boundary conditions are

$$u \to u_\infty, T \to T_\infty, \omega \to \omega_\infty \tag{6.12}$$

Introducing the stream function ψ,

$$u = \frac{\partial \psi}{\partial y} \quad v = -\frac{\partial \psi}{\partial x} \tag{6.13}$$

the continuity Eq. (6.2) is automatically satisfied, and the momentum equation in terms of the stream function becomes

$$\frac{\partial \psi}{\partial y}\frac{\partial^2 \psi}{\partial x \partial y} - \frac{\partial \psi}{\partial x}\frac{\partial^2 \psi}{\partial y^2} = v\frac{\partial^3 \psi}{\partial y^3} \tag{6.14}$$

A similarity solution for Eq. (6.14) does not exist unless the injection velocity $v_{\rm w}$ is proportional to $x^{1/2}$, and the incoming mass fraction of the sublimable substance, ω_∞, is equal to the saturation mass fraction corresponding to the incoming temperature T_∞ (Kurosaki 1974). Zhang et al. (1996) obtained a local nonsimilarity solution by defining the following similarity variables:

$$\begin{aligned} \xi &= \frac{x}{L}, \quad \eta = y\sqrt{\frac{u_\infty}{2vL\xi}}, \quad f = \frac{\psi}{\sqrt{2vu_\infty L\xi}} \\ \theta &= \frac{k(T - T_\infty)}{q''_{\rm w}\sqrt{2vL\xi/u_\infty}}, \qquad \varphi = \frac{\rho h_{\rm sv} D(\omega - \omega_\infty)}{q''_{\rm w}\sqrt{2vL\xi/u_\infty}} \end{aligned} \tag{6.15}$$

Figure 6.4 shows the typical dimensionless temperature and mass fraction profiles obtained by local nonsimilarity solution. The effect of injection velocity at the surface due to sublimation is reflected by

$$B = \frac{q''_{\rm w}}{\rho h_{\rm sv} v}\sqrt{\frac{2vL}{u_\infty}} \tag{6.16}$$

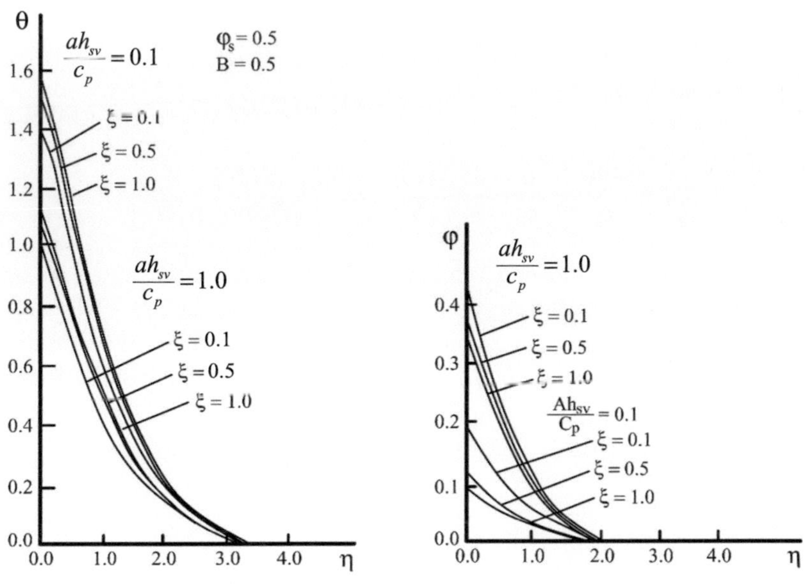

Fig. 6.4 Temperature and mass fraction distributions (Zhang et al. 1996)

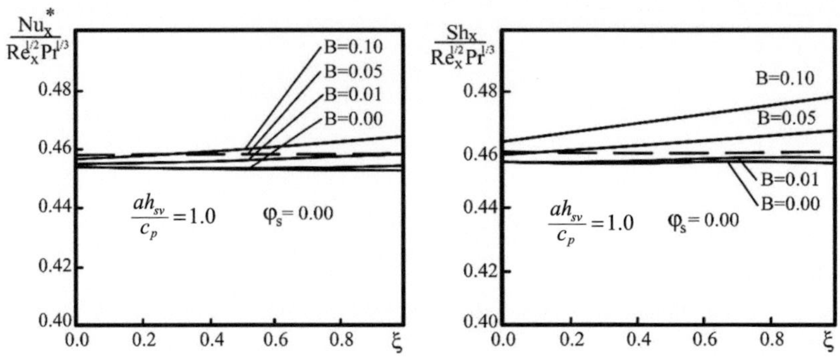

Fig. 6.5 Nusselt number based on convection and Sherwood number (Zhang et al. 1996)

and the effect of the mass fraction of the sublimable substance in the incoming flow is represented by

$$\varphi_{\mathrm{s}} = \frac{\rho h_{\mathrm{sv}} D(\omega_{\mathrm{sat},\infty} - \omega_{\infty})}{q''_{\mathrm{w}}\sqrt{2\nu L/u_{\infty}}} \tag{6.17}$$

where $\omega_{sat,\infty}$ is the saturation mass fraction corresponding to the incoming temperature:

$$\omega_{\mathrm{sat},\infty} = aT_{\infty} + b \tag{6.18}$$

It can be seen from Fig. 6.4 that the dimensionless temperature and mass fraction at different ξ are also different, which is the further evidences that a similarity solution does not exist.

The local Nusselt number based on the *total* heat flux at the bottom of the flat plate is

$$Nu_x = \frac{h_{\mathrm{w}}x}{k} = \frac{[q''_{\mathrm{w}}/(T_{\mathrm{w}} - T_{\infty})]x}{k} = \frac{Re_x^{1/2}}{\sqrt{2}\theta(\xi,0)} \tag{6.19}$$

and the Nusselt number based on convective heat transfer is

$$Nu_x^* = \frac{h_x x}{k} = \frac{-x}{T_{\mathrm{w}} - T_{\infty}}\left(\frac{\partial T}{\partial y}\right)_{y=0} = -\frac{\theta'(\xi,0)}{\sqrt{2}\theta(\xi,0)}Re_x^{1/2} \tag{6.20}$$

The Sherwood number is

$$Sh_x = \frac{h_{\mathrm{m}}x}{D} = \frac{-x}{\omega_{\mathrm{w}} - \omega_{\infty}}\frac{\partial \omega}{\partial y}\bigg|_{y=0} = -\frac{\varphi'(\xi,0)}{\sqrt{2}\varphi\theta(\xi,0)}Re_x^{1/2} \tag{6.21}$$

Figure 6.5 shows the effect of blowing velocity on the Nusselt number based on convective heat transfer and the Sherwood number for $\varphi_{\mathrm{sat},\infty} = 0$; i.e., the mass fraction of sublimable substance is equal to the saturation mass fraction corresponding to the incoming temperature. It can be seen that the effect of blowing velocity on mass transfer is stronger than that on heat transfer.

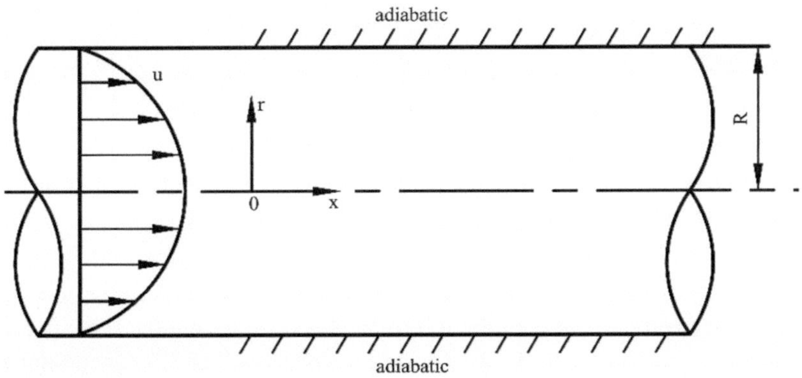

Fig. 6.6 Sublimation in an adiabatic tube

6.2.2 Sublimation Inside an Adiabatic Tube

In addition to the external sublimation discussed in the preceding subsection, internal sublimation is also very important. Sublimation inside an adiabatic and externally heated tube will be analyzed in this and the next subsections. The physical model of the problem under consideration is shown in Fig. 6.6 (Zhang and Chen 1990). The inner surface of a circular tube with radius R is coated with a layer of sublimable material which will sublime when gas flows through the tube. The fully developed gas enters the tube with a uniform inlet mass fraction of the sublimable substance, ω_0, and a uniform inlet temperature, T_0. Since the outer wall surface is adiabatic, the latent heat of sublimation is supplied by the gas flow inside the tube; this in turn causes the change in gas temperature inside the tube. It is assumed that the flow inside the tube is incompressible laminar flow with constant properties. In order to solve the problem analytically, the following assumptions are made:

1. The entrance mass fraction ω_0 is assumed to be equal to the saturation mass fraction at the entry temperature T_0.
2. The saturation mass fraction can be expressed as a linear function of the corresponding temperature.
3. The mass transfer rate is small enough that the transverse velocity components can be neglected.

The fully developed velocity profile in the tube is

$$u = 2\bar{u}\left[1 - \left(\frac{r}{R}\right)^2\right] \tag{6.22}$$

where $\bar{u}$ is the mean velocity of the gas flow inside the tube.

Neglecting axial conduction and diffusion, the energy and mass transfer equations are (Faghri et al. 2010):

$$ur\frac{\partial T}{\partial x} = \alpha\frac{\partial}{\partial r}\left(r\frac{\partial T}{\partial r}\right) \tag{6.23}$$

$$ur\frac{\partial \omega}{\partial x} = D\frac{\partial}{\partial r}\left(r\frac{\partial \omega}{\partial r}\right) \tag{6.24}$$

which are subjected to the following boundary conditions:

$$T = T_0 \quad x = 0 \tag{6.25}$$

$$\omega = \omega_0, \quad x = 0 \tag{6.26}$$

$$\frac{\partial T}{\partial r} = \frac{\partial \omega}{\partial r} = 0, \quad r = 0 \tag{6.27}$$

$$-k\frac{\partial T}{\partial r} = \rho D h_{sv}\frac{\partial \omega}{\partial r}, \quad r = R \tag{6.28}$$

Equation (6.28) implies that the latent heat of sublimation is supplied as the gas flows inside the tube. Another boundary condition at the tube wall is obtained by setting the mass fraction at the wall as the saturation mass fraction at the wall temperature (Kurosaki 1973). According to the second assumption, the mass fraction and temperature at the inner wall have the following relationship:

$$\omega = aT + b, \quad r = R \tag{6.29}$$

where a and b are constants.

The following nondimensional variables are then introduced:

$$\begin{aligned} &\eta = \frac{r}{R} \quad \xi = \frac{x}{RPe} \quad Le = \frac{\alpha}{D} \quad Re = \frac{2\bar{u}R}{\nu} \\ &Pe = \frac{2\bar{u}R}{\alpha} \quad \theta = \frac{T - T_f}{T_0 - T_f} \quad \varphi = \frac{\omega - \omega_f}{\omega_0 - \omega_f} \end{aligned} \tag{6.30}$$

where T_f and ω_f are temperature and mass fraction of the sublimable substance, respectively, after heat and mass transfer are fully developed. Equations (6.23)–(6.29) then become

$$\eta(1 - \eta^2)\frac{\partial \theta}{\partial \xi} = \frac{\partial}{\partial \eta}\left(\eta\frac{\partial \theta}{\partial \eta}\right) \tag{6.31}$$

$$\eta(1 - \eta^2)\frac{\partial \varphi}{\partial \xi} = \frac{1}{Le}\frac{\partial}{\partial \eta}\left(\eta\frac{\partial \varphi}{\partial \eta}\right) \tag{6.32}$$

$$\theta = \varphi = 1, \quad \xi = 0 \tag{6.33}$$

$$\frac{\partial \theta}{\partial \eta} = \frac{\partial \varphi}{\partial \eta} = 0, \quad \eta = 0 \tag{6.34}$$

$$-\left(\frac{ah_{sv}}{c_p}\right)\frac{\partial \theta}{\partial \eta} = \frac{1}{Le}\frac{\partial \varphi}{\partial \eta}, \quad \eta = 1 \tag{6.35}$$

$$\varphi = \left(\frac{ah_{sv}}{c_p}\right)\theta, \quad \eta = 1 \tag{6.36}$$

The heat and the mass transfer Eqs. (6.31) and (6.32) are independent, but their boundary conditions are coupled by Eqs. (6.35) and (6.36). The solution of Eqs. (6.31) and (6.32) can be obtained via separation of variables. It is assumed that the solution of θ can be expressed as a product of the function of η and a function of ξ, i.e.,

$$\theta = \Theta(\eta)\Gamma(\xi) \tag{6.37}$$

Substituting Eq. (6.37) into Eq. (6.31), the energy equation becomes

$$\frac{\Gamma'}{\Gamma} = \frac{\dfrac{\mathrm{d}}{\mathrm{d}\eta}\left(\dfrac{\mathrm{d}\Theta}{\mathrm{d}\eta}\right)}{\eta(1-\eta^2)\Theta} = -\beta^2 \tag{6.38}$$

where β is the eigenvalue for the energy equation.

Equation (6.38) can be rewritten as the following two ordinary differential equations:

$$\Gamma' + \beta^2\Gamma = 0 \tag{6.39}$$

$$\frac{\mathrm{d}}{\mathrm{d}\eta}\left(\frac{\mathrm{d}\Theta}{\mathrm{d}\eta}\right) + \beta^2\eta(1-\eta^2)\Theta = 0 \tag{6.40}$$

The solution of Eq. (6.39) is

$$\Gamma = C_1 \mathrm{e}^{-\beta^2\xi} \tag{6.41}$$

The boundary condition of Eq. (6.40) at $\eta = 0$ is

$$\Theta'(0) = 0 \tag{6.42}$$

The dimensionless temperature is then

$$\theta = C_1\Theta(\eta)\mathrm{e}^{-\beta^2\xi} \tag{6.43}$$

Similarly, the dimensionless mass fraction is

$$\varphi = C_2\Phi(\eta)\mathrm{e}^{-\gamma^2\xi} \tag{6.44}$$

where γ is the eigenvalue for the conservation of species equation and $\Phi(\eta)$ satisfies

$$\frac{\mathrm{d}}{\mathrm{d}\eta}\left(\frac{\mathrm{d}\Phi}{\mathrm{d}\eta}\right) + Le\gamma^2\eta(1-\eta^2)\Phi = 0 \tag{6.45}$$

and the boundary condition of Eq. (6.45) at $\eta = 0$ is

$$\Phi'(0) = 0 \tag{6.46}$$

Substituting Eqs. (6.43)–(6.44) into Eqs. (6.35)–(6.36), one obtains

$$\beta = \gamma \tag{6.47}$$

$$-\left(\frac{Ah_{sv}}{c_p}\right)\frac{\Theta(1)}{\Phi(1)} = Le\frac{\Theta'(1)}{\Phi'(1)} \tag{6.48}$$

To solve Eqs. (6.40) and (6.45) using the Runge–Kutta method, it is necessary to specify two boundary conditions for each. However, there is only one boundary condition for each: Eqs. (6.42) and (6.46), respectively. Since both Eqs. (6.40) and (6.45) are homogeneous, one can assume that the other boundary conditions are $\Theta(0) = \Phi(0) = 1$ and solve for Eq. (6.40) and (6.45) numerically. It is necessary to point out that the eigenvalue, β, is still unknown at this point and must be obtained by Eq. (6.48). There will be a series of β which satisfy Eq. (6.48), and for each value of the β_n, there is one set of corresponding Θ_n and Φ_n functions $(n = 1, 2, 3, \ldots)$.

If we use any one of the eigenvalue β_n and corresponding eigenfunctions—Θ_n and Φ_n—in Eqs. (6.43) and (6.44), the solutions of Eq. (6.31) and (6.32) become

$$\theta = C_1 \Theta_n(\eta) e^{-\beta_n^2 \xi} \tag{6.49}$$

$$\varphi = C_2 \Phi_n(\eta) e^{-\beta_n^2 \xi} \tag{6.50}$$

which satisfy all boundary conditions except those at $\xi = 0$. In order to satisfy boundary conditions at $\xi = 0$, one can assume that the final solutions of Eqs. (6.31) and (6.32) are

$$\theta = \sum_{n=1}^{\infty} G_n \Theta_n(\eta) e^{-\beta_n^2 \xi} \tag{6.51}$$

$$\varphi = \sum_{n=1}^{\infty} H_n \Phi_n(\eta) e^{-\beta_n^2 \xi} \tag{6.52}$$

where G_n and H_n can be obtained by substituting Eqs. (6.51) and (6.52) into Eq. (6.33), i.e.,

$$1 = \sum_{n=1}^{\infty} G_n \Theta_n(\eta) \tag{6.53}$$

$$1 = \sum_{n=1}^{\infty} H_n \Phi_n(\eta) \tag{6.54}$$

Due to the orthogonal nature of the eigenfunctions Θ_n and Φ_n, expressions of G_n and H_n can be obtained by

$$G_n = \frac{\int_0^1 \eta(1-\eta^2)\Theta_n(\eta)d\eta + \left[\frac{\Theta_n(1)}{\Phi_n(1)}\right] \int_0^1 \eta(1-\eta^2)\Phi_n(\eta)d\eta}{\int_0^1 \eta(1-\eta^2)\left\{\Theta_n^2(\eta) + (Ah_{sv}/c_p)\left[\frac{\Theta_n(1)}{\Phi_n(1)}\right]^2 \Phi_n^2(\eta)\right\}d\eta} \tag{6.55}$$

$$H_n = \frac{Ah_{sv}}{c_p}\frac{\Theta_n(1)}{\Phi_n(1)} G_n \tag{6.56}$$

The Nusselt number due to convection and Sherwood number are

$$Nu = \frac{-k\frac{\partial T}{\partial r}\big|_{r=R}}{\bar{T} - T_w}\frac{2R}{k} = -\frac{2}{\bar{\theta} - \theta_w}\sum_{n=1}^{\infty} G_n e^{-\beta_n^2 \xi}\Theta_n'(1) \tag{6.57}$$

$$Sh = \frac{-D\frac{\partial \omega}{\partial r}\big|_{r=R}}{\bar{\omega} - \omega_w}\frac{2R}{D} = -\frac{2}{\bar{\varphi} - \varphi_w}\sum_{n=1}^{\infty} H_n e^{-\beta_n^2 \xi}\Phi_n'(1) \tag{6.58}$$

where $\bar{T}$ and $\bar{\omega}$ are mean temperature and mean mass fraction in the tube.

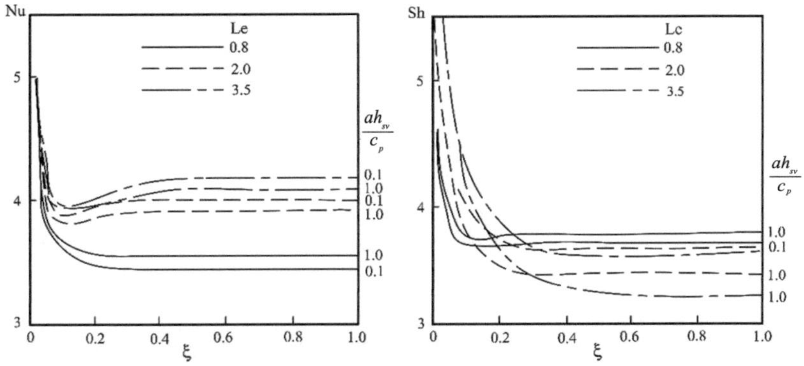

Fig. 6.7 Nusselt and Sherwood numbers for sublimation inside an adiabatic tube (Zhang and Chen 1990)

Figure 6.7 shows heat and mass transfer performance during sublimation inside an adiabatic tube. For all cases, both Nusselt and Sherwood numbers become constant when ξ is greater than a certain number, thus indicating that heat and mass transfer in the tube have become fully developed. The length of the entrance flow increases with increasing Lewis number. While the fully developed Nusselt number increases with increasing Lewis number, the Sherwood number decreases with increasing Lewis number, because a larger Lewis number indicates larger thermal diffusivity or low mass diffusivity. The effect of (Ah_{sv}/c_p) on Nusselt and Sherwood numbers is relatively insignificant: both Nusselt and Sherwood numbers increase with increasing (Ah_{sv}/c_p) for $Le < 1$, but increasing (Ah_{sv}/c_p) for $Le > 1$ results in decreasing Nusselt and Sherwood numbers.

6.2.3 Sublimation with Chemical Reaction

During combustion involving a solid fuel, the solid fuel may burn directly, or it may be sublimated before combustion. In the latter case—which will be discussed in this subsection—gaseous fuel diffuses away from the solid–vapor surface. Meanwhile, the gaseous oxidant diffuses toward the solid–vapor interface. Under the right conditions, the mass flux of vapor fuel and the gaseous oxidant meet and the chemical reaction occurs at a certain zone known as the flame. The flame is usually a very thin region with a color dictated by the temperature of combustion.

Figure 6.8 shows the physical model of the problem under consideration (Kaviany 2001). The concentration of the fuel is highest at the solid fuel surface and decreases as the location of the flame is approached. The gaseous fuel diffuses away from the solid fuel surface and meets the oxidant as it

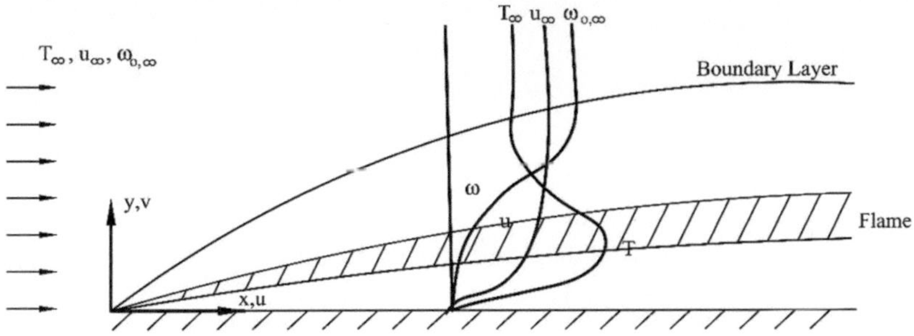

Fig. 6.8 Sublimation with chemical reaction

flows parallel to the solid fuel surface. Combustion occurs in a thin reaction zone where the temperature is the highest, and the latent heat of sublimation is supplied by combustion. The combustion of solid fuel through sublimation can be modeled as a steady-state boundary-layer type flow with sublimation and chemical reaction.

To model the problem, the following assumptions are made:

1. The fuel is supplied by sublimation at a steady rate.
2. The Lewis number is unity, so the thermal and concentration boundary layers have the same thickness.
3. The buoyancy force is negligible.

The conservations of mass, momentum, energy, and species of mass in the boundary layer are

$$\frac{\partial(\rho u)}{\partial x}+\frac{\partial(\rho v)}{\partial y}=0 \tag{6.59}$$

$$u\frac{\partial u}{\partial x}+v\frac{\partial u}{\partial y}=\frac{\partial}{\partial y}\left(v\frac{\partial u}{\partial y}\right) \tag{6.60}$$

$$\frac{\partial}{\partial x}(\rho c_p uT)+\frac{\partial}{\partial y}(\rho c_p vT)=\frac{\partial}{\partial y}\left(k\frac{\partial T}{\partial y}\right)+\dot{m}_o'''h_{c,o} \tag{6.61}$$

$$\frac{\partial}{\partial x}(\rho u\omega_o)+\frac{\partial}{\partial y}(\rho v\omega_o)=\frac{\partial}{\partial y}\left(\rho D\frac{\partial \omega_o}{\partial y}\right)-\dot{m}_o''' \tag{6.62}$$

where $\dot{m}_o'''$ is rate of oxidant consumption (kg/m^3-s). $h_{c,o}$ is the heat released by combustion per unit mass consumption of the oxidant (J/kg), which is different from the combustion heat defined in Chap. 3. ω_o is mass fraction of the oxidant in the gaseous mixture.

The corresponding boundary conditions of Eqs. (6.59)–(6.62) are

$$u\to u_\infty,\quad T\to T_\infty,\quad \omega_o\to\omega_{o,\infty}\ \text{at}\ y\to\infty \tag{6.63}$$

$$u=0,\quad v=\frac{\dot{m}_f''}{\rho},\quad \frac{\partial\omega_o}{\partial y}=0\ \text{at}\ y=0 \tag{6.64}$$

where $\dot{m}_f''$ is the rate of solid fuel sublimation per unit area (kg/m^2-s) and ρ is the density of the mixture.

The shear stress at the solid fuel surface is

$$\tau_w=\mu\frac{\partial u}{\partial y},\quad y=0 \tag{6.65}$$

The heat flux at the solid fuel surface is

$$q_w''=-k\frac{\partial T}{\partial y},\quad y=0 \tag{6.66}$$

The exact solution of the heat and mass problem described by Eqs. (6.59)–(6.62) can be obtained using conventional numerical simulation, which is very complex. However, it is useful here to introduce the results obtained by Kaviany (2001) using analogy between momentum and heat transfer. Multiplying Eq. (6.62) by $h_{c,o}$ and adding the result to Eq. (6.61), one obtains

$$\begin{aligned}\frac{\partial}{\partial x}\left[\rho u(c_{p}T+\omega_{o}h_{c,o})\right]+\frac{\partial}{\partial y}\left[\rho v(c_{p}T+\omega_{o}h_{c,o})\right]\\ =\frac{\partial}{\partial y}\left[k\frac{\partial T}{\partial y}+\rho Dh_{c,o}\frac{\partial\omega_{o}}{\partial y}\right]\end{aligned}\tag{6.67}$$

Considering the assumption that Lewis number is unity, i.e., $Le=\alpha/D=1$, Eq. (6.67) can be rewritten as

$$\begin{aligned}\frac{\partial}{\partial x}\left[\rho u(c_{p}T+\omega_{o}h_{c,o})\right]+\frac{\partial}{\partial y}\left[\rho v(c_{p}T+\omega_{o}h_{c,o})\right]\\ =\frac{\partial}{\partial y}\left[\rho\alpha\frac{\partial}{\partial y}(c_{p}T+\omega_{o}h_{c,o})\right]\end{aligned}\tag{6.68}$$

which can be viewed as an energy equation with quantity $c_pT+\omega_oh_{c,o}$ as a dependent variable.

Since $\partial\omega_o/\partial y=0$ at $y=0$, i.e., the solid fuel surface is not permeable for the oxidant, Eq. (6.66) can be rewritten as

$$q''_{w}=-\rho\alpha\frac{\partial}{\partial y}(c_{p}T+\omega_{o}h_{c,o}),\quad y=0\tag{6.69}$$

Analogy between surface shear stress and the surface energy flux yields

$$\begin{aligned}q''_{w}&=\frac{\tau_{w}}{u_{\infty}}\left[(c_{p}T+\omega_{o}h_{c,o})_{w}-(c_{p}T+\omega_{o}h_{c,o})_{\infty}\right]\\ &=\frac{\tau_{w}}{u_{\infty}}\left[c_{p}(T_{w}-T_{\infty})+h_{c,o}(\omega_{o,w}-\omega_{o,\infty})\right]\end{aligned}\tag{6.70}$$

The energy balance at the surface of the solid fuel is

$$-q''_{w}=\dot{m}''_{f}h_{sv}+q''_{\ell}\tag{6.71}$$

where the two terms on the right-hand side of Eq. (6.71) represent the latent heat of sublimation, and the sensible heat required to raise the surface temperature of the solid fuel to sublimation temperature and heat loss to the solid fuel.

Combining Eqs. (6.70) and (6.71) yields the rate of sublimation on the solid fuel surface

$$\dot{m}''_{f}=Z\frac{\tau_{w}}{u_{\infty}}\tag{6.72}$$

where Z is transfer driving force or transfer number defined as

$$Z=\frac{c_{p}(T_{\infty}-T_{w})+h_{c,o}(\omega_{o,\infty}-\omega_{o,w})}{h_{sv}+q''_{\ell}/m''_{f}}\tag{6.73}$$

By using the friction coefficient—

$$C_f = \frac{\tau_w}{\rho u_\infty^2/2} \tag{6.74}$$

Equation (6.72) becomes

$$\dot{m}''_f = \frac{C_f}{2}\rho u_\infty Z \tag{6.75}$$

The surface blowing velocity of the gaseous fuel is then

$$v_w = \frac{\dot{m}''_f}{\rho} = \frac{C_f}{2}u_\infty Z \tag{6.76}$$

where the friction coefficient C_f can be obtained from the solution of boundary-layer flow over a flat plate with blowing on the surface (Kaviany 2001; Kays et al. 2004). The similarity solution of the boundary-layer flow problem exists only if blowing velocity satisfies $v_w \propto x^{-1/2}$. In this case, one can define a blowing parameter as

$$B = \frac{(\rho v)_w}{(\rho u)_\infty} Re_x^{1/2} \tag{6.77}$$

Combination of Eqs. (6.76) and (6.77) yields

$$B = \frac{Z}{2} Re_x^{1/2} C_f \tag{6.78}$$

Glassman and Yetter (2008) recommended an empirical form of Eq. (6.78) based on numerical and experimental results:

$$B = \frac{\ln(1+Z)}{2.6Z^{0.15}} \tag{6.79}$$

Example 6.1

Air with a temperature of 27 °C flows at 1 m/s over a 1-m long solid fuel surface with a temperature of 727 °C. The concentration of the oxidant at the solid fuel surface is 0.1, and the heat released per unit mass of the oxidant consumed is 12,000 kJ/kg. The latent heat of sublimation for the solid fuel is 1500 kJ/kg. Neglect the sensible heat required to raise the surface temperature of the solid fuel to sublimation temperature and heat loss to the solid fuel. Estimate the average blowing velocity due to sublimation on the fuel surface.

Solution The mass fractions of the oxygen at the solid fuel surface and in the incoming air are, respectively, $\omega_{o,w} = 0.1$ and $\omega_{o,\infty} = 0.21$. The specific heat of gas, approximately taken as specific heat of air at $T_{ave} = (T_w + T_\infty)/2 = 377$ °C, is $c_p = 1.063$ kJ/kg-K. The combustion heat per unit oxidant consumed is $h_{c,o} = 12,000\,\text{kJ/kg}$. The latent heat of sublimation is $h_{sv} = 1500\,\text{kJ/kg}$. The density at the wall and the incoming temperatures are, respectively, $\rho_w = 0.3482\,\text{kg/m}^3$ and $\rho_\infty = 1.1614\,\text{kg/m}^3$. The viscosity at T_{ave} is $\nu = 60.21 \times 10^{-6}\,\text{m}^2/\text{s}$.
The transfer driving force can be obtained from Eq. (6.73), i.e.,

$$Z = \frac{c_p(T_\infty - T_w) + h_{c,o}(\omega_{o,\infty} - \omega_{o,w})}{h_{sv}}$$
$$= \frac{1.063 \times (27 - 727) + 12000 \times (0.21 - 0.1)}{1500} = 0.3839$$

The blowing parameter obtained from Eq. (6.79) is

$$B = \frac{\ln(1+Z)}{2.6Z^{0.15}} = \frac{\ln(1+0.5257)}{2.6 \times 0.5257^{0.15}} = 0.1443$$

The blowing velocity at the surface is obtained from Eq. (6.77):

$$v_w = \frac{\rho_\infty}{\rho_w} B u_\infty Re_x^{-1/2} = \frac{\rho_\infty}{\rho_w} B (u_\infty \nu)^{1/2} x^{-1/2}$$

which can be integrated to yield the average blowing velocity:

$$\bar{v}_w = \frac{2\rho_\infty}{\rho_w} B (u_\infty \nu L)^{1/2}$$
$$= \frac{2 \times 1.1614}{0.3482} \times 0.1443 \times \left(1 \times 60.21 \times 10^{-6} \times 1\right)^{1/2} = 0.007469\,\text{m/s}$$

6.3 Chemical Vapor Deposition (CVD)

6.3.1 Introduction

CVD is widely used to fabricate semiconductor devices. It depends on the availability of a volatile gaseous chemical that can be converted to solid film through some thermally activated chemical reaction. Chemical vapor deposition can be used to produce a large variety of thin films with different precursors. It is very crucial that the chemical reaction takes place on the substrate surface only, so that a thin film can be deposited onto the substrate. If undesired chemical reactions occur in the gas phase, the solid particles can be formed which may fall onto the substrate or coat the chamber walls. To avoid the undesired chemical reaction, the substrate surface temperature, deposition time, pressure, and surface specificity should be carefully selected. The chemical reaction during a CVD process is usually accomplished in several steps. The path of chemical reactions can be altered by changing the substrate temperature. For example, when titanium tetrabromide ($TiBr_4$) is used as a precursor to deposit titanium film, the chemical reaction is accomplished in the following steps (Mazumder and Kar 1995):

$$TiBr_4(g) \rightarrow TiBr_2(s) + Br_2(g)$$

$$3TiBr_2(s) \rightarrow 2TiBr(s) + TiBr_4(g)$$

$$4TiBr(g) \rightarrow 3Ti(s) + TiBr_4(g)$$

Table 6.1 Overall chemical reaction of CVD processes (including LCVD)

Thin films	Overall reaction	Temperature of reaction (°C)	References
Al_2O_3	$Al(l) + H_2O(g) = AlO(g) + H_2(g)$ $AlO(g) + H_2O(g) = Al_2O_3(s) + H_2(g)$	1230–1255	Powell et al. (1966)
C	$C_xH_y(g) = xC(s) + (y/2)H_2(g)$	700–1450	Taylor et al. (2004)
GaAs	GaCI(g) + (1/4)As4(g) = GaAs(s) + HCI(g)		Sivaram (1995)
GaAs	GaAs(g) + HCI(g) =GaCI(g) + 1/4(As_4(g)) + 1/2(H_2(g))	700–850	Sivaram (1995)
	$Ga(CH_3)_3 + AsH_3 = GaAs + 3CH_4$ $Al(CH_3)_3 + AsH_3 = AlAs + 3CH$	500–800	Ueda (1996)
GaN	$Ga(g) + NH_3 = GaN(s) + (3/2)H_2\ (g)$	650	Elyukhin et al. (2002)
Ge(s)	$GeH_4 = Ge(s) + 2H_2$		Herring (1990)
Si	$SiH_4(g) = Si(s) + 2H_2(g)$	>600 (polysilicon) >850–900 (single crystal)	Herring (1990)
SiC	$Si(CH_3)_4(g) = SiC(s) + 3CH_4(g)$	700–1450	Sun et al. (1998)
SiO_2	$SiH_4 + O_2 = SiO_2 + 2H_2$		Sivaram (1995)
	$SiH_4 + 2N_2O = SiO_2 + 2H_2O + 2N_2$	800	Sivaram (1995)
	$SiH_2Cl_2 + 2N_2O = SiO_2 + 2HCI + 2N_2$	>900	Sivaram (1995)
TiO_2	$TiCl_4(g) + O_2(g) = TiO_2(S) + 2Cl_2(g)$		Jakubenas et al. (1997)
TiN	$TiCl_4(g) + 2H_2(g) + (1/2)N_2(g) = TiN$ $(s) + 4HCl(g)$	900	Mazumder and Kar (1995)

The mechanisms of chemical reactions for many CVD processes are not clear, so the chemical reactions occurring in a CVD process are often represented by a single overall chemical reaction equation. Table 6.1 summarizes some examples of the overall chemical reactions occurring in CVD processes (including LCVD).

CVD reactors may operate at atmospheric reduced pressure (APCVD)—which varies from 0.1 to 1 atm—or at low pressure (LPCVD). The typical pressure for LPCVD is 10^{-3} atm. A wide variety of CVD reactors have been developed for its various applications; some of them are illustrated in Fig. 6.9 (Jensen et al. 1991; Mahajan 1996). The horizontal reactor shown in Fig. 6.9a is one of the most established configurations: a rectangular duct. The wafers to be coated are placed on a heated susceptor that is tilted by about 3° in order to ensure uniformity of deposition (Mahajan 1996). The horizontal reactor is primarily used in CVD research and epitaxial growth of silicon semiconductors (Jensen et al. 1991). In the vertical reactor shown in Fig. 6.9b, the precursors are injected into a slowly rotating susceptor on which CVD takes place (Evans and Greif 1987). The barrel reactor shown in Fig. 6.9c is frequently used for large volume production of silicon epitaxial wafers. The wafers sit in shallow pockets on a slightly tapered, slowly rotating heated susceptor. In the CVD reactors shown in Figs. 6.9a–c, the activation energy for chemical reaction is supplied directly to the susceptors, and the walls are either unheated or cooled. The CVD reactor shown in Fig. 6.9d, however, is a hot wall tubular reactor that is heated from outside; it is commonly used to deposit polycrystalline silicon and other dielectric films. The reactor operates at a low pressure (0.1 to 10 Torr) and is nearly at isothermal condition, with temperatures ranging from 300 to 900 °C (Jensen

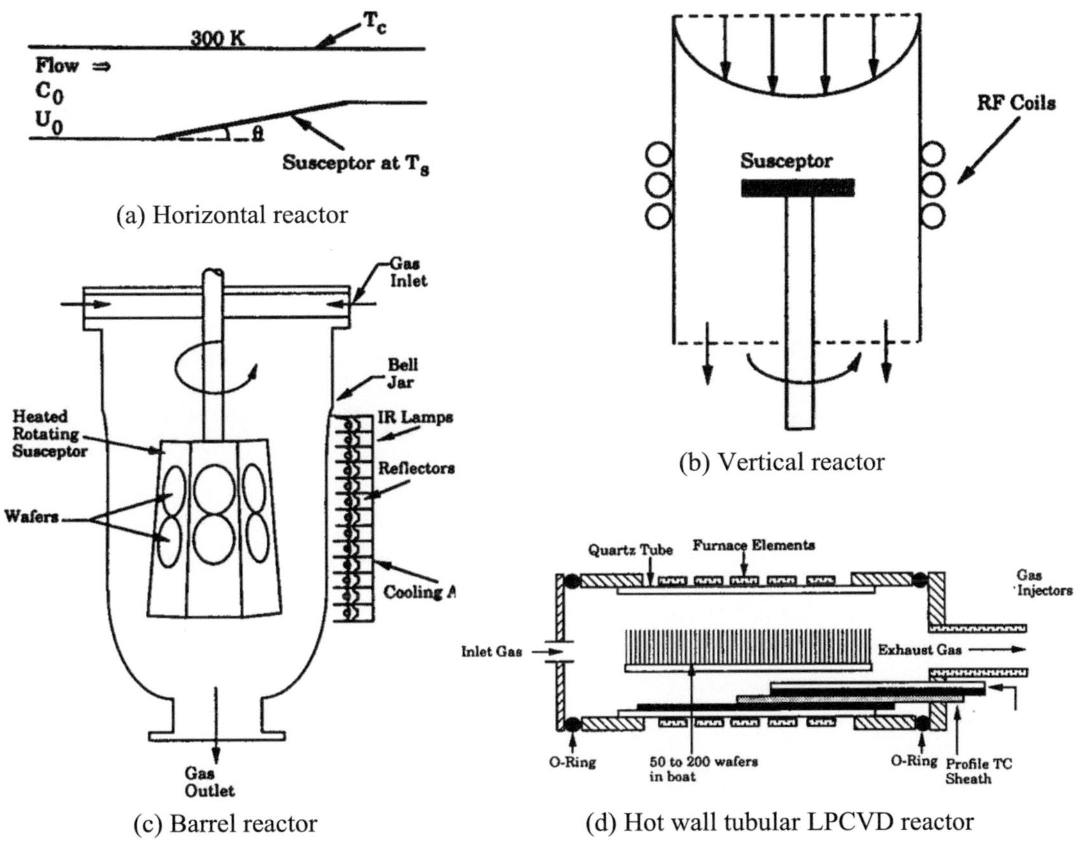

Fig. 6.9 Common CVD reactors (Mahajan 1996; Reprinted with permission from Elsevier)

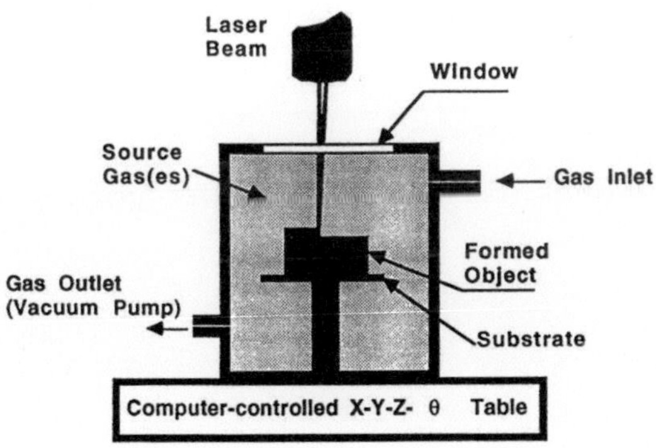

Fig. 6.10 SALD system (Marcus et al. 1993)

et al. 1991). In addition to gaseous precursors discussed above, the precursor for CVD can also be liquid as reported by Versteeg et al. (1995).

Figure 6.10 shows a reaction chamber for the Selective Area Laser Deposition (SALD) process (Marcus et al. 1993). In contrast to conventional CVD, in which the entire susceptor is heated, only a very small spot on the substrate is heated by a directed laser beam. Scanning of the substrate surface is accomplished by a movable table. After the first layer of the solid is deposited, consecutive layers can be deposited to build the three-dimensional part based on the CAD design. The pressure inside the chamber is usually under 1 atm, and the temperature of the spot under laser irradiation can range from 700 to 1500 °C. Successful deposition of various ceramic and metallic materials using various gaseous precursors has been reported.

6.3.2 Governing Equations of CVD

Since the velocity of the precursors is generally very low and the characteristic length is also very small, the corresponding Reynolds number is under 100 and the Grashof number governing natural convection is under 10^6. Therefore, the transport phenomena in the CVD process are laminar in nature. The temperature in a reactor varies significantly (typically from 300 to 900 K), so the Boussinesq approximation is no longer appropriate. It is necessary to use the compressible model for transport phenomena in CVD processes.

The governing equations for the CVD process can be obtained by simplifying the generalized governing equations in Chap. 3. The following assumptions can be made to obtain the governing equations:

1. The reference frame is stationary.
2. The body force $\mathbf{X}$ is gravitational force, which is the same for all components in the precursors.
3. Dilute approximation is valid because the partial pressure of the reactant is much lower than that of the carrier gas.
4. The deposited film is very thin (from nanometers to microns), and its effect on the flow field can be neglected.

The continuity equation is

$$\frac{D\rho}{Dt} + \rho\nabla \cdot \mathbf{V} = 0 \tag{6.80}$$

where the precursor gases are treated as a compressible fluid mixture.

The momentum equation is

$$\rho\frac{D\mathbf{V}}{Dt} = \nabla \cdot \tau' + \rho\mathbf{g} \tag{6.81}$$

where the stress tensor is

$$\tau' = -p\mathbf{I} + 2\mu\mathbf{D} - \frac{2}{3}\mu(\nabla \cdot \mathbf{V})\mathbf{I} \tag{6.82}$$

The energy equation is

$$\rho c_\mathrm{p}\frac{DT}{Dt} = \nabla \cdot (k\nabla T) + T\beta\frac{Dp}{Dt} \tag{6.83}$$

where the effect of viscous dissipation and the Dufour effect have been neglected.

The conservation of species mass in terms of the mass fraction is

$$\rho \frac{D\omega_i}{Dt} = -\nabla \cdot \mathbf{J}_i + \dot{m}_i''', \quad i = 1, 2, \ldots N - 1 \tag{6.84}$$

where ω_i is the mass fraction of the i-th component in the gaseous precursor. The mass flux $\mathbf{J}_i$ includes mass fluxes due to ordinary diffusion driven by the concentration gradient and thermal (Soret) diffusion. The production rate of the i-th species, $\dot{m}_i'''$, can be obtained by analyzing the chemical reaction. If the number of chemical reactions taking place in the system is N_g, the mass production rate is (Mahajan 1996)

$$\dot{m}_i''' = \sum_{j=1}^{N_g} a_{ij}^g M_i \Re_j^g \tag{6.85}$$

where a_{ij}^g is the stoichiometric coefficient for the i-th component in the j-th chemical reaction in the gas phase, and $\Re_j^g$ is the net reaction rate of the j-th chemical reaction in the gas phase.

The density of the gas is related to the pressure and temperature by the ideal gas law:

$$\rho = \frac{p}{R_g T} \tag{6.86}$$

The boundary conditions for the governing equations of a CVD process depend on the geometric configuration of the reactor. It is generally assumed that the nonslip condition is applicable to all solid walls. The normal velocity on a solid wall is zero (no penetration) for all walls except the susceptor where chemical reaction takes place. The total net mass flux of all species can lead to a normal velocity component on the susceptor as (Mahajan 1996)

$$v_n = \frac{1}{\rho} \sum_{i=1}^{N} \sum_{j=1}^{N_s} a_{ij}^s M_i \Re_j^s \tag{6.87}$$

where a_{ij}^s is the stoichiometric coefficient for the i-th component in the j-th chemical reaction on the susceptor surface, and $\Re_j^g$ is the net reaction rate of the j-th chemical reaction on the susceptor surface.

The net surface reaction rate $\Re_j^g$ is a product of the sticking coefficient γ_j (fraction of product that can be stuck on the substrate) and the effusive flux of the j-th species (Mahajan 1996), i.e.,

$$\Re_j^s = \gamma_j \frac{x_j p_j}{\sqrt{2\pi M_j RT}} \tag{6.88}$$

The growth rate of the deposit on the susceptor is

$$\frac{\mathrm{d}\delta}{\mathrm{d}t} = M_f \sum_{i=1}^{N} \sum_{j=1}^{N_s} a_{ij}^s \Re_j^s \alpha_{i,f} \tag{6.89}$$

where M_f is the molecular mass of the deposited film, and $\alpha_{i,f}$ is the number of film atoms in the i-th species.

Table 6.2 Transport properties of the common gas(es) for CVD (Mahajan 1996)

Properties	Gas(es)	c_0	c_1	c_2
μ^a (N-s/m^2)	TMGa	-1.15×10^{-6}	3.35×10^{-8}	-6.68×10^{-12}
	AsH_3	-4.32×10^{-7}	5.94×10^{-8}	-1.46×10^{-11}
	H_2	2.63×10^{-6}	2.22×10^{-8}	-5.19×10^{-12}
	N_2	4.93×10^{-6}	4.55×10^{-8}	-1.08×10^{-11}
	SiH_4	1.47×10^{-6}	3.66×10^{-8}	-6.81×10^{-12}
k^a (W/m-K)	TMGa	-3.52×10^{-3}	3.85×10^{-5}	-3.84×10^{-8}
	AsH_3	-7.16×10^{-3}	6.53×10^{-5}	-3.47×10^{-9}
	H_2	5.77×10^{-2}	4.43×10^{-4}	-7.54×10^{-8}
	N_2	8.15×10^{-3}	6.24×10^{-5}	-4.48×10^{-9}
	SiH_4	-2.12×10^{-2}	1.45×10^{-4}	-1.31×10^{-8}
c_p^a (kJ/Kg-K)	TMGa	5.40×10^{2}	1.60	0
	AsH_3	2.45×10^{2}	1.08×10^{0}	-4.24×10^{-4}
	H_2	1.44×10^{4}	-2.61×10^{-1}	8.67×10^{-4}
	N_2	1.03×10^{3}	4.58×10^{-3}	1.34×10^{-4}
	SiH_4	4.74×10^{2}	3.26	-1.08×10^{-3}
D_{12}^b (m^2/s)	SiH_4, N_2	-9.64×10^{-1}	6.25×10^{-3}	8.50×10^{-6}
	SiH_4, H_2	−2.90	2.06×10^{-2}	2.81×10^{-5}
	N_2, H_2	−3.20	2.44×10^{-2}	3.37×10^{-5}
	TMGa, H_2	−1.87	1.64×10^{-2}	3.13×10^{-5}
	TMGa, N_2	-4.17×10^{-1}	2.89×10^{-3}	4.93×10^{-6}
	AsH_3, H_2	−2.26	1.73×10^{-2}	2.80×10^{-5}
	AsH_3, N_2	-6.15×10^{-1}	4.57×10^{-3}	7.49×10^{-6}
	TMGa, AsH_3	-2.26×10^{-1}	1.27×10^{-3}	3.18×10^{-6}
$k_{12}^{T\,c}$	H_2,SiH_4	-2.74×10^{-1}	−1.70	-6.35×10^{-3}
	N_2,SiH_4	-5.15×10^{-2}	−1.69	-4.94×10^{-3}
	H_2,N_2	-2.71×10^{-1}	−1.61	-9.15×10^{-3}
	TMGa, H_2	1.32	−1.54	-3.57×10^{-3}
	TMGa, N_2	6.36×10^{-1}	−1.58	-3.36×10^{-3}
	AsH_3, H_2	8.86×10^{-1}	−1.57	-4.35×10^{-3}
	AsH_3, N_2	3.09×10^{-1}	−1.55	-4.06×10^{-3}
	TMGa, AsH_3	1.94×10^{-1}	−1.79	-1.91×10^{-3}

[a] $\mu, k, c_p = c_0 + c_1 T + c_2 T^2$; [b] $D_{12} = D_{21} = (c_0 + c_1 T + c_2 T^2)/p$; [c] $k_{12}^T = -k_{12}^T = c_0 x_1 x_2 [1 + c_1 \exp(c_2 T)]$, for $x_1 \to 0$. T is absolute temperature (K); Reprinted with permission from Elsevier

6.3.3 Transport Properties

The thermophysical properties of various gaseous precursors are necessary to utilize the transport models outlined above. Since the temperature varies significantly throughout a CVD system, the transport phenomena must be modeled using variable thermal physical properties. This requires knowledge of the dependence of the thermophysical properties on temperature. The thermophysical properties of the commonly used gas(es) in CVD are tabulated in Table 6.2.

The viscosity and thermal conductivity of some precursors that are not readily available can be estimated using the method recommended by Bird et al. (2006). The viscosity is

$$\mu = 2.6693 \times 10^{-6} \frac{\sqrt{MT}}{\sigma^2 \Omega_\mu} \tag{6.90}$$

Table 6.3 Dependence of collision integral Ω_μ on dimensional temperature $k_b T/\varepsilon$ (Bird et al. 2006)[a]

$k_b T/\varepsilon$	Ω_μ	$\Omega_{D,12}$	$k_b T/\varepsilon$	Ω_μ	$\Omega_{D,12}$	$k_b T/\varepsilon$	Ω_μ	$\Omega_{D,12}$
0.30	2.840	2.649	1.70	1.249	1.141	4.2	0.9598	0.8748
0.35	2.676	2.468	1.75	1.235	1.128	4.3	0.9551	0.8703
0.40	2.531	2.314	1.80	1.222	1.117	4.4	0.9506	0.8659
0.45	2.401	2.182	1.85	1.209	1.105	4.5	0.9462	0.8617
0.50	2.284	2.066	1.90	1.198	1.095	4.6	0.9420	0.8576
0.55	2.178.	1.965	1.95	1.186	1.085	4.7	0.9380	0.8537
0.60	2.084	1.877	2.00	1.176	1.075	4.8	0.9341	0.8499
0.65	1.999	1.799	2.10	1.156	1.058	4.9	0.9304	0.8463
0.70	1.922	1.729	220	1.138	1.042	5.0	0.9268	0.8428
0.75	1.853	1.667	2.30	1.122	1.027	6.0	0.8962	0.8129
0.80	1.790	1.612	2.40	1.107	1.013	7.0	0.8727	0.7898
0.85	1.734	1.562	2.50	1.0933	1.0006	8.0	0.8538	0.7711
0.90	1.682	1.517	2.60	1.0807	0.9890	9.0	0.8380	0.7555
0.95	1.636	1.477	2.7	1.0691	0.9782	10.0	0.8244	0.7422
1.00	1.593	1.440	2.8	1.0583	0.9682	12.0	0.8018	0.7202
1.05	1.554	1.406	2.9	1.0482	0.9588	14.0	0.7836	0.7025
1.10	1.518	1.375	3.0	1.0388	0.9500	16.0	0.7683	0.6878
1.15	1.485	1.347	3.1	1.0300	0.9418	18.0	0.7552	0.6751
1.20	1.455	1.320	3.2	1.0217	0.9340	20.0	0.7436	0.6640
1.25	1.427	1.296	3.3	1.0139	0.9267	25.0	0.7198	0.6414
1.30	1.401	1.274	3.4	1.0066	0.9197	30.0	0.7010	0.6235
1.35	1.377	1.253	3.5	0.9996	0.9131	35.0	0.6854	0.6088
1.40	1.355	1.234	3.6	0.9931	0.9068	40.0	0.6723	0.5964
1.45	1.334	1.216	3.7	0.9868	0.9008	50.0	0.6510	0.5763
1.50	1.315	1.199	3.8	0.9809	0.8952	75.0	0.6140	0.5415
1.55	1.297	1.183	3.9	0.9753	0.8897	100.0	0.5887	0.5180
1.60	1.280	1.168	4.0	0.9699	0.8845			
1.65	1.264	1.154	4.1	0.9647	0.8796			

[a]Reproduced with permissions from John Wiley and Sons, Limited

where M is the molecular mass and σ is the collision diameter ($\AA = 10^{-10}$ m) of the molecule that can be estimated by

$$\sigma = 0.841\bar{v}_{\mathrm{c}}^{1/3} \tag{6.91}$$

or

$$\sigma = 1.166\bar{v}_{\mathrm{b,liq}}^{1/3} \tag{6.92}$$

where $\bar{v}_{\mathrm{c}}$ and $\bar{v}_{\mathrm{b,liq}}$ are the specific volumes (cm^3/mol) of the precursor at critical point, and of the saturated liquid at normal boiling point, respectively.

The collision integral Ω_μ in Eq. (6.90) is a slowly varying function of dimensionless temperature, $k_{\mathrm{b}}T/\varepsilon$, and is tabulated in Table 6.3. k_{b} is the Boltzmann constant and ε is a characteristic energy of interaction between molecules which can be estimated by

$$\frac{\varepsilon}{k_{\mathrm{b}}} = 0.77T_{\mathrm{c}} \tag{6.93}$$

or

$$\frac{\varepsilon}{k_{\mathrm{b}}} = 1.15T_{\mathrm{b}} \tag{6.94}$$

where T_{c} and T_{b} are critical temperature and normal boiling point, respectively.

The thermal conductivity of the polyatomic gas is related to its viscosity by

$$k = \left(c_{\mathrm{p}} + \frac{5}{4}R_{\mathrm{g}}\right)\mu \tag{6.95}$$

where R_{g} is the gas constant.

For a mixture of different gases, as is usually the case in CVD processes, the viscosity and the thermal conductivity of the mixture are related to those of the individual components by

$$\mu = \sum_{i=1}^{N} \frac{x_i\mu_i}{\sum_{j=1}^{N} x_i\phi_{ij}} \tag{6.96}$$

$$k = \sum_{i=1}^{N} \frac{x_i k_i}{\sum_{j=1}^{N} x_i\phi_{ij}} \tag{6.97}$$

where

$$\phi_{ij} = \frac{1}{\sqrt{8}}\left(1 + \frac{M_i}{M_j}\right)^{-1/2}\left[1 + \left(\frac{\mu_i}{\mu_j}\right)^{1/2}\left(\frac{M_j}{M_i}\right)^{1/4}\right]^2 \tag{6.98}$$

The specific heat of the gaseous mixture is related to those of the individual components by

$$c_{\mathrm{p}} = \sum_{i=1}^{N} x_i c_{\mathrm{p},i} \tag{6.99}$$

For applications that involve unknown mass diffusivity, it can be estimated by

$$D_{12} = 1.8583 \times 10^{-7} \frac{\sqrt{T^3\left(M_1^{-1} + M_2^{-1}\right)}}{p\sigma_{12}^2 \Omega_{D,12}} \tag{6.100}$$

where the unit for pressure is atm and

$$\sigma_{12} = \frac{1}{2}(\sigma_1 + \sigma_2) \tag{6.101}$$

$\Omega_{D,12}$ is a function of $k_b T/\varepsilon_{12}$ that can be obtained from Table 6.3 using

$$\varepsilon_{12} = \sqrt{\varepsilon_1 \varepsilon_2} \tag{6.102}$$

The concentration of the reactant is usually much lower than that of the carrier gas(es). When the reactant is a single gas diluted by the carrier gas, the diffusivity of the reactant to the carrier gaseous mixture is of interest. If the reactant is defined as component 1 in the precursor, and the carrier gases are components 2 through N, the diffusivity of the reactant—1—to the carrier gas mixture—m—can be obtained by

$$\frac{1 - x_1}{D_{1m}} = \sum_{j=2}^{N} \frac{x_j}{D_{1j}} \tag{6.103}$$

Example 6.2

Titanium tetrachloride ($TiCl_4$) is used as a reactant to deposit titanium nitride, TiN (Zhang and Faghri 2000). The boiling point of $TiCl_4$ is 136.4 °C, and its liquid specific volume at normal boiling point is 109.7 cm^3/mol. Its specific heat at 900 K is 560.92 J/kg-k. Estimate its viscosity and thermal conductivity at 900 K.

Solution The molecular mass of $TiCl_4$ is $M = 189.71$ kg/kmol. The collision diameter of $TiCl_4$ is obtained from Eq. (6.92):

$$\sigma = 1.166\bar{v}_{b,liq}^{1/3} = 1.166 \times 109.7^{1/3} = 5.582\,\text{Å}$$

The dimensionless temperature

$$\frac{k_b T}{\varepsilon} = \frac{T}{1.15 T_b} = \frac{900}{1.15(136.4 + 273.15)} = 1.9108$$

can be used to find the collision integral from Table 6.3; the result is $\Omega_\mu = 1.1946$.

The viscosity can be obtained from Eq. (6.90), i.e.,

$$\begin{aligned}\mu &= 2.6693 \times 10^{-6} \frac{\sqrt{MT}}{\sigma^2 \Omega_\mu} \\ &= 2.6693 \times 10^{-6} \times \frac{\sqrt{189.71 \times 900}}{5.582^2 \times 1.1946} = 2.96 \times 10^{-5}\,\text{kg/s-m}\end{aligned}$$

The gas constant of $TiCl_4$ is

$$R_g = \frac{R_u}{M} = \frac{8.3143 \times 10^3}{189.71} = 43.8 \text{J/kg - K}$$

The thermal conductivity of $TiCl_4$ can be obtained from Eq. (6.97), i.e., $k = \left(c_p + \frac{5}{4}R_g\right)\mu = (560.92 + \frac{5}{4} \times 43.8) \times 2.96 \times 10^{-5} = 0.01826 \text{W/m} - \text{K}$

6.3.4 Laser Chemical Vapor Deposition (LCVD)

During LCVD, the spot on the substrate under laser irradiation is at a very high temperature (1200 K or higher). Temperature gradients in the source gases will cause natural convection in the chamber. The concentration of the gas mixture near the hot spot on the substrate is affected by the chemical reaction taking place on the substrate. Concentration differences in the chamber become another force driving natural convection in the chamber. For the case of LCVD by a stationary laser beam, Lee et al. (1995) concluded that the effect of natural convection on the thin-film deposition rate was negligible and that the heat and mass transfer in the gases were dominated by diffusion. In the SALD process, a laser beam scans the substrate and induces chemical reaction; the resulting product forms a line on the substrate. These lines, formed by multiple laser scans, are subsequently interwoven to form a part layer. To thoroughly understand the effects of various physical phenomena—including natural convection—on the SALD process, natural convection during LCVD with a moving laser beam was investigated by Zhang (2003).

The physical model of LCVD under consideration is illustrated in Fig. 6.11. A substrate made of Incoloy 800 with a thickness of h is located in the bottom of a chamber. Before the vapor deposition starts, the chamber is evacuated and then filled with a mixture of H_2, N_2, and $TiCl_4$. A laser beam moves along the surface of the substrate with a constant velocity, u_b. The initial temperature of the substrate, T_i, is below the chemical reaction temperature. Vapor deposition starts when the surface temperature reaches the chemical reaction temperature. The chemical reaction that occurs on the top substrate surface absorbs part of the laser energy and consumes the $TiCl_4$. A concentration difference is thereby established and becomes the driving force for mass transfer. The physical model of the LCVD process includes natural convection, heat transfer in the substrate and gases, and chemical reaction, as well as mass transfer in the gases.

The laser beam travels with a constant velocity u_b along the surface of the substrate, constituting a typical moving heat source problem. If the substrate is sufficiently large in comparison with the diameter of the laser beam, which has an order of magnitude of 10^{-3} m, a quasi-steady state occurs.

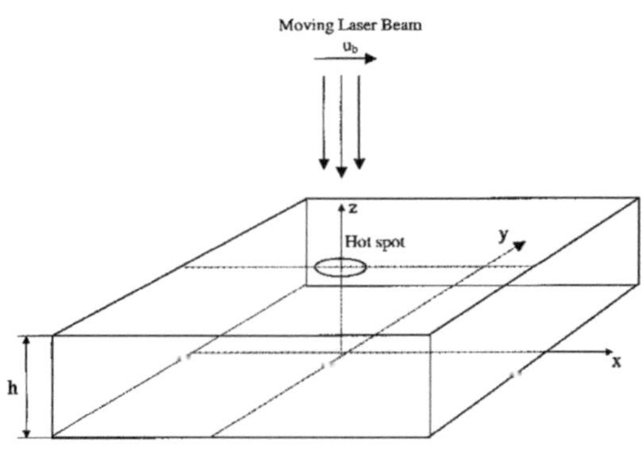

Fig. 6.11 Physical model of Laser chemical vapor deposition

The system appears to be in steady state from the standpoint of an observer located in and traveling with the laser beam. By simulating LCVD with a moving laser beam in the moving coordinate system, the computational time will be substantially shortened, thereby enabling numerical simulation for a significant number of cases.

Heat transfer in the substrate and gases is modeled as one problem with different thermal properties in each region. In the substrate region, the velocity is set to zero in the numerical solution. The advantage of modeling the heat and mass transfer in the substrate and the gases as one problem is that the temperature distribution in the substrate and gases can be obtained by solving one equation. This eliminates the iteration procedure needed to match the boundary condition at the substrate–gas interface. Since the model geometry is symmetric about the xz plane, only half of the problem needs to be investigated. For a coordinate system moving with the laser beam, as shown in Fig. 6.11, the laser beam is stationary but the substrate and the chamber move with a velocity $-u_b$. The heat and mass transfer in the substrate and gases is governed by Eqs. (6.80)–(6.84), with buoyancy forces due to temperature and concentration gradients accounted for, but the Soret effect neglected (Zhang 2003). For the substrate region, the thermal properties are those of Incoloy 800, the substrate material. For the gaseous region, the thermal properties are determined by the individual thermal properties of H_2, N_2, and $TiCl_4$ as well as their molar fractions [see Eq. (6.96)–(6.97)]. The mass diffusivity of $TiCl_4$ in the gas mixture is determined by the Stefan–Maxwell equation, using the binary diffusivity of $TiCl_4$ with respect to all other species, which is calculated using the hard sphere model.

The heat flux at the substrate surface due to laser beam irradiation and chemical reaction is expressed as

$$q'' = \frac{2P\alpha_a}{\pi r_0^2}\exp\left[-\frac{2(x^2+y^2)}{r_0^2}\right] - \varepsilon\sigma(T^4 - T_\infty^4) + \rho_{\mathrm{TiN}}\Delta H_R u_b \frac{d\delta}{dx}, \quad z = h \tag{6.104}$$

where ΔH_R is chemical reaction heat, and $d\delta/dt$ is the deposition rate. For a chemical reaction in the order of unity, the deposition rate is expressed as

$$\frac{d\delta}{dx} = -\frac{\gamma_{\mathrm{TiN}} K_0}{u_b \rho_{\mathrm{TiN}}}\exp\left(-\frac{E}{R_u T_s}\right)\omega_s \tag{6.105}$$

where ω_s represents the concentration of $TiCl_4$ at the surface of the substrate. The constant K_0 in Eq. (6.105) is defined as $K_0 = (\omega_{H_2})_i(\omega_{N_2})_i^{1/2}K_0'$.

The coefficient γ_{TiN} in Eq. (6.105) is a sticking coefficient defined as

$$\gamma_{\mathrm{TiN}} = \begin{cases} 1 & T < T_m \\ 1 + (T_m - T_s)/(T_M - T) & T_m \le T \le T_M \\ 0 & T > T_M \end{cases} \tag{6.106}$$

where T_s is the surface temperature of the substrate, T_m is a threshold temperature below which the product of the chemical reaction can fully stick to the substrate, and T_M is another threshold temperature above which no product of chemical reaction can be stuck on the substrate. If the surface temperature is between T_m and T_M, the product of chemical reaction can only be partially stuck on the substrate. The values of T_m and T_M are chosen as 1473 and 1640 K, respectively (Conde et al. 1992).

The boundary conditions of the velocities are

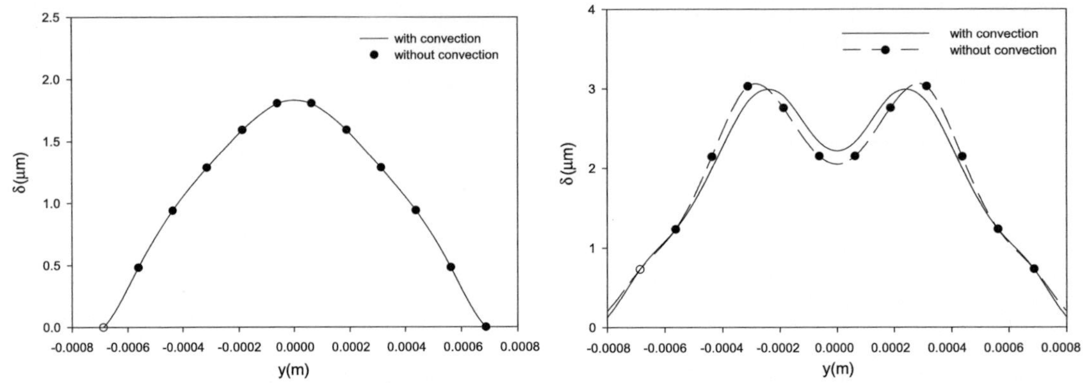

(a) P=300W, u_b=1.2mm/s (b) P=360W, u_b=1.2mm/s

Fig. 6.12 Comparison of cross sections (Zhang 2003)

$$u = -u_b, \quad v = w = 0, \quad |x| \to \infty \tag{6.107}$$

$$v = \frac{\partial u}{\partial y} = \frac{\partial w}{\partial y} = 0, \quad y = 0 \tag{6.108}$$

$$u = -u_b, \quad v = w = 0, \quad y \to \infty$$

$$u = -u_b, \quad v = w = 0, \quad z = 0, \infty \tag{6.110}$$

The governing equations are discretized using the finite volume method (Patankar 1980); the SIMPLEC algorithm (Van Doormaal and Raithby 1984) was employed to handle the linkage between velocity and pressure. The results show that the effect of natural convection on the shape of the deposited film is negligible for the laser power of P = 300 W (see Fig. 6.12a). When the laser power is increased to 360 W, the effect of natural convection on the shape of the cross section becomes important, although the cross-sectional area is almost unchanged (see Fig. 6.12b). A groove is observed on the top of the deposited film for P = 360 W due to a low sticking coefficient.

Zhang (2004) presented a parametric study on shape and cross-sectional area of the thin film produced by LCVD with a moving laser beam. The effect of natural convection on the LCVD process is neglected because it has little effect on the shapes of the deposited film, and it has no effect on the cross-sectional area of the thin film. The effects of laser scanning velocity, laser power, and radius of the laser beam on the shapes of the deposited film were investigated. The results showed that a groove could be observed on the top of the film in conjunction with higher laser power and lower scanning velocity. The cross-sectional area, calculated by

$$A_c = \frac{2}{r_0^2} \int_0^\infty \delta \mathrm{d}y \tag{6.111}$$

at different processing parameters, is shown in Fig. 6.13. It decreases with increasing scanning velocity. It also increases with increasing laser power and decreasing laser beam radius. The following empirical correlation on the dimensionless cross sectional area is obtained (Zhang 2004):

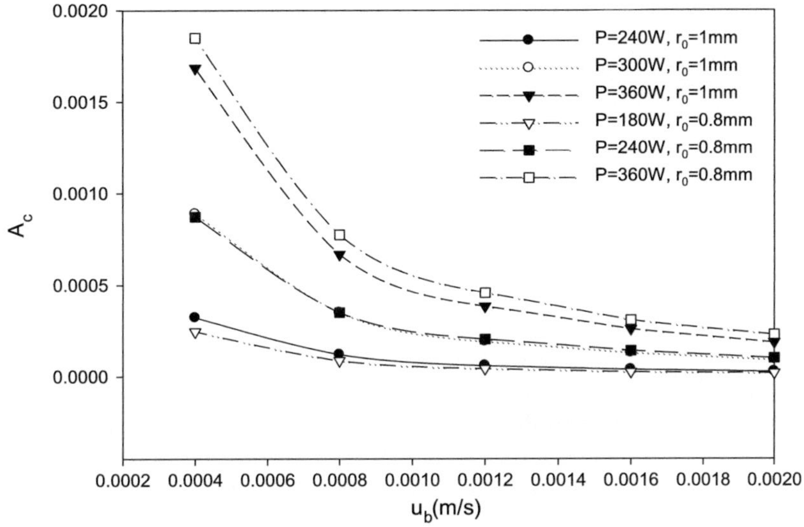

Fig. 6.13 Dimensionless cross-sectional area versus scanning velocity (Zhang 2004)

$$A_c = a_0 + a_1 \left(\frac{Bi}{Pe} \right)^{1.35} \tag{6.112}$$

where

$$a_0 = \begin{cases} -2.136 \times 10^{-5} + 6.126 \times 10^{-6} Bi & r_0 = 1.0\,\text{mm} \\ -9.206 \times 10^{-5} + 9.352 \times 10^{-5} Bi & r_0 = 0.8\,\text{mm} \end{cases} \tag{6.113}$$

$$a_1 = \begin{cases} 8.343 \times 10^{-7} + 1.423 \times 10^{-5} Bi & r_0 = 1.0\,\text{mm} \\ -2.587 \times 10^{-5} + 4.372 \times 10^{-5} Bi & r_0 = 0.8\,\text{mm} \end{cases} \tag{6.114}$$

The Biot number and Peclet number are defined as

$$Bi = \frac{\alpha_a P}{\pi r_0 k_s (T_c - T_i)} \tag{6.115}$$

$$Pe = \frac{u_b r_0}{\alpha_s} \tag{6.116}$$

where α_a is the absorptivity of the laser beam on the substrate surface, P is the laser power, r_0 is the radius of the laser beam, T_c is the chemical reaction temperature, T_i is the initial temperature of the gases, and u_b is the laser scanning velocity. The thermal conductivity k_s and the thermal diffusivity α_s in Eqs. (6.115)–(6.115) are those of the substrate (Incoloy) at chemical reaction temperature.

Problems

6.1. A subcooled solid is exposed to its superheated vapor as shown in Fig. 6.2a. The temperature at the left surface of the solid is T_0, which is below the interfacial temperature. Depending on the direction of the overall heat flux at the interface, both sublimation and deposition are possible. Derive the criteria for sublimation and deposition.

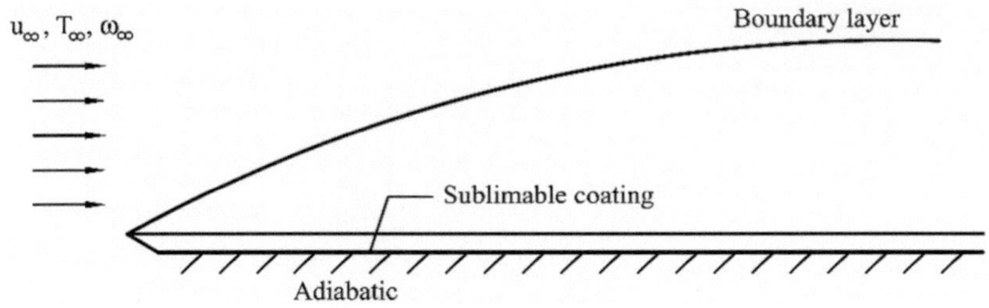

Fig. P6.4

6.2. Superheated vapor is brought into contact with a cold surface at a temperature of T_0, and deposition takes place on the cold surface. Find the deposition rate by solving transient conduction in the deposited solid phase.

6.3. The inner surface of a circular tube with radius R is coated with a layer of sublimable material, and the outer wall of the tube is kept at a constant temperature T_w. The fully developed gas enters the tube with a uniform inlet mass fraction of the sublimable substance ω_0 that equals the saturation mass fraction corresponding to the inlet temperature T_0. The thermal and mass diffusivities are assumed to be the same, i.e., Le = 1. Find the local Nusselt number based on convective heat flux and the total heat flux at the wall, and the local Sherwood number.

6.4. A gas with a velocity of u_∞, a concentration of a sublimable substance, ω_∞, and a temperature of T_∞ flows parallel to a flat plate coated with sublimable materials; the back of the flat plate is adiabatic (see Fig. P6.4). Specify the governing equations and the corresponding boundary conditions of the sublimation problem.

6.5. Suppose the blowing velocity on the surface of the flat plate satisfies $v_w \propto x^{-1/2}$. Introduce appropriate similarity variables to the governing equations in Problem 6.4 and reduce the governing equations into a set of ordinary differential equations.

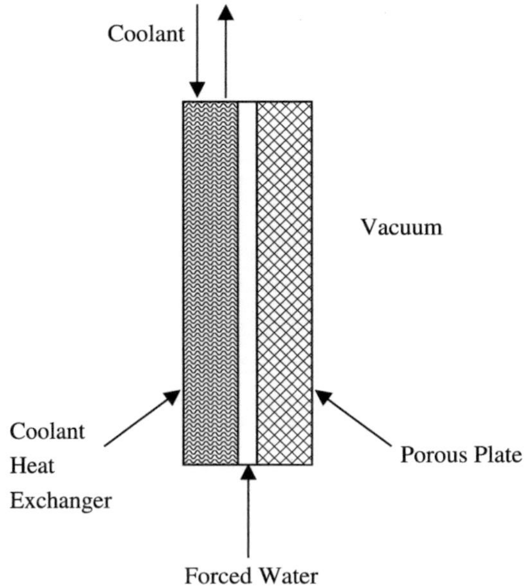

Fig. P6.8

6.6. Write a computer program to solve for the ordinary differential equations of Problem 6.5, and obtain the local Nusselt number and Sherwood number.

6.7. Air with a temperature of 27 °C flows at 1 m/s over a 1-m-long solid fuel surface at a temperature of 527 °C. The average blowing velocity due to sublimation of the solid fuel is 0.01 m/s, and the heat released per unit mass of the oxidant consumed is 10,000 kJ/kg. The latent heat of sublimation for the solid fuel is 1350 kJ/kg. The sensible heat required to raise the surface temperature of the solid fuel to sublimation temperature, and heat loss to the solid fuel, can be neglected. Estimate the mass fraction of the oxidant at the solid fuel surface.

6.8. Manned spacecraft and space suits reject excess thermal energy by sublimating water into the vacuum of space. The sublimator consists of a porous plate exposed to vacuum on one side and feed water on the other side. The feed water seeps into the porous plate, where it then freezes. In Fig. P6.8, the sublimator also has a separate coolant heat exchanger that interfaces with the feed water. Describe in detail how the process of sublimation keeps the astronaut or spacecraft cool.

6.9. The precursor for SALD of TiN film is a mixture of titanium tetrachloride, nitrogen, and hydrogen. The total pressure in the chamber is 207 torr and the partial pressure of titanium tetrachloride is 7 torr. The partial pressures of N_2 and H_2 are the same. Estimate the viscosity and thermal conductivity of the precursors at 900 K.

6.10. Estimate the binary mass diffusivity of titanium tetrachloride to nitrogen gas in the gaseous mixture described in Problem 6.9. The collision diameter of a nitrogen molecule is $\sigma = 3.681$ Å, and the characteristic energy of interaction between molecules satisfies $\varepsilon/k_b = 91.5$ K.

6.11. What is the binary mass diffusivity of titanium tetrachloride to the hydrogen gas in the gaseous mixture described in Problem 6.9? The collision diameter of the hydrogen molecule is $\sigma =$ 2.915 Å, and the characteristic energy of interaction between hydrogen molecules satisfies $\varepsilon/k_b = 38.0\,\text{K}$.

6.12. Find the mass diffusivity of $TiCl_4$ to a mixture of N_2 and H_2 used in SALD of TiN. The total pressure of the precursors is 207 torr, and the partial pressure of $TiCl_4$ is 7 torr. The partial pressures of both N_2 and H_2 are at 100 torr.

6.13. In an LCVD of TiN using a mixture of $TiCl_4$, N_2, and H_2, a laser beam with a power of 350 W and a radius of 1 mm scans at a velocity of 1.5 mm/s. The absorptivity of the laser beam on the substrate surface is 0.23. The initial gas temperature is 338 K, and the chemical reaction temperature is 1173 K. The thermal conductivity and thermal diffusivity of the substrate at 1173 K are 24.5 W/m-K and $4.677 \times 10^{-6}\,\text{m}^2/\text{s}$, respectively. Estimate the dimensional cross-sectional area of the deposited film.

6.14. Laser chemical vapor infiltration (LCVI) is an additive manufacturing process that uses gas precursors and powder particles to build three-dimensional parts. The powder particles are bounded together through LCVD on the particle surface by decomposition of gas precursors. A nearly homogeneous distribution of the heat flux on the particle surface can be assumed due

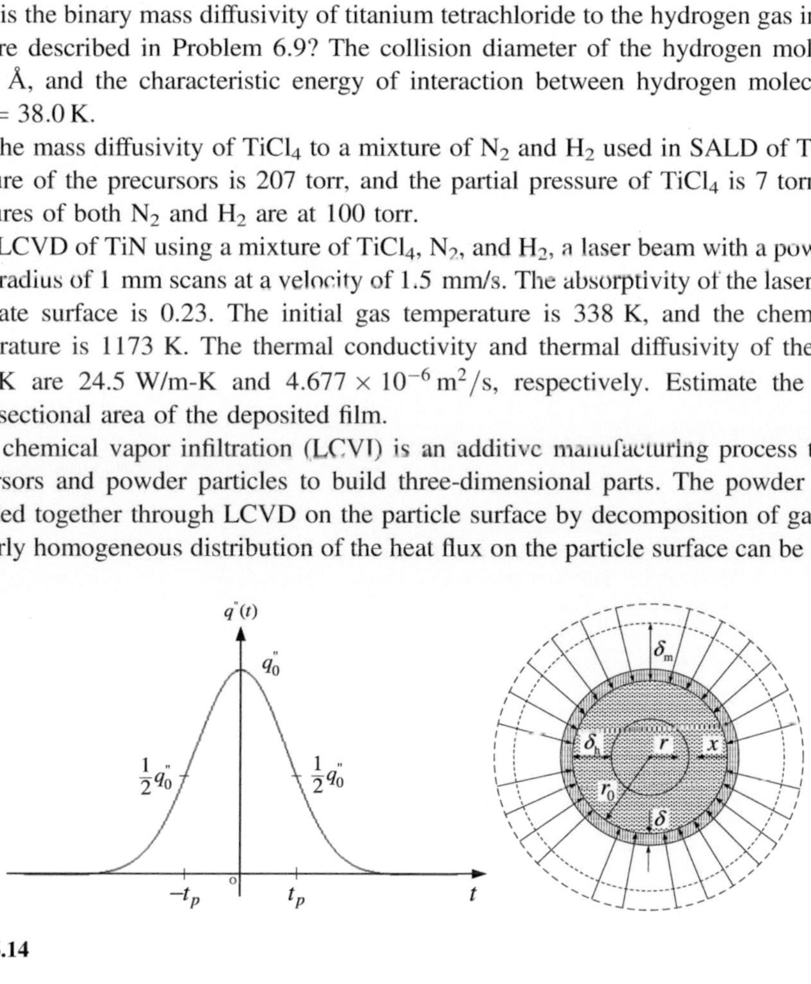

Fig. P6.14

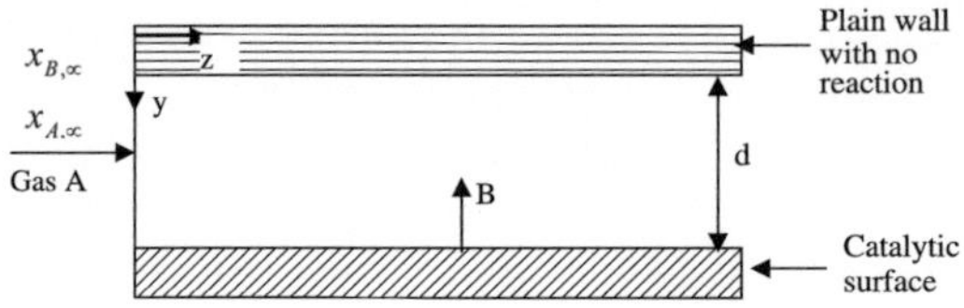

Fig. P6.16

to multiple scattering of the radiation. Figure P6.14 illustrates a metal powder particle made of Incoloy 800 that is surrounded by a mixture of H_2, N_2, and $TiCl_4$ and subjected to temporal Gaussian heat flux from a nanosecond pulsed laser beam. Because of symmetry of the spherical particle and the assumption of uniform heat flux distribution, the problem can be simplified to be one-dimensional in the *r*-direction. Assuming that natural convection can be neglected, give the governing equations and boundary conditions to obtain the deposited film thickness by considering heat transfer in the particle, chemical reaction on the particle surface, and mass transfer in the gas phase. The origin of time is chosen to be when the heat flux is at its maximum; thus the time-dependent heat flux is expressed as $q''(t) = q''_0 \exp[-\ln\left(2t^2/t_p^2\right]$, where q''_0 is the maximum heat flux and t_p is the half-width of the laser pulse at half maximum (see Fig. P6.14).

6.15. Solve heat transfer in the powder particle and mass transfer in the gas phase in Problem 6.14 using integral solution (see Chap. 5) and outline the procedure to obtain the deposited film thickness.

6.16. Model a simplified catalytic reaction in Fig. P6.16 in which gas A enters the reactor and is convected to B. Assume that at the bottom surface a reaction 2A $\rightarrow$ B is being carried out steadily, irreversibly, and instantaneously in an isothermal process. Assume ideal gas and obtain the local mass flux rate of A to B.

6.17. Repeat Problem 6.16, but account for the finite reaction kinetics at the catalytic surface. All other assumptions remain the same. Assume the rate at which A disappears at the catalyst surface is proportional to the molar concentration of A in the gas at the surface (first-order surface reaction).

References

Bauerle, D. (1996). *Laser processing and chemistry*. New York, NY: Springer.

Bird, R. B., Stewart, W. E., & Lightfoot, E. N. (2006). *Transport phenomena* (2nd ed.). New York: Wiley. (Revised).

Conde, O., Kar, A., & Mazumder, J. (1992). Laser chemical vapor deposition of TiN Dot: A comparison of theoretical and experimental results. *Journal of Applied Physics, 72,* 754–761.

Eckert, E. R. G., & Goldstein, R. J. (1976). *Measurement in heat transfer*. New York, NY: McGraw-Hill.

Elyukhin, V. A., Garcia-Salgado, G., & Pena-Sierra, R. (2002). Thermodynamic model for low temperature metalorganic chemical vapor deposition of GaN. *Journal of Applied Physics, 91,* 9091–9094.

Evans, G., & Greif, R. (1987). A numerical model of the flow and heat transfer in a rotating disk chemical vapor deposition reactor. *Journal of Heat Transfer, 109,* 928–935.

Faghri, A., Zhang, Y., & Howell, J. R. (2010). *Advanced heat and mass transfer*. Columbia, MO: Global Digital Press.

Glassman, I., & Yetter, R. A. (2008). *Combustion* (4th ed.). Burlington, MA: Elsevier.

Herring, R. B. (1990). Silicon epitaxy, In W. C. O'Mara, R. B. Herring, & L. P. Hunt (Eds.), *Handbook of semiconductor silicon technology* (pp. 258–336). Park Ridge, NJ: Noyes Publications.

Jakubenas, K. J., Birmingham, B., Harrison, S., Crocker, J., Shaarawi, M.S., Tompkins, J. V., et al. (1997). Recent development in SALD and SALDVI. In *Proceedings of 7th International Conference on Rapid Prototyping*, San Francisco, CA.

Jensen, K. F., Einset, E. O., & Fotiadis, D. I. (1991). Flow phenomena in chemical vapor deposition of thin films. *Annual Review of Fluid Mechanics, 23,* 197–232.

Kaviany, M. (2001). *Principles of convective heat transfer* (2nd ed.). New York: Springer.

Kays, W. M., Crawford, M. E., & Weigand, B. (2004). *Convective heat transfer* (4th ed.). New York, NY: McGraw-Hill.

Kurosaki, Y. (1973). Coupled heat and mass transfer in a flow between parallel flat plate (Uniform heat flux). *Journal of the Japan Society of Mechanical Engineers, Part B, 39,* 2512–2521. (in Japanese).

Kurosaki, Y. (1974). Coupled heat-mass transfer of a flat plate with uniform heat flux in a laminar parallel flow. *Journal of the Japan Society of Mechanical Engineers, Part B, 40,* 1066–1072. (in Japanese).

Lee, Y. L., Tompkins, J. V., Sanchez, J. M., & Marcus, H. L. (1995). Deposition rate of silicon carbide by selected area laser deposition. *Proceedings of Solid Freeform Fabrication Symposium, 1995,* 433–439.

Mahajan, R. L. (1996). Transport phenomena in chemical vapor-deposition systems. In *Advances in heat transfer*. San Diego, CA: Academic Press.

Marcus, H. L., Zong, G., & Subramanian, P. K. (1993). Residual stresses in laser processed solid freeform fabrication, residual stresses in composites. In E. V. Barrera & I. Dutta (Eds.), *Measurement, modeling and effect on thermomechanical properties* (pp. 257–271). TMS.

Mazumder, J., & Kar, A. (1995). *Theory and application of laser chemical vapor deposition*. New York, NY: Plenum Publishing Co.

Patankar, S. V. (1980). *Numerical heat transfer and fluid flow*. Washington, DC: Hemisphere.

Powell, C., Blocher, M., & Oxley, J. (1966). *Vapor deposition*. New York: Wiley.

Sivaram, S. (1995). *Chemical vapor deposition: Thermal and plasma deposition of electronic materials*. Bordrecht, Netherlands: Kluwer Academic Publishers.

Sun, L., Jakubenas, K. J., Crocker, J. E., Harrison, S., Shaw, L. L., & Marcus, H. L. (1998). In situ thermocouples in micro-components fabricated using SALD/SALDVI techniques: II evaluation of processing parameters. *Materials and Manufacturing Processes, 13,* 883–907.

Taylor, C. A., Wayne, M. F., & Chiu, W. K. S. (2004). Microstructural characterization of thin carbon films deposited from hydrocarbon mixtures. *Surface & Coatings Technology, 182,* 131–137.

Ueda, O. (1996). *Reliability and degradation of III-V optical devices*. Boston: Artech House Inc.

Van Doormaal, J. P., & Raithby, G. D. (1984). Enhancements of the simple method for predicting incompressible fluid flows. *Numerical Heat Transfer, 7,* 147–163.

Versteeg, V. A., Avedisian, C. T., & Raj, R. (1995). Metalorganic chemical vapor deposition by pulsed liquid injection using an ultrasonic nozzle: Titanium dioxide on sapphire from titanium (IV) isopropoxide. *Journal of the American Ceramic Society, 78,* 2763–2768.

Zhang, Y. (2003). Quasi-steady state natural convection in laser chemical vapor deposition with a moving laser beam. *Journal of Heat Transfer, 125,* 429–437.

Zhang, Y. (2004). A simulation-based correlation of cross-sectional area of the thin film produced by laser chemical vapor deposition with a moving laser beam. *Journal of Manufacturing Science and Engineering, 126,* 796–800.

Zhang, Y., & Chen, Z. Q. (1990). Analytical solution of coupled laminar heat-mass transfer inside a tube with adiabatic external wall. In *Proceedings of the 3rd National Interuniversity Conference on Engineering Thermophysics* (pp. 341–345). Xi'an Jiaotong University Press, Xi'an, China.

Zhang, Y., Chen, Z. Q., & Chen, M. (1996). Local non-similarity solution of coupled heat-mass transfer of a flat plate with uniform heat flux in a laminar parallel flow. *Journal of Thermal Science, 5,* 112–116.

Zhang, Y., & Faghri, A. (2000). Thermal modeling of selective area laser deposition of titanium nitride on a finite slab with stationary and moving laser beams. *International Journal of Heat and Mass Transfer, 43,* 3835–3846.

Condensation 7

7.1 Introduction

Condensation occurs when a saturated or superheated vapor—pure or multicomponent—comes into contact with an object, such as a wall or other contaminant, which has a temperature below the saturation temperature. In most applications involving the condensation of a vapor, heat is transferred to a solid wall adjacent to the vapor. If the solid wall temperature is below the equilibrium saturation temperature at the system pressure, a liquid droplet embryo may form at this solid–vapor interface. This condensation is referred to as *heterogeneous nucleation* of a liquid droplet embryo. Heterogeneous liquid droplet nucleation is nucleation of a vapor droplet embryo at the interface of a metastable vapor phase and another phase; this second phase is usually solid and is naturally held at a lower temperature than the vapor. A metastable vapor is one that is supercooled below its equilibrium saturation temperature at the system pressure (see Chap. 2). Figure 7.1 is a flowchart schematic showing the different modes by which a liquid droplet embryo can form. It can be seen from this figure that the embryos can also be formed homogeneously. *Homogeneous nucleation* of a liquid droplet occurs entirely within a supercooled vapor. The liquid droplet is completely surrounded by supercooled vapor and is not attached to a lower temperature wall, as is the case in the heterogeneous process.

The heterogeneous condensation occurs when vapor condenses on the cooled surface as a thin film or as droplets, depending on whether the surface is wettable or nonwettable with the condensate. If the surface is wettable with the condensate, a continuous condensate film forms on the surface and filmwise condensation occurs (see Fig. 7.2a). Conversely, if the surface is not wettable with the condensate, a series of condensate droplets form on the surface (i.e., dropwise condensation occurs) (see Fig. 7.2b). Dropwise condensation is preferred over filmwise condensation, because its resulting overall heat transfer coefficient is higher by as much as an order of magnitude. This difference occurs because the cooler wall surface always has an area that is not covered by condensate. The resistances encountered at the liquid–vapor interface, along with the conduction resistance through the liquid itself, are removed and the heat transfer is increased significantly. Therefore, in industrial applications, it is wise to introduce conditions that promote dropwise condensation. The dropwise condensation can be promoted by taking one or more of the following steps:

1. Introduce a nonwetting agent into the vapor that will eventually deposit on the cooling surface to break up wetting conditions.

A. Faghri and Y. Zhang, *Fundamentals of Multiphase Heat Transfer and Flow*,
https://doi.org/10.1007/978-3-030-22137-9_7

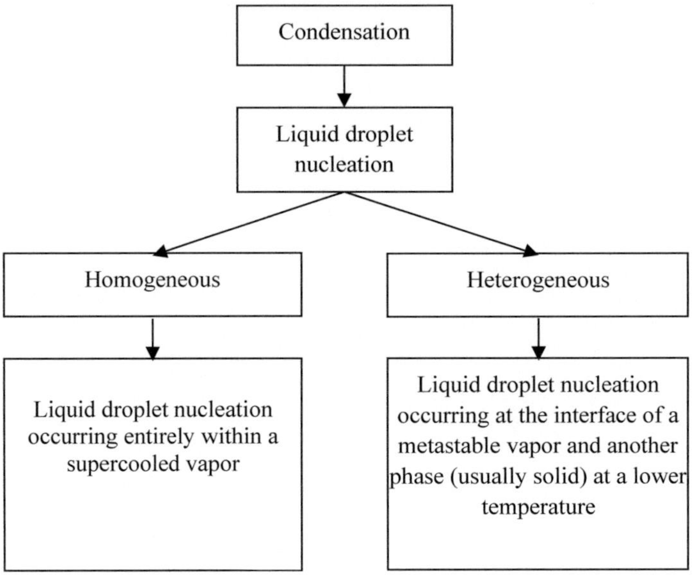

Fig. 7.1 Flowchart of the different modes of condensation

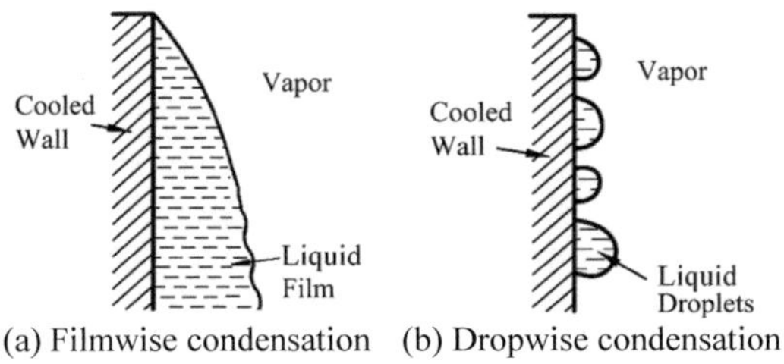

Fig. 7.2 Heterogeneous condensation. **a** Filmwise condensation, **b** dropwise condensation

2. Apply grease or waxy products that are poor wetting agents to the cool wall surface in order to promote nonwetting conditions.
3. Permanently coat the wall surface with a low surface energy or noble metal.

If a tiny amount of sufficiently supercooled contaminant is introduced to the vapor, condensate can form on the contaminant, even if no cold wall is present. This is an example of homogeneous condensation that is different from the heterogeneous condensation mentioned above, in that it relies on a solid, liquid, or even vapor contaminant to initiate the condensation process. This type of condensation produces a mist-like quality and is depicted in Fig. 7.3a. Two more examples of homogeneous condensation are shown in Fig. 7.3b, c. When cold liquid is introduced into vapor through a nozzle, liquid droplets are formed; vapor condenses on the surface of these droplets suspended in the vapor phase (see Fig. 7.3b). When vapor bubbles are introduced into the bulk liquid, as shown in Fig. 7.3c, the vapor bubbles surrounded by the cold liquid shrink and eventually collapse due to condensation.

In many industrial applications, the saturated vapor to be condensed is in fact a miscible binary vapor mixture. In a multicomponent vapor, the saturation temperature is referred to as the *dew point*. As will be described below, this binary mixture does condense differently than a pure vapor and has

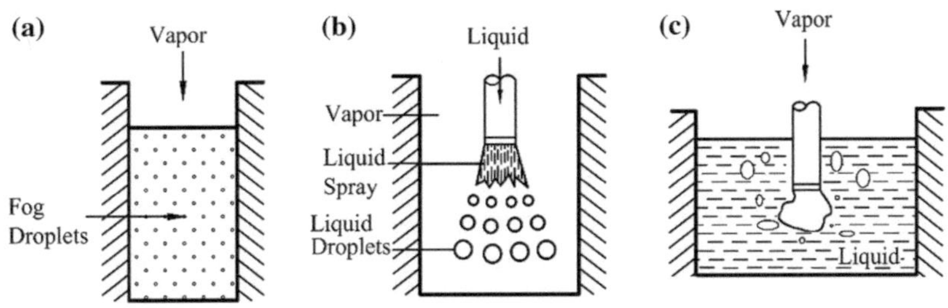

Fig. 7.3 Homogeneous condensation: **a** condensation on small contaminant particles in the vapor mixture, **b** condensation on liquid droplets, and **c** condensation of vapor bubbles

lower heat transfer and condensation rates (Fujii 1991). Figure 7.4a shows the transverse distributions of temperature T and mass fractions ω_{1v} and ω_{2v} in the condensate film, vapor boundary layer, and bulk vapor of a steady condensation process consisting of a binary miscible vapor mixture in contact with a vertical cooled wall; the phase equilibrium diagram of this condensation process is shown in Fig. 7.4b. The dew point line reflects the points at which the binary vapor/gas begins to condense. The boiling point line reflects the points at which the binary liquid begins to boil or when the binary vapor is completely condensed. When a binary vapor mixture is cooled below its saturation temperature by contact with a cold wall, the less volatile component 2 (with higher saturation temperature) condenses more than the volatile component 1. In other words, as the vapor mixture cools, the component with the higher saturation temperature at the system saturation pressure will condense first. If it is assumed that the bulk vapor mixture is kept at a constant density, the volatile component (with lower saturation temperature) must become very dense (while remaining in its vapor form) at the liquid–vapor interface. Meanwhile, the less volatile component condenses into the liquid phase first. This can be seen in Fig. 7.4a, where the volatile component increases in mass at the interface, while the less volatile component is at its lowest mass at the interface. Gravity or drag forces constantly flush the surface of the wall, so the dense mass of the volatile component is removed by convection forces. However, due to this buildup of the volatile component at the interface, a larger resistance to heat flow is produced in a multicomponent system than in a pure vapor condensation process. Due to the difference in saturation properties of the binary mixture components, the temperature drop across the interface is larger than for a pure vapor condensation process. Therefore, the interfacial resistance for the condensation of a binary vapor mixture usually cannot be neglected, as is sometimes the case for the condensation of pure vapors. This interfacial resistance will, however, be developed in Sect. 7.2.3 for a pure vapor. Figure 7.4b illustrates condensation of a binary mixture of miscible vapors on a phase equilibrium chart. Initially, the vapor is at T_∞, and it is placed in contact with a cooler wall. The vapor temperature decreases until it reaches T_{sat}, at which point the condensation process begins. The dew point line on the diagram is the saturation temperature for various concentrations of the two components in the mixture. The less volatile component in the mixture (component 2 in this case) will condense faster than the other. At the temperature of the interface between the liquid and the vapor phases, the mass fraction of each phase can be read off the tie line between the dew point and boiling point lines. For example, the mass fractions of the component 1 at the interface ($y = \delta$) are $\omega_{1v\delta}$ and $\omega_{1\ell\delta}$ from the vapor and liquid sides, respectively. As can be seen, the condensate film will be much heavier with the less volatile component, and the vapor will have a higher concentration of the more volatile component.

Other applications involve the condensation of vapors of partially or completely immiscible liquids such as water or organic compounds. During condensation of vapors of some immiscible liquids, the condensate will form in a combination of dropwise and filmwise liquid. One component will

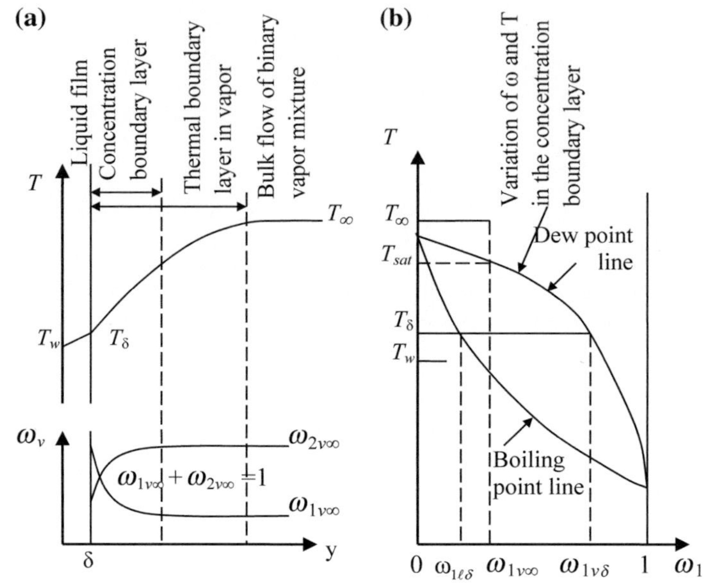

Fig. 7.4 Temperature and mass fraction in the condensation of a binary miscible vapor mixture: **a** T and ω distribution in the condensate film and vapor boundary layer; **b** variation of T and ω on a diagram of phase equilibrium (Fujii 1991; Reproduced with kind permission of Springer Science and Business Media)

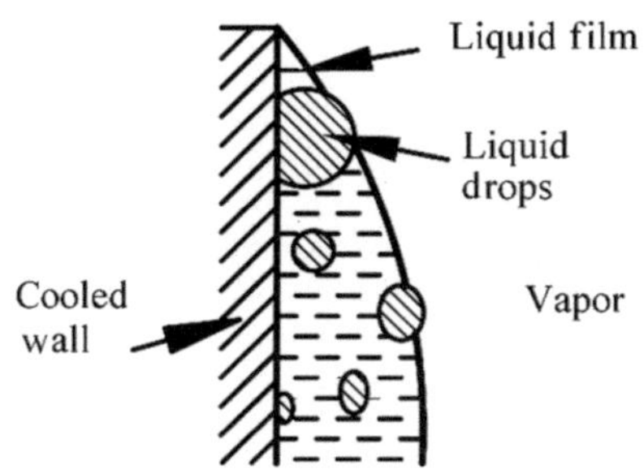

Fig. 7.5 Condensation of immiscible fluids

condense as a liquid film with droplets of the other component floating within or on top of it. This mixture can be seen in Fig. 7.5 and is caused by the different surface tension forces of the two components in relation to the vapor and solid wall. Figures 7.6 and 7.7 show the phase diagrams for binary mixtures with a miscibility gap and with completely immiscible liquids, respectively. The point where the dew point lines meet the immiscible liquid regions is the eutectic point, which occurs at the eutectic temperature for a given pressure. Condensation may occur in several forms for these cases. If the wall temperature is greater than the eutectic temperature or if the condensate film is sufficiently thick, the interface temperature must also be above the eutectic temperature. This case is similar to the condensation of miscible liquids: a homogenous condensate film will form, and the concentrations of each component in the vapor and liquid phases can be read off the tie line corresponding to the interface temperature. When the wall temperature is less than the eutectic temperature and the condensate film is thin, the interface temperature will be equal to the eutectic temperature. As condensation occurs, two immiscible liquid phases form. These liquid phases are in equilibrium with the remaining vapor, which will have the eutectic composition.

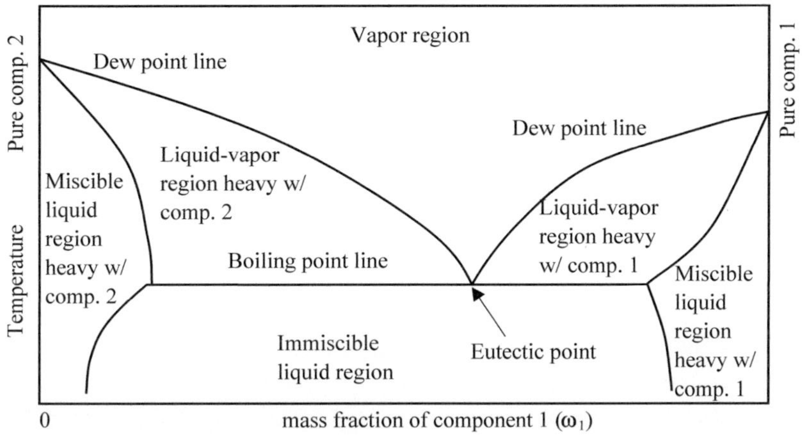

Fig. 7.6 Phase diagram of liquids with a miscibility gap

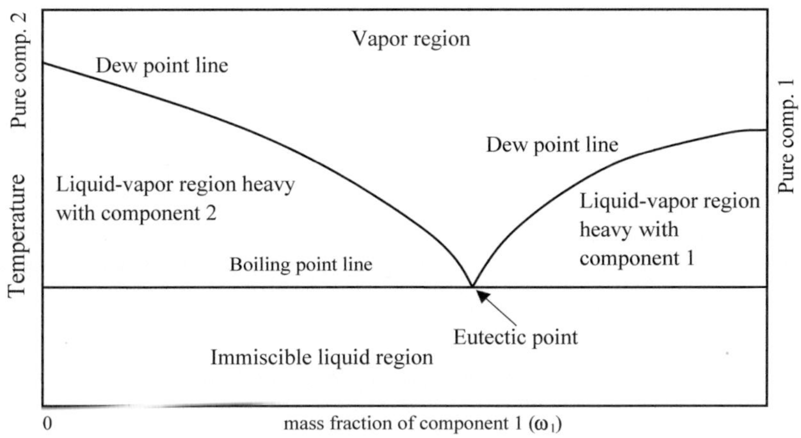

Fig. 7.7 Phase diagram of completely immiscible pure liquid

In Sect. 7.2, the mechanisms of dropwise condensation will be discussed, starting with the critical equilibrium radius for the initial formation of a liquid droplet embryo. This discussion will show that the critical equilibrium radius is dependent on the contact angle of the liquid droplet with the wall surface and on the interfacial surface tensions found in the liquid droplet. A detailed discussion on heat transfer resistance in the condensation process will be presented, with emphasis on the dropwise condensation process. Section 7.3 begins with an introduction to the filmwise condensation problem, which is followed by the generalized governing equations for laminar film condensation of a binary vapor. The classical Nusselt analysis of laminar film condensation on a vertical wall is discussed, and the generalized condensation process for a flowing vapor reservoir for both laminar and turbulent film condensation will be addressed. Turbulent film condensation in a tube with both cocurrent and countercurrent vapor flow is also presented, as are empirical correlations for filmwise condensation configurations. Section 7.3 closes with a discussion of the effect of noncondensable gases on the condensation process. Filmwise condensation in porous media will be discussed in Chap. 12.

7.2 Dropwise Condensation

7.2.1 Dropwise Condensation Formation

Several theories have been proposed to explain the mechanism of dropwise condensation. The first model, proposed by Eucken (1937), has been supported by many experimental studies, such as that by McCormick and Baer (1963). It states that liquid droplets form only heterogeneously at nucleation sites; if they are formed with a radius exceeding that of equilibrium, they will continue to grow and then join with surrounding droplets. Once the mass of the condensate reaches a critical point, it will be removed from the surface by gravitational forces or by drag forces produced by the surrounding gas. As droplets are removed, the surface is wiped clean of condensate and the process restarts at the nucleation sites. This periodic cleansing constitutes the advantage of dropwise condensation over filmwise condensation, as there is no resistance to heat transfer through the condensate when the condensate layer is removed, and thus, the heat transfer rate increases greatly.

The second approach postulates that between drops there exists a thin and unstable liquid film on a solid surface. As the condensation process continues and the thin film grows thicker, the film reaches a critical thickness—estimated to be on the order of 1 μm—at which point it breaks up into droplets. Condensation then continues in the dry areas between the recently formed droplets, and on top of the already formed droplets. The majority of new condensate does occur on the wall surface because there is less resistance to heat conduction than if the new condensate formed on the already existing droplets. These new condensate droplets are then drawn into the neighboring droplets by surface tension effects, producing a new thin film. This film will then grow and rupture at the critical thickness, and the process will repeat continuously.

Dropwise condensation takes place if the condensate cannot wet the surface. The wettability can be characterized by a contact angle (see Fig. 7.8) defined by the following Young's equation:

$$\cos\theta = \frac{\sigma_{sv} - \sigma_{s\ell}}{\sigma_{\ell v}} \tag{7.1}$$

where σ_{sv}, $\sigma_{s\ell}$, and $\sigma_{\ell v}$ are surface tensions between solid–vapor, solid–liquid, and liquid–vapor interfaces. When the contact angle, θ, is greater than 90°, the condensate cannot wet the surface and dropwise condensation occurs. The criterion for dropwise condensation is the critical surface tension σ_{cr} which is characteristic of the surface alone. If the surface tension between the liquid–vapor interface $\sigma_{\ell v}$ is greater than σ_{cr}, dropwise condensation occurs (Shafrin and Zisman 1960). Critical surface tensions for selected solid surfaces are given in Table 7.1. It can be seen that the critical surface tension for all solids listed in the table is below the surface tension of water at 1 atm ($\sigma_{\ell v} = 58.91 \times 10^{-3}\,\mathrm{N/m}$). Therefore, a metal surface, on which film condensation usually occurs, can be coated with another substance that has lower critical surface tension, such as one of those in Table 7.1, to promote dropwise condensation.

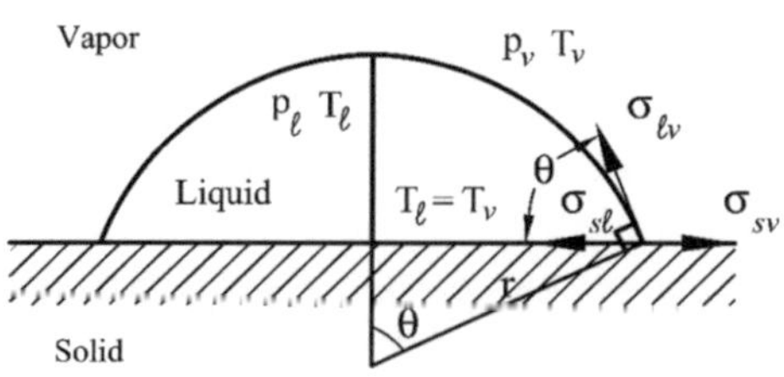

Fig. 7.8 Contact angle and equilibrium of surface tensions of a liquid droplet embryo

Table 7.1 Critical surface tension for selected solid surfaces (Shafrin and Zisman 1960)[a]

Solid surface	σ_{cr} (10^{-3} N/m)
Kel-F®	31
Nylon	46
Platinum with perfluorobutyric acid monolayer	10
Platinum with perfluorolauric acid monolayer	6
Polyethylene	31
Polystyrene	33
Polyvinyl chloride	39
Teflon®	18

[a]Reprinted with permission from American Chemical Society

7.2.2 Critical Droplet Radius for Spontaneous Growth and Destruction

As mentioned before, upon formation in their nucleation sites, droplets grow only if they form with a radius that exceeds the equilibrium radius. The analysis that leads to the definition of the critical equilibrium radius is presented below.

A good place to start this derivation lies in the minimum Gibbs free energy principle. The Gibbs free energy, G, arises from one of the Legendre transformations for closed systems states, which states that

$$G = G(T,p) = E - TS + pV \tag{7.2}$$

where T, p, V, E, and S are the system temperature, pressure, volume, internal energy, and entropy, respectively. Equation (7.2) can be differentiated to obtain

$$\mathrm{d}G = \mathrm{d}E - T\mathrm{d}S - S\mathrm{d}T + p\mathrm{d}V + V\mathrm{d}p \tag{7.3}$$

Assuming the only work term is of the $p\mathrm{d}V$ type, and neglecting potential and kinetic terms, the fundamental thermodynamic relationship is (see Sect. 2.2)

$$\mathrm{d}E \leq T\mathrm{d}S - p\mathrm{d}V \tag{7.4}$$

Substituting Eq. (7.4) into Eq. (7.3) yields

$$\mathrm{d}G \leq -S\mathrm{d}T + V\mathrm{d}p \tag{7.5}$$

Assuming that the system pressure and temperature are held at fixed values, the above expression reduces to

$$\mathrm{d}G \leq 0 \tag{7.6}$$

Along with the assumption that the system pressure and temperature are fixed, G becomes the availability of the system, Ψ:

$$\Psi = G = E + p_0 V - T_0 S \tag{7.7}$$

where p_O and T_O are the reservoir pressure and temperature, respectively. Availability is defined as the maximum amount of work one can extract from a system as it comes into equilibrium with a large reference environment of pressure p_O and temperature T_O. It follows from this analysis that for equilibrium to occur in the system, $d\Psi = 0$. Furthermore, stable equilibrium corresponds to a minimum value of Ψ.

Carey (2016) considered a system with supersaturated vapor, adjacent to a vertical wall, that has an initial temperature and pressure of T_v and p_v. The initial availability Ψ_0 of this system is given by

$$\Psi_0 = m_{\text{total}} g_v + (A_{sv})_I \sigma_{sv} \tag{7.8}$$

where m_{total} is the total mass of the system, and g_v is the specific Gibbs free energy of the vapor phase which is a function of T_v and P_v. The second term on the right-hand side represents the contribution of the work done by surface tension at the solid–vapor interface on the availability [see Eq. (2.209)], and $(A_{sv})_I$ is the initial surface area shared by the solid and vapor of the system.

Now, consider that the wall temperature is lowered and droplets begin to form on the surface of the wall. The total availability of the system is the sum of the availabilities of the liquid Ψ_ℓ, the vapor Ψ_v, and the interface Ψ_I. The expressions for each are as follows:

$$\Psi_\ell = m_\ell [g_\ell + (p_v - p_\ell) v_\ell] \tag{7.9}$$

where the second term in the brackets on the right-hand side corrects for the difference between the vapor and liquid, and

$$\Psi_v = (m_{\text{total}} - m_\ell) g_v \tag{7.10}$$

$$\Psi_I = A_{\ell v}\sigma_{\ell v} + (A_{sv})_f \sigma_{sv} + A_{s\ell}\sigma_{s\ell} \tag{7.11}$$

where the three terms on the right-hand side represent the work done by surface tension at the liquid–vapor, solid–vapor, and solid–liquid interfaces. $(A_{sv})_f$ is the difference between the initial surface area shared by the solid–vapor interface and the surface area shared by the solid–liquid interface.

$$(A_{sv})_f = (A_{sv})_I - A_{s\ell} \tag{7.12}$$

The areas are calculated directly from Fig. 7.8:

$$A_{\ell v} = 2\pi r^2 (1 - \cos\theta) \tag{7.13}$$

$$A_{s\ell} = \pi r^2 (1 - \cos^2\theta) \tag{7.14}$$

and the volume of the liquid droplet is

$$V_\ell = \frac{\pi r^3}{3}(2 - 3\cos\theta + \cos^3\theta) \tag{7.15}$$

The change of the total availability of the system due to formation of the condensate embryo is then

$$\begin{aligned}\Delta\Psi &= (\Psi_\ell + \Psi_v + \Psi_I) - \Psi_0 \\ &= [m_\ell g_\ell + V_\ell(p_v - p_\ell) + (m_{total} - m_\ell)g_v \\ &\quad + A_{\ell v}\sigma_{\ell v} + (A_{sv})_f\sigma_{sv} + A_{s\ell}\sigma_{s\ell}] - [m_{total}g_v - (A_{sv})_I\sigma_{sv}] \\ &= m_\ell(g_\ell - g_v) + V_\ell(p_v - p_\ell) \\ &\quad + A_{\ell v}\sigma_{\ell v} + [(A_{sv})_f - (A_{sv})_I]\sigma_{sv} + A_{s\ell}\sigma_{s\ell}\end{aligned} \tag{7.16}$$

Substituting Eq. (7.12) into Eq. (7.16), the change of availability becomes

$$\Delta\Psi = m_\ell(g_\ell - g_v) + V_\ell(p_v - p_\ell) + A_{\ell v}\sigma_{\ell v} + A_{s\ell}(\sigma_{s\ell} - \sigma_{sv}) \tag{7.17}$$

Substituting Young's equation, Eq. (7.1), into Eq. (7.17), one obtains

$$\Delta\Psi = m_\ell(g_\ell - g_v) + V_\ell(p_v - p_\ell) + (A_{\ell v} - A_{s\ell}\cos\theta)\sigma_{\ell v} \tag{7.18}$$

Substituting Eqs. (7.13) and (7.14) into Eq. (7.18), the following expression is obtained for a change in availability of the system:

$$\begin{aligned}\Delta\Psi &= \Psi - \Psi_0 \\ &= m_\ell(g_\ell - g_v) + V_\ell(p_v - p_\ell) + 4\pi r^2\sigma_{\ell v}F\end{aligned} \tag{7.19}$$

where

$$F = \frac{2 - 3\cos\theta + \cos^3\theta}{4} \tag{7.20}$$

Considering that the pressure in the liquid droplet is related to the pressure in the vapor phase by $p_\ell - p_v = 2\sigma_{\ell v}/r$ and substituting Eq. (7.15) into Eq. (7.19), one obtains

$$\Delta\Psi = \Psi - \Psi_0 = m_\ell(g_\ell - g_v) + \frac{4}{3}\pi r^2\sigma_{\ell v}F \tag{7.21}$$

The mass of the embryo is

$$m_\ell = \frac{V_\ell}{v_\ell} = \frac{4}{3v_\ell}\pi r^3 F \tag{7.22}$$

Substituting Eq. (7.22) into Eq. (7.21), one obtains

$$\Delta\Psi = \Psi - \Psi_0 = \frac{4}{3v_\ell}\pi r^3 F(g_\ell - g_v) + \frac{4}{3}\pi r^2\sigma_{\ell v}F \tag{7.23}$$

If the embryo droplet has the exact equilibrium radius, r_e, for which the liquid droplet is in thermodynamic and mechanical equilibrium with the surrounding vapor,

$$g_{\ell,e} = g_{v,e} \tag{7.24}$$

Therefore, the change of availability at equilibrium becomes

$$\Delta\Psi_{\mathrm{e}} = \frac{4}{3}\pi r_{\mathrm{e}}^2 \sigma_{\ell \mathrm{v}} F \tag{7.25}$$

The change of availability near the equilibrium radius can be obtained by expanding Eq. (7.21) in the form of a Taylor series, i.e.,

$$\Delta\Psi = \Delta\Psi_{\mathrm{e}} + \left.\frac{\partial \Delta\Psi}{\partial r}\right|_{\mathrm{e}} (r - r_{\mathrm{e}}) + \frac{1}{2}\left.\frac{\partial^2 \Delta\Psi}{\partial r^2}\right|_{\mathrm{e}} (r - r_{\mathrm{e}})^2 + \cdots \tag{7.26}$$

The derivative of $\Delta\Psi$ with respect to r can be found from Eq. (7.23), i.e.,

$$\frac{\partial \Delta\Psi}{\partial r} = \frac{4}{3v_\ell}\pi r^3 F\left(\frac{\partial g_\ell}{\partial r} - \frac{\partial g_{\mathrm{v}}}{\partial r}\right) + \frac{4}{v_\ell}\pi r^2 F(g_\ell - g_{\mathrm{v}}) + \frac{8}{3}\pi r \sigma_{\ell \mathrm{v}} F \tag{7.27}$$

The second-order derivative can be found by differentiating Eq. (7.27), i.e.,

$$\begin{aligned} \frac{\partial^2 \Delta\Psi}{\partial r^2} = {} & \frac{4}{3v_\ell}\pi r^3 F\left(\frac{\partial^2 g_\ell}{\partial r^2} - \frac{\partial^2 g_{\mathrm{v}}}{\partial r^2}\right) + \frac{8}{v_\ell}\pi r^2 F\left(\frac{\partial g_\ell}{\partial r} - \frac{\partial g_{\mathrm{v}}}{\partial r}\right) \\ & + \frac{8}{v_\ell}\pi r F(g_\ell - g_{\mathrm{v}}) + \frac{8}{3}\pi \sigma_{\ell \mathrm{v}} F \end{aligned} \tag{7.28}$$

Since the pressure and temperature of the vapor phase, p_{v} and T_{v}, are fixed, the Gibbs free energy for the vapor phase remains the same near the equilibrium:

$$g_{\mathrm{v}} = g_{\mathrm{v,e}} \tag{7.29}$$

i.e.,

$$\frac{\partial g_{\mathrm{v}}}{\partial r} = 0 \tag{7.30}$$

$$\frac{\partial^2 g_{\mathrm{v}}}{\partial r^2} = 0 \tag{7.31}$$

While the temperature of the liquid droplet is the same as the vapor temperature ($T_\ell = T_{\mathrm{v}}$), the pressure in the liquid droplet is related to the pressure in the vapor phase by $p_\ell = p_{\mathrm{v}} + 2\sigma_{\ell \mathrm{v}}/r$. Therefore, the Gibbs free energy for the liquid phase is

$$g_\ell = g_{\ell,\mathrm{e}} + \mathrm{d}g_\ell \tag{7.32}$$

where

$$\mathrm{d}g_\ell = v_\ell \mathrm{d}p_\ell = -\frac{2v_\ell \sigma_{\ell \mathrm{v}}}{r^2}\mathrm{d}r \tag{7.33}$$

i.e.,

$$\frac{\partial g_\ell}{\partial r} = -\frac{2v_\ell \upsilon_{\ell \mathrm{v}}}{r^2} \tag{7.34}$$

and

$$\frac{\partial^2 g_\ell}{\partial r^2} = \frac{4 v_\ell \sigma_{\ell v}}{r^3} \tag{7.35}$$

Substituting Eqs. (7.30), (7.31), (7.34), and (7.35) into Eqs. (7.27) and (7.28), one obtains

$$\frac{\partial \Delta \Psi}{\partial r} = \frac{4}{v_\ell} \pi r^2 F (g_\ell - g_v) \tag{7.36}$$

$$\frac{\partial^2 \Delta \Psi}{\partial r^2} = -8\pi F \sigma_{\ell v} + \frac{8}{v_\ell} \pi r F (g_\ell - g_v) \tag{7.37}$$

At equilibrium, Eqs. (7.36) and (7.37) become

$$\left. \frac{\partial \Delta \Psi}{\partial r} \right|_e = 0 \tag{7.38}$$

$$\left. \frac{\partial^2 \Delta \Psi}{\partial r^2} \right|_e = -8\pi F \sigma_{\ell v} \tag{7.39}$$

Substituting Eqs. (7.38) and (7.39) into Eq. (7.26), the change of availability near the equilibrium radius becomes

$$\Delta \Psi = \frac{4}{3} \pi r_e^2 \sigma_{\ell v} F - 4\pi \sigma_{\ell v} F (r - r_e)^2 + \cdots \tag{7.40}$$

It can be seen from the above equation that $\Delta\Psi$ is maximum at $r = r_e$, and, therefore, it is shown once again that at the equilibrium radius the droplet is in unstable equilibrium (see Fig. 7.9). However, to maintain equilibrium $d\Psi = dG < 0$, i.e., the system tends naturally to achieve the lowest Gibbs free energy value. Therefore, as the droplet increases in radius, the availability of the system decreases and the droplet is in equilibrium. Also, it should be pointed out that if a droplet forms with a radius smaller than equilibrium it spontaneously destroys itself. This can also be seen by the minimum Gibbs free energy principle. If the liquid droplet forms with a radius below the equilibrium radius and tries to grow, $\Delta\Psi > 0$; therefore, the system is increasing in availability, which is highly

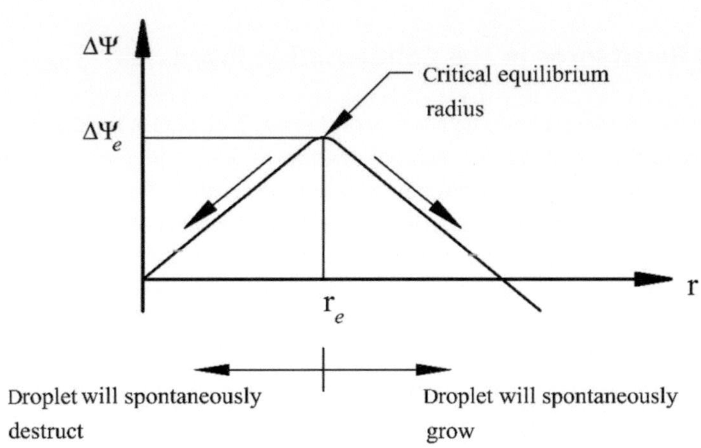

Fig. 7.9 Variation of the system availability with droplet radius

unstable. For the availability to decrease and approach equilibrium, the liquid droplet would need to continuously decrease in size until it disappears. This phenomenon can also be seen in Fig. 7.9. The unstable equilibrium radius is at the top of the curve. If a droplet with such a radius loses one molecule, it will need to continuously decrease in size to meet the $\Delta\Psi < 0$ criterion. If the droplet gains one molecule, it will need to continuously increase in size to meet the $\Delta\Psi < 0$ criterion.

An expression to determine the minimum equilibrium radius of a newly formed droplet that will not spontaneously disappear was developed in Sect. 2.4.3; the result is expressed by Eq. (2.250), i.e.,

$$r_{\min} = \frac{2\sigma v_\ell}{R_g T \ln[p_v/p_{\text{sat}}(T)]} \quad (7.41)$$

Considering the Clapeyron equation(2.158) with $T_v = T_0$, $T = T_w$, and $T_v T_w \approx T_w^2$, the expression for minimum equilibrium radius size becomes

$$r_{\min} = \frac{D_{\min}}{2} = \frac{2 v_\ell \sigma T_w}{h_{\ell v}(T_v - T_w)} \quad (7.42)$$

where $h_{\ell v}$ is the latent heat of energy for the vapor-to-liquid conversion, and σ is the surface tension of the condensing fluid.

Example 7.1

Saturated steam at 1 atm is in contact with a cold wall at 95 °C. Estimate the minimum equilibrium size for dropwise condensation.

Solution The saturated vapor temperature at 1 atm is $T_v = 100\,°\text{C}$, and the latent heat of vaporization is $h_{\ell v} = 2251.2\,\text{kJ/kg}$. The surface tension is $\sigma = 58.91 \times 10^{-3}\,\text{N/m}$. The specific volume of liquid water, determined at $T_{\text{ave}} = (T_v + T_w)/2 = 97.5\,°\text{C}$, is $v_\ell = 1.041 \times 10^{-3}\,\text{m}^3/\text{kg}$. The minimum radius of the liquid droplet is obtained from Eq. (7.42):

$$r_{\min} = \frac{2 v_\ell \sigma T_w}{h_{\ell v}(T_v - T_w)} = \frac{2 \times 1.041 \times 10^{-3} \times (95 + 273.15)}{2251.2 \times 10^3 \times (100 - 95)} = 6.8 \times 10^{-8}\,\text{m} = 0.068\,\mu\text{m}$$

7.2.3 Thermal Resistances in the Condensation Processes

The condensation process must overcome a series of thermal resistances for the heat and mass transfer to occur. These resistances include the thermal resistance found in the vapor, thermal resistance encountered during the phase change from vapor to liquid, resistance caused by capillary depression of the equilibrium saturation temperature at the interface, thermal resistance found in the liquid phase, and thermal resistance found at the wall where heat is conducted from the surface into the wall. The mode of conduction into the wall depends on whether dropwise condensation or filmwise condensation occurs, as will be discussed below. In short, when filmwise condensation occurs uniformly along the surface, the heat flux can simply be found from a straight application of Fourier's law of conduction into a solid. However, if dropwise condensation occurs, conduction into the wall is constricted around the individual droplets and cannot occur uniformly over the solid wall. However,

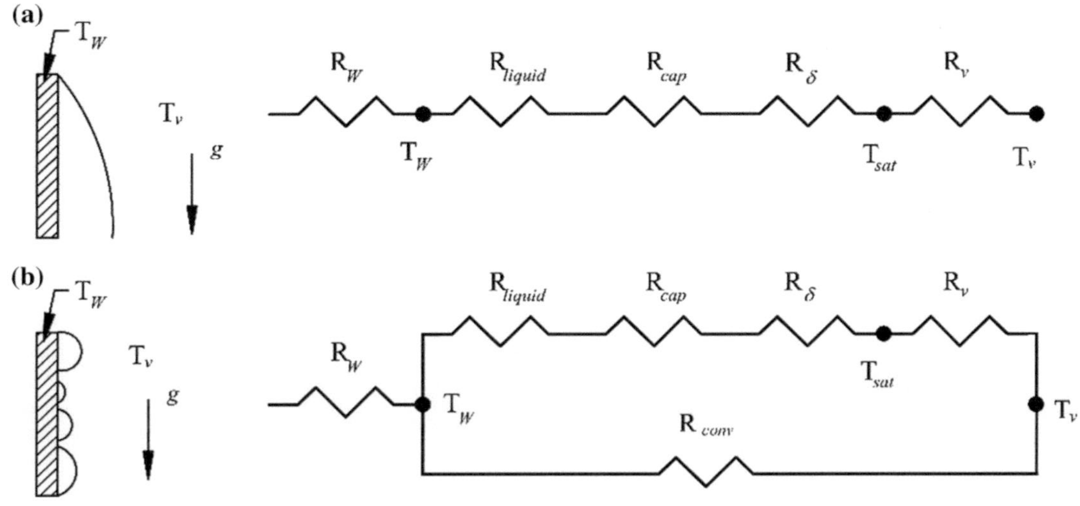

Fig. 7.10 Schematic of the resistance to heat flow in the condensation process: **a** filmwise condensation; **b** time-averaged dropwise condensation

some resistances can be neglected in relation to others, except in special cases. These individual resistances and their importance to the overall resistance will be discussed in this section.

Figure 7.10 shows the resistance to heat flow associated with both filmwise and dropwise condensations, in which R_w is resistance resulting from conduction of heat through the cold wall, R_{liquid} is resistance resulting from heat conduction through liquid film or a droplet, R_{cap} is the resistance resulting from capillary depression of the equilibrium saturation temperature, R_δ is interfacial thermal resistance, R_v is resistance resulting from heat transfer in the vapor phase, and R_{conv} is the convective thermal resistance for the area not covered by the droplet in the dropwise conduction. It should be pointed out that the dropwise condensation is a highly transient process where each resistance value would be fluctuating in time. Therefore, the thermal resistances for dropwise condensation are time-averaged values. In addition, the resistance network approach shown in Fig. 7.10b is valid for the situations that liquid droplets form only heterogeneously at nucleation sites and there is no thin and unstable liquid film between liquid droplets.

Overall, the thermal resistance associated with vapor will be the same for both dropwise and filmwise condensation, as is also the case for the interfacial resistance and capillary depression resistance. The conduction resistances found in dropwise and filmwise condensation are different; i.e., heat is conducted differently through liquid droplets as compared to a liquid film. Conduction through an individual liquid droplet is a function of the size of the droplet (an expression that takes into account all droplets found on the wall surface will also be presented), while the conduction through a thin liquid film is a function of the film thickness and thus a function of position on the wall. The thin liquid film will be discussed in great detail in the next section.

It is assumed in this discussion that the wall temperature is held at a constant value, T_w, and therefore, the overall temperature drop for the area covered by the droplet is as follows:

$$\Delta T_{total} = T_v - T_w = \Delta T_v + \Delta T_\delta + \Delta T_{cap} + \Delta T_{droplet} \tag{7.43}$$

where the temperature differences are for the *vapor*, interface δ, capillary depression of the equilibrium saturation temperature cap, and conduction through the droplet, respectively.

7.2.3.1 Resistance in the Vapor

The thermal resistance found in the vapor can usually be ignored except in special cases, because it is usually an order of magnitude less than the other resistances found in the condensation process. This low contribution to the overall resistance is the result of the vapor's ability to mix extremely well if either free or forced convection is present; this in turn allows the heat and mass transfer toward the cooler surface to easily occur. However, if the vapor is superheated, this mixing process is severely limited, and the resistance of the vapor phase would have to be considered in the form of convection through vapor. In that case, the controlling temperature difference would be $(T_v - T_{sat})$, i.e., the temperature difference between the bulk superheated vapor and the saturated temperature at the liquid–vapor interface. Other cases in which the thermal resistance of the vapor would have to be taken into account include the condensation of vapor mixtures (described above for binary mixtures) and vapor mixtures that include an inert gas (noncondensable gas). The inert gas effectively insulates the convection of heat through a vapor. However, in this discussion, the thermal resistance of the vapor will be ignored because it is usually negligible except for the above cases.

7.2.3.2 Interfacial Resistance

The next resistance encountered in the thermal path from vapor to wall is that found at the vapor–liquid interface. The high heat transfer coefficients associated with the condensation process make it possible to achieve a large heat transfer rate with a small temperature drop (Faghri 2016). This is necessary because the temperature drop at the vapor–liquid interface in a condensation process is very small. This resistance is found in both filmwise and dropwise condensation, and the expressions are identical.

The heat flux at the interface can be obtained by Eq. (4.156), i.e.,

$$q''_\delta = \left(\frac{2\alpha}{2-\alpha}\right)\left(\frac{h_{\ell v}^2}{T_v v_{\ell v}}\right)\sqrt{\frac{M_v}{2\pi R_u T_v}}\left(1 - \frac{p_v v_{\ell v}}{2h_{\ell v}}\right)(T_v - T_\ell) \tag{7.44}$$

where α is an accommodation coefficient.

The corresponding heat transfer coefficient across the interface is obtained by Eq. (4.157), i.e.,

$$h_\delta = \frac{q''_\delta}{(T_v - T_\ell)} = \left(\frac{2\alpha}{2-\alpha}\right)\left(\frac{h_{\ell v}^2}{T_v v_{\ell v}}\right)\sqrt{\frac{M_v}{2\pi R_u T_v}}\left(1 - \frac{p_v v_{\ell v}}{2h_{\ell v}}\right) \tag{7.45}$$

For most systems, the second term in the last parentheses is very small compared to unity and, therefore, can be neglected to obtain the following:

$$h_\delta = \frac{q''_\delta}{(T_v - T_\ell)} = \left(\frac{2\alpha}{2-\alpha}\right)\left(\frac{h_{\ell v}^2}{T_v v_{\ell v}}\right)\sqrt{\frac{M_v}{2\pi R_u T_v}} \tag{7.46}$$

The temperature drop across the interface of a single liquid droplet—assuming it is hemispherical—is

$$\Delta T_\delta = \frac{q_d}{h_\delta(\pi D^2/2)} \tag{7.47}$$

where $\pi D^2/2 = A_{\ell v}$ is the surface area of the hemispherical liquid droplet.

7.2.3.3 Resistance Due to the Capillary Depression of the Equilibrium Saturation Temperature

According to Graham and Griffith (1973), a resistance exists due to the slight depression of the equilibrium interface temperature below that of the normal saturation temperature for a droplet of diameter D. Assuming that a droplet forms with the minimum equilibrium diameter and grows spontaneously to any nonstable diameter D, we can replace the temperature difference $T_{sat} - T_w$ with ΔT_{cap} and the minimum droplet diameter D_{min} with the actual size of the droplet D in Eq. (7.42) to get the following expression for the temperature drop across the capillary depression:

$$\Delta T_{cap} = \frac{4 v_\ell \sigma T_{sat}}{h_{\ell v} D} \tag{7.48}$$

Combining Eq. (7.48) with the minimum equilibrium droplet size expression, Eq. (7.42), one obtains

$$\Delta T_{cap} = \frac{(T_{sat} - T_w) D_{min}}{D} \tag{7.49}$$

7.2.3.4 Resistance Due to Conduction through the Droplet

The conduction of heat through either a liquid droplet or liquid film is usually the controlling factor in resistance to heat flow. This is due directly to the fact that the largest temperature drop in the condensation process occurs in the liquid film or droplet, even though the conduction path is relatively short in comparison to the other heat flow lengths in the condensation process. This leads to high resistances and low heat transfer coefficients. Graham and Griffith (1973) also developed an expression for the conduction of heat through a single droplet of diameter D from the liquid–vapor interface to the wall. As the heat flux travels through the droplet from interface to wall, the planar area normal to the heat flux varies due to the droplet's spherical shape. Also, the distance that the heat flux has to travel through the droplet depends on where it entered the droplet at the liquid–vapor interface. Therefore, Graham and Griffith (1973) took into account these variations in the development of the heat flux through a single hemispherical droplet. This heat flux can be written in terms of the temperature drop through the droplet due to conduction as follows:

$$\Delta T_{droplet} = \frac{q_d (D/2)}{4\pi k_\ell (D/2)^2} = \frac{q_d}{2 k_\ell \pi D} \tag{7.50}$$

7.2.4 Heat Transfer Coefficient for Dropwise Condensation

By substituting the expressions for temperature drops through the interface, Eq. (7.47); capillary depression, Eq. (7.49); and liquid droplets, Eq. (7.50), into Eq. (7.43) and neglecting the temperature drop in the vapor phase, the overall temperature drop is obtained:

$$\Delta T_{total} = T_{sat} - T_w = \frac{2 q_d}{h_\delta \pi D^2} + (T_{sat} - T_w)\frac{D_{min}}{D} + \frac{q_d}{2 k_\ell \pi D} \tag{7.51}$$

The heat flux through a single droplet can then be given by rearranging Eq. (7.51), i.e.,

$$q_{\rm d} = \left(\frac{\pi D^2}{2}\right) \Delta T_{\rm total} \frac{(1 - D_{\rm min}/D)}{(1/h_\delta + D/4k_\ell)} \tag{7.52}$$

To obtain an expression for the total heat transfer through all of the droplets, one must integrate over the total number of droplets and the whole size distribution of the droplets. The size distribution of the droplets can be described using a number distribution function $n''(D)$ that the number of droplets with sizes between D and $D + \mathrm{d}D$ per unit area of surface is $n''(D)\,\mathrm{d}D$. Therefore, to obtain the expression for total heat flux, we multiply $q_{\rm d}$ by the number density $n''(D)\,\mathrm{d}D$ of droplets of size D and integrate over the whole range of droplet size (Carey 2016):

$$q'' = \frac{\pi \Delta T_{\rm total}}{2} \int_{D_{\rm min}}^{D_{\rm max}} n''(D) D^2 \frac{(1 - D_{\rm min}/D)}{(1/h_\delta + D/4k_\ell)} \mathrm{d}D \tag{7.53}$$

The total heat transfer coefficient for the area covered by the liquid droplets can be found for the dropwise condensation process:

$$h_{\rm drops} = \frac{\pi}{2} \int_{D_{\rm min}}^{D_{\rm max}} n''(D) D^2 \frac{(1 - D_{\rm min}/D)}{(1/h_\delta + D/4k_\ell)} \mathrm{d}D \tag{7.54}$$

In dropwise condensation, the type of conduction that occurs in the wall is a direct result of the constricted heat flow around and between the large droplets on the surface of the wall. In filmwise condensation, conduction from the liquid–solid interface would be obtained by applying Fourier's law through the liquid film. However, in dropwise condensation, the resistance to thermal transport found in the liquid droplets is much higher than the resistance found in the liquid-free area between the droplets. Conduction through the walls would prefer to initiate at the bare areas. Therefore, the droplets effectively "squeeze" or "constrict" the heat flow toward the small bare areas. Mikic (1969) developed an expression for the resistance encountered in this constriction:

$$R_{\rm cons} = \left(\frac{1}{3\pi k_\ell}\right) \int_{D_{\rm min}}^{D_{\rm max}} \frac{n''(D) D^2 \mathrm{d}D}{[1 - f(D)]} \tag{7.55}$$

where $f(D)$ is the fraction of the surface area covered by droplets with diameters greater than D. As can be seen above, this expression is a total expression for resistance, because it takes into consideration the full range and number of droplets.

Finally, if this resistance is combined and assumed to be in series with all the other resistances described in previous sections, the total heat transfer coefficient, including conduction through the wall (in this development, the *constriction* conduction mode was used due to the dropwise condensation discussion), is found to be

$$h_{\rm total} = \left(\frac{1}{h_{\rm drops}} + R_{\rm cons}\right)^{-1} \tag{7.56}$$

This theoretical model cannot be used to predict the heat transfer coefficient for dropwise condensation unless we know the droplet size distribution function $f_{\rm n}(D)$. Determination of droplet size is very difficult, especially for droplets smaller than 10 μm, which are major contributors to the heat transfer in dropwise condensation.

Among the possible combinations of fluids and surfaces, steam and well-promoted copper surfaces have been investigated extensively. Griffith (1983) recommended the following correlation for prediction of the heat transfer coefficient for dropwise condensation of steam:

$$h = \begin{cases} 51104 + 2044T_{sat} & 22\,°\mathrm{C} < T_{sat} < 100\,°\mathrm{C} \\ 255510 & T_{sat} > 100\,°\mathrm{C} \end{cases} \tag{7.57}$$

Example 7.2

On a clear winter night, dropwise condensation occurred on the inner surface of window glass at a temperature of 5 °C. The air temperature in the room is 25 °C, and the relative humidity in the room is 80%. Estimate the condensation rate using Griffith's correlation.

Solution The saturation pressure of the water vapor at 25 °C is $p_{sat} = 0.03596\,\mathrm{bar}$. The partial pressure of the water vapor in an air with 80% humidity is $p_{water} = 0.8 \times 0.03596 = 0.02877\,\mathrm{bar}$. Therefore, the dew point is the saturation temperature corresponding to p_{water}, i.e., T_{sat} = 22.1 °C. The latent heat of vaporization at this temperature is $h_{\ell v} = 2448.8\,\mathrm{kJ/kg}$.

Assuming Eq. (7.57) is valid for dropwise condensation, the heat transfer coefficient is

$$h = 51104 + 2044T_{sat} = 51104 + 2044 \times 22.1 = 96276.4\,\mathrm{W/°C\,m^2}$$

The condensation rate is then

$$\dot{m}'' = \frac{q''}{h_{\ell v}} = \frac{h(T_{sat} - T_w)}{h_{\ell v}} = \frac{96276.4 \times (22.1 - 5)}{2448.8 \times 10^3} = 0.672\,\mathrm{kg/m^2s}$$

7.3 Filmwise Condensation

7.3.1 Regimes of Filmwise Condensation

As in the case of external heterogeneous dropwise condensation, filmwise condensation occurs when a cold wall surface is in contact with a vapor near saturation conditions. Filmwise condensation on a vertical surface occurs when the liquid phase fully wets the surface, whereas in dropwise condensation, the liquid incompletely wets the solid surface.

The condensation process begins with vapor condensing directly on the wall surface. However, in contrast to dropwise condensation, after the wall is initially wetted, it remains covered by a thin film of condensate. After that point, condensation occurs only at the liquid–vapor interface. Therefore, the condensation rate is directly a function of the rate at which heat is transported across the liquid film from the liquid–vapor interface to the wall.

Figure 7.11 shows three distinct regimes of filmwise condensation on a vertical wall. These regimes are proceeding in order from the top of the wall (x = 0): laminar, wavy, and turbulent. The Reynolds number is defined as $Re_\delta = 4\Gamma/\mu_\ell$, where Γ is mass flow rate of condensate per unit width. At the top of the wall, where the film is thinnest, the laminar regime exists. As the condensation

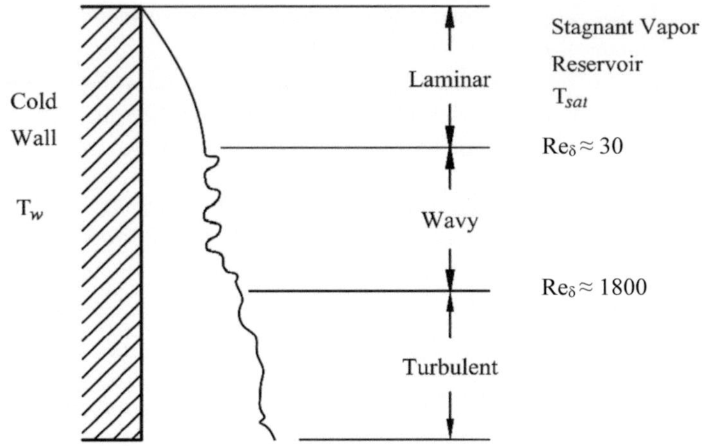

Fig. 7.11 Flow regimes of film condensate on a vertical wall

process proceeds, more and more condensate appears on the surface and the liquid condensate is pulled downward by gravity. As the condensate moves downward, the film becomes thicker. The first sign of transition to a nonlaminar regime appears as a series of regular ripples or waves of condensate. This regime is called the wavy regime and is considered neither laminar nor turbulent. It is characterized by consistent, regular series of waves in time. Finally, if the wall is long enough, the film thickness becomes so great that irregular ripples in both time and space will appear which is identified as turbulent flow regime.

The laminar regime was first rigorously analyzed by Nusselt (1916). Because many simplifying assumptions were made, this analysis provided a closed-form solution. This classical analysis was a very good building block for later studies that gradually chipped away at the assumptions made by Nusselt by employing numerical methods. The classical laminar flow condensation analysis will be presented in Sect. 7.3.2, followed by some of the later studies that improved Nusselt's model. The improvements presented here will include the consideration of noncondensable vapors in the condensation process and the effect of vapor flow (Nusselt assumed a stagnant vapor reservoir). The wavy and turbulent regimes are obviously much more difficult to solve than the laminar, and numerical methods are required to obtain an acceptable solution. However, the reasoning behind these regimes will be presented because the overall heat transfer rate from the vapor reservoir to the cooled wall is *dominated* by contributions from the wavy and turbulent sections. In fact, most industrial applications require that the walls in surface condensers are of a sufficient length and have the surface modified in order to guarantee wavy and turbulent regimes of operation.

7.3.2 Laminar Film Condensation of a Binary Vapor Mixture

Knowledge of binary vapors is of the utmost practical importance, because pure vapor is rarely present in everyday life or in industrial applications. Consider the steady laminar film condensation of a binary vapor on a smooth, vertical, and cooled wall. The bulk vapor mixture flows downwards and parallels the vapor boundary layer and condensate film flows, which are also flowing downwards due to the effects of gravity. The physical model of this condensation of a binary vapor is shown schematically in Fig. 7.12, where x is the distance measured along the flat plate starting from the leading edge and y is the normal distance from the plate. This figure depicts the condensate film

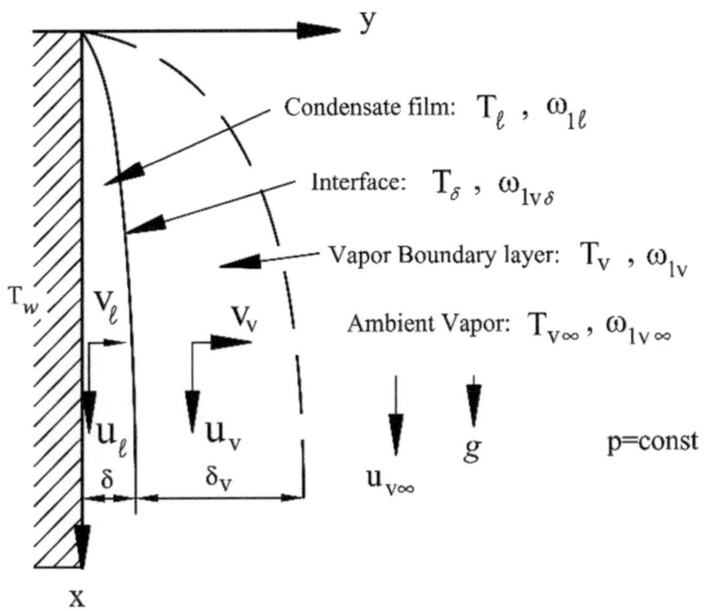

Fig. 7.12 Physical model and coordinate system for condensation of a binary vapor mixture (Fujii 1991; Reproduced with kind permission of Springer Science and Business Media)

boundary layer directly adjacent to the wall with thickness $\delta = \delta(x)$. Directly adjacent to this condensate film is the binary vapor boundary layer, with thickness $\delta_v = \delta_v(x)$, which develops between the condensate film and bulk vapor flow. The velocity components of the x- and y-directions are u and v, respectively. The temperature is denoted by T, and the pressure of the system is a constant p. The mass fraction of a component is denoted by ω. The subscripts ℓ, v, δ, w, and ∞ denote the liquid, vapor, interface, wall, and bulk conditions, respectively. Finally, the subscript 1 denotes the component of the binary vapor with the lower boiling point.

The following assumptions were made to describe the laminar film condensation (Fujii 1991):

1. T_w, T_δ, $\omega_{1v\delta}$, $T_{v\infty}$, $\omega_{1v\infty}$, $u_{v\delta}, u_{\ell\delta}$, and $u_{v\infty}$ are independent of x.
2. The condensate film and binary vapor boundary layers both develop from the leading edge of the vertical surface, $x = 0$.
3. Condensation takes place only at the vapor–liquid interface. In other words, no condensation takes place within the binary vapor boundary layer in the form of a mist or fog.
4. Both temperature and velocity are continuous at the vapor–liquid interface.
5. The condensate is miscible.
6. The physical properties of the system are assumed to be constant with respect to concentration and temperature except in the case of buoyancy terms.
7. The density of the condensate liquid is assumed to be much greater than that of the binary vapor ($\rho_\ell \gg \rho_v$).
8. The vapor mixture can be treated as an ideal gas, and the thermal diffusion is negligible.

Before the governing equations are presented, an explanation is needed for the physical conditions that occur during this process. When a binary vapor makes contact with a cooled vertical wall, the less volatile component of the mixture (the component with the higher boiling point) begins to condense first and in greater quantity. Since the system must maintain the total mass concentration to maintain equilibrium, the volatile component becomes much denser at the liquid–vapor interface. During this process, the bulk vapor with a constant mass concentration is steadily feeding to the liquid–vapor interface. Therefore, the vapor boundary layer develops with a very high concentration of the volatile component at the liquid–vapor interface, which is quickly diluted to the concentration of the bulk vapor.

The governing equations for the laminar film condensation of a binary vapor mixture can be given by taking the above assumptions into account and using boundary-layer analysis.

For the condensate film, the continuity, momentum, and energy equations are:

$$\frac{\partial u_\ell}{\partial x} + \frac{\partial v_\ell}{\partial y} = 0 \tag{7.58}$$

$$u_\ell \frac{\partial u_\ell}{\partial x} + v_\ell \frac{\partial u_\ell}{\partial y} = \nu_\ell \frac{\partial^2 u_\ell}{\partial y^2} + g - \frac{1}{\rho_\ell}\frac{\mathrm{d}p}{\mathrm{d}x} \tag{7.59}$$

$$u_\ell \frac{\partial T_\ell}{\partial x} + v_\ell \frac{\partial T_\ell}{\partial y} = \alpha_\ell \frac{\partial^2 T_\ell}{\partial y^2} \tag{7.60}$$

For the vapor boundary layer, the continuity, momentum, energy, and species equations are:

$$\frac{\partial u_\mathrm{v}}{\partial x} + \frac{\partial v_\mathrm{v}}{\partial y} = 0 \tag{7.61}$$

$$u_\mathrm{v} \frac{\partial u_\mathrm{v}}{\partial x} + v_\mathrm{v} \frac{\partial u_\mathrm{v}}{\partial y} = \nu_\mathrm{v} \frac{\partial^2 u_\mathrm{v}}{\partial y^2} + g\left(1 - \frac{\rho_{\mathrm{v}\infty}}{\rho_\mathrm{v}}\right) \tag{7.62}$$

$$u_\mathrm{v} \frac{\partial T_\mathrm{v}}{\partial x} + v_\mathrm{v} \frac{\partial T_\mathrm{v}}{\partial y} = \alpha_\mathrm{v} \frac{\partial^2 T_\mathrm{v}}{\partial y^2} + Dc_{\mathrm{p}12} \frac{\partial \omega_{1\mathrm{v}}}{\partial y}\frac{\partial T_\mathrm{v}}{\partial y} \tag{7.63}$$

$$u_\mathrm{v} \frac{\partial \omega_{1\mathrm{v}}}{\partial x} + v_\mathrm{v} \frac{\partial \omega_{1\mathrm{v}}}{\partial y} = D \frac{\partial^2 \omega_{1\mathrm{v}}}{\partial y^2} \tag{7.64}$$

where ν is the kinematic viscosity, ρ is the density, k is the thermal conductivity, and D is the diffusivity between components 1 and 2. The second term on the right-hand side of Eq. (7.63) represents the contribution of mass concentration gradient to the energy balance, which is usually negligible.

The isobaric specific heat difference of the binary vapor, $c_{\mathrm{p}12}$, is a weighted average of the isobaric specific heats of the two individual components; it is nondimensional as follows:

$$c_{\mathrm{p}12} = \frac{c_{\mathrm{p}1\mathrm{v}} - c_{\mathrm{p}2\mathrm{v}}}{c_{\mathrm{p}1\mathrm{v}}\omega_{1\mathrm{v}} + c_{\mathrm{p}2\mathrm{v}}\omega_{2\mathrm{v}}} = \frac{c_{\mathrm{p}1\mathrm{v}} - c_{\mathrm{p}2\mathrm{v}}}{c_{\mathrm{pv}}} \tag{7.65}$$

This isobaric specific heat term is assumed to be constant even though it has a larger fluctuation than the other physical properties, because overall the second term on the right-hand side of Eq. (7.63) is much smaller than the first.

Other definitions of the terms in Eqs. (7.58)–(7.64) are as follows, beginning with the mass fractions $\omega_{1\mathrm{v}}$ and $\omega_{2\mathrm{v}}$ in the vapor phase:

$$\omega_{1\mathrm{v}} = \frac{\rho_{1\mathrm{v}}}{\rho_\mathrm{v}} \tag{7.66}$$

$$\omega_{2v} = \frac{\rho_{2v}}{\rho_v} \tag{7.67}$$

where

$$\rho_v = \rho_{1v} + \rho_{2v} \tag{7.68}$$

Therefore, it follows from Eqs. (7.66)–(7.68) that a relationship between ω_{1v} and ω_{2v} can be written as follows:

$$\omega_{1v} + \omega_{2v} = 1 \tag{7.69}$$

Note that only the continuity equation for component 1 of the vapor is given. However, it should be immediately recognized from Eq. (7.69) that if the mass fraction of one component is known at any point in space, then the mass fraction of the other components can easily be determined.

The partial pressures of the system are determined by the following simple expressions:

$$\frac{p_1}{p} = \left(1 + \frac{M_1 \omega_{2v}}{M_2 \omega_{1v}}\right)^{-1} \tag{7.70}$$

$$\frac{p_2}{p} = \left(1 + \frac{M_2 \omega_{1v}}{M_1 \omega_{2v}}\right)^{-1} \tag{7.71}$$

where M_1 and M_2 are the molecular masses of components 1 and 2, respectively.

The boundary equations for the above-generalized governing equations at the surface of the cold wall are as follows:

$$u_\ell = 0, \quad y = 0 \tag{7.72}$$

$$v_\ell - 0, \quad y = 0 \tag{7.73}$$

$$T_\ell = T_w, \quad y = 0 \tag{7.74}$$

These boundary conditions represent no slip at the wall and the continuity of temperature at the wall.

The boundary conditions at locations far from the cold wall are

$$u_v = u_{v\infty}, \quad y \to \infty \tag{7.75}$$

$$T_v = T_{v\infty}, \quad y \to \infty \tag{7.76}$$

$$\omega_{1v} = \omega_{1v\infty}, \quad y \to \infty \tag{7.77}$$

These boundary conditions represent conditions in the constant bulk binary vapor. Finally, the boundary conditions that exist at the liquid–vapor interface ($y = \delta$) are given as follows:

$$u_{\ell\delta} = u_{v\delta} = u_\delta \tag{7.78}$$

$$\left(\mu_\ell \frac{\partial u_\ell}{\partial y}\right)_\delta = \left(\mu_\mathrm{v} \frac{\partial u_\mathrm{v}}{\partial y}\right)_\delta \tag{7.79}$$

$$\left[\rho_\ell\left(u_\ell \frac{\mathrm{d}\delta}{\mathrm{d}x} - v_\ell\right)\right]_\delta = \left[\rho_\mathrm{v}\left(u_\mathrm{v} \frac{\mathrm{d}\delta}{\mathrm{d}x} - v_\mathrm{v}\right)\right]_\delta = \dot{m}'' = \dot{m}_1'' + \dot{m}_2'' \tag{7.80}$$

$$T_{\ell\delta} = T_{\mathrm{v}\delta} = T_\delta \tag{7.81}$$

$$\left(k_\ell \frac{\partial T_\ell}{\partial y}\right)_\delta = h_{\ell\mathrm{v}}\dot{m}'' + \left(k_\mathrm{v} \frac{\partial T_\mathrm{v}}{\partial y}\right)_\delta \tag{7.82}$$

$$\omega_{1\mathrm{v}} = \omega_{1\mathrm{v}\delta} \tag{7.83}$$

where μ is the dynamic viscosity, and $h_{\ell\mathrm{v}}$ is the latent heat of vaporization. Finally, $\dot{m}'' = \dot{m}_1'' + \dot{m}_2''$ is the condensation mass flux perpendicular to the vertical plate; it is a dependent of the location along the x-axis. Equations (7.78), (7.81), and (7.83) are the boundary conditions that refer to the continuity of velocity, temperature, and mass fraction at the liquid–vapor interface, respectively. Equation (7.79) recognizes that the shear stresses in the liquid and vapor layers are equal at the liquid–vapor interface. Equation (7.82) is an energy balance at the interface, which relates the heat transferred to and from the interface with that released by the latent heat. Equation (7.80) is the overall condensation mass continuity at the interface. The mass flux of the noncondensable component 2, $\dot{m}_{2x}''$, at interface in Eq. (7.80), is not zero because the noncondensable component 2 may dissolve into the condensate.

The mass fluxes of the vapor in the binary vapor system are [see Eq. (1.55)]

$$\dot{m}_1'' = -\rho D_{12}\frac{\partial \omega_1}{\partial y} + \omega_1(\dot{m}_1'' + \dot{m}_2'') \tag{7.84}$$

$$\dot{m}_2'' = -\rho D_{21}\frac{\partial \omega_2}{\partial y} + \omega_2(\dot{m}_1'' + \dot{m}_2'') \tag{7.85}$$

and the molar fluxes of the vapor in the binary vapor system are [see Eq. (1.56)]

$$\dot{n}_1'' = -D_{12}\frac{\partial c_1}{\partial y} + x_1\dot{n}_\mathrm{T}'' \tag{7.86}$$

$$\dot{n}_2'' = -D_{21}\frac{\partial c_2}{\partial y} + x_2\dot{n}_\mathrm{T}'' \tag{7.87}$$

where c_i is molar concentration (kmol/m^3) of component i (i = 1, 2). If both components can be treated as ideal gases, the molar fraction is identical to the partial pressure obtained from Eqs. (7.70) and (7.71). $\dot{n}_T'' = \dot{n}_1'' + \dot{n}_2''$ is the total molar flux of components 1 and 2. Equations (7.86) and (7.87) are often expressed in terms of partial pressure, i.e.,

$$\dot{n}_1'' = -\frac{D}{R_\mathrm{u}T}\frac{\partial p_1}{\partial y} + \dot{n}_T''\frac{p_1}{p} \tag{7.88}$$

$$\dot{n}_2'' = -\frac{D}{R_\mathrm{u}T}\frac{\partial p_2}{\partial y} + \dot{n}_T''\frac{p_2}{p} \tag{7.89}$$

where R_u is the universal gas constant. Equations (7.86)–(7.89) are valid in the binary vapor boundary layer ($\delta \le y \le \delta + \delta_\mathrm{v}$).

Also, the mass fraction of component 1 in the condensate film can be found from the following expression:

$$\omega_{1\ell} = \frac{\dot{m}''_{1x}}{\dot{m}''_{1x} + \dot{m}''_{2x}} \tag{7.90}$$

If the noncondensable component cannot be dissolved into the liquid film, $\omega_{1\ell}$ will be unity. Effects of noncondensable gas on film condensation will be discussed in Sect. 7.3.7.

7.3.3 Filmwise Condensation in a Stagnant Pure Vapor Reservoir

7.3.3.1 Laminar Flow Regime

The classical analysis of laminar film condensation on a vertical or inclined wall was performed by Nusselt (1916). The physical conditions of laminar film condensation are shown in Fig. 7.13. As is the case in any heat transfer analysis, the objective of this section is to find the heat transfer coefficient and the Nusselt number for the laminar flow regime in film condensation on a vertical surface. The classical Nusselt analysis requires many assumptions in order to achieve a closed-form solution of the boundary-layer-type momentum equations. These assumptions include the following:

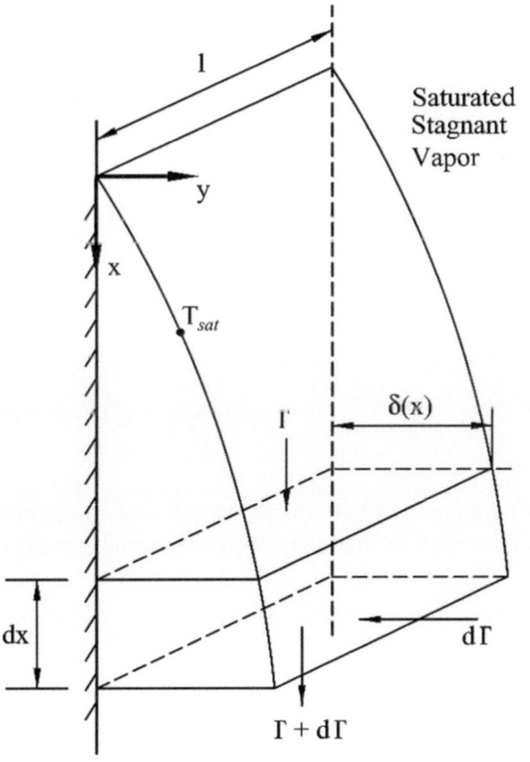

Fig. 7.13 Overview of the control volume under consideration in the Nusselt analysis

1. The flow is laminar and steady.
2. Fluid properties are constant.
3. Subcooling of the liquid is negligible in the energy balance, i.e., all condensation occurs at the saturation temperature corresponding to the pressure in the liquid film near the wall.
4. Inertia and convection effects are negligible in the boundary-layer momentum and energy equations, respectively.
5. The vapor is assumed stagnant, and therefore, shear stress is considered to be negligible at the liquid–vapor interface.

Since the vapor phase is stationary, the governing equations for the vapor phase are no longer needed. The continuity equation for the liquid phase is the same as Eq. (7.58). The momentum equation in the liquid phase is Eq. (7.59), in which the pressure gradient may be approximated in terms of conditions outside the liquid film, as required by the boundary-layer approximations. It follows that pressure in the liquid film satisfies

$$\frac{\mathrm{d}p_\ell}{\mathrm{d}x} = \frac{\mathrm{d}p_\mathrm{v}}{\mathrm{d}x} = \rho_\mathrm{v} g \tag{7.91}$$

Substituting Eq. (7.91) into Eq. (7.59), the momentum equation becomes

$$\rho_\ell \left(u \frac{\partial u}{\partial x} + v \frac{\partial u}{\partial y} \right) = \mu_\ell \frac{\partial^2 u}{\partial y^2} + g(\rho_\ell - \rho_\mathrm{v}) \tag{7.92}$$

The left-hand side term represents the inertia effects in the slender film region, while the right-hand terms are the effects due to friction and the sinking effect. As stated by Assumption #4 of the above list, the inertia term is negligible in this analysis. Therefore, the above equation simplifies to

$$\frac{\partial^2 u}{\partial y^2} = \frac{g}{\mu_\ell}(\rho_\mathrm{v} - \rho_\ell) \tag{7.93}$$

Integrating this equation twice with respect to y and using the boundary conditions of nonslip at the wall ($u = 0$ at $y = 0$) and zero shear at the liquid–vapor interface ($\partial u/\partial y = 0$ at $y = \delta$, Assumption #5) yields the following:

$$u(x, y) = \frac{(\rho_\ell - \rho_\mathrm{v}) g}{\mu_\ell} \left(y\delta - \frac{y^2}{2} \right) \tag{7.94}$$

It can be seen from this equation that the downward velocity component is directly dependent on both the x- and y-coordinates due to the varying (increasingly thicker in the x-direction) film thickness.

The mass flow rate per unit width of surface, Γ, of this liquid film at any point can be found by integrating the above velocity profile across the liquid film thickness and multiplying by the liquid density, i.e.,

$$\Gamma = \rho_\ell \int_0^\delta u\,\mathrm{d}y = \frac{\rho_\ell (\rho_\ell - \rho_\mathrm{v}) g \delta^3}{3\mu_\ell} \tag{7.95}$$

where it can now be seen that the mass flow rate is a function of the x-coordinate.

Recalling that the heat transfer across the liquid film is by conduction only, and assuming that no subcooling exists at the liquid–vapor interface (Assumption #3), Fourier's law can be used to obtain the heat flux across the film thickness:

$$q'' = \frac{k_\ell (T_{\text{sat}} - T_{\text{w}})}{\delta} \tag{7.96}$$

where T_{sat} and T_{w} are the saturation temperature and wall temperature, respectively.

The heat transfer rate per unit width for the control volume shown in Fig. 7.13 is given as

$$\text{d}q' = \frac{k_\ell \Delta T}{\delta} \text{d}x \tag{7.97}$$

where $\Delta T = T_{\text{sat}} - T_{\text{w}}$.

Since no subcooling in the liquid film exists, the latent heat effects of condensation dominate the process. Therefore, it can be said that

$$\text{d}q' = h_{\ell\text{v}}\, \text{d}\Gamma \tag{7.98}$$

where $\text{d}\Gamma$ is found by differentiating the expression for mass flow rate per unit surface, Eq. (7.95). It is found to be

$$\text{d}\Gamma = \frac{\rho_\ell (\rho_\ell - \rho_{\text{v}}) g \delta^2}{\mu_\ell} \text{d}\delta \tag{7.99}$$

This expression and the expression for heat flow $\text{d}q'$ across the control volume are then substituted into the energy equation, Eq. (7.98), and the following is found:

$$\delta^3 \frac{\text{d}\delta}{\text{d}x} = \frac{k_\ell \mu_\ell \Delta T}{\rho_\ell (\rho_\ell - \rho_{\text{v}}) g h_{\ell\text{v}}} \tag{7.100}$$

Finally, δ can be found by integrating the above and using the boundary condition that $\delta = 0$ at $x = 0$:

$$\delta = \left[\frac{4 k_\ell \mu_\ell x \Delta T}{\rho_\ell (\rho_\ell - \rho_{\text{v}}) g h_{\ell\text{v}}} \right]^{1/4} \tag{7.101}$$

The local heat transfer coefficient h_x is found to be

$$h_x = \frac{q''}{T_{\text{sat}} - T_{\text{w}}} = \frac{k_\ell}{\delta} = k_\ell^{3/4} \left[\frac{\rho_\ell (\rho_\ell - \rho_{\text{v}}) g h_{\ell\text{v}}}{4 \mu_\ell x \Delta T} \right]^{1/4} \tag{7.102}$$

The local Nusselt number, Nu_x, of the laminar film condensation process on a vertical plate is found to be

$$Nu_x = \frac{h_x x}{k_\ell} = \left[\frac{\rho_\ell (\rho_\ell - \rho_{\text{v}}) g h_{\ell\text{v}} x^3}{4 k_\ell \mu_\ell \Delta T} \right]^{1/4} \tag{7.103}$$

It is also desirable to obtain the average heat transfer coefficient and Nusselt number for a plate of length L. The mean heat transfer coefficient can be defined as

$$\bar{h} = \frac{1}{L}\int_0^L h_x(x)\mathrm{d}x \tag{7.104}$$

Substituting the expression for h_x, Eq. (7.102), into Eq. (7.104) and integrating, the following expressions are obtained:

$$\overline{Nu}_{\mathrm{L}} = \frac{\bar{h}L}{k_\ell} = 0.943\left[\frac{\rho_\ell(\rho_\ell - \rho_{\mathrm{v}})gh_{\ell\mathrm{v}}L^3}{\mu_\ell k_\ell \Delta T}\right]^{1/4} \tag{7.105}$$

The local and average heat transfer coefficient can also be nondimensionalized in term of Reynolds number Re_δ, defined as

$$Re_\delta = \frac{4\Gamma}{\mu_\ell} \tag{7.106}$$

Substituting Eq. (7.95) into Eq. (7.106), the Reynolds number for *laminar* film condensation becomes

$$Re_\delta = \frac{4\rho_\ell(\rho_\ell - \rho_{\mathrm{v}})g\delta^3}{3\mu_\ell^2} \tag{7.107}$$

Substituting Eqs. (7.106) and (7.101) into Eqs. (7.103) and (7.105), the following nondimensional correlations are obtained:

$$\frac{h_x}{k_\ell}\left[\frac{\mu_\ell^2}{\rho_\ell(\rho_\ell - \rho_{\mathrm{v}})g}\right]^{1/3} = 1.1Re_\delta^{-1/3} \tag{7.108}$$

$$\frac{\bar{h}}{k_\ell}\left[\frac{\mu_\ell^2}{\rho_\ell(\rho_\ell - \rho_{\mathrm{v}})g}\right]^{1/3} = 1.47Re_\delta^{-1/3} \tag{7.109}$$

where the bracket term on the left-hand side, together with its exponent (1/3), is the characteristic length.

The above Nusselt analysis assumes no subcooling in the liquid condensate. This assumption can be removed by making an energy balance at the interface that takes subcooling of the liquid into account, as follows:

$$\frac{\mathrm{d}q'}{\mathrm{d}x} = \frac{k_\ell \Delta T}{\delta} = h_{\ell\mathrm{v}}\frac{\mathrm{d}\Gamma}{\mathrm{d}x} + \frac{\mathrm{d}}{\mathrm{d}x}\int_0^\delta \rho_\ell c_{\mathrm{p}\ell}u(T_{\mathrm{sat}} - T)\mathrm{d}y \tag{7.110}$$

The final term of the right-hand side takes into account a temperature gradient across the liquid condensate film. Substituting the velocity profile in Eq. (7.94) and using a linear temperature profile,

$$\frac{T_{\mathrm{sat}} - T}{T_{\mathrm{sat}} - T_{\mathrm{w}}} = 1 - \frac{y}{\delta} \tag{7.111}$$

to evaluate Eq. (7.110), the energy balance can be written as

$$\frac{k_\ell \Delta T_\ell}{\delta} = h'_{\ell \mathrm{v}} \frac{\mathrm{d}\Gamma}{\mathrm{d}x} \tag{7.112}$$

where

$$h'_{\ell \mathrm{v}} = h_{\ell \mathrm{v}} \left\{ 1 + \frac{3}{8} \left[\frac{c_{\mathrm{p}\ell}(T_{\mathrm{sat}} - T_{\mathrm{w}})}{h_{\ell \mathrm{v}}} \right] \right\} \tag{7.113}$$

Equation (7.112) is identical to the energy balance in Eq. (7.97) and (7.98) except that $h_{\ell \mathrm{v}}$ has been replaced by $h'_{\ell \mathrm{v}}$.

Rohsenow (1956) improved this analysis even more by including the effects of convection in the liquid along with liquid subcooling and thereby developing the following:

$$h'_{\ell \mathrm{v}} = h_{\ell \mathrm{v}} \left\{ 1 + 0.68 \left[\frac{c_{\mathrm{p}\ell}(T_{\mathrm{sat}} - T_{\mathrm{w}})}{h_{\ell \mathrm{v}}} \right] \right\} \tag{7.114}$$

Finally, if it is desirable to include the effects of vapor superheat in the above analysis, the latent heat obtained by Eqs. (7.113) or (7.114) can be further modified by adding $c_{\mathrm{pv}}(T_{\mathrm{v}} - T_{\mathrm{sat}})$.

The above analysis can also be applied for condensation outside a vertical tube if $\delta / D \ll 1$, where D is the diameter of the tube. For an inclined wall with an inclination angle of θ (the angle between the wall and the vertical direction), the component of the gravitational acceleration along the inclined wall is $g \cos\theta$. It follows from Eqs. (7.102) and (7.105) that $h_x \propto (\cos\theta)^{1/4}$ and $\bar{h} \propto (\cos\theta)^{1/4}$.

7.3.3.2 Wavy Condensate Regime

Wavy flows of thin liquid films have higher heat transfer coefficients than smooth thin films, due to mixing action and an increase in the interfacial surface area. It has been shown by experiments that even during well defined laminar film flow, the film surface (liquid–vapor interface) can be wavy. These waves can occur on film flow over a rough wall or a polished wall. This wave formation can even occur when the vapor reservoir is stagnant; however, the waves can be more pronounced when the vapor reservoir has an average velocity, which leads to a higher shear stress at the liquid–vapor interface. These small disturbances can amplify under specific circumstances, i.e., when the film reaches a critical thickness and produces full waves that are regular with respect to time and therefore are not considered turbulent. These waves can lead to an improvement in the heat transfer coefficient by as much as 50%, compared to that of the above laminar Nusselt analysis. The mechanisms of heat transfer enhancement by waves include enlargement of liquid–vapor interfacial area and decrease in mean film thickness.

The flow regime of the liquid condensate is determined by the Reynolds number defined by Eq. (7.107). While experimental observations of the condensate indicate that the laminar flow regime usually becomes wavy in the general vicinity of $Re_\delta = 30$, Brauer (1956) suggested that the Reynolds number for onset of waves is related to the Archimedes number by

$$Re_\delta > 9.3Ar_\ell^{1/5} \tag{7.115}$$

where Archimedes number is defined as

$$Ar_\ell = \frac{\rho_\ell \sigma^{3/2}}{\mu_\ell^2 g^{1/2} (\rho_\ell - \rho_v)^{3/2}} \tag{7.116}$$

The wavy laminar film remains so until approximately $Re_\delta = 1800$, where it becomes turbulent. In heat transfer analysis of the wavy film regime, there is no one theory reliable for calculating the heat transfer across the wavy film. Many numerical analyses have been performed, and different coefficients have been found for the heat transfer coefficient and corresponding Nusselt number. Kutateladze (1982) gave the following correlation for the mean Nusselt number of film condensation on a vertical plate where wave effects are present:

$$\bar{h} = \frac{k_\ell}{(v_\ell^2/g)^{1/3}} \frac{Re_\delta}{Re_\delta^{1.22} - 5.2}, \quad 30 \le Re_\delta \le 1800 \tag{7.117}$$

To use Eq. (7.117) to determine the heat transfer coefficient, it is first necessary to know the Reynolds number, which depends on the mass flow rate of condensate per unit width, Γ, as indicated by Eq. (7.107). The mass flow rate of the condensate can be determined by

$$\Gamma = \frac{q'}{h'_{\ell v}} = \frac{\bar{h}L(T_{sat} - T_w)}{h'_{\ell v}} \tag{7.118}$$

Substituting Eq. (7.118) into Eq. (7.106), one obtains

$$Re_\delta = \frac{4\bar{h}L(T_{sat} - T_w)}{h'_{\ell v}\mu_\ell} \tag{7.119}$$

Rearranging Eq. (7.119) yields

$$\bar{h} = \frac{Re_\delta \, h'_{\ell v} v_\ell}{4L(T_{sat} - T_w)} \tag{7.120}$$

Combining Eq. (7.117) and (7.120) yields the Reynolds number for film condensation with waves as follows:

$$Re_\delta = \left[4.81 + \frac{3.7Lk_\ell(T_{sat} - T_w)}{\mu_\ell h'_{\ell v}} \left(\frac{g}{v_\ell^2}\right)^{1/3}\right]^{0.82} \tag{7.121}$$

After the Reynolds number is determined by Eq. (7.121), the heat transfer coefficient can be determined by Eq. (7.117).

7.3.3.3 Turbulent Film Regime

If one follows the film condensate farther down the vertical wall, the film reaches a critical thickness where the waves become irregular with respect to both time and space. This is where the turbulent film regime exists and can exist even with a stagnant vapor reservoir. It follows that this regime occurs when the Reynolds number, Re_δ, exceeds a critical value that is much higher than that of the wavy regime.

As stated in the previous section, the condensate film becomes turbulent at approximately $Re_\delta = 1800$. The heat transfer rate in this turbulent regime is much larger than that in laminar or wavy flow. The turbulent region in film condensation, as in any turbulent flow, is extremely difficult to model, and all accurate results come from empirical correlations or from detailed numerical modeling. For turbulent flow of condensation on a vertical plate, Labuntsov (1957) recommended the following empirical correlation:

$$\frac{h_x}{k_\ell}\left[\frac{\mu_\ell^2}{\rho_\ell(\rho_\ell - \rho_v)g}\right]^{1/3} = 0.023Re_\delta^{0.25}Pr_\ell^{-0.5}, \quad Pr_\ell \geq 10 \tag{7.122}$$

Chun and Seban (1971) obtained the following experimental correlation for local turbulent heat transfer coefficient for evaporation of water from a vertical wall, which is applicable for local heat transfer coefficient for condensation:

$$\frac{h_x}{k_\ell}\left(\frac{v_\ell^2}{g}\right)^{1/3} = 3.8 \times 10^{-3}Re^{0.4}Pr_\ell^{-0.65}, \quad Re > 5800Pr_\ell^{-1.06} \tag{7.123}$$

Butterworth (1983) obtained the average heat transfer coefficient for film condensation that covers laminar, wavy laminar, and turbulent flow by combining Eqs. (7.109), (7.117), and (7.122) as follows

$$\bar{h} = \frac{k_\ell}{(v_\ell^2/g)^{1/3}} \frac{Re_\delta}{8750 + 58Pr_\ell^{-0.5}(Re_\delta^{0.75} - 253)} \tag{7.124}$$

The Reynolds number, Re_δ, is needed in order to use Eq. (7.124) to determine the heat transfer coefficient for turbulent film condensation. Equation (7.120) was obtained by energy balance, and it is valid for all film condensation regimes. Combining Eqs. (7.124) and (7.120), the Reynolds number for turbulent flow is obtained:

$$Re_\delta = \left[\frac{0.069Lk_\ell Pr_\ell^{0.5}(T_{sat} - T_w)}{\mu_\ell h'_{\ell v}}\left(\frac{g}{v_\ell^2}\right)^{1/3} - 151Pr_\ell^{0.5} + 253\right]^{4/3} \tag{7.125}$$

which can be used together with Eq. (7.124) to determine the heat transfer coefficient for turbulent film condensation.

Example 7.3

Saturated steam at 1 atm condenses on a vertical wall with a height of L = 1 m and width of b = 1.5 m. The surface temperature of the vertical wall is 80 °C. What are the average heat transfer and condensation rates?

Solution The saturation temperature of steam at 1 atm is T_{sat} = 100 °C. The vapor density at this temperature is $\rho_v = 0.5974\,\text{kg/m}^3$, and the latent heat of vaporization is $h_{\ell v} = 2251.2\,\text{kJ/kg}$. The liquid properties evaluated at $T_f = (T_{sat} + T_w)/2 = 90\,°\text{C}$ are $\rho_\ell = 965.3\,\text{kg/m}^3$, $c_{p\ell} = 4.206\,\text{kJ/kg K}$, $\mu_\ell = 0.315 \times 10^{-3}\,\text{kg/ms}$, $k_\ell = 0.675\,\text{W/m K}$, and $v_\ell = \mu_\ell/\rho_\ell = 0.326 \times 10^{-6}\,\text{m}^2/\text{s}$.

The revised latent heat of vaporization is

$$h'_{\ell v} = h_{\ell v}\left\{1 + 0.68\left[\frac{c_{p\ell}(T_{sat} - T_w)}{h_{\ell v}}\right]\right\}$$
$$= 2251.2 \times \left\{1 + 0.68\left[\frac{4.206 \times (100 - 80)}{2251.2}\right]\right\} = 2308.4\,\text{kJ/kg}$$

Assuming the film condensation is laminar (as will be verified later), the heat transfer coefficient can be obtained from Eq. (7.105), i.e.,

$$\bar{h} = 0.943\left[\frac{\rho_\ell(\rho_\ell - \rho_v)gk_\ell^3 h_{\ell v}}{\mu_\ell \Delta T L}\right]^{1/4}$$
$$= 0.943 \times \left[\frac{965.3 \times (965.3 - 0.5974) \times 9.8 \times 0.675^3 \times 2308.4 \times 10^3}{0.315 \times 10^{-3}(100 - 80) \times 1}\right]^{1/4}$$
$$= 5340.2\,\text{W/m}^2\,\text{K}$$

The heat transfer rate is then

$$q = \bar{h}Lb(T_{sat} - T_w) = 5340.2 \times 1 \times 1.5 \times (100 - 80) = 1.602 \times 10^5\,\text{W}$$

The condensation rate is

$$\dot{m} = \frac{q}{h'_{\ell v}} = \frac{1.602 \times 10^5}{2308.4 \times 10^3} = 0.0694\,\text{kg/s}$$

The assumption of laminar film condensation is now checked by obtaining the Reynolds number defined in Eq. (7.106), i.e.,

$$Re_\delta = \frac{4\Gamma}{\mu_\ell} = \frac{4\dot{m}}{\mu_\ell b} = \frac{4 \times 0.0694}{0.315 \times 10^{-3} \times 1.5} = 588$$

which is greater than 30 and below 1800. This means that the assumption of laminar film condensation is invalid, and it is necessary to consider the effect of waves on the film condensation. It should be kept in mind the above Reynolds number of 588 is obtained by assuming laminar film condensation. For film condensation with wavy effects, the Reynolds number should be obtained from Eq. (7.121), i.e.,

$$Re_\delta = \left[4.81 + \frac{3.7Lk_\ell(T_{sat} - T_w)}{\mu_\ell h'_{\ell v}}\left(\frac{g}{\nu_\ell^2}\right)^{1/3}\right]^{0.82}$$
$$= \left[4.81 + \frac{3.7 \times 1 \times 0.675 \times (100 - 80)}{0.315 \times 10^{-3} \times 2308.4 \times 10^3}\left(\frac{9.8}{(0.326 \times 10^{-6})^2}\right)^{1/3}\right]^{0.82}$$
$$= 730.9$$

which confirms that the film condensation on the lowest section of the plate is in the wavy regime. The heat transfer coefficient is obtained from Eq. (7.117), i.e.,

$$\bar{h} = \frac{k_\ell}{\left(v_\ell^2/g\right)^{1/3}} \frac{Re_\delta}{Re_\delta^{1.22} - 5.2}$$
$$= \frac{0.675}{\left[(0.326 \times 10^{-6})^2/9.8\right]^{1/3}} \frac{730.9}{730.9^{1.22} - 5.2} = 7160\,\mathrm{W/m^2\,K}$$

The heat transfer rate is then

$$q = \bar{h}Lb(T_{\mathrm{sat}} - T_{\mathrm{w}}) = 7160 \times 1 \times 1.5 \times (100 - 80) = 2.15 \times 10^5\,\mathrm{W}$$

The condensation rate is

$$\dot{m} = \frac{q}{h'_{\ell \mathrm{v}}} = \frac{2.15 \times 10^5}{2308.4 \times 10^3} = 0.0931\,\mathrm{kg/s}$$

which is much higher than the condensation rate obtained by assuming laminar film condensation near the bottom of the plate.

7.3.4 Effects of Vapor Motion

7.3.4.1 Laminar Condensate Flow

It was assumed in Sect. 7.3.3 that the vapor reservoir is stagnant. This assumption was made in order to simplify the analysis of the heat transfer across a thin condensing film. In most real systems, however, the effect of vapor motion must be taken into account. This vapor motion can be due to free (natural effects) or forced (mechanical effects) convection processes. The analysis is conducted in the same way in both cases except that with vapor motion the shear stress at the liquid–vapor interface cannot be assumed to be zero.

In the analysis below, the vapor will be considered as having a downward motion. The analysis follows the same outline as the Nusselt analysis presented above for a case where the vapor reservoir is stagnant, producing zero interfacial shear. In other words, referring to Fig. 7.12, the shear stress at the $y = \delta$ location of the interfacial control volume will now have a finite real value of shear stress. The boundary-layer momentum equation for liquid is once again given as

$$\rho_\ell \left(u \frac{\partial u}{\partial x} + v \frac{\partial u}{\partial y} \right) = -\frac{\mathrm{d}p_\ell}{\mathrm{d}x} + \mu_\ell \frac{\partial^2 u}{\partial y^2} + \rho_\ell g \tag{7.126}$$

where $\mathrm{d}p/\mathrm{d}x$ has a value different from the Nusselt analysis, because another pressure gradient imposed by the motion of the adjacent vapor exists along with the hydrostatic pressure gradient in the liquid, i.e.,

$$\frac{\mathrm{d}p_\ell}{\mathrm{d}x} = \rho_\mathrm{v} g + \left(\frac{\mathrm{d}p}{\mathrm{d}x}\right)_\mathrm{v} \tag{7.127}$$

where ρ_v is the vapor density and the subscript v denotes vapor motion. For the sake of convenience, this superimposed pressure gradient can be combined into a fictitious density, ρ_v^*, denoted as

$$\rho_v^* = \rho_v + \frac{1}{g}\left(\frac{dp}{dx}\right)_v \tag{7.128}$$

Substituting Eq. (7.128) into Eqs. (7.127) and (7.126) produces

$$\rho_\ell\left(u\frac{\partial u}{\partial x} + v\frac{\partial u}{\partial y}\right) = \mu_\ell\frac{\partial^2 u}{\partial y^2} + g(\rho_\ell - \rho_v^*) \tag{7.129}$$

The left-hand side of Eq. (7.129) represents the inertia effects of the slender film region, while the right-hand side gives the effects of friction and the pressure gradient in the liquid. It is assumed once again that the inertia terms are negligible compared to the other terms in this analysis. Therefore, the above equation simplifies to

$$\frac{\partial^2 u}{\partial y^2} = \frac{g}{\mu_\ell}(\rho_v^* - \rho_\ell) \tag{7.130}$$

Integrating Eq. (7.130) twice with respect to y, using the nonslip boundary conditions at the wall ($u = 0$ at $y = 0$), and using the new boundary condition of a finite shear stress at the liquid–vapor interface ($\partial u/\partial y = \text{const}$ at $y = \delta$), the following is obtained:

$$u(x,y) = \frac{(\rho_\ell - \rho_v^*)g}{\mu_\ell}\left(y\delta - \frac{y^2}{2}\right) + \frac{\tau_\delta y}{\mu_\ell} \tag{7.131}$$

where τ_δ is the shear stress at the interface.

In this equation, it can be seen once again that the downward velocity component depends directly on both the x- and y-coordinates due to the varying (increasingly thick in the x-direction) film thickness. In addition, u is now also a function of the interfacial shear stress.

The mass flow rate per unit width of surface, Γ, of this liquid film at any point can now be found by integrating the velocity profile across the liquid film thickness and multiplying by the liquid density –

$$\Gamma = \rho_\ell\int_0^\delta u\,dy = \frac{\rho_\ell(\rho_\ell - \rho_v^*)g\delta^3}{3\mu_\ell} + \frac{\tau_\delta\rho_\ell\delta^2}{2\mu_\ell} \tag{7.132}$$

where it can now be seen that the mass flow rate is also a function of the x-coordinate.

The previously described procedure for Nusselt analysis is followed verbatim, using Eq. (7.132) for the mass flux across the control volume. Since it is still assumed that heat transfer across the liquid film is by conduction only, and that no subcooling exists at the liquid–vapor interface, heat flux across the film thickness can be obtained using Fourier's law:

$$q'' = \frac{k_\ell(T_{sat} - T_w)}{\delta} \tag{7.133}$$

where T_{sat} and T_w are the saturation temperature and wall temperature, respectively ($\Delta T = T_{sat} - T_w$). Rewriting Eq. (7.133) in terms of heat transfer rate for the control volume shown in Fig. 7.13, the heat flow is given as

$$\mathrm{d}q' = \frac{k_\ell \Delta T}{\delta}\mathrm{d}x \tag{7.134}$$

Since no subcooling of the liquid film exists, the latent heat effects of condensation dominate the process. Thus,

$$\mathrm{d}q' = h_{\ell \mathrm{v}}\,\mathrm{d}\Gamma \tag{7.135}$$

where $\mathrm{d}\Gamma$ can be found by differentiating the above expression for mass flow rate per unit surface area, with the result

$$\mathrm{d}\Gamma = \left(\frac{\rho_\ell(\rho_\ell - \rho_\mathrm{v}^*)g\delta^2}{\mu_\ell} + \frac{\tau_\delta \rho_\ell \delta}{\mu_\ell}\right)\mathrm{d}\delta \tag{7.136}$$

Substituting this expression and the expression for heat flow across the control volume, $\mathrm{d}q$, into the conservation of mass and energy equation, the following is found:

$$\frac{\mathrm{d}\delta}{\mathrm{d}x} = \frac{k_\ell \mu_\ell \Delta T}{\rho_\ell(\rho_\ell - \rho_\mathrm{v}^*)g h_{\ell \mathrm{v}}\delta^3 + \tau_\delta \rho_\ell h_{\ell \mathrm{v}}\delta^2} \tag{7.137}$$

Finally, δ can be found by integrating the above and using the boundary condition that $\delta = 0$ at $x = 0$.

$$\delta^4 + \frac{4\tau_\delta \delta^3}{3(\rho_\ell - \rho_\mathrm{v}^*)g} = \left[\frac{4k_\ell \mu_\ell x \Delta T}{\rho_\ell(\rho_\ell - \rho_\mathrm{v}^*)g h_{\ell \mathrm{v}}}\right] \tag{7.138}$$

This equation can be nondimensionalized by using the following dimensionless numbers (Rohsenow et al. 1956):

$$\delta^* = \frac{\delta}{L_\mathrm{F}} \tag{7.139}$$

$$x^* = \left(\frac{x}{L_\mathrm{F}}\right)\frac{4c_{p\ell}\Delta T}{Pr_\ell h_{\ell \mathrm{v}}} \tag{7.140}$$

$$\tau_\delta^* = \frac{\tau_\delta}{L_\mathrm{F}(\rho_\ell - \rho_\mathrm{v}^*)g} \tag{7.141}$$

where

$$L_\mathrm{F} = \left[\frac{\mu_\ell^2}{\rho_\ell(\rho_\ell - \rho_\mathrm{v}^*)g}\right]^{1/3} \tag{7.142}$$

is the characteristic length of the nondimensional problem. Equation (7.138), which relates film thickness to the vertical location on the plate, can be rewritten as follows:

$$x^* = (\delta^*)^4 + \frac{4}{3}(\delta^*)^3\tau_\delta^* \tag{7.143}$$

It directly follows that the mean Nusselt number and Reynolds number for laminar flow with finite vapor shear can be expressed as follows:

$$\overline{Nu_{L_F}} = \frac{h_L}{k_\ell}\left[\frac{\mu_\ell^2}{\rho_\ell(\rho_\ell - \rho_v^*)g}\right]^{1/3} = \frac{4}{3}\frac{(\delta^*)^3}{x^*} + \frac{2(\delta^*)^2\tau_\delta^*}{x^*} \tag{7.144}$$

$$Re_\delta = \frac{4\Gamma}{\mu_\ell} = \frac{4}{3}(\delta^*)^3 + 2(\delta^*)^2\tau_\delta^* \tag{7.145}$$

From the above, it can be seen that in order to solve for the average Nusselt number or Reynolds number, the dimensionless numbers x^* and δ^* are needed. These numbers are directly dependent on each other; therefore, an iterative procedure is required to solve for the heat transfer parameters for each and every point. Additional complications follow from the fact that δ depends on position x and either the variation of heat removal rate across the film with respect to x or the varying temperature drop across the film thickness. In other words, unless the problem statement is oversimplified, the only easy way to solve this problem is numerically.

For the case in which the gravitational force is negligible compared with the interfacial shear force imposed by the cocurrent vapor flow, Butterworth (1981) recommended the following correlation for local heat transfer coefficient:

$$Nu_x^* = 1.41 Re_\delta^{-1/2}(\tau_\delta^+)^{1/2} \tag{7.146}$$

where the modified local Nusselt number is defined as

$$Nu_x^* = \frac{h_x}{k}\left[\frac{\mu_\ell^2}{\rho_\ell(\rho_\ell - \rho_v)g}\right]^{1/3} \tag{7.147}$$

and the dimensionless interfacial shear stress is

$$\tau_\delta^+ = \frac{\rho_\ell \tau_\delta}{[\rho_\ell(\rho_\ell - \rho_v)\mu_\ell g]^{2/3}} \tag{7.148}$$

Butterworth (1981) also recommended the following expression for the cases where both gravity and interfacial vapor shear are significant:

$$h = (h_{\text{shear}}^2 + h_{\text{grav}}^2)^{1/2} \tag{7.149}$$

where h_{grav} is heat transfer coefficient for gravity-dominated film condensation—determined with Eqs. (7.103) or (7.108)—and h_{shear} is heat transfer coefficient for shear-dominated film condensation, Eq. (7.146).

Lin and Faghri (1998) developed a model for predicting the condensation heat transfer coefficient for annular flow in rotating stepped wall heat pipes. The theoretical result is compared with experimental data. The effect of vapor shear drag on the condensation heat transfer is discussed (see Problem 7.24).

7.3.4.2 Turbulent Condensate Flow

The velocity and pressure in turbulent flow experience random fluctuations, and they can be expressed

$$u = \bar{u} + u' \tag{7.150}$$

$$v = \bar{v} + v' \tag{7.151}$$

$$p = \bar{p} + p' \tag{7.152}$$

where the bar notation denotes the mean value averaged over time and the prime notation denotes fluctuations from the mean.

The boundary-layer equation for forced turbulent flow along a planar surface is

$$\bar{u}\frac{\partial \bar{u}}{\partial x} + \bar{v}\frac{\partial \bar{u}}{\partial y} = -\frac{1}{\rho}\frac{\partial \bar{p}}{\partial x} + \frac{1}{\rho}\frac{\partial}{\partial y}\left(\mu\frac{\partial \bar{u}}{\partial y} - \rho\overline{u'v'}\right) \tag{7.153}$$

The momentum fluctuation term $-\rho\overline{u'v'}$ is the turbulent stress and can be obtained from:

$$-\rho\overline{u'v'} = \rho\varepsilon\frac{\partial \bar{u}}{\partial y} \tag{7.154}$$

where ε is an empirical function known as the momentum eddy diffusivity that is a flow property, not a fluid property. A close look at Eq. (7.153) shows that the shear stress expression has changed from the normal laminar flow form. The apparent shear stress expression for turbulent flow can be written as follows:

$$\tau_{\text{app}} = \mu\frac{\partial \bar{u}}{\partial y} - \rho\overline{u'v'} \tag{7.155}$$

Substituting Eq. (7.154) into Eq. (7.155) gives

$$\tau_{\text{app}} = \mu\frac{\partial \bar{u}}{\partial y} + \rho\varepsilon\frac{\partial \bar{u}}{\partial y} = \rho(v+\varepsilon)\frac{\partial \bar{u}}{\partial y} \tag{7.156}$$

which is the shear stress expression for turbulent flow consisting of both laminar and turbulent portions. If the following is defined

$$\varepsilon^{+} = 1 + \frac{\varepsilon}{v} \tag{7.157}$$

Equation (7.156) can be simplified to

$$\tau_{\text{app}} = \rho\varepsilon^{+}v\frac{\partial \bar{u}}{\partial y} \tag{7.158}$$

which shows that when $\varepsilon^{+} = 1$, the problem simplifies to the laminar case. Further, it can be said that when $\varepsilon/v \gg 1$, then $\varepsilon^{+} = \varepsilon/v$, and therefore Eq. (7.158) becomes simplified to represent the fully turbulent case. The turbulent regime in film condensation with vapor motion is very difficult to model, so empirical correlations are often used to predict the heat transfer coefficient.

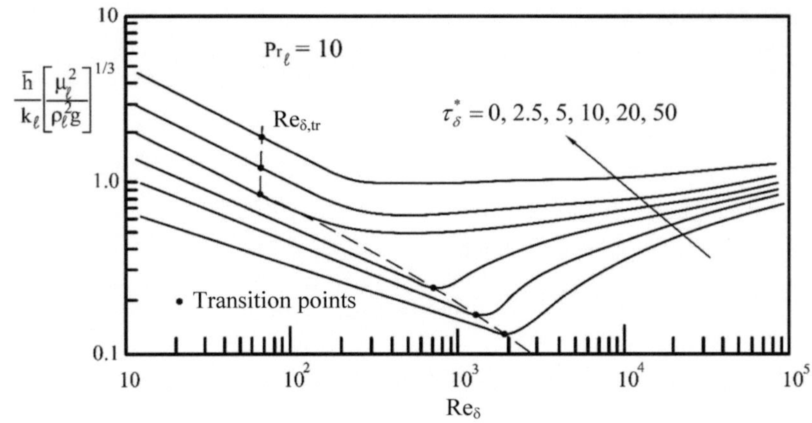

Fig. 7.14 Variation of the mean film condensation heat transfer coefficient with Reynolds number and τ_δ^* as predicted by Rohsenow et al. (1956)

Rohsenow et al. (1956) extended Seban's (1954) falling film condensation analysis for turbulent flow from zero vapor shear to finite vapor shear cases to arrive at the following expression

$$\bar{h} = 0.065 Pr_\ell^{1/2} \left(\tau_\delta^*\right)^{1/2} k_\ell \left(\frac{g}{\nu_\ell^2}\right)^{1/3}, \quad Re_\delta > Re_{\delta,\mathrm{tr}} \tag{7.159}$$

where $Re_{\delta,\mathrm{tr}}$ is the Reynolds number at transition point and can be obtained from:

$$Re_{\delta,\mathrm{tr}} = 1800 - 246\left(1 - \frac{\rho_\mathrm{v}}{\rho_\ell}\right)^{1/3} \tau_\delta^* + 0.667\left(1 + \frac{\rho_\mathrm{v}}{\rho_\ell}\right)\left(\tau_\delta^*\right)^3 \tag{7.160}$$

Equation (7.159) may be used to predict the average heat transfer coefficient beyond the transition point into turbulent flow. It is found to agree well with a relationship obtained by Carpenter and Colburn (1951).

Figure 7.14 shows the variation of the average film condensation heat transfer coefficient with Reynolds number and the nondimensional shear stress τ_δ^* ; the figure shows both the laminar and turbulent flow regimes. For the cases with low nondimensional shear stress values, the transition from laminar to turbulent film flow takes place at approximately $Re_\delta = 1100$, after which the heat transfer rate rises sharply. It appears that wavy flow at these low shear stress values does not contribute to any change in the flow's average heat transfer coefficient and Reynolds number. However, higher shear stress numbers allow for a more gradual change (shallow gradient) from laminar to fully turbulent flow, in turn allowing a wavy flow to exist. This shallow gradient shows neither the abrupt steep rise of a turbulent flow nor the consistent downward gradient characteristic of Nusselt flow. Instead, it is a combination of the two types of flow.

7.3.5 Turbulent Film Condensation

We will now consider turbulent flow for a specific application—turbulent condensate flow in a circular tube, as shown in Fig. 7.15. Turbulent film condensation occurs at the inner surface of the circular tube. While the liquid condensate flows downward due to gravity, the vapor flows either

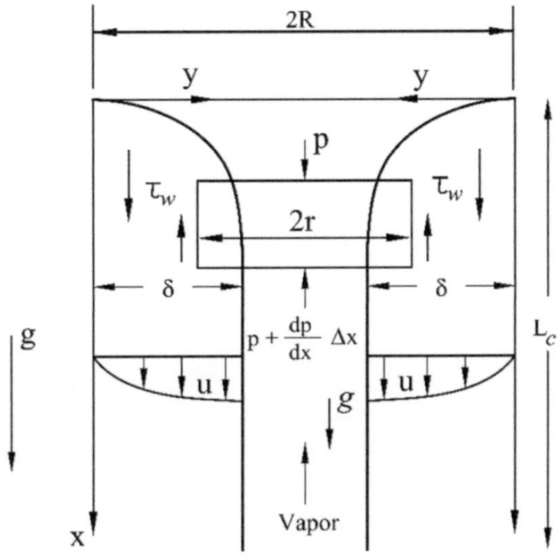

Fig. 7.15 Physical model of the condensation phenomena in contact with flowing vapor

downward (cocurrent vapor flow) or upward (countercurrent vapor flow). Faghri (1986) proposed a method of predicting the average film thickness, the local heat transfer coefficient, and the overall heat transfer coefficient for turbulent film condensation in a tube with interfacial shear stress caused by cocurrent and countercurrent vapor flow. In a fashion similar to that of Nusselt condensation, the inertia terms are neglected and the only forces included are body, pressure, and viscous forces. This particular model takes into account the decrease in the stream flow rate due to condensation.

To obtain an expression of the shear stress in the liquid film, a control volume with radius r and height Δx as shown in Fig. 7.15 is considered. For the case of countercurrent flow, a force balance results in

$$p\left(\pi r^2\right) + \rho_v g V_v + \rho_\ell g V_\ell = \left(p + \frac{dp}{dx}\Delta x\right)\pi r^2 + \tau(2\pi r \Delta x) \tag{7.161}$$

where the volume of the vapor and the liquid portion of the control volume are as follows:

$$V_v = (R - \delta)^2 \pi \Delta x \tag{7.162}$$

$$V_\ell = (R - y)^2 \pi \Delta x - V_v \tag{7.163}$$

Substituting Eqs. (7.162) and (7.163) into Eq. (7.161) and dividing through by Δx to solve for the shear stress, the following is obtained:

$$\tau = \frac{R - y}{2}\left(\rho_\ell g - \frac{dp}{dx}\right) - (\rho_\ell - \rho_v) g \frac{(R - \delta)^2}{2(R - y)} \tag{7.164}$$

The shear stress at the liquid–vapor interface can be found by letting $y = \delta$ in Eq. (7.164), i.e.,

$$\tau_\delta = \frac{R - \delta}{2}\left(\rho_v g - \frac{dp}{dx}\right) \tag{7.165}$$

Substituting Eq. (7.165) into Eq. (7.164), the shear stress at any radius, τ, can be related to the shear stress at the liquid film surface, τ_δ, by

$$\tau = \frac{R - y}{R - \delta}\tau_\delta + (\rho_\ell - \rho_\mathrm{v})g\left[\frac{2R(\delta - y) - \delta^2 + y^2}{2(R - y)}\right] \tag{7.166}$$

Assuming the tube radius is much greater than the condensate film thickness forming on the inner surface, the curvature of the liquid film can be neglected and the resulting analysis would be applicable to condensation between two flat plates. Taking this into account, Eq. (7.165) reduces to

$$\tau_\delta = \frac{R}{2}\left(\rho_\mathrm{v} g - \frac{\mathrm{d}p}{\mathrm{d}x}\right) \tag{7.167}$$

If it is further assumed that $\rho_\ell \gg \rho_\mathrm{v}$, Eq. (7.166) would reduce to

$$\tau = \tau_\delta + \rho_\ell g(\delta - y) \tag{7.168}$$

which can be written into a generalized form that includes the case of cocurrent flow, i.e.,

$$\tau = \pm\tau_\delta + \rho_\ell g(\delta - y) \tag{7.169}$$

where the + sign denotes downward vapor flow (cocurrent flow) and the − sign denotes upward vapor flow (countercurrent flow).

The shear stress at the wall, $y = 0$, is

$$\tau_\mathrm{w} = \pm\tau_\delta + \rho_\ell g\delta \tag{7.170}$$

The velocity profile in the liquid film can be found from the following differential equation when all axial terms and the curvature are neglected:

$$\frac{\mathrm{d}}{\mathrm{d}y}\left[(\nu_\ell + \varepsilon)\frac{\mathrm{d}u}{\mathrm{d}y}\right] + g = 0 \tag{7.171}$$

where ε is the momentum eddy diffusivity and is a time-measured flow property that adjusts the viscosity term for turbulent flow.

Assuming that the wall is impermeable and that the interfacial shear stress is known, the boundary conditions for this problem are given as

$$u = 0, \quad y = 0 \tag{7.172}$$

$$-\mu\frac{\partial u}{\partial y} = \pm\tau_\delta, \quad y = \delta \tag{7.173}$$

Integrating Eq. (7.171) twice with respect to y and applying boundary conditions specified by Eqs. (7.172) and (7.173), the velocity profile in the liquid film is obtained:

$$u = \int_0^y \frac{[g(\delta \quad y) \pm \tau_\delta/\rho_\ell]}{\nu_\ell + \varepsilon}\mathrm{d}y \tag{7.174}$$

The liquid Reynolds number obtained by the following expression:

$$Re_\delta = \frac{4\Gamma}{\mu_\ell} = 4\rho_\ell \int_0^\delta \frac{u\mathrm{d}y}{\mu_\ell} \tag{7.175}$$

To generalize the problem statement, the following nondimensional variables are defined:

$$\begin{gathered}\delta^+ = \frac{\delta u_\mathrm{f}}{\nu_\ell},\ \delta^* = \delta\left(\frac{\nu_\ell^2}{g}\right)^{-1/3},\ y^+ = \frac{y u_\mathrm{f}}{\nu_\ell},\ \tau_\delta^* = \frac{\tau_\delta(\nu_\ell g)^{-2/3}}{\rho_\ell}\\ u^+ = \frac{u}{u_\mathrm{f}},\ x^+ = \frac{x u_\mathrm{f}}{\nu_\ell},\ \varepsilon^+ = 1 + \frac{\varepsilon}{\nu_\ell},\ D^+ = \frac{D u_\mathrm{f}}{\nu_\ell}\end{gathered} \tag{7.176}$$

where u_f is the fractional velocity, defined as

$$u_\mathrm{f} = \left(\frac{\tau_\mathrm{w}}{\rho_\ell}\right)^{1/2} \tag{7.177}$$

Applying these nondimensional variables to Eqs. (7.170), (7.174), and (7.175), their nondimensional forms are obtained as follows:

$$u_\mathrm{f}^3 \mp \tau_\delta^+ u_\mathrm{f}(\nu_\ell g)^{2/3} - (\nu_\ell g)\delta^+ = 0 \tag{7.178}$$

$$u^+ = \int_0^{y^+} \frac{(1 - g\nu_\ell y^+/u_\mathrm{f}^3)}{\varepsilon^+}\mathrm{d}y^+ \tag{7.179}$$

$$Re_\delta = 4\int_0^\delta u^+\mathrm{d}y^+ \tag{7.180}$$

It should be noted that if $\tau_\delta^+ = 0$ and $u_\mathrm{f}^3 = \nu_\ell g \delta^+$, then Eq. (7.178) would reduce to the nonsheared film case (classical Nusselt analysis).

We will now consider the thermal side of the problem. An energy balance can be written for the case of constant heat flux at the wall (q''_w). This energy balance also assumes that heat transfer across the liquid film is dominated by conduction, so the convective terms can therefore be neglected.

$$\rho_\ell c_{\mathrm{p}\ell}\left(\frac{\nu_\ell}{Pr} + \frac{\varepsilon}{Pr_t}\right)\frac{\mathrm{d}T}{\mathrm{d}y} = q''_\mathrm{w} = \mu_\ell h_{\ell\mathrm{v}}\frac{\mathrm{d}(\Gamma/\mu_\ell)}{\mathrm{d}x} \tag{7.181}$$

where Pr_t is the turbulent Prandtl number, which will be discussed thoroughly toward the end of this subsection.

Equation (7.181) can be nondimensionalized to obtain

$$\frac{\mathrm{d}(\Gamma/\mu_\ell)}{\mathrm{d}x^+} = N_\mathrm{T}\left\{\int_0^{\delta^+}\left[\frac{Pr}{Pr_\mathrm{t}}\left(\frac{Pr_\mathrm{t}}{Pr} - 1 + \varepsilon^+\right)\right]^{-1}\mathrm{d}y^+\right\}^{-1} \tag{7.182}$$

where

$$N_{\mathrm{T}} = \frac{(T_{\mathrm{sat}} - T_{\mathrm{w}})c_{p\ell}}{h_{\ell\mathrm{v}} Pr} \tag{7.183}$$

Finally, the local heat transfer coefficient can be found directly from Eq. (7.181):

$$h_x = \frac{q''_{\mathrm{w}}}{T_{\mathrm{sat}} - T_{\mathrm{w}}} = \left\{ \int_0^{\delta} \left[\rho_\ell c_{p\ell} \left(\frac{\nu_\ell}{Pr} + \frac{\varepsilon}{Pr_{\mathrm{t}}} \right) \right]^{-1} \mathrm{d}y \right\}^{-1} \tag{7.184}$$

Nondimensionalizing Eq. (7.184) as a Nusselt number, the following is obtained:

$$Nu_x = \frac{h_x \delta}{k_\ell} = \delta^+ \left\{ \int_0^{\delta^+} \left[\frac{Pr}{Pr_{\mathrm{t}}} \left(\frac{Pr_{\mathrm{t}}}{Pr} - 1 + \varepsilon^+ \right) \right]^{-1} \mathrm{d}y^+ \right\}^{-1} \tag{7.185}$$

An average heat transfer coefficient is desirable in many practical applications. A modified Nusselt number Nu_x^+ is related to Nu_x by

$$Nu_x^+ = \frac{h_x}{k_\ell} \left(\frac{\nu_\ell}{g} \right)^{1/3} = Nu_x \frac{u_{\mathrm{f}}}{\delta^+} (\nu_\ell g)^{-1/3} \tag{7.186}$$

The average modified Nusselt number is then found from the following relation:

$$\overline{Nu^+} = \int_0^1 Nu^+ \mathrm{d}\left(\frac{x^+}{L^+} \right) \tag{7.187}$$

The dimensionless shear stress at the interface can be written as

$$\tau_\delta^* = (\nu_\ell g)^{-2/3} u_{\mathrm{f}}^2 (u_{\mathrm{v}}^+ + u_{\ell,\delta}^+) \left[\frac{f_{\mathrm{E}}}{2} \left(\frac{\rho_{\mathrm{v}}}{\rho_\ell} \right) (u_{\mathrm{v}}^+ + u_{\ell,\delta}^+) + \frac{\mathrm{d}(\Gamma/\mu_\ell)}{\mathrm{d}x^+} \right] \tag{7.188}$$

where f_{E} is the friction factor for vapor flow, which is different for upflows and downflows. The friction factor for vapor flow can be obtained by modifying the friction factor for single-phase flow, f, to accommodate the two-phase nature of the flow (Asali et al. 1985). It is also different for ripple ($Re_\delta \le 75$; the film is covered by long crested ripples with a steep front) and roll wave ($Re_\delta > 75$) regimes, i.e.,

$$f_{\mathrm{E}} = \begin{cases} f[1 + 0.045(Mg^+ - 5.9)] & Re_\delta \le 75 \\ f[1 + 0.045 Re_{\mathrm{v}}^{-0.2}(Mg^+ - 5.9)] & Re_\delta > 75 \end{cases} \tag{7.189}$$

where

$$Mg^{+} = \begin{cases} 0.78\, Re_{\delta}^{0.6}\left(\dfrac{\nu_\ell}{\nu_v}\right)\left(\dfrac{\rho_\ell}{\rho_v}\dfrac{\tau_\delta}{\tau_c}\right)^{1/2} & Re_\delta \le 75 \\ 0.50\, Re_{\delta}^{0.7}\left(\dfrac{\nu_\ell}{\nu_v}\right)\left(\dfrac{\rho_\ell}{\rho_v}\dfrac{\tau_\delta}{\tau_c}\right)^{1/2} & Re_\delta > 75 \end{cases} \tag{7.190}$$

The characteristic stress is given by

$$\frac{\tau_\delta}{\tau_c} = \left(\frac{1}{3} + \frac{2}{3}\frac{\tau_w}{\tau_\delta}\right)^{-1} \tag{7.191}$$

To calculate the velocity distribution and the heat transfer coefficient, a definition is required for ε_m and Pr_t from an appropriate turbulence model. In modeling of ε_m, it is customary to divide the flow into two regions—the inner region, where the turbulent transport is dominated by the wall, and another wave-like region that is directly adjacent. Faghri (1986) used a combination of the Szablewski (1968) and Van Driest models to obtain the following expression:

$$\varepsilon_m^{+} = \frac{1}{2} + \frac{1}{2}\left\{1 + 0.64 y^{+2}\frac{\tau}{\tau_w}\left[1 - \exp\left(-\frac{y^{+}}{A^{+}}\sqrt{\frac{\tau}{\tau_w}}\right)\right]^2\left[\exp\left(-1.66\left(1 - \frac{\tau}{\tau_w}\right)\right)\right]^2\right\} \tag{7.192}$$

where $A^{+} = 25.1$ and $\tau/\tau_w = 1 - y^{+}(g\nu_\ell)/u_f^3$.

This profile represents the eddy diffusivity in the inner layer closest to the wall ($0 \le y^{+} \le 0.6\delta^{+}$), where the influence of the wall is important. In the outer layer ($0.6\delta^{+} < y^{+} \le \delta^{+}$), the eddy viscosity is assumed to be constant, with a continuous transition to the inner layer.

Finally, because the turbulent transport near the liquid–vapor interface is quite different from that near the wall, the turbulent Prandtl number, Pr_t, cannot be assumed to be constant. Faghri (1986) used the following expression (Habib and Na 1974) for the analysis of turbulent transfer in pipes:

$$Pr_t = \frac{1 - \exp(-y^{+}/A^{+})}{\left[1 - \exp\left(-y^{+}\sqrt{Pr}/B^{+}\right)\right]} \tag{7.193}$$

where

$$B^{+} = \sum_{i=1}^{5} c_i (\log_{10} Pr)^{i-1} \tag{7.194}$$

and $c_1 = 34.96$; $c_2 = 28.79$; $c_3 = 33.95$; $c_4 = 6.3$; $c_5 = -1.186$.

The solution procedure begins with guessing an initial value of τ_δ^* for the initial values of δ_0^{+} and the initial vapor flow, which specifies the initial value of $Re_v = 4\dot{m}_v/(\pi D \mu_v)$. Based on the values of τ_δ^* and δ_0^{+}, the fractional velocity, u_f, is then obtained by solving Eq. (7.192). The dimensionless eddy diffusivity, ε^{+}, is obtained from Eq. (7.188). Equations (7.179) and (7.180) are integrated numerically to obtain the velocity profile and the liquid Reynolds number. An updated dimensionless shear stress at the interface can be obtained from Eq. (7.188). The process is repeated until the Re_ℓ values between two consecutive iterations differ by less than 0.5%. The local convective heat transfer coefficient can be obtained from Eqs. (7.186) and (7.187). The above procedure can be repeated for different x until heat transfer coefficients are obtained at all locations.

Heat transfer in the condenser sections of conventional and annular two-phase closed thermosyphon tubes has been studied analytically by Faghri et al. (1989). The method involved extending

Nusselt theory to include the variation of the shear at the vapor–liquid film interface. Harley and Faghri (1994) presented a transient two-dimensional condensation in a thermosyphon that accounts for conjugate heat transfer through the wall and the falling condensate film. The complete transient two-dimensional conservation equations are solved for the vapor flow and pipe wall, and the liquid film was modeled using a quasi-steady-state Nusselt-type solution.

7.3.6 Other Filmwise Condensation Configurations

Other configurations for external condensation have been investigated over the years. One of the most popular is external condensation on a vertical cylinder. If the thickness of the liquid film is smaller than the diameter of the cylinder by at least an order of magnitude, the heat transfer and Nusselt number expressions for a vertical cylinder are the same as those for a vertical plate. Nusselt analysis for laminar film condensation for a vertical plate can also be applied to condensation on a plate at an angle (with respect to the vertical) by replacing g in Eq. (7.105) with $g\cos\theta$:

$$\overline{Nu_{\mathrm{L}}} = \frac{\bar{h}_{\mathrm{L}}L}{k_\ell} = 0.943\left[\frac{\rho_\ell(\rho_\ell-\rho_{\mathrm{v}})g\cos\theta h'_{\ell\mathrm{v}}L^3}{\mu_\ell k_\ell(T_{\mathrm{sat}}-T_{\mathrm{w}})}\right]^{1/4} \tag{7.195}$$

When the cold surface is curved as in a horizontal cylinder or sphere, the tangential direction of the gravity varies along the condensate film. The empirical correlations for other configurations also have almost the same forms as that of a vertical plate, except the leading coefficient and the characteristic length differ. For example, for laminar film condensation on a horizontal cylinder, the leading coefficient changes from 0.943 to 0.729 and the characteristic length changes from L to D, the diameter of the cylinder, i.e.,

$$\overline{Nu_{\mathrm{D}}} = \frac{\bar{h}_{\mathrm{D}}D}{k_\ell} = 0.729\left[\frac{D^3 h'_{\ell\mathrm{v}}g(\rho_\ell-\rho_{\mathrm{v}})}{k_\ell\nu_\ell(T_{\mathrm{sat}}-T_{\mathrm{w}})}\right]^{1/4} \tag{7.196}$$

which can be obtained by applying Nusselt analysis (see Problem 7.15).

For laminar film condensation on a sphere, the average heat transfer coefficient can also be obtained by Nusselt analysis (see Problem 7.18).

$$\overline{Nu_{\mathrm{D}}} = \frac{\bar{h}_{\mathrm{D}}D}{k_\ell} = 0.815\left[\frac{D^3 h'_{\ell\mathrm{v}}g(\rho_\ell-\rho_{\mathrm{v}})}{k_\ell\nu_\ell(T_{\mathrm{sat}}-T_{\mathrm{w}})}\right]^{1/4} \tag{7.197}$$

Another expression of great importance in heat exchanger design is that of a vertical column of n horizontal tubes. The average heat transfer coefficient for all n cylinders can be obtained by:

$$\overline{Nu_{\mathrm{D}}} = \frac{\bar{h}_{\mathrm{D,n}}D}{k_\ell} = 0.729\left[\frac{D^3 h'_{\ell\mathrm{v}}g(\rho_\ell-\rho_{\mathrm{v}})}{nk_\ell\nu_\ell(T_{\mathrm{sat}}-T_{\mathrm{w}})}\right]^{1/4} \tag{7.198}$$

which indicates that the average heat transfer coefficient for all n cylinders, $\bar{h}_{\mathrm{D,n}}$, is related to the average heat transfer coefficient for the first tube of this array, $\bar{h}_{\mathrm{D}}$, by

$$\bar{h}_{\mathrm{D,n}} = \frac{\bar{h}_{\mathrm{D}}}{n^{1/4}} \tag{7.199}$$

This calculation would lead to a lower average heat transfer coefficient for an array than that found for a single tube. In practice, however, it has been found that these calculations significantly underestimate the heat transfer capabilities of this system, due to splashing effects that occur as the condensate (in the form of sheet or droplets) falls from one tube to a tube underneath.

For condensation on an upward-facing horizontal surface of a finite size, the condensate in the central region flows toward the edge where it is spilled (Bejan 1991). For condensation over a long horizontal strip with a width of L, the average heat transfer can be obtained by

$$\overline{Nu_{\mathrm{L}}} = \frac{\bar{h}L}{k_\ell} = 1.079\left[\frac{\rho_\ell(\rho_\ell - \rho_{\mathrm{v}})gh'_{\ell\mathrm{v}}L^3}{\mu_\ell k_\ell(T_{\mathrm{sat}} - T_{\mathrm{w}})}\right]^{1/5} \tag{7.200}$$

The heat transfer coefficient for film condensation over an upward-facing horizontal circular disk with a diameter of D is

$$\overline{Nu_{\mathrm{D}}} = \frac{\bar{h}D}{k_\ell} = 1.368\left[\frac{\rho_\ell(\rho_\ell - \rho_{\mathrm{v}})gh'_{\ell\mathrm{v}}D^3}{\mu_\ell k_\ell(T_{\mathrm{sat}} - T_{\mathrm{w}})}\right]^{1/5} \tag{7.201}$$

Example 7.4

Saturated acetone at $T_{\mathrm{sat}} = 60\,°\mathrm{C}$ condenses on the outside of a horizontal copper tube with a diameter of $D = 3.0\,\mathrm{cm}$. The outer surface temperature of the copper tube is $T_{\mathrm{w}} = 40\,°\mathrm{C}$. Find the heat transfer coefficient and the rate of condensation per unit length of the tube.

Solution The vapor properties are evaluated at saturation temperature $T_{\mathrm{sat}} = 60\,°\mathrm{C}$. The vapor density at this temperature is $\rho_{\mathrm{v}} = 2.37\,\mathrm{kg/m^3}$, and the latent heat of vaporization is $h_{\ell\mathrm{v}} = 517\,\mathrm{kJ/kg}$. The liquid properties evaluated at $T_{\mathrm{f}} = (T_{\mathrm{sat}} + T_{\mathrm{w}})/2 = 50\,°\mathrm{C}$ are $\rho_\ell = 756.0\,\mathrm{kg/m^3}$, $c_{\mathrm{p}\ell} = 2255\,\mathrm{J/kg\,K}$, $\mu_\ell = 0.248 \times 10^{-3}\,\mathrm{kg/m\,s}$, $k_\ell = 0.172\,\mathrm{W/m\,K}$, and $\nu_\ell = \mu_\ell/\rho_\ell = 0.328 \times 10^{-6}\,\mathrm{m^2/s}$.

The revised latent heat of vaporization is

$$\begin{aligned} h'_{\ell\mathrm{v}} &= h_{\ell\mathrm{v}}\left\{1 + 0.68\left[\frac{c_{\mathrm{p}\ell}(T_{\mathrm{sat}} - T_{\mathrm{w}})}{h_{\ell\mathrm{v}}}\right]\right\} \\ &= 517 \times \left\{1 + 0.68\left[\frac{2.255 \times (60 - 40)}{517}\right]\right\} = 547.7\,\mathrm{kJ/kg} \end{aligned}$$

The heat transfer coefficient can be obtained from Eq. (7.196), i.e.,

$$\begin{aligned} \bar{h} &= 0.729\left[\frac{(\rho_\ell - \rho_{\mathrm{v}})gk_\ell^3 h'_{\ell\mathrm{v}}}{\nu_\ell \Delta T D}\right]^{1/4} \\ &= 0.729 \times \left[\frac{(756.0 - 2.37) \times 9.8 \times 0.172^3 \times 547.7 \times 10^3}{0.328 \times 10^{-6}(60 - 40) \times 0.03}\right]^{1/4} \\ &= 2331.3\,\mathrm{W/m^2K} \end{aligned}$$

The heat transfer rate per unit width is then

$$q' = \bar{h}\pi D(T_{sat} - T_w) = 2331.3 \times \pi \times 0.03 \times (60 - 40) = 4394.4\,\text{W}$$

The condensation rate per unit width is

$$\dot{m}' = \frac{q'}{h'_{\ell v}} = \frac{4394.4}{547.7 \times 10^3} = 0.00802\,\text{kg/s m}$$

The correlations that we discussed thus far have been limited to cases in which the flow of the condensate is driven by gravity. When the vapor is forced to flow over the cooler surface, the vapor interacts with the condensate and drags liquid in the vapor flow direction. For condensation on the outside of a horizontal tube in cross-flow, the heat transfer coefficient is affected by both free-steam velocity of vapor, u_∞, and gravitational force (Shekriladze and Gomelauri 1966), so that

$$\overline{Nu_D} = \frac{\bar{h}D}{k_\ell} = 0.64 Re_D^{1/2}\left[1 + \left(1 + 1.69\frac{g h'_{\ell v}\mu_\ell D}{u_\infty^2 k_\ell (T_{sat} - T_w)}\right)^{1/2}\right]^{1/2} \tag{7.202}$$

where $Re_D = u_\infty D/\nu_\ell$ is based on the viscosity of liquid. Equation (7.202) is valid for a Reynolds number up to 10^6.

For laminar film condensation on a horizontal flat plate in a parallel stream of saturated vapor, the average heat transfer coefficient is

$$\overline{Nu_L} = \frac{\bar{h}L}{k_\ell} = 0.872 Re_L^{1/2}\left[\frac{1.508}{(1 + Ja/Pr_\ell)^{3/2}} + \frac{Pr_\ell}{Ja}\left(\frac{\rho_v \mu_v}{\rho_\ell \mu_\ell}\right)^{1/2}\right]^{1/3} \tag{7.203}$$

where L is the length of the flat plate along the vapor flow direction, $Re_L = u_\infty L/\nu_\ell$ is based on the viscosity of liquid, and $Ja = c_{p\ell}(T_{sat} - T_w)/h_{\ell v}$ is the Jakob number. Equation (7.203) is valid for $\rho_\ell \mu_\ell/\rho_v \mu_v = 10 \sim 500$ and $Ja/Pr_\ell = 0.01 \sim 1$.

7.3.7 Effects of Noncondensable Gas

The physical model and governing equations for binary vapor (condensable vapor and noncondensable gas) have been discussed in Sect. 7.3.2; heat and mass transfer in the vapor phase must be studied, in addition to heat transfer in the liquid film. There exists a boundary layer in the vapor phase ($\delta < y < \delta + \delta_v$), in which the partial pressure of the condensable vapor, p_v, decreases from a constant value $p_{v,\infty}$ at $y = \delta + \delta_v$ to the value $p_{v,\delta}$ at $y = \delta$, where the vapor is condensing to a liquid. The condensing vapor must diffuse through this vapor boundary layer to the liquid–vapor interface. The partial pressure of the noncondensable gas, p_g, on the other hand, increases from its reservoir value, $p_{g,\infty}$, to the value $p_{g,\delta}$ at the liquid–vapor interface. At any point in space and time, the summation of the partial pressures of this binary system must equal the constant total pressure.

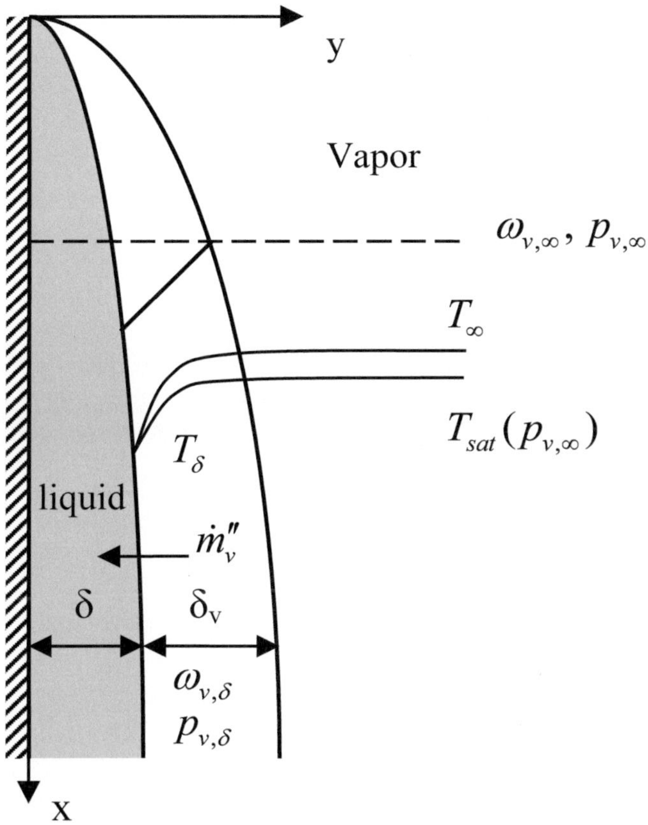

Fig. 7.16 Mass transfer in the equivalent laminar film

$$p_v + p_g = p \tag{7.204}$$

The partial pressure of the condensing gas decreases as it approaches the phase interface, and its corresponding saturation temperature $T_{sat}(p_v)$ also falls. Depending on the noncondensable gas content, the temperature at the interface can be much lower than if no such gas were present. The temperature difference across the interface would also be lower as a result, which would lead to a lower overall heat transfer coefficient. This clearly demonstrates the benefit of removing as much noncondensable gas from the system as possible. However, systematic purity cannot always be achieved, and the noncondensable gas content must be taken into account.

The mass transfer in the vapor boundary layer can be rigorously analyzed by solving the boundary-layer type of governing equations discussed in Sect. 7.3.2. Hewitt et al. (1994) presented an alternative model based on the concept of equivalent laminar film—a layer in the vapor phase in which the mass fraction of the condensable vapor, ω_v, varies linearly from $\omega_{v,\delta}$ at the liquid–vapor interface to $\omega_{v,\infty}$ at $y = \delta + \delta_v$ (see Fig. 7.16). Meanwhile, the partial pressure of the condensable vapor changes from $p_{v,\delta}$ to $p_{v,\infty}$ in the equivalent laminar film. Assuming that the noncondensable gas cannot be dissolved into the liquid condensate, the molar flux of the condensable vapor at any point within the equivalent laminar layer can be obtained by [see Eq. (7.86)]

$$\dot{n}_v'' = -D_{vg}\frac{\partial c_v}{\partial y} + \dot{n}_v'' x_v \tag{7.205}$$

which can be rearranged to obtain

$$\dot{n}_v'' = -\frac{D_{vg}}{1 - c_v/c_T}\frac{\partial c_v}{\partial y} \tag{7.206}$$

where c_T is total molar concentration of vapor and noncondensable gas.

Integrating Eq. (7.206) over the equivalent laminar layer and considering $\dot{n}_v''$ and c_T are constants, one obtains

$$\dot{n}_v'' \int_{\delta}^{\delta+\delta_v} dy = -D_{vg} c_T \int_{c_{v,\delta}}^{c_{v,\infty}} \frac{1}{c_T - c_v} dc_v \tag{7.207}$$

Equation (7.207) can be rearranged to yield

$$\dot{n}_v'' = \frac{c_T D_{vg}}{\delta_v}\ln\left(\frac{c_T - c_{v,\delta}}{c_T - c_{v,\infty}}\right) = c_T h_{m,G}\ln\left(\frac{c_T - c_{v,\delta}}{c_T - c_{v,\infty}}\right) \tag{7.208}$$

where

$$h_{m,G} = \frac{D_{vg}}{\delta_v} \tag{7.209}$$

is mass transfer coefficient (m/s). Equation (7.209) is valid under the assumption of a linear boundary layer. Since the thickness of equivalent laminar film, δ_v, is still unknown at this point, the mass transfer coefficient can be approximately related to the heat transfer coefficient, h_G, through the Lewis equation, i.e.,

$$h_{m,G} = \frac{1}{\rho_{vg}}\frac{h_G}{c_{p,vg}} \tag{7.210}$$

where $c_{p,vg}$ is the constant-pressure specific heat of the vapor–gas mixture.

If the condensable vapor and noncondensable gas can be treated as ideal gas, the molar flux in Eq. (7.208) can also be expressed in term of mass flux, $\dot{m}_v''$, and partial pressure of the condensable vapor, p_v, in the mixture (see Problem 7.29)

$$\dot{m}_v'' = \rho_{vg} h_{m,G}\ln\frac{p - p_{v,\delta}}{p - p_{v,\infty}} \tag{7.211}$$

The energy balance across a differential control volume at the liquid–vapor interface, as shown in Fig. 7.17, is (Stephan 1992).

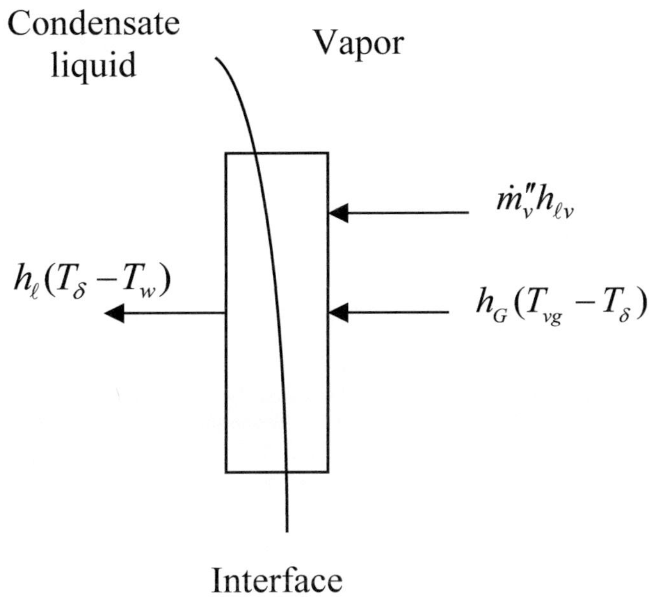

Fig. 7.17 Energy balance at the liquid–vapor interface for film condensation on a vertical plate including the effects of noncondensable gases

$$h_\ell(T_\delta - T_w) = \dot{m}_v'' h_{\ell v} + h_G(T_{vg} - T_\delta) \tag{7.212}$$

where h_G is the heat transfer coefficient from the vapor–gas mixture to the liquid–vapor interface, and where T_δ is the temperature at the liquid–vapor interface.

Substituting Eqs. (7.210) and (7.211) into the energy balance Eq. (7.212) and using $h_{vg} = \xi h_G$, which corrects for the fact that vapor does not flow along the wall but stops at the liquid–vapor interface (where ξ is a correction factor), the following is obtained (Stephan 1992):

$$T_\delta - T_w = \frac{h_G}{h_\ell}\left[\frac{h_{\ell v}}{c_{p,vg}} \ln \frac{p - p_{v,\delta}}{p - p_{v,\infty}} + \xi\left(T_{vg} - T_\delta\right)\right] \tag{7.213}$$

In the case of small inert gas content, the above equation will reduce to

$$T_\delta - T_w = \frac{h_G}{h_\ell}\left[\frac{h_{\ell v}}{c_{p,vg}} \ln \frac{p - p_{v,\delta}}{p - p_{v,\infty}}\right] \tag{7.214}$$

In the best-case scenario where there is no noncondensable gas in the vapor, the heat flux across the liquid film would be as follows:

$$q'' = \frac{k_\ell}{\delta}(T_{sat} - T_w) \tag{7.215}$$

However, if a noncondensable gas is present, the temperature drop across the film would be lower and the heat flux would be as follows:

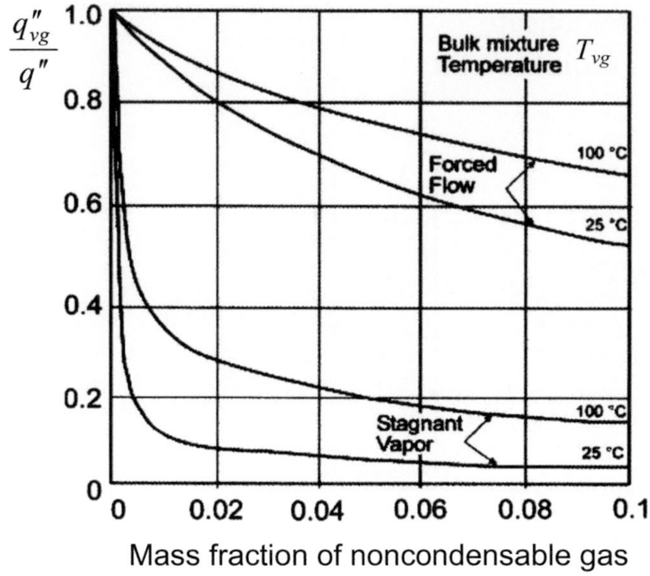

Fig. 7.18 Falling film condensation of steam with noncondensable gas (Collier and Thome 1994; Reprinted with permission from Oxford University Press)

$$q''_{vg} = \frac{k_\ell}{\delta}(T_\delta - T_w) \tag{7.216}$$

where q''_{vg} is the heat flux across the liquid film in the presence of a noncondensable gas. The ratio of the heat fluxes obtained by Eqs. (7.216) and (7.215) is

$$\frac{q''_{vg}}{q''} = \frac{T_\delta - T_w}{T_{sat} - T_w} \leq 1 \tag{7.217}$$

Substituting the expression for temperature difference across a liquid film in the presence of a noncondensable gas—Eq. (7.214)—into the above ratio, the following is obtained:

$$\frac{q''_{vg}}{q''} = \frac{h_G h_{\ell v}}{(T_{sat} - T_w) h_\ell c_{vg,\infty}} \ln \frac{p - p_{v,\delta}}{p - p_{v,\infty}} \tag{7.218}$$

or substituting $\dot{m}_v$ from Eq. (7.211):

$$\frac{q''_{vg}}{q''} = \frac{\dot{m}''_v h_{\ell v}}{(T_{sat} - T_w) h_\ell} \tag{7.219}$$

It can be seen from the above expression that for large ($T_{sat} - T_w$), the velocity or mass flow rate $\dot{m}_v$, must be made sufficiently large to acquire a large heat transfer coefficient for the heat transfer from the vapor–gas mixture to the liquid–vapor interface. This must be done in order to make q''_{vg} not too small and therefore remove the undesirable effects of the noncondensable gas as much as possible.

Figure 7.18 shows the effects of noncondensable gas (air) in saturated steam on falling film condensation under both stagnant and forced convection conditions. It can be seen that the effect of noncondensable gas on condensation in a stagnant vapor is very significant. For forced convective condensation, the effect of noncondensable gas is still significant, but is much weaker than its effect on the stagnant vapor condensation. Therefore, the effect of noncondensable gas on condensation can be minimized by allowing vapor to flow.

Example 7.5

A mixture of 90% steam and 10% air at 100 °C and 1 atm flows across a horizontal cylinder. The diameter of the cylinder is 0.1 m, and the velocity of the mixture is 30 m/s. The condensation rate on the cylinder is $\dot{m}''_v = 0.02\,\text{kg/m}^2\text{s}$. The properties of the mixture are $\rho_{vg} = 0.944\,\text{kg/m}^3$, $\mu_{vg} = 8.2 \times 10^{-6}\,\text{N s/m}^2$, $D_{vg} = 3.64 \times 10^{-5}\,\text{m}^2/\text{s}$, respectively. What is the temperature at liquid–vapor interface? If the temperature of the tube is 80 °C, what is the percentage of heat transfer reduction due to the existence of noncondensable gas?

Solution The mass fraction of the steam is $\omega_{v,\infty} = 0.9$. The molecular mass of water and air is $M_v = 18.02\,\text{kg/kmol}$ and $M_g = 28.96\,\text{kg/kmol}$. The partial pressure of the steam can be obtained from Eq. (7.70), i.e.,

$$\begin{aligned} p_{v,\infty} &= p\left[1 + \frac{M_v(1-\omega_{v,\infty})}{M_g\omega_{v,\infty}}\right]^{-1} \\ &= 1.013 \times 10^5 \times \left[1 + \frac{18.02 \times (1-0.9)}{28.96 \times 0.9}\right]^{-1} = 0.9475 \times 10^5\,\text{kPa} \end{aligned}$$

The Reynolds number of the mixture is

$$Re_D = \frac{\rho_{vg}u_\infty D}{\mu_{vg}} = \frac{0.944 \times 30 \times 0.1}{8.2 \times 10^{-6}} = 34{,}536$$

The Schmidt number of the mixture is

$$Sc = \frac{\mu_{vg}}{\rho_{vg}D_{vg}} = \frac{8.2 \times 10^{-6}}{0.944 \times 3.64 \times 10^{-5}} = 0.239$$

The empirical correlation for forced convective heat transfer across a cylinder is (see Appendix D)

$$\overline{Nu}_D = 0.3 + \frac{0.62Re_D^{1/2}Pr^{1/3}}{[1+(0.4/Pr)^{2/3}]^{1/4}} \times \left[1 + \left(\frac{Re_D}{282{,}000}\right)^{5/8}\right]^{4/5}$$

Analogy between mass and heat transfer gives us

$$\begin{aligned} \overline{Sh}_D &= 0.3 + \frac{0.62Re_D^{1/2}Sc^{1/3}}{[1+(0.4/Sc)^{2/3}]^{1/4}}\left[1 + \left(\frac{Re_D}{282{,}000}\right)^{5/8}\right]^{4/5} \\ &= 0.3 + \frac{0.62 \times 34{,}536^{1/2} \times 0.239^{1/3}}{[1+(0.4/0.239)^{2/3}]^{1/4}}\left[1 + \left(\frac{34536}{282000}\right)^{5/8}\right]^{4/5} = 108.1 \end{aligned}$$

The mass transfer coefficient is therefore

$$h_{m,G} = \frac{\overline{Sh}_D D_{vg}}{D} = \frac{108.1 \times 3.64 \times 10^{-5}}{0.1} = 0.03935\,\text{m/s}$$

The partial pressure of the vapor at the liquid–vapor interface, $p_{v,\delta}$, can be obtained from Eq. (7.211)

$$\begin{aligned} p_{v,\delta} &= p - (p - p_{v,\infty}) \exp\left(\frac{\dot{m}''_v}{\rho_{vg} h_{m,G}}\right) \\ &= 1.013 \times 10^5 - (1.013 \times 10^5 - 0.9475 \times 10^5) \exp\left(\frac{0.02}{0.944 \times 0.03935}\right) \\ &= 0.900 \times 10^5\,\text{Pa} \end{aligned}$$

The interfacial temperature, T_δ, is the saturation temperature corresponding to the above partial pressure. It can be found from the Clapeyron–Clausius Eq. (2.158).

$$\ln \frac{p_{v,\delta}}{p} = -\frac{h_{\ell v}}{R_g}\left(\frac{1}{T_\delta} - \frac{1}{T_{sat}}\right) \tag{7.220}$$

where $h_{\ell v} = 2251.2\,\text{kJ/kg}$ and $R_g = 0.4615\,\text{kJ/kg K}$ for water at 1 atm. Equation (7.220) can be rearranged to obtain

$$\begin{aligned} T_\delta &= \left(\frac{1}{T_{sat}} - \frac{R_g}{h_{\ell v}} \ln \frac{p_{v,\delta}}{p}\right)^{-1} \\ &= \left(\frac{1}{373.15} - \frac{0.4615}{2251.2} \ln \frac{0.900 \times 10^5}{1.013 \times 10^5}\right)^{-1} = 369.80\,\text{K} = 96.65\,^\circ\text{C} \end{aligned}$$

The ratio of heat fluxes with and without noncondensable gas can be obtained from Eq. (7.217)

$$\frac{q''_{vg}}{q''} = \frac{T_\delta - T_w}{T_{sat} - T_w} = \frac{96.65 - 80.0}{100.0 - 80.0} = 0.8325$$

In other words, the heat transfer is decreased by 16.75% due to the presence of 10% of non-condensable gas.

Problems

7.1. For a liquid–vapor phase-change problem, explain how the interface affects heat transfer. Describe the conditions under which the interfacial heat transfer resistance is not negligible.

7.2. Perform an energy balance for a control volume that includes a liquid–vapor interface, and show that the conservation of mass at the liquid–vapor interface in film condensation is [see Eq. (7.80)]

$$\left[\rho_\ell\left(u_\ell \frac{d\delta}{dx} - v_\ell\right)\right]_I = \left[\rho_v\left(u_v \frac{d\delta}{dx} - v_v\right)\right]_I, \quad y = \delta$$

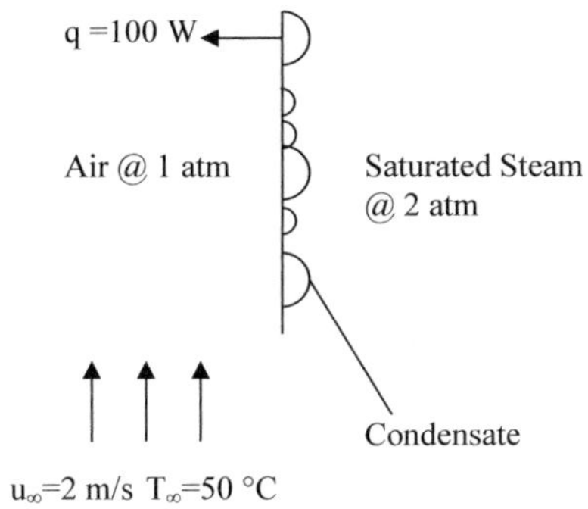

Fig. P7.4

7.3. A beverage can with diameter of 65 mm and a height of 120 mm is removed from a refrigerator with a temperature of 4 °C and placed in a room at temperature of 30 °C and relative humidity of 75%. What is the condensation rate if the condensation can be assumed to be dropwise? If the relative humidity is changed to 50%, how will condensation rate change?

7.4. A square vertical plate 0.5 m long separates saturated steam at 2 atm from 50 °C air blowing parallel to the plate at 2 m/s (see Fig. P7.4). 100 W is transferred from the plate to the air by convective cooling. What is the condensation rate assuming dropwise condensation?

7.5. A bathroom mirror is initially at 20 °C. As a bath is drawn, the room temperature increases to 30 °C and the relative humidity increases to 90%. Calculate the heat transfer rate for a single droplet of condensate 1 mm in diameter. Discuss which thermal resistances are most significant.

7.6. Consider the dropwise condensation of saturated vapor at 30 °C on a vertical plate that is maintained at 20 °C. The equilibrium contact angle of the plate is $\theta_c = 90°$, while the receding and advancing contact angles are $\theta_r = 75°$ and $\theta_a = 105°$, respectively. If the maximum adhesion forces on the droplet is given as:

$$F_\sigma = \frac{\sigma_{\ell v} D}{\frac{\pi}{\theta_a - \theta_r} - \frac{\theta_a - \theta_r}{\pi}} [\sin(\theta_a - \theta_r) + \cos(\theta_a - \theta_r) + 1]$$

where D is the droplet diameter.

(a) Calculate the maximum stable diameter of a droplet attached to the surface of the plate (*ignore droplet deformation and assume that the volume of the droplets is* $V_d = (2/3)\pi(D/2)^3$, $\rho_\ell = 1000\,\text{kg/m}^3$, g = 9.81 m/s^2, and $\sigma_{\ell v} = 0.0725\,\text{N/m}$).

(b) Droplet distribution function, $n''(D)$, represents the number of droplets of that diameter D per unit area at any given time. If the droplet distribution function, $n''(D)$, is given as

$$n''(D) = \frac{2}{3\pi D_{max}} D^{-1} \quad (\#/\text{unit area})$$

Find the fraction of the surface covered by liquid droplets (*ignore droplet deformation and assume that the interfacial area between a droplet and the plate is* $A_d = \pi(D/2)^2$). The minimum sustainable droplet diameter is $D_{min} = 4\sigma_{\ell v} v_\ell T_v / [h_{\ell v}(T_v - T_w)]$ with $v_\ell = 1.003 \times 10^{-3}\,\text{m}^3/\text{kg}$ and $h_{\ell v} = 2429.77\,\text{kJ/kg}$.

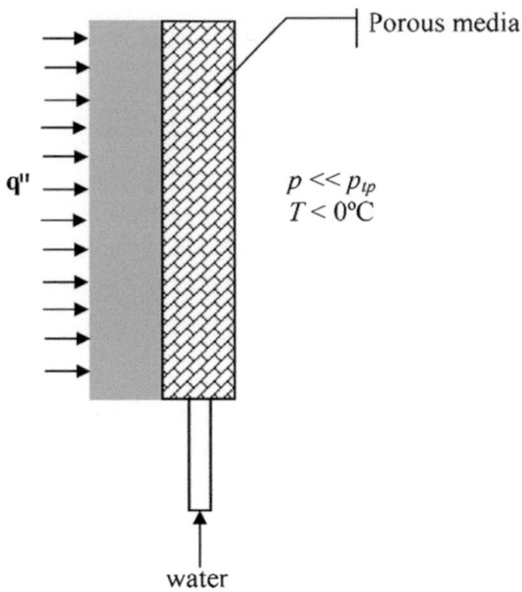

Fig. P7.7

c) If the convective heat transfer coefficient between the flat plate and gas is given as $h = 25$ W/m^2 K, calculate the rate of heat transfer through the vapor phase per unit area of the flat plate.

7.7. In very low pressure environments, sensible cooling is typically not sufficient due to very low conductivity of the gas. To be able to reject high rates of heat, a cooling scheme as shown in Fig. P7.7 is proposed. The porous media is connected to a liquid reservoir on the bottom. One side of the surface is open to environment, and the other side is connected to the heat source. Explain how this system works, and outline the heat transfer mechanism.

7.8. Many assumptions are made in Nusselt analysis. Discuss the reasonability of these assumptions and their expected effects on heat transfer.

7.9. For film condensation of steam at atmospheric pressure, find the distance from the top of a vertical plate at which a laminar flow would become wavy. What is the average heat transfer coefficient for this plate? The heat removal rate can be assumed to be 500 kW/m^2.

7.10. Assuming that the plate in the previous problem has a length of 5 cm and recognizing that this exceeds the maximum laminar length found in the previous problem, find the average heat transfer coefficient for the plate by using both the Kutateladze correlation (which assumes that wavy flow exists along the length at some point) and the classical Nusselt analysis. Compare the two answers.

7.11. For a vertical pipe at 25 °C with a length of 2 m in a saturated steam environment of 100 °C, find the average heat transfer coefficient of an external turbulent film condensation. What is the corresponding mass flow rate?

7.12. Assume that an interfacial shear stress of 0.5 N/m^2 exists in the film condensation on a vertical plate with the physical properties of Problem 7.7. Find the corresponding average heat transfer coefficient for this configuration. The imposed pressure gradient that drives the flow is 2000 N/m^3.

7.13. The effect of vapor flow on film condensation was considered in Sect. 7.3.4 by introducing a pressure gradient imposed by the motion of the adjacent vapor and a shear stress at the interface. A more rigorous analysis requires consideration of the vapor flow using boundary-layer theory. Write the complete governing equations for both liquid and vapor flows, and give the corresponding boundary conditions.

7.14. A pure vapor at its saturation temperature, T_{sat}, flows downward across a horizontal cylindrical tube with a wall temperature of T_w ($T_w < T_{sat}$). The downward vapor velocity is u_∞. Write the governing equations and corresponding boundary conditions for the convective condensation problem. The pressure gradient in the vapor flow can be determined by assuming potential vapor flow.

7.15. When a stagnant pure vapor condenses on the outer surface of a horizontal cylinder, the condensate generated from the top of the cylinder flows along its surface and eventually drips from the bottom of the cylinder. The driving force of the liquid condensate flow is the tangential component of the gravitational force along the surface of the cylinder. Perform Nusselt analysis to get the *average* heat transfer coefficient, and compare your result with Eq. (7.196).

7.16. Condensation on the outer surface of a vertical cylinder is often estimated using a correlation for condensation on a vertical plate. This treatment is valid only if the thickness of the liquid film is much smaller than the radius of the vertical cylinder. For condensation over a thin vertical cylinder, the effect of curvature of the cylindrical surface on the condensation must be

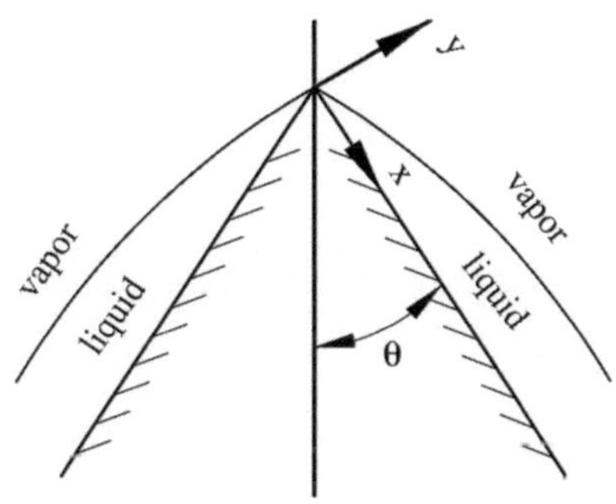

Fig. P7.17

considered. This requires that the governing equations be written in the cylindrical coordinate system. Obtain the film thickness for film condensation on the outer surface of a cylinder using Nusselt's theory.

7.17. A saturated vapor condenses on the outside surface of a constant-temperature cone (see Fig. P7.17). Neglecting vapor shear stress at the liquid–vapor interface and the subcooling effect, obtain a single differential equation relating local film thickness to the temperature difference, the fluid properties, and the distance from the apex of the cone (hint: make assumptions similar to those made for the Nusselt Theory).

7.18. During laminar film condensation on the outer surface of a sphere, the condensation starts from the top of the sphere and the condensate eventually drips from the bottom of the sphere. When the condensate flows down along the spherical surface, both the driving force and the width of the liquid film vary. Derive the average heat transfer coefficient for this case, and compare your result with Eq. (7.197).

7.19. For laminar film condensation occurring on a vertical surface, write down an appropriate set of governing equations and boundary conditions that could be solved to predict the variation of the film thickness and heat transfer coefficient with x for the following different boundary conditions.

a. The cold surface temperature T_w is constant.
b. There is a constant heat flux at the cold surface.
c. The back of the cold surface is subjected to convective cooling with a coolant temperature of T_∞ and heat transfer coefficient of h_c. The thickness of the cold wall can be neglected.

7.20. The process of condensation inside a vertical tube is complex. Build a reasonable model; write governing equations and boundary conditions for the model.

7.21. The effect of liquid subcooling was neglected in the Nusselt analysis. Starting from the energy equation for liquid film with convection terms accounted for, show that the energy balance in the liquid film is Eq. (7.110). If the temperature profile in the liquid can be assumed to be linear like Eq. (7.111), show that the Nusselt analysis is still valid provided that the latent heat, $h_{\ell v}$, is replaced by $h'_{\ell v}$ as defined in Eq. (7.113).

7.22. A pure vapor flows into a horizontal channel formed by two parallel plates (see Fig. P7.22) with uniform velocity (u_∞) and temperature (T_∞). The bottom plate is maintained at a tem-

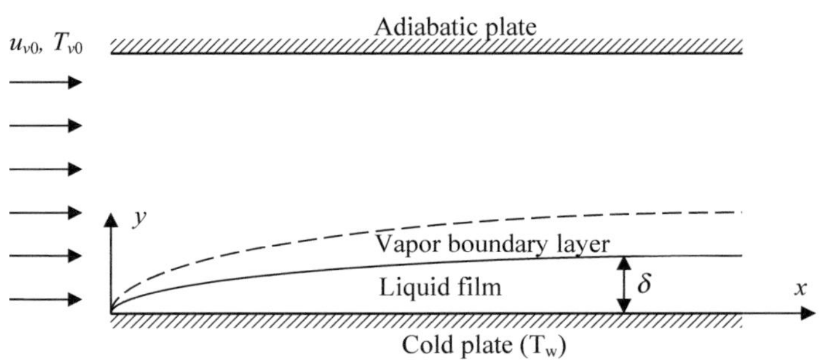

Fig. P7.22

perature, T_w, below the saturation temperature, while the upper plate is adiabatic. Condensation occurs on the surface of the bottom plate, and the condensate is dragged by the vapor and flows to the positive x-direction. The thickness of the liquid film is much smaller than the distance between the two plates ($\delta \ll H$). Both velocity and temperature distribution in the liquid film can be assumed to be linear. The shear stress at the liquid–vapor interface can be estimated by $\tau_\delta = 0.5f\rho_v(u_v - u_\delta)^2 + \dot{m}''_\delta(u_v - u_\delta)$, where the friction coefficient $f = CRe^n$ (C and n are constants), u_δ is the axial velocity at interface, and $\dot{m}''_\delta$ is rate of condensation. Obtain the ordinary differential equation that governs the liquid film thickness.

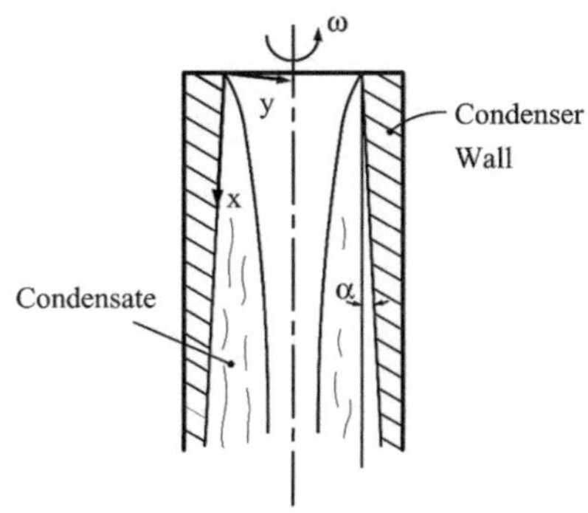

Fig. P7.24

7.23. A condenser is made of a vertical array of 200 horizontal tubes, each 0.5 m long and with diameters of 10 mm. If the condenser is meant to condense saturated steam at 0.5 bars and the cooling tubes are at 25 °C, what is the average condensation rate for a single tube of the condenser?

7.24. The condenser section of a rotating heat pipe is shown in Fig. P7.24. The liquid condensate is removed by centrifugal force, and the effect of gravity can be neglected. The temperature of the condenser wall uniformly equals T_w, and the vapor is saturated. The effect of vapor flow on the film condensation can be neglected. Assuming the circumferential velocity and temperature gradient are negligible, find the velocity profile in the liquid condensate.

7.25. Heat conduction in the condenser wall was neglected in Problem 7.24 by assuming the entire condenser wall is at a uniform temperature. If the outer surface of the condenser wall is maintained at T_w, repeat the above problem while considering heat conduction in the condenser wall. Heat conduction in the axial direction can be neglected for simplicity.

7.26. In a rotating heat pipe, thermal energy transport relies on the evaporation and condensation of a

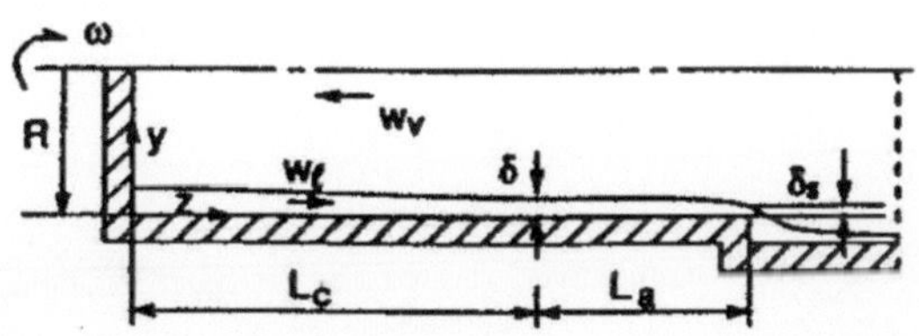

Fig. P7.26

small amount of working fluid, whereas the condensate return relies on the centrifugal force. In the cylindrical section, including condenser and adiabatic zones with a step at the position connecting the evaporator, the condensate return is maintained by centrifugal forces acting on the liquid film with varying film thickness, which results in a hydrostatic pressure change in the film. The coordinate system for annular flow is shown in Fig. P7.26. It can be assumed that (1) the condensate film flow is laminar, (2) the liquid film thickness is much smaller than the heat pipe's inner radius, (3) the radial heat flux is uniform, (4) inertial and convective effects in

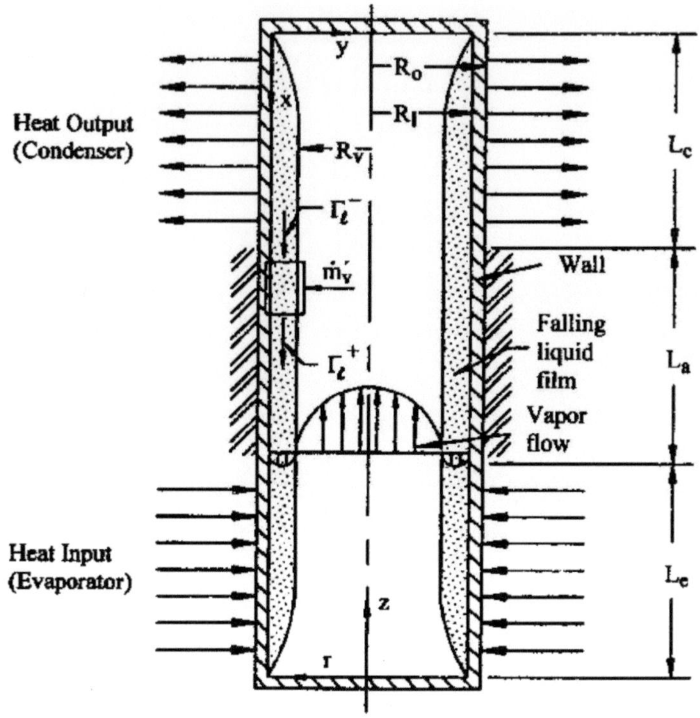

Fig. P7.27

the liquid are neglected, and (5) the gravitational force is negligible compared to the centrifugal force. Obtain the liquid film thickness along the axial direction.

7.27. Under an ideal situation, no liquid accumulates at the bottom of the thermosyphon (see Fig. P7.27). It is assumed that (1) the vapor condensation is filmwise with no interfacial waves, (2) liquid subcooling is negligible, (3) inertial effects in the liquid are negligible, (4) convection

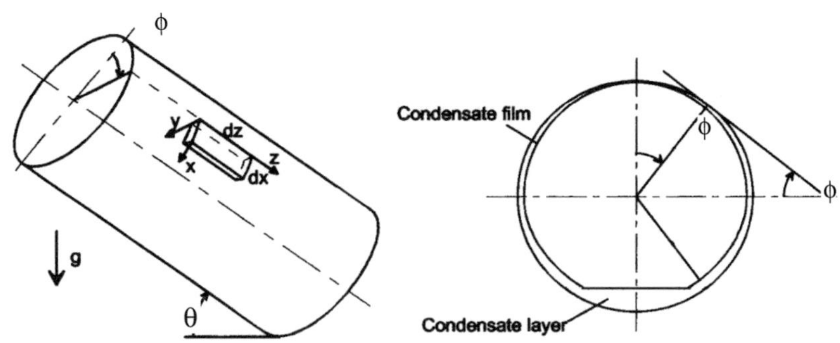

(a) configuration (b) liquid film inside the tube

Fig. P7.28

effects in the liquid are negligible, (5) the liquid film thickness is much smaller than the vapor space radius, and (6) the vapor density is much smaller than the liquid density. Specify the governing equations for flow and heat transfer in the liquid film in the cylindrical coordinate system. Analyze heat transfer in the liquid film using the Nusselt-type solution procedure.

7.28. The reflux condensation in a small diameter tube is shown in Fig. P7.28. It is assumed that the shear stress on the surface of the liquid film is negligible, and all other assumptions for the Nusselt analysis are applicable. Derive the differential equation for the liquid film thickness distribution, and specify the corresponding boundary conditions. You do not need to consider the liquid layer at the lower portion of the cross section.

7.29. The mass flux at the liquid–vapor interface for condensation of condensable vapor and non-condensable gas can be obtained from Eq. (7.208). Show that the mass flux can also be obtained from Eq. (7.211) if the mixture of condensable vapor and noncondensable gas can be treated as ideal gas.

7.30. Consider a system using an annular pipe with its inner wall made of a porous material (Fig. P7.30). In this system, steam flows through a porous pipe and cooling liquid flows through the annular region between a solid outer pipe and the porous inner pipe. The steam condenses onto the inner wall of the porous material, because the cooling liquid keeps the temperature of the porous wall below that of the steam's saturation temperature. A small pressure drop is maintained between the steam and the cooling flow in order to drive the condensate through the porous material to the annular region, where it mixes with the cooling

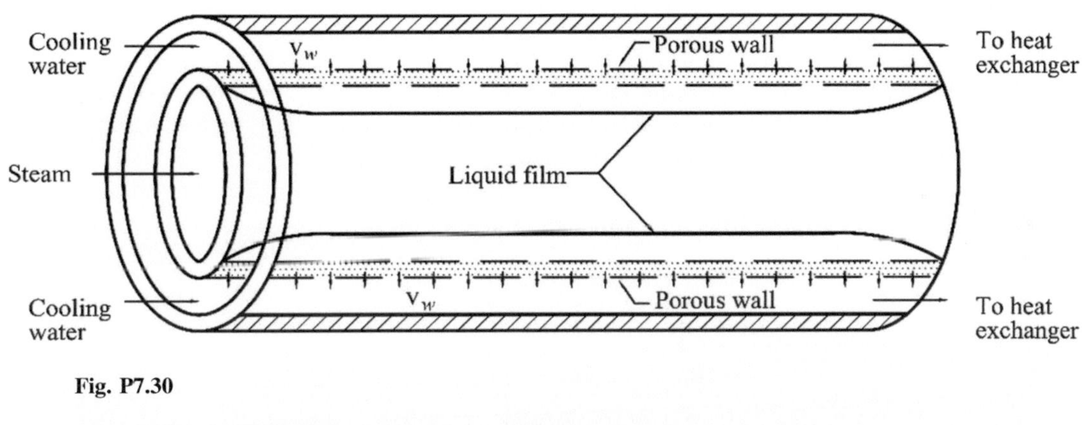

Fig. P7.30

liquid and is swept away to a heat exchanger. Assuming the suction velocity at wall $v \cong v_w$ = const, extend the Nusselt analysis for laminar film condensation inside a tube with a constant-temperature porous wall. The shear stress at the liquid–vapor interface decreases due to condensation. Assuming that the properties of the fluid remain constant and that the curvature of the condensate film can be neglected, develop the dimensionless velocity and temperature distributions using the following nondimensional variables:

$$y^* = -\frac{yv_w}{\alpha_\ell}, u^* = -\frac{u_\ell}{v_w}, T^* = \frac{T - T_w}{T_{sat} - T_w}, N_2 = \frac{\tau_\delta \alpha_\ell}{\mu_\ell v_w^2}, \delta^* = \frac{-v_w \delta}{\alpha_\ell} \quad (7.221)$$

7.31. Consider the use of an annular pipe as a condenser for microgravity application, with vapor shear as the driving force for condensate removal as shown in Fig. P7.31. Here, saturated vapor enters the annular region at $x = 0$, while the inner and outer walls are held at constant

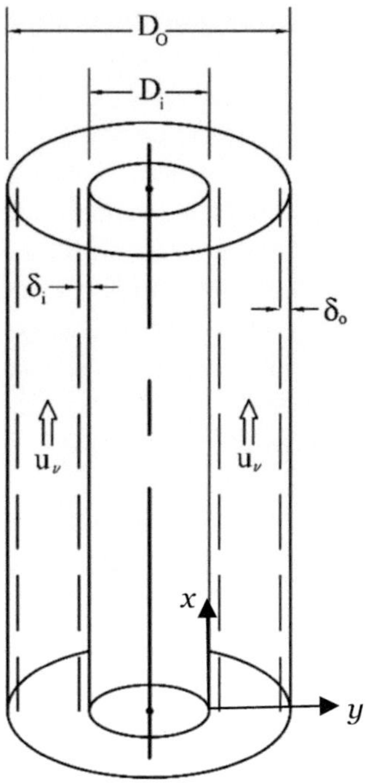

Fig. P7.31

temperatures $T_{w,i}$ and $T_{w,o,}$ respectively. A pressure drop occurs along the x-axis due to the friction at the walls. Also, the vapor temperature, density, and viscosity were all assumed to be constant. It is assumed that the film thicknesses, δ_i at inner wall and δ_o at outer wall, are small compared to the inner diameter of the annulus, D_i. Neglect inertia term and convection effect in the liquid films and obtain the following expression for the inner and outer walls, respectively:

$$\frac{\mathrm{d}\,Re_{\ell,i}}{\mathrm{d}x}=\frac{k_\ell\left(T_{sat}-T_{w,i}\right)}{h_{\ell v}\mu_\ell\delta_i}=\frac{Ja_i}{Pr_\ell\delta_i}=\frac{N_{1i}}{\delta_i} \tag{7.222}$$

$$\frac{\mathrm{d}\,Re_{\ell,o}}{\mathrm{d}x}=\frac{k_\ell\left(T_s-T_{w,o}\right)}{h_{\ell v}\mu_\ell\delta_o}=\frac{Ja_o}{Pr_\ell\delta_o}=\frac{N_{1o}}{\delta_o} \tag{7.223}$$

where Ja is the Jacob number $c_{p\ell}(T_{sat}-T_w)/h_{\ell w}$ and N_1 is the ratio of the Jacob number to the liquid Prandtl number.

7.32. Prove the following expression for Problem 7.31 using the mass balance between the liquid and vapor:

$$Re_{v,h}=Re_{v,e,h}-4\frac{\mu_\ell}{\mu_v}\frac{K^*}{1+K^*}Re_{\ell,i}-4\frac{\mu_\ell}{\mu_v}\frac{1}{1+K^*}Re_{\ell,o} \tag{7.224}$$

where

$$K^*=\frac{D_i}{D_o} \tag{7.225}$$

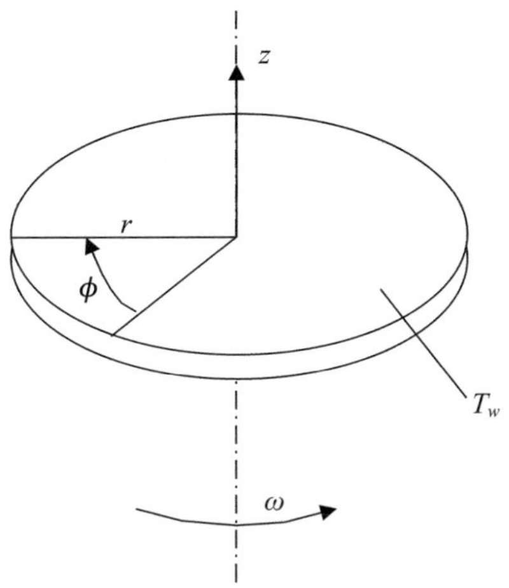

Fig. P7.33

7.33. One method to create artificial gravity for condensate removal involves use of a centrifugal field that is, a rotating disk. Consider a cooled rotating disk in a large quiescent body of pure saturated vapor, as shown in Fig. P7.33. The liquid forms a continuous film on the disk, and the fluid in this film will move radially outward due to the centrifugal force. Develop the conservation equations for mass, momentum in the r-, ϕ-, and z-directions and energy for an incompressible, steady, and constant-property liquid including the associated boundary conditions.

References

Asali, J. C., Hanratty, T. J., & Andreuss, P. (1985). Interfacial drag and film height for vertical annular flow. *AIChE Journal, 31*(6), 895–902.

Bejan, A. (1991). Film condensation on a upward facing plate with free edges. *International Journal of Heat and Mass Transfer, 34,* 578–582.

Brauer, H. (1956). Stromung und Warmeubergang bei Reiselfilmen. *VDI Forschung, 22,* 1–40.

Butterworth, D. (1981). *Simplified methods for condensation on a vertical surface with vapor shear* (UKAEA Report). AERE-R9683.

Butterworth, D. (1983). Film condensation of pure vapor. *Heat exchanger handbook, chapter 2.6.2*. Washington, DC: Hemisphere.

Carey, V. P. (2016). *Liquid-vapor phase-change phenomena: An introduction to the thermophysics of vaporization and condensation processes in heat transfer equipment* (3rd ed.). New York, NY: Taylor & Francis.

Carpenter, F. S., & Colburn, A. P. (1951). The effect of vapor velocity on condensation inside tubes, in *Proceedings of General Discussion of Heat Transfer* (pp. 20–26). Institute of Mechanical Engineers and American Society of Mechanical Engineers.

Chun, K. R., & Seban, R. A. (1971). Heat transfer to evaporating liquid films. *Journal of Heat Transfer, 93,* 391–396.

Collier, J. G., & Thome, J. R. (1994). *Convection boiling and condensation* (3rd ed.). Oxford: Oxford University Press.

Eucken, A. (1937). Energie- und Stoffaustausch an Grenzflächen. *Naturwissenschaften, 25,* 209–219.

Faghri, A. (1986). Turbulent film condensation in a tube with cocurrent and countercurrent vapor flow, in *AIAA Paper No*. 86-1354.

Faghri, A. (2016). *Heat pipe science & technology* (2nd ed.). Columbia, MO: Global Digital Press.

Faghri, A., Chen, M. M., & Morgan, M. (1989). Heat transfer characteristics in two-phase closed conventional and concentric annular thermosyphons. *Journal of Heat Transfer, 111*(3), 611–618.

Fujii, T. (1991). *Theory of laminar film condensation*. New York, NY: Springer.

Graham, C., & Griffith, P. (1973). Drop size distribution and heat transfer in dropwise condensation. *International Journal of Heat and Mass Transfer, 16,* 337–346.
Griffith, P. (1983). Dropwise condensation. In E. U. Schlunder (Ed.), *Heat exchange design handbook, chapter 2.6.5* (Vol. 2). New York, NY: Hemisphere Publishing.
Habib, I. S., & Na, T. P. (1974). Prediction of heat transfer pipe flow with constant wall temperature. *Journal of Heat Transfer, 96,* 253–254.
Harley, C., & Faghri, A. (1994). Transient two-dimensional gas-loaded heat pipe analysis. *Journal of Heat Transfer, 116,* 716–723.
Hewitt, G. F., Bott, T. R., & Shires, G. L. (1994). *Process heat transfer*. Boca Raton, FL: CRC Press.
Kutateladze, S. S. (1982). Semi-empirical theory of film condensation of pure vapors. *International Journal of Heat and Mass Transfer, 25,* 653–660.
Labuntsov, D. A. (1957). Heat transfer in film condensation of pure steam on vertical surfaces and horizontal tubes. *Teploenergetika, 4,* 72–80.
Lin, L., & Faghri, A. (1998). Condensation in a rotating stepped wall heat pipe with hysteretic annular flow. *Journal of Thermophysics and Heat Transfer, 12*(1), 94–99.
McCormick, J. L., & Baer, E. (1963). On the mechanism of heat transfer in dropwise condensation. *Journal of Colloid Science, 18,* 208–216.
Mikic, B. B. (1969). On mechanism of dropwise condensation. *International Journal of Heat and Mass Transfer, 12,* 1311–1323.
Nusselt, W. (1916). Die Oberflächenkondensation des Wasserdampfes. *Z. Vereins deutscher Ininuere, 60,* 541–575.
Rohsenow, W. M. (1956). Heat transfer and temperature distribution in laminar film condensation. *Transactions of the ASME, 78,* 1645–1648.
Rohsenow, W. M., Webber, J. H., & Ling, T. (1956). Effect of vapor velocity on laminar and turbulent film condensation. *Transactions of the ASME, 78,* 1637–1643.
Seban, R. (1954). Remarks on film condensation with turbulent flow. *Transactions of the ASME, 76,* 299–303.
Shafrin, E. G., & Zisman, W. A. (1960). Constitutive relations in the wetting of low energy surfaces and the theory of the retraction method of preparing monolayers. *Journal of Physical Chemistry, 64,* 519–524.
Shekriladze, I. G., & Gomelauri, V. I. (1966). Theoretical study of laminar film condensation of flow vapour. *International Journal of Heat and Mass Transfer, 9,* 581–591.
Stephan, K. (1992). *Heat transfer in condensation and boiling*. Berlin: Springer.
Szablewski, W. (1968). Turbulent Parallelstromunjen. *Zeitshr. Ang. Math Mech., 48,* 35.

8 Evaporation

8.1 Introduction

Evaporation is a liquid-to-vapor transformation process that occurs across a liquid–vapor interface; it is different from boiling that occurs at a solid–liquid (heating) interface (see Fig. 8.1). Another distinction between evaporation and boiling is that there are no vapor bubbles formed during the evaporation process. Evaporation processes generally occur from liquid films, drops, and jets. Films may flow on a heated or adiabatic surface as a result of gravity or vapor shear. Drops may evaporate from a heated substrate, or they may be suspended in a gas mixture. Jets may be cylindrical in shape or elongated (ribbon-like).

In film evaporators, a liquid film is evaporated in order to: (1) cool the surface on which it flows, (2) cool the liquid itself, or (3) increase the concentration of some component(s) in the liquid. Heat may be transferred to the film surface by conduction or convection. The design of film evaporators is most frequently constrained by the extent of the liquid superheat, since nucleate boiling must be prevented. Nucleate boiling can lead to product deterioration because elevated temperatures increase chemical reaction rates. Also, the accumulation of deposits on walls (fouling) as a result of nucleate boiling hinders heat transfer. To avoid nucleate boiling, liquid superheating should not exceed 3–40 K, depending on the liquid. This chapter deals with evaporation with no nucleate boiling.

In addition to the capability for low liquid superheating, film evaporators may be designed to achieve short residence time (i.e., contact time between the liquid and the evaporator wall), which further reduces the tendency for chemical reactions to occur. The tendency for fouling is also reduced in more rapidly flowing films. These features are ideal for heat-sensitive fluids such as those in food processing and polymer devolatilization (Alhusseini et al. 1998).

A thin liquid film in an evaporator can be produced by several arrangements, two of which are falling and climbing films (see Fig. 8.2). In falling film evaporators, the liquid is evenly distributed around the opening of a vertical channel—usually a tube—and falls under gravitational force as its surface evaporates to a vapor or gas mixture. If the vapor is drawn to the bottom, the flow is concurrent as shown in Fig. 8.2a and the shear stress of the vapor on the liquid film increases its speed downward. When the vapor is drawn to the top, the flow is countercurrent as shown in Fig. 8.2b. A climbing film evaporator Fig. 8.2c is designed so that liquid is fed into the bottom, and vapor shear pushes the liquid film along the wall to the top. In all of these cases, the lack of a hydrostatic pressure drop eliminates the drop of the corresponding saturation temperature along the evaporator length, thereby increasing the uniformity in temperature. A more complicated example,

A. Faghri and Y. Zhang, *Fundamentals of Multiphase Heat Transfer and Flow*, https://doi.org/10.1007/978-3-030-22137-9_8

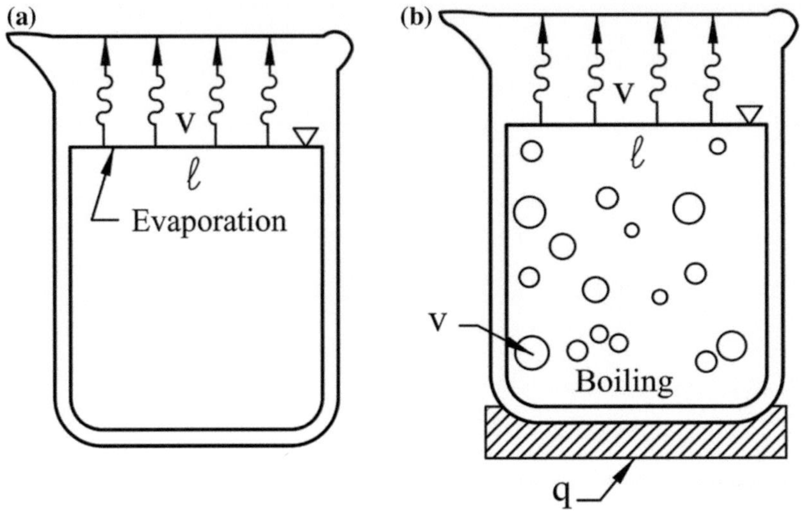

Fig. 8.1 Boiling and evaporation: **a** evaporation occurring at a liquid–vapor interface, and **b** boiling occurring at a solid–liquid interface

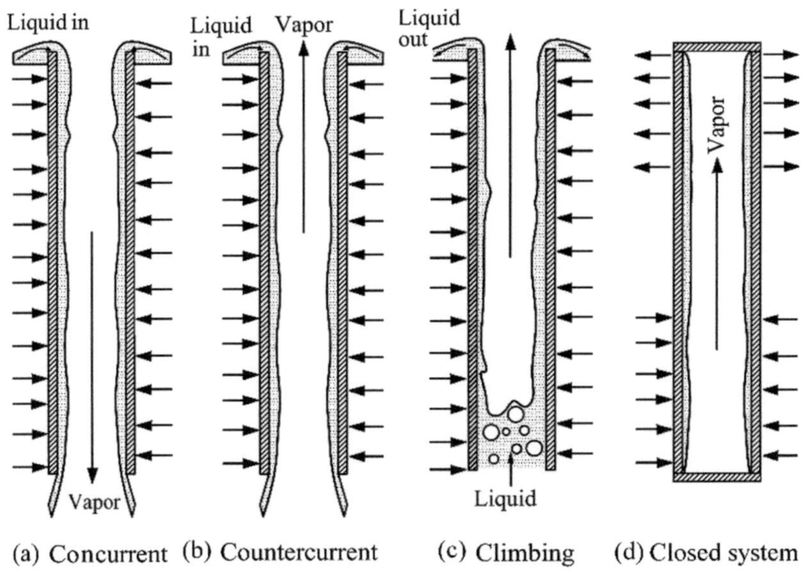

Fig. 8.2 Examples of film evaporators

shown in Fig. 8.2d, is a two-phase closed thermosyphon where evaporation takes place in the evaporator section and liquid and vapor flow in opposite directions.

The residence time of a liquid in a falling film evaporator may be increased with a countercurrent configuration. If this configuration is chosen, however, the likelihood of entrainment, flooding, or liquid blocking should be considered. Unlike the concurrent configuration, countercurrent flow is more prone to entrainment, which occurs when liquid is carried upward by the vapor (Faghri 2016). Liquid blocking—when liquid bridges the channel and blocks the vapor path—can also be more likely in a countercurrent configuration.

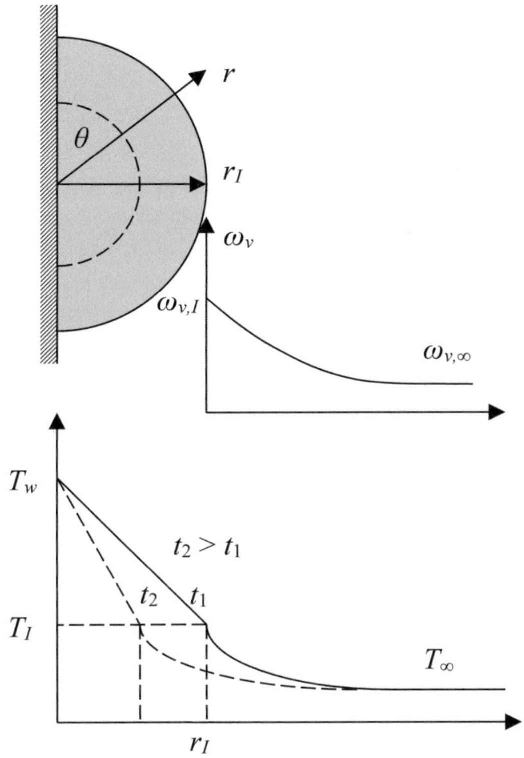

Fig. 8.3 Evaporation from a liquid droplet on a heated wall

In addition to the film evaporation described above, evaporations from liquid droplets attached to a heated wall or surrounded by hot gas can also find their application in irrigation of crops, firefighting, and combustion. Figure 8.3 shows evaporation from a liquid droplet attached to a heated wall; its application can be found in surface spray cooling, where liquid droplets are sprayed onto a hot surface. Heat is conducted through the droplet to the liquid–vapor interface, where an abrupt temperature drop takes place due to evaporation. The mass fraction of the vapor component in the gas mixture, ω_v, is greatest next to the interface; by diffusion and convection, the mass fraction decreases to its bulk level with increasing distance from the wall. The size of the liquid drop and the temperature distribution at two different times $(t_2 > t_1)$ are shown in Fig. 8.3. As time goes on, the liquid droplet becomes smaller (as indicated by the dashed line), while the temperatures at the heated wall and interface remains unchanged.

When a liquid droplet is surrounded by a hot gas mixture, evaporation takes place on the surface of the droplet, as shown in Fig. 8.4. Such processes are used in bulk cooling as well as spray fuel injection in diesel engines. The time required for such a droplet to completely evaporate as a member of a dispersed phase will determine the overall heat transfer (cooling effect). The partial pressure, p_v, and the mass fraction, ω_v, of the vapor in the gas mixture are the greatest adjacent to the surface of the droplet, and diffusion and convection cause the partial pressure and mass fraction to decrease with increasing distance from the interface. The temperature profile in the drop is dominated by transient conduction, but convection takes place outside the drop. Evaporation on the drop surface can be dominated by heat transfer or mass diffusion, depending on the size and thermophysical properties of the drop, as well as external flow condition.

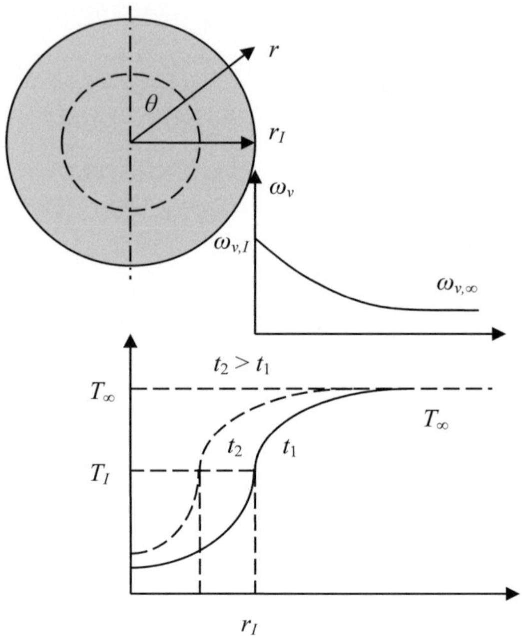

Fig. 8.4 Evaporation from a liquid droplet suspended in vapor–gas mixture

The criteria for evaporation in pure vapor or gas mixtures are discussed in Sect. 8.2. Section 8.3 includes a discussion of evaporation from a liquid film on a horizontal adiabatic wall, as well as evaporation from a vertical falling film with waves on an adiabatic surface. The contributions of the waves are considered by introducing additional corrections on viscosity, thermal conductivity, and mass diffusivity. Evaporation from a vertical falling film on a heated surface is presented in Sect. 8.4; this includes the classical Nusselt evaporation, i.e., evaporation from a laminar film with a smooth surface, and laminar films with waves, as well as turbulent film. Section 8.5 discusses direct-contact evaporation from liquid drops surrounded by hot gas and hot immiscible liquid, as well as direct evaporation of a liquid jet surrounded by hot gas.

8.2 Classification and Criteria

The latent heat of vaporization acts as a heat sink during phase change. It is supplied either by migration through the liquid, as in *heterogeneous evaporation*, or directly to the interface, as in *direct-contact evaporation*. In the former, the heat must migrate by conduction (and in some cases convection) through the liquid to the interface. Figure 8.5 illustrates the difference between heterogeneous and direct-contact evaporation. When a beaker of water is placed on a hot plate, as shown in Fig. 8.5a, only heterogeneous evaporation can occur, because heat must migrate from the bottom of the beaker to the surface of the water in order for evaporation to take place. The liquid's temperature is highest at its point of contact with the bottom of the beaker. Adjacent to the bottom is a layer through which heat passes only by conduction. Convection takes place throughout most of the depth, causing the temperature to decrease slightly. Just below the surface, a thin layer exists in which the temperature drops abruptly as the interface is approached. The surface temperature of the liquid can be assumed to equal the saturation temperature corresponding to the partial pressure of the vapor at the surface. Evaporation from a liquid film or a droplet attached to a heated wall is heterogeneous evaporation.

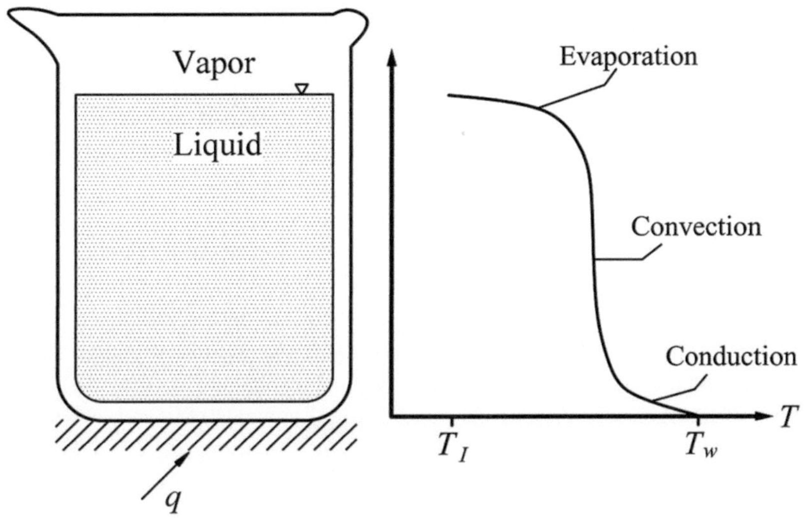

(a) Water in a beaker on a heated surface

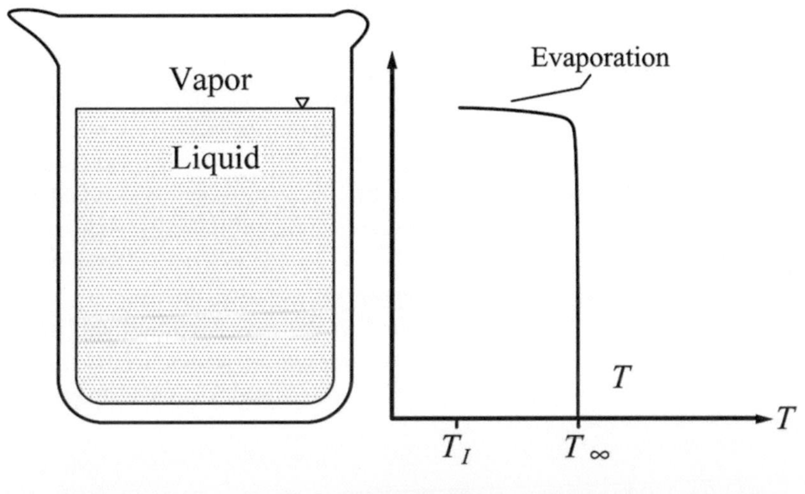

(b) Water in a beaker surrounded by constant temperature air with humidity < 100%

Fig. 8.5 Comparison of **a** heterogeneous and **b** direct-contact evaporation

Figure 8.5b shows an example of *direct-contact evaporation*, where a beaker of water sits on a table in a room of temperature T_∞ with a relative humidity <100%. Since the heat required for the evaporation is supplied at the surface, there is no temperature gradient through the depth of the liquid except just below the interface. Evaporation of a liquid droplet suspended in a hot gas mixture is another example of direct-contact evaporation, because the latent heat of evaporation is supplied to the surface of the droplet by the surrounding hot gas mixture. Evaporation of a liquid film or droplet attached to an adiabatic wall is direct evaporation because the latent heat is provided by the gas mixture, not the wall. Evaporation from a liquid surface is a mass transfer process, where the rate at which the molecules leave the liquid exceeds the rate at which the molecules move back into the liquid from the vapor. Therefore, the vapor pressure at the interface, which is the saturation pressure corresponding to the liquid surface temperature, $p_{sat}(T_I)$—if the effects of capillary and disjoining pressures are negligible—must exceed the vapor pressure, p_v. The difference between the saturation pressure and the vapor pressure is negligibly small when a liquid is evaporating to its pure vapor. Consequently, the temperature at the interface is equal to the saturation temperature corresponding to the local vapor pressure.

If evaporation is from a liquid to a mixture of vapor and noncondensable gas, the vapor generated at the interface must be diffused to the main stream of the mixture. In this case, evaporation will be further limited by the mass diffusion process. However, if the effect of vapor diffusion is negligible, the temperature at the interface equals the saturation temperature corresponding to the *partial* pressure of the vapor in the mixture. In this case, evaporation can occur only if saturation pressure corresponding to the interfacial temperature exceeds the partial pressure in the vapor. For example, a beaker of water is in thermal equilibrium with its surroundings at 20 °C and 50% relative humidity. The saturation pressure of water at 20 °C is 2.3 kPa, and the vapor partial pressure in air at 50% humidity is half of that, 1.15 kPa. Since the liquid saturation pressure exceeds the vapor partial pressure, evaporation occurs.

Therefore, heterogeneous evaporation requires that the liquid is superheated to a temperature above the saturation temperature corresponding to the partial pressure of the vapor, $T_\ell > T_{sat}(p_v)$. For direct-contact evaporation, the temperature of the vapor–gas mixture in contact with the liquid must exceed the saturation temperature corresponding to the partial pressure of the vapor. Let us reuse the example of a beaker of water in thermal equilibrium with its surroundings at 20 °C and 50% relative humidity. Since the gas temperature of 20 °C is greater than the saturation temperature of 8.8 °C, corresponding to the partial pressure of the vapor, evaporation occurs.

The conditions under which evaporation to a mixture of vapor and noncondensable gas can occur are demonstrated in a *T–s* diagram, as shown in Fig. 8.6. The saturated liquid and vapor lines represent the temperature and entropy conditions necessary for equilibrium between liquid and vapor phases. The saturated vapor line and the isotherm $T = T_I$ intersect at a point *I* from which the interface saturation pressure isobar $p = p_I$ is drawn. For evaporation to occur, the interface saturation pressure p_I must be greater than the vapor partial pressure p_v. If the temperature of the vapor is above the interfacial temperature (see point *a*), the latent heat of evaporation is supplied by the gas; evaporation under this condition is referred to as *hot evaporation*. The liquid for hot evaporation can be either

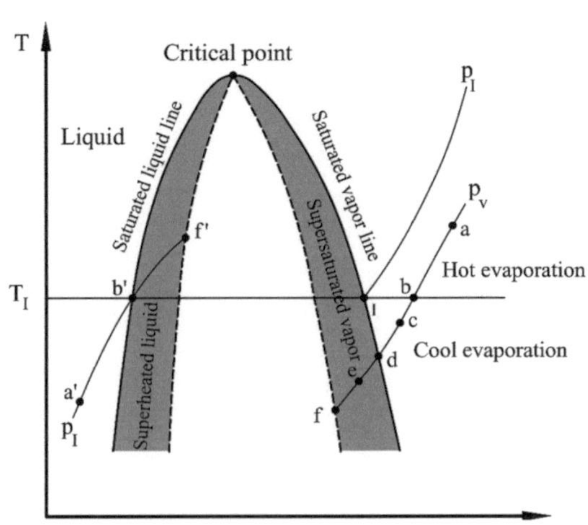

Fig. 8.6 Evaporation domains for a vapor–gas mixture

saturated or subcooled, as represented by curve $a'b'$ in Fig. 8.6. The above example with the beaker of water, shown in Fig. 8.5b, occurs at point a because the vapor temperature (20 °C) is greater than the interfacial temperature, which is at the saturation temperature corresponding to the partial pressure of the vapor (8.8 °C). At point b, which represents a case where the relative humidity is 100%, the gas temperature equals the saturation temperature and hot evaporation ceases.

At point c, the temperature of the gas is below the saturation temperature, so the gas cannot provide the latent heat of vaporization. For evaporation occurs at point c, heat must be supplied by a superheated liquid, represented by curve $b'f'$ in Fig. 8.6. As discussed in Chap. 2, the superheated liquid is in a metastable state. If the temperature of the liquid is higher than that of point f', vapor bubbles will appear in the liquid. Evaporation occurring under this condition is referred to as *cool evaporation* because the interface is warmer than the gas. At the intersection of isobar $p = p_v$ and the saturated liquid line (point d), the vapor is saturated. As a result, cool evaporation does not occur under equilibrium condition. Practically speaking, evaporation is still possible, but it will result in supersaturated vapor, which is in a metastable state as discussed in Chap. 2. If the vapor–gas mixture is already supersaturated, as represented by point e, evaporation is possible provided that sufficient superheat exists in the liquid. Evaporation to the vapor–gas mixture at point f will be impossible, because it represents the maximum possible supersaturation of the vapor–gas mixture.

8.3 Evaporation from Liquid Film on an Adiabatic Wall

8.3.1 Evaporation from Horizontal Films

Evaporation from thin liquid films occurs in many industrial and natural processes, including drying, evaporative cooling, and sweating. Figure 8.7 shows the physical model of horizontal thin film evaporation under consideration (Carey 2016). A film wets a horizontal surface over which flows a gas of ambient temperature T_∞ and mass fraction of water ω_∞ at a velocity of u_∞. Since the solid surface underneath the liquid is adiabatic, the latent heat of vaporization is provided by the vapor–gas mixture flowing above the liquid film. The heat source (vapor–gas mixture) and the evaporating liquid are in direct contact, making this scenario an example of the direct-contact evaporation. The liquid is evaporated and the latent heat is absorbed in the evaporation process. The resulting vapor injects into the boundary layer and is removed by the gas flow. The boundary layer becomes thicker, and a free stream of the gas flow is displaced from the surface being cooled. While phase change is the dominant mechanism of heat transfer at lower gas temperatures, vapor injection becomes more important at higher gas temperatures that correspond to higher evaporation rates.

Due to diffusion and convection, the temperature rises and the concentration falls from their values at the film surface to their ambient values at the edge of the boundary layer. A nonslip boundary

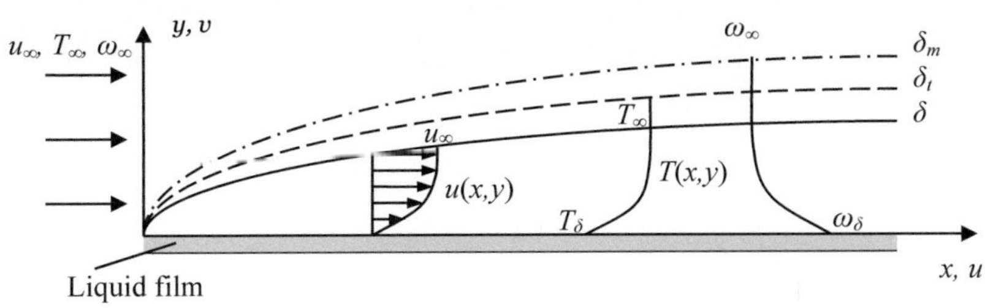

Fig. 8.7 Evaporation from a thin liquid film on a horizontal surface

condition exists at the film surface, and the film itself is considered stationary with respect to the gas. Assuming a steady state, constant density, and incompressible flow, the continuity equation is

$$\frac{\partial u}{\partial x}+\frac{\partial v}{\partial y}=0 \tag{8.1}$$

With no pressure gradients and constant viscosity assumed, the boundary-layer momentum equation is written as

$$u\frac{\partial u}{\partial x}+v\frac{\partial u}{\partial y}=\tilde{\nu}\frac{\partial^2 u}{\partial y^2} \tag{8.2}$$

where $\sim$ on top of the viscosity signifies the mass-averaged properties of the mixture.

Neglecting viscous dissipation and assuming constant thermal diffusivity and specific heats, the boundary-layer energy balance is

$$u\frac{\partial T}{\partial x}+v\frac{\partial T}{\partial y}=\tilde{\alpha}\frac{\partial^2 T}{\partial y^2} \tag{8.3}$$

where the axial conduction on the x-direction has been neglected because heat transfer occurred mainly in the y-direction.

The mass fraction of water is also accounted for by assuming a constant mass diffusivity:

$$u\frac{\partial \omega}{\partial x}+v\frac{\partial \omega}{\partial y}=D\frac{\partial^2 \omega}{\partial y^2} \tag{8.4}$$

As $y \to \infty$, the boundary conditions can be taken directly from Fig. 8.7:

$$u \to u_\infty, \quad T \to T_\infty, \quad \omega \to \omega_\infty \tag{8.5}$$

The boundary conditions are now laid out at the film surface $y = 0$. There is a nonslip condition, and the mass flux $\dot{m}''$ evaporates normal to the surface.

$$u=0, \quad v=\frac{\dot{m}''}{\tilde{\rho}}\omega=\omega_\delta, \quad T=T_\delta=T_{\text{sat}}(\omega_\delta, p_\infty) \tag{8.6}$$

where $T_{\text{sat}}(\omega_\delta, p_\infty)$ is the adiabatic saturation temperature of the liquid at the liquid surface. For an air–water system, it is equal to the wet-bulb temperature that can be measured by a thermometer covered in water-soaked cloth over which air is passed.

An energy balance at the interface accounts for the heat of vaporization and mass flux of the evaporating fluid. With the nonslip boundary condition, the only mode of heat transfer is conduction:

$$\dot{m}''h_{\text{lv}}=-q''=\tilde{k}\left(\frac{\partial T}{\partial y}\right)_{y=0} \tag{8.7}$$

Substituting Eq. (8.7) into Eq. (8.6), the velocity component in the y-direction becomes

$$v|_{y=0}=\frac{\tilde{k}}{\rho h_{\text{lv}}}\left(\frac{\partial T}{\partial y}\right)_{y=0} \tag{8.8}$$

The mass flux at the interface is the result of both diffusion and convection. Therefore, the mass balance at the interface is written as

$$\dot{m}'' = -\rho D\left(\frac{\partial \omega}{\partial y}\right)_{y=0} + \rho\omega_\delta v|_{y=0} \tag{8.9}$$

Substituting Eq. (8.7) into Eq. (8.9), conservation of mass at the interface becomes

$$-\tilde{\rho} D\left(\frac{\partial \omega}{\partial y}\right)_{y=0} + \tilde{\rho}\omega_\delta v|_{y=0} = \frac{\tilde{k}}{h_{\ell v}}\left(\frac{\partial T}{\partial y}\right)_{y=0} \tag{8.10}$$

Equations (8.1)–(8.2) are the same as the laminar boundary-layer equations for forced convection over a flat plate with blowing on the liquid surface. While this problem can be easily solved numerically using a commercial package (e.g., ANSYS Fluent), Eqs. (8.1)–(8.4) can be converted to a set of ordinary differential equations that can be solved using a shooting method. By defining a stream function ψ as follows,

$$u = \frac{\partial \psi}{\partial y} \quad v = -\frac{\partial \psi}{\partial x} \tag{8.11}$$

the continuity equation (8.1) is automatically satisfied and the momentum equation (8.2) in terms of the stream function becomes

$$\frac{\partial \psi}{\partial y}\frac{\partial^2 \psi}{\partial x \partial y} - \frac{\partial \psi}{\partial x}\frac{\partial^2 \psi}{\partial y^2} = \tilde{\nu}\frac{\partial^3 \psi}{\partial y^3} \tag{8.12}$$

The similarity solution of the problem can be obtained by introducing the following similarity variables:

$$\eta = y\sqrt{\frac{u_\infty}{\tilde{\nu} x}}, \quad f = \frac{\psi}{\sqrt{\tilde{\nu} u_\infty x}}, \quad \theta = \frac{T - T_\delta}{T_\infty - T_\delta}, \quad \varphi = \frac{\omega - \omega_\delta}{\omega_\infty - \omega_\delta} \tag{8.13}$$

and the local heat transfer coefficient can be evaluated by

$$h_x = \frac{\tilde{k}}{T_\infty - T_\delta}\left(\frac{\partial T}{\partial y}\right)_{y=0} = \tilde{k}\sqrt{\frac{u_\infty}{\tilde{\nu} x}}\theta'(0) \tag{8.14}$$

The local Nusselt number is then

$$Nu_x = \frac{h_x x}{k} = \theta'(0) Re_x^{\frac{1}{2}} \tag{8.15}$$

The local mass transfer coefficient can be evaluated by

$$h_{mx} = \frac{D}{\omega_\infty - \omega_\delta}\left(\frac{\partial \omega}{\partial y}\right)_{y=0} = D\sqrt{\frac{u_\infty}{\tilde{\nu} x}}\varphi'(0) \tag{8.16}$$

and the local Sherwood number is

$$Sh_x = \frac{h_m x}{D} = \varphi'(0) Re_x^{\frac{1}{2}} \tag{8.17}$$

where

$$Re_x = \frac{u_\infty x}{\tilde{\nu}} \tag{8.18}$$

Although Eqs. (8.15) and (8.17) have the same form as the case of single-phase forced convective heat or mass transfer over a flat plate, the solutions of $\theta'(0)$ and $\varphi'(0)$ depend on Prandtl number ($Pr = \tilde{\nu}/\alpha$), Schmidt number ($Sc = \tilde{\nu}/D$), Jakob number [$Ja = c_p(T_\infty - T_{sat})/h_{\ell v}$], and ω_∞. Consequently, simple empirical correlations similar to the case of single-phase forced convective heat or mass transfer over a flat plate are very difficult to obtain.

Since the blowing velocity at the surface of the liquid film, determined by Eq. (8.8), is usually much lower than the incoming vapor–gas mixture velocity, u_∞, one can assume that the blowing velocity at the liquid surface is negligible. The heat and mass transfer problems then cease to be a conjugated problem. The solutions of $\theta'(0)$ and $\varphi'(0)$ are then (Faghri et al. 2010):

$$\theta'(0) = 0.332 Pr^{1/3} \tag{8.19}$$

$$\varphi'(0) = 0.332 Sc^{1/3} \tag{8.20}$$

Substituting Eqs. (8.19) and (8.20) into Eqs. (8.15) and (8.17) yields

$$Nu_x = \frac{h_x x}{k} = 0.332 Re_x^{1/2} Pr^{1/3} \tag{8.21}$$

$$Sh_x = \frac{h_m x}{k} = 0.332 Re_x^{1/2} Sc^{1/3} \tag{8.22}$$

The average heat and mass transfer characteristics are often relevant for practical applications. The average Nusselt and Sherwood numbers based on average heat and mass transfer coefficients for a liquid film with a length of L are obtained as

$$\overline{Nu} = \frac{\bar{h} L}{k} = 0.664 Re_L^{1/2} Pr^{1/3} \tag{8.23}$$

$$\overline{Sh} = \frac{\bar{h}_m L}{k} = 0.664 Re_L^{1/2} Sc^{1/3} \tag{8.24}$$

where

$$Re_L = \frac{u_\infty L}{\tilde{\nu}} \tag{8.25}$$

The mass fraction of water at the liquid surface can be obtained by substituting Eqs. (8.19) and (8.20) into Eq. (8.10), i.e.,

$$\omega_\delta = \frac{\omega_\infty + (Sc/Pr)^{2/3} Ja}{1 + (Sc/Pr)^{2/3} Ja} \tag{8.26}$$

The applications of the above empirical correlations can be demonstrated by the following example.

Example 8.1

Air at 1 atm with 30 °C and 55% humidity flows over a 0.8 m-long wet flat plate with a velocity of 1 m/s. Estimate the rates of heat transfer to and evaporation from the wet flat plate. Estimate the evaporation-induced blowing velocity on the surface, and compare it with the incoming velocity of the humid air.

Solution Assuming the binary (air–water) mixture can be modeled as an ideal gas, the total density of the mixture is

$$\rho = \rho_1 + \rho_2 = \frac{p_1 M_1}{R_u T} + \frac{(p - p_1) M_2}{R_u T} \tag{8.27}$$

where M_1 and M_2 are the molecular masses of water and air, respectively, which can be found from Table 1.1 as $M_1 = 18.015\,\text{kg/kmol}$ and $M_2 = 28.97\,\text{kg/kmol}$.
The mass fraction of the water in the mixture is then

$$\omega = \frac{\rho_1}{\rho} = \frac{p_1 M_1}{p_1 M_1 + (p - p_1) M_2} \tag{8.28}$$

The saturation pressure corresponding to the air temperature of $T_\infty = 30\,°\text{C}$ is $p_{1,\text{sat}} = 4.246\,\text{kPa}$. Since the relative humidity in the binary mixture is 55%, the partial pressure of the water in the mixture is $p_1 = 0.55 p_{1,\text{sat}} = 2.335\,\text{kPa}$. Therefore, the mass fraction of water at the locations far from the flat plate is

$$\omega_\infty = \frac{2.335 \times 18.015}{2.335 \times 18.015 + (101.325 - 2.335) \times 28.97} = 0.0144$$

The properties of the moist air can be evaluated using those for pure air in Table B.1, because the mass fraction of the water is very low, i.e., $\rho = 1.165\,\text{kg/m}^3$, $c_p = 1.006\,\text{kJ/kg K}$, $\nu = 0.16 \times 10^{-4}\,\text{m}^2/\text{s}$, $k = 0.026\,\text{W/m K}$, $\alpha = 2.23 \times 10^{-3}\,\text{m}^2/\text{s}$, and $Pr = 0.72$. The mass diffusivity of water in air can be found from Table B.49 as $D_{12} = 0.26 \times 10^{-4}\text{m}^2/\text{s}$, so the Schmidt number is $Sc = \nu/D = 0.62$.
The surface temperature of the flat plate, T_δ, is the saturation temperature corresponding to the partial pressure of the water. If the partial pressure at the surface can be treated as unchanged (an approximation which will be corrected later), the surface temperature is the saturation temperature corresponding to $p_1 = 2.335\,\text{kPa}$, i.e., $T_\delta = 20\,°\text{C}$. The latent heat of evaporation for water at 20 °C is $h_{\ell v} = 2463.8\,\text{kJ/kg}$. The Jakob number is

$$Ja = \frac{c_p (T_\infty - T_\delta)}{h_{\ell v}} = \frac{1.006 \times (30 - 20)}{2463.8} = 0.00408$$

The mass fraction of water at the surface of the flat plate must satisfy Eq. (8.26), i.e.,

$$\omega_\delta = \frac{\omega_\infty + (Sc/Pr)^{2/3} Ja}{1 + (Sc/Pr)^{2/3} Ja} = \frac{0.0144 + (0.62/0.72)^{2/3} \times 0.00408}{1 + (0.62/0.72)^{2/3} \times 0.00408} = 0.0180$$

which is based upon the assumption that the partial pressure of water is the same throughout the boundary layer. With the mass fraction obtained above, the partial pressure of the water at the surface can be corrected using Eq. (8.28), i.e.,

$$p_{1,\delta} = \frac{\omega_\delta M_2 p}{M_1 - \omega_\delta (M_1 - M_2)} = \frac{0.0180 \times 28.97 \times 101.325}{18.015 - 0.018 \times (18.015 - 28.97)} = 2.901\,\text{kPa}$$

The saturation temperature corresponding to the above partial pressure is $T_\delta = 23.49\,°\text{C}$. The mass fraction with this new surface temperature is found to be $\omega_\delta = 0.0168$.
The Reynolds number is

$$Re_L = \frac{u_\infty L}{\nu} = \frac{1 \times 0.8}{0.16 \times 10^{-4}} = 5 \times 10^4$$

The average Nusselt and Sherwood numbers can be obtained with Eqs. (8.23)–(8.24):

$$\overline{Nu} = 0.664\, Re_L^{1/2} Pr^{1/3} = 0.664 \times (5 \times 10^4)^{1/2} \times 0.72^{1/3} = 133.08$$
$$\overline{Sh} = 0.664\, Re_L^{1/2} Sc^{1/3} = 0.664 \times (5 \times 10^4)^{1/2} \times 0.62^{1/3} = 126.60$$

The average heat and mass transfer coefficients can be found as

$$\bar{h} = \frac{Nuk}{L} = \frac{133.08 \times 0.026}{0.8} = 4.33\,\text{W/m}^2\,\text{K}$$
$$\bar{h}_m = \frac{ShD}{L} = \frac{126.6 \times 0.26 \times 10^{-4}}{0.8} = 0.00411\,\text{m/s}$$

The rates of heat transfer and evaporation are

$$q' = \bar{h} L (T_\infty - T_\delta) = 4.33 \times 0.8 \times (30 - 23.49) = 22.55\,\text{W/m}$$
$$\dot{m}' = \bar{h}_m L \rho (\omega_\infty - \omega_\delta)$$
$$= 0.00411 \times 0.8 \times 1.165 \times (0.0180 - 0.0144) = 1.379 \times 10^{-5}\,\text{kg/s m}$$

The average blowing velocity due to evaporation is

$$\bar{v}_\delta = \frac{\dot{m}'}{\rho L} = \frac{1.379 \times 10^{-5}}{1.165 \times 0.8} = 1.480 \times 10^{-5}\,\text{m/s}$$

which is very low, compared with the velocity in the x-direction.

8.3.2 Evaporation from Vertical Falling Films

In falling film evaporators, the flowing gas provides the latent heat of vaporization, and the gas cools as it passes through the evaporator. If the gas flows downward, it grows denser as it cools. The buoyancy force creates a *chimney effect*, by means of which the gas exits the bottom and more gas enters the top. An experimental study of this flow was conducted by Yan et al. (1991) and followed by a numerical simulation by Yan and Lin (1991), which is presented below. Figure 8.8 shows the physical model of evaporation from a vertical falling film on two parallel walls with a spacing of $2b$ ($b \ll L$). While the outer sides of the wall are insulated, the liquid falling films on the inner walls evaporate into a gas that flows through the channel. Assuming there is no pressure gradient and the inertial term is negligible, the momentum balance in the assumed laminar liquid film with negligible inertia is

$$\frac{\partial}{\partial y}\left[\left(\mu_\ell + \mu'_\ell\right)\frac{\partial u_\ell}{\partial y}\right] + \rho_\ell g = 0 \tag{8.29}$$

where μ'_ℓ is a correction to the viscosity in response to presumed surface waves on the film.

Considering conduction through the liquid and sensible heat loss, the energy balance is

$$\rho_\ell c_{\mathrm{p},\ell} u_\ell \frac{\partial T_\ell}{\partial x} = \frac{\partial}{\partial y}\left[\left(k_\ell + k'_\ell\right)\frac{\partial T_\ell}{\partial y}\right] \tag{8.30}$$

where k'_ℓ is a correction to the conductivity to account for the effect of the surface waves.

Natural convection occurs in the gas region, and the continuity equation is

$$\frac{\partial}{\partial x}(\rho u)_\mathrm{g} + \frac{\partial}{\partial y}(\rho v)_\mathrm{g} = 0 \tag{8.31}$$

The momentum equation in the gas region is considered only in the axial direction, and constant density over the cross-section is assumed

$$(\rho u)_\mathrm{g}\frac{\partial u_\mathrm{g}}{\partial x} + (\rho v)_\mathrm{g}\frac{\partial u_\mathrm{g}}{\partial y} = -\frac{\mathrm{d}p_\mathrm{dyn}}{\mathrm{d}x} + \frac{\partial}{\partial y}\left[\left(\mu_\mathrm{g} + \mu'_\mathrm{g}\right)\frac{\partial u_\mathrm{g}}{\partial y}\right] + (\rho_\mathrm{g} - \rho_0)g \tag{8.32}$$

where p_dyn is dynamic pressure defined as

$$p_\mathrm{dyn} = p - \frac{1}{2}\rho u_0^2 \tag{8.33}$$

The energy equation includes energy storage by the gas, conduction through the gas, and convection of the vapor component through the gas.

$$\begin{aligned}(\rho c_\mathrm{p} u)_\mathrm{g}\frac{\partial T_\mathrm{g}}{\partial x} + (\rho c_\mathrm{p} v)_\mathrm{g}\frac{\partial T_\mathrm{g}}{\partial y} \\ = \frac{\partial}{\partial y}\left[\left(k_\mathrm{g} + k'_\mathrm{g}\right)\frac{\partial T_\mathrm{g}}{\partial y}\right] + \rho_\mathrm{g}\left(\alpha_\mathrm{g} + \alpha'_\mathrm{g}\right)\left(c_\mathrm{p,v} - c_\mathrm{p,g}\right)\frac{\partial T_\mathrm{g}}{\partial y}\frac{\partial \omega}{\partial y}\end{aligned} \tag{8.34}$$

where the diffusivity is corrected to reflect the effect of surface waves. The last term on the right-hand side of Eq. (8.34) represents the energy transport through interspecies diffusion.

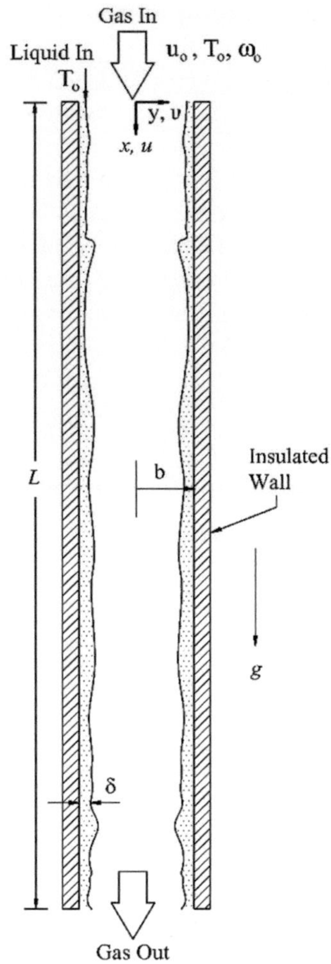

Fig. 8.8 Evaporation of a thin liquid film on a vertical adiabatic surface

A concentration balance is also required:

$$(\rho u)_g \frac{\partial \omega}{\partial x} + (\rho v)_g \frac{\partial \omega}{\partial y} = \frac{\partial}{\partial y}\left[\rho_g (D + D') \frac{\partial \omega}{\partial y}\right] \tag{8.35}$$

The boundary conditions at the inlet are now specified with uniform temperature, velocity, and concentration:

$$x = 0: \quad u = u_0 \quad T = T_0 \quad \omega = \omega_0 \quad p_{dyn} = -\rho u_0^2/2 \tag{8.36}$$

At the center of the channel, symmetric conditions must be satisfied, i.e.,

$$y = 0: \quad \frac{\partial u}{\partial y} = 0, \quad \frac{\partial T}{\partial y} = 0, \quad \frac{\partial \omega}{\partial y} = 0 \tag{8.37}$$

At the gas–liquid interface, the velocity, temperature, and concentration are at their interface values:

$$y = b - \delta: \quad u = u_\delta, \quad T = T_\delta, \quad \omega = \omega_\delta \tag{8.38}$$

With the nonslip condition, the shear stresses act equally and in opposite directions:

$$\tau_\delta = \left[(\mu + \mu')\frac{\partial u}{\partial y}\right]_\ell = \left[(\mu + \mu')\frac{\partial u}{\partial y}\right]_{\mathrm{g}} \tag{8.39}$$

The mass balance that accounts for the mass flux convected from the interface is

$$\dot{m}'' = \frac{\rho(D + D')}{1 - \omega_\delta}\left(\frac{\partial \omega}{\partial y}\right)_\delta \tag{8.40}$$

An energy balance that accounts for the heat required to release the mass from the liquid film is

$$\left[(k + k')\frac{\partial T}{\partial y}\right]_\ell = \left[(k + k')\frac{\partial T}{\partial y}\right]_{\mathrm{g}} + \dot{m}'' h_{\ell \mathrm{v}} \tag{8.41}$$

The mass fraction at the interface is assumed to be that for thermodynamic equilibrium

$$\omega_\delta = \frac{M_{\mathrm{v}} p_\delta}{M_{\mathrm{g}}(p - p_\delta) + M_{\mathrm{v}} p_\delta} \tag{8.42}$$

where M_{v} and M_{g} are the molecular masses of vapor and gas, respectively.

At the channel wall, there is also a nonslip condition with the liquid, and the wall is adiabatic

$$y = b: \quad u_\ell = 0, \quad \frac{\partial T_\ell}{\partial y} = 0 \tag{8.43}$$

At the channel exit, the dynamic pressure falls to zero:

$$x = 0: \quad p_{\mathrm{dyn}} = 0 \tag{8.44}$$

Corrections to the viscosity due to presumed surface waves on the film are (Yan and Lin 1991):

$$\mu'_\ell = c_\ell \rho_\ell \delta'^2 f d_\ell \tag{8.45}$$

$$\mu'_g = c_g \rho_g \delta'^2 f d_g \tag{8.46}$$

where the values of c_ℓ and c_g are 0.1 and 10, respectively. The amplitude, δ' (m), and frequency, f (Hz), of the wave are estimated by the following empirical correlations:

$$\delta' = \begin{cases} 4.1236 \times 10^{-4} We^{0.5505} & We \le 0.038357 \\ 1.5340 \times 10^{-4} We^{0.2472} & We > 0.038357 \end{cases} \tag{8.47}$$

$$f = \begin{cases} 86.70 We^{0.371327} & We \le 0.0264443 \\ 71.79 We^{0.319413} & We > 0.0264443 \end{cases} \tag{8.48}$$

where We is Weber number, defined as

$$We = \frac{(Re_\delta/4)^{5/3}}{\Omega} \tag{8.49}$$

and the Reynolds number, Re_δ, and the surface tension parameter, Ω, are

$$Re_\delta = \frac{4\Gamma}{\mu_\ell} \tag{8.50}$$

$$\Omega = 0.6736219 \frac{\sigma}{\rho_\ell \nu_\ell^{4/3}} \tag{8.51}$$

The damping functions d_ℓ and d_g in Eqs. (8.45) and (8.46) are obtained by

$$d_\ell = 1 - \exp\left(-\frac{b-y}{\delta}\right) \tag{8.52}$$

$$d_g = \exp\left[-\frac{5(b-\delta-y)}{\delta}\right] \tag{8.53}$$

The correction on thermal diffusivity, α', can be estimated by introducing the wave Prandtl number, $Pr' = \nu'/\alpha'$. Similarly, the correction on mass diffusivity is estimated by introducing the wave Schmidt number, $Sc' = D'/\alpha'$. It is recommended that the values of wave Prandtl and Schmidt numbers be unity and that the problem be solved numerically.

Figure 8.9 shows predicted the axial velocity, temperature, and mass fraction for the evaporation of a falling ethanol liquid film. It can be seen from Fig. 8.9a that the inlet velocity of the gas stream is nearly uniform at the inlet and that the velocity profile becomes distorted with the maximum velocity shifted away from the central line. As can be seen from Fig. 8.9b, the temperatures in the liquid and gas near the entrance (x/L = 0.039 and 0.187) monotonically increase with y, which means that the direction of heat transfer is from channel wall to the liquid–vapor interface, and then to the gas phase. At downstream (x/L = 0.468 and 0.93), the liquid and gas temperatures near the liquid–vapor interface is higher than the interfacial temperature; i.e., the latent heat of vaporization is supplied from

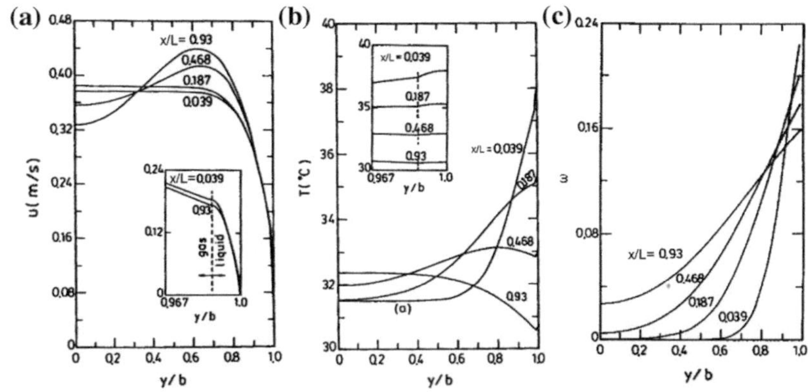

Fig. 8.9 Predicted axial **a** velocity, **b** temperature, and **c** mass fraction profile for evaporation of ethanol film ($T_{\ell,i} = 40\,°C$, $\Gamma = 0.02\,kg/m\,s$; Yan and Lin 1991; Reprinted with permission from Elsevier)

both liquid and gas phases. The mass fraction of the ethanol in the central region of the channel increases with increasing x due to evaporation (see Fig. 8.9c).

Experimental results using ethanol revealed that a larger inlet film temperature increases the liquid film cooling; this was to be expected because the temperature differential between the film and gas was increased. Lowering the liquid mass flow rate also increases the rates of cooling of liquid film; this is most likely due to longer residence times in the channel. It was also found that the exit temperature of the film could drop below the ambient temperature, which can be expected due to the volatility of ethanol.

Yan et al. (1991) performed an experimental study of the evaporative cooling of falling liquid film in an adiabatic vertical channel. Their experimental data yielded an empirical correlation for the mean heat flux, $\bar{q}''$, of water film evaporation:

$$\bar{q}'' = 0.0113 T_{\ell,0}^{3.15} b^{1/3} \tag{8.54}$$

where $T_{\ell,0}$ is the liquid film inlet temperature. It can be seen from Eq. (8.54) that the heat flux is not a function of the mass flow rate of the liquid film.

Example 8.2

Falling water film evaporation occurs on the inner sides of two parallel plates separated by a distance of $2b = 3\,cm$. The length of the parallel plates is 1 m, and the inlet temperature of the liquid film is 40 °C. The mass flow rate per unit width is $\Gamma = 0.04\,kg/s\,m$. Find the liquid film temperature at the outlet of the vertical channel.

Solution The specific heat of water should be evaluated at its mean temperature. Since the exit temperature is unknown, the inlet temperature is approximated as the mean temperature (which will be verified later). The specific heat at inlet temperature (40 °C) is $c_{p\ell} = 4.179\,kJ/kg\,°C$. The mean heat flux on the liquid film can be obtained from Eq. (8.54), i.e.,

$$\bar{q}'' = 0.0113 T_{\ell,0}^{3.15} b^{1/3} = 0.0113 \times (40)^{3.15} (0.015)^{1/3} = 310.17\,W/m^2$$

The mean heat flux is related to the liquid film temperature by

$$\bar{q}'' = \Gamma c_{p\ell}(T_{\ell,0} - T_{\ell,L})/L \tag{8.55}$$

where the length of the parallel plates $L = 1$ m. The liquid film temperature at outlet is then

$$T_{\ell,L} = T_{\ell,0} - \frac{\bar{q}''L}{\Gamma c_{p\ell}} = 40 - \frac{310.2 \times 1}{0.04 \times 4.179 \times 10^3} = 38.1\,°\text{C}$$

Since the temperature change in the liquid film is insignificant, the assumption that the inlet temperature was equivalent to the mean temperature was reasonable.

Example 8.3

Humid air with a temperature of T_0 and mass fraction of water of ω_0 enters at the top of a vertical channel composed of two parallel plates separated by a distance of b (Debbissi et al. 2003; see Fig. 8.10). The inlet velocity and pressure of the humid air are respectively u_0 and p_0. There is an extremely thin liquid film on the inner side of the left wall, and the outer side of the wall is insulated. The right wall is dry and maintained at a constant temperature, T_w. Since the humid air is cooled and its humidity increases as it flows through the vertical channel, the downward flow is accelerated by the chimney effect. Specify the governing equations and the corresponding boundary conditions of the problem.

Solution Since the liquid film is extremely thin, its thickness can be neglected and one can assume that the gas–vapor mixture flows in the entire channel ($0 < y < b$). Since the flow is mainly in the x-direction, and the velocity component in the y-direction is several orders of magnitude smaller than that in the x-direction, one can assume that the flow in the channel is boundary-layer type; i.e., diffusion of momentum and mass in the x-direction can be neglected. Thus, the continuity equation is

$$\frac{\partial \rho u}{\partial x} + \frac{\partial \rho v}{\partial y} = 0 \tag{8.56}$$

The momentum equation in the x-direction is

$$u\frac{\partial u}{\partial x} + v\frac{\partial u}{\partial y} = -\frac{1}{\rho}\frac{\mathrm{d}p_{\mathrm{dyn}}}{\mathrm{d}x} - g\beta(T - T_0) + \beta_{\mathrm{m}}g(\omega - \omega_0) + \nu\frac{\partial^2 u}{\partial y^2} \tag{8.57}$$

where β and β_{m} are the coefficients of thermal expansion and concentration expansion, respectively.

Neglecting viscous dissipation and assuming constant thermal diffusivity and specific heats, the energy balance is

$$u\frac{\partial T}{\partial x} + v\frac{\partial T}{\partial y} = \alpha\frac{\partial^2 T}{\partial y^2} + \frac{c_{\mathrm{pv}} - c_{\mathrm{pg}}}{c_{\mathrm{p}}}D_{\mathrm{vg}}\frac{\partial T}{\partial y}\frac{\partial \omega}{\partial y} \tag{8.58}$$

where the second term on the right-hand side represents energy transport through interspecies diffusion, and D_{vg} is binary mass diffusivity from vapor to gas.

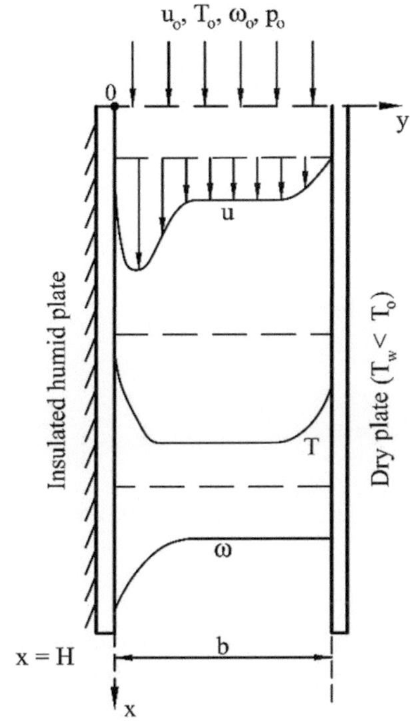

Fig. 8.10 Evaporation into humid air between parallel plates

The mass fraction of vapor in the mixture, ω, satisfies

$$u\frac{\partial \omega}{\partial x}+v\frac{\partial \omega}{\partial y}=\frac{1}{\rho}\frac{\partial}{\partial y}\left(\rho D_{\mathrm{vg}}\frac{\partial \omega}{\partial y}\right) \tag{8.59}$$

which is applicable in cases where the mass diffusivity is not a constant.

The mass flow rate at any location x equals the mass flow rate at the inlet (x = 0) plus accumulated evaporation from the left wall (0, x), i.e.,

$$\int_0^b \rho u \mathrm{d}y=\rho_0 u_0 b+\int_0^x \rho v|_{y=0}\mathrm{d}x \tag{8.60}$$

Since this is a boundary-layer problem, we will need to specify boundary conditions at the inlet, and the left and right walls.

The boundary conditions at the inlet are similar to those of Eq. (8.36), i.e.,

$$x = 0: \quad u = u_0 \quad T = T_0 \quad \omega = \omega_0 \quad p_{\mathrm{dyn}} = -\rho u_0^2/2 \tag{8.61}$$

The velocity components in the x-direction at both walls should be zero, as required by the nonslip condition:

$$u = 0, \quad y = 0, b \tag{8.62}$$

The velocity component in the y-direction at the left wall is determined by the evaporation rate, i.e.,

$$v = \frac{\dot{m}''}{\rho} = -\frac{q''}{\rho h_{\ell \mathrm{v}}} = \frac{k}{\rho h_{\ell \mathrm{v}}} \frac{\partial T}{\partial y}\bigg|_{y=0}, \quad y = 0 \tag{8.63}$$

while the velocity in the y-direction at the right wall is zero, i.e.,

$$v = 0, \quad y = b \tag{8.64}$$

The latent heat of evaporation is supplied by the humid air flow through the vertical channel; therefore, the boundary condition of the energy equation at the left wall is

$$-k\frac{\partial T}{\partial y}\bigg|_{y=0} = \frac{\rho h_{\ell \mathrm{v}} D_{\mathrm{vg}}}{1 - \omega|_{y=0}} \frac{\partial \omega}{\partial y}\bigg|_{y=0}, \quad y = 0 \tag{8.65}$$

The temperature of the right wall is constant, i.e.,

$$T = T_{\mathrm{w}}, \quad y = b \tag{8.66}$$

The mass fraction of the vapor at the left wall is

$$\omega|_{y=0} = \frac{p_{\mathrm{v}} M_{\mathrm{v}}}{p_{\mathrm{v}} M_{\mathrm{v}} + (p - p_{\mathrm{v}}) M_{\mathrm{g}}} \tag{8.67}$$

where p_{v} is the partial pressure of the vapor at the left wall, which equals the saturation pressure corresponding to the temperature of the left wall.

Since the right wall is dry, there is no mass transfer at the right wall, i.e.,

$$\frac{\partial C}{\partial y} = 0, \quad y = b \tag{8.68}$$

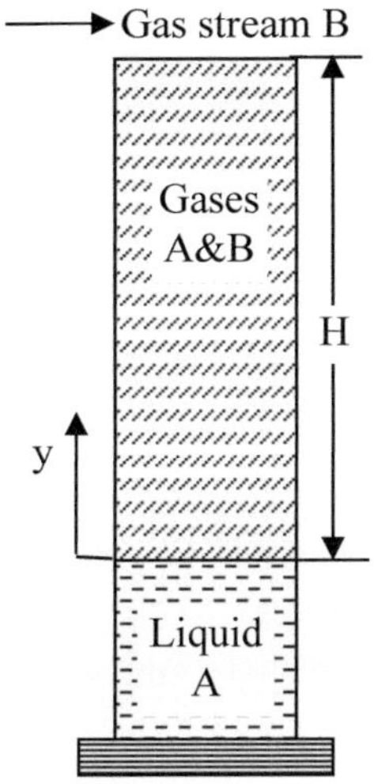

Fig. 8.11 Evaporation of liquid A into a stagnant gas B

Example 8.4

Determine the rate at which volatile liquid A evaporates into a stagnant gas B in a tube configuration as shown in Fig. 8.11, where the total pressure and temperature remain constant.

Solution This is the classic Stefan's diffusion problem. The following assumptions are made to solve this problem:

1. Steady state.
2. 1D problem.
3. Constant total pressure and temperature.
4. Liquid A is permeable to gas B.
5. Ideal gas for gas mixture.
6. Flat equilibrium condition at the liquid–vapor interface.
7. The mass diffusivity D_{AB} is a constant throughout the system.

Based on the above six assumptions, the continuity and species equations are reduced to the following form

$$\frac{d\rho v}{dy} = 0 \tag{8.69}$$

$$\rho v \frac{d\omega_A}{dy} = D_{AB} \frac{d\left(\rho \frac{d\omega_A}{dy}\right)}{dy} \tag{8.70}$$

where ω_A is the mass fraction of species A in the gas mixture and v is the velocity in the y-direction. It should be noted that energy and momentum equations are unnecessary since temperature and pressure are constant. However, mixture density is not constant. It is customary to solve this problem in terms of mole fraction, for simplicity, rather than mass fraction. The appropriate boundary conditions are
At vapor/liquid interface

$$y = 0, \quad \omega_A = \omega_{A,\delta}, \quad v_B = 0 \tag{8.71}$$

At the top of the tube

$$y = H, \quad \omega_A = \omega_{A,H} \tag{8.72}$$

The velocity of the gas mixture, v, is related to the velocity of species, i.e.,

$$v = \omega_A v_A + \omega_B v_B \tag{8.73}$$

At the interface, $v_B = 0$, since liquid is impermeable to gas B. The mixture velocity at the interface is

$$v = \omega_A v_A \tag{8.74}$$

The rate of evaporation of species A at the interface is

$$\dot{m}''_A = \rho_A v_A = \rho_A v - \rho D_{AB} \frac{d\omega_A}{dy} \tag{8.75}$$

Substituting Eq. (8.74) into Eq. (8.75) yields the fully related equation for evaporation at the interface

$$\dot{m}''_A = \rho_A v_A = \frac{-\rho D_{AB} \frac{d\omega_A}{dy}}{1 - \omega_A} \tag{8.76}$$

For a one-dimensional, steady-state problem

$$\frac{d\dot{m}''_A}{dy} = 0 \tag{8.77}$$

Substituting Eq. (8.76) into Eq. (8.77) yields

$$\frac{d}{dx}\left(\frac{-\rho D_{AB}}{1 - \omega_A}\frac{d\omega_A}{dy}\right) = 0 \tag{8.78}$$

$$\frac{-\rho D_{\mathrm{AB}}}{1-\omega_{\mathrm{A}}}\frac{\mathrm{d}\omega_{\mathrm{A}}}{\mathrm{d}y} = \dot{m}''_{\mathrm{A}} = \text{constant} \tag{8.79}$$

Since the gas mixture is assumed to be an ideal gas and isothermal, the total pressure is the sum of the partial pressures of components A and B

$$p = p_{\mathrm{A}} + p_{\mathrm{B}} \tag{8.80}$$

where $p = \rho R_{\mathrm{u}} T/M$, $p_{\mathrm{A}} = \rho_{\mathrm{A}} R_{\mathrm{u}} T/M_{\mathrm{A}}$, and $p_{\mathrm{B}} = \rho_{\mathrm{B}} R_{\mathrm{u}} T/M_{\mathrm{B}}$.
Further simplification reduces Eq. (8.80) to

$$\frac{1}{M} = \frac{\rho_{\mathrm{A}}}{\rho}\frac{1}{M_{\mathrm{A}}} + \frac{\rho_{\mathrm{B}}}{\rho}\frac{1}{M_{\mathrm{B}}} \tag{8.81}$$

Using Eq. (8.81) and definitions, with some simplification, the mass fraction is

$$\omega_{\mathrm{A}} = \frac{p_{\mathrm{A}}(M_{\mathrm{A}}/M_{\mathrm{B}})}{p - p_{\mathrm{A}}(1 - M_{\mathrm{A}}/M_{\mathrm{B}})} \tag{8.82}$$

Similarly, the mixture density is given by

$$\rho = \frac{pM}{R_{\mathrm{u}}T} = \frac{M_{\mathrm{B}}}{R_{\mathrm{u}}T}\left[p - p_{\mathrm{A}}\left(1 - \frac{M_{\mathrm{A}}}{M_{\mathrm{B}}}\right)\right] \tag{8.83}$$

Substituting Eqs. (8.82) and (8.83) into Eq. (8.79) yields

$$-\left(\frac{pM_{\mathrm{A}}}{R_{\mathrm{u}}T}\right)\left(\frac{D_{\mathrm{AB}}}{p - p_{\mathrm{A}}}\right)\frac{\mathrm{d}p_{\mathrm{A}}}{\mathrm{d}y} = \dot{m}''_{\mathrm{A}} = \text{constant} \tag{8.84}$$

Integrating Eq. (8.84) and applying the boundary conditions yields

$$\ln\left(\frac{p - p_{\mathrm{A}}}{p - p_{\mathrm{A},\delta}}\right) = \frac{y}{H}\ln\left(\frac{p - p_{\mathrm{A,H}}}{p - p_{\mathrm{A},\delta}}\right) \tag{8.85}$$

where $p_{\mathrm{A},\delta}$ and $p_{\mathrm{A,H}}$ are the partial pressures of species A at the liquid–vapor interface and at the top of the tube, respectively. Substituting Eq. (8.85) into Eq. (8.84) yields the rate of evaporation of species A

$$\dot{m}''_{\mathrm{A}} = \frac{pM_{\mathrm{A}}D_{\mathrm{AB}}}{R_{\mathrm{u}}TH}\ln\left(\frac{p - p_{\mathrm{A,H}}}{p - p_{\mathrm{A},\delta}}\right) \tag{8.86}$$

where temperature T and pressure p are assumed to be uniform within the gas mixture.

8.4 Falling Film Evaporation on a Heated Wall

As mentioned previously, falling film evaporators are used in both mass transfer and heat transfer applications. In general, the film's mass flow rate decreases as evaporation occurs. If the film is being concentrated (as in fruit juice concentration), evaporation is also accompanied by an increase in viscosity. With the decrease in mass flow rate and the corresponding increase in viscosity, the Reynolds number decreases and the flow tends to progress from turbulent (if that ever existed) to wavy laminar, and then to laminar flow.

8.4.1 Classical Nusselt Evaporation

The simplest evaporation heat transfer coefficient calculation uses the Nusselt treatment of falling film evaporation, as shown in Fig. 8.12. A flat plate with constant temperature T_w is placed vertically, and a liquid film flows downward due to gravity. Evaporation takes place on the surface of the liquid film because the temperature of the vertical plate is above the saturation temperature of the vapor corresponding to the partial pressure of the vapor in the mixture. Since the heat must be transferred from the heat source (vertical wall) to the liquid–vapor interface, where evaporation takes place, this is an example of the heterogeneous evaporation. The following assumptions are made:

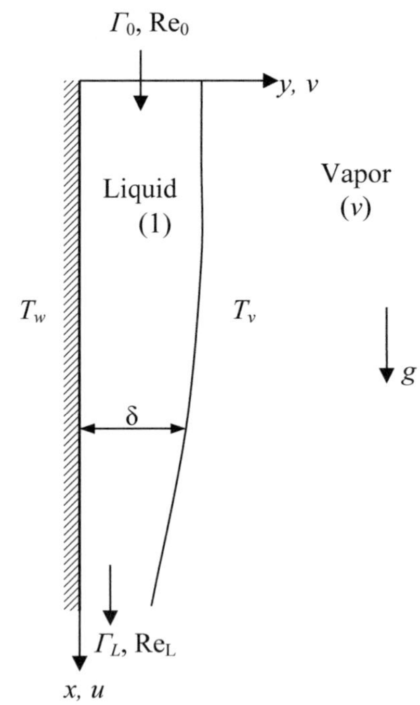

Fig. 8.12 Schematic of Nusselt evaporation

1. The vapor is quiescent.
2. Inertia in the laminar liquid film is negligible.
3. Heat transfer through the liquid film is by conduction only; i.e., convective terms in the energy equation are negligible.
4. Constant wall temperature.
5. The process occurs at steady state.
6. A nonslip condition exists at the wall.

Proceeding from the assumptions of negligible inertia and constant fluid properties, the momentum balance is

$$\mu_\ell \frac{d^2 u_\ell}{dy^2} = -(\rho_\ell - \rho_v)g \tag{8.87}$$

Integrating Eq. (8.87) twice, with respect to y and using the boundary conditions of nonslip at the wall ($u = 0$ at $y = 0$) and zero shear at the liquid–vapor interface ($\partial u/\partial y = 0$ at $y = \delta$), the following velocity profile is obtained:

$$u_\ell = \frac{(\rho_\ell - \rho_v)g}{\mu_\ell}\left(\delta y - \frac{y^2}{2}\right) \tag{8.88}$$

The mass flow rate per unit width is defined by integrating the velocity profile from the wall to the interface.

$$\Gamma = \int_0^\delta \rho_\ell u dy \tag{8.89}$$

The liquid velocity profile, Eq. (8.88), is substituted into Eq. (8.89), and the result is integrated from the wall to the interface. The result is then rearranged to obtain liquid film thickness:

$$\delta = \left[\frac{3\mu_\ell \Gamma}{\rho_\ell(\rho_\ell - \rho_v)g}\right]^{1/3} \tag{8.90}$$

Since heat transfer through the film is by conduction only, the heat transfer coefficient is $h_x = k_\ell/\delta$. Considering Eq. (8.90), the local heat transfer coefficient becomes

$$h_x = k_\ell \left[\frac{\rho_\ell(\rho_\ell - \rho_v)g}{3\mu_\ell \Gamma}\right]^{1/3} \tag{8.91}$$

Note that the film thickness δ drops out of the equation, and the heat transfer coefficient h is now a function of the mass flow rate per unit width, Γ.

The local energy balance across the liquid film is

$$h_{\ell v}\frac{d\Gamma}{dx} = -h_x(T_w - T_v) \tag{8.92}$$

Combining Eqs. (8.91) and (8.92), one obtains,

$$\Gamma^{1/3}d\Gamma = -\frac{k_\ell}{h_{\ell v}}\left[\frac{\rho_\ell(\rho_\ell - \rho_v)g}{3\mu_\ell}\right]^{1/3}(T_w - T_v)dx \tag{8.93}$$

Integrating from the inlet to some arbitrary length *L*,

$$\int_{\Gamma_0}^{\Gamma_L}\Gamma^{1/3}d\Gamma = -\int_0^L\frac{k_\ell}{h_{\ell v}}\left[\frac{\rho_\ell(\rho_\ell - \rho_v)g}{3\mu_\ell}\right]^{1/3}(T_w - T_v)dx \tag{8.94}$$

A relation for Γ at outlet is obtained in terms of the inlet mass flow rate, Γ_0, for the case of constant wall and vapor temperatures:

$$\Gamma_L^{4/3} = \Gamma_0^{4/3} - \frac{4}{3}\frac{k_\ell}{h_{\ell v}}\left[\frac{\rho_\ell(\rho_\ell - \rho_v)g}{3\mu_\ell}\right]^{1/3}(T_w - T_v)L \tag{8.95}$$

Now, the mass flow rate per unit width is related to the Reynolds number by

$$Re = \frac{4\Gamma}{\mu_\ell} \tag{8.96}$$

The local heat transfer coefficient from Eq. (8.91) can be expressed in terms of Reynolds number as follows:

$$\frac{h_x}{k_\ell}\left[\frac{\mu_\ell^2}{\rho_\ell(\rho_\ell - \rho_v)g}\right]^{1/3} = \left(\frac{4}{3}\right)^{1/3}\frac{1}{Re^{1/3}} = 1.10Re^{-1/3} \tag{8.97}$$

Equation (8.97) for local heat transfer coefficient is valid for both laminar condensation and evaporation at both constant wall temperature and constant heat flux at wall.

To calculate the average heat transfer coefficient, $\bar{h}$, the energy flux over the entire length $0 < x < L$ is balanced with the mass flux from the interface:

$$\bar{h} = \frac{h_{\ell v}(\Gamma_0 - \Gamma_L)}{L(T_w - T_v)} = \frac{\mu_\ell h_{\ell v}(Re_0 - Re_L)}{4L(T_w - T_v)} \tag{8.98}$$

Equation (8.98) is valid for laminar, wavy laminar, and turbulent flows with constant wall temperature.

Equation (8.92), which reflects the local energy balance across the liquid film, can be written in terms of Reynolds number as

$$\frac{\mathrm{d}Re}{h_x} = -\frac{4(T_\mathrm{w} - T_\mathrm{v})}{\mu_\ell h_{\ell \mathrm{v}}}\mathrm{d}x \tag{8.99}$$

Integrating Eq. (8.99) in the interval of (0, *L*), one obtains

$$\int_{Re_0}^{Re_L} \frac{\mathrm{d}Re}{h_x} = -\frac{4(T_\mathrm{w} - T_\mathrm{v})L}{\mu_\ell h_{\ell \mathrm{v}}} \tag{8.100}$$

Combining Eqs. (8.98) and (8.100), the average heat transfer coefficient is expressed as

$$\bar{h} = -\frac{(Re_0 - Re_L)}{\int_{Re_0}^{Re_L} 1h_x dRe} \tag{8.101}$$

Equation (8.101) for $\bar{h}$ is valid for laminar, wavy laminar, and turbulent flow.

Substituting Eq. (8.97) into Eq. (8.101), an expression for the average heat transfer coefficient for laminar flow in terms of Reynolds number is obtained:

$$\frac{\bar{h}}{k_\ell}\left[\frac{\mu_\ell^2}{\rho_\ell(\rho_\ell - \rho_\mathrm{v})g}\right]^{1/3} = \left(\frac{4}{3}\right)^{4/3}\frac{(Re_0 - Re_L)}{\left(Re_0^{4/3} - Re_L^{4/3}\right)} \tag{8.102}$$

In order to use Eq. (8.102) to obtain the average heat transfer coefficient, it is necessary to know Reynolds number at $x = L$, Re_L. It can be obtained by rewriting the relation for Γ, Eq. (8.95), in terms of Reynolds number, as follows:

$$Re_L^{4/3} = Re_0^{4/3} - 4\left(\frac{4}{3}\right)^{4/3}\frac{k_\ell(T_\mathrm{w} - T_\mathrm{v})L}{\mu_\ell^{4/3}h_{\ell \mathrm{v}}}\left[\frac{\rho_\ell(\rho_\ell - \rho_\mathrm{v})g}{3\mu_\ell}\right]^{1/3} \tag{8.103}$$

The above analysis is valid for falling film evaporation on a vertical wall maintained at a constant temperature. Falling film evaporation from a vertical wall with constant heat flux is discussed in Example 8.5.

Example 8.5

A liquid film flows downward and evaporates on a vertical plate subjected to a constant heat flux. Determine the average heat transfer coefficient.

Solution The physical model of the problem under consideration is similar to that shown in Fig. 8.12 except that the thermal boundary condition is changed. The local heat transfer coefficient for falling film evaporation can still be calculated by Eq. (8.91) or (8.97), because no assumption about boundary conditions was used to obtain them.
For the case of constant heat flux, the energy balance accounts for the heat addition required for evaporation, i.e.,

$$h_{\ell v}\frac{d\Gamma}{dx} = -q''_w \tag{8.104}$$

Since q''_w is not a function of x, Eq. (8.104) can be integrated in the interval of (0, L) to yield

$$\Gamma = \Gamma_0 - \frac{q''_w}{h_{\ell v}}L \tag{8.105}$$

Equation (8.105) can be rewritten in terms of Reynolds number as

$$Re_L = Re_0 - \frac{4q''_w}{\mu_\ell h_{\ell v}}L \tag{8.106}$$

which can be used to obtain the Reynolds number at $x = L$. The average heat transfer coefficient can be obtained by

$$\bar{h} = \frac{h_{\ell v}(\Gamma_0 - \Gamma_L)}{L(\bar{T}_w - T_v)} = \frac{\mu_\ell h_{\ell v}(Re_0 - Re_L)}{4L(\bar{T}_w - T_v)} \tag{8.107}$$

which is similar to Eq. (8.98) except that the wall temperature is the average value along the wall ($0 < x < L$), i.e.,

$$\bar{T}_w = \int_0^L T_w dx \tag{8.108}$$

For the case with constant heat flux, the energy balance across the liquid film in terms of Reynolds number can still be represented by Eq. (8.99). Integrating Eq. (8.99) in the interval of (0, L) and considering the fact that T_w is not a constant, one obtains

$$\int_{Re_0}^{Re_L} \frac{dRe}{h_x} = -\frac{4(\bar{T}_w - T_v)L}{\mu_\ell h_{\ell v}} \tag{8.109}$$

Combining Eqs. (8.107) and (8.109) to eliminate $(\bar{T}_w - T_v)$ yields the same equation as Eq. (8.101), which eventually leads to Eq. (8.102). Therefore, Eq. (8.102) is also valid for the case with constant heat flux heating, while the Reynolds number at $x = L$ must be calculated by Eq. (8.106).

8.4.2 Laminar Falling Film with Surface Waves

Wavy flows of thin liquid films have higher heat transfer coefficients than smooth thin films. This effect is due to the former's greater interfacial surface area and mixing action. Faghri and Seban (1985) analyzed a system in which a liquid film at an initial nondimensional temperature of zero flowed down a vertical wall and evaporated at the free liquid–vapor interface. The laminar liquid film thickness varied sinusoidally, and the Reynolds number ranged between 35 and 472. The energy equation for a two-dimensional situation using the same vertical wall shown in Fig. 8.12 is as follows:

$$\frac{\partial T}{\partial t}+u\frac{\partial T}{\partial x}+v\frac{\partial T}{\partial y}=\alpha\left(\frac{\partial^2 T}{\partial x^2}+\frac{\partial^2 T}{\partial y^2}\right) \tag{8.110}$$

The continuity equation for this system is

$$\frac{\partial u}{\partial x}+\frac{\partial v}{\partial y}=0 \tag{8.111}$$

Faghri and Seban (1985) assumed a quasi-parabolic velocity profile in the x-direction as follows:

$$u=3\bar{u}(x,t)\left(\eta-\frac{\eta^2}{2}\right) \tag{8.112}$$

where the nondimensional coordinate η is defined as

$$\eta=\frac{y}{\delta} \tag{8.113}$$

and $\bar{u}$ is the local mean velocity. Equation (8.111) can be rearranged and integrated with respect to y to obtain the velocity component v:

$$v=-\int_0^y \frac{\partial u}{\partial x}\mathrm{d}y=-3\left[\frac{\partial \bar{u}}{\partial x}\delta\left(\frac{\eta^2}{2}-\frac{\eta^3}{6}\right)-\bar{u}\frac{\partial \delta}{\partial x}\left(\frac{\eta^2}{2}-\frac{\eta^3}{3}\right)\right] \tag{8.114}$$

It can be seen that Eq. (8.112) has been substituted in order to find an expression for the v component of the velocity where the local mean velocity and film thickness δ are still unknown as functions of the wavy surface. To determine these two variables, the following mass balance can be written as a function of time and space:

$$\frac{\partial \delta}{\partial t}=-\frac{\partial}{\partial x}\int_0^\delta u\mathrm{d}y=-\frac{\partial}{\partial x}(\bar{u}\delta) \tag{8.115}$$

The film thickness can be written in terms of its average value, $\bar{\delta}$, and a local amplitude ψ.

$$\delta = \bar{\delta}(1 + \varphi) \tag{8.116}$$

Assuming that the wave formation has a periodic characteristic with a velocity c, the following relations can be written for the fluctuation of local mean velocity and film thickness:

$$\frac{\partial \bar{u}}{\partial t} = -c\frac{\partial \bar{u}}{\partial x} \tag{8.117}$$

$$\frac{\partial \delta}{\partial t} = -c\frac{\partial \delta}{\partial x} \tag{8.118}$$

Combining Eqs. (8.115)–(8.118) and integrating, the following expression is obtained:

$$(c - \bar{u})(1 + \varphi) = [c - \bar{u}_0] \tag{8.119}$$

where $\bar{u}_0$ is the average velocity for the average film thickness $\bar{\delta}$. For very small local amplitudes, $\phi < 1$, $\bar{u}$ may be approximated by expanding Eq. (8.119) and neglecting the third-order terms.

$$\bar{u} = \bar{u}_0 + (c - \bar{u}_0)\varphi - (c - \bar{u}_0)\varphi^2 \tag{8.120}$$

Substituting Eqs. (8.116) and (8.120) into Eq. (8.114), one obtains the velocity profile in terms of the wavy surface parameters:

$$\begin{aligned} v = -3\bar{u}_0\bar{\delta}\frac{\partial \phi}{\partial x}\Bigg\{&\left(\frac{c}{\bar{u}_0} - 1\right)(1 - 2\phi)\frac{\delta}{\bar{\delta}}\left(\frac{\eta^2}{2} - \frac{\eta^3}{6}\right) \\ &-\left[1 + \left(\frac{c}{\bar{u}_0} - 1\right)\phi - \left(\frac{c}{\bar{u}_0} - 1\right)\phi^2\right]\left(\frac{\eta^2}{2} - \frac{\eta^3}{3}\right)\Bigg\} \end{aligned} \tag{8.121}$$

Assuming the wave is sinusoidal,

$$\varphi = A\sin\left[\left(\frac{2\pi}{\lambda}\right)(x - ct)\right] \tag{8.122}$$

and introducing the nondimensional variable ξ defined as

$$\xi = \left(\frac{2\pi}{\lambda}\right)(x - ct) \tag{8.123}$$

where A is magnitude of the wave, and λ is the wavelength, Eq. (8.110) is transformed from the (t, x, y) to the (ξ, η) coordinate system,

$$C_1\frac{\partial T}{\partial \xi} + C_2\frac{\partial T}{\partial \eta} = C_3\frac{\partial^2 T}{\partial \eta^2} + C_4\frac{\partial^2 T}{\partial \eta \partial \xi} + C_5\frac{\partial^2 T}{\partial \xi^2} \tag{8.124}$$

where the constants are as follows:

$$C_1 = \frac{2\pi}{\lambda}(u - c) \tag{8.125}$$

$$\begin{aligned} C_2 = {} & \frac{2\pi\eta}{\lambda}\frac{\bar{\delta}}{\delta}(c - u)A\cos\xi - \left(\frac{2\pi}{\lambda}\right)^2 \frac{\bar{\delta}}{\delta}\alpha\eta A\sin\xi \\ & - 2\left(\frac{2\pi}{\lambda}\right)^2 \alpha\left(\frac{\bar{\delta}}{\delta}\right)\eta A^2\cos^2\xi + \frac{v}{\delta} \end{aligned} \tag{8.126}$$

$$C_3 = \frac{\alpha}{\delta^2} + \alpha\left(\frac{2\pi}{\lambda}\right)^2 \left(\frac{\bar{\delta}}{\delta}\right)^2 \eta^2 A^2 \cos^2\xi \tag{8.127}$$

$$C_4 = -2\alpha\left(\frac{2\pi}{\lambda}\right)^2 \eta\frac{\bar{\delta}}{\delta}A\cos\xi \tag{8.128}$$

$$C_5 = \alpha\left(\frac{2\pi}{\lambda}\right)^2 \tag{8.129}$$

These coefficients are evaluated with $\bar{u}$ from Eqs. (8.112) and (8.120), and v from Eq. (8.121).

In the evaporating film with a fluid with a high latent heat of vaporization, the average film thickness over a wave length, $\bar{\delta}$, will not vary substantially with distance, x. Furthermore, the temporal average temperature profile will not vary with x far from where heating begins. Therefore, the boundary condition for Eq. (8.124) that describes evaporation on a wavy film is approximated as below.

The equality of temperatures at corresponding points in the period gives

$$T(0, \eta) = T(2\pi, \eta) \tag{8.130}$$

Boundary conditions for dimensionless temperature at the heated vertical wall and wavy surface are, respectively,

$$T(\xi, 0) = 1 \tag{8.131}$$

$$T(\xi, 1) = 0 \tag{8.132}$$

Therefore, the energy equation with the above boundary conditions can be solved to obtain the wavy condensate film's temperature profile. With this profile, the heat flux at the wall can be found from

$$q''_{\mathrm{w}} = -\frac{k}{\delta}\frac{\partial T}{\partial \eta}\bigg|_{\eta=0} \tag{8.133}$$

and the heat flux normal to the wavy surface ($\eta = 1$) is as follows:

$$\frac{\mathbf{q}''_{\delta}}{k} = \left[\frac{2\pi}{\lambda}\left(\frac{\bar{\delta}}{\delta}A\cos\xi\right)\mathbf{i} - \frac{1}{\delta}\mathbf{j}\right]\frac{\partial T}{\partial \eta}\bigg|_{\eta=1} \tag{8.134}$$

Table 8.1 Comparison of wavy film analysis with experimental data (Faghri and Seban 1985)

	References	Rogovan et al. (1969)		Kapitza and Kapitza (1975)		Rogovan et al. (1969)		Rogovan et al. (1969)			
1	T (°C)	20		25		20		20			
2	$4\Gamma/\mu$	35		92		268		472			
3	$\bar{\delta}\,(10^{-4}\ \text{ft})$	4.132		5.96		9.41		11.5			
4	$\bar{w}_0$ (ft/s)	0.279		0.383		0.836		1.164			
5	$2\pi/\lambda\ (\text{ft}^{-1})$	191		241		159		136			
6	$c/\bar{u}_0$	2.12		1.93		1.65		1.55			
7	A	0.30		0.52		0.55		0.40			(0.55)
8	ν/α	7.2	1.7	6.2	1.7	7.2	1.7	7.2	1.7	7.0	1.7
9	$\bar{h}\bar{\delta}/k\vert_{\text{w}}$	1.059	1.057	1.309	1.296	1.567	1.543	1.393	1.372	1.724	1.694
10	$\bar{h}\bar{\delta}/k\vert_{\delta}$	1.058	1.056	1.314	1.300	1.611	1.580	1.433	1.405	1.816	1.772
11	$1/\sqrt{1-A^2}$	1.05		1.17		1.19		1.09		1.19	
12	$\bar{h}\bar{\delta}/k\vert_{\text{w}}$ [a]	1.049	1.048	1.171	1.169	1.199	1.199	1.092	1.092	1.199	1.199
13	$\bar{h}\bar{\delta}/k\vert_{\delta}$ [a]	1.049	1.048	1.171	1.169	1.197	1.197	1.091	1.091	1.197	1.197

[a]Conduction. Reprinted with permission from Elsevier

where **i** and **j** are unit vectors in the x- and y-directions, respectively. The average heat flow over a period $0 < \xi < 2\pi$ is found by integrating Eqs. (8.133) and (8.134). For a sinusoidal wavy layer, such as the one considered here, this average is identical to the average obtained for the actual surface length. The average heat flux at the wall ($\eta = 0$) is approximated by

$$\frac{\bar{q}''_{\text{w}}}{k} = \frac{1}{2\pi}\int_0^{2\pi} \frac{1}{\delta}\frac{\partial T}{\partial \eta}\bigg|_0 \mathrm{d}\xi \tag{8.135}$$

and the average normal heat flux at the wavy surface is as follows:

$$\frac{\bar{q}''_{\delta}}{k} = \frac{1}{2\pi}\int_0^{2\pi} \frac{1}{\delta}\frac{\partial T}{\partial \eta}\bigg|_{\eta=1} \left[1 + \left(\frac{\partial \delta}{\partial x}\right)^2\right]^{1/2} \mathrm{d}\xi \tag{8.136}$$

The calculations were made using 40 increments in both ξ and η for the value of $2\pi/\lambda$, A, and $c/\bar{u}_0$. Table 8.1 lists these experimental determinations made by the cited authors. Rows 1 through 7 give the experimental conditions and measurements. Row 8 gives the Prandtl numbers for which the calculations were made. One, on the order of 7, corresponds to the experimental conditions, and the other, 1.7, was selected as a comparison to show the effect of Prandtl number. Row 9 gives the calculated average Nusselt number at the wall, and Row 10 gives the average Nusselt number at the outer edge of the layer. These values should be the same, and the difference indicates the failure of the calculation to satisfy the energy balance. This difference is small; it increases with the Reynolds number, reflecting

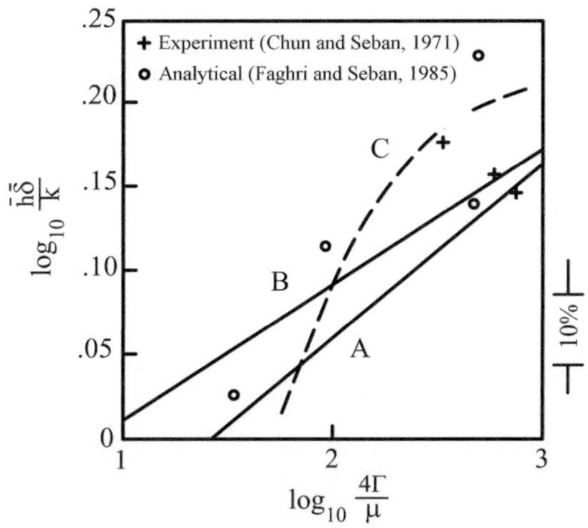

Fig. 8.13 Average Nusselt number for evaporation of a wavy falling film as a function of the Reynolds number: Curve A, Kutateladze (1982); Curve B, Kutateladze (1963); Curve C, Hirschburg and Florschuetz (1982); and ο for $f^+ = 0.65$ (Faghri and Seban 1985; Reprinted with permission from Elsevier)

some increase in truncation error. Row 11 gives the average Nusselt number as evaluated for unidimensional conduction. On this basis, the local Nusselt number is $h\delta/k = 1$, and then

$$\frac{\bar{h}}{k} = \frac{1}{2\pi}\int_0^{2\pi} \frac{d\xi}{\delta} \tag{8.137}$$

For the sinusoidal wave,

$$\begin{aligned}\frac{\bar{h}}{k} &= \frac{1}{2\pi\bar{\delta}}\int_0^{2\pi} \frac{d\xi}{1 + A\sin\xi} \\ &= \frac{1}{2\pi\bar{\delta}}\left[\int_0^{\pi} \frac{d\xi}{1 + A\sin\xi} + \int_0^{\pi} \frac{d\xi}{1 - A\sin\xi}\right] = \frac{1}{\bar{\delta}\sqrt{1 - A^2}}\end{aligned} \tag{8.138}$$

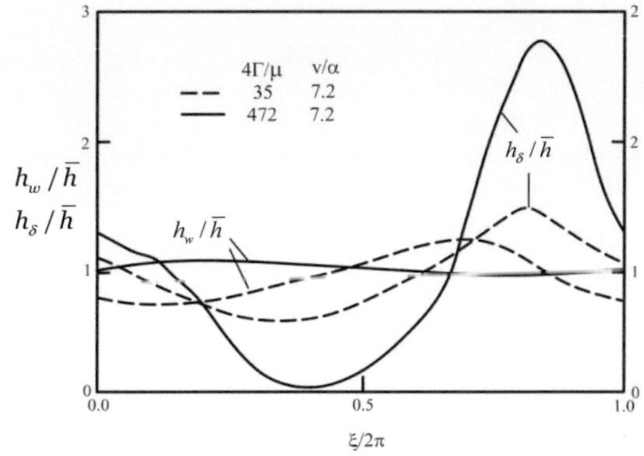

Fig. 8.14 Local Nusselt number for evaporation of wavy falling film (Faghri and Seban 1985; Reprinted with permission from Elsevier)

The difference between the calculated average Nusselt numbers is due to the contribution of convection and to the two-dimensional nature of the conduction that exists because of the variation in layer thickness. The result for $w = v = c = 0$ corresponds to the two-dimensional conduction solution for the wave, and the Nusselt numbers for such a calculation are shown in Rows 12 and 13. These rows should be identical, and the truncation error in the calculation creates a discrepancy between them, as it does in the cases in which there is fluid motion. These values are essentially the same as Row 11, for unidimensional conduction, and this correspondence shows that two-dimensional conduction effects are negligible. Therefore, the difference between the Nusselt numbers in Rows 9 and 11 reflects only the effect of convection.

Figure 8.13 shows the values of the Nusselt number $(\bar{h}\bar{\delta}/k)$ as given by Row 9 of Table 8.1 for a Prandtl number of about 7. The value of 472 for $(4\Gamma/\mu)$ is also shown for an arbitrarily higher value of $A = 0.55$, because the value of A probably should not decrease as the Reynolds number increases. Figure 8.13 also contains lines A—to show the specification of Zazuli given by Kutateladze (1963)—and B—to show that of Kutateladze (1982). Curve C is that of Hirschburg and Florschuetz (1982) for the intermediate wave solution designated by them as $f^+ = 0.65$, which best fits the average of the data. Three data points from Chun and Seban (1971) for evaporation of water with a Prandtl number of about 5.6 are shown by plus (+) symbols, and four numerical results of Faghri and Seban (1985) are shown by circles (°) in the same figure.

Figure 8.14 shows the variation of the local heat transfer coefficient as a function of ξ, normalized with respect to the average value over a period. The results are for the Prandtl number of 7.2 and the Reynolds numbers of 35 and 472 (the same portrayal for the Prandtl number of 1.7 is not much different). The ratio $h_w/\bar{h}$ for the wall varies considerably for the low Reynolds number, but not very much for the high Reynolds number. For the surface, the ratio $h_\delta/\bar{h}$ varies substantially for both cases. The numerical results for the Nusselt numbers of Faghri and Seban (1985) are on the order of, but tend to be higher than, the correlation equations of Kutateladze (1963, 1982). Comparison of Rows 9 and 11 in Table 8.1 reveals the considerable effects of convection and two-dimensional conduction.

For practical purposes, Chun and Seban (1971) suggested that the following correlation based on wavy laminar film condensation is valid for wavy laminar falling film evaporation:

$$h_x = 0.876\left(\frac{Re}{4}\right)^{0.11} h_{\text{Nusselt}} \tag{8.139}$$

An empirical correlation for the local heat transfer coefficient for laminar wavy flow is obtained by combining Eq. (8.139) with Eq. (8.97),

$$\frac{h_x}{k_\ell}\left[\frac{\mu_\ell^2}{\rho_\ell(\rho_\ell - \rho_v)g}\right]^{1/3} = 0.828Re^{-0.22}, \quad Re > Re_{\text{wavy}} \tag{8.140}$$

Chun and Seban (1971) suggested the following empirical correlation to predict the onset of wavy laminar flow:

$$Re \geq Re_{\text{wavy}} = 2.43Ka^{-1/11} \tag{8.141}$$

where Ka is Kapitza number, a ratio of surface tension to viscous force:

$$Ka = \frac{\mu_\ell^4 g}{(\rho_\ell - \rho_v)\sigma^3} \tag{8.142}$$

The average heat transfer coefficient is often relevant to the practical application and can be obtained by using Eq. (8.101), which is also valid for laminar flow with waves. Substituting Eq. (8.140) into Eq. (8.101), one obtains an empirical correlation of the average heat transfer coefficient as follows:

$$\frac{\bar{h}}{k_\ell}\left[\frac{\mu_\ell^2}{\rho_\ell(\rho_\ell - \rho_v)g}\right]^{1/3} = \frac{(Re_0 - Re_L)}{\left(Re_0^{1.22} - Re_L^{1.22}\right)} \tag{8.143}$$

Substituting Eq. (8.98) into Eq. (8.143), an equation correlating Re_L, the Reynolds number at $x = L$, is obtained:

$$Re_L^{1.22} = Re_0^{1.22} - 4\frac{k_\ell L(T_w - T_v)}{\mu_\ell h_{\ell v}}\left[\frac{\rho_\ell(\rho_\ell - \rho_v)g}{\mu_\ell^2}\right]^{1/3} \tag{8.144}$$

Example 8.6

A uniformly distributed liquid water film at 0.01 kg/s is introduced at the top of a tube with an inner diameter of 4 cm and a height of 2 m. The inner surface temperature of the tube is uniformly 105 °C, and the saturation temperature of the steam is 100 °C. Find the evaporation rate and the average heat transfer coefficient.

Solution The properties of the liquid water and the steam can be approximated as those at the saturation temperature: $k_\ell = 0.68\,\text{W/m}\,^\circ\text{C}$, $\rho_\ell = 158.77\,\text{kg/m}^3$, $\mu_\ell = 2.79\times10^{-4}\,\text{N s/m}^2$, $\sigma = 58.910^{-4}\,\text{N/m}$, $\rho_v = 0.596\,\text{kg/m}^3$, and $h_{\ell v} = 2251.2\,\text{kJ/kg}$. It is assumed that the liquid film thickness is much smaller than the radius of the tube. Therefore, the falling film analysis for evaporation over a vertical flat plate can be used.

The Kapitza number is obtained from Eq. (8.142):

$$Ka = \frac{\mu_\ell^4 g}{(\rho_\ell - \rho_v)\sigma^3} = \frac{(2.79\times10^{-4})^4 \times 9.8}{(958.77 - 0.596)\times(58.9\times10^{-3})^3} = 3.032\times10^{-13}$$

Therefore, the Reynolds number for wavy film is

$$Re_{\text{wavy}} = 2.43Ka^{-1/11} = 2.43\times(3.03\times10^{-13})^{-1/11} = 33.4$$

The inlet mass flow rate per unit perimeter is

$$\Gamma_0 = \frac{\dot{m}_0}{\pi D} = \frac{0.01}{\pi\times0.04} = 7.95\times10^{-2}\,\text{kg/s m}$$

The Reynolds number at the inlet is

$$Re_0 = \frac{4\Gamma_0}{\mu_\ell} = \frac{4 \times 7.95 \times 10^{-2}}{2.79 \times 10^{-4}} = 1139.8 > Re_{\text{wavy}}$$

which is also under the transition Reynolds number for turbulent, $Re_{\text{turb}} = 5840\,Pr^{-1.05} = 1862$ [see Eq. (8.145)]. This means that the liquid film is in the laminar wavy regime.The Reynolds number at outlet can be obtained from Eq. (8.144), i.e.,

$$\begin{aligned} Re_L^{1.22} &= Re_0^{1.22} - 4\frac{k_\ell L(T_w - T_v)}{\mu_\ell h_{\ell v}}\left[\frac{\rho_\ell(\rho_\ell - \rho_v)g}{\mu_\ell^2}\right]^{1/3} \\ &= 1139.8^{1.22} - 4 \times \frac{0.68 \times 2 \times (105 - 100)}{2.79 \times 10^{-4} \times 2251.2 \times 10^3} \\ &\quad \times \left[\frac{958.77 \times (958.77 - 0.596) \times 9.8}{(2.79 \times 10^{-4})^2}\right]^{1/3} \\ &= 3257.5 \end{aligned}$$

Thus,

$$Re_L = 757.6$$

which is still in the wavy regime so the entire surface is wavy.

The mass flow rate of the liquid at the outlet is

$$\dot{m}_L = \dot{m}_0 \frac{Re_L}{Re_0} = 0.01 \times \frac{757.6}{1139.8} = 6.64 \times 10^{-3}$$

The evaporation rate is therefore

$$\dot{m}_v = \dot{m}_0 - \dot{m}_L = 0.01 - 6.64 \times 10^{-3} = 3.36 \times 10^{-3}\,\text{kg/m}$$

The heat transfer coefficient can be obtained from Eq. (8.143), i.e.,

$$\begin{aligned} \bar{h} &= k_\ell \left[\frac{\rho_\ell(\rho_\ell - \rho_v)g}{\mu_\ell^2}\right]^{1/3} \frac{(Re_0 - Re_L)}{(Re_0^{1.22} - Re_L^{1.22})} \\ &= 0.68 \times \left[\frac{958.77 \times (958.77 - 0.596) \times 9.8}{(2.79 \times 10^{-4})^2}\right]^{1/3} \\ &\quad \times \frac{1139.8 - 757.6}{1139.8^{1.22} - 757.6^{1.22}} = 6017\,\text{W/m}^2\,{}^\circ\text{C} \end{aligned}$$

8.4.3 Turbulent Falling Film

Stephan (1992) presented an empirical relationship between Reynolds number and Prandtl number that could be used to determine whether a falling film is completely turbulent:

$$Re \geq Re_{\text{turb}} = 5840 Pr^{-1.05} \tag{8.145}$$

In order to model turbulent flow, the following development from Seban and Faghri (1976) proceeds with assumptions similar to those made in the classical Nusselt laminar analysis. The momentum equation is

$$\frac{\mathrm{d}}{\mathrm{d}y}(\nu_\ell + \varepsilon_\mathrm{M})\frac{\mathrm{d}u}{\mathrm{d}y} + g = 0 \tag{8.146}$$

where ε_M is the turbulent eddy diffusivity.

Defining the following dimensionless variables,

$$u_\tau = \sqrt{\tau_\mathrm{w}/\rho_\ell},\ u^+ = u/u_\tau,\ y^+ = y u_\tau/\nu_\ell \tag{8.147}$$

the momentum equation becomes

$$\frac{\mathrm{d}}{\mathrm{d}y^+}\left(\frac{\varepsilon_\mathrm{M}}{\nu_\ell} + 1\right)\frac{\mathrm{d}u^+}{\mathrm{d}y^+} + \frac{g\nu_\ell}{u_\tau^3} = 0 \tag{8.148}$$

which is subject to the following boundary conditions:

$$u^+ = 0, \quad y^+ = 0 \tag{8.149}$$

$$\frac{\mathrm{d}u^+}{\mathrm{d}y^+} = 0, \quad y^+ = \delta^+ \tag{8.150}$$

Integrating Eq. (8.148) twice and using Eqs. (8.149) and (8.150) to determine the integral constant, the dimensionless velocity becomes

$$u^+ = \int_0^{y^+} \frac{(1 - y^+/\delta^+)}{(1 + \varepsilon_\mathrm{M}/\nu_\ell)}\mathrm{d}y^+ \tag{8.151}$$

The energy equation is

$$u\frac{\partial T}{\partial x} = \frac{\partial}{\partial y}(\varepsilon_\mathrm{H} + \alpha)\frac{\partial T}{\partial y} \tag{8.152}$$

where ε_H is turbulent eddy thermal diffusivity. Equation (8.152) is subject to the following boundary conditions:

$$T = T_{\text{sat}} \quad x = 0 \tag{8.153}$$

$$\frac{\partial T}{\partial y} = -\frac{q''_{\rm w}}{k_\ell}, \quad y = 0 \tag{8.154}$$

$$T = T_{\rm sat}, \quad y = \delta \tag{8.155}$$

By defining dimensionless temperature as

$$T^+ = \frac{T\rho_\ell c u_\tau}{q''_{\rm w}} \tag{8.156}$$

and using the dimensionless variables from Eq. (8.147), the energy equation becomes

$$u^+ \frac{\partial T}{\partial x^+} = \frac{\partial}{\partial y^+}\left(\frac{\alpha}{\nu_\ell} + \frac{\varepsilon_{\rm H}}{\nu_\ell}\right)\frac{\partial T^+}{\partial y^+} \tag{8.157}$$

The solution requires choosing a turbulence model for the specification of $\varepsilon_{\rm M}/\nu$ and of $\varepsilon_{\rm H}/\varepsilon_{\rm M}$. The detailed turbulence model and solution procedure can be found in Seban and Faghri (1976). The predicted result agrees very well with the experimental result of Chun and Seban (1971) for constant heat flux at wall. To predict local heat transfer in turbulent falling film flow on a wall heated under constant heat flux, the following empirical correlation, recommended by Chun and Seban (1971), can be used:

$$\frac{h}{k_\ell}\left[\frac{\mu_\ell^2}{\rho_\ell(\rho_\ell - \rho_{\rm v})g}\right]^{1/3} = 3.8 \times 10^{-3} Re^{0.4} Pr^{0.65} \tag{8.158}$$

The average heat transfer coefficient can be obtained by substituting Eq. (8.158) into Eq. (8.101), i.e.,

$$\frac{\overline{h}}{k_\ell}\left[\frac{\mu_\ell^2}{\rho_\ell(\rho_\ell - \rho_{\rm v})g}\right]^{1/3} = 2.28 \times 10^{-3} \frac{(Re_0 - Re_L)}{\left(Re_0^{0.6} - Re_L^{0.6}\right)} Pr^{0.65} \tag{8.159}$$

As was demonstrated in Example 8.5, Eq. (8.159) is valid for cases with either constant wall temperature or constant heat flux. However, the methods for determining the Reynolds number at $x = L$ will vary for different thermal boundary conditions on the wall. For cases with constant wall temperature, the Reynolds number at $x = L$, Re_L, is obtained by substituting Eq. (8.98) into Eq. (8.159), i.e.,

$$Re_L^{0.6} = Re_0^{0.6} - 9.12 \times 10^{-3} \frac{k_\ell L(T_{\rm w} - T_{\rm v})}{\mu_\ell h_{\ell \rm v}}\left[\frac{\rho_\ell(\rho_\ell - \rho_{\rm v})g}{\mu_\ell^2}\right]^{1/3} \tag{8.160}$$

For the case with constant heat flux, Eq. (8.106) can be used to determine the Reynolds number at $x = L$, Re_L.

While gravitational force drives the falling film evaporation discussed here, liquid flow in a liquid film can also be driven by centrifugal force. Rahman and Faghri (1992) analyzed the processes of heating and evaporation in a thin liquid film adjacent to a horizontal disk rotating about a vertical axis at a constant angular velocity. The fluid emanates axisymmetrically from a source at the center of the disk and then is carried downstream by inertial and centrifugal forces. Closed-form analytical solutions were derived for fully developed flow and heat transfer. Simplified analyses were also presented for developing heat transfer in a fully developed flow field. Rice et al. (2005) provided a detailed analysis for evaporation from a thin liquid film on a rotating disk including conjugate effects.

8.4.4 Surface Spray Cooling

When liquid drops are sprayed onto a hot surface, heat conduction causes the liquid to superheat and evaporate into a gas (see Fig. 8.3). The cooling capacity of the process depends on the time required for an average-sized drop to cool. If the shape of the droplet can be assumed to be hemispherical during the evaporation process, the energy balance at the interface can be written as

$$\rho_\ell h_{\ell v} 2\pi r_I^2 \frac{dr_I}{dt} = k_\ell \int_0^{\pi/2} \frac{\partial T_\ell(r_I,\theta)}{\partial r} 2\pi r_I^2 \cos\theta d\theta - h2\pi r_I^2 (T_\infty - T_{sat}) \tag{8.161}$$

where r_I is the radius of the droplet. Equation (8.161) can be simplified as

$$\rho_\ell h_{\ell v} \frac{dr_I}{dt} = k_\ell \int_0^{\pi/2} \frac{\partial T_\ell(r_I,\theta)}{\partial r} \cos\theta d\theta - h(T_\infty - T_{sat}) \tag{8.162}$$

Analytical solution of Eq. (8.162) is very difficult because the temperature distribution in the liquid droplet is two-dimensional in nature. A scale analysis similar to that in Lock (1994) is presented here to estimate the time required for droplet with an initial radius of R_i to evaporate completely.

The scale of the radius of the droplet, r_I, is R_i. Thus the scale of the temperature gradient at the interface is

$$\frac{\partial T(r_I,\theta)}{\partial r} \sim \frac{T_w - T_{sat}}{R_i} \tag{8.163}$$

The scale analysis of Eq. (8.162) yields

$$\rho_\ell h_{\ell v} \frac{R_i}{t_f} \sim k_\ell \frac{T_w - T_{sat}}{R_i} + h(T_\infty - T_{sat}) \tag{8.164}$$

which can be rearranged as

$$\frac{\rho_\ell h_{\ell v}}{k_\ell} \frac{R_i^2}{t_f} \sim (T_w - T_{sat}) + \frac{hR_i}{k_\ell}(T_\infty - T_{sat}) \tag{8.165}$$

Since the liquid droplet is very small, hR_i/k_ℓ is very small, so the second term on the right-hand side of Eq. (8.164) is much smaller than the first. Equation (8.164) can be simplified by neglecting the second term on the right-hand side, i.e.,

$$\rho_\ell h_{\ell v}\frac{R_i^2}{k_\ell t_f}\sim(T_w-T_{sat}) \tag{8.166}$$

The physical significance of Eq. (8.166) is that the latent heat of evaporation is balanced primarily by conduction in the liquid droplet, while the effect of convection on the surface of the droplet is negligible.

The time that it takes to completely evaporate the droplet with an initial radius of R_i is therefore estimated by

$$t_f\sim\frac{\rho_\ell h_{\ell v}R_i^2}{k_\ell(T_w-T_{sat})} \tag{8.167}$$

For the cases where the substrate heat flux beneath the droplet, q''_w, is known, Eq. (8.166) should be rewritten in terms of heat flux. The scale of the heat flux at the heating surface is

$$q''_w\sim\frac{k_\ell(T_w-T_{sat})}{R_i} \tag{8.168}$$

Combining Eqs. (8.167) and (8.168) yields

$$t_f\sim\frac{\rho_\ell h_{\ell v}R_i}{q''_w} \tag{8.169}$$

which indicates that a smaller drop will provide a larger cooling effect.

While Eqs. (8.167) and (8.169) provide the order of magnitude of the time required to completely evaporate the droplet, quantitative estimation of the evaporation time is often also desirable. A simple approximate analysis will be done below by estimating the conduction in the liquid droplet by the following correlation:

$$q_d=k_\ell\bar{A}\frac{T_w-T_{sat}}{\bar{\delta}} \tag{8.170}$$

where $\bar{A}$ and $\bar{\delta}$ are the average cross-sectional area of heat conduction and the average path length of the conduction, respectively. For a hemispherical droplet, the contact area between the droplet and the heated wall is $A_w=\pi r_I^2$ and the interfacial area of the droplet is $A_I=2\pi r_I^2$. Thus, we can take the average conduction area as

$$\bar{A}=\frac{1}{2}(A_w+A_I)=\frac{3}{2}\pi r_I^2 \tag{8.171}$$

The average path length for conduction is

$$\bar{\delta}=\frac{V}{\bar{A}}=\frac{(2/3)\pi r_I^3}{(3/2)\pi r_I^2}=\frac{4}{9}r_I \tag{8.172}$$

Therefore, the conduction in the liquid droplet becomes

$$q_{\mathrm{d}} = \frac{27}{8}\pi k_\ell r_I (T_{\mathrm{w}} - T_{\mathrm{sat}}) \tag{8.173}$$

Replacing the first term on the right-hand side of Eq. (8.162) with Eq. (8.173), and dropping the second term on the right-hand side of Eq. (8.162) (see the above scale analysis), the energy balance for the hemispherical droplet becomes

$$\rho_\ell h_{\ell \mathrm{v}} 2\pi r_I^2 \frac{dr_I}{dt} = \frac{27}{8}\pi k_\ell r_I (T_{\mathrm{w}} - T_{\mathrm{sat}}) \tag{8.174}$$

which can be rearranged as

$$r_I \frac{dr_I}{dt} = \frac{27}{16} \frac{k_\ell (T_{\mathrm{w}} - T_{\mathrm{sat}})}{\rho_\ell h_{\ell \mathrm{v}}} \tag{8.175}$$

which is subject to the following initial conditions:

$$r_I = R_i, \quad t = 0 \tag{8.176}$$

Integrating Eq. (8.175) and considering its initial condition, Eq. (8.176), one obtains the transient radius of the liquid droplet:

$$r_I = \sqrt{R_i^2 - \frac{27}{8} \frac{k_\ell (T_{\mathrm{w}} - T_{\mathrm{sat}}) t}{\rho_\ell h_{\ell \mathrm{v}}}} \tag{8.177}$$

The time required for the droplet to evaporate completely can be obtained by letting r_I in Eq. (8.177) equal zero, i.e.,

$$t_{\mathrm{f}} = \frac{8 \rho_\ell h_{\ell \mathrm{v}} R_i^2}{27 k_\ell (T_{\mathrm{w}} - T_{\mathrm{sat}})} \tag{8.178}$$

which agrees with the results obtained by scale analysis, Eq. (8.167).

The above analysis is otherwise simplistic because it assumes that the shape of the droplet is hemispherical. In fact, the shape of the droplet depends on the velocity with which it impacts the wall as well as the wettability of the liquid droplet on the wall. If the contact area between the droplet and the wall is larger (due to higher impacting velocity or good wettability), the time for conduction through the liquid is shorter and the life of the drop will be shorter as well.

Example 8.7

A hemispherical liquid water droplet with an initial radius of 20 μm is attached to a heated wall and exposed to pure water vapor at 1 atm. The temperature of the heated wall is 180 °C. Estimate the time required for the liquid droplet to evaporate completely.

Solution The saturation temperature corresponding to 1 atm is T_{sat} = 100 °C. The properties of the liquid water can be approximately evaluated at the saturation temperature: $k_\ell = 0.68\,\mathrm{W/m\,°C}$, $\rho_\ell = 958.77\,\mathrm{kg/m^3}$, and $h_{\ell \mathrm{v}} = 2251.2\,\mathrm{kJ/kg}$.

The time required for the liquid droplet to evaporate completely can be obtained from Eq. (8.178), i.e.,

$$t_{\mathrm{f}} = \frac{8\rho_\ell h_{\ell\mathrm{v}} R_i^2}{27k_\ell(T_{\mathrm{w}} - T_{\mathrm{sat}})}$$
$$= \frac{8 \times 958.77 \times 2251.2 \times 10^3 \times (20 \times 10^{-6})^2}{27 \times 0.68 \times (180 - 100)} = 0.0047\,\mathrm{s} = 4.7\,\mathrm{ms}$$

8.5 Direct-Contact Evaporation

Direct-contact evaporation processes are the most effective cooling strategies when the coolant is mixed with the medium it will cool. Since direct-contact evaporation requires no container to separate the coolant from the medium to be cooled, there are no corresponding thermal resistances.

8.5.1 Evaporation of a Liquid Droplet in a Hot Gas

Figure 8.4 describes the situation in which a liquid drop evaporates to a surrounding hot gas. Since the liquid droplet is surrounded by the hot gas, it is assumed that the liquid droplet is spherical in shape and the evaporation is a spherically symmetric problem. Neglecting radiation effect, the energy equation for a liquid droplet is

$$\frac{\partial T_\ell}{\partial t} = \frac{\alpha_\ell}{r^2}\frac{\partial}{\partial r}\left(r^2 \frac{\partial T_\ell}{\partial r}\right) \tag{8.179}$$

where the thermophysical properties of the liquid are assumed to be temperature-independent. The initial temperature of the droplet is

$$T_\ell(r,t) = T_i, \quad t = 0 \tag{8.180}$$

The symmetric condition at the center of the liquid droplet requires that

$$\frac{\partial T_\ell}{\partial r} = 0, \quad r = 0 \tag{8.181}$$

Since the evaporation takes place on the surface of the liquid droplet, its surface temperature equals the saturation temperature corresponding to the partial pressure of the vapor at the interface, i.e.,

$$T(r,t) = T_{\mathrm{sat}}, \quad r = r_I \tag{8.182}$$

An energy balance at the surface of the liquid droplet accounts for the latent heat exchanged through conduction inside the drop and convection away from the droplet; that energy balance is

$$\rho_\ell h_{\ell\mathrm{v}} \frac{\mathrm{d}r_I}{\mathrm{d}t} = k_\ell \frac{\partial T_\ell}{\partial r} - h(T_\infty - T_{\mathrm{sat}}), \quad r = r_I \tag{8.183}$$

where h is the heat transfer coefficient between the gas and the liquid droplet.

If the liquid droplet is evaporating to the vapor–gas mixture, the mass balance at the interface can be written as

$$\rho_\ell \frac{\mathrm{d}r_I}{\mathrm{d}t} = -\rho_{\mathrm{g}} h_{\mathrm{m}} (\omega_I - \omega_\infty) \tag{8.184}$$

where h_{m} is the mass transfer coefficient, ρ_{g} is the density of the mixture, and ω is the mass fraction of the evaporating component in the mixture.

Analytical solution of the problem defined by Eqs. (8.179)–(8.184) will yield a very complex expression of the final results. Since the liquid droplet is usually very small, one can assume that the heat convected to the droplet surface is balanced by the latent heat of evaporation, so heat conduction to the liquid droplet can be neglected. This assumption is valid because the bulk temperature of the droplet will quickly approach the saturation temperature (Lock 1994), i.e.,

$$\frac{\mathrm{d}r_I}{\mathrm{d}t} = -\frac{h(T_\infty - T_{\mathrm{sat}})}{\rho_\ell h_{\ell \mathrm{v}}}, \quad r = r_I \tag{8.185}$$

which is subject to the following initial condition:

$$r_I = R_i, \quad t = 0 \tag{8.186}$$

Integrating Eq. (8.185) and considering its initial condition, Eq. (8.186), the transient radius of the liquid droplet is obtained.

$$r_I = R_i - \frac{h(T_\infty - T_{\mathrm{sat}})t}{\rho_\ell h_{\ell \mathrm{v}}} \tag{8.187}$$

In arriving at Eq. (8.187), the heat transfer coefficient was assumed to be constant throughout the entire evaporation process. The time required to evaporate the droplet completely, t_{f}, can be obtained by setting r_I equal to zero in Eq. (8.187), i.e.,

$$t_{\mathrm{f}} = \frac{\rho_\ell h_{\ell \mathrm{v}} R_i}{h(T_\infty - T_{\mathrm{sat}})} \tag{8.188}$$

The time required to evaporate the droplet completely can also be estimated from the mass balance Eq. (8.184). By following a procedure similar to that which obtained Eq. (8.188), one obtains

$$t_{\mathrm{f}} = \frac{\rho_\ell R_i}{\rho_{\mathrm{g}} h_{\mathrm{m}} (\omega_I - \omega_\infty)} \tag{8.189}$$

Combining Eqs. (8.188)–(8.189), one obtains

$$\omega_I - \omega_\infty = \frac{h}{h_{\mathrm{m}}} \frac{Ja_{\mathrm{g}}}{\rho_{\mathrm{g}} c_{p_{\mathrm{g}}}} \tag{8.190}$$

where Ja_g is the Jakob number, defined as

$$Ja_g = \frac{c_{pg}(T_\infty - T_{sat})}{h_{\ell v}} \tag{8.191}$$

The above analysis is based on the assumption that conduction to the liquid droplet is negligible. Evaporation of a liquid droplet with consideration of the effect of conduction heat loss will be investigated in Example 8.8.

Example 8.8

Heat conduction in a liquid droplet can be considered using an integral approximate solution with a parabolic temperature distribution in the liquid droplet. Find the transient radius of the droplet, with conduction in the liquid droplet considered.

Solution Integrating Eq. (8.179) in the interval of (0, r_I) with respect to r, one obtains

$$\int_0^{r_I} r^2 \frac{\partial T_\ell}{\partial t} \mathrm{d}r = \alpha_\ell r_I^2 \frac{\partial T_\ell}{\partial r}\bigg|_{r=r_I} \tag{8.192}$$

The left-hand side of Eq. (8.192) can be rewritten using Leibnitz's rule, i.e.,

$$\frac{\mathrm{d}}{\mathrm{d}t}\int_0^{r_I} r^2 T_\ell \mathrm{d}r - r_I^2 T_{sat}\frac{\mathrm{d}r_I}{\mathrm{d}t} = \alpha_\ell r_I^2 \frac{\partial T_\ell}{\partial r}\bigg|_{r=r_I} \tag{8.193}$$

which is the integral equation of the heat conduction problem.

Assuming the temperature distribution in the liquid droplet is parabolic profile, and using Eqs. (8.181)–(8.182) to eliminate two of the three constants, the temperature distribution in the liquid droplet becomes (Dombrovsky and Sazhin 2003)

$$T_\ell(r,t) = T_c(t) + [T_{sat} - T_c(t)]\left(\frac{r}{r_I}\right)^2 \tag{8.194}$$

where $T_c(t)$ represents the temperature in the center of the liquid droplet, which is still unknown at this point.

Substituting Eq. (8.194) into Eq. (8.193), one obtains

$$\frac{\mathrm{d}(\Delta T)}{\mathrm{d}t} = -15\Delta T\left(\frac{\alpha_\ell}{r_I^2} + \frac{1}{r_I}\frac{\mathrm{d}r_I}{\mathrm{d}t}\right) \tag{8.195}$$

where $\Delta T = T_{sat} - T_c(t)$.

Substituting Eq. (8.194) into Eq. (8.183), the energy balance at interface becomes

$$\rho_\ell h_{\ell v}\frac{dr_I}{dt}=\frac{2k_\ell(T_{sat}-T_c)}{r_I}-h(T_\infty-T_{sat}),\quad r=r_I$$

which can be rearranged to

$$\frac{dr_I}{dt}=\frac{2k_\ell}{\rho_\ell h_{\ell v}r_I}\left[\Delta T-\frac{1}{4}\frac{k_g}{k_\ell}Nu(T_\infty-T_{sat})\right] \tag{8.196}$$

where $Nu=2hr_I/k_g$ is the known Nusselt Number.

Equations (8.195)–(8.196) are a set of ordinary differential equations with r_I and ΔT as unknowns. They can be solved numerically with the following initial conditions at $t=0$:

$$r_I=R_i$$

$$\Delta T=T_{sat}-T_i$$

Elperin et al. (2005) numerically investigated simultaneous heat and mass transfer during evaporation and condensation on the surface of a stagnant droplet in the presence of admixtures that contain noncondensable soluble gas. The equations for Stefan velocity and the transient radius of the droplet were obtained by using conservation of species mass at droplet surface with absorption and evaporation at the surface accounted for. Their results showed that droplet evaporation rate, droplet temperature, interfacial absorbate concentration, and rate of mass transfer are highly interdependent.

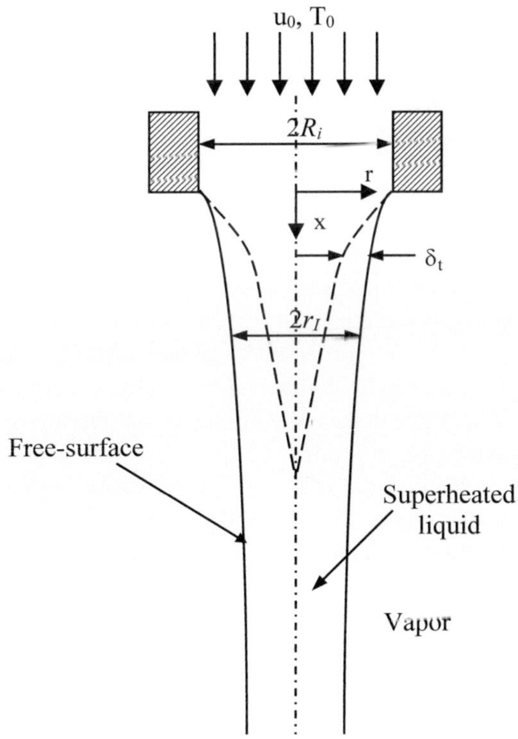

Fig. 8.15 Evaporation from a jet surrounded by a hot gas (Lock 1994; Reproduced by permission from Oxford University Press)

8.5.2 Evaporation of a Liquid Jet in a Pure Vapor

The physical model of the problem under consideration is shown in Fig. 8.15 (Lock 1994). A liquid jet flows out from a nozzle with a radius of r_0 and is surrounded by pure vapor at saturation temperature. At the exit of the nozzle, the velocity and temperature are uniformly T_0 and u_0, respectively. It is further assumed that the velocity in the jet remains uniformly equal to u_0 as it continues flowing; then, the jet can be treated as slug flow. This is possible when the friction between the liquid jet and the surrounding vapor is negligible. The temperature of the jet will be affected by the hot gas as soon as it exits the nozzle. Because evaporation occurs on the surface of the liquid jet, its surface temperature is equal to the saturation temperature corresponding to the vapor pressure. The energy equation for the liquid jet is

$$u_0 \frac{\partial T_\ell}{\partial x} = \frac{\alpha_\ell}{r} \frac{\partial}{\partial r}\left(r \frac{\partial T_\ell}{\partial r} \right) \tag{8.197}$$

where the thermophysical properties have been assumed to be constants. The initial temperature of the jet is

$$T_\ell(r, x) = T_0, \quad t = 0 \tag{8.198}$$

The boundary conditions at the center and surface of the liquid jet are

$$\frac{\partial T_\ell}{\partial r} = 0, \quad r = 0 \tag{8.199}$$

$$T(r, x) = T_{\text{sat}}, \quad r = r_I(x) \tag{8.200}$$

where $r_I(x)$ is the radius of the liquid jet, which equals r_0 at $x = 0$ but decreases with increasing x.

The energy balance at the surface of the liquid jet is

$$\rho_\ell h_{\ell v} u_0 \frac{\mathrm{d} r_I}{\mathrm{d} x} = k_\ell \frac{\partial T_\ell}{\partial r}, \quad r = r_I(x) \tag{8.201}$$

which demonstrates that superheated liquid supplies the latent heat of vaporization. Since the vapor temperature equals the saturation temperature, there is no convection on the surface of the liquid jet.

As the liquid jet moves downstream, a thermal boundary layer, $r_0 - \delta_\text{t} < r < r_0$, develops in the liquid jet. The evaporation of the jet can be divided into two stages: the early stage when $\delta_\text{t} \ll r_0$ and a later stage when $\delta_\text{t} = r_0$. These two stages are similar to the thermal entrance problem and fully developed flow for forced convection in a tube.

During the first stage, the effect of the curvature of the cylindrical liquid jet can be neglected because the thermal boundary layer is much thinner than the radius of the tube. The solution of the temperature distribution in the thermal boundary layer can be obtained by using the solution of heat conduction in a semi-infinite solid, i.e.,

$$T(y, x) = T_{\text{sat}} + (T_0 - T_{\text{sat}}) \operatorname{erf}\left(\frac{y}{2\sqrt{\alpha_\ell x / u_0}} \right) \tag{8.202}$$

where the y-coordinate originates at the jet's surface and points toward its center. It is related to r by $y = r_I - r$. Substituting Eq. (8.202) into the energy balance equation at the interface, Eq. (8.201), one obtains

$$\frac{\mathrm{d}r_I}{\mathrm{d}x} = -\frac{k_\ell (T_0 - T_{\mathrm{sat}})}{\rho_\ell h_{\ell \mathrm{v}} \sqrt{\pi \alpha_\ell u_0 x}} \tag{8.203}$$

The variation in the radius of the liquid jet during the first stage can be obtained by integrating Eq. (8.203), i.e.,

$$r_I = R_i - \frac{2k_\ell (T_0 - T_{\mathrm{sat}})}{\rho_\ell h_{\ell \mathrm{v}}} \sqrt{\frac{x}{\pi \alpha_\ell u_0}} \tag{8.204}$$

Evaporation during the second stage ($\delta_{\mathrm{t}} = r_I$), which occurs at the lower portion of the jet, is much slower than the first stage and ceases after the superheat in the liquid jet vanishes. The final radius of the liquid jet, $r_{I,\mathrm{f}}$, can be obtained by a simple energy balance:

$$\rho_\ell c_{p\ell} \pi R_i^2 (T_0 - T_{\mathrm{sat}}) = \rho_\ell h_{\ell \mathrm{v}} \pi (R_i^2 \quad r_{I,\mathrm{f}}^2) \tag{8.205}$$

Rearranging Eq. (8.205), one obtains the final radius of the liquid jet as

$$r_{I,\mathrm{f}} = R_i \sqrt{1 - Ja} \tag{8.206}$$

where

$$Ja = \frac{c_{p_\ell} (T_0 - T_{\mathrm{sat}})}{h_{\ell \mathrm{v}}} \tag{8.207}$$

is Jakob number.

Problems

8.1. A liquid drop is on the top of a substrate at temperature T_{s}. Suppose the liquid drop is disk-like and the initial thickness is δ_0. Find the time required to completely evaporate the liquid drop. Assume that heat transfer across the disk-like liquid drop is one-dimensional conduction.

8.2. A small amount of water is contained in a tray, as shown in Fig. P8.2. The bottom of the tray is maintained at a temperature, T_{w}, which is above the ambient temperature, T_∞. The initial bulk temperature of the water in the tray, T_{b}, is between T_{w} and T_∞. The rate of evaporation at the surface of the water is $\dot{m}_{\mathrm{evp}}$ (kg/s). Find the bulk temperature of the water as a function of time.

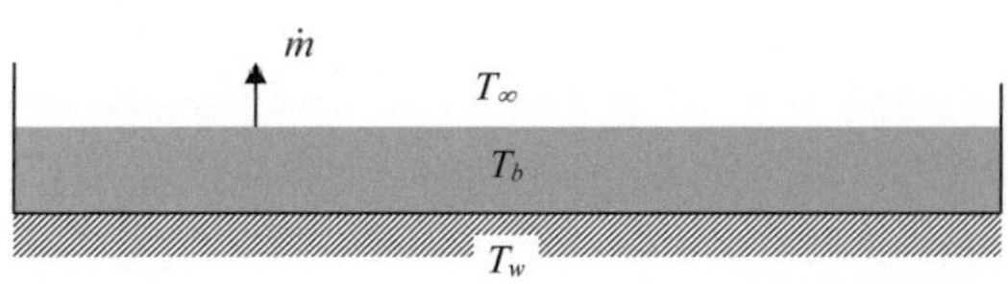

Fig. P8.2

8.3. A thin layer of liquid film sits on top of a heated surface at temperature T_0 (see Fig. P8.3). The top surface of the liquid film is at temperature T_δ ($T_\delta < T_0$). The mass flux on the top surface of the liquid film due to evaporation is $\dot{m}''_\delta$. It is suggested that the liquid layer could be divided into upper and lower regions with linear temperature distribution in each region. Build an appropriate model to determine the location and temperature of the boundary between the two regions.

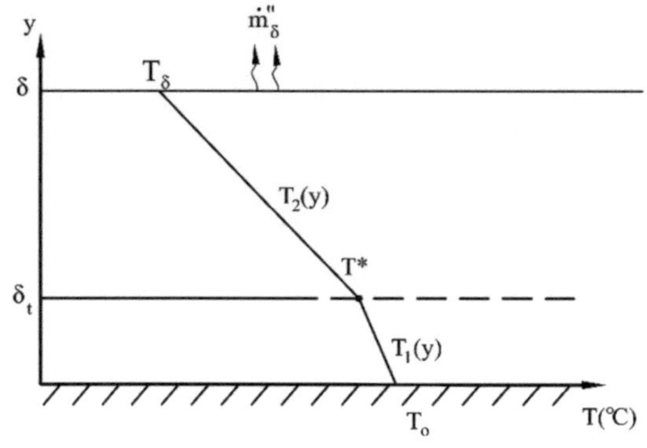

Fig. P8.3

8.4. Solve Stefan's diffusion problem (see Example 8.2) in terms of molar concentration by assuming the molar fractions of species A at liquid–vapor interface, $x_{A,\delta}$, and at the top of the tube, $x_{A,H}$, are known. The mixture concentration can be assumed constant.

8.5. Do the Stefan diffusion problem (Problem 8.4) as quasi-steady rather than steady condition, and obtain the change in height with time in terms of molar concentration of mixture and initial height H.

8.6. Evaporation from a horizontal thin film is described by three partial differential equations (8.12), (8.3), and (8.4). Convert these equations to ordinary differential equations using similarity variables defined in Eq. (8.13), and obtain their boundary conditions from Eqs. (8.5)–(8.10).

8.7. Evaporation from a horizontal thin liquid film is described by Eqs. (8.1)–(8.6), (8.8), and (8.10). To solve the problem using an integral method, the profiles of velocity, temperature, and concentration need to be determined. If these profiles can be assumed to be second-degree polynomial functions, find velocity, temperature, and concentration distribution in the boundary layers.

8.8. Solve the evaporation problem using integral approximate method. For the sake of simplicity, assume $Pr = Sc = 1$ so that the velocity, thermal, and concentration boundary-layer thicknesses are the same. Use the velocity, temperature, and concentration distributions in the boundary layers obtained from the previous problem. The blowing velocity at the surface due to evaporation can be neglected.

8.9. Sweat cooling is a technique used to protect an intensely heated porous wall by evaporating a liquid injected from the pores to the wall surface. In order to cool a 0.5-m-long porous stainless surface, water is injected to the surface at a rate that is just sufficient to keep the surface wet. Dry air at 600 K and 1 atm flows over the plate at a velocity of 10 m/s. Find the rate of water supply needed to maintain the wall temperature at 400 K.

8.10. Falling water film evaporation occurs on the inner sides of two parallel plates with a length of 1 m, and they are separated by a distance of $2b = 3\,\text{cm}$. The inlet and outlet liquid film temperatures are 50 and 45 °C, respectively. Find the mass flow rate of the liquid film per unit width, Γ.

8.11. Air enters the bottom of a vertical channel formed by two parallel walls (see Fig. P8.11). A liquid water film is present on the left wall, which is maintained at a constant temperature. The right wall is adiabatic and has no liquid film on it. The liquid film on the left wall is so thin that its thickness can be neglected. The left wall can be considered as a surface where the concentration of the vapor equals the saturation concentration of the vapor at its partial pressure. Assuming that the air flow in the vertical channel is laminar and the air inlet velocity is uniform, find the governing equations and boundary conditions to describe this conjugate heat and mass transfer process.

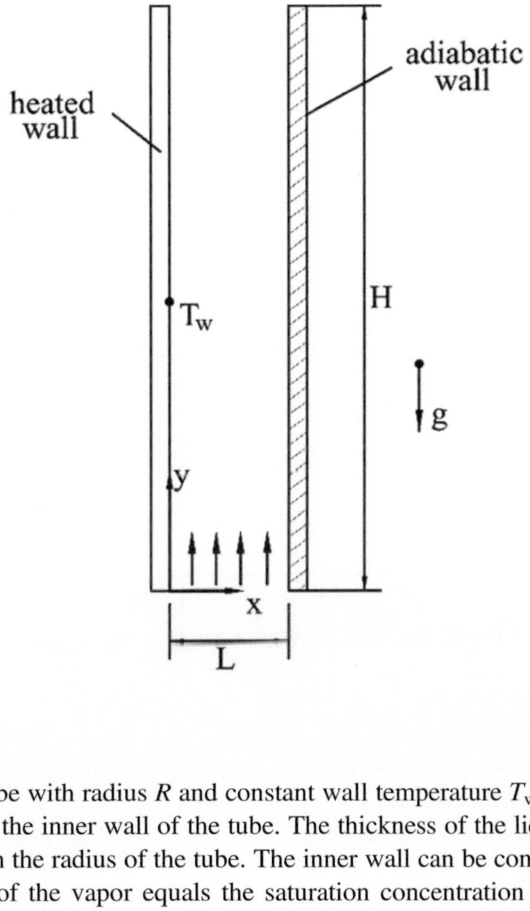

Fig. P8.11

8.12. Air flows into a tube with radius R and constant wall temperature T_w. Evaporation occurs on a thin liquid film on the inner wall of the tube. The thickness of the liquid film can be neglected in comparison with the radius of the tube. The inner wall can be considered as a surface where the concentration of the vapor equals the saturation concentration of the vapor at its partial pressure. The air flow at the inlet of the tube is fully developed, but the temperature and concentration distributions at the inlet are uniform. Specify the governing equations and corresponding boundary conditions of the problem.

8.13. Solve the energy and species equations of the above problem to obtain the temperature and concentration distributions in the tube. The Schmidt number, Sc, can be assumed to be equal to one. Obtain the local convective heat transfer coefficient for the system.

8.14. A two-phase thermosyphon, shown in Fig. P8.14, is needed to deliver a steady-state heat rate of $\dot{Q} = 1\,\text{kW}$ at an operating temperature of $T_v = 333$ K. An acceptable wall superheat is $\Delta T = 5$ K. The inside diameter of the circular channel is $D = 0.1$ m, and the working fluid is water. Use classical Nusselt analysis to determine how long the evaporator should be to avoid dryout.

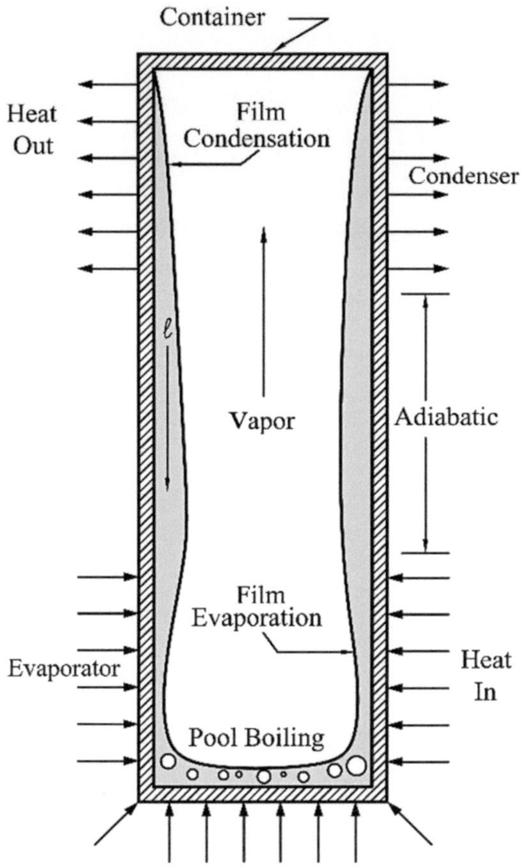

Fig. P8.14

8.15. Derive Eq. (8.114) from Eqs. (8.111)–(8.112), using the transformation of variables $(x, y) \rightarrow (\xi, \eta)$ where $\xi = x$ and $\eta = y/\delta(x)$.

8.16. Show that Eq. (8.110) can be transformed into Eq. (8.124) by using the variables defined in Eqs. (8.113), (8.116), and (8.122)–(8.123).

8.17. In a falling liquid film evaporation experiment, the temperature of the heating wall is 106 °C and the vapor pressure is at one atmosphere. The inlet mass flow rate of the liquid film per unit width is $\Gamma_0 = 2.24 \times 10^{-2}\,\text{kg/s m}$. Is the film flow laminar, laminar with waves, or turbulent? If the height of the vertical plate is 0.1 m, determine the average heat transfer coefficient. Please also check the overall energy balance by using Eq. (8.98).

8.18. A uniformly distributed liquid water film at 0.06 kg/s is introduced at the top of a tube with an inner diameter of 4 cm and height of 3 m. The inner surface temperature of the tube is uniformly 105 °C, and the saturation temperature of the steam is 100 °C. Is the liquid film at the inlet and outlet laminar, laminar with waves, or turbulent? Find the evaporation rate and the average heat transfer coefficient.

8.19. Falling film evaporation on a vertical wall is analyzed in great detail in Sect. 8.4. How would you apply the solutions of falling film evaporation on a vertical heated wall to predict heat transfer of falling film evaporation on an inclined heating wall? The inclination angle between the heating wall and the vertical direction is θ.

8.20. You are asked to select the most weight-efficient evaporant for a particular evaporator. The temperature at which the evaporator will operate has not been set. Go through Appendix B.26–B.48, and rank each working fluid based on its heat of vaporization ($h_{\ell v}$). Also, for each working fluid, note what the operating temperature of the evaporator would be if the operating pressure was 1 atm. Which evaporants stand out in the different temperature ranges?

8.21. Consider evaporation for a constant-temperature wall rotating disk with controlled liquid jet impingement at its center (see Fig. P8.21). Rotational forces cause the film to spread across the disk and then off the outer edge. The flow field is similar to a laminar, steady falling film, except that rotation replaces gravity as the driving force and the system must be solved in cylindrical coordinates. Obtain the heat transfer coefficient for the thin liquid film evaporation on the rotating disk by neglecting inertia and convective terms.

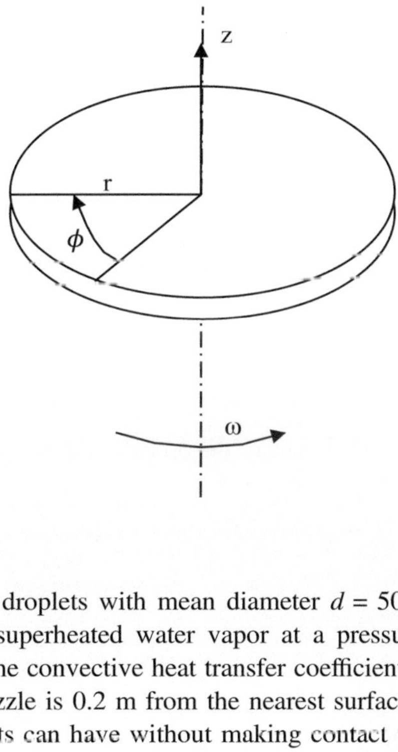

Fig. P8.21

8.22. A spray of liquid water droplets with mean diameter d = 50 μm at a temperature 373 K is injected into a tank of superheated water vapor at a pressure p = 1.013 × 10^5 Pa held at constant T_∞ = 423 K. The convective heat transfer coefficient on the surface of the droplet is 4000 W/m^2-K. If the nozzle is 0.2 m from the nearest surface of the tank, what is the maximum velocity the droplets can have without making contact with the wall?

8.23. The analysis presented in Sect. 8.5.1 assumed that the heat transfer coefficient on the surface of the liquid droplet was a constant. When a liquid droplet is spread into the hot gas, the following empirical correlation can be used to estimate the heat transfer coefficient: $Nu = 2 + 0.6 Re^{1/2} Pr^{1/3}$. Since the radius of the droplet is changing in the evaporation process, the heat transfer coefficient is not a constant. How will you modify Eq. (8.188) when heat transfer coefficient is not a constant?

8.24. Slurry droplets and hot air are brought into contact in a slurry-drying process. During the early stage, the slurry droplet can be considered as a porous core covered by pure water (see Fig. P8.24). Evaporation takes place on the surface of the slurry droplet, the temperature of which can be assumed to be uniform all the time. The first phase of evaporation is referred to as evaporation of the pure water around the porous core. Build an appropriate model to predict the transient radius and temperature of the slurry droplet during the first phase of evaporation.

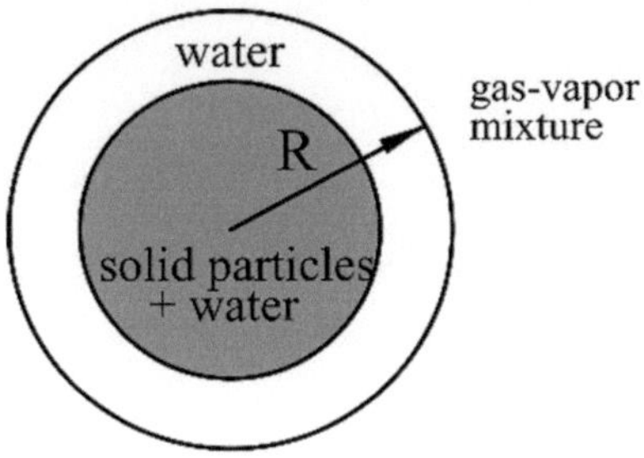

Fig. P8.24

8.25. A stationary liquid fuel droplet with an initial temperature of T_0 is introduced into hot gas at a temperature of T_∞. The initial pressures in the liquid droplet and the hot gas are p_0 and p_∞ respectively. The initial mass fraction of the fuel vapor in the hot gas is ω_∞. Evaporation takes place on the surface of the liquid fuel droplet, which is assumed to be in spherical shape throughout the process. Since the temperature of the fuel droplet is below the temperature of the hot gas, a laminar downward motion of hot gas is formed near the liquid fuel droplet. Write the governing conservation equations and corresponding boundary conditions for liquid and gas flow, and heat and mass transfer in both liquid fuel droplet and hot gas. The spherical coordinate system with origin located at the center of the liquid fuel droplet should be used.

References

Alhusseini, K. A., Tuzla, K., & Chen, J. C. (1998). Falling film evaporation of single component liquids. *Journal of Heat Transfer, 41,* 1623–1632.

Carey, V. P. (2016). *Liquid-vapor phase-change phenomena: An introduction to the thermophysics of vaporization and condensation processes in heat transfer equipment* (3rd ed.). New York, NY: Taylor & Francis.

Chun, K. R., & Seban, R. A. (1971). Heat transfer to evaporating liquid films. *Journal of Heat Transfer, 93,* 391–396.

Debbissi, C., Orfi, J., & Nasrallah, S. B. (2003). Evaporation of water by free or mixed convection into humid air and superheated steam. *International Journal of Heat and Mass Transfer, 46,* 4703–4715.

Dombrovsky, L. A., & Sazhin, S. S. (2003). A simplified non-isothermal model for droplet heating and evaporation. *International Communications in Heat and Mass Transfer, 30,* 787–796.

Elperin, T., Fominykh, A., & Krasovitov, B. (2005). Modeling of simultaneous gas absorption and evaporation of large droplet. In *Proceedings of 2005 International Mechanical Engineering Congress and Exposition.* Orlando, FL, November 5–11, 2005.

Faghri, A. (2016). *Heat pipe science & technology* (2nd ed.). Columbia, MO: Global Digital Press.

Faghri, A., & Seban, R. (1985). Heat transfer in wavy liquid films. *International Journal of Heat and Mass Transfer, 28,* 506–508.

Faghri, A., Zhang, Y., & Howell, J. R. (2010). *Advanced heat and mass transfer.* Columbia, MO: Global Digital Press.

Hirschburg, R. F., & Florschuetz, L. W. (1982). Laminar wavy film flow, parts I and II. *Journal of Heat Transfer, 104,* 452–464.
Kapitza, P. L., & Kapitza, K. P. (1975). Wavy flow of thin layers of viscous fluid. In *Collected papers of P. L. Kapitza* (Vol. 2, pp. 662–709). New York: Pergamon.
Kutateladze, S. S. (1963). *Fundamentals of heat transfer*. New York: Academic Press Inc.
Kutateladze, S. S. (1982). Semi-empirical theory of film condensation of pure vapors. *International Journal of Heat and Mass Transfer, 25,* 653–660.
Lock, G. S. H. (1994). *Latent heat transfer.* Oxford, UK: Oxford Science Publications, Oxford University.
Rahman, M. M., & Faghri, A. (1992). Analysis of heating and evaporation from a liquid film adjacent to a horizontal rotating disk. *International Journal of Heat and Mass Transfer, 35,* 2644–2655.
Rice, J., Faghri, A., & Cetegen, B. M. (2005). Analysis of a free surface flow for a controlled liquid impinging jet over a rotating disk including conjugate effects, with and without evaporation. *International Journal of Heat and Mass Transfer, 48,* 5192–5204.
Rogovan, I. A., Olevski, V. M., & Runova, N. G. (1969). Measurement of the parameters of film type wavy flow on a vertical plane. *Theoretical Foundations of Chemical Engineering, 3,* 164.
Seban, R. A., & Faghri, A. (1976). Evaporation and heating with turbulent falling liquid films. *Journal of Heat Transfer, 98,* 315–318.
Stephan, K. (1992). *Heat transfer in condensation and boiling*. New York: Springer.
Yan, W. M., & Lin, T. F. (1991). Evaporative cooling of a liquid film through interfacial heat and mass transfer in a vertical channel—II. Numerical study. *International Journal of Heat and Mass Transfer, 34,* 1113–1124.
Yan, W. M., Lin, T. F., & Tsay, Y. L. (1991). Evaporative cooling of a liquid film through interfacial heat and mass transfer in a vertical channel—I. Experimental study. *International Journal of Heat and Mass Transfer, 34,* 1105–1111.

9 Boiling

9.1 Introduction

When a solid surface is immersed in a liquid and the solid surface temperature, T_w, exceeds the saturation temperature, $T_{sat}(p_\ell)$, for the liquid at that pressure, vapor bubbles can form, grow, and detach from the solid surface. This phase-change process is called *boiling* or more precisely *heterogeneous boiling*, and the energy transport involved is classified as convective heat transfer with phase change. Two types of boiling are commonly distinguished: *pool boiling* and *flow* or *forced convective boiling*. In the former case, the bulk liquid is quiescent while the liquid near the heating surface moves due to free convection and the mixing induced by bubble growth and detachment. In the latter case, bulk liquid motion driven by some external means is superimposed on the motion that also occurs in pool boiling. Vapor bubbles within the liquid phase are the primary visual characteristics that distinguish boiling from evaporation. The influence of bubbles is also the primary cause of differences in the thermodynamic and hydrodynamic analyses of these two phase-change modes. Figure 9.1 compares, in schematic fashion, the temperature profile of a quiescent volume of fluid experiencing evaporation by natural convection with that of one experiencing boiling. As the figure indicates, boiling requires a much larger temperature difference between the bulk liquid and the heating surface than evaporation does. A second distinction that arises from the figure is that the thermal boundary layer is less precise in the case of boiling; this may be attributed to the mixing action caused by the release of bubbles from the heating surface.

In addition to the pool and forced convection classifications, boiling can also be categorized according to the initial temperature of the liquid. Figure 9.2 shows pool boiling for different initial liquid temperatures. *Subcooled boiling* occurs if the liquid temperature starts below the saturation temperature. In this case, bubbles formed at the heater surface experience condensation as they rise through the cooler bulk liquid and may collapse before reaching the free surface. Saturated boiling, on the other hand, occurs if the temperature of the liquid equals the saturation temperature. In this case, bubbles form at the heating surface, travel intact through the liquid, and escape at the free surface.

Section 9.2 introduces the pool boiling curve and the various regimes (free convection, nucleate boiling, transition boiling, and film boiling) of which it is composed. After this introduction, the four regimes and the characteristic points of the curve are discussed in detail. Section 9.3 discusses nucleate boiling, including nucleation, bubble dynamics and detachment, nucleation side density, numerical simulation of bubble growth and merger, and heat transfer analysis. The critical heat flux is presented in Sect. 9.4, followed by discussions on transition boiling and minimum heat flux in Sect. 9.5. This chapter is closed by the discussion of film boiling (Sect. 9.6), including film boiling

A. Faghri and Y. Zhang, *Fundamentals of Multiphase Heat Transfer and Flow*, https://doi.org/10.1007/978-3-030-22137-9_9

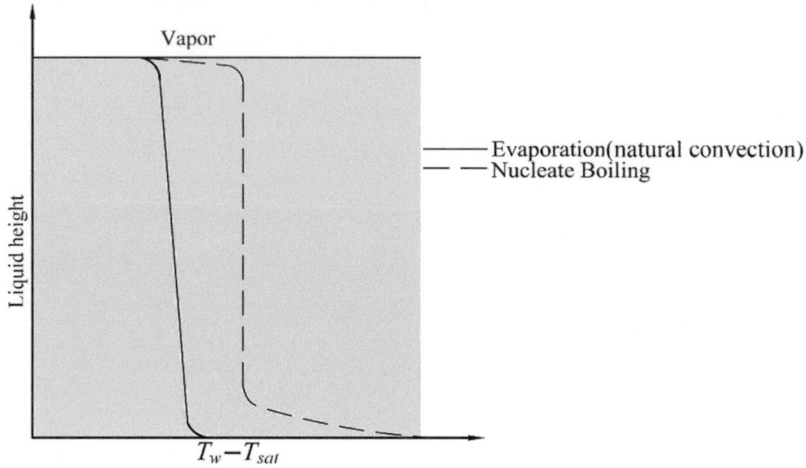

Fig. 9.1 Schematic comparison of evaporation and nucleate boiling temperature profiles

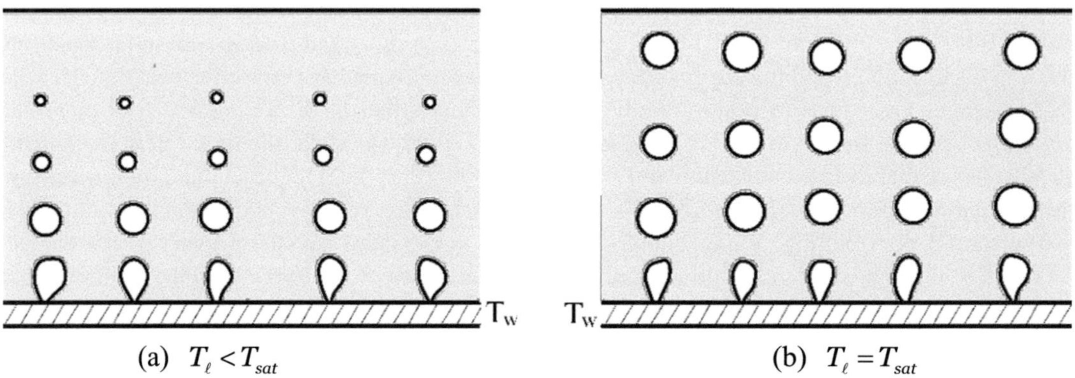

Fig. 9.2 Effects of liquid temperature: **a** subcooled boiling, **b** saturated boiling (□ vapor, ■ liquid)

laminar boundary-layer analysis and correlations, direct numerical simulations, and Leidenfrost phenomena. Forced convection boiling in both macro- and microtubes, which involves liquid–vapor two-phase flow with transitions between several characteristic flow patterns, is addressed in Chap. 10 under the heading of Two-Phase Flow. Boiling in porous media is discussed in Chap. 12.

9.2 Pool Boiling Regimes

The classical pool boiling curve is a plot of heat flux, q'', versus excess temperature, $\Delta T = T_w - T_{sat}$. As the value of the excess temperature increases, the curve traverses four different regimes: (1) natural or free convection, (2) nucleate boiling, (3) transition boiling, and (4) film boiling. Different experimental methods may be used to define the pool boiling curve; constant-temperature control and constant heat flux control are the two most commonly cited.

A typical boiling curve for saturated pool boiling of water heated by a platinum wire at atmospheric pressure for a temperature-controlled environment is shown in Fig. 9.3. When the excess temperature ΔT is less than 5 °C, no bubbles form. Instead, heat is transferred from the solid surface

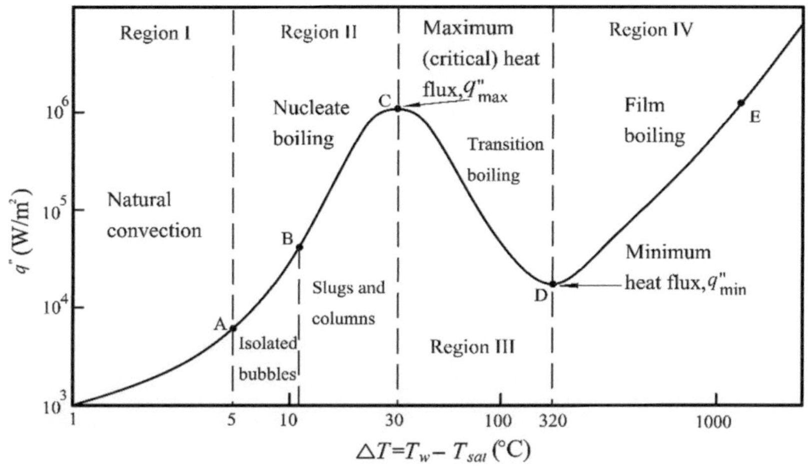

Fig. 9.3 Pool boiling curve for saturated water in a temperature-controlled environment

to the bulk liquid via natural convection. Heat transfer coefficients in this regime can be calculated using the semi-empirical correlations for natural convection.

When the excess temperature increases beyond 5 °C, the system enters the nucleate boiling regime–point A on Fig. 9.3. Vapor bubbles are generated at certain preferred locations on the heater surface called *nucleation sites*; these are often microscopic cavities or cracks on the solid surface. Nucleation occurs repeatedly from the same sites, indicating a causal link between bubble formation and some surface feature as well as the cyclical nature of the bubble-forming process. Small cavities and surface cracks act as sites for bubble generation because: (1) The contact area between the liquid and heating surface increases relative to a perfectly smooth surface, so liquid trapped in these areas vaporizes first; and (2) The presence of trapped gases in such cracks creates liquid–vapor interfaces, which serve as sites where transfer of energy in the form of latent heat from the liquid to the vapor phase takes place. Once a vapor bubble has been initiated at a nucleation site, under the right conditions the bubble grows to a certain required diameter, detaches from the heating surface, and rises to the liquid free surface. If the excess temperature remains at the low end of the nucleate boiling regime, shown between points A and B of Fig. 9.3, each bubble generated can grow and detach from the surface independently–that is, without interaction between bubbles. As the bubble-generating process occurs at the active nucleation sites, the surface area between these sites retains the liquid–solid contact that characterizes the natural convection regime. Convection remains the primary mechanism of heat transfer in this so-called "isolated bubble" regime. As we shall see, however, the character of this convection is markedly different from that of the natural convection encountered at lower excess temperatures ($\Delta T < 5$ °C for water).

As the excess temperature increases beyond point B in Fig. 9.3, additional nucleation sites become active and more bubbles are generated. The higher density of bubbles leads to their interaction with each other. Bubbles from separate sites now merge to form columns and slugs of vapor, thus decreasing the overall contact area between the heating surface and the saturated liquid. Consequently, the slope of the boiling curve begins to decrease and the heat flux eventually reaches a maximum value, q''_{max}, referred to as the critical heat flux. The critical heat flux, which marks the upper limit of the nucleate boiling regime, reaches a value of approximately 10^6 W/m^2 for water at an excess temperature of about $\Delta T_c = 30$ °C. The nucleate boiling regime is most desirable for many industrial applications because of its high heat flux at relatively low levels of excess temperature (5 °C $\leq \Delta T \leq$ 30 °C for water). However, under certain circumstances, there is a need to avoid nucleate boiling, such as wicked heat pipes (Faghri 2016).

As the temperature increases beyond the critical heat flux point, the rate of bubble generation exceeds the rate of bubble detachment from the heater surface. Bubbles from an increasing number of sites merge to form continuous vapor films over portions of the surface, further decreasing the contact area between the heating surface and the saturated liquid. These vapor films are not stable, however, they can detach from the surface, leading to restoration of contact with the liquid and resumption of nucleate boiling. Under these unstable conditions, the surface temperature may fluctuate rapidly, so the excess temperature shown on the horizontal coordinate of Fig. 9.3 between points C and D should be regarded as an average value. Since the boiling in this regime combines unstable film with partial nucleate boiling types, it is referred to as the region of transition boiling. When the excess temperature becomes high enough to sustain a stable vapor film, the heat flux reaches its minimum value, q''_{min}. This point, known as the Leidenfrost temperature, marks the upper temperature limit of the transition boiling regime.

At temperatures above the Leidenfrost temperature, the bulk liquid and the heating surface are completely separated by a stable vapor film, so boiling in this regime is known as film boiling. The phase change in film boiling occurs at a liquid–vapor interface, instead of directly on the surface, as in the case of nucleate boiling. Thermal energy from the heating surface reaches the liquid–vapor interface by convection in the vapor film as well as by direct radiation to the interface. In the film boiling regime, the surface heat flux becomes a monotonically increasing function of the excess temperature, because radiation heat transfer from the solid surface to the liquid plays a significant role at high surface temperature. Pool boiling continues in this regime until the surface temperature reaches the maximum allowable temperature of the heating surface (2042 K for platinum, for example). Beyond that point, the heating surface can melt in a potentially catastrophic failure. If a protective insulation is provided, however, as in the case of refractory metals, for example, it is possible for film boiling heat flux to exceed the critical heat flux, q''_{max}.

The boiling curve presented in Fig. 9.3 and described above assumes that the surface temperature is independently controlled and that the heat flux is the dependent variable (temperature-controlled environment). However, direct control over the surface temperature is not always possible. For example, when electric heating provides thermal energy to the solid, the controllable parameter is heat flux. Surface temperature then becomes the dependent variable, and heat flux becomes the independent variable. If the experiment of gradually increasing the added thermal energy is repeated using constant heat flux instead of constant-temperature control, the resulting boiling curve matches that of controlled temperature up to the critical heat flux, q''_{max} When the surface heat flux is increased slightly above the critical heat flux, however, portion C-D-E of the boiling curve is bypassed and the surface temperature increases abruptly to point E of the stable film boiling regime (Nukiyama 1934). This abrupt increase in surface temperature is undesirable because T_E usually exceeds the melting point of the solid material. The critical heat flux can thus serve as the warning point above which burnout can occur; consequently, the critical point is sometimes referred to as the *burnout point*. If the pool boiling curve is defined by decreasing the controlled heat flux from an initial point in the film boiling regime (point E, for example), a characteristic identical to Fig. 9.3 appears down to the point of minimum heat flux. After that point, a continued decrease of q'' yields a second hysteresis path that leads immediately to the nucleate boiling regime between points A and B. In this case, the transition boiling regime and a portion of the nucleate boiling regime are bypassed. Therefore, the transition regime that is observed when temperature control is used to define the pool boiling curve is unavailable in the controlled heat flux case.

The characteristics of the boiling curve for forced-convection surface boiling are examined in detail first by Bergles and Rohsenow (1964). In the region of low wall superheat, the heat transfer can be predicted by available correlations for forced convection. Bergles and Rohsenow (1964) indicated that the curves for forced-convection surface boiling cannot be based on data for saturated pool boiling but must rather be based on actual forced-convection data.

9.3 Nucleate Boiling

As was indicated in the previous section, nucleate boiling is the most desirable regime in the majority industrial applications (with few exceptions) because relatively high rates of heat transfer can be achieved with relatively low-temperature differentials. Because vapor bubbles exert a major influence on the hydrodynamic and heat transfer characteristics of the nucleate boiling regime, a thorough understanding of how bubbles form and how they influence heat transfer is essential. Three stages of bubble production, each of which will be discussed in the following subsections, are defined as follows: (1) initiation or nucleation, (2) growth, and (3) detachment. The section then ends with a summary of modeling approaches that have been developed to describe heat transfer in nucleate boiling.

9.3.1 Nucleation and Inception

The nucleation process is commonly divided in two categories: heterogeneous nucleation and homogeneous nucleation. Heterogeneous nucleation, or bubble initiation, in typical industrial applications is characterized by the cyclic formation of vapor bubbles at preferred sites on a heated wall. Any surface, regardless of how highly polished, contains surface irregularities–microcracks, defects, cavities, pits, scratches, or boundaries between solid crystals–that can trap small gas pockets. These serve as nucleation sites for vapor bubbles. The trapped gas and/or vapor, known as the bubble embryo, grows by acquiring mass via evaporation from the nearby liquid. It grows until it reaches a critical size, at which point the forces tending to cause separation of bubbles from the heating surface overcome the adhesive forces. When the force imbalance favors the separating forces, the bubble is released from the surface. This process, by which solid surface imperfections with entrapped gases promote the formation of bubble embryos, is known as heterogeneous nucleation.

Pure liquids that have been thoroughly degassed may still experience bubble formation at molecular vapor clusters within the bulk liquid, (i.e., away from any solid surface) by a process known as homogeneous nucleation. However, the degree of superheat required for homogeneous nucleation is substantially larger than that for heterogeneous nucleation. The common observation of bubble formation at the low superheat indicated by point A of Fig. 9.3, coupled with the cyclic formation of bubbles at selected points on heater surfaces, confirms that heterogeneous nucleation is by far the more common process.

Since the trapped gases and/or vapors are central to the nucleation event, a consideration of the thermodynamics of vapor bubbles immersed in a liquid is appropriate. We will find that it justifies the expectation that the presence of gas or vapor pockets enhances bubble formation.

We begin by considering a vapor bubble, including the possibility of noncondensable gas, immersed in its own liquid phase. Recall from Chap. 2 (see Sect. 2.4.3) that Laplace–Young equation must hold in order for the vapor/gas and liquid to be in equilibrium:

$$p_g + p_v - p_\ell = \frac{2\sigma}{R_b} \tag{9.1}$$

where R_b is the radius of the vapor bubble. Equation (9.1) indicates that the pressure of the vapor and gas must be greater than the pressure of the liquid to maintain a balance of forces in the vapor-interface-liquid system. The physical implications of this imbalance of pressures across the boundary separating vapor and liquid phases can be visualized at a molecular level. The higher vapor pressure in the bubble causes an increased number of molecules to strike the interface where they are absorbed by the liquid phase. To maintain the mass balance that equilibrium requires, there must be a corresponding increase in the number of molecules emitted through the interface from the liquid. Under the given constraints, this can be accomplished only by increasing the temperature of the liquid side of the interface. Consequently, the liquid near the bubble's surface must be superheated to a temperature above the saturation temperature that corresponds to the prevailing bulk liquid pressure. To put these qualitative observations on a quantitative basis, consider that the vapor pressure for a curved bubble interface can be expressed as (Sect. 2.4.3)

$$p_v = p_{v,sat}(T) \exp\left[\frac{-2\sigma\rho_v}{p_{v,sat}(T) R_b \rho_\ell}\right] \tag{9.2}$$

where $p_{v,sat}(T)$ is the normal saturation pressure corresponding to temperature, T. Since $2\sigma\rho_v/[p_{v,sat}(T)R_b\rho_\ell] \ll 1$, Eq. (9.2) can be approximated as

$$p_v \doteq p_{v,sat}(T)\left[1 - \frac{2\sigma\rho_v}{p_{v,sat}(T) R_b \rho_\ell}\right] \tag{9.3}$$

Combining Eqs. (9.1) and (9.3) yields

$$p_g + p_{v,sat}(T) - p_\ell \doteq \frac{2\sigma}{R_b}\left(1 + \frac{\rho_v}{\rho_\ell}\right) \tag{9.4}$$

Recalling the fundamental relationship between pressure and temperature differences in the two-phase region, the Clapeyron equation,

$$\frac{dp}{dT} = \frac{h_{\ell v}}{T(1/\rho_v - 1/\rho_\ell)} \tag{9.5}$$

and assuming that the fluid obeys the ideal gas law, i.e.,

$$p = \rho R_g T \tag{9.6}$$

and that $\rho_\ell \gg \rho_v$, we arrive at

$$\frac{dp}{p} = \frac{h_{\ell v}}{R_g T^2} dT \tag{9.7}$$

For a pressure change from p_ℓ to $p_{v,sat}(T)$, the corresponding saturation temperature changes from T_{sat} to T. So by integrating Eq. (9.7) between these limits, we obtain

$$\ln\frac{p_{\mathrm{v,sat}}(T)}{p_\ell}=\frac{h_{\ell\mathrm{v}}}{R_{\mathrm{g}}T_{\mathrm{sat}}}\frac{T-T_{\mathrm{sat}}}{T} \tag{9.8}$$

Substituting Eq. (9.4) into Eq. (9.8) gives

$$\Delta T=T-T_{\mathrm{sat}}=\frac{R_{\mathrm{g}}T_{\mathrm{sat}}T}{h_{\ell\mathrm{v}}}\ln\left[1+\frac{2\sigma}{p_\ell R_{\mathrm{b}}}\left(1+\frac{\rho_{\mathrm{v}}}{\rho_\ell}\right)-\frac{p_{\mathrm{g}}}{p_\ell}\right] \tag{9.9}$$

Considering that $2\sigma/(p_\ell R_{\mathrm{b}})-p_{\mathrm{g}}/p_\ell\ll 1$ and again that $\rho_\ell\gg\rho_{\mathrm{v}}$, Eq. (9.9) can be further simplified as

$$\Delta T=T-T_{\mathrm{sat}}=\frac{R_{\mathrm{g}}T_{\mathrm{sat}}T}{p_\ell h_{\ell\mathrm{v}}}\left(\frac{2\sigma}{R_{\mathrm{b}}}-p_{\mathrm{g}}\right) \tag{9.10}$$

For the case with no noncondensable gas $(p_{\mathrm{g}}=0)$, Eq. (9.10) can be rearranged to yield the critical vapor bubble radius:

$$R_{\mathrm{b}}=\frac{2\sigma T_{\mathrm{sat}}}{h_{\ell\mathrm{v}}\rho_{\mathrm{v}}\Delta T} \tag{9.11}$$

This critical radius is necessary for a bubble to exist at ΔT above the saturation temperature that corresponds to the prevailing vapor pressure. Any bubble with a radius less than the size given by Eq. (9.11) will collapse, but a bubble with a radius equal to R_{b} will grow in a spontaneous fashion. It should be noted from Eq. (9.10) that the size of the equilibrium vapor nucleus becomes smaller as the superheat (ΔT) increases. This identifies the basic mechanism for the increase of q'' with rising excess temperature between points A and C of Fig. 9.3, namely, the increased bubble density that results from the activation of more and more nucleation sites at ever-smaller surface imperfections.

Equation (9.10) shows that the presence of noncondensable gas reduces the superheat required to generate bubbles. The subject of superheat was presented in Sect. 2.5.4. The bubble growth is spontaneous if the excess temperature is greater than the value given in Eq. (9.10). Any real surface always has some small cavities with very small inclined angles where liquid can only partially fill the cavities, so opportunities for the formation of gas or vapor pockets are common. Equation (9.10) gives the superheat required to initiate nucleate boiling from a pre-existing nucleus with a radius of R_{b}. Griffith and Wallis (1960) suggested that it is also applicable for nucleate boiling from a surface with a microcavity whose mouth radius is R_{b}.

Alternative studies by Mizukami (1975), Nishio (1985), and Wang and Dhir (1993a) suggested that the superheat required to initiate nucleate boiling is determined by the instability of vapor nuclei in the cavity. If the curvature of the liquid–vapor interface, K, increases with increasing vapor volume ($\mathrm{d}K/\mathrm{d}V^*>0$, where V^* is dimensionless vapor volume), the vapor nucleus is stable. On the other hand, the vapor bubble embryo or nucleus is unstable if K decreases with increasing vapor volume. For the nucleate boiling to initiate from the wall, the wall superheat must be sufficient to ensure the vapor nuclei are unstable.

Figure 9.4 shows the dependence of the nondimensional modified curvature of the interface on the dimensionless volume of the vapor nucleus in a spherical cavity with a mouth angle $\theta_{\mathrm{m}}=30°$. The noncondensable gas is not present and the results at different contact angles are shown. The dimensionless volume of the vapor nucleus is defined as $V^*=V_{\mathrm{b}}/V_{\mathrm{c}}$ where V_{b} is the volume of the nucleus and V_{c} is the volume of the nucleus at point A, where the curvature is zero. It can be seen from Fig. 9.4 that the nucleus is stable between points A and D, since $\mathrm{d}K/\mathrm{d}V^*>0$. When the contact

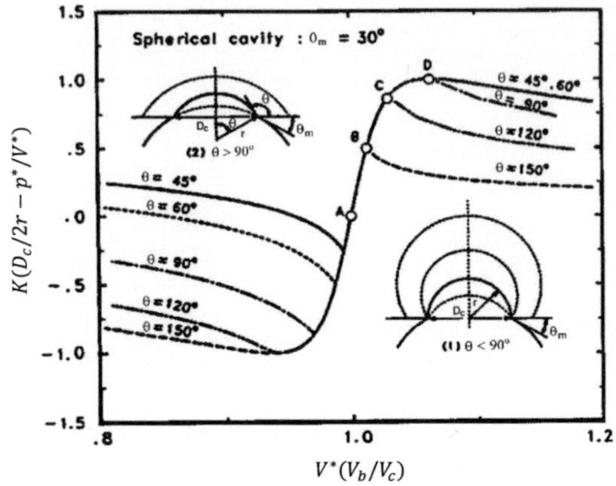

Fig. 9.4 Dimensionless modified curvature versus dimensionless volume of vapor bubble nucleus in a spherical cavity (Wang and Dhir 1993a)

angle is less than 90°, the maximum curvature, K_{max}, is 1. For the cases in which $\theta > 90°$, the maximum curvature becomes $\sin\theta$. Nucleation occurs beyond the point where the curvature reaches its maximum; the corresponding superheat is (Wang and Dhir 1993a)

$$\Delta T = T - T_{sat} = \frac{4\sigma T_{sat}}{p_v h_{\ell v} D_c} K_{max} \tag{9.12}$$

where D_c is the diameter of the cavity mouth.

Example 9.1

A steam bubble with a radius of 5 μm is surrounded by liquid water at 120 °C. Will this bubble grow or collapse?

Solution The liquid pressure is $p_\ell = 1$ atm $= 1.013 \times 10^5$ Pa. The saturation temperature at this pressure is $T_{sat} = 100\,°C = 373.15$ K. The properties of water at this temperature are $\sigma = 58.9 \times 10^{-3}$ N/m, $h_{\ell v} = 2251.2$ kJ/kg, and $\rho_v = 0.5974$ kg/m^3. Therefore, the critical bubble radius can be determined from Eq. (9.11), i.e.,

$$\begin{aligned} R_b &= \frac{2\sigma T_{sat}}{h_{\ell v}\rho_v \Delta T} \\ &= \frac{2 \times 58.9 \times 10^{-3} \times 373.15}{2251.2 \times 10^3 \times 0.5974 \times (120 - 100)} \\ &= 1.63 \times 10^{-6}\ \text{m} = 1.63\ \mu\text{m} \end{aligned}$$

Since the radius of the bubble, 5 μm, is greater than the critical bubble radius, the bubble will grow.

Figure 9.5 shows an idealized model of a surface crevice containing a vapor pocket as it expands under the influence of evaporation. The inner surface of the cone is wettable and the contact angle of the liquid is greater than the cone angle ϕ ($\phi < \theta \leq 90°$). The vapor–liquid interface is idealized as spherical in shape. As the vapor pocket expands upward in the surface crevice, its interface takes a

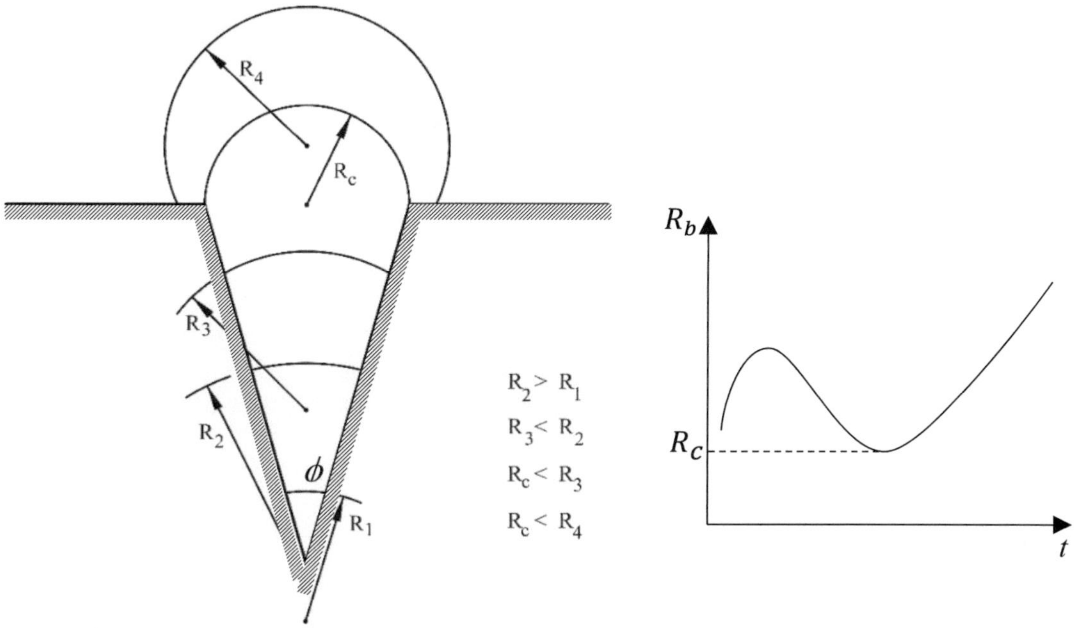

Fig. 9.5 Variation of bubble radius as the bubble grows within and emerges from an idealized cavity

concave shape and the radius of curvature, R_b, increases first $(R_2 > R_1)$ as the interface moves toward the lip of the cavity. Once the incipient bubble reaches the lip of the interface, the radius of curvature begins to decrease $(R_3 < R_2)$. It reaches its minimum value (approximately) as the liquid–vapor interface becomes a hemisphere with a radius equal to that of the cavity mouth, R_c. Beyond this point, any further evaporation pushes the liquid–vapor interface out of the surface cavity, and the interface radius of curvature begins to increase again $(R_4 < R_c)$. Consideration of this sequence of radius variation, along with Eq. (9.10), indicates that the superheat required to form a vapor bubble from this cavity can be determined by substituting $R_b = R_c$ into Eq. (9.10). In normal circumstances, of course, this variation deviates from the idealized model in Fig. 9.5 and is a function of cavity angle ϕ and contact angle θ. A liquid-filled cavity does not promote nucleation.

Bankoff (1958) proposed that an advancing liquid front will fill the upper portion of the cavity while trapping gas under it, if the contact angle θ is greater than the wedge angle ϕ. However, an advancing gas or vapor front will not be able to displace the liquid if $\theta > \pi - \phi$. Wang and Dhir (1993a) proposed a gas/vapor entrapment criterion by minimizing Helmholtz free energy of a system including the liquid–gas interface in a cavity.

In reality, the shape of a vapor bubble is not hemispherical, and there is a microlayer (a very thin liquid layer, see Fig. 9.6) that is responsible for more than half of heat transfer in some cases. Plesset and Sadhal (1979) provided the average liquid film thickness $\bar{\delta}$ in terms of bubble lift time t_0 and liquid kinematic viscosity by the following equation

$$\bar{\delta} = \frac{8}{7}(3\nu_\ell t_0)^{1/2} \tag{9.13}$$

The contribution of the microlayer to the overall heat transfer obviously depends on the fraction of area occupied by the microlayer. Utaka et al. (2013) investigated the structure of the microlayer for nucleate pool boiling of water and ethanol using a laser extinction method. They suggested that the

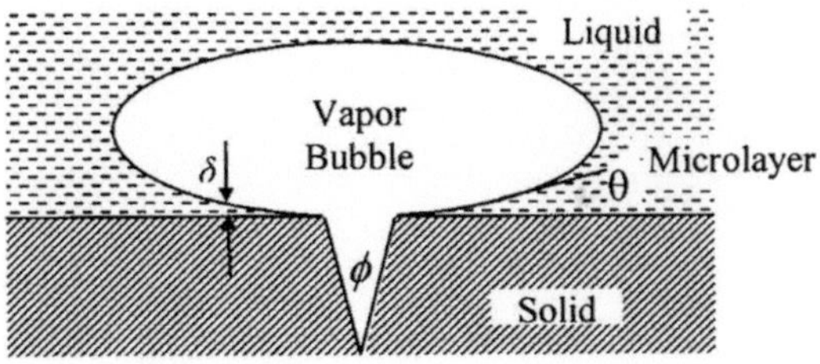

Fig. 9.6 Liquid microlayer under a vapor bubble at a nucleation site

thickness of initially formed microlayer increased linearly with increasing distance between the incipient bubble site and the measurement position.

9.3.2 Bubble Dynamics

The second stage of bubble production is the growth stage. In real applications, the conditions under which growth occurs can be complex. The temperature field of the liquid near the heating surface must be nonuniform for heat to be transferred within the liquid. Several forces influence bubble growth, including the inertia of the surrounding liquid, shear forces at the interface, surface tension, and the pressure difference between the vapor and the liquid. The energy exchanged includes heat transfer by conduction and convection as well as latent heat of vaporization. The process is by definition transient, the terms in the mass, momentum, and energy equations constantly shifting over time. So we cannot expect that modeling the process of bubble growth will yield simple solutions.

The growth of a bubble can be classified into two groups (both of them will be discussed below): one in which the bubble grows in a quiescent superheated liquid of infinite extent (homogeneous); another in which the bubble is attached and grows on a heated wall (heterogeneous). Finally, the dynamics of bubble growth within a superheated liquid droplet will also be discussed in this subsection.

9.3.2.1 Homogeneous Bubble Growth

Bubble growth within a superheated liquid will be considered first. Once a growing vapor bubble's radius reaches the critical vapor bubble radius, it grows spontaneously. During the early stage, when the bubble radius is small, the Laplace–Young equation indicates that the pressure differential across the interface is at its maximum value. The resulting high interfacial velocity leads to significant inertia terms in the momentum equation. Meanwhile, the temperature of the interface is close to the superheat temperature of the surrounding liquid, so heat transfer into the vapor bubble experiences the highest driving temperature differential that occurs at any time during the process. As a result, early bubble growth tends to be limited by inertia, or the exchange of momentum between the vapor and liquid phases.

In the physical model shown in Fig. 9.7, the instantaneous radius and the liquid–vapor interface velocity at time t are $R(t)$ and dR/dt, respectively. For an incompressible, radially symmetric inviscid flow, the continuity equation in the spherical coordinate system is

$$\frac{1}{r^2}\frac{\partial(r^2 u)}{\partial r} = 0 \tag{9.14}$$

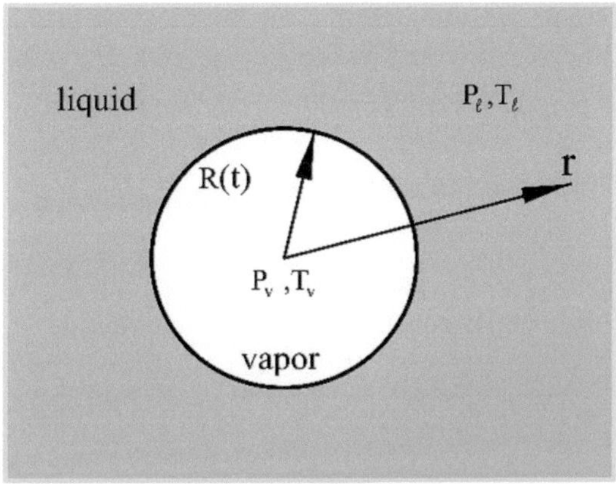

Fig. 9.7 Homogeneous nucleation of a vapor bubble in superheated liquid

where u is the radial velocity in the liquid phase. Assuming $\rho_v \ll \rho_\ell$, this velocity at the liquid–vapor interface is equal to the growth rate, i.e.,

$$u(R,t) = \frac{dR}{dt} \tag{9.15}$$

Integrating Eq. (9.14) over the interval of (R, r), where r is an arbitrary location in the liquid phase, one obtains

$$u(r,t) = \frac{dR}{dt}\left(\frac{R}{r}\right)^2 \tag{9.16}$$

The momentum equation for the surrounding liquid is

$$\rho_\ell\left(\frac{\partial u}{\partial t} + u\frac{\partial u}{\partial r}\right) = -\frac{\partial p}{\partial r} \tag{9.17}$$

Substituting Eq. (9.16) into Eq. (9.17) yields

$$\frac{\rho_\ell}{r^2}\left[2R\left(\frac{dR}{dt}\right)^2 + R^2\frac{d^2R}{dt^2}\right] - \frac{2\rho_\ell R^4}{r^5}\left(\frac{dR}{dt}\right)^2 = -\frac{\partial p}{\partial r} \tag{9.18}$$

Integrating Eq. (9.18) from the bubble surface ($r = R$) to infinity ($r \to \infty$) yields

$$R\frac{d^2R}{dt^2} + \frac{3}{2}\left(\frac{dR}{dt}\right)^2 = \frac{1}{\rho_\ell}[p_\ell(R) - p_\ell(\infty)] \tag{9.19}$$

Substituting Eq. (9.1) with $p_g = 0$ into Eq. (9.19) gives

$$R\frac{d^2R}{dt^2} + \frac{3}{2}\left(\frac{dR}{dt}\right)^2 = \frac{1}{\rho_\ell}\left[p_v - p_\ell(\infty) - \frac{2\sigma}{R}\right] \tag{9.20}$$

which is referred to as Rayleigh's equation. During the inertia-controlled stage of bubble growth, the surface tension term $2\sigma/R$ is negligible compared to $p_v - p_\ell(\infty)$. The pressure difference $p_v - p_\ell(\infty)$ can be evaluated by using the Clausius–Clapeyron equation, i.e.,

$$p_v - p_\ell(\infty) = \frac{\rho_v h_{\ell v}[T_\infty - T_{sat}[p_\ell(\infty)]]}{T_{sat}[p_\ell(\infty)]} \tag{9.21}$$

Substituting Eq. (9.21) into Eq. (9.20) and neglecting the surface tension term $2\sigma/R$ in Eq. (9.20), one can obtain an ordinary differential equation for the bubble radius:

$$R\frac{d^2R}{dt^2} + \frac{3}{2}\left(\frac{dR}{dt}\right)^2 = \frac{\rho_v h_{\ell v}}{\rho_\ell}\frac{T_\infty - T_{sat}[p_\ell(\infty)]}{T_{sat}[p_\ell(\infty)]} \tag{9.22}$$

which is subject to the following initial condition:

$$R = 0, \quad t = 0 \tag{9.23}$$

The instantaneous bubble radius can be obtained by solving Eq. (9.22):

$$R(t) = \left\{\frac{2}{3}\frac{\rho_v h_{\ell v}}{\rho_\ell}\frac{T_\infty - T_{sat}[p_\ell(\infty)]}{T_{sat}[p_\ell(\infty)]}\right\}^{\frac{1}{2}} t \tag{9.24}$$

which indicates that the radius of the bubble is a linear function of time during the early stage of the bubble's growth.

During the later stage of bubble growth, the inertial force becomes insignificant as the pressure difference $p_v - p_\ell$ (remember Laplace–Young's equation) diminishes and the interfacial motion slows. At the same time, the temperature difference between the vapor inside the bubble and the interface diminishes. With a lower level of interfacial force, heat transfer becomes the process that limits bubble growth and the energy equation for the liquid phase is:

$$\frac{\partial T_\ell}{\partial t} + u\frac{\partial T_\ell}{\partial r} = \frac{\alpha_\ell}{r^2}\frac{\partial}{\partial r}\left(r^2\frac{\partial T_\ell}{\partial r}\right) \tag{9.25}$$

where the radial velocity, u, is obtained from Eq. (9.16). The initial and the boundary conditions for Eq. (9.25) are

$$T(r,0) = T_\ell(\infty),\ T(R,t) = T_{sat}(p_v),\ T(\infty,t) = T_\ell(\infty) \tag{9.26}$$

The energy balance at the liquid–vapor interface is

$$k_\ell \left.\frac{\partial T}{\partial r}\right|_{r=R} = \rho_v h_{\ell v}\frac{dR}{dt} \tag{9.27}$$

The instantaneous bubble radius is

$$R(t) = 2C_R\sqrt{\alpha_\ell t} \tag{9.28}$$

where C_R is a constant.

For a large Jakob number (corresponding to low pressure), C_{R} can be obtained by (Plesset and Zwick 1954)

$$C_{\mathrm{R}} = \sqrt{\frac{3}{\pi} Ja} \tag{9.29}$$

and the Jakob number is defined as

$$Ja = \frac{\rho_\ell c_{p\ell} [T_\ell(\infty) - T_{\mathrm{sat}}(p_{\mathrm{v}})]}{\rho_{\mathrm{v}} h_{\ell\mathrm{v}}} \tag{9.30}$$

For a small Jakob number (corresponding to high pressure)

$$C_{\mathrm{R}} = \sqrt{\frac{Ja}{2}} \tag{9.31}$$

Equation (9.28) indicates that the bubble radius is proportional to the square root of time.

Bubble growth in the two limiting cases described above has been analyzed. The inertia-controlled bubble growth model ignored heat transfer, which is only valid in the early stage of bubble growth. In the heat transfer-controlled bubble growth model, on the other hand, the effect of inertia on bubble growth was neglected; this assumption is appropriate only during the late stage of bubble growth. A comprehensive bubble growth model must include both inertia and heat transfer and provide a smooth transition between the two regimes. For bubble growth in an initially uniformly superheated liquid, Mikic et al. (1970) obtained the following expression of the relationship between bubble radius and time:

$$R^+ = \frac{2}{3}\left[(t^+ + 1)^{\frac{3}{2}} - (t^+)^{\frac{3}{2}} - 1\right] \tag{9.32}$$

where R^+ and t^+ are nondimensional radius and time, respectively, defined as

$$R^+ = \frac{RA}{B^2} \qquad t^+ = \frac{tA^2}{B^2} \tag{9.33}$$

and

$$A = \left\{\frac{b[T_\ell(\infty) - T_{\mathrm{sat}}(p_\ell(\infty))] h_{\ell\mathrm{v}} \rho_{\mathrm{v}}}{\rho_\ell T_{\mathrm{sat}}(p_\ell(\infty))}\right\}^{\frac{1}{2}} \tag{9.34}$$

$$B = \left(\frac{12\alpha_\ell}{\pi}\right)^{\frac{1}{2}} Ja \tag{9.35}$$

$$Ja = \frac{\rho_\ell c_{p\ell} [T_\ell(\infty) - T_{\mathrm{sat}}(p_\ell(\infty))]}{\rho_{\mathrm{v}} h_{\ell\mathrm{v}}} \tag{9.36}$$

where b in Eq. (9.34) is a geometric parameter with a value of 2/3 for a perfect spherical bubble. Equation (9.32) provides a generalized solution for the bubble radius because it can be reduced to

Eq. (9.24) for small t^+ and to Eq. (9.28) for larger t^+. Prosperetti and Plesset (1978) developed a more advanced model for vapor bubble growth in a superheated liquid.

Example 9.2

A vapor bubble is initiated and grows in liquid water at 120 °C. Find the time at which the bubble sizes predicted by the inertia-controlled model and the heat transfer-controlled model are the same. What is the bubble size obtained by both models?

Solution The saturation temperature at 1 atm is $T_{sat} = 100\,°\text{C} = 373.15$ K. The properties of water at this temperature are $\sigma = 58.9 \times 10^{-3}$ N/m, $h_{\ell v} = 2251.2$ kJ/kg, $\rho_\ell = 958.77$ kg/m^3, $\rho_v = 0.5974$ kg/m^3, $c_{p\ell} = 4.216$ kJ/kg K, and $k_\ell = 0.68$ W/m K. The thermal diffusivity is $\alpha_\ell = k_\ell/(\rho_\ell c_{p\ell}) = 1.68 \times 10^{-7}$ m^2/s.

The Jakob number is obtained from Eq. (9.30), i.e.,

$$Ja = \frac{\rho_\ell c_{p\ell}[T_\ell(\infty) - T_{sat}(p_v)]}{\rho_v h_{\ell v}} = \frac{958.77 \times 4.216 \times (120 - 100)}{0.5974 \times 2251.2} = 60.11$$

and C_R for the heat transfer-controlled model can be obtained from Eq. (9.29):

$$C_R = \sqrt{\frac{3}{\pi}} Ja = \sqrt{\frac{3}{\pi}} \times 60.11 = 58.7$$

The bubble sizes in the inertia-controlled model and heat transfer-controlled model can be obtained from Eqs. (9.24) and (9.28), respectively. The time at which the bubble sizes obtained by both models are the same can be found by equalizing Eqs. (9.24) and (9.28):

$$\left\{\frac{2}{3}\frac{\rho_v h_{\ell v}}{\rho_\ell}\frac{T_\infty - T_{sat}[p_\ell(\infty)]}{T_{sat}[p_\ell(\infty)]}\right\}^{1/2} t = 2C_R\sqrt{\alpha_\ell t}$$

i.e.,

$$\begin{aligned} t &= 4C_R^2\alpha_\ell \Big/ \left\{\frac{2}{3}\frac{\rho_v h_{\ell v}}{\rho_\ell}\frac{T_\infty - T_{sat}[p_\ell(\infty)]}{T_{sat}[p_\ell(\infty)]}\right\} \\ &= 4 \times 58.7^2 \times 1.68 \times 10^{-7} \Big/ \left\{\frac{2}{3} \times \frac{0.5974 \times 2251.2 \times 10^3}{958.77}\frac{120 - 100}{373.15}\right\} \\ &= 4.62 \times 10^{-5}\ \text{s} \end{aligned}$$

The bubble radius obtained is

$$\begin{aligned} R &= 2C_R\sqrt{\alpha_\ell t} = 2 \times 58.7 \times \sqrt{1.68 \times 10^{-7} \times 4.62 \times 10^{-5}} \\ &= 3.27 \times 10^{-4}\ \text{m} = 0.327\ \text{mm} \end{aligned}$$

9.3.2.2 Heterogeneous Bubble Growth

The bubble growth discussed so far is limited to bubbles in an extensive, uniformly superheated liquid. For a vapor bubble attached to a heating surface, growth occurs in a nonuniform temperature field. The criteria for initiation and growth of a bubble near a wall were analyzed by Han and Griffith (1965). They assumed that a surface cavity, which serves as a nucleation site, has a hemispherical vapor cap on it. As a result, the bubble can grow only if the thermal layer adjacent to the nucleation site is sufficiently thick. A relation between cavity size and surface temperature elevation was derived on the basis of an assumed relation between the thermal layer thickness and the nucleus size at incipience. For saturation conditions, they found the maximum and minimum cavity sizes that can be active under a constant temperature:

$$(R_c)_{\max,\min} = \frac{\delta}{3}\left\{1 \pm \left[1 - \frac{12\sigma T_{sat}}{\delta\rho_v h_{\ell v}(T_w - T_{sat})}\right]^{1/2}\right\} \tag{9.37}$$

where δ is the thermal layer thickness.

Howell and Siegel (1966) examined bubble characteristics on single nucleation sites of known diameter (0.1–1 mm) on highly polished surfaces in pool boiling (see Fig. 9.8). Since the bubbles were large and slow-growing, only surface tension and buoyancy forces were important at departure. If the vapor bubble extends through the liquid thermal layer ($R_c > \delta$) as shown in Fig. 9.8a, evaporation occurs from the portion of the nucleus surface within the thermal layer while condensation occurs over the remaining portion of the nucleus. For the vapor bubble to grow, the amount of evaporative heat transfer must be greater than that of the condensation, which leads to the following criterion:

$$T_w - T_{sat} > \frac{4\sigma T_{sat}}{\rho_v h_{\ell v}\delta}, \quad R_c > \delta \tag{9.38}$$

If the nucleus is wholly contained in the thermal layer, as shown in Fig. 9.8b, the temperature required for the bubble to grow is

$$T_w - T_{sat} > \frac{2\sigma T_{sat}}{\rho_v h_{\ell v} R_c}\frac{1}{1 - R_c/(2\delta)}, \quad R_c < \delta \tag{9.39}$$

For the growth rate of a vapor bubble attached to the wall, Mikic et al. (1970) suggested that Eqs. (9.32)–(9.36) are applicable, provided that the geometric parameters $b = \pi/7$ and liquid temperature $T_\ell(\infty)$ are replaced by the heating surface temperature T_w.

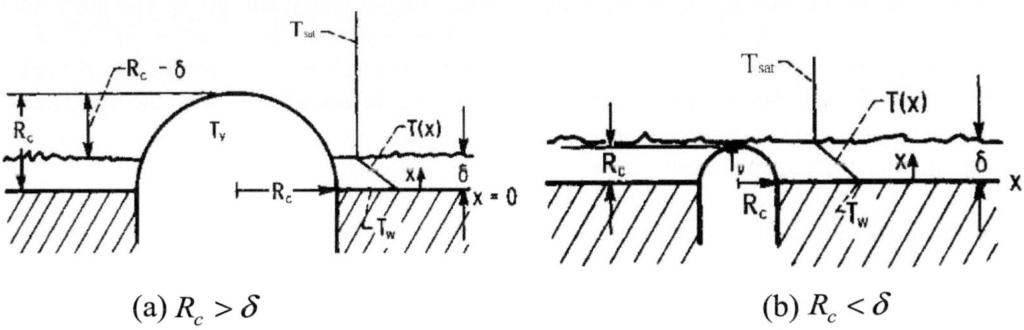

Fig. 9.8 Criteria for growth of hemispherical bubble nucleus (Howell and Siegel 1966)

As noted before, the microlayer plays a very important role during bubble growth near a heated surface (Fig. 9.6). Van Stralen et al. (1975a) proposed the following relation for bubble growth using microlayer theory, which is applicable to both inertia and heat transfer-controlled region for pure liquids:

$$R(t) = \frac{R_1(t)R_2(t)}{R_1(t) + R_2(t)} \tag{9.40}$$

where

$$R_1(t) = 0.8165\sqrt{\frac{\rho_v h_{\ell v}(T_w - T_{sat})\exp\left[-(t/t_d)^{1/2}\right]}{\rho_\ell T_{sat}}}t \tag{9.41}$$

$$\begin{aligned} R_2(t) = {} & 1.954\left\{R^* \exp\left[-\left(\frac{t}{t_d}\right)^{1/2}\right] + \frac{T_\infty - T_{sat}}{T_w - T_{sat}}\right\}Ja(\alpha_\ell t)^{1/2} \\ & + 0.373Pr_\ell^{-1/6}\left\{\exp\left[-\left(\frac{t}{t_d}\right)^{1/2}\right]\right\}^{1/2} Ja(\alpha_\ell t)^{1/2} \end{aligned} \tag{9.42}$$

$$R^* = 1.3908\frac{R_2(t_d)}{Ja(\alpha_\ell t_d)^{1/2}} - 0.1908Pr_\ell^{-1/6} \tag{9.43}$$

R^* can be determined if the departure time t_d is known. Van Stralen et al. (1975b) compared the above equation with experimental bubble growth rate data for water over a range of temperature, and obtained good agreement.

9.3.2.3 Bubble Growth within Superheated Liquid Droplets

Bubble growth within superheated liquid droplets is important in applications such as preparation of emulsified liquids, fuel coolant interactions in postulated nuclear reactor accidents, and heat exchangers. In these cases, droplets of the volatile liquid are dispersed in another stagnant liquid. Heating the field liquid leads to superheating of the droplets, homogeneous nucleation, and bubble growth.

Two immiscible liquids of different volatility are mixed together such that one liquid is dispersed in the form of droplets within the other, so that the droplets may be heated by direct contact across the liquid/liquid interface. Heat can be supplied by conduction, convection, nucleate boiling, or film boiling.

When nucleation sites are not present, single-phase conduction or convection can exist beyond the normal saturation state of liquid 1; this is termed superheating. This state can exist as long as the liquid does not come into contact with a vapor phase with which it is in equilibrium. The upper limit to the temperature of the liquid at a given pressure before a phase change occurs is called the superheat limit. At the superheat limit, homogeneous nucleation will begin to occur inside liquid 1 and form a vapor droplet. These vapor bubbles are in metastable equilibrium with the surrounding liquid, and subsequent growth of the initial bubbles will complete the phase transition. The two steps in phase change related to superheated liquid droplets are (Avedisian 1986):

1. The initial stage, during which microscopic bubbles form in the droplet
2. A second or bubble growth stage, in which the initial bubble grows as the liquid droplet vaporizes

Bubble growth in a volatile droplet suspended in an immiscible nonvolatile liquid was developed by Avedisian and Suresh (1985) by solving the coupled energy and momentum equations for the

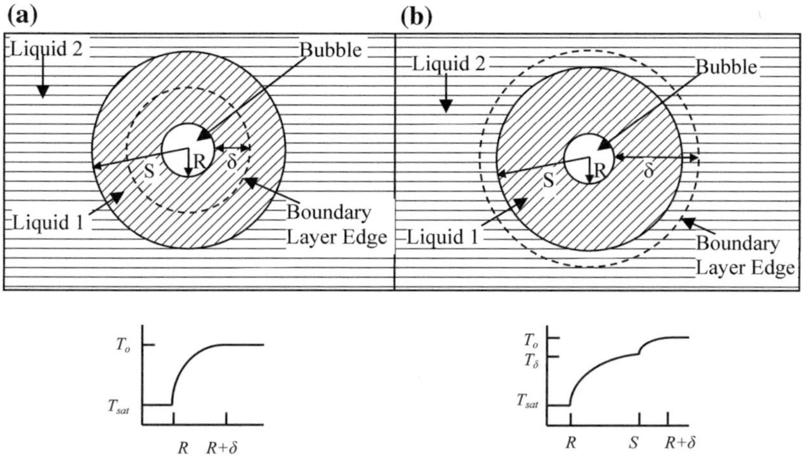

Fig. 9.9 Schematic illustration of bubble growth in a droplet suspended in an immiscible fluid. **a** Early stage where thermal boundary layer (δ) is within the droplet; **b** later stage when boundary layer extends into liquid 2 (Avedisian and Suresh 1985; reproduced with permission from Elsevier)

temperature fields in both liquids. A numerical solution for a two-phase droplet modeled as a vapor bubble growing from the center of the liquid was presented. When the properties of the two liquids are very different, the bubble growth rate can experience a significant change when the thermal boundary layer extends into nonvolatile liquid, as shown in Fig. 9.9.

Consider the following conditions for superheat bubble growth. At $t = 0$, a vapor bubble appears in liquid 1. The initial temperatures of liquids 1 and 2 are the same, and there is no motion between the droplet and the liquid. The bubble begins to grow due to a perturbation caused by a slight reduction in ambient pressure or an increase in ambient temperature. Continuous growth of the bubble completely consumes liquid 1 until the mass of the droplet is entirely vaporized and only a vapor bubble in liquid 2 is left. The radius of the final bubble is

$$R_\mathrm{f} = (1 - \varepsilon)^{-1/3} S_\mathrm{o} \tag{9.44}$$

where ε is the density ratio, and S_o is the initial overall drop radius. To formulate an equation of motion for a spherical bubble growing from the center of a spherical droplet, assume a vapor bubble growing from the center of a volatile liquid suspended in a nonvolatile liquid, as shown in Fig. 9.9. The velocity, pressure, and temperature in the two liquids are governed by the continuity, momentum, and energy equations for one-dimensional and unsteady conditions, given as

$$\frac{1}{r^2}\frac{\partial}{\partial r}\left(r^2 v_i\right) = 0 \tag{9.45}$$

$$\frac{\partial v_i}{\partial t} + v_i \frac{\partial v_i}{\partial r} = -\frac{1}{\rho}\frac{\partial p_i}{\partial r} + v_i\left[\frac{1}{r^2}\frac{\partial}{\partial r}\left(r^2\frac{\partial v_i}{\partial r}\right) - \frac{2v_i}{r^2}\right] \tag{9.46}$$

$$\frac{\partial T_i}{\partial t} + v_i \frac{\partial T_i}{\partial r} = \alpha_i \frac{1}{r^2}\frac{\partial}{\partial r}\left(r^2\frac{\partial T_i}{\partial r}\right) \tag{9.47}$$

where $i = 1$ and 2, v_i is the radial velocity, p_i is the pressure within liquid i, T_i is the temperature within liquid i, and α_i is the thermal diffusivity of liquid i.

The boundary and initial conditions are:

$$T_1(r,0) = T_2(r,0) = T_o \tag{9.48}$$

$$T_1(R,t) = T_v(t) \tag{9.49}$$

$$T_1(S,t) = T_2(S,t) \tag{9.50}$$

$$k_1 \frac{\partial T_1}{\partial r}\bigg|_{S,t} = k_2 \frac{\partial T_2}{\partial r}\bigg|_{S,t} \tag{9.51}$$

$$T_2(\infty,t) = T_o \tag{9.52}$$

$$R(0) = R_o \tag{9.53}$$

$$\dot{R}(0) = 0 \tag{9.54}$$

where k_1 and k_2 are the thermal conductivity of liquid 1 and 2, respectively, R is the bubble radius and S is the droplet radius. The interfacial energy balance around the bubble yields

$$k_1 \frac{\partial T_1}{\partial r}\bigg|_{R,t} = \rho_v h_{\ell v} \dot{R} \tag{9.55}$$

where $\dot{R}$ is the velocity of the bubble wall. T_o and R_o are the initial conditions of the limit of superheat and initial unstable bubble radius. These terms, defined by the critical nucleus state for homogeneous nucleation, are intrinsic properties of liquid 1 and are functions of ambient pressure for a given nucleation rate.

Radial velocity is continuous in this problem, because there is no mass transfer across the liquid's interface. The velocity can be found by integrating Eq. (9.45) and applying a mass balance around the bubble.

$$v_i = \varepsilon \frac{R^2}{r^2} \dot{R} \tag{9.56}$$

where ε is the density ratio $(1 - \rho_v/\rho_1)$. Equation (9.56) is the same as bubble growth in an infinite medium.

The Rayleigh equation for this problem is obtained by integrating Eq. (9.46) over r twice: from R to S and again from S to ∞.

$$\int_R^\beta \left[\frac{\partial v_i}{\partial t} + v_i \frac{\partial v_i}{\partial r} = -\frac{1}{\rho}\frac{\partial p_i}{\partial r} + v_i \left(\frac{1}{r^2}\frac{\partial}{\partial r}\left(r^2 \frac{\partial v_i}{\partial r}\right) - \frac{2v_i}{r^2}\right)\right] \mathrm{d}r \tag{9.57}$$

where $\beta = S$ or ∞. When $r = R$,

$$p_v - p_{R1} = \frac{2\sigma_1}{R} - 2\mu_1 \frac{\partial v_1}{\partial r}\bigg|_{r=R} \tag{9.58}$$

and when $r = S$

$$p_{S1} - p_{S2} = \frac{2\sigma_{12}}{S} - 2(\mu_2 - \mu_1)\frac{\partial v_1}{\partial r}\bigg|_{r=S} \tag{9.59}$$

The equation of motion for a bubble growing in the center of a droplet is found by combining Eqs. (9.56) and (9.55), integrating twice, and substituting Eqs. (9.58) and (9.59):

$$\begin{aligned} \left[R\ddot{R} + 2\dot{R}^2\right]\left(1 - \bar{\varepsilon}\frac{R}{S}\right) - \varepsilon\frac{\dot{R}^2}{2}\left(1 - \bar{\varepsilon}\frac{R^4}{S^4}\right) \\ + 4v_1\frac{\dot{R}}{R}\left(1 - \bar{\mu}\frac{R^3}{S^3}\right) \\ = \frac{p_v - p_o}{\rho_1 \varepsilon} - \frac{2\sigma_1}{\varepsilon\rho_1 R}\left[1 + \frac{\sigma_{12}}{\sigma_1}\frac{R}{S}\right] \end{aligned} \tag{9.60}$$

where $\bar{\varepsilon}$ is the density ratio $(1 - \rho_2/\rho_1)$, $\bar{\mu}$ is the viscosity ratio $(1 - \mu_2/\mu_1)$, σ_{12} is the liquid 1/liquid 2 interfacial tension, and p_o is pressure at infinity in liquid 2.

Figure 9.10 shows the evolution of pressure, temperature and average radius for a bubble growing in an n-octane droplet at $Ja = 10\ [\rho_1 c_{p1}(T_o - T_{sat})/(\rho_v h_{\ell v})]$, where $\bar{R} = R/S_o$ and $\tau = t\alpha_1/S_o^2$. The early stages of bubble growth in a droplet are similar to growth in a semi-infinite medium. The initial state corresponds to the homogeneous nucleation limit and the asymptotic temperatures and pressures correspond to saturation conditions.

Sideman and Isenberg (1967) developed an analytical, closed form solution for boiling a droplet in another immiscible liquid with the following assumptions:

1. No radial convection
2. Constant and uniform temperature of liquid 1
3. Inviscid and uniform translatory motion of liquid 2 around liquid 1 droplet

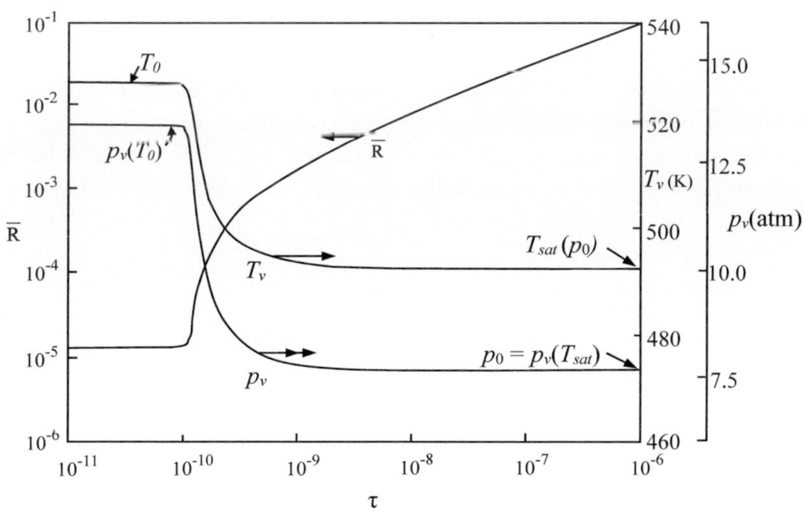

Fig. 9.10 Early time variation of vapor pressure, vapor temperature and radius of a bubble growing in a superheated n-octane droplet. Initial conditions correspond to the kinetic limit of superheat of octane at the indicated pressure (Avedisian and Suresh 1985; reproduced with permission from Elsevier)

This closed form solution is valid only after boundary-layer penetration of liquid 2. During early bubble growth, the assumptions are not valid because the temperature field is transient and the boundary layer is entirely in liquid 1. The analytical solution for boiling of a droplet in another immiscible liquid is given as

$$\bar{S} = (1-\varepsilon)^{-1/3}\left\{1-\varepsilon\left[\left(\frac{9}{2\pi}\right)^{1/2}(1-\varepsilon)JaPe^{1/2}\tau - 1\right]^2\right\}^{1/3} \tag{9.61}$$

where $\bar{S} = S/S_o$, Pe is the Peclet number $(U_\infty S_o/\alpha_2)$, and Ja is the Jakob number.

9.3.3 Bubble Detachment

A vapor bubble attached to a heating surface is subjected to different forces. While surface tension and liquid inertia tend to prevent such a bubble from detaching, buoyancy and dynamic forces act directly on the bubble to lift it from the heating surface. When the bubble grows large enough, the detachment-oriented forces will dominate and the bubble will detach from the heating surface. The inertial force is a strong function of liquid superheat, which, as indicated by Eq. (9.12), is inversely proportional to the size of the cavity. Therefore, a bubble formed from a small cavity grows faster than one formed from a large cavity.

The bubble size at departure can be found by performing a force balance on the bubble. For a small cavity size ($r_c < 10\,\mu\text{m}$), the bubble size at departure is dominated by a balance between buoyancy and liquid inertial forces. When the cavity size is large, the bubble grows more slowly, thus the inertial force becomes unimportant. Instead, a force balance between buoyancy and surface tension dominates the bubble size at departure. Fritz (1935) was the first to correlate the bubble departure diameter by balancing buoyancy and surface tension forces, and he proposed the following equation:

$$D_b = 0.0208\theta\sqrt{\frac{2\sigma}{g(\rho_\ell - \rho_v)}} \tag{9.62}$$

where θ is the contact angle measured in degrees. For a detailed discussion of different types of contact angles and definitions including apparent, advancing and receding see Sect. 4.3. Further experimental investigations indicated that Eq. (9.62) can provide a correct length scale for the bubble diameter at departure, but significant deviation of the bubble diameter at departure has been reported. Cole and Rohsenow (1969) proposed much better correlations for the bubble diameter at departure:

For water:

$$D_b = 1.5\times10^{-4}\sqrt{\frac{2\sigma}{g(\rho_\ell - \rho_v)}}(Ja^*)^{\frac{5}{4}} \tag{9.63}$$

For other fluid:

$$D_b = 4.65\times10^{-4}\sqrt{\frac{2\sigma}{g(\rho_\ell - \rho_v)}}(Ja^*)^{\frac{5}{4}} \tag{9.64}$$

where

$$Ja^* = \frac{\rho_\ell c_{p\ell} T_{sat}}{\rho_v h_{\ell v}} \tag{9.65}$$

At high heat flux, the bubble departure diameter can be obtained from (Gorenflo et al. 1986)

$$D_b = C_1 \left(\frac{Ja^4 h_{\ell v}^2}{g} \right)^{1/3} \left[1 + \left(1 + \frac{2\pi}{3Ja} \right)^{1/2} \right]^{4/3} \tag{9.66}$$

where the constant C_1 obtained by fitting experimental data for boiling of R-12, R-22, and propane are 14.7, 16.0, and 2.78, respectively.

The bubble release frequency f_b is related to the waiting time between detachment of the one bubble and initiation of the next bubble, t_w, and the growth time, t_g, of a bubble before its detachment. Attempts to obtain bubble release frequency by the prediction of t_w and t_g were rarely successful because (a) these models generally did not consider evaporation from both the base and surface of bubbles; (b) cavity size variations significantly alter the growth time; (c) bubble activity, fluid flow, and heat transfer near a nucleation site can influence both bubble growth and waiting time; and (d) bubble shapes always change during growth (Kandilikar et al. 1999).

While the bubble release frequency depends distinctly upon the bubble diameter at departure, which differs from site to site, it is a constant for each individual nucleation site. It is recommended that the correlation have the form of $f_b D_b^n = \text{const}$, where n is the exponent. Ivey (1967) suggested that the value of exponent n for inertia-controlled bubble growth is 2. When heat transfer controls bubble growth, the value of n is equal to 1/2. Malenkov (1971) proposed the most comprehensive correlation for the product of bubble release frequency and bubble diameter at departure:

$$f_b D_b = \frac{V_b}{\pi \left(1 - \frac{1}{1 + V_b \rho_v h_{\ell v} / q''} \right)} \tag{9.67}$$

where V_b is the bubble departure velocity and is calculated by

$$V_b = \sqrt{\frac{D_b g (\rho_\ell - \rho_v)}{2(\rho_\ell + \rho_v)} + \frac{2\sigma}{D_b (\rho_\ell + \rho_v)}} \tag{9.68}$$

Eastman (1984) developed an analysis related to dynamics of bubble departure that is presented here. During nucleate boiling, the forces that hold the bubble to the wall (negative) are larger than the forces that pull the bubble from the wall (positive) when the bubble is very small. As the bubble grows, the positive forces grow faster than the negative forces, until the total force becomes positive and pulls the bubble away from the surface. The negative forces are surface tension and drag force. The positive forces are buoyancy (for an upward-facing surface), internal pressure, and inertia. These forces and directions are shown in Fig. 9.11.

The five forces that act on a bubble during growth from a heated wall are: internal pressure, surface tension, buoyancy, drag, and inertia. Analyses show that buoyancy is related to the gravitational

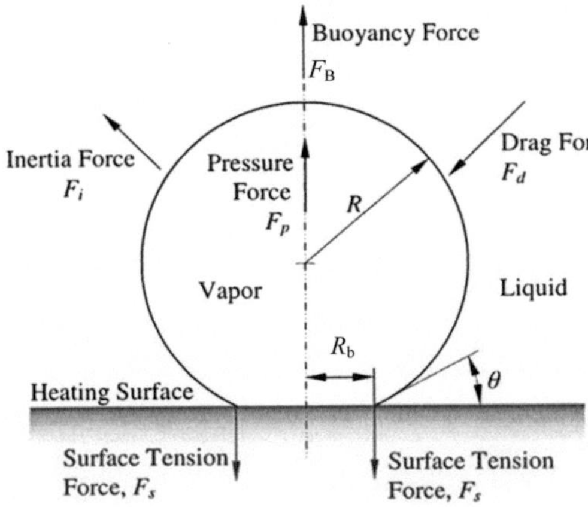

Fig. 9.11 Forces acting on a vapor bubble growing on a heating surface (Eastman 1984)

acceleration and the inertial force is dependent on the deceleration of the bubble. The bubble departure can be determined by a force balance:

$$F_{\mathrm{d}} + F_{\mathrm{s}} = F_{\mathrm{i}} + F_{\mathrm{p}} + F_{\mathrm{B}} \tag{9.69}$$

where F_i and F_p are liquid inertial force and pressure force, respectively. The surface tension force, F_s, is caused by the attraction of the liquid to the surface that acts around the perimeter of the bubble base. The surface tension force is proportional to the fluid surface tension, σ, and the advancing contact angle, θ, i.e.,

$$F_{\mathrm{s}} = 2\pi R_{\mathrm{b}}\sigma \sin\theta \tag{9.70}$$

where R_b is the base radius. The base contact angle goes to 90° as the bubble nears departure.

When the bubble is growing in a viscous fluid, it will be subject to drag force. In realistic applications, this force is negligible for most fluids. A very rough estimate of this force was made by Keshock and Siegel (1962), and this approximation will be used here. In estimating drag force, it is also assumed that the bubble is spherical and growing away from the wall at a velocity equal to the change of its radius with time, i.e., d*R*/d*t*. The drag force can be calculated as:

$$F_{\mathrm{d}} = C_{\mathrm{D}}\frac{\rho_\ell}{2}\left(\frac{\mathrm{d}R}{\mathrm{d}t}\right)^2 \pi R^2 \tag{9.71}$$

where drag coefficient C_D can be calculated by using the following experimental correlation

$$C_{\mathrm{D}} = \frac{45}{Re} \tag{9.72}$$

where

$$Re = \frac{2\rho_\ell}{\mu_\ell} R \frac{\mathrm{d}R}{\mathrm{d}t} \tag{9.73}$$

For a spherical bubble, which is submerged in a stagnant fluid, the buoyancy force is equal to the weight of the fluid displaced:

$$F_{\mathrm{B}} = \frac{4\pi R^3}{3}(\rho_\ell - \rho_{\mathrm{v}})g \tag{9.74}$$

Vapor inertia is negligible for most of the bubble growth; however, the inertia for the liquid surrounding the bubble needs to be accounted for. The liquid around the bubble is moved by the growth of the bubble. The affected mass of the fluid is that occupied by 11/16 of the bubble volume (Han and Griffith 1965). The growth rate of the bubble decreases after the initial unbinding. The inertia of the liquid works against the decrease in velocity and tries to pull the vapor away from the surface. According to Newton's second law, the inertial force can be approximated as follows, based on the above argument:

$$F_i = \frac{11}{6}\pi R^3 \rho_\ell \frac{\mathrm{d}^2 R}{\mathrm{d}t^2} \tag{9.75}$$

The pressure force results from the contribution of the dynamic excess vapor pressure and capillary pressure. It can be written as follows:

$$F_{\mathrm{p}} = \left(\frac{2\sigma}{R} + p_{\mathrm{v}}\right)\pi R_{\mathrm{b}}^2 \tag{9.76}$$

where the vapor pressure is given below

$$p_{\mathrm{v}} - p_\ell = \frac{2\sigma}{R} \tag{9.77}$$

The departure radius is assumed to be the radius at which the net force changed signs from negative to positive. Equations (9.70)–(9.77) were used to calculate the departure radius for saturated water for 1 and 0.229 g, respectively. The forces acting on a bubble versus its radius are shown in Figs. 9.12 and 9.13. Analysis shows that buoyancy is directly related to the local gravitational acceleration. The inertial force is independent of gravity but dependent on the deceleration of the bubble.

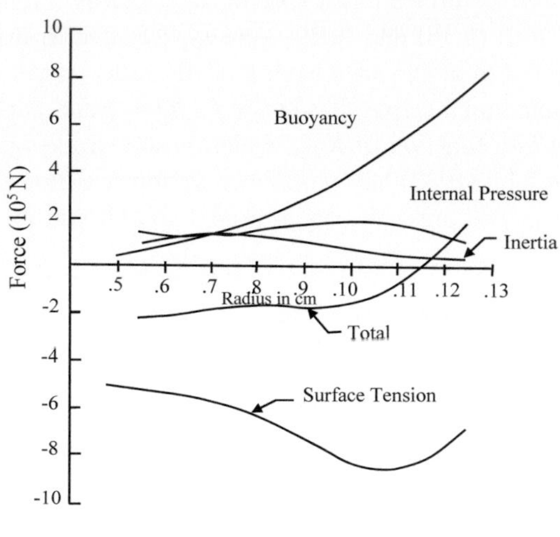

Fig. 9.12 Forces acting on a bubble in saturated water with 1-g acceleration (Eastman 1984)

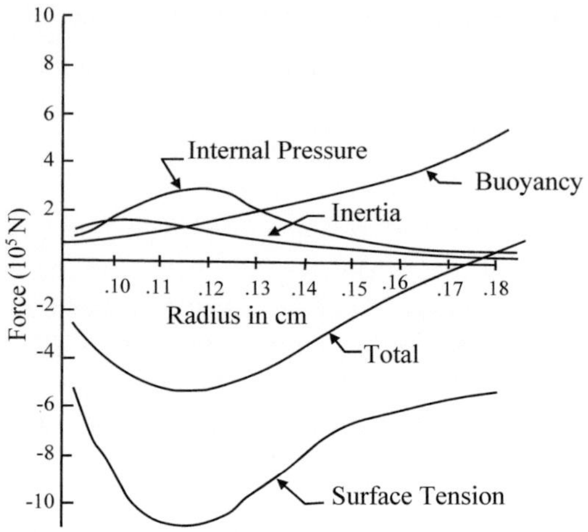

Fig. 9.13 Forces acting on a bubble in saturated water under 0.229-g acceleration (Eastman 1984)

9.3.4 Nucleate Site Density

The knowledge of distribution of nucleation sites is an important factor in determining the boiling characteristics of a surface under specific operating conditions. The number density of sites, or total number of active sites per unit area, is a function of contact angle, cavity half angle, and heat flux (or superheat) (Fig. 9.6), i.e.,

$$N''_{\text{a}} = f(\theta, \phi, \Delta T, \text{fluid properties}) \tag{9.78}$$

Equation (9.11) indicated that for a given local heat flux or superheat, a cavity will be active if $R_{\min}$ is greater than R_{b},

$$R_{\min} \geq \frac{2\sigma T_{\text{sat}}}{h_{\ell v}\rho_{\text{v}}\Delta T} \tag{9.79}$$

Obviously, each cavity on a real surface has a specific $R_{\min}$ that is a function of geometry and the contact angle. Considering Eqs. (9.78) and (9.79), one expects that as the wall superheat increases, $R_{\min}$ decreases and the number of active sites having cavity radii greater than $R_{\min}$ increases.

Lorenz et al. (1974) counted total active sites/cm^2 on a #240 (sand paper) finished copper surface for different working fluids as a function of $R_{\min}$, which is shown in Fig. 9.14.

Kocamustafaogullari and Ishii (1983) have correlated various existing experimental data of N''_{a} for water on a variety of surfaces and pressure ranges from 1 to 198 atm by

$$N''_{\text{a}} = D_{\text{d}}^2\left[\left(\frac{D_{\text{c}}}{D_{\text{d}}}\right)^{-0.44} F\right]^{1/4.4} \tag{9.80}$$

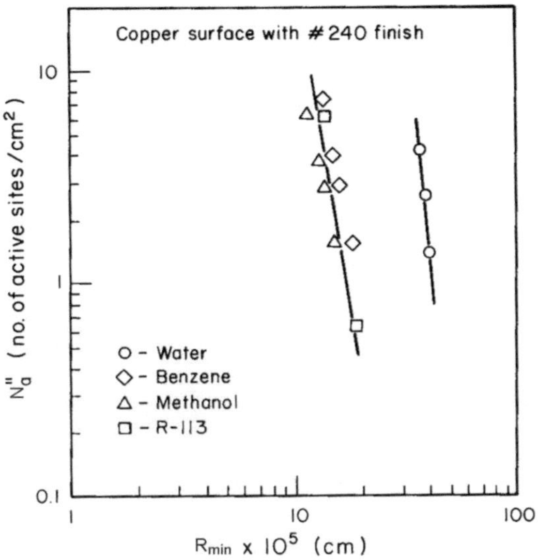

Fig. 9.14 Number density of active sites for boiling on a copper surface (Lorenz et al. 1974)

where

$$F = 2.157 \times 10^{-7}\left(\frac{\rho_\ell - \rho_v}{\rho_v}\right)^{-3.2}\left[1 + 0.0049\left(\frac{\rho_\ell - \rho_v}{\rho_v}\right)\right]^{4.13} \tag{9.81}$$

$$D_c = 4\sigma[1 + (\rho_\ell/\rho_v)]/p_\ell \cdot \{\exp[h_{\ell v}(T_v - T_{sat})/(R_g T_v T_{sat})] - 1\} \tag{9.82}$$

$$D_d = 0.02080\theta\sqrt{\frac{\sigma}{g(\rho_\ell - \rho_v)}} \cdot 0.0012\left(\frac{\rho_\ell - \rho_v}{\rho_v}\right)^{0.9} \tag{9.83}$$

where R_g is the gas constant for the vapor.

Wang and Dhir (1993a, b) have studied number density for boiling of water at 1 atm on a mirror-finished copper surface, and they provided a mechanistic approach for relating the cavities that are present on the surface to the cavities that actually nucleate.

9.3.5 Bubble Growth and Merger

Although nucleate boiling has been a subject of extensive studies, a fundamental understanding of the dynamics of bubble growth, departure, and merger is still lacking. Most existing analytical models did not consider how the flow and temperature field are influenced by the bubbles motion. While early efforts to develop numerical simulations of bubble growth during partial nucleate boiling did not consider the merger of the vapor bubbles (e.g., Lee and Nydahl 1989; Welch 1998), single and multiple bubble mergers during nucleate boiling have been numerically simulated, and the results compared favorably with experimental observations (Son et al. 2002; Mukherjee and Dhir 2004; Dhir 2005, 2018). The numerical simulation of the bubble merger process on a single nucleation site during pool boiling, performed by Son et al. (2002), will be introduced in this subsection.

The transport phenomena in the nucleate boiling occur at different scales, and for this reason, Son et al. (2002) divided the computational domain into macro- and microregions, as shown in Fig. 9.15.

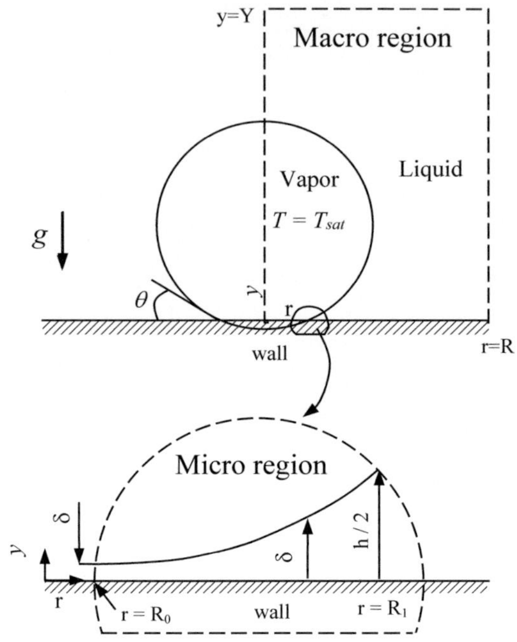

Fig. 9.15 Macro- and microregion in numerical simulation (Son et al. 2002)

The macroregion includes the vapor bubble and the surrounding liquid, while the microregion is the thin-film region beneath the bubble. Evaporation in the microregion was solved using lubrication theory. The phase change and macroregion were solved using a level set formulation.

Assuming laminar and constant properties, the conservation of mass in the microlayer is

$$\frac{\partial \delta}{\partial t} = v_\ell - \frac{q''}{\rho_\ell h_{\ell v}} \tag{9.84}$$

where the liquid velocity normal to the film surface, v_ℓ, is obtained by integrating the continuity equation in the cylindrical coordinate system, i.e.,

$$v_\ell = -\frac{1}{r}\frac{\partial}{\partial r}\int_0^\delta r u_\ell \mathrm{d}y \tag{9.85}$$

Neglecting inertial and gravity effects, the momentum equation in the thin film is

$$\frac{\partial p_\ell}{\partial r} = \mu_\ell \frac{\partial^2 u_\ell}{\partial y^2} \tag{9.86}$$

Assuming conduction is the mechanism for heat transfer across the thin film, the heat flux is

$$q'' = k_\ell \frac{T_\mathrm{w} - T_\delta}{\delta} \tag{9.87}$$

where T_δ is the temperature at the liquid–vapor interface. The evaporating heat flux can also be written using a modified Clausis-Clapeyron equation:

$$q'' = \left(\frac{2}{\pi R_\mathrm{g} T_\mathrm{v}}\right)^{1/2} \frac{\rho_\mathrm{v} h_{\ell\mathrm{v}}^2}{T_\mathrm{v}} \left[T_\delta - T_\mathrm{v} + \frac{(p_\ell - p_\mathrm{v}) T_\mathrm{v}}{\rho_\ell h_{\ell\mathrm{v}}}\right] \tag{9.88}$$

The pressures in the liquid and vapor are related by momentum balance at interface (Chap. 5)

$$p_\ell = p_\mathrm{v} - \sigma K - p_\mathrm{d} + \frac{q''^2}{\rho_\mathrm{v} h_{\ell\mathrm{v}}^2} \tag{9.89}$$

where surface tension, σ, is a function of temperature, and K is curvature:

$$K = \frac{1}{r}\frac{\partial}{\partial r}\left[r \frac{\partial \delta}{\partial r} \Big/ \sqrt{1 + \left(\frac{\partial \delta}{\partial r}\right)^2}\right] \tag{9.90}$$

and $p_\mathrm{d} = A\delta^{-3}$ is the disjoining pressure.

The combination of Eqs. (9.84)–(9.88) yields a fourth order ordinary differential equation in the following form:

$$\delta'''' = f(r, t, \delta, \delta', \delta'', \delta''') \tag{9.91}$$

which is subject to the following boundary conditions:

$$\delta = \delta_0;\ \delta' = \delta''' = 0 \quad \text{at } r = R_0 \tag{9.92}$$

$$\delta = h/2;\ \delta' = \tan\theta;\ \delta'' = 0 \quad \text{at } r = R_1 \tag{9.93}$$

where δ_0 is the nonevaporating film thickness for which Son et al. (2002) used $\delta_0 = 6 \times 10^{-10}$ m, $h/2$ is the distance to the first computational node for the level set function from the wall, and θ is the apparent contact angle. R_1 is obtained from the solution of the macroregion.

As for the macroregion, the level set formulation is used to accommodate the effect of liquid–vapor phase change. In this methodology, the interface separating the two phases is captured by a level set function, ϕ, which is defined as a signed distance from the interface: a negative sign for the vapor phase and a positive sign for the liquid phase. The governing equations for the macroregion are

$$\begin{aligned}\rho\left(\frac{\partial \mathbf{V}}{\partial t} + \mathbf{V}\cdot\nabla\mathbf{V}\right) &= -\nabla p + \rho\mathbf{g} - \rho\mathbf{g}\beta(T - T_\mathrm{sat}) \\ &\quad - \sigma K \nabla H + \nabla\cdot\mu\nabla\mathbf{V} + \nabla\cdot\mu\nabla\mathbf{V}^T\end{aligned} \tag{9.94}$$

$$\begin{aligned}\rho c_{p\ell}\left(\frac{\partial T}{\partial t} + \mathbf{V}\cdot\nabla T\right) &= \nabla\cdot k\nabla T \quad \text{for } H > 0 \\ T &= T_\mathrm{sat}(p_\mathrm{v}) \quad \text{for } H = 0\end{aligned} \tag{9.95}$$

$$\nabla \cdot \mathbf{V} = \frac{\mathbf{m}''}{\rho^2} \nabla \rho + \dot{V}_{\text{micro}} \tag{9.96}$$

where the properties are evaluated by $\rho = \rho_{\text{v}} + (\rho_\ell - \rho_{\text{v}})H$, $\mu^{-1} = \mu_{\text{v}}^{-1} + (\mu_\ell^{-1} - \mu_{\text{v}}^{-1})H$, and $k^{-1} = k_\ell^{-1}H$. Due to the discontinuity at the liquid–vapor interface, the step function H is defined by

$$H = \begin{cases} 0 & \phi \leq -1.5h \\ 0.5 + \phi/(3h) + \sin[2\pi\phi/(3h)]/(2\pi) & |\phi| \leq 1.5h \\ 1 & \phi > 1.5h \end{cases} \tag{9.97}$$

where h is grid spacing. The source term included in Eq. (9.96) is obtained by conservation of mass and energy at the interface,

$$\mathbf{m}'' = \rho(\mathbf{V}_\delta - \mathbf{V}) = \frac{k \nabla T}{h_{\ell \text{v}}} \tag{9.98}$$

and the rate of volume production from microlayer $\dot{V}_{\text{micro}}$ is obtained from

$$\dot{V}_{\text{micro}} = \int_{R_0}^{R_1} \frac{k_\ell (T_{\text{w}} - T_\delta)}{\rho_{\text{v}} h_{\ell \text{v}} \delta \Delta V_{\text{micro}}} r dr \tag{9.99}$$

where ΔV_{micro} is the vapor side control volume near the microregion.

The level set function, ϕ, is advanced by

$$\frac{\partial \phi}{\partial t} = -\mathbf{V}_\delta \cdot \nabla \phi \tag{9.100}$$

and reinitialized by solving

$$\frac{\partial \phi}{\partial t} = \frac{\phi_0}{\sqrt{\phi_0^2 + h^2}} (1 - |\nabla \phi|) \tag{9.101}$$

The boundary conditions for the macroregion are

$$u = v = 0, \quad T = T_{\text{w}}, \quad \frac{\partial \phi}{\partial y} = -\cos\theta \text{ at } y = 0 \tag{9.102}$$

$$u = \frac{\partial v}{\partial r} = \frac{\partial T}{\partial r} = \frac{\partial \phi}{\partial r} \text{ at r} = 0, R \tag{9.103}$$

$$\frac{\partial u}{\partial y} = \frac{\partial v}{\partial y} = \frac{\partial \phi}{\partial y} = 0, \quad T = T_{\text{sat}}, \text{ at } y = Y \tag{9.104}$$

The grid size used by Son et al. (2002) was 26 μm; the time step was 8×10^{-3} ms, which satisfies $\Delta t \leq h/(|u| + |v|)$ because of the explicit treatment of the convection terms.

Figure 9.16 shows the bubble growth pattern during one cycle for ΔT = 10 K and waiting time of 1.28 ms. The computational domain was 2.5 × 10 mm^2 ($R \times Y$). It can be seen that the bubble merger occurred twice during one cycle at t = 1.4 ms and t = 4.5 ms. The merged bubble departed from the surface at t = 5.5 ms, with no more bubble merger during this cycle. Son et al. (2002) also compared the predicted bubble merger pattern with the experimental observations and the agreement was very good.

The bubble dynamics and heat transfer associated with lateral bubble merger during transition from partial to fully developed nucleate boiling were studied numerically by Mukherjee and Dhir (2004). Multiple bubble mergers in a line and in a plane were simulated, and the bubble dynamics and wall heat transfer are compared to those for a single bubble. The results show that merger of multiple bubbles significantly increases the overall wall heat transfer, which is caused by trapping the liquid layer between the bubble bases during merger and the drawing of cooler liquid toward the wall during the post-merger contraction. While these numerical investigations provided insights about bubble formation, growth, and mergers in nucleate boiling, the results are as good as the knowledge of the distribution of the nucleation sites.

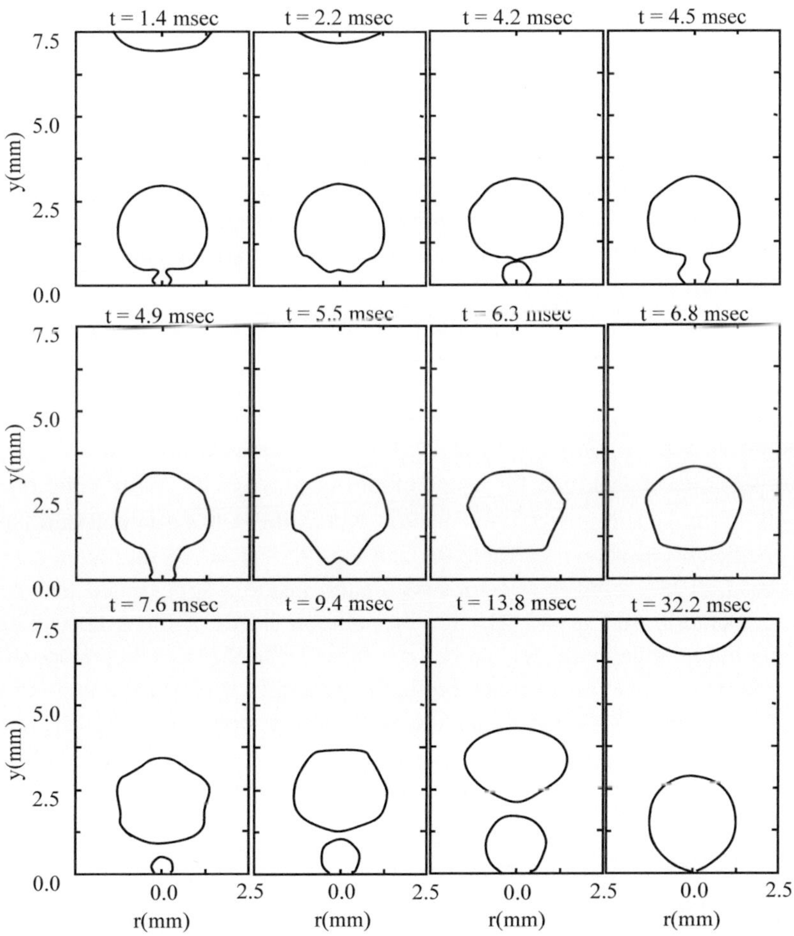

Fig. 9.16 Bubble growth and demerger pattern for $\Delta T = 10\,\text{K}$ and waiting time of 1.28 ms (Son et al. 2002)

9.3.6 Heat Transfer in Nucleate Boiling

In general, heat flux and heat transfer coefficient during evaporation and nucleate boiling can be correlated with the driving temperature difference ($T_w - T_{sat}$) according to the following equations:

$$q'' = c_1(T_w - T_{sat})^m \tag{9.105}$$

and since $q'' = h(T_w - T_{sat})$ one can get

$$h = c_2(T_w - T_{sat})^{m-1} = c_2(T_w - T_{sat})^n \tag{9.106}$$

$$h = c_3 q''^{(m-1)/m} = c_3 q''^P \tag{9.107}$$

The values of m, n and P are shown in Table 9.1 and are almost constant with weak dependencies on liquid properties and heated surfaces. However, c_1, c_2 and c_3 depend greatly upon the properties of liquid and heated surface material and geometry. For evaporation, one can use free convection correlation and therefore

$$h = c_2(T_w - T_{sat})^{1/4} \quad \text{Laminar flow} \tag{9.108}$$

$$h = c_2(T_w - T_{sat})^{1/3} \quad \text{Turbulent flow} \tag{9.109}$$

A typical temperature distribution during nucleate boiling under 1 atm over a horizontal plate at 109.1 °C is shown in Fig. 9.17. It can be seen that the difference between wall temperature and the liquid temperature, $T_w - T_\ell$, is significantly greater than the difference between the liquid temperature and saturation temperature, $T_\ell - T_{sat}$. The driving force of heat transfer is $T_w - T_{sat}$, which is used to define the heat transfer coefficient. It is found that the heat transfer coefficient is approximately proportional to the third power of this temperature difference, i.e.,

$$h = c_2(T_w - T_{sat})^3 \tag{9.110}$$

Figure 9.18 shows the comparison of equations shown in Table 9.1 with experimental results for evaporation and nucleate pool boiling of water at 100 °C from a horizontal surface. Clearly, there are two distinct and separate ranges, one for evaporation (up to $q'' = 10^4$ W/m^2) and one for nucleate boiling $\left(q'' > 10^4 \text{ W/m}^2\right)$. The power law equation works for both regions a and b with different exponents and coefficients as shown in Table 9.1 and Fig. 9.18 (P = 3/4 for region a and P = 1/5 for region b). Cao et al. (1989) used $m = 1$ for evaporation inside a spot heated heat pipe with high working fluid temperature and $m = 1.215$ for evaporation inside a block-heated heat pipe with moderate working fluid temperature. As can be seen from Table 9.1, n = 3 corresponds to $P = 0.75$ in Eq. (9.107). The range of P for nucleate boiling is generally between 0.6 and 0.8, which corresponds to $2.5 \le m \le 5$ or $1.5 \le n \le 4$ (e.g., the well-known Rohsenow Eq. (9.115) for nucleate

Table 9.1 Heat transfer comparison of evaporation and nucleate pool boiling[a]

		$q'' = c_1(T_w - T_{sat})^m$	$h = c_2(T_w - T_{sat})^n$	$h = c_3 q''^P$
Evaporation	Laminar	m = 5/4	n = 1/4	P = 1/5
	Turbulent	m = 4/3	n = 1/3	P = 1/4
Nucleate boiling		m = 4	n = 3	P = 3/4

[a] c_1, c_2, and c_3 are constants, not functions of $\Delta T = T_w - T_{sat}$

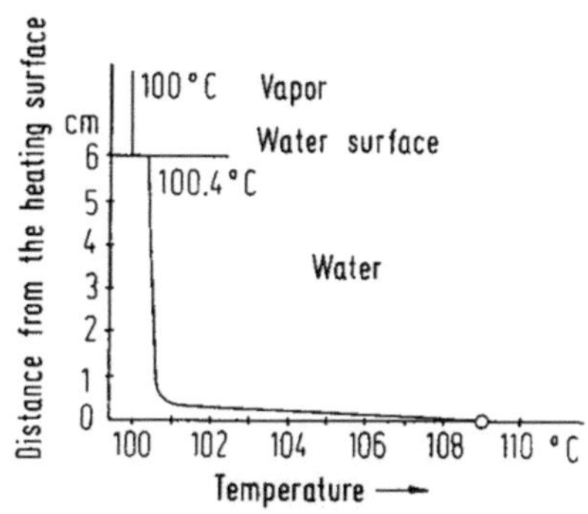

Fig. 9.17 Temperature profile above a heating surface during nucleate boiling (Stephan 1992; reprinted with kind permission of Springer Science and Business Media)

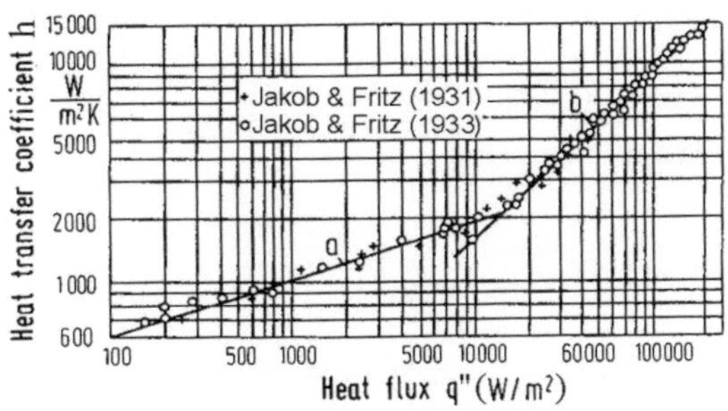

Fig. 9.18 Evaporation and nucleate boiling of water at 100 °C on a heated surface (Stephan 1992; reprinted with kind permission of Springer Science and Business Media)

boiling has $m = 3$ and $n = 2$). For low-boiling point liquids such as helium, the value of P can be as small as 0.5 ($m = 2$ and $n = 1$).

Heat transfer in nucleate boiling processes has been described through a variety of heat transport models, all of which have limitations of some sort because they were developed under ideal conditions. For nucleate boiling on an upward-facing horizontal surface in the isolated bubble region, heat transfer mechanisms include (1) transient heat conduction from the heating surface to the adjacent liquid, (2) natural convection on inactive areas of the heating surface, and (3) evaporation from the microlayer underneath the vapor bubble (Forster and Greif 1959). The liquid adjacent to the heating surface becomes superheated, is pushed outward by the vapor bubble, and then mixes with the bulk liquid. An expression of the heat flux during partial nucleate boiling is (Kandilikar et al. 1999)

$$q'' = \frac{C_1^2}{2}\sqrt{\pi(k\rho c_p)_\ell f_b D_b^2 N_a''}\Delta T + \left(1 - \frac{C_1^2}{2} N_a \pi D_b^2\right)\bar{h}_{nc}\Delta T + N_a'' \frac{\pi}{4} D_b^2 \bar{h}_{evp}\Delta T \tag{9.111}$$

where the three terms on the right-hand side represent the contributions of transient conduction, natural convection, and microlayer evaporation. In order to use Eq. (9.111) to calculate heat transfer, it is necessary to know the proportionality constant C_1 for bubble area influence, the bubble diameter at departure D_b, the bubble release frequency f_b, the number density of the nucleate site N_a'', and average heat transfer coefficient of natural convection and evaporation, $\bar{h}_{nc}$ and $\bar{h}_{evp}$.

Mikic and Rohsenow (1969) justified Eq. (9.111) without the last term by using empirical correlations for several of these parameters. Mikic and Rohsenow's (1969) correlation was revisited by Bergles (2005), who concluded that their correlation is the ultimate pool-boiling model because all of the parameters can be obtained by various empirical correlations and only one experimental observation—the cavity size distribution. The problem is that it is impossible to get closer than one empirical constant since the nucleation site distribution is unknown a priori. The third term on the right-hand side of Eq. (9.111) was introduced by Judd and Hwang (1976) to account for microlayer evaporation at the base of the bubble. The above nucleate boiling model considers only isolated bubbles. The analysis breaks down when bubbles coalesce and become slugs and columns. In addition, the above model includes expressions of bubble departure diameter D_b and frequency f_b, which are only correlations for most cases. There may be a high degree of uncertainty associated with the resulting correlations for heat transfer. Complete and reliable mathematical models for heat transfer in nucleate boiling have yet to be developed. Consequently, empirical correlations based on experimental data are widely used to predict heat transfer in nucleate boiling.

Jakob and Linke (1935) proposed one of the earliest and most popular models. Their model assumes that bubble growth and departure cause the convective heat transfer of liquid from the solid surface. This reasoning allowed Rohsenow (1952) to construct the correlation for nucleate boiling in the following single-phase forced convection type:

$$Nu = \frac{1}{C_{s,\ell}} Re^{1-m} Pr_\ell^{1-n} \tag{9.112}$$

where $C_{s,\ell}$ is an empirical constant that accounts for the effect of different combinations of heating surface and liquid. The Nusselt number is defined as

$$Nu = \frac{h}{k_\ell}\left[\frac{\sigma}{(\rho_\ell - \rho_v)g}\right]^{1/2} = \frac{q''}{k_\ell(T_w - T_{sat})}\left[\frac{\sigma}{(\rho_\ell - \rho_v)g}\right]^{1/2} \tag{9.113}$$

where the bracket term at the right-hand side, together with its exponent (1/2) is the characteristic length.

The Reynolds number using the same characteristic length is

$$Re = \left(\frac{q''}{\rho_\ell h_{\ell v}}\right)\left[\frac{\sigma}{(\rho_\ell - \rho_v)g}\right]^{1/2}\frac{1}{\nu_\ell} \tag{9.114}$$

where $q''/(\rho_\ell h_{\ell v})$ is superficial velocity of the liquid.

The heat flux in nucleate boiling, q'', is proportional to $(T_w - T_{sat})^3$; therefore, the exponent m in Eq. (9.112) is chosen to be 1/3. The exponent n is, however, dependent on the specific combination of the surface and the liquid. Substituting Eqs. (9.113) and (9.114) into Eq. (9.112), the following empirical correlation is obtained (Rohsenow 1952):

$$q'' = \mu_\ell h_{\ell v}\left[\frac{g(\rho_\ell - \rho_v)}{\sigma}\right]^{1/2}\left[\frac{c_{p\ell}(T_w - T_{sat})}{C_{s,\ell} h_{\ell v} Pr_\ell^n}\right]^3 \tag{9.115}$$

which applies to clean surfaces and is insensitive to the shape and orientation of the surface, which is obviously an approximation. It is still by far the most widely used empirical correlation for heat transfer in nucleate boiling. The coefficient $C_{s,\ell}$ and the exponent n depend on the liquid–surface combination, and their representative values are shown in Table 9.2. It is necessary to point out that the error of heat flux estimated by Eq. (9.115) can be as high as $\pm 100\%$.

If the heat flux is specified and it is necessary to estimate the temperature of the heated surface, one can rearrange Eq. (9.115) to obtain

$$T_{\rm w} - T_{\rm sat} = \frac{h_{\ell v}}{c_{p\ell}} Pr_\ell^n C_{s,\ell} \left[\frac{q''}{\mu_\ell h_{\ell v}} \left(\frac{\sigma}{g(\rho_\ell - \rho_v)} \right)^{1/2} \right]^{1/3} \tag{9.116}$$

under these circumstances, the error is within $\pm 25\%$.

In addition to Rohsenow's empirical correlation presented here, there are numerous other empirical correlations for heat transfer in nucleate boiling. Cooper (1984) proposed an empirical correlation for the heat transfer coefficient for pool nucleate boiling as a function of reduced pressure and surface roughness:

$$h = 55 p_{\rm r}^{0.12 - 0.4343 \ln R_{\rm p}} (-0.4343 \ln p_{\rm r})^{-0.55} M^{-0.5} q''^{0.67} \tag{9.117}$$

where $p_{\rm r}$ is reduced pressure defined as $p_{\rm r} = p/p_{\rm c}$, $R_{\rm p}$ is the surface roughness in μm (it can be set to 1.0 μm for unknown roughness), and M is molecular mass of the fluid. Equation (9.117) is valid for reduced pressure between 0.001 and 0.9, and molecular masses ranging from 2 to 200. For horizontal copper cylinders, Cooper (1984) recommended that the boiling heat transfer coefficient can be obtained by multiplying the results obtained from Eq. (9.117) by 1.7. However, Thome (2016) pointed out that this correction is unnecessary for the boiling of a refrigerant on copper tubes.

While Cooper (1984) attempted to develop a correlation that is valid for all fluids, Gorenflo (1993) proposed a fluid-specific correlation that includes the effect of the surface roughness, i.e.,

$$h = h_0 F_{\rm PF} (q''/q_0'')^{nf} (R_{\rm p}/R_{\rm p0})^{0.133}, \quad 0.0005 < p_{\rm r} < 0.95 \tag{9.118}$$

Table 9.2 Values of $C_{s\ell}$ and n for various combinations of surfaces and fluids for Eqs. (9.115) and (9.116) (Rohsenow 1952)

Surface material	Surface finish	Fluid	$C_{s\ell}$	n
Brass		Water	0.0060	1.0
Chromium		Benzene	0.0101	1.7
Chromium		Ethyl alcohol	0.0027	1.7
Copper		Carbon tetrachloride	0.0130	1.7
Copper		Isopropanol	0.0130	1.7
Copper	Lapped	n-Pentane	0.0049	1.7
Copper	Polished	n-Pentane	0.0154	1.7
Copper	Polished	Water	0.0130	1.0
Copper	Scored	Water	0.0068	1.0
Nickel		Water	0.0060	1.0
Platinum		Water	0.0130	1.0
Stainless steel	Chemically etched	Water	0.0130	1.0
Stainless steel	Ground and polished	Water	0.0060	1.0
Stainless steel	Mechanically polished	Water	0.0130	1.0
Stainless steel	Teflon pitted	Water	0.0058	1.0

where $p_r = p/p_{cr}$ is the reduced pressure and h_0 is a reference heat transfer coefficient at fixed reference conditions: $p_{r0} = 0.1$ $R_{p0} = 0.4 \mu m$, and $q''_0 = 2.0 \times 10^4$ W/m^2; it is different for different fluids and is listed in Table 9.3. For the fluid that is not listed, h_0 can be obtained by the experimental results or other correlations. The pressure correction factor is

$$F_{PF} = \begin{cases} 1.73 p_r^{0.27} + [6.1 + 0.68/(1 - p_r)] p_r^2, & \text{for water} \\ 1.2 p_r^{0.27} + 2.5 p_r + p_r/(1 - p_r), & \text{for all fluids except water and helium} \end{cases} \tag{9.119}$$

and the exponent in Eq. (9.118), nf, decreases with increasing p_r:

$$nf = \begin{cases} 0.9 - 0.3 p_r^{0.3}, & \text{for water} \\ 0.9 - 0.3 p_r^{0.15}, & \text{for all fluids except water and helium} \end{cases} \tag{9.120}$$

Table 9.3 Critical pressure and reference heat transfer coefficient (Gorenflo 1993)

Fluid	Chemical formula	Molecular mass (kg/kmol)	P_{crit} (bar)	h_0 (W/m^2 K)
Acetone	$(CH_3)_2CO$	58.08	47.0	3950
Argon (on Cu)	Ar	39.95	49.0	8200
Argon (on Pt)	Ar	39.95	49.0	6700
Ammonia	NH_3	17.03	113.0	7000
Benzene	C_6H_6	78.11	48.9	2750
n-Butane	$CH_3CH_2CH_2CH_3$	58.12	38.0	3600
n-Butanol	$C_4H_{10}O$	74.12	49.6	2600
i-Butanol	$C_4H_{10}O$	74.12	43.0	4500
Carbon Dioxide (at triple point)	CO_2	44.01	73.8	5100
Chloromethane	CH_3Cl	50.49	66.8	4400
Diphenyl	$C_{12}H_{10}O_2S$	154.2	38.5	2100
Ethane	C_2H_6	30.07	48.8	4500
Ethanol	$C_2F_3Cl_3$	46.07	63.8	4400
Freon®-11	$CFCl_3$	137.4	44.0	2800
Freon®-12	$C_2F_3Cl_3$	120.9	41.6	4000
Freon®-13	$CClF_3$	104.5	38.6	3900
Freon®-22	CHF_2Cl	86.47	49.9	3900
Freon®-23	CHF_3	70.02	48.7	4400
Freon®-113	$C2F3Cl_3$	187.4	34.1	2650
Freon®-114	CCl_2FCF_3	170.9	32.6	2800
Freon®-115	C_2ClF_5	154.5	31.3	4200
Freon®-123	$CHCl_2CF_3$	152.9	36.7	2600
Freon®-134a	CF_3CH_2F	102.0	40.6	4500
Freon®-152a	CHF_2CH_3	66.05	45.2	4000
Freon®-502	CHF_2Cl/C_2ClF_5	111.6	40.8	3300

(continued)

Table 9.3 (continued)

Fluid	Chemical formula	Molecular mass (kg/kmol)	P_{crit} (bar)	h_0 (W/m^2 K)
n-Heptane	C_7H_{16}	100.2	27.3	3200
n-Hexane	C_6H_{14}	86.18	29.7	3300
Hydrogen	H_2	2.02	12.97	24,000
Methane	CH_4	16.04	46.0	7000
Neon	Ne	20.18	26.5	20,000
Nitrogen (on Cu)	N_2	28.02	34	10,000
Nitrogen (on Pt)	N_2	28.02	34	7000
Oxygen (on Cu)	O_2	32.00	50.5	9500
Oxygen (on Pt)	O_2	32.00	50.5	7200
n-Pentane	C_5H_{12}	72.15	33.7	3400
i-Pentane	C_5H_{12}	72.15	33.3	3400
Propane	$CH_3CH_2CH_3$	44.10	42.4	4000
n-Propanol	C_3H_8O	60.1	51.7	3800
i-Propanol	C_3H_8O	60.1	47.6	3000
RC-318	C_4F_8	200.0	28.0	4200
Sulfur hexafluoride	SF_6	146.1	37.6	3700
Tetrafluoromethane	CF_4	88.00	66.8	4400
Toluene	C_7H_8	92.14	41.1	2650
Water	H_2O	18.02	220.6	5600

The surface roughness R_p is in μm. For the cases with unknown surface roughness, R_p is set as 0.4 μm.

The effects of various physical parameters, such as surface wettability, surface roughness, heater geometry, system pressure, liquid subcooling, and gravity on the rate of nucleate boiling heat transfer are reviewed in detail by Dhir (2018).

9.4 Critical Heat Flux

In nucleate boiling, heat flux increases and reaches a maximum value with increasing surface temperature. Further increase in the surface temperature results in decreasing heat flux because the transition from nucleate boiling to film boiling takes place. The maximum heat flux that can be obtained by nucleate boiling is referred to as critical heat flux (CHF). In the case of controlled heat flux, a slight increase of heat flux beyond the CHF can cause the surface temperature to rise to a value exceeding the surface material's maximum allowable temperature. This in turn can cause severe damage or meltdown of the surface. The CHF is also referred to as burnout heat flux for this reason. The value of the critical heat flux is affected by hydrodynamic instabilities, wetting criteria, heat capacity, heater geometry, heating method, and pressure.

The CHF phenomenon has received considerable attention in the past, and different mechanisms have been proposed to interpret its cause. Heat transfer in nucleate boiling involves (1) heat conduction through liquid from the heating surface to the liquid–vapor interface, (2) evaporation at the liquid–vapor interface, and (3) the escape of vapor from the heating surface. The critical heat flux phenomena occur when either generation or escape of vapor is restricted.

Significant efforts have been made over the years in predicting critical heat flux. Kutateladze (1948) developed the following pioneering semi-empirical criterion for critical heat flux on a flat plate using a dimensional analysis based on similarity between critical heat flux and column flooding:

$$\frac{q''_{\max}}{\rho_v h_{\ell v}\left[\sigma g(\rho_\ell - \rho_v)/\rho_v^2\right]^{1/4}} = K \tag{9.121}$$

where K is a constant that is equal to 0.16 for pool boiling on a large horizontal flat surface.

Based on the assumption that CHF proceeds from the vapor escape limit, Zuber (1959) proposed one of the earliest successful models. This model was later refined by Lienhard and Dhir (1973). In the late stage of nucleate boiling, vapor bubbles join together to form vapor columns and slugs. The vapor columns and slugs are subject to the Helmholtz instability (Sect. 4.6.1), which occurs when small disturbances in the interface between counter-flowing liquid and vapor are amplified. As a result, the columns and slugs become distorted and block the liquid from wetting the heating surface. A vapor blanket forms over part of or the entire heating surface and separates the heating surface and the liquid.

Zuber (1959) postulated that the CHF is reached when the Helmholtz instability appears in the interface of the large vapor columns leaving the heating surface. He suggested a correlation that has the same form as Eq. (9.121), except the constant $K = \pi/24 = 0.131$. Eq. (9.121) is often referred to as the Kutateladze–Zuber equation.

Zuber (1959) assumed that the vapor columns were in a square array. To determine the centerline spacing of the vapor columns, Lienhard and Dhir (1973) used the most dangerous wavelength in the two-dimensional wave pattern for the Taylor instability of the interface between a horizontal semi-infinite liquid region above a layer of vapor, λ_D. The diameter of each column is assumed to be $\lambda_D/2$(see Fig. 9.19).

For vertical liquid and vapor flow, the critical Helmholtz velocity is [see Eq. (4.186)]

$$u_c = |\bar{u}_\ell - \bar{u}_v| = \sqrt{\frac{\sigma\alpha(\rho_\ell + \rho_v)}{\rho_\ell \rho_v}} \tag{9.122}$$

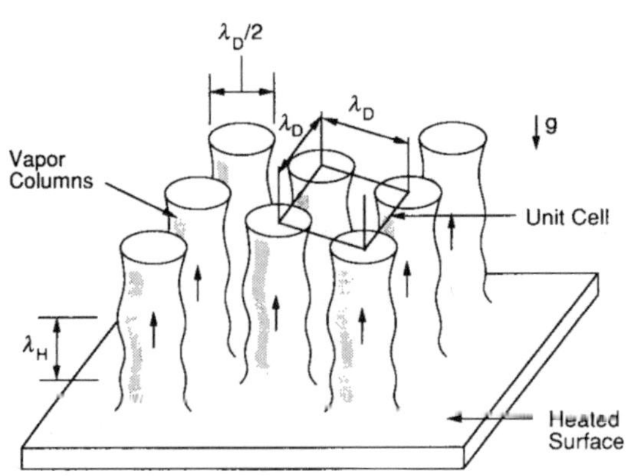

Fig. 9.19 Formation of vapor column prior to CHF (Carey 2016)

where the wave number α and the wavelength λ_D have the following relationship:

$$\alpha = \frac{2\pi}{\lambda_D} \tag{9.123}$$

Substituting Eq. (9.123) into Eq. (9.122) and considering $\rho_\ell \gg \rho_v$, the critical Helmholtz velocity becomes

$$u_c = \sqrt{\frac{2\pi\sigma}{\rho_v \lambda_D}} \tag{9.124}$$

Since the density of the vapor is significantly lower than that of the liquid, the upward vapor column velocity is much higher than the downward liquid velocity. Therefore, the critical Helmholtz velocity can be approximated as

$$u_c \doteq u_v = \frac{q''_{max}}{\rho_v h_{\ell v}} \frac{A_s}{A_c} \tag{9.125}$$

where A_s/A_c is the ratio of the total surface area to the area of the surface occupied by the vapor column. Since the spacing of the centerline of the vapor column is λ_D and the diameter of the vapor column is $\lambda_D/2$, the area ratio is

$$\frac{A_s}{A_c} = \frac{\lambda_D^2}{\pi(\lambda_D/2)^2/4} = \frac{16}{\pi} \tag{9.126}$$

Substituting Eq. (9.126) into Eq. (9.125), the critical Helmholtz velocity becomes

$$u_c = \frac{16 q''_{max}}{\pi \rho_v h_{\ell v}} \tag{9.127}$$

Combining Eqs. (9.124) and (9.127), the critical heat flux is obtained:

$$q''_{max} = \frac{\pi \rho_v h_{\ell v}}{16} \sqrt{\frac{2\pi\sigma}{\rho_v \lambda_D}} \tag{9.128}$$

The most dangerous wavelength, λ_D, as obtained by the Raleigh–Taylor instability analysis, was given by Eq. (5.219), i.e.,

$$\lambda_D = 2\pi \sqrt{\frac{3\sigma}{(\rho_\ell - \rho_v)g}} \tag{9.129}$$

Substituting Eq. (9.129) into Eq. (9.128), the critical heat flux becomes

$$q''_{max} = 0.149 \rho_v h_{\ell v} \left[\frac{\sigma g(\rho_\ell - \rho_v)}{\rho_v^2}\right]^{1/4} \tag{9.130}$$

which is identical to the Kutateladze-Zuber Eq. (9.121) except that the constant is 0.149—14% higher than that of Zuber model. Lienhard and Dhir (1973) showed that the constant of 0.149 could provide better agreement with the available experimental data. The critical heat fluxes predicted by Eq. (9.130) reasonably agreed with the experimental data for water, hydrocarbons, cryogenic liquids, and halogenated refrigerants. The experimentally measured CHF for the boiling of liquid metal, on the other hand, is two to four times higher than that predicted by Eq. (9.130). This discrepancy is due to a strong conduction-convection mechanism of liquid metal that the above model did not account for. Equation (9.130) indicates that the critical heat flux is proportional to $g^{1/4}$, which means the critical heat flux will be reduced under microgravity conditions.

Equation (9.130) is applicable to saturated boiling. The critical heat flux increases with subcooling of the liquid according to Zuber et al. (1961) by the following correlation:

$$\frac{q''_{\max}}{q''_{\max,\text{sat}}} = 1 + \frac{5.3(T_{\text{sat}} - T_\ell)}{\rho_v h_{\ell v}} \left(k_\ell \rho_\ell c_{p\ell}\right)^{1/2} \left[\frac{\sigma(\rho_\ell - \rho_v)g}{\rho_v^2}\right]^{-1/8} \left[\frac{(\rho_\ell - \rho_v)g}{\sigma}\right]^{1/4} \tag{9.131}$$

Equation (9.130) is applicable to an infinite horizontal heater surface, since there is no characteristic length in the expression. In order to use Eq. (9.130), which was obtained by assuming equality in spacing between vapor columns and the most dangerous wavelength, λ_D, the length scale of the heater surface, L, must be greater than $2\lambda_D$. Considering Eq. (9.129), this criterion can be written in terms of confinement number:

$$Co = \frac{\sqrt{\sigma/[(\rho_\ell - \rho_v)g]}}{L} \leq \frac{1}{4\pi\sqrt{3}} \tag{9.132}$$

Lienhard and Lienhard (2011) provides detailed physical descriptions and historical perspectives on both nucleate boiling and CHF. There has been some debate regarding how surface conditions affect the value for critical heat flux. Some experimental results (in accordance with Zuber's theory) show that there is no relationship between the two. To be more specific, surface-roughening techniques such as scoring and sandblasting have been found not to greatly influence the critical heat flux. In contrast, some experiments have shown that certain surface conditions (such as oxidation and deposition) do seem to increase the critical heat flux by increasing wettability of the fluid. The critical heat flux is significantly reduced on essentially nonwetting surfaces. In general, if the surface condition can directly increase the wettability of the fluid, then the value of the critical heat flux may be increased. In other circumstances, no general conclusion will apply in every instance.

Many different CHF mechanisms are presented in literature over the years for the main purpose of improving the accuracy of predictive CHF models including instability, bubble interface, dry spot, and dryout. These efforts also discussed the effect of contact angle, surface size, surface roughness, surface orientation, and pressure on CHF. The effects of various physical parameters such as surface wettability, surface finish, heater geometry and orientation on CHF are reviewed in detail by Dhir (2018). Liang and Mudawar (2018a) provided a detailed review of CHF for pool boiling for flat surfaces including the discussion of various CHF models and their associated mechanism and parametric trends. Liang and Mudawar (2018b) assessed different models and correlations for CHF based on a new consolidated CHF database consisting of 800 data points from 37 sources and includes 14 working fluids, pressures from 0.0016 to 5.2 MPa, orientation angles from 0° to 180°, and contact angles from 0° to 113°. They concluded that Eq. (9.130) is one of the best correlations for pool boiling on horizontal, upward-facing surfaces in the entire range of pressures.

Example 9.3

An electrically heated 5 mm platinum wire boils water at atmospheric conditions. Find the maximum heat flux and surface temperature that can be attained in the nucleate boiling regime.

Solution The saturation temperature of water at 1 atm is $T_{sat} = 100\,°\mathrm{C}$ and the properties are as follows: $h_{\ell v} = 2251.2\,\mathrm{kJ/kg}$, $\sigma = 58.9 \times 10^{-3}\,\mathrm{N/m}$, $\rho_\ell = 958.77\,\mathrm{kg/m^3}$, $\rho_v = 0.5974\,\mathrm{kg/m^3}$, $c_{p\ell} = 4.22$ kJ/kg K, $Pr_\ell = 1.76$, and $\mu_\ell = 279 \times 10^{-6}$ N s/m^2.
The critical heat flux can be found from Eq. (9.130), i.e.,

$$q''_{max} = 0.149\rho_v h_{\ell v}\left[\frac{\sigma g(\rho_\ell - \rho_v)}{\rho_v^2}\right]^{1/4}$$

$$= 0.149 \times 0.5974 \times 2251.2 \times 10^3 \times \left[\frac{58.9 \times 10^{-3} \times 9.8 \times (958.77 - 0.5974)}{0.5974^2}\right]^{1/4}$$

$$= 1257.3\,\mathrm{kW/m^2}$$

The surface temperature can be estimated from Eq. (9.116). For water boiling on the platinum surface, we have $C_{s,f} = 0.013$ and $n = 1$, i.e.,

$$T_w = T_{sat} + \frac{h_{\ell v}}{c_{p\ell}}\Pr_\ell^n C_{s,\ell}\left[\frac{q''}{\mu_\ell h_{\ell v}}\left(\frac{\sigma}{g(\rho_\ell - \rho_v)}\right)^{1/2}\right]^{1/3}$$

$$= 100 + \frac{2251.2}{4.22} \times 1.76^1$$

$$\times\, 0.013 \times \left[\frac{1257.3}{279 \times 10^{-6} \times 2251.2} \times \left(\frac{58.9 \times 10^{-3}}{9.8 \times (958.77 - 0.5974)}\right)^{1/2}\right]^{1/3}$$

$$= 120.9\,°\mathrm{C}.$$

9.5 Transition Boiling and Minimum Heat Flux

9.5.1 Transition Boiling

The transition boiling regime is probably the least studied region of the boiling curve because: (1) transition boiling is less important, practically speaking, than nucleate boiling, and (2) the mechanism of transition is very complicated and experimental investigation is very difficult. In this region, the liquid periodically contacts the heating surface creating large amounts of vapor and forcing the liquid away from the surface. The combined result is an unstable vapor film or blanket. Then the film collapses, allowing the liquid to contact the heating surface again. It is a region with alternate nucleate and film boiling, and is bypassed if the boiling curve experiment is performed with a controlled heat flux setting. It is only obtainable by experiments with controlled surface temperature, which are very rare in practical applications. The boiling curve in Fig. 9.3 shows that the excess temperature ΔT for transition boiling at one atmosphere is about 30–120 °C. The limited conditions conducive to this regime in turn limit its practical appeal to many investigators. However, investigation of transition boiling can help the experimenter to avoid burnout caused by excessive surface temperature. It is also possible to help reducing the heat flux fluctuations that result from unstable dry regions in contact with the heating surface.

Witte and Lienhard (1982) reported the existence of a hysteresis in transition boiling and suggested that there are two distinct boiling curves: the *nucleate-transition boiling* curve, obtained by increasing the excess temperature from nucleate boiling, and the *film-transition boiling* curve, created by decreasing excess temperature from film boiling (see Fig. 9.20). If the liquid wets the heating surface very well, which is generally presumed to be the case, the boiling curve does not progress immediately into the transition regime once the excess temperature exceeds its value corresponding to the critical heat flux. Witte and Lienhard (1982) suggested that the low contact angle made it easier for the liquid to spread over the surface, which promotes vigorous nucleation instead of the expected transition boiling. The curve also deviated from the conventional pool boiling curve when film boiling was initiated first, and the curve was then traversed by decreasing the excess temperature. As the point of minimum heat flux is approached, vapor is not produced as rapidly. The vapor layer becomes less stable, and the film breaks down. The lower the contact angle (i.e., the better the wetting characteristics of the liquid), the higher the temperature or heat flux at which the curve departs from film boiling. In this case, the boiling curve departs from film boiling without first reaching the minimum heat flux.

Recent experimental work in transition boiling does not support the existence of a hysteresis, because there was no evidence of two distinctive boiling curves. In an experimental work by Hohl et al. (1996), the entire boiling curve for FC-72 under steady and transient conditions was measured. The heater surface was a horizontal circular flat plate of 34 mm in diameter, coated with a 20-μm nickel layer. The surface temperature was controlled via feedback control. The typical boiling curve is shown in Fig. 9.21. As is evident, the boiling curves are identical regardless of the direction of the temperature change. The contradicting evidence with regard to the existence of hysteresis in the

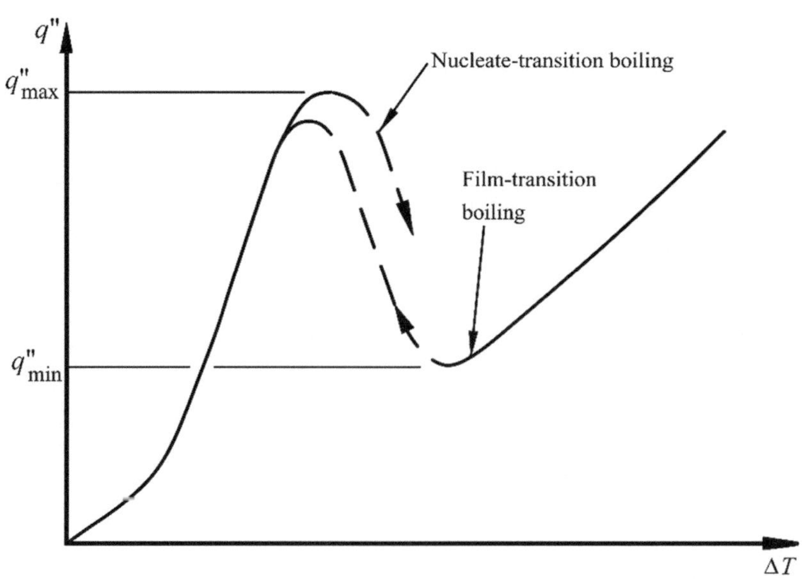

Fig. 9.20 Transition boiling curves

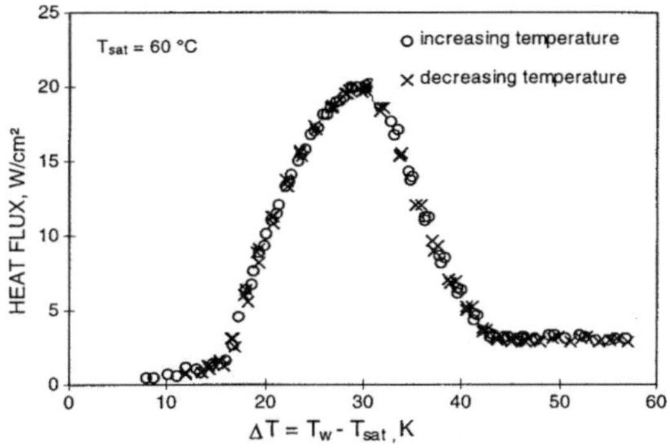

Fig. 9.21 Boiling curves of saturated FC 72 (Hohl et al. 1996)

boiling curve could be due to different geometric configurations and the temperature control system. Another possible cause is contamination, since some combinations of liquid and solid surface can affect wettability. At this time, there is no universal criterion for the existence of hysteresis in transition boiling.

Heat transfer in the transition boiling regime is very challenging to predict because it is the most complex regime in pool boiling. Contradiction about the existence of a hysteresis makes accurately predicting heat transfer even more difficult. Transition boiling can be viewed as a combination of unstable nucleate boiling and unstable film boiling at different locations, or at different times at the same location. Therefore, the heat flux for transition boiling can be expressed as (Berenson 1962)

$$q'' = Fq''_\ell + (1 - F)q''_v \tag{9.133}$$

where F is the average proportion of the heating surface in contact with liquid at any given moment. It can be estimated by the following correlation:

$$F = \exp\left(-2.2\frac{\Delta T}{\Delta T_{CHF}} + 2\right) \tag{9.134}$$

where ΔT_{CHF} is the excess temperature at critical heat flux (CHF), q''_{max}.

The heat flux during the vapor contact can reasonably be assumed to be equal to the minimum heat flux, i.e.,

$$q''_v = q''_{min} \tag{9.135}$$

The heat flux during the liquid contact, q''_ℓ, is time-dependent. Its time-averaged value can be correlated as

$$\frac{q''_\ell}{q''_{max}} = \frac{1 - 0.18q''_{min}/q''_{max}}{0.82\Delta T/\Delta T_{CHF}} \tag{9.136}$$

The minimum heat flux, q''_{min}, in the above correlations needs to be determined in the next subsection.

9.5.2 Minimum Heat Flux

The minimum heat flux point in the boiling curve is the boundary between the transition boiling regime and the film boiling regime. The minimum heat flux, $q''_{\min}$, is reached when the heat flux is equal to the minimum vapor formation rate that can sustain a stable vapor film over the heating surface. In controlled heat flux pool boiling, nucleate boiling is re-established when the heat flux falls below the minimum heat flux (see Fig. 9.3). The minimum heat flux is referred to as the Leidenfrost point. The vapor in the vapor film can be released by regular generation of vapor bubbles at some point and time. The minimum heat flux is related to the vapor bubble release rate and can be estimated by (Zuber 1959)

$$q''_{\min} = e_{\mathrm{b}} n''_{\mathrm{b}} f_{\min} \tag{9.137}$$

where e_{b} represents the energy per vapor bubble, n''_{b} is the number of vapor bubbles released per unit area and per release cycle, and $f_{\min}$ is the minimum number of cycles per second needed to compensate for normal collapse rate.

The energy per bubble is assumed to be equal to the latent heat carried away by the bubble with a radius equal to $\lambda_{\mathrm{D}}/4$ and released at the node of the Taylor wave:

$$e_{\mathrm{b}} = \frac{4\pi}{3}\left(\frac{\lambda_{\mathrm{D}}}{4}\right)^3 \rho_{\mathrm{v}} h_{\ell\mathrm{v}} \tag{9.138}$$

Zuber (1959) postulated that the vapor generated at the liquid–vapor interface was released at the node and antinode of a two-dimensional Taylor wave pattern, or once per half cycle, i.e.,

$$n''_{\mathrm{b}} = \frac{2}{\lambda_{\mathrm{D}}^2} \tag{9.139}$$

The bubble release frequency, $f_{\min}$, is

$$f_{\min} = C_1 \beta_{\max} \tag{9.140}$$

where $\beta_{\max}$ is the frequency of the most rapidly growing disturbance obtained by an interfacial stability analysis:

$$\beta_{\max} = \left[\frac{4(\rho_\ell - \rho_{\mathrm{v}})^3 g^3}{27(\rho_\ell + \rho_{\mathrm{v}})^3 \sigma}\right] \tag{9.141}$$

The minimum heat flux can be obtained by substituting Eqs. (9.138)–(9.140) into Eq. (9.137) and using Eq. (9.129) to obtain λ_{D}, i.e.,

$$q''_{\min} = C\rho_{\mathrm{v}} h_{\ell\mathrm{v}} \left[\frac{g\sigma(\rho_\ell - \rho_{\mathrm{v}})}{(\rho_\ell + \rho_{\mathrm{v}})^2}\right]^{1/4} \tag{9.142}$$

where $C = C_1(\pi^2/12)(4/3)^3$ is a new constant. Berenson (1961) recommended $C = 0.09$ by fitting the experimental data for pool boiling so that Eq. (9.142) becomes

$$q''_{min} = 0.09\rho_v h_{\ell v}\left[\frac{g\sigma(\rho_\ell - \rho_v)}{(\rho_\ell + \rho_v)^2}\right]^{1/4} \quad (9.143)$$

Equation (9.143) provides minimum heat flux for a horizontal surface; it is accurate within approximately 50% for most fluids at moderate pressure and is less accurate for higher pressure. Lienhard and Witte (1985) provided a much more detailed analysis on this subject. In a series of quenching experiments with smooth/structure hydrophilic/hydrophobic surfaces, the effects of surface wettability and surface structures on the minimum heat flux behavior were observed by Jo et al. (2018).

Example 9.4

Estimate the minimum heat flux for pool boiling of water at one atmosphere. When pressure increases, how will the minimum heat flux change?

Solution The properties of water at 1 atm can be found in Example 9.3. The minimum heat flux for pool boiling of water can be obtained from Eq. (9.143), i.e.,

$$\begin{aligned} q''_{min} &= 0.09\rho_v h_{\ell v}\left[\frac{g\sigma(\rho_\ell - \rho_v)}{(\rho_\ell + \rho_v)^2}\right]^{1/4} \\ &= 0.09 \times 0.5974 \times 2251.2 \times \left[\frac{9.8 \times 58.9 \times 10^{-3} \times (958.77 - 0.5974)}{(957.85 + 0.5974)^2}\right]^{1/4} \\ &= 18.95\,\text{kW/m}^2 \end{aligned}$$

To estimate the effect of pressure on the minimum heat flux, it is necessary to simplify Eq. (9.143). Considering the fact that $\rho_\ell \gg \rho_v$, Eq. (9.143) can be simplified as

$$q''_{min} = 0.09\rho_v h_{\ell v}\left(\frac{g\sigma}{\rho_\ell}\right)^{1/4}$$

Since the vapor density, ρ_v, is a strong function of pressure, i.e., it increases significantly with increasing pressure, it is expected that the minimum heat flux q''_{min} also increases with increasing pressure.

9.6 Film Boiling

9.6.1 Film Boiling Analysis

When the excess temperature increases beyond the minimum heat flux point, boiling enters the film boiling regime, characterized by a continuous vapor film covering the entire surface. The major thermal resistance is in this vapor layer. Unlike the vapor film in the transition boiling regime that in the film boiling regime is stable. Since the liquid is separated from the heating surface by a stable vapor film, heat transfer in this regime can be obtained by analyzing evaporation at the liquid–vapor interface.

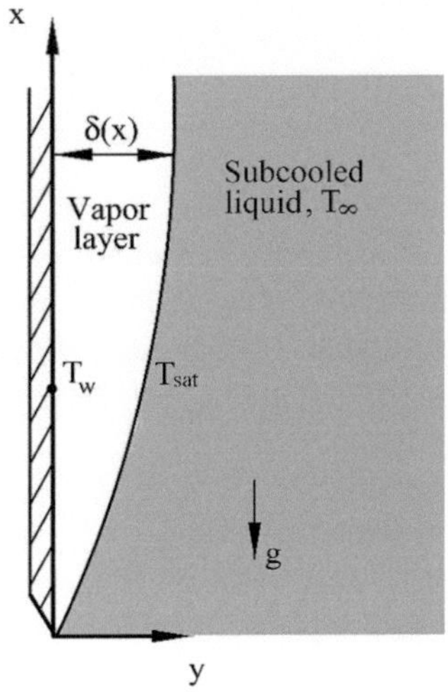

Fig. 9.22 Film boiling on a vertical surface

Figure 9.22 shows film boiling from a heated vertical flat plate in a motionless pool of subcooled liquid. Vapor generated at the liquid–vapor interface flows upward due to buoyancy force. Drag force from the vapor causes the adjacent liquid to flow upward as well. Therefore, a velocity boundary-layer forms in the liquid phase adjacent to the vapor film. Successful solution of the film boiling problem requires solutions of vapor and liquid flow as well as heat transfer in the vapor phase and at the interface. Since the liquid phase is subcooled, there is also heat transfer in the liquid. The vapor layer thickness is usually small compared to the length scale of the vertical flat plate, so boundary-layer approximations are applicable to the vapor film. It is further assumed that the vapor flow is laminar and two-dimensional. The continuity, momentum, and energy equations in the vapor layer are (Faghri et al. 2010):

$$\frac{\partial u_v}{\partial x} + \frac{\partial v_v}{\partial y} = 0 \tag{9.144}$$

$$u_v \frac{\partial u_v}{\partial x} + v_v \frac{\partial u_v}{\partial y} = \nu_v \frac{\partial^2 u_v}{\partial y^2} + \frac{g(\rho_\ell - \rho_v)}{\rho_v} \tag{9.145}$$

$$u_v \frac{\partial T_v}{\partial x} + v_v \frac{\partial T_v}{\partial y} = \alpha_v \frac{\partial^2 T_v}{\partial y^2} \tag{9.146}$$

Vapor flow drags the liquid adjacent to the vapor film, but the rest of the liquid phase is stationary. Therefore, boundary-layer theory can be applied to describe the motion of the liquid, i.e.,

$$\frac{\partial u_\ell}{\partial x} + \frac{\partial v_\ell}{\partial y} = 0 \tag{9.147}$$

$$u_\ell \frac{\partial u_\ell}{\partial x} + v_\ell \frac{\partial u_\ell}{\partial y} = \rho_\ell g \beta_\ell (T_\ell - T_\infty) + \nu_\ell \frac{\partial^2 u_\ell}{\partial y^2} \tag{9.148}$$

There will also be a thermal boundary layer in the liquid phase, and its temperature satisfies

$$u_\ell \frac{\partial T_\ell}{\partial x} + v_\ell \frac{\partial T_\ell}{\partial y} = \alpha_\ell \frac{\partial^2 T_\ell}{\partial y^2} \tag{9.149}$$

The boundary conditions at the heated wall ($y = 0$) are

$$y = 0: \quad u_v = v_v = 0, \quad T = T_w \tag{9.150}$$

The boundary conditions at the liquid–vapor interface ($y = \delta$) are necessary to couple the fluid flows in the liquid and vapor phases. The velocities for both phases are the same due to the nonslip condition, i.e.

$$y = \delta: \quad u_v = u_\ell \tag{9.151}$$

The shear stresses calculated from the liquid and vapor phases are the same as required by force balance, i.e.,

$$\left(\mu \frac{\partial u}{\partial y}\right)_v = \left(\mu \frac{\partial u}{\partial y}\right)_\ell, \quad y = \delta \tag{9.152}$$

The mass balance at the liquid–vapor interface can be obtained by applying conservation of mass for a control volume shown in Fig. 9.23, i.e.,

$$\left(\rho u \frac{\mathrm{d}\delta}{\mathrm{d}x} - \rho v\right)_v = \left(\rho u \frac{\mathrm{d}\delta}{dx} - \rho v\right)_\ell \tag{9.153}$$

The temperature at the liquid–vapor interface is equal to the saturation temperature:

$$T_v = T_\ell = T_{sat}, \quad y = \delta \tag{9.154}$$

The energy balance at the liquid–vapor interface is

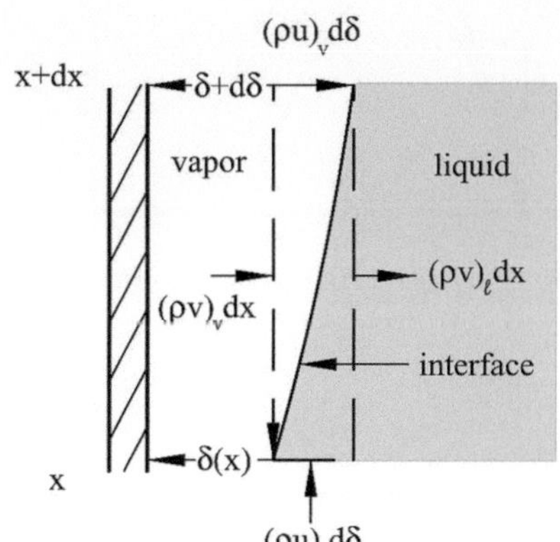

Fig. 9.23 Mass balance at the liquid–vapor interface

$$-k_{\mathrm{v}}\frac{\partial T_{\mathrm{v}}}{\partial y}+q''_{\mathrm{wR}}=\rho_{\mathrm{v}}\left(u_{\mathrm{v}}\frac{\mathrm{d}\delta}{\mathrm{d}x}-v_{\mathrm{v}}\right)-k_{\ell}\frac{\partial T_{\ell}}{\partial y} \tag{9.155}$$

where q''_{wR} is heat flux due to radiation from the heat wall to the interface.

The boundary condition in the liquid far from the heated surface is

$$u_{\ell}\to 0,\quad y\to\infty \tag{9.156}$$

For a case in which the liquid is saturated ($T_{\ell}=T_{\mathrm{sat}}$), the first term on the right-hand side of Eq. (9.148), representing buoyancy force in the liquid phase, will disappear. The energy Eq. (9.149) for the liquid phase will not be needed. Bromley (1950) was the first to provide a theoretical heat transfer prediction for film boiling from a horizontal tube using an analysis and assumptions similar to Nusselt's analysis for condensation (Sect. 7.3.3). He considered two types of boundary conditions at the liquid and vapor interfaces. One type of boundary condition assumes the liquid surrounding tube is stagnant (zero velocity at liquid/vapor interface) and the other type assumes liquid moves completely freely with vapor (zero shear interface at liquid/vapor). Dhir and Lienhard (1973) did the similar analysis for spheres.

Koh (1962) used similarity solution for the film boiling in a saturated liquid by converting the set of partial differential equations and boundary equations [Eqs. (9.144)–(9.156)] to a set of three ordinary differential equations with their corresponding boundary conditions using the following dimensionless independent variables:

$$\eta=\left(\frac{g}{4\nu_{\mathrm{v}}^{2}}\frac{\rho_{\ell}-\rho_{\mathrm{v}}}{\rho_{\mathrm{v}}}\right)^{1/4}\frac{y}{x^{1/4}} \tag{9.157}$$

$$\xi=\left(\frac{g}{4\nu_{\ell}^{2}}\frac{\rho_{\ell}-\rho_{\mathrm{v}}}{\rho_{\mathrm{v}}}\right)^{1/4}\frac{y}{x^{1/4}} \tag{9.158}$$

The velocity profile in the vapor and liquid, as well as the temperature profile in the vapor layer—obtained by Koh (1962) for the case of $Pr_{\mathrm{v}}=1.0$, $(\rho_{\mathrm{v}}\mu_{\mathrm{v}})/(\rho_{\ell}\mu_{\ell})=0.01$—is shown in Fig. 9.24.

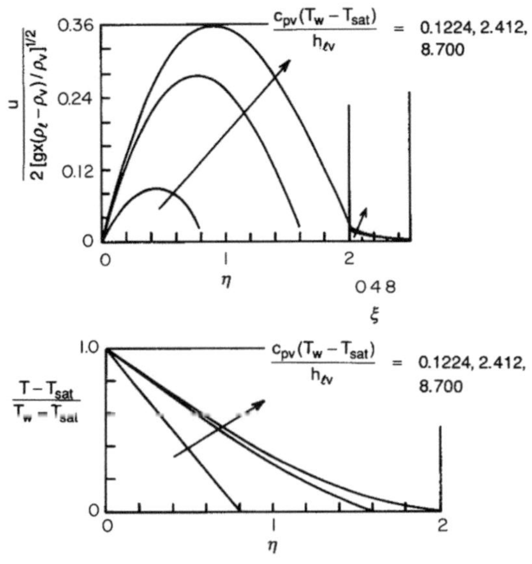

Fig. 9.24 Dimensionless velocity and temperature profiles ($Pr_{\mathrm{v}}=1.0$, $[(\rho_{\mathrm{v}}\mu_{\mathrm{v}})/(\rho_{\ell}\mu_{\ell})]=0.01$; Koh 1962)

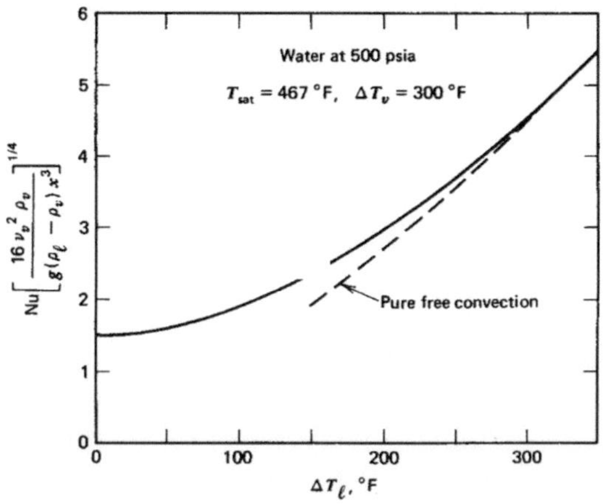

Fig. 9.25 Effect of subcooling on Nusselt number for film boiling from a vertical plate (Sparrow and Cess 1962)

The effect of subcooling on boiling from a vertical plate was studied by Sparrow and Cess (1962). Figure 9.25 shows the effect of subcooling on Nusselt number for film boiling of water with $T_{sat} =$ 467 °F from a vertical plate at $T_w = 767$ °F. It can be seen that liquid subcooling can have a significant effect on the Nusselt number. When the liquid subcooling is very large, the Nusselt number approaches that for the cases of free convection.

Forced convection laminar film boiling from a horizontal constant-temperature plate for the cases of saturated and subcooled liquids is solved by liquid and vapor boundary-layer equations by Cess and Sparrow (1961a, b). The result for saturated film boiling is presented below:

$$\frac{1}{2}K^{1/2} = NuRe^{-1/2}\left(\frac{\mu_v}{\mu_\ell}\right)\left[1+\pi^{1/2}NuRe^{-1/2}\left(\frac{\mu_v}{\mu_\ell}\right)\right]^{1/2} \tag{9.159}$$

where

$$K = \left(\frac{\rho_v\mu_v}{\rho_\ell\mu_\ell}\right)\frac{Pr_v h_{\ell v}}{c_{p,v}(T_w - T_{sat})} \tag{9.160}$$

$$Nu = \frac{hx}{k} \tag{9.161}$$

$$Re = \frac{u_\infty x}{\nu_\ell} \tag{9.162}$$

where x is distance from leading edge, and u_∞ is liquid forced velocity far from the plate. Equation (9.159) is valid for $10^{-3} \le K \le 1$. Gravity is negligible in the above analysis since forced convection was the driving force and therefore the result for $u_\infty = 0$ is not meaningful. Liquid subcooling will increase both heat transfer and plate shear.

The thermophysical properties of both liquid and vapor in the above analysis were assumed to be independent from the temperature. The influence of constant properties' assumption was examined by Nishikawa et al. (1976) by numerically solving the two-phase boundary-layer equations. They confirmed that, except at high system pressure, the results based on the constant properties' assumption reasonably agreed with the results based on variable properties. Nakayama (1986) also provided additional analysis related to forced convective film subcooled boiling for cases of constant surface temperature.

The analysis developed above do not account for heat transfer due to radiation from the heating surface across the vapor layer to the liquid. When the surface temperature is very high, the radiation across the vapor film contributes more and more to the overall heat transfer rate from the heating surface to the liquid. The rate of vapor production increases and the vapor layer thickens when radiation is significant. A detailed analysis of the radiation heat transfer must be performed in order to accurately predict the film boiling heat transfer.

Since the vapor is virtually transparent to infrared radiation emitted by the heating surface and liquid, the vapor can be treated as a nonparticipating medium; this assumption significantly simplifies the modeling of radiation heat transfer. The vapor films in film boiling are usually thin, allowing the radiation heat transport to be modeled as the radiation exchange between two parallel plates. Heat transfer due to radiation acts independently from the heat transfer from conduction and convection. The total heat transfer can be found by adding the contribution from radiation directly to the contributions from conduction and convection. Since the heating surface and the interface are treated as two parallel plates with different temperatures, the contribution from radiation can be obtained by

$$q''_{\mathrm{rad}} = \frac{\sigma_{\mathrm{SB}}(T_{\mathrm{w}}^4 - T_{\mathrm{sat}}^4)}{\frac{1}{\varepsilon_{\mathrm{s}}} + \frac{1}{\varepsilon_{\mathrm{I}}} - 1} \tag{9.163}$$

where σ_{SB} is the Stefan–Boltzmann constant, and ε_{s} and ε_{I} are the emissivity of the heating surface and the interface, respectively.

The effective heat transfer coefficient due to radiation is then obtained by

$$h_{\mathrm{rad}} = \frac{q''_{\mathrm{rad}}}{T_{\mathrm{w}} - T_{\mathrm{sat}}} = \frac{\sigma_{\mathrm{SB}}(T_{\mathrm{w}}^2 + T_{\mathrm{sat}}^2)(T_{\mathrm{w}} + T_{\mathrm{sat}})}{\frac{1}{\varepsilon_{\mathrm{s}}} + \frac{1}{\varepsilon_{\mathrm{I}}} - 1} \tag{9.164}$$

The overall heat transfer coefficient for film boiling can be obtained by (Bromley 1950)

$$h = h_{\mathrm{con}}\left(\frac{h_{\mathrm{con}}}{h}\right)^{1/3} + h_{\mathrm{rad}} \tag{9.165}$$

where h_{con} is the convective heat transfer coefficient. Equation (9.165) is not convenient for determining the overall heat transfer coefficient h, since h appears on both sides of the equation. Bromley (1950) further recommended that Eq. (9.165) could be simplified to

$$h = h_{\mathrm{con}} + 0.75h_{\mathrm{rad}} \tag{9.166}$$

The difference between the overall heat transfer coefficient obtained by Eqs. (9.165) and that obtained by Eq. (9.166) is within 5%.

The average Nusselt number for film boiling on a horizontal cylinder (Bromley 1950) or sphere (Lienhard and Dhir 1973) is given by

$$\overline{Nu} = \frac{\bar{h}D}{k_{\mathrm{v}}} = C\left[\frac{\rho_{\mathrm{v}}g(\rho_{\ell} - \rho_{\mathrm{v}})h'_{\ell\mathrm{v}}D^3}{\mu_{\mathrm{v}}k_{\mathrm{v}}}\right]^{\frac{1}{4}} \tag{9.167}$$

$$h'_{\ell v} = h_{\ell v}\left[1 + \frac{0.4c_{pv}(T_w - T_{sat})}{h_{\ell v}}\right] \tag{9.168}$$

The correlation constants C for horizontal cylinders and spheres are 0.62 and 0.67, respectively. The vapor properties are evaluated at the film temperature, $T_f = (T_w + T_{sat})/2$. Eq. (9.167) is valid for $0.8 < \lambda_c/D \le 8$ where $\lambda_c = 2\pi\{\sigma/[(\rho_\ell - \rho_v)g]\}^{1/2}$ is critical wavelength, or as long as the diameter of the horizontal cylinder or sphere is much greater than the vapor film thickness.

Another useful geometric configuration is film boiling over a horizontal, large flat plate, with the dense liquid overlaying a less-dense vapor. Vapor leaves the film in the form of detaching bubbles, which determines the heat transfer rate. Berenson (1961) performed an analysis of film boiling on a horizontal heat surface in which he assumed that the bubbles are released at the node of the Taylor wave. The correlation resulting from the Taylor instability analysis is given by

$$h = 0.425\left\{\left[\frac{k_v^3 g\rho_v(\rho_\ell - \rho_v)h'_{\ell v}}{\mu_v(T_w - T_{sat})}\right]\left[\frac{g(\rho_\ell - \rho_v)}{\sigma}\right]^{1/2}\right\}^{1/4} \tag{9.169}$$

which is in good agreement with experimental results obtained by using n-pentane and carbon tetrachloride.

Klimenko (1981) correlated the film boiling from a horizontal plate

$$Nu = \begin{cases} 0.19[Ga(\rho_\ell/\rho_v - 1)]^{1/3} Pr_v^{1/3} f_1 & Ga(\rho_\ell/\rho_v - 1) < 10^8 \text{ (Laminar)} \\ 0.0086[Ga(\rho_\ell/\rho_v - 1)]^{1/2} Pr_v^{1/3} f_2 & Ga(\rho_\ell/\rho_v - 1) > 10^8 \text{ (Turbulent)} \end{cases} \tag{9.170}$$

where

$$f_1 = \begin{cases} 1 & Ja_v \ge 0.71 \\ 0.89 Ja_v^{-1/3} & Ja_v < 0.71 \end{cases} \tag{9.171}$$

$$f_2 = \begin{cases} 1 & Ja_v \ge 0.5 \\ 0.71 Ja_v^{-1/2} & Ja_v < 0.5 \end{cases} \tag{9.172}$$

where

$$Nu = \frac{h\lambda_c}{k} \tag{9.173}$$

$$Ja_v = \frac{c_{p,v}(T_w - T_{sat})}{h_{\ell v}} \tag{9.174}$$

$$Ga = \frac{g\lambda_c^3}{2v_v^2} \tag{9.175}$$

are Nusselt, Jakob, and Galileo numbers, respectively. All vapor properties are evaluated at $(T_w + T_{sat})/2$. Equation (9.170) is valid for $7 \times 10^4 < Ga(\rho_\ell/\rho_v - 1) < 3 \times 10^8$, $0.69 < Pr_v < 3.45$, $0.14 < Ja_v < 32.26$, $0.0045 < p/p_{crit} < 0.98$ and up to 21.7 g. The effect of plate size can be accounted for by

$$\frac{h}{h_\infty} = \begin{cases} 1, & L/\lambda_c > 5 \\ 2.9(\lambda_c/L)^{0.67}, & L/\lambda_c < 5 \end{cases} \tag{9.176}$$

where L is the minimum dimension of the horizontal plate.

Example 9.5

Film boiling at 1 atm occurs on the surface of a 2-cm-diameter sphere made out of polished copper. The surface temperature of the sphere is $T_w = 360$ °C. The emissivity of the sphere surface is $\varepsilon_w = 0.05$, but the liquid–vapor interface can be treated as a black body. Find the heat transfer coefficients due to convection and radiation, and the overall heat transfer coefficient.

Solution The liquid density and latent heat are evaluated at $T_{sat} = 100\,°\text{C}$, i.e., $\rho_\ell = 958.77\,\text{kg/m}^3$, $h_{\ell v} = 2251.2\,\text{kJ/kg}$. The properties of the vapor must be evaluated at the mean temperature of the vapor, $(T_w + T_{sat})/2 = 230\,°\text{C}$, i.e., $\rho_v = 0.4381\,\text{kg/m}^3$, $\mu_v = 1.715 \times 10^{-5}\,\text{N s/m}^2$, $k_v = 0.0341\,\text{W/m K}$, and $c_{pv} = 1.986\,\text{kJ/kg K}$.
The corrected latent heat is obtained from Eq. (9.168):

$$\begin{aligned} h'_{\ell v} &= h_{\ell v}\left[1 + \frac{0.4c_{pv}(T_w - T_{sat})}{h_{\ell v}}\right] \\ &= 2251.2 \times \left[1 + \frac{0.4 \times 1.986 \times (360 - 100)}{2251.2}\right] \\ &= 2457.7\,\text{kJ/kg} \end{aligned}$$

The Nusselt number for film boiling can be obtained from Eq. (9.167), i.e.,

$$\begin{aligned} \overline{\text{Nu}} &= 0.67\left[\frac{\rho_v g(\rho_\ell - \rho_v)h'_{\ell v}D^3}{\mu_v k_v}\right]^{1/4} \\ &= 0.67 \times \left[\frac{0.4381 \times 9.8 \times (958.77 - 0.4381) \times 2457.7 \times 10^3 \times 0.02^3}{1.715 \times 10^{-5} \times 0.0341}\right]^{1/4} \\ &= 408.6 \end{aligned}$$

The heat transfer coefficient due to film boiling is

$$\bar{h}_{con} = \frac{k_v\overline{Nu}}{D} = \frac{0.0341 \times 408.6}{0.02} = 696.7\,\text{W/m}^2\,\text{K}$$

The radiative heat transfer coefficient is

$$\begin{aligned} \bar{h}_{rad} &= \frac{\sigma_{SB}(T_w^2 + T_{sat}^2)(T_w + T_{sat})}{1/\varepsilon_w + 1/\varepsilon_I - 1} \\ &= \frac{5.67 \times 10^{-8} \times (633^2 + 373^2) \times (633 + 373)}{1/0.05 + 1/1.0 - 1} \\ &= 1.54\,\text{W/m K} \end{aligned}$$

The overall heat transfer coefficient can be obtained from Eq. (9.166), i.e.,

$$\bar{h} = \bar{h}_{\text{con}} + 0.75\bar{h}_{\text{rad}} = 696.7 + 0.75 \times 1.54 = 697.8\,\text{W/m K}.$$

9.6.2 Direct Numerical Simulation of Film Boiling

In additional to the above analytical results about film boiling, numerical solutions on this topic were also reported in the recent literature (Esmaeeli and Tryggvason 2004). Son and Dhir (1997) simulated film boiling for both two-dimensional and axisymmetric flows using a moving body-fitted coordinate system, which is limited to the modest deformation of the phase boundary. This limitation was later overcome by Juric and Tryggvason (1998) using a front-tracking method and by Son and Dhir (1998) using a level set method. The subcooled film boiling was solved by Banerjee and Dhir (2001) using a level set method. The Volume of Fluid (VOF) method was employed to simulate film boiling by Welch and Wilson (2000). In addition to the above work that use one method, the front-tracking method was also combined with the finite difference method (Esmaeeli and Tryggvason 2001, 2003) and level contour technique (Shin and Juric 2002). The direct numerical simulation of film boiling conducted by Esmaeeli and Tryggvason (2004) will be presented here.

The continuity, momentum, and energy equations for each phase in the boiling flow are

$$\frac{\partial \rho}{\partial t} + \nabla \cdot (\rho \mathbf{V}) = 0 \tag{9.177}$$

$$\frac{\partial \rho \mathbf{V}}{\partial t} + \nabla \cdot (\rho \mathbf{V}\mathbf{V}) = -\nabla p + \rho \mathbf{g} + \nabla \cdot \mu[\nabla \mathbf{V} + (\nabla \mathbf{V})^T] \tag{9.178}$$

$$\frac{\partial \rho c_{\text{p}} T}{\partial t} + \nabla \cdot (\rho c_{\text{p}} \mathbf{V} T) = \nabla \cdot (k \nabla T) \tag{9.179}$$

where the viscosity dissipation has been neglected in Eq. (9.179). At the liquid–vapor interface, the following jump conditions must be satisfied:

$$\rho_\ell(\mathbf{V}_\ell - \mathbf{V}_I) \cdot \mathbf{n} = \rho_{\text{v}}(\mathbf{V}_{\text{v}} - \mathbf{V}_{\text{I}}) \cdot \mathbf{n} = \dot{m}'' \tag{9.180}$$

$$\dot{m}''(\mathbf{V}_{\text{v}} - \mathbf{V}_\ell) = (\boldsymbol{\tau}_{\text{v}} - \boldsymbol{\tau}_\ell) \cdot \mathbf{n} - (p_{\text{v}} - p_\ell)\mathbf{I} \cdot \mathbf{n} + \sigma K \mathbf{n} \tag{9.181}$$

$$\dot{m}'' h_{\ell \text{v}} = q_I'' = k_{\text{v}} \left.\frac{\partial T}{\partial n}\right|_{\text{v}} - k_\ell \left.\frac{\partial T}{\partial n}\right|_\ell \tag{9.182}$$

where $\mathbf{V}_\ell$ and $\mathbf{V}_{\text{v}}$ are the fluid velocities at the liquid and vapor sides of the interface, and $\mathbf{V}_{\text{I}}$ is the interfacial velocity.

The above momentum and energy equations and their jump conditions can be rewritten as the following "single-field" form:

$$\frac{\partial \rho \mathbf{V}}{\partial t} + \nabla \cdot (\rho \mathbf{V}\mathbf{V}) = -\nabla p + \rho \mathbf{g} + \nabla \cdot \mu(\nabla \mathbf{V} + \nabla \mathbf{V}^T) + \sigma \int_I \delta(\mathbf{r} - \mathbf{r}_I) K \mathbf{n}_I \mathrm{d}A_I \tag{9.183}$$

$$\frac{\partial \rho c_p T}{\partial t} + \nabla \cdot (\rho c_p \mathbf{V} T) = \nabla \cdot (k \nabla T) - \left[1 - (c_v - c_\ell)\frac{T_{sat}}{h_{\ell v}}\right] \int_I \delta(\mathbf{r} - \mathbf{r}_I)\mathbf{q}''_I dA_I \tag{9.184}$$

where δ is a two- or three-dimensional delta function, $\mathbf{r}$ is the position vector and $\mathbf{r}_I$ is the position of the interface. The subscript I in Eqs. (9.183) and (9.184) represents interface.

Although both liquid and vapor can be treated as incompressible flows, the incompressible assumption near the interface is invalid. In order to rewrite the continuity equation and the jump condition into a single-field equation, the velocity field can be rewritten as

$$\mathbf{V} = \mathbf{V}_v H + \mathbf{V}_\ell (1 - H) \tag{9.185}$$

where H is a Heaviside function that has a value of one in vapor but zero in the liquid phase. The gradient of the Heaviside function can be expressed in terms of the interfacial properties:

$$\nabla H = \int_I \delta(\mathbf{r} - \mathbf{r}_I)\mathbf{n}_I dA_I \tag{9.186}$$

which is zero everywhere except at the interface.

The divergence of the velocity defined in Eq. (9.185) is

$$\nabla \cdot \mathbf{V} = H \nabla \cdot \mathbf{V}_v + (1 - H)\nabla \cdot \mathbf{V}_\ell + (\mathbf{V}_v - \mathbf{V}_\ell) \cdot \nabla H \tag{9.187}$$

Substituting Eq. (9.186) into Eq. (9.187) and considering liquid and vapor are incompressible ($\nabla \mathbf{V}_v = \nabla \mathbf{V}_\ell = 0$), one obtains

$$\nabla \cdot \mathbf{V} = \int_I \delta(\mathbf{r} - \mathbf{r}_I)(\mathbf{V}_v - \mathbf{V}_\ell) \cdot \mathbf{n}_I dA_I \tag{9.188}$$

Eliminating $\mathbf{V}_I$ in Eq. (9.180) yields

$$(\mathbf{V}_v - \mathbf{V}_\ell) \cdot \mathbf{n} = \dot{m}'' \left(\frac{1}{\rho_v} - \frac{1}{\rho_\ell}\right) = \frac{q''}{h_{\ell v}} \left(\frac{1}{\rho_v} - \frac{1}{\rho_\ell}\right) \tag{9.189}$$

Substituting Eq. (9.189) into Eq. (9.188), the single-field continuity equation becomes

$$\nabla \cdot \mathbf{V} = \frac{q''}{h_{\ell v}} \left(\frac{1}{\rho_v} - \frac{1}{\rho_\ell}\right) \int_I \delta(\mathbf{r} - \mathbf{r}_I) q''_I dA_I \tag{9.190}$$

Esmaeeli and Tryggvason (2004) solved the single-field governing Eqs. (9.183), (9.184), and (9.190) using a second-order space-time accurate front-tracking/finite difference method on a staggered grid. They obtained a numerical solution of film boiling on a horizontal surface where a thin vapor film separates the liquid and the heated surface. The domain size is $\lambda_{d3} \times \lambda_{d3} \times 2\lambda_{d3}$ ($\lambda_{d3} = 2\pi\sqrt{6\sigma/(\rho_\ell - \rho_v)g}$ is the most dangerous three-dimensional wavelength) and the grid resolution is $96 \times 96 \times 192$. Phase change takes place at the liquid–vapor interface. Evolution of a liquid vapor interface and velocity field during film boiling is shown in Fig. 9.26. The properties of the liquid and vapor correspond to the properties of water at $p_{sat} = 169$ bar, with the exception of the

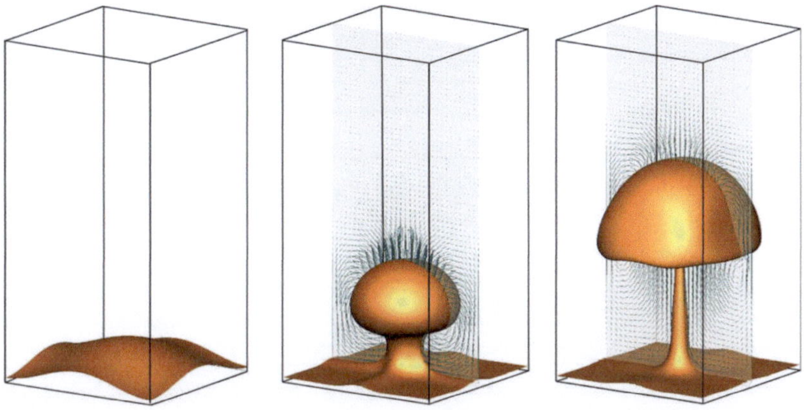

(a) t = 0 (b) t = 8.38 (c) t = 16.76

Fig. 9.26 Film boiling on a heated horizontal surface ($Pr = 4.2$, $Gr = 17.85$, $Ja = 0.064$, $\rho_v/\rho_\ell = 0.21$, $\mu_v/\mu_\ell = 0.386$, $c_v/c_\ell = 1.83$; Esmaeeli and Tryggvason 2004; reprinted with permission from Elsevier)

Grashof number. Initially, a quiescent liquid pool rests on a hot horizontal plate blanketed by a thin vapor film, as shown in Fig. 9.26a. The initial temperatures of liquid and vapor are at the saturation temperature. As the horizontal plate is heated, liquid vapor phase change takes place at the interface and a large vapor bubble is formed as is shown in Fig. 9.26b. As the heating continues, the neck of the bubble shrinks and the vapor bubble rises due to buoyancy effect (see Fig. 9.26c).

9.6.3 Leidenfrost Phenomena

A drop of liquid is introduced to a solid surface of constant temperature. If the temperature of the solid is around the boiling point of the liquid, the drop will rapidly boil and evaporate. However, if the solid is held at a temperature much higher than the boiling point of the liquid, a thin film of vapor forms between the solid and the liquid (see Fig. 9.27). A drop that floats on its own vapor in this way is called a *Leidenfrost drop*, named after the German physician, Johann Gottlob Leidenfrost, who first reported the phenomenon in 1756. The thin vapor film keeps the liquid and solid from coming into direct contact with each other thus inhibiting bubble nucleation. The film also acts as an insulator for the liquid above it, thereby slowing the evaporation process such that, for example, a droplet of water with radius of 1 mm can float for an entire minute over a metallic surface at 200 °C to evaporate completely.

A variety of technological applications involves imposing liquid drops on a hot surface, therefore making the Leidenfrost phenomenon an important topic of study. Examples of such applications include spray cooling of hot metal during metallurgical production, film cooling of a rocket nozzle, mist flow heat transfer in evaporators, fuel droplet vaporization in fuel-injected engines, and reflooding of a nuclear reactor core after an accidental loss of coolant.

The Leidenfrost effect can be easily observed by placing a few drops of water on a hot frying pan (nonstick). When the pan reaches the Leidenfrost temperature, the water droplets will bead up and "dance" across the surface due to the immediate vaporization of the bottom of the liquid drop. Physicists have hypothesized that firewalkers (people who walk barefooted across hot beds of coal) used the Leidenfrost phenomenon to their advantage to impress crowds. However, after several

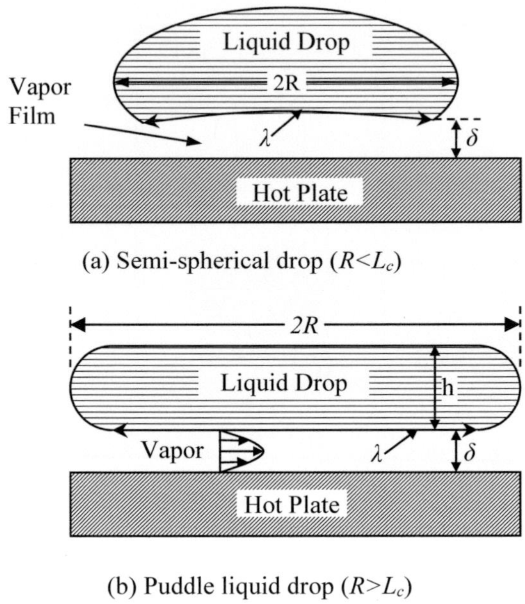

Fig. 9.27 Schematic of a Leidenfrost drop hovering over a solid hot surface with radius R and vapor film thickness δ

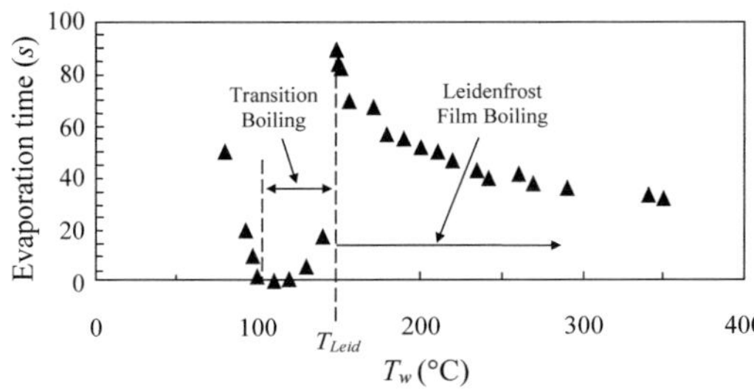

Fig. 9.28 Lifetime of a millimetric water droplet of radius R_o = 1 mm deposited on a Duralumin plate of temperature T [reused with permission from Biance et al. (2003), Copyright 2003, American Institute of Physics]

injurious experiments, they lost faith in their argument and decided instead that the entertainers completed this feat due to a combination of other hidden conditions (such as a layer of insulating ash).

To explain the Leidenfrost phenomenon, the lifetime τ of a water droplet and an n-heptane droplet are shown as a function of the solid plate temperature in Figs. 9.28 and 9.29, respectively. The lifetime of the water droplet decreases to around 200 ms at temperatures below 100 °C. As the plate temperature increases from 100 to 150 °C, the vapor film begins to form and a considerable increase in the droplet's lifetime is observed. The temperature at which the longest droplet lifetime is recorded is known as the *Leidenfrost temperature*. This temperature, and correspondingly the maximum droplet lifetime, is dependent on a number of factors including solid roughness, liquid purity, and the manner in which the drop is placed on the solid. At temperatures greater than the Leidenfrost temperature, the lifetime of the droplet gradually decreases. The same trend is true for n-heptane for which the Leidenfrost temperature is 200 °C.

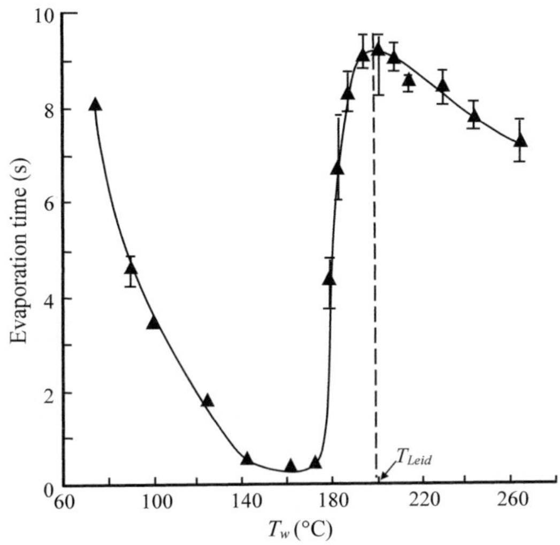

Fig. 9.29 Lifetime of an n-heptane droplet of radius $R_o = 0.75$ mm deposited on a stainless steel surface of temperature T (Chandra and Avedisian 1991)

Leidenfrost drops are considered nonwetting and exhibit levitation characteristics. The capillary length L_c for a liquid drop is defined as

$$L_c = \sqrt{\frac{\sigma}{\rho g}} \tag{9.191}$$

where σ is the liquid surface tension and ρ is the liquid density. If the radius R is smaller than the capillary length, the drop is spherical except for a flattened bottom (Fig. 9.27a). Contact is taken as the region of the drop interface parallel to the solid surface. The relationship between the contact size λ, radius R, and capillary length L_c is found by combining the correlation established by Mahadevan et al. (1999) with the geometric Hertz relation

$$\lambda \sim \frac{R^2}{L_c} \tag{9.192}$$

If the radius R is larger than the capillary length, the drop forms a puddle flattened by gravity (Fig. 9.27b). In this case, the contact size λ is related only to the radius R:

$$\lambda \sim R \tag{9.193}$$

The puddle's thickness h is found by balancing the surface tension and the hydrostatic force, yielding

$$h = 2L_c \tag{9.194}$$

It has been observed that the radius R of a Leidenfrost drop of water must be on the order of 1 cm. In cases where the radius R exceeded 1 cm, a bubble of vapor rises at the center and bursts at the upper interface due to a Rayleigh-Taylor instability at the lower interface (Biance et al. 2003). The largest radius a droplet of water can have without bubbles forming is named the critical radius R_c and it is linearly related to the capillary length L_c and to the puddle height according to the following two equations

$$R_c = 3.84 L_c \tag{9.195}$$

$$R_c = 1.92\,\mathrm{h} \tag{9.196}$$

Biance et al. (2003) presented a simple dimensional scaling and analysis by combining force and energy balance to show the lifetime of a droplet for $R < L_c$ and $R > L_c$, which is presented below. The thickness of the vapor film δ varies with time due to the evaporation process of the liquid drop on top of it.

For a puddle, $R > L_c$ (Fig. 9.27b), an energy balance yields the following equation for the evaporation rate:

$$\frac{\mathrm{d}m}{\mathrm{d}t} = \frac{k_v}{h_{\ell v}} \frac{\Delta T}{\delta} \pi \lambda^2 \tag{9.197}$$

where $\pi\lambda^2$ is the contact zone surface area, k_v is the vapor thermal conductivity, $\Delta T/\delta$ is the temperature gradient, m is the mass of the liquid drop and $\Delta T = T_\infty - T_{sat}$.

Assuming fully developed vapor flow between the liquid drop and the solid surface (Fig. 9.27b), a force balance yields the following solution

$$\frac{\mathrm{d}m}{\mathrm{d}t} = \rho_v \frac{2\pi\delta^3}{3\mu_v} \Delta p \tag{9.198}$$

where Δp is the pressure drop imposed by the liquid drop and μ_v is the gas viscosity.

By equating Eqs. (9.197) and (9.198) and using λ from Eq. (9.193) with $\Delta p = 2\rho_\ell g L_c$, one obtains the following equation for vapor film thickness for $R > L_c$:

$$\delta = \left(\frac{3 k_v \Delta T \mu_v}{4 h_{\ell v} \rho_v \rho_\ell g L_c} \right)^{1/4} R^{1/2} \tag{9.199}$$

For small drops, $R < L_c$ (Fig. 9.25a), contact size is found using Eq. (9.192), and the pressure drop acting on the vapor film is the Laplace pressure $2\sigma/R$. Equations (9.200) and (9.201) predict the forms of evaporation rate and vapor film thickness for small drops, respectively:

$$\frac{\mathrm{d}m}{\mathrm{d}t} \sim \frac{k_v}{h_{\ell v}} \frac{\Delta T}{R} R^2 \tag{9.200}$$

$$\delta \sim \left(\frac{k_v \Delta T \mu_v \rho_\ell g}{h_{\ell v} \rho_v \sigma^2} \right)^{1/3} R^{4/3} \tag{9.201}$$

For both small drops and puddles, the film thickness δ increases monotonically with the drop radius.

To determine the lifetime of a Leidenfrost drop of a particular radius, we must assume that the radius and thickness are related by the quasi-steady conditions developed above [Eqs. (9.197)–(9.201)]. For a puddle $R > L_c$ assuming $m = \pi R^2 h \rho_\ell$,

$$R(t) = R_o \left(1 - \frac{1}{\tau} \right)^2 \tag{9.202}$$

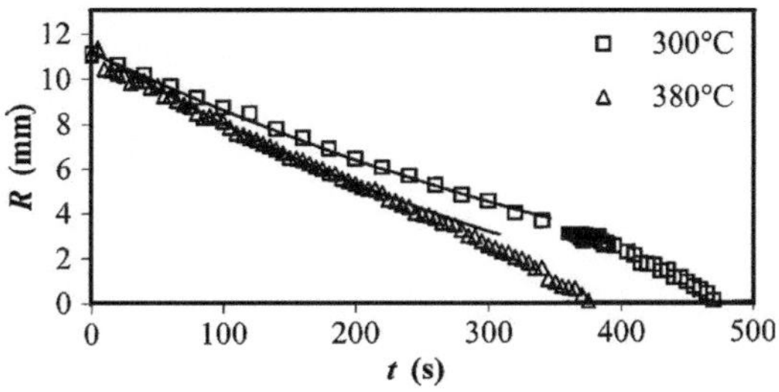

Fig. 9.30 Radius of a water droplet placed on a duralumin plate and filmed from above as a function of time. The fitted lines are predicted by Eq. (9.202) [reused with permission from Biance et al. (2003), Copyright 2003, American Institute of Physics]

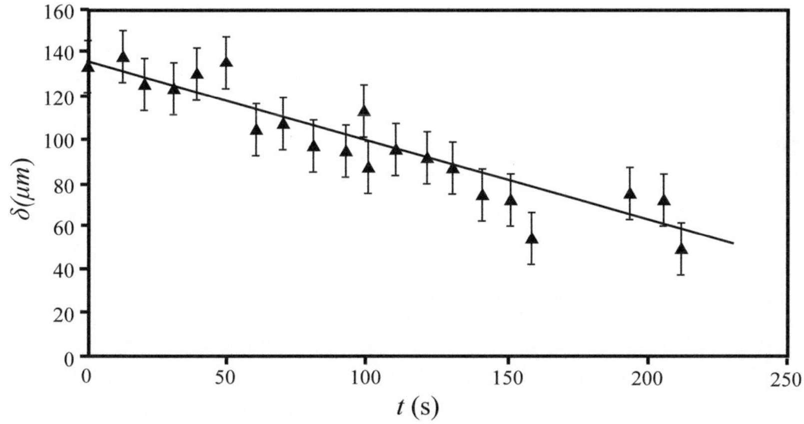

Fig. 9.31 Thickness of the vapor film of a water droplet placed on a duralumin plate and filmed from above as a function of time. The fitted line is predicted by Eq. (9.204) [Reused with permission from Anne-Laure Biance et al. (2003), Copyright 2003, American Institute of Physics]

where R_o is the radius at $t = 0$ and the lifetime τ is

$$\tau = 2\left(\frac{4\rho_\ell L_c h_{\ell v}}{k_v \Delta T}\right)^{3/4}\left(\frac{3\mu_v}{\rho_v g}\right)^{1/4} R_o^{1/2} \tag{9.203}$$

Equation (9.202) is the line in Fig. 9.30 and shows good agreement with the experimental data for water droplets for two different wall temperatures. The time dependence of the film thickness is derived from the above equations and gives the following linear relationship, which decreases as the time approaches the lifetime of the drop (as shown in Fig. 9.31):

$$\delta(t) = \left(\frac{3k_v \Delta T \mu_v R_o^2}{4h_{\ell v}\rho_v \rho_\ell g L_c}\right)^{1/4}\left(1 - \frac{t}{\tau}\right) \tag{9.204}$$

Similarly, for smaller drops, $R < L_c$, evaporation occurs over the whole drop surface as described earlier. Therefore, assuming $m = 4\pi R^3 \rho_\ell / 3$, the time dependence of the radius is

$$R(t) = R_o\left(1 - \frac{t}{\tau}\right)^{1/2} \tag{9.205}$$

where

$$\tau \sim \frac{\rho_\ell h_{\ell v}}{k_v \Delta T} R_o^2 \tag{9.206}$$

Equation (9.205) fits well with the trend shown in Fig. 9.30, where the radius variations occur more quickly toward the end of the drop's life. A comparison of Eqs. (9.203) and (9.206) reveals that the lifetime of a small drop is slightly more dependent on changes in temperature and much more on radius than the lifetime of a puddle. Bernardin and Mudawar (2002) presented a theoretically based physical model for Leidenfrost point which was constructed around vapor bubble nucleation, growth, interference criteria, and surface cavity size characterization including evaluation with an extensive data base.

The above analysis was performed using a simple, one-dimensional analysis of evaporation of a Leidenfrost droplet. A numerical solution of the two-dimensional analysis for two boundary conditions was performed by Nguyen and Avedisian (1987). The first case is a horizontal surface maintained at a constant temperature, and in the second case, the surface is insulated while the surrounding gas is heated.

In the two-dimensional analysis, vapor flows out of the droplet in the radial direction and the streamlines bend due to the presence of the wall. This results in the creation of a pressure field that lifts the droplet off the surface. The droplet continues to levitate until it has completely evaporated.

For both cases, assume the droplet shape to be spherical (which is valid for droplets in most industrial sprays and droplets with diameters of 100 μm or less). Also, take the evaporation process to be quasi-steady and properties to be constant. Radiative effects, spatial nonuniformities of temperature in the liquid, internal liquid motion, and buoyancy induced flow are all neglected.

The continuity, momentum, energy, and species equations for the steady case in the vapor phase are

$$\nabla \cdot \mathbf{V} = 0 \tag{9.207}$$

$$\mathbf{V} \cdot \nabla \mathbf{V} = -\frac{1}{\rho_v} \nabla p + \frac{\mu_v}{\rho_v} \nabla^2 \mathbf{V} \tag{9.208}$$

$$\mathbf{V} \cdot \nabla T = \alpha_v \nabla^2 T \tag{9.209}$$

$$\mathbf{V} \cdot \nabla \omega_i = D \nabla^2 \omega_i \tag{9.210}$$

where i = 1 is the droplet and i = 2 is the inert ambient, such that $\omega_1 + \omega_2 = 1$.

Performing a force balance and neglecting the effects of droplet acceleration caused by variations in levitation height, the weight of the droplet is found to be equal to the net force acting on the droplet due to viscous stress and pressure, as follows

$$\int \mathbf{j} \cdot \tau' dA = \mathbf{g}(\rho_\ell - \rho_v) \cdot \mathbf{V} \tag{9.211}$$

where τ' is the total stress tensor and $\mathbf{j}$ is the unit vector in the vertical direction.

The following boundary conditions are used:

Solid surface

$$\begin{cases} \mathbf{V} = 0 \\ \dfrac{\partial V_n}{\partial n} = \dfrac{\partial \omega_1}{\partial n} = 0 \\ T = T_w \quad (\text{case } 1) \\ \dfrac{\partial T}{\partial n} = 0 \quad (\text{case } 2) \end{cases} \tag{9.212}$$

Droplet surface

$$\begin{cases} T = T_{sat} \\ \omega = \omega_1 s \end{cases} \tag{9.213}$$

Ambient

$$\begin{cases} \mathbf{V} \to 0 \\ \omega_1 \to 0 \\ T \to T_\infty \end{cases} \tag{9.214}$$

Equations (9.212)–(9.214) were classified in the bispherical coordinate system, which is made up of a family of spheres (β = constant) that are each orthogonal to a family of spindle-shaped surfaces (α = constant).

The droplet evaporation time is obtained by the following mass balance:

$$\frac{4}{3}\pi\frac{d}{dt}\left(\rho_\ell R^3\right) = \rho_v \int u_\beta dA \tag{9.215}$$

where a droplet is the sphere $\beta = \beta_o$ and the solid surface corresponds to $\beta = 0$.

Variation of the evaporation time and evolution of levitation height based on the above two-dimensional analysis are presented in Figs. 9.32 and 9.33 for *n*-heptane. Of particular interest are

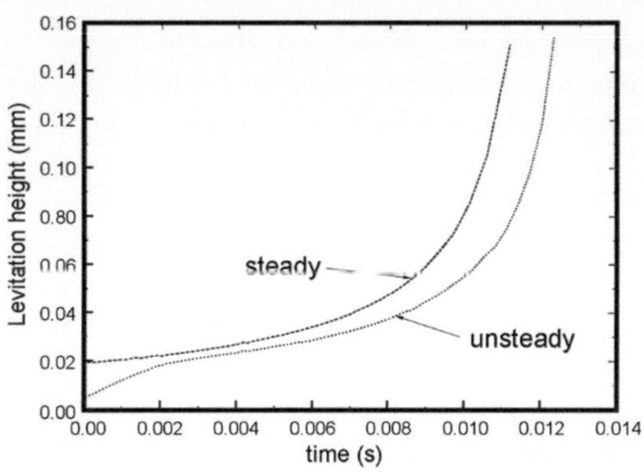

Fig. 9.32 Evolution of levitation height for n-heptane evaporating above an isothermal surface. Surface temperature is 600 K, ambient and initial droplet temperature is 300 K and initial droplet radius of 25 μm. Note that the droplet moves away from the surface during evaporation (Nguyen and Avedisian 1987; reprinted with permission from Elsevier).

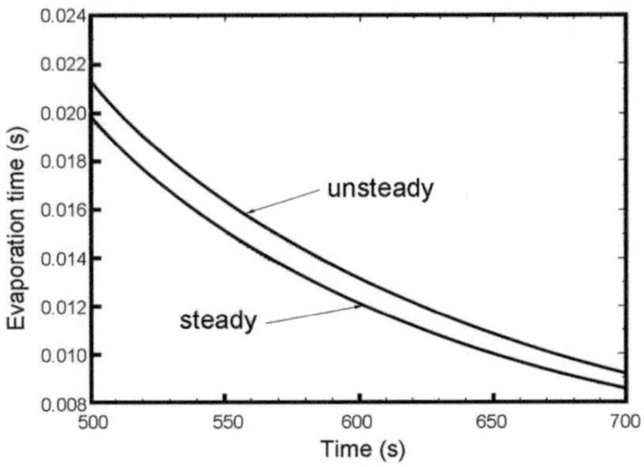

Fig. 9.33 Variation of droplet lifetime of n-heptane due to changes in surface temperature of the isothermal solid plate, initial droplet and ambient temperature is 300 K and the initial droplet radius is 25 μm (Nguyen and Avedisian 1987; reprinted with permission from Elsevier)

simulations shown in Fig. 9.32 of a Leidenfrost droplet moving away from the surface as it evaporates. This unusual effect has formed the basis of recent work on Leidenfrost drops flying off a surface for various industrial applications. Other characteristics of interest include evaporation at porous surfaces, and the ability of a drop to bounce when thrown at a solid hot surface, which require further study.

Problems

9.1 In saturated pool boiling at 1.013×10^5 Pa, what are the liquid temperatures at which vapor bubbles with diameters of 0.1 and 1 mm can exist and grow?

9.2 For vapor bubble growth in an initially uniformly superheated liquid, the relationship between the bubble radius and time is expressed by Eq. (9.32). At the early stage of bubble growth, the relationship between the bubble radius and time is expressed by Eq. (9.24). Show that Eq. (9.32) can be reduced to Eq. (9.24) for the early stage of bubble growth.

9.3 At the late stage of bubble growth, the relationship between the bubble radius and time is expressed by Eq. (9.28). Show that the general relationship between the bubble radius and time expressed by Eq. (9.32) can be reduced to Eq. (9.28) for the late stage of bubble growth.

9.4 A vapor bubble is initiated at $t = 0$ and grows in liquid water at 120 °C. Find the bubble size at time $t = 0.1$ ms using different models and compare the results.

9.5 Estimate the bubble departure diameter for water at 1 atm. Redo the problem for water at 2 atm and discuss the effect of pressure on the bubble departure diameter and frequency.

9.6 Use an example from industry to explain why it is very important to keep heat flux below critical heat flux.

9.7 A mechanically polished thin-wall stainless tube with a diameter of 3.5 mm and a length of 100 mm is immersed into nearly saturated water at 1 atm. The electric current passes through the stainless tube to supply the heat. Estimate the heat transfer coefficient when the electric power is 100 W.

9.8 A mechanically polished stainless steel pan with a diameter of 30 cm is filled with water and placed on top of the heating unit. The top surface of the water is exposed to ambient air at 1 atm pressure. Find the inner surface temperature of the pan that is necessary to generate vapor at a rate of 2.3 kg/h.

9.9 If the heat flux in Problem 9.8 is 50% of the critical heat flux, find (a) the rate of vapor generation, and (b) the inner surface temperature at the bottom of the pan.

9.10 Estimate the maximum heat flux for nucleate boiling of water at $p = 1.013 \times 10^5$Pa on the moon, where the gravitational acceleration is $g/6$.

9.11 Estimate the minimum heat flux for pool boiling of water at two atmospheric pressures. Compare your results with those in Example 6.4 and discuss the effect of pressure on the minimum heat flux.

9.12 The critical heat flux of water at 1 atm is $q''_{\max} = 1258.7\,\mathrm{kW/m^2}$, which occurs at $\Delta T = 30\,^\circ\mathrm{C}$. Boiling enters the transition boiling regime when the excess temperature is increased further. Estimate the heat flux when the excess temperature is equal to 50 °C.

9.13 If the convection terms in Eqs. (9.145) and (9.146) can be neglected, find the vapor film thickness and heat transfer coefficient for film boiling over a vertical flat plate. The liquid adjacent to the vapor layer can be treated as stationary and the effect of radiation from the heating surface to the interface can be neglected.

9.14 The film boiling of a saturated liquid on a vertical plate is described by Eqs. (9.144)–(9.156). Simplify these equations to a set of ordinary differential equations and corresponding boundary conditions using the similarity variables defined in Eqs. (9.157) and (9.158) as well as $\psi_v = 4\nu_v\left(\frac{g}{4\nu_v^2}\frac{\rho_\ell - \rho_v}{\rho_v}\right)^{1/4} x^{3/4} f(\eta)$, $\psi_\ell = 4\nu_\ell\left(\frac{g}{4\nu_\ell^2}\frac{\rho_\ell - \rho_v}{\rho_v}\right)^{1/4} x^{3/4} F(\xi)$, $\theta = \frac{T - T_{sat}}{T_w - T_{sat}}$, where ψ_v and ψ_ℓ are stream functions for the vapor and liquid phases ($u_v = \partial\psi_v/\partial y$, $v_v = -\partial\psi_v/\partial x$, $u_\ell = \partial\psi_\ell/\partial y, v_\ell = -\partial\psi_\ell/\partial x$). The vapor film thickness must satisfy the energy balances in the vapor control volume showing in Fig. P9.14.

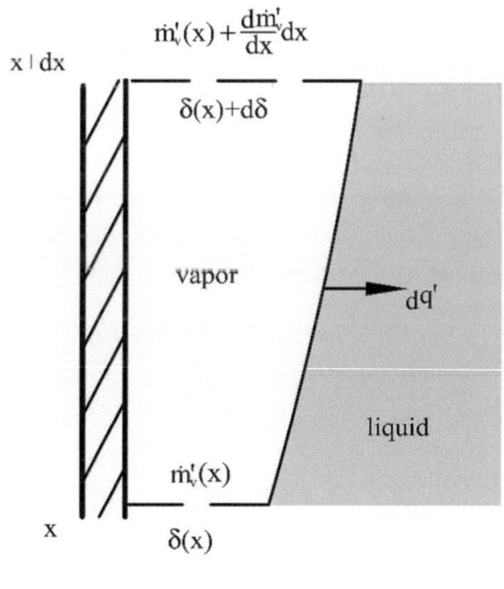

Fig. P9.14

9.15 Write a computer program to solve the film boiling problem described by a set of ordinary differential equations and corresponding boundary conditions obtained in Problem 9.14 and compare your results with those in Fig. 9.24.

9.16 Perform a scale analysis of the governing equations and corresponding boundary conditions for subcooled boiling, Eqs. (9.144)–(9.156), and identify the dominant nondimensional parameters for boiling of subcooled liquid on a vertical plate.

9.17 Define appropriate similarity variables for subcooled boiling and reduce the problem to a set of ordinary differential equations. Obtain the necessary boundary conditions for the ordinary differential equations to mathematically close the problem.

9.18 Film boiling occurs on an electrically heated 1.27 mm platinum wire placed horizontally in the water at atmospheric condition. The surface temperature of the wire is 754 °C. The emissivity of the platinum wire and the liquid vapor interface can be assumed as $\varepsilon_s = \varepsilon_i = 1$. Find the power dissipated per unit length of the wire.

9.19 Film boiling at 1 atm occurs on the surface of a 2-cm diameter sphere made out of polished copper. The surface temperature of the sphere is T_w = 800 °C. The emissivity of the sphere surface is $\varepsilon_w = 0.05$, but the liquid–vapor interface can be treated as a black body. Find the heat transfer coefficients due to convection, radiation, and the overall heat transfer coefficient. Compare your results with those of Example 9.5 and discuss the effect of radiation at different temperatures.

9.20 Specify the governing equations and corresponding boundary conditions for film boiling of saturated liquid on an isothermal cylinder with diameter D and temperature T_w. The effect of radiation from the cylinder surface to the liquid–vapor interface can be neglected.

9.21 If the inertia terms in the governing equations for both vapor and liquid phases are negligible and $\mu_v/\mu_\ell \doteq 0$, obtain the vapor film thickness for film boiling in Problem 9.20.

9.22 The heat transfer coefficient for boiling or condensation depends on the properties of the fluid, a characteristic length L, the temperature difference ΔT, buoyancy force $(\rho_\ell - \rho_v)g$, latent heat of vaporization $h_{\ell v}$, and surface tension, σ:

$$h = h[k, \mu, c_p, \rho, L, \Delta T, (\rho_\ell - \rho_v)g, h_{\ell v}, \sigma] \tag{9.216}$$

Using Buckingham's Π theorem show that the number of nondimensional variables is 5, as opposed to the 10 dimensional variables in Eq. (9.216). Furthermore, use dimensional analysis to show that

$$Nu = f(Gr, Ja, Pr, Bo) \tag{9.217}$$

where

$$Gr = \frac{\rho g(\rho_\ell - \rho_v)L^3}{\mu^2}, \quad Ja = \frac{c_p \Delta T}{h_{\ell v}}, \quad Bo = \frac{(\rho_\ell - \rho_v)gL^2}{\sigma} \tag{9.218}$$

are the Grashof number, Jakob number, and Bond number, respectively.

9.23 Block-heated heat pipes are normally used to transport energy from one place to another for the purpose of energy conservation or electronic component cooling. The evaporator and condenser segments are normally separated, which is similar to conventional heat pipes except for the localized heating. The working temperature is comparatively low or moderate, which means that radiation heat transfer is not important in the analysis of these heat pipes. Consider the heat pipe shown in Fig. P9.23. It has an evaporator section of length LE, a wall thickness δ, an outside radius R, and a block-heated area of width WH and length LH. An examination of the reported data shows that a power law boiling relation is appropriate for relating the heat flux to the evaporating temperature drop in a heat pipe, i.e., $q''_e = a(T - T_s)^b$. Assume the wall thickness is much smaller than the radius of the heat pipe and that variation of temperature along the thickness direction of the heat pipe is negligible. Specify the governing equation and corresponding boundary conditions to describe temperature distribution in the heat pipe wall.

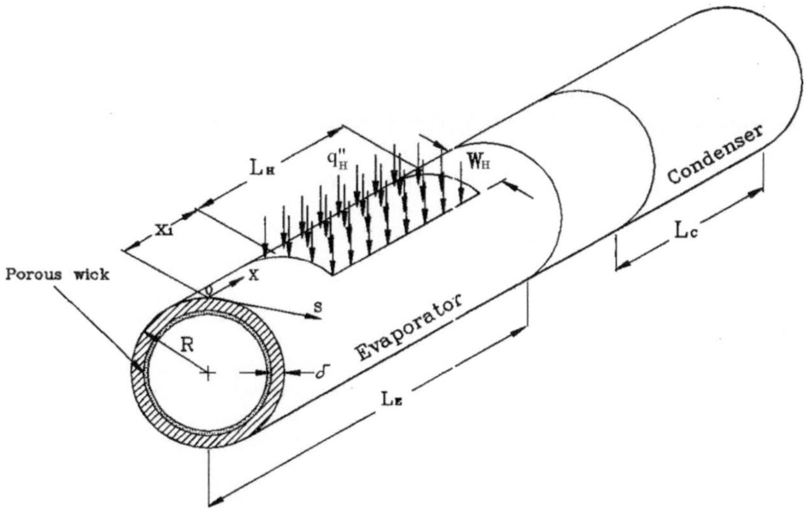

Fig. P9.23

References

Avedisian, C. T. (1986). Bubble growth in superheated liquid droplets. In *Encyclopedia of fluid mechanics*. Houston, TX: Gulf Publishing Company.

Avedisian, C. T., & Suresh, K. (1985). Analysis of non-explosive bubble growth within a superheated liquid droplet suspended in an immiscible liquid. *Chemical Engineering Science, 40,* 2249–2259.

Banerjee, D., & Dhir, V. K. (2001). Study of subcooled film boiling on a horizontal disc: Part I—Analysis. *Journal of Heat Transfer, 123,* 271–284.

Bankoff, S. G. (1958). Entrapment of gas in the spreading of liquid over a rough surface. *AIChE Journal, 4,* 24–26.

Berenson, P. J. (1961). Film boiling heat transfer from a horizontal surface. *Journal of Heat Transfer, 83,* 351–356.

Berenson, P. J. (1962). Experiments on pool-boiling heat transfer. *International Journal of Heat and Mass Transfer, 5,* 985–999.

Bergles, A. E. (2005). Bora Mikic and pool boiling. In *Proceedings of the 2005 ASME Summer Heat Transfer Conference*, San Francisco, CA.

Bergles, A. E., & Rohsenow, W. M. (1964). The determination of forced convection surface-boiling heat transfer. *Journal of Heat Transfer, 86,* 365–372.

Bernardin, J. D., & Mudawar, I. (2002). A cavity activation and bubble growth model of the Leidenfrost point. *Journal of Heat Transfer, 124,* 864–874.

Biance, A. L., Clanet, C., & Quere, D. (2003). Leidenfrost drops. *Physics of Fluids, 15,* 1632–1637.

Bromley, L. A. (1950). Heat transfer in stable film boiling. *Chemical Engineering Progress, 46,* 221–227.

Cao, Y., Faghri, A., & Mahefkey, E. T. (1989). The thermal performance of heat pipes with localized heat input. *International Journal of Heat and Mass Transfer, 32,* 1279–1287.

Carey, V. P. (2016). *Liquid-vapor phase-change phenomena: An introduction to the thermophysics of vaporization and condensation processes in heat transfer equipment* (3rd ed.). New York, NY: Taylor & Francis.
Cess, R. D., & Sparrow, E. M. (1961a). Film boiling in a forced-convection boundary-layer flow. *Journal of Heat Transfer, 83,* 370–376.
Cess, R. D., & Sparrow, E. M. (1961b). Subcooled forced-convection film boiling on a flat plate. *Journal of Heat Transfer, 83,* 377–379.
Chandra, S., & Avedisian, C. T. (1991). On the collision of a droplet with a solid surface. *Proceedings of Royal Society of London A, 432,* 13–41.
Cole, R., & Rohsenow, W. M. (1969). Correlations for bubble departure diameters for boiling of saturated liquid. *Chemical Engineering Progress, 65,* 211–213.
Cooper, M. G. (1984). Heat flow rates in saturated nucleate boiling—A wide ranging examination using reduced properties. In *Advances in heat transfer* (Vol. 16, pp. 157–239). Princeton, NJ: Academic Press.
Dhir, V. (2005). Mechanistic prediction of nucleate boiling heat transfer—Achievable or a hopeless task. In M. Jakob (Eds.), *Lecture in 2005 ASME Summer Heat Transfer Conference*, San Francisco, CA.
Dhir, V. K. (2018). Nucleate pool boiling. In F. Kulacki (Ed.), *Handbook of thermal science and engineering*. Cham: Springer.
Eastman, R. E. (1984). Dynamics of bubble departure. In *AIAA 1984 Thermophysics* (pp. 1–5.), AIAA-1984-1707.
Esmaeeli, A., & Tryggvason, G. (2001). Direct numerical simulations of boiling flows. In *Proceedings of the Fourth International Conference on Multiphase Flow* (ICMF-2001), New Orleans, LA.
Esmaeeli, A., & Tryggvason, G. (2003). Computations of explosive boiling in microgravity. *Journal of Scientific Computing, 19,* 163–182.
Esmaeeli, A., & Tryggvason, G. (2004). Computations of film boiling. Part I: Numerical method. *Int. J. Heat Mass Transfer, 47,* 5451–5461.
Faghri, A. (2016). *Heat pipe science and technology* (2nd ed.). Columbia, MO: Global Digital Press.
Faghri, A., Zhang, Y., & Howell, J. R. (2010). *Advanced heat and mass transfer*. Columbia, MO: Global Digital Press.
Fritz, W. (1935). Maximum volume of vapor bubbles. *Physikalishce Zeitschrift, 36,* 379–384.
Forster, D. E., & Greif, R. (1959). Heat transfer to a boiling liquid-mechanism and correlation. *Journal of Heat Transfer, 81,* 43–53.
Gorenflo, D. (1993). Pool boiling. In *VDI heat atlas*. Düsseldorf: VDI-Verlag.
Gorenflo, D., Knabe, V., & Beiling, V. (1986). Bubble density on surface with nucleate boiling—Its influence on heat transfer and burnout heat flux at elevated saturation pressure. In *Proceedings of the 8th International Heat Transfer Conference* (Vol. 4, pp. 1995–2000), San Francisco, CA.
Griffith, P., & Wallis, J. D. (1960). The role of surface conditions in nucleate boiling. *The Chemical Engineering Progress Symposium Series, 56*(30), 49–63.
Han, Y. Y., & Griffith, P. (1965). The mechanism of heat transfer in nucleate pool boiling. I-bubble initiation, growth and departure. *International Journal of Heat and Mass Transfer, 8,* 887–904.
Hohl, R., Auracher, H., Blum, J., & Marquardt, W. (1996). Pool boiling heat transfer experiments with controlled wall temperature transients. In *2nd European Thermal Science and 14th UIT National Heat Transfer Conference*, Rome (pp. 1647–1652).
Howell, J. R., & Siegel, R. (1966). Incipience, growth, and detachment of boiling bubbles in saturated water from artificial nucleation site of known geometry and size. In *Proceedings of the 3rd International Heat Transfer Conference*, Chicago, IL (Vol. 4, pp. 12–23).
Ivey, H. J. (1967). Relationship between bubble frequency, departure diameter and rise velocity in nucleate boiling. *International Journal of Heat and Mass Transfer, 10,* 1023–1040.
Jakob, M., & Linke, W. (1935). Heat transmission in the evaporation of liquids at vertical and horizontal surfaces. *Physik Z., 36,* 267–280.
Judd, R. L., & Hwang, K. S. (1976). A comprehensive model for nucleate boiling heat transfer including microlayer evaporation. *Journal of Heat Transfer, 98,* 623–629.
Juric, D., & Tryggvason, G. (1998). Computations of boiling flow. *International Journal of Multiphase Flow, 24,* 387–410.
Jo, H. J., Yeom, H., Yoon, D. S., Duarte, J. P., & Corradini, M. L. (2018). Minimum heat flux (MHF) Behavior with different surface characteristics including structured surfaces and different surface energies. *International Journal of Heat and Mass Transfer, 127,* 414–421.
Kandilikar, S. G., Dhir, V. K., & Shoji, M. (1999). *Handbook of phase change: Boiling and condensation*. Philadelphia, PA: Taylor and Francis
Keshock, E. G., & Siegel, R. (1962). *Forces acting on bubbles in nucleate boiling under normal and reduced gravity conditions*. NASA TN D-2299.
Klimenko, V. V. (1981). Film boiling on a horizontal plate-new correlation. *International Journal of Heat and Mass Transfer, 24,* 69–79.

Kocamustafaogullari, G., & Ishii, M. (1983). Interfacial area and nucleation site density in boiling systems. *International Journal of Heat and Mass Transfer, 26,* 1377–1387.

Koh, J. C. Y. (1962). Analysis of film boiling on vertical surfaces. *Journal of Heat Transfer, 84,* 55–62.

Kutateladze, S. S. (1948). On the transition to film boiling under natural convection. *Kotloturbostroenie, 3,* 10–12.

Lee, R. C., & Nydahl, J. E. (1989). Numerical calculation of bubble growth in nucleate boiling from inception through departure. *Journal of Heat Transfer, 111,* 474–479.

Liang, G., & Mudawar, I. (2018a). Pool boiling critical heat flux (CHF)—Part 1: Review of mechanisms, models, and correlations. *International Journal of Heat and Mass Transfer, 117,* 1352–1367.

Liang, G., & Mudawar, I. (2018b). Pool boiling critical heat flux (CHF)—Part 2: Assessment of models and correlations. *International Journal of Heat and Mass Transfer, 117,* 1368–1383.

Lienhard, J. H., & Dhir, V. K. (1973). *Extended hydrodynamic theory of the peak and minimum heat fluxes*. NASA CR-2270.

Lienhard, J. H. (2011). *A heat transfer textbook* (4th ed.). Mineola, NY: Dover Civil and Mechanical Engineering.

Lienhard, J. H., & Witte, L. C. (1985). An historical review of the hydrodynamic theory of boiling. *Reviews in Chemical Engineering, 3,* 187–280.

Lorenz, J. J., Mikic, B. B., & Rohsenow, W. M. (1974). The effect of surface conditions on boiling characteristics. In *Proceedings of 5th International Heat Transfer Conference* (Vol. 1, pp. 35–39).

Mahadevan, L., & Pomeau, Y. (1999). Rolling droplets. *Physics of Fluids, 11,* 2449–2453.

Malenkov, I. G. (1971). The frequency of vapor bubble separation as function of bubble size. *Fluid Mechanics: Soviet Research, 1,* 36–42.

Mikic, B. B., & Rohsenow, W. M. (1969). A new correlation of pool-boiling data including the effect of heating surface characteristics. *Journal of Heat Transfer, 91,* 245–250.

Mikic, B. B., Rohsenow, W. M., & Griffith, P. (1970). On bubble growth rate. *International Journal of Heat and Mass Transfer, 13,* 657–666.

Mizukami, K. (1975). Entrapment of vapor in re-entrant cavities. *Letters in Heat and Mass Transfer, 2,* 279–284.

Mukherjee, A., & Dhir, V. K. (2004). Study of lateral merger of vapor bubble during nucleate pool boiling. *Journal of Heat Transfer, 126,* 1023–1039.

Nakayama, A. (1986). Subcooled forced convection film boiling on plane and axisymmetric bodies in the presence of pressure gradient. *AIAA Journal, 24,* 230–236.

Nguyen, T. K., & Avedisian, C. T. (1987). Numerical solution for film evaporation of a spherical liquid droplet on an isothermal and adiabatic surface. *International Journal of Heat and Mass Transfer, 30,* 1497–1509.

Nishikawa, K., Ito, T., & Matsumoto, K. (1976). Investigation of variable thermophysical property problem concerning pool film boiling from vertical plate with prescribed uniform temperature. *International Journal of Heat and Mass Transfer, 19,* 1173–1181.

Nishio, A. (1985). Stability of pre-existing vapor nucleus in uniform temperature field. *Transactions of JSME, Series B, 54–303,* 1802–1807.

Nukiyama, S. (1934). The maximum and minimum values of heat Q transmitted from metal to boiling water under atmospheric pressure. *Journal of Japanese Society of Mechanical Engineering, 37* (pp. 367–374) (translated in *International Journal of Heat and Mass Transfer, 9*, 1419–1433 (1966)).

Prosperetti, A., & Plesset, M. S. (1978). Vapor bubble growth in a superheated liquid. *Journal of Fluid Mechanics, 85,* 349–368.

Plesset, M. S., & Sadhal, S. S. (1979). An analytical estimate of the microlayer thickness in nucleate boiling. *Journal of Heat Transfer, 101,* 180–182.

Plesset, M. S., & Zwick, S. A. (1954). The growth of vapor bubble in superheated liquids. *Journal of Applied Physics, 25,* 493–500.

Rohsenow, W. M. (1952). A method for correlating heat-transfer data for surface boiling of liquids. *Transactions of ASME, 74,* 969–976.

Shin, S., & Juric, D. (2002). Modeling three-dimensional multiphase flow using a level contour reconstruction method for front tracking without connectivity. *Journal of Computational Physics, 180,* 427–470.

Sideman, S., & Isenberg, J. (1967). Direct contact heat transfer with change of phase: Bubble growth in three-phase systems. *Desalination, 2,* 207–214.

Son, G., & Dhir, V. K. (1997). Numerical simulation of saturated film boiling on a horizontal surface. *Journal of Heat Transfer, 119,* 525–533.

Son, G., & Dhir, V. K. (1998). Numerical simulation of film boiling near critical pressures with a level set method. *Journal of Heat Transfer, 120,* 183–192.

Son, G., Ramanujapu, N., & Dhir, V. K. (2002). Numerical simulation of bubble merger process on a single nucleation site during pool nucleate boiling. *Journal of Heat Transfer, 124,* 51–62.

Sparrow, E. M., & Cess, R. D. (1962). The effect of subcooled liquid on laminar film boiling. *Journal of Heat Transfer, 84,* 149–156.

Stephan, K. (1992). *Heat transfer in condensation and boiling*. Berlin: Springer.

Thome, J. R. (2016). *The heat transfer engineering data book III*. Ulm: Wieland-Werke AG.

Utaka, Y., Kashiwabara, Y., & Ozak, M. (2013). Microlayer structure in nucleate boiling of water and ethanol at atmospheric pressure. *International Journal of Heat and Mass Transfer, 57*(1), 222–230.

Van Stralen, S. J. D., Sohal, M. S., Cole, R., & Sluyter, W. M. (1975a). Bubble growth rates in pure and binary systems: Combined effects of relaxation and evaporation microlayers. *International Journal of Heat and Mass Transfer, 18,* 453–467.

Van Stralen, S. J. D., Cole, R., Sluyter, W. M., & Sohal, M. S. (1975b). Bubble growth rates in nucleate boiling of water at subatmospheric pressures. *International Journal of Heat and Mass Transfer, 18,* 655–669.

Wang, C. H., & Dhir, V. K. (1993a). On the gas entrapment and nucleation density during pool boiling of saturated water. *Journal of Heat Transfer, 115,* 670–679.

Wang, C. H., & Dhir, V. K. (1993b). Effect of surface wettability on active nucleation site density during pool boiling of water on a vertical surface. *Journal of Heat Transfer, 115,* 659–669.

Welch, S. W. J. (1998). Direct simulation of vapor bubble growth. *International Journal of Heat and Mass Transfer, 41,* 1655–1666.

Welch, S. W. J., & Wilson, J. J. (2000). A volume of fluid based method for fluid flows with phase change. *Journal of Computational Physics, 160,* 662–682.

Witte, L. C., & Lienhard, I. H. (1982). On the existence of two 'transition' boiling curves. *International Journal of Heat and Mass Transfer, 25,* 771–779.

Zuber, N. (1959). *Hydrodynamic aspects of boiling heat transfer*. USAEC Report AECU-4439.

Zuber, N., Tribus, M., & Westwater, J. W. (1961). The Hydrodynamic crisis in pool boiling of saturated and subcooled liquids. In *International Development in Heat Transfer: Proceedings of 1961–62 International Heat Transfer Conference*, Boulder, CO (pp. 230–236).

10 Two-Phase Flow and Heat Transfer

10.1 Introduction

Two-phase flow refers to the interactive flow of two distinct phases—each phase representing a mass or volume of matter—with common interfaces in a channel. Multiphase flow can occur in a single-component or multicomponent system (Table 1.1). Possible phase combinations include: (1) solid–liquid, where solid particles are mostly dispersed in the liquid; (2) solid–gas, where the solid particles are carried by a stream of gas; (3) liquid–vapor (gas), where the volume fraction of one phase relative to the other results in different flow regimes; and (4) a combination of the above. While each of these modes represents a significant area of two-phase flow, liquid–vapor (gas) flow is by far the most common in various industries and thus has been investigated in greater depth. Therefore, the major emphasis of this chapter is on liquid–vapor (gas) flow.

Each regime in liquid–vapor (gas) two-phase flow has a characteristic flow behavior that can substantially affect both pressure drop and heat transfer. In the case of a single-component two-phase flow, such as forced convective condensation or evaporation, continuous mass transfer occurs between the vapor and liquid phases. Examples of liquid–gas flow processes include distillation, fractionation, flashing, spray-drying, stripping, and absorption. There are also two-component liquid–gas systems where the gas is noncondensable; these include air–water flow in aeration, deaeration, and humidification or dehumidification processes. A biomedical engineering example is the development of artificial lungs where absorption of oxygen or desorption of CO_2 from blood is required.

Section 10.2 presents the important parameters for liquid–vapor (gas) two-phase flow, and flow patterns in vertical and horizontal tubes. The various two-phase flow models, as well as the prediction of pressure drop and void fraction, are presented in Sect. 10.3. The two-phase flow regimes and heat transfer characteristics for forced convective condensation and boiling are presented in Sects. 10.4 and 10.5, respectively. The chapter is closed by discussions of two-phase flow, condensation, and boiling heat transfer in micro/miniature channels in Sect. 10.6.

A. Faghri and Y. Zhang, *Fundamentals of Multiphase Heat Transfer and Flow*, https://doi.org/10.1007/978-3-030-22137-9_10

10.2 Flow Patterns of Two-Phase Flow

Among the kinds of two-phase flows summarized in the preceding section, liquid–vapor (gas) flow is the most complex because the interfaces are deformable and the vapor or gas phase is more common. Furthermore, the interfacial configurations in two-phase flow are also very complicated due to heat and mass transfer and can vary over a wide range. The interfacial distribution in the liquid–vapor (gas) flow can be classified into a number of categories known as flow patterns or flow regimes. Flow regimes and flow regime transitions for vertical and horizontal tubes will be discussed. The flow patterns that will be presented in this section apply to two-phase flow in straight tubes only and are not applicable for two-phase flow in bends or coils. In addition, the flow patterns that will be addressed in this section are for cocurrent flow only and are not valid for countercurrent flow. However, one can easily obtain useful two-phase flow information in open literature concerning various configurations in geometry and/or countercurrent flow.

Before introducing the flow patterns, it is useful to establish notations and to develop some fundamental concepts and relationships.

10.2.1 Concepts and Notation

Since liquid–vapor two-phase flow with phase change will be the focus in this section, one phase is designated liquid and the other as vapor. The vapor or holdup *void fraction* represents the time-averaged volumetric fraction of vapor in a two-phase mixture, i.e.,

$$\alpha = \frac{\int_{V_{\mathrm{v}}} \mathrm{d}V}{\int_{V} \mathrm{d}V} = \frac{V_{\mathrm{v}}}{V_{\mathrm{v}} + V_{\ell}} \tag{10.1}$$

It should be noted that $\alpha = \varepsilon_{\mathrm{v}}$.

For liquid–vapor two-phase flow in a pipe, the volume of each phase consists of the cross-sectional area of the flow tube covered by that phase times a differential length element. Since the differential length is common in both phases, the void fraction can be considered the time-averaged area fraction:

$$\alpha = \frac{\Delta z \int_{A_{\mathrm{v}}} \mathrm{d}A}{\Delta z \int_{A} \mathrm{d}A} = \frac{A_{\mathrm{v}}}{A_{\mathrm{v}} + A_{\ell}} \tag{10.2}$$

The density of the two-phase mixture is defined as the average mass per unit volume:

$$\rho = \frac{\int_{V_{\ell}} \rho_{\ell} \mathrm{d}V + \int_{V_{\mathrm{v}}} \rho_{\mathrm{v}} \mathrm{d}V}{V_{\ell} + V_{\mathrm{v}}} \tag{10.3}$$

If the density of each phase is constant, the following relationship can be obtained by employing Eq. (10.1):

$$\rho = (1 - \alpha)\rho_{\ell} + \alpha\rho_{\mathrm{v}} \tag{10.4}$$

The phase velocity is the mean velocity of each phase and is defined as the volumetric flow rate of that phase through its cross-sectional area. This area-averaged velocity should also be considered as time-averaged velocity to eliminate random fluctuations, i.e.,

$$\langle w_\ell \rangle^\ell = \frac{Q_\ell}{A_\ell} \tag{10.5}$$

$$\langle w_v \rangle^v = \frac{Q_v}{A_v} \tag{10.6}$$

where Q_ℓ and Q_v are the volumetric flow rates of the liquid and vapor phases. The phase velocities defined in Eqs. (10.5) and (10.6) are intrinsic-averaged velocity and will be represented by w_ℓ and w_v in future references for ease of notations. The superficial velocity, or volumetric flux, of each phase is defined as the volumetric flow rate of that phase divided by the total cross-sectional flow area in question, i.e.,

$$j_\ell = \frac{Q_\ell}{A_\ell + A_v} \tag{10.7}$$

$$j_v = \frac{Q_v}{A_\ell + A_v} \tag{10.8}$$

The superficial velocity defined in Eqs. (10.7) and (10.8) is equal to the velocity that each phase would have if it were to flow alone in the channel at its specified mass flow rate. The relationship between the superficial velocity and the phase velocity of the liquid phase can be obtained by combining Eqs. (10.5) and (10.7) and using the definition of the void fraction in Eq. (10.2), i.e.,

$$j_\ell = w_\ell (1 - \alpha) \tag{10.9}$$

The relationship between the superficial velocity and the phase velocity of the vapor phase can be obtained in a similar manner:

$$j_v = w_v \alpha \tag{10.10}$$

Since the void fraction in two-phase flow always ranges between zero and one, it can be seen from Eqs. (10.9) and (10.10) that the phase velocities for each phase are greater than the corresponding volumetric flux. The total volumetric flux of the two-phase mixture, j, can be expressed as

$$j = j_\ell + j_v \tag{10.11}$$

The volumetric flow fraction, β, is defined as the volumetric flow rate of the vapor divided by the total volumetric flow rate:

$$\beta = \frac{j_v}{j_\ell + j_v} = \frac{Q_v}{Q_\ell + Q_v} \tag{10.12}$$

which is a particularly convenient quantity for the experimentalist since the volumetric flow rates can be readily calculated or measured.

It has been observed experimentally that the one-dimensional phase velocity of the vapor is normally greater than the one-dimensional phase velocity of the liquid in flowing two-phase systems. A *slip ratio* is defined as the ratio of the phase velocity of the vapor to that of the liquid, i.e.,

$$S = \frac{w_v}{w_\ell} \tag{10.13}$$

For homogeneous flow, the velocities for the liquid and vapor phases are identical and therefore the slip ratio $S = 1$, i.e.,

$$w_v = w_\ell \tag{10.14}$$

Substituting Eqs. (10.9) and (10.10) into Eq. (10.14) yields

$$\frac{j_v}{\alpha} = \frac{j_\ell}{1 - \alpha} \tag{10.15}$$

The void fraction for homogeneous flow is then

$$\alpha = \frac{j_v}{j_\ell + j_v} \tag{10.16}$$

which indicates that $\alpha = \beta$ for homogeneous flow.

The quality is defined as the vapor (gas) content of the two-phase flow. It is a parameter that identifies the dryness or the wetness of the two-phase system.

$$x = \frac{\dot{m}_v}{\dot{m}_\ell + \dot{m}_v} \tag{10.17}$$

where $\dot{m}$ is the mass flow rate of the liquid or vapor (gas) phase, obtained by

$$\dot{m}_\ell = \rho_\ell w_\ell A_\ell \tag{10.18}$$

$$\dot{m}_v = \rho_v w_v A_v \tag{10.19}$$

The *superficial mass flux* or *mass velocity* of liquid and vapor is defined as

$$G_\ell = \frac{\dot{m}_\ell}{A} = \rho_\ell j_\ell = \rho_\ell w_\ell (1 - \alpha) \tag{10.20}$$

$$G_v = \frac{\dot{m}_v}{A} = \rho_v j_v = \rho_v w_v \alpha \tag{10.21}$$

The total mass flux in two-phase flow is then

$$G = G_\ell + G_v = \rho_v j_v + \rho_\ell j_\ell = \rho_v w_v \alpha + \rho_\ell w_\ell (1 - \alpha) \tag{10.22}$$

The quality, x, can also be defined in terms of mass flux as

$$x = \frac{G_v}{G} \tag{10.23}$$

Substituting Eq. (10.23) into Eqs. (10.20) and (10.21), one can relate the superficial velocity to the quality as

$$j_\ell = \frac{G(1-x)}{\rho_\ell} \tag{10.24}$$

$$j_\mathrm{v} = \frac{Gx}{\rho_\mathrm{v}} \tag{10.25}$$

10.2.2 Vertical Tubes

Two-phase flow in a vertical tube tends to be more symmetric, since gravity acts equally in the circumferential directions. The gravitational force plays a more dominant role in the liquid phase—and therefore in the whole of the two-phase flow—in a vertical channel. Figure 10.1 shows the flow patterns encountered in vertical upward flow, where the patterns are defined as follows (Hewitt 1998; Thome 2016):

- **Bubbly flow**: Vapor bubbles are dispersed in a continuous liquid phase. The size of the bubbles varies widely but is generally small compared to the diameter of the tube.
- **Slug or plug flow**: As the bubble size increases and the bubbles start to consolidate, plugs can form. The plugs are often *bullet-shaped* in an upward flow and may be separated by regions occupied by liquid with a dispersion of small bubbles. These bullet-shaped plugs are commonly referred to as "Taylor bubbles" after the Taylor instability.
- **Churn flow**: As the slug bubbles grow larger, they start to break up, leading to a more random and unstable flow. Although the vapor continuously flows upward, the liquid phase may experience intermittent upward and downward motion, because the shear force from the vapor phase can just balance the imposed pressure gradient and the downward gravitational force. This oscillatory

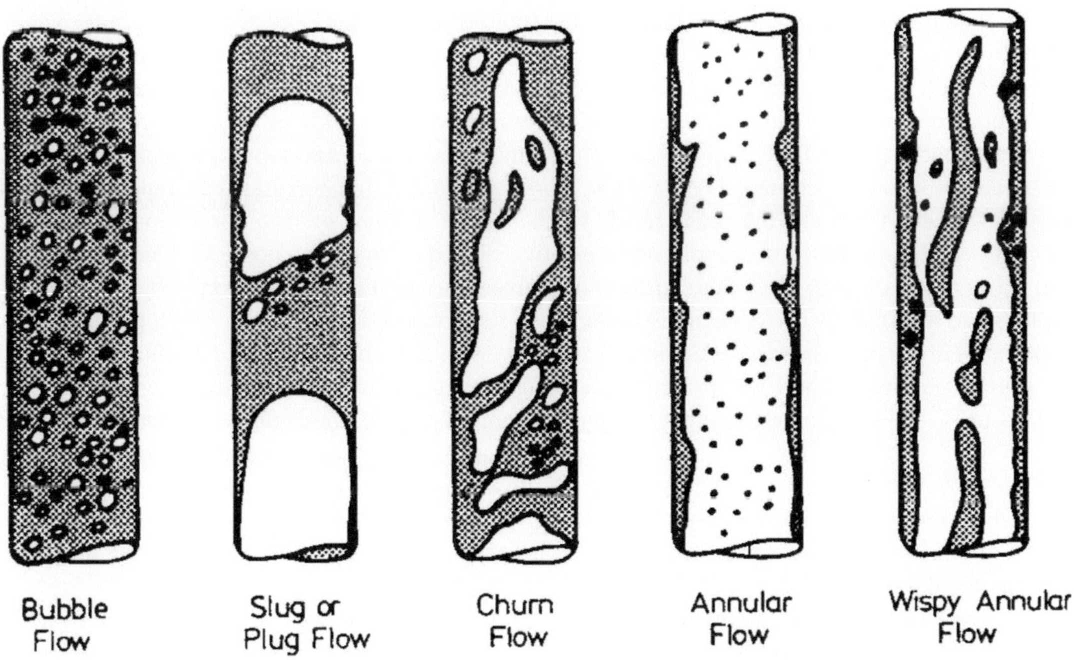

Fig. 10.1 Flow regimes in vertical upward cocurrent two-phase flow (▦ liquid □ vapor) (Hewitt 1998)

pattern is termed churn flow, and it is an intermediate regime between slug flow and annular flow. When the diameter of the vertical tube is small, the flow pattern can change directly from slug flow to annular flow without going through the churn flow pattern.

- **Annular flow**: At relatively high quality, the thin liquid layer flows along the inner wall of the tube and the central core of the flow consists of the vapor (gas) phase. Annular flow results when the interfacial shear of the high-velocity gas or vapor on the liquid film becomes dominant over gravity. Liquid is then expelled from the center of the tube to form a film on the tube wall. Since the vapor core velocity is much higher than the liquid velocity, the vapor core may ripple the liquid layer and cause waves on the liquid film. It is also possible that some of the liquid phase may be entrained as small droplets in the gas core, or that some bubbles may be entrained in the liquid film.
- **Wispy annular flow**: At a high liquid flow rate, the concentration of the liquid drops in the vapor (gas) core increases. The merging of these liquid droplets can lead to large lumps, breaks, or wisps of liquid in the gas core.

In different regimes of two-phase flow, the pressure drop and heat transfer characteristics are significantly different, so it is necessary to identify the conditions corresponding to the flow regimes described above. A number of approaches have been developed to allow two-phase flow regime prediction; these include analytical and numerical models and the use of experimental and operational data that have been collected over the decades by various investigators. The models based on experimental data have resulted in semi-empirical correlations that have had some degree of success in predicting two-phase flow. The analytical models that are based on physical concepts are often used to derive one-dimensional continuity, momentum, and energy balances for simple boundary conditions and steady-state systems. In transient systems and more complex systems where boundary conditions are not exact, numerical models based on physical concepts (Chap. 3) are used to predict two-phase flow. This information and/or experimental details have been transformed into readily available and simple tools that can develop a number of graphic diagrams, referred to as flow maps. These diagrams represent the flow in terms of important and useful parameters providing information about flow rate and operating parameters such as heat transfer coefficient and pressure drop. Based on these data, flow regimes representing different fractions of liquid and gas can be observed. Then, for a given set of operating conditions, the expected flow regime can quickly be determined.

The accuracy of each flow map depends on the methods used to develop it. For example, a flow map developed from experimental data for one working fluid may not necessarily be representative of another working fluid. On the other hand, when a large amount of data for various fluids and operating conditions are used, a more representative flow map might be possible. One of the limitations of experimental or operational data is the subjectivity in the data as reported by an observer. Furthermore, it is not always easy to distinguish one flow regime from another. Discrepancies in the reporting of flow regimes can translate into errors in a given flow map. Conversely, a flow map based on accurate physical models will provide a better representation of the actual flow regimes.

Two-phase flow is generally a complex phenomenon; therefore, to obtain a reasonable analytical prediction, a number of approximations must be made while solving the momentum and energy balances. Generally, the pressure drop across the flow channel and/or heat transfer coefficients are calculated to determine the flow characteristics at a given point and allow prediction of the flow regime. To that end, prediction methods based on heat transfer coefficients and pressure drop correlations are developed

The flow regime map is usually given in terms of superficial velocities or some other generalized flow parameters for liquid and vapor flow. The different regimes are separated by lines that represent

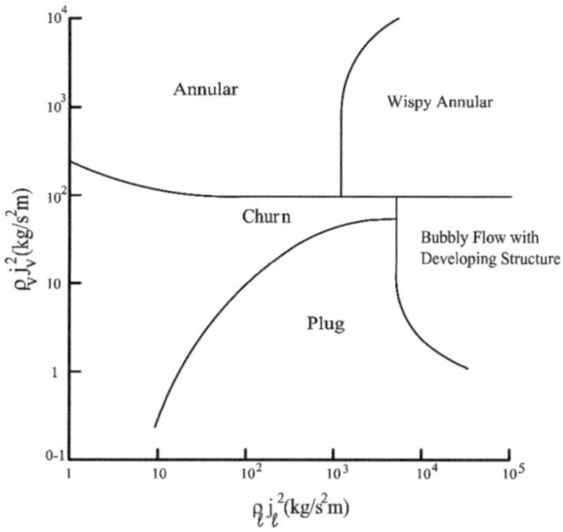

Fig. 10.2 Flow regime map obtained by Hewitt and Roberts (1969) for vertical upward two-phase flow

the conditions for transition between flow regimes. A flow regime map proposed by Hewitt and Roberts (1969), which represents a fairly wide range of experimental data for upward two-phase flow, is shown in Fig. 10.2. The horizontal and vertical coordinates of the map are the superficial momenta of the liquid and vapor phases, respectively. It can be seen that the conditions for different flow regimes and the boundaries that separate them can be expressed as a combination of the superficial momentum fluxes of the liquid and vapor phases. The boundaries between different flow regimes are the criteria that identify points of flow regime transition and are often of interest in practical applications. In addition to the flow map developed by Hewitt and Roberts (1969), there are many flow maps available in the literature. Since the identification of the flow pattern, and even the name of the flow regime, is subjective and may differ from one observer to another, it is very challenging to present every correlation and flow pattern for all of the different flow regimes and configurations. A very detailed review of the boundaries between the different flow regimes, and the corresponding empirical correlations, can be found in Thome (2016).

The flow regimes discussed above are applicable to vertical upward two-phase flow. Although downward two-phase flow finds applications in many areas, especially in condensation, it has received very little attention from investigators, and so there is very little flow pattern data for this case. The regimes for vertical downward two-phase flow are similar to those shown in Fig. 10.1, except that the downward-acting shear force, gravitational force, and imposed pressure gradient eliminate the churn flow regime. The limited research shows that the annular flow regime is dominant in vertical downward flow, which is different from upward flow. A flow pattern map based on low-pressure air–water flow data proposed by Golan and Stenning (1969) is shown in Fig. 10.3.

Example 10.1

Determine the flow regime for vertical upward flow of 0.8 kg/s of R134a in a vertical 0.1 m diameter tube at −20 °C and 25% quality.

Solution The mass flow rate is $\dot{m} = 0.8\,\mathrm{kg/s}$, and the diameter of the tube is $D = 0.1\,\mathrm{m}$. Therefore, the mass flux of the R11 in the tube is

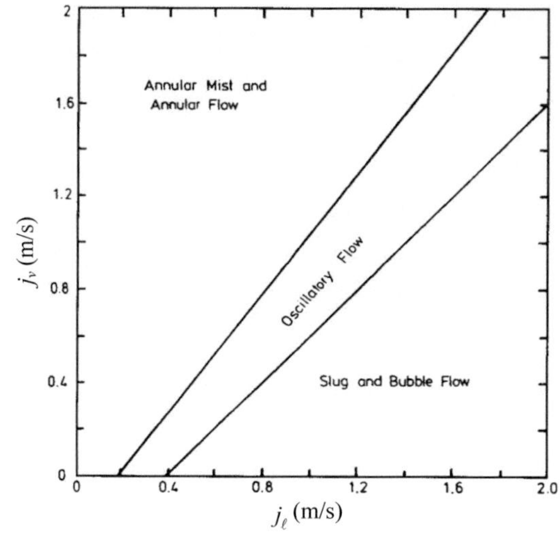

Fig. 10.3 Flow regime map for vertical downward two-phase flow (Golan and Stenning 1969)

$$G = \frac{\dot{m}}{A} = \frac{4\dot{m}}{\pi D} = \frac{4 \times 0.8}{\pi \times 0.1^2} = 101.86\,\text{kg/m}^2\text{-s}$$

The densities of liquid and vapor R-134 at $T = -20$ °C can be found from the Table B.34 as $\rho_\ell = 1358\,\text{kg/m}^3$ and $\rho_\text{v} = 6.785\,\text{kg/m}^3$. The superficial velocities of the vapor and liquid are obtained from Eqs. (10.24) and (10.25), i.e.,

$$j_\ell = \frac{G(1-x)}{\rho_\ell} = \frac{101.86 \times (1 - 0.25)}{1358} = 0.0562\,\text{m/s}$$

$$j_\text{v} = \frac{Gx}{\rho_\text{v}} = \frac{101.86 \times 0.25}{6.785} = 3.75\,\text{m/s}$$

The superficial momenta of the liquid and vapor phases are, respectively:

$$\rho_\ell j_\ell^2 = 1358 \times 0.0562^2 = 4.30\,\text{kg/s}^2\text{-m}$$

$$\rho_\text{v} j_\text{v}^2 = 6.785 \times 3.75^2 = 95.4\,\text{kg/s}^2\text{-m}$$

It can be found from Fig. 10.2 that the flow regime is churn flow.

10.2.3 Horizontal Tubes

Horizontal two-phase flow exhibits flow patterns different from those in vertical two phase flow, because gravity acts perpendicularly to the flow direction. Two-phase flow in horizontal tubes is more complex than vertical two-phase flow, because the flow is usually not axisymmetric, due to the effect of gravity. The flow patterns encountered in horizontal two-phase flow are defined as follows (Hewitt 1998; Thome 2016) and shown in Fig. 10.4:

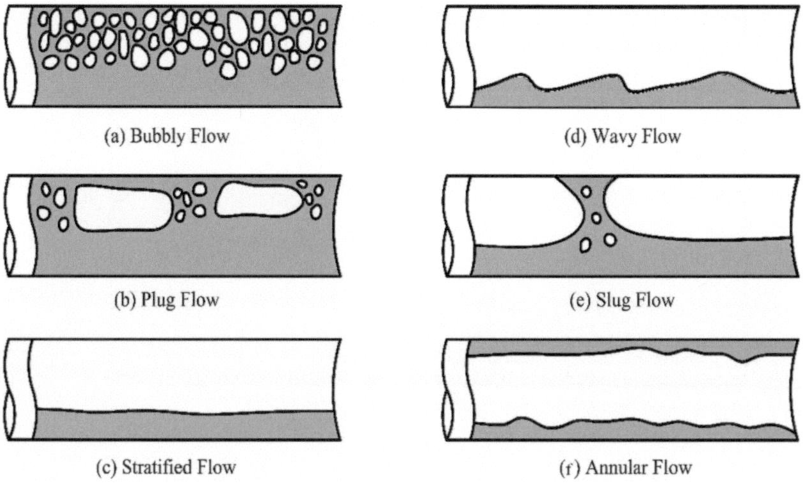

Fig. 10.4 Flow regimes in horizontal two-phase flow (■ liquid, □ vapor)

- **Dispersed bubble flow**: The vapor (gas) phase appears as distinct bubbles in a continuous liquid phase. The bubbles tend to rise to the top of the flow due to buoyancy effects. When the liquid velocity is high, the bubbles may be more uniformly distributed in the liquid, as in vertical flow.
- **Plug flow**: An increase of quality results in larger vapor (gas) bubble size and the formation of bullet-shaped plugs, which tend to remain at the top of the flow channel due to buoyancy force.
- **Stratified flow**: The liquid and vapor (gas) velocities in this regime are relatively slow, and the quality is relatively high. The liquid flows along the bottom of the tube due to gravity while the vapor (gas) phase flows at the top of the tube. It is also possible that a very thin layer of the liquid forms at the top of the flow channel.
- **Stratified wavy flow**: As the vapor (gas) velocity increases in the stratified flows, the shear forces of the vapor (gas) flow over the liquid cause ripples on the top of the liquid phase and result in the formation of waves on the liquid–vapor (gas) interface. The waves climb up the sides of the tube, and the liquid layer at the bottom of the channel starts to stretch thin.
- **Slug flow**: The amplitude of the waves increases as the liquid flow rate increases. The crests can span the entire tube, and a bridge starts to develop, separating the slugs from one another. However, a substantial liquid phase remains, and gravity pulls it to the bottom of the flow channel. The top of the flow channel is still wetted by a relatively thin film of liquid.
- **Annular-dispersed flow**: Similar to the flow pattern in vertical two-phase flow, the liquid layer flows near the inner wall of the tube and the vapor (gas) flows in the central core. However, the liquid layer at the bottom is thicker than that on the top of the channel due to the effect of gravitational force.

The most widely used flow pattern map for horizontal two-phase flow, proposed by Taitel and Dukler (1976), is shown in Fig. 10.5. This map is based on a semi-theoretical method, and it is computationally more difficult to use than other flow maps. The horizontal coordinate of the Taitel and Dukler (1976) map is the Martinelli parameter:

$$X = \left[\frac{(\mathrm{d}p_F/\mathrm{d}z)_\ell}{(\mathrm{d}p_F/\mathrm{d}z)_\mathrm{v}}\right]^{1/2} \tag{10.26}$$

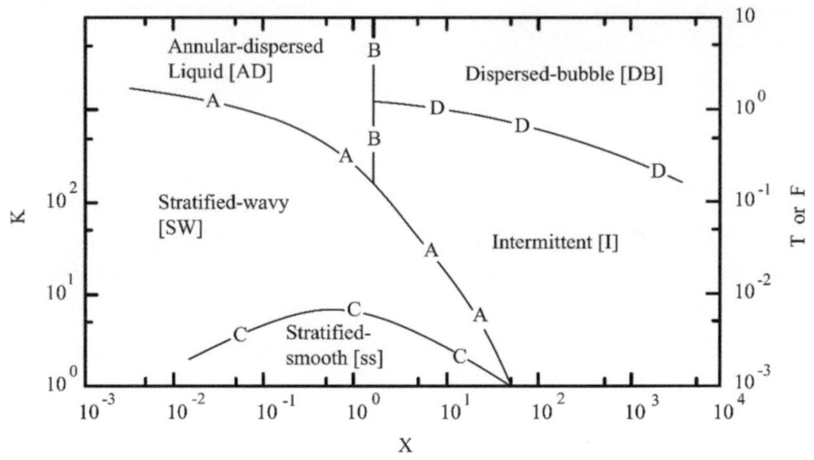

Fig. 10.5 Flow regime map for horizontal two-phase flow (Taitel and Dukler 1976)

where $(dp_F/dz)_\ell$ and $(dp_F/dz)_v$ are the pressure gradients for the liquid and vapor phases, flowing along the channel. The vertical coordinates of the Taitel and Dukler (1976) map are defined as

$$F = \sqrt{\frac{\rho_v}{\rho_\ell - \rho_v}} \frac{j_v}{\sqrt{Dg\cos\theta}} \tag{10.27}$$

$$K = \left[\frac{\rho_v j_v^2}{(\rho_\ell - \rho_v)Dg\cos\theta} \frac{Dj_\ell}{\nu_\ell}\right]^{1/2} \tag{10.28}$$

$$T = \left[\frac{(dp/dz)_\ell}{(\rho_\ell - \rho_v)g\cos\theta}\right]^{1/2} \tag{10.29}$$

where D is the tube diameter and θ is the angle of inclination of the channel to the horizontal. It can be seen that determination of the flow regime using Taitel and Dukler's (1976) map requires the pressure gradient for the liquid and vapor phases flowing along the channel, which should be determined using appropriate flow models.

It should be pointed out that the flow maps presented in Figs. 10.2, 10.3, and 10.4 were obtained for adiabatic two-phase flow; however, the transition boundaries between different flow regimes depend on the heat flux. Nevertheless, these flow maps are often used to determine the flow patterns for evaporation and condensation inside tubes, for which external heating or cooling is required. Application of these flow maps to forced convective boiling or condensation inside a tube may not yield reliable results.

10.3 Two-Phase Flow Models

A number of models have been proposed to calculate pressure drop and heat transfer in two-phase flow. These models provide a simple and accurate representation of the flow regimes. The models that have been extensively used by most researchers in two-phase are the homogeneous model and the separated flow model. The latter is a simple version of multifluid model; it allows two phases to have different properties and one-dimensional velocities, while the conservation equations are written for the combined flow. The major difference between these two models is the way phase velocity is addressed in each. The homogeneous model lumps both phases together to provide a homogeneous flow, and the behavior of the homogeneous flow is then determined. In the separated flow model, on the other hand, the flow of each phase is determined independently and the effects of the two phases are then summed.

The homogeneous flow model provides an easier approach to determining flow properties and behaviors, but it underestimates the pressure drop, particularly in a moderate pressure range. Furthermore, the homogeneous model is less accurate when velocity and flow conditions for both phases are more dispersed. A separated flow model, on the other hand, is somewhat more complex but tends to produce more accurate results. The one-dimensional models for homogeneous flow and separated flow are presented in Sects. 10.3.1 and 10.3.2, respectively.

10.3.1 Homogeneous Flow Model

The central assumption of the homogeneous flow model is that the two phases travel at equal velocities and mix well; therefore, they can be treated as if there is only one phase. This model works better for two-phase flow near the critical point, where the differences between the properties of the liquid and vapor are insignificant, or when the mass velocity of the two-phase flow is very high so that the flow regime is either bubbly or misty flow. In arriving at the homogeneous model for two-phase flow, area averaging is performed for both phases and the resultant governing equations become one-dimensional in nature. The density of the homogeneous mixture satisfies the relation

$$\frac{1}{\rho} = \frac{x}{\rho_v} + \frac{1-x}{\rho_\ell} \tag{10.30}$$

i.e.,

$$\rho = \frac{\rho_v \rho_\ell}{\rho_\ell x + \rho_v (1-x)} \tag{10.31}$$

The mass flow rates of the liquid and vapor phases defined in Eqs. (10.18) and (10.19) become

$$\dot{m}_\ell = \rho_\ell w A_\ell \tag{10.32}$$

$$\dot{m}_v = \rho_v w A_v \tag{10.33}$$

Substituting Eqs. (10.32) and (10.33) into Eq. (10.2) yields

$$\alpha = \frac{\dot{m}_v/\rho_v}{\dot{m}_v/\rho_v + \dot{m}_\ell/\rho_\ell} \tag{10.34}$$

Considering the definition of quality in Eq. (10.17), the void fraction becomes

$$\alpha = \frac{x}{x + (1-x)\rho_v/\rho_\ell} \tag{10.35}$$

The total mass flux in the channel becomes

$$G = \frac{\dot{m}_\ell + \dot{m}_v}{A} = \frac{\rho_\ell w A_\ell + \rho_v w A_v}{A} = \rho w \tag{10.36}$$

The governing equations for the homogeneous model include continuity, momentum, and energy equations, which are listed below (see Example 3.12):

$$A\frac{\partial \rho}{\partial t} + \frac{\partial}{\partial z}(GA) = 0 \tag{10.37}$$

$$A\frac{\partial G}{\partial t} + \frac{\partial (G^2 A/\rho)}{\partial z} = -\frac{\partial (pA)}{\partial z} - \rho g \cos\theta A - \tau_w P \tag{10.38}$$

$$\begin{aligned} &\frac{\partial}{\partial t}\left[\rho\left(h + \frac{w^2}{2} + gz\cos\theta\right)\right] + \frac{1}{A}\frac{\partial}{\partial z}\left[\rho w A\left(h + \frac{w^2}{2} + gz\cos\theta\right)\right] \\ &= \frac{P}{A}q''_w + q''' + \frac{\partial p}{\partial t} \end{aligned} \tag{10.39}$$

where the angle brackets, "⟨⟩," which represent the area average, have been dropped for ease of notation. P is perimeter, p is pressure, q''_w is heat flux at the wall, and q''' is internal heat generation per unit volume within the fluid.

Substituting Eq. (10.36) into Eq. (10.39), the energy equation becomes

$$\begin{aligned} &\frac{\partial (\rho h)}{\partial t} + \frac{1}{A}\frac{\partial}{\partial z}(GAh) \\ &= \frac{P}{A}q''_w + q''' - \frac{1}{A}\frac{\partial}{\partial z}\left(\frac{G^3 A}{2\rho^2}\right) \\ &\quad - gG\cos\theta - \frac{\partial}{\partial t}\left(\frac{G^2}{2\rho}\right) + \frac{\partial p}{\partial t} \end{aligned} \tag{10.40}$$

For steady-state two-phase flow in a circular tube with constant cross-sectional area, the momentum Eq. (10.38) reduces to

$$-\frac{dp}{dz} = \frac{4\tau_w}{D} + \frac{\partial (G^2/\rho)}{\partial z} + \rho g \cos\theta \tag{10.41}$$

The three terms on the right-hand side of Eq. (10.41) represent pressure drops due to friction, dp_F/dz, acceleration, dp_a/dz, and gravity, dp_g/dz. Thus, Eq. (10.41) can be rewritten as

$$-\frac{dp}{dz} = -\frac{dp_F}{dz} - \frac{dp_a}{dz} - \frac{dp_g}{dz} \tag{10.42}$$

where the pressure drop due to friction, dp_F/dz, must be obtained using appropriate correlations.

The energy equation (including mechanical and thermal energy) for steady-state two-phase flow in a circular tube with constant cross-sectional area can be obtained by simplifying Eq. (10.40),

$$\frac{dh}{dz} = \frac{4q''_w}{GD} + \frac{q'''}{G} - \frac{G^2}{2}\frac{d}{dz}\left(\frac{1}{\rho^2}\right) - g\cos\theta \tag{10.43}$$

The enthalpy of the two-phase mixture can be expressed as $h = h_\ell + x(h_v - h_\ell)$. For a two-phase flow system with condensation or evaporation, where kinetic and potential energy as well as internal heat generation can be neglected, Eq. (10.43) reduces to

$$\frac{dx}{dz} = \frac{4q''_w}{GDh_{\ell v}} \tag{10.44}$$

10.3.2 Separated Flow Model

Compared with the homogeneous model, the separated flow model has been used more widely, because it provides a better prediction of flow behavior with a manageable level of complexity. The separated flow model assumes that each phase displays different properties and flows at different velocities. It is a simpler version of the two-fluid model discussed in Chap. 3 because it is assumed that only velocities differ between the two phases, while the conservation equations are written only for the combined flow. In addition, the pressure across any given cross section of a channel carrying a multiphase flow is assumed to be the same for both phases (Hewitt 1998).

The mass flow rates of the liquid and vapor phases are obtained from Eqs. (10.18) and (10.19). The cross-sectional area of the channel occupied by liquid and vapor can be obtained by rearranging Eqs. (10.18) and (10.19), i.e.,

$$A_\ell = \dot{m}_\ell / (\rho_\ell w_\ell) \tag{10.45}$$

$$A_v = \dot{m}_v / (\rho_v w_v) \tag{10.46}$$

The void fraction of the two-phase flow can be obtained by substituting Eqs. (10.45) and (10.46) into Eq. (10.2), i.e.,

$$\alpha = \frac{\dot{m}_v / (\rho_v w_v)}{\dot{m}_v / (\rho_v w_v) + \dot{m}_\ell / (\rho_\ell w_\ell)} \tag{10.47}$$

Considering the definition of quality in Eq. (10.17), the void fraction becomes

$$\alpha = \frac{x}{x + (1 - x)\frac{\rho_v w_v}{\rho_\ell w_\ell}} \tag{10.48}$$

It can be seen that when the vapor velocity, w_v, is greater than the liquid velocity, w_ℓ (which is often the case for vertical upward and horizontal cocurrent flow), the homogeneous model overpredicts the void fraction. On the other hand, when the vapor velocity, w_v, is lower than the liquid velocity, w_ℓ, as occurs in vertical downward flow, the homogeneous model underpredicts the void fraction.

The governing equations for steady-state two-phase flow in a channel for separated flow model have been given in Example 3.13. More generalized governing equations that are applicable to transient flow are presented here. The continuity equations for the vapor and liquid phases are, respectively

$$\frac{\partial}{\partial t}(\rho_v \alpha A) + \frac{\partial}{\partial z}(\rho_v w_v \alpha A) = \dot{m}'''_v \tag{10.49}$$

$$\frac{\partial}{\partial t}[\rho_\ell(1-\alpha)A]+\frac{\partial}{\partial z}[\rho_\ell w_\ell(1-\alpha)A]=\dot{m}_\ell''' \tag{10.50}$$

where $\dot{m}_\mathrm{v}'''$ and $\dot{m}_\ell'''$ are the mass production rates of vapor and liquid due to phase change in the two-phase flow system. Conservation of mass requires that the summation of $\dot{m}_\mathrm{v}'''$ and $\dot{m}_\ell'''$ equals zero, so the continuity equation for the two-phase system can be obtained by adding Eqs. (10.49) and (10.50) together, i.e.,

$$\frac{\partial}{\partial t}(\rho A)+\frac{\partial}{\partial z}[\rho_\mathrm{v}w_\mathrm{v}\alpha A+\rho_\ell w_\ell(1-\alpha)A]=0 \tag{10.51}$$

where ρ is the density of the two-phase mixture from Eq. (10.4). Considering the definition of mass flux in Eqs. (10.20)–(10.22), the continuity equation can be rewritten as

$$\frac{\partial}{\partial t}(\rho A)+\frac{\partial}{\partial z}(GA)=0 \tag{10.52}$$

The momentum equations for the vapor and liquid phases are

$$\begin{aligned}&\frac{\partial}{\partial t}(\rho_\mathrm{v}w_\mathrm{v}\alpha A)+\frac{\partial}{\partial z}\left(\rho_\mathrm{v}w_\mathrm{v}^2\alpha A\right)\\&=-\alpha A\frac{\partial p}{\partial z}-g\rho_\mathrm{v}\alpha A\cos\theta-\tau_{\mathrm{w,v}}P_{\mathrm{w,v}}+F_{\ell,\mathrm{v}}\end{aligned} \tag{10.53}$$

$$\begin{aligned}&\frac{\partial}{\partial t}[\rho_\ell w_\ell(1-\alpha)A]+\frac{\partial}{\partial z}\left[\rho_\ell w_\ell^2(1-\alpha)A\right]\\&=-(1-\alpha)A\frac{\partial p}{\partial z}-g\rho_\ell(1-\alpha)A\cos\theta-\tau_{\mathrm{w},\ell}P_{\mathrm{w},\ell}+F_{\mathrm{v},\ell}\end{aligned} \tag{10.54}$$

where $\tau_{\mathrm{w,v}}$ and $\tau_{\mathrm{w},\ell}$ are shear stresses at the wall for vapor and liquid, respectively. $P_{\mathrm{w,v}}$ and $P_{\mathrm{w},\ell}$ are the portion of the perimeter that is in contact with vapor and liquid, respectively. $F_{\ell,\mathrm{v}}$ and $F_{\mathrm{v},\ell}$ are the interactive forces between the liquid and vapor phases, which satisfy $F_{\ell,\mathrm{v}}=-F_{\mathrm{v},\ell}$ as required by Newton's third law. Assuming constant wall shear stress τ_w around the periphery of the channel, the momentum equation of the two-phase system can be obtained by adding Eqs. (10.53) and (10.54), i.e.,

$$\begin{aligned}&A\frac{\partial \dot{m}''}{\partial t}+\frac{\partial}{\partial z}\left[\rho_\mathrm{v}w_\mathrm{v}^2\alpha A+\rho_\ell w_\ell^2(1-\alpha)A\right]\\&=-A\frac{\partial p}{\partial z}-g\rho A\cos\theta-\tau_\mathrm{w}P\end{aligned} \tag{10.55}$$

Substituting Eqs. (10.20) and (10.21) into Eq. (10.55), the momentum equation becomes

$$A\frac{\partial G}{\partial t}+\frac{\partial}{\partial z}\left\{A\left[\frac{G_\mathrm{v}^2}{\rho_\mathrm{v}\alpha}+\frac{G_\ell^2}{\rho_\ell(1-\alpha)}\right]\right\}=-A\frac{\partial p}{\partial z}-g\rho A\cos\theta-\tau_\mathrm{w}P \tag{10.56}$$

Considering that the mass flux of the vapor phase can be expressed as $G_v = xG$ (Eq. 10.23) and the mass flux of the liquid is $G_\ell = G - G_v = (1-x)G$, the momentum equation in the separated flow model becomes

$$\frac{\partial G}{\partial t} + \frac{1}{A}\frac{\partial}{\partial z}\left\{GA\left[\frac{x^2}{\rho_v \alpha} + \frac{(1-x)^2}{\rho_\ell(1-\alpha)}\right]\right\} = -\frac{\partial p}{\partial z} - g\rho\cos\theta - \frac{\tau_w P}{A} \tag{10.57}$$

The energy equations for the vapor and liquid phases are

$$\begin{aligned}&\frac{\partial}{\partial t}\left[\rho_v \alpha A\left(h_v + \frac{w_v^2}{2} + gz\cos\theta\right)\right] + \frac{\partial}{\partial z}\left[\rho_v w_v \alpha A\left(h_v + \frac{w_v^2}{2} + gz\cos\theta\right)\right]\\ &= P_v q''_w + q'''\alpha A + \alpha A\frac{\partial p}{\partial t} + q'''_{\ell,v}\end{aligned} \tag{10.58}$$

$$\begin{aligned}&\frac{\partial}{\partial t}\left[\rho_\ell(1-\alpha)A\left(h_\ell + \frac{w_\ell^2}{2} + gz\cos\theta\right)\right]\\ &\quad + \frac{\partial}{\partial z}\left[\rho_\ell w_\ell(1-\alpha)A\left(h_\ell + \frac{w_\ell^2}{2} + gz\cos\theta\right)\right]\\ &= P_\ell q''_w + q'''(1-\alpha)A + (1-\alpha)A\frac{\partial p}{\partial t} + q'''_{v,\ell}\end{aligned} \tag{10.59}$$

where the heat flux at the surface of the channel, q''_w, is assumed to be the same for the perimeter of the channel, whether it is in contact with liquid or vapor. $q'''_{\ell,v}$ and $q'''_{v,\ell}$ are the interphase heat transfers that satisfy $q'''_{\ell,v} = -q'''_{v,\ell}$. The energy equation for the two-phase mixture is then obtained by adding Eqs. (10.58) and (10.59), i.e.,

$$\begin{aligned}&A\frac{\partial}{\partial t}\left[\rho_v\alpha\left(h_v + \frac{w_v^2}{2} + gz\cos\theta\right) + \rho_\ell(1-\alpha)\left(h_\ell + \frac{w_\ell^2}{2} + gz\cos\theta\right)\right]\\ &+ \frac{\partial}{\partial z}\left\{A\left[\rho_v w_v\alpha\left(h_v + \frac{w_v^2}{2} + gz\cos\theta\right) + \rho_\ell w_\ell(1-\alpha)\left(h_\ell + \frac{w_\ell^2}{2} + gz\cos\theta\right)\right]\right\}\\ &= Pq''_w + q'''A + A\frac{\partial p}{\partial t}\end{aligned} \tag{10.60}$$

Substituting Eqs. (10.20) and (10.21) into Eq. (10.60), the energy equation becomes

$$\begin{aligned}&A\frac{\partial}{\partial t}\left[\rho_v\alpha\left(h_v + \frac{G_v^2}{2\rho_v^2\alpha^2} + gz\cos\theta\right) + \rho_\ell(1-\alpha)\left(h_\ell + \frac{G_\ell^2}{2\rho_\ell^2(1-\alpha)^2} + gz\cos\theta\right)\right]\\ &+ \frac{\partial}{\partial z}\left\{A\left[G_v\left(h_v + \frac{G_v^2}{2\rho_v^2\alpha^2} + gz\cos\theta\right) + G_\ell\left(h_\ell + \frac{G_\ell^2}{2\rho_\ell^2(1-\alpha)^2} + gz\cos\theta\right)\right]\right\}\\ &= Pq''_w + q'''A + A\frac{\partial p}{\partial t}\end{aligned} \tag{10.61}$$

Since $G_v = xG$ and $G_\ell = G - G_v = (1-x)G$, Eq. (10.61) can be modified as

$$
\begin{aligned}
& A\frac{\partial}{\partial t}[\rho_{\mathrm{v}}h_{\mathrm{v}}\alpha+\rho_{\ell}h_{\ell}(1-\alpha)]+\frac{\partial}{\partial z}\{GA[xh_{\mathrm{v}}+(1-x)h_{\ell}]\} \\
& = Pq_{\mathrm{w}}''+q'''A-\frac{\partial}{\partial z}\left\{\frac{G^3A}{2}\left[\frac{x^3}{\rho_{\mathrm{v}}^2\alpha^2}+\frac{(1-x)^3}{\rho_{\ell}^2(1-\alpha)^2}\right]\right\}-GAg\cos\theta \\
& -A\frac{\partial}{\partial t}\left\{\frac{G^2}{2}\left[\frac{x^2}{2\rho_{\mathrm{v}}\alpha}+\frac{(1-x)^2}{2\rho_{\ell}(1-\alpha)}\right]\right\}+A\frac{\partial p}{\partial t}
\end{aligned}
\tag{10.62}
$$

For steady-state two-phase flow in a circular tube with constant cross-sectional area (A = constant), the continuity equation (10.52) results in $dG/dz = 0$. The momentum Eq. (10.57) reduces to

$$
-\frac{\mathrm{d}p}{\mathrm{d}z}=\frac{4\tau_{\mathrm{w}}}{D}+G\frac{\mathrm{d}}{\mathrm{d}z}\left[\frac{x^2}{\rho_{\mathrm{v}}\alpha}+\frac{(1-x)^2}{\rho_{\ell}(1-\alpha)}\right]+g\rho\cos\theta \tag{10.63}
$$

The three terms on the right-hand side of Eq. (10.63) represent pressure drops due to friction, $\mathrm{d}p_F/\mathrm{d}z$, acceleration, $\mathrm{d}p_a/\mathrm{d}z$, and gravity, $\mathrm{d}p_g/\mathrm{d}z$. To predict the pressure drop of two-phase flow using the separated flow model, empirical correlations for friction and void fraction are needed (see Eq. 10.63).

The energy equation in the separated flow model for steady-state two-phase flow in a circular tube with constant cross-sectional area can be obtained by simplifying Eq. (10.62), i.e.,

$$
\frac{\mathrm{d}}{\mathrm{d}z}[xh_{\mathrm{v}}+(1-x)h_{\ell}]=\frac{4q_{\mathrm{w}}''}{GD}+\frac{q'''}{G}-\frac{G^2}{2}\frac{\mathrm{d}}{\mathrm{d}z}\left[\frac{x^3}{\rho_{\mathrm{v}}^2\alpha^2}+\frac{(1-x)^3}{\rho_{\ell}^2(1-\alpha)^2}\right]-g\cos\theta \tag{10.64}
$$

For a two-phase flow system with condensation or evaporation, where kinetic and potential energy, as well as internal heat generation, can be neglected, Eq. (10.64) also reduces to Eq. (10.44). Therefore, the energy equations for the homogeneous model and the separated flow model are the same if kinetic and potential energy, as well as internal heat generation, can be neglected.

10.3.3 Frictional Pressure Drop

The pressure drop is very important for the design of two-phase devices, because it dictates the pump power that is required to drive the flow. As shown in Eqs. (10.41) and (10.63), the pressure drop includes three parts: frictional, accelerational, and gravitational pressure drops. The accelerational and gravitational pressure drop terms in Eqs. (10.41) and (10.63) can be calculated based on physical properties and the void fraction. The frictional pressure drop for steady-state two-phase flow in a circular tube can be calculated by

$$
-\frac{\mathrm{d}p_F}{\mathrm{d}z}=\frac{4\tau_{\mathrm{w}}}{D} \tag{10.65}
$$

The frictional pressure drop for two-phase flow must be accounted for by using an empirical correlation, which can be based on either the homogeneous or separated flow model. While the former assumes that both liquid and gas move at the same velocity (slip ratio S is 1) and is also referred to as zero-slip model, the latter allows liquid and gas to move at different velocities.

10.3.3.1 Correlations Based on the Homogeneous Model

Since the homogeneous model treats the two phases in the mixture as a pseudo-single-phase flow, the frictional pressure gradient can be calculated by (Beattie and Whalley 1982)

$$-\frac{dp_F}{dz} = \frac{4\tau_w}{D} = \frac{2f\dot{m}''^2}{D\rho} \tag{10.66}$$

where f is the two-phase friction factor, which can be determined by empirical correlation for single-phase flow. For many two-phase homogeneous flow considerations, particularly in the annular and bubble regimes, the following equation is proposed to determine the homogeneous friction factor, f:

$$\frac{1}{\sqrt{f}} = 3.48 - 4\log_{10}\left[\frac{2\kappa}{D} + \frac{9.35}{(Re\sqrt{f})}\right] \tag{10.67}$$

where the term κ/D represents the surface roughness/diameter ratio. The Reynolds number can be calculated based on a homogeneous flow using

$$Re = \frac{GD}{\mu} \tag{10.68}$$

The homogeneous viscosity of the two-phase mixture is obtained by

$$\frac{1}{\mu} = \frac{x}{\mu_v} + \frac{1-x}{\mu_\ell} \tag{10.69}$$

10.3.3.2 Correlations Based on the Separated Flow Model

The frictional pressure gradient of two-phase flow can be related to that of either the vapor or liquid phase flowing alone in the channel (Lockhart and Martinelli 1949; Chisholm 1967). The frictional pressure gradients of the vapor or liquid phase flow in the channel, with their actual flow rate and properties, can be defined as

$$-\left(\frac{dp_F}{dz}\right)_v = \frac{2f_v G^2 x^2}{D\rho_v} \tag{10.70}$$

$$-\left(\frac{dp_F}{dz}\right)_\ell = \frac{2f_\ell G^2 (1-x)^2}{D\rho_\ell} \tag{10.71}$$

where f_v and f_ℓ are, respectively, the friction factors for the vapor and liquid phases with their actual mass flux flowing in the channel alone.

Similarly, the frictional pressure gradient in the channel—with the same total mass flow rate of the two-phase flow, but with the properties of the vapor or liquid phase—can be defined as

$$-\left(\frac{dp_F}{dz}\right)_{vo} = \frac{2f_{vo} G^2}{D\rho_v} \tag{10.72}$$

$$-\left(\frac{dp_F}{dz}\right)_{\ell o} = \frac{2f_{\ell 0}G^2}{D\rho_\ell} \tag{10.73}$$

where f_{v0} is the vapor friction factor if the vapor phase with mass flux, G, occupies the entire channel, whereas $f_{\ell 0}$ is the liquid fraction factor if the channel is taken by liquid phase with mass flux G alone.

Through the standard equations and charts for the single-phase flow, the friction factors defined in Eqs. (10.70)–(10.73) can be related to the respective Reynolds numbers:

$$Re_v = \frac{GxD}{\mu_v} \tag{10.74}$$

$$Re_\ell = \frac{G(1-x)D}{\mu_\ell} \tag{10.75}$$

$$Re_{vo} = \frac{GD}{\mu_v} \tag{10.76}$$

$$Re_{\ell o} = \frac{GD}{\mu_\ell} \tag{10.77}$$

The relationships between the frictional factor and the Reynolds number are different for laminar and turbulent flow.

$$f = \begin{cases} \frac{16}{Re} & Re < 2000 \\ 0.079Re^{-0.25} & Re > 2000 \end{cases} \tag{10.78}$$

The frictional pressure gradient of the two-phase flow can be related to those defined in Eqs. (10.70)–(10.73) through pressure drop multipliers defined as

$$\phi_v^2 = \frac{dp_F/dz}{(dp_F/dz)_v} \tag{10.79}$$

$$\phi_\ell^2 = \frac{dp_F/dz}{(dp_F/dz)_\ell} \tag{10.80}$$

$$\phi_{v0}^2 = \frac{dp_F/dz}{(dp_F/dz)_{v0}} \tag{10.81}$$

$$\phi_{\ell 0}^2 = \frac{dp_F/dz}{(dp_F/dz)_{\ell 0}} \tag{10.82}$$

Two commonly used parameters in two-phase flow investigations are the Martinelli parameter, X, which was defined in Eq. (10.26), and the Chisholm parameter, Y,

$$Y = \left[\frac{(dp_F/dz)_{\ell 0}}{(dp_F/dz)_{v0}}\right]^{1/2} \tag{10.83}$$

Parameter X, the Martinelli parameter, is a ratio of pressure drops of single-phase flow terms. As can be seen from Eqs. (10.79)–(10.82), the pressure drop in two-phase flow can be determined if any one of the four multipliers is known. A generalized method to determine the frictional pressure gradient multiplier was proposed by Lockhart and Martinelli (1949), who related the frictional multipliers ϕ_v and ϕ_ℓ to the Martinelli parameter X as shown in Fig. 10.6. It can be seen that the trends for ϕ_v and ϕ_ℓ are different because ϕ_v increases with increasing X, but ϕ_ℓ decreases with increasing X. The multiplier curves also depend on whether the liquid–phase-alone flow and the vapor-phase-alone flow are laminar or turbulent. There are four curves for ϕ_v and ϕ_ℓ, and each corresponds to the combination of laminar (viscous) and turbulent flow for the vapor- or liquid-phase-alone flows in the channel. For example, $\phi_{\ell,\mathrm{vt}}$ represents the multiplier in the liquid alone pressure drop for cases where the liquid–phase flowing alone in the channel is laminar (viscous) but the vapor phase flowing alone in the channel is turbulent. Chisholm (1967) correlated the curves of Lockhart and Martinelli (1949) and recommended the following relationships:

$$\phi_\ell^2 = 1 + \frac{C}{X} + \frac{1}{X^2} \tag{10.84}$$

$$\phi_v^2 = 1 + CX + X^2 \tag{10.85}$$

where C is a dimensionless constant that depends on the combination of the natural and the phase-alone flows. The value of the constant C recommended by Chisholm (1967) can be found in Table 10.1. The correlation by Lockhart and Martinelli (1949) can provide a good prediction if $\mu_\ell/\mu_v > 1000$ and $\dot{m}'' < 100$ kg/m^2-s. Alternative correlations should be used when the two-phase flow falls outside these ranges.

For cases where $\mu_\ell/\mu_v > 1000$ and $G > 100\,\mathrm{kg/m^2}$-s, the following correlation proposed by Chisholm (1973a) should be used:

$$\phi_{\ell 0}^2 = 1 + (Y^2 - 1)\left[Bx^{(2-n)/2}(1-x)^{(2-n)/2} + x^{2-n}\right] \tag{10.86}$$

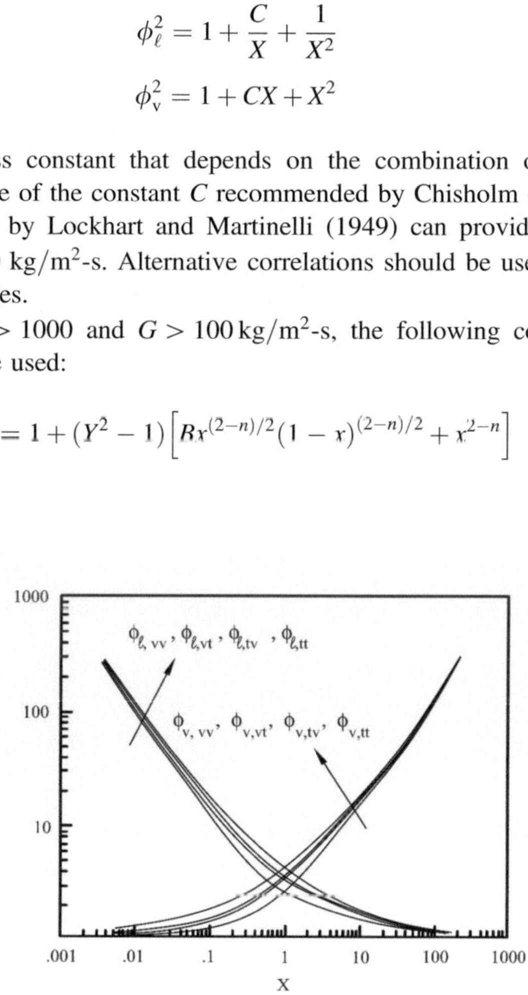

Fig. 10.6 Lockhart–Martinelli correlations for pressure drop (Hewitt 1998)

Table 10.1 Value of C in Eqs. (10.84) and (10.85)

Liquid	Vapor	Subscripts	C
Turbulent	Turbulent	tt	20
Viscous	Turbulent	vt	12
Turbulent	Viscous	tv	10
Viscous	Viscous	vv	5

where n is the exponent in the friction factor–Reynolds number relationship ($f\,Re^n$ = constant). According to Eq. (10.78), n equals 1 for laminar flow and 0.25 for turbulent flow. The parameter B is given by

$$B = \begin{cases} \dfrac{55}{\sqrt{G}} & 0<Y<9.5 \\ \dfrac{520}{Y\sqrt{G}} & 9.5<Y<28 \\ \dfrac{15000}{Y^2\sqrt{G}} & Y > 28 \end{cases} \tag{10.87}$$

For cases where $\mu_\ell/\mu_v < 1000$, the following correlation developed by Friedel (1979) using a database of 25,000 points can provide a better prediction:

$$\phi_{\ell 0}^2 = C_1 + \frac{3.24 C_2}{Fr^{0.045} We^{0.035}} \tag{10.88}$$

where

$$C_1 = (1-x)^2 + x^2 \left(\frac{\rho_\ell}{\rho_v}\right)\left(\frac{f_{v0}}{f_{\ell 0}}\right) \tag{10.89}$$

$$C_2 = x^{0.78}(1-x)^{0.24}\left(\frac{\rho_\ell}{\rho_v}\right)^{0.91}\left(\frac{\mu_v}{\mu_\ell}\right)^{0.19}\left(1-\frac{\mu_v}{\mu_\ell}\right)^{0.7} \tag{10.90}$$

$$Fr = \frac{G^2}{gD\rho^2} \tag{10.91}$$

$$We = \frac{G^2 D}{\rho\sigma} \tag{10.92}$$

10.3.3.3 Bounds on Two-Phase Flow

The advantage of the pressure drop correlations based on the separated flow model is that it is applicable for all flow patterns. This flexibility is accompanied by low accuracy. Awad and Muzychka (2005a) developed rational bounds for two-phase pressure gradients. The lower bound of the friction pressure drop is

$$\left(\frac{dp}{dz}\right)_{F,lower} = \frac{0.158G^{1.75}(1-x)^{1.75}\mu_\ell^{0.25}}{D^{1.25}\rho_\ell}\left[1+\left(\frac{x}{1-x}\right)^{0.7368}\left(\frac{\rho_\ell}{\rho_v}\right)^{0.4211}\left(\frac{\mu_v}{\mu_\ell}\right)^{0.1053}\right]^{2.375} \tag{10.93}$$

where D is the diameter of the tube.

The upper bound of the friction pressure drop is

$$\left(\frac{dp}{dz}\right)_{F,upper} = \frac{0.158G^{1.75}(1-x)^{1.75}\mu_\ell^{0.25}}{D^{1.25}\rho_\ell}\left[1+\left(\frac{x}{1-x}\right)^{0.4375}\left(\frac{\rho_\ell}{\rho_v}\right)^{0.25}\left(\frac{\mu_v}{\mu_\ell}\right)^{0.0625}\right]^{4} \tag{10.94}$$

An acceptable prediction of pressure drop can be obtained by averaging the maximum and minimum values, i.e.,

$$\begin{aligned}\left(\frac{dp}{dz}\right)_{F,ave} &= \frac{0.079G^{1.75}(1-x)^{1.75}\mu_\ell^{0.25}}{D^{1.25}\rho_\ell} \\ &\cdot\left\{\left[1+\left(\frac{x}{1-x}\right)^{0.7368}\left(\frac{\rho_\ell}{\rho_v}\right)^{0.4211}\left(\frac{\mu_v}{\mu_\ell}\right)^{0.1053}\right]^{2.375}\right. \\ &\left.+\left[1+\left(\frac{x}{1-x}\right)^{0.4375}\left(\frac{\rho_\ell}{\rho_v}\right)^{0.25}\left(\frac{\mu_v}{\mu_\ell}\right)^{0.0625}\right]^{4}\right\}\end{aligned} \tag{10.95}$$

Figure 10.7 shows the lower and upper bounds, and the average fraction pressure gradient versus mass flux. The experimental results by Bandel (1973) for R-12 flow at $x = 0.3$ and $T_{sat} = 0\,°C$ in a

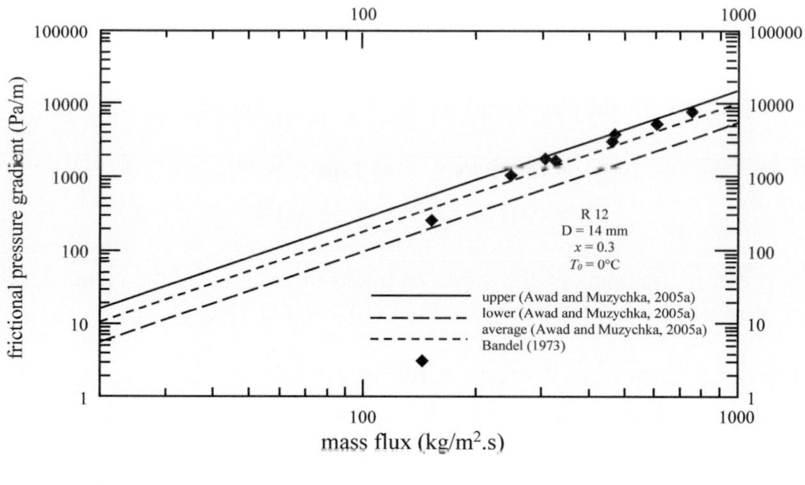

Fig. 10.7 Pressure gradient versus mass flux (Awad and Muzychka 2005a)

smooth horizontal tube at $D = 14\,\text{mm}$ are also plotted in Fig. 10.7 for comparison. Equation (10.95) could predict the pressure drop with the root mean square error of 26.4%.

Example 10.2

Use the Taitel–Dukler flow map to determine the flow regime for a flow of 8 kg/s of water–steam at 15% quality and 180 °C in a 0.1 m ID horizontal tube.

Solution The thermophysical properties of water at p = 10 bar can be found from Table B.48 in Appendix B, i.e., $\rho_\ell = 887.31\ \text{kg/m}^3$, $\rho_\text{v} = 5.1597\,\text{kg/m}^3$, $\mu_\ell = 1493 \times 10^{-7}\,\text{N}-\text{s/m}^2$, and $\mu_\text{v} = 149 \times 10^{-7}$ N-s/m^2.

The mass flux of the two-phase flow in the horizontal tube is

$$G = \frac{\dot{m}}{A} = \frac{4\dot{m}}{\pi D^2} = \frac{4 \times 8}{\pi \times 0.1^2} = 1018.6\,\text{kg/m}^2\text{-s}$$

The superficial velocities of the vapor and liquid are

$$j_\text{v} = \frac{Gx}{\rho_\text{v}} = \frac{1018.6 \times 0.15}{5.1597} = 29.61\,\text{m/s}$$

$$j_\ell = \frac{G(1-x)}{\rho_\ell} = \frac{1018.6 \times (1-0.15)}{887.31} = 0.976\,\text{m/s}$$

The Reynolds numbers of the vapor and liquid phases are obtained from Eqs. (10.74) and (10.75), i.e.,

$$Re_\text{v} = \frac{GxD}{\mu_\text{v}} = \frac{1018.6 \times 0.15 \times 0.1}{149 \times 10^{-7}} = 1.025 \times 10^6$$

$$Re_\ell = \frac{G(1-x)D}{\mu_\ell} = \frac{1018.6 \times (1-0.15) \times 0.1}{1493 \times 10^{-7}} = 5.799 \times 10^5$$

The fraction coefficients for the vapor and liquid phases are determined from Eq. (10.78), i.e.,

$$f_\text{v} = 0.079\,Re_\text{v}^{-0.25} = 2.48 \times 10^{-3}$$

$$f_\ell = 0.079\,Re_\ell^{-0.25} = 2.86 \times 10^{-3}$$

The frictional pressure gradients of the vapor or liquid phase flow in the channel, with its actual flow rate and properties, can be found from Eqs. (10.70) and (10.71), i.e.,

$$-\left(\frac{\text{d}p_F}{\text{d}z}\right)_\text{v} = \frac{2f_\text{v}G^2x^2}{D\rho_\text{v}} = \frac{2 \times 2.48 \times 10^{-3} \times 1018.6^2 \times 0.15^2}{0.1 \times 5.1597} = 224.4\,\text{Pa/m}$$

$$-\left(\frac{\text{d}p_F}{\text{d}z}\right)_\ell = \frac{2f_\ell G^2(1-x)^2}{D\rho_\ell}$$

$$= \frac{2 \times 2.86 \times 10^{-3} \times 1018.6^2 \times 0.85^2}{0.1 \times 887.31} = 48.3\,\text{Pa/m}$$

Therefore, the Martinelli parameter can be obtained from Eq. (11.26), i.e.,

$$X = \left[\frac{(\mathrm{d}p_F/\mathrm{d}z)_\ell}{(\mathrm{d}p_F/\mathrm{d}z)_\mathrm{v}}\right]^{1/2} = \left[\frac{48.3}{224.4}\right]^{1/2} = 0.464$$

It can be seen from the Taitel and Dukler (1976) flow map that actual flow regime depends on the values of F and K, which can be found from Eqs. (10.27) and (10.28), i.e.,

$$F = \sqrt{\frac{\rho_\mathrm{v}}{\rho_\ell - \rho_\mathrm{v}}\frac{j_\mathrm{v}}{\sqrt{Dg}}} = \sqrt{\frac{5.1597}{887.31 - 5.1597}\frac{29.61}{\sqrt{0.1 \times 9.8}}} = 2.29$$

$$K = \left[\frac{\rho_\mathrm{v} j_\mathrm{v}^2}{(\rho_\ell - \rho_\mathrm{v})Dg}\frac{\rho_\ell D j_\ell}{\mu_\ell}\right]^{1/2}$$

$$= \left[\frac{5.1597 \times 29.61^2}{(887.31 - 5.1597) \times 0.1 \times 9.8}\frac{887.31 \times 0.1 \times 0.976}{1493 \times 10^{-7}}\right]^{1/2} = 1742.2$$

The flow regime is annular-dispersed liquid (AD), as indicated by Fig. 10.5.

Example 10.3

Determine the pressure gradient due to friction in Example 10.2.

Solution Since the ratio of the viscosity is

$$\frac{\mu_\ell}{\mu_\mathrm{v}} = \frac{1493 \times 10^{-7}}{149 \times 10^{-7}} = 10 < 1000$$

Equation (10.88) developed by Friedel (1979), should be used. In addition to the properties found in Example 10.2, the surface tension is $\sigma = 42.19 \times 10^{-3}\,\mathrm{N/m}$.
The Reynolds numbers of the vapor or liquid phases obtained from Eqs. (10.76) and (10.77) are

$$Re_{\mathrm{v}0} = \frac{GD}{\mu_\mathrm{v}} = \frac{1018.6 \times 0.1}{149 \times 10^{-7}} = 7.289 \times 10^6$$

$$Re_{\ell 0} = \frac{GD}{\mu_\ell} = \frac{1018.6 \times 0.1}{1493 \times 10^{-7}} = 6.821 \times 10^5$$

The fraction coefficients for the vapor and liquid phases determined from Eq. (10.78) are

$$f_{\mathrm{v}0} = 0.079\,Re_{\mathrm{v}0}^{-0.25} = 1.52 \times 10^{-3}$$

$$f_{\ell 0} = 0.079\,Re_{\ell 0}^{-0.25} = 2.75 \times 10^{-3}$$

The frictional pressure gradients when the liquid flows in the channel with the total mass flow rate of the two-phase flow, found from Eq. (10.73), is

$$-\left(\frac{\mathrm{d}p_F}{\mathrm{d}z}\right)_{\ell 0} = \frac{2 f_{\ell 0} G^2 (1 - x)^2}{D\rho_\ell}$$

$$= \frac{2 \times 2.75 \times 10^{-3} \times 1018.6^2 \times 0.85^2}{0.1 \times 887.31} = 46.45\,\mathrm{Pa/m}$$

The homogeneous density of the two-phase system is obtained from Eq. (10.31):

$$\rho = \frac{\rho_v \rho_\ell}{\rho_\ell x + \rho_v (1-x)} = \frac{887.31 \times 5.1597}{887.31 \times 0.15 + 5.1597 \times 0.85} = 33.30\,\mathrm{kg/m^3}$$

All the parameters necessary to use Eq. (10.88) are obtained from Eqs. (10.89)–(10.92):

$$C_1 = (1-x)^2 + x^2 \left(\frac{\rho_\ell}{\rho_v}\right)\left(\frac{f_{v0}}{f_{\ell 0}}\right)$$

$$= (1-0.15)^2 + 0.15^2 \times \left(\frac{887.31}{5.1597}\right) \times \left(\frac{1.52 \times 10^{-3}}{2.75 \times 10^{-3}}\right) = 2.86$$

$$C_2 = x^{0.78}(1-x)^{0.24}\left(\frac{\rho_\ell}{\rho_v}\right)^{0.91}\left(\frac{\mu_v}{\mu_\ell}\right)^{0.19}\left(1-\frac{\mu_v}{\mu_\ell}\right)^{0.7}$$

$$= 0.15^{0.78}(1-0.15)^{0.24}\left(\frac{887.3}{5.1597}\right)^{0.91}\left(\frac{149 \times 10^{-7}}{1493 \times 10^{-7}}\right)^{0.19}\left(1-\frac{149 \times 10^{-7}}{1493 \times 10^{-7}}\right)^{0.7}$$

$$= 14.21$$

$$Fr = \frac{G^2}{gD\rho^2} = \frac{1018.6^2}{9.8 \times 0.1 \times 33.30^2} = 954.76$$

$$We = \frac{G^2 D}{\rho\sigma} = \frac{1018.6^2 \times 0.1}{33.30 \times 42.19 \times 10^{-3}} = 7.385 \times 10^4$$

Therefore, the pressure multiplier is

$$\phi_{\ell 0}^2 = C_1 + \frac{3.24 C_2}{Fr^{0.045} We^{0.035}} = 2.86 + \frac{3.24 \times 14.21}{954.76^{0.045} \times (7.385 \times 10^4)^{0.035}} = 25.7$$

The pressure gradient due to friction is obtained from Eq. (10.82), i.e.,

$$-\frac{\mathrm{d}p_F}{\mathrm{d}z} = -\phi_{\ell 0}^2 \left(\frac{\mathrm{d}p_F}{\mathrm{d}z}\right)_{\ell 0} = 25.7 \times 46.45 = 1.19 \times 10^3\ \mathrm{Pa/m}$$

10.3.4 Void Fraction

Correlations for the frictional pressure drop were discussed in the preceding subsection. To obtain the total pressure drop, accelerational and gravitational pressure drops must also be determined. As can be seen from Eqs. (10.41) and (10.63), knowledge of the void fraction is required for determination of the accelerational and gravitational pressure drops for both the homogeneous and separated flow models. In the case of a horizontal circular tube, the gravitational pressure drop term becomes zero but the accelerational pressure drop terms are still present. Therefore, correlations for the void fraction in the two-phase flow will be discussed.

When the phase velocities differ for the liquid and vapor phases, the slip ratio, defined in Eq. (10.13), is frequently used in lieu of the void fraction. Substituting Eqs. (10.20) and (10.21) into Eq. (10.13), one obtains

$$S = \frac{G_v \rho_\ell (1 - \alpha)}{G_\ell \rho_v \alpha} \tag{10.96}$$

which can be simplified by using Eq. (10.23), i.e.,

$$S = \frac{\rho_\ell x (1 - \alpha)}{\rho_v (1 - x) \alpha} \tag{10.97}$$

The slip ratio can also be related to the volumetric flow rate by substituting Eqs. (10.5) and (10.6) into Eq. (10.13), i.e.,

$$S = \frac{Q_v A_\ell}{Q_\ell A_v} \tag{10.98}$$

Substituting Eq. (10.2) into Eq. (10.98), one obtains

$$S = \frac{Q_v (1 - \alpha)}{Q_\ell \alpha} \tag{10.99}$$

The relationships between the void fraction and slip ratio can be obtained by rearranging Eqs. (10.97) and (10.99):

$$\alpha = \frac{1}{1 + \frac{1-x}{x} \frac{\rho_v}{\rho_\ell} S} \tag{10.100}$$

$$\alpha = \frac{Q_v}{S Q_\ell + Q_v} \tag{10.101}$$

It can be observed from Eqs. (10.97) and (10.99), (10.100), and (10.101) that the void fraction may be determined if the slip ratio is known, and vice versa. The void fraction in the homogeneous model obtained by Eq. (10.35) is a special case of Eq. (10.100) where the slip ratio $S = 1$. The homogeneous model is the better choice in a two-phase application for bubbly or mist flow, since the liquid–vapor interface is very small and the slip ratio is very close to unity. For other flow patterns where the velocity of the vapor phase is significantly higher than the velocity of the liquid phase, the homogeneous model tends to significantly overpredict the void fraction. Another case in which the homogeneous model will provide good results is near-critical point ($\rho_\ell \approx \rho_v$), in which case $\alpha \approx x$.

To predict the void fraction for cases where the velocities of the two phases differ significantly, correlations based on the separated flow model should be employed. The earliest void fraction correlation is based on the separated flow model proposed by Lockhart and Martinelli (1949), shown in Fig. 10.6. The relationship between the void fraction and the Martinelli parameter, X, is fitted by the following equation:

$$\alpha = \frac{\phi_{\ell,tt} - 1}{\phi_{\ell,tt}} \tag{10.102}$$

where $\phi_{\ell,tt}$ is the frictional multiplier for turbulent flow in both vapor and liquid phases, obtained from Eq. (10.84) with the appropriate value of C from Table 10.1, i.e.,

$$\phi_{\ell,\mathrm{tt}}^2 = 1 + \frac{20}{X} + \frac{1}{X^2} \tag{10.103}$$

Butterworth (1975) recommended the following simpler correlation to replace Eq. (10.102):

$$\alpha = \left[1 + 0.28X^{0.71}\right]^{-1} \tag{10.104}$$

The void fraction correlation proposed by Lockhart and Martinelli (1949) is the most widely used correlation. However, it tends to overpredict the void fraction at high mass flow rate. A number of alternative correlations have been proposed to overcome this problem. The correlation proposed by Premoli et al. (1971) is worthwhile to introduce here because it covers a wide range of data. The correlation of Premoli et al. (1971) given in terms of the slip ratio is

$$S = 1 + E_1\left[\left(\frac{y}{1 + yE_2}\right) - yE_2\right]^{\frac{1}{2}} \tag{10.105}$$

where

$$y = \frac{\beta}{1 - \beta} \tag{10.106}$$

$$E_1 = 1.578\, Re_{\ell 0}^{-0.19}\left(\frac{\rho_\ell}{\rho_\mathrm{v}}\right)^{0.22} \tag{10.107}$$

$$E_2 = 0.0273\, We\, Re_{\ell 0}^{-0.51}\left(\frac{\rho_\ell}{\rho_\mathrm{v}}\right)^{-0.08} \tag{10.108}$$

where β is the volumetric flow fraction of vapor flow defined by Eq. (10.12). The Weber number in Eqs. (10.107) and (10.108) is defined as

$$We = \frac{G^2 D}{\rho_\ell \sigma} \tag{10.109}$$

The void fraction can be obtained from Eq. (10.100) after the slip ratio is obtained from Eq. (10.105).

Butterworth (1975) compared different void fraction correlations for two-phase flow and suggested that the void fractions can be expressed in the following form

$$\alpha = \frac{1}{1 + c\left(\frac{1-x}{x}\right)^q \left(\frac{\rho_\mathrm{v}}{\rho_\ell}\right)^r \left(\frac{\mu_\ell}{\mu_\mathrm{v}}\right)^s} \tag{10.110}$$

where the values of the constant and exponents are given in Table 10.2. Chisholm (1973b) presented a simple correlation in terms of slip-ratio

$$S = \left[1 - x\left(1 - \frac{\rho_\ell}{\rho_\mathrm{v}}\right)\right]^{0.5} \tag{10.111}$$

Table 10.2 Constants and exponents for different void fraction models

Models	c	q	r	s
Homogeneous model	1	1	1	0
Zivi (1964)	1	1	0.67	0
Turner (1966)	1	0.72	0.40	0.08
Lockhart–Martinelli (1949)	0.28	0.64	0.36	0.07
Thome (1964)	1	1	0.89	0.18
Baroczy (1965)	1	0.74	0.65	0.13

Awad and Muzychka (2005b) compared these models for two-phase flow of R-22 at $T_{sat} = 50\,°C$ in a smooth tube of D = 10 mm, and the results are shown in Fig. 10.8. It can be seen that no two models can provide the same prediction on the void fraction.

Based on the assumption that both liquid and vapor are turbulent (both Re_ℓ and Re_v are greater than 2000), Awad and Muzychka (2005b) developed rational bounds for two-phase void fraction. The lower bound of the void fraction is

$$\alpha_{lower} = \left\{ 1 + \left[\left(\frac{1-x}{x} \right)^{0.875} \left(\frac{\rho_v}{\rho_\ell} \right)^{0.5} \left(\frac{\mu_\ell}{\mu_v} \right)^{0.125} \right]^{16/19} \right\}^{-1} \tag{10.112}$$

The upper bound of the void fraction is

$$\alpha_{upper} = \left\{ 1 + 0.28 \left[\left(\frac{1-x}{x} \right)^{0.875} \left(\frac{\rho_v}{\rho_\ell} \right)^{0.5} \left(\frac{\mu_\ell}{\mu_v} \right)^{0.125} \right]^{0.71} \right\}^{-1} \tag{10.113}$$

By averaging the lower and upper bounds, an empirical correlation for void fraction in two-phase flow can be obtained.

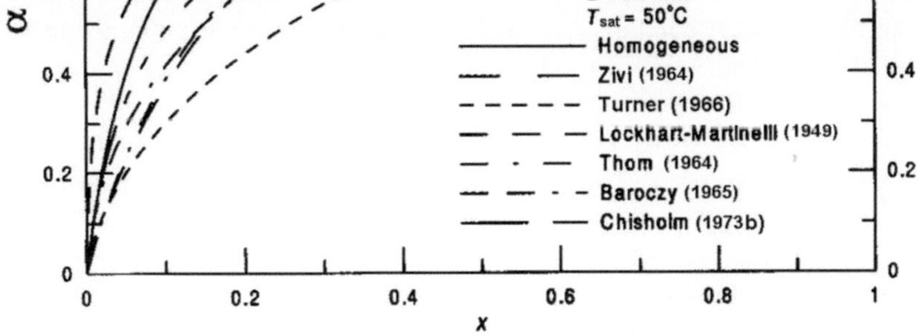

Fig. 10.8 Comparison of void fraction models for two-phase flow (Awad and Muzychka 2005b)

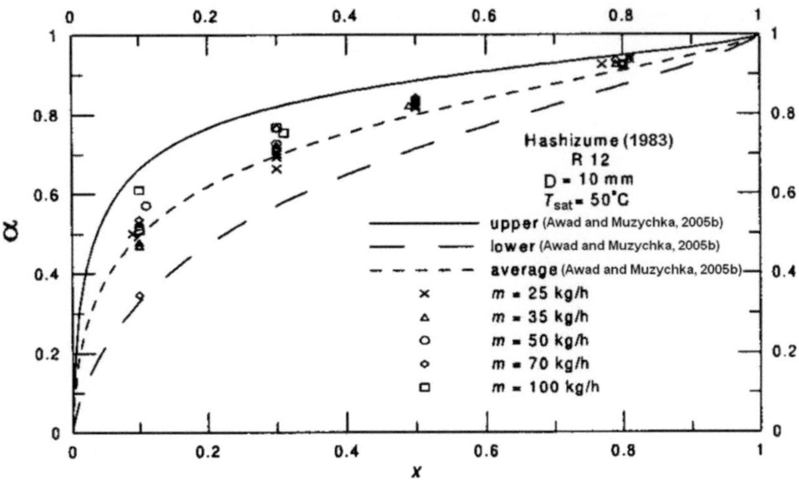

Fig. 10.9 Void fraction versus quality (Awad and Muzychka 2005b)

$$
\begin{aligned}
\alpha_{\text{ave}} = & \frac{1}{2}\left\{1+\left[\left(\frac{1-x}{x}\right)^{0.875}\left(\frac{\rho_{\text{v}}}{\rho_\ell}\right)^{0.5}\left(\frac{\mu_\ell}{\mu_{\text{v}}}\right)^{0.125}\right]^{16/19}\right\}^{-1} \\
& +\frac{1}{2}\left\{1+0.28\left[\left(\frac{1-x}{x}\right)^{0.875}\left(\frac{\rho_{\text{v}}}{\rho_\ell}\right)^{0.5}\left(\frac{\mu_\ell}{\mu_{\text{v}}}\right)^{0.125}\right]^{0.71}\right\}^{-1}
\end{aligned}
\tag{10.114}
$$

Figure 10.9 shows comparisons of the void fraction with Hashizume's (1983) experimental data for R-12 at $T_{\text{sat}} = 50\,°\text{C}$ in a smooth horizontal tube at $D = 10\,\text{mm}$. The root mean square error is 9.04%.

10.4 Forced Convective Condensation in Tubes

10.4.1 Two-Phase Flow Regimes

Convective condensation in tubes occurs in the condensers of air-conditioning systems and refrigerators. Depending on the tube's geometric configuration, condensation within can be classified as condensation in vertical or horizontal tubes. Condensation in a vertical tube is relatively simple, because gravity acts parallel to the flow direction and, consequently, an annular liquid film forms on the inner surface of the tube. Therefore, the flow pattern for condensation in a vertical tube is limited to annular flow, and an analytical solution can easily be obtained (see Chap. 7). Condensation in a horizontal tube, on the other hand, is much more complicated, because gravity acts perpendicularly to the flow direction and various flow patterns are possible. When condensation occurs in a horizontal tube, different flow regimes are observed at different positions along its length. Two possible sequences of flow regimes for convective condensation in a horizontal tube are possible, depending on the mass flow rate (Hewitt 1998).

Figure 10.10a shows the sequence of flow regimes that arise during condensation in a horizontal circular tube with higher mass flow rate. Superheated vapor enters the horizontal tube, which has a temperature below the saturation temperature of the vapor. The flow at this point in the tube is single-phase vapor flow, since no condensation occurs near the inlet. After the vapor cools and

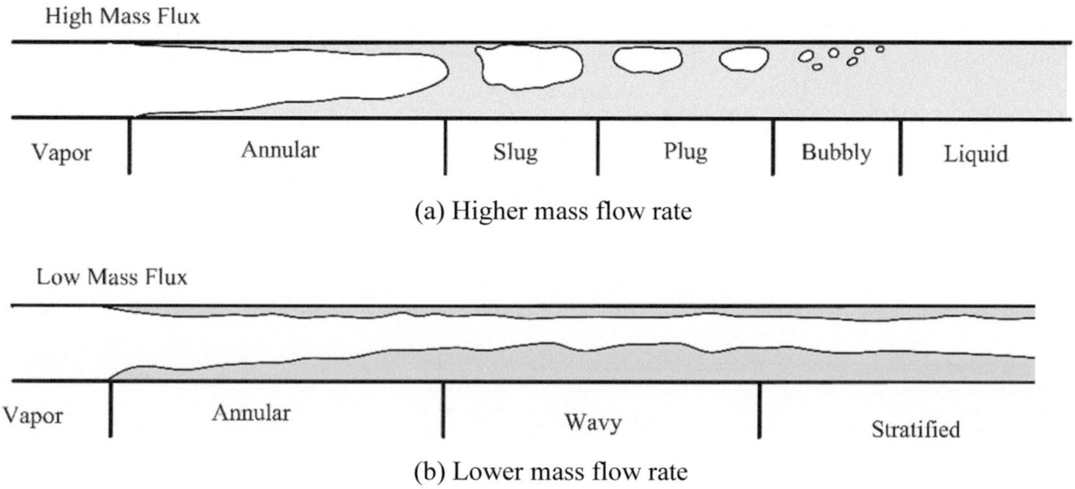

Fig. 10.10 Flow patterns for condensation inside horizontal tubes

becomes saturated, condensation starts to occur on the inner wall of the tube. The flow pattern at the beginning of condensation is annular, because the velocity of the vapor is much higher than that of the condensate. The dominant force in the annular flow regime is shear stress at the liquid–vapor interface, with gravity playing a less important role. As condensation continues, the velocity of the vapor phase decreases and the dominant force shifts from shear force at the interface to gravitational force. Liquid accumulates at the bottom of the tube, while condensation takes place mainly at the top portion of the tube where the liquid film is thin. The void fraction in the tube decreases as the vapor condenses downstream, and the flow patterns change to slug and plug flows. The size of the vapor plugs further decreases downstream, and the two-phase flow enters a brief bubbly flow regime before the completion of condensation. Near the outlet of the horizontal tube, the quality reduces to zero and the flow in the tube becomes single-phase liquid flow.

The sequence of flow regimes for convective condensation in a horizontal tube at a lower mass flow rate is shown in Fig. 10.10b. Superheated vapor enters the tube and the flow is single-phase vapor near the inlet, as in the high mass flow rate case. Another similarity is that the flow regime is annular flow at the commencement of condensation. As condensation continues, the condensate at the top portion of the tube flows to the bottom portion due to gravitational force. Since the mass flow rate is not high enough to produce slug or plug flow, two-phase flow in the horizontal tube enters a stratified flow regime where liquid flows in the bottom portion of the tube and vapor flows in the top portion. Unlike the case of high flow rate, the vapor at lower flow rate never completely condenses. Breber et al. (1980) used an extensive collection of data from the literature to demonstrate that the flow pattern of Taitel and Dukler (1976) is applicable to condensation in horizontal tubes. Experimental data from more than 700 condensation flow regime observations in the literature, including water, R-12 and R-113, inner-tube diameters of 4.8–50.8 mm, and a wide range of operating parameters, were used for the comparison. With the exception of data from the smallest inner diameter tube (4.8 mm), agreement was good with the flow regime transition boundaries as predicted by Taitel and Dukler.

A flow map for condensation in a horizontal tube was suggested by Tandon et al. (1982) and is shown in Fig. 10.11. The dimensionless vapor velocity is defined as

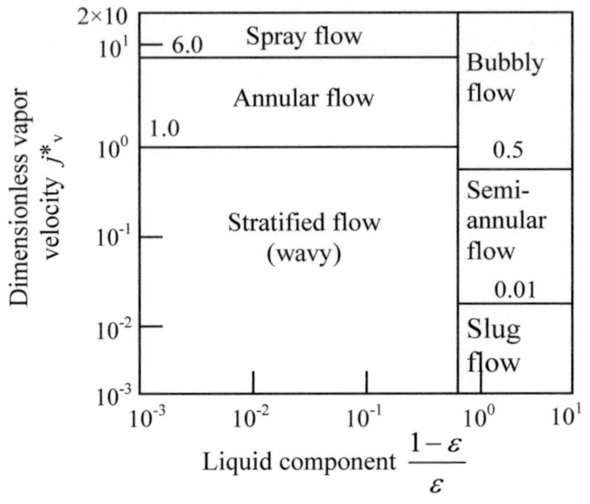

Fig. 10.11 Flow pattern map for condensation in a horizontal tube (Tandon et al. 1982)

Table 10.3 Flow regimes of convective condensation in a horizontal tube

$(1-\varepsilon)/\varepsilon$	j_v^*	Flow regimes
$(1-\varepsilon)/\varepsilon \le 0.5$	$j_v^* \le 1$	Stratified flow
	$1 \le j_v^* \le 6$	Annular-dispersed flow
	$j_v^* > 6$	Spray flow
$(1-\varepsilon)/\varepsilon > 0.5$	$j_v^* \le 0.01$	Slug flow
	$0.01 \le j_v^* \le 0.5$	Semi-annular flow
	$j_v^* > 0.5$	Bubbly flow

$$j_v^* = \frac{xG}{\sqrt{gD\rho_v(\rho_\ell - \rho_v)}} \tag{10.115}$$

where D is the diameter of the tube. The abscissa is the relative cross-sectional area of the tube occupied by the liquid and vapor, $A_\ell/A_v = (1-\varepsilon)/\varepsilon$. The flow regimes shown in Fig. 10.11 can be summarized in Table 10.3. The most frequently occurring flow regimes for condensation in a horizontal tube are the annular-dispersed flow and the stratified flow. The dispersed-annular flow can also occur in condensation in a vertical tube if the vapor velocity is high enough to allow the effect of gravity to disregarded.

10.4.2 Heat Transfer Predictions

Heat transfer characteristics for convective condensation in a tube depend strongly on the flow regimes and the different structures of their liquid–vapor interfaces. Heat transfer in stratified and annular regimes has been extensively studied, and different correlations based on theoretical analysis and experimental data have been proposed.

Gravitational effects can be neglected for the high mass flux region, ($G > 400\,\text{kg/m}^2\text{-s}$), so that annular flow with convective condensation in a horizontal tube can be analyzed with the method applied to the vertical tube in Chap. 7. Furthermore, a detailed analysis of annular flow condensation in a miniature tube presented in Sect. 10.6.2 under the title "Annular Flow Condensation in a Miniature Tube" is also applicable to conventional single tube.

For practical applications, the following empirical correlation, proposed by Shah (1979) from a fit of 474 experimental data points, is recommended:

$$Nu = 0.023\, Re_{\ell 0}^{0.8}\, Pr_{\ell}^{0.4}\left[(1-x)^{0.8} + \frac{3.8x^{0.76}(1-x)^{0.04}}{(p_{\mathrm{sat}}/p_{\mathrm{c}})^{0.38}}\right] \tag{10.116}$$

where $Re_{\ell 0}$ is defined by Eq. (10.77) and Pr_ℓ is the liquid Prandtl number. Equation (10.116) was obtained by modifying the Dittus–Boelter equation for single-phase forced convection in a tube. It is applicable to the following conditions: $0.002 \le p_{\mathrm{sat}}/p_{\mathrm{c}} \le 0.44$, $21\,°\mathrm{C} \le T_{\mathrm{sat}} \le 310\,°\mathrm{C}$, $3\,\mathrm{m/s} \le w_{\mathrm{v}} \le 300\,\mathrm{m/s}$, $0 \le x \le 1$, $10.8\,\mathrm{kg/m^2} - \mathrm{s} \le G \le 1599\,\mathrm{kg/m^2\text{-}s}$, $Re_{\ell 0} \ge 350$, and $\mathrm{Pr}_\ell > 0.5$. Moser et al. (1998) verified Eq. (10.116) using 1197 data points for 6 different kinds of refrigerants and concluded that the deviation of Eq. (10.116) is 14.37%.

At lower vapor velocity, interfacial shear force becomes less important and gravity becomes the dominant force in convective condensation in the tube. As shown in Fig. 10.12, due to gravitational force, the condensate in the top portion of the tube flows into the bottom portion, and the two-phase flow lies in the stratified regime. The stratified regime is very common for low mass flow rates or short tubes.

The average heat transfer coefficient for condensation in a tube with stratified flow is

$$\bar{h} = \frac{\theta_{\mathrm{strat}}}{\pi} h_{\mathrm{d}} + \left(1 - \frac{\theta_{\mathrm{strat}}}{\pi}\right) h_{\mathrm{b}} \tag{10.117}$$

where h_{d} and h_{b} are heat transfer coefficients of the drainage condensate and the condensate at the bottom. Since condensation occurs primarily at the top portion of the tube where the liquid layer is

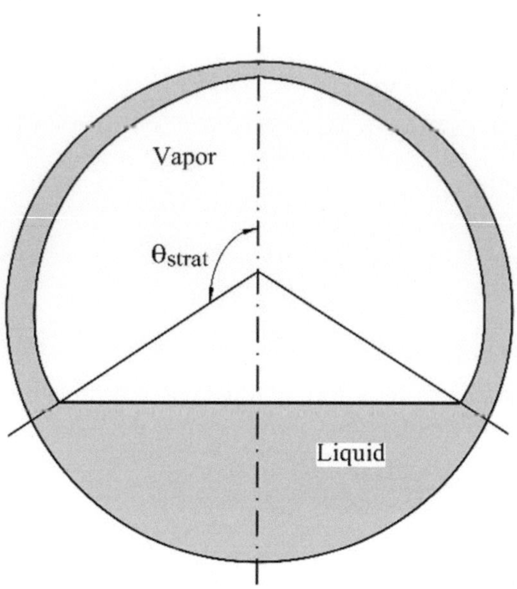

Fig. 10.12 Cross section of stratified flow

relatively thin, the condensation occurring on the surface of the stratified liquid can be neglected ($h_b \ll h_d$). Thus, Eq. (10.117) is simplified as

$$\bar{h} = \frac{\theta_{\text{strat}}}{\pi} h_d \tag{10.118}$$

Since condensation occurring in the drainage condensate is film condensation, one can expect that the heat transfer coefficient can be obtained using a correlation similar to the Nusselt solution.

$$h_d = F(\theta_{\text{strat}}) \left[\frac{g\rho_\ell(\rho_\ell - \rho_v) k_\ell^3 h'_{\ell v}}{\mu_\ell D (T_{\text{sat}} - T_w)} \right]^{1/4} \tag{10.119}$$

where the function $F(\theta_{\text{strat}})$ can be obtained by (Jaster and Kosky 1976)

$$F(\theta_{\text{strat}}) = 0.728 \frac{\pi}{\theta_{\text{strat}}} \left[\frac{\pi}{\theta_{\text{strat}}} + \frac{\sin 2(\pi - \theta_{\text{strat}})}{2\pi} \right]^{3/4} \tag{10.120}$$

The void fraction, α, is related to the stratification angle, θ_{strat}, by

$$\alpha = \frac{\pi}{\theta_{\text{strat}}} + \frac{\sin 2(\pi - \theta_{\text{strat}})}{2\pi} \tag{10.121}$$

Substituting Eqs. (10.120) and (10.121) into Eq. (10.119) yields

$$h_d = 0.728 \frac{\pi}{\theta_{\text{strat}}} \alpha^{3/4} \left[\frac{g\rho_\ell(\rho_\ell - \rho_v) k_\ell^3 h'_{\ell v}}{\mu_\ell D} \right]^{1/4} \tag{10.122}$$

The average heat transfer coefficient can be obtained by substituting Eq. (10.122) into Eq. (10.118), i.e.,

$$\bar{h} = 0.728 \alpha^{3/4} \left[\frac{g\rho_\ell(\rho_\ell - \rho_v) k_\ell^3 h'_{\ell v}}{\mu_\ell D} \right]^{1/4} \tag{10.123}$$

The condition for which Eq. (10.123) applies is $Re_{v0} \le 3.5 \times 10^4$. The void fraction in Eq. (10.123) can be obtained by

$$\alpha = \left[1 + \frac{1-x}{x} \left(\frac{\rho_v}{\rho_\ell} \right)^{2/3} \right]^{-1} \tag{10.124}$$

In addition to Eqs. (10.116) and (10.123), Dobson and Chato (1998) proposed the empirical correlations that took into account the effect of different flow regimes. For annular flow and $\dot{m}'' \ge 500\ \text{kg/m}^2\text{-s}$, Dobson and Chato (1998) recommended the following correlation:

$$Nu = \frac{hD}{k_\ell} = 0.023 Re_\ell^{0.8} Pr_\ell^{0.4} \left(1 + \frac{2.22}{X_{tt}^{0.89}} \right) \tag{10.125}$$

where X_{tt} is the Martinelli parameter for turbulent flow in both liquid and vapor phases, i.e.,

$$X_{\mathrm{tt}} = \left(\frac{1-x}{x}\right)^{0.9}\left(\frac{\rho_{\mathrm{v}}}{\rho_{\ell}}\right)^{0.5}\left(\frac{\mu_{\ell}}{\mu_{\mathrm{v}}}\right)^{0.1} \tag{10.126}$$

When the flow pattern is stratified wavy flow, the empirical correlation is

$$Nu = \frac{hD}{k_{\ell}} = \frac{0.23\,Re_{\mathrm{v0}}^{0.12}}{1+1.11\,X_{\mathrm{tt}}^{0.58}}\left[\frac{Ga_{\ell}\,\mathrm{Pr}_{\ell}}{Ja_{\ell}}\right]^{0.25} + \left(1-\frac{\theta_{\mathrm{strat}}}{\pi}\right)Nu_{\mathrm{strat}} \tag{10.127}$$

where the first term on the right-hand side of Eq. (10.127) accounts for the contribution of the film condensation on the top portion of the tube. The contribution of the stratified liquid is accounted for by the second term on the right-hand side of Eq. (10.127). Ga_{ℓ} is the Galileo number for the liquids defined as

$$Ga_{\ell} = \frac{gD^3}{\nu_{\ell}^2} \tag{10.128}$$

The angle from the top of the tube to the stratified liquid layer in the bottom θ_{strat} is estimated by

$$\theta_{\mathrm{strat}} = \pi - \arccos(2\alpha - 1) \tag{10.129}$$

where the void fraction is approximated by Eq. (10.100), with the slip ratio $S = (\rho_{\ell}/\rho_{\mathrm{v}})^{1/3}$, i.e.,

$$\alpha = \frac{x}{x+(1-x)(\rho_{\mathrm{v}}/\rho_{\ell})^{2/3}} \tag{10.130}$$

which is referred to as the Zivi void fraction equation (Zivi 1964).

The Nusselt number for forced convective heat transfer of the stratified liquid, Nu_{strat}, in Eq. (10.127) is

$$Nu_{\mathrm{strat}} = 0.0195\,Re_{\ell}^{0.8}Pr_{\ell}^{0.4}\left(1.376+\frac{c_1}{X_{\mathrm{tt}}^{c_2}}\right)^{1/2} \tag{10.131}$$

where

$$c_1 = \begin{cases} 4.172+5.48Fr_{\ell}-1.564Fr_{\ell}^2 & Fr_{\ell}\le 0.7 \\ 7.242 & Fr_{\ell} > 0.7 \end{cases} \tag{10.132}$$

and

$$c_2 = \begin{cases} 1.773-0.169Fr_{\ell} & Fr_{\ell}\le 0.7 \\ 1.655 & Fr_{\ell} > 0.7 \end{cases} \tag{10.133}$$

The transition from annular flow to stratified flow is characterized by the modified Froude transition number, which is defined as

$$Fr_{so} = c_3 Re_{\ell}^{c_4}\left(\frac{1+1.09X_{\mathrm{tt}}^{0.039}}{X_{\mathrm{tt}}}\right)^{1.5} \tag{10.134}$$

where $c_3 = 0.025$ and $c_4 = 1.59$ for $Re_\ell \le 1250$ and $c_3 = 1.26$ and $c_4 = 1.04$ for $Re_\ell > 1250$. Dobson and Chato (1998) suggested that Eq. (10.127) is applicable when $\dot{m}'' < 500$ kg/m^2-s and $Fr_{s0} < 20$.

Example 10.4

Steam condenses at 80 °C as it flows inside a horizontal tube with an inner diameter of 5 cm. The mass flow rate is 1.2 kg/s, and the quality is $x = 0.75$. Find the heat transfer coefficient.

Solution The properties of the water and steam at 80 °C can be found from Table B.48 in Appendix B: $p_{sat} = 47.359\,\text{kPa}$, $h_{\ell v} = 2308.9\,\text{kJ/kg}$, $\rho_\ell = 971.82\,\text{kg/m}^3$, $\rho_v = 0.2932\,\text{kg/m}^3$, $\mu_\ell = 3510 \times 10^{-7}\,\text{N-s/m}^2$, $\mu_v = 113 \times 10^{-7}\text{N-s/m}^2$, $k_\ell = 0.669\,\text{W/m-K}$ and $\text{Pr}_\ell = \mu_\ell c_{p\ell}/k_\ell = 2.2$. The mass flow rate is $\dot{m} = 1.2\,\text{kg/s}$, and the diameter of the tube is $D = 0.05$ m. Therefore, the mass flux of the steam–water mixture in the tube is

$$G = \frac{\dot{m}}{A} = \frac{4\dot{m}}{\pi D} = \frac{4 \times 1.2}{\pi \times 0.05^2} = 611.15\,\text{kg/(m}^2\text{s)}$$

Since $G > 500\,\text{kg/m}^2$-s, the heat transfer coefficient can be obtained from Eq. (10.125). The Martinelli parameter for turbulent flow in both the liquid and vapor phases, X_{tt}, is obtained from Eq. (10.126):

$$X_{tt} = \left(\frac{1-x}{x}\right)^{0.9} \left(\frac{\rho_v}{\rho_\ell}\right)^{0.5} \left(\frac{\mu_\ell}{\mu_v}\right)^{0.1}$$
$$= \left(\frac{1-0.75}{0.75}\right)^{0.9} \left(\frac{0.2932}{971.82}\right)^{0.5} \left(\frac{3510 \times 10^{-7}}{113 \times 10^{-7}}\right)^{0.1} = 0.0091$$

The liquid Reynolds numbers are obtained from Eq. (10.75), i.e.,

$$Re_\ell = \frac{G(1-x)D}{\mu_\ell} = \frac{611.15 \times (1-0.75) \times 0.05}{3510 \times 10^{-7}} = 2.176 \times 10^4$$

The Nusselt number is obtained from Eq. (10.125), i.e.,

$$Nu = \frac{hD}{k_\ell} = 0.023\,Re_\ell^{0.8}\,Pr_\ell^{0.4}\left(1 + \frac{2.22}{X_{tt}^{0.89}}\right)$$
$$= 0.023 \times (2.176 \times 10^4)^{0.8} \times 2.2^{0.4} \times \left(1 + \frac{2.22}{0.0091^{0.89}}\right)$$
$$= 1.363 \times 10^4$$

The heat transfer coefficient is then

$$h = \frac{Nu k_\ell}{D} = \frac{1.363 \times 10^4 \times 0.669}{0.05} = 1.824 \times 10^5\,\text{W/m}^2\text{-K}$$

10.5 Forced Convective Boiling in Tubes

10.5.1 Regimes in Horizontal and Vertical Tubes

Forced convective boiling in tubes finds application in evaporators for air conditioners and refrigerators, as well as in boilers. The two most common configurations in practical applications are convective boiling in horizontal tubes and upward flow boiling in vertical tubes. The flow regimes and heat transfer correlations for these two configurations are different and will be discussed separately.

The typical sequence of flow regimes for upward flow forced convective boiling in a uniformly heated vertical tube ($q'' = \text{const}$) is shown in Fig. 10.13. Since gravity acts parallel to the flow direction, the upward flow boiling in the tube is axisymmetric. The gravitational force plays a more dominant role while the liquid–vapor interfacial shear force is less important. If the fluid enters the tube as subcooled liquid and leaves the tube as superheated vapor, all of the flow regimes for horizontal tubes discussed in Sect. 10.2 will be encountered in the interim. The void fraction increases from zero at the inlet of the tube to one at the outlet of the tube. Since the density of the vapor is significantly lower than that of the liquid, the density of the two-phase mixture significantly decreases along the flow direction. To maintain constant mass flux along the flow direction, the mean velocity must increase substantially to match the significant increase in vapor-phase velocity. The growing disparity between vapor and liquid velocities along the flow direction will result in changing flow patterns along the flow direction.

Subcooled liquid enters the vertical tube and prior to initiation of any bubbles (Zone A), the heat transfer mechanism is single-phase forced convection. When the rising wall temperature exceeds the saturation temperature, vapor bubbles are generated and dispersed in the continuous liquid phase. While the fluid temperature in Zone B (subcooled boiling) is still below saturation temperature, the fluid temperature in Zone C (saturated nucleate boiling) is equal to the saturation temperature. The vapor bubbles in the bubbly flow region rise due to buoyancy force and the external pressure difference. As the vaporization process continues, more liquid is converted to vapor, the void fraction increases, then the flow regime progresses from bubbly to slug, churn, and then to annular (Zones D through F). When the liquid film on the inner surface of the tube is completely evaporated, the flow pattern becomes drop or mist flow (Zone G) and the wall temperature abruptly increases. When the last droplet of the liquid is vaporized, the flow in the tube becomes single phase (Zone H). Under constant heat flux conditions, both wall and fluid temperatures increase linearly in Zone H.

Figure 10.14 shows the typical sequence of flow regimes for forced convective boiling in a horizontal tube. Two-phase flow with boiling in a horizontal tube is no longer axisymmetric, because the gravitational force acts perpendicularly to the flow direction. When the subcooled liquid enters the horizontal tube, because no bubbles are present, the initial heat transfer mechanism is single-phase forced convection. After vapor is generated, the bubbles are dispersed in the continuous liquid phase and tend to rise to the top portion of the tube due to buoyancy effects. As boiling continues in the horizontal tube, the flow regime changes to plug, annular, and mist flow. As with the vertical flow case, the flow becomes single phase in the horizontal tube after the last droplet of the liquid is vaporized.

Kattan et al. (1998a, b, c) studied forced convective boiling in horizontal tubes, and proposed a new flow map, shown in Fig. 10.15. The boundary to separate different flow regimes can be found in Thome and El Hajal (2003) or Thome (2016). In contrast to the Taitel–Dukler flow map in Fig. 10.5, which is given in generic form, the Kattan–Thome–Favrat flow map is valid for one fluid (R134a) at a given tube diameter (D = 12 mm).

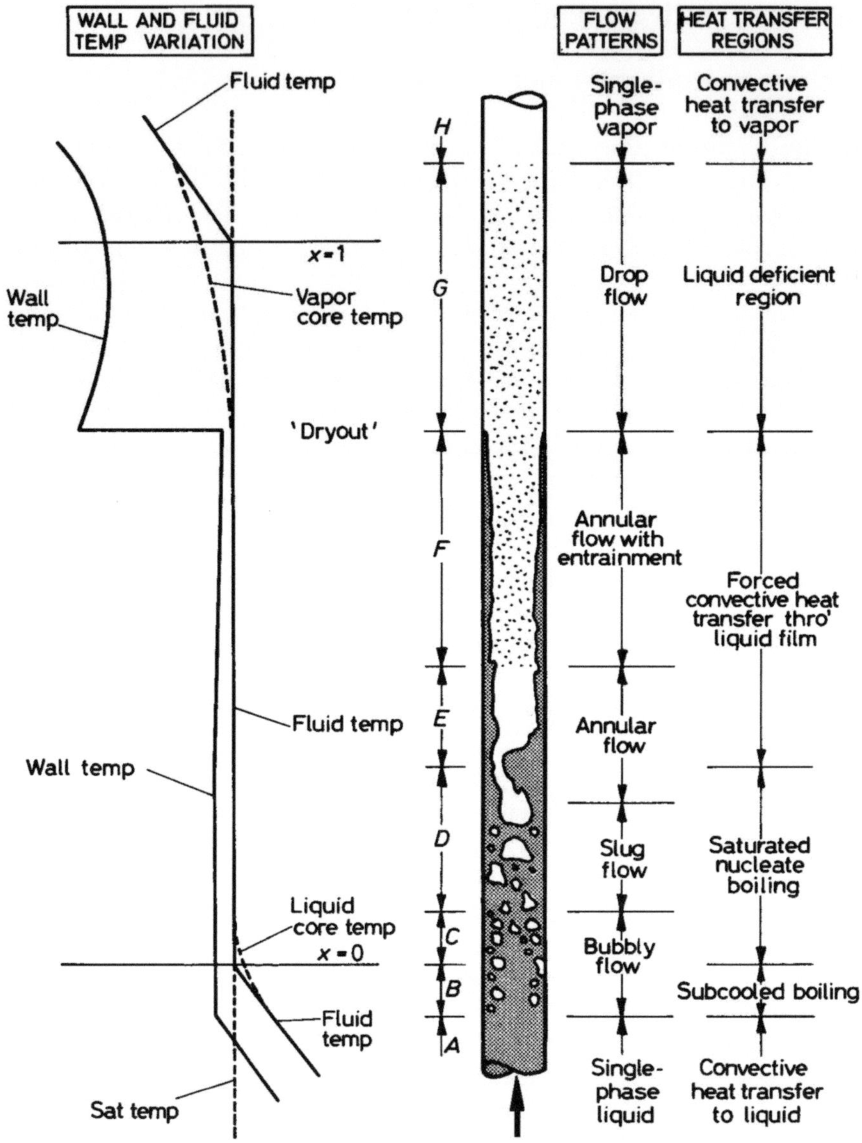

Fig. 10.13 Regimes for convective boiling in a vertical tube (Collier and Thome 1994; Reprinted with permission from Oxford University Press)

Many techniques have been tried to increase heat transfer coefficients in heat transfer devices. For single-phase applications, these methods concentrate mostly on disruption and destabilization of the thermal boundary layer, where two-phase flow applications rely on heat transfer enhancers or surface modification to increase boiling near the surface. However, these techniques all result in a pressure drop. One way in which heat transfer may be increased and not experience a severe pressure drop is through electrohydrodynamics (EHD). For two-phase flow, the difference in the specific dielectric

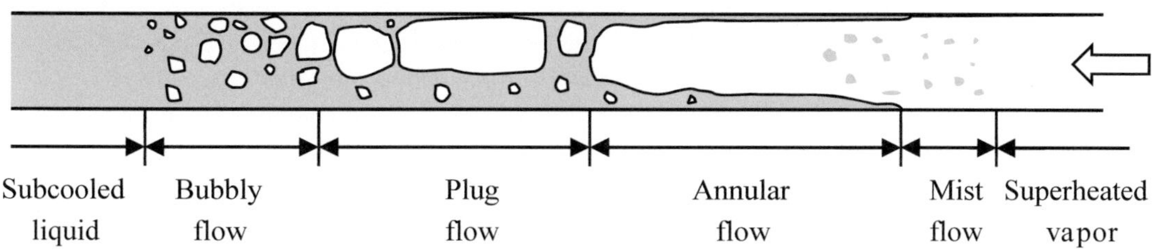

Fig. 10.14 Flow regimes for convective boiling in a horizontal tube (■ liquid, □ vapor)

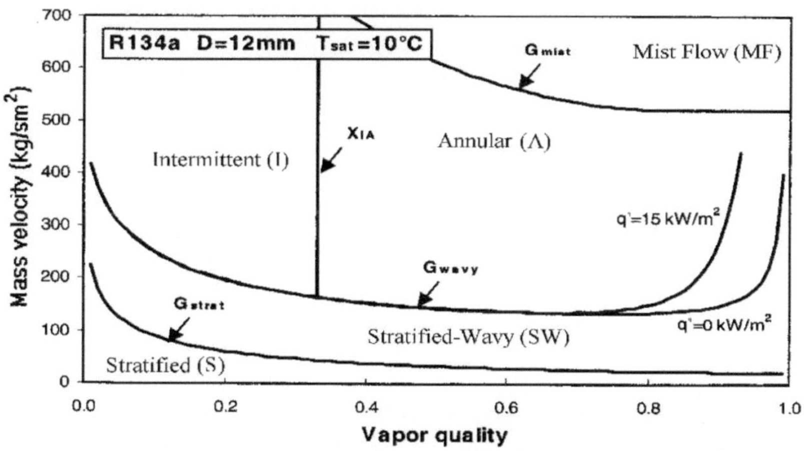

Fig. 10.15 Kattan–Thome–Favrat flow pattern map for forced convective boiling in horizontal tube (Thome and El Hajal 2003; Reproduced by permissions of Routledge/Taylor & Francis Group, LLC)

constant in an electric field can contribute additional body forces, which may lead to an increase in heat transfer. Experimental and analytical studies both show a decrease in the thermal boundary layer, increased convection, and interfacial instabilities that result in high heat transfer rates.

Existing research on the effects of EHD on convective boiling has confirmed that EHD can significantly increase the heat transfer rates. The optimum enhancement has been attributed to the relative influence of EHD forces and inertial forces on determining the flow regime. At low qualities, EHD effects cause a flow regime redistribution that significantly influences the heat transfer. Its effect diminishes as the inertial forces become the dominant force in determining the flow pattern. Also, some studies show that the liquid extraction phenomena may promote early partial dryout, which negatively affects heat transfer.

Modeling of electric field action in two-phase flow is difficult, because fluid-dynamics conservation equations are coupled with Maxwell's and Ohm's laws and must be solved together.

An electric field in two-phase flow adds additional body forces in the fluid as well as at the liquid–vapor interface. The electric force, $\mathbf{F_e}$, applied to the field is given as

$$\mathbf{F_e} = \rho_e \mathbf{E} - \frac{1}{2}E^2 \nabla \varepsilon + \frac{1}{2}\nabla \left[\rho E^2 \left(\frac{\partial \varepsilon}{\partial \rho}\right)_T\right] \qquad (10.135)$$

where ρ_e and ρ are the charge density and mass density, respectively, E is the electric field strength, and ε is the permittivity. The three terms on the right-hand side in the above equation are the electrophoretic, dielectophoretic, and electrostrictive components of the force, respectively.

The electrophoretic force (Coulomb force) results from the net free charge in the fluid or injected form electrodes. Phase interactions are typically associated with this component. The dielectrophoretic force arises from inhomogeneity in the permittivity of the dielectric fluid due to nonuniform electric fields, temperature gradients, and phase differences. The electrostrictive force results from inhomogeneous electric field strength and the variation of the dielectric constant with temperature and density.

To examine electroconvective problems, the force, $\mathbf{F_e}$, must be added to the right-hand side of the Navier–Stokes equation as a body force. Some other electrodynamic equations needed in the problem are

$$\mathrm{div}(\varepsilon\mathbf{E}) = \rho_e \tag{10.136}$$

$$\mathbf{E} = -\mathrm{grad}\varphi \tag{10.137}$$

$$\mathrm{div}\mathbf{J} = 0 \tag{10.138}$$

where ρ_e is the charge density, φ is the electric potential, and $\mathbf{J}$ is the current density given by

$$\mathbf{J} = \rho_e\mathbf{V} + \sigma_E\mathbf{E} + \frac{\partial(\varepsilon\mathbf{E})}{\partial t} \tag{10.139}$$

where charge convection, Ohm's law, and displacement current are accounted for, respectively. Magnetic field effects can be neglected, since the electrical current is low. The most important coupling mechanisms in the flow, temperature, and electrostatic fields are free charge convection, electrical body force $\mathbf{F}_e$, and variations of ε and σ with temperature. The presence and displacement of vapor masses can change the electrostatic field distribution.

The primary factor causing flow is the variation in the dielectric constant, and therefore the dielectrophoretic effect. A two-phase flow pattern map for annular channels under a DC applied voltage and the application to EHD convective boiling analysis was presented by Cotton et al. (2005). In the Cotton et al. (2005) study, the difference in dielectric constant for liquid was $\varepsilon_\ell = 9.5$ and for vapor $\varepsilon_v = 1.09$ at the average saturation temperature of 24 °C for HFC-134a. Given the magnitude for conductivity in typical dielectrics, additional electrical energy consumed by the electric field is considered negligible.

A dimensional analysis can be used for the equations of EHD in two-phase flow. The two dimensionless EHD numbers that will result from the analysis of the electric body force are the EHD number or conductive Rayleigh number

$$E_{\mathrm{hd}} = \frac{J_o L^3}{\rho_o v^2 \mu_c A} \tag{10.140}$$

where I_o is the outlet current, L is the length, ρ_o is the outlet density, v is the kinematic viscosity, μ_c is the ion mobility, and A is the area. The Masuda number or dielectric Rayleigh number is given as

$$M_{\mathrm{d}} = \frac{\varepsilon_o E_o^2 T_o (\partial\varepsilon_s/\partial T)_\rho L^2}{2\rho_o v^2} \tag{10.141}$$

where ε_o is the outlet permittivity, E_o is the outlet electric field strength, T_o is the outlet temperature, and ε_s is the dielectric constant.

The combined effects of electric and forced convection must be considered when $E_{hd}/Re^2 \sim 1$ and/or $M_d/Re^2 \sim 1$. If $E_{hd}/Re^2 \ll 1$ or $M_d/Re^2 \ll 1$, electric convection effects can be neglected. If $E_{hd}/Re^2 \gg 1$ or $M_d/Re^2 \gg 1$, forced convection effects may be neglected. This is analogous to buoyancy-driven flow, and the EHD numbers can be compared to the Grashof number in the absence of forced convection. This order of magnitude helps determine the range and extent of EHD influence on the flow and needs to be identified to determine the voltage levels required to affect heat transfer through migration of the liquid. Based on dimensionless analysis (Chang and Watson 1994) it is expected that $E_{hd}/Re_1^2 \sim \geq 0.1$ is sufficient to define the minimum condition above which electric fields significantly influence the flow of liquid.

An experimental study by Cotton et al. (2005) was conducted on the tube side boiling heat transfer of HFC-134a in a single-pass countercurrent flow heat exchanger in an electric field. The mechanics of EHD-induced flow and heat transfer have been investigated for various flow conditions by applying 0–8 kV to a concentric inner electrode. A theoretical Steiner-type two-phase flow pattern map was developed for flow boiling in the annular channel under DC high voltage. Flow regimes (Fig. 10.16) encountered in the convective boiling process were reconstructed experimentally and compared with the EHD map. Results show that when $M_d \sim Re^2$, EHD interfacial forces have a strong influence on the flow pattern, which is considered to be the primary mechanism affecting the increase in pressure drop and the augmentation or even suppression of heat transfer.

10.5.2 Bubble Lift-off Size

The bubble lift-off size, which is the bubble size when it detaches from the heating surface, can be different from bubble departure size, which is bubble size when it departs from the nucleate site. The bubble departure size for pool boiling has been discussed in detail in Chap. 9. For forced convective boiling, the bubble departure is usually analyzed by force balance along the flow direction, while the force balance normal to the flow direction plays a crucial role on bubble lift-off size. Situ et al. (2005) analyzed the force balance during the bubble growth to predict the bubble lift-off size; their work will be presented here.

Figure 10.17a shows an active nucleation site in upward forced convection subcooled boiling. The bubble is nucleated at the nucleation site and grows until it reaches to the departure size, at which point it departs from the nucleate site. The departed bubble slides along the heating surface and vaporization continue at the inner surface of the bubble. Since the liquid is subcooled, condensation may take place at the outer surface of the bubble if the size of the bubble is greater than the superheated liquid layer. Depending on the overall effects of vaporization and condensation, the bubble may either grow or shrink. The bubble eventually lifts-off from the heating surface at some distance from the nucleate site.

The force balance of the bubble at the nucleate site is shown in Fig. 10.17b. The momentum equations along the x- and y-directions are

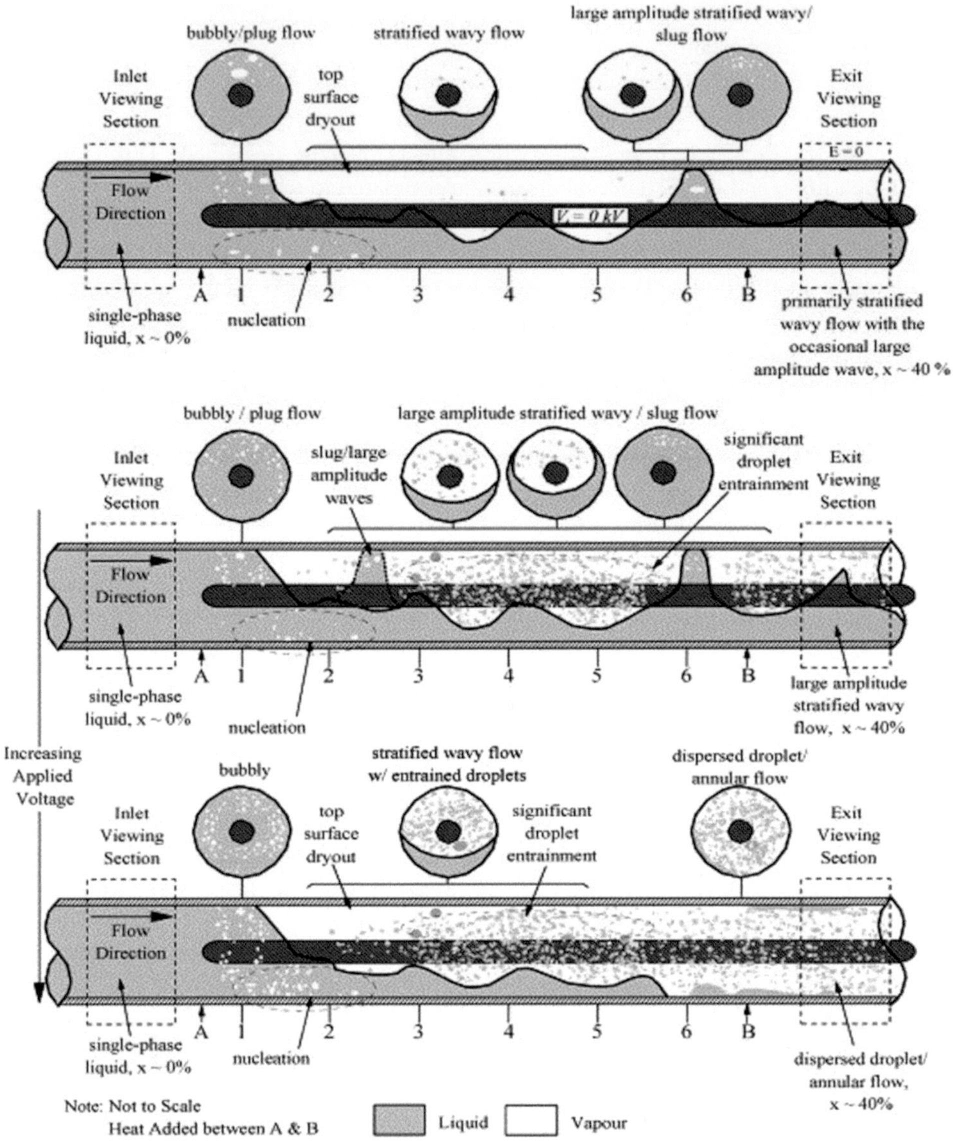

Fig. 10.16 Proposed reconstructed flow pattern from surface temperature measurements and inlet and outlet flow observations for increasing DC voltage levels: **a** 0 applied voltage, **b** moderate voltage, and **c** high voltage ($\dot{m}'' = 100\,\mathrm{kg/m^2}$-s, $q'' = 10\,\mathrm{kW/m^2}$, and $x_{\mathrm{in}} = 0\%$; Cotton et al. 2005; Reprinted with permission from Elsevier)

$$F_{\mathrm{sx}} + F_{\mathrm{dux}} + F_{\mathrm{sl}} = \rho_{\mathrm{v}} V_{\mathrm{b}} \frac{\mathrm{d}u_{\mathrm{v}}}{\mathrm{d}t} \tag{10.142}$$

$$F_{\mathrm{sy}} + F_{\mathrm{duy}} + F_{\mathrm{p}} + F_{\mathrm{g}} + F_{\mathrm{qs}} = \rho_{\mathrm{v}} V_{\mathrm{b}} \frac{\mathrm{d}v_{\mathrm{v}}}{\mathrm{d}t} \tag{10.143}$$

where F_{s} is the surface tension force, F_{du} is the unsteady drag force, F_{sl} is the shear lift force, F_{p} is the pressure force, F_{g} is the gravity force, and F_{qs} is the quasi-steady force, V_{b} is bubble volume, and u_{v} and v_{v} are bubble velocity in x- and y-directions, respectively.

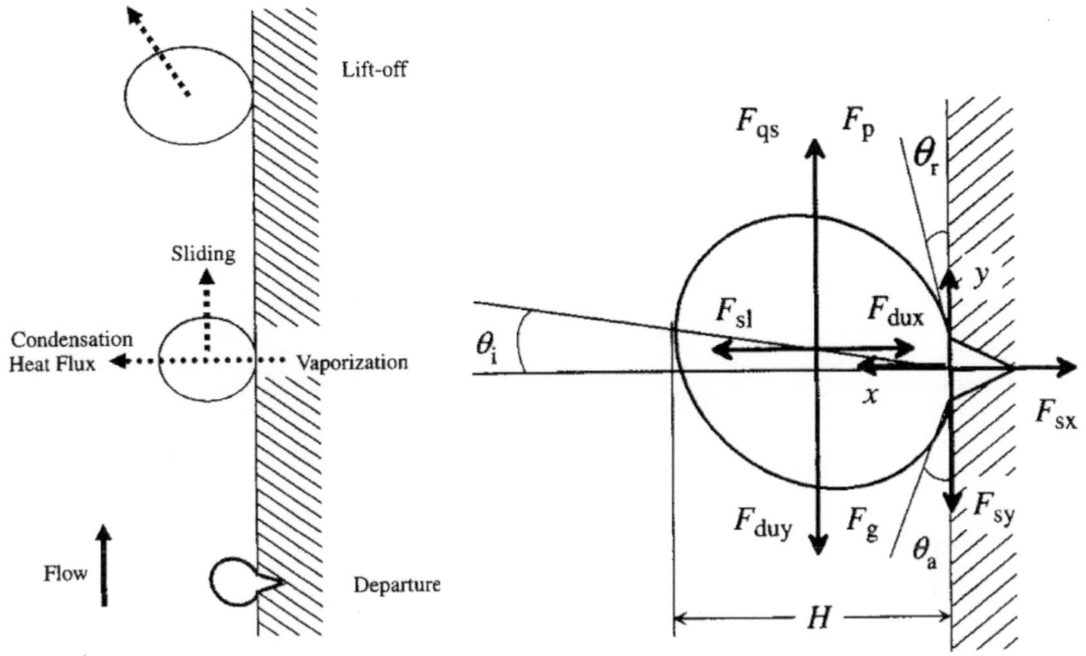

(a) bubble nucleation phenomena (b) force balance at a nucleation site

Fig. 10.17 Physical model of bubble lift-off (Situ et al. 2005; Reprinted with permission from Elsevier)

The surface tension force in the x- and y-directions are (Klausner et al. 1993)

$$F_{\mathrm{sx}} = -D_{\mathrm{w}}\sigma \frac{\pi}{\theta_{\mathrm{a}} - \theta_{\mathrm{r}}}(\cos\theta_{\mathrm{r}} - \cos\theta_{\mathrm{a}}) \tag{10.144}$$

$$F_{\mathrm{sy}} = -1.25 D_{\mathrm{w}}\sigma \frac{\pi(\theta_{\mathrm{r}} - \theta_{\mathrm{a}})}{\pi^2 - (\theta_{\mathrm{r}} - \theta_{\mathrm{a}})^2}(\sin\theta_{\mathrm{a}} + \sin\theta_{\mathrm{r}}) \tag{10.145}$$

where D_{w} is bubble contact diameter on the heating surface, and θ_{a} and θ_{r} are advancing and receding contact angles (see Fig. 10.17b).

The unsteady drag force (also referred to as growth force) is

$$F_{\mathrm{du}} = \frac{\mathrm{d}(\rho_\ell V_\ell u_{\mathrm{b}})}{\mathrm{d}t} = \rho_\ell \left(V_\ell \frac{\mathrm{d}^2 H}{\mathrm{d}t^2} + \frac{\mathrm{d}H}{\mathrm{d}t}\frac{\mathrm{d}V_\ell}{\mathrm{d}t} \right) \tag{10.146}$$

where V_ℓ is the volume of the virtual added mass (Chen et al. 2003)

$$V_\ell = \frac{11}{12}\pi R_{\mathrm{b}}^3 \tag{10.147}$$

and H is bubble height measured from the heating wall, $u_{\mathrm{b}} = \mathrm{d}H/\mathrm{d}t$ is the bubble front velocity in the x-direction. For a spherical bubble, H is the diameter of the bubble and the bubble front velocity becomes $u_{\mathrm{b}} = 2\mathrm{d}R_{\mathrm{b}}/\mathrm{d}t$. The unsteady drag force for spherical bubble can be expressed as

$$F_{\mathrm{du}} = \rho_{\ell}\pi R_{\mathrm{b}}^{2}\left(\frac{11}{12}\dot{R}_{\mathrm{b}}^{2} + \frac{11}{6}R_{\mathrm{b}}\ddot{R}_{\mathrm{b}}\right) \tag{10.148}$$

where $\dot{R}_{\mathrm{b}} = \mathrm{d}R_{\mathrm{b}}/\mathrm{d}t$ and $\ddot{R}_{\mathrm{b}} = \mathrm{d}^{2}R_{\mathrm{b}}/\mathrm{d}t^{2}$.

The components of the unsteady drag force in the x- and y-directions are

$$F_{\mathrm{dux}} = F_{\mathrm{du}}\cos\theta_{\mathrm{i}} \tag{10.149}$$

$$F_{\mathrm{duy}} = F_{\mathrm{du}}\sin\theta_{\mathrm{i}} \tag{10.150}$$

The transient radius of the vapor bubble can be obtained from Zuber's (1961) bubble growth model (Chap. 9)

$$R_{\mathrm{b}}(t) = \frac{2b}{\sqrt{\pi}}Ja\sqrt{\alpha_{\ell}t} \tag{10.151}$$

where $b = 1.73$ (Zeng et al. 1993) and the Jakob number is defined as

$$Ja = \frac{\rho_{\ell}c_{\ell}(T_{\mathrm{w}} - T_{\mathrm{sat}})}{\rho_{\mathrm{v}}h_{\ell\mathrm{v}}} \tag{10.152}$$

While Eq. (10.151) can provide a good estimate for saturated boiling, the bubble radius will depend on effective Jakob number, Reynolds number, and Prandtl number (Kocamustafaogullari and Ishii 1983)

$$R_{\mathrm{b}}(t) = f(Ja_{\mathrm{e}}, Re, Pr, t) \tag{10.153}$$

where

$$Ja_{\mathrm{e}} = \frac{\rho_{\ell}c_{\ell}\Delta T_{e}}{\rho_{\mathrm{v}}h_{\ell\mathrm{v}}} \tag{10.154}$$

and

$$\Delta T_{\mathrm{e}} = S(T_{\mathrm{w}} - T_{\mathrm{sat}}) \tag{10.155}$$

where S is the suppression factor.

The shear lift force is (Mei and Klausner 1994)

$$F_{\mathrm{sl}} = \frac{1}{2}C_{\mathrm{l}}\rho_{\ell}\pi R_{\mathrm{b}}^{2}v_{\mathrm{r}}^{2} \tag{10.156}$$

where $v_{\mathrm{r}} = v_{\ell} - v_{\mathrm{v}}$ is the relative velocity between the bubble center mass and liquid phase. The shear lift coefficient is (Klausner et al. 1993)

$$C_{\mathrm{l}} = 3.877G_{s}^{1/2}(Re_{\mathrm{b}}^{-2} + 0.014G_{\mathrm{s}}^{2})^{1/4} \tag{10.157}$$

where

$$G_s = \left|\frac{dv_\ell}{dx}\right|\frac{R_b}{v_r} \tag{10.158}$$

and

$$Re_b = \frac{2R_b v_r}{v_\ell} \tag{10.159}$$

is the bubble Reynolds number.

The pressure and gravity forces are

$$F_p = \rho_\ell g V_b \tag{10.160}$$

$$F_g = -\rho_v g V_b \tag{10.161}$$

The quasi-steady drag force is (Klausner et al. 1993)

$$F_{qs} = 6\pi\rho_\ell v_\ell v_r R_b \left\{\frac{2}{3} + \left[\left(\frac{12}{Re_b}\right)^{0.65} + 0.862\right]^{-1.54}\right\} \tag{10.162}$$

Figure 10.18 shows the force balance in the x-direction, normal to the flow direction, at the moment of bubble lift-off. Since the contact area between the bubble and the heating wall becomes zero at lift-off, the effect of surface tension can be neglected. The unsteady drag force (growth force) is also negligible since the inclination angle is zero at lift-off. The force balance in the x-direction is

$$F_{du} + F_{sl} = 0 \tag{10.163}$$

Substituting Eqs. (10.148) and (10.156) into Eq. (10.163) yields

$$\rho_\ell \pi R_b^2\left(\frac{11}{12}\dot{R}_b^2 + \frac{11}{6}R_b\ddot{R}_b\right) + \frac{1}{2}C_l\rho_\ell\pi R_b^2 v_r^2 = 0 \tag{10.164}$$

Substituting Eq. (10.151) into Eq. (10.164), one obtains

$$\frac{\alpha_\ell}{t_{lo}} = \frac{3\pi C_l v_r^2}{22b^2 Ja^2} \tag{10.165}$$

where t_{lo} is the time at which a bubble lifts-off, which can be obtained from Eq. (10.151), i.e.,

$$t_{lo} = \frac{\pi r_{lo}^2}{4b^2 Ja^2 \alpha_\ell} \tag{10.166}$$

Combining Eqs. (10.165) and (10.166), a dimensionless lift-diameter is obtained

$$D_{lo}^* = \frac{4\sqrt{22/3b^2}}{\pi} Ja^2 Pr_\ell^{-1} \tag{10.167}$$

where

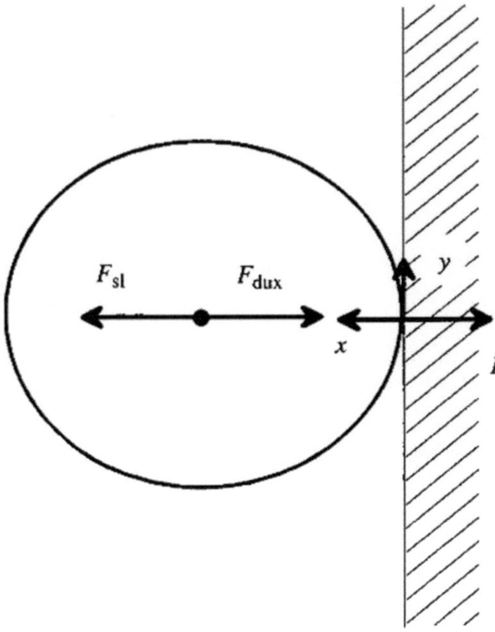

Fig. 10.18 Force balance of a vapor bubble at lift-off (Situ et al. 2005; Reprinted with permission from Elsevier)

$$D_{lo}^* = \sqrt{C_l}\, Re_b = \sqrt{C_l}\left(\frac{v_r D_{lo}}{v_\ell}\right) \tag{10.168}$$

For forced convective subcooled boiling, the Jakob number in Eq. (10.167) should be replaced by the effective Jakob number defined in Eq. (10.154). The bubble lift-off size predicted by (10.167) agreed well with the experimental data within an averaged deviation of $\pm 35.2\%$.

10.5.3 Heat Transfer Predictions

The different flow regimes have significant effects on the heat transfer characteristics of convective boiling in a tube. While heat transfer for subcooled liquid and superheated vapor can be easily handled by correlations for single-phase heat transfer, the intermediate heat transfer mechanism is complicated by phase change from liquid to vapor. After boiling is initiated in the tube, vapor bubbles are generated at certain nucleate sites while the rest of the inner surface of the tube remains in contact with the liquid. Under these conditions, the heat transfer mechanism is a combination of two parallel processes: single-phase convection in the liquid and nucleate boiling.

The overall heat transfer coefficient for convective boiling in an upward vertical tube can be written as (Chen 1963)

$$h = Fh_\ell + Sh_b \tag{10.169}$$

where h_ℓ and h_b are heat transfer coefficients for single-phase convection of the liquid and nucleate boiling, respectively. F and S in Eq. (10.169) are dynamic factors that modify the contributions of single-phase liquid convection and nucleate boiling, respectively

The single-phase heat transfer coefficient for liquid alone can be obtained by using the Dittus–Boelter/McAdams equation:

$$h_\ell = 0.023\left(\frac{k_\ell}{D}\right) Re_\ell^{0.8} Pr_\ell^{0.4} \tag{10.170}$$

The contribution of nucleate boiling is determined by using the correlation proposed by Forster and Zuber (1955) for pool boiling:

$$h_\mathrm{b} = 0.00122\left[\frac{k_\ell^{0.79} c_{\mathrm{p},\ell}^{0.45} \rho_\ell^{0.49}}{\sigma^{0.5} \mu_\ell^{0.29} h_\mathrm{fg}^{0.24} \rho_\mathrm{v}^{0.24}}\right] [T_\mathrm{w} - T_\mathrm{sat}(p_\ell)]^{0.24} [p_\mathrm{sat}(T_\mathrm{w}) - p_\ell]^{0.75} \tag{10.171}$$

The convective boiling factor F can be obtained by a regression of experimental data (Chen 1963):

$$F = \begin{cases} 1 & X_\mathrm{tt}^{-1} \le 0.1 \\ 2.35(0.213 + X_\mathrm{tt}^{-1})^{0.736} & X_\mathrm{tt}^{-1} > 0.1 \end{cases} \tag{10.172}$$

where X_tt is the Lockhart–Martinelli parameter obtained by Eq. (10.126).

The nucleate boiling suppression factor S is

$$S = \frac{1}{1 + 2.53 \times 10^{-6} Re_{TP}^{1.17}} \tag{10.173}$$

where

$$Re_\mathrm{TP} = Re_\ell F^{1.25} \tag{10.174}$$

is the local two-phase Reynolds number.

Example 10.5

Calculate the flow boiling two-phase heat transfer coefficient for 300 kg/(sm^2) of water at 1 atm and 20% quality in an upward vertical tube with a diameter of 2 cm and a wall temperature of 140 °C.

Solution The saturation temperature at $p_\ell = 1\,\mathrm{atm} = 1.013 \times 10^5\,\mathrm{Pa}$ is $T_\mathrm{sat} = 100\,°\mathrm{C}$. The properties of water at this temperature are $\sigma = 58.9 \times 10^{-3}\,\mathrm{N/m}$, $h_{\ell v} = 2251.2\,\mathrm{kJ/kg}$, $\rho_\ell = 958.77\,\mathrm{kg/m^3}$, $\rho_\mathrm{v} = 0.5974\,\mathrm{kg/m^3}$, $\mu_\ell = 2790 \times 10^{-7}\,\mathrm{N\text{-}s/m^2}$, $\mu_\mathrm{v} = 121 \times 10^{-7}\mathrm{N\text{-}s/m^2}$, $c_{p\ell} = 4.216\,\mathrm{kJ/kg°C}$, $k_\ell = 0.68\,\mathrm{W/m\text{-}°C}$, and $Pr_\ell = \mu_\ell c_{p\ell}/k_\ell = 1.73$.

The mass flux of $\dot{m}'' = 300\,\mathrm{kg/s\text{-}m^2}$ yields a liquid Reynolds number of

$$Re_\ell = \frac{G(1-x)D}{\mu_\ell} = \frac{300 \times (1 - 0.2) \times 0.02}{2790 \times 10^{-7}} = 1.720 \times 10^4$$

The single-phase heat transfer coefficient for liquid alone can be obtained from Eq. (10.170), i.e.,

$$h_\ell = 0.023\left(\frac{k_\ell}{D}\right)Re_\ell^{0.8}Pr_\ell^{0.4}$$
$$= 0.023 \times \left(\frac{0.68}{0.02}\right) \times (1.72 \times 10^4)^{0.8} \times 1.73^{0.4}$$
$$= 9.481 \times 10^4 \text{ W/m}^2\text{-K}$$

The saturation pressure corresponding to the wall temperature is $p_{\text{sat}}(T_{\text{w}}) = 3.61 \times 10^5$ Pa. The contribution of nucleate boiling is determined from Eq. (10.171):

$$h_{\text{b}} = 0.00122\left[\frac{k_\ell^{0.79}c_{\text{p},\ell}^{0.45}\rho_\ell^{0.49}}{\sigma^{0.5}\mu_\ell^{0.29}h_{\text{fg}}^{0.24}\rho_{\text{v}}^{0.24}}\right][T_{\text{w}} - T_{\text{sat}}(p_\ell)]^{0.24}[p_{\text{sat}}(T_{\text{w}}) - p_\ell]^{0.75}$$
$$= 0.00122 \times \left[\frac{0.68^{0.79} \times (4.216 \times 10^3)^{0.45} \times 958.77^{0.49}}{(58.9 \times 10^{-3})^{0.5} \times (2790 \times 10^{-7})^{0.29} \times (2251.2 \times 10^3)^{0.24} \times 0.5974^{0.24}}\right]$$
$$\times [140 - 100]^{0.24} \times \left[3.61 \times 10^5 - 1.013 \times 10^5\right]^{0.75}$$
$$= 4.639 \times 10^4 \text{ W/m}^2\text{-K}$$

The Martinelli parameter for turbulent flow in both liquid and vapor phases, X_{tt}, is obtained from Eq. (10.126):

$$X_{\text{tt}} = \left(\frac{1-x}{x}\right)^{0.9}\left(\frac{\rho_{\text{v}}}{\rho_\ell}\right)^{0.5}\left(\frac{\mu_\ell}{\mu_{\text{v}}}\right)^{0.1}$$
$$= \left(\frac{1-0.2}{0.2}\right)^{0.9}\left(\frac{0.5974}{958.77}\right)^{0.5}\left(\frac{2790 \times 10^{-7}}{121 \times 10^{-7}}\right)^{0.1} = 0.119$$

The convective boiling factor F is obtained from Eq. (10.172), i.e.,

$$F = 2.35(0.213 + X_{\text{tt}}^{-1})^{0.736} = 2.35 \times (0.213 + 0.119^{-1})^{0.736} = 11.47$$

The local two-phase Reynolds number is

$$Re_{\text{TP}} = Re_\ell F^{1.25} = 1.720 \times 10^4 \times 11.47^{1.25} = 3.63 \times 10^5$$

The nucleate boiling suppression factor S is obtained from Eq. (10.173), i.e.,

$$S = \frac{1}{1 + 2.53 \times 10^{-6}Re_{\text{TP}}^{1.17}} = \frac{1}{1 + 2.53 \times 10^{-6} \times (3.63 \times 10^5)^{1.17}} = 0.11$$

Therefore, the heat transfer coefficient is obtained from Eq. (10.169) as

$$h = Fh_\ell + Sh_{\text{b}}$$
$$= 11.47 \times 9.481 \times 10^4 + 0.11 \times 4.639 \times 10^4 = 1.09 \times 10^6 \text{ W/m}^2\text{-K}$$

Based on more than 10,000 experimental data points for different fluids, including water, refrigerants, and cryogents, Kandlikar (1990, 1991) proposed the following generalized heat transfer correlation for convective boiling in both vertical and horizontal tubes:

$$h = \max[h_{\mathrm{NBD}}, \quad h_{\mathrm{CBD}}] \tag{10.175}$$

where h_{NBD} and h_{CBD} are the nucleate boiling dominant and convective boiling dominant heat transfer coefficients, and they are obtained by

$$h_{\mathrm{NBD}} = \left[0.6683 Co^{-0.2} f_2(Fr_{\ell o}) + 1058\, Bo^{0.7} F_{\mathrm{f}\ell}\right](1-x)^{0.8} h_{\ell o} \tag{10.176}$$

$$h_{\mathrm{CBD}} = \left[1.136 Co^{-0.9} f_2(Fr_{\ell o}) + 667.2\, Bo^{0.7} F_{\mathrm{f}\ell}\right](1-x)^{0.8} h_{\ell o} \tag{10.177}$$

where Co is the convective number, Bo is the boiling number, and $Fr_{\ell o}$ is the Froude number.

$$Co = \left(\frac{\rho_{\mathrm{v}}}{\rho_\ell}\right)^{0.5} \left(\frac{1-x}{x}\right)^{0.8} \tag{10.178}$$

$$Bo = \frac{q''_{\mathrm{w}}}{AGh_{\ell \mathrm{v}}} \tag{10.179}$$

$$Fr_{\ell o} = \frac{G^2}{\rho_\ell^2 g D} \tag{10.180}$$

The Froude number multiplier, $f_2(Fr_{\ell o})$, is

$$f_2(Fr_{\ell o}) = \begin{cases} 1 & \text{vertical or horizontal tubes } (Fr_{\ell o} \geq 0.04) \\ (25 Fr_{\ell o})^{0.3} & \text{horizontal tubes } (Fr_{\ell o} < 0.04) \end{cases} \tag{10.181}$$

For liquids with a Prandtl number between 0.5 and 2000, a range that covers most fluids except liquid metal, the single-phase all-liquid heat transfer coefficient $h_{\ell o}$ in Eqs. (10.176) and (10.177) is

$$h_{\ell o} = \begin{cases} \dfrac{k_\ell}{D} \dfrac{(Re_{\ell o} - 1000) Pr_\ell (f/2)}{1 + 12.7(Pr_\ell^{2/3} - 1)(f/2)^{0.5}} & 2300 \leq Re_{\ell o} \leq 10^4 \\ \dfrac{k_\ell}{D} \dfrac{Re_{\ell o} Pr_\ell (f/2)}{1.07 + 12.7(Pr_\ell^{2/3} - 1)(f/2)^{0.5}} & 10^4 \leq Re_{\ell o} \leq 5 \times 10^6 \end{cases} \tag{10.182}$$

where the friction factor f in Eq. (10.182) is given by

$$f = (1.58 \ln Re_{\ell o} - 3.28)^{-2} \tag{10.183}$$

The fluid-surface parameter $F_{\mathrm{f}\ell}$ depends on the combination of the liquid and tube material used. For stainless steel tubing, $F_{\mathrm{f}\ell}$ is taken as 1 regardless of the type of fluid. For copper tubing, $F_{\mathrm{f}\ell}$ can be obtained from Table 10.4. The above correlation is valid for $0.001 \leq x \leq 0.95$.

Under constant heat flux conditions, the wall temperature will sharply increase when the heat transfer mechanism inside the tube suddenly changes from two-phase to single-phase vapor heat transfer. Since the heat transfer inside the tube always begins with the single-phase liquid, then changes to convective boiling, and finally to single-phase vapor, the critical heat flux (CHF) phenomenon always occurs at a point near the exit of the tube. For this reason, CHF for convective boiling inside the tube is a local phenomenon. There are numerous empirical correlations available in

Table 10.4 $F_{f\ell}$ for copper tube (Kandlikar 1990, 1991)

Fluid	$F_{f\ell}$	Fluid	$F_{f\ell}$
Water	1.00	R-114	1.24
R-11	1.30	R-134a	1.63
R-12	1.50	R-152a	1.10
R-13B1	1.31	Nitrogen	4.70
R-22	2.20	Neon	3.50
R-113	1.30		

the literature. The most generalized model to predict CHF for forced convective boiling is recommended by Katto and Ohno (1984).

$$q''_{\max} = q''_{\text{co}}\left(1 + K_K \frac{h_{\ell,\text{sat}} - h_{\text{in}}}{h_{\ell\text{v}}}\right) \tag{10.184}$$

For $\gamma = \rho_{\text{v}}/\rho_\ell < 0.15$

$$q''_{\text{co}} = \begin{cases} q''_{\text{co1}} & q''_{\text{co1}} < q''_{\text{co2}} \\ \min\left(q''_{\text{co2}}, q''_{\text{co3}}\right) & q''_{\text{co1}} > q''_{\text{co2}} \end{cases} \tag{10.185}$$

$$K_K = \max(K_{K1}, K_{K2}) \tag{10.186}$$

For $\gamma = \rho_{\text{v}}/\rho_\ell > 0.15$

$$q''_{\text{co}} = \begin{cases} q''_{\text{co1}} & q''_{\text{co1}} < q''_{\text{co5}} \\ \min\left(q''_{\text{co4}}, q''_{\text{co5}}\right) & q''_{\text{col}} > q''_{\text{co5}} \end{cases} \tag{10.187}$$

$$K_K = \begin{cases} K_{K1} & K_{K1} > K_{K2} \\ \min(K_{K2}, K_{K3}) & K_{K1} \le K_{K2} \end{cases} \tag{10.188}$$

where

$$q''_{\text{co1}} = C_K G h_{\ell\text{v}} We_k^{-0.043} \left(\frac{L}{D}\right)^{-1} \tag{10.189}$$

$$q''_{\text{co2}} = 0.10 G h_{\ell\text{v}} \gamma^{1.33} We_k^{-1/3} \left[\frac{1}{1 + 0.0031(L/D)}\right] \tag{10.190}$$

$$q''_{\text{co3}} = 0.098 G h_{\ell\text{v}} \gamma^{1.33} We_k^{-0.433} \left[\frac{(L/D)^{0.27}}{1 + 0.0031(L/D)}\right] \tag{10.191}$$

$$q''_{\text{co4}} = 0.0384 G h_{\ell\text{v}} \gamma^{0.6} We_k^{-0.173} \left[\frac{1}{1 + 0.280 We_k^{-0.233}(L/D)}\right] \tag{10.192}$$

$$q''_{co5} = 0.234 G h_{\ell v} \gamma^{0.513} We_k^{-0.433} \left[\frac{(L/D)^{0.27}}{1 + 0.0031(L/D)} \right] \tag{10.193}$$

$$K_{K1} = \frac{1.043}{4 C_K We_k^{-0.043}} \tag{10.194}$$

$$K_{K2} = \left(\frac{5}{6}\right) \frac{0.0124 + D/L}{\gamma^{1.33} We_k^{-1/3}} \tag{10.195}$$

$$K_{K3} = 1.12 \frac{1.52\, We_k^{-0.233} + D/L}{\gamma^{0.6} We_k^{-0.173}} \tag{10.196}$$

$$\gamma = \frac{\rho_v}{\rho_\ell} \tag{10.197}$$

$$We_k = \frac{G^2 L}{\rho_\ell \sigma} \tag{10.198}$$

$$C_K = \begin{cases} 0.25 & L/D < 50 \\ 0.25 + 0.009[L/D - 50] & 50 \le L/D \le 150 \\ 0.34 & L/D > 150 \end{cases} \tag{10.199}$$

The critical heat flux predicted using Eq. (10.184) agreed reasonably well with a variety of fluids, including water, ammonia, benzene, ethanol, helium, hydrogen, nitrogen, R-12, R-21, R-22, R-113, and potassium.

Heat transfer coefficients for boiling in both vertical and horizontal tubes are often measured from experiments using electrical heating that result in axially and circumferentially uniform heat flux. While this approach can give reasonable boundary conditions for boiling in a vertical tube, Thome (2016) pointed out that electrical heating for boiling in a horizontal tube is not preferred, because circumferential conduction in the tube wall from the hot, dry-wall condition at the top to the colder, wet-wall condition at the bottom yields unknown boundary conditions. Therefore, countercurrent hot water heating that can provide reasonable boundary conditions is preferred.

10.6 Two-Phase Flow and Heat Transfer in Micro- and Minichannels

10.6.1 Two-Phase Flow Patterns

The need for reliable, high-performance, price-competitive electronic devices, most notably electronic chips, has created demand for comparably small heat transfer devices capable of removing the required heat load within a limited temperature range. Closed two-phase devices, such as miniature/microheat pipes and capillary-pumped loops, have been and are being used successfully for this application. Typically, these use a spreading strategy and feature a large number of small circular channels arranged in parallel rows in a rectangular body. Whatever configuration is used, the heat energy removed at the chip is transported away and rejected from the system by condensation at a remote location (Begg et al. 1999; Zhang et al. 2001). Therefore, it is necessary to gain a fundamental understanding of two-phase flow and phase change heat transfer in miniature/microchannels. Another area of application is in the refrigeration industry, since miniaturization of the heat exchanger can

(a) reduce plant size, (b) lower materials costs, and (c) reduce fluid inventory (Vlasie et al., 2004). Reducing refrigerant charge in the refrigeration system can also minimize possible leakage.

The transport phenomena occurring in different scales can vary significantly. In general, the scales of two-phase flow and heat transfer in channels can be classified according to the hydraulic diameter of the channels, although it also depends on the properties of the fluid and flow conditions. One classification is based on the hydraulic diameter of the channels:

- Macrochannels: $D_h > 6\,\text{mm}$
- Minichannels: $200\,\mu\text{m} < D_h \le 6\,\text{mm}$
- Microchannels: $10\,\mu\text{m} < D_h \le 200\,\mu\text{m}$
- Transitional channels: $100\,\text{nm} < D_h \le 10\,\mu\text{m}$
- Nanochannels: $D_h \le 100\,\text{nm}$.

Flow regimes for two-phase flow in both horizontal and vertical tubes have been studied intensively, as outlined in the preceding section. However, most flow regime studies on two-phase flow were performed in conventionally sized (macroscale) tubes with hydraulic diameters greater than 6 mm. Unlike in conventionally sized passages, in which surface tension effects are limited, surface tension in miniature/microchannels can have significant effects on flow pattern transitions, on overall hydrodynamics, and in particular on the thin films that are believed to be the dominant mechanism controlling the heat transfer characteristics. The flow regime maps for condensation and boiling in conventionally sized tubes may not be relevant to flow in miniature circular tubes, where the surface tension has a significant effect on the hydrodynamics. At present, the data regarding basic flow patterns for two-phase flow, with or without heat transfer in miniature/microcircular tubes, is still very limited. It follows that little is known about the mechanisms associated with transition in these flows.

The classifications based on the size of the channel do not necessarily address the threshold where the characteristics of two-phase flow and heat transfer depart from the macroscale descriptions. For example, Barnea et al. (1983) found that the effect of surface tension is important for stratified-slug transition in horizontal flow in small diameter (4–12 mm) tubes. The best threshold criterion for distinguishing the microscale from the macroscale is bubble growth. With its diameter confined by the channel, a bubble can grow only in length rather than diameter. This criterion should be a function of geometry, size, and fluid properties and does not exist currently.

Kawaji and Chung (2004) presented a thorough review of adiabatic two-phase flow in minichannels and microchannels, and suggested that the transition from minichannels to microchannels occurred between 100 and 250 μm. The flow pattern for an air–water system in circular and semi-triangular minichannels with hydraulic diameters between 1.09 and 1.49 mm was investigated by Triplett et al. (1999). They identified five flow patterns for all test sections: bubbly, slug, churn, slug-annular, and annular flow (see Fig. 10.19). Since the gravitational force no longer dominates the two-phase flow in minichannels, the stratified flow and stratified wavy flow that appeared in the horizontal conventional channels shown in Fig. 10.4 were not observed in minichannels. Other than that, the flow patterns in the minichannels and conventionally sized channels are similar.

The studies of flow patterns in microchannels are still quite limited at this time. The flow patterns for an air–water system in 25 and 100 μm microchannels were studied by Feng and Serizawa (1999), Serizawa and Feng (2001), and Serizawa et al. (2000). They identified five flow patterns as shown in Fig. 10.20: dispersed bubbly flow, gas slug flow, liquid-ring flow, liquid lump flow, and liquid droplet flow. Chung and Kawaji (2004) investigated in detail nitrogen gas–water two-phase flow patterns in microchannels with diameters between 50 and 500 μm; they identified slug flow, liquid-ring flow, gas core flow with a serpentine liquid film, and semi-annular flow patterns; dispersed bubbly and droplet flows were not observed. Bubbly flow does not appear in Chung and Kawaji (2004) because it requires bubbles smaller than the channel diameter and would occur at extremely low gas flow rates in adiabatic systems. Another flow pattern that can be observed in microchannels but not minichannels is *liquid-ring flow,* which occurs when the liquid bridge between two consecutive gas slugs becomes unstable at higher flow rates.

10.6.2 Flow Condensation

10.6.2.1 Flow Patterns in Two-Phase Condensation in Miniature and Capillary Tubes

Condensation in miniature or microchannels finds its applications in electronics cooling, microscale energy, and Microelectro-Mechanical Systems (MEMS). Figure 10.21 shows patterns of flow

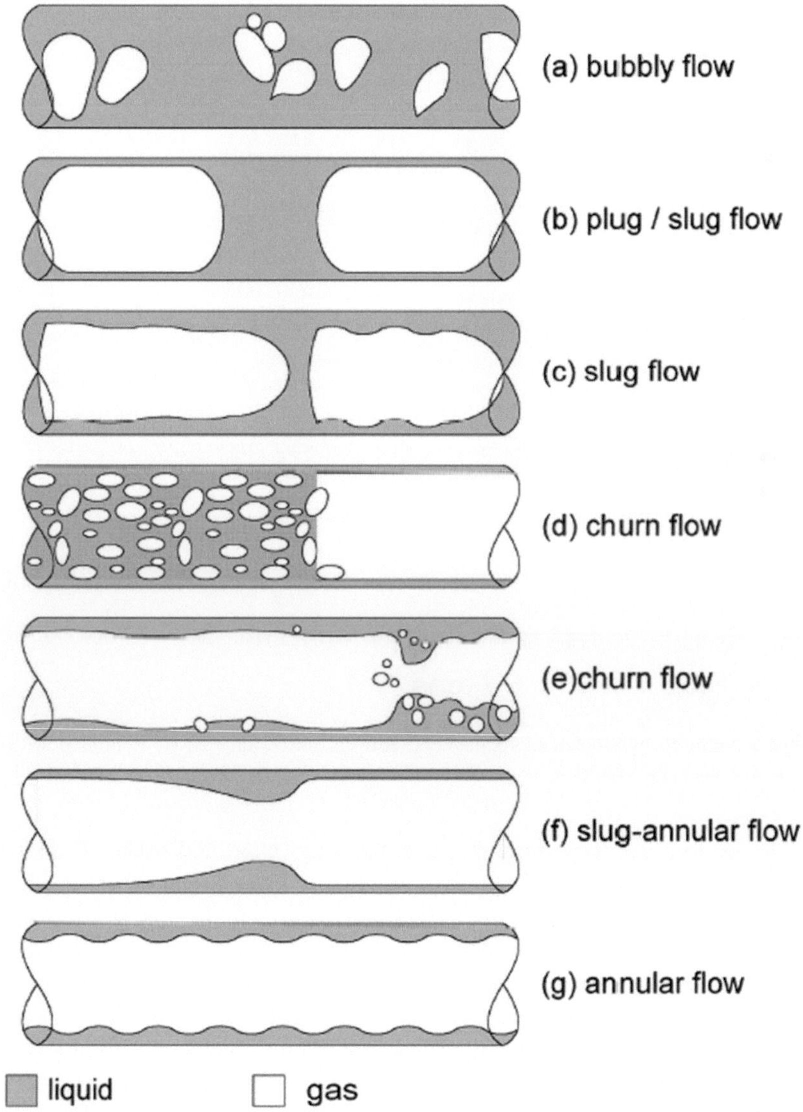

Fig. 10.19 Flow patterns in horizontal minichannels (Triplett et al. 1999; Reprinted with permission from Elsevier)

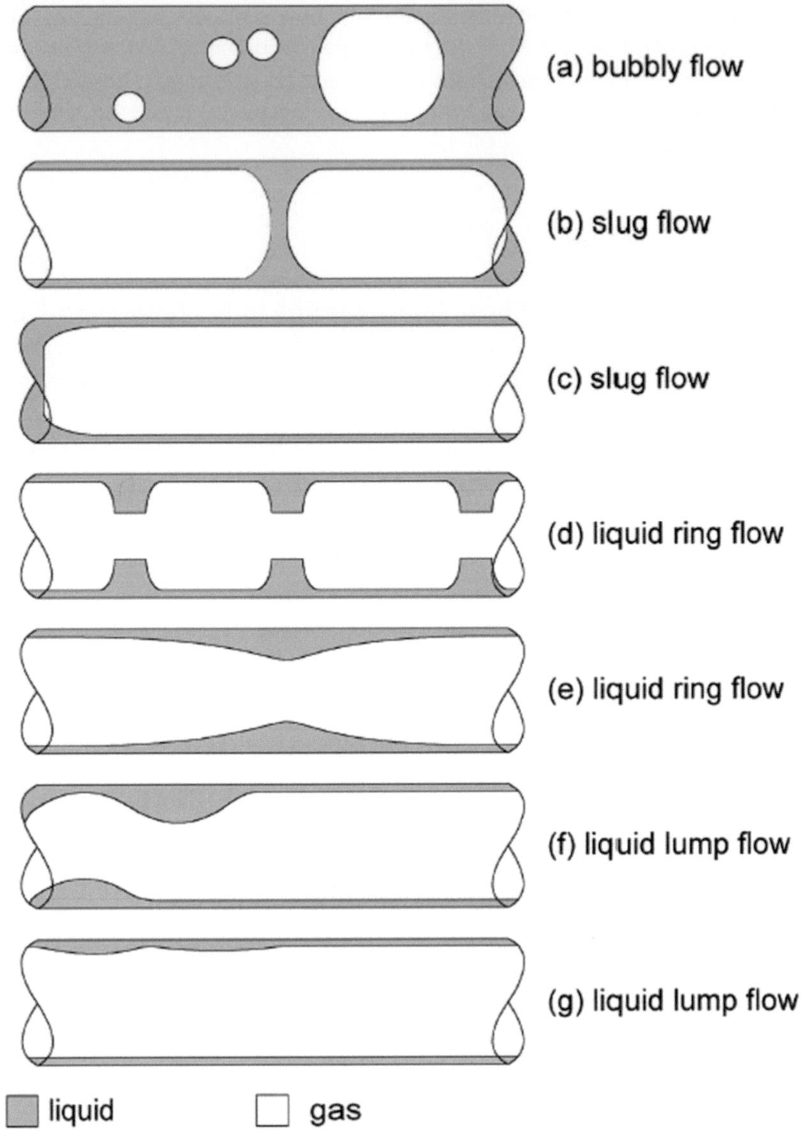

Fig. 10.20 Flow patterns in horizontal microchannels (Kawaji and Chung 2004; Reprinted with permission from Routledge/Taylor & Francis Group, LLC)

particularly relevant to a capillary tube, i.e., annular, slug, plug, and bubble (Tabatabai and Faghri 2001). At the beginning of the two-phase flow, an annular layer forms on the inside of the tube. As vapor or gas velocity decreases, it causes ripples to form on the liquid surface, leading to the formation of collars. The collar can also result from condensation of the vapor on the liquid film in the tube. Eventually, the collars grow to form a bridge. The relative size of the gap formed as a result of bridging establishes slug, plug, and bubble regimes. In a large tube, the bridge does not form because of high gravitational pull on the liquid film. The recognition of and accuracy in reporting on various flow regimes and their respective operating conditions is a major consideration in developing a flow map. Some investigators have reported more intermittent (slug and plug) regimes than others. For example, Dobson and Chato (1998) reported annular, wavy, wavy–annular, and mist–annular regimes

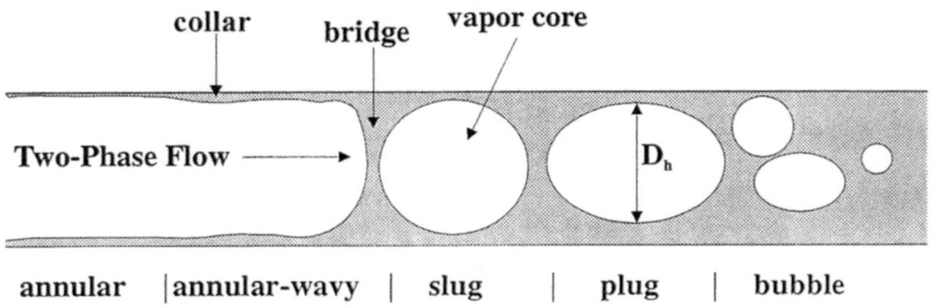

Fig. 10.21 Flow pattern of two-phase condensation flow in a capillary horizontal tube (Tabatabai and Faghri 2001)

for condensation of refrigerants; Soliman (1974) identified annular, semi-annular, semi-annular-wavy, spray annular, annular wavy, and spray regimes for condensation of refrigerants. In general, there is bound to be some subjectivity in regime reporting that reflects the accuracy of measurement, visualization, and regime identification techniques. This fact becomes even more critical in tubes of smaller diameter where flow regimes may be more difficult to observe.

In general, film or interface instability can be used as a criterion for flow transitions. Rabas and Minard (1987) suggested two forms of flow instabilities occurring inside horizontal tubes with complete condensation. The two forms are distinguished by a transition Froude number. It is suggested that the first instability results from the low vapor flow rate associated with a stratified exit condition and from vapor flowing into the tube exit, which causes condensate chugging or water hammer instability. The second instability results from a high vapor flow rate, which produces an inadequate distribution of the vapor and blockage of the tube exit, in turn causing large subcooled condensate temperature variations.

Instability can also occur in small diameter tubes, due mainly to capillary blocking where the liquid film bridges the tube to form a plug. In an integro-differential approach by Teng et al. (1999), capillary blocking was investigated in a thermosyphon condenser tube with an axisymmetric viscous annular condensate film with a vapor core. It was found that at low relative vapor velocities, surface tension was responsible for film instability in capillary tubes. At high relative vapor velocities, on the other hand, hydrodynamic force was responsible for the instability. Additionally, liquid bridges are maintained by buoyant motion of the vapor bubbles. Tabatabai and Faghri (2001) proposed a detailed flow map that emphasizes the importance of surface tension in two-phase flow in horizontal miniature and microchannels. Tabatabai and Faghri's (2001) flow map for two-phase flow in horizontal tubes plots the ratio of vapor to liquid superficial velocity as defined by

$$SV = \frac{j_{\mathrm{v}}}{j_{\ell}} \tag{10.200}$$

versus the ratio of pressure drop due to surface tension and shear:

$$\Delta p_{\mathrm{ratio}} = \frac{(\mathrm{d}p/\mathrm{d}z)_{\mathrm{surface\ tension}}}{(\mathrm{d}p/\mathrm{d}z)_{\ell,\mathrm{shear}} + (\mathrm{d}p/\mathrm{d}z)_{\mathrm{g,shear}}} \tag{10.201}$$

is shown in Fig. 10.22. A transition boundary based on a force balance that includes shear, buoyancy, and surface tension forces is also proposed. The flow map is compared to a number of existing experimental data sets totaling 1589 data points. Comparison of the proposed map and model with previous models shows substantial improvement and accuracy in determining surface tension-dominated regimes. Furthermore, the proposed flow map shows how each regime transition boundary is affected by surface tension. Tabatabai and Faghri (2001) presented a detailed methodology for calculating various pressure drops.

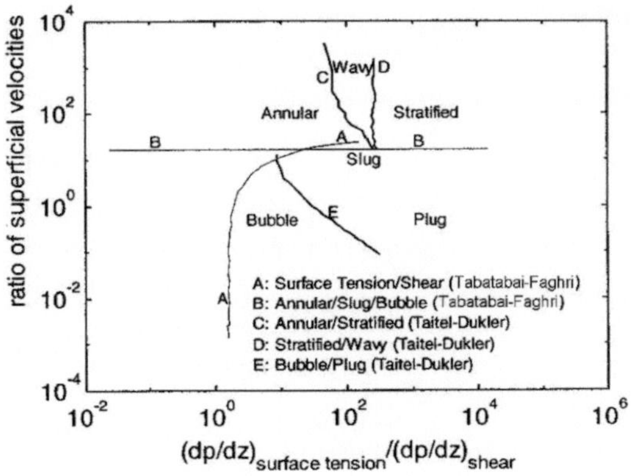

Fig. 10.22 Flow map for condensation in miniature/micro channels (Tabatabai and Faghri 2001)

It is very difficult to measure the local heat transfer coefficient of flow condensation in a single small diameter tube, because of complexities involved in controlling the flow conditions, heat flux, and vapor quality. Shin and Kim (2005) performed an experimental study of flow condensation of R-134a inside circular channels with diameters of 0.493, 0.691, and 1.067 mm and square channels with hydraulic diameters of 0.494, 0.658, and 0.972 mm. The mass flux varied from 100 to 600 kg/m^2-s, and the heat flux varied from 5 to 20 kW/m^2. They compared their experimental results with the existing empirical correlations for conventional size, including Eq. (10.116), and found that the existing correlations generally underpredict the heat transfer coefficient at lower mass flux regions. As the mass flux increases, the Nusselt number predicted by Eq. (10.116) agrees well with the experimental results.

Since the heat transfer capacity of a single miniature/microchannel is very low, multiple channels inevitably will be necessary for practical applications. Riehl and Ochterbeck (2002) studied convective condensation heat transfer of methanol in square channels with dimensions of 0.5, 0.75, 1.0, and 1.5 mm; the corresponding channel numbers are 14, 12, 10, and 8, respectively. The experiments were performed at two different saturation temperatures: 45 and 55 °C. The experimental results were correlated in the following form:

$$Nu = We^{-Ja}\, Re\, Pr^{Y} \tag{10.202}$$

where

$$We = \frac{\rho_\ell w^2 L}{\sigma} \tag{10.203}$$

$$Ja = \frac{c_{p\ell}(T_{\text{sat}} - T_{\text{w}})}{h_{\ell v}} \tag{10.204}$$

$$Re = \frac{GD_h}{\mu_\ell} \tag{10.205}$$

$$Y = \begin{cases} 1.3 & Re \leq 65 \\ \dfrac{0.5D_h - 1}{2D_h} & Re > 65 \end{cases} \tag{10.206}$$

where w is working fluid velocity (m/s), and L is the channel length. Equation (10.202) correlated over 95% of experimental results with a relative error of less than 25%.

10.6.2.2 Annular Flow Condensation in a Miniature Tube

Analytical methods for condensation in a vertical tube are found to be applicable to the annular flow patterns in both horizontal tube condensation and a reduced-gravity environment. Because of the symmetry in both upward and downward flow, the analysis of condensation in a vertical tube is a problem that has been treated extensively. Many of the studies use the same type of one-dimensional analysis as the classic Nusselt approach, but modified to account for pressure drop in the vapor generated by both friction at the liquid/vapor interface, and momentum exchange between the vapor and liquid arising from the condensation mass flux. The objective in this section is to model annular film condensation in miniature circular tubes where capillary phenomena can conceivably result in blocking of the tube cross section with liquid at some distance from the condenser entrance. A physical and mathematical model of annular film condensation in a miniature tube was developed by Begg et al. (1999). The physical description illustrated in Fig. 10.23a. Vapor condenses inside a

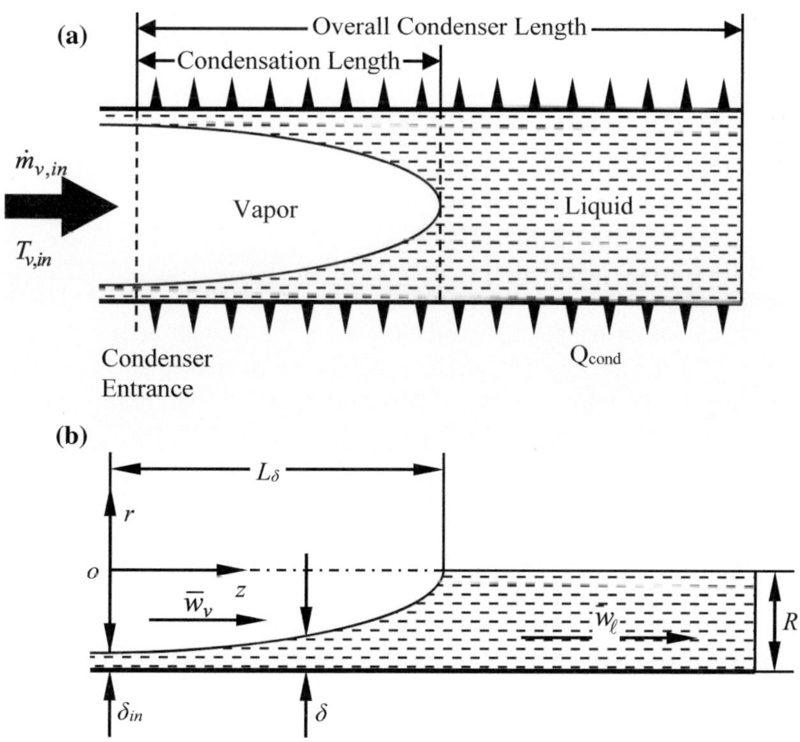

Fig. 10.23 Description of the physical model for annular film condensation in a miniature tube: **a** structure of two-phase flow for complete condensation, and **b** coordinate system and conventions for film condensation model (Begg et al. 1999)

tube, forming a stationary liquid–vapor interface. At some point downstream of the inlet, all the incoming vapor is condensed, and only liquid flows in the tube cross section. This can occur due to surface tension effects that are usually neglected in film condensation models for tubes with conventional larger diameters.

The model condensation in miniature tube differs from traditional models for film condensation in conventionally sized tubes, due to the following features:

1. The disjoining pressure and the interfacial resistance can affect both the liquid and vapor flow for extremely small channels and film thickness. Therefore, the terms containing the disjoining pressure and interfacial resistance should be included in the model.
2. The effect of the surface tension on the fluid flow is included in the model and the two principal radii of the liquid–vapor interface curvature are used in the Laplace–Young equation.
3. Liquid subcooling is accounted for in the model, since it can result in a significant variation of the wall temperature under convective cooling conditions.
4. The liquid momentum conservation equation in cylindrical coordinates is utilized and the corresponding expression for the shear stress at the liquid–vapor interface is accounted for.

A steady-state mathematical model of condensation leading to complete condensation was developed by Begg et al. (1999). It includes coupled vapor and liquid flows with shear stresses at the liquid free surface resulting from the vapor–liquid frictional interaction and surface tension gradient. The model is based on the following simplifying assumptions:

1. The vapor is saturated and there is no temperature gradient in the vapor in the radial direction.
2. Heat transport in the thin film is due to conduction in the radial direction only.
3. Inertia terms can be neglected for the viscous flow in the liquid films with low Reynolds numbers.
4. Force on liquid due to surface tension is much greater than the gravitational force, thus the gravitational body force is neglected. Therefore, the liquid is distributed onto the walls in an axisymmetric film.
5. The solid tube wall is infinitely thin, so that its thermal resistance in the radial direction, as well as the axial heat conduction, can be neglected.

The cylindrical coordinate system used is shown in Fig. 10.23b. Both the vapor and liquid flow along the z-coordinate. The physical situation should be described, taking into consideration the vapor compressibility and the vapor temperature variation along the channel. Also, the second principal radius of curvature of the liquid–vapor interface should be accounted for, although it is usually neglected in modeling of film flows in tubes of larger diameters. Within the assumptions considered above, the mass and energy balances for the liquid film shown in Fig. 10.23a yield

$$\int_{R-\delta}^{R} r w_\ell(r) \mathrm{d}r = \frac{1}{2\pi\rho_\ell}\left(\dot{m}_{\ell,\mathrm{in}} - \frac{Q(z)}{h_{\ell \mathrm{v}}}\right) \tag{10.207}$$

$\dot{m}_{\ell,\mathrm{in}}$ is the liquid mass flow rate at the condenser inlet, and $Q(z)$ is the rate of heat through a given cross section, due to phase change for $z > 0$, and is defined as follows:

$$Q(z) = 2\pi R \int_0^z q''_{\mathrm{w}} \mathrm{d}z - \left(c_{\mathrm{p},\ell}\dot{m}_\ell \overline{T}_\ell - c_{\mathrm{p},\ell,\mathrm{in}}\dot{m}_{\ell,\mathrm{in}}\overline{T}_{\ell,\mathrm{in}}\right) \tag{10.208}$$

$q''_{\mathrm{w}}(z)$ is the heat flux at the solid–liquid interface due to heat conduction through a cylindrical film with a thickness of δ balanced with the enthalpy change. Note that $T_\delta > T_{\mathrm{w}}$ and therefore $Q(z)$ in Eq. (10.207) will be a negative quantity. The heat flux at the wall is

$$q''_{\mathrm{w}} = k_\ell \frac{T_{\mathrm{w}} - T_\delta}{R \ln[R/(R-\delta)]} \tag{10.209}$$

where T_δ is the local temperature of the liquid–vapor interface. From Eqs. (10.208) and (10.209), we obtain the following equation for the heat rate rejected per unit length of the tube:

$$\frac{dQ}{\mathrm{d}z} = 2\pi k_\ell \frac{T_{\mathrm{w}} - T_\delta}{\ln[R/(R-\delta)]} - \frac{\mathrm{d}}{\mathrm{d}z}\left(c_{\mathrm{p},\ell}\dot{m}_\ell \overline{T}_\ell\right) \tag{10.210}$$

where $\overline{T}_\ell$ is the area-averaged liquid temperature for a given z location. For consideration of subcooling in the condensed liquid, $\overline{T}_\ell$ is found from an area average given by

$$\overline{T}_\ell = \frac{2\int_{R-\delta}^{R} rT_\ell(r)\mathrm{d}r}{R^2 - (R-\delta)^2} \tag{10.211}$$

where $T_\ell(r)$ is the assumed liquid film temperature profile given by the temperature distribution in a cylindrical wall.

$$T_\ell(r) = T_\delta + \frac{T_{\mathrm{w}} - T_\delta}{\ln\frac{R}{R-\delta}} \ln\frac{r}{R-\delta} \tag{10.212}$$

Substituting Eq. (10.212) into Eq. (10.211) and integrating results in an expression for $T_\ell(r)$, the derivate of $\overline{T}_\ell$ in the axial is approximated by

$$\begin{aligned} \frac{d\overline{T}_\ell}{\mathrm{d}z} &= \frac{\mathrm{d}T_\delta}{\mathrm{d}z} + \left(\frac{\mathrm{d}T_{\mathrm{w}}}{\mathrm{d}z} - \frac{\mathrm{d}T_\delta}{\mathrm{d}z}\right)\left[\ln\frac{R-\delta}{R}\right]^{-1} \frac{2}{R^2 - (R-\delta)^2} \\ &\times \left[\left(\ln\frac{R}{R-\delta} - \frac{1}{2}\right)\frac{R^2}{2} + \frac{(R-\delta)^2}{4}\right] \end{aligned} \tag{10.213}$$

where the change of film thickness in the axial direction is assumed to be negligibly small. This assumption may need to be reconsidered; however, the reduction in mathematical complexity resulting from its use is significant.

The axial momentum conservation for viscous flow in a liquid film in which the inertia terms are assumed to be negligible is

$$\frac{1}{r}\frac{\partial}{\partial r}\left(r\frac{\partial w_\ell}{\partial r}\right) = \frac{1}{\mu_\ell}\left(\frac{\mathrm{d}p_\ell}{\mathrm{d}z} + \rho_\ell g \sin\varphi\right) \tag{10.214}$$

where φ is the inclination angle. The boundary conditions for the last equation are the nonslip condition at $r = R$ and shear stresses at the liquid–vapor interface due to the frictional liquid–vapor interaction, $\tau_{\ell,\mathrm{v}}$, and the surface tension gradient related to the interfacial temperature gradient along the channel.

$$w_\ell\big|_{r=R} = 0 \tag{10.215}$$

$$\left.\frac{\partial w_\ell}{\partial r}\right|_{(r=R-\delta)} = \frac{1}{\mu_\ell}\left[-\tau_{\ell,\mathrm{v}} - \frac{\mathrm{d}\sigma}{\mathrm{d}T}\frac{\mathrm{d}T_\delta}{\mathrm{d}z}\right] \equiv E \tag{10.216}$$

where T_δ is the local liquid–vapor interface temperature, and the term with $\mathrm{d}\sigma/\mathrm{d}T$ is due to the Marangoni effect.

Taking into account the effect of the condensation process on the shear stress term, an expression from Munoz-Cobol et al. (1996) for annular filmwise condensation in vertical tubes with noncondensable gases is used for $\tau_{\ell,\mathrm{v}}$.

$$\tau_{\ell,\mathrm{v}} = \tau_{\ell,\mathrm{v}0} a'/[\exp(a') - 1] \tag{10.217}$$

where $\tau_{\ell,\mathrm{v}0}$ is defined as the interfacial shear stress in the absence of phase change and is given by

$$\tau_{\ell,,\mathrm{v}0} = 0.5 f_{\ell,\mathrm{v}} \rho_\mathrm{v} \left(\bar{w}_\mathrm{v} - w_{\ell,\delta}\right)^2 \tag{10.218}$$

where ρ_v is the vapor density, w_v is the axial vapor velocity, and a' is the ratio of the local condensation mass flow rate to the vapor mass flux rebounding from the interface, as approximated by

$$a' = -\frac{\mathrm{d}Q}{\mathrm{d}z} / \left[\pi R h_{\ell\mathrm{v}} \rho_\mathrm{v} f_{\ell,\mathrm{v}} \left(\bar{w}_\mathrm{v} - w_{\ell,\delta}\right)\right] \tag{10.219}$$

$$f_{\ell,\mathrm{v}} = f_\mathrm{v}\left(1 + 360\frac{\delta}{2R}\right) \tag{10.220}$$

and $f_{\ell,\mathrm{v}}$ is the interfacial friction factor. From an experimental study for a simulated change of phase, the vapor friction factor, f_v, is given as follows:

$$f_\mathrm{v} = 16[1.2337 - 0.2337\exp(-0.0363\,Re_\mathrm{r})] \times [\exp(1.2Ma)]/Re \tag{10.221}$$

Re_r is found using the vapor suction velocity at the liquid–vapor interface due to condensation, $v_{\mathrm{v},\delta}$, which is defined through the condensing mass flux.

$$v_{\mathrm{v},\delta} = \frac{\mathrm{d}Q}{\mathrm{d}z}\frac{1}{2\pi(R-\delta)h_{\ell\mathrm{v}}\rho_\mathrm{v}} \tag{10.222}$$

Solving Eqs. (10.214)–(10.216), the velocity profile is expressed as follows:

$$w_\ell = -\frac{1}{\mu_\ell}\frac{\mathrm{d}p_\ell}{\mathrm{d}z}\left[\frac{1}{4}\left(R^2 - r^2\right) + \frac{(R-\delta)^2}{2}\ln\frac{r}{R}\right] + E(R-\delta)\ln\frac{r}{R} \tag{10.223}$$

Substituting Eq. (10.223) into Eq. (10.207), we obtain the following equation for the axial pressure gradient in the liquid:

$$\frac{dp_\ell}{dz} = \rho_\ell g \sin\varphi + \mu_\ell \left[\frac{1}{2\pi\rho_\ell}\left(\frac{Q}{h_{\ell v}} - \dot{m}_{\ell,in}\right) + E(R-\delta)F\right] \times \left[\frac{R^4}{16} + \frac{(R-\delta)^2}{2}\left(F + \frac{(R-\delta)^2}{8} - \frac{R^2}{4}\right)\right]^{-1} \tag{10.224}$$

where

$$F = \frac{(R-\delta)^2}{2}\left(\ln\frac{R}{R-\delta} + \frac{1}{2}\right) - \frac{R^2}{4} \tag{10.225}$$

The pressure difference between the vapor and liquid phases is due to capillary effects and disjoining pressure, p_d,

$$p_v - p_\ell = \sigma\left\{\frac{d^2\delta}{dz^2}\left[1 + \left(\frac{d\delta}{dz}\right)^2\right]^{-3/2} + \frac{1}{R-\delta}\cos\left(\text{atan}\frac{d\delta}{dz}\right)\right\} - p_d \tag{10.226}$$

The term with cosine on the right-hand side of this equation is due to the second principal radius of interfacial curvature. Introducing an additional variable

$$\frac{d\delta}{dz} = \Delta \tag{10.227}$$

Equation (10.226) can be rewritten as follows:

$$\frac{d\Delta}{dz} = \left[1 + (\Delta)^2\right]^{3/2}\left(\frac{p_v - p_\ell + p_d}{\sigma} - \frac{\cos(\text{atan}\,\Delta)}{R-\delta}\right) \tag{10.228}$$

The integral equations of mass conservation for the vapor and liquid flows take the following form:

$$\rho_v A_v \bar{w}_v(z) = \bar{w}_{v,in}\rho_{v,in}A_{v,in} + Q(z)/h_{\ell v} \tag{10.229}$$

$$\rho_\ell A_\ell \bar{w}_\ell(z) = \bar{w}_{\ell,in}\rho_{\ell,in}A_{\ell,in} - Q(z)/h_{\ell v} \tag{10.230}$$

$A_v = \pi(R-\delta)^2$ is the cross-sectional area of the vapor channel, and $\bar{w}_{v,in}$ is the average vapor velocity at $z = 0$.

The compressible quasi-one-dimensional momentum equation for the vapor flow is modified to account for nonuniformity of the vapor cross-sectional area of the liquid–vapor interface, following Faghri (2016).

$$\frac{dp_v}{dz} = \rho_v g \sin\varphi + \frac{1}{A_v}\left[\frac{d}{dz}\left(-\beta_v \rho_v w_v^2 A_v\right) - f_v \rho_v w_v^2 \pi(R-\delta) + 2\pi(R-\delta)\rho_v v_{v,\delta}^2 \sin(\text{atan}\,\Delta)\right] \tag{10.231}$$

with $\beta_v = 1.33$ for small radial Reynolds numbers.

The perfect gas law is employed to account for the compressibility of the vapor flow,

$$\rho_{\rm v} = \frac{p_{\rm v}}{R_g T_{\rm v}} \tag{10.232}$$

Therefore,

$$\frac{{\rm d}\rho_{\rm v}}{{\rm d}z} = \frac{1}{R_g}\left(\frac{{\rm d}p_{\rm v}}{{\rm d}z}\frac{1}{T_{\rm v}} - \frac{p_{\rm v}}{T_{\rm v}^2}\frac{{\rm d}T_{\rm v}}{{\rm d}z}\right) \tag{10.233}$$

The saturated vapor temperature and pressure are related by the Clausius–Clapeyron equation, which can be written in the following form:

$$\frac{{\rm d}T_{\rm v}}{{\rm d}z} = \frac{{\rm d}p_{\rm v}}{{\rm d}z}\frac{R_{\rm g}T_{\rm v}^2}{p_{\rm v}h_{\ell{\rm v}}} \tag{10.234}$$

The seven first-order differential equations, Eqs. (10.210), (10.224), (10.227), (10.228), (10.231), (10.233), and (10.234), include the following seven variables: $\delta, \Delta, p_\ell, Q, p_{\rm v}, \rho_{\rm v}$, and $T_{\rm v}$. Therefore, seven boundary conditions are set forth at $z = 0$.

$$\delta = \delta_{\rm in} \tag{10.235}$$

$$\Delta = 0 \tag{10.236}$$

$$p_\ell = p_{\rm v,in} - \frac{2\sigma}{R - \delta_{\rm in}} + p_{\rm d} \tag{10.237}$$

$$Q = 0 \tag{10.238}$$

$$p_{\rm v} \equiv p_{\rm v,in} = p_{\rm v,sat}(T_{\rm v,in}) \tag{10.239}$$

$$\rho_{\rm v,in} = \frac{p_{\rm v,in}}{R_{\rm g}T_{\rm v,in}} \tag{10.240}$$

$$T_{\rm v} = T_{\rm v,in} \tag{10.241}$$

In the boundary condition given by Eq. (10.235), $\delta_{\rm in}$ is defined from the condition that in the adiabatic zone, just before the entrance of the condenser, the liquid and vapor pressure gradients should be equal. To find $\delta_{\rm in}$, Eqs. (10.224) and (10.231) should be solved for the case of $Q = 0$, $v_{{\rm v},\delta} = 0$, ${\rm d}A_{\rm v}/{\rm d}z = 0$, and ${\rm d}T_\delta/{\rm d}z = 0$. The boundary condition, Eq. (10.238), directly follows from Eq. (10.208). There are also parameters $\dot{m}_{\ell,{\rm in}}$ and $\bar{w}_{\rm v,in}$, and an additional variable, T_δ, involved in this problem. They will be considered using additional algebraic equations. The parameter $\dot{m}_{\ell,{\rm in}}$ should be found using a constitutive condition at the entrance of the condenser

$$\dot{m}_{\ell,{\rm in}} = \dot{m}_{\rm t} - Q_{\rm in}/h_{\ell{\rm v}} \tag{10.242}$$

where $Q_{\rm in}$ is the total heat load into the condenser. Also $\bar{w}_{\rm v,in} = Q_{\rm in}/(h_{\ell{\rm v}}\rho_{\rm v}A_{\rm v,in})$.

The liquid–vapor interface temperature, T_δ, differs from the saturated bulk vapor temperature because of the interfacial resistance and effects of curvature on saturation pressure over liquid films. The interfacial resistance is defined as (Chap. 4)

$$q''_{\delta} = -\left(\frac{2\alpha}{2-\alpha}\right)\frac{h_{\ell v}}{\sqrt{2\pi R_g}}\left[\frac{p_v}{\sqrt{T_v}} - \frac{(p_{sat})_{\delta}}{\sqrt{T_{\delta}}}\right] \tag{10.243}$$

where p_v and $(p_{sat})_{\delta}$ are the saturation pressures corresponding to T_v and T_{δ}, the temperatures associated with the thin liquid film interface, respectively. The following two algebraic equations should be solved to determine T_{δ} for every point along the z-direction. The relation between the saturation vapor pressure over the thin condensing film, $(p_{sat})_{\delta}$, affected by the surface tension, and the normal saturation pressure corresponding to T_{δ}, $p_{sat}(T_{\delta})$, is given by the following equation (Chap. 4):

$$(p_{sat})_{\delta} = p_{sat}(T_{\delta})\exp\left[\frac{(p_{sat})_{\delta} - p_{sat}(T_{\delta}) - \sigma K + p_d}{\rho_{\ell}R_g T_{\delta}}\right] \tag{10.244}$$

where K is the local curvature of the liquid–vapor interface defined by the term in outer brackets in Eq. (10.226). Notice that under steady conditions, q''_{δ} is due to heat conduction through the liquid film. It follows from Eq. (10.209) and (10.243)

$$T_{\delta} = T_w + \frac{R}{k_{\ell}(R-\delta)}\ln\frac{R}{R-\delta}\left(\frac{2\alpha}{2-\alpha}\right) \times \frac{h_{\ell v}}{\sqrt{2\pi R_g}}\left[\frac{p_v}{\sqrt{T_v}} - \frac{(p_{sat})_{\delta}}{\sqrt{T_{\delta}}}\right] \tag{10.245}$$

Equations (10.244) and (10.245) determine the interfacial temperature, T_{δ}, and pressure, $(p_{sat})_{\delta}$, for a given vapor pressure, $p_v = p_{v,sat}(T_v)$, temperature of the solid–liquid interface, T_w, and the liquid film thickness, δ. Note that T_w is the local temperature of the wall and can vary along the condenser depending on the cooling conditions.

For the case of variable wall temperature, T_w becomes an additional variable. If the convective heat transfer coefficient at the outer tube wall, h_o, and the cooling liquid temperature, T_{∞}, are known, the local wall temperature can be defined using an energy balance

$$h_o(T_w - T_{\infty}) = \frac{1}{2\pi R}\frac{dQ}{dz} \tag{10.246}$$

To solve this problem, the first-order ordinary differential equation, Eq. (10.246), must be added to the seven previously specified. The additional boundary condition is given by

$$z = 0, \quad T_w = T_{w,in} \tag{10.247}$$

Equations (10.210), (10.224), (10.227), (10.228), (10.231), (10.233), and (10.234) with corresponding boundary conditions have been solved using the standard Runge–Kutta procedure by Begg et al. (1999). Algebraic Eqs. (10.244) and (10.245), with two unknowns, $(p_{sat})_{\delta}$ and T_{δ}, have been solved numerically for every point on z using Wegstein's iteration method.

The length of the two-phase zone is given by the numerical solution. At the end of the two-phase zone, $d\Delta/dz$ became infinitely large, and the film thickness increased dramatically, causing the solution to break down.

Two cases of heat-load-in are presented, one representing complete condensation and the other incomplete condensation. Complete condensation is defined as the condensation of all incoming vapor in a filmwise manner, in which the vapor flow terminates at a well-defined location, forming a steady meniscus-like interface. The condition of complete condensation requires the overall energy balance to be satisfied. Thus, the energy convected into the tube, or heat-load-in, Q_{in}(W), must be equal to the cumulative heat rate rejected from the condensing vapor in the tube, referred to simply as Q(W). This occurs over a distance from the inlet to the tube known as the condensation length, L_δ (m). Incomplete condensation is said to exist for a heat-load-in greater than Q_{in}(W) for complete condensation.

Figure 10.24a shows variation of the liquid film thickness along the condenser with a vapor inlet temperature of 363 K and a constant wall temperature of 340 K for two cases of heat-load-in, $Q_{in} = 10$ W and $Q_{in} = 12$ W. The total mass flow rate of $\dot{m}_t = 1.00 \times 10^{-6}$ (kg/s) is the same for both cases. The film thickness decreases in the downstream direction before converging due to the

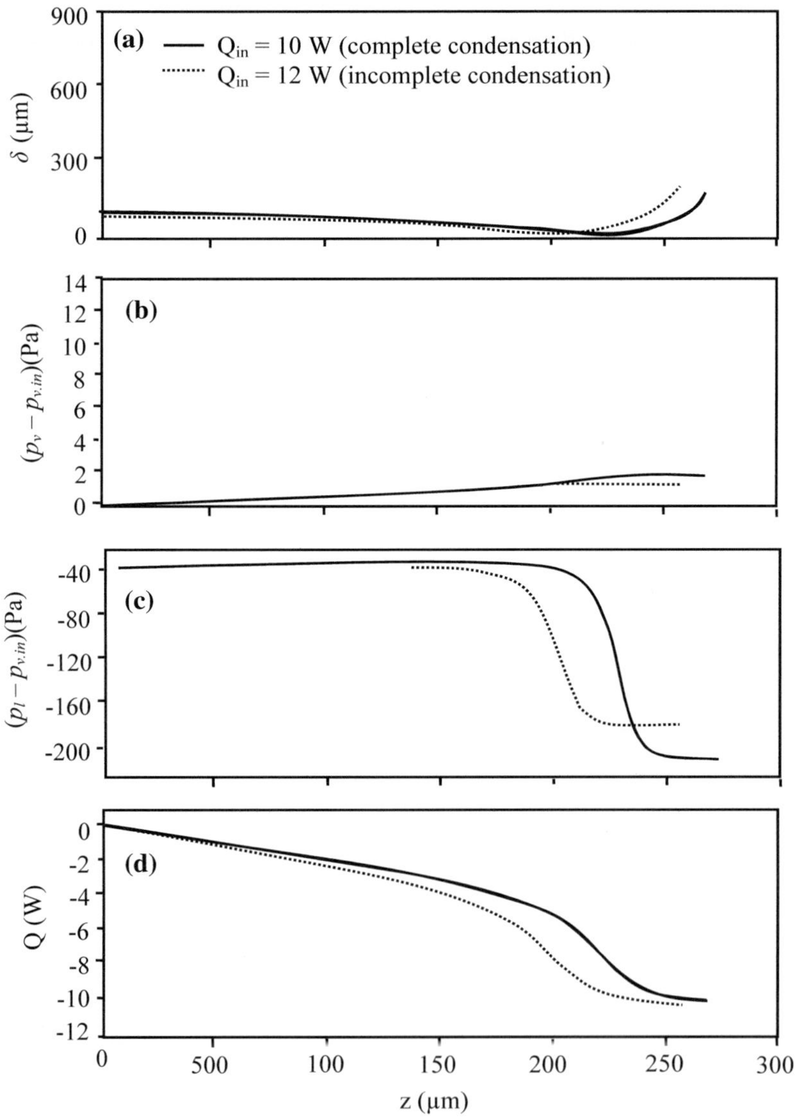

Fig. 10.24 Annular film condensation in a circular tube with constant wall temperature boundary condition. $R = 1.55$ mm, $T_w = 363$ K, $\dot{m} = 0.01$ g/s. **a** Film thickness versus position, **b** vapor pressure versus position, **c** liquid pressure versus position, **d** cumulative heat rejected versus position (Begg et al. 1999)

capillary forces acting at the liquid–vapor interface. Condensation is more intensive in the region where the film thickness is at the minimum. The values of δ_{in} will be different for each case of heat-load-in even though the total mass flow rate and thermal boundary conditions are the same. At the inlet, the mass flow rates of liquid and vapor adjust themselves to satisfy the energy balance. The vapor pressure drop across the condensation length is insignificant compared to the liquid pressure drop as shown in Fig. 11.24b, c.

Figure 10.24d shows the cumulative sum of heat removed from the vapor by condensation. For the case of Q_{in} = 10 W, representing complete condensation, all the incoming vapor is condensed and a location is calculated for the beginning of single-phase (liquid) flow in the tube. Thus, all the heat transfer and associated pressure drop occur upstream of this location. Although the incomplete condensation case represents a valid numerical solution for equations modeling axisymmetric film condensation, it must be considered with reference to an overall energy balance to provide a meaningful physical interpretation. For the case of incomplete condensation, Q_{in} = 12 W, not all of the heat-load-into the tube is rejected at the termination of the calculation. At the termination of the calculation, the final value of Q at $z = L_\delta$ is seen to be less than Q_{in}. Thus, the overall energy balance is not satisfied.

During complete condensation, the values of the dependent variables (δ_{in}, p_ℓ, p_v, Q) are the steady field solution of the two-phase flow and are independent of time as calculated and as interpreted physically. The simple physical interpretation is that the calculated liquid film profile is in fact stationary and goes to the radius of the tube at L_δ. Obviously, the vapor mass flow rate becomes zero at this point also. However, during incomplete condensation, the numerical solution is interpreted as representing a time-averaged steady-state two-phase flow field where the vapor mass flow rate does not terminate at the calculated condensation length. For values of Q_{in} greater than but close to the maximum heat-load-in, the incomplete condensation solution predicts that two-phase flow will exist downstream of the point calculated for L_δ. By implication, the flow field will no longer be steady in an instantaneous sense.

Observations from a flow visualization experiment of water vapor condensing in a horizontal glass tube, by Begg et al. (1999), confirm the existence and qualitative features of annular film condensation leading to the complete condensation phenomenon in small diameter (d < 3.5 mm) circular tubes.

10.6.3 Flow Evaporation and Boiling

10.6.3.1 Onset of Nucleate Boiling in Microchannel Flow

The first onset of the vapor bubble marks the transition from the single-phase flow in the microchannel to two-phase flow, and for this reason, the prediction of onset of nucleate boiling (ONB) is very important. Liu et al. (2005) thoroughly reviewed the existing works on ONB in subcooled boiling. Hsu (1962) proposed the ONB heat flux correlation based on the minimum superheat criterion for pool boiling

$$q''_{ONB} = \frac{k_\ell h_{\ell v} \rho_v (T_W - T_{sat})^2}{12.8 \sigma T_{sat}} \tag{10.248}$$

The effect of contact angle on the ONB heat flux was considered by Davis and Anderson (1966), who proposed the following correlation

$$q''_{\mathrm{ONB}} = \frac{k_\ell h_{\ell v} \rho_v (T_w - T_{sat})^2}{8(1 + \cos\theta)\sigma T_{sat}} \tag{10.249}$$

Both Eqs. (10.248) and (10.249) are obtained based on pool boiling, but they are also used to predict ONB in conventional channels. Liu et al. (2005) proposed an analytical model for ONB in microchannels (grooves) as shown in Fig. 10.25, which will be presented here.

The physical model of ONB in a microchannel is shown in Fig. 10.26. It is assumed that (1) the bubble shape is a truncated sphere, (2) the liquid temperature is not affected by the bubble nucleus due to its small size, and (3) the vapor and liquid are in equilibrium under saturation. The condition required for the bubble nucleus to grow is that the liquid temperature at $y = y_b$ is greater than the superheat requirement (see Fig. 10.26b).

The superheat equation for the bubble nucleus can be obtained from equilibrium theory

$$T_b - T_{sat} = \frac{2\sigma T_b}{h_{\ell v} \rho_v R_b} \tag{10.250}$$

where T_b is the liquid temperature required for the bubble nucleus to grow and T_{sat} is the saturation temperature corresponding to the liquid pressure p_ℓ.

Since the bubble shape is a truncated sphere, we have (Fig. 10.26)

$$y_b = R_b(1 + \cos\theta) \tag{10.251}$$

$$r_c = R_b \sin\theta \tag{10.252}$$

Substituting Eq. (10.251) into Eq. (10.250) and solving for T_b, one obtains

$$T_b = T_{sat} \Big/ \left(1 - \frac{2\sigma C}{h_{\ell v} \rho_v y_b}\right) \tag{10.253}$$

where $C = 1 + \cos\theta$.

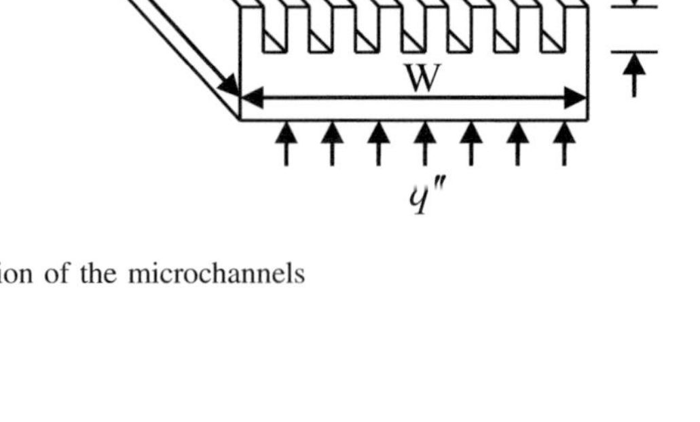

Fig. 10.25 Geometric configuration of the microchannels

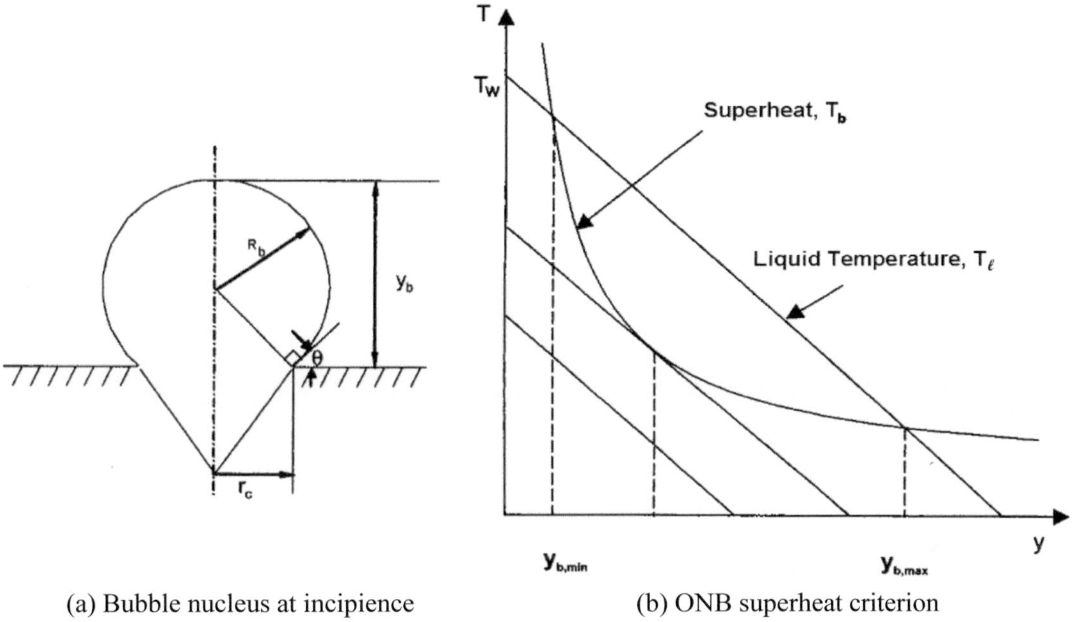

Fig. 10.26 Onset of nucleate boiling in microchannel (Liu et al. 2005; Reprinted with permission from Elsevier)

The liquid temperature near the wall of the microchannel can be assumed as

$$T_\ell(y) = T_{\mathrm{w}} - q''_{\mathrm{w}} y / k_\ell \tag{10.254}$$

where q''_{w} is the effective wall heat flux at the inner wall of the microchannels.

If a constant heat flux, q'', is applied to the bottom, the liquid temperature at the outlet is related to that at the inlet by

$$T_{\ell,\mathrm{out}} = T_{\ell,\mathrm{in}} + \frac{q'' WL}{\rho_\ell c_\ell u_0 (n w_{\mathrm{c}} H_{\mathrm{c}})} \tag{10.255}$$

where u_0 is the liquid inlet velocity, and n is the number of microchannels. If the convective heat transfer coefficient is uniform along the channel surfaces and flow in the channel is fully developed, the channel wall temperature can be obtained by

$$T_{\mathrm{w}} = T_\ell + \frac{q''_{\mathrm{w}}}{\overline{Nu} k_\ell / D_{\mathrm{h}}} \tag{10.256}$$

where $\overline{Nu}$ is the Nusselt number for fully developed flow in a three-sided (bottom and side walls) rectangular channel (Shah and London 1978)

$$\overline{Nu} = 8.235(1 - 1.883/\alpha + 3.767/\alpha^2 - 5.814/\alpha^3 + 5.361/\alpha^4 - 2/\alpha^5) \tag{10.257}$$

where α is the aspect ratio of the microchannel.

The effective wall heat flux, q''_{w}, is related to the applied heat flux by

$$q''_{\mathrm{w}} = \left(\frac{\alpha}{1+2\eta\alpha} \frac{w_{\mathrm{c}}+w_{\mathrm{w}}}{H_{\mathrm{c}}} \right) q'' \tag{10.258}$$

where η is the fin efficiency.

The condition for nucleation boiling to occur is $T_\ell(y_\mathrm{b}) \geq T_\mathrm{b}$. The necessary condition for ONB can be obtained from Eqs. (10.253) and (10.254), i.e.,

$$T_{\mathrm{w}} - \frac{q''_{\mathrm{w}}}{k_\ell} y_{\mathrm{b}} = T_{\mathrm{sat}} \Big/ \left(1 - \frac{2\sigma C}{h_{\ell\mathrm{v}} \rho_{\mathrm{v}} y_{\mathrm{b}}} \right) \tag{10.259}$$

which can be rearranged as an equation of y_b,

$$\frac{q''_{\mathrm{w}}}{k_\ell} y_{\mathrm{b}}^2 - \left(T_{\mathrm{w}} + \frac{2\sigma C}{\rho_{\mathrm{v}} h_{\ell\mathrm{v}}} \frac{q''_{\mathrm{w}}}{k_\ell} \right) y_{\mathrm{b}} + \frac{2\sigma C}{h_{\ell\mathrm{v}} \rho_{\mathrm{v}} y_{\mathrm{b}}} T_{\mathrm{w}} = 0 \tag{10.260}$$

For the roots of Eq. (10.260) to be real, the following condition must be satisfied

$$\left(T_{\mathrm{w}} + \frac{2\sigma C}{\rho_{\mathrm{v}} h_{\ell\mathrm{v}}} \frac{q''_{\mathrm{w}}}{k_\ell} \right)^2 - 4 \frac{q''_{\mathrm{w}}}{k_\ell} \frac{2\sigma C}{h_{\ell\mathrm{v}} \rho_{\mathrm{v}} y_{\mathrm{b}}} T_{\mathrm{w}} \geq 0 \;. \tag{10.261}$$

which can be rearranged to yield the superheat criterion

$$\sqrt{T_{\mathrm{w}}} - \sqrt{T_{\mathrm{sat}}} \geq \sqrt{\frac{2\sigma C}{\rho_{\mathrm{v}} h_{\ell\mathrm{v}}} \frac{q''_{\mathrm{w}}}{k_\ell}} \tag{10.262}$$

Equation (10.262) can be used to determine if nucleate boiling will occur in the microchannels under specified parameters. If the liquid inlet conditions are prescribed but the heat flux is allowed to vary, the ONB heat flux can be obtained by

$$\begin{aligned} &\sqrt{T_{\ell,\mathrm{in}} + \frac{q'' WL}{\rho_\ell c_\ell u_0 (n w_{\mathrm{c}} H_{\mathrm{c}})} + \frac{\left(\frac{\alpha}{1+2\eta\alpha} \frac{w_{\mathrm{c}}+w_{\mathrm{w}}}{H_{\mathrm{c}}} \right) q''_{\mathrm{ONB}}}{\overline{Nu}\, k_\ell / D_{\mathrm{h}}}} - \sqrt{T_{\mathrm{sat}}} \\ &= \sqrt{\frac{2\sigma C}{\rho_{\mathrm{v}} h_{\ell\mathrm{v}}} \frac{\left(\frac{\alpha}{1+2\eta\alpha} \frac{w_{\mathrm{c}}+w_{\mathrm{w}}}{H_{\mathrm{c}}} \right) q''_{\mathrm{ONB}}}{k_\ell}} \end{aligned} \tag{10.263}$$

If nucleate boiling is to be avoided or delayed for some application, one can increase the liquid inlet velocity u_0, or reduce the liquid inlet temperature, $T_{\ell,\mathrm{in}}$. In this case, Eq. (10.263) can be used to obtain the minimum liquid velocity or maximum liquid inlet velocity to avoid nucleate boiling.

The wall superheat can be obtained from inequality Eq. (10.262) as

$$T_{\mathrm{w}} - T_{\mathrm{sat}} = \frac{2\sigma C}{\rho_{\mathrm{v}} h_{\ell\mathrm{v}}} \frac{q''_{\mathrm{w}}}{k_\ell} + 2\sqrt{T_{\mathrm{sat}}} \sqrt{\frac{2\sigma C}{\rho_{\mathrm{v}} h_{\ell\mathrm{v}}} \frac{q''_{\mathrm{w}}}{k_\ell}} \tag{10.264}$$

Substituting Eq. (10.258) into Eq. (10.264), the wall superheat becomes

$$T_{\mathrm{w}} - T_{\mathrm{sat}} = \frac{2\sigma C}{\rho_{\mathrm{v}} h_{\ell\mathrm{v}} k_{\ell}} \left(\frac{\alpha}{1+2\eta\alpha} \frac{w_{\mathrm{c}} + w_{\mathrm{w}}}{H_{\mathrm{c}}} \right) q'' + 2\sqrt{T_{\mathrm{sat}}} \sqrt{\frac{2\sigma C}{\rho_{\mathrm{v}} h_{\ell\mathrm{v}} k_{\ell}} \left(\frac{\alpha}{1+2\eta\alpha} \frac{w_{\mathrm{c}} + w_{\mathrm{w}}}{H_{\mathrm{c}}} \right) q''} \tag{10.265}$$

10.6.3.2 Flow Evaporation in Minichannel Heat Sink with Axial Grooves

New advanced cooling technologies are needed to meet the demands for dissipating extremely high heat fluxes (over 300 W/cm^2) from electronic components. The performance characteristics of a flat high heat flux evaporative minichannel heat sink with axial capillary grooves on its inner surface, as shown in Fig. 10.27, were modeled by Khrustalev and Faghri (1996). The small axial grooves allow this heat sink to withstand high heat fluxes with a small pressure drop, low thermal resistance, and a comparatively large effective length. To evaluate the advantages of this design compared to existing designs, the maximum heat flux, thermal resistance, and the pressure drop on the element will be estimated for some independent operational parameters, such as the saturated vapor outlet temperature $T_{\mathrm{v,o}}$, the geometry of the grooved surfaces, and the effective lengths of the evaporators under consideration.

The cross sections of the heat sink and the coordinate system are shown in Fig. 10.27. Introduced by a pump, the liquid enters the heat sink and, under the influence of an extremely high heat flux, boils on the wall at a short distance from the entrance. While the vapor phase appears to be moving along the z-coordinate, the vapor and liquid cocurrent flows become separated, with the liquid flowing mostly in the grooves. The separation of the phases in a small, horizontal, flat rectangular channel with smooth walls (19.05 × 3.18 mm^2) has been experimentally investigated by Wambsganss et al. (1991). They have observed the annular flow pattern for a water–air mixture starting with an air velocity of about 10 m/s. Wilmarth and Ishii (1994) examined two test sections with gap widths of 1 and 2 mm with channel widths of 20 and 15 mm, respectively. In their experiments, the annular flow regime was observed with air velocity of about 5 m/s. For the heat sink with a flow channel cross section of 7 × 1 mm^2, and the heat flux of 100 W/cm^2 on both side walls, it can be estimated that the velocity of the vapor reaches 5 m/s at the distance of about 3.3 mm from the entrance (for T_{v} = 100 °C). This means that for such a heat sink, the flow boiling length preceding the annular flow pattern would be very small, even with smooth walls. Since surface tension tends to pull the liquid into the grooves, the flow boiling length should be even smaller, and the annular regime transforms into the separated cocurrent two-phase flow for the considered heat sink. Surface tension seems to be a stabilizing factor for the liquid flow in a groove; it predominates over the destabilizing dynamic vapor–liquid interaction when the Weber number is less than unity. An expression for the Weber number—including the empirically adjusted characteristic size for a capillary-grooved surface with vapor flow over it—has been given by Vasiliev et al. (1984). Their expression is based on the analysis of the experimental data obtained from the gravity-assisted heat pipe experiments with several fluids, including water.

$$We \equiv \frac{\rho_{\mathrm{v}} \bar{w}_{\mathrm{v}}^2 D_{\mathrm{h},\ell}^2}{2\pi\sigma W} < 1 \tag{10.266}$$

where W is the half-width of the groove. Khrustalev and Faghri (1997) confirmed the importance of the entrainment limitation through experimental results obtained for a miniature heat sink with axial grooves. Therefore, with the inequality given by Eq. (10.266) satisfied, the emphasis is on the mode where vapor and liquid flows are separated with the liquid flowing exclusively in the grooves. Since

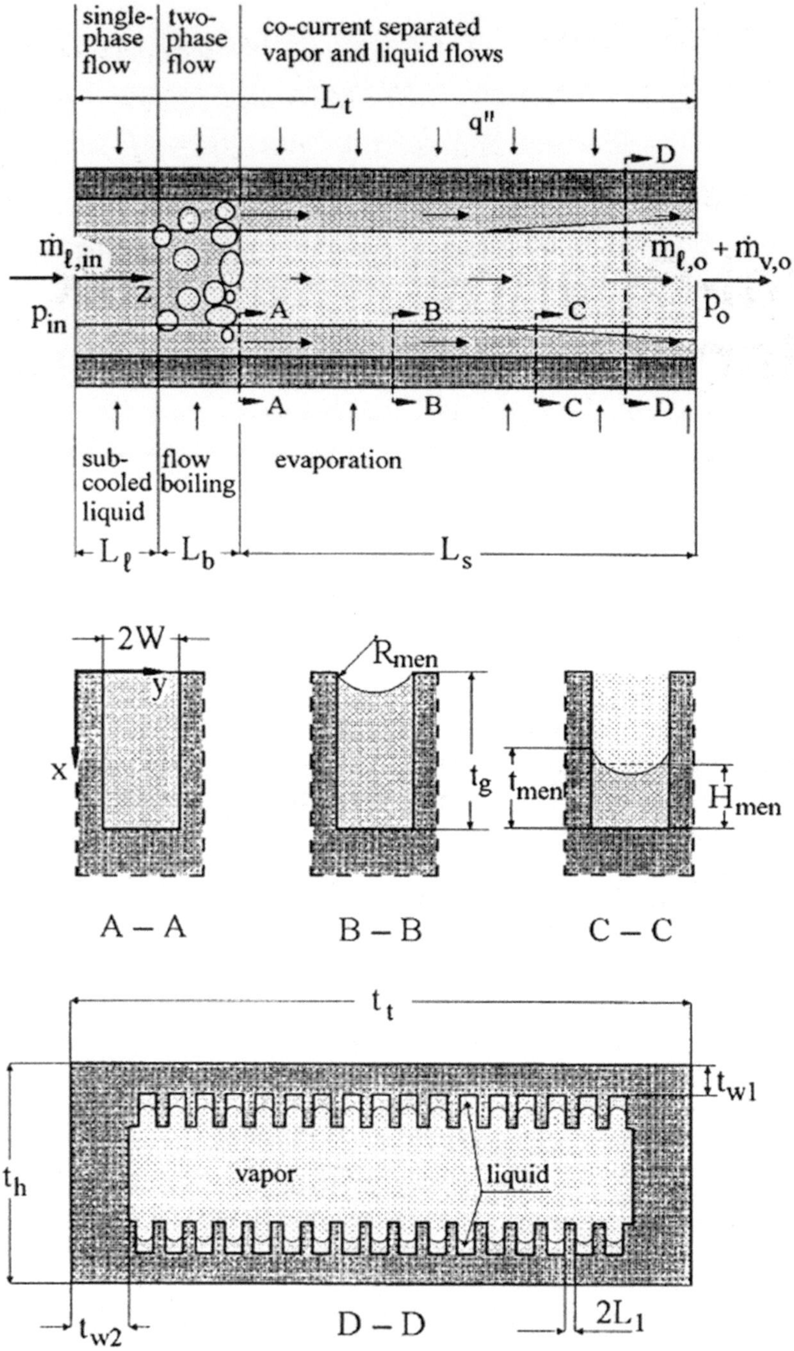

Fig. 10.27 Schematic of the flat miniature evaporator with axial capillary grooves (not to scale; Khrustalev and Faghri 1996)

the inlet liquid is supposed to be saturated, and the cross section of the vapor channel is very small, it is assumed in this analysis that $L_\ell = L_b = 0$ (Fig. 10.27), which means that only the part of the evaporator with separated vapor and liquid flows is modeled.

Another significant feature of this evaporator is that the formation of vapor bubbles in the liquid, which can occur with high heat fluxes, does not obstruct the liquid flow in the grooves (unlike in heat pipes) due to the concurrency of the liquid and vapor flows. Perturbations in the liquid flow caused by a growing vapor bubble are compensated by the dynamic vapor–liquid interaction during the bubble's explosion into the vapor space, with the subsequent return of the accelerated liquid droplets to the liquid flow. Moreover, for comparatively moderate heat fluxes, the liquid in the grooves does not boil along most of the effective length, because the liquid superheat is not sufficient for boiling due to the intensive evaporation of the liquid from the surfaces of the thin films attached to the liquid–vapor menisci corners. The analysis also attempts to predict the onset of nucleate boiling using the methodology by Khrustalev and Faghri (1997).

Another essential feature of this evaporator is that the liquid flow in the grooves is strongly influenced by the vapor flow. The vapor–liquid frictional interaction can be the most important factor of those affecting the hydraulic resistance of the liquid flow. Another factor of importance to the liquid flow is the axial gradient of the capillary pressure, due to the variation of the main radius of the liquid–vapor meniscus along the groove. This variation takes place not far from the entrance region of the evaporator (see Fig. 10.27), where the grooves are nearly filled with liquid. Farther from the entrance, starting from the point on z where the meniscus main radius R_{men} reaches its minimum value $R_{men,min}$, the effective meniscus height H_{men} decreases along the z-coordinate, so that at some point closer to the evaporator outlet, the liquid–vapor meniscus interface nearly touches the bottom of the groove. At this interval along z, the liquid flows in a groove and is influenced by the pressure gradient in the vapor and the vapor–liquid frictional interaction at the interface. However, the surface tension indirectly controls the liquid distribution by collecting the occasional liquid droplets from the walls of the grooves.

With a heat flux close to the maximum, the effective meniscus height, H_{men}, can become very small at the outlet of the evaporator; therefore, this region, which can be called the "thin-film flow region," requires a special treatment. Under close examination, it is evident that the bottom of a capillary groove rarely is rectangular, a shape that is essential for modeling the "thin-film flow region" (but can be practically insignificant for other regions). This depends upon the manufacturing process used. When the electro-discharge machining process (EDM) is used to manufacture grooves on the inner evaporator surface, the groove bottom is rounded, and the liquid cross section has a very complicated form. This reality should be kept in mind, even though the present analysis and its model are simplified by use of a rectangular strip to approximate the liquid cross-sectional shape in the "thin-film flow region." Simultaneously, the axial capillary pressure gradient is assumed to be negligible in this region in comparison to the vapor pressure gradient, which is consistent with the description of the previously discussed region with a larger H_{men}. With extremely high heat fluxes, dryout of the evaporator can occur for two reasons: (1) hydrodynamic limitations of the fluid transport (the corresponding heat flux is denoted as the maximum, q''_{max}), or (2) the beginning of vapor film boiling (the corresponding heat flux is usually referred to as the critical heat flux, CHF). The present model can predict q''_{max} but CHF should be determined experimentally.

At any axial location, the following mass conservation equation must hold over the cross section of the evaporator modeled under steady-state conditions:

$$\dot{m}_\ell(z) = \dot{m}_{\ell,o} + \dot{m}_{v,o} - \dot{m}_v(z) \tag{10.267}$$

where $\dot{m}_v = \bar{w}_v \rho_v A_v$, $\dot{m}_\ell = N \bar{w}_\ell \rho_\ell A_\ell$, N is the number of grooves, and the subscript "o" denotes the evaporator outlet.

For laminar liquid flow in a groove, the momentum conservation equation is

$$\frac{dp_\ell}{dz} + \rho_\ell g \sin\phi = -f_\ell \frac{2\rho_\ell \bar{w}_\ell^2}{D_{h,\ell}} \equiv -f_\ell \frac{2\dot{m}_\ell^2}{D_{h,\ell}\rho_\ell A_\ell^2 N^2} \tag{10.268}$$

where ϕ is the inclination angle. The inertia terms in Eq. (10.268) are neglected, since Khrustalev and Faghri (1994) demonstrated that this simplification is true even when the liquid cross-sectional area varies along the flow.

For vapor flow in a channel with a small Mach number ($Ma \ll 1$), the momentum conservation equation is (Faghri 2016)

$$\frac{d}{dz}(p_v + \rho_v \beta_v \bar{w}_v^2) + \rho_v g \sin\phi = -f_v \frac{2\rho_v \bar{w}_v^2}{D_{h,v}} \tag{10.269}$$

The Laplace–Young equation relates the interfacial radius of curvature to the pressure difference between the liquid and vapor. In differential form, the equation is

$$\frac{d}{dz}\left(\frac{\sigma}{R_{men}}\right) = \frac{dp_v}{dz} - \frac{dp_\ell}{dz} \tag{10.270}$$

The mass and energy conservation balances relate the vapor mass flow rate with the heat flux at the inside surface of the vapor channel, q''_v:

$$\frac{d\dot{m}_v}{dz} = \frac{2q''_v(z)}{h_{\ell v}}(t_t - 2t_{w2}) \tag{10.271}$$

where

$$q''_v = h_e(\overline{T}_w - T_v)$$

h_e is the local effective heat transfer coefficient describing the evaporative heat transfer from the wall of the evaporator to the saturated vapor, through the liquid films and the fin between the grooves. $\overline{T}_w$ is the local mean wall temperature. Note that $q''_v = q''$ for cases in which axial conduction in the wall is insignificant and where q'' is the heat flux on the outer surfaces of the heat sink.

These conservation equations should be solved simultaneously for the inlet interval of z where R_{men} varies and $t_{men} = t_g$ (note that H_{men} is not constant). However, for the outlet z-interval at the heat sink outlet (see Fig. 10.27), the liquid recedes into the grooves, and t_{men} varies with z (as does H_{men}) while $R_{men} = R_{men,min}$. At this "outlet" interval, H_{men} is one of the unknown variables, while the pressure gradients in the liquid and vapor are equal: $dp_\ell/dz = dp_v/dz$ (see Eq. 10.270). For the "outlet" interval, taking the equality of the pressure gradients into consideration, the following transcendental equation for H_{men} can be obtained from Eqs. (10.267) and (10.268):

$$\dot{m}_{\ell,o} + \dot{m}_{v,o} - \dot{m}_v(z) = \left[-\frac{D_{h,\ell}\rho_\ell A_\ell^2 N^2}{2f_\ell}\left(\frac{dp_v}{dz} + \rho_\ell g \sin\phi\right)\right]^{1/2} \tag{10.272}$$

which should be solved for H_{men} at every point on z since $D_{h,\ell}$, A_ℓ, f_ℓ and dp_v/dz depend on H_{men}. According to Eq. (10.271), the vapor mass flow rate at the outlet is

$$\dot{m}_{v,o} = 2 \int_{2}^{L_s} q''_v(z)(t_t - 2t_{w2}) \frac{dz}{h_{\ell v}} \tag{10.273}$$

The coefficients, f_v, f_ℓ, $D_{h,\ell}$, $D_{h,v}$, and others appearing in the governing equations, can be found in Khrustalev and Faghri (1996) and will not be repeated here.

For the given temperature of the vapor at the outlet, $T_{v,o}$, the boundary conditions for Eqs. (10.269) and (10.271) at the right-hand end of the evaporator ($z = L_s$) are

$$p_v = p_{sat}(T_{v,o}) \tag{10.274}$$

$$\dot{m}_v = \dot{m}_{v,o} \tag{10.275}$$

Taking into consideration that $h_{\ell v}$ varies along the z-coordinate due to the axial vapor temperature gradient, $m_{v,o}$—estimated initially by Eq. (10.273) at $T_{v,o}$—should be adjusted so as to satisfy the following condition:

$$\dot{m}_v|_{z=0} = 0 \tag{10.276}$$

When predicting the maximum heat flux, the liquid flow rate at the outlet of the evaporator is infinitesimal: $\dot{m}_{\ell,o} = \varepsilon \dot{m}_{v,o}$ where ε is a very small number set equal to 10^{-3} in the numerical experiments. In this case, the heat flux, q'', should be chosen so as to satisfy the following boundary condition at the beginning of the evaporator (nearly planar liquid–vapor interface):

$$\frac{1}{R_{men}|_{z=0}} = \frac{\cos \theta_{men,min}}{t_h/2 - t_{w1} - t_g} \tag{10.277}$$

which implies that the liquid and vapor flows become separated at $z = 0$ under the influence of the capillary forces. In cases of a given heat flux below the maximum heat flux, $\dot{m}_{\ell,o}$ should be chosen so as to satisfy Eq. (10.277).

The numerical procedure begins at $z = L_s$ and moves toward the left-hand end of the vapor zone. At some point on z—which can be denoted as z_1—the condition $t_{men} = t_g$ is reached, and, starting from this point, Eqs. (10.268)–(10.271) should be solved for p_ℓ, p_v, R_{men} and $\dot{m}_v$ with the following boundary conditions (at $z = z_1$):

$$p_\ell = p_v|_{z=z_1} - \frac{\sigma}{R_{men,min}} \tag{10.278}$$

$$p_v = p_v|_{z=z_1} \tag{10.279}$$

$$R_{men} = R_{men,min} \tag{10.280}$$

$$\dot{m}_v = \dot{m}_v|_{z=z_1} \tag{10.281}$$

Khrustalev and Faghri (1996) solved the above problem numerically, and an example of the characteristics of the copper–water evaporator for the following operating parameters was presented (see Fig. 10.28): $W = 0.063$ mm, $L_1 = 0.02$ mm, $L_t = 80$ mm, $t_g = 0.229$ mm, $t_h = 2.58$ mm, $t_t = 7.75$ mm, $t_{w1} = 0.5$ mm, $\phi = 0$, $\theta_{men,min} = 33°$, $\alpha = 0.05$, $T_{v,o} = 353$ K, and $q'' = 65\ \mathrm{W/cm^2}$. A grooved structure featuring such a thin fin was reported by Plesch et al. (1991). In most numerical experiments, the overall meniscus height at the outlet, $t_{men,o}$ was larger than $t_{men,min}$, and in all cases

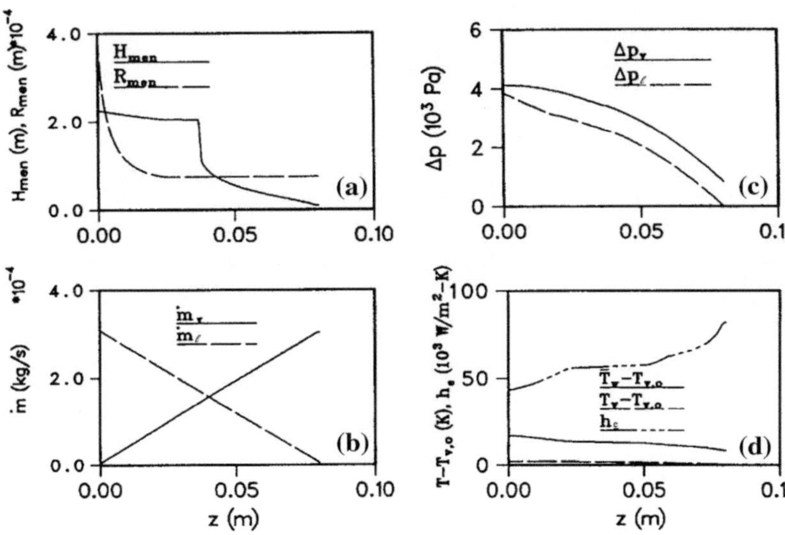

Fig. 10.28 Performance characteristics of the miniature evaporator along the axial direction: **a** effective height and radius of the liquid–vapor meniscus; **b** vapor and liquid mass flow rates; **c** vapor and liquid pressure variations; and **d** vapor and wall temperature variations, and effective heat transfer coefficient (Khrustalev and Faghri 1996)

$H_{\text{men,o}}$ was larger than the roughness, R_a. This means that the concept of the "thin-film region" did not practically affect the results obtained.

10.6.3.3 Boiling Heat Transfer in a Miniature Axially Grooved Rectangular Channel

Different configurations of miniature heat sinks are needed for the varied applications. The most important characteristics of miniature heat sinks are listed below (Khrustalev and Faghri 1997):

1. Heated length and external thickness
2. Uniformity of temperature over the surface
3. Effective heat transfer coefficients (or thermal resistance)
4. Maximum attainable heat fluxes on the surface
5. Pressure drop and mass flow rate
6. Stability of operation
7. Operating temperature versus absolute pressure range.

Obviously, heat sinks with low thermal resistance, and those of sufficient length and capability to withstand high heat fluxes, would be more practical for various technical applications. For this reason, it would be useful to enhance the inner surfaces of the miniature heat sinks. However, the mechanisms of heat transfer and those leading to dryout phenomena in two-phase miniature channels with enhanced surfaces are not completely understood, due to their complexity, especially for the case of discrete, in-line sources addressed by Maddox and Mudawar (1989) and Gersey and Mudawar (1993). While axially grooved evaporators have been used in heat pipes (Plesch et al. 1991; Faghri 2016), Khrustalev and Faghri (1997) attempted to foster a better understanding of the performance of miniature two-phase heat sinks in a minichannel with forced convection.

The goal of the experimental investigation presented by Khrustalev and Faghri (1997) was to obtain and analyze the performance characteristics of a flat copper–water miniature two-phase heat sink with small axial grooves on the inner walls, as shown in Fig. 10.29. These capillary grooves

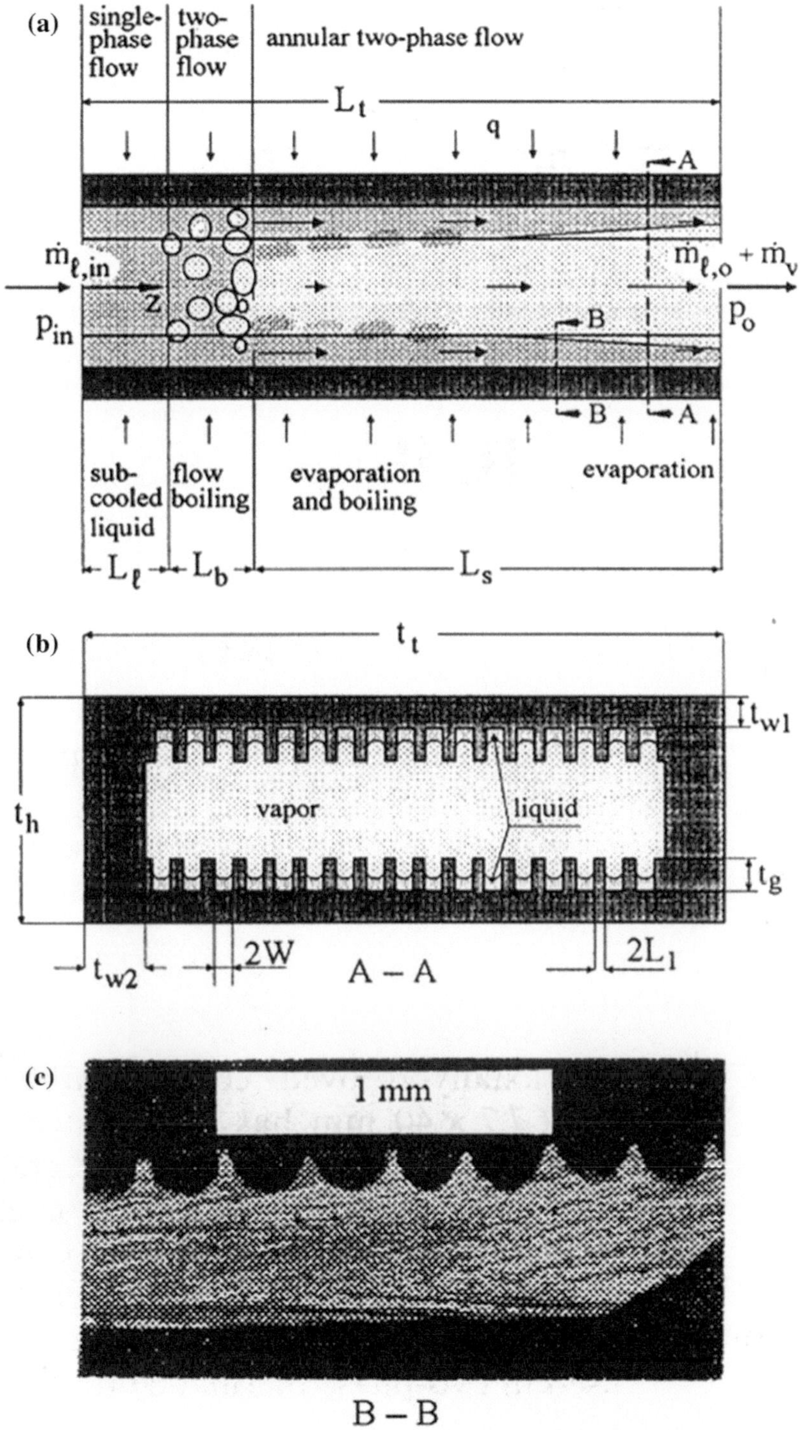

Fig. 10.29 Boiling heat transfer in the miniature axially grooved heat sink (Khrustalev and Faghri 1997)

provide significant heat transfer enhancement during evaporation and boiling of the liquid, and accelerate rewetting of the inner surface after occasional local dryout. Moreover, since the liquid can flow along the axial grooves under the influence of the capillary pressure and the pressure gradient along the flow channel, the heated length can be larger for the considered configuration than it would be for miniature channels with smooth walls.

The heat pipe-type (i.e., pure evaporative) regime has been modeled by Khrustalev and Faghri (1996; see the preceding subsubsection) for moderate heat loads ($q'' \le 65\,\mathrm{W/cm^2}$). Because surface tension tends to pull the liquid into the grooves, the flow boiling length should be even smaller than that for the case with smooth walls, so that the annular regime transforms into separated cocurrent two-phase flow. Surface tension seems to be a stabilizing factor for liquid flow in a groove, and it predominates over the destabilizing dynamic vapor–liquid interaction when the Weber number is small. As mentioned above, the liquid flows in the grooves under the influence of the capillary pressure gradient and the vapor pressure gradient, which can be expressed by the Laplace–Young equation in differential form.

The test module shown in Fig. 10.30 was manufactured from a plate of oxygen-free copper using an electro-discharge machining process (EDM). The module consisted of a rectangular cross-sectional channel with an axially grooved inner surface on the side walls and six thick-film resistors soldered onto the outer surfaces (three resistors per side). The test section had the following dimensions: $t_t = 7.7\,\mathrm{mm}$, $t_h = 2.6\,\mathrm{mm}$, $L_t = 50\,\mathrm{mm}$, and $t_{w1} = t_{w2} = 0.5\,\mathrm{mm}$. There were 27 grooves on each of the two inner surfaces of the side with the shape and dimensions shown in the microphotograph of Fig. 10.29c. The half-width of each groove, W, was about 0.11 mm. The ends of the test module were inserted into copper tubes with crimped ends and soldered so that 44 mm of the section length was available for installation of the resistors. The surface area of a resistor soldered onto the section wall was $6 \times 12\ \mathrm{mm}^2$. The gap between the resistors was about 1 mm, so that the maximum effective heated length was about 40 mm. As a result of the manufacturing process, the test section had two longitudinal fins with cross sections of $1 \times 1\ \mathrm{mm}^2$ along the centerlines of the 2.6 mm wide lateral walls. Nine copper–constantan thermocouple beads were soldered directly to the test section 10.5 mm

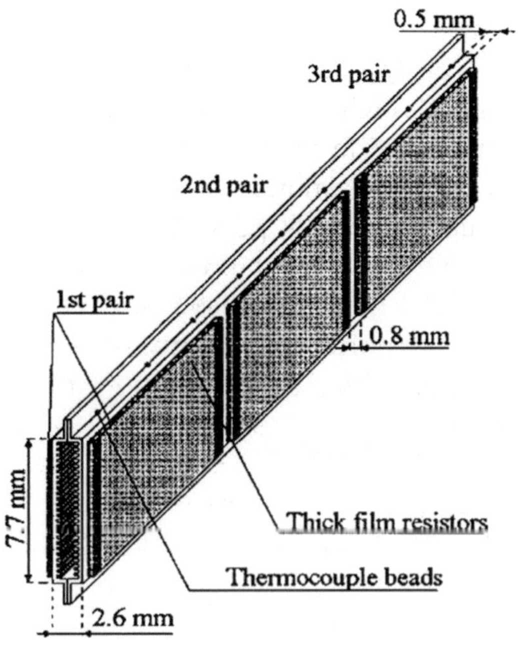

Fig. 10.30 Test section with six thick-film resistors (Khrustalev and Faghri 1997)

apart, in the corner between the section wall and the fin (Fig. 10.32). This allowed for the observation of local fluctuations in the wall temperature along the heat sink. The thermocouple beads were covered with a high-temperature, low thermal conductivity cement layer. The test section and the connecting copper tubes were thoroughly insulated to prevent heat losses into the environment. Heat losses from the heat sink were estimated to be less than 1% of the total heat load.

The heat flux on the wall of the heat sink was restricted by dryout, which led to a drastic increase in the heated wall temperature. Khrustalev and Faghri (1997) noted that the entrainment of the liquid in the grooves by the high-velocity vapor flow is one of the most important mechanisms for the comparatively long miniature channel. The entrainment of the liquid begins when the dynamic impact of the vapor flow is comparable with the stabilizing effect of the capillary pressure. The latter is usually described by the condition $We = 1$, where the Weber number is defined in Eq. (10.266). Therefore, for uniform heat flux, and neglecting the effect of subcooling, it follows from Eq. (10.266) and the condition $We = 1$ that

$$q''_{\mathrm{crit}} = \frac{(\pi \sigma W \rho_{\mathrm{v}}/2)^{1/2}}{L_{\mathrm{eff}} D_{\mathrm{h},\ell}} h_{\ell \mathrm{v}} [t_{\mathrm{h}} - 2(t_{\mathrm{w1}} + t_{\mathrm{g}})] \tag{10.282}$$

assuming that the grooves are filled with liquid prior to the beginning of entrainment. Table 10.5 compares the critical heat fluxes predicted by Eq. (10.282) and those obtained experimentally. For the short length of 13 mm, Eq. (10.282) failed to predict the experimental value of q''_{crit} because in that case, dryout occurred due to the traditionally recognized mechanism whereby the intense vapor bubble formation began on the heated surface before the entrainment could do so. For larger heated lengths, however, the prediction was accurate. The predicted critical heat flux increased with the saturation pressure. To better determine the critical mechanisms inducing dryout, more experimental data with different geometries of the grooved surfaces are needed.

For the effective heat transfer coefficients measured, the values were comparatively high (up to 100,000 W/m-K) and of the same order of magnitude as those predicted by the evaporative model by Khrustalev and Faghri (1996). The effective heat transfer coefficients did not increase with the heat flux for $q'' \leq 45$ W/cm^2. This means that with moderate heat flux, the impact of evaporation on heat transfer prevailed over the impact of boiling across the heated length. Actually, the heat sink worked in the mixed boiling–evaporation regime. With small mass flow rates (even with a vapor quality less than 0.3), instabilities and dryout occurred with heat fluxes less than 50 W/cm^2. This may have been because the grooves were too shallow to provide the necessary mass flow rate across the entire heated length. With larger mass flow rates, operation was stable due to higher pressure gradients in the channel. The critical heat flux decreased with the heated length. Since the total heat load was proportional to the heated length, it was assumed that dryout was induced mainly by the entrainment of the liquid in the grooves with the vapor flow. However, the mechanisms of dryout in this heat sink should be further investigated.

Table 10.5 Comparison of the experimental q''_{crit} with the prediction by Eq. (10.282)

Effective heated length (mm)	$p_{o,\mathrm{sat}}$(kPa)	q''_{crit}, W/cm^2 Experimental	q''_{crit} (W/cm^2) Equation (10.282)
40	101	84	87
26	101	108	134
13	101	130	269
13	20	–	138

10.6.3.4 Boiling in Micro/Miniature Channels

For two-phase flow in a microchannel, there is no stratified flow; therefore, the orientation of the channel has little effect on the flow pattern. The flow patterns for boiling in microchannels in order of appearance from the inlet are: bubbly flow, elongated bubbles, annular, mist, and flows with partial dryout. Since the diameter of the channel is so small, the bubbly flow regime is very short, as the bubbles grow to the size of the channel very quickly. Once the bubbles grow to the size of the channel, the flow pattern becomes an elongated bubble flow, followed by an annular flow regime. Cyclic dryout is also possible in the elongated bubble flow regime, which is also referred to as the confined bubble regime (Vlasie et al. 2004).

Thome et al. (2004) developed a three-zone flow boiling model for the evaporation of elongated bubbles in microchannels (Fig. 10.31). The vapor bubble is quickly nucleated and grows to the size of the channel to become an elongated bubble. At any given location in the microchannel that is passed by a liquid slug, an elongated bubble then follows, leaving liquid film behind. Evaporation of the thin film makes the elongated bubble grow longer and move forward. If the thin film dries out before the next liquid slug arrives, a vapor plug (dry zone) passes. This cycle repeats itself so any given point in the microchannel is passed by liquid slugs, elongated bubbles, and vapor slugs. Thome et al. (2004) and Dupont and Thome (2004) developed a three-zone heat transfer model based on a homogeneous flow assumption, i.e., the liquid and vapor were assumed to have the same velocity. The model illustrated the importance of strong cyclic variation in the heat transfer coefficient as well as the strong dependency of the heat transfer on bubble frequency, the minimum film thickness at dryout, and the liquid film's formation thickness. They concluded that the heat transfer coefficient in the elongated film region is several orders of magnitude larger than that in the liquid slug region, while heat transfer in the dry zone is nearly negligible. The time-averaged overall heat transfer coefficient strongly depends on the relative lengths of these three zones.

Vlasie et al. (2004) thoroughly reviewed the available empirical correlations for heat transfer in flow boiling in small diameter channels. Lazarek and Black (1982) studied flow boiling of R-113 in 123- and 246-mm-long circular vertical tubes with internal diameters of 3.1 mm. They recommended the following correlation:

$$Nu_{\mathrm{D}} = 30\, Re_{\mathrm{D}}^{0.857}\, Bo^{0.714} \tag{10.283}$$

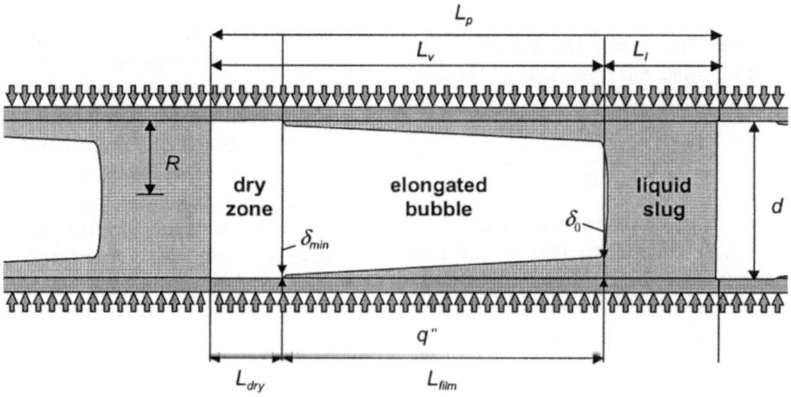

Fig. 10.31 Three-zone heat transfer model for elongated bubble flow in microchannels (Thome et al. 2004; Reprinted with permission from Elsevier)

where

$$Bo = \frac{q''}{h_{\ell v} G} \tag{10.284}$$

is the boiling number. Equation (10.283) is obtained by analyzing experimental data performed in the following ranges: $14\,\mathrm{kW/m^2} < q'' < 380\,\mathrm{kW/m^2}$, $125\,\mathrm{kg/m^2\text{-}s} < \dot{m}'' < 725\,\mathrm{kg/m^2\text{-}s}$, and $130\,\mathrm{kPa} < p < 140\,\mathrm{kPa}$. In arriving at Eq. (10.283), the effect of quality on heat transfer is neglected.

Kew and Cornwell (1997) performed experiments of boiling of R-141 in a 500-mm-long, small diameter tube ($D = 1.39$ to $3.69\,\mathrm{mm}$). While the results for 3.69- and 2.87-mm tubes follow trends similar to those observed in conventionally sized tubes, the results for the 1.39-mm tube depart from these trends significantly. To account for the effect of quality on boiling heat transfer, Kew and Cornwell (1997) proposed a modified Lazarek and Black correlation as follows:

$$Nu_D = 30\, Re_D^{0.857}\, Bo^{0.714} (1 - x)^{-0.143} \tag{10.285}$$

Despite the fact that there can be three different flow regimes—isolated bubble, confined bubble, and annular flow—in flow boiling in small diameter tubes, nucleate boiling correlations, such as Eqs. (10.283) and (10.285), often yield better results.

Tran et al. (1996) studied nucleate boiling of R-12 in a circular channel with diameter of 2.46 mm and rectangular channel of $1.70 \times 4.06\ \mathrm{mm}^2$ and proposed the following correlation

$$h = 8.4 \times 10^5 (Bo^2 We_\ell)^{0.3} (\rho_\ell/\rho_v)^{-0.4} \tag{10.286}$$

where the boiling number, Bo, is defined in Eq. (10.284) and the Weber number is

$$We_\ell = \frac{G^2 D}{\rho_\ell \sigma} \tag{10.287}$$

Equation (10.286) is obtained within the following range: $3.6\,\mathrm{kW/m^2} < q'' < 129\,\mathrm{kW/m^2}$, $44\,\mathrm{kg/m^2\text{-}s} < G < 832\,\mathrm{kg/m^2\text{-}s}$, and $510\,\mathrm{kPa} < p < 820\,\mathrm{kPa}$.

Yu et al. (2002) experimentally investigated flow boiling of water in a horizontal tube of 2.98-mm inner diameter at a system pressure of 200 kPa. The inlet temperature is 80 °C, and the mass flux ranges from 50 to 200 kg/m^2-s. They proposed the following correlations

$$h = 6.4 \times 10^6 (Bo^2 We_\ell)^{0.27} (\rho_\ell/\rho_v)^{-0.2} \tag{10.288}$$

To enhance boiling heat transfer on the conventional surface, one can use special surface geometry to promote nucleate boiling by reducing the boiling inception superheat, increasing heat transfer coefficient, and increasing the critical heat flux. The early studies focused on application of rough surface to promote nucleation. However, the effect of roughness will diminish due to aging effects. In order to overcome the aging effect, one can utilize the microchannels with re-entrant cavities as shown in Fig. 10.32 (Koşar et al. 2005). The heat flux versus average surface temperature at different mass flux is shown in Fig. 10.33. It can be seen that there is an abrupt change in slope near 100 °C, where transition from single-phase flow to boiling flow occurs. As the effective heat flux increases to a certain value, the surface temperature significantly increases, which indicates a critical heat flux condition.

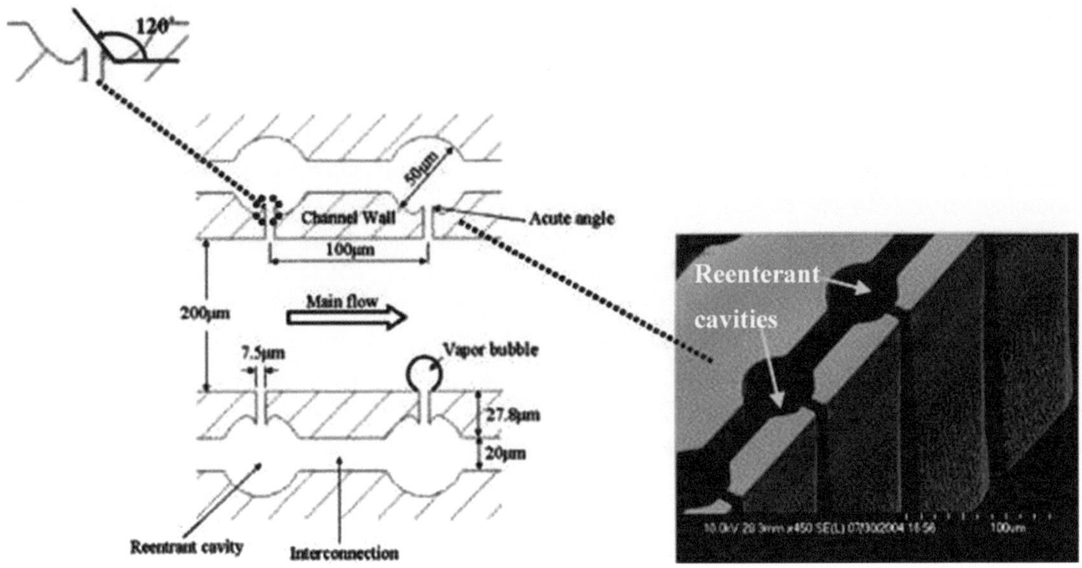

Fig. 10.32 Microchannels with re-entrant cavities (Koşar et al. 2005; Reprinted with permission from Elsevier)

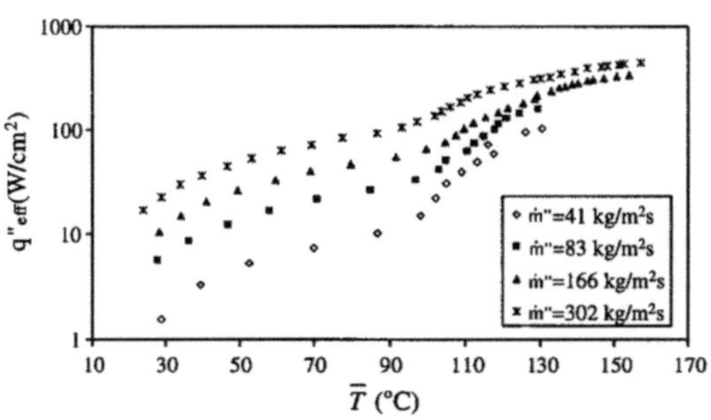

Fig. 10.33 Effective heat flux versus average surface temperature (Koşar et al. 2005; Reprinted with permission from Elsevier)

For nucleate boiling in five parallel rectangular channels with a hydraulic diameter of $D_h = 0.233\,mm$, Koşar et al. (2005) proposed the following correlation

$$h = 1.068 q''^{0.64} \tag{10.289}$$

For convective boiling heat transfer in the same geometric configuration, they suggested:

$$h = 4.068 \times 10^4\, Re_{\ell o}^{0.857} (1 - x_e)^{0.8} \left(\frac{1 - x_e}{x_e}\right)^{0.02} \tag{10.290}$$

where x_e is the quality at exit.

Qu and Mudawar (2004) studied the CHF phenomena in microchannels and found that when there is a significant compressible volume upstream of the heating section, an oscillating flow may lead to CHF. Another phenomenon observed by Qu and Mudawar (2004) that causes CHF is excursive

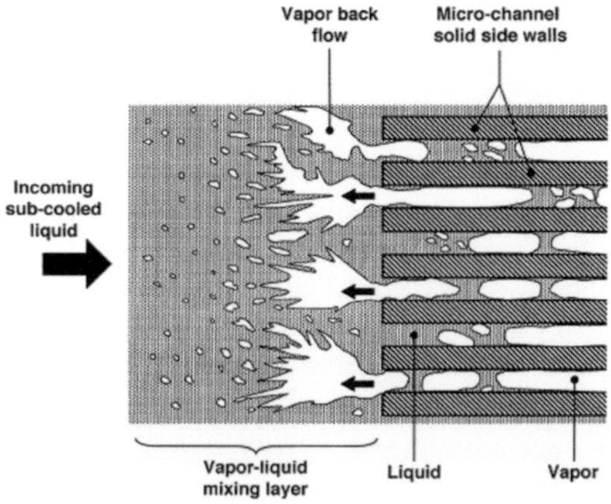

Fig. 10.34 Vapor back-flow near CHF (Qu and Mudawar 2004; Reprinted with permission from Elsevier)

instability caused by back-flow of the vapor, as shown in Fig. 10.34. The vapor flows backward from the individual microchannels into the upstream shallow plenum and forms a thick intermittent vapor layer. This layer breaks up into many small vapor bubbles, which propagate further upstream even deeper in the plenum. There, they mix with the incoming liquid. Bergles and Kandlikar (2005) suggested that all CHF data for parallel microchannels are subject to the parallel channel/excursive instability. The incoming flow is simply absorbed by the compressible volume, and the stagnated flow in the headed channel may lead to CHF before the flow through the channel is resumed. Therefore, the upstream compressible volume leads to an oscillating flow in the headed channel, which causes the CHF.

Qu and Mudawar (2004) correlated 42 data points from their experimental results for water in microchannels with $D_h = 380\,\mu m$, as well as data for R-113 in circular tube with $D = 510\ \mu m$ and 2.54 mm from their previous study (Bowers and Mudawar 1994). They recommended the following empirical correlation for critical heat flux in microchannels:

$$\frac{q''_{max}}{Gh_{\ell v}} = 33.43\left(\frac{\rho_v}{\rho_\ell}\right)^{1.11} We^{-0.21}\left(\frac{L}{D_h}\right)^{-0.36} \tag{10.291}$$

where

$$We = \frac{GL}{\sigma\rho_\ell} \tag{10.292}$$

is the Weber number based on the channel heated length L.

It should be pointed out that the CHF obtained from Eq. (10.291) represents a hydrodynamic CHF, which is caused by excursive instability. It is different from the CHF obtained from a stabilized single channel, in which case the CHF is caused by the usual dryout mechanism.

As noted above, boiling heat transfer in micro- and minichannels has attracted significant interest in recent years. However, the complex natures of convective flow boiling and two-phase flow in micro- and minichannels are still not well understood. Additional semi-empirical correlations that are physically based and cover a wide range of conditions in terms of size, fluid properties, and flow conditions are needed.

Problems

10.1. Express slip ratio in terms of the total cross-sectional area, total volume flow rate, void fraction, and vapor velocity.

10.2. Redo Problem 10.1 using the mass flux instead of the total volume flow rate.

10.3. Express volumetric flow fraction in terms of the total cross-sectional area, total volume flow rate, and gas superficial velocity.

10.4. Consider liquid–vapor two-phase flow. Find the expression for mass flux as a function of the quality, the individual phase velocities, and the densities.

10.5. Express total volumetric flux in terms of the individual mass flow rates, the total cross-sectional area, and the phase densities.

10.6. Determine the flow regime for upward flow of 10 kg/s of R-11 in a vertical 0.1 m diameter tube at 10 bar and 25% quality.

10.7. For a horizontal miniature tube of 1 mm inner diameter, determine the flow regime for 1.5×10^{-3} kg/s of water–steam at 1 atm and 10% quality.

10.8. For a horizontal circular tube with inner diameter of 5 cm ($D = 0.05$ m), determine the flow regime for 1.5×10^{-3} kg/s of steam-water flow at 1 atm and 59% quality.

10.9. The flow of 0.1 kg/s of air and 5 kg/s of water at 20 °C and 1 atm at a given point in a horizontal channel is in the bubbly regime. Describe qualitatively how the two-phase flow should change to achieve stratified flow.

10.10. Determine the pressure drop due to friction for a 1 kg/s flow of steam–water in a 10 cm inner diameter tube at 2 atm and 40% quality.

10.11. Calculate the void fraction using the homogeneous model for a two-phase water–steam flow with a quality of 35% at 15 bars.

10.12. Specify the governing equations for *adiabatic* two-phase flow in a circular tube of constant cross-sectional area using (1) the homogeneous model and (2) the separated flow model.

10.13. A capillary tube-suction line heat exchanger consists of a capillary tube and a suction line welded together as shown in Fig. P10.13. While evaporation takes place in the capillary tube, the suction line flow is single phase. Specify the governing equations for fluid flow and heat transfer in the system using the homogeneous model.

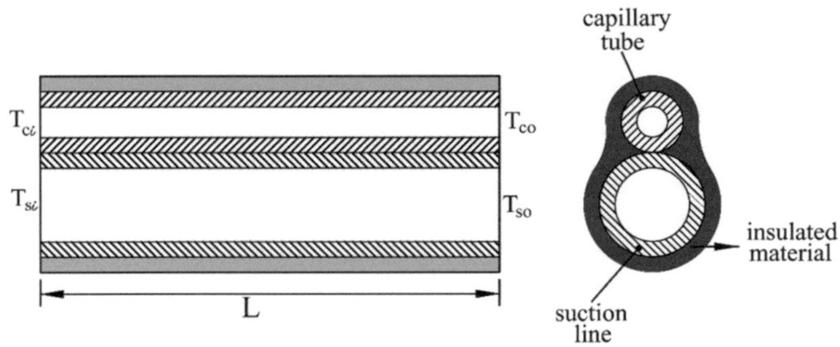

Fig. P10.13

10.14. A mixture of steam–water flows in a 5-cm inner diameter vertical tube at 10 kg/s, 5 atm, and 10% quality. What is the void fraction?

10.15. A mixture of steam–water flows in a 5-cm inner diameter vertical tube at 1 atm. The superficial velocities of the liquid and vapor are 1 and 100 m/s, respectively. Determine the void fraction using Eq. (10.105).

10.16. Steam condenses at 80 °C as it flows inside a horizontal tube with an inner diameter of 5 cm. The mass flow rate is 0.5 kg/s, and the quality is $x = 0.75$. Find the heat transfer coefficient.

10.17. Calculate the flow boiling two-phase heat transfer coefficient for a 400 kg/sm^2 of R22 at a saturation temperature of −20 °C and 20% quality in an upward vertical tube with an inner diameter of 1 cm and a wall temperature of 0 °C.

10.18. Saturated liquid water at 1 atm enters a vertical tube evaporator (circular cross section) with a constant heat flux of q'' (see Fig. P10.18). The tube inner diameter is D, and the thickness of the tube can be neglected (i.e., assume that the outer diameter of the tube is equal to the inner diameter).

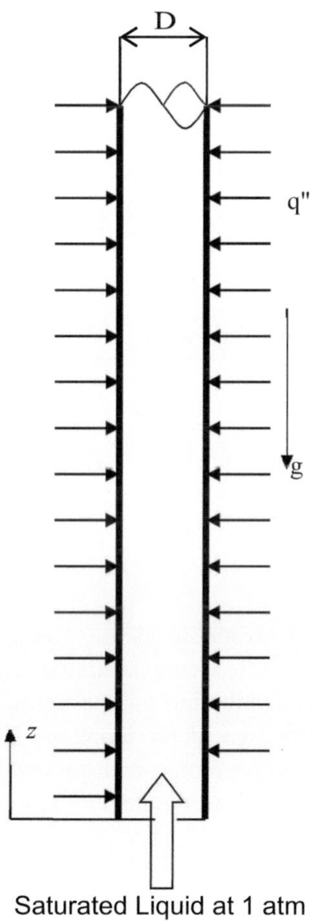

Fig. P10.18

a. Determine the quality of mixture as a function of distance from the inlet if the kinetic and potential energy can be neglected.
b. Assuming homogeneous flow model is valid, determine the void fraction as a function of distance from the inlet.
c. Develop an expression for (but not evaluate) the gravitational pressure drop across the tube for a given length, *L*.
d. If the mass flow rate is 0.05 kg/s, heat flux is 100 kW/m^2, and the inner diameter of the tube is 2 cm, determine the minimum length of the tube to completely evaporate the liquid.

10.19. A stainless steel heat exchanger (Fig. P10.19) is being used as a flow boiler. The evaporant flows through 10 tubes with a liquid counterflow. Find the heat transfer coefficient when the quality, *x*, is 0.5. The evaporant is ammonia at 200 K, and the evaporant passage effective diameter is 1 mm. The wall heat flux is 5000 W/m^2.

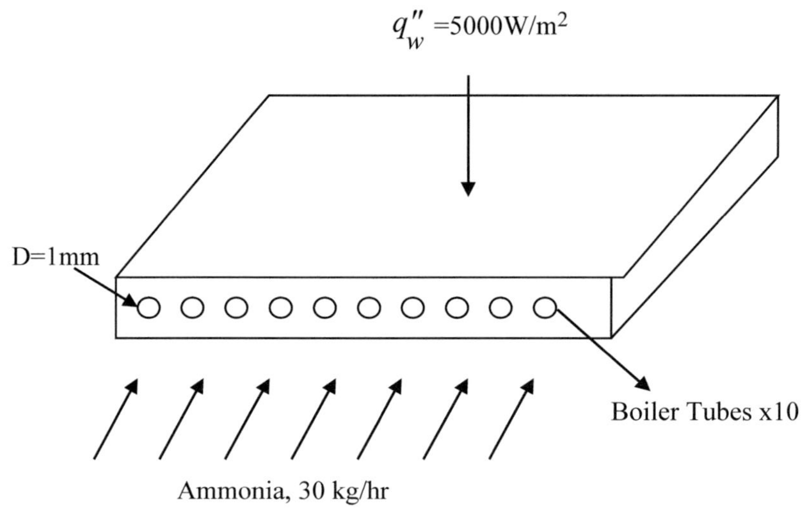

Fig. P10.19

10.20. You are given a flow boiler that is fed with saturated liquid ammonia at 200 K. 100% saturated vapor exits the boiler. The boiler has 547 passages, each of which is 1.25 mm high and 1.25 mm wide. The mass flow rate of the ammonia is 11.45 kg/hr (see Fig. P10.20). Determine the phase velocity at the beginning and the end of the flow boiler. What is the percentage increase in phase velocity from the beginning to the end of the boiler?

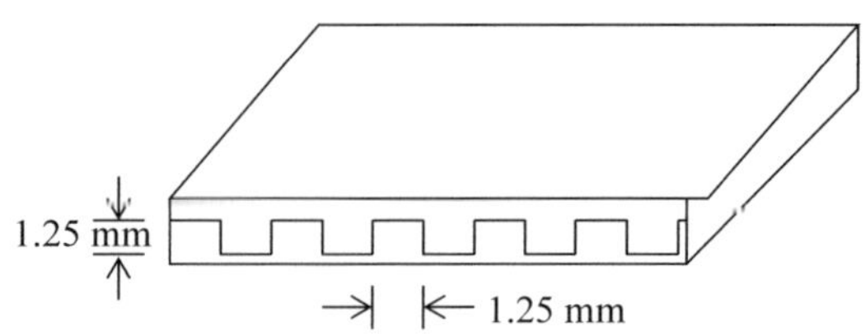

Fig. P10.20

10.21. In an electronic cooling system, a row of parallel tubes with diameters of D are embedded in a horizontal plate with thickness W (Fig. P10.21). A mixture of liquid and vapor with void fraction of α flows in the tube and the flow pattern is stratified. The upper surface of the plate is heated at constant heat flux while the bottom of the plate is insulated. The temperature of both liquid and vapor inside the tube is T_{sat}, but the heat transfer coefficients for liquid and vapor are different. Since the structure is symmetric, it is sufficient to consider only one cell with a width of S. Specify the governing equations and corresponding boundary conditions for heat conduction in the plate. You can use either the Cartesian or the cylindrical coordinate system.

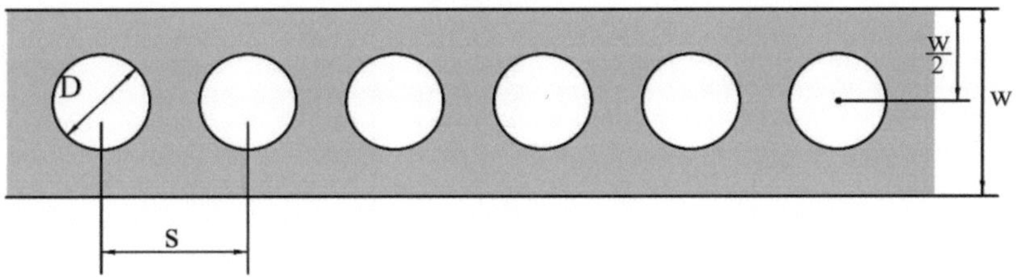

Fig. P10.21

10.22. The spacing between the tubes in Problem 10.21 is a critical parameter for the performance of the electronic cooling system. While small spacing allows more tubes per unit width, it also increases the conductance resistance from the top of the plate to the bottom of the plate. An optimal spacing will minimize the maximum temperature in the system. Write a computer program to solve the above conduction problem for the following parameters: D = 1 mm, W = 2 mm, $\alpha = 0.5$, $q'' = 10^6\ \mathrm{W/m^2}$, $h_\ell = 10^4\ \mathrm{W/m^2K}$, $h_v = 100\ \mathrm{W/m^2\text{-}K}$. Perform a parametric study to find the optimal spacing for the system in Problem 10.21.

10.23. A pulsating heat pipe (PHP) is a long miniature tube bent into many turns with the evaporator and condenser sections located at these turns (see Sect. 1.5.4). Since the diameter of the PHP is very small (less than 5 mm), the working fluid charged in the PHP forms vapor plugs and liquid slugs due to capillary action. Specify the governing equation to describe the oscillatory motion of an arbitrary liquid slug with a length of L in the PHP. The effects of surface tension and friction must be taken into account.

10.24. Evaporation and condensation on the thin film left behind by the liquid slug are the mechanisms that sustain the pressure difference between the two ends of the liquid slug. If the vapor phase satisfies the ideal gas law, obtain the energy equation for the vapor plug. Note that the mass of the vapor plug is not constant due to evaporation and condensation.

10.25. For a miniature passage with a very large length to width ratio, a fully developed flow in both vapor and liquid is very typical. A cross section of a typical miniature passage with axial rectangular grooves is shown in Fig. P10.25a. The liquid flows in the capillary microgrooves and the vapor flows in the central part of the passage free of liquid. Fluid flow in the grooved passage can be analyzed by considering a computational domain as shown in Fig. P10.25b. The vapor and liquid flow along the z-coordinate (concurrently or countercurrently). It can be assumed that (1) the fluid flow is laminar with fully developed velocity profiles, (2) the curvature of the liquid–vapor interface is constant, and (3) symmetry conditions are applicable at the three exterior boundaries of the vapor domain. Develop a steady-state mathematical model for the liquid and vapor flows that includes coupled vapor and liquid flows with shear stresses at the liquid free surface due to the vapor–liquid frictional interaction.

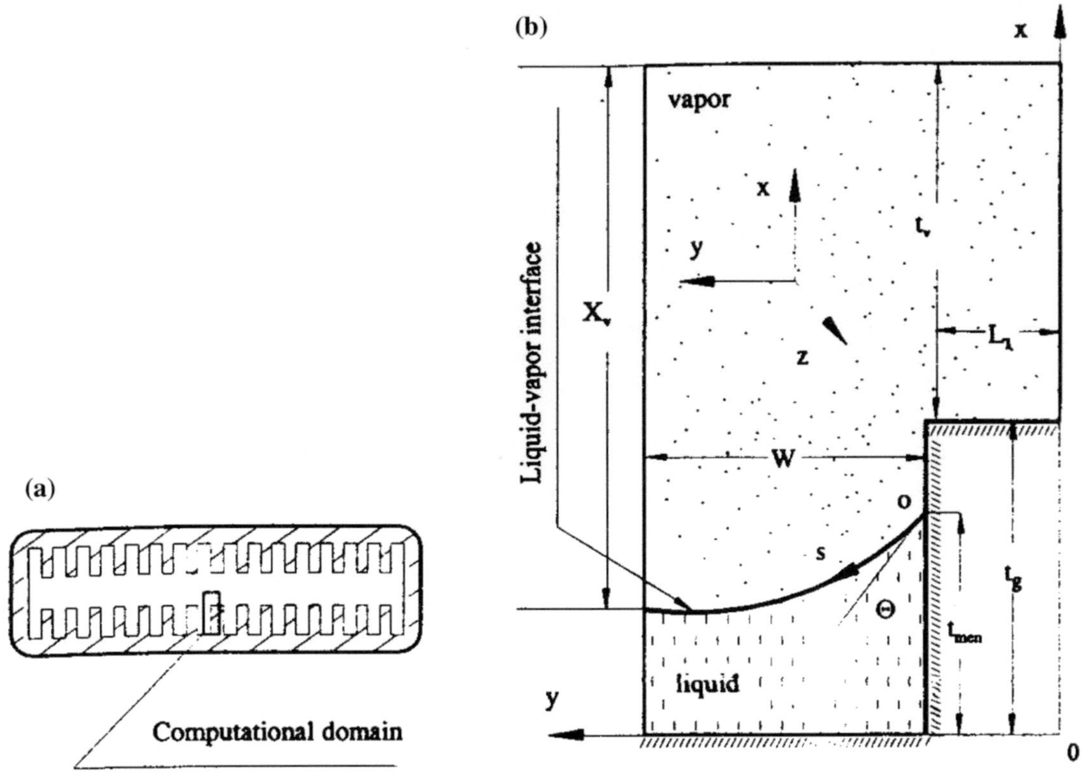

Fig. P10.25

References

Awad, M. M., & Muzychka, Y. S. (2005a). Bounds on two-phase flow: Part I fractional pressure gradient in circular tubes. In *Proceedings of International Mechanical Engineering Congress and Exposition*, Orlando, FL (DVD).

Awad, M. M., & Muzychka, Y. S. (2005b). Bounds on two-phase flow: Part II void fraction in circular tubes. In *Proceedings of International Mechanical Engineering Congress and Exposition*, Orlando, FL (DVD).

Bandel, J. (1973). *Druckverlust ünd Wärmeübergang bei der Verdampfung siedender Kältemittel im durchströmten waagerechten Rohr*, Doctoral Dissertation, Universität Karlsruhe.

Baroczy, C. J. (1965). Correlation of liquid fraction in two-phase flow with applications to liquid metals. *Chemical Engineering Progress Symposium Series, 61,* 179–191.

Barnea, D., Luninski, Y., & Taitel, Y. (1983). Flow pattern in horizontal and vertical two phase flow in small diameter pipes. *Canadian Journal of Chemical Engineering, 61,* 617–620.

Beattie, D. R. H., & Whalley, P. B. (1982). A simple two-phase frictional pressure drop calculation method. *International Journal of Multiphase Flow, 8,* 83–87.

Begg, E., Khrustalev, D., & Faghri, A. (1999). Complete condensation of forced convection two-phase flow in a miniature tube. *Journal of Heat Transfer, 121,* 904–915.

Bergles, A. E., & Kandlikar, S. G. (2005). On the nature of critical heat flux in microchannels. *Journal of Heat Transfer, 127,* 102–107.

Bowers, M. B., & Mudawar, I. (1994). High flux boiling in low flow rate, low pressure drop mini-channel and microchannel heat sinks. *International Journal of Heat and Mass Transfer, 37,* 321–332.

Breber, G., Palen, J. W., & Taborek, J. (1980). Prediction of horizontal tubeside condensation of pure components using flow regime criteria. *Journal of Heat Transfer, 102,* 471–476.

Butterworth, D. (1975). A comparison of some void-fraction relationships for cocurrent gas-liquid flow. *International Journal of Multiphase Flow, 1*(6), 845–850.

Chang, J. S., & Watson, A. (1994). Electromagnetic hydrodynamics. *IEEE Transactions on Dielectrics and Electrical Insulation, 1,* 871–895.

Chen, J. C. (1963). A correlation for boiling heat transfer to saturated fluids in convective flow. In *ASME preprint 63-HT-34, Presented at 6th National Heat Transfer Conference,* Boston, MA.

Chen, Y., Groll, M., Mertz, R., & Kulenovic, R. (2003). Force analysis for isolated bubbles growing from smooth and evaporator tubes. *Transaction of the Institute of Fluid-Flow Machinery, 112,* 57–74.

Chisholm, D. (1967). A theoretical basis for the Lockhart-Martinelli correlation for two-phase flow. *International Journal of Heat and Mass Transfer, 10,* 1767–1778.

Chisholm, D. (1973a). Pressure gradients due to friction during the flow of evaporating two-phase mixtures in smooth tubes and channels. *International Journal of Heat and Mass Transfer, 16,* 347–348.

Chisholm, D. (1973b). Void fraction during two-phase flow. *Journal of Mechanical Engineering Science, 15,* 235–236.

Chung, P. M. Y., & Kawaji, M. (2004). The effect of channel diameter on adiabatic two-phase flow characteristics in microchannels. *International Journal of Multiphase Flow, 30,* 735–761.

Collier, J. G., & Thome, J. R. (1994). *Convective boiling and condensation* (3rd ed.). Oxford, UK: Oxford University Press.

Cotton, J., Robinson, A. J., Shoukri, M., & Chang, J. S. (2005). A two-phase flow pattern map for annular channels under a DC applied voltage and the application to electrohydrodynamic convective boiling analysis. *International Journal of Heat and Mass Transfer, 48,* 5563–5579.

Davis, E. J., & Anderson, G. H. (1966). The incipience of nucleate boiling in forced convection flow. *AIChE Journal, 12,* 774–780.

Dobson, M. K., & Chato, J. C. (1998). Condensation in smooth horizontal tubes. *Journal of Heat Transfer, 120,* 193–213.

Dupont, V., & Thome, J. R. (2004). Evaporation in microchannels: Influence of the channel diameter on heat transfer. *Microfluidics and Nanofluidics, 1,* 119–127.

Faghri, A. (2016). *Heat pipe science and technology* (2nd ed.). Columbia, MO: Global Digital Press.

Feng, Z. P., & Serizawa, A. (1999). Visualization of two-phase flow patterns in an ultra-small tube. In *Proceedings of the 18th Multiphase Flow Symposium of Japan* (pp. 33–36).

Forster, H. K., & Zuber, N. (1955). Dynamics of vapor bubbles and boiling heat transfer. *AIChE Journal, 1,* 531–535.

Friedel, L. (1979). Improved friction pressure drop correlations for horizontal and vertical two-phase pipe flow. In *European Two-Phase Flow Group Meeting*, Paper E2, Ispra, Italy.

Gersey, C. O., & Mudawar, I. (1993). Orientation effects on critical heat flux from discrete, in-line heat sources in a flow channel. *Journal of Heat Transfer, 115,* 973–985.

Golan, L. P., & Stenning, A. H. (1969). Two-phase vertical flow maps. *Proceedings of the Institute of Mechanical Engineers, 184*(3C), 110–116.

Hashizume, K. (1983). Flow pattern, void fraction and pressure drop of refrigerant two-phase flow in a horizontal pipe, Part I: Experimental data. *International Journal of Multiphase Flow, 9,* 399–410.

Hewitt, G. F. (1998). Multiphase fluid flow and pressure drop. In *Heat exchanger design handbook* (Vol. 2). New York, NY: Begell House.

Hewitt, G. F., & Roberts, D. N. (1969). Studies of two-phase flow patterns by simultaneous X-ray and flash photography. AERE-M 2159, HMSO.

Hsu, Y. Y. (1962). On the size range of active nucleation cavities on a heating surface. *Journal of Heat Transfer, 84,* 207–216.

Jaster, H., & Kosky, P. G. (1976). Condensation heat transfer in a mixed flow regime. *International Journal of Heat and Mass Transfer, 19,* 95–99.

Kandlikar, S. G. (1990). A general correlation for two-phase flow boiling heat transfer coefficient inside horizontal and vertical tubes. *Journal of Heat Transfer, 112,* 219–228.

Kandlikar, S. G. (1991). Development of a flow boiling map for subcooled and saturated flow boiling of different fluids in circular tubes. *Journal of Heat Transfer, 113,* 190–200.

Kattan, N., Thome, J. R., & Favrat, D. (1998a). Flow boiling in horizontal tubes: Part 1—Development of adiabatic two-phase flow pattern map. *Journal of Heat Transfer, 120,* 140–147.

Kattan, N., Thome, J. R., & Favrat, D. (1998b). Flow boiling in horizontal tubes: Part 2—New heat transfer data for five refrigerants. *Journal of Heat Transfer, 120,* 148–155.

Kattan, N., Thome, J. R., & Favrat, D. (1998c). Flow boiling in horizontal tubes: Part 3—Heat transfer model based on flow pattern. *Journal of Heat Transfer, 120,* 156–165.

Katto, Y., & Ohno, H. (1984). An improved version of the generalized correlation of critical heat flux for the forced convection boiling in uniformly heated vertical tubes. *International Journal of Heat and Mass Transfer, 27,* 1641–1648.

Kawaji, M., & Chung, P. M. Y. (2004). Adiabatic gas-liquid flow in microchannels. *Microscale Thermophysical Engineering, 8,* 239–257.

Kew, P. A., & Cornwell, K. (1997). Correlations for prediction of boiling heat transfer in small diameter channels. *Applied Thermal Engineering, 17,* 705–715.

Khrustalev, D., & Faghri, A. (1994). Thermal analysis of a micro heat pipe. *Journal of Heat Transfer, 116,* 189–198.

Khrustalev, D., & Faghri, A. (1996). High flux evaporative mini-channel heat sink with axial capillary grooves. *Journal of Enhanced Heat Transfer, 3,* 221–232.

Khrustalev, D., & Faghri, A. (1997). Boiling heat transfer in miniature axially-grooved rectangular channel with discrete heat sources. *Journal of Enhanced Heat Transfer, 4,* 163–174.

Klausner, J. F., Mei, R., Bernhard, D. M., & Zeng, L. Z. (1993). Vapor bubble departure in forced convection boiling. *International Journal of Heat and Mass Transfer, 36,* 651–662.

Kocamustafaogullari, G., & Ishii, M. (1983). Interfacial area and nucleation site density in boiling systems. *International Journal of Heat and Mass Transfer, 26,* 1377–1387.

Koşar, A., Kuo, C. J., & Peles, Y. (2005). Boiling heat transfer in rectangular microchannels with reentrant cavities. *International Journal of Heat and Mass Transfer, 48,* 4867–4886.

Lazarek, G. M., & Black, S. H. (1982). Evaporative heat transfer, pressure drop and critical heat flux in a small vertical tube with R-113. *International Journal of Heat and Mass Transfer, 25,* 945–960.

Liu, D., Lee, P. S., & Garimella, S. V. (2005). Prediction of the onset of nucleate boiling in microchannel flow. *International Journal of Heat and Mass Transfer, 48,* 5134–5149.

Lockhart, R. W., & Martinelli, R. C. (1949). Proposed correlation of data for isothermal two-phase, two-component flow in pipes. *Chemical Engineering Progress Symposium Series, 45,* 39–48.

Maddox, D. E., & Mudawar, I. (1989). Single- and two-phase convective heat transfer from smooth and enhanced microelectronic heat sources in a rectangular channel. *Journal of Heat Transfer, 111,* 1045–1052.

Mei, R., & Klausner, J. F. (1994). Shear lift force on spherical bubbles. *International Journal of Heat Fluid Flow, 15,* 62–65.

Moser, K. W., Webb, R. L., & Na, B. (1998). A new equivalent reynolds number model for condensation in smooth tubes. *Journal of Heat Transfer, 120,* 410–417.

Munoz-Cobol, J. L., Herranz, L., Sancho, J., Takachenko, I., & Verdu, G. (1996). Turbulent vapor condensation with noncondensable gases in vertical tubes. *International Journal of Heat and Mass Transfer, 39,* 3249–3260.

Plesch, D., Bier, W., Seidel, D., & Schubert, K. (1991). Miniature heat pipes for heat removal from microelectronic circuits. In *Micromechanical Sensors, Actuators, and Systems*, DSC-32 (pp. 303–313). New York: ASME.

Premoli, A., Francesco, D., & Prina, A. (1971). A dimensionless correlation for determining the density of two-phase mixtures. *Thermotecnica, 25,* 17–26.

Qu, W., & Mudawar, I. (2004). Measurement and correlation of critical heat flux in two-phase micro-channel heat sinks. *International Journal of Heat and Mass Transfer, 47,* 2045–2059.

Rabas, T. J., & Minard, P. G. (1987). Two types of flow instabilities occurring inside horizontal tubes with complete condensation. *Heat Transfer Engineering, 8,* 40–49.

Riehl, R. R., & Ochterbeck, J. M. (2002). Experimental investigation of the convective condensation heat transfer in microchannel flows. In *The 9th Brazilian Congress of Thermal Engineering and Sciences*, Caxambu, MG, Paper CIT02-0495.

Serizawa, A., & Feng, Z. P. (2001). Two-phase flow in micro-channels. In *Proceedings of the 4th International Conference on Multiphase Flow*, New Orleans, LA.

Serizawa, A., Feng, Z., & Kawara, Z. (2000). Two-phase flow in microchannels. *Experimental Thermal and Fluid Science, 26,* 703–714.

Shah, M. M. (1979). A general chart correlation for heat transfer during condensation inside pipes. *International Journal of Heat and Mass Transfer, 22,* 547–556.

Shah, R. K., & London, A. L. (1978). Laminar flow forced convection in ducts. In *Advances in Heat Transfer*, Suppl. 1. San Diego, CA: Academic Press.

Shin, J. S., & Kim, M. H. (2005). An experimental study of flow condensation heat transfer inside circular and rectangular mini-channels. *Heat Transfer Engineering, 26,* 36–44.

Situ, R., Hibiki, T., Ishii, M., & Mori, M. (2005). Bubble lift-off size in forced convective subcooled boiling flow. *International Journal of Heat and Mass Transfer, 48,* 5536–5548.

Soliman, H. M. (1974). *Analytical and experimental studies of flow patterns during condensation inside horizontal tubes*, Ph.D. Dissertation, Kansas State University, Manhattan, KS.

Tabatabai, A., & Faghri, A. (2001). A new two-phase flow map and transition boundary accounting for surface tension effects in horizontal miniature and micro tubes. *Journal of Heat Transfer, 123,* 958–968.

Taitel, Y., & Dukler, A. E. (1976). A model for predicting flow regime transitions in horizontal and near horizontal gas-liquid flow. *AIChE Journal, 22,* 47–55.

Tandon, T. N., Varma, H. K., & Gupta, C. P. (1982). A new flow regime map for condensation inside horizontal tubes. *Journal of Heat Transfer, 104,* 763–768.

Teng, H., Cheng, P., & Zhao, T. S. (1999). Instability of condensate film and capillary blocking in small-diameter-thermosyphon condensers. *International Journal of Heat and Mass Transfer, 42,* 3071–3083.

Thome, J. R. S. (1964). Prediction of pressure drop during forced circulation boiling of water. *International Journal of Heat and Mass Transfer, 7,* 709–724.

Thome, J. R. (2016). *The heat transfer engineering data book III.* Ulm: Wieland-Werke AG.

Thome, J. R., Dupont, V., & Jacobi, A. M. (2004). Heat transfer model for evaporation in microchannels. Part I: Presentation of model. *International Journal of Heat and Mass Transfer, 47,* 3375–3385.

Thome, J. R., & El Hajal, J. (2003). Two-phase flow pattern map for evaporation in horizontal tubes: Latest version. *Heat Transfer Engineering, 24,* 3–10.

Tran, T. N., Wambsganns, M. W., & France, D. M. (1996). Small circular-and rectangular-channel boiling with two refrigerants. *International Journal of Multiphase Flow, 22,* 485–498.

Triplett, K. A., Ghiaasiaan, S. M., Abdel-Khalik, S. I., & Sadowski, D. L. (1999). Gas-liquid two-phase flow in microchannels—Part I: Two-phase flow pattern. *International Journal of Multiphase Flow, 25,* 377–394.

Turner, J. M. (1966). *Annular two-phase flow*. Ph.D. Dissertation, Dartmouth College, Hanover, NH.

Vasiliev, L. L, Grakovich, L. P., & Khrustalev, D. K. (1984). Gravity-assisted heat pipes for solar energy collectors. In *Proceedings of the 5th International Heat Pipe Conference* (Vol. 3, pp. 19–24). Tsukuba, Japan, Pre-prints.

Vlasie, C., Macchi, H., Guilpart, J., & Agostini, B. (2004). Flow boiling in small diameter channels. *International Journal of Refrigeration, 27,* 191–201.

Wambsganss, M. W., Jendrzejczyk, J. A., & France, D. M. (1991). Two-phase flow patterns and transitions in a small, horizontal, rectangular channel. *International Journal of Multiphase Flow, 17,* 327–342.

Wilmarth, T., & Ishii, M. (1994). Two-phase flow regimes in narrow rectangular vertical and horizontal channels. *International Journal of Heat and Mass Transfer, 37,* 1749–1758.

Yu, W., France, D. M., Wambsgans, M. W., & Hull, J. R. (2002). Two-phase pressure drop, boiling heat transfer, and critical heat flux to water in a small-diameter horizontal tube. *International Journal of Multiphase Flow, 28,* 927–941.

Zeng, L. Z., Klausner, J. F., Bernhard, D. M., & Mei, R. (1993). A unified model for the prediction of bubble detachment diameters in boiling systems - II. Flow boiling. *International Journal of Heat and Mass Transfer, 36,* 2271–2279.

Zhang, Y., Faghri, A., & Shafii, M. B. (2001). Capillary blocking in forced convective condensation in horizontal miniature channels. *Journal of Heat Transfer, 123,* 501–511.

Zivi, S. M. (1964). Estimation of steady-state void fraction by means of the principle of minimum energy production. *Journal of Heat Transfer, 86,* 247–252.

Zuber, N. (1961). The dynamics of vapor bubbles in nonuniform temperature fields. *International Journal of Heat and Mass Transfer, 2,* 83–98.

11 Fluid-Particle Flow and Heat Transfer

11.1 Introduction

Dispersed flow is a multiphase flow where one phase is dispersed within another continuous phase. The dispersed phase can be droplets, bubbles, or solid particles, while the continuous phase can be either liquid or gas. Condensation and evaporation on the liquid droplet surfaces were discussed in Chaps. 7 and 8, while dispersed flows that involve gas or vapor bubbles were discussed in Chaps. 9 and 10. Both phases in the dispersed liquid-gas flow are fluids so that either phase can be dispersed or continuous, depending on the volumetric flow rates. In other words, the dispersed phase has a lower volumetric flow rate, while the continuous phase has a higher volumetric flow rate. On the other hand, the solid phase in a gas/liquid–solid two-phase flow is always a dispersed phase and the continuous phase can only be a gas or liquid. The foci of this chapter are the cases where solid particles are carried by gas or liquid as well as the interaction between the particles and the carrier fluids.

A gas–solid flow is a flow of discrete particles in a continuous gas phase, and it includes a granular flow, fluidized bed, and pneumatic transport. For the cases of granular flow where particles move down a chute or inclined wall, the dominant force for the particle movement is gravity and the forces between particles due to collisions among particles, or forces between the wall and particles due to collisions between them. The interaction between the particles and gas in this case is negligible. A fluidized bed is consisted of a vertical vessel containing solid particles and gas introduced from the bottom of the vessel to suspend the solid particles and implement various physical and/or chemical processes. Pneumatic transport or pneumatic conveying is an approach that uses tubes or ducts to transport a mixture of solid materials and gas (air). While a pneumatic tube can be used at a drive-through window of a bank to exchange articles between the customer and teller, flowing powdery materials such as pulverized coal and cement can also be conveyed via pneumatic transport.

Liquid–solid particle flow is referred to as slurry flow, and it involves the transport of solid particles carried by a liquid. For example, coal can be transported via slurry pipelines that use water as carrier liquid mixed with pulverized coal. In cases where the density of solid particles is higher than that of liquid, the particles at the bottom will slow down and form a sludge layer—a phenomenon referred to as sedimentation. The most common type of sedimentation is responsible for the formation of riverbanks caused by the natural dispersion of mud and sand from the water. When nanosized particles are dispersed, stable nanofluid can be formed.

Section 11.2 presents the size distribution of particles with both discrete and continuous distributions to be discussed. Section 11.3 discusses the interaction between particles in the absence of fluid. A computer simulation will also be presented for particle packing based on a raindrop algorithm

A. Faghri and Y. Zhang, *Fundamentals of Multiphase Heat Transfer and Flow*, https://doi.org/10.1007/978-3-030-22137-9_11

and dynamics modeling. Section 11.4 discusses fluid-particle interactions that include continuity equations, momentum, and heat transfer between particles and fluid. Section 11.5 discusses the gas-particle system that includes a fluidized bed, pneumatic transport, and coupled heat and mass transfer with chemical reaction. This chapter closes with discussions of liquid–solid flow, including liquid–solid transport in ducts, nanofluids, microencapsulated phase-change materials, and self-assembly of submicron particles suspended in water.

11.2 Size Distribution of Particles

11.2.1 Discrete Size Distributions

Size distribution of particles in a dispersed flow plays an important role in the fluid-solid two-phase flow. Different sizes of particles can be characterized by their diameters (for spherical particles) or equivalent diameter (for nonspherical particles). The equivalent diameter for nonspherical particle can be defined based on an equivalent sphere with the same volume or the same surface area:

$$d_{\mathrm{pv}} = \left(\frac{6V_{\mathrm{p}}}{\pi}\right)^{1/3} \tag{11.1}$$

$$d_{\mathrm{ps}} = \left(\frac{S}{\pi}\right)^{1/2} \tag{11.2}$$

The equivalent diameter based on an equivalent sphere with the same volume, d_{pv}, is often used to calculate drag force, while the equivalent diameter with the same surface area, d_{ps} is useful to calculate heat and mass transfer between the particle and fluid. Table 11.1 shows the ranges of some typical particle sizes for different applications (Guo 2002). The size distribution can be either monodisperse distribution where all particles are close to a single size, or, more commonly, polydisperse where particle sizes are distributed over a wide range (Crowe et al. 2012).

For a system with n_{t} particles with sizes ranging from $d_{\min}$ to $d_{\max}$, the range of size, $d_{\max} - d_{\min}$, can be divided into a number N of equal or unequal intervals and the size of the i-th interval is Δd_i. Thus, it follows that:

Table 11.1 Typical ranges of particle sizes (Guo 2002)

Substances	The range of particle sizes (μm)
Smoke from tobacco	0.01–1
Cooking fumes	0.03–1
Bacteria	0.3–50
Fog	2–80
Soot	1–100
Coal ash	1–200
Pulverized coal	3–600
Spray	6–6000
Raindrop	600–6000

$$\sum_{i=1}^{N} \Delta d_i = d_{\max} - d_{\min} \tag{11.3}$$

The particle size in each interval can be represented by diameter at midpoint of the interval. It is imperative that Δd_i is large enough to contain enough particles and small enough to obtain a meaningful distribution. The discrete number frequency in the ith interval for particle diameter d_i, $\tilde{f}_n(d_i)$, can be obtained by counting the number of particles in the i-th interval, n_i, and dividing by the total number of particles in the system, n_t:

$$\tilde{f}_n(d_i) = \frac{n_i}{n_t}. \tag{11.4}$$

The discrete number frequency of particle diameters in the entire range is shown in Fig. 11.1 which means that:

$$\sum_{i=1}^{N} \tilde{f}_n(d_i) = 1 \tag{11.5}$$

The frequency function $f_n(d_i)$ is defined as

$$f_n(d_i) = \frac{\tilde{f}_n(d_i)}{\Delta d_i} = \frac{n_i}{n_t \Delta d_i} \tag{11.6}$$

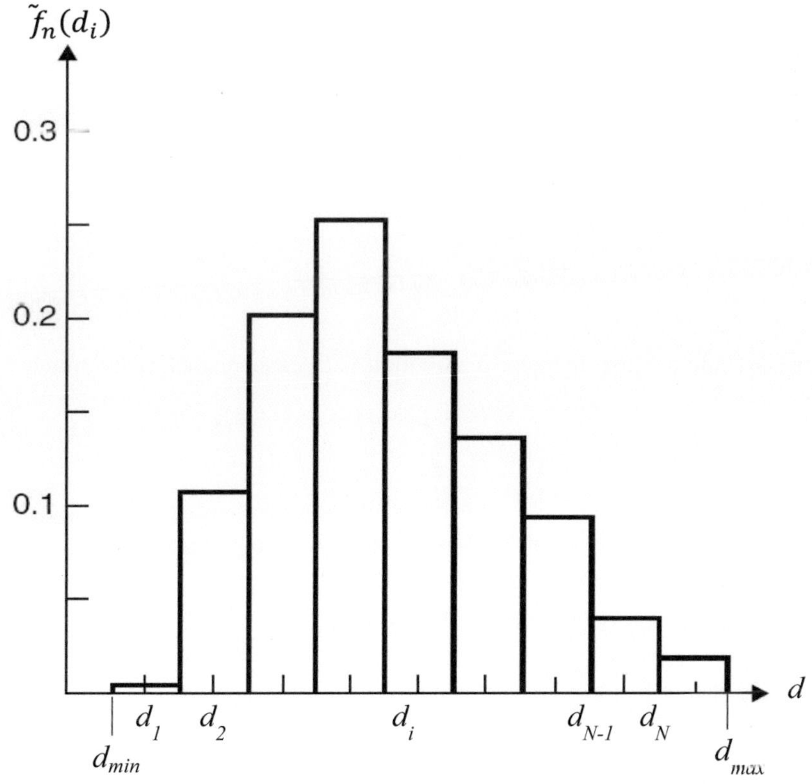

Fig. 11.1 Discrete number frequency of particle diameters

While the discrete number frequency $\tilde{f}_n(d_i)$ is dimensionless, the frequency function, $f_n(d_i)$, has a unit of m^{-1} in the SI unit system.

Substituting Eqs. (11.6)–(11.5) yields:

$$\sum_{i=1}^{N} f_n(d_i)\Delta d_i = 1 \tag{11.7}$$

The accumulative distribution that includes a number fraction of all particles ranging from d_{min} to d_i can be obtained by

$$F_n(d_i) = \sum_{j=1}^{i} f_n(d_j)\Delta d_j \tag{11.8}$$

The number-averaged particle diameter for the entire system is

$$\overline{d}_n = \sum_{i=1}^{N} d_i \tilde{f}_n(d_i) \tag{11.9}$$

The width of the distribution of size can be measured by the number variance:

$$\sigma_n^2 = \frac{N}{N-1}\sum_{i=1}^{N} (d_i - \overline{d}_n)^2 \tilde{f}_n(d_i) \tag{11.10}$$

If the number of intervals is very large, $N/(N-1) \approx 1$ and Eq. (11.10) becomes:

$$\sigma_n^2 = \sum_{i=1}^{N} (d_i - \overline{d}_n)^2 \tilde{f}_n(d_i) \tag{11.11}$$

or equivalently:

$$\sigma_n^2 = \sum_{i=1}^{N} d_i^2 \tilde{f}_n(d_i) - \overline{d}_n^2 \tag{11.12}$$

The standard deviation of the diameter distribution is the square root of the number variance:

$$\sigma_n = \sqrt{\sigma_n^2} = \left[\sum_{i=1}^{N} d_i^2 \tilde{f}_n(d_i) - \overline{d}_n^2\right]^{1/2} \tag{11.13}$$

A low number variance or standard deviation indicates that the particle diameters in the system are close to the number-averaged particle diameter, but a high number variance or standard deviation shows that the particle sizes are spread out over a wide range.

11.2.2 Continuous Size Distributions

When the diameter interval becomes very small, the discrete frequency function defined in Eq. (11.6) becomes:

$$f_n(d) = \lim_{\Delta d \to 0} \frac{\tilde{f}_n(d)}{\Delta d} \tag{11.14}$$

which is a continuous function (see Fig. 11.2). The continuous accumulative distribution that includes number fraction of all particles ranged from d_{min} to d can be obtained by

$$F_n(d) = \int_0^d f_n(\lambda) d\lambda \tag{11.15}$$

where the lower limit has been changed from d_{min} to 0 for simplification. The geometric mean diameter or number median diameter, μ_g, is defined as the particle diameter at which the accumulative distribution is 0.5, i.e.,

$$F_n(\mu_g) = 0.5 \tag{11.16}$$

It follows that

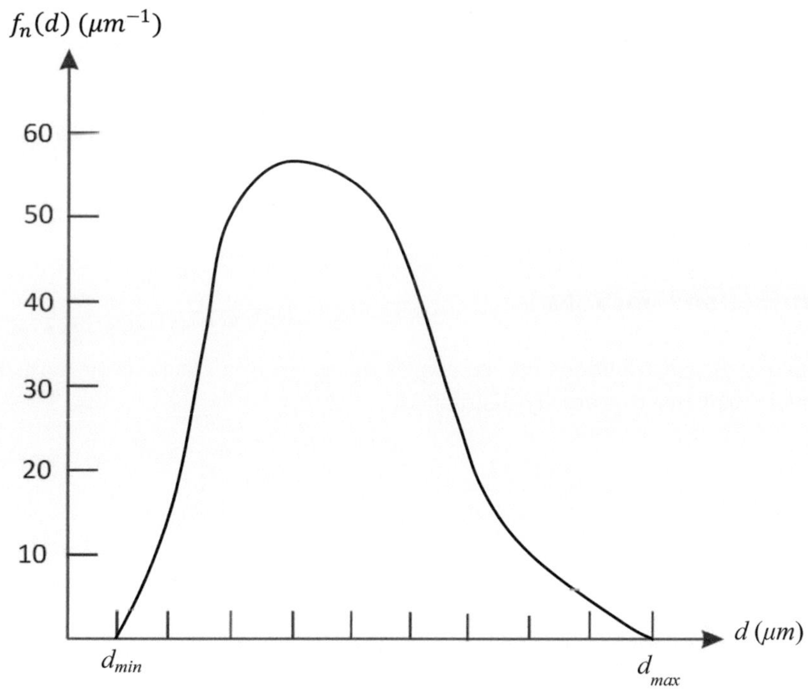

Fig. 11.2 Continuous number distribution of particle diameter

$$\int_{0}^{d_{\max}} f_{\mathrm{n}}(\lambda)\mathrm{d}\lambda = 1 \tag{11.17}$$

which is similar to Eq. (11.5) for discrete number frequency.

The mean number diameter of the particles for the entire system is

$$\mu_{\mathrm{n}} = \int_{0}^{d_{\max}} \lambda f_{\mathrm{n}}(\lambda)\mathrm{d}\lambda \tag{11.18}$$

The spread of the particle sizes can be described by the variance of the frequency distribution:

$$\sigma_{\mathrm{n}}^2 = \int_{0}^{d_{\max}} (\lambda - \mu_{\mathrm{n}})^2 f_{\mathrm{n}}(\lambda)\mathrm{d}\lambda \tag{11.19}$$

which can be equivalently expressed as

$$\sigma_{\mathrm{n}}^2 = \int_{0}^{d_{\max}} \lambda^2 f_{\mathrm{n}}(\lambda)\mathrm{d}\lambda - \mu_{\mathrm{n}}^2 \tag{11.20}$$

The standard deviation of the continuous number distribution is

$$\sigma_{\mathrm{n}} = \sqrt{\sigma_{\mathrm{n}}^2} = \left[\int_{0}^{d_{\max}} \lambda^2 f_{\mathrm{n}}(\lambda)\mathrm{d}\lambda - \mu_{\mathrm{n}}^2\right]^{1/2} \tag{11.21}$$

The distribution can be considered as monodisperse if $\sigma_{\mathrm{n}}/\mu_{\mathrm{n}}$ is less than 0.1.

11.2.3 Distribution Functions

The continuous number distributions can be approximated using a Gaussian or normal distribution where the particle diameter is normally distributed:

$$f_{\mathrm{n}}(d) = \frac{1}{\sqrt{2\pi}\sigma_{\mathrm{n}}} \exp\left[-\frac{1}{2}\left(\frac{d - \mu_{\mathrm{n}}}{\sigma_{\mathrm{n}}}\right)^2\right]\frac{1}{d} \tag{11.22}$$

or a lognormal distribution where the log of the particle diameter is normally distributed:

$$f_{\mathrm{n}}(d) = \frac{1}{\sqrt{2\pi}\sigma_0} \exp\left[-\frac{1}{2}\left(\frac{\ln d - \mu_0}{\sigma_0}\right)^2\right]\frac{1}{d} \tag{11.23}$$

where μ_0 and σ_0 are the mean and standard deviation of the *lognormal* distribution.

The cumulative distribution function for the lognormal distribution is

$$F_n(d) = \frac{1}{\sqrt{2\pi}\sigma_0} \int_0^d \exp\left[-\frac{1}{2}\left(\frac{\ln\lambda - \mu_0}{\sigma_0}\right)^2\right] \frac{d\lambda}{\lambda} \tag{11.24}$$

which can be integrated to yield

$$F_n(d) = \frac{1}{2}\left[1 + \operatorname{erf}\left(\frac{\ln d - \mu_0}{\sigma_0}\right)\right] \tag{11.25}$$

Substituting Eq. (11.25) into Eq. (11.16), one obtains

$$\mu_0 = \ln\mu_g \tag{11.26}$$

which indicates that the mean deviation of the lognormal distribution is a natural log of the geometric mean diameter or median number diameter. The standard deviation of the lognormal distribution σ_0 is related to the geometric standard deviation σ_g by:

$$\sigma_0 = \ln\sigma_g \tag{11.27}$$

Thus, the lognormal distribution, Eq. (11.23), and cumulative distribution, Eq. (11.25), can be rewritten as

$$f_n(d) = \frac{1}{\sqrt{2\pi}\sigma_0} \exp\left[-\frac{1}{2}\left(\frac{\ln(d/\mu_g)}{\ln\sigma_g}\right)^2\right] \frac{1}{d} \tag{11.28}$$

$$F_n(d) = \frac{1}{2}\left[1 + \operatorname{erf}\left(\frac{\ln(d/\mu_g)}{\ln\sigma_g}\right)\right] \tag{11.29}$$

The geometric mean diameter and geometric standard deviation are related to the number mean diameter and standard deviation by:

$$\mu_g = \frac{\mu_n^2}{\sqrt{\mu_n^2 + \sigma_n^2}} \tag{11.30}$$

$$\sigma_g = \exp\left\{\left[\ln\left(1 + \frac{\sigma_n^2}{\mu_n^2}\right)\right]^{1/2}\right\} \tag{11.31}$$

11.3 Interaction Between Dry Particles

11.3.1 Random Packing of Spherical Particles

Random packing of particles is very useful for many applications in physical, chemical, and engineering. For the case of spherical particles, the problem may be stated as follows: given spheres with radii distributed according to a prescribed probability density, and given that they are packed together by some rules to be specified, what is the nature of the resultant heap (Visscher and Bolsterli 1972)?

A common technique in the chemical industry for obtaining extended solid–fluid interfacial areas or good fluid mixing is to pass the fluid through a fixed bed of solid particles. This type of solid particle bed can be found in catalytic reactors and packed filters; the designs of these units are based upon the characteristics of those sphere packing systems. In addition, knowledge on the random packing of spheres will lead to further understanding of the heat transfer phenomena and fluid flow in powder-based additive manufacturing processes, such as the selective laser sintering (SLS).

Experimental investigations and numerical simulations on random spheres packing have been intensively carried out. Experimental work has identified some important variables, and it is well known that the mean packing density (fraction of space filled by the particles) depends strongly on the methods of packing, and the maximum packing density for truly random experimental packing appears to be 0.6366 (Scott and Kilgour 1969). However, the experimental results could not offer much insight into the details of the packing process, and thus, it is necessary to use numerical simulation to assist these studies. Computer simulation can be used to model structures of materials ranging from manufacturing raw materials to developing advanced products. The impact of particle properties on their packing structures is of prime importance to the entire packing process and is always essential for fabrication. A better understanding of packing is beneficial to optimize and to improve industrial applications. There are various particle packing algorithms; moreover, existing packing simulation techniques can be grouped into three types: (1) sequential addition, (2) nonsequential rearrangement, and (3) dynamic simulation.

An example of a simple algorithm via sequential addition is to simulate the successive packing of a container with spheres dropping in the gravitational field (Shi and Zhang 2008). A sphere that has made three contacts with other packed spheres is in its final position if it is stable. This method describes the actual process of packing generation in experiments without considering possible interactive forces among particles and vibration; thus, the method that allows stability based solely on gravity yields loose packing with a density of around 0.58 in the case of equal spheres. With systems of hard spheres of equal or different sizes, there have been many theoretical studies of random sphere packing. Most studies have been concerned with maximizing the packing density. Jodrey and Tory (1981) successfully simulated equal spheres packing with a density of 0.637, which is the maximum value of the experimental result. They then improved the algorithm and obtained a packing density between 0.642 and 0.649 (Jodrey and Tory 1985). By using the force-biased algorithm, Moscinski et al. (1989) obtained a maximum packing density of 0.67.

The discrete element method (DEM) gives us an opportunity to study the dynamics of a system by investigating the behavior of individual particles. There are two kinds of DEM: rigid body DEM, and nonrigid body DEM. Rigid body DEM assumes that collisions between particles occur instantly. Its advantage is that it can simulate a larger amount of distinct components, but the detail of the collision process cannot be captured. Conversely, nonrigid body DEM considers the duration of the collision and incorporates both elastic and cohesive forces during the contact of colliding bodies, thereby allowing us to model the collision realistically. However, due to the higher complexity of the nonrigid body approach, it takes more computational time. In any DEM simulation, the time step must be very small to avoid an unrealistic large contact force between the colliding bodies, which leads to longer CPU times.

Computer simulations of particle packing using the raindrop algorithm (Shi and Zhang 2008) and DEM simulation of fine cohesive particles (Jia et al. 2012; Tayeb et al. 2016a) will be discussed in the rest of this section

11.3.2 Simulation of Random Packing Using the Raindrop Algorithm

Computer simulation starts with randomly choosing a sphere according to a pregenerated continuous particle-size distribution and then dropping the sphere into a dimension-specified box. Its final position is obtained by using dropping and rolling rules that are derived from similar physical processes of spheres dropping into the gravitational field to minimize its gravity potential (Shi and Zhang 2008).

Initially, the sphere is given a coordinate of (x, y, z), where (x, y) is chosen randomly, and z is well above the top of the defined box (see Fig. 11.3). Each sphere moves along a linear trajectory (constant (x, y), decreasing z) until its surface contacts the floor of the box or another deposited sphere. If it contacts the floor, its final position will be determined. If not, the rolling rules are employed as it rolls down in a vertical plane on its contacting sphere until it makes contact with another sphere. At this point, the generated sphere has already made contact with two deposited spheres, but it is in an unstable position; thus, its gravitational potential can be reduced further. The generated particle will roll downward retaining its contact with the two spheres it connected with on the way down until it hits the third particle. In this case, a stable status will be examined to ensure that the generated sphere is in its stable position (with a possible minimum gravity potential). If the sphere is stable, the final position of the sphere is determined and the model will turn to generate a new sphere. Otherwise, it rolls on the double contact that continues its steep descent or it will connect with a sphere that is in the most unstable position. During the above-mentioned procedure, anytime the sphere contacts the floor, it stops. More details about the algorithm can be found in Shi and Zhang (2008) and Zhou et al. (2009). In addition, periodic boundaries are employed on the sides of the box to eliminate the wall effects. By using these boundaries, any sphere that leaves the box through one of its four surrounding vertical walls or corners will immediately reenter through the opposite wall or corner.

The packing structures with different particle-size distributions can be quantitatively described by the packing density and the coordination number. The packing density is defined as the volume of solids divided by the total volume of the arrangement (solids and voids). In other words, it is the total volume of spheres divided by the volume of box housing these spheres. The coordination number is the number of spheres in contact with any given sphere. The average coordination number, which is defined as the summation of the coordination number of each sphere divided by the number of

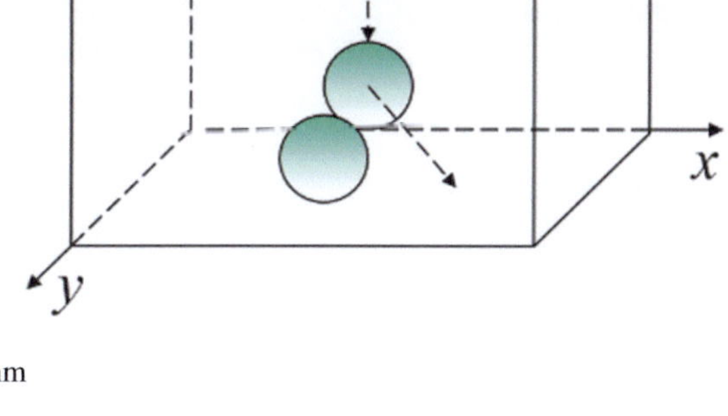

Fig. 11.3 Raindrop algorithm

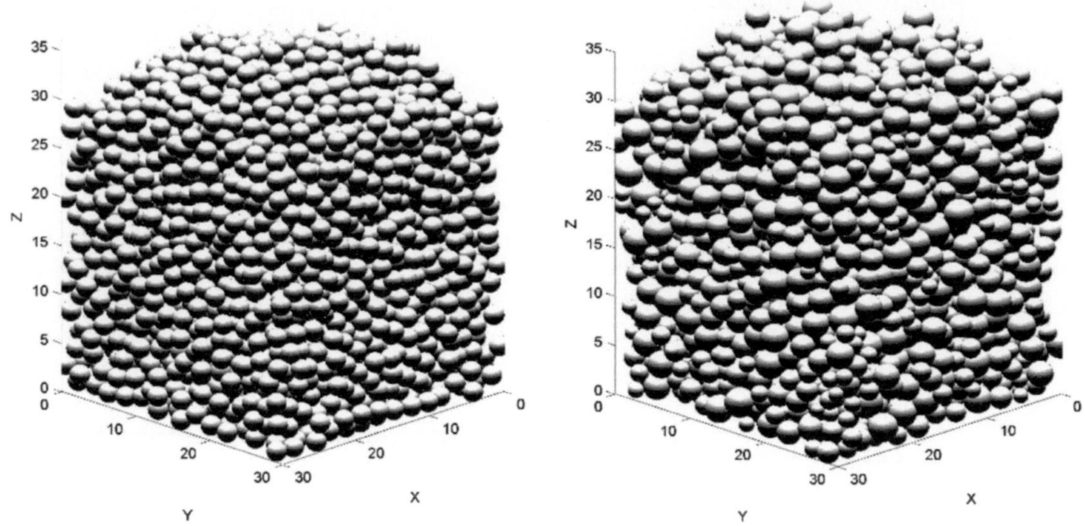

Fig. 11.4 Packing of spherical particles with **a** equal size distribution and **b** Gaussian distribution

spheres, is usually used to investigate the packing structures. Mechanical vibration can be used to increase the packing density in experiments, but the coordination number is hard to obtain since it is not easy to record the contacting points for each sphere in such experiments.

Shi and Zhang (2008) simulated packing particles with different size distributions and representative results are shown in Fig. 11.4. Although the equal size powder particles are impossible in a manufacturing process, it is still interesting to study the packing problem with monosized particles because such a study helps verify the present model. It can also be used to determine the dimensions of packing volume, which can eliminate the effect of packing volume on packing structure in the simulation. The radius of the particle is set as the unit of dimension; thus, all radii discussed here are dimensionless parameters. Five different size packing volumes were studied for monosized packing. This study revealed that the packing density and coordination number become stable once the packing volume reaches the value of $30 \times 30 \times 35$. The stable packing density is 0.578, and the coordination number is 5.97, which are close to the values of 0.58 and 6.0 reported by Troy et al. (1973). To verify the effects of random functions on the results, the program was run five times to obtain packing structures under the same volume of $30 \times 30 \times 35$. The deviations were 0.0007 and 0.01 for the packing density and the coordination number, respectively. This means that the randomicity of the program has little effect on the results as long as the packing volume is sufficiently large.

In reality, the system generally consists of different sized particles distributed over a certain range because of the limitations of the processes used to produce different powders. For a normal distribution, the particle-size distribution can be represented by Eq. (11.22). A serious problem with normal distribution is that it predicts a finite fraction of particles with sizes less than zero. Furthermore, the largest particle size does not have a limit and is infinite. Hence, it is necessary to determine a size range for the Gaussian distribution in simulation. The deviation of Gaussian distribution, σ_n, is obviously a key parameter that affects the packing structure; moreover, the limited range will have its influence on the structure. The results are given in Tables 11.2 with a mean radius of 1. The three-dimensional (3D) structure is illustrated in Fig. 11.4b for the Gaussian distribution case.

It can be seen from Table 11.2 that with a broader radius range of particles, higher packing density is obtained, but the average coordination number is decreased. The higher packing density is obtained

Table 11.2 Packing density and coordination number with normal distribution

Dimensionless radius range	Standard deviation, σ_n	1	2	3	4	5
0.34–3.0	Packing density	0.600	0.594	0.593	0.592	0.591
	Coordination number	5.71	5.66	5.62	5.57	5.64
0.50–2.0	Packing density	0.592	0.592	0.591	0.591	0.593
	Coordination number	5.80	5.83	5.84	5.83	5.81
0.67–1.5	Packing density	0.585	0.586	0.586	0.587	0.584
	Coordination number	5.89	5.88	5.89	5.89	5.89

because the smaller particles can fit into the gaps between the larger particles. On the other hand, even though the large particles have higher coordination numbers considering the many small particles around them, the larger particles also prevent adjacent small particles from contacting with more particles; thus, the coordination numbers of small particles are low, which results in a low average coordination number. For narrow radius ranges, the packing density slightly increases, while the coordination number is just slightly reduced compared to the results of equal size distribution, and there is little effect of deviation on packing structure. For applications in which a higher packing density is desirable, such as selective laser sintering (SLS), powder particles with Gaussian distribution should be limited to a narrow radius range.

11.3.3 Analysis of Cohesive Microsized Particle Packing Structure

Although particle packing with different size distributions has been studied using dropping and rolling rules as described in the preceding section, these simulations did not consider the interparticle force and were created solely from the geometric point of view. The effects of various forces, such as contact force, gravity, or the van der Waals force, were not considered in the simulation.

When performing dynamic modeling of particle collisions, the contact forces among the colliding particles need to be calculated first. This contact force is a result of the deformation of the colliding particles and can be decomposed into two components: normal viscoelastic force and tangential frictional force. The former can be calculated by the linear-spring model (Francesco and Alberto 2004) or a more accurate nonlinear Hertz model (Dintawa et al. 2008), and the latter by the Mindlin–Deresiewicz theory (Remond 2010). One important aspect when modeling the tangential frictional force between contacted particles is to clarify two distinct conditions: static friction and sliding friction. According to the Mindlin–Deresiewicz theory (Remond 2010), a maximum tangential displacement can be determined once the normal deformation is known. If the tangential displacement is less than or equal to the maximum tangential displacement, the friction is static friction; otherwise, the friction becomes sliding friction. The tangential displacement stops growing once it reaches its maximum value.

Micrometer-scale particle packing, ranging from 50 to 100 μm, was studied by Jia et al. (2012) using the soft-sphere DEM method. The advantage of the soft-sphere DEM method is that many-particle simultaneous collisions can be modeled. The difference between the models of millimeter-scale and micrometer-scale is that in the millimeter scale, van der Waals force is not considered, but in micrometer scale, the van der Waals force needs to be incorporated into the model. In the process of the collision, the coefficient of restitution depends on impact velocity and should not be a constant. Therefore, a damping coefficient is employed in the present simulation to consider the collision duration. The present model is based on a nonlinear-spring method that is derived from the Mindlin–Deresiewicz theory (Remond 2010), which is more accurate for modeling the tangential frictional force.

There are two kinds of motions for each particle: translation and rotation. The equations describing these motions for particle i are as follows:

$$m_i \frac{d^2\mathbf{r}_i}{dt^2} = \mathbf{F}_i \tag{11.32}$$

$$I_i \frac{d\boldsymbol{\omega}_i}{dt} = \mathbf{T}_i \tag{11.33}$$

where m_i is the mass of the particle, $\boldsymbol{r}_i$ is the position of the particle, ω_i is the angular velocity, I_i is the inertial moment that is equal to $0.4m_iR_i^2$, and R_i is the particle radius. $\boldsymbol{F}_i$ and $\boldsymbol{T}_i$ are, respectively, the resultant force and torque acting on the particle i:

$$\mathbf{F}_i = \sum_j (\mathbf{F}_{ij}^{\mathrm{n}} + \mathbf{F}_{ij}^{\mathrm{t}} + \mathbf{F}_{ij}^{\mathrm{v}}) + m_i\mathbf{g} \tag{11.34}$$

$$\mathbf{T}_i = \sum_j (\mathbf{T}_{ij}^{\mathrm{t}} + \mathbf{T}_{ij}^{\mathrm{r}}) \tag{11.35}$$

where $\boldsymbol{F}_{ij}^{\mathrm{n}}$ is the normal contact force, $\boldsymbol{F}_{ij}^{\mathrm{t}}$ is the tangential frictional force, $\boldsymbol{F}_{ij}^{\mathrm{v}}$ is the van der Waals force, $\boldsymbol{T}_{ij}^{\mathrm{t}}$ is the torque caused by the tangential contact force $\boldsymbol{F}_{ij}^{\mathrm{t}}$, and $\boldsymbol{T}_{ij}^{\mathrm{r}}$ is the torque caused by the rolling friction. The subscripts ij denote the force (torque) acting on particle i by particle j.

According to the nonlinear Hertz theory (Tibor and Fritz 2005), the normal contact force acting on particle i by particle j is calculated by:

$$\mathbf{F}_{ij}^{\mathrm{n}} = \left[\frac{2}{3}E\sqrt{\bar{R}}\xi_{\mathrm{n}}^{3/2} - \gamma E\sqrt{\bar{R}}\sqrt{\xi_{\mathrm{n}}}(\mathbf{v}_{ij}.\mathbf{n}_{ij})\right]\mathbf{n}_{ij} \tag{11.36}$$

where $E = Y/(1-\sigma^2)$ with Y denoting Young's modulus and σ is Poisson's ratio. $\bar{R}$ is an effective radius that equals $R_iR_j/(R_i + R_j)$ with $\mathbf{R_i}$ and $\mathbf{R}_i$ being the vectors running from the center of particle i and j to the contact point of the two particles, respectively. γ is the damping coefficient. $\mathbf{v}_{ij}$ is the velocity of particle i relative to the velocity of particle j, $\xi_{\mathrm{n}} = R_i + R_j - |R_{ij}|$ is the deformation between the two particles with R_{ij} being the distance between the spherical centers of particle i and j. Finally, $\mathbf{n}_{ij}$ is the unit vector whose direction is from the center of particle j to that of particle i.

Based on the Mindlin–Deresiewicz theory (Remond 2010), the tangential contact force $\boldsymbol{F}_{ij}^{\mathrm{t}}$ is given as:

$$\mathbf{F}_{ij}^{\mathrm{t}} = \mu\left|\mathbf{F}_{ij}^{\mathrm{n}}\right|\left[1 - \left(1 - \frac{|\boldsymbol{\xi}_{\mathbf{t}}|}{|\xi_{\max}|}\right)^{1.5}\right]\mathbf{t}_{ij} \tag{11.37}$$

where μ is sliding friction coefficient, $\left|\boldsymbol{F}_{ij}^{\mathrm{n}}\right|$ is the magnitude of the normal contact force, and ξ_{t} is the tangential displacement which is determined as:

$$\boldsymbol{\xi}_{\mathbf{t}} = \int_{t_0}^{\mathrm{t}} \mathbf{v}_{\mathbf{t}}dt \tag{11.38}$$

where t_0 is the time when the two particles barely touch without deformation, and $\mathbf{v}_t$ is the tangential relatively velocity:

$$\mathbf{v}_t = (\mathbf{v}_i - \mathbf{v}_j) \cdot \mathbf{t}_{ij} + (\boldsymbol{\omega}_i \times \mathbf{R}_i - \boldsymbol{\omega}_j \times \mathbf{R}_j) \tag{11.39}$$

where $\mathbf{t}_{ij}$ is the unit vector along the tangential direction, and $\boldsymbol{\omega}_i$ and $\boldsymbol{\omega}_j$ are angular velocities of particles i and j, respectively. If the tangential displacement is larger than the magnitude of the maximum tangential displacement $|\xi_{\max}|$, sliding friction occurs between the two particles. The maximum tangential displacement $\xi_{\max}$ is calculated as:

$$\boldsymbol{\xi}_{\mathbf{max}} = \mu \frac{2-\sigma}{2-2\sigma} \boldsymbol{\xi}_{\mathbf{n}} \tag{11.40}$$

The van der Waals force is expressed in the following form:

$$\mathbf{F}_{ij}^{\mathrm{v}} = -\frac{\mathrm{Ha}}{6} \times \frac{64 R_i^3 R_j^3 (h + R_i + R_j)}{(h^2 + 2R_i h + 2R_j h)^2 (h^2 + 2R_i h + 2R_j h + 4R_i R_j)^2} \mathbf{n_{ij}} \tag{11.41}$$

where Ha is the Hamaker constant that depends on particle material property, and h is the distance between the centers of particles i and j minus the sum of the radius of the two particles. It should be pointed out that a minimum value $h_{\min}$ for h is chosen in order to prevent numerical instability (e.g., $h_{\min}$ = 10 nm). In other words, if h is smaller than $h_{\min}$, h is forced to be equal to $h_{\min}$.

The torque acting on particle i due to the rolling friction between particle i and j is:

$$\mathbf{T}_{ij}^{\mathrm{r}} = -\mu_{\mathrm{r}} R_i \left| \mathbf{F}_{ij}^{\mathrm{n}} \right| \boldsymbol{\omega}_i \tag{11.42}$$

where μ_{r} is rolling friction coefficient. The torque $\boldsymbol{T}_{ij}^{\mathrm{t}}$ caused by the tangential contact force $\boldsymbol{F}_{ij}^{\mathrm{t}}$ is given by:

$$\mathbf{T}_{ij}^{\mathrm{t}} = \mathbf{R}_i \times \mathbf{F}_{ij}^{\mathrm{t}} \tag{11.43}$$

The Verlet method is employed due to its simplicity and effectiveness. The computer code is written in FORTRAN, and simulations are carried out on a PC.

Initially, there are 4500 (15 × 15 × 20) particles assigned in a rectangular space, i.e., 15 cell (layers) lengths in each of the x- and y-directions and 20 cell lengths in the z-direction. To make sure no overlap exists, the cell lengths are slightly larger than the largest particle diameter. The particles start to fall down due to gravity and collide with each other. Periodic boundary conditions are applied to the horizontal directions. The normal size distributions following Eq. (11.22) are considered, and 15 cases are simulated for three ranges of particle diameters: 50–60 μm (the mean diameter is 55 μm), 50–80 μm (the mean diameter is 65 μm), and 50–100 μm (the mean diameter is 75 μm) and five deviations: 1, 2, 3, 4, and 5 μm.

Table 11.3 shows the particle material properties including Young's modulus, Poisson's ratio, density, Hamaker constant, friction coefficients between the particles, and the damping coefficient that is used to calculate the normal contact force between colliding particles. In the DEM simulations, the time step must be small enough to avoid any unrealistically large deformation of the colliding particles. If the deformation is unrealistically large, the contact force acting on the colliding particles will also be unrealistically large. The contact force also depends on particle size. When dealing with Gaussian particles, the time step not only depends on the smallest particle but also depends on the size

Table 11.3 Parameters used in simulation of cohesive particle packing

Parameter	Value
Young's modulus, Y	10^7 N/m^2
Poisson's ratio, σ	0.3
Particle density, ρ	2500 kg/m^3
Hamaker constant, Ha	6.5×10^{-20} J
Sliding friction coefficient μ	0.3
Rolling friction coefficient, μ_r	0.002
Damping coefficient, γ	2.0×10^{-5} s

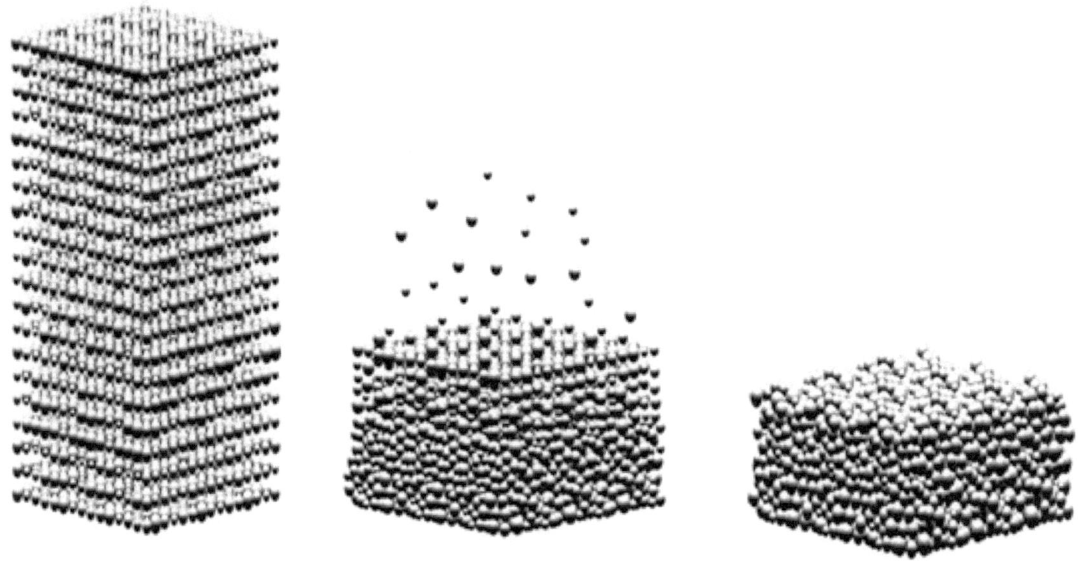

(a) Initial (0 s) (b) Middle (0.05 s) (c) Final (0.1 s)

Fig. 11.5 Gaussian particles packing process

difference of the colliding particles. The larger the size difference is, the smaller the time step should be. The time step used in the DEM simulation is 9×10^{-8} s. Figure 11.5 shows the Gaussian particles packing process for the particle diameters ranging from 50 to 100 μm with a particle diameter mean value of 75 μm and a diameter deviation of 5 μm.

Figure 11.6 shows the coordination number and porosity of Gaussian particles granular structures. It can be observed from Fig. 11.6 that as the particle mean diameter decreases, the coordination number decreases and the porosity increases. On the other hand, just the opposite occurs as the particle diameter deviation increases, i.e., the coordination number increases and the porosity decreases. The reason is that as the particle mean diameter decreases, the mean magnitude of the van

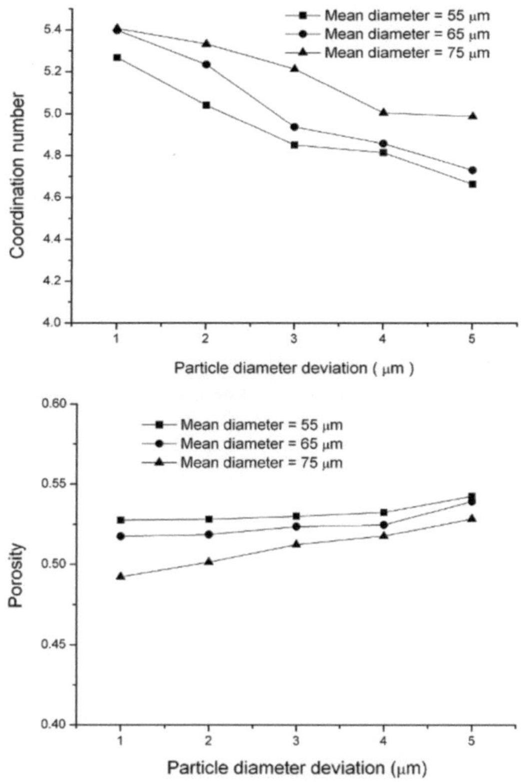

Fig. 11.6 Coordination number and porosity of Gaussian granular structures

der Waals force increases, making the packing structure looser. As the particle diameter deviation increases, the variety of particle size increases, making the packing structure looser.

The radial distribution function (RDF), $g(r)$, is a measure of how particle number density changes at a given distance from a reference particle:

$$g(r) = \frac{\mathrm{d}N(r)}{4\pi r^2 \mathrm{d}r\rho} \tag{11.44}$$

where $N(r)$ is the number of particles in the spherical space with a radius of r around the reference point, and ρ is the number of particles per volume.

Figure 11.7 shows the RDFs for the Gaussian particle packing cases (d = 50–100 μm). The units of the distance r are the radius of the smallest particles. Notably, the RDF peak value for the case with a particle diameter deviation equal to 1 μm is larger than RDF peak value in the other four cases. As the deviation increases, the radial distance where RDF reaches the peak value also increases. The physical meaning of this is that as the deviation increases, there are more contacts between small and large particles in the final granular packing structures.

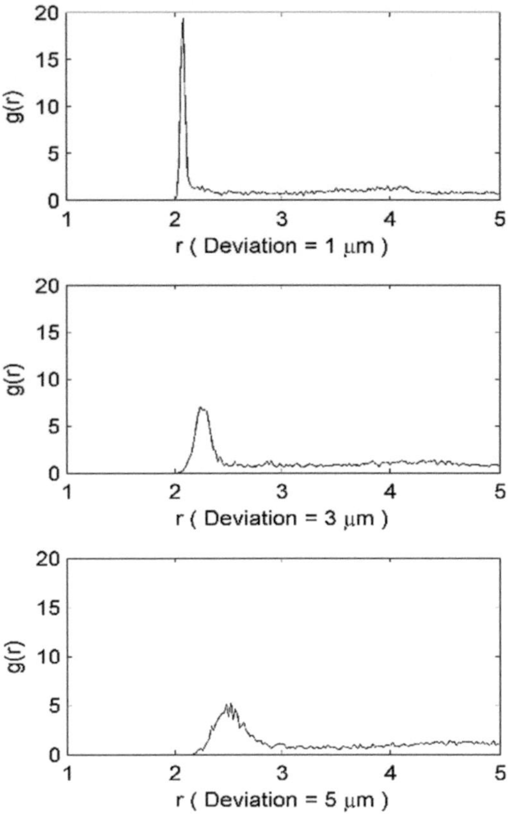

Fig. 11.7 Radial distribution function of Gaussian granular structures (d = 50–100 μm)

11.4 Fluid-Particle Interactions

In a fluid-particle system, the behavior of an individual solid particle can be influenced by the interaction between the fluid and solid particle, as well as the interaction with other solid particles. In the case of dispersed flow, the momentum response time (see Eq. 11.99) is longer than the mean free time (the average time between collisions), and the behavior of the particles is governed by the interaction between fluid and particles. On the other hand, when the momentum response time is shorter than the mean free time, the behavior of the particles is affected by the interaction between particles; thus, the fluid-particle flow becomes a dense flow.

There are two types of approaches that can be used to model fluid-particle flow: (1) the Euler–Lagrange approach, and (2) the Euler–Euler approach. Under the Euler–Lagrange approach, the continuous phase is described using the Euler approach, while the particles are described using the Lagrange approach. Under the Euler–Euler approach, the carrier fluid and particles are described using the multifluid model (see Sect. 3.4.2). The focus of this section is the interactions between fluid and particles under the Euler–Lagrange approach. It is assumed that the fluid phase is multicomponent while the solid phase is a single component.

11.4.1 Continuity Equation

The continuity equation in the multifluid model, Eq. (3.264), can be applied to the fluid phase:

$$\frac{\partial(\rho_f \varepsilon_f)}{\partial t} + \nabla \cdot (\rho_f \varepsilon_f \mathrm{V}) = (\dot{m}'''_{pf} - \dot{m}'''_{fp}) \tag{11.45}$$

where ρ_f and ε_f are the density and volume fraction of the continuous fluid phase, respectively. $\dot{m}'''_{pf}$ represents mass transfer per unit volume from the particle to the fluid phase, and $\dot{m}'''_{fp}$ represents the mass transfer per unit volume from the fluid phase to the particle phase. The difference between $\dot{m}'''_{pf}$ and $\dot{m}'''_{fp}$ represents the *net* mass transfer per unit volume from the particles to the fluid phase. The mass transfer between phases can be due to phase change (sublimation and deposition or melting and solidification), dissolution or chemical reaction.

The continuity equation for the i-th particle is

$$\frac{\mathrm{d}m_i}{\mathrm{d}t} = -\pi d_p^2 \rho_f w \tag{11.46}$$

where w is the outward radial velocity in the fluid phase and $\rho_f w$ represents mass flux over the particle surface from the solid to fluid phase due to phase change or chemical reaction.

11.4.2 Momentum Transfer

The momentum equation in the multifluid model, Eq. (3.272), can be applied to the fluid phase:

$$\begin{aligned}&\frac{\partial}{\partial t}(\varepsilon_f \rho_f \mathrm{V}) + \nabla \cdot (\varepsilon_f \rho_f \mathrm{VV}) \\ &\quad = -\nabla \cdot (\varepsilon_f p_{\rho gh}) - \mathrm{g} \cdot \mathrm{r}\nabla(\varepsilon_f \rho_f) + \nabla \cdot (\varepsilon_f \boldsymbol{\tau}) - \mathrm{F_d} + (\dot{m}'''_{pf}\mathbf{V}_\mathrm{I} - \dot{m}'''_{fp}\mathbf{v}_{p,\mathrm{I}})\end{aligned} \tag{11.47}$$

where ρ_f is the average fluid density, $\mathbf{V}$ is the average velocity of the fluid, $p_{\rho gh} = p - \rho_f \mathbf{g} \cdot \mathbf{r}$ is the dynamic pressure, $\mathbf{F}_d$ is the drag force, and $\mathbf{g}$ is the gravitational acceleration. The last term on the right-hand side of Eq. (11.47) accounts for the momentum transfer with particles; $\mathbf{V}_\mathrm{I}$ and $\mathbf{v}_{p,I}$ are the intrinsic phase-averaged velocity of the fluid and particle at the fluid-particle interface.

The motions of the solid particles can be both translation and rotation and can be described by the following momentum and moment of momentum equations (Tayeb et al. 2016b):

$$m_i \dot{\mathbf{v}}_{p,i} = \mathbf{F}^n_{ij} + \mathbf{F}^t_{ij} + \mathbf{F}_{coh} + m_i \mathbf{g} + \mathbf{F}_d + \mathbf{F}_b \tag{11.48}$$

$$I_i \dot{\boldsymbol{\omega}}_i = \mathbf{T}^r_{ij} + \mathbf{T}^t_{ij} \tag{11.49}$$

where m_i is the mass of the i-th particle, $\mathbf{v}_{p,i}$ is the velocity of the i-th particle, I_i is the moment of the particle's inertia, and $\boldsymbol{\omega}_i$ is the angular velocity of the i-th particle. $\mathbf{F}^n_{ij}$ and $\mathbf{F}^t_{ij}$ are the normal and tangential force due to contact, $\mathbf{F}_{coh}$ is the cohesive force between two particles, $m_i\mathbf{g}$ is a gravitational force, $\mathbf{F}_d$ is the drag force due to relative fluid motion, and $\mathbf{F}_b$ is the buoyant force. $\mathbf{T}^n_{ij}$ is the torque due to rolling friction, and $\mathbf{T}^t_{ij}$ is the torque due to tangential force.

The drag force accounts for the force acting on the particle due to the relative motion between particle and fluid. For spherical rigid particles, the drag coefficient can be obtained as

$$\mathbf{F}_{\mathrm{d}} = \frac{1}{8}\pi d_{\mathrm{p}}^2 \rho_{\mathrm{f}} C_{\mathrm{D}} \left(\mathbf{V} - \mathbf{v}_{\mathrm{p}}\right)\left|\mathbf{V} - \mathbf{v}_{\mathrm{p}}\right| \tag{11.50}$$

where C_{D} is the drag coefficient

$$C_{\mathrm{D}} = \begin{cases} \frac{24}{Re}\left(1 + \frac{1}{6} Re^{3/2}\right) & Re \leq 1000 \\ 0.424 & Re > 1000 \end{cases} \tag{11.51}$$

where the Reynolds number is

$$Re = \frac{\rho_{\mathrm{f}} d_{\mathrm{p}} \left|\mathbf{V} - \mathbf{v}_{\mathrm{p}}\right|}{\mu_{\mathrm{f}}} \tag{11.52}$$

The drag model that is suitable for particles in the near submicron range is (Ounis et al. 1991):

$$\mathrm{F}_{\mathrm{d}} = \frac{18\mu_{\mathrm{f}}}{d_{\mathrm{p}}^2 C_{\mathrm{D}}}\left(\mathbf{V} - \mathbf{v}_{\mathrm{p}}\right) \tag{11.53}$$

where C_{D} is the drag coefficient

$$C_{\mathrm{D}} = 1 + \frac{2\lambda}{d_{\mathrm{p}}}\left[1.257 + 0.4 \exp\left(-\frac{1.1 d_{\mathrm{p}}}{2\lambda}\right)\right] \tag{11.54}$$

where λ is the molecular mean free path of the fluid (for liquid water, it is 2.5×10^{-10} m).

For spherical particles, the buoyant force is

$$\mathbf{F}_{\mathrm{b}} = -\frac{1}{6}\rho_{\mathrm{f}} \pi d_{\mathrm{p}}^3 \mathbf{g} \tag{11.55}$$

A contact force model is implemented to account for the interaction between particles and between particle and wall. The model used is the Gran–Hertz–History model where a nonlinear relationship is assumed between overlap distance during the collision and normal contact force. Other forces that are included are viscous damping force and cohesive force. Tangential force during collision is also taken into account.

The normal contact force during collision $\mathbf{F}_{ij}^{\mathrm{n}}$ is given by (Dou et al. 2014):

$$\mathbf{F}_{ij}^{\mathbf{n}} = \left[Y_{\mathrm{n}}\delta_{\mathrm{n}} - c_{\mathrm{n}}\left(\mathbf{v}_{ij} \cdot \hat{\mathbf{n}}_{ij}\right)\right]\hat{\mathbf{n}}_{ij} \tag{11.56}$$

where

$$\begin{aligned} Y_{\mathrm{n}} &= 4E_{\mathrm{eff}}\sqrt{\bar{\mathrm{R}}\delta_{\mathrm{n}}}/3, \quad c_{\mathrm{n}} = 2e_{\mathrm{eff}}\sqrt{5\mathrm{B}_{\mathrm{n}} m_{\mathrm{eff}}/6} \\ e_{\mathrm{eff}} &= \ln(e)/\sqrt{\ln^2(e) + \pi^2}, \quad B_{\mathrm{n}} = 2E_{\mathrm{eff}}\sqrt{\bar{\mathrm{R}}\delta_{\mathrm{n}}} \end{aligned} \tag{11.57}$$

where $\hat{\mathbf{n}}_{ij}$ is the unit vector pointing from particle i to particle j, e is the coefficient of restitution of the particles, $\bar{R} = R_i R_j / \left(R_i + R_j\right)$ represents the geometric mean radius of the i-th and j-th particle, $\mathbf{v}_{ij}$ represents the velocity of the particle i relative to the velocity of the particle j, $E_{\mathrm{eff}} = \left[\left(1 - \sigma_1^2\right)/E_1 + \left(1 - \sigma_2^2\right)/E_2\right]^{-1}$ is the effective elastic modulus computed by each particle's elastic modulus and Poisson ratio, $\delta_n = R_i + R_j - |R_{ij}|$ is the displacement in the normal direction, and $m_{eff} = m_i m_j / (m_i + m_j)$ is the effective masses of the particles.

The contact force in the tangential direction is given as (Kloss et al. 2012):

$$\mathbf{F}_{ij}^{t} = -\min\left[\mu_s \left|\mathbf{F}_{ij}^{\mathbf{n}}\right|, Y_t(\delta_t \cdot \hat{\mathbf{t}}_{ij}) - c_t(\mathbf{v}_t \cdot \hat{\mathbf{t}}_{ij})\right]\hat{\mathbf{t}}_{ij} \tag{11.58}$$

where

$$\begin{aligned} &Y_t = 8S_{eff}\sqrt{\bar{R}\delta_n}, c_t = 2e_{eff}\sqrt{5B_t m_{eff}/6}, B_t = 8S_{eff}\sqrt{\bar{R}\delta_n}, \\ &S_{eff} = 1/[2(2-\sigma_1)(1+\sigma_1)/E_1 + 2(2-\sigma_2)(1+\sigma_2)/E_2] \end{aligned} \tag{11.59}$$

where δ_t is the tangential displacement vector given by $\delta_t = \int_{t_0}^{t} v_t dt$, $\hat{t}_{ij}$ is the unit vector along the tangential direction, t_0 is the time when the two particles barely contact without deformation, and t is the time of the collision. The tangential relative velocity is given by $v_t = [(v_i - v_j)\cdot\hat{t}_{ij}]\hat{t}_{ij} + (\omega_i \times R_i - \omega_j \times R_j)$, where $\boldsymbol{\omega}_i$ and $\boldsymbol{\omega}_j$ are the angular velocities of particles i or j, and $\mathbf{R}_i$ or $\mathbf{R}_j$ is the radial vector.

For the cohesive force, the Johnson–Kendall–Roberts (JKR) model (Johnson et al. 1971) based on the Hertz elastic theory is used:

$$F_{coh} = \gamma\pi(D - R_i - R_j)(D + R_i - R_j)(D - R_i + R_j)(D + R_i + R_j)/4D^2 \tag{11.60}$$

where γ is the surface energy density and D is the distance between the centers of the i-th and j-th particles.

The torque due to rolling friction and tangential force is given by (Zhou et al. 1999):

$$\mathbf{T}_{ij}^{t} = \mathbf{R}_i \times \mathbf{F}_{ij}^{t} \tag{11.61}$$

$$\mathbf{T}_{ij}^{r} = \mu_r \bar{R} Y_n \delta_n \frac{\boldsymbol{\omega}_{ij}\cdot\hat{\mathbf{t}}_{ij}}{|\boldsymbol{\omega}_{ij}|}\hat{\mathbf{t}}_{ij} \tag{11.62}$$

where μ_r is the rolling friction coefficient, and $\omega_{ij} = \omega_i - \omega_j$ is the relative angular velocity.

Example 11.1

The terminal velocity (or settling velocity) of a particle is the highest velocity that a particle can achieve when it falls through a stagnant fluid, either gas or liquid. Determine the terminal velocity of a particle with a density ρ_p and diameter d_p as it falls through gas with a density of ρ_g.

Solution The forces acting on the particle includes downward gravity, F_g, and upward drag force, F_d, and buoyant force F_b. The terminal velocity is reached when the forces acting on the particle are balanced:

$$F_g = F_d + F_b \tag{11.63}$$

The gravity force is

$$F_{\mathrm{g}} = m_{\mathrm{p}}g = \rho_{\mathrm{p}}\frac{1}{6}\pi d_{\mathrm{p}}^3 g \tag{11.64}$$

The drag force and buoyant forces can be obtained from Eqs. (11.50) and (11.55):

$$F_{\mathrm{d}} = \frac{1}{8}\pi d_{\mathrm{p}}^2 \rho_{\mathrm{g}} C_{\mathrm{D}} U_{\mathrm{t}}^2 \tag{11.65}$$

$$F_{\mathrm{b}} = \rho_{\mathrm{g}}\frac{1}{6}\pi d_{\mathrm{p}}^3 g \tag{11.66}$$

Substituting Eqs. (11.64)–(11.66) into Eq. (11.63) and solving for the terminal velocity yields

$$U_{\mathrm{t}} = \left[\frac{4gd_{\mathrm{p}}}{3C_{\mathrm{D}}}\left(\frac{\rho_{\mathrm{p}} - \rho_{\mathrm{g}}}{\rho_{\mathrm{g}}}\right)\right]^{1/2} \tag{11.67}$$

Equation (11.67) cannot be directly used to obtain the terminal velocity because the drag coefficient C_{D} depends on the terminal velocity. The terminal velocity will be different for different correlations of drag coefficients. Haider and Levenspiel (1989) proposed the following correlation of drag coefficient for spherical and nonspherical particles:

$$\begin{aligned} C_{\mathrm{D}} = &\frac{24}{Re}\left\{1 + [8.1716\exp(-4.0655\psi)]Re^{0.0964 + 0.5565\psi}\right\} \\ &+ \frac{73.69Re \cdot \exp(-5.0748\psi)}{Re + 5.378\exp(6.2122\psi)} \end{aligned} \tag{11.68}$$

where ψ is the particle sphericity:

$$\psi = \frac{A_{\text{sphere}}}{A_{\text{particle}}} \tag{11.69}$$

where A_{sphere} is the surface area of an equivalent spherical particle with the same volume, and A_{particle} is the surface area of the particle (any shape). The sphericity must be less than or equal to 1. Luckos and Koekemoer (2014) reported that the sphericities of pulverized coal and char particles are 0.793 and 0.749, respectively.

Haider and Levenspiel (1989) then developed the following explicit terminal velocity of the falling sphere and nonspherical particles:

$$U_{\mathrm{t}}^* = \left[\frac{18}{d_{\mathrm{p}}^{*2}} + \frac{2.3348 - 1.7439\psi}{d_{\mathrm{p}}^{*0.5}}\right]^{-1}, \quad 0.5 \le \psi \le 1 \tag{11.70}$$

where U_t^* and d_p^* are the dimensionless terminal velocity and dimensionless particle diameter:

$$U_t^* = U_t \left[\frac{\rho_g^2}{g\mu_g(\rho_p - \rho_g)} \right]^{1/3} \tag{11.71}$$

$$d_p^* = d_p \left[\frac{g\rho_g(\rho_p - \rho_g)}{\mu_g^2} \right]^{1/3} \tag{11.72}$$

Equation (11.70) is very easy to use because, unlike implicit Eq. (11.67), terminal velocity only appears on the left-hand side.

11.4.3 Heat Transfer

The energy equation for the fluid phase is:

$$\begin{aligned} &\frac{\partial}{\partial t}(\rho_f c_{pf} \alpha_f T_f) + \nabla \cdot (\rho_f c_{pf} \alpha_f \mathbf{V} T_f) \\ &= \nabla \cdot (\alpha_f k_f \nabla T_f) + q''' + G(T_s - T_f) \end{aligned} \tag{11.73}$$

where c_{pf} is the mean specific heat of the fluid, k_f is the mean thermal conductivity of the fluid, q''' is the source term due to phase change or chemical reaction, G is the coupling factor between the particle and fluid, and T_s is the mean temperature of the particles.

Assuming the particle is small enough for the temperature in the particle to be treated as uniform (lumped capacitance model), the particle temperature, $T_{p,i}$, satisfies (ANSYS Fluent Theory Guide 2017):

$$m_i c_{p,i} \frac{dT_{p,i}}{dt} = h_{p,i} A_{p,i} (T_f - T_{p,i}) + \varepsilon_{p,i} A_{p,i} \sigma (\theta_R^4 - T_{p,i}^4) \tag{11.74}$$

where $h_{p,i}$ is the convective heat transfer coefficient between the i-th particle and the fluid and can be obtained using the following empirical correlation (Li and Mason 2000):

$$Nu_i = \frac{h_{p,i} d_{p,i}}{k_f} = \begin{cases} 2 + 0.6\varepsilon_f^n Re^{1/2} \Pr^{1/3}, & Re \le 200 \\ 2 + 0.5\varepsilon_f^n Re^{1/2} \Pr^{1/3} + 0.02\varepsilon_f^n Re^{0.8} \Pr^{1/3}, & 200 < Re \le 1500 \\ 2 + 0.000045\varepsilon_f^n Re^{1.8}, & Re > 1500 \end{cases} \tag{11.75}$$

where the Reynolds number, Re, can be calculated from Eq. (11.52), and $n = 3.5$.

The second term on the right-hand side of Eq. (11.74) is to account for the absorption/emission of radiation at the particle surface and it is significant only if the fluid phase is gas. $\varepsilon_{p,i}$ is the emissivity of the particle surface and θ_R is the radiation temperature:

$$\theta_R = \left(\frac{G_R}{4\sigma} \right)^{1/4} \tag{11.76}$$

where G is the incident radiation

$$G_R = \int_{\Omega=4\pi} I d\Omega \tag{11.77}$$

where I is the radiation intensity, and Ω is the solid angle.

11.5 Gas-Particle Systems

11.5.1 Fluidized Bed

In a fluidized bed, solid particles are placed in a vertical vessel with holes at the bottom where pressurized gas is introduced to cause the particles to flow, as shown in Fig. 11.8. The mixture of gas and particles in a fluidized bed behaves like a fluid that has the ability to flow and form "bubbles" (regions with very low particle density). The interaction between the fluid and particles can be promoted by a large interfacial surface area per unit bed volume and high relative velocities between the gas and the dispersed solid particles. The motions of particles in a fluidized bed are dominated by the interaction between fluid and particles as well as particle–particle and particle–wall collisions.

11.5.1.1 Gas Velocity for Fluidization

In order to suspend the particles in a fluidized bed, the pressure drop across the entire height must be able to balance the buoyed weights of all particles:

$$\Delta p_{\rm bed} = (\rho_p - \rho_g)(1 - \bar{\varepsilon}_g)gH \tag{11.78}$$

where $\bar{\varepsilon}_g$ is the volume fraction of gas averaged over the entire bed volume, and H is the height of the fluidized bed. It should be pointed out that the effects of wall friction and particle accelerations, which are flow pattern-dependent, are not taken into account in Eq. (11.78).

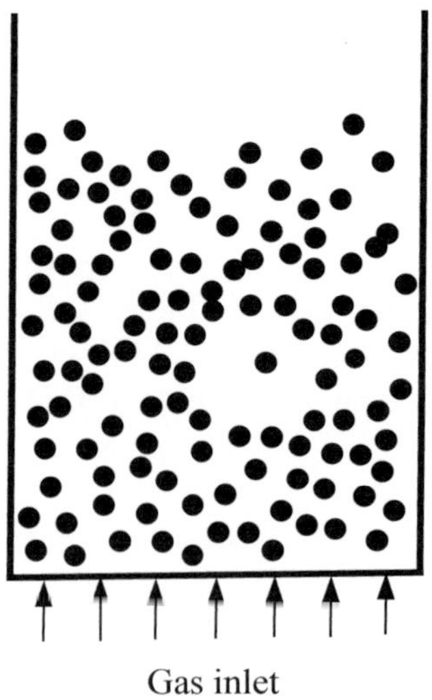

Fig. 11.8 Diagram of a fluidized bed

The pressure difference calculated from Eq. (11.78) is required for fluidization. Before the onset of the fluidization, the particles are loosely packed, and the system is a fixed bed, while gas flows in between the gaps among particles. Assuming all particles are spherical with an average diameter of d_p, the pressure difference can be obtained from the Ergun equation:

$$\Delta p_{FB} = \frac{150H\mu_g U_g(1-\varepsilon_{g0})^2}{\psi^2 d_p^2 \varepsilon_{g0}^3} + \frac{1.75H\rho_g U_g^2(1-\varepsilon_{g0})}{\psi d_p \varepsilon_{g0}^3} \tag{11.79}$$

where ψ is the particle sphericity (see Eq. 11.69), U_g is the superficial velocity of the gas, and ε_{g0} is the volume fraction of the particles in a fixed state. The velocity at which transition from a fixed bed to fluidized bed takes place is defined as the *minimum fluidization velocity*, U_{mf}, and it can be obtained by combining Eqs. (11.78) and (11.79):

$$Re_{mf} = \frac{\rho_g U_{mf} d_p}{\mu_g} = \sqrt{C_1^2 + C_2 Ar} - C_1 \tag{11.80}$$

where $C_1 = 33.7$ and $C_2 = 0.0408$, and Ar is Archimedes number:

$$Ar = \frac{g\rho_g(\rho_p - \rho_g)d_p^3}{\mu_g^2} \tag{11.81}$$

The bed voidage at minimum fluidization, ε_{mf}, is related to the bed height, H_{mf}, by the following relationship:

$$\varepsilon_{mf} = 1 - \frac{M_p}{\rho_p A H_{mf}} \tag{11.82}$$

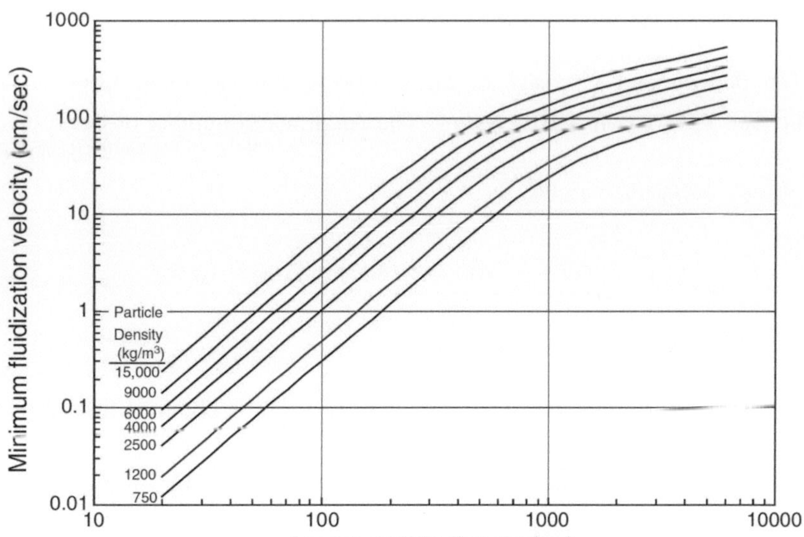

Fig. 11.9 Minimum fluidization velocity versus particle diameter for particles with different densities fluidized by air at atmospheric temperature and pressure (Michaelides et al. 2016)

where M_p is the total mass of all particles in the bed. Figure 11.9 shows the variation of minimum fluidization velocity with particle diameters on a log-log scale plot at different particle material densities. Clearly, the minimum fluidization velocity increases as the average particle diameter and particle density increases.

Equation (11.80) is valid for a dense phase system where the solid concentration is greater than 0.2. Abrahamsen and Geldart (1980) compared several empirical correlations for minimum fluidization velocity and suggested that the correlation proposed by Baeyens (1973) could give the best prediction for the entire range of solid fraction:

$$U_{mf} = \frac{0.0009(\rho_p - \rho_g)^{0.934} g^{0.934} d_p^{1.8}}{\mu_g^{0.87} \rho_g^{0.066}} \tag{11.83}$$

The voidage at the minimum fluidization velocity depends on the particle Reynolds number and can be obtained as follows:

$$\varepsilon_{mf}^3 = 0.091 \frac{1 - \varepsilon_{mf}}{\psi^3}, \quad Re_{mf} \le 20 \tag{11.84}$$

$$\varepsilon_{mf} = \left(\frac{0.071}{\psi^3}\right)^{1/3}, \quad Re_{mf} > 20 \tag{11.85}$$

In order to have particles circulate in the fluidized bed, rather than being conveyed pneumatically, the superficial gas velocity must be less than the terminal velocity obtained from Eq. (11.70).

Example 11.2

The pulverized coal particles with a diameter of 100 μm are fluidized in a vessel by air at 600 K. The density and sphericity of the pulverized coal particles are 800 kg/m^3 and 0.793, respectively. Determine the minimum fluidization velocity using Eqs. (11.80) and (11.83). What is the highest superficial gas velocity for the fluidized bed to function?

Solution The density and viscosity of air at 600 K can be found from Table B.1: $\rho_g = 0.5804\,\text{kg/m}^3$ and $\mu_g = 3.058 \times 10^{-5}\,\text{Ns/m}^2$. The Archimedes number of the system is

$$Ar = \frac{g\rho_g(\rho_p - \rho_g)d_p^3}{\mu_g^2} = \frac{9.81 \times 0.5804 \times (800 - 0.5804) \times (100 \times 10^{-6})^3}{(3.058 \times 10^{-5})^2} = 4.867$$

The Reynolds number at minimum fluidization velocity can be obtained from Eq. (11.80):

$$Re_{mf} = \sqrt{C_1^2 + C_2 Ar} - C_1 = \sqrt{33.7^2 + 0.0408 \times 4.867} - 33.7 = 2.946 \times 10^{-3}$$

The minimum fluidization velocity is

$$U_{\mathrm{mf}} = \frac{\mu_{\mathrm{g}} Re_{\mathrm{mf}}}{\rho_{\mathrm{g}} d_{\mathrm{p}}} = \frac{3.058 \times 10^{-5} \times 2.946 \times 10^{-3}}{0.5804 \times 100 \times 10^{-6}} = 1.552 \times 10^{-3}\,\mathrm{m/s}$$

Alternatively, the minimum fluidization velocity can also be determined from Eq. (11.83):

$$\begin{aligned} U_{mf} &= \frac{0.0009(\rho_{\mathrm{p}} - \rho_{\mathrm{g}})^{0.934} g^{0.934} d_{\mathrm{p}}^{1.8}}{\mu_{\mathrm{g}}^{0.87} \rho_{\mathrm{g}}^{0.066}} \\ &= \frac{0.0009(800 - 0.5804)^{0.934} 9.81^{0.934} (100 \times 10^{-6})^{1.8}}{(3.058 \times 10^{-5})^{0.87} 0.5804^{0.066}} = 2.162 \times 10^{-3}\,\mathrm{m/s} \end{aligned}$$

The fluidized bed will function when the superficial gas velocity is higher than the minimum fluidization velocity but lower than the terminal velocity. To calculate the terminal velocity, the dimensionless particle diameter can be obtained from Eq. (11.72):

$$\begin{aligned} d_{\mathrm{p}}^{*} &= d_{\mathrm{p}} \left[\frac{g \rho_{\mathrm{g}} (\rho_{\mathrm{p}} - \rho_{\mathrm{g}})}{\mu_{\mathrm{g}}^{2}} \right]^{1/3} \\ &= 100 \times 10^{-6} \left[\frac{9.81 \times 0.5804(800 - 0.5804)}{(3.058 \times 10^{-5})^{2}} \right]^{1/3} = 1.698 \end{aligned}$$

The dimensionless terminal velocity from Eq. (11.70) is

$$\begin{aligned} U_{\mathrm{t}}^{*} &= \left[\frac{18}{d_{\mathrm{p}}^{*2}} + \frac{2.3348 - 1.7439\psi}{d_{\mathrm{p}}^{*0.5}} \right]^{-1} \\ &= \left[\frac{18}{1.698^{2}} + \frac{2.3348 - 1.7439 \times 0.793}{1.698^{0.5}} \right]^{-1} = 0.1434 \end{aligned}$$

The terminal velocity can be obtained from Eq. (11.71):

$$\begin{aligned} U_{\mathrm{t}} &= U_{\mathrm{t}}^{*} \left[\frac{g \mu_{\mathrm{g}} (\rho_{\mathrm{p}} - \rho_{\mathrm{g}})}{\rho_{\mathrm{g}}^{2}} \right]^{1/3} \\ &= 0.1434 \left[\frac{9.81 \times 3.058 \times 10^{-5} (800 - 0.5804)}{0.5804^{2}} \right]^{1/3} \\ &= 0.0626\,\mathrm{m/s} \end{aligned}$$

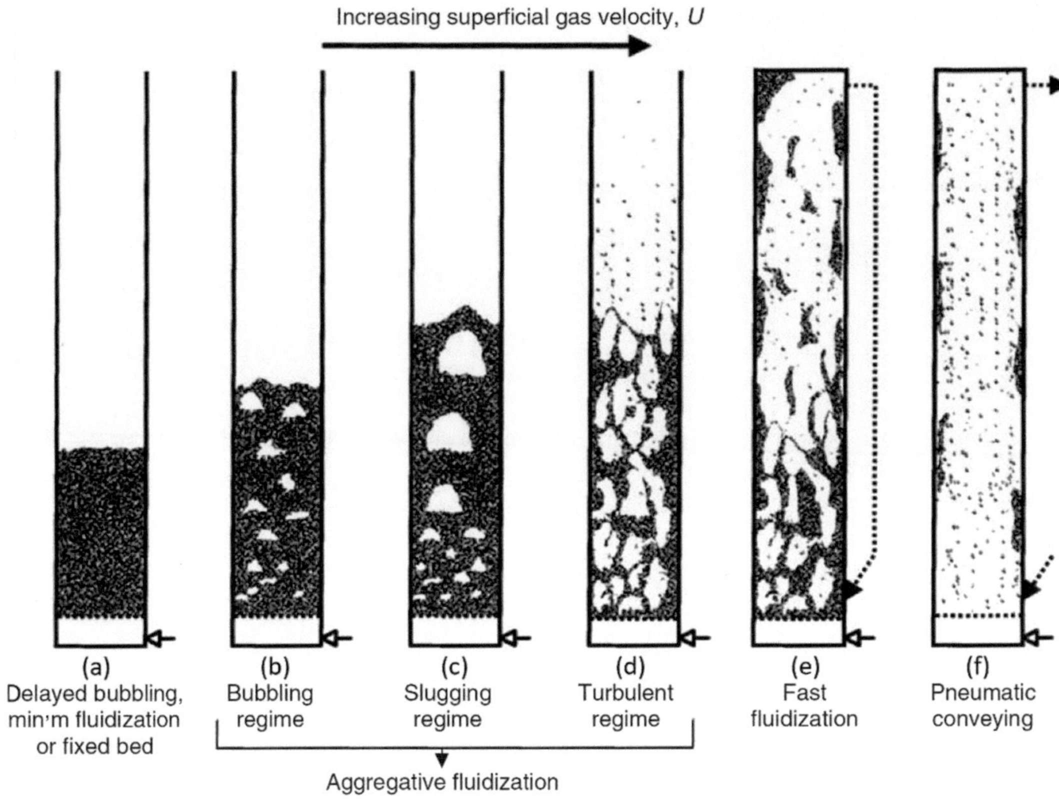

Fig. 11.10 Flow patterns for gas–solid fluidization (Michaelides et al. 2016)

11.5.1.2 Flow Patterns

Figure 11.10 shows the evolution of flow patterns in a gas–solid fluidized bed. As the superficial velocity of the gas U_g increases, the following flow patterns can be observed (Michaelides et al. 2016):

- **Bubble-free bed expansion**. Once the superficial velocity exceeds the minimum fluidization velocity U_{mf}, the particles start to become loose and the volume of the bed increases. No bubbles are in the fluidized bed and the flow pattern is referred to as particulate or homogeneous fluidization.
- **Bubbling fluidization**. As the superficial velocity further increases, some small regions of voids start to appear; these voids are referred to as "bubbles," which are similar to the gas/vapor bubbles in the liquid as discussed in Chap. 10. These bubbles are surrounded by the dense phase of the particle–gas mixture that behaves like a non-Newtonian fluid. The bubbles coalesce as they rise up through the bed.
- **Slug flow**. Once the bubbles grow into a size that exceeds half of the diameter of the vessel, the flow pattern becomes a slug flow. The slugs are bullet-shaped with a blunt head and normally formed in the center of the vessel that is away from the wall (see Fig. 11.10c). The slug breaks at the top surface of the bed, which causes the top surface of the bed to fall sharply.
- **Turbulent fluidization**. As the superficial velocity further increases for the slug flow, the slugs break down and the flow enters into turbulent fluidization. The voids in this regime are much

smaller than that in the slug flow; their shapes are irregular and unstable. For the applications that require strong interaction between gas and particles, such as chemical reactions, turbulent fluidization is very favorable.

- **Fast fluidization and dense suspension up-flow.** Particles are conveyed by the gas phase to be ejected at the top of the vessel, and new particles must be supplied at the bottom. This is now a dense phase pneumatic conveying system and a fluidized "bed" no longer exist.

It follows from the above description that only Fig. 11.10b–e represent the actual flow pattern in a fluidized bed because Fig. 11.10a is an expansion bed and Fig. 11.10f is a dense phase pneumatic conveying system. The following criteria are often used to distinguish the different flow regimes:

11.5.1.3 Minimum Bubbling Velocity, U_{mb}

When the particles are large, and the interparticle cohesive forces are weak relative to the weight–buoyancy forces, bubbling starts as soon as the superficial velocity exceeds the minimum fluidization velocity, $U_{\rm mf}$. For the cases that interparticle cohesive forces are dominant, the minimum bubbling velocity can be estimated by (Abrahamsen and Geldart 1980):

$$U_{\rm mb} = 2.07\exp(0.716\chi)\frac{d_{\rm p}\rho_{\rm g}^{0.06}}{\mu_{\rm g}^{0.347}} \tag{11.86}$$

where χ is the mass fraction of fine particles with diameters smaller than 45 μm. If the minimum bubbling velocity obtained from Eq. (11.86) is lower than the minimum fluidization velocity obtained from Eq., (11.80), the minimum fluidization velocity should be used as the bubbling velocity. The ratio of minimum bubbling velocity and minimum fluidization velocity can be obtained by combining Eqs. (11.86) and (11.83):

$$\frac{U_{\rm mb}}{U_{\rm mf}} = \frac{2300\rho_{\rm g}^{0.13}\mu_{\rm g}^{0.52}}{d_{\rm p}^{0.8}(\rho_{\rm p}-\rho_{\rm g})^{0.93}}\exp(0.716\chi) \tag{11.87}$$

The voidage at minimum bubbling velocity can be obtained from

$$\frac{1-\varepsilon_{\rm mb}}{1-\varepsilon_{\rm mf}} = \left(\frac{U_{\rm mb}}{U_{\rm mf}}\right)^{0.22} \tag{11.88}$$

11.5.1.4 Minimum Slugging Velocity, U_{ms}

Slug flow requires that the aspect ratio $H_0/D_{\rm h} \geq 2$, where H_0 is the height of the powder bed at a fixed state, and $D_{\rm h}$ is the hydraulic diameter of the vessel. The minimum slugging velocity can be obtained from (Geldart 1986):

$$U_{\rm ms} = U_{\rm mf} + 0.0016(60D_{\rm h}^{0.175} - H_{\rm mf})^2 + 0.07(gD_{\rm h})^{1/2} \tag{11.89}$$

where $H_{\rm mf}$ is the height of the bed at minimum fluidization velocity, and all length units are in centimeters.

Agu et al. (2017) studied the onset of bubbling and slugging in a fluidized bed using a dual-plane electrical capacitance tomography system and stated that the average voidage at the minimum slugging condition could be obtained from the following relationship:

$$\frac{1-\varepsilon_{\mathrm{ms}}}{1-\varepsilon_{\mathrm{mf}}}=\left[1.206+0.604\left(\frac{U_{\mathrm{ms}}}{U_{\mathrm{mf}}}\right)^{-10}\right]^{-1} \tag{11.90}$$

11.5.1.5 Critical Velocity for Turbulent Fluidization, U_c

The critical velocity for turbulent fluidization is an experimentally determined superficial gas velocity, at which the standard deviation of the pressure fluctuation reaches the maximum (Michaelides et al. 2016). Ellis et al. (2004) experimentally studied the hydrodynamics of gas–solid turbulent fluidized beds for fluid-cracking catalyst particles in columns of diameter 0.26, 0.61, and 1.56 m. They recommended that the Reynolds number based on the critical velocity could be obtained from the following correlation:

$$Re_{\mathrm{c}}=\frac{\rho_{\mathrm{g}}U_{\mathrm{c}}d_{\mathrm{p}}}{\mu_{\mathrm{g}}}=\frac{(H/D)^{0.43}Ar^{0.74}}{\sqrt{4.2(H/D)^{0.86}+3.1Ar^{0.33}}} \tag{11.91}$$

11.5.1.6 Transport Velocity for Fast Fluidization, U_{tr}

At even higher velocity, the system is no longer a bed because there is not any apparent upper bed surface and a large number of particles are in suspension. The particles rise in the core of the vessel and fall back into clusters near the wall of the vessel, causing the flow regime to enter fast fluidization. The transport velocity is at a minimum velocity below which fast fluidization cannot occur. The following empirical correlation can be used to estimate the transport velocity (Perales et al. 1991):

$$Re_{\mathrm{tr}}=\frac{\rho_{\mathrm{g}}d_{\mathrm{p}}U_{\mathrm{tr}}}{\mu_{\mathrm{g}}}=1.45Ar^{0.484},\quad 20<Ar<50,000 \tag{11.92}$$

11.5.1.7 Dense Suspension Up-flow (DSU) or Dilute Pneumatic Conveying (DPC)

When the superficial gas velocity is very high, the time-averaged mass flux of the particles is upward over the entire cross section, which initiates an upward net particle circulation flux, $G_{\mathrm{s}}\left(\mathrm{kg/m^2s}\right)$. Either DSU or DPC can occur depending on whether the volume fraction of the particle is higher than 7%. Kim et al. (2004) studied the flow behavior and flow regime transitions in a circulating fluidized bed riser. They recommended that the onsets of DSU and DPC can be estimated using the following criteria:

$$\frac{U_{\mathrm{DSU}}}{\varepsilon_{\mathrm{A}}V_{\mathrm{t}}}=22.8\left[\frac{G_{\mathrm{s}}}{\rho_{\mathrm{p}}(1-\varepsilon_{\mathrm{A}})U_{\mathrm{t}}}\right]^{0.59}Ar^{-0.20},\quad \varepsilon_{\mathrm{A}}>7\% \tag{11.93}$$

$$\frac{U_{\mathrm{DPC}}}{\varepsilon_{\mathrm{A}}V_{\mathrm{t}}}=1160\left[\frac{G_{\mathrm{s}}}{\rho_{\mathrm{p}}(1-\varepsilon_{\mathrm{A}})U_{\mathrm{t}}}\right]^{-0.90}Ar^{-0.20},\quad \varepsilon_{\mathrm{A}}<7\% \tag{11.94}$$

where ε_{A} is the cross-sectional averaged voidage, and U_{t} is the terminal velocity of particles at a mean size that can be estimated using Eq. (11.70). For a fluidized bed to function properly, the superficial gas velocity should be lower than the onset velocity of DSU or DPC.

11.5.1.8 Heat Transfer

Many applications require adding or extracting thermal energy from the fluidized bed. There are several intensive heat transfer processes in a fluidized bed: (1) internal heat transfer between the gas and particles, (2) internal heat transfer between different points in the fluidized bed, and (3) external heat transfer between a heat transfer area and the gas-particle mixture. The heat transfer between the gas and particles can be calculated using Eq. (11.75). Since the thermal response time of the particles —the time it takes for a particle to respond to a temperature change of the gas (see Eq. 11.101)—is much shorter than the characteristic time in a fluidized bed, the local temperatures of gas and particles are practically equal. Therefore, the focus here will be the heat transfer between a heat transfer surface and the fluidized bed.

The heat transfer between the fluidized bed and a heat transfer area can be calculated using Newton's law of cooling:

$$Q = hA(T_{\mathrm{b}} - T_{\mathrm{w}}) \tag{11.95}$$

where T_{w} and A are the temperature and the area of the heat transfer surface, and T_{b} is the bulk temperature of the fluidized bed. The heat transfer coefficient depends on the particle size, superficial gas velocity, and the thermophysical properties of particles and the gas. The overall heat transfer coefficient can be expressed as:

$$h = h_{\mathrm{p}} + h_{\mathrm{g}} + h_{\mathrm{r}} \tag{11.96}$$

where h_{p} and h_{g} are the contributions to the heat transfer coefficients from particle-phase contact and gas-phase contact with the heat transfer surface, and h_{r} is the radiative heat transfer coefficient. Figure 11.11 provides a schematic of the particle and gas convective heat transfer coefficient in a dense, bubbling fluidized bed at different particle sizes. Fluidization of the bed takes place when the superficial gas velocity is greater than the minimum fluidization velocity (Eq. 11.80 or 11.83) and lower than the terminal velocity (Eq. 11.70). The particle convective heat transfer to the surface

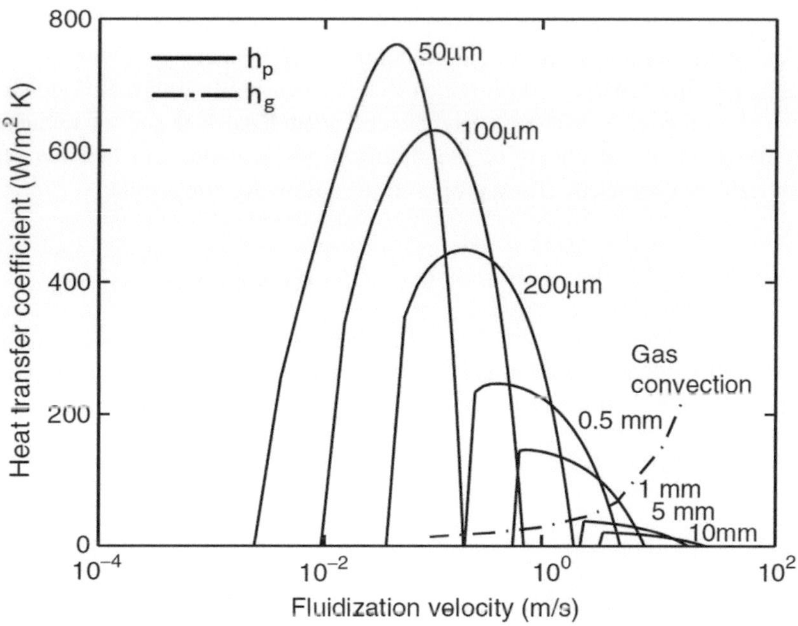

Fig. 11.11 Heat transfer coefficients versus fluidization velocity in a dense, bubbling fluidized bed (Michaelides et al. 2016)

initially increases with increasing gas velocity until reaches a peak value, after which the particle heat transfer coefficient decreases due to a simultaneous increase in voidage. On the other hand, the gas convective heat transfer coefficient monotonically increases with increasing gas velocity. When the velocity is greater than its optimal value, the main contributor toward heat transfer is gas convection due to a decrease in contributions from the particles.

The optimum velocity corresponding to maximum h_p can be obtained from the following Todes' correlation (Todes 1965):

$$Re_{opt} = \frac{Ar}{18 + 5.22Ar^{0.5}} \tag{11.97}$$

The two convection terms on the right hand of Eq. (11.96) can be lumped together to define an overall convective heat transfer coefficient:

$$h_c = h_p + h_g \tag{11.98}$$

In addition to the gas velocity, the particle size also plays a significant role in the overall convective heat transfer coefficient. The overall convective heat transfer coefficient at optimal gas velocity can be obtained from Zabrodsky (1966) as:

$$h_{c,max} = 35.7 d_p^{-0.36} \rho_p^{0.2} k_g^{0.6} \tag{11.99}$$

Assuming that the particle surface is gray and nondiffusive and the gas is nonprecipitating, the radiative heat transfer coefficient can be determined from:

$$h_r = \varepsilon\sigma(T_b + T_w)(T_b^2 + T_w^2) \tag{11.100}$$

where ε is emissivity, which should not be confused with voidage.

While the heat transfer surfaces can take many different shapes, the most common approach is to insert heat transfer tubes with heating or cooling fluids flowing inside the tube. Numerous empirical correlations about heat transfer between an inserted heat transfer tube and fluidized bed are available in the literature, but the results from these correlations often do not agree with each other and the difference could be in the order of 100% (Chen 2003; Chen et al. 2005). Borodulya et al. (1991) employed a two-zone model of heat transfer between a fluidized bed and an immersed surface to correctly take into account the effects of the fluidized gas pressure and bed temperature on the convective heat transfer coefficient. They proposed the following correlation:

$$\frac{h_c d_p}{k_g} = 0.74 Ar^{0.1} \left(\frac{\rho_p}{\rho_g}\right)^{0.14} \left(\frac{c_{pp}}{c_{pg}}\right)^{0.24} (1-\varepsilon)^{2/3} + 0.46 (Re_p Pr_g) \frac{(1-\varepsilon)^{2/3}}{\varepsilon} \tag{11.101}$$

where ε is the overall voidage. Equation (11.101) is valid under the following range of parameters: $0.1 < d_p < 4\,\text{mm}, 0.1 < p < 10\,\text{MPa}$, $140 < Ar < 1.1 \times 10^7$ for both horizontal and vertical tubes.

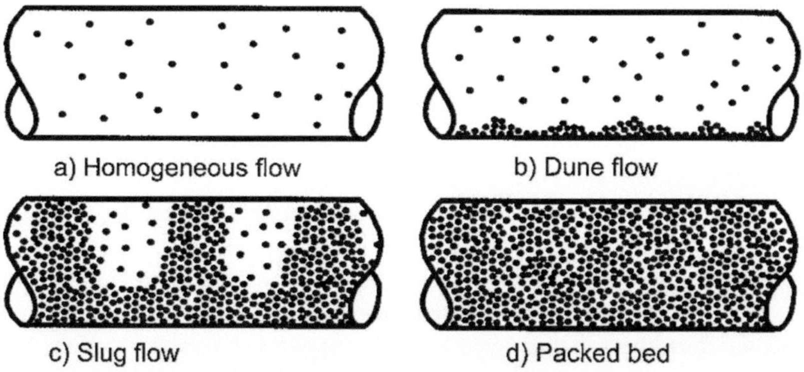

Fig. 11.12 Flow pattern in horizontal pneumatic transport (Crowe et al. 2012)

11.5.2 Pneumatic Transport

11.5.2.1 Flow Patterns

The pneumatic transport is a simple, flexible, reliable, and safe approach to the transport of solid material products (particles) over a long distance. However, multiple components are in a pneumatic conveying system (Michaelides et al. 2016): a gas mover (fan, centrifugal blower, rotary lobe blower, etc.), a product feeder (rotary feeder, screw feeder, and blow tank), conveying line, and a gas–solid separator (cyclone, baghouse, etc.). A conveying line is a tube/duct that transports the mixture of solid particles; thus, a pressure drop in the conveying line is very important for designing and operating the pneumatic conveying system. The pressure drop is significantly affected by the flow patterns.

The flow pattern in a pneumatic transport system depends on many factors, including orientation of the tube (vertical vs. horizontal), particulate loading, shape and size distributions of the particles, and Reynolds number. The flow patterns encountered in horizontal pneumatic transport flow of solid–gas are defined as follows (Crowe et al. 2012) and are shown in Fig. 11.12:

- **Homogeneous flow**. The gas velocity is sufficiently high, and the particles are fully mixed with the gas by turbulent mixing. The higher the gas velocity is, the more uniformly distributed the particles in the gas are. The particle loading for homogeneous flow is the lowest among the four patterns showing in Fig. 11.12 and the pressure drop is also the lowest.
- **Dune flow**. If the gas velocity is reduced to below a certain threshold, referred to as *saltation velocity*, particles will start to deposit at the bottom of the tube to form dunes, which are ripples formed by the gas flow. Both particulate loading and the pressure drop for dune flow are higher than that of the homogeneous flow.
- **Slug flow**. As the gas velocity is reduced further, more particles will be deposited at the bottom of the tube and the peaks of the dunes will touch the top of the tube. There will be regions completely filled by the settled particles and other regions with particles that are suspended between the completely filled regions. The particle loading for the slug flow is much higher than that of the homogeneous and dune flow.
- **Packed bed**. When the gas velocity is even lower, the entire tube will be filled with settled particles, and the only the gas flow will be in the porous media formed by the packed particles. Pneumatic transport is no longer possible because there is no particle flow.

The behavior of a pneumatic transport system is significantly affected by the solid loading defined as:

$$\mu_s = \frac{\dot{m}_s}{\dot{m}_g} = \frac{\dot{m}_s}{\rho_g A U_g} \tag{11.102}$$

where $\dot{m}_s$ and $\dot{m}_g$ are mass flow rates of solid particles and gas, respectively. Moreover, ρ_g, A, and U_g represent the density of the gas, cross-sectional area of the tube, and superficial velocity of the gas, respectively.

A pneumatic conveying system can be classified as either a dilute-phase conveying system or a dense phase conveying system. The dilute-phase conveying system requires that the gas velocity in all segments be higher than the saltation velocity. The maximum particle loading for dilute-phase conveying is usually less than 10, and the gauge pressure of the system is below one bar. In a dense phase conveying system, the particles are conveyed in nonsuspension modes (e.g., slug flow).

The saltation solid loading that separates dilute and dense phase conveying can be obtained from the following correlation (Michaelides et al. 2016):

$$\mu_{\text{saltation}} = \alpha Fr_{ss}^{\beta} \tag{11.103}$$

where Fr_{ss} is the Froude number at saltation velocity

$$Fr_{ss} = \frac{U_{gs}}{\sqrt{gD}} \tag{11.104}$$

and α and β are dimensionless coefficients:

$$\alpha = 10^{-(1.4d_p + 1.96)} \tag{11.105}$$

$$\beta = 1.1d_p + 2.5 \tag{11.106}$$

where d_p is the particle diameter in *millimeters*. The saltation velocity can be obtained by substituting Eqs. (11.102) and (11.104) into Eq. (11.103):

$$U_{gs} = \left[\frac{\dot{m}_s\left(\sqrt{gD}\right)^{\beta}}{A\rho_g\alpha}\right]^{1/(\beta+1)} \tag{11.107}$$

which is referred to as Rizk's correlation (Rizk 1976) and is valid for the pneumatic transport of coarse particles whose diameters are greater than 0.5 mm and the particle specific gravity ($\rho_p g$) is less than 3.

Matsumoto et al. (1977) suggested that the saltation velocities of coarse and fine particles behaved differently, and the critical diameter that separates the coarse and fine particles can be expressed as:

$$\frac{d_{p,\text{crit}}}{D} = 1.39\left(\frac{\rho_p}{\rho_g}\right)^{-0.74} \tag{11.108}$$

The saltation solid loading can be obtained from:

$$\mu_s = \begin{cases} 0.373\left(\frac{\rho_p}{\rho_g}\right)^{1.06}\left(\frac{Fr_p}{10}\right)^{-3.7}\left(\frac{Fr_{ss}}{10}\right)^{3.61} & d_p \geq d_{p,crit} \\ 55\left(\frac{d_p}{D}\right)^{1.43}\left(\frac{Fr_{ss}}{10}\right)^{4.0} & d_p < d_{p,crit} \end{cases} \tag{11.109}$$

where

$$Fr_p = \frac{U_t}{\sqrt{gd_p}} \tag{11.110}$$

where U_t is the terminal velocity of a single particle in the gas (see Eq. 11.70).

When coarse particles are transported in tubes with a small diameter, choking may occur. The system then exhibits slugging behavior, and solid particles will not flow (Chong and Leung 1986). The choking velocity in a vertical pneumatic conveying can be obtained from (Yousifi and Gau 1974):

$$\frac{U_c}{\sqrt{gd_p}} = 32Re_p^{-0.06}\mu_s^{0.28} \tag{11.111}$$

where

$$Re_p = \frac{\rho_g U_t d_p}{\mu_g} \tag{11.112}$$

Substituting Eq. (11.102) into Eq. (11.111), an explicit expression for choking velocity can be obtained as:

$$U_c = \left[32\sqrt{gd_p}Re_p^{-0.06}\left(\frac{\dot{m}_s}{A\rho_g}\right)^{0.28}\right]^{1/1.28} \tag{11.113}$$

While the saltation and choking velocities for coarse particles are almost identical, the saltation velocity is about 3–5 times larger than the choking velocity for fine particles.

Example 11.3

A pneumatic conveying line with an inner tube diameter of 0.2 m is being used to transport solid particles with a mean diameter of 2.5 mm and a density of 800 kg/m^3. The conveying rate of the particles is 5 kg/s. Assuming the system is at 300 K, determine the saltation velocity using Rizk's correlation and Matsumoto correlation. What is choking velocity for the pneumatic conveying line?

Solution The density and viscosity of air at 300 K can be found from Table B.1: $\rho_g = 1.1614 kg/m^3$ and $\mu_g = 1.846 \times 10^{-5} Ns/m^2$. The dimensionless coefficients for Rizk's correlation are:

$$\alpha = 10^{-(1.4d_p + 1.96)} = 10^{-(1.4\times 2.5 + 1.96)} = 3.467 \times 10^{-6}$$

$$\beta = 1.1d_p + 2.5 = 1.1 \times 2.5 + 2.5 = 5.25$$

The saltation velocity can be obtained from Eq. (11.107):

$$U_{gs} = \left[\frac{\dot{m}_s(\sqrt{gD})^{\beta}}{A\rho_g\alpha}\right]^{1/(\beta+1)}$$
$$= \left[\frac{5\times(\sqrt{9.81\times0.2})^{5.25}}{(1/4)\times\pi\times0.2^2\times1.1614\times3.467\times10^{-6}}\right]^{1/(5.25+1)} = 21.80\,\text{m/s}$$

The critical diameter that separates coarse and fine particles for the Matsumoto correlation is:

$$d_{p,crit} = 1.39D\left(\frac{\rho_p}{\rho_g}\right)^{-0.74} = 1.39\times0.2\left(\frac{800}{1.1614}\right)^{-0.74} = 2.2\times10^{-3}\text{m} = 2.2\,\text{mm}$$

The dimensionless particle diameter for the terminal velocity can be obtained from Eq. (11.72):

$$d_p^* = d_p\left[\frac{g\rho_g(\rho_p-\rho_g)}{\mu_g^2}\right]^{1/3}$$
$$= 2.5\times10^{-3}\left[\frac{9.81\times1.1614(800-1.1614)}{(1.846\times10^{-5})^2}\right]^{1/3} = 747.29$$

The dimensionless terminal velocity from Eq. (11.70) is

$$U_t^* = \left[\frac{18}{d_p^{*2}} + \frac{2.3348-1.7439\psi}{d_p^{*0.5}}\right]^{-1}$$
$$= \left[\frac{18}{747.29^2} + \frac{2.3348-1.7439\times1}{747.29^{0.5}}\right]^{-1} = 46.19$$

The terminal velocity can be obtained from Eq. (11.71):

$$U_t = U_t^*\left[\frac{g\mu_g(\rho_p-\rho_g)}{\rho_g^2}\right]^{1/3}$$
$$= 46.19\left[\frac{9.81\times1.846\times10^{-5}(800-1.1614)}{1.1614^2}\right]^{1/3}$$
$$= 21.95\,\text{m/s}$$

The Froude number based on terminal velocity is:

$$Fr_p = \frac{U_t}{\sqrt{gd_p}} = \frac{21.95}{\sqrt{9.81\times2.5\times10^{-3}}} = 140.2$$

Since $d_p > d_{p,crit}$, the saltation solid loading satisfies:

$$\mu_{saltation} = \frac{\dot{m}_s}{\rho_g A U_{gs}} = 0.373\left(\frac{\rho_p}{\rho_g}\right)^{1.06}\left(\frac{Fr_p}{10}\right)^{-3.7}\left(\frac{Fr_{ss}}{10}\right)^{3.61} \tag{11.114}$$

Substituting Eq. (11.104) into Eq. (11.114) and solving for saltation velocity yields

$$U_{gs} = \left[\left(\frac{\rho_p}{\rho_g}\right)^{-1.06}\left(\frac{Fr_p}{10}\right)^{3.7}\frac{\dot{m}_s\left(10\sqrt{gD}\right)^{3.61}}{0.373(\pi/4)D^2\rho_g}\right]^{1/4.61} \tag{11.115}$$

which can be used to calculate the saltation velocity:

$$U_{gs} = \left[\left(\frac{800}{1.1614}\right)^{-1.06}\left(\frac{140.2}{10}\right)^{3.7}\frac{5\left(10\sqrt{9.81\times 0.2}\right)^{3.61}}{0.373(\pi/4)\times 0.2^2\times 1.1614}\right]^{1/4.61} = 19.34\,\mathrm{m/s}$$

which is slightly lower than that obtained from Rizk's correlation.

To calculate the choking velocity, the Reynolds number based on terminal velocity and particle diameter is:

$$Re_p = \frac{\rho_g U_t d_p}{\mu_g} = \frac{1.1614\times 21.95\times 2.5\times 10^{-3}}{1.846\times 10^{-5}} = 3.45\times 10^3$$

The choking velocity is then obtained from Eq. (11.113):

$$\begin{aligned}U_c &= \left[32\sqrt{gd_p}Re_p^{-0.06}\left(\frac{\dot{m}_s}{A\rho_g}\right)^{0.28}\right]^{1/1.28}\\ &= \left[32\sqrt{9.81\times 2.5\times 10^{-3}}\left(3.45\times 10^3\right)^{-0.06}\left(\frac{5}{(\pi/4)\times 0.2^2\times 1.1614}\right)^{0.28}\right]^{1/1.28}\\ &= 7.05\,\mathrm{m/s}\end{aligned}$$

It can be seen that the saltation velocity is about three times larger than the choking velocity.

11.5.2.2 Pressure Drop for Dilute-Phase Conveying

The pressure drop in a straight tube of a steady-state dilute-phase conveying system includes static and frictional pressure drops

$$\Delta p = \Delta p_s + \Delta p_f \tag{11.116}$$

where the static pressure drop can be obtained from:

$$\Delta p_s = \rho_p(1-\varepsilon)\Delta Zg + \rho_g\varepsilon\Delta Zg \tag{11.117}$$

where ΔZ is the increase in elevation from the inlet to the exit. ΔZ is zero for the horizontal tube; it is equal to the length of the tube serving as a vertical tube where the inlet is at the bottom, and it is negative for a vertical tube with an inlet at the top. The voidage can be obtained from:

$$\varepsilon = 1 - \frac{\dot{m}_s}{A\rho_p U_p} \tag{11.118}$$

where

$$U_{\rm p} = U_{\rm g} - U_{\rm t} \tag{11.119}$$

The frictional pressure drop can be due to either gas or solid particles:

$$\Delta p_{\rm f} = \Delta p_{\rm g} + \Delta p_{\rm p} \tag{11.120}$$

where the pressure drop due to gas can be obtained from correlations for single-phase gas flowing alone in a tube:

$$\Delta p_{\rm g} = \lambda_{\rm g} \frac{L}{D} \frac{\rho_{\rm g} U_{\rm g}^2}{2} \tag{11.121}$$

where the friction factor for gas depends on the behavior of the gas phase alone and can be obtained from (Plasynski et al. 1994; Gao et al. 2013)

$$\lambda_{\rm g} = \begin{cases} 0.3164 Re^{-0.25} & 2320 < Re < 10^5 \\ 0.0056 + 0.5 Re^{-0.32} & Re > 10^5 \end{cases} \tag{11.122}$$

The pressure drop due to solid can be calculated similarly to gas:

$$\Delta p_{\rm p} = \mu_{\rm s} \lambda_{\rm z} \frac{L}{D} \frac{\rho_{\rm g} U_{\rm g}^2}{2} \tag{11.123}$$

where the friction factor for particles can be obtained from (Weber 1982; 1991):

$$\lambda_{\rm z} = K \mu_{\rm s}^{\rm a} Fr^{\rm b} Fr_{\rm s}^{\rm c} \left(\frac{D}{d_{\rm p}} \right)^d \tag{11.124}$$

where

$$Fr = \frac{U_{\rm g}^2}{gD}, \quad {\rm Fr_s} = \frac{U_{\rm t}^2}{gD} \tag{11.125}$$

and the other parameters in Eq. (11.124) can be estimated from Table 11.4 if experimental data for the materials are not available.

The total frictional pressure drop can be obtained by substituting Eqs. (11.121) and (11.123) into Eq. (11.120):

$$\Delta p_{\rm f} = \left(\lambda_{\rm g} + \mu_{\rm s} \lambda_{\rm z} \right) \frac{L}{D} \frac{\rho_{\rm g} U_{\rm g}^2}{2} \tag{11.126}$$

Table 11.4 Parameters for frictional factor for the particle, Eq. (11.124)

Materials	K	a	b	c	d
Fine powders ($d_p < 0.5\,\text{mm}$)	2.1	–0.3	–1	0.25	0.1
Coarse powders ($d_p > 0.5\,\text{mm}$)	0.082	–0.3	–0.86	0.25	0.1

Example 11.4

Solid particles with a mean diameter of 2.5 mm and a density of 800 kg/m^3 are pneumatically conveyed in a 15-m-long horizontal tube with an inner diameter of 0.2 m at a temperature of 300 K. The conveying rate of the particle is 5 kg/s and the superficial gas velocity is 25 m/s. What is the pressure loss in this pneumatic conveying tube?

Solution The Reynolds number is

$$Re = \frac{\rho_g U_g D}{\mu_g} = \frac{1.1614 \times 25 \times 0.2}{1.846 \times 10^{-5}} = 3.146 \times 10^5$$

The friction factor for gas phase is

$$\lambda_g = 0.0056 + 0.5 Re^{-0.32} = 0.0056 + 0.5 \times (3.146 \times 10^5)^{-0.32} = 0.0143$$

The solid loading is

$$\mu_s = \frac{\dot{m}_s}{\dot{m}_g} = \frac{\dot{m}_s}{\rho_g A U_g} = \frac{5}{1.1614 \times (1/4) \times \pi \times 0.2^2 \times 25} = 5.48$$

The Froude numbers for the solid phase friction factor are:

$$Fr = \frac{U_g^2}{gD} = \frac{25^2}{9.81 \times 0.2} = 318.6$$

$$Fr_s = \frac{U_t^2}{gD} = \frac{21.95^2}{9.81 \times 0.2} = 245.6$$

Since the particle diameter is greater than 0.5 mm, the friction factor for particles is

$$\lambda_z = K \mu_s^a Fr^b Fr_s^c \left(\frac{D}{d_p}\right)^d$$
$$= 0.082 \times 5.48^{-0.3} \times 318.6^{-0.86} \times 245.6^{0.25} \left(\frac{0.2}{2.5 \times 10^{-3}}\right)^{0.1} = 0.00212$$

The pressure drop for the tube is

$$\Delta p_f = (\lambda_g + \mu_s \lambda_z) \frac{L}{D} \frac{\rho_g U_g^2}{2}$$
$$= (0.0143 + 5.48 \times 0.00212) \frac{15}{0.2} \frac{1.1614 \times 25^2}{2} = 706.14 Pa$$

11.5.2.3 Pressure Drop for Dense Phase Conveying

The particles in the dense phase conveying system are transported by two different modes: (1) the slug flow, and (2) the fluidized dense-phase flow. For the slug flow (see Fig. 11.8c), the slugs move in a wave-like fashion. The particles at the bottom of the tube are dragged by the gas to join the front (left) of the slug, while the particles on the end (right) of the slug are dropped to the bottom of the tube. This process is repeated at the next slug and the particles move forward one slug at a time. For slug flow, the stresses between the moving slugs of particles and the pipe wall dominate the solid-phase frictional losses (Michaelides et al. 2016).

Fluidized dense phase flow is appropriate for fine particles with good air retention capability. It was suggested that Eq. (11.126) can also be applied to fluidized dense phase conveying, but the solid frictional factor λ_z must be determined differently. Jones and Williams (2003) proposed the following solid friction factor:

$$\lambda_z = 83\mu_s^{-0.9} Fr^{-1} \tag{11.127}$$

which is independent of particle properties based on experimental data from PFA, iron powder, copper ore, and democrat flour.

Gao et al. (2013) proposed the following correlation for solids friction factor λ_z based on extensive bench-scale data derived from the horizontal dense-phase pneumatic conveying of pulverized coal:

$$\lambda_z = \begin{cases} 0.005 + 0.63Fr^{-0.95} & 16 < Fr \leq 174.2 \\ 0.016Fr^{-0.1} & 174.2 < Fr < 1600 \end{cases} \tag{11.128}$$

where the Froude number is defined in Eq. (11.125).

11.5.3 Multicomponent Gas-Particle Flows

Due to the unique characteristics of coherency, monochromaticity, and collimation, lasers have been widely used in various areas, such as etching and ablation of polyimide, ablation of biological tissues, and interaction with composite materials, to name a few. Numerous work on the simulation of laser material processing have been published in the past decades, but the study of simultaneous fluid flow and heat and mass transfer in a coated medium induced by laser heating is scant. Afrin et al. (2016) investigated the effects of laser irradiation on heat transfer and mass destruction of a urethane paint-coated substrate via a 3D numerical simulation. The paint began to decompose through a chemical reaction when the paint's hottest spot reached a threshold temperature of 560 K. As a result, combustion products CO_2, H_2O, and NO_2 were produced and ejected into the gas phase. In addition, chromium (III) oxide, which is buried (as pigment) in the paint, was ejected as particles from the paint into the surrounding gaseous phase.

The entire process of laser irradiation to a paint coated on a steel substrate includes: (1) heat conduction in the paint and substrate, (2) chemical reaction in the paint, and (3) convective heat and mass transfer of reactant and products in a multicomponent gaseous phase. The solid domain encompasses two layers: The bottom layer is stainless steel (AISI 304L), and the top layer is the paint that is a homogeneous mixture of binder (40%) and pigment (60%). Only the heat conduction equation is solved for this domain. For stainless steel, both thermal conductivity and heat capacity are temperature-dependent. For the gaseous phase, the governing equations are similar to those presented in Sect. 11.4.

The urethane-based paint is considered because it is an industrial standard. In the past two decades, it has mostly replaced acrylic paints. The binder for this paint is polyurethane ($C_3H_7NO_2$) and the pigment is chromium (III) oxide (Cr_2O_3). The overall chemical reaction occurring in the paint is: $4C_3H_7NO_2 + 19O_2 = 12CO_2 + 14H_2O + 4NO_2$. The chemical reaction is assumed to be a complete-type at a zero-order level. As mentioned previously, polyurethane will start decomposing when the temperature at the top surface of the paint reaches the threshold temperature. The produced gas species (CO_2, H_2O, and NO_2) then diffuse into the gaseous domain. Meanwhile, the pigment in the paint will be ejected as particles into the gaseous domain from the paint surface. The reaction rate is approximated by the Arrhenius equation:

$$k_c = \left(RT/2\pi M_{O_2}\right)^{1/2} e^{-E/RT} \tag{11.129}$$

where E is the activation energy, and M_{O2} is the molar weight of oxygen. Alternatively, the reaction rate can also be defined as the slope of the concentration–time plot for a species divided by the stoichiometric coefficient of that species. For consistency, a negative sign is added if the species is a reactant. Thus, the reaction rate constant can be represented as (Petrucci et al. 2006):

$$\begin{aligned} k_c &= -\Delta[C_3H_7NO_2]/\Delta t = -4\Delta[O_2]/19\Delta t = 4\Delta[CO_2]/12\Delta t \\ &= 4\Delta[H_2O]/14\Delta t = 4\Delta[NO_2]/4\Delta t \end{aligned} \tag{11.130}$$

Therefore, the reactant-consuming rate and the products generating rate can be estimated by Eq. (11.130).

Since it is computationally exhaustive to capture all particles' dynamic behaviors during the whole computational process due to the presence of a large number of particles, a concept of the *parcel* which is a collection of real particles (pigment) is adopted to deal with solid particle dynamics in a fluid flow. In this sense, all the particles in one parcel share the same particle properties, i.e., size, velocity, temperature, etc. To facilitate the parcel generating mechanism, a field activation type of particle injection model that mimics the particle generation process is developed with the idea of introducing many local "injectors." Specifically, mass destruction is considered as parcel injection from "injectors" which are buried at the center of the cells in the solid that are adjacent to the coupled boundary between the solid and gaseous domains. The fields associated with those cells, namely temperature and oxygen concentration, determine whether injectors start or stop to work. In other words, when the temperature on the paint surface exceeds the threshold value and the oxygen concentration is sufficiently high, the parcel will be generated and ejected into the gaseous domain according to the chemical reaction rate. The number of particles in single parcel generated in each time step is estimated by

$$N = \Delta V_{rem}\rho^N = \Delta V_{rem}\frac{\omega_{pg}/\left(\rho_{pg}\frac{4}{3}\pi\bar{r}^3\right)}{\omega_{bd}/\rho_{bd} + \omega_{pg}/\rho_{pg}} \tag{11.131}$$

where ΔV_{rem} is the volume removed from an individual "injector," ρ^N is number density of particles buried in paint, ρ_{pm} and ρ_{bd} are the density of pigment and binder, respectively, and $\bar{r}$ is the mean radius of particles in a parcel.

The parcels' motion can be determined by Newton's second law of motion. It should be noticed that the result should account for gravity and the drag force due to the velocity difference from the gaseous phase. The drag force is estimated in consideration of particle's submicron size as follows:

$$\mathrm{F}_{drag} = \left[\frac{3}{4}m\mu_c\left(\frac{24}{C_c}\right)/\rho_p d_p^2\right](\mathrm{U} - \mathrm{U_p}) \tag{11.132}$$

where μ_c is the molecular viscosity of the fluid, $\mathbf{U}_p$ is the velocity of the parcel, and C_c is the Cunningham correction to Stokes' drag law.

In addition, the simulation will be automatically terminated when the maximum temperature in the stainless metal reaches to its melting point (1670 K) since the current work only focuses on the

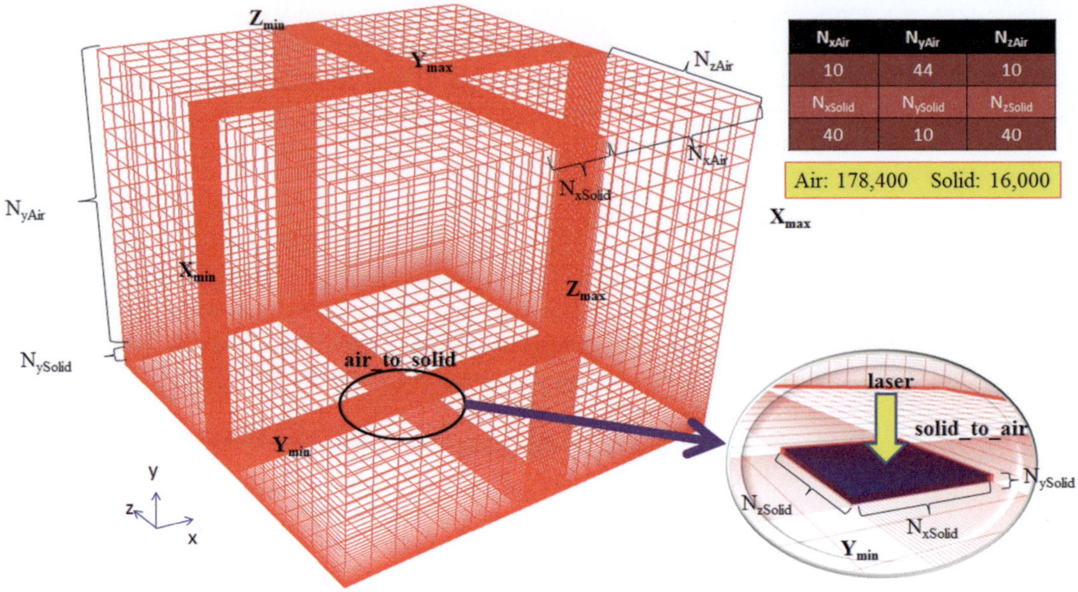

N_xAir	N_yAir	N_zAir
10	44	10
N_xSolid	N_ySolid	N_zSolid
40	10	40

Air: 178,400 Solid: 16,000

Fig. 11.13 Illustration of mesh arrangement for laser interaction with a painted stainless steel substrate (Afrin et al. 2016)

physical process of paint removal before the phase change of stainless steel occurs. The simulation is performed by using the OpenFOAM-2.3.0 framework and incorporating the recently developed solver.

Figure 11.13 illustrates the physical and geometric model of the problem under consideration. A solid (paint + stainless steel) with a size of $l \times w \times h$ (length × width × height) in the x-, z-, and y-directions is placed at the bottom of the simulation domain that has a size of $L \times W \times H$. The dimensions used in the simulation are 250 × 250 × 200 mm for the gaseous domain and 30 × 30 × 1.35 mm for the solid domain. The thickness is 0.150 mm for the paint, and the rest of the 1.2 mm is stainless steel (AISI 304L). The material absorptivities are assumed to be constants: 0.8 for the paint and 0.1 for the stainless steel (AISI 304L). The total laser irradiation time on the paint is set to be 10 s. The initial temperature of the entire domain is 300 K. The particles are assumed to have a monosized distribution with a diameter of 0.75 μm. The initial velocity of each ejected parcel is zero. The initial flow is static with a temperature of 300 K, while the pressure is 1 atm. All the boundaries, except the interface with the solid region, are considered as open to the surrounding air.

Figure 11.14 shows the interaction between the ejected parcels and gaseous phase. It can be seen that parcels (white dots) are lifted by the surrounding air as shown in Fig. 11.14a, and the entire velocity field is found to be symmetric-like. Figure 11.14b shows a similar phenomenon, but with more parcels at a higher velocity. However, asymmetric velocity appears in Fig. 11.14c due to the momentum transfer between the parcels and gaseous phase, which is the leading factor affecting the entire velocity field in the simulation domain. In addition, from the perspective of momentum conservation, the momentum exchange will lower the gas velocity and accelerate the parcels' velocity if the drag force is larger than gravity. However, once the number of the parcel is too large, the momentum of the gaseous phase will be completely dragged down as a result of this intensive momentum exchange. In other words, the gas velocity will be pulled downward the ground. Figure 11.14d shows the velocity at the zone close to the laser heating spot is directed toward the ground. As a result, the parcels on the paint will be pulled to a place nearby as shown. A comparison between

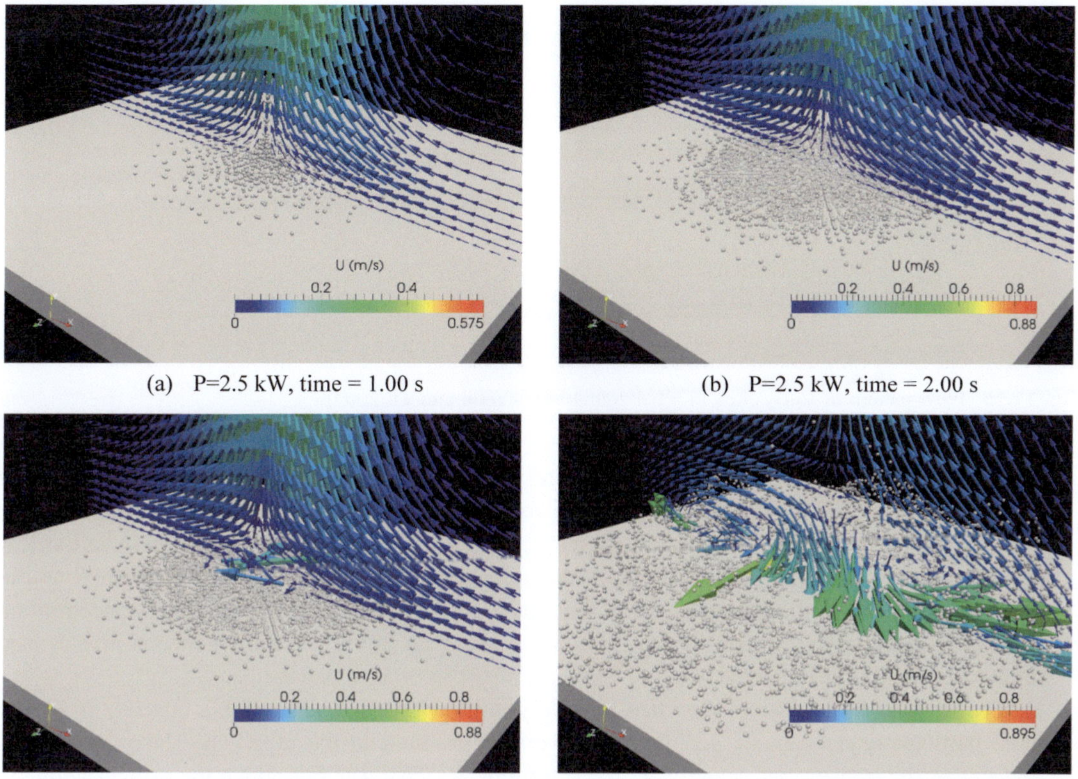

(a) P=2.5 kW, time = 1.00 s

(b) P=2.5 kW, time = 2.00 s

(c) P=2.5 kW, time = 2.20 s

(d) P=2.5 kW, time = 2.77 s

Fig. 11.14 State of parcel flow and gaseous phase at different times for laser interaction with a painted stainless steel substrate (Afrin et al. 2016)

Fig. 11.14c and d shows that the velocity does not have a significant increase during the period of 0.57 s due to a significant increase in the number of parcels, which certainly consumes a large amount of momentum held by the gaseous phase.

11.6 Liquid-Particle Systems

11.6.1 Liquid–Solid Transport in Ducts

11.6.1.1 Flow Patterns

The flow of suspension in liquid is referred to as slurry flow. Particles with different sizes ranging from a few microns to tens of millimeters can be transported via a slurry flow. Transportation of large fragments, like rocks, chunks of coal, clay, and ice, via slurry flow is referred to as *hydrotransport*. Similar to the gas-particle flow that was discussed in the previous section, the slurry flow can either be homogeneous or heterogeneous, depending on the properties of the fragments and liquid, as well as the flow velocity. The flow pattern is very important, and it affects the performance of the flow slurry transport, including pressure loss and erosion. The flow patterns that can be encountered in horizontal slurry flow are shown in Fig. 11.15 and are defined as follows (Michaelides et al. 2016; Crowe et al. 2012; Yeoh and Tu 2009):

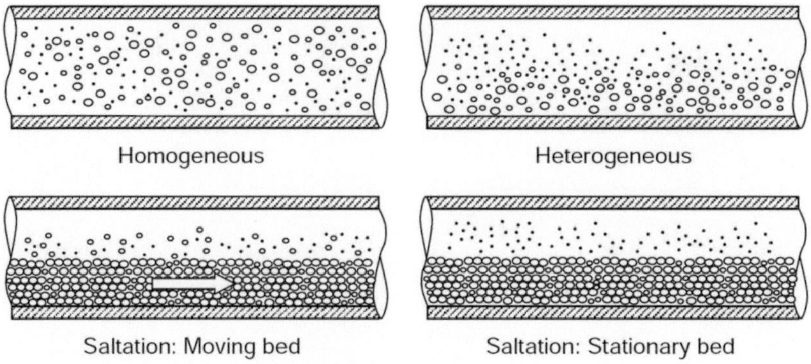

Fig. 11.15 Flow pattern of slurry flow in a horizontal tube (Yeoh and Tu 2009)

- **Homogeneous flow**. The particles are sufficiently small, and the liquid velocity is higher than the settling velocity, U_s, so that the particles are fully mixed with liquid by turbulent mixing. The homogeneous flow occurs when the particles' sizes are in the order of 10 μm or less and at low particle density. The slurry usually exhibits the behavior of non-Newtonian fluid, but the flow can be described using a homogeneous model (see Sect. 3.4.3). The variation of particle concentration from the top to the bottom of the tube should be less than 20%. The homogeneous flow is also referred to as nonsettling flow.
- **Heterogeneous flow** takes place when relatively large and dense particles are present, and these large particles tend to create sediment as a vertical gradient in the tube. The slurry velocity is lower than the transitional velocity, U_{m1}, or settling velocity, U_s, but above the critical velocity, U_c, or transitional velocity, U_{m2}. Although the particles are not uniformly distributed, the majority of the particles are still fully suspended. The velocities of the solid particles are different from that of the liquid; thus, the homogeneous model can no longer be applied.
- **Saltation flow**. When the particles form a packed bed at the bottom of the tube, saltation flow can be observed. The saltation flow can be subdivided into a moving bed or stationary bed regimes:

  a. **Moving bed**. The large and dense particles accumulate at the bottom of the tube and the particles in the bed can still move or slide along the bottom of the tube due to the shear force from the moving liquid. The upper portion of the flow is still a heterogeneous flow. The bottom portion moves much slower than the top portion, which results in a much lower overall mass flow rate. Saltation flow with moving bed occurs when the slurry velocity is below the critical velocity, U_c, or transitional velocity, U_{m2}, but above the transitional velocity, U_{m3}.
  b. **Stationary bed**. The slurry flow velocity is too low to drive all deposited particles to move so that a stationary layer is formed at the bottom of the tube. The flow in the stationary bed behaves like the liquid flow in a porous medium. Saltation flow in a stationary bed occurs when the slurry velocity is below the transitional velocity, U_{m3}, but above the transitional velocity, U_{m4}. The stationary bed should be avoided because it can become highly unsteady or can even plug the tube if the stationary bed becomes too thick.

- **Blockage**. If the slurry velocity is below the transitional velocity, U_{m4}, the tube will be blocked, and the system becomes liquid flow through a porous medium.

In contrast to nonsettling flow for homogeneous flow, heterogeneous flow and saltation flow are considered to be settling flow. The flow pattern in the slurry flow is influenced by many parameters,

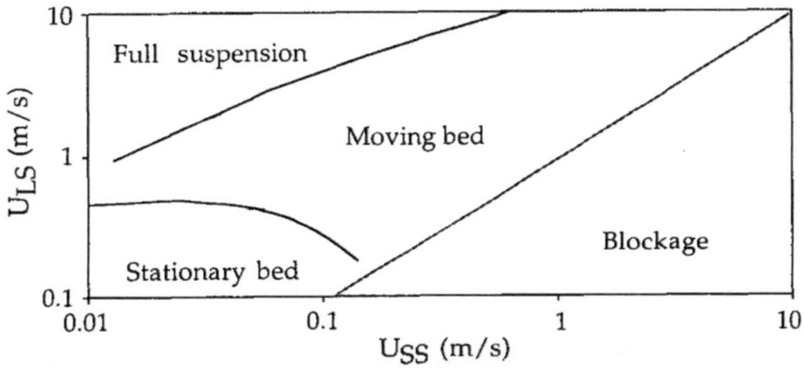

Fig. 11.16 Flow pattern regime map for slurry flow ($\rho_p = 1240\text{kg/m}^3, d_p = 3\,\text{mm}, D = 50\,\text{mm}$) from Doron and Barnea (1996)

including tube diameter, inclination angle, superficial velocities of liquid and particles, as well as physical properties of the particles and liquid. Therefore, a flow map in term of dimensionless parameters has not been successful, and existing flow maps are usually presented in terms of dimensional parameters for a specific combination of particles and liquid under limited particle sizes and tube diameters. Figure 11.16 shows a flow regime map in terms of superficial velocities of liquid, U_{LS}, and particles, U_{SS}, which was created by Doron and Barnea (1996) based on a three-layer model. The superficial velocities of particles and liquid are equivalent to their volumetric flow rates. The full suspension regime in Fig. 11.16 includes both homogeneous and heterogeneous flows discussed above. While this flow map illustrates the trends of flow regime transitions, it is only valid for parameters shown in the figure caption. More flow maps under different particle sizes, densities, and tube diameters can be found in Doron and Barnea (1996).

The homogeneous and heterogeneous flows are separated by the settling velocity, U_s, which can be estimated from Newitt et al. (1955).

$$U_s = 38.7(gDU_t)^{1/3} \tag{11.133}$$

where U_t is the terminal velocity of the particle with size d_p in the liquid.

In order for the slurry flow system to work normally, one would expect that the slurry flow velocity is above U_c or U_{m2} so that the full suspension flow, including homogeneous or heterogeneous flow, can be maintained. The critical velocity is the slurry velocity below in which a layer of sliding particles will accumulate on the bottom of a circular tube. Numerous efforts to determine the critical velocity have been made in the last several decades (Michaelides et al. 2016). The following correlations can be used to estimate the critical velocity (Wasp et al. 1977):

$$\frac{U_c}{[2gD(\rho_p - \rho_\ell)/\rho_\ell]^{0.5}} = 3.399\varepsilon_p^{0.2156}\left(\frac{d_p}{D}\right) \tag{11.134}$$

where ε_p is the volume fraction of particles.

The slurry flow velocity below U_c is not desirable; thus, the transitional velocities U_{m3} and U_{m4} are not important and will not be discussed here.

Example 11.5

Solid particles with a mean diameter of 3 mm and a density of 1240 kg/m^3 are transported in a tube with an inner diameter of 50 mm. The carrier liquid is water at 20 °C and the volume fraction of the particle is 10%. Determine the settling velocity and critical velocity.

Solution The density and viscosity of water at 20 °C can be found from Table B.48: $\rho_\ell =$ 999kg/m^3 and $\mu_\ell = 1.0016 \times 10^{-3}$Ns/m^2. To calculate the terminal velocity, the dimensionless particle diameter can be obtained from Eq. (11.72) is:

$$d_p^* = d_p \left[\frac{g\rho_\ell(\rho_p - \rho_\ell)}{\mu_\ell^2}\right]^{1/3}$$
$$= 3 \times 10^{-3} \left[\frac{9.81 \times 999(1240 - 999)}{(1.0016 \times 10^{-3})^2}\right]^{1/3} = 39.91$$

The dimensionless terminal velocity from Eq. (11.70) is

$$U_t^* = \left[\frac{18}{d_p^{*2}} + \frac{2.3348 - 1.7439\psi}{d_p^{*0.5}}\right]^{-1}$$
$$= \left[\frac{18}{39.91^2} + \frac{2.3348 - 1.7439 \times 1}{39.91^{0.5}}\right]^{-1} = 9.54$$

The terminal velocity obtained can be from Eq. (11.71):

$$U_t = U_t^* \left[\frac{g\mu_\ell(\rho_p - \rho_\ell)}{\rho_\ell^2}\right]^{1/3}$$
$$= 9.54 \left[\frac{9.81 \times 1.0015 \times 10^{-3}(1240 - 999)}{999^2}\right]^{1/3} = 0.0705\,\text{m/s}$$

The settling velocity is:

$$U_s = 38.7(gDU_t)^{1/3} = 38.7 \times (9.81 \times 0.05 \times 0.0705)^{1/3} = 12.61\,\text{m/s}$$

The critical velocity can be obtained from Eq. (11.134):

$$U_c = 3.399\varepsilon_p^{0.2156}\left(\frac{d_p}{D}\right)\left[\frac{2gD(\rho_p - \rho_\ell)}{\rho_\ell}\right]^{0.5}$$
$$= 3.399 \times 0.1^{0.2156}\left(\frac{0.003}{0.05}\right)\left[\frac{2 \times 9.81 \times 0.05 \times (1240 - 999)}{999}\right]^{0.5}$$
$$= 0.0249\ \text{m/s}$$

which is much lower than the settling velocity.

11.6.1.2 Pressure Drop

The pressure drop in steady-state slurry in a straight tube with a uniform cross section includes static and frictional pressure drops

$$\Delta p = \Delta p_s + \Delta p_f \tag{11.135}$$

where the static pressure drop can be obtained from:

$$\Delta p_s = \rho_p \varepsilon_p \Delta Z g + \rho_\ell (1 - \varepsilon_p) \Delta Z g \tag{11.136}$$

For a tube with a diameter of D and length of L, the friction pressure loss is

$$\Delta p_f = f \frac{4L}{D} \frac{\rho_m U_m^2}{2} \tag{11.137}$$

where the density of the slurry can be obtained from:

$$\rho_m = \varepsilon_p \rho_p + (1 - \varepsilon_p) \rho_\ell \tag{11.138}$$

The Fanning friction factor is:

$$f = \frac{\tau_w}{\rho_m U_m^2 / 2} \tag{11.139}$$

which depends on the flow regimes (laminar vs. turbulent) and type of slurry (Newtonian vs. non-Newtonian).

Assuming the slurry is incompressible, the stress, τ, and strain rate, $\dot{\gamma}$, are related by Newton's law of viscosity (see Eq. 1.15):

$$\tau = \mu_m \dot{\gamma} \tag{11.140}$$

where μ_m is the apparent viscosity, which is a constant for a Newtonian fluid.

For the cases of low particle concentration, the slurry can be considered Newtonian. The following correlation recommended by Churchill (1977) can be applied to both laminar and turbulent flow:

$$f = 2 \left[\left(\frac{8}{Re} \right)^{12} + (A + B)^{-1.5} \right]^{1/12} \tag{11.141}$$

where

$$A = \left[2.457 \ln \left(\frac{1}{C} \right) \right]^{16}, \; B = \left(\frac{37,530}{Re} \right)^{16}, \; C = \left(\frac{7}{Re} \right)^{0.9} + 0.27 \frac{\kappa}{D} \tag{11.142}$$

where κ is the surface roughness (m), and the Reynolds number is defined based on the velocity and properties of the slurry mixture:

$$Re = \frac{\rho_m U_m D}{\mu_m} \tag{11.143}$$

where the apparent viscosity of the slurry can be estimated by (Thomas 1965)

$$\mu_m/\mu_\ell = 1 + 2.5\varepsilon_p + 10.05\varepsilon_p^2 + 0.00273\exp(16.6\varepsilon_p) \tag{11.144}$$

For the cases with high solid particle concentration, the apparent viscosity μ_m is no longer a constant, and it is a function of the magnitude of shear stress, τ, or strain rate, $\dot{\gamma}$; thus, the slurry must be treated as non-Newtonian. There are many types of non-Newtonian fluid models such as the power-law model, Bingham plastic model, and Casson model, to name a few. The friction factors for different non-Newtonian fluid will take different forms. Many slurries can be treated as Bingham plastic fluid so that the pressure drop based on the Bingham plastic model will be discussed here.

Bingham plastic fluids show Newtonian behavior but require a finite yield stress before it begins to flow (Buckingham 1921); thus, the stress–strain rate relationship is:

$$\tau = \tau_b + \mu_b\dot{\gamma} \tag{11.145}$$

where τ_b and μ_b are yield stress and limiting viscosity, respectively. These two parameters depend on the volume fraction and size distribution of the particles. There are no reliable correlations for these two properties (Bird et al. 1983); thus, the most reliable approach is to measure them experimentally.

For the laminar flow of Bingham plastic fluid in a circular tube with a diameter of D or radius of $R = D/2$, the strain rate is

$$\frac{\partial u}{\partial r} = \begin{cases} 0, & \tau < \tau_b, \quad 0 < r < r_c \\ -(\tau - \tau_b)/\mu_b, & \tau \geq \tau_b, \quad r_c \leq r \leq R \end{cases} \tag{11.146}$$

where r_c is the radius of the unsheared core at which the shear stress is equal to the yield stress, τ_b. The velocity in the unsheared core is uniformly equal to a plug velocity:

$$u = U_p = \frac{\tau_b}{2\mu_b r_c}(R - r_c)^2, 0 < r < r_c \tag{11.147}$$

The velocity in the sheared annular region ($r_c \leq r \leq R$) is

$$u = \frac{R - r}{\mu_b}\left[\frac{\tau_w}{2}\left(1 + \frac{r}{R}\right) - \tau_b\right], \; r_c \leq r \leq R \tag{11.148}$$

The Fanning friction factor can be obtained from (see Problem 11.28):

$$f_L = \frac{16}{Re_b}\left[1 + \frac{1}{6}\frac{\mathrm{He}_b}{Re_b} - \frac{1}{3}\frac{\mathrm{He}_b^4}{f_L^3 Re_b^7}\right] \tag{11.149}$$

where

$$Re_b = \frac{\rho_m U_m D}{\mu_b} \tag{11.150}$$

$$He_b = \frac{D^2 \rho_m \tau_b}{\mu_b^2} \tag{11.151}$$

are the Bingham Reynolds number and Hedstrom number, respectively. Equation (11.149) is implicit for friction factor f_L so that iteration is necessary to obtain friction factor. Swamee and Aggarwal (2011a) proposed the following explicit expression for the Fanning friction factor:

$$f_L = \frac{16}{Re_b}\left[1 + \left(\frac{He_b}{6.2218 Re_b}\right)^{0.958}\right] \tag{11.152}$$

For turbulent flow, the friction factor can be obtained from (Darby et al. 1992):

$$f_T = 10^a Re_b^{-0.193} \tag{11.153}$$

where

$$a = -1.47\left[1 + 0.146 \exp(-2.9 \times 10^{-5} He_b)\right] \tag{11.154}$$

Unlike Newtonian fluid where the critical Reynolds number that distinguishes laminar and turbulent flows can be well defined, the critical Reynolds number for Bingham plastic fluid is much more difficult to predict. The transition from laminar to turbulent flow occurs gradually. Moreover, two critical Reynolds numbers must be considered: a lower critical Reynolds number, Re_{cr1}, below which the flow is laminar, and an upper critical Reynolds number, Re_{cr2}, above which flow is fully turbulent. While the upper critical Reynolds number depends on flow disturbance and is difficult to define precisely, the lower critical Reynolds number can be obtained from (Swamee and Aggarwal 2011b):

$$Re_{cr1} = 2100\left(1 + \frac{He_b}{3600}\right)^{0.35} \tag{11.155}$$

Without knowing the flow regime, a single correlation that covers the entire flow regime will be beneficial. Darby and Melson (1981) combined the Fanning friction factors for laminar and turbulent flow and recommended the following empirical correlation:

$$f = \left(f_L^m + f_T^m\right)^{1/m} \tag{11.156}$$

where

$$m = 1.7 + \frac{40,000}{Re_b} \tag{11.157}$$

The Fanning factor for laminar flow can be obtained from Eq. (11.149) or more conveniently from Eq. (11.152), and the turbulent friction factor can be obtained from Eq. (11.153).

11.6.2 Nanofluids

Nanofluids are a new class of nanotechnology-based heat transfer fluids that are devised by stably suspending a small amount (1 vol.% or less) of nanoparticles, nanofibers, or nanotubes with a length on the order of 1–50 nm in the traditional heat transfer fluids (Choi 2009). Thermophysical properties, nonequilibrium heat and mass transfer, and nanofluid flow in a two-fluid microchannel heat exchanger will be discussed.

11.6.2.1 Thermophysical Properties

The effective density and capacitance of the nanofluid are calculated using the classical models for very dilute suspensions are available in the literature and are expressed as:

$$\rho_{\mathrm{nf}} = \left(1 - \varepsilon_{\mathrm{p}}\right)\rho_{\mathrm{f}} + \varepsilon_{\mathrm{p}}\rho_{\mathrm{p}} \tag{11.158}$$

$$\left(\rho c_{\mathrm{p}}\right)_{\mathrm{nf}} = \left(1 - \varepsilon_{\mathrm{p}}\right)\left(\rho\, c_{\mathrm{p}}\right)_{\mathrm{f}} + \varepsilon_{\mathrm{p}}\left(\rho\, c_{\mathrm{p}}\right)_{\mathrm{p}} \tag{11.159}$$

where ε_{p} is nanoparticle volume fraction, while subscript *p* and *f* stand for nanoparticle and base fluid.

The thermal conductivity of nanofluid is affected by many factors, including particle size, volume concentration, particle shape, and temperature. It was reported by early researchers that nanofluids exhibited very high thermal conductivity even when the concentrations of suspended nanoparticles were low. In response to the claim of anomalous thermal conductivity of nanofluids, an International Nanofluid Property Benchmark Exercise (INPBE) was conducted by 34 organizations worldwide to measure the thermal conductivity of identical samples of the nanofluids (Buongiorno et al. 2009). They found that the thermal conductivity enhancement afforded by the tested nanofluids increased with increasing particle loading, particle aspect ratio, and decreasing base fluid thermal conductivity. They also suggested that the effective medium theory was found to be in good agreement with the experimental data and that no anomalous enhancement of thermal conductivity was achieved in the nanofluids tested in their exercise. They suggested that, for well-dispersed noninteracting spherical particles with negligible thermal resistance at the particle/fluid interface, the nanofluid thermal conductivity k_{eff} is related to the thermal conductivity of the base fluid, k_{f}, by Maxwell's model:

$$\frac{k_{\mathrm{eff}}}{k_{\mathrm{f}}} = \frac{k_{\mathrm{p}} + 2k_{\mathrm{f}} + 2\varepsilon_{\mathrm{p}}(k_{\mathrm{p}} - k_{\mathrm{f}})}{k_{\mathrm{p}} + 2k_{\mathrm{f}} - \varepsilon_{\mathrm{p}}(k_{\mathrm{p}} - k_{\mathrm{f}})} \tag{11.160}$$

where k_{p} is the thermal conductivity of the nanoparticle material.

Koo and Kleinstreuer (2004, 2005) suggested that Maxwell's model can be improved by considering the effects of Brownian motion, and the thermal conductivity of the nanofluid can be calculated according to following equations:

$$k_{\mathrm{eff}} = k_{\mathrm{f}}\frac{k_{\mathrm{p}} + 2k_{\mathrm{f}} + 2\varepsilon_{\mathrm{p}}(k_{\mathrm{p}} - k_{\mathrm{f}})}{k_{\mathrm{p}} + 2k_{\mathrm{f}} - \varepsilon_{\mathrm{p}}(k_{\mathrm{p}} - k_{\mathrm{f}})} + 5 \times 10^4 \beta\varepsilon_{\mathrm{p}}(\rho c_{\mathrm{p}})_{\mathrm{f}}\sqrt{\frac{k_{\mathrm{b}}T}{d_{\mathrm{p}}\rho_{\mathrm{p}}}}f(T, \varepsilon_{\mathrm{p}}) \tag{11.161}$$

where the second term on the right-hand side can account for the effects of Brownian motion. d_{p} and ρ_{p} are the mean diameter and a density of nanoparticles, respectively. k_{b} is Boltzmann constant. The β-function is to account for the fraction of the liquid volume that travels with a nanoparticle. For water-based nanofluids, it can be obtained from

$$\beta = \begin{cases} 0.0137(100\varepsilon_{\mathrm{p}})^{-0.8229} & \varepsilon_{\mathrm{p}} < 1\%, \mathrm{CuO} \\ 0.0011(100\varepsilon_{\mathrm{p}})^{-0.7272} & \varepsilon_{\mathrm{p}} > 1\%, \mathrm{CuO} \\ 0.0017(100\varepsilon_{\mathrm{p}})^{-0.0841} & \varepsilon_{\mathrm{p}} > 1\%, \mathrm{Al_2O_3} \end{cases} \tag{11.162}$$

The *f*-function is to account for the effect of particle interactions, which is a linear function of the temperature:

$$f(T, \varepsilon_p) = (-6.04\varepsilon_p + 0.4705)T + (1722.3\varepsilon_p - 134.63) \quad (11.163)$$

which is valid for $1\% < \varepsilon_p < 4\%$ and $300K < T < 325K$.

Thermal conductivity of the nanofluids has been extensively studied in the last two decades and numerous empirical correlations based on analytical, numerical, and experimental studies have been proposed. Angayarkanni and Philip (2015) reviewed the latest development on the thermal properties of the nanofluids and over 30 different correlations for effective thermal conductivity were summarized.

Xuan and Roetzel (2000) suggested that the effective viscosity of the nanofluid can be obtained from:

$$\mu_{eff} = \frac{\mu_f}{(1 - \varepsilon_p)^{2.5}} \quad (11.164)$$

Koo and Kleinstreuer (2005) studied the laminar nanofluid flow in microheat sinks and suggested that the Brownian motion affected the viscosity in a similar way like its effect on nanofluid thermal conductivity. Therefore, the effective viscosity of the nanofluid can be obtained from

$$\mu_{eff} = \frac{\mu_f}{(1 - \alpha)^{2.5}} + 5 \times 10^4 \beta \varepsilon_p \rho_f \sqrt{\frac{k_b T}{d_p \rho_p}} f(T, \varepsilon_p) \quad (11.165)$$

11.6.2.2 Coupling Between Nanoparticles and Base Fluid

Motivated by the application of nanofluid in the oscillating heat pipe and other two-phase devices, heat conduction across a thin nanofluid film subject to periodic heat flux heating on one side and specified temperature on the other side was investigated by Zhang and Ma (2008). When the nanofluid layer is heated, heat is transferred to the base fluid first and then the nanoparticles are heated by the base fluid. On the other hand, when the nanofluid layer is cooled, the base fluid is cooled first and the nanoparticles are then cooled by interaction with the base fluid. Therefore, it is expected that there will be a delay from the variation of the temperature of the base fluid to that of the nanoparticles. The nonequilibrium between the nanoparticles and the base fluid can be modeled using a two-temperature model. The energy equations of the nanoparticles (*s*) and the base fluid (*f*) can be, respectively, expressed as:

$$\varepsilon_p (\rho c_p)_s \frac{\partial T_s}{\partial t} = G(T_f - T_s) \quad (11.166)$$

$$(1 - \varepsilon_p)(\rho c_p)_f \frac{\partial T_f}{\partial t} = k_{eff} \nabla^2 T_f + G(T_s - T_f) \quad (11.167)$$

where G is the coupling factor; ε_p is the volume fraction of the nanoparticle; T_s is the temperature of nanoparticles; T_f is the temperature of the base fluid, and k_{eff} is the effective thermal conductivity of the nanofluids.

The two-temperature model represented by Eqs. (11.166) and (11.167) is commonly applied to nonequilibrium heat conduction in porous media and passive thermal control systems for electronics. In the two-temperature model, one of the most important properties is the coupling factor, G, that indicates the intensity of the heat exchange between the nanoparticle and the base fluid. Kang et al.

(2012) addressed nonequilibrium heat transfer in a nanofluid, and the coupling factor was obtained by a molecular dynamics simulation.

The heat exchange between the nanoparticles and base fluid can be obtained by:

$$Q = GV_{\mathrm{p}}(T_{\mathrm{s}} - T_{\mathrm{f}}) \tag{11.168}$$

where V_{p} is the volume of the nanoparticle. If the amount of the heat exchange, the volume of the nanoparticles, and the temperature difference between nanoparticles and the base fluid are known, the coupling factor G can be obtained from Eq. (11.168).

A nonequilibrium molecular dynamics (NEMD) simulation is used to calculate the coupling factor. In the simulation, the nanoparticle atom with a minimum velocity is chosen first, followed by the base fluid atom that has the maximum velocity. These two velocities change with the system still resting in the energy conservation mode. This procedure was performed every few steps in the simulation causing the temperatures of the nanoparticles and the base fluid to increase and decrease, respectively, which caused a temperature difference between the nanoparticles and the base fluid. As a result, heat flux from the nanoparticles to the base fluid was generated. When this flux attained the kinetic energy exchange that results from the artificial velocity exchange between the nanoparticle and base fluid atoms, the system reached dynamic equilibrium and the temperatures of the nanoparticles and the base fluid became steady. Therefore, the coupling factor was calculated by Eq. (11.168). For the 5.5% vol. nanofluid with a 5.384 nm nanoparticle diameter, the resulting coupling factor was 4.34×10^{15} W/m^3-K.

Zhang and Ma (2008) suggested that the coupling factor can be approximately calculated by the following equation:

$$G = 6\mathrm{Nu}\varepsilon_{\mathrm{p}}k_{\mathrm{eff}}/d_{\mathrm{p}}^2 \tag{11.169}$$

where the Nusselt number Nu is approximately 2 for the case of pure conduction and the effective conductivity k_{eff} is 0.171 W/m-K. The resulting coupling factor is 3.87×10^{15} W/m^3-K. Compared with the result obtained by the MD simulation, it is 10.8% lower. This is still valid because 2 is the minimum possible value for the Nusselt number. From the molecular dynamics simulation result, it is suggested that the Nusselt number should be approximately 2.2.

The influences of the nanoparticle volume fraction and diameter on the coupling factor were then studied. The coupling factors of the nanofluid that contains a 5.384-nm-diameter nanoparticle with vol. 2.48, 3.16, 4.1, and 5.46% were obtained using the molecular dynamics simulation, and the results are shown in Fig. 11.17a. The coupling factor linearly increases with increasing nanoparticle volume fraction, which is consistent with Eq. (11.169). Figure 11.17b shows the coupling factor for different nanoparticle diameters but also has the same volume fraction of 5.46% vol. using molecular dynamics simulation. The coupling factor linearly decreases with increasing nanoparticle diameter. This phenomenon can, once again, be explained by Eq. (11.169).

11.6.2.3 Nanofluid Flow in a Two-*Fluid Microchannel Heat Exchanger*

Microchannel heat exchangers (MCHEs) became increasingly important to the current future of different industries due to their dependence on the successful cooling of electronic components in telecommunication, robotics, metrology, and automotive industries. One of the special types of MCHEs is the counterflow microchannel heat exchangers (CFMCHEs), which are more attractive than single-layered microheat exchangers and have better thermal performance. This type of MCHE can be used for cooling or heating a specific fluid using a secondary fluid. Enhancement of thermal efficiency of such systems is very essential from the energy saving perspective. Different factors such as geometry, velocity, number of channels, and thermal properties of both solid and fluid affect the

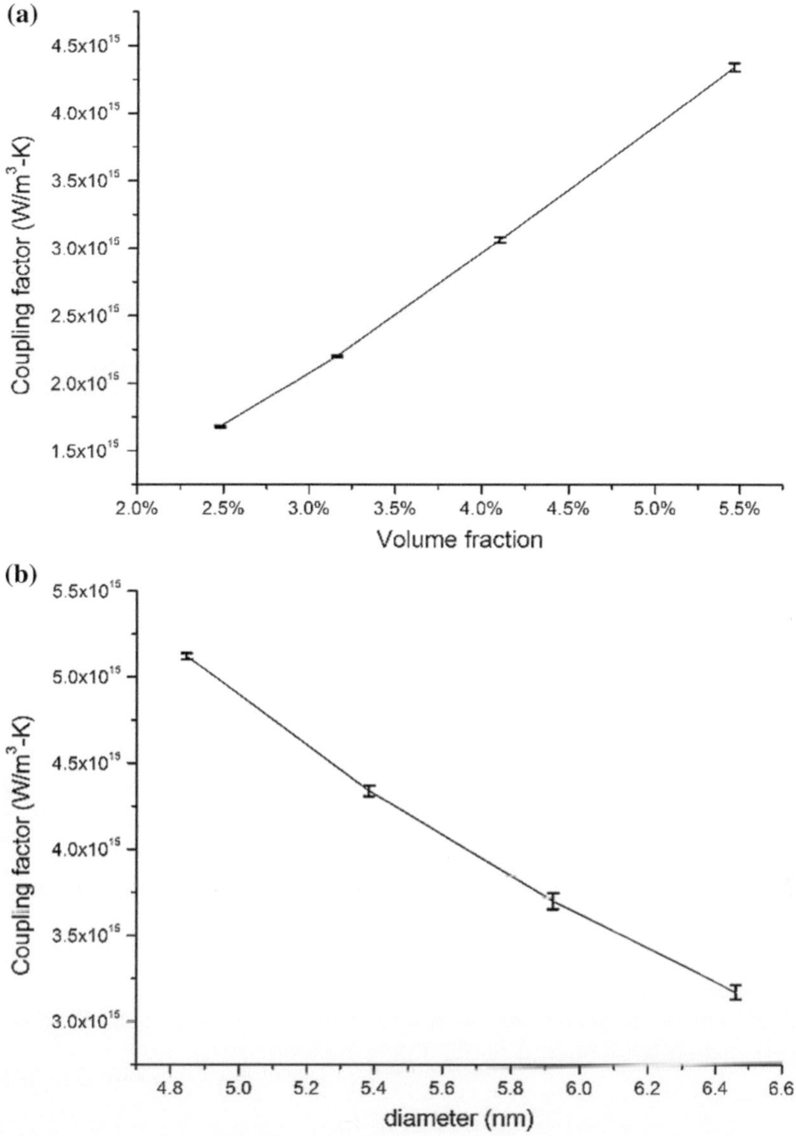

Fig. 11.17 Effect of **a** volume fraction and **b** nanoparticle diameter on coupling factor

thermal performance of MCHEs. The thermal and hydrodynamic performances of MCHEs can be improved by finding an optimum configuration of the system and by improving the heat transfer properties of the working fluid.

A counterflow microchannel heat exchanger (Fig. 11.18) composed of many channels is the usual device for the heating/cooling of a specific fluid. Mohammadian et al. (2014) carried out a numerical study with Al_2O_3–water nanofluid as the working fluid of hot or cold channels. Conjugated heat transfer with and without Al_2O_3–water nanofluid in a CFMCHE had been considered. The nanoparticles are ultrafine (<100 nm) so that the single-phase approach proposed by Xuan and Roetzel (2000) was used for modeling the nanofluid. To simplify the analysis, only one symmetrical part of the CFMCHE is considered. The cold fluid flows through the top channels, which takes away

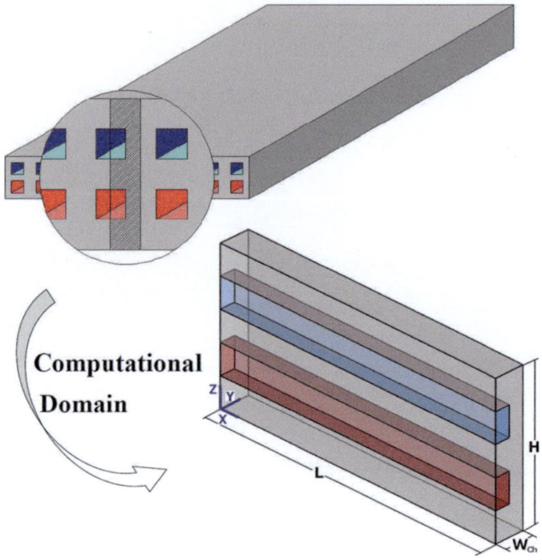

Fig. 11.18 Schematic of counterflow microchannel heat exchanger with Al_2O_3–water nanofluid (Mohammadian et al. 2014)

heat from the bottom channels where the hot fluid is flowing. The inlet temperatures of fluid at the bottom and top channels are $T_h = 325\,\text{K}$ and $T_c = 300\,\text{K}$,, respectively. The width, total length, and height of each channel are 40 μm, 10 mm, and 40 μm, respectively. The thickness of solid material between the two adjacent channels and between hot and cold channels is 40 μm each. Since the temperature fields in the hot and cold channels are different, using nanofluids as a coolant or heating agent can lead to different thermal efficiencies. Two different combinations of working fluids were studied: (i) water in the cold channel and nanofluid in the hot channel, and (ii) nanofluid in the cold channel and water in the hot channel.

The nanofluid flow is assumed to be incompressible and the inlet Reynolds number is less than 480; hence, flow is assumed to be laminar. The continuity, momentum, and energy equations for a steady flow with temperature-dependent material properties are used. The effective thermal conductivity of the nanofluid was obtained by the following correlation:

$$k_{\text{eff}} = k_f \left[\frac{\left(d_p \left(R_b + \frac{d_p}{k_p} \right)^{-1} + 2k_f \right) - 2\varepsilon_p \left(k_f - d_p \left(R_b + \frac{d_p}{k_p} \right)^{-1} \right)}{\left(d_p \left(R_b + \frac{d_p}{k_p} \right)^{-1} + 2k_f \right) + \varepsilon_p \left(k_f - d_p \left(R_b + \frac{d_p}{k_p} \right)^{-1} \right)} \right] + 5 \times 10^4 \beta \varepsilon_p (\rho c_p)_f \sqrt{\frac{\kappa_b T}{d_p \rho_p}} f(T, \varepsilon_p) \tag{11.170}$$

which is based on Eq. (11.161) with the effects of Kapitza resistance considered. The value of Kapitza resistance is $R_b = 4 \times 10^{-8}\,\text{km}^2/\text{W}$. The effective viscosity is calculated using Eq. (11.165).

Major parameters of heat transfer and hydrodynamic performance observed in this study are the effectiveness and pumping power. The heat exchanger effectiveness is the ratio of the actual heat transfer to the maximum possible heat that can be transferred:

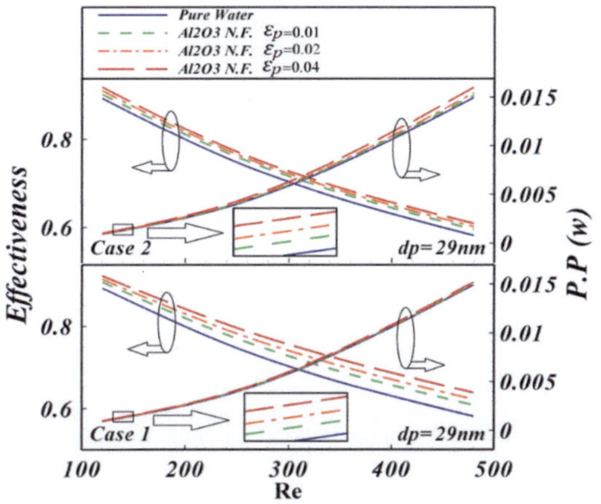

Fig. 11.19 Effect of Reynolds number on thermal effectiveness and pumping power at different volume fraction of nanoparticles

$$\varepsilon = \frac{C_h(T_{hi} - T_{ho})}{C_{min}(T_{hi} - T_{ci})} = \frac{C_c(T_{co} - T_{ci})}{C_{min}(T_{hi} - T_{ci})} \tag{11.171}$$

where C_c and C_h are cold and hot heat capacity rates, respectively. C_{min} is equal to C_c or C_h—whichever is smaller. T_{hi}, T_{ci}, T_{ho}, and T_{ho} are the temperature at hot and cold channel inlets and the bulk temperature at the outlet of hot and cold channels.

The pumping power required to circulate hot and cold fluids in a CFMCHE is:

$$P.P = \dot{V}_h(P_{hi} - P_{ho}) + \dot{V}_c(P_{ci} - P_{co}) \tag{11.172}$$

where $\dot{V}$ is the volumetric flow rate.

Figure 11.19 shows the results of the effectiveness and pumping power obtained for the CFMCHE using different Reynolds numbers and volume fractions. Clearly, increasing Reynolds number and adding low volume fractions of nanoparticles (0.01–0.04) to the base fluid leads to a significant increase in effectiveness; with increasing the volume fraction, the enhancement increases. In addition to the advantage of increasing thermal performance as nanoparticles are added to the base fluid. However, there is a drawback associated with the increase in the pumping power in the device with volume fraction. It is clear from Fig. 11.19 that increasing volume fraction of nanoparticles, which is responsible for larger heat transfer performance, also lead to higher pumping power but as expected the increase is small, especially for lower particle volume fractions. Notably, the low range of volume fractions and subsequent small increases in viscosity will not cause a noticeable penalty on pressure drop or pumping power.

11.6.3 Nanoencapsulated Phase-Change Material (PCM) Slurry

The nanoencapsulated phase-change material slurry is an effective coolant that disperses nanoencapsulated phase-change material (NEPCM) in a heat transfer fluid, such as water. NEPCM particles are composed of a core of phase-change material (PCM; e.g., paraffin wax) with a wall around it, which can be made of cross-linked polymer. The wall material is usually 14–20% of the total capsule

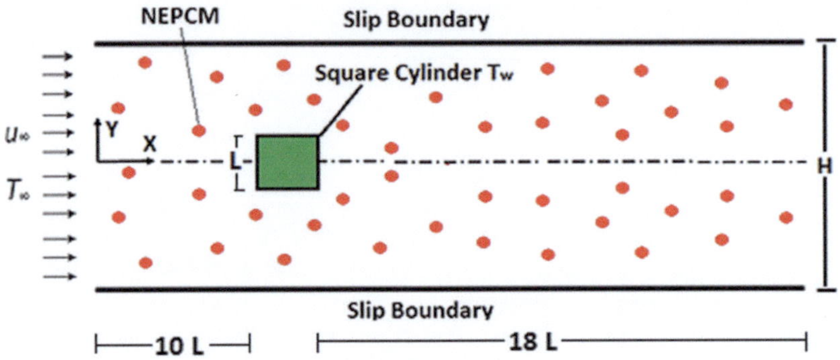

Fig. 11.20 Schematic of flow around a square cylinder and computational domain (Seyf et al. 2014)

mass and is sufficiently flexible to accommodate the volume changes that accompany solid–liquid phase change (Sabbah et al. 2011). The latent heat of the NEPCM particles provides an approach to energy storage and enhanced heat transfer when the particles experience the phase change. The high effective thermal conductivity of the mixture due to microconvection induced by nanoparticles is coupled with the large latent heat of NEPCM, achieving high heat fluxes with little temperature variation. Therefore, NECPM coolants are well suited for use in high-performance compact heat exchangers and heat sinks used in electronic equipment.

Seyf et al. (2014) studied the effect of NEPCM slurry on unconfined flow and heat transfer over a square isothermal cylinder. The system of interest in this work is a square cylinder with a wall temperature of T_w and side L placed symmetrically between the upper and lower boundaries as shown in Fig. 11.20. The cylinder is exposed to a free stream with constant temperature, T_∞, and velocity, U_∞. The ratio of the height of computational domain to side of block (*H*/*L*) is 25 to prevent the misleading effects of the presence of the lower and upper boundaries on the flow and heat transfer near the block. To reduce the effect of outlet and inlet boundary conditions on flow patterns in the vicinity of block, the upstream and downstream distances are chosen as 18 and 10 times that of the side of cylinder, respectively. As shown in Fig. 11.20, an NEPCM slurry flows over the square cylinder and its temperature increases as it moves over the block and reaches the melting temperature of the PCM. There is no mass transfer between the carrier fluid and capsules because when PCM melts inside the capsules; the melted PCM does not mix with the carrier fluid because it remains contained in the capsules. The carrier fluid exhibits small temperature change when the PCM melts. The NEPCM slurry consists of n-octadecane as nanoparticles and water. The problem is described using the two-dimensional incompressible Navier–Stokes and energy equations in the Cartesian coordinate system with temperature-dependent thermophysical properties.

The slurry consists of water and n-octadecane phase-change particles with diameter, melting point, density, specific heat, thermal conductivity, and a latent heat of 100 nm, 296.15 K, 815 kg/m^3, 2000 J/kgK, 0.18 W/mK, and 2.44×10^5 J/kg, respectively. The dynamic viscosity of slurry is calculated using following equation:

$$\mu_{\text{slurry}} = \mu_{\text{water}}\left(1 - C - 1.16C^2\right)^{-2.5} \tag{11.173}$$

where C is volume concentration of NEPCM, and μ_{water} is the water viscosity that is a function of temperature.

The static thermal conductivity of the suspension or thermal conductivity of slurry at rest (no shear rate) can be obtained from Eq. (11.160). Due to particle–particle, particle–liquid, and particle–wall interaction, the effective thermal conductivity of slurry under motion increases. These interactions play a major role in the enhancement of the thermal conductivity of mixture. The effective thermal conductivity of slurry flow is specified by the following correlation:

$$\begin{aligned} &k_{\text{slurry}} = k_{\text{water}}\left(1 + BCPe_{\text{p}}^{\text{m}}\right) \\ &B = 3, \quad m = 1.5, \qquad Pe_{\text{p}} < 0.67 \\ &B = 1.8, \quad m = 0.18, \quad 0.67 < Pe_{\text{p}} < 250 \\ &B = 3, \quad m = \tfrac{1}{11}, \qquad Pe_{\text{p}} > 250 \end{aligned} \tag{11.174}$$

The particle Peclet number is defined as

$$Pe_{\text{p}} = \frac{e\, d_{\text{p}}^2}{\alpha_{\text{water}}} \tag{11.175}$$

where α_{water} is thermal diffusivity of water. The shear rate is a function of all the spatial coordinates and corresponding velocities. The magnitude of the shear rate can be calculated using the following equation:

$$e = \sqrt{\frac{1}{2}\left[2\left(\frac{\partial u}{\partial y} + \frac{\partial v}{\partial x}\right)^2 + \left(2\frac{\partial u}{\partial x}\right)^2 + \left(2\frac{\partial v}{\partial y}\right)^2\right]} \tag{11.176}$$

The effective thermal conductivity correlation given above shows that the thermal conductivity is strongly dependent on (i) the shear rate, which can be increased by decreasing the conduction dimension or increasing slurry flow rate; and (ii) particle size is due to the interactions (drag force, lift force, and virtual mass) between the liquid and the particles. Slurry density and specific heat are given by

$$\rho_{\text{slurry}} = (1 - \phi)\rho_{\text{water}} + \phi\, \rho_{\text{p}} \tag{11.177}$$

$$c_{\text{p,slurry}} = (1 - \phi)c_{\text{p,water}} + \phi\, c_{\text{p,p}} \tag{11.178}$$

$$\phi = \frac{C\, \rho_{\text{p}}}{\rho_{\text{water}} + C\left(\rho_{\text{p}} - \rho_{\text{water}}\right)} \tag{11.179}$$

where $c_{\text{p,water}}$, ρ_{water}, ϕ, and ρ_{p} are the specific heat and density of water, mass concentration, or loading fraction of slurry and density of NEPCM, respectively. It is assumed that melting takes place in a range of temperature between T_{m1} and T_{m2}, and the following sine profile is used to represent the NEPCM particle equivalent specific heat for $T_{\text{m1}} \le T \le T_{\text{m2}}$:

$$c_{\text{p,p}} = c_{\text{p,pcm}} + \left[\frac{\pi}{2}\left(\frac{h_{\text{m}}}{T_{\text{mr}}} - c_{\text{p,pcm}}\right) \sin \pi\left(\frac{T - T_{\text{m1}}}{T_{\text{mr}}}\right)\right] \tag{11.180}$$

where $T_{\text{mr}} = T_{\text{m2}} - T_{\text{m1}}$ is a range of melting temperature. For temperatures higher or lower than melting range the value of the specific heat of particle is equal to $c_{\text{p,pcm}}$.

The two-dimensional governing equations are discretized using the finite volume method with the SIMPLE algorithm to handle the linkage between velocity and pressure. A collocated grid was used to discretize the solution domain, i.e., the pressure, temperature, velocity components, and all properties are stored on the main grid, which is at the center of the control volume. The convection–diffusion terms in the momentum and energy equations are discretized by a QUICK scheme. The resulting systems of equations for all dependent variables are solved using a point implicit linear equation in conjunction with an algebraic multigrid (AMG) method. The typical number of iterations to get a converged solution with a residual of less than 10^{-7} for all variables is about 5500.

Figures 11.21 and 11.22 show streamlines and temperature contour around the cylinder for *Re* = 45 for water and NEPCM slurry with *C* = 0.3. It can be seen that the flow is attached to the block until it reaches the trailing edge of the upper surface, and then, it separates from the cylinder and forms two symmetric recirculation cells. At a constant inlet volumetric flow rate, due to the higher viscosity of slurry compared to pure water, the flow of fluid over cylinder becomes slower so the shape of the core of these cells changes when the NEPCM slurry is employed as working fluid. The cores of these cells are slightly oval for the cases of water and circular for the cases of the NEPCM slurry. Since the inlet Reynolds number was kept as a constant, the inlet velocity was higher for the slurry cases which cause higher velocity gradient and velocity magnitude in the computational domain as shown in Fig. 11.21; hence, the recirculation zones became larger for NEPCM slurry cases. On the other hand, NEPCM particles increased the viscosity of mixture which had a significant effect on the size of the recirculation zone behind the block. The positive enhancement in velocity balances the negative effect of the viscosity increase; hence, the size of recirculation zones remains nearly constant compared to the cases of pure water.

Figure 11.22 indicates that using slurry as the working fluid causes the isotherms to become more condensed near the surface hence the overall heat transfer from wall increases. It is worth noting that a higher volume fraction of nanoparticles causes the thickness of the thermal boundary layer near the surface to decrease, which exhibits the large temperature gradient and hence higher total heat transfer coefficient; this is due to the latent heat of NEPCM particles, which acts as a heat sink, reduces the thickness of the thermal boundary layer on the surface.

Figure 11.23 shows the heat transfer coefficient along the surfaces for *Re* = 45 at different volume fractions of NEPCM. The heat transfer coefficient is low at point A; then, it gradually increases along

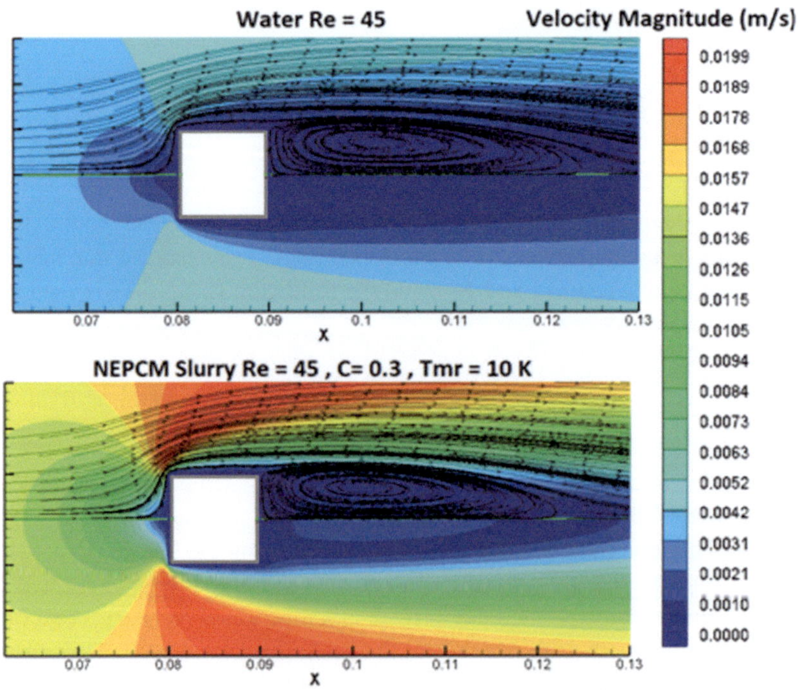

Fig. 11.21 Streamlines and velocity contours of water and slurry at *Re* = 45 (Seyf et al. 2014)

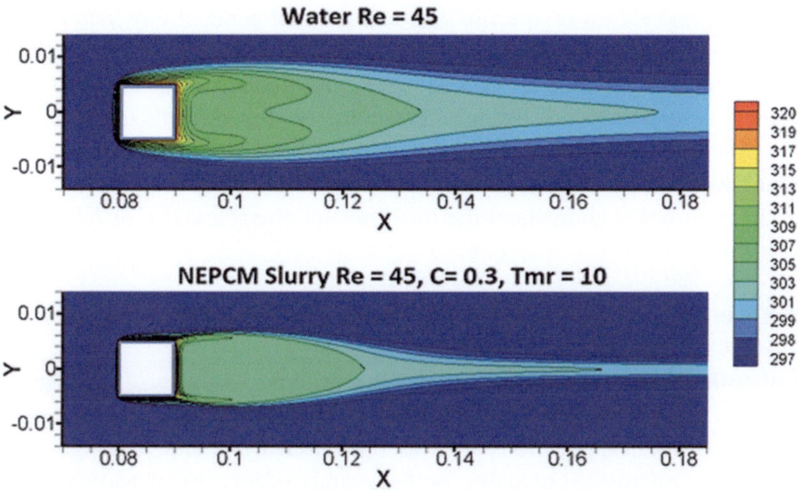

Fig. 11.22 Temperature contours of water and slurry at Re = 45 (Seyf et al. 2014)

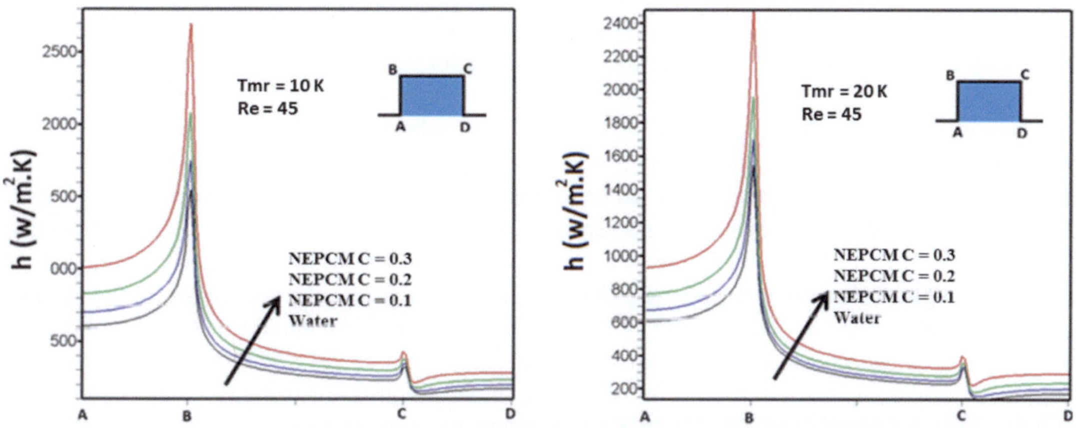

Fig. 11.23 Local heat transfer coefficients along the edge of the square cylinder in cross-flow of a nanofluid (Seyf et al. 2014)

the vertical surface until point B due to the expected bending of the isotherm and streamlines contours to negotiate the corner. From point B to near point C, the heat transfer coefficient decreases, and at point C, it suddenly increases and reaches the second maxima that are always lower than the value seen at point B; beyond this point, it again decreases up to the point D (the rear stagnation point) where the local heat transfer coefficient is minimum due to the very low velocity of the fluid. The heat transfer coefficient over the rear surface of the cylinder is lower than the upper and front surfaces as the temperature contours are less dense behind the square cylinder. Furthermore, due to the high-temperature gradient at corners of the block, the heat transfer coefficient reaches a peak in these locations.

Figure 11.23 indicates that increasing volume fraction of NEPCM increases the heat transfer coefficient over the block. This is due to the fact that adding NEPCM particles remarkably enhances the heat capacity of the slurry caused by the large latent heat of the NEPCM. The maximum enhancement occurs at point B due to both the specific heat enhancement and the bending of the isotherm and streamlines contours. It is worth noting that to achieve a constant Reynolds number, the inlet velocity is adjusted for different cases of NEPCM slurries. It is clear that as volume fractions of nanoparticles increase, the values of specific heat and thermal conductivity increase, but the density decreases. To keep the Reynolds number constant, the overall interaction of these parameters is such that the inlet velocity should increase; hence, the velocity gradient increases and consequently the temperature gradient around the block changes depending on the viscosity and specific heat of slurry. On the other hand, with the increasing volume fraction of particles, the viscosity of slurry increases which may tend to decrease the velocity gradient near the block. Therefore, these opposing behaviors should be considered to better understand the heat transfer characteristics of NEPCM.

11.6.4 Self-assembly of Submicron Particles Suspended in Liquid

Self-assembly of colloids and microscale particles has become a subject of great interest due to the widespread advancement of microscale technologies. Fabrication of microscale structures that is essential for many industrial applications such as optics, catalysis and printing technology can be achieved by assembling microsized particles. Self-assembly of microscale particles induced by evaporation is a simple yet widely used microfabrication technique. A complete numerical analysis of such phenomena can be of great importance by providing a deeper understanding of the assembling process. CFD-DEM is a robust and efficient method for simulating interactions among a fluid and large number of particles. In CFD-DEM, particle to particle interaction is computed by the discrete element method (DEM) which is a highly appreciated Lagrangian approach for particle interaction simulation. The fluid phase is governed by Navier–Stokes equation, and finite volume method is used for the CFD part.

Tayeb et al. (2016b) investigated the self-assembly of submicron particles suspended in a water film. The size of the CFD domain is 6 × 6 × 6 mm. Liquid water filled the container to a height of 0.5 mm with 4500 particles each with a radius of 5 μm, which are initially suspended in liquid water film without touching each other. To prevent the influence of side walls, the periodic boundary condition is applied in lateral directions. The governing equations for the fluid and particles can be found in Sect. 11.4. The time step for the DEM case is 10^{-8} s whereas the time step for the CFD case is 10^{-5} s, which means that for each CFD iteration, DEM runs for 1000 iterations. A smaller time step for DEM is applied to prevent any unrealistic overlap between particles during collisions. The entire CFD domain is kept at a temperature of 373.15 K initially with the exception of the bottom wall which is kept at a temperature of 378.15 K at all times.

Figure 11.24 shows the packing structure of the particles after they settled to the bottom of the container. It can be seen that packing configuration tends to form a pattern with the particles

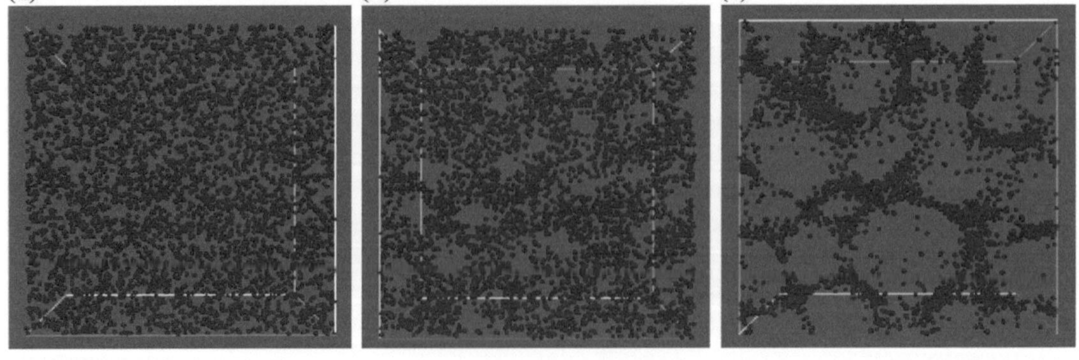

Fig. 11.24 Arrangement of submicron particles suspended in a water film at **a** 1 s, **b** 1.9 s, and **c** 2.0 s (Tayeb et al. 2016b)

agglomerated in a special way. This clustering of particles continues as the water continues to evaporate. The liquid water film is allowed to completely evaporate while the particles at the bottom of the container are left alone. Interesting patterns are seen to emerge as the liquid water film evaporates. Tayeb et al. (2016b) also analyzed the resulting packing structure in terms of the range of coordination number and radial distribution function which also indicate the self-assembly of the particles.

Problems

11.1. The size distribution of particles in a system is given in Table P11.1. Find discrete number frequency, $\tilde{f}_n(d_i)$, number frequency function, $f_n(d_i)$, and accumulative distribution, $F_n(d_i)$ in each size range. What are the number-averaged particle diameter and the standard deviation for the system?

Table P11.1 Size distribution of a particle system

Range	Range of diameters (μm)	Representative diameter in range, d_i	Number of particles in range, n_i
1	0–0.5	0.25	17
2	0.5–1	0.75	99
3	1–2	1.50	429
4	2–4	3.0	225
5	4–6	5.0	40
6	6–10	8.0	13

11.2. In addition to particle size distribution based on particle diameter discussed in Sect. 11.2, another approach is to use particle mass or volume. Under this approach, discrete mass frequency, $\tilde{f}_m(d_i)$; mass frequency function, $f_m(d_i)$; accumulative distribution, $F_m(d_i)$; mass-averaged particle diameter, $\bar{d}_m$, and standard deviation, σ_m, can be defined. Repeat the above problem for distribution based on the masses of particles.

11.3. Derive the relationship between continuous frequency function, $f_n(d)$, and the continuous frequency function, $f_m(d)$.

11.4. Show that Eq. (11.19) is equivalent to Eq. (11.20).

11.5. The Sauter mean diameter (SMD) is defined as the diameter of a sphere with the same volume-to-surface area ratio as the particle of interest: $SD = D_{32} = d_{pv}^3/d_{ps}^2$ where d_{pv} and d_{ps} are equivalent diameters based on the same volume and area. They can be obtained from Eqs. (11.1) and (11.2). What is the SMD of a nonspherical particle with volume V_p and the surface area S?

11.6. For a system with n_t particles whose diameters are ranged from d_{min} to d_{max}, the Sauter mean diameter can be defined as the diameter of the particle with the same volume-to-surface area as the entire system. What is the Sauter mean diameter for the system defined in Problem 11.1?

11.7. The three close-packed cubic arrangements of particles are shown in Fig. P11.7 (Kaviany 2013). Find the packing density and porosity of each arrangement. What are the coordinate numbers for these three arrangements?

11.8. When fibers with a uniform diameter are closely packed in parallel, the cross-sectional view is like packing two-dimensional circular disks. Find the maximum packing density that can be achieved in a two-dimensional packing of disks with a uniform diameter.

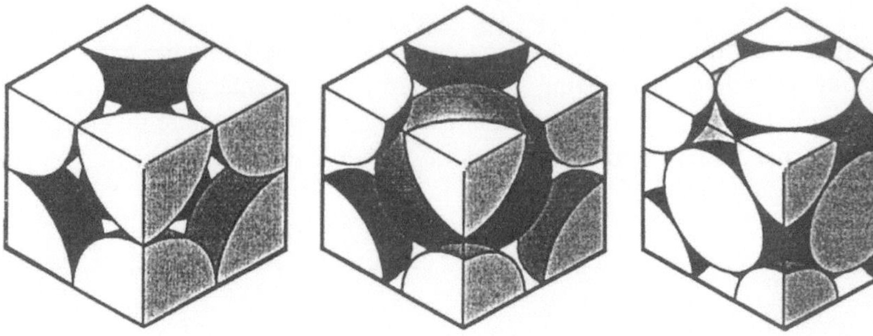

Figure P11.7

11.9. In a selective laser sintering (SLS) process, a bimodal mixture of monosized particles with 50 μm and 1 μm diameters is used. Assuming the large particles exhibit simple cubic packing (see Fig. P11.7a) and the fine particles exhibit loose random packing, what is the porosity for this system? If the large particles are made of copper and the fine particles are aluminum, what is the density of the powder mixture?

11.10. If the large particle in Problem 11.9 is changed to 100 μm in diameter and body-centered cubic packing (see Fig. P11.7b), what are the porosity and the density of the system?

11.11. Estimate the van der Waals force between two particles with an equal size of 1, 10, 20, 50, 100, 500, and 1000 μm and compare the magnitudes of the van der Waals forces with the gravity of one particle for each case. The distance between surfaces of two particles in all cases is 10 nm and the Hamaker constant is 6.5×10^{-20} J. The density of the particle material is 2500 kg/m^3.

11.12. Estimate the terminal velocity of pulverized coal particles of 500 μm in the air at 27 °C and 1 atm.

11.13. When a small solid particle falls in a viscous fluid, the falling velocity of the particle will be very slow and inertia forces of the fluid are negligible. In this case, the Reynolds number will be very low ($Re \ll 1$) and the flow is termed creeping flow or Stokes flow. What are the drag coefficient and terminal velocity for creeping flow?

11.14. A spherical particle with diameter d_p is released in a viscous fluid at time $t = 0$ and starts to fall. Assuming $Re \ll 1$, what is the transient falling velocity of the particle?

11.15. What is the voidage at minimum fluidization velocity for the fluidized bed in Example 11.2?

11.16. The sand is packed in a vessel and air at 27 °C and 1 atm is introduced from the holes in the bottom to fluidize the sand. If the sand can be treated as nonspherical particles with an equivalent diameter of 5 mm and sphericity of 0.7, what is the voidage of the bed at minimum fluidization velocity? The density of sand is 1602kg/m^3.

11.17. Coffee beans can be roasted in a fluidized bed roaster where hot air is introduced from the bottom. The coffee bean can be treated as a triaxial ellipsoid with the lengths of three axes of $a = 8.19$ mm, $b = 6.11$ mm, and $c = 4.60$ mm, respectively. The density of coffee bean is 740 kg/m^3 and the hot air temperature is 500 K for medium-dark roasts. What are the equivalent spherical diameter and sphericity of the coffee bean? If it is desired to operate the fluidized bed in the bubbling fluidization regime, what is the lowest superficial gas velocity?

11.18. A fluidized bed with a diameter of 0.203 m is loaded with spherical glass beads with a mean diameter of 125 μm and density of 1.04 g/cm^3. The height of the glass bead bed at the fixed state is 0.5 m. The glass beads and air are in thermal equilibrium, and their temperature is

300 K. If the superficial gas velocity is 0.16 m/s and the voidage is 0.43, what is the regime of fluidization? A heat transfer tube with a diameter of 2.9 cm, the emissivity of 0.8, and temperature 350 K is submerged horizontally in the fluidized bed to heat the fluidized bed. Estimate the convective and radiative heat transfer coefficient. The specific heat of the glass beads is 0.84 kJ/kg-K.

11.19. Repeat the above problem for a fluidized bed temperature of 500 K and heat transfer tube temperature of 550 K. Discuss the contribution of radiative heat transfer at different temperatures.

11.20. A pneumatic conveying line with an inner tube diameter of 0.2 m is being used to transport solid particles with a mean diameter of 1.5 mm and density of 800 kg/m^3. The conveying rate of the particle is 5 kg/s and the temperature of the system is at 300 K. Determine the saltation velocity using the Matsumoto correlation and choking velocity.

11.21. Solid particles with a mean diameter of 4 mm and density of 945 kg/m^3 are pneumatically conveyed in a 15-m-long horizontal tube with an inner diameter of 0.25 m at a temperature of 300 K. The conveying rate of the particle is 3.8 kg/s and the superficial gas velocity is 25 m/s. What is the pressure loss in this pneumatic conveying tube?

11.22. Solid particles with a mean diameter of 3 mm and density of 800 kg/m^3 are pneumatically conveyed upward in a 15-m-long *vertical* tube with an inner diameter of 0.2 m at a temperature of 300 K. The conveying rate of the particle is 3 kg/s and the superficial gas velocity is 25 m/s. What is the total pressure loss in this pneumatic conveying tube?

11.23. Pulverized coal is transported via a *dense* phase conveying system with an inner diameter of 42 mm at a temperature of 300 K. It is assumed that the particles are spherical with an average diameter of 61 μm and the density of the coal particle is 1400 kg/m^3. The mass flow rate of the coal particle is 3 kg/s, and the superficial gas velocity is 10 m/s. What is the pressure loss per unit length, $\Delta p/L$?

11.24. Solid particles with a mean diameter of 3 mm and density of 1240 kg/m^3 are transported in a tube with an inner diameter of 50 mm. The carrier liquid is water at 20 °C and the volume fraction of the particle is 10%. When the slurry flows at a velocity of 15 m/s, what is the superficial velocity of the liquid and particles?

11.25. Solid particles with a mean diameter of 3 mm and density of 1240 kg/m^3 are transported in a tube with an inner diameter of 50 mm and a length of 15 m. The velocity of the slurry is 15 m/s and the roughness of the tube is 0.0015 mm. The carrier liquid is water at 20 °C and the volume fraction of the particle is 10%. If it is assumed that the slurry can be approximated as a Newtonian fluid, what is the pressure loss?

11.26. Coal slurry pipeline is an effective approach to transport coal from where it is mined to where it is consumed. The pulverized coal and water are mixed at about 1:1 by weight. The coal particle diameter is 75 μm and the density is 1400 kg/m^3. The inner diameter is 150 mm and the length of the pipeline is 25 m. If the system is operated at 20 °C, what is the volume fraction of the coal particle in the mixture? What is the settling velocity and critical velocity?

11.27. If the coal slurry in the above problem can be approximated as a Newtonian fluid and the velocity of the slurry is 10 m/s, what is the pressure loss over the entire pipe? The roughness of the pipe is 0.002 mm.

11.28. Show that the Fanning friction factor for a Bingham plastic fluid laminar flow in a circular tube can be obtained from Eq. (11.149).

11.29. A limestone slurry is transported through a horizontal pipe with a diameter of 150 mm and length of 20 m. The slurry can be considered to be a Bingham plastic fluid with a density of 1530 kg/m^3, a yield stress of 65 N/m^2 and a limiting viscosity of 0.15 Pa-s. If the volumetric flow rate of the limestone slurry is 0.02 m^3/s, what is the pressure loss?

11.30. Drilling mud considered to be a Bingham plastic fluid has a density of 1200 kg/m^3, a yield stress of 10 N/m^2 and a limiting viscosity of 0.035 Pa-s and is being pumped through a 50-m-long pipe with a diameter of 0.1 m at a mass flow rate of 40 kg/s. If the flow laminar or turbulent? What is the pressure loss?

11.31. The behaviors of some slurries can be better described using the power-law model that the stress–strain relation is $\tau = K_p \dot{\gamma}^n$, where K_p and n are the flow consistency index and flow behavior index, respectively. What is the Fanning friction factor for laminar flow in a circular tube?

11.32. Calculate the thermal conductivity of a CuO–water nanofluid with volume fraction of nanoparticles at 0.8% using Eqs. (11.160), (11.161) and (11.170) and discuss the effects of Brownian motion thermal interfacial resistance on thermal conductivity. The diameter of the nanoparticles is 50 μm and the temperature of the nanofluid is 27 °C. The thermal conductivity of bulk CuO is 33 W/m-K.

11.33. What is the effect of Brownian motion on the viscosity in Problem 11.32?

11.34. Obtain a single equation with a base fluid temperature, T_f, as the sole unknown by combining Eqs. (11.166) and (11.167).

References

Abrahamsen, A. R., & Geldart, D. (1980). Behaviour of gas fluidized beds of fine powders. *Powder Technology, 26,* 35–46.

Afrin, N., Mao, Y., Zhang, Y., Chen, J. K., Ritter, R., Lampson, A., et al. (2016). Multicomponent gas-particle flow and heat/mass transfer induced by a localized laser irradiation on a urethane-coated stainless steel substrate. *Frontiers in Heat and Mass Transfer, 7,* 7.

Agu, C. E., Tokheim, L. A., Eikeland, M., & Moldestad, B. M. E. (2017). Determination of onset of bubbling and slugging in a fluidized bed using a dual-plane electrical capacitance tomography system. *Chemical Engineering Journal, 328,* 997–1008.

Angayarkanni, S. A., & Philip, J. (2015). Review on thermal properties of nanofluids: Recent developments. *Advances in Colloid and Interface Science, 225,* 146–176.

ANSYS Fluent Theory Guide. ANSYS, Inc., 2017.

Baeyens, J. (1973). *Heat transfer in gas fluidized beds* (Ph.D. Thesis) University of Bradford.

Bird, R. B., Dai, G. C., & Yarusso, B. L. (1983). The rheology and flow of viscoplastic materials. *Reviews in Chemical Engineering, 1,* 1–70.

Borodulya, V. A., Teplitsky, A. P., Sorokin, V. V., Markevich, I. I., Hassan, A. F., & Yeryomenko, T. P. (1991). Heat transfer between a surface and a fluidized bed: Consideration of pressure and temperature effects. *International Journal of Heat and Mass Transfer, 34,* 47–53.

Buckingham, E. (1921). On plastic flow through capillary tubes. *ASTM Proceedings, 21,* 1154–1156.

Buongiorno, J., Venerus, D. C., & Zhou, S.-Q. (2009). A benchmark study on the thermal conductivity of nanofluids. *Journal of Applied Physics, 106,* 094312.

Chen, J. C. (2003). Heat transfer. In W.C. Yang, (Ed.), *Handbook of fluidization and fluid-particle systems*. New York: Marcel Dekker, Inc.

Chen, J. C., Grace, J. R., & Golriz, (2005). Heat transfer in fluidized beds: Design methods. *Powder Technology, 150,* 123–132.

Choi, S. U. S. (2009). Nanofluids: From vision to reality through research. *Journal of Heat Transfer, 131*(3), 033106.

Chong, Y. O., & Leung, L. S. (1986). Comparison of choking velocity correlations in vertical pneumatic conveying. *Powder Technology, 47,* 43–50.

Churchill, S. W. (1977). Friction-factor equation spans all fluid-flow regimes. *Chemical Engineering, 7,* 91–92.

Crowe, C. T., Schwartzkopf, J. D., Sommerfeld, M., & Tsuji, T. (2012). *Multiphase flows with droplets and particles* (2nd ed.). Boca Raton: CRC Press.

Darby, R., & Melson, J. (1981). How to predict the friction factor for the flow of bingham plastics. *Chemical Engineering, 88,* 59–61.

Darby, R., Mun, R., & Boger, V. (1992). Prediction Friction Loss in Slurry Pipes. *Chemical Engineering*, September, 116–119.

Dintawa, E., Tijskens, E., & Ramon, H. (2008). On the accuracy of the hertz model to describe the normal contact of soft elastic spheres. *Granular Matter, 10,* 209–221.

Doron, P., & Barnea, (1996). Flow pattern maps for solid-liquid flow in pipes. *International Journal of Multiphase Flow, 22*(2), 273–283.

Dou, X., Mao, Y., & Zhang, Y. (2014). Effects of contact force model and size distribution on microsized granular packing. *Journal of Manufacturing Science and Engineering, 136*(2), 021003.

Ellis, N., Bi, H. T., Lim, C. J., & Grace, J. R. (2004). Hydrodynamics of turbulent fluidized beds of different diameters. *Powder Technology, 141,* 124–136.

Francesco, P. D. M., & Alberto, D. R. (2004). Analytical solution for the problem of frictional-elastic collisions of spherical particles using the linear model. *Chemical Engineering Science, 59,* 3461–3475.

Gao, X., Lu, W., Lu, H., Gong, X., Xie, K., Liu, H., et al. (2013). Pressure drop prediction for horizontal dense-phase pneumatic conveying of pulverized coal associated with feeding to gasifier. *Chemical Engineering Research and Design, 91*(12), 2509–2514.

Geldart, D. (1986). *Gas fluidization technology*. New York: Wiley.

Guo, L. J. (2002). *Two-phase and multiphase flow dynamics*. Xi'an, China: Xi'an Jiaotong Press. (in Chinese).

Haider, A., & Levenspiel, O. (1989). Drag coefficient and terminal velocity of spherical and nonspherical particles. *Powder Technology, 58,* 63–70.

Jia, T., Zhang, Y., Chen, J. K., & He, Y. L. (2012). Dynamic simulation of granular packing of fine cohesive particles with different size distributions. *Powder Technology, 218,* 76–85.

Jodrey, W. S., & Tory, E. M. (1981). Computer simulation of isotropic, homogeneous, dense random packing of equal spheres. *Powder Technology, 30,* 111–118.

Jodrey, W. S., & Tory, E. M. (1985). Computer simulation of close random packing of equal spheres. *Physical Review A, 32,* 2347–2351.

Johnson, K. L., Kendall, K., & Roberts, A. D. (1971). Surface energy and the contact of elastic solids. *Proceedings of the Royal Society of London, Series A: Mathematical and Physical Sciences, 324*(1558), 301–313.

Jones, M. G., & Williams, K. C. (2003). Solids friction factors for fluidized dense-phase conveying. *Particulate Science and Technology, 21*(1), 45–56.

Kang, H., Zhang, Y., Yang, M., & Li, L. (2012). Nonequilibrium molecular dynamics simulation of coupling between nanoparticles and base-fluid in a nanofluid. *Physics Letters A, 376*(4), 521–524.

Kaviany, M. (2013). *Principles of heat transfer in porous media* (2nd ed.). New York: Springer Verlag.

Kim, S. W., Kirbas, G., Bi, H., Lim, C. J., & Grace, J. R. (2004). Flow behavior and regime transition in a high-density circulating fluidized. *Chemical Engineering Science, 59,* 3955–3963.

Kloss, C., Goniva, C., Hager, A., Amberger, S., & Pirker, S. (2012). Models, algorithms and validation for OpenSource DEM and CFD-DEM. *Progress in Computational Fluid Dynamics, 12*(2), 140–152.

Koo, J., & Kleinstreuer, C. (2004). A new thermal conductivity model for nanofluids. *Journal of Nanoparticle Research, 6,* 577–588.

Koo, J., & Kleinstreuer, C. (2005). Laminar nanofluid flow in micro heat-sinks. *International Journal of Heat and Mass Transfer, 48,* 2652–2661.

Li, J., & Mason, D. J. (2000). A computational investigation of transient heat transfer in pneumatic transport of granular particles. *Powder Technology, 112*(3), 273–282.

Luckos, A., & Koekemoer, A. (2014). On the sphericity of coal and char particles. *South African Journal of Chemical Engineering, 19*(3), 62–71.

Matsumoto, S., Kikuta, M., & Maeda, S. (1977). Effect of particle size on the minimum transport velocity for horizontal pneumatic conveying of solids. *Journal of Chemical Engineering of Japan, 10*(2), 273–279.

Michaelides, E., Crowe, C. T., & Schwarzkopf, J. D. (2016). *Multiphase flow handbook* (2nd ed.). Boca Raton: CRC Press.

Mohammadian, S. K., Seyf, H. R., & Zhang, Y. (2014). Performance augmentation and optimization of aluminum oxide-water nanofluid flow in a two-fluid microchannel heat exchanger. *Journal of Heat Transfer, 136*(2), 021701.

Moscinski, J., Bargie, M., Rycerz, Z. A., & Jacobs, P. W. M. (1989). The force-biased algorithm for the irregular close packing of equal hard spheres. *Molecular Simulation, 3,* 201–212.

Newitt, D. M., Richardson, J. F., Abbott, M., & Turtle, R. B. (1955). Hydraulic conveying of solids in horizontal pipes. *Transactions. Institute of Chemical Engineers, 33,* 93–113.

Ounis, H., Ahmadi, G., & McLaughlin, J. B. (1991). Brownian diffusion of submicrometer particles in the viscous sublayer. *Journal of Colloid and Interface Science, 143*(1), 266–277.

Perales, J. F., Coll, T., Llop, M. F., Puigjaner, L., Arnaldos, J., & Casal, J. (1991). On the transition from bubbling to fast fluidization regimes. In P. Basu, M. Hasatani, & M. Horio (Eds.), *Circulating Fluidized bed technology III* (pp. 73–78). Pergamon Press, Oxford.

Petrucci, R. H., Harwood, W. S., Herring, G. E. Madura, J. (2006). *General chemistry: Principles and modern application* (9th edn). Prentice Hall.

Plasynski, S. I., Klingzing, G. E., & Mathur, M. P. (1994). High-pressure vertical pneumatic transport investigation. *Powder Technology, 79,* 95–109.

Remond, S. (2010). DEM simulation of small particles clogging in the packing of large beads. *Physica A, 389,* 4485–4496.

Rizk, F. (1976). Pneumatic conveying at optimal operating conditions and a solution of Barth's equation. In *Proceedings of the BHRA Fluid Engineering (PNEUMOTRANSPORT 3),* Cranfield, U.K., Paper D4.

Sabbah, R., Seyed-Yagoobi, J., & Al-Hallaj, S. (2011). Heat transfer characteristics of liquid flow with micro-encapsulated phase change material: Numerical study. *Journal of Heat Transfer, 133*(12), 121702–121710.

Scott, G. D., & Kilgour, D. M. (1969). The density of random close packing of spheres. *Journal of Physics D, 2,* 863–866.

Seyf, H. R., Wilson, M., Zhang, Y., & Ma, H. B. (2014). Flow and heat transfer of nanoencapsulated phase change material slurry past a unconfined square cylinder. *Journal of Heat Transfer, 136*(5), 051902.

Shi, Y., & Zhang, Y. (2008). Simulation of random packing of spherical particles with different size distributions. *Applied Physics A-Mater, 92*(3), 621–626.

Swamee, P. K., & Aggarwal, N. (2011a). Explicit equations for laminar flow of herschel-bulkley fluids. *Canadian Journal of Chemical Engineering, 89,* 1426–1433.

Swamee, P. K., & Aggarwal, N. (2011b). Explicit equations for laminar flow of bingham plastic fluids. *Journal of Petroleum Science and Engineering, 76*(3–4), 178–184.

Tayeb, R., Dou, X., Mao, Y., & Zhang, Y. (2016a). Analysis of cohesive micro-sized particle packing structure using history-dependent contact models. *Journal of Manufacturing Science and Engineering, 138*(4), 041005.

Tayeb, R., Mao, Y., & Zhang, Y., (2016b) Numerical Investigation of evaporation induced self-assembly of sub-micron particles suspended in water, In *ASME 2016* 5th *Micro/Nanoscale Heat & Mass Transfer International Conference,* Biopolis, Singapore, January 4–6, 2016.

Thomas, D. G. (1965). Transport characteristics of suspensions: VIII. A note on the viscosity of newtonian suspensions of uniform spherical particles. *Journal of Colloid Science, 20,* 267–277.

Tibor, G. J., & Fritz, E. (2005). Three-dimensional discrete element simulations in hoppers and silos. *Powder Technology, 158,* 58–68.

Todes, O. M. (1965). *Applications of fluidized beds in the industry, part 11(4).* Leningrad: Znanie.

Tory, E. M., Church, B. H., Tam, M. K., & Ratner, M. (1973). Simulated random packing of equal spheres. *Canadian Journal of Chemical Engineering, 51,* 484–493.

Visscher, W. M., & Bolsterli, M. (1972). Random packing of equal and unequal spheres in two and three dimensions. *Nature, 329,* 504–507.

Wasp, E. L., Kenny, J. P., & Gandhi, R. L. (1977). *Solid liquid flow–slurry pipeline transportation* (1st ed.). Clausthal, Germany: Trans-Tech Publications.

Weber, M. (1982). Correlation analyses in the design of pneumatic transport plant. *Bulk Solids Handling, 2,* 231–233.

Weber, M. (1991). Friction of the air and the air/solid mixture in pneumatic conveying. *Bulk Solids Handling, 11,* 99–102.

Xuan, Y., & Roetzel, W. (2000). Conceptions for heat transfer correlations of nanofluids. *International Journal of Heat and Mass Transfer, 43,* 3701–3707.

Yeoh, G. H., & Tu, J. (Eds.). (2009). *Computational techniques for multiphase flows.* Burlington, MA: Elsevier.

Yousifi, Y., & Gau, G. (1974). Aerodynamique de l'ecoulement vertical de suspensions concentrees gaz-solides—l. Regimes d'ecoulement et stabilite aerodynamique. *Chemical Engineering Science, 29*(9), 1939–1946.

Zabrodsky, S. S. (1966). *Hydrodynamics and heat transfer in fluidized beds.* Cambridge, Massachusetts: MIT Press.

Zhang, Y., & Ma, H. B. (2008). Nonequilibrium heat conduction in a nanofluid layer with periodic heat flux. *International Journal of Heat and Mass Transfer, 51,* 4862–4874.

Zhou, Y. C., Wright, B. D., Yang, R. Y., Xu, B. H., & Yu, A. B. (1999). Rolling friction in the dynamic simulation of sandpile formation. *Physica A-Statistical Mechanics and Its Applications, 269*(2), 536–553.

Zhou, J., Zhang, Y., & Chen, J. K. (2009). Numerical simulation of random packing of spherical particles for powder-based additive manufacturing. *Journal of Manufacturing Science and Engineering, 131*(3), 031004.

12 Flow and Heat Transfer in Porous Media

12.1 Introduction

Fluid flow and heat transfer in porous media are applicable to a wide range of fields, including mechanical, chemical, environmental, and petroleum engineering, as well as geology. Porous media can be found naturally in rocks and sand beds and also can be fabricated as wicks and catalytic pellets. They are an essential component in high-technology devices such as fuel cells and heat pipes. A fundamental formulation of the governing equations within a porous media will be presented here. A porous medium (Nield and Bejan 2017) is a solid matrix with voids or pores, which are continuously connected. The voids are filled with one or more fluids that can pass through the medium because the voids are interconnected. The void fraction in the solid matrix is most frequently referred to as the porosity, ε. Conversely, $1-\varepsilon$ is the volume fraction of the solid matrix, ε_{sm}, that makes the porous medium. If $g(x, y, z, t)$ is a geometric function that is identically equal to 1 in a void, and 0 in the solid matrix, the porosity is defined as:

$$\varepsilon = \frac{1}{\Delta V} \int_{\Delta V} g \mathrm{d}V. \tag{12.1}$$

where ΔV is an elemental volume of the porous zone. When there is more than one phase present in a porous medium, there is a porosity of a single phase, ε_k, which is simply the volume fraction of a particular phase in an elemental volume.

$$\varepsilon_k = \frac{1}{\Delta V} \int_{\Delta V_k} g \mathrm{d}V. \tag{12.2}$$

The subscript, *k*, pertains only to phase *k*. The volume fraction of the solid matrix in an elemental volume is represented by ε_{sm}. Therefore,

$$1 = \varepsilon + \varepsilon_{sm} = \sum_{k=1}^{\Pi-1} \varepsilon_k + \varepsilon_{sm} \tag{12.3}$$

To model a porous zone, it is imperative that one be familiar with the different length scales of the zone. The first dimension is the particle or void length scale, *d*, and the second dimension, *L*, is the

A. Faghri and Y. Zhang, *Fundamentals of Multiphase Heat Transfer and Flow*, https://doi.org/10.1007/978-3-030-22137-9_12

system or porous zone length scale. If d is on the order of L, such as in a very thin porous layer on a heat transfer surface, where the bulk flow is normal to the surface, the porous zone can be directly modeled with minimal assumptions. However, this situation is usually not the case and more often $d \ll L$. When $d \ll L$, direct simulation of transport characteristics in an individual pore is not practical; therefore, the local mean velocity is used. Because of this difference in scale, the following intrinsic and extrinsic phase averages (see Sect. 3.4.1) are often used to describe a porous system:

$$\langle \Phi_k \rangle^k = \frac{1}{\Delta V_k} \int_{\Delta V_k} \Phi_k \mathrm{d}V \tag{12.4}$$

$$\langle \Phi_k \rangle = \frac{1}{\Delta V} \int_{\Delta V_k} \Phi_k \mathrm{d}V \tag{12.5}$$

Figure 12.1a shows how the flow in an individual pore can be very difficult to model when a larger scale is needed, and therefore a volume-averaged velocity, seen in Fig. 12.1b, is often more useful. For analyses in a porous system, one of two velocities are often used: the intrinsic average velocity, $\langle \mathbf{V}_k \rangle^k$, which is the volume-averaged velocity over the volume filled with a particular phase or the extrinsic average velocity, $\langle \mathbf{V}_k \rangle$, which is the extrinsic averaged velocity over an entire volume including a solid. These two velocities are related by the Dupuit–Forchheimer relationship:

$$\langle \mathbf{V}_k \rangle = \varepsilon_k \langle \mathbf{V}_k \rangle^k \tag{12.6}$$

The governing equations for the fluid phase in a porous medium can now be developed. It is assumed that the porous medium is saturated with the fluid phase and that the relative velocity, $\mathbf{V}_{\mathrm{f,rel}}$, only refers to the relative motion of a fluid phase f with respect to the solid matrix. In this section, the solid matrix is assumed to be stationary, therefore $\mathbf{V}_{\mathrm{f}} = \mathbf{V}_{\mathrm{f,rel}}$. The conservation laws will be derived for a porous zone by using a volume-averaging technique for a single fluid occupying the pores. The momentum equations will be derived in a manner in which the viscous interactions between a fluid and the solid matrix are modeled with a property of the porous wick called the permeability, $\mathbf{K}$. The energy equation will be derived in such a way that local thermal equilibrium between the solid matrix and the fluid can be assumed or local thermal nonequilibrium conditions can be assumed.

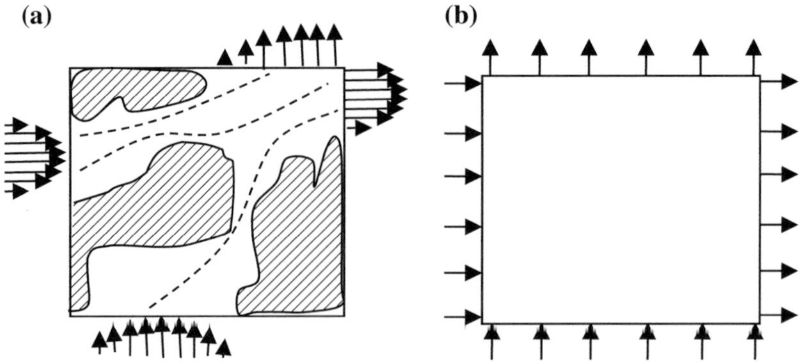

Fig. 12.1 Comparison of velocity distributions in an elemental volume in a porous medium: **a** actual velocity and **b** volume-averaged velocity

Section 12.2 presents single-phase flow and heat transfer in porous media. Multiphase transport in porous media, including multifluid model (MFM) and multiphase mixture model (MMM), are discussed in Sect. 12.3. Section 12.4 discusses boiling in porous media, including boiling on a wicked surface, boiling in porous media heated from below, and an analysis of film boiling in porous media. Section 12.5 is related to condensation in porous media, and finally, Sect. 12.6 presents melting and solidification in porous media.

12.2 Fluid Flow and Heat Transfer in Porous Media

12.2.1 Conservation of Mass

The fluid-saturated porous media can be considered as a two-phase system that contains fluid (f) and solid matrix (sm). The continuity equation for the fluid phase (f) in the porous media can be obtained by considering Eq. (3.264) ($k \rightarrow \mathrm{f}$) without mass transfer:

$$\frac{\partial}{\partial t}\left(\varepsilon\langle\rho_{\mathrm{f}}\rangle^{\mathrm{f}}\right)+\nabla\cdot\left(\varepsilon\langle\rho_{\mathrm{f}}\rangle^{\mathrm{f}}\langle\mathbf{V}_{\mathrm{f}}\rangle^{\mathrm{f}}\right)=0 \tag{12.7}$$

where $\langle\mathbf{V}_{\mathrm{f}}\rangle^{\mathrm{f}}$ is the intrinsic phase-averaged velocity. The right-hand side is zero, because there is no phase change between the porous matrix and the fluid. If the fluid phase in the porous media can be assumed as incompressible, the volume-averaged density is equivalent to the density of a single phase, $\langle\rho_{\mathrm{f}}\rangle^{\mathrm{f}}=\rho_{\mathrm{f}}$ and Eq. (12.7) can be simplified as

$$\frac{\partial}{\partial t}(\varepsilon\rho_{\mathrm{f}})+\nabla\cdot(\rho_{\mathrm{f}}\langle\mathbf{V}_{\mathrm{f}}\rangle)=0 \tag{12.8}$$

where $\langle\mathbf{V}_{\mathrm{f}}\rangle=\varepsilon\langle\mathbf{V}_{\mathrm{f}}\rangle^{\mathrm{f}}$ is the extrinsic phase-averaged velocity.

For an incompressible flow in a porous media of uniform porosity, the continuity equation is

$$\nabla\cdot\langle\mathbf{V}_{\mathrm{f}}\rangle=\nabla\cdot\left(\varepsilon\langle\mathbf{V}_{\mathrm{f}}\rangle^{\mathrm{f}}\right)=0 \tag{12.9}$$

12.2.2 Conservation of Momentum

Since a macroscopic approach is the most feasible way to model transport, there needs to be a way to model the bulk resistance to the flow caused by the porous zone. In 1856, Henry Darcy experimentally measured the resistance to a steady, one-dimensional, gravitationally driven flow through an unconsolidated, uniform, rigid, and isotropic solid matrix. He came up with a relationship for the pressure gradients/resistance as a function of the dynamic viscosity, μ_{f}, the extrinsic phase-averaged velocity, and the permeability, K, now known as Darcy's law.

$$\rho\mathbf{g}-\frac{1}{\varepsilon}\nabla(\varepsilon p)=\frac{\mu_{\mathrm{f}}}{K}\langle\mathbf{V}_{\mathrm{f}}\rangle \tag{12.10}$$

where the permeability, K, is the resistance to the flow, which is analogous to the thermal conductivity in Fourier's law of heat conduction. It has units of meters squared.

Darcy's law can be expanded into vectorial form for an anisotropic solid matrix.

$$\rho \mathbf{g} - \frac{1}{\varepsilon}\nabla(\varepsilon p) = \frac{\mu_f}{\mathbf{K}}\langle \mathbf{V}_f \rangle \tag{12.11}$$

where the permeability, **K**, becomes a second-order symmetric tensor. When the porous medium is isotropic (having equal resistance in all directions), the permeability reduces to a scalar (single value), K. Darcy's law is valid for $Re < 1$ or the creeping flow regime, where the viscous forces dominate. The Reynolds number of the porous media is defined by the mean volumetric velocity and the average characteristic length scale of the voids, $\bar{d}$.

$$Re = \frac{\rho_f |\langle \mathbf{V}_f \rangle| \bar{d}}{\mu_f} \tag{12.12}$$

In this flow regime, the wall effects are confined to within one or two particle diameters of the wall. Also, in the region where the flow enters the porous zone, the entrance region is confined to approximately three particle diameters. Therefore, the flow becomes very closely uniform in one main direction, and the walls bounding the porous zone have minimal effect.

As the Reynolds number increases from unity to about 10, the drag smoothly transitions from linear to nonlinear. This linear to nonlinear transition of drag in a porous zone represents the flow moving from the creeping flow to the laminar flow regime. To account for this nonlinearity, when $Re > 10$, Ward (1964) came up with a quadratic drag term dependent of $\rho/\mathbf{K}^{1/2}$.

$$\rho \mathbf{g} - \frac{1}{\varepsilon}\nabla(\varepsilon p) = \frac{\mu_f}{\mathbf{K}}\langle \mathbf{V}_f \rangle + \frac{C_f}{\mathbf{K}^{1/2}}\rho_f |\langle \mathbf{V}_f \rangle| \langle \mathbf{V}_f \rangle \tag{12.13}$$

where C_f is a dimensionless drag constant, initially thought by Ward (1964) to be a universal constant, 0.55. Since the drag smoothly transitions from linear to nonlinear when $0 < Re < 10$, the significance of the quadratic drag term can be formed as a linear function of Re.

$$\rho \mathbf{g} - \frac{1}{\varepsilon}\nabla(\varepsilon p) = \frac{\mu_f}{\mathbf{K}}\langle \mathbf{V}_f \rangle + \frac{C_f}{\mathbf{K}^{1/2}}\rho_f |\langle \mathbf{V}_f \rangle| \langle \mathbf{V}_f \rangle \frac{(Re - 1)}{9} \tag{12.14}$$

It was later found to vary based on the nature of the porous medium. From the differential momentum equation, Eq. (3.63), the advective term is $\rho_f(\mathbf{V}_f \cdot \nabla)\mathbf{V}_f$. From a scaling analysis, the velocity of the fluid is on the order of the volume-weighted average velocity over the porosity, for a single phase, $\mathbf{V}_f \sim \langle \mathbf{V}_f \rangle / \varepsilon$. Therefore, the advective term is on the order of $\rho_f \langle \mathbf{V}_f \rangle^2 / (\varepsilon^2 L)$. A scaling analysis between the advective term and the quadratic drag term gives a ratio proportional to $\mathbf{K}^{1/2}/(C_f \varepsilon^2 L)$, where L is the macroscopic characteristic length scale. The ratio of these two terms is usually very small; therefore, the advective term often can be assumed to have a negligible effect on the momentum equations.

An extension of the Stokes drag force on a sphere in an infinite domain is to include the effects of the neighboring spheres, as with the case of packed beds made of spheres. The Stokes drag force is the drag exerted by a single sphere in an infinite domain on a flow with a $Re \ll 1$ (negligible inertia). This was accomplished by superimposing the Stokes flow on the Darcy flow, i.e.,

$$\rho \mathbf{g} - \frac{1}{\varepsilon}\nabla(\varepsilon p) = \frac{\mu_{\mathrm{f}}}{\mathbf{K}}\langle \mathbf{V}_{\mathrm{f}}\rangle - \mu_{\mathrm{f}}'\nabla^2\langle \mathbf{V}_{\mathrm{f}}\rangle \tag{12.15}$$

where the effective viscosity denoted by $\mu_{\mathrm{f}}' = \mu_{\mathrm{f}}'(\mu_{\mathrm{f}}, \varepsilon, \Im)$ is a function of the dynamic viscosity, the porosity, and the tortuosity $\Im$ of the porous media. The tortuosity is a measure of the connectivity of void space in a porous zone. Equation (12.15) is called the Brinkman equation, and it is usually noted that it is valid for high porosity, $\varepsilon > 0.8$. When the porosity is low, the stresses felt on a fluid in one pore are communicated to the fluid in another pore mainly by pressure because the solid matrix prevents direct viscous interaction of the fluid in separate pores. Even though this equation is only valid for higher porosities, the Laplacian operator of $\langle \mathbf{V}_{\mathrm{f}}\rangle$ is needed when the boundaries of the porous media affect the flow field.

An empirical momentum equation can be heuristically obtained if Darcy's law, Eq. (12.10), is combined with the inertial drag component, Eq. (12.13), and Brinkman's equation, Eq. (12.15).

$$\rho \mathbf{g} - \frac{1}{\varepsilon}\nabla(\varepsilon p) = -\mu_{\mathrm{f}}'\nabla^2\langle \mathbf{V}_{\mathrm{f}}\rangle + \frac{\mu_{\mathrm{f}}}{\mathbf{K}}\langle \mathbf{V}_{\mathrm{f}}\rangle + \frac{C_{\mathrm{f}}}{\mathbf{K}^{1/2}}\rho_{\mathrm{f}}|\langle \mathbf{V}_{\mathrm{f}}\rangle|\langle \mathbf{V}_{\mathrm{f}}\rangle \tag{12.16}$$

However, for a more complete understanding, this equation is compared to a volume-averaged momentum equation derived in Sect. 3.4.2. If the fluid phase is incompressible, i.e., $\langle \rho_{\mathrm{f}}\rangle^{\mathrm{f}} = \rho_{\mathrm{f}}$, the volume-averaged momentum equation for the fluid phase, Eq. (3.272), becomes

$$\begin{aligned} &\frac{\partial}{\partial t}\left(\varepsilon\rho_{\mathrm{f}}\langle \mathbf{V}_{\mathrm{f}}\rangle^{\mathrm{f}}\right) + \nabla\cdot\left(\varepsilon\rho_{\mathrm{f}}\langle \mathbf{V}_{\mathrm{f}}\rangle^{\mathrm{f}}\langle \mathbf{V}_{\mathrm{f}}\rangle^{\mathrm{f}}\right) \\ &= \nabla\cdot\left(\varepsilon\langle \boldsymbol{\tau}_{\mathrm{f}}'\rangle^{\mathrm{f}}\right) + \varepsilon\rho_{\mathrm{f}}\mathbf{g} + \langle \mathbf{F}_{\mathrm{sm,f}}\rangle - \nabla\cdot\left(\varepsilon\rho_{\mathrm{f}}\langle \hat{\mathbf{V}}_{\mathrm{f}}\hat{\mathbf{V}}_{\mathrm{f}}\rangle^{\mathrm{f}}\right) \end{aligned} \tag{12.17}$$

where $\langle \mathbf{V}_{\mathrm{f}}\mathbf{V}_{\mathrm{f}}\rangle^{\mathrm{f}}$ in Eq. (3.272) is replaced by $\langle \mathbf{V}_{\mathrm{f}}\rangle^{\mathrm{f}}\langle \mathbf{V}_{\mathrm{f}}\rangle^{\mathrm{f}} + \langle \hat{\mathbf{V}}_{\mathrm{f}}\hat{\mathbf{V}}_{\mathrm{f}}\rangle^{\mathrm{f}}$ and $\mathbf{X}_{\mathrm{f}}$ is replaced by $\mathbf{g}$, i.e., gravity is the only body force. Since the flow in a porous media is usually laminar because of the characteristic diameter, the product of the velocity deviations, $\langle \hat{\mathbf{V}}_{\mathrm{f}}\hat{\mathbf{V}}_{\mathrm{f}}\rangle^{\mathrm{f}}$, is also small and therefore neglected henceforth. The third term on the right-hand side of Eq. (12.17) represents the interaction between the solid matrix and the fluid phase.

For an incompressible fluid, the volume average of the stress tensor is

$$\langle \boldsymbol{\tau}_{\mathrm{f}}'\rangle^{\mathrm{f}} = -\langle p_{\mathrm{f}}\rangle^{\mathrm{f}}\mathbf{I} + \mu_{\mathrm{f}}\left[\nabla\langle \mathbf{V}_{\mathrm{f}}\rangle^{\mathrm{f}} + \left(\nabla\langle \mathbf{V}_{\mathrm{f}}\rangle^{\mathrm{f}}\right)^{\mathrm{T}}\right] \tag{12.18}$$

Since the local static pressure is needed in the equation of state for an ideal gas, or for the temperature at a liquid vapor interface, the local pressure is more accurately described as the intrinsic phase average pressure, not the extrinsic phase-averaged pressure. If the deviation of pressure within a fluid volume element is small, $\hat{p}_k \approx 0$, then the volume-averaged pressure is

$$\langle p_{\mathrm{f}}\rangle = \varepsilon\langle p_{\mathrm{f}}\rangle^{\mathrm{f}} = \varepsilon p. \tag{12.19}$$

Substituting Eq. (12.18) into Eq. (12.17) and considering Eq. (12.19), the momentum equation becomes

$$\begin{aligned}\frac{\partial}{\partial t}\left(\varepsilon\rho_{\rm f}\langle\mathbf{V}_{\rm f}\rangle^{\rm f}\right)+\nabla\cdot\left(\varepsilon\rho_{\rm f}\langle\mathbf{V}_{\rm f}\rangle^{\rm f}\langle\mathbf{V}_{\rm f}\rangle^{\rm f}\right)&=-\nabla(\varepsilon p)\\+\mu_{\rm f}\nabla^2\left(\varepsilon\langle\mathbf{V}_{\rm f}\rangle^{\rm f}\right)+\varepsilon\rho_{\rm f}\mathbf{g}+\langle\mathbf{F}_{\rm sm,f}\rangle\end{aligned}\tag{12.20}$$

Using the continuity equation, Eq. (12.8), and assuming constant porosity, the momentum equation becomes

$$\begin{aligned}\rho_{\rm f}\left[\frac{1}{\varepsilon}\frac{\partial\langle\mathbf{V}_{\rm f}\rangle}{\partial t}+\frac{\langle\mathbf{V}_{\rm f}\rangle}{\varepsilon^2}\nabla\cdot(\langle\mathbf{V}_{\rm f}\rangle)\right]\\=-\nabla p+\frac{\mu_{\rm f}}{\varepsilon}\nabla^2\langle\mathbf{V}_{\rm f}\rangle+\rho_{\rm f}\mathbf{g}+\frac{1}{\varepsilon}\langle\mathbf{F}_{\rm sm,f}\rangle\end{aligned}\tag{12.21}$$

The empirical momentum relationships are heuristically related to the volume-averaged momentum equation through reasonable observations. In the second term on the right-hand side of Eq. (12.21), the value of $\mu_{\rm f}/\varepsilon$ represents $\mu_{\rm f}'$ in Eq. (12.16). The fourth term on the right-hand side of Eq. (12.21) is replaced by the Darcian relationship and the quadratic drag term. The Darcian drag term comes from the deformation stress caused by the solid–fluid interaction, and the quadratic drag term comes from the pressure coefficient associated with the fluid flowing around the solid matrix.

$$\begin{aligned}\rho_{\rm f}\left[\frac{1}{\varepsilon}\frac{\partial\langle\mathbf{V}_{\rm f}\rangle}{\partial t}+\frac{\langle\mathbf{V}_{\rm f}\rangle}{\varepsilon^2}\nabla\cdot\langle\mathbf{V}_{\rm f}\rangle\right]&=-\nabla p\\+\frac{\mu_{\rm f}}{\varepsilon}\nabla^2\langle\mathbf{V}_{\rm f}\rangle+\rho_{\rm f}\mathbf{g}-\frac{\mu_{\rm f}}{\mathbf{K}}\langle\mathbf{V}_{\rm f}\rangle-\frac{\rho_{\rm f}C_{\rm f}}{\mathbf{K}^{1/2}}|\langle\mathbf{V}_{\rm f}\rangle|\langle\mathbf{V}_{\rm f}\rangle\end{aligned}\tag{12.22}$$

This equation reduces to Eq. (12.16) if the inertia terms are negligible. In Eq. (12.22), the viscous interaction between the fluid and the solid matrix is replaced by Darcy loss and inertial loss terms. The shear losses between the fluid and the solid matrix are accounted for without having any information on the characteristics of the flow in each pore.

The permeability, K, can be obtained by empirical relations, which are well tabulated for various types of porous media. For packed beds of spherical particles of diameter d, the permeability is

$$K=\frac{d^2\varepsilon^3}{150(1-\varepsilon)^2}\tag{12.23}$$

Further details on permeability and porosity of various porous materials, including porous wicks in heat pipes, can be found in Kaviany (2013) and Faghri (2016). For convenience, the porous zone properties for a variety of applications are presented in Table 12.1, including K, ε, $\Delta A_{\rm sm}/\Delta V$ and $r_{\rm eff}$.

Table 12.1 Properties of porous media materials (Nield and Bejan 2017; Faghri 2016)

Solid matrix	Porosity	Permeability K (m^2)	Effective pore radius r_{eff} (m)	$\frac{\Delta A_{sm}}{\Delta V}$ (m^{-1})
Black slate powder	0.57–0.66	4.9×10^{-14}–1.2×10^{-13}		7×10^5–8.9×10^5
Brick	0.12–0.34	4.8×10^{-15}–2.2×10^{-13}		
Cigarette		1.1×10^{-9}		
Cigarette filters	0.17–0.49			
Coal	0.02–0.12			
Concrete (ordinary mixes)	0.02–0.07			
Copper powder (hot-compacted)	0.09–0.34	3.3×10^{-10}–1.5×10^{-9}		
Corkboard		2.4×10^{-11}-5.1×10^{-11}		
Fiberglass	0.88–0.93			$(5.6–7.7) \times 10^4$
Hair (on mammals)	0.95–0.99			
Hair felt		8.3×10^{-10}–1.2×10^{-9}		
Leather	0.56–0.59	9.5×10^{-14}–1.2×10^{-13}		$(1.2–1.6) \times 10^6$
Limestone (dolomite)	0.04–0.10	2×10^{-15}–4.5×10^{-14}		
Sand	0.37–0.50	2.0×10^{-11}–1.8×10^{-10}		$(1.5–2.2) \times 10^4$
Screen, SST, 200 mesh	0.733	5.2×10^{-11}	58×10^{-6}	
Screen, nickel, 50 mesh, sintered	0.625	6.63×10^{-10}	305×10^{-6}	
Felt, sintered, SST	0.916	5.46×10^{-10}	94×10^{-6}	
Felt, Nickel, A30	0.815	3.06×10^{-11}	120×10^{-6}	
Powder, sintered, nickel	0.597		58×10^{-6}	
Beads, monel, 70–80 mesh	0.4	7.75×10^{-11}	96.9×10^{-6}	
Sandstone ("oil sand")	0.08–0.38	5×10^{-16}–3×10^{-12}		
Silica grains	0.65			
Silica powder	0.37–0.49	1.3×10^{-14}–5.1×10^{-14}		$(6.8–8.9) \times 10^5$
Soil	0.43–0.54	2.9×10^{-13}–1.4×10^{-11}		
Wire crimps	0.68–0.76	3.8×10^{-9}–1.0×10^{-8}		$(2.9–4.0) \times 10^3$

Example 12.1

A cylindrical tube has a porous zone inside it. The porous zone consists of N smaller cylindrical tubes within it as shown in Fig. 12.2. The large cylinder has a radius of R_0, while the smaller internal cylinders all have a radius of r_0. Derive the permeability from the momentum equation derived by the volume averaging technique. Assume that the flow is incompressible, a fully developed flow within each tube and that the volume-averaged velocity, $\langle \mathbf{V}_k \rangle$, is constant throughout the larger cylinder.

Solution There are two parts in solving this problem, (1) knowing the frictional losses in each pore and (2) then substituting these losses into the simplified porous media momentum equation, Eq. (12.10). Since the flow is steady, incompressible and one-dimensional, continuity is already specified by having a uniform flow through the cylinder. Also the volume-averaged momentum equation, Eq. (3.267) repeated below with $k \rightarrow$ f, can be simplified by:

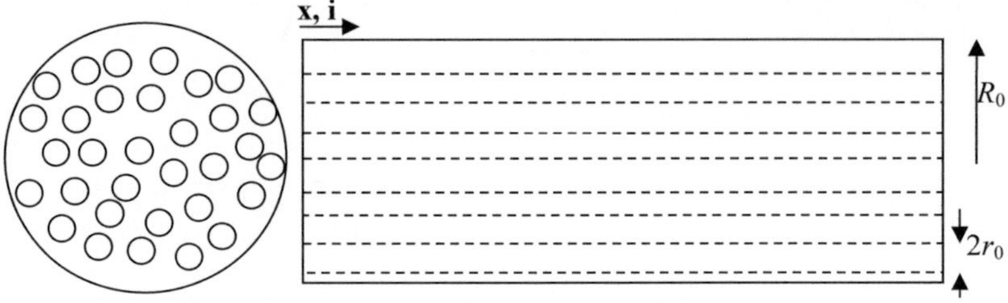

Fig. 12.2 Cross-sectional and side view of a cylinder with cylindrical pores

$$\frac{\partial}{\partial t}\left(\varepsilon\rho_{\mathrm{f}}\cancelto{0}{\langle\mathbf{V}_{\mathrm{f}}\rangle^{\mathrm{f}}}\right)+\nabla\cdot\left(\varepsilon\rho_{\mathrm{f}}\cancelto{0}{\langle\mathbf{V}_{\mathrm{f}}\mathbf{V}_{\mathrm{f}}\rangle^{\mathrm{f}}}\right)=\nabla\cdot\left(\varepsilon\langle\boldsymbol{\tau}\rangle^{\mathrm{f}}\right)-\frac{1}{\Delta V}\int_{\Delta A_{\mathrm{sm}}}\cancelto{0}{\rho_{\mathrm{f}}\mathbf{V}_{\mathrm{f}}}\left(\mathbf{V}_{\mathrm{f}}-\mathbf{V}_{\mathrm{I}}\right)\cdot\mathbf{n}\mathrm{d}A$$

$$+\frac{1}{\Delta V}\int_{A_k}\cancelto{0}{\boldsymbol{\tau}\cdot\mathbf{n}}\mathrm{d}A_k+\varepsilon\rho_{\mathrm{f}}\mathbf{X}_{\mathrm{f}}$$

With constant properties, and the assumed fully developed profile, the above equation becomes

$$\varepsilon\frac{\partial\langle p_{\mathrm{f}}\rangle^{\mathrm{f}}}{\partial x}=\frac{2\mu_{\mathrm{f}}}{\Delta V}\int\limits_{\Delta A_{\mathrm{sm}}}(\mathbf{D}_{\mathrm{f}}\cdot\mathbf{n})\mathrm{d}A \tag{12.24}$$

where $\mathbf{D}_{\mathrm{f}}\cdot\mathbf{n}$ is the rate of deformation in the normal direction of the pore surface. The fully developed velocity profile in a single tube is:

$$\mathbf{V}_{\mathrm{f}}\cdot\mathbf{i}=\frac{(r_0^2-r^2)}{4\mu}\left(-\frac{\mathrm{d}p_{\mathrm{f}}}{\mathrm{d}x}\right) \tag{12.25}$$

where $\mathbf{i}$ is the unit vector in the x-direction. From the velocity profile, we can obtain the rate of deformation, $\mathbf{D}_{\mathrm{f}}\cdot\mathbf{n}$, with respect to the smaller cylindrical tubes surface.

$$\mathbf{D}_{\mathrm{f}}\cdot\mathbf{n}=\frac{1}{2}\frac{\partial\mathbf{V}_{\mathrm{f}}\cdot\mathbf{i}}{\partial r}=\frac{r_0}{4\mu}\frac{\mathrm{d}p_{\mathrm{f}}}{\mathrm{d}x} \tag{12.26}$$

Integrating this value over all of the pore wall areas gives the total frictional loss.

$$\frac{2\mu_{\mathrm{f}}N}{\Delta V}\int\limits_{\Delta A_{\mathrm{sm}}}(\mathbf{D}_f\cdot\mathbf{n})\mathrm{d}A=N\left(\frac{r_0}{R_0}\right)^2\frac{\mathrm{d}p_{\mathrm{f}}}{\mathrm{d}x} \tag{12.27}$$

The extrinsic average velocity is equal to the area average velocity.

$$\langle\mathbf{V}_{\mathrm{f}}\rangle\cdot\mathbf{i}=\frac{2N}{R_0^2}\int\limits_0^{r_0}\mathbf{V}_{\mathrm{f}}\cdot\mathbf{i}r\mathrm{d}r=\frac{Nr_0^4}{8\mu R_0^2}\left(-\frac{\mathrm{d}p_{\mathrm{f}}}{\mathrm{d}x}\right) \tag{12.28}$$

Now the frictional loss can be converted to a function of $\langle \mathbf{V}_{\mathrm{f}} \rangle$ and substituted into the original momentum equation, Eq. (12.24).

$$\varepsilon \frac{\mathrm{d}\langle p_{\mathrm{f}} \rangle^{\mathrm{f}}}{\mathrm{d}x} = \frac{-8\mu_{\mathrm{f}} \langle \mathbf{V}_{\mathrm{f}} \rangle \cdot \mathbf{i}}{r_0^2} \tag{12.29}$$

The porosity is:

$$\varepsilon = \frac{N r_0^2}{R_0^2} \tag{12.30}$$

To maintain the form of Darcy's law,

$$-\frac{\mathrm{d}\langle p_{\mathrm{f}} \rangle^{\mathrm{f}}}{\mathrm{d}x} = \frac{\mu}{K} \langle \mathbf{V}_{\mathrm{f}} \rangle \cdot \mathbf{i} = \frac{8 R_0^2 \mu}{N r_0^4} \langle \mathbf{V}_{\mathrm{f}} \rangle \cdot \mathbf{i} \tag{12.31}$$

Therefore, the permeability is

$$K = \frac{N r_0^4}{8 R_0^2} \tag{12.32}$$

Example 12.2

When using Darcy's law in a porous media, the effect of a no-slip boundary condition cannot be accounted for because of the order of the equation. The shear stress caused by a wall (Fig. 12.3) can be captured using Brinkman's equation. Use the analysis of a semi-infinite, fully developed velocity in a porous zone in no gravitational field to determine the criteria when the wall effects should be incorporated in the governing equations. Develop a solution using a no-slip wall condition, and use this solution to determine the significance of the wall for different porous properties. Also, find the volume average velocity profile of a flow through a porous media bounded by parallel plates under a given pressure gradient. Use no-slip boundary conditions, and compare the mean flow rate to the mean flow rate using Darcy's law.

Solution Brinkman's equation in a one-dimensional porous zone is:

$$-\frac{\mathrm{d}p}{\mathrm{d}x} = \mu_{\mathrm{f}} \left(\frac{\langle u_{\mathrm{f}} \rangle}{K} - \frac{1}{\varepsilon} \frac{\partial^2 \langle u_{\mathrm{f}} \rangle}{\partial y^2} \right) \tag{12.33}$$

where $\langle u_{\mathrm{f}} \rangle$ is the extrinsic phase-averaged velocity in the x-direction and the effective viscosity is assumed to be $\mu' = \mu/\varepsilon$. With boundary conditions at the wall and at the free stream of:

$$y = 0, \quad u_{\mathrm{f}} = 0, \langle u_{\mathrm{f}} \rangle = 0 \tag{12.34}$$

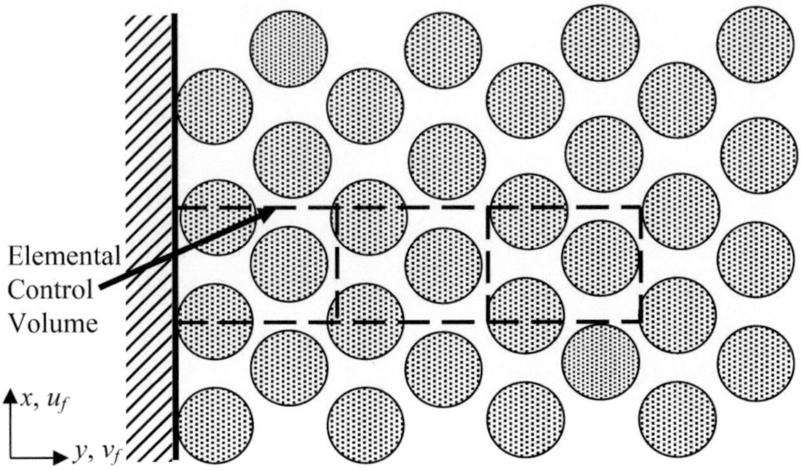

Fig. 12.3 Porous region near a wall

$$y = \infty, \quad \frac{\partial \langle \mathbf{V}_\mathrm{f} \rangle}{\partial y} = 0 \tag{12.35}$$

Assume that the volume average velocity at the wall is zero (knowing it may not be correct). The solution to the semi-infinite Brinkman's equation is:

$$-\frac{\mu_\mathrm{f} \langle u_\mathrm{f} \rangle}{K \frac{\mathrm{d}p}{\mathrm{d}x}} = \left(1 - \exp\left(-y\sqrt{\frac{\varepsilon}{K}}\right)\right) \tag{12.36}$$

which is plotted in Fig. 12.4.

It can be seen in Fig. 12.4 that the wall effects are confined to a distance of approximately $6\sqrt{K/\varepsilon}$. The average pore length in the y-direction is about $\sqrt{K/\varepsilon}$; therefore, the wall effects will only be noticeable in cells that are within approximately six pore lengths from the wall. For computational considerations, if the grid is not resolved within six average pore lengths, then there is no advantage to using Brinkman's equation over Darcy's law.

To quantify the overall size of the solution to determine what scale is large enough such that wall effects can be neglected, we must solve Brinkman's equation for a bounded flow between two parallel plates $(0 \leq y \leq h)$.

$$-\frac{\mathrm{d}p}{\mathrm{d}x} = \mu_\mathrm{f}\left(\frac{\langle u_\mathrm{f} \rangle}{K} - \frac{1}{\varepsilon}\frac{\partial^2 \langle u_\mathrm{f} \rangle}{\partial y^2}\right) \tag{12.37}$$

The boundary conditions are:

$$y = 0, \ \langle u \rangle = 0 \tag{12.38}$$

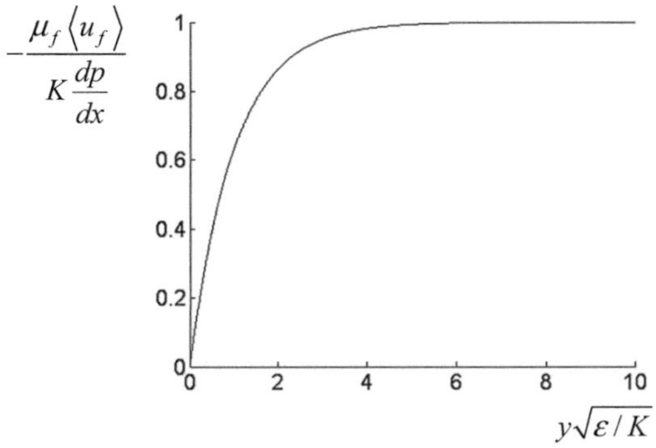

Fig. 12.4 Solution to Brinkman's equation in a semi-infinite domain

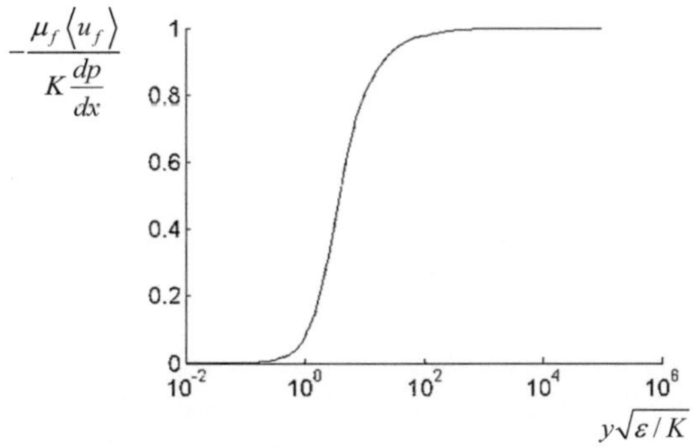

Fig. 12.5 Nondimensional average velocity versus nondimensional distance

$$y = h,\ \langle u \rangle = 0 \tag{12.39}$$

The general solution to this equation after applying boundary conditions in Eqs. (12.38) and (12.39) is:

$$\langle u_{\mathrm{f}} \rangle = -\frac{K}{\mu_{\mathrm{f}}}\frac{\mathrm{d}p}{\mathrm{d}x}\left(1 - A\exp\left(\sqrt{\frac{\varepsilon}{K}}y\right) - B\exp\left(-\sqrt{\frac{\varepsilon}{K}}y\right)\right) \tag{12.40}$$

where the constants A and B are:

$$A = \left[\exp\left(-\sqrt{\frac{\varepsilon}{K}}h\right) - 1\right]\left[\exp\left(-\sqrt{\frac{\varepsilon}{K}}h\right) - \exp\left(\sqrt{\frac{\varepsilon}{K}}h\right)\right]^{-1} \tag{12.41}$$

$$B = \left[1 - \exp\left(\sqrt{\frac{\varepsilon}{K}}h\right)\right]\left[\exp\left(-\sqrt{\frac{\varepsilon}{K}}h\right) - \exp\left(\sqrt{\frac{\varepsilon}{K}}h\right)\right]^{-1} \tag{12.42}$$

By analyzing Fig. 12.5, it can be noted that the wall effects are insignificant when the overall length of the system η is greater than 10^3, and therefore, Darcy's law is a good approximation. However, when η is less than 10^3, Brinkman's equation should be applied. If η is on the order of 1, the volume-averaging approach is not good, and the local velocities should be modeled.

12.2.3 Energy Equation

The derivation of the energy equation in a porous medium is similar to that of continuity and momentum equations. The volume-averaged energy Eq. (3.285) ($k \rightarrow$ f and no mass transfer) is valid for both fluid phase and the solid matrix. Assuming the fluid phase is incompressible, and neglecting the viscous dissipation, the energy equation for the fluid phase becomes

$$\frac{\partial}{\partial t}\left(\varepsilon\rho_{\mathrm{f}}\langle h_{\mathrm{f}}\rangle^{\mathrm{f}}\right)+\nabla\cdot\left(\varepsilon\rho_{\mathrm{f}}\langle \mathbf{V}_{\mathrm{f}}\rangle^{\mathrm{f}}\langle h_{\mathrm{f}}\rangle^{\mathrm{f}}\right)=-\nabla\cdot\langle \mathbf{q}_{\mathrm{f}}''\rangle+\langle q_{\mathrm{f}}'''\rangle+\left\langle q_{\mathrm{sm,f}}'''\right\rangle \tag{12.43}$$

where $\langle \mathbf{V}_{\mathrm{f}} h_{\mathrm{f}}\rangle^{\mathrm{f}}$ in Eq. (3.285) is replaced by $\langle \mathbf{V}_{\mathrm{f}}\rangle^{\mathrm{f}}\langle h_{\mathrm{f}}\rangle^{\mathrm{f}}$ in Eq. (12.43), i.e., the product of the deviations were neglected. $\left\langle q_{\mathrm{sm,f}}'''\right\rangle$ is heat transfer from solid matrix to the fluid phase.

By applying the continuity eqiation (12.8), Eq. (12.43) becomes

$$\varepsilon\rho_{\mathrm{f}}\frac{\partial\langle h_{\mathrm{f}}\rangle^{\mathrm{f}}}{\partial t}+\varepsilon\rho_{\mathrm{f}}\langle \mathbf{V}_{\mathrm{f}}\rangle^{\mathrm{f}}\cdot\nabla\langle h_{\mathrm{f}}\rangle^{\mathrm{f}}=-\nabla\cdot\langle \mathbf{q}_{\mathrm{f}}''\rangle+\langle q_{\mathrm{f}}'''\rangle+\left\langle q_{\mathrm{sm,f}}'''\right\rangle \tag{12.44}$$

The volume-averaged energy equation in the solid matrix is:

$$\varepsilon_{\mathrm{sm}}\langle\rho_{\mathrm{sm}}\rangle^{\mathrm{sm}}\frac{\partial\langle h_{\mathrm{sm}}\rangle^{\mathrm{sm}}}{\partial t}=-\nabla\cdot\langle \mathbf{q}_{\mathrm{sm}}''\rangle+\langle q_{\mathrm{f}}'''\rangle-\left\langle q_{\mathrm{sm,f}}'''\right\rangle \tag{12.45}$$

where the convection term is dropped since the solid matrix is in the solid state and has no velocity. The thermal boundary conditions at the fluid-solid interface, ΔA_{sm}, are:

$$T_{\mathrm{f}}=T_{\mathrm{sm}} \tag{12.46}$$

$$\mathbf{q}_{\mathrm{f}}''\cdot\mathbf{n}_{\mathrm{f}}=\mathbf{q}_{\mathrm{sm}}''\cdot\mathbf{n}_{\mathrm{sm}} \tag{12.47}$$

These boundary conditions between the fluid and the solid matrix are what make up the heat transfer term between the fluid and solid, $\left\langle q_{\mathrm{sm,f}}'''\right\rangle$. Equations (12.46) and (12.47) are very general, and no assumptions about the thermal equilibrium between the fluid and the solid matrix have been made. If the fluid and the solid matrix are considered to be in local thermal equilibrium, $\langle T_{\mathrm{f}}\rangle^{\mathrm{f}}=\langle T_{\mathrm{sm}}\rangle^{\mathrm{sm}}=T$, the energy equations for the solid and the fluid can be added to find the total porous media energy equation:

$$\begin{aligned}&\varepsilon\rho_{\mathrm{f}}\frac{\partial\langle h_{\mathrm{f}}\rangle^{\mathrm{f}}}{\partial t}+\varepsilon_{\mathrm{sm}}\langle\rho_{\mathrm{sm}}\rangle^{\mathrm{sm}}\frac{\partial\langle h_{\mathrm{sm}}\rangle^{\mathrm{sm}}}{\partial t}+\varepsilon\rho_{\mathrm{f}}\langle \mathbf{V}_{\mathrm{f}}\rangle^{\mathrm{f}}\cdot\nabla\langle h_{\mathrm{f}}\rangle^{\mathrm{f}}\\&=-\nabla\cdot\langle \mathbf{q}_{\mathrm{f}}''\rangle-\nabla\cdot\langle \mathbf{q}_{\mathrm{sm}}''\rangle+\langle q_{\mathrm{f}}'''\rangle+\langle q_{\mathrm{sm}}'''\rangle\end{aligned} \tag{12.48}$$

The energy equation can be further simplified by assuming that the enthalpy is only a function of temperature and constant specific heat:

$$(\rho c_{\mathrm{p}})_{\mathrm{eff}}\frac{\partial T}{\partial t}+(\rho c_{\mathrm{p}})_{\mathrm{f}}\langle \mathbf{V}_{\mathrm{f}}\rangle\cdot\nabla T=\nabla\cdot k_{\mathrm{eff}}\nabla T+q_{\mathrm{eff}}''' \tag{12.49}$$

The effective heat capacity, thermal conductivity, and heat generation rates per unit volume, denoted by the subscript "eff" are defined by

$$(\rho c_p)_{\text{eff}} = \varepsilon_f(\rho c_p)_f + \varepsilon_{sm}(\rho c_p)_{sm} \tag{12.50}$$

$$k_{\text{eff}} = \varepsilon k_{\text{eff,f}} + \varepsilon_{sm} k_{\text{eff,sm}} \tag{12.51}$$

$$q'''_{\text{eff}} = \varepsilon \langle q'''_f \rangle^f + \varepsilon_{sm} \langle q'''_{sm} \rangle^{sm} \tag{12.52}$$

where $k_{\text{eff,f}}$ and $k_{\text{eff,sm}}$ are, respectively, the effective thermal conductivity of the fluid and solid matrix in the *porous media*, and both of them depend on porosity and pore structure in the porous media. When the thermal conductivities of the fluid and the solid matrix are close to each other, $k_f \sim k_{sm}$, one can assume that $k_{\text{eff,f}} \approx k_f$ and $k_{\text{eff,sm}} \approx k_{sm}$, and use Eq. (12.51) to evaluate the effective thermal conductivity. In a case where thermal conductivities of the fluid and the solid matrix differ significantly, the following equation should be used (Hadley 1986):

$$\begin{aligned}\frac{k_{\text{eff}}}{k_f} &= (1-\alpha_0)\frac{\varepsilon f_0 + (1-\varepsilon f_0)k_{sm}/k_f}{1-\varepsilon(1-f_0)+\varepsilon(1-f_0)k_{sm}/k_f} \\ &+ \alpha_0 \frac{2(1-\varepsilon)(k_{sm}/k_f)^2 + (1+2\varepsilon)k_{sm}/k_f}{(2+\varepsilon)k_{sm}/k_f + 1 - \varepsilon}\end{aligned} \tag{12.53}$$

where

$$f_0 = 0.8 + 0.1\varepsilon \tag{12.54}$$

$$\log \alpha_0 = \begin{cases} -4.898\varepsilon & 0 \le \varepsilon \le 0.0827 \\ -0.405 - 3.154(\varepsilon - 0.0827) & 0.0827 \le \varepsilon \le 0.298 \\ -1.084 - 6.778(\varepsilon - 0.298) & 0.298 \le \varepsilon \le 0.580 \end{cases} \tag{12.55}$$

The above energy equation, Eq. (12.49), is only valid when the fluid and the solid matrix are assumed to be in local thermal equilibrium. While local thermal equilibrium is an often used hypothesis for heat transfer in porous media, it is well recognized that local equilibrium between different phases in a system of solid and fluid cannot be achieved when thermal properties for different phases differ widely or during rapid heating or cooling. The fluid and solid-matrix energy equations need to be kept separate, and this situation is referred to as *local thermal nonequilibrium*. In this situation, heat transfer from the solid matrix to the fluid phase, $\langle q'''_{sm,f} \rangle$, in Eqs. (12.44) and (12.45) can be modeled as a convective boundary condition, with a heat transfer coefficient of h_c. When Fourier's law of heat conduction is applied, the energy equations for the fluid and solid matrix are

$$\begin{aligned}&(\rho c_p)_f\left(\varepsilon\frac{\partial T_f}{\partial t} + \langle \mathbf{V}_f \rangle \cdot \nabla T_f\right) \\ &= \nabla \cdot (\varepsilon k_{\text{eff,f}} \nabla T_f) + \varepsilon \langle q'''_f \rangle^f + h_{f,sm}\frac{\Delta A_{sm}}{\Delta V}(T_{sm} - T_f)\end{aligned} \tag{12.56}$$

$$\begin{aligned}\varepsilon_{sm}(\rho c_p)_{sm}\frac{\partial T_{sm}}{\partial t} &= \nabla \cdot (\varepsilon_{sm} k_{\text{eff,sm}} \nabla T_{sm}) + \varepsilon_{sm}\langle q'''_{sm} \rangle \\ &- h_c \frac{\Delta A_{sm}}{\Delta V}(T_{sm} - T_f)\end{aligned} \tag{12.57}$$

Equations (12.56) and (12.57) are more accurate representations of the thermal field in a porous zone. However, they require an empirical relationship for the heat transfer coefficient between the solid matrix and the fluid phase.

12.2.4 Species

The volume-averaged species Eq. (3.289) can be readily applied to the fluid phase in the porous media:

$$\frac{\partial\left(\varepsilon\langle\rho_{\mathrm{f},i}\rangle^{\mathrm{f}}\right)}{\partial t}+\nabla\cdot\left(\varepsilon\langle\rho_{\mathrm{f},i}\rangle^{\mathrm{f}}\langle\mathbf{V}_{\mathrm{f}}\rangle^{\mathrm{f}}\right)=-\nabla\cdot\langle\mathbf{J}_{\mathrm{f},i}\rangle+\left\langle\dot{m}_{\mathrm{f},i}'''\right\rangle \tag{12.58}$$

Applying the continuity Eq. (12.8) and the definition of the mass fraction, Eq. (1.43), the averaged species equation becomes

$$\varepsilon\rho_{\mathrm{f}}\frac{\partial\langle\omega_{\mathrm{f},i}\rangle^{\mathrm{f}}}{\partial t}+\rho_{\mathrm{f}}\mathbf{v}_{\mathrm{f}}\cdot\nabla\langle\omega_{\mathrm{f},i}\rangle^{\mathrm{f}}=-\nabla\cdot\langle\mathbf{J}_{\mathrm{f},i}\rangle+\left\langle\dot{m}_{\mathrm{f},i}'''\right\rangle \tag{12.59}$$

12.3 Multiphase Transport in Porous Media

12.3.1 Overview

The framework for multiphase transport in a porous zone (see Fig. 12.6) is already laid out through the single-phase transport in porous media. When the porous medium is saturated with both liquid and vapor, the governing equations for liquid and vapor as well as solid matrix must be specified. In addition, the solid-liquid-vapor interactions play an important role in heat and mass transfer applications. The governing equation for the liquid and vapor phase must also consider possible phase change between liquid and vapor. Chapter 4 is entirely dedicated to problems of involving solid-liquid-vapor interfacial phenomena. Problems that directly solve the solid-liquid-vapor phenomena are typically applied to a single pore.

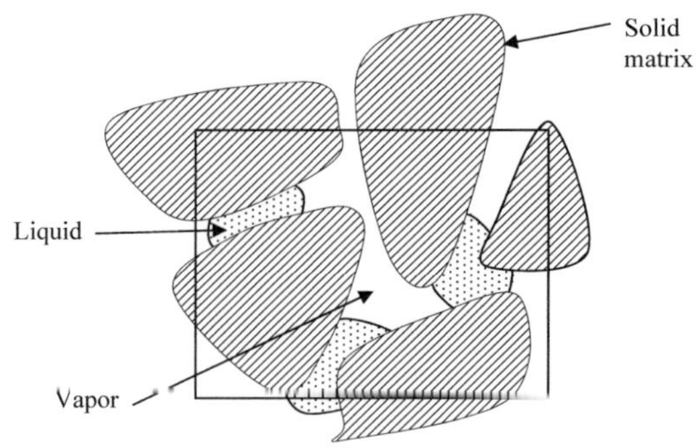

Fig. 12.6 Typical elemental volume with periodic dispersion of phases

The premise for both the multifluid model (MFM) and the multiphase mixture model (MMM) for solving multiphase porous media are the volume-averaged Navier–Stokes equations. The major assumption for these models used in a porous medium is that the flow is considered noninertial. One of the benefits of a noninertial flow is that it behaves very well and is in the laminar flow regime. Therefore, the deviatoric values from the volume-averaged equation are neglected without compromising the accuracy of the solution. This approximation is reasonable as the pore size is very small, which makes viscous effects dominate. The MFM solves both phases separately, along with the details of each phase, such as temperature and species mass fraction. The MMM adds the governing equations for each phase and converts the flow variables into their mass-averaged values counterparts.

12.3.2 Multifluid Model (MFM)

In this model, there are numerous liquid–vapor interfacial regions within each elemental pore; therefore, the average change of these interfacial characteristics is considered as a model for this type of problem. The transport equations for a single-component substance coexisting in the vapor phase, v, and the liquid phase, ℓ, are given. In these equations, the liquid and vapor are considered to be in local thermal equilibrium. However, the liquid and vapor phases are in local thermal nonequilibrium with the solid. Whitaker (1977) developed governing equations for the periodic dispersion of phases within a porous zone, as seen in Fig. 12.6.

The continuity Eq. (12.8) for the vapor and liquid phases becomes

$$\frac{\partial}{\partial t}(\varepsilon_{\mathrm{v}}\rho_{\mathrm{v}}) + \nabla\cdot(\rho_{\mathrm{v}}\langle\mathbf{V}_{\mathrm{v}}\rangle) = \langle\dot{m}'''_{\ell\mathrm{v}}\rangle \tag{12.60}$$

$$\frac{\partial\varepsilon_\ell}{\partial t} + \nabla\cdot\langle\mathbf{V}_\ell\rangle = -\frac{\langle\dot{m}'''_{\ell\mathrm{v}}\rangle}{\rho_\ell} \tag{12.61}$$

where the density in the liquid phase is assumed constant. The volumetric mass transfer term, $\langle\dot{m}'''_{\ell\mathrm{v}}\rangle$, with units of kg/m^3 s, represents the evaporation rate when it has a positive value and the condensation rate when it has a negative value.

When gravity is the only body force, the momentum Eq. (12.20), for the vapor and liquid phases becomes

$$\begin{aligned}&\rho_{\mathrm{v}}\left[\frac{1}{\varepsilon_{\mathrm{v}}}\frac{\partial\langle\mathbf{V}_{\mathrm{v}}\rangle}{\partial t} + \frac{\langle\mathbf{V}_{\mathrm{v}}\rangle}{\varepsilon_{\mathrm{v}}}\cdot\nabla\left(\frac{\langle\mathbf{V}_{\mathrm{v}}\rangle}{\varepsilon_{\mathrm{v}}}\right)\right]\\ &= -\nabla\langle p_{\mathrm{v}}\rangle^{\mathrm{v}} + \frac{\mu_{\mathrm{v}}}{\varepsilon_{\mathrm{v}}}\nabla^2\langle\mathbf{V}_{\mathrm{v}}\rangle - \frac{\mu_{\mathrm{v}}}{\mathbf{K}_{\mathrm{v}}}\langle\mathbf{V}_{\mathrm{v}}\rangle - \frac{C_{\mathrm{f}}}{\mathbf{K}^{1/2}}\rho_{\mathrm{v}}|\langle\mathbf{V}_{\mathrm{v}}\rangle|\langle\mathbf{V}_{\mathrm{v}}\rangle + \rho_{\mathrm{v}}\mathbf{g}\end{aligned} \tag{12.62}$$

$$\begin{aligned}&\rho_\ell\left[\frac{1}{\varepsilon_\ell}\frac{\partial\langle\mathbf{V}_\ell\rangle}{\partial t} + \frac{\langle\mathbf{V}_\ell\rangle}{\varepsilon_\ell}\cdot\nabla\left(\frac{\langle\mathbf{V}_\ell\rangle}{\varepsilon_\ell}\right)\right]\\ &= -\frac{\mu_\ell}{\mathbf{K}_\ell}\langle\mathbf{V}_\ell\rangle + \frac{\mu_\ell}{\varepsilon_\ell}\nabla^2\langle\mathbf{V}_\ell\rangle - \left[C_\varepsilon\nabla\varepsilon_\ell + C_{\mathrm{T}}\nabla\langle T_{\mathrm{f}}\rangle^{\mathrm{f}} + \nabla\langle p_{\mathrm{v}}\rangle^{\mathrm{v}}\right] + \rho_\ell\mathbf{g}\end{aligned} \tag{12.63}$$

The first two terms in the bracket on the right-hand side of Eq. (12.63) represent the pressure gradient due to the change in capillary pressure, p_{c}, as a function of volume fraction of liquid and temperature. The two coefficients in Eq. (12.63) are defined as

$$C_\varepsilon = \frac{\partial p_c}{\partial \varepsilon_\ell} \tag{12.64}$$

$$C_T = \frac{\partial p_c}{\partial \langle T_{\ell v} \rangle^{\ell v}} \tag{12.65}$$

where the capillary pressure is defined as

$$p_c = \frac{2\sigma_{\ell v}}{r_{\text{eff}}} = \langle p_v \rangle^v - \langle p_\ell \rangle^\ell \tag{12.66}$$

and r_{eff} is the effective radius of curvature of the interface between the vapor and the liquid. For more information on interfacial phenomena, the reader is referred to Chap. 4. The effective radius is a function of the volume fraction of the liquid and the pore geometry.

The energy equation for the fluid (liquid and vapor) phase, where $\langle T_\ell \rangle^\ell = \langle T_v \rangle^v = \langle T_f \rangle^f$, is

$$\begin{aligned}
&\left(\varepsilon_\ell \rho_\ell c_{p\ell} + \varepsilon_v \langle \rho_v \rangle^v c_{pv}\right) \frac{\partial \langle T_f \rangle^f}{\partial t} + \left(\rho_\ell c_{p\ell} \mathbf{v}_\ell + \langle \rho_v \rangle^v c_{pv} \mathbf{v}_v\right) \cdot \nabla \langle T_f \rangle^f \\
&= \nabla \cdot \left[\left(\varepsilon_\ell k_{\text{eff},\ell} + \varepsilon_v k_{\text{eff},v}\right) \nabla \langle T_{\ell v} \rangle^{\ell v}\right] + h_{\text{sm},\ell}\left(\langle T_{\text{sm}} \rangle^{\text{sm}} - \langle T_f \rangle^f\right) \frac{\Delta A_{\text{sm},\ell}}{\Delta V} \\
&+ h_{\text{sm},v}\left(\langle T_{\text{sm}} \rangle^{\text{sm}} - \langle T_f \rangle^f\right) \frac{\Delta A_{\text{sm},v}}{\Delta V}
\end{aligned} \tag{12.67}$$

where $h_{c,\text{sm},\ell}$ and $h_{c,\text{sm},v}$ are convective heat transfer coefficients from solid matrix to liquid and vapor phases, respectively. The energy equation for the solid matrix is

$$\begin{aligned}
&\left(\varepsilon_{\text{sm}} \rho_{\text{sm}} c_{p,\text{sm}}\right) \frac{\partial \langle T_{\text{sm}} \rangle^{\text{sm}}}{\partial t} = \nabla \cdot \left(\varepsilon_{\text{sm}} k_{\text{eff,sm}} \nabla \langle T_{\text{sm}} \rangle^{\text{sm}}\right) \\
&- h_{\text{sm},\ell}\left(\langle T_{\text{sm}} \rangle^{\text{sm}} - \langle T_f \rangle^f\right) \frac{\Delta A_{\text{sm},\ell}}{\Delta V} - h_{\text{sm},v}\left(\langle T_{\text{sm}} \rangle^{\text{sm}} - \langle T_f \rangle^f\right) \frac{\Delta A_{\text{sm},v}}{\Delta V}
\end{aligned} \tag{12.68}$$

The volume fractions of all the phases must satisfy the following constraint:

$$\varepsilon_\ell + \varepsilon_v + \varepsilon_{\text{sm}} = 1 \tag{12.69}$$

The equation of state for the vapor phase, assuming it is an ideal gas, is

$$\langle p_v \rangle^v = \langle \rho_v \rangle^v R_g \langle T_f \rangle^f \tag{12.70}$$

The thermodynamic relation for saturation pressure is

$$\langle p_v \rangle^v = p_0 \exp\left\{-\left[\frac{2\sigma_{\ell v}}{r_{\text{eff}} \rho_\ell R_g \langle T_f \rangle^f} + \frac{h_{lv}}{R_g}\left(\frac{1}{\langle T_f \rangle^f} - \frac{1}{T_0}\right)\right]\right\} \tag{12.71}$$

where T_0 is saturation temperature corresponding to a reference pressure p_0.

The above governing equations for multiphase transport in porous media are based upon the assumption that the liquid and vapor phases are in thermal equilibrium. Gray (2000) addressed macroscale equilibrium conditions for liquid-gas two-phase flow in porous media. Duval et al. (2004) used the method of volume averaging to derive a three-temperature macroscopic model that considered local thermal nonequilibrium among the three phases. A closed form of the evaporation rate at the macroscopic level can be obtained depending on the macroscopic temperatures and the effective properties. Many multiphase problems have a clear liquid–vapor interface within the porous zone. These problems can be dealt with by solving the phases as separate solutions with boundary conditions between the phases. Example 12.3 demonstrates this class of problem.

Example 12.3

A heat pipe uses a thermal cycle in which a liquid is evaporated from one end, and the vapor is condensed at the other end. The liquid motion is caused by capillary action through a wick, which is a porous zone (see Fig. 12.7). The maximum pressure drop that the wick can handle is based on the effective radius, r_{eff}, by $\Delta p_{\max} = 2\sigma/r_{\mathrm{eff}}$. If the effective radius is 2.5×10^{-5} m, calculate whether the wick has enough pumping capability to compensate for 10 kW/m^2 of heat entering the evaporator and leaving the condenser. The adiabatic section, L_a, has a length of 0.5 m, while the evaporator and condenser sections have a length of 0.1 m ($L_e = L_c$). The permeability, K, of the wick is 5×10^{-9} m^2, while the wick thickness, δ_{wick}, is 0.005 m. The porosity of the wick is 0.3. The working fluid is water, and the only effect the vapor phase has on the wick is that the pressure applied to the interface is 1 atmosphere. Assume that the inertial losses in the wick are negligible and that the fluid flow in the wick is one-dimensional. Also assume that the wick is thin enough that conduction through the wick can be considered one-dimensional and that the flow of the liquid has a negligible effect on the temperature distribution in the wick.

Solution Because the saturation temperature of water at 1 atm is 100 °C, the density of the liquid is $\rho_v = 958.77\,\mathrm{kg/m^3}$, the dynamic viscosity is 0.000279 Ns/m^2, and the latent heat of vaporization is $h_{\ell v} = 2251.2\,\mathrm{kJ/kg}$. The surface tension is $\sigma = 0.05891\,\mathrm{N/m}$.

Since the volume-averaged velocity is uniform across the wick, the volume-averaged continuity equation, Eq. (3.264), can be integrated over the thickness of the wick and reduced to:

$$\rho_\ell \delta_{\mathrm{wick}} \frac{d\bar{u}}{dx} = \dot{m}''_{v\ell} \tag{12.72}$$

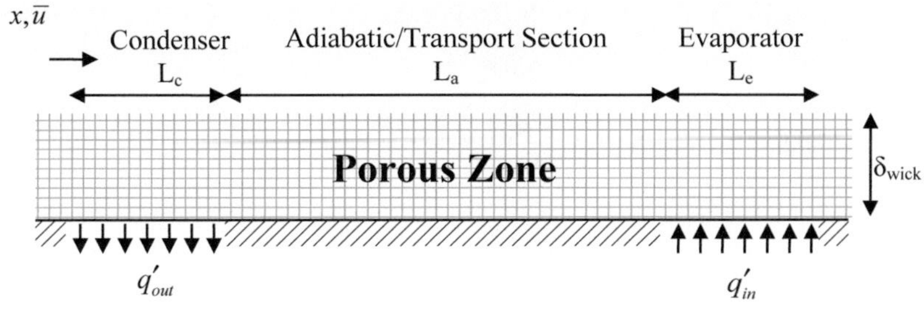

Fig. 12.7 Evaporator, condenser, and adiabatic sections of a wick in a heat pipe

where $\dot{m}''_{v\ell}$ is the condensation mass flux (negative for evaporation). The mass flux is related to the heat flux by the latent heat:

$$\dot{m}''_{v\ell} = \frac{q''}{h_{\ell v}} \tag{12.73}$$

where q'' is the heat flux. Therefore,

$$\rho_\ell \delta_{\text{wick}} \frac{d\bar{u}}{dx} = \frac{q''}{h_{\ell v}} \tag{12.74}$$

Multiplying both sides by dx, and then integrating, the volume-averaged velocity become:

$$\bar{u} = \frac{q''x}{\rho\delta_{\text{wick}} h_{\text{lv}}}, \quad x < L_e \tag{12.75}$$

$$\bar{u} = \frac{q''L_e}{\rho\delta_{\text{wick}} h_{\text{lv}}}, \quad L_e < x < L_e + L_a \tag{12.76}$$

$$\bar{u} = \frac{q''}{\rho\delta_{\text{wick}} h_{\text{lv}}}(2L_e + L_a - x), \quad x > L_e + L_a \tag{12.77}$$

Darcy's law can now be applied:

$$\frac{dp_\ell}{dx} = -\frac{\mu}{K}\bar{u} \tag{12.78}$$

The overall pressure drop in the liquid needed can be obtained by integrating Darcy's law with respect to x,

$$\Delta p_\ell = \frac{\mu}{K}\frac{q''x^2}{2\rho\delta_{\text{wick}} h_{\ell v}}, \quad x < L_e \tag{12.79}$$

$$\Delta p_\ell = \frac{\mu}{K}\frac{q''}{\rho\delta_{\text{wick}} h_{\ell v}}\left(L_e x - \frac{L_e^2}{2}\right), \quad L_e < x < L_e + L_a \tag{12.80}$$

$$\Delta p_\ell = \frac{\mu}{K}\frac{q''}{\rho\delta_{\text{wick}} h_{\ell v}}\left(L_e(2x - L_e - L_a) + L_a\left(x - \frac{L_a}{2}\right) - \frac{x^2}{2}\right), \quad x > L_e + L_a \tag{12.81}$$

The pressure vs. distance along the wick is plotted in Fig. 12.8. The maximum difference in pressure with a value of 3102.3 Pa is at the end of the evaporator. The maximum pressure drop the wick can handle is

$$\Delta p_{\max} = \Delta p_v + \Delta p_\ell = \frac{2\sigma}{r_{\text{eff}}} = \frac{2 \times (0.05891\ \text{N/m})}{2.5 \times 10^{-5}\ \text{m}} = 4712.8\ \text{Pa}$$

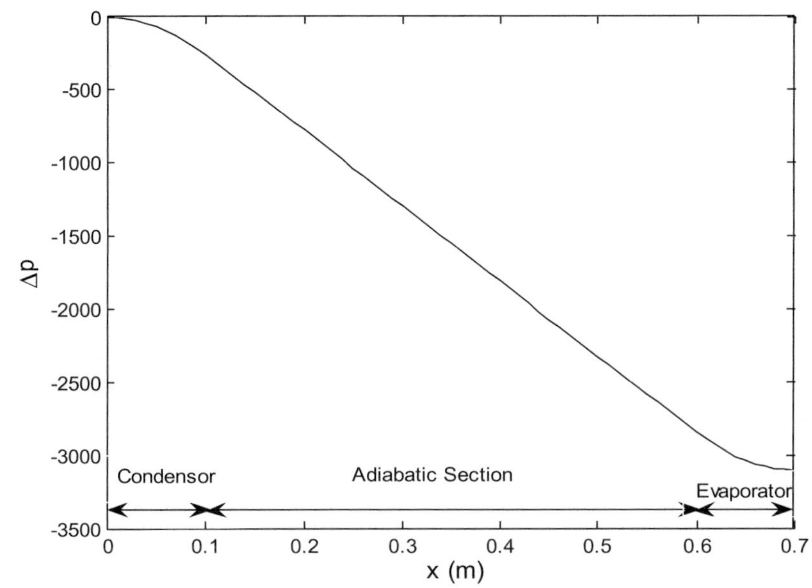

Fig. 12.8 Liquid pressure drop (Pa) in the wick section of a heat pipe

Since the maximum pressure drop the wick can handle is greater than the calculated pressure drop in the liquid, assuming the pressure drop in the vapor is zero, the wick has enough pumping capacity.

The basic equations for the MFM model can also be laid out in terms of phase saturation. The volume-averaged continuity equation for phase k in terms of phase saturation is:

$$\frac{\partial}{\partial t}(\varepsilon s_k \rho_k) + \nabla \cdot \left(\varepsilon s_k \rho_k \langle \mathbf{V}_k \rangle^k\right) = \dot{m}_k''' \tag{12.82}$$

The porosity of the porous medium is ε, the phase saturation is s_k, the intrinsic phase average velocity is $\langle \mathbf{V}_k \rangle^k$, and the volumetric mass production due to the phase change from all other phases to phase k is $\dot{m}_k'''$. The summation of the phase saturation is unity $\sum_{k=1}^{\Pi} s_k = 1$. The phase saturation is the fraction of a pore occupied by phase k; therefore, the product of the phase saturation and the porosity is the total volume fraction of that phase, ε_k.

$$\varepsilon_k = \varepsilon s_k \tag{12.83}$$

The summation of the total volume fraction of each phase and the solid matrix is unity, $\sum_{k=1}^{\Pi} \varepsilon_k + \varepsilon_{sm} = 1$. The momentum equation in a porous medium for phase k is Darcian if inertia and macroscopic shear effects (shear stresses between pores) are neglected.

$$\varepsilon s_k \langle \mathbf{V}_k \rangle^k = -\frac{K K_{rk}}{\mu_k} (\nabla p_k - \rho_k \mathbf{g}) \tag{12.84}$$

Equation (12.84) can be directly inserted into the continuity equation to get a Laplace-type equation for the pressure of phase k. The volume-averaged species equation for the species mass fraction, $\langle \omega_{k,i} \rangle^k$, is:

$$\frac{\partial}{\partial t}\left(\varepsilon s_k \rho_k \langle \omega_{k,i} \rangle^k\right) + \nabla \cdot \left(\varepsilon s_k \rho_k \langle \mathbf{V}_k \rangle^k \langle \omega_{k,i} \rangle^k\right) = -\langle \nabla \cdot \mathbf{J}_{k,i} \rangle + \dot{m}'''_{k,i} \tag{12.85}$$

The diffusive mass flux of species i is $\mathbf{J}_{k,i}$, and the species generation rate due to the phase change or chemical reaction is $\dot{m}'''_{k,i}$. The products of chemical reactions in phase k are also in phase k; therefore, the sum of the species generation term is simply the mass production due to the phase change:

$$\sum_{i=1}^{N} \dot{m}'''_{k,i} = \dot{m}'''_k \tag{12.86}$$

Also, the summation of the species generation for one component overall the phases is simply the species generation rate due to the chemical reactions, $\dot{m}'''_i$, only because phase-change processes do not change the species.

$$\sum_{k=1}^{\Pi} \dot{m}'''_{k,i} = \dot{m}'''_i \tag{12.87}$$

The final equation is the energy equation for phase k, and it is:

$$\begin{aligned} &\frac{\partial}{\partial t}\left(\varepsilon s_k \rho_k \langle h_k \rangle^k\right) + \nabla \cdot \left(\varepsilon s_k \rho_k \langle \mathbf{V}_k \rangle^k \langle h_k \rangle^k\right) \\ &= -\langle \nabla \cdot \mathbf{q}''_{k,i} \rangle + \dot{q}'''_k + \dot{m}'''_k \langle h_k \rangle^k + \dot{q}'''_{k,\mathrm{E}} \end{aligned} \tag{12.88}$$

The second term on the right-hand side represents the total volumetric heat transfer rate from all other phases to phase k, and the third term on the right-hand side is the heat added through the mass production of phase k through phase change. The last term is the heat generated in phase k to an external heat source such as radiation or electrical current. The summation of the second and third terms over all the phases is zero, because of the interfacial energy balance.

$$\sum_{k=1}^{\Pi} \left(\dot{q}'''_k + \dot{m}'''_k \langle h_k \rangle^k\right) = 0 \tag{12.89}$$

Example 12.4

Transport phenomena in the gas diffusion layer of a proton exchange membrane fuel cell (PEMFC) can be simplified into a one-dimensional steady-state problem, as shown in Fig. 12.9 (Nam and Kaviany 2003). Gaseous oxygen is supplied from the channel and diffused into the gas diffusion layer. Water vapor enters at the boundary between the catalyst layer and the diffusion layer and

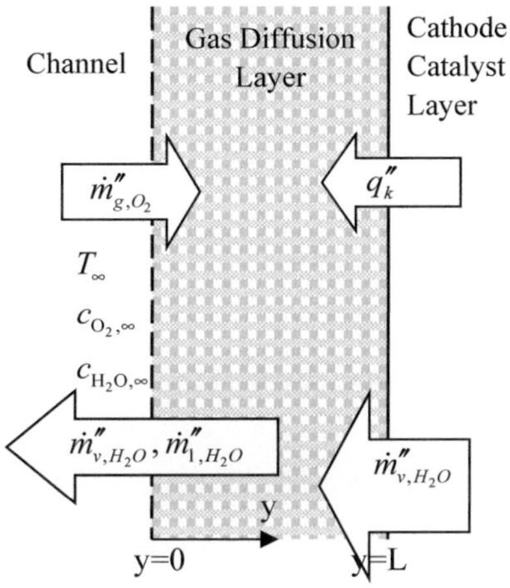

Fig. 12.9 One-dimensional model of a gas diffusion layer

condenses into liquid water in the gas diffusion layer. The condensate is then depleted to the channel. The gas diffusion layer also receives chemical reaction heat from its interface with the catalyst layer. Specify the governing equations and corresponding boundary conditions for transport phenomena in the gas diffusion layer.

Solution Assuming that different phases in the gas diffusion layer are in thermal equilibrium and that the transport phenomena in the gas diffusion layer are steady state, the energy equation, Eq. (3.285), is

$$\frac{\mathrm{d}}{\mathrm{d}y}\left(\langle k_{\mathrm{eff}}\rangle\frac{\mathrm{d}\langle T\rangle}{\mathrm{d}y}\right) = h_{\ell\mathrm{v}}\dot{m}'''_{\mathrm{v}\ell,\mathrm{H_2O}} \tag{12.90}$$

where $\langle k_{\mathrm{eff}}\rangle$ is effective thermal conductivity, $\langle T\rangle$ is volume-averaged temperature, $\dot{m}'''_{\mathrm{v}\ell,\mathrm{H_2O}}$ is the volumetric condensation rate of water (m^3/kg s).
The boundary condition at the interface between the channel and the gas diffusion layer is at a prescribed temperature, i.e.,

$$\langle T\rangle = 0, \quad y = 0 \tag{12.91}$$

At the interface between the catalyst and gas diffusion layers, we have

$$\langle k\rangle\frac{\partial\langle T\rangle}{\partial y} = q''_k, \quad y = L \tag{12.92}$$

The species conservation equation, Eq. (3.289), for oxygen uses a Fickian diffusion term and an advection transport term. Fickian diffusion assumes that the diffusive transport of species is driven by the concentration gradient of that species.

$$\frac{\mathrm{d}}{\mathrm{d}y}\left(-cM_{\mathrm{O_2}}\langle D_{\mathrm{m,O_2}}\rangle\frac{\mathrm{d}x_{\mathrm{O_2}}}{\mathrm{d}y} + x_{\mathrm{O_2}}M_{\mathrm{O_2}}\sum_i\frac{J_i}{M_i}\right) = 0 \tag{12.93}$$

where $c = p(R_g T)$ is total molar concentration, x_i is molar fraction of species i, M_i is molecular mass for species i, and $\langle D_{m,O_2} \rangle$ is the effective mass diffusivity of oxygen.
The boundary conditions for Eq. (12.93) are

$$c_{O_2} = c_{O_2,\infty}, \quad y = 0 \tag{12.94}$$

$$-cM_{O_2} \langle D_{m,O_2} \rangle \frac{dx_{O_2}}{dy} = J_{O_2}, \quad y = L \tag{12.95}$$

where $c_{O_2} = cx_{O_2}$ is the mole concentration of oxygen.
The species conservation equation for water vapor also has a diffusive and advection transport term, plus a term representing the condensation rate.

$$\frac{d}{dy}\left(-cM_{H_2O} \langle D_{m,H_2O} \rangle \frac{dx_{H_2O}}{dy} + x_{H_2O} M_{H_2O} \sum_i \frac{J_i}{M_i}\right) = -\dot{m}'''_{v\ell,H_2O} \tag{12.96}$$

subject to the following boundary conditions

$$c_{H_2O} = c_{H_2O,\infty}, \quad y = 0 \tag{12.97}$$

$$cM_{H_2O} \langle D_{m,H_2O} \rangle \frac{dx_{H_2O}}{dy} = J_{H_2O}, \quad y = L \tag{12.98}$$

where $c_{H_2O} = cx_{H_2O}$ is the mole concentration of water vapor.
The volumetric condensation rate, $\dot{m}'''_{v\ell,H_2O}$, is obtained by

$$\dot{m}'''_{v\ell,H_2O} = M_{H_2O} \gamma \frac{p_{v,H_2O} - p_{sat,H_2O}(T)}{R_g T} \tag{12.99}$$

where γ is the volumetric condensation coefficient, and p_{v,H_2O} and $p_{sat,H_2O}(T)$ are partial pressure of the water vapor and saturation pressure of water at temperature T, respectively.
The liquid flow in the gas diffusion layer is driven by capillary force, assuming that the gas-phase pressure is constant, and the Reynolds number is less than unity, i.e.,

$$\frac{d}{dy}\left[-\rho_\ell \frac{KK_{r\ell}}{\mu_\ell} \left(\frac{dp_c}{dS}\right) \frac{dS}{dy}\right] = \dot{m}'''_{v\ell,H_2O} \tag{12.100}$$

where $S = (s - s_{ir})/(1 - s_{ir})$ is reduced water saturation (s is saturation defined as the fraction of pore space occupied by liquid water, s_{ir} is irreducible saturation), K is permeability, and $K_{r\ell} = S^3$ is relative permeability. The relative permeability accounts for the increased resistance to the flow in the liquid when the liquid saturation decreases.

12.3.3 Multiphase Mixture Model (MMM)

The general procedure to relate the MFM to the MMM is to sum each equation over all of the phases. The introductions of the mass-averaged density and velocity, as well as the relative mobility, are given where appropriate. The MMM was initially developed by Wang and Beckermann (1993) and was applied to model multiphase flow in fuel cells.

The continuity equation summed over two fluid phases, k and j, is:

$$\frac{\partial}{\partial t}\left[\varepsilon\left(s_k\rho_k+s_j\rho_j\right)\right]+\nabla\cdot\left[\varepsilon\left(s_k\rho_k\langle\mathbf{V}_k\rangle^k+s_j\rho_j\langle\mathbf{V}_j\rangle^j\right)\right]=0 \tag{12.101}$$

Defining mixture density and the mixture velocity

$$\bar{\rho}=s_k\rho_k+s_j\rho_j \tag{12.102}$$

$$\bar{\rho}\bar{\mathbf{V}}=\varepsilon s_k\rho_k\langle\mathbf{V}_k\rangle^k+\varepsilon s_j\rho_j\langle\mathbf{V}_j\rangle^j \tag{12.103}$$

the continuity equation can be rewritten as

$$\frac{\partial}{\partial t}(\varepsilon\bar{\rho})+\nabla\cdot(\bar{\rho}\bar{\mathbf{V}})=0 \tag{12.104}$$

The momentum equation can be summed over both phases.

$$\varepsilon s_k\rho_k\langle\mathbf{V}_k\rangle^k+\varepsilon s_j\rho_j\langle\mathbf{V}_j\rangle^j=-K\left[\frac{K_{rk}}{\nu_k}(\nabla p_k-\rho_k\mathbf{g})+\frac{K_{rj}}{\nu_j}(\nabla p_j-\rho_j\mathbf{g})\right] \tag{12.105}$$

A capillary pressure that relates the pressure in phase k to the pressure in phase j is introduced:

$$p_c=p_j-p_k \tag{12.106}$$

The capillary pressure is often expressed as the Leverett function (Leverett 1941). This function relates the capillary pressure to the wetting phase saturation, s_w. A phase is said to wet the porous material if the contact angle that phase makes with it is less than 90°. Therefore, if $\theta_k<90°$, the wetting phase is phase k, $s_w=s_k$; if $\theta_j<90°$, then phase j is the wetting phase, $s_w=s_j$.

$$p_c=\sigma\cos\theta\left(\frac{\varepsilon}{K}\right)^{1/2}\left[1.417(1-s_w)-2.120(1-s_w)^2+1.263(1-s_w)^3\right] \tag{12.107}$$

This specific function was developed to describe the capillary pressure in soils engineering; however, its use has been extended to other technology such as fuel cells. The reason for the over usage of the Leverett function is the lack of functions to describe the capillary pressure for other types of porous media. With the definition of the mixture velocity and the capillary pressure, the mixture momentum equation can be rewritten as:

$$\bar{\rho}\bar{\mathbf{V}}=-K\left[\left(\frac{K_{rk}}{\nu_k}+\frac{K_{rj}}{\nu_j}\right)\nabla p_j-\frac{K_{rk}}{\nu_k}\nabla p_c-\left(\frac{\rho_k K_{rk}}{\nu_k}+\frac{\rho_j K_{rj}}{\nu_j}\right)\mathbf{g}\right] \tag{12.108}$$

Note that the capillary pressure gradient can be reduced into its relative components.

$$\nabla p_c = \frac{\partial p_c}{\partial s_k}\nabla s_k + \frac{\partial p_c}{\partial \sigma_{jk}}\sum_{i=1}^{N-1}\frac{\partial \sigma_{jk}}{\partial \langle \omega_{k,i}\rangle^k}\nabla \langle \omega_{k,i}\rangle^k + \frac{\partial p_c}{\partial \sigma_{jk}}\frac{\partial \sigma_{jk}}{\partial T}\nabla T \tag{12.109}$$

Now the mixture kinematic viscosity, $\bar{\nu}$, and relative mobility, λ_k, are introduced.

$$\bar{\nu} = \left(\frac{K_{rk}}{\nu_k} + \frac{K_{rj}}{\nu_j}\right)^{-1} \tag{12.110}$$

$$\lambda_k = \frac{K_{rk}}{\nu_k}\bar{\nu} \tag{12.111}$$

The momentum Eq. (12.108) can be rewritten as:

$$\bar{\rho}\bar{\mathbf{V}} = -\frac{K}{\bar{\nu}}\left[\nabla p_j - \lambda_k \nabla p_c - \left(\lambda_k \rho_k + \lambda_j \rho_j\right)\mathbf{g}\right] \tag{12.112}$$

A definition of the mixture pressure is introduced so that:

$$\nabla \bar{p} = \nabla p_j - \lambda_k \nabla p_c \tag{12.113}$$

A density correction factor, γ_ρ, is also introduced.

$$\gamma_\rho = \frac{1}{\bar{\rho}}\left(\rho_k \lambda_k + \rho_j \lambda_j\right) \tag{12.114}$$

The momentum Eq. (12.112) can be written in the form

$$\bar{\rho}\bar{\mathbf{V}} = -\frac{K}{\bar{\nu}}\left[\nabla \bar{p} - \gamma_\rho \rho \mathbf{g}\right] \tag{12.115}$$

In order to simplify the MMM derivation for the species and energy equations, a *diffusive phase*-mass flux is introduced. This term is analogous to the diffusion mass flux in multicomponent mixtures, but refers to each phase, rather than each component in a phase. This value relates the actual mass flux of phase k to the mixture mass flux.

$$\mathbf{j}_k = \varepsilon s_k \rho_k \langle \mathbf{V}_k \rangle^k - \lambda_k \bar{\rho}\bar{\mathbf{V}} \tag{12.116}$$

From this relation, it can be shown that the *diffusive phase*-mass flux is:

$$\mathbf{j}_k = -\mathbf{j}_j = \frac{\lambda_k \lambda_j}{\bar{\nu}}\left[K\nabla p_c + \left(\rho_k - \rho_j\right)\mathbf{g}\right] \tag{12.117}$$

This relation will be useful and will be used henceforth. The mixture species equation for species i is obtained by adding the species i equation for phase k and phase j.

$$\begin{aligned}&\frac{\partial}{\partial t}\left(\varepsilon s_k \rho_k \langle \omega_{k,i}\rangle^k + \varepsilon s_j \rho_j \langle \omega_{j,i}\rangle^j\right) + \nabla \cdot \left(\varepsilon s_k \rho_k \langle \mathbf{V}_k\rangle^k \langle \omega_{k,i}\rangle^k + \varepsilon s_j \rho_j \langle \mathbf{V}_j\rangle^j \langle \omega_{j,i}\rangle^j\right)\\ &= -\nabla \cdot \left\langle \mathbf{J}_{k,i} + \mathbf{J}_{j,i}\right\rangle + \dot{m}'''_{k,i} + \dot{m}'''_{j,i}\end{aligned} \tag{12.118}$$

From Eq. (12.87), the summation of the species production over all the phases is the species production due to chemical reaction. The mixture mass fraction is:

$$\bar{\rho}\bar{\omega}_i = s_k\rho_k\langle\omega_{k,i}\rangle^k + s_j\rho_j\langle\omega_{j,i}\rangle^j \tag{12.119}$$

The correction factor for species advection, γ_i, is introduced.

$$\gamma_i = \frac{1}{\bar{\omega}_i}\left(\lambda_k\langle\omega_{k,i}\rangle^k + \lambda_j\langle\omega_{j,i}\rangle^j\right) \tag{12.120}$$

The mixture species equation is:

$$\begin{aligned}&\frac{\partial}{\partial t}(\varepsilon\bar{\rho}\bar{\omega}_i) + \nabla\cdot(\gamma_i\bar{\rho}\bar{\mathbf{V}}\bar{\omega}_i)\\&= -\nabla\cdot\langle\mathbf{J}_{k,i} + \mathbf{J}_{j,i}\rangle - \nabla\cdot\left(\mathbf{j}_k\left(\langle\omega_{k,i}\rangle^k - \langle\omega_{j,i}\rangle^j\right)\right) + \dot{m}_i'''\end{aligned} \tag{12.121}$$

It should be noted that Eqs. (12.104), (12.105), and (12.121) are also applicable to multiphase and not necessary to two phase as developed here.

It is important to point out that the mixture species equation still contains the species mass fraction of each phase. Therefore, the species mass fraction in phase k and phase j must be related to the mixture mass fraction through thermodynamic equilibrium. Expanding all the terms in Eq. (12.119) yields:

$$(s_k\rho_k + s_j\rho_j)\bar{\omega}_i = s_k\rho_k\langle\omega_{k,i}\rangle^k + s_j\rho_j\langle\omega_{j,i}\rangle^j \tag{12.122}$$

Since there are two phases presented, the phase saturations add to unity. Therefore, the phase saturation of phase k can be calculated by:

$$s_k = \frac{\rho_j\left(\bar{\omega}_i - \langle\omega_{j,i}\rangle^j\right)}{\rho_k\left(\bar{\omega}_i - \langle\omega_{k,i}\rangle^k\right) + \rho_j\left(\langle\omega_{j,i}\rangle^k - \bar{\omega}_i\right)} \tag{12.123}$$

It is important to note that when the saturation of phase k is calculated in this manner, the one-phase continuity equation is not solved; instead, all N species equations are solved. For more discussion on this approach to the calculation of the phase saturation, refer to the comparison of MFM and MMM models below.

The last transport equation that must be examined is the energy equation. The energy equations of both phases are added together plus an unsteady and heat conduction term represents the influence of the solid matrix on the energy equation.

$$\begin{aligned}&\frac{\partial}{\partial t}\left((1-\varepsilon)\rho_{\text{sm}}\langle h_{\text{sm}}\rangle^{\varepsilon\text{m}} + \varepsilon\left[s_k\rho_k\langle h_k\rangle^k + s_j\rho_j\langle h_j\rangle^j\right]\right)\\&+\nabla\cdot\left(\varepsilon\left[s_k\rho_k\langle\mathbf{V}_k\rangle^k\langle h_k\rangle^k + s_j\rho_j\langle\mathbf{V}_j\rangle^j\langle h_j\rangle^j\right]\right) = -\nabla\cdot\left\langle\mathbf{q}_{k,i}'' + \mathbf{q}_{j,i}'' + \mathbf{q}_{\text{sm},i}''\right\rangle + \dot{q}_{\text{E}}'''\end{aligned} \tag{12.124}$$

The external heat generation is represented by $\dot{q}_{\text{E}}'''$, and it is the summation of the external heat generation over all of the phases. The mixture enthalpy is

$$\bar{\rho}\bar{h} = s_k \rho_k \langle h_k \rangle^k + s_j \rho_k \langle h_j \rangle^j \tag{12.125}$$

The correction factor for energy advection, γ_h, is also introduced:

$$\gamma_h = \frac{1}{\bar{h}} \left(\lambda_k \langle h_k \rangle^k + \lambda_j \langle h_j \rangle^j \right) \tag{12.126}$$

An assumption that corresponds with the mixture enthalpy is that each phase is in thermodynamic equilibrium, $\langle T_k \rangle^k = \langle T_j \rangle^j$. If Fourier's law governs the heat conduction and effective thermal conductivity can be used, the energy equation can be rewritten using an effective thermal conductivity and the *diffusive phase*-mass flux as:

$$\begin{gathered} \frac{\partial}{\partial t}((1-\varepsilon)\rho_{\mathrm{sm}} \langle h_{\mathrm{sm}} \rangle^{\mathrm{sm}} + \varepsilon \bar{\rho}\bar{h}) + \nabla \cdot (\gamma_h \bar{\rho} \bar{\mathbf{V}} \bar{h}) \\ = \nabla \cdot (k_{\mathrm{eff}} \nabla T) - \nabla \cdot \left(\mathbf{j}_k \left(\langle h_k \rangle^k - \langle h_j \rangle^j \right) \right) + \dot{q}_{\mathrm{E}}''' \end{gathered} \tag{12.127}$$

where the effective thermal conductivity is related to the phase saturation, porosity, geometry of the porous medium (tortuosity, $\Im$) and the conductivity of the porous media and both phases, $k_{\mathrm{eff}} = f\left(k_{\mathrm{sm}}, k_k, k_j, \varepsilon, s_k, \Im\right)$.

It is important to note that the heat generation and consumption of chemical reactions and latent heat are all embedded in this governing equation.

12.3.4 Comparison of MFM and MMM Models

The major advantage of MFM over MMM is that MFM has the potential to capture the details of each phase, including interface. Furthermore, the nonequilibrium thermodynamics for temperature and species concentration in each phase can be obtained by MFM. Another benefit is the potential to account for the pore size distribution in a porous media and model how the saturation of each pore is not equivalent. The pore size distribution is not well researched in the literature. This is why the Leverett function of capillary pressure is commonly used, even though this function applies to soil engineering and probably is not the best choice for other applications.

In the MFM model, the momentum equation can be directly inserted into the continuity equation, therefore eliminating the intrinsic velocity of each phase. It also can be directly inserted into the energy and species equations. Also, the phase saturation of phase j is known if the saturation of phase k is known. Similarly, only $N-1$ species equations are needed in each phase because the species balance is maintained through continuity. Therefore, the main flow variables in the MFM model are $s_k, p_j, \langle \omega_{k,i} \rangle^k, \langle \omega_{k,i} \rangle^k, \langle \omega_{j,i} \rangle^j, \langle T_k \rangle^k, \langle T_j \rangle^j$.

The number of equations and the number of unknowns using the MFM model for two phases when the energy equation is needed are $5 + 2 \times (N-1)$. These equations are two continuity equations, three energy equations, and $N-1$ species equations for each phase. Applying the appropriate boundary conditions, this is a well-posed problem. In the MMM model, applying the momentum equation directly into the continuity equation to eliminate the velocity, the main flow variables are $\bar{\rho}, \bar{p}, \bar{\omega}_i, \bar{h}$.

The variables are the mixture density, the mixture pressure, $N-1$ species mass fraction, and the temperature, i.e., the number of variables is $3 + (N-1)$. The main equations are one mixture continuity equation, $N-1$ mixture mass fraction equations, and one energy equation, i.e., the number of equations = $2+(N-1)$. One more equation needs to be solved in order for this approach to be well posed, since there is one more dependent variable than equations. The first approach to this problem is to solve the continuity equation for one of the phases. Using the MMM model variables, the continuity equation for phase k is:

$$\frac{\partial}{\partial t}(\varepsilon s_k \rho_k) + \nabla \cdot (\lambda_k \bar{\rho} \bar{\mathbf{V}}) = -\nabla \cdot \mathbf{J}_k + \dot{m}_k''' \tag{12.128}$$

The other approach, which is suggested by Wang and Cheng (1996) and used in Wang and Wang (2003) to model fuel cells, is to solve species equations for all N species and not to solve the continuity equation for phase k. It is very important to note that if this approach is taken, the mixture species mass fraction as well the species mass fraction in each phase must add up to unity.

$$\sum_{i=1}^{N} \bar{\omega}_i = \sum_{i=1}^{N} \langle \omega_{k,i} \rangle^k = \sum_{i=1}^{N} \langle \omega_{j,i} \rangle^j = 1 \tag{12.129}$$

If these criteria are met, the saturation calculated by Eq. (12.123) is correct. Assuming the relations in Eq. (12.129) are upheld, then the species equation for Nth component can be written in terms of all the other components.

$$\begin{aligned} &\frac{\partial}{\partial t}(\varepsilon \bar{\rho}) - \sum_{i=1}^{N-1} \frac{\partial}{\partial t}(\varepsilon \bar{\rho} \bar{\omega}_i) + \nabla \cdot (\bar{\rho} \bar{\mathbf{V}}) + \sum_{i=1}^{N-1} \nabla \cdot (\gamma_i \bar{\rho} \bar{\mathbf{V}} \bar{\omega}_i) \\ &= \sum_{i=1}^{N-1} \nabla \cdot \langle \mathbf{J}_{k,i} + \mathbf{J}_{j,i} \rangle + \sum_{i=1}^{N-1} \nabla \cdot \left(\mathbf{j}_k \left(\langle \omega_{k,i} \rangle^k - \langle \omega_{j,i} \rangle^j \right) \right) - \sum_{i=1}^{N-1} \dot{m}_i''' \end{aligned} \tag{12.130}$$

This equation is simply the continuity equation subtracted by $N-1$ species equations. Mathematically, this equation is not a new (independent) equation and therefore solving for the Nth species equation still leaves the system of equations ill-posed. This leads us to using the continuity equation for phase k to complete the system of equations.

The formulation of MMM reduces the total number of equations, which, at first glance, would seem to decrease the computational time. However, when looking at the number of terms that these equations incorporate, it is not very different from the MFM. For example, examining the species equation in the MFM model for Π phases and D_x dimensions, and comparing it to the MMM model for the same number of phases and dimensions can show how many terms are solved (Table 12.2; Rice 2006).

For two-phase, two-dimensional systems, the MFM model solves a total of ten terms, while the MMM model solves a total of 11 terms. So the computational time difference between the MMM and the MFM models are probably not as dramatic as it first appears. Since the total number of terms

Table 12.2 Comparison of MFM and MMM, total terms computed in species equation

Model	First-order time	First-order space (advection)	Second-order space (diffusion)	Total
MFM	Π	ΠD_x	ΠD_x	$\Pi(2D_x+1)$
MMM	1	$(1+\Pi)D_x$	ΠD_x	$D_x(2\Pi+1)+1$

solved is very close, the computational time is probably limited by the convergence rate more than anything else.

Unsaturated flow theory comes from MFM, with additional assumptions. In unsaturated flow theory (UFT), only the liquid phase is considered, and the gas pressure is assumed to be equal to the hydrostatic pressure of the gas. If the capillary pressure is assumed to be only a function of liquid saturation, and there is no mass transfer considered, then Richard's equation is valid. Richard's equation is:

$$\frac{\partial}{\partial t}(\varepsilon s_k \rho_k) + \nabla \cdot \left[K \frac{K_{rk}}{\nu_k} \left(\frac{dp_c}{ds} \nabla s_k + (\rho_k - \rho_j)\mathbf{g} \right) \right] = 0 \qquad (12.131)$$

12.4 Boiling in Porous Media

12.4.1 Boiling in a Wicked Surface

A wicked surface is formed by coating a surface with a thin porous layer. It can be applied as the wick structures of a heat pipe as well as form a modified surface for enhanced heat transfer. Boiling in porous media is more complicated than that from plain surfaces due to the presence of capillary structures. Also, heat transfer and vapor formation may change with different types of wick structures and working fluids. The emphasis in this subsection is the thin porous layers. However, boiling in porous media in general is discussed in Sect. 12.4.2. As a result, different heat transfer modes may exist in the wicks. For a thin layer of wick structure saturated with liquid as it is applied in heat pipes, four basic modes of heat transfer and vapor formation are shown schematically in Fig. 12.10 (Faghri 2016).

Mode 1: Conduction–convection. The whole wick is filled with liquid; conduction occurs across the liquid layer and evaporation takes place from its surface. No boiling occurs within the wick. However, natural convection may take place within some thick wicks under a gravitational field. This is the situation for nonmetallic working fluids under a low heat flux and metallic working fluids under a low or moderately high heat flux. The heat transfer across the wick can be calculated by a conduction model with sufficient accuracy.

Mode 2: Receding liquid. As heat flux is increased, the evaporation at the liquid surface intensifies. The capillary or body forces available in the heat pipe may not be capable of driving enough liquid back to the evaporator zone. As a result, the liquid layer begins to recede into the wick structure. Until the liquid is completely depleted, the heat transfer across the liquid layer is still by conduction, and liquid vaporization takes place at the liquid–vapor interface. No boiling occurs within the wick structure. This is applicable to both nonmetallic and metallic working fluids.

Mode 3: Nucleate boiling. For some nonmetallic working fluids, a large temperature difference across the wick may cause nucleate boiling within the wick. Bubbles grow at the heated wall and escape to the liquid surface and burst rapidly. As noted before, nucleate boiling has significance for heat transfer improvements in many applications; except some cases, such as wicked heat pipes which is one of the limits of operation. For wicked heat pipes that require capillary force to sustain liquid circulation, nucleate boiling in the wick indeed represents a heat transfer limit for the following reasons: (a) large bubbles bursting at the liquid surface may disrupt the menisci established at the liquid–vapor interface and eliminate the capillary force circulating the liquid condensate; and (b) vapor bubbles generated at the evaporator section may block the liquid return from the condenser section.

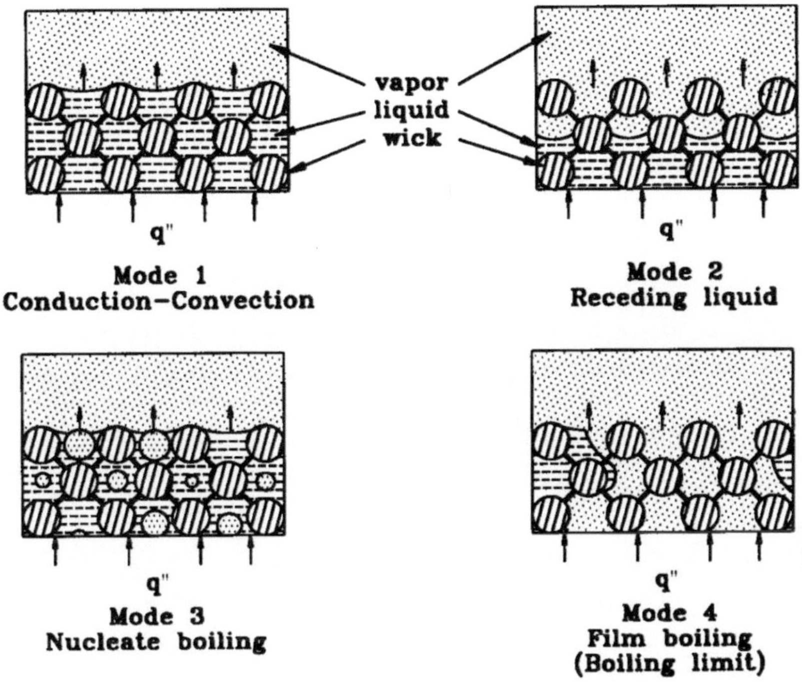

Fig. 12.10 Modes of heat transfer in wicks (Faghri 2016)

Mode 4: Film boiling. As the temperature difference across the wick increases, a large quantity of bubbles is generated at the heated wall. Before escaping to the surface, these bubbles coalesce and form a layer of vapor adjacent to the heated wall. This layer prevents the liquid from reaching the wall surface. As a result, the wall temperature will increase rapidly and the heat pipe may burn out. This heat transfer limit is similar to film boiling in pool boiling heat transfer and is a common heat transfer limit for wicked heat pipes.

Heat transfer from wicked surfaces has been extensively studied. Since the wick provides additional sites for nucleation, heat transfer in the wick is more complicated than boiling heat transfer from plain surfaces. In addition, the superheat ΔT for boiling incipience in a wick should be lower than that for a smooth surface. Marto and Lepere (1982) experimentally studied pool boiling heat transfer from various porous metallic coatings and enhanced surfaces. Figure 12.11shows their experimental results for a porous coating called the high flux surface. It can be seen from the figure that boiling heat transfer from a surface covered by a porous coating requires a much lower wall superheat due to the presence of additional nucleation sites.

Ferrell and Alleavitch (1970) experimentally studied the heat transfer from a horizontal surface covered with beds of Monel® beads. The bed depth ranged from 3.2 to 25.4 mm, and water was the working fluid at atmospheric pressure. The experimental results are shown in Fig. 12.12. They concluded that the heat transfer mechanism at lower ΔT was conduction through the saturated wick–liquid matrix to the liquid–vapor interface, which corresponds to Mode 1 in Fig. 12.10. At higher superheats, nucleate boiling occurs, and the experimental data deviate from the curve predicted by the conduction model. The experimental data intercept the boiling curves from the plane surface and Rohsenow's pool boiling correlation (Rohsenow et al. 1985) at a higher ΔT. The superheat at which nucleate boiling occurs is lower than that of pool boiling.

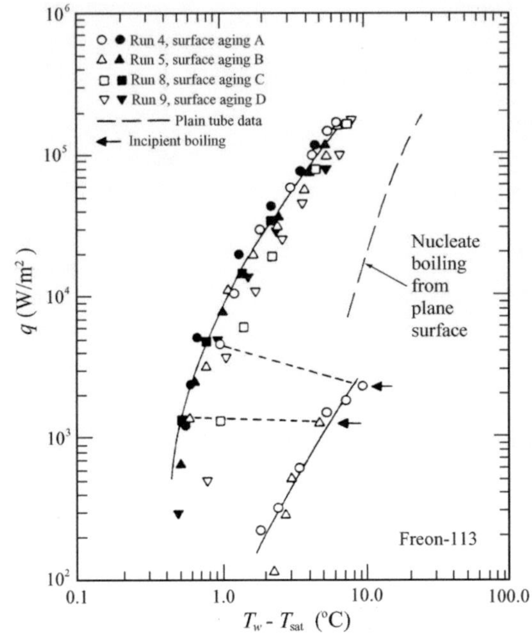

Fig. 12.11 Comparison of high heat flux surface to plain tube for R-113 (Marto and Lepere 1982)

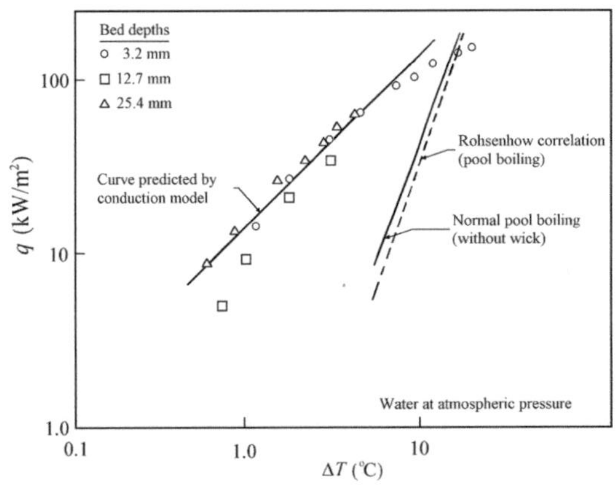

Fig. 12.12 Heat transfer from a submerged wick (Ferrell and Alleavitch 1970)

Analysis of boiling in the wick involves the formation of bubbles (nucleation) as well as their subsequent growth and motion. Some vapor nuclei or small bubbles always exist within the wick structure, but the superheat is essential for these bubbles to grow. A small, trapped, hemispherical vapor bubble with effective radius R_b in the vicinity of the wall–wick interface is shown in Fig. 12.13. The bubble, as well as the liquid adjacent to the wall, is at the wall temperature T_w. The vapor pressure inside the vapor bubble is $p_{b,v}$, while the liquid pressure adjacent to the bubble is p_ℓ. The temperature and pressure of the vapor adjacent to the meniscus are T_v and p_v. At equilibrium,

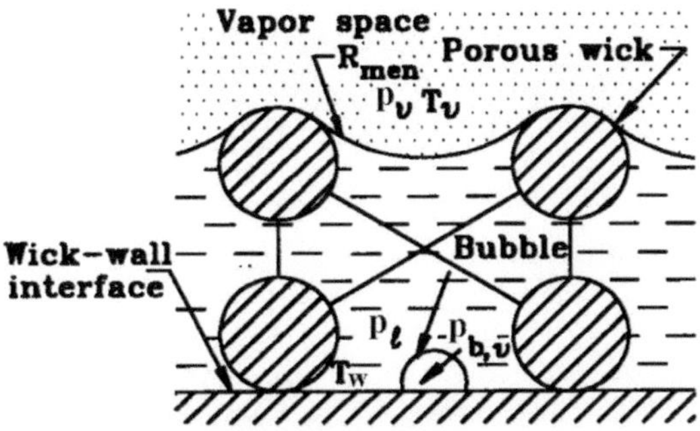

Fig. 12.13 Bubble formation at the wall–wick interface (Faghri 2016)

$$p_{b,v} - p_\ell = \frac{2\sigma}{R_b} \tag{12.132}$$

The relation between the liquid pressure and the vapor-space pressure is

$$p_v - p_\ell = \frac{2\sigma}{R_{men}} \tag{12.133}$$

where R_{men} is the radius of the liquid–vapor meniscus. Combining of Eqs. (12.132) and (12.133) yields

$$p_{b,v} - p_v = 2\sigma\left(\frac{1}{R_b} - \frac{1}{R_{men}}\right) \tag{12.134}$$

The vapor pressure in the bubble, $p_{v,b}$, is related to the saturation pressure corresponding to the wall temperature, $p_{sat}(T_w)$, by

$$p_{b,v} = p_{sat}(T_w)\left(1 - \frac{2\sigma\rho_v}{p_{sat}(T_w)R_b\rho_\ell}\right) \tag{12.135}$$

Combining Eqs. (12.134) and (12.135), one obtains

$$p_{sat}(T_w) - p_v = 2\sigma\left(\frac{1}{R_b} - \frac{1}{R_{men}}\right) + \frac{2\sigma\rho_v}{R_b\rho_\ell} \tag{12.136}$$

Assuming the vapor adjacent to the meniscus is at saturation state, and applying the Clausius–Clapeyron equation between (p_v, T_v) and ($p_{sat}(T_w)$, T_w), a critical temperature difference is obtained:

$$\Delta T_{crit} = T_w - T_v = \frac{R_g T_v T_w}{h_{\ell v}} \ln\left[1 + \frac{2\sigma}{p_v}\left(\frac{1}{R_b} - \frac{1}{R_{men}}\right) + \frac{2\sigma\rho_v}{p_v R_b \rho_\ell}\right] \tag{12.137}$$

If

$$\frac{2\sigma}{p_v}\left(\frac{1}{R_b}-\frac{1}{R_{men}}\right)+\frac{2\sigma\rho_v}{p_v R_b \rho_\ell}<1$$

Equation (12.137) can be rewritten as

$$\Delta T_{crit}=T_w-T_v=\frac{R_g T_v T_w}{h_{\ell v}}\left[\frac{2\sigma}{p_v}\left(\frac{1}{R_b}-\frac{1}{R_{men}}\right)+\frac{2\sigma\rho_v}{p_v R_b \rho_\ell}\right] \tag{12.138}$$

considering

$$\frac{2\sigma\rho_v}{p_v R_b \rho_\ell}\ll\frac{2\sigma}{p_v}\left(\frac{1}{R_b}-\frac{1}{R_{men}}\right)$$

and assuming $T_v T_w \approx T_v^2$, a commonly used critical superheat for bubble formation is obtained (Faghri 2016):

$$\Delta T_{crit}=\frac{2\sigma T_v}{h_{\ell v}\rho_v}\left(\frac{1}{R_b}-\frac{1}{R_{men}}\right) \tag{12.139}$$

which indicates that the required superheat in a wick is lower than that for a plain surface.

12.4.2 Boiling in Porous Media

When a thick porous medium saturated with liquid sits on a horizontal impermeable surface heated from below, boiling may occur when the horizontal surface temperature reaches the saturation temperature of the liquid. The experiments conducted by Sondergeld and Turcotte (1977) revealed that no significant superheat is required to initiate boiling in porous media. The heat transfer mode is similar to boiling in a wicked surfaces—discussed in the preceding subsection—except that evaporation on the top of the porous medium does not have much effect on the formation and growth of the vapor bubbles. In addition, the vapor bubbles cannot escape from the top of the porous medium as they can in the case of boiling on a wicked surface. Consequently, the entire porous medium can be divided into two regions: a two-phase region and a liquid region (see Fig. 12.14; Ramesh and Torrance 1990). The two-phase region is isothermal at the saturation temperature, and heat transfer in the two-phase region is achieved by counterpercolation of liquid and vapor. If, on the other hand, the temperature in the liquid region is below saturation temperature, convection may take place before and after boiling is initiated. Since the density of the liquid overlaying the two-phase region is higher than that of the two-phase region itself, the interface between liquid and the two-phase region may become unstable due to buoyancy and gravitational instabilities. The gravitational instability differs from the classical Rayleigh–Taylor instability because of heat and mass transfer across the interface.

When the superheat on the surface is not very high, the vapor bubbles generated at the horizontal heating surface move upward while the liquid flows downward; this motion is similar to nucleate boiling in bulk liquid. The horizontal heating surface can be kept wet, and the vaporization process can be sustained. As the superheat increases and the critical heat flux is exceeded, the vapor bubbles join together to form a vapor film that covers the horizontal surface and prevents it from coming into direct contact with the liquid. In this case, a vapor region is present and the entire porous medium is divided into three layers.

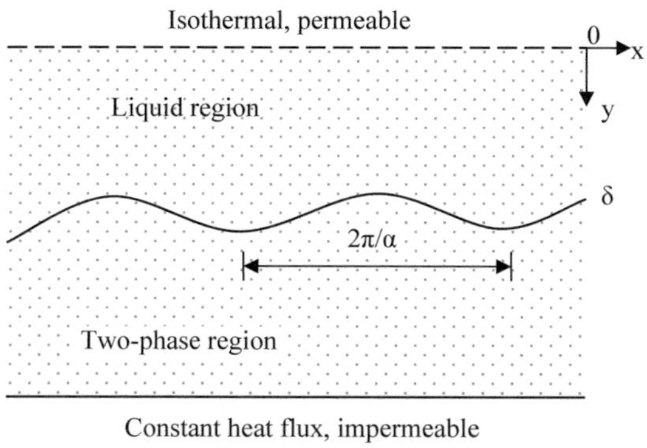

Fig. 12.14 Boiling in a porous medium

The characteristics of boiling in porous media are often described using the Bond number

$$Bo = \frac{g(\rho_\ell - \rho_v)K/\varepsilon}{\sigma} \tag{12.140}$$

where K and ε are permeability and porosity, respectively. Fukusako et al. (1986) conducted an experimental study of the boiling of water, R-11 and R-113 saturated in packed beds of spherical beads with diameters ranging from 1.0 to 16.5 mm. The height of the packed bed ranged from 10 to 300 mm. The results for boiling of R-11 in an 80-mm-high packed bed with glass beads with different diameters are shown in Fig. 12.15. The corresponding Bond number obtained by Kaviany (2013) is also shown in Fig. 12.15. When there are no glass beads present ($Bo \to \infty$), Fukusako et al.'s (1986) experimental results are consistent with the conventional pool boiling curve discussed in Sect. 9.2. The variation of the heat flux follows the order of nucleate, transient, and film boiling. The existences

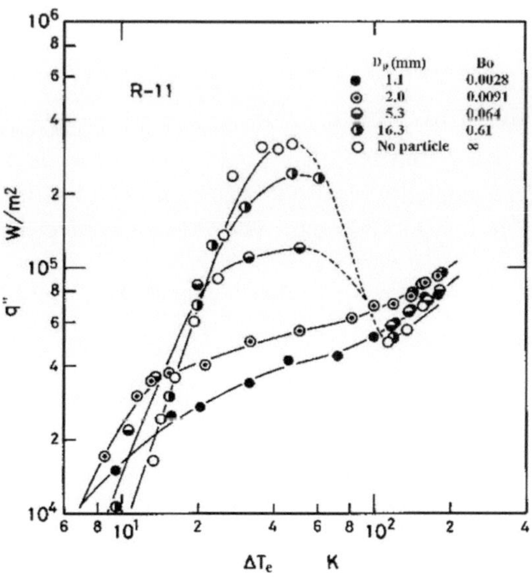

Fig. 12.15 Effect of particle size (Bond number) on boiling curve for R-11 in a packed bed with glass beads (Fukusako et al. 1986)

of maximum heat flux (transition from nucleate to transient boiling) and minimum heat flux (transition from transient to film boiling) are apparent. For boiling in a packed bed with large glass beads (D_p = 16.3 mm, Bo = 0.61), the variation of heat flux follows a similar trend except that the maximum heat flux is lower. As the diameters of the glass beads decrease, the maximum heat flux continuously decreases, while the minimum heat flux is almost unaffected. As a result, the heat flux becomes a monotonic function of ΔT_e when the diameter of the glass beads is less than 2.0 mm or Bo < 0.01. In this case, the transition from nucleate to film boiling occurred without going through a maximum heat flux.

For transition boiling in porous media, Fukusako et al. (1986) recommended the following empirical correlation based on the experimental results from their own and other results in the literature:

$$Nu_\ell = \frac{q''D_p}{\Delta T_e k_{\mathrm{eff},\ell}} = 0.075\left(\frac{D_p}{\sqrt{\sigma/[g(\rho_\ell-\rho_v)]}}\right)^{0.9}\left(\frac{h_{\ell v}}{c_{\mathrm{eff},\ell}\Delta T_e}\right)^m Pr_\ell^{2.37}\left(\frac{k_{\mathrm{eff},\ell}}{k_\ell}\right)^n \tag{12.141}$$

where $k_{\mathrm{eff},\ell} = \varepsilon k_\ell + (1-\varepsilon)k_{\mathrm{sm}}$ and $c_{\mathrm{eff},\ell} = \varepsilon c_\ell + (1-\varepsilon)c_{\mathrm{sm}}$ are thermal conductivity and specific heat of the porous media saturated with liquid. The exponents in Eq. (12.141) are

$$m = 1.3\left(\frac{D_p}{\sqrt{\sigma/[g(\rho_\ell-\rho_v)]}}\right)^{0.6} Pr_\ell^{-0.8} \tag{12.142}$$

and

$$n = -0.59 Pr_\ell^{0.3} \tag{12.143}$$

Fukusako et al. (1986) showed that Eq. (12.141) correlates with most experimental results obtained by them and other researchers within ±40%.

For film boiling in porous media, Fukusako et al. (1986) recommended the following correlation:

$$Nu_v = \frac{q''D_p}{\Delta T_e k_{\mathrm{eff},v}} = 4.10(GrPr_v)^{0.25}\left(\frac{h_{\ell v}}{c_{\mathrm{eff},v}\Delta T_e}\right)^{0.04}\left(\frac{k_{\mathrm{eff},v}}{k_v}\right)^{-0.95}\left(\frac{D_p}{H_p}\right)^{0.15} \tag{12.144}$$

where $k_{\mathrm{eff},v} = \varepsilon k_v + (1-\varepsilon)k_{\mathrm{sm}}$ and $c_{\mathrm{eff},v} = \varepsilon c_v + (1-\varepsilon)c_{\mathrm{sm}}$ are thermal conductivity and specific heat of the porous media saturated with vapor. Equation (12.144) was obtained for the packed bed with height, H_p, ranging from 10 to 300 mm. The Grashof number is defined as:

$$Gr = \frac{(\rho_\ell-\rho_v)\rho_v g D_p^3}{\mu_v^2} \tag{12.145}$$

They also obtained an empirical correlation for minimum heat flux as follows

$$\frac{q_{\min}^2 D_p}{\rho_v \sigma h_{\ell v}^2} = 9.21\times 10^3\left(\frac{D_p}{\sqrt{\sigma/[g(\rho_\ell-\rho_v)]}}\right)^{0.65}\left(\frac{\rho_v}{\rho_\ell}\right)^{0.5}\left(\frac{k_{\mathrm{eff},v}}{k_v}\right)^{0.1} \tag{12.146}$$

12.4.3 Film Boiling in Porous Media

Film boiling of liquid saturated in a porous medium at an initial temperature of $T_\infty < T_{sat}$ next to a vertical, impermeable heated wall at a temperature of $T_w > T_{sat}$ is analyzed (see Fig. 12.16; Cheng and Verma 1981; Nield and Bejan 2017). Vapor generated at the liquid-vapor interface flows upward due to buoyancy force. The liquid adjacent to the vapor layer is dragged upward by the vapor. The temperature at the liquid–vapor interface is at the saturation temperature. There are velocity and thermal boundary layers in the liquid phase adjacent to the vapor film. The solution of the film boiling problem requires solutions of vapor and liquid flow, as well as heat transfer in both the vapor and liquid phases. It is assumed that boundary-layer approximations are applicable to the vapor film and to convection heat transfer in the liquid phase. It is further assumed that the vapor flow is laminar, two-dimensional; Darcy's law is applicable in both the vapor and liquid phases. The continuity, momentum, and energy equations in the vapor film are

$$\frac{\partial u_v}{\partial x} + \frac{\partial v_v}{\partial y} = 0 \tag{12.147}$$

$$u_v = -\frac{K}{\mu_v}(\rho_\ell - \rho_v)g \tag{12.148}$$

$$u_v \frac{\partial T_v}{\partial x} + v_v \frac{\partial T_v}{\partial y} = \alpha_{mv} \frac{\partial^2 T_v}{\partial y^2} \tag{12.149}$$

where α_{mv} is thermal diffusivity of the porous medium saturated with the vapor.

The governing equations for the liquid boundary layer are

$$\frac{\partial u_\ell}{\partial x} + \frac{\partial v_\ell}{\partial y} = 0 \tag{12.150}$$

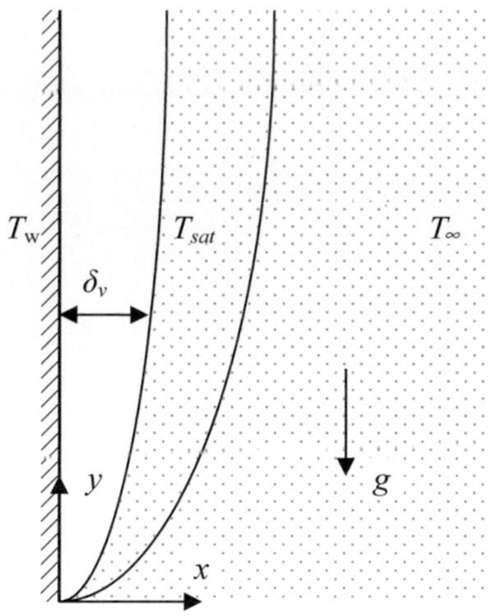

Fig. 12.16 Film boiling in porous media

$$u_\ell = \frac{K}{\mu_\ell} \rho_\infty g \beta_\ell (T_\ell - T_\infty) \tag{12.151}$$

$$u_\ell \frac{\partial T_\ell}{\partial x} + v_\ell \frac{\partial T_\ell}{\partial y} = \alpha_{m\ell} \frac{\partial^2 T_\ell}{\partial y^2} \tag{12.152}$$

where $\alpha_{m\ell}$ is thermal diffusivity of the porous medium saturated with the liquid.

The boundary conditions at the heated wall ($y = 0$) are

$$v_v = 0, \quad y = 0 \tag{12.153}$$

$$T = T_w, \quad y = 0 \tag{12.154}$$

It should be pointed out that u_v is not equal to zero at the heating surface under Darcy's law, i.e., slip occurs at the surface. The boundary condition in the liquid that is far from the heated surface is

$$u_\ell = 0, \quad y \to \infty \tag{12.155}$$

$$T_\ell = T_\infty, \quad y \to \infty \tag{12.156}$$

The mass balance at the liquid–vapor interface is [see Eq. (9.153)]:

$$\left(\rho u \frac{d\delta}{dx} - \rho v\right)_v = \left(\rho u \frac{d\delta}{dx} - \rho v\right)_\ell, \quad y = \delta_v \tag{12.157}$$

The temperature at the liquid–vapor interface is equal to the saturation temperature:

$$T_v = T_\ell = T_{sat}, \quad y = \delta_v \tag{12.158}$$

The above film boiling problem can be solved using a similarity solution. The results obtained by Cheng and Verma (1981) are shown in Fig. 12.17. The dimensionless parameters used in Fig. 12.17 are defined as

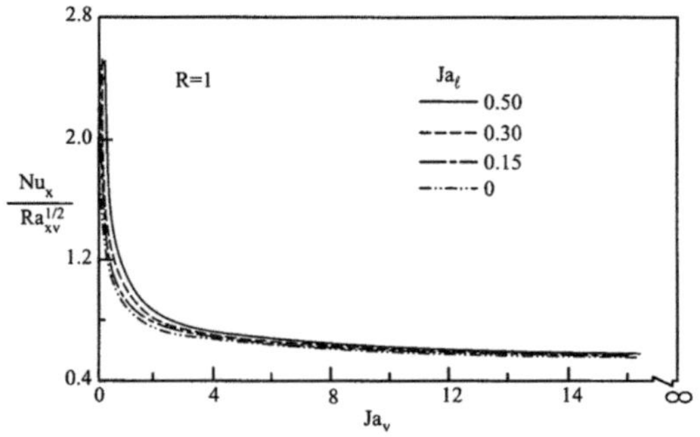

Fig. 12.17 Heat transfer for film boiling on a vertical wall in porous media (Cheng and Verma 1981; Reprinted with permission from Elsevier)

$$Nu_x = \frac{h_x x}{k_{\mathrm{mv}}} = \frac{q''_{\mathrm{w}} x}{k_{\mathrm{mv}}(T_{\mathrm{w}} - T_{\mathrm{sat}})}, Ra_{xv} = \frac{(\rho_\ell - \rho_{\mathrm{v}})gKx}{\mu_{\mathrm{v}}\alpha_{\mathrm{mv}}}$$
$$R = \frac{\rho_{\mathrm{v}}}{\rho_\ell}\left[\frac{\mu_\ell \alpha_{\mathrm{mv}}(\rho_\ell - \rho_{\mathrm{v}})c_{\mathrm{p}\ell}}{\mu_{\mathrm{v}}\alpha_{\mathrm{m}\ell}\rho_\ell \beta_\ell h_{\ell\mathrm{v}}}\right], \quad \mathrm{Ja}_{\mathrm{v}} = \frac{c_{\mathrm{p}\ell}(T_{\mathrm{sat}} - T_\infty)}{h_{\ell\mathrm{v}}}, \tag{12.159}$$
$$Ja_\ell = \frac{c_{\mathrm{pv}}(T_{\mathrm{w}} - T_{\mathrm{sat}})}{h_{\ell\mathrm{v}}}$$

where Jakob numbers Ja_{v} and Ja_ℓ, measure the degrees of superheat in the vapor and subcooling in the liquid. For all cases shown in Fig. 12.17, the effect of liquid subcooling on the heat transfer is insignificant. The effect of vapor superheat on heat transfer is significant when Ja_{v} is less than 2. The following asymptotic result can be obtained from Fig. 12.17:

$$Nu_x = 0.5642 Ra_{xv}^{1/2}, \quad Ja_{\mathrm{v}} \to \infty \tag{12.160}$$

12.5 Film Condensation in Porous Media

12.5.1 Overview

Film condensation occurs when the temperature of a vertical, impermeable, and wettable wall next to a porous medium saturated with vapor falls below saturation temperature (Fig. 12.18). In addition to gravity-driven downward liquid flow, the liquid also infiltrates the vapor region due to capillary force. The latter will create a two-phase region between the liquid film and the vapor region, where both condensate and vapor are present. It is assumed that the vapor temperature is equal to the saturation temperature; i.e., there is no superheat in the vapor phase. The temperature in the liquid region is below saturation temperature, while the temperature in the two-phase region is at saturation temperature. The discussion in this section is limited to the case where the thickness of the liquid film is much greater than the diameter of the pore size. This is referred to as a thick-film region, and the local volume average is applicable (Kaviany 2013). If the liquid film thickness is less than or comparable to

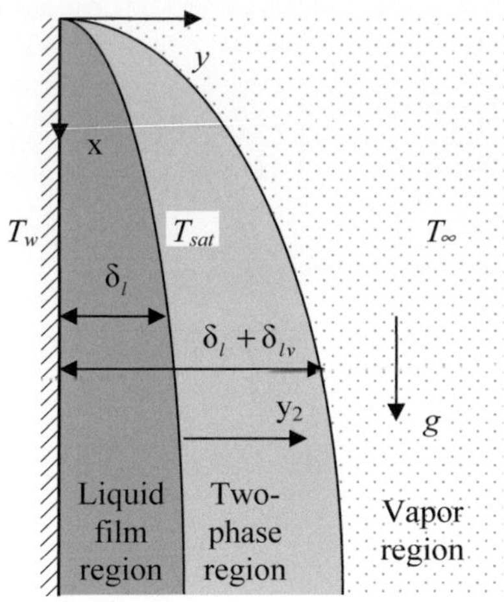

Fig. 12.18 Film condensation in a porous medium

the pore size, the local volume average will no longer be applicable and a direct simulation at the pore level must be performed.

The dominant forces in the condensation process are gravitational and capillary forces, and the latter dictates the thickness of the two-phase region, $\delta_{\ell v}$. The ratio of gravity and capillary forces is measured by Bond number:

$$Bo = \frac{g(\rho_\ell - \rho_v)K/\varepsilon}{\sigma} \tag{12.161}$$

where K and ε are, respectively, permeability and porosity. When $Bo \simeq 1$, condensation in a porous medium is dominated by both gravitational and capillary force. The condensation is dominated by capillary force when $Bo < 1$. When gravity dominates, Bond number will be greater than 1, and there will be no two-phase region, in which case the analysis will be significantly simplified.

In the following two subsections, an analysis of gravity-dominated condensation will be discussed first, followed by a discussion of the effect of surface tension on the condensation process.

12.5.2 Gravity-Dominated Film Condensation

When condensation is dominated by gravity, the effect of surface tension is negligible, and consequently, no two-phase region exists. Condensation along an inclined wall in a porous medium (see Fig. 12.19) will be discussed in this subsection. A porous medium saturated with dry vapor at its saturation temperature, T_{sat}, is bounded by an inclined impermeable wall with a temperature T_w ($T_w < T_{sat}$). Since the wall temperature is below saturation temperature, film condensation occurs on the inclined wall and the condensate flows downward due to gravity. It is assumed that the condensation is gravity-dominated and, therefore, that the liquid and vapor are separated by a sharp interface, not a two-phase region. In addition, the following assumptions are made by Cheng (1981) in order to obtain an analytical solution:

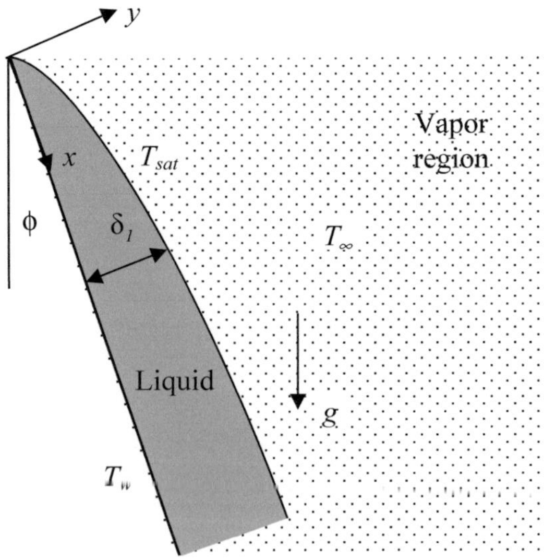

Fig. 12.19 Gravity dominated film condensation on an inclined wall in a porous medium

1. The condensate film is very thin compared to the length of the inclined wall ($\delta_\ell \ll L$) so that boundary-layer assumption is valid.
2. The properties for the porous medium, liquid, and vapor are independent from temperature.
3. The inclination angle, ϕ, is small enough for the gravity component in the normal direction of the surface to be negligible.
4. Darcy's law is valid for both liquid and vapor phases.
5. The saturation temperature, T_{sat}, is constant.

Under these assumptions, the continuity, momentum, and energy equations for the liquid layer are

$$\frac{\partial u_\ell}{\partial x} + \frac{\partial v_\ell}{\partial y} = 0 \tag{12.162}$$

$$u_\ell = \frac{K}{\mu_\ell}(\rho_\ell - \rho_v) g \cos\phi \tag{12.163}$$

$$u_\ell \frac{\partial T_\ell}{\partial x} + v_\ell \frac{\partial T_\ell}{\partial y} = \alpha_\ell \frac{\partial^2 T_\ell}{\partial y^2} \tag{12.164}$$

The boundary conditions at the wall are

$$v_\ell = 0, \quad y = 0 \tag{12.165}$$

$$T_\ell = T_w, \quad y = 0 \tag{12.166}$$

At the interface, the boundary conditions are

$$T_\ell = T_{sat}, \quad y = \delta_\ell \tag{12.167}$$

$$\dot{m}'' = \rho_\ell \left(u_\ell \frac{d\delta_\ell}{dx} - v_\ell \right), \quad y = \delta_\ell \tag{12.168}$$

$$\dot{m}'' h_{\ell v} = -k_{m\ell} \frac{\partial T}{\partial y}, \quad y = \delta_\ell \tag{12.169}$$

where $\dot{m}''$ is mass flux of condensate across the interface, and $k_{m\ell}$ is thermal conductivity of the porous medium saturated with liquid. Combining of Eqs. (12.168) and (12.169) yields

$$\rho_\ell h_{\ell v} \left(u_\ell \frac{d\delta_\ell}{dx} - v_\ell \right) = -k_{m\ell} \frac{\partial T}{\partial y}, \quad y = \delta_\ell \tag{12.170}$$

Introducing stream function

$$u_\ell = \frac{\partial \psi}{\partial y}, \quad v_\ell = -\frac{\partial \psi}{\partial x} \tag{12.171}$$

and the following similarity variables:

$$\eta = \sqrt{Ra_{\ell x}}\frac{y}{x} \tag{12.172}$$

$$f(\eta) = \frac{\psi}{\alpha_\ell \sqrt{Ra_{\ell x}}} \tag{12.173}$$

$$\theta(\eta) = \frac{T_\ell - T_{\text{sat}}}{T_{\text{w}} - T_{\text{sat}}} \tag{12.174}$$

where

$$Ra_{\ell x} = \frac{(\rho_\ell - \rho_{\text{v}})g\cos\phi K x}{\mu_\ell \alpha_\ell} \tag{12.175}$$

the governing equations and the corresponding boundary conditions become

$$f' = 1 \tag{12.176}$$

$$2\theta'' + f\theta' = 0 \tag{12.177}$$

$$f(0) = 0 \tag{12.178}$$

$$\theta(0) = 1 \tag{12.179}$$

$$\theta(\eta_\delta) = 0 \tag{12.180}$$

$$Ja_\ell \theta'(\eta_\delta) = -\frac{1}{2}f(\eta_\delta) \tag{12.181}$$

where

$$\eta_\delta = \sqrt{Ra_{\ell x}}\frac{\delta_\ell}{x} \tag{12.182}$$

is the dimensionless liquid film thickness and

$$Ja_\ell = \frac{c_{p\ell}(T_{\text{sat}} - T_{\text{w}})}{h_{\ell\text{v}}} \tag{12.183}$$

is Jakob number that measures the degree of subcooling at the wall.

Integrating Eq. (12.176) and considering Eq. (12.178), one obtains

$$f = \eta \tag{12.184}$$

which can be substituted into Eqs. (12.177) and (12.181) to get

$$2\theta'' + \eta\theta' = 0 \tag{12.185}$$

$$Ja_\ell \theta'(\eta_\delta) = -\frac{1}{2}\eta_\delta \tag{12.186}$$

The solution of Eq. (12.185) with Eqs. (12.179) and (12.180) as boundary conditions is

$$\theta(\eta) = 1 - \frac{\text{erf}(\eta/2)}{\text{erf}(\eta_\delta/2)} \tag{12.187}$$

where the dimensionless film thickness can be obtained by substituting Eq. (12.187) into Eq. (12.186):

$$Ja_\ell = \sqrt{\frac{\pi\eta_\delta}{2}}\exp\left(\frac{\eta_\delta^2}{4}\right)\text{erf}\left(\frac{\eta_\delta}{2}\right) \tag{12.188}$$

The heat flux at the wall is

$$q''_w = -k_{m\ell}\left(\frac{\partial T_\ell}{\partial y}\right)_{y=0} = \frac{k_{m\ell}(T_{sat} - T_w)\sqrt{Ra_{\ell x}}}{x}\theta'_\ell(0) \tag{12.189}$$

and the local Nusselt number is

$$Nu_x = \frac{q''_w x}{k_{m\ell}(T_{sat} - T_w)} = \frac{Ra_{\ell x}^{1/2}}{\sqrt{\pi}\text{erf}(\eta_\delta/2)} \tag{12.190}$$

where η_δ is function of Jakob number, Ja_ℓ, as indicated by Eq. (12.188). Cheng (1981) recommended that Eq. (12.190) can be approximated using

$$Nu_x = \left(\frac{1}{2Ja_\ell} + \frac{1}{\pi}\right)^{1/2} Ra_{\ell x}^{1/2} \tag{12.191}$$

In practical application, the average Nusselt number is often of the interest. It can be obtained by integrating Eq. (12.191):

$$\overline{Nu} = \frac{\bar{h}L}{k_{m\ell}} = \left(\frac{1}{Ja_\ell} + \frac{2}{\pi}\right)^{1/2} Ra_{\ell L}^{1/2} \tag{12.192}$$

12.5.3 Effect of Surface Tension on Condensation

The analysis in the preceding subsection is valid for gravity-dominated condensation in porous media ($Bo \gg 1$). When the condensation is gravity–capillary forces dominated ($Bo \sim 1$) or capillary force dominated ($Bo \ll 1$), there will be a two-phase region that is saturated by a mixture of liquid and vapor, as shown in Fig. 12.18. The fraction of liquid in the pore space is defined as saturation:

$$\gamma_\ell = \frac{\varepsilon_\ell}{\varepsilon} \tag{12.193}$$

where ε_ℓ and ε are volume fraction of the liquid and porosity in the porous media. The effect of surface tension on condensation in porous media was studied by Majumdar and Tien (1990) and their work will be briefly described below.

The continuity equation for the two-phase region is

$$\rho_\ell\left(\frac{\partial u_\ell}{\partial x}+\frac{\partial v_\ell}{\partial y_2}\right)+\rho_v\left(\frac{\partial u_v}{\partial x}+\frac{\partial v_v}{\partial y_2}\right)=0 \tag{12.194}$$

where y_2 is measured from the interface between liquid and the two-phase region (see Fig. 8.29). The mass fluxes for liquid and vapor are governed by Darcy's law, as follows:

$$\dot{m}''_\ell=-\frac{KK_{r\ell}}{v_\ell}\nabla p_\ell \tag{12.195}$$

$$\dot{m}''_v=-\frac{KK_{rv}}{v_v}\nabla p_v \tag{12.196}$$

where $K_{r\ell}$ and K_{rv} are relative permeabilities (dimensionless) for liquid and vapor phases, respectively. Their products with the permeability K represent the permeability for liquid and vapor flow in the porous media.

Since the density of the vapor is much lower than that of the liquid, the pressure variation in the liquid phase is very insignificant. Also, the change of capillary pressure is due mainly to the change of liquid pressure. Consequently, the vapor flow is negligible compared to the liquid flow, so that Eq. (12.194) is reduced to

$$\frac{\partial u_\ell}{\partial x}+\frac{\partial v_\ell}{\partial y_2}=0 \tag{12.197}$$

The velocity components in the x- and y-directions are

$$u_\ell=K_{r\ell}u_D=\frac{KK_{r\ell}(\rho_\ell-\rho_v)g}{\mu_\ell} \tag{12.198}$$

$$v_\ell=-\frac{KK_{r\ell}}{\mu_\ell}\frac{\partial p_\ell}{\partial y_2}=\frac{KK_{r\ell}}{\mu_\ell}\frac{\partial p_{cap}}{\partial y_2} \tag{12.199}$$

where $u_D=K(\rho_\ell-\rho_v)g/\mu_\ell$ is Darcian velocity. The capillary pressure can be written as

$$p_{cap}=\frac{\sigma}{\sqrt{K/\varepsilon}}f(s) \tag{12.200}$$

where $f(s)$ is a Leverett function (see Sect. 12.3.3):

$$f(s)=1.417(1-s)-2.120(1-s)^2+1.263(1-s)^3 \tag{12.201}$$

and s is dimensionless saturation defined as

$$s=\frac{\gamma_\ell-\gamma_{\ell i}}{1-\gamma_{\ell i}} \tag{12.202}$$

where $\gamma_{\ell i}$ is irreducible saturation, below which liquid flow will not occur. The relative permeability in Eqs. (12.198) and (12.199) is obtained from

$$K_{r\ell} = s^3 \tag{12.203}$$

Substituting Eqs. (12.198)–(12.203) into Eq. (12.197), the following form of the dimensionless continuity equation is obtained:

$$3\frac{\partial s}{\partial x} + \left[\frac{\sigma/\sqrt{K/\varepsilon}}{(\rho_\ell - \rho_v)g}\right]\left[(3f' + sf'')\left(\frac{\partial s}{\partial x}\right)^2 + sf'\frac{\partial^2 s}{\partial y_2^2}\right] = 0 \tag{12.204}$$

which is subjected to the following boundary conditions:

$$s = 0, \quad x = 0 \tag{12.205}$$

$$s = 1, \quad y_2 = 0 \tag{12.206}$$

$$s = 0, \quad y_2 \to \infty \tag{12.207}$$

Introducing the following similarity variable:

$$\eta = y_2\left[\frac{(\rho_\ell - \rho_v)g}{(\sigma/\sqrt{K/\varepsilon})x}\right]^2 \tag{12.208}$$

Equations (12.204)–(12.207) are transformed to

$$s'' = \frac{3\eta s' - 2(3f' + sf'')s'^2}{2sf'} \tag{12.209}$$

$$s = 1, \quad \eta = 0 \tag{12.210}$$

$$s = 0, \quad \eta \to \infty \tag{12.211}$$

The numerical solution of Eq. (12.209) was obtained using the Runge–Kutta method, and it is concluded that $\eta = 1.296$ can be chosen to determine the thickness of the two-phase region.

For the liquid region, the model in the preceding subsection overpredicts the heat transfer coefficient because of slip in the velocity at the wall. They presented a new model based on the following assumptions: (1) the boundary-layer assumption applies in the liquid film, (2) convection terms in the energy equation are negligible, (3) subcooling in the liquid is negligible, and (4) the fluid properties are constants. The governing equations for the liquid film in dimensionless form are

$$\frac{\partial u_\ell^+}{\partial x^+} + \frac{\partial v_\ell^+}{\partial y^+} = 0 \tag{12.212}$$

$$\frac{\partial^2 u_\ell^+}{\partial y^{+2}} + 1 - u_\ell^+ = 0 \tag{12.213}$$

$$\frac{\partial^2 \theta_\ell}{\partial y^{+2}} = 0 \tag{12.214}$$

where the dimensionless variables are defined as

$$u_\ell^+ = \frac{u_\ell}{u_\mathrm{D}}, v_\ell^+ = \frac{v_\ell}{u_\mathrm{D}}, x^+ = \frac{x}{\sqrt{K}}, y^+ = \frac{y}{\sqrt{K}}, \delta_\ell^+ = \frac{\delta_\ell}{\sqrt{K}}, \theta = \frac{T - T_\mathrm{w}}{T_\mathrm{sat} - T_\mathrm{w}} \tag{12.215}$$

Compared with Eq. (12.163) in Cheng's (1981) model, Eq. (12.213) allows for nonslip conditions at the wall. The boundary conditions at the wall are

$$u_\ell^+ = v_\ell^+ = \theta = 0, \quad y^+ = 0 \tag{12.216}$$

The boundary conditions at the interface between the liquid film and the two-phase region require that the velocity and shear stress in these two regions match, which makes the solution of the condensation problem very challenging. Majumdar and Tien (1990) proposed three models to handle the boundary condition at the interface between the liquid and the two-phase regions; two of them are discussed below.

Model 1. At the interface between the liquid and the two-phase region, the shear stress is zero, i.e., $\partial u_\ell^+/\partial y^+ = 0$ at $\tilde{y} = \delta_\ell^+$, which is the same as in classical Nusselt analysis. The velocity profile in the liquid layer is

$$u_\ell^+ = 1 - \cosh y^+ + \tanh \delta_\ell^+ \sinh y^+ \tag{12.217}$$

The liquid layer thickness can be obtained from an energy balance at the interface, and the result is

$$(1 - \mathrm{sech}^2\delta_\ell^+)\frac{\mathrm{d}\delta_\ell^+}{\mathrm{d}x^+} + (1 - \mathrm{sech}\delta_\ell^+)\frac{0.373}{\sqrt{Box^+}} = \frac{Ja_\ell}{\delta_\ell^+ Ra_\mathrm{K}} \tag{12.218}$$

where

$$Ra_K = \frac{K^{3/2}(\rho_\ell - \rho_\mathrm{v})g}{\mu_\ell \alpha_\mathrm{e}} \tag{12.219}$$

is the Rayleigh number based on permeability, and $\alpha_\mathrm{e} = k_\mathrm{eff}/(\rho_\ell c_{p\ell})$ is effective thermal diffusivity. Analytical solution of Eq. (12.218) is not possible, and it must be solved numerically.

Model 2. This model also employs Eq. (12.213) to obtain the velocity in the liquid layer, except that the boundary condition at $y^+ = \delta_\ell^+$ is changed to $u_\ell^+ = 1$. Although it is not as rigorous as Model 1, it is an improvement over Cheng (1981) because it uses a nonslip condition at the wall. The velocity profile in the liquid layer is

$$u_\ell^+ = 1 - \cosh y^+ + \coth \delta_\ell^+ \sinh y^+ \tag{12.220}$$

and the overall energy balance at the interface is

$$\left(1 - \frac{1}{1 + \cosh \delta_\ell^+}\right)\frac{\mathrm{d}\delta_\ell^+}{\mathrm{d}x^+} + \frac{0.373}{\sqrt{Box^+}} = \frac{Ja_\ell}{\delta_\ell^+ Ra_\mathrm{K}} \tag{12.221}$$

which also needs to be solved numerically.

In the liquid region, Eq. (12.214) will yield a linear temperature profile, and the local Nusselt number is

$$Nu_x = \frac{x^+}{\delta_\ell^+} \tag{12.222}$$

Figure 12.20 shows comparison between the results predicted from Models 1 and 2, together with the experimental results by White and Tien (1987) for constant porosity media. The parameter R in the figure is defined as

$$R = \frac{Ra_K}{Bo} = \frac{\sigma\sqrt{K\varepsilon}}{\mu_\ell \alpha_e} \tag{12.223}$$

which reflects the ratio of surface tension force and viscous force. It can be seen from Fig. 12.20a that for aluminum foam, the agreement between Model 1 and experimental results is very good, but Model 2 significantly overpredicts the heat transfer rates. This is expected because Model 2 is not as rigorous as Model 1. The slight overprediction of Model 1 may be attributed to the fact that shear

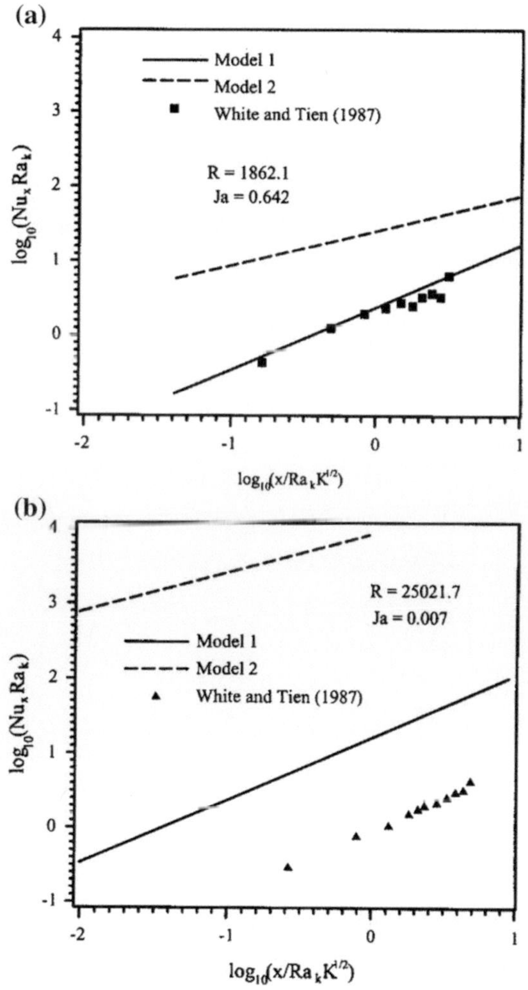

Fig. 12.20 Comparison of results of models 1 and 2 with experiments: **a** aluminum foam metal, **b** polyurethane foam (Majumdar and Tien 1990)

stress at the interface is neglected. For polyurethane foam, on the other hand, Model 1 also significantly overpredicts heat transfer rate. This overprediction is due to the fact that surface tension drag at the liquid region interface is neglected in Model 1. Since R in Fig. 12.20b is one order of magnitude higher than that in Fig. 12.20a, it is expected that surface tension plays a more significant role in Fig. 12.20b.

12.6 Melting and Solidification in Porous Media

Melting and solidification in porous media can find application in a wide range of problems, including freezing and thawing of the soil in a cold region, latent heat thermal energy storage, and selective laser sintering of metal powders. Applications in life science include cryopreservation of tissue, human embryos, and organs. While the early studies about phase change in porous media treated melting and solidification as conduction-controlled, the effect of natural convection is often very important. Convection-controlled melting and solidification of PCMs saturated in porous media will be discussed in this section.

12.6.1 One-Region Problem

Melting in an enclosure filled with a porous medium and solid PCM studied by Jany and Bejan (1987) will be presented here (see Fig. 12.21). It is assumed that all of the walls of the two-dimensional rectangular cavity are impermeable and that the initial temperature of the porous medium and PCM, T_i, is at the melting point of the PCM, T_m. At time $t = 0$, the temperature of the left side of the cavity is increased to T_w, while the right side, top, and bottom of the cavity remain insulated. The following assumptions are made:

1. Liquid flow is very slow, so Darcy's law is applicable.
2. The PCM and porous medium are in thermal equilibrium.
3. The porous medium is isotropic.

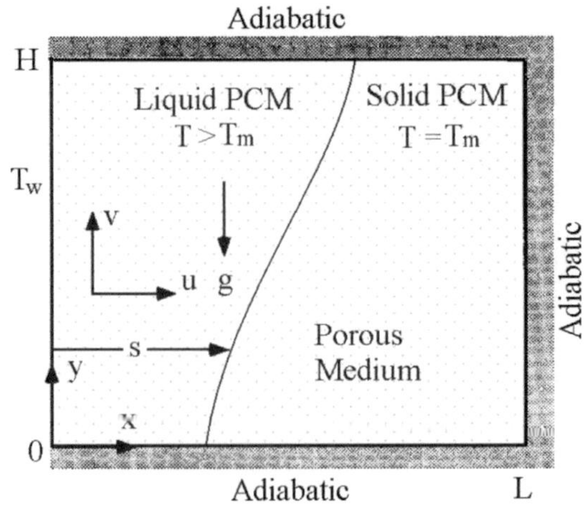

Fig. 12.21 Convection-controlled melting in a porous medium

4. Constant thermal properties for both PCM and porous medium.

Since the temperature in the solid region is equal to the melting point, this is a one-region melting problem. The governing equations for the liquid region are

$$\frac{\partial\langle u_\ell\rangle}{\partial x}+\frac{\partial\langle v_\ell\rangle}{\partial y}=0 \tag{12.224}$$

where $\langle u_\ell\rangle$ and $\langle v_\ell\rangle$ are the components of extrinsic phase-averaged velocity in the x- and y-directions [see Eq. (12.10)], and they can be obtained by Darcy's law:

$$\langle u_\ell\rangle=-\frac{K}{\mu_\ell}\frac{\partial\langle p_\ell\rangle^\ell}{\partial x} \tag{12.225}$$

$$\langle v_\ell\rangle=-\frac{K}{\mu_\ell}\left\{\frac{\partial\langle p_\ell\rangle^\ell}{\partial y}+\rho_\ell g[1-\beta(\langle T\rangle-T_{\mathrm{m}})]\right\} \tag{12.226}$$

The energy equation is

$$\Upsilon\frac{\partial\langle T\rangle}{\partial t}+\langle u_\ell\rangle\frac{\partial\langle T\rangle}{\partial x}+\langle v_\ell\rangle\frac{\partial\langle T\rangle}{\partial y}=\alpha_{\mathrm{m}}\left(\frac{\partial^2\langle T\rangle}{\partial x^2}+\frac{\partial^2\langle T\rangle}{\partial y^2}\right) \tag{12.227}$$

where

$$\Upsilon=\varepsilon+(1-\varepsilon)\frac{(\rho c_{\mathrm{p}})_{\mathrm{sm}}}{(\rho c_{\mathrm{p}})_\ell} \tag{12.228}$$

is the ratio of the heat capacity of the porous medium saturated with PCM over the heat capacity of the PCM. ε is the volume fraction of PCM, $(\rho c_{\mathrm{p}})_\ell$ and $(\rho c_{\mathrm{p}})_{\mathrm{sm}}$ are heat capacities of the liquid PCM and porous matrix. The thermal diffusivity of the porous medium saturated with the PCM is $\alpha_{\mathrm{m}}=k_{\mathrm{eff}}/(\rho c_{\mathrm{p}})_\ell$.

Equations (12.224)–(12.227) are subjected to the following boundary conditions:

$$\langle u_\ell\rangle=0,\quad \langle T\rangle=T_{\mathrm{w}},\quad x=0 \tag{12.229}$$

$$\langle u_\ell\rangle=0,\quad \langle T\rangle=T_{\mathrm{m}},\quad x=s \tag{12.230}$$

$$\langle v_\ell\rangle=0,\quad \frac{\partial\langle T\rangle}{\partial y}=0,\quad y=0,H \tag{12.231}$$

The energy balance at the solid–liquid interface is (Jany and Bejan 1987)

$$\frac{\partial s}{\partial t}=-\frac{\alpha_{\mathrm{m}}c_{\mathrm{p}\ell}}{h_{s\ell}}\left(\frac{\partial\langle T\rangle}{\partial x}-\frac{\partial s}{\partial y}\frac{\partial\langle T\rangle}{\partial y}\right),\quad x=s \tag{12.232}$$

where the second term in the parentheses at the right-hand side represents the effect of the inclination of the interface.

To change $\langle u_\ell\rangle$ and $\langle v_\ell\rangle$ into one dependent function, a stream function, ψ, is defined as

$$\langle u_\ell \rangle = \frac{\partial \psi}{\partial y}, \quad \langle v_\ell \rangle = -\frac{\partial \psi}{\partial x} \tag{12.233}$$

Introducing the following dimensionless variables

$$\begin{gathered} X = \frac{x}{H}, Y = \frac{y}{H}, Fo = \frac{\alpha_{\mathrm{m}} t}{H^2}, S = \frac{s}{H}, \langle U_\ell \rangle = \frac{\langle u_\ell \rangle H}{\alpha_{\mathrm{m}}}, \langle V_\ell \rangle = \frac{\langle v_\ell \rangle H}{\alpha_{\mathrm{m}}} \\ \langle \theta \rangle = \frac{\langle T \rangle - T_{\mathrm{m}}}{T_{\mathrm{w}} - T_{\mathrm{m}}}, \Psi = \frac{\psi}{\alpha_{\mathrm{m}}}, Ra = \frac{g\beta KH(T_{\mathrm{w}} - T_{\mathrm{m}})}{\nu_\ell \alpha_{\mathrm{m}}}, Ste = \frac{c_{p\ell}(T_{\mathrm{w}} - T_{\mathrm{m}})}{h_{s\ell}} \end{gathered} \tag{12.234}$$

where Ra and Ste are the Rayleigh and Stefan numbers, respectively, the governing equations and boundary conditions become

$$\frac{\partial^2 \Psi}{\partial x^2} + \frac{\partial^2 \Psi}{\partial y^2} = -Ra \frac{\partial \theta}{\partial X} \tag{12.235}$$

$$\Upsilon \frac{\partial \theta}{\partial Fo} + U \frac{\partial \theta}{\partial X} + V \frac{\partial \theta}{\partial Y} = \frac{\partial^2 \theta}{\partial x^2} + \frac{\partial^2 \theta}{\partial y^2} \tag{12.236}$$

$$U = \frac{\partial \Psi}{\partial Y} = 0, \quad \theta = 1, \quad X = 0 \tag{12.237}$$

$$U = \frac{\partial \Psi}{\partial Y} = 0, \quad \theta = 0, \quad X = S \tag{12.238}$$

$$V = -\frac{\partial \Psi}{\partial X} = 0, \quad \frac{\partial \theta}{\partial Y} = 0, \quad Y = 0, 1 \tag{12.239}$$

$$\frac{\partial S}{\partial Fo} = -Ste \left(\frac{\partial \theta}{\partial X} - \frac{\partial S}{\partial Y} \frac{\partial \theta}{\partial Y} \right), \quad x = S \tag{12.240}$$

where $\langle \rangle$ for phase averaging has been dropped in the dimensionless variables for ease of notation.

The dimensionless governing equations were solved numerically using a finite difference method by Jany and Bejan (1987). Propagations of the melting fronts in a square cavity $(L/H = 1)$ for different Rayleigh numbers are shown in Fig. 12.22. It can be seen that both the melting velocity and the shape of the melting front are strongly affected by natural convection. Heat transfer from the heating wall is measured by overall Nusselt number, defined as

$$\overline{Nu} = \frac{\bar{h} H}{k_{\mathrm{eff}}} = -\frac{1}{(T_{\mathrm{w}} - T_{\mathrm{m}})} \int_0^H \frac{\partial T}{\partial x} \mathrm{d}y = -\int_0^1 \left(\frac{\partial \theta}{\partial X} \right)_{X=0} \mathrm{d}Y \tag{12.241}$$

Figure 12.23 shows Nusselt number as a function of dimensionless time, SteFo, at different Rayleigh numbers for a square cavity. The result for $Ra = 0$ represents Nusselt number for pure conduction solution. It can be seen that the effect of natural convection is pronounced when the Rayleigh number is greater than 50. For all cases where $Ra > 0$, there is a knee point beyond which the Nusselt number drops significantly. The knee point represents the time at which the top of the melting front arrives at the right wall ($X = 1$) of the cavity. The Nusselt number decreases sharply after the knee point because the height of the PCM being molten decreases after the top of the melting front reaches to the right wall.

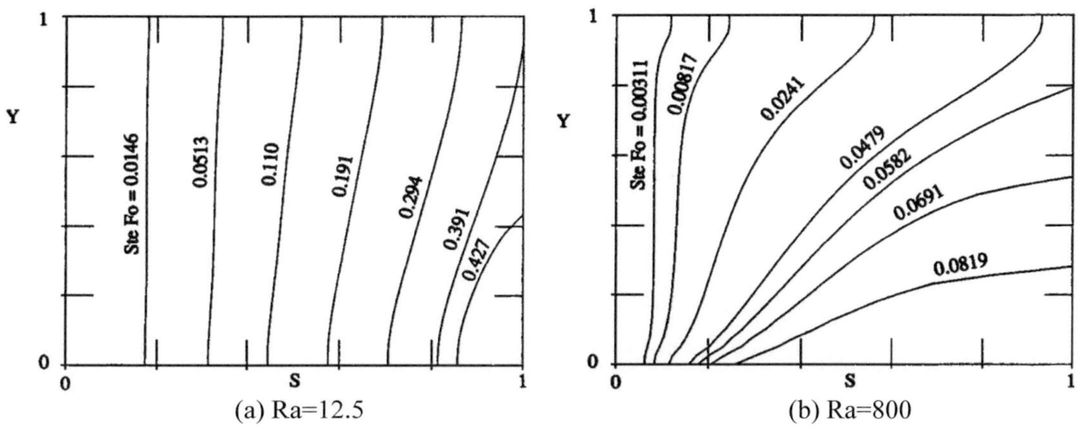

Fig. 12.22 Propagation of the melting fronts (Jany and Bejan 1987)

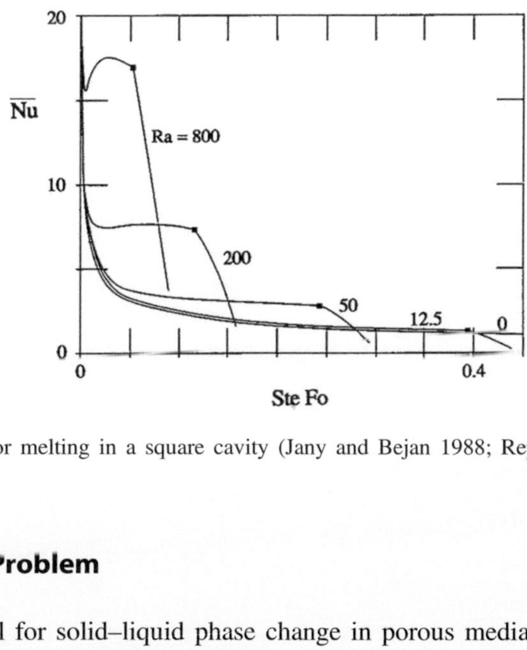

Fig. 12.23 Nusselt number for melting in a square cavity (Jany and Bejan 1988; Reprinted with permission from Elsevier)

12.6.2 Two-Region Problem

A more generalized model for solid–liquid phase change in porous media was proposed by Beckermann and Viskanta (1988). Instead of using Darcy's law, which is valid only at very low velocity, the effect of inertia terms was included in the model. This treatment allows for nonslip conditions at the wall of the cavity, e.g., $v = 0$ at $x = 0$ (under Darcy's law, slip occurs at the wall). The generalized governing equations that are applicable to the entire computational domain, including both liquid and solid phases, are obtained by using an enthalpy model that can be applied to a two-region problem (melting of a subcooled solid or solidification of a superheated liquid).

Beckermann and Viskanta (1988) developed their model based on volume-averaged governing equations (see Sect. 12.3). Figure 12.24 shows the illustration of a phase-change process in a porous medium. The volume fraction of pore space in the porous medium is defined as

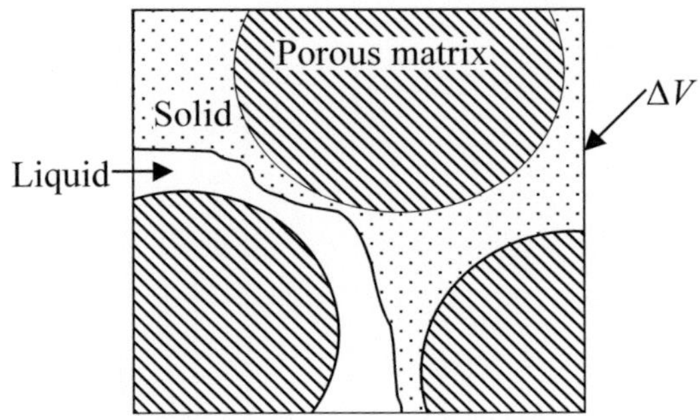

Fig. 12.24 Control volume for volume averaging

$$\varepsilon_{\mathrm{p}} = \Delta V_{\mathrm{p}}/\Delta V \tag{12.242}$$

where ΔV_{p} is the volume of porous space and ΔV is the total volume. The porous space can be occupied by either liquid (ℓ) or solid (s) phase of the PCM. The liquid volume fraction in the porous space is defined as

$$\gamma = \Delta V_{\ell}/\Delta V_{\mathrm{p}} \tag{12.243}$$

The volume fraction of the liquid in the porous media is therefore

$$\varepsilon_{\ell} = \Delta V_{\ell}/\Delta V = \varepsilon_{\mathrm{p}}\gamma \tag{12.244}$$

Although the PCM under consideration is single component and its melting point is well defined, the melting front can have a finite thickness because the phase change can be simultaneously inhabited by solid and liquid in the pore. A two-phase region, similar to a mushy zone for a binary system, exists between the solid and liquid phases. The liquid fraction, γ, in this two-phase region varies between 0 and 1, while the average temperature in this two-phase region is equal to the melting point, T_{m}. It was assumed that the liquid fraction and the temperature in the two-phase region were related by

$$\gamma = \frac{T - T_{\mathrm{m}} + \Delta T}{2\Delta T}, \quad T_{\mathrm{m}} - \Delta T \le T \le T_{\mathrm{m}} + \Delta T \tag{12.245}$$

Viskanta (1988) assumed that: (1) the flow and heat transfer is two-dimensional and laminar, (2) the properties of both porous matrix and PCM are homogeneous and isotropic, (3) the porous matrix and the PCM are at thermal equilibrium, (4) the velocities of the porous matrix and solid phase are zero, (5) the flow is incompressible and the Boussinesq approximation (density change is considered only in the buoyancy term) applies, (6) the thermal physical properties are constants, (7) the dispersion flux due to velocity fluctuation is negligible, and (8) the densities of the liquid and solid phases are identical.

The continuity, momentum, and energy equations are

$$\nabla \cdot \mathbf{V} = 0 \tag{12.246}$$

$$\frac{\rho_\ell}{\varepsilon_\ell}\left[\frac{\partial \mathbf{V}}{\partial t}+\frac{1}{\varepsilon_\ell}(\mathbf{V}\cdot\nabla)\mathbf{V}\right]=-\nabla p+\frac{\mu_\ell}{\varepsilon_\ell}\nabla^2\mathbf{V} - \left(\frac{\mu_\ell}{K}+\frac{\rho_\ell C_{\mathrm{f}}}{K^{1/2}}|\mathbf{V}|\right)\mathbf{V}-\rho_\ell \mathbf{g}\beta(T-T_{\mathrm{ref}}) \tag{12.247}$$

$$(\rho c_{\mathrm{p}})_{\mathrm{eff}}\frac{\partial T}{\partial t}+\rho_\ell c_{\mathrm{p}\ell}(\mathbf{V}\cdot\nabla T)=-\nabla\cdot(k_{\mathrm{eff}}\nabla T)-\varepsilon_{\mathrm{p}}\rho_\ell h_{\mathrm{s}\ell}\frac{\partial \gamma}{\partial t} \tag{12.248}$$

where $\mathbf{V}=\langle \mathbf{V}_\ell\rangle^\ell$ is intrinsic phase-averaged velocity. For the porous medium consisting of spherical beads with diameter d_{p}, the permeability can be obtained using the following Kozeny–Carman relation:

$$K=\frac{d_{\mathrm{p}}^2\varepsilon_\ell^3}{175(1-\varepsilon_\ell)^2} \tag{12.249}$$

which indicates that the permeability is zero in the solid region. The Forchheimer inertia coefficient C_{f} is taken as 0.55 (Ward 1964). The effective heat capacity $(\rho c_{\mathrm{p}})_{\mathrm{eff}}$ in Eq. (12.249) is

$$(\rho c_{\mathrm{p}})_{\mathrm{eff}}=\varepsilon_{\mathrm{p}}\rho\left[\gamma c_{\mathrm{p}\ell}+(1-\gamma)c_{\mathrm{ps}}\right]+(1-\varepsilon_{\mathrm{p}})(\rho c_p)_{\mathrm{sm}} \tag{12.250}$$

where $(\rho c_{\mathrm{p}})_{\mathrm{sm}}$ is heat capacity of the porous matrix. The effective thermal conductivity of the porous medium saturated with PCM is estimated by (Veinberg 1967)

$$k_{\mathrm{eff}}+\varepsilon k_{\mathrm{eff}}^{1/3}\frac{k_{\mathrm{sm}}-k_{\ell\mathrm{s}}}{k_{\ell\mathrm{s}}^{1/3}}-k_{\mathrm{sm}}=0 \tag{12.251}$$

where $k_{\ell\mathrm{s}}$ is the average thermal conductivity of the PCM:

$$k_{\ell\mathrm{s}}=\gamma k_\ell+(1-\gamma)k_{\mathrm{s}} \tag{12.252}$$

Beckermann and Viskanta (1988) applied the above model to solve solid–liquid phase-change problems as shown in Fig. 12.25. A rectangular cavity is filled with a rigid porous matrix that is saturated with PCM. The left wall temperature is held at T_{h}, which is above the melting point of the PCM, T_{m}, while the right wall is kept at a temperature T_{c}, below the melting point. The top and bottom of the cavity are adiabatic. The initial temperature of the porous medium and PCM, T_{i}, is equal to T_{c} for a melting problem or T_{h} for a solidification problem.

The governing equations were nondimensionalized and numerical simulations for both melting and solidification were performed by Beckermann and Viskanta (1988). They also performed experiments on melting and solidification in a system containing spherical glass beads ($d_{\mathrm{p}}=6$ mm, $\varepsilon_{\mathrm{p}}=0.385$) and gallium as a PCM ($T_{\mathrm{m}}=29.78$ °C) in a 3.81 × 4.76 cm^2 ($L\times H$) rectangular cavity. Figure 12.26 shows comparisons between numerical and experimental results for melting and solidification. It can be seen that the agreement between numerical and experimental results is very good for both melting and solidification. Under the conditions studied by Beckermann and Viskanta (1988), the effect of natural convection on the melting and solidification was insignificant because the interfaces shown in Fig. 12.26 are nearly planar and vertical. The reason for the weak effect of natural convection is that the Rayleigh numbers—as defined in Eq. (12.234)—for the cases shown in

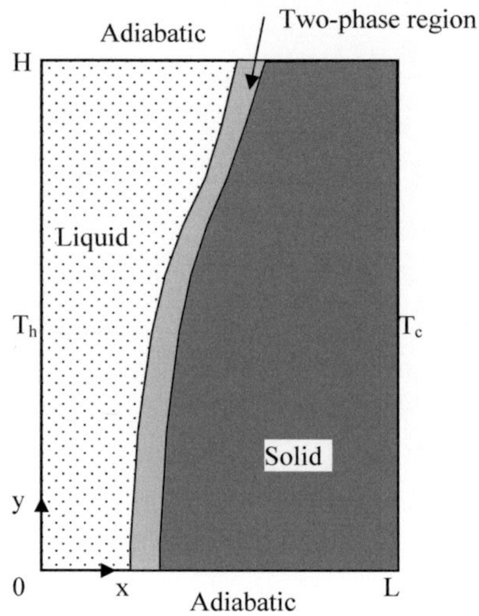

Fig. 12.25 Solid–liquid phase change in a rectangular cavity filled with porous media (Beckermann and Viskanta 1988; Reproduced with permission from Elsevier)

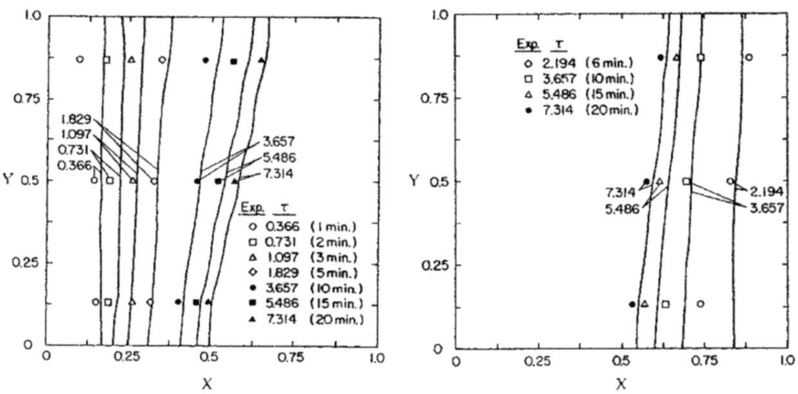

(a) Melting (T_h=45 °C, T_c=20 °C) (b) Solidification (T_h=40 °C, T_c=20 °C)

Fig. 12.26 Comparison between numerical and experimental results (Beckermann and Viskanta 1988; Reproduced with permission from Elsevier)

Fig. 12.26a, b are, respectively, 9.22 and 11.52 (Nield and Bejan 2017), which are comparable to Fig. 12.22a.

Oosthuizen (1988) investigated natural convection and heat transfer in an enclosure containing a porous medium saturated with the PCM (see Fig. 12.27a). The left wall of the cavity is maintained at a temperature T_h, which is above the melting point of the PCM, while the right wall temperature is kept at T_c, a temperature below the melting point. The problem under the configuration shown in Fig. 12.27a can reach steady state, at which point the solid–liquid interface ceases to move. At steady state, the location of the interface is not affected by the latent heat of the PCM, $h_{s\ell}$, because there is no phase change in the steady state. The location and shape of the interface are affected by natural convection in the liquid region as well as conduction in the solid region. Oosthuizen (1988) solved the

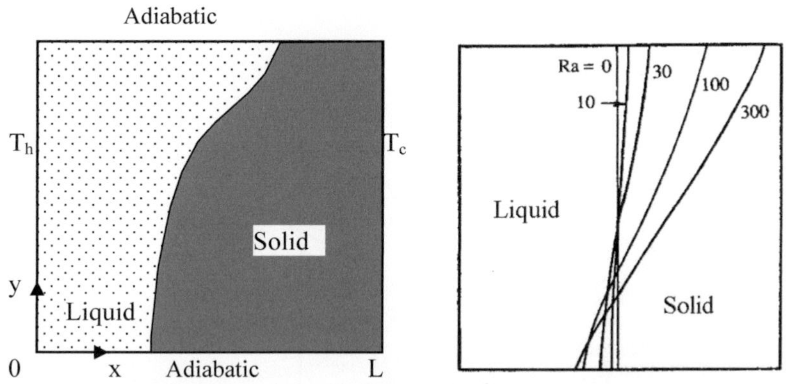

(a) Physical model (b) Effect of Ra on the freezing front

Fig. 12.27 Steady-state convection and heat transfer in a porous medium with differentially heated sidewalls (Oosthuizen 1988)

phase-change problem shown in Fig. 12.27a by using a finite element method. The effects of Rayleigh number ($Ra = g\beta KH(T_h - T_c)/(\nu_\ell \alpha_m) = 0$–$500$), aspect ratio ($H/L = 0.5$–$2$), and thermal conductivity ratio of the liquid and solid region ($k_{\mathrm{eff},\ell}/k_{\mathrm{eff,s}} = 1-3$) on the fluid flow and heat transfer were investigated. The effect of the Rayleigh number on the solid–liquid interface for $k_{\mathrm{eff},\ell}/k_{\mathrm{eff,s}} = 1-3$ and $\theta_c = (T_m - T_c)/(T_h - T_c) = 0.5$ is shown in Fig. 12.27b. In the absence of natural convection ($Ra = 0$), the phase change is conduction-controlled and the location of the interface is located at $L/2$. After the Rayleigh number is greater than 30, the interface becomes inclined and the total volume of the liquid region increases with increasing Rayleigh number.

Problems

12.1. A capillary porous structure with partial heating and evaporation on its upper surface is shown in Fig. P12.1. The entire porous structure is saturated with liquid from the bottom surface ($y = 0$), which is connected to a pool for liquid supply. A constant heat flux, q_0'', is applied over part of the upper surface ($0 < x < L_{xf}$), which is impermeable to the fluid. The rest of the upper surface ($L_{xf} < x < L_x$) exposes liquid in the pores of the capillary structure to the vapor space above. Heat is transferred from the left portion of the upper surface to the right portion of the upper surface, where evaporation takes place. The left and right sidewalls are adiabatic and impermeable. At steady state, the liquid is drawn from the bottom of the porous structure and flows to the right portion of the upper surface due to evaporation. If the liquid velocity at the bottom surface and the vapor velocity at the upper surface are uniform, determine the liquid velocity at the bottom, v_ℓ, and the vapor velocity at the upper surface, v_v.

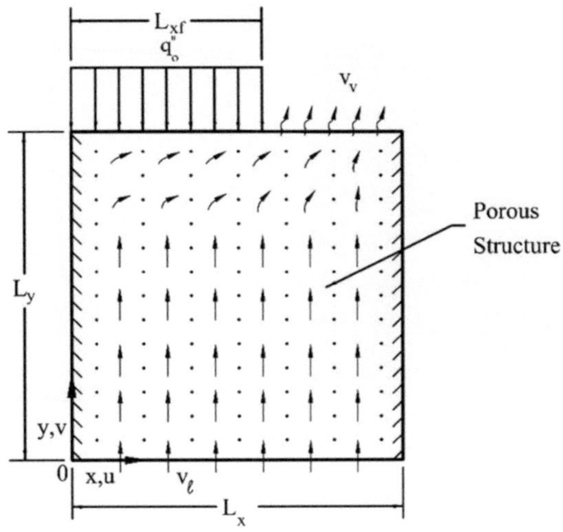

Fig. P12.1

12.2. The area-averaged velocity components in the *x*- and *y*-directions in Problem 12.1 can be estimated by Darcy's law:

$$u = -\frac{K}{\mu}\frac{\partial p}{\partial x} \quad v = -\frac{K}{\mu}\frac{\partial p}{\partial y}$$

where *K* is permeability. The continuity equation is

$$\frac{\partial u}{\partial x} + \frac{\partial v}{\partial y} = 0$$

Determine the analytical solution for the pressure and velocity distributions in the porous structure.

12.3. Specify the energy equation and the corresponding boundary conditions for Problem 12.2; then nondimensionalize them. Write a computer program to obtain dimensionless temperature distribution in the porous media.

12.4. For a porous bed made of spherical balls packed in a body-centered cubic (BCC) formation, use the continuity Eq. (3.51) and momentum Eqs. (3.72)–(3.74), to develop the boundary conditions and numerically solve the flow field, and determine the permeability of the packed bed by plotting the product of $\mu\mathbf{v}$ versus dp/dx. Assume that the elemental cube's sides all have a length of 10^{-3} m and that the spheres all have a radius of 4.6×10^{-4} m. (Note: the spheres are slightly compressed against each other, and therefore, their radii are slightly larger than if they were only contacting one another at a single point). Model the top right quadrant of an elementary cubic cell, Fig. P12.4. Numerically solve the problem for different fluids and determine the permeability.

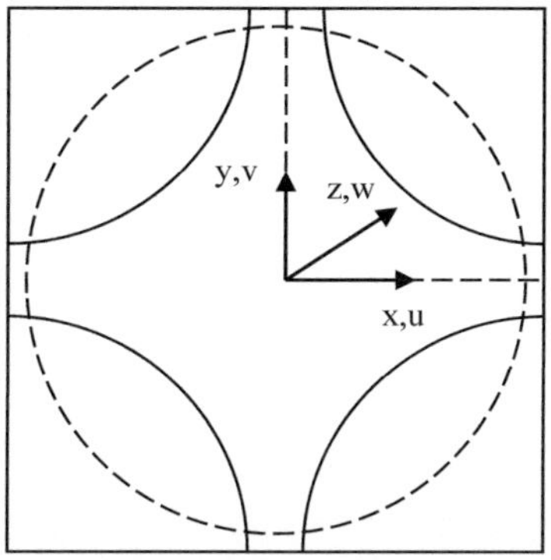

Fig. P12.4

12.5. For most heat pipes, the wick structures are very thin, so the liquid flow in the wick can be simplified to one-dimensional axial flow. Since the liquid velocity and its gradient in the axial direction are very small, the liquid flow in the wick satisfies Darcy's law, i.e., $w_\ell = -\frac{K}{\mu_\ell}\frac{\mathrm{dp}_\ell}{\mathrm{d}z}$, where K is permeability. For an axially grooved wick structure as shown in Fig. P12.5, determine the permeability analytically.

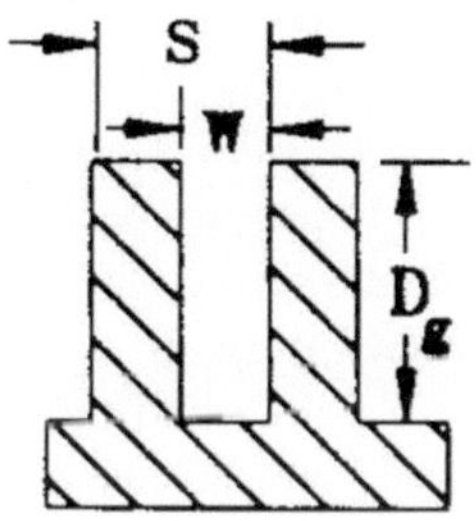

Fig. P12.5

12.6. The designer of a coffee maker wants to make a filtration system in which the water is heated and filtered at the same time. The filter can be modeled as a porous zone within a tube, and the heat can be modeled as a uniform heat flux applied to the tube's outer wall. Set up the governing equations and boundary conditions so that the designer can get a temperature and pressure field, and thus know when the liquid may start to boil.

12.7. A conventional heat pipe with multiple heat sources, as shown in Fig. P12.7, is divided into three regions: vapor-flow region, liquid–wick region, and solid wall. The effect of gravity can be neglected so that the fluid flow and heat transfer can be treated as two-dimensional. Develop conservation governing equations and corresponding boundary conditions for all three regions. Furthermore, develop interfacial mass and energy balance between regions.

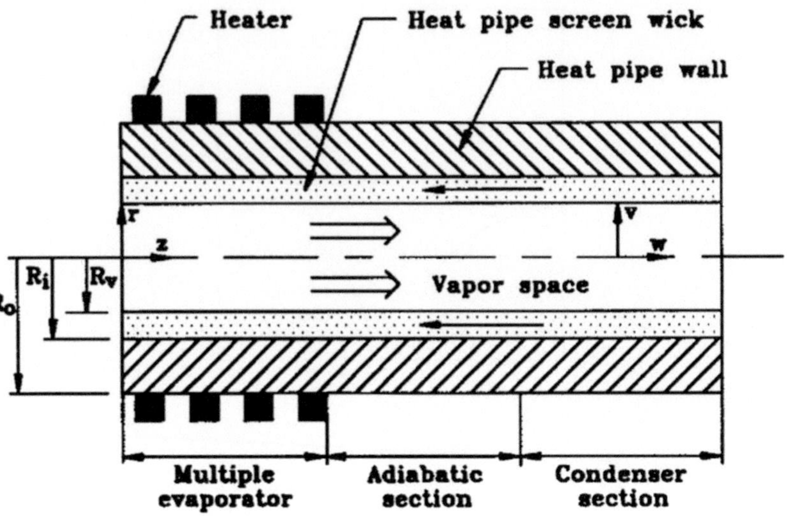

Fig. P12.7

12.8. The boundary of a porous zone is subjected to a constant heat flux. The porous zone is in local thermal nonequilibrium. Specify the thermal boundary conditions for the fluid phase and the thermal boundary condition for the solid phase. Also discuss the physical significance of your boundary conditions.

12.9. Consider the direct–methanol fuel cell (DMFC) with a liquid feed presented in Fig. P12.9. Assume that the methanol solution is passing over the fuel cell at a very high velocity, and that solution has a constant concentration at the top of the anode gas diffusion layer. The oxygen supplied to the cathode comes from air. Assume that the catalyst layers are very thin; therefore, their thickness can be neglected and the reaction rates can be considered to be a surface reaction rate. The membrane can be considered saturated at all times. Develop the governing equations and boundary conditions for the transport of liquid and gas through the fuel cell as well as the electric potentials of the membrane and gas diffusion layers for a one-dimensional steady-state simulation.

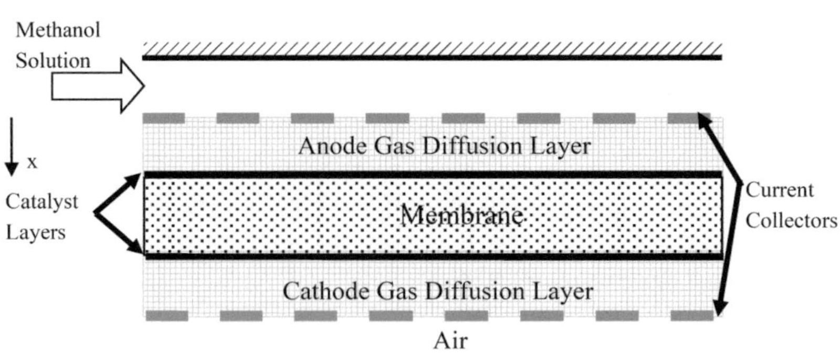

Fig. P12.9

12.10. The multiphase transport model in porous media can also be used to model two-phase flow in microchannel heat exchangers. The transport phenomena in a microchannel heat exchanger can be characterized as unidirectional flow combined with three-dimensional conduction. Set up the governing equations for this system, while leaving the terms caused by phase interaction and solid–fluid interaction unclosed. The energy equation shall be written in terms of internal energy, and the solid, liquid, and vapor equations shall be set up separately.

12.11. In a fluidized bed, particles are supported by an upward flow of fluid around them as shown in Fig. P12.11. Formulate the problem as multifluid model and deduce the fluid flux and pressure gradient needed to cause fluidization in a bed with void fraction ε_f by neglecting inertia. Assume one-dimensional, incompressible, steady flow with negligible interparticle forces and no phase change.

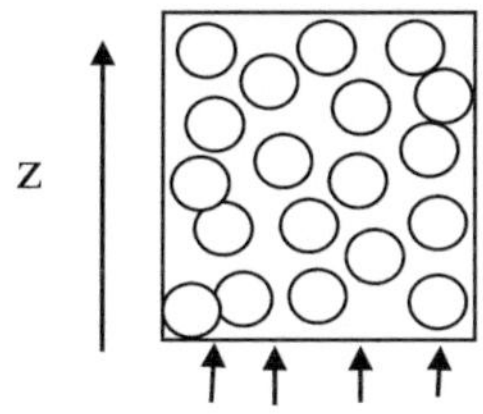

Fig. P12.11

12.12. A horizontal elliptical tube with a major axis 2a and minor axis 2b is embedded in a porous medium filled with a dry saturated vapor, as shown in Fig. P12.12. A dry saturated vapor flows downward with uniform vertical velocity U_∞. Since the temperature of the elliptical tube, T_w, is below the saturation temperature of the vapor, T_{sat}, condensation takes place at the surface of the elliptical tube. Specify the governing equation and corresponding boundary conditions for this problem.

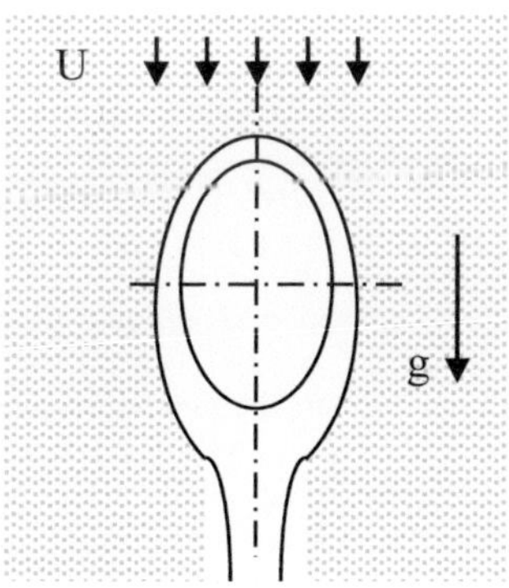

Fig. P12.12

12.13. Gas-loaded heat pipes have been applied in many diverse fields and are useful when the temperature of a device must be held constant while a variable heat load is dissipated. The physical configuration and coordinate system of a gas-loaded heat pipe are shown in Fig. P12.13. Develop the appropriate conservation equations for transient, compressible, two-species flow for mass, momentum, energy, and species in vapor space, wick, and wall including initial and boundary condition for a two-dimensional transient gas-loaded heat pipe.

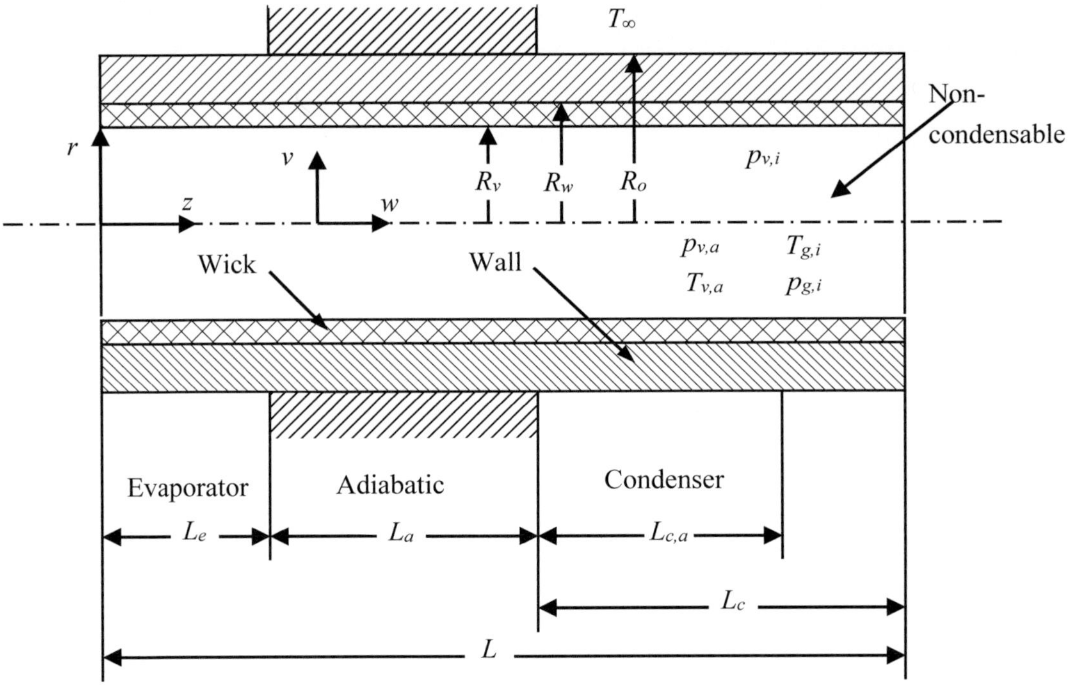

Fig. P12.13

References

Beckermann, C., & Viskanta, R. (1988). Natural convection solid/liquid phase change in porous media. *International Journal of Heat and Mass Transfer, 31,* 35–46.

Cheng, P. (1981). Film condensation along an inclined surface in a porous medium. *International Journal of Heat and Mass Transfer, 24,* 983–990.

Cheng, P., & Verma, A. K. (1981). The effect of subcooling liquid on film boiling about a vertical heated surface in a porous medium. *International Journal of Heat and Mass Transfer, 24,* 1151–1160.

Duval, F., Fichot, F., & Quintard, M. (2004). A local thermal non-equilibrium model for two-phase flows with phase-change in porous media. *International Journal of Heat and Mass Transfer, 47,* 613–639.

Fukusako, S., Komoriya, T., & Seki, N. (1986). An experimental study of transition and film boiling heat transfer in liquid-saturated porous bed. *Journal of Heat Transfer, 108,* 117–124.

Faghri, A. (2016). *Heat pipe science and technology* (2nd ed.). Columbia, MO: Global Digital Press.

Ferrell, J. K., & Alleavitch, J. (1970). Vaporization heat transfer in capillary wick structures. *Chemical Engineering Progress Symposium Series, 66*(102), 82–91.

Gray, W. G. (2000). Macroscale equilibrium conditions for two-phase flow in porous media. *International Journal of Multiphase Flow, 26,* 467–501.

Hadley, G. R. (1986). Thermal conductivity of packed metal powders. *International Journal of Heat and Mass Transfer, 29,* 909–920.

Jany, P., & Bejan, A. (1987). *Melting in the presence of natural convection in a rectangular cavity filled with porous medium* (Report DU-AB-6). Department of Mechanical Engineering and Materials Science, Duke University.

Jany, P., & Bejan, A. (1988) Scaling theory of melting with natural convection in an enclosure. *International Journal of Heat and Mass Transfer, 31*(6), 1221–1235.

Kaviany, M. (2013). *Principles of heat transfer in porous media* (2nd ed.). New York: Springer.

Leverett, M. C. (1941). Capillary behavior in porous solids. *Transaction of AIME, 142,* 152–169.

Majumdar, A., & Tien, C. L. (1990). Effects of surface tension on film condensation in a porous medium. *Journal of Heat Transfer, 112,* 751–757.

Marto, P. J., & Lepere, V. J. (1982). Pool boiling heat transfer from enhanced surfaces to dielectric fluids. *Journal of Heat Transfer, 104,* 292–299.

Nam, J.-H., & Kaviany, M. (2003). Effective mass diffusivity and water saturation distribution in single- and two-layer PEMFC diffusion medium. *International Journal of Heat and Mass Transfer, 46,* 4595–4611.

Nield, D. A., & Bejan, A. (2017). *Convection in porous media* (5th ed.). New York: Springer.

Oosthuizen, P. H. (1988). The effects of free convection on steady state freezing in a porous medium filled cavity. *ASME HTD, 96*(1), 321–327.

Ramesh, P. S., & Torrance, K. E. (1990). Stability of boiling in porous media. *International Journal of Heat and Mass Transfer, 33,* 1895–1990.

Rice, J. (2006). *Personal communication*. Storrs, CT: University of Connecticut.

Rohsenow, W. M., Hartnett, J. P., & Ganic, E. N. (Eds.). (1985). *Handbook of heat transfer fundamentals*. New York: McGraw-Hill.

Sondergeld, C. H., & Turcotte, D. L. (1977). An Experimental study of two- phase convection in a porous medium with applications to geological problems. *Journal of Geophysical Research, 82,* 2045–2053.

Veinberg, A. K. (1967). Permeability, electrical conductivity, dielectric constant and thermal conductivity of a medium with spherical and ellipsoidal inclusions. *Soviet Physics Doklady, 11,* 593–595.

Viskanta, R. (1988). Heat transfer during melting and solidification of metals. *Journal of Heat Transfer, 110,* 1205–1219.

Wang, C. Y., & Beckermann, C. (1993). A two-phase mixture model of liquid-gas flow and heat transfer in capillary porous media, Part I: Model formulation. *International Journal of Heat and Mass Transfer, 36,* 2747–2758.

Wang, C. Y., & Cheng, P. (1996). A multiphase mixture model for multiphase multicomponent transport in capillary porous media Part I: Model development. *International Journal of Heat and Mass Transfer, 39,* 3607–3618.

Wang, Z. H., & Wang, C. Y. (2003). Mathematical modeling of liquid-feed direct methanol feed cells. *Journal of the Electrochemical Society, 140,* A508–A519.

Ward, J. C. (1964). Turbulent flow in porous media. *ASCE Journal of the Hydraulics Division, 90*(HY5), 1–12.

Whitaker, S. (1977). Simultaneous heat, mass and momentum transfer in porous media: A theory of drying. *Advances in Heat Transfer, 13,* 119–203.

White, S. M., & Tien, C. L. (1987). An experimental investigation of film condensation in porous structures. In *Presented at the 6th International Heat Pipe Conference*, Grenoble, France.

Appendix A: Constants, Units, and Conversion Factors

See Tables A.1, A.2 and A.3

Table A.1 Physical constants

Universal gas constant	$R_u = 8314.34\,\text{J/(kmol K)}$
Boltzmann constant	$k_b = 1.38054 \times 10^{-23}\,\text{J/K}$
Stefan–Boltzmann constant	$\sigma_{SB} = 5.67 \times 10^{-8}\,\text{W/(m}^2\,\text{K}^4)$
Atmospheric pressure	$p_{atm} = 1.013 \times 10^5\,\text{Pa}$
Gravitational acceleration	$g = 9.807\,\text{m/s}^2$

Table A.2 Prefixes

Factor	Prefix	Symbol	Factor	Prefix	Symbol
10^{18}	Exa	E	10^{-1}	Deci	d
10^{15}	Peta	P	10^{-2}	Centi	c
10^{12}	Tera	T	10^{-3}	Milli	m
10^{9}	Giga	G	10^{-6}	Micro	μ
10^{6}	Mega	M	10^{-9}	Nano	n
10^{3}	Kilo	k	10^{-12}	Pico	p
10^{2}	Hecto	h	10^{-15}	Femto	f
10	Deka	da	10^{-18}	Atto	a

A. Faghri and Y. Zhang, *Fundamentals of Multiphase Heat Transfer and Flow*, https://doi.org/10.1007/978-3-030-22137-9

Table A.3 Conversion factors

Physical quantity		Conversion factor	
Acceleration	ft/s^2	0.30480	m/s^2
	m/s^2	3.2808	ft/s^2
Area	ft^2	0.092903	m^2
	cm^2	0.15500	in^2
	m^2	10.764	ft^2
Density	lbm/in^3	1728.0	lbm/ft^3
	lbm/ft^3	16.018	kg/m^3
	kg/m^3	0.062428	lbm/ft^3
	g/m^3	62.428	lbm/ft^3
Energy	Btu	1.0551	kJ
	Btu	0.0002930	kW h
	Btu	0.25200	kcal
	ft lbf	0.0012851	Btu
	kW h	3412.8	Btu
	kcal	3.9683	Btu
	kJ	0.94782	Btu
	HP h	2544.4	Btu
Energy flux	Btu/h ft^2	3.1546	W/m^2
	kcal/h m^2	0.36867	Btu/h ft^2
	cal/s cm^2	13272.0	Btu/h ft^2
	W/cm^2	3170.0	Btu/h ft^2
	W/m^2	0.8598	kcal/h m^2
	W/m^2	0.31700	Btu/h ft^2
Enthalpy	Btu/lbm	2324.4	J/kg
	J/kg	0.00043021	Btu/lbm
Force	lbf	32.1740	lbm ft/s^2
	lbf	4.448	N
	kgf	2.2046	lbf
	kgf	9.80665	N
	N	0.22481	lbf
Heat transfer coefficient	Btu/h ft^2 °F	5.6782	W/m^2 K
	kcal/h m^2 °C	0.20482	Btu/h-ft^2 °F
	W/m^2 K	0.17611	Btu/h-ft^2 °F
	W/m^2 K	0.8598	kcal/h m^2 °C
Kinematic viscosity (ν) Thermal diffusivity (α) Mass diffusivity (D)	ft^2/h	2.5807×10^{-5}	m^2/s
	ft^2/s	0.092905	m^2/s
	m^2/s	10000	cm^2/s (stokes)
	m^2/s	38750	ft^2/h
	m^2/s	10.764	ft^2/s
Length	in	25.4	mm
	ft	0.3048	m
	in	0.08333	ft
	mm	0.039370	in
	m	3.2808	ft

(continued)

Table A.3 (continued)

Physical quantity		Conversion factor	
Mass	lbm	0.45359	kg
	kg	2.2046	lbm
Mass flow rate	lbm/h	0.45359	kg/h
	lbm/s	3600	lbm/h
	kg/s	7936.6	lbm/h
	kg/h	2.2046	lbm/h
Power	Btu/s	1.055	kW
	Btu/h	0.293	W
	W	3.412	Btu/h
	W	9.48×10^{-4}	Btu/s
	HP	0.746	kW
	HP	0.707	Btu/s
Pressure	psi	6895	Pa
	atm	1.013×10^{5}	Pa
	atm	14.696	psi
	bar	10^{5}	Pa
	torr	1.000	mmHg
	torr	133.32	Pa
	psi	27.68	in H_2O
	ft-H_2O	0.4335	psi
Specific heat, specific entropy	Btu/lbm °F	4.1868	kJ/kg K
	kcal/kg °C	1.000	Btu/lbm °F
	kJ/kg K	0.23885	Btu/lbm °F
Temperature	°F	T(°C) = 5/9[T(°F) − 32]	°C
	°C	T(°F) = 9/5 T(°C) + 32	°F
	°C	T(K) = T(°C) + 273.15	K
	°F	T(°R) = T(°F) + 459.67	°R
Thermal conductivity	Btu/h ft °F	1.7307	W/m K
	cal/cm s °C	418.68	W/m K
	W/m K	0.5778	Btu/h ft °F
Velocity	ft/s	0.30480	m/s
	km/h	0.27778	m/s
	mile/h	1.609	km/h
Viscosity	kg/s m	1	N s/m^2
	posi	0.1	N s/m^2
	lbm/s ft	1.4882	N s/m^2
	lbm/h ft	4.1338×10^{-4}	N s/m^2
	N s/m^2	0.67195	lbm/s ft
	N s/m^2	2419.08	lbm/h ft
Volume	L	1	dm^3
	L	0.001	m^3
	ft^3	0.02832	m^3
	in^3	16.39	cm^3
	yd^3	0.7646	m^3
	m^3	35.313	ft^3

(continued)

Table A.3 (continued)

Physical quantity		Conversion factor	
Volume	Gal (U.S.)	3.785	L
	Gal (IMP)	4.546	L
	Pint (U.S.)	0.4732	L
	Pint (IMP)	0.5683	L
Volume flow rate	ft^3/min	4.7196×10^{-4}	m^3/s
	ft^3/s	0.0028318	m^3/s
	m^3/s	2118.8	ft^3/min
Volumetric heat generation rate	Btu/h ft^3	10.35	W/m^3
	W/m^3	0.0966	Btu/h ft^3

Appendix B: Transport Properties

List of Properties Tables

A. Faghri and Y. Zhang, *Fundamentals of Multiphase Heat Transfer and Flow*, https://doi.org/10.1007/978-3-030-22137-9

Table B.1 Air at 1 atm (Bergman and Lavine 2017)

T Temp. (K)	ρ Density (kg/m^3)	c_p Specific heat (kJ/Kg-K)	μ Viscosity (10^{-7} N s/m^2)	ν Kinematic viscosity (10^{-6} m^2/s)	*k* Thermal conductivity (10^{-3} W/m K)	α Thermal diffusivity (10^{-6} m^2/s)	*Pr* Prandtl number
100	3.5562	1.032	71.1	2.00	9.34	2.54	0.786
150	2.3364	1.012	103.4	4.426	13.8	5.84	0.758
200	1.7458	1.007	132.5	7.59	18.1	10.3	0.737
250	1.3947	1.006	159.6	11.44	22.3	15.9	0.72
300	1.1614	1.007	184.6	15.89	26.3	22.5	0.707
350	0.995	1.009	208.2	20.92	30.0	29.9	0.700
400	0.8711	1.014	230.1	26.41	33.8	38.3	0.690
450	0.7740	1.021	250.7	32.39	37.3	47.2	0.686
500	0.6964	1.030	270.1	38.79	40.7	56.7	0.684
550	0.6329	1.040	288.4	45.57	43.9	66.7	0.683
600	0.5804	1.051	305.8	52.69	46.9	76.9	0.685
650	0.5356	1.063	322.5	60.21	49.7	87.3	0.690
700	0.4975	1.075	338.8	68.10	52.4	98	0.695
750	0.4643	1.087	354.6	76.37	54.9	109	0.702
800	0.4354	1.099	369.8	84.93	57.3	120	0.709
850	0.4097	1.110	384.3	93.80	59.6	131	0.716
900	0.3868	1.121	398.1	102.9	62.0	143	0.720
950	0.3666	1.131	411.3	112.2	64.3	155	0.723
1000	0.3482	1.141	424.4	121.9	66.7	168	0.726
1100	0.3166	1.159	449	141.8	71.5	195	0.728
1200	0.2902	1.175	473	162.9	76.3	224	0.728
1300	0.2679	1.189	496	185.1	82	238	0.719
1400	0.2488	1.207	530	213	91	303	0.703
1500	0.2322	1.23	557	240	100	350	0.685
1600	0.2177	1.248	584	268	106	390	0.688
1700	0.2049	1.267	611	298	113	435	0.685
1800	0.1935	1.286	637	329	120	482	0.683
1900	0.1833	1.307	663	362	128	534	0.677
2000	0.1741	1.337	689	396	137	589	0.672
2100	0.1658	1.372	715	431	147	646	0.667
2200	0.1582	1.417	740	468	160	714	0.655
2300	0.1513	1.478	766	506	175	783	0.647
2400	0.1448	1.558	792	547	196	869	0.63
2500	0.1389	1.665	818	589	222	960	0.613
3000	0.1135	2.726	955	841	486	1570	0.536

Table B.2 Carbon dioxide (CO_2) at 1 atm (Bergman and Lavine 2017)

T Temp. (K)	ρ Density (kg/m^3)	c_p Specific heat (kJ/kg K)	μ Viscosity (10^{-7} N s/m^2)	ν Kinematic viscosity (10^{-6} m^2/s)	k Thermal conductivity (10^{-3} W/m K)	α Thermal diffusivity (10^{-6} m^2/s)	Pr Prandtl number
280	1.9022	0.830	140	7.36	15.20	9.63	0.765
300	1.7730	0.851	149	8.40	16.55	11.00	0.766
320	1.6609	0.872	156	9.39	18.05	12.50	0.754
340	1.5618	0.891	165	10.60	19.70	14.20	0.746
360	1.4743	0.908	173	11.70	21.20	15.80	0.741
380	1.3961	0.926	181	13.00	22.75	17.60	0.737
400	1.3257	0.942	190	14.30	24.30	19.50	0.737
450	1.1782	0.981	210	17.80	28.30	24.50	0.728
500	1.0594	1.020	231	21.80	32.50	30.10	0.725
550	0.9625	1.050	251	26.10	36.60	36.20	0.721
600	0.8826	1.080	270	30.60	40.70	42.70	0.717
650	0.8143	1.100	288	35.40	44.50	49.70	0.712
700	0.7564	1.130	305	40.30	48.10	56.30	0.717
750	0.7057	1.150	321	45.50	51.70	63.70	0.714
800	0.6614	1.170	337	51.00	55.10	71.20	0.716

Table B.3 Helium (He) at 1 atm (Bejan 2013)

T Temp. (K)	ρ Density (kg/m^3)	c_p Specific heat (kJ/kg K)	μ Viscosity (10^{-6} N s/m^2)	ν Kinematic viscosity (10^{-6} m^2/s)	k Thermal conductivity (10^{-3} W/m K)	α Thermal diffusivity (10^{-6} m^2/s)	Pr Prandtl number
4.22	16.900	9.78	1.25	0.0739	0.011	0.00064	1.15
7	7.530	5.71	1.76	0.234	0.014	0.00321	0.73
10	5.020	5.41	2.26	0.449	0.018	0.00642	0.70
20	2.440	5.25	3.58	1.470	0.027	0.0209	0.70
30	1.620	5.22	4.63	2.860	0.034	0.0403	0.71
60	0.811	5.20	7.12	8.800	0.053	0.125	0.70
100	0.487	5.20	9.78	20.10	0.074	0.291	0.69
200	0.244	5.19	15.1	62.20	0.118	0.932	0.67
300	0.162	5.19	19.9	122.0	0.155	1.830	0.67
600	0.0818	5.19	32.2	396.0	0.251	5.940	0.67
1000	0.0487	5.19	46.3	946.0	0.360	14.20	0.67

Table B.4 Hydrogen (H_2) at 1 atm (Bergman and Lavine 2017)

T Temp. (K)	ρ Density (kg/m^3)	c_p Specific heat (kJ/kg K)	μ Viscosity (10^{-7} N s/m^2)	ν Kinematic viscosity (10^{-6} m^2/s)	k Thermal conductivity (10^{-3} W/m K)	α Thermal diffusivity (10^{-6} m^2/s)	Pr Prandtl number
100	0.24255	11.230	42.1	17.4	67	24.6	0.707
150	0.16156	12.600	56.0	34.7	101	49.6	0.699
200	0.12115	13.540	68.1	56.2	131	79.9	0.704
250	0.09693	14.060	78.9	81.4	157	115	0.707
300	0.08078	14.310	89.6	111	183	158	0.701
350	0.06924	14.430	98.8	143	204	204	0.700
400	0.06059	14.480	108.2	179	226	258	0.695
450	0.05386	14.500	117.2	218	247	316	0.689
500	0.04848	14.520	126.4	261	266	378	0.691
550	0.04407	14.530	134.3	305	285	445	0.685
600	0.04040	14.550	142.4	352	305	519	0.678
700	0.03463	14.610	157.8	456	342	676	0.675
800	0.03030	14.700	172.4	569	378	849	0.670
900	0.02694	14.830	186.5	692	412	1030	0.671
1000	0.02424	14.990	201.3	830	448	1230	0.673
1100	0.02204	15.170	213.0	966	488	1460	0.662
1200	0.02020	15.370	226.2	1120	528	1700	0.659
1300	0.01865	15.590	238.5	1279	568	1955	0.655
1400	0.01732	15.810	250.7	1447	610	2230	0.650
1500	0.01616	16.020	262.7	1626	655	2530	0.643
1600	0.01520	16.280	273.7	1801	697	2815	0.639
1700	0.01430	16.580	284.9	1992	742	3130	0.637
1800	0.01350	16.960	296.1	2193	786	3435	0.639
1900	0.01280	17.490	307.2	2400	835	3730	0.643
2000	0.01210	18.250	318.2	2630	878	3975	0.661

Table B.5 Nitrogen (N_2) at 1 atm (Bergman and Lavine 2017)

T Temp. (K)	ρ Density (kg/m^3)	c_p Specific heat (kJ/kg K)	μ Viscosity (10^{-7} N s/m^2)	ν Kinematic viscosity (10^{-6} m^2/s)	k Thermal conductivity (10^{-3} W/m K)	α Thermal diffusivity (10^{-6} m^2/s)	Pr Prandtl number
100	3.4388	1.070	68.8	2.00	9.58	2.6	0.768
150	2.2594	1.050	100.6	4.45	13.9	5.86	0.759
200	1.6883	1.043	129.2	7.65	18.3	10.4	0.736
250	1.3488	1.042	154.9	11.48	22.2	15.8	0.727
300	1.1233	1.041	178.2	15.86	25.9	22.1	0.716
350	0.9625	1.042	200.0	20.78	29.3	29.2	0.711
400	0.8425	1.045	220.4	26.16	32.7	37.1	0.704
450	0.7485	1.050	239.6	32.01	35.8	45.6	0.703
500	0.6739	1.056	257.7	38.24	38.9	54.7	0.700
550	0.6124	1.065	274.7	44.86	41.7	63.9	0.702

(continued)

Table B.5 (continued)

T Temp. (K)	ρ Density (kg/m^3)	c_p Specific heat (kJ/kg K)	μ Viscosity (10^{-7} N s/m^2)	ν Kinematic viscosity (10^{-6} m^2/s)	k Thermal conductivity (10^{-3} W/m K)	α Thermal diffusivity (10^{-6} m^2/s)	Pr Prandtl number
600	0.5615	1.075	290.8	51.79	44.6	73.9	0.701
700	0.4812	1.098	321.0	66.71	49.9	94.4	0.706
800	0.4211	1.220	349.1	82.9	54.8	116	0.715
900	0.3743	1.146	375.3	100.3	59.7	139	0.721
1000	0.3368	1.167	399.9	118.7	64.7	165	0.721
1100	0.3062	1.187	423.2	138.2	70.0	193	0.718
1200	0.2807	1.204	445.3	158.6	75.8	224	0.707
1300	0.2591	1.219	466.2	179.9	81.0	256	0.701

Table B.6 Oxygen (O_2) at 1 atm (Bergman and Lavine 2017)

T Temp. (K)	ρ Density (kg/m^3)	c_p Specific heat (kJ/kg K)	μ Viscosity (10^{-7} N s/m^2)	ν Kinematic viscosity (10^{-6} m^2/s)	k Thermal conductivity (10^{-3} W/m K)	α Thermal diffusivity (10^{-6} m^2/s)	Pr Prandtl number
100	3.9450	0.9620	76.4	1.94	9.25	2.44	0.796
150	2.5850	0.9210	114.8	4.44	13.8	5.80	0.766
200	1.9300	0.9150	147.5	7.64	18.3	10.4	0.737
250	1.5420	0.9150	178.6	11.58	22.6	16.0	0.723
300	1.2840	0.9200	207.2	16.14	26.8	22.7	0.711
350	1.1000	0.9290	233.5	21.23	29.6	29.0	0.733
400	0.9620	0.9420	258.2	26.84	33.0	36.4	0.737
450	0.8554	0.9560	281.4	32.90	36.3	44.4	0.741
500	0.7698	0.9720	303.3	39.40	41.2	55.1	0.716
550	0.6998	0.9880	324.0	46.30	44.1	63.8	0.726
600	0.6414	1.0030	343.7	53.59	47.3	73.5	0.729
700	0.5498	1.0310	380.8	69.26	52.8	93.1	0.744
800	0.4810	1.0540	415.2	86.32	58.9	116	0.743
900	0.4275	1.0740	447.2	104.6	64.9	141	0.740
1000	0.3848	1.0900	477.0	124.0	71.0	169	0.733
1100	0.3498	1.1030	505.5	144.5	75.8	196	0.736
1200	0.3206	1.1150	532.5	166.1	81.9	229	0.725
1300	0.2960	1.1250	588.4	188.6	87.1	262	0.721

Table B.7 Water (H_2O) vapor at 1 atm (Bergman and Lavine 2017)

T Temp. (K)	ρ Density (kg/m^3)	c_p Specific heat (kJ/kg K)	μ Viscosity (10^{-7} N s/m^2)	ν Kinematic viscosity (10^{-6} m^2/s)	k Thermal conductivity (10^{-3} W/m K)	α Thermal diffusivity (10^{-6} m^2/s)	Pr Prandtl number
380	0.5863	2.060	127.1	21.68	24.6	20.4	1.060
400	0.5542	2.014	134.4	24.25	26.1	23.4	1.040
450	0.4902	1.980	152.5	31.11	29.9	30.8	1.010
500	0.4405	1.985	170.4	38.68	33.9	38.8	0.998
550	0.4005	1.997	188.4	47.04	37.9	47.4	0.993
600	0.3652	2.026	206.7	56.60	42.2	57.0	0.993
650	0.3380	2.056	224.7	66.48	46.4	66.8	0.996
700	0.3140	2.085	242.6	77.26	50.5	77.1	1.000
750	0.2931	2.119	260.4	88.84	54.9	88.4	1.000
800	0.2739	2.152	278.6	101.7	59.2	100.0	1.010
850	0.2579	2.186	296.9	115.1	63.7	113.0	1.020

Table B.8 Volume expansion coefficients for liquids (Mills and Coimbra 2015)

Liquid	T (K)	$\beta \times 10^3$ (1/K)	Liquid	T (K)	$\beta \times 10^3$ (1/K)
Ammonia	293	2.45	Hydrogen	20.3	15.1
Engine oil (SAE 50)	273	0.70	Mercury	273	0.18
	430	0.70		550	0.18
Ethylene glycol $C_2H_4(OH)_2$	273	0.65	Nitrogen	70	4.9
	373	0.65		77.4	5.7
Refrigerant-22	250	2.27		80	5.9
	260	2.41		90	7.2
	270	2.58		100	9.0
	280	2.78		110	12
	290	3.03		120	24
	300	3.35	Oxygen	89	2.0
	310	3.75	Sodium	366	0.27
	320	4.30	Therminol® 60	230	0.79
	330	5.09		250	0.75
	340	6.34		300	0.70
	350	8.64		350	0.70
Refrigerant-134a	230	2.00		400	0.76
	240	2.09		450	0.84
	250	2.20		500	0.96
	260	2.32		550	1.1
	270	2.47			
	280	2.65			
	290	2.86			
	300	3.13			
	310	3.48			
	320	3.95			
	330	4.61			
	340	5.60			
	350	7.32			
Glycerin $C_3H_5(OH)_3$	280	0.47			
	300	0.48			
	320	0.50			

Table B.9 Density and volume expansion coefficients of water (Mills and Coimbra 2015)

T (K)	ρ (kg/m^3)	$\beta \times 10^6$ (1/K)	T (K)	ρ (kg/m^3)	$\beta \times 10^6$ (1/K)
273.15	999.8679	−68.05	320.00	989.12	436.7
274.00	999.9190	−51.30	330.00	984.25	504.0
275.00	999.9628	−32.74	340.00	979.43	566.0
276.00	999.9896	−15.30	350.00	973.71	624.4
277.00	999.9999	1.16	360.00	967.12	697.9
278.00	999.9941	16.78	370.00	960.61	728.7
279.00	999.9727	31.69	373.15	957.85	750.1
280.00	999.9362	46.04	380.00	953.29	788
285.00	999.5417	114.1	390.00	945.17	841
290.00	998.8281	174.0	400.00	937.21	896
295.00	997.8332	227.5	450.00	890.47	1129
300.00	996.5833	276.1	500.00	831.26	1432
310.00	993.4103	361.9			

Table B.10 Aluminum

Aluminum, Al, T_m = 933 K (Rohsenow et al. 1998)			
T Temp. (K)	ρ Density (kg/m^3)	c Specific heat (kJ/kg K)	k Thermal conductivity (W/m K)
100	2732	0.481	300
150	2726	0.683	250
200	2719	0.797	237
250	2710	0.859	235
300	2701	0.902	237
400	2681	0.949	240
600	2639	1.042	231
800	2591	1.134	218

Table B.11 Aluminum alloy, 2024-T6

Aluminum Alloy, 2024-T6, T_m = 775 K (Bergman and Lavine 2017) (4.5% Cu, 1.5% Mg, 0.6% Mn)			
T Temp. (K)	ρ Density (kg/m^3)	c Specific heat (kJ/kg K)	k Thermal conductivity (W/m K)
100		0.473	65
200		0.787	163
300	2770	0.875	177
400		0.925	186
600		1.042	186

Table B.12 Cartridge brass

Cartridge brass, T_m = 1188 K (Bergman and Lavine 2017)			
T Temp. (K)	ρ Density (kg/m^3)	c Specific heat (kJ/kg K)	k Thermal conductivity (W/m K)
100			75
200		0.360	95
300	8530	0.380	110
400		0.395	137
600		0.425	149

Table B.13 Copper

Copper, Cu, T_m = 1358 K (Rohsenow et al. 1998)			
T Temp. (K)	ρ Density (kg/m^3)	c Specific heat (kJ/kg K)	k Thermal conductivity (W/m K)
100	9009	0.254	480
150	8992	0.323	429
200	8973	0.357	413
250	8951	0.377	406
300	8930	0.386	401
400	8884	0.396	393
600	8787	0.431	379
800	8642	0.448	366
1000	8568	0.446	352
1200	8458	0.480	339

Table B.14 Fused silica

Silicon dioxide, SiO, T_m = 1883 K (Bergman and Lavine 2017)			
T Temp. (K)	ρ Density (kg/m^3)	c Specific heat (kJ/kg K)	k Thermal conductivity (W/m K)
100			0.69
200			1.14
300	2220	0.745	1.38
400		0.905	1.51
600		1.040	1.75
800		1.105	2.17
1000		1.155	2.87
1200		1.195	4.00

Table B.15 Inconel® X-750

Inconel X-750, T_m = 1665 K (Bergman and Lavine 2017) (73% Ni, 15% Cr, 6.7% Fe)			
T Temp. (K)	*ρ* Density (kg/m³)	*c* Specific heat (kJ/kg K)	*k* Thermal conductivity (W/m K)
100			8.7
200		0.372	10.3
300	8510	0.439	11.7
400		0.473	13.5
600		0.510	17.0
800		0.546	20.5
1000		0.626	24.0
1200			27.6
1500			33.0

Table B.16 Iron

Iron, Fe, T_m = 1810 K (Rohsenow et al. 1998)			
T Temp. (K)	*ρ* Density (kg/m³)	*c* Specific heat (kJ/kg K)	*k* Thermal conductivity (W/m K)
100	7900	0.216	134
150	7890	0.324	104
200	7880	0.384	94
250	7870	0.422	87
300	7860	0.450	80
400	7830	0.491	70
600	7760	0.555	55
800	7690	0.692	43
1000	7650	1.034	32
1200	7620		28
1400	7520		31
1600	7420		
1800	7420		

Table B.17 Molybdenum

Molybdenum, Mo, T_m = 2892 K (Rohsenow et al. 1998)			
T Temp. (K)	*ρ* Density (kg/m³)	*c* Specific heat (kJ/kg K)	*k* Thermal conductivity (W/m K)
100	10,260	0.140	180
150	10,250	0.196	149
200	10,250	0.223	143
250	10,250	0.241	140
300	10,240	0.248	138
400	10,220	0.261	134
500	10,210	0.268	130
600	10,190	0.274	126
800	10,160	0.280	118
1000	10,120	0.292	112
1200	10,080		105
1400	10,040		100

Table B.18 Nickel

Nickel, Ni, T_m = 1728 K (Rohsenow et al. 1998)			
T Temp. (K)	ρ Density (kg/m^3)	c Specific heat (kJ/kg K)	k Thermal conductivity (W/m K)
100	8960	0.323	165
150	8940	0.329	120
200	8930	0.383	105
250	8910	0.416	98
300	8900	0.444	91
400	8860	0.490	80
600	8780	0.590	66
800	8690	0.530	68
1000	8610	0.556	72
1200	8510	0.582	76
1400	8410		80
1600	8320		

Table B.19 Niobium

Niobium, Nb, T_m = 2740 K (Rohsenow et al. 1998)			
T Temp. (K)	ρ Density (kg/m^3)	c Specific heat (kJ/kg K)	k Thermal conductivity (W/m K)
100	8600	0.202	55
150	8590	0.238	53
200	8580	0.254	53
250	8570	0.263	53
300	8570	0.268	54
400	8550	0.272	55
500	8530	0.277	57
600	8510	0.281	58
800	8470	0.290	61
1000	8430	0.298	64
1200	8380	0.307	68
1400	8340		71

Table B.20 Plain carbon steel

Plain Carbon Steel, T_m = 1480 °C (Bergman and Lavine 2017)			
T Temp. (K)	ρ Density (kg/m^3)	c Specific heat (kJ/kg K)	k Thermal conductivity (W/m K)
300	7854	0.434	60.5
400		0.487	56.7
600		0.559	48.0
800		0.685	39.2
1000		1.169	30.0

Table B.21 Stainless steel 304

Stainless Steel 304, T_m = 1670 K (Bergman and Lavine 2017)			
T Temp. (K)	ρ Density (kg/m^3)	c Specific heat (kJ/kg K)	k Thermal conductivity (W/m K)
100		0.272	9.2
200		0.402	12.6
300	7900	0.477	14.9
400		0.515	16.6
600		0.557	19.8
800		0.582	22.6
1000		0.611	25.4
1200		0.640	28.0
1500		0.682	31.7

Table B.22 Tantalum

Tantalum, Ta, T_m = 3252 K (Rohsenow et al. 1998)			
T Temp. (K)	ρ Density (kg/m^3)	c Specific heat (kJ/kg K)	k Thermal conductivity (W/m K)
100	16,490	0.108	59
150	16,480	0.125	58
200	16,460	0.132	58
250	16,450	0.137	57
300	16,440	0.141	58
400	16,410	0.145	58
500	16,370	0.148	59
600	16,340	0.149	59
800	16,270	0.152	59
1000	16,200	0.160	60
1200	16,130		61
1400	16,060		62

Table B.23 Titanium

Titanium, Ti, T_m = 1953 K (Rohsenow et al. 1998)			
T Temp. (K)	ρ Density (kg/m^3)	c Specific heat (kJ/kg K)	k Thermal conductivity (W/m K)
100	4510	0.295	31
150	4515	0.406	27
200	4520	0.464	25
250	4515	0.501	23
300	4510	0.525	21
400	4490	0.555	20
600	4470	0.597	19
800	4440	0.627	19
1000	4410	0.652	21
1200	4380		22
1400	4350		24
1600	4320		

Table B.24 Tungsten

Tungsten, W, T_m = 3660 K (Rohsenow et al. 1998)			
T Temp. (K)	ρ Density (kg/m^3)	c Specific heat (kJ/kg K)	k Thermal conductivity (W/m K)
100	19,310	0.089	208
150	19,300	0.113	192
200	19,290	0.125	185
250	19,280	0.131	180
300	19,270	0.135	174
400	19,240	0.137	159
500	19,220	0.139	146
600	19,190	0.140	137
800	19,130	0.144	125
1000	19,080	0.148	118
1200	19,020		112
1400	18,950		108

Table B.25 Phase change materials (PCMs)

PCMs	Chemical formula	T_m melting point (°C)	$h_{s\ell}$ latent heat (kJ/kg)	ρ_s Solid density (kg/m^3)	ρ_ℓ Liquid density (kg/m^3)	μ_ℓ Liquid viscosity (10^{-3} N s/m^2)	k_s Solid thermal conductivity (W/m K)	k_ℓ Liquid thermal conductivity (W/m K)	$c_{p,s}$ Solid specific heat (kJ/kg K)	$c_{p,\ell}$ Liquid specific heat (kJ/kg-K)	β Liquid thermal expansion coefficient (10^{-4}1/K)
n-Tetradecane[a]	$C_{14}H_{30}$	5.5	226	825	771			0.15			
n-Hexadecane[b]	$C_{16}H_{34}$	18.2	228.9	833	774			0.1505	1.80	2.31	
n-Octadecane[c]	$C_{18}H_{38}$	27.5	244	814[a]	774[a]	3.9	0.358	0.152	2.15	2.18	8.5
n-Eicosane[b]	$C_{20}H_{42}$	36.40	247.3	815	780			0.150	1.92	2.46	8.5
Gallium[d]	Ga	29.78	80.16	6095	6093	1.81	33.5	32.0	0.340[c]	0.3815	1.2
Aluminum[c]	Al	660.4	395	2702	2380[e]	1.3	238	94.03	1.076	1.08	1.2
Water[f]	H_2O	0	333.7[g]	920	1000	1.75	1.88	0.569	2.04	4.23	−0.6805
Acetic acid[a]	CH_3COOH	16.7	187	1214	1050	1.31		0.18	2040	1960	
Sodium hydrogen phosphate dodecahydrate[a]	$Na_2HPO_4 \cdot 12H_2O$	36	280	1520	1446		0.514	0.476	1690	1940	4.35

[a]Hale et al. (1971), [b]Humphries and Griggs (1977), [c]Bennon and Incropera (1988); [d]Brent et al. (1988), [e]Iida and Guthrie (1988), [f]Bergman and Lavine (2017), [g]Cengel et al. (2019)

Table B.26 Thermophysical properties at saturation for acetone

Acetone, $(CH_3)_2CO$, Molecular mass: 58.1, (T_{sat} = 56.25 °C; T_m = −93.15 °C; Reay et al. 2014; Faghri 2016)

T Temp. (°C)	p_v Saturation pressure (10^5 Pa)	$h_{\ell v}$ Latent heat (kJ/kg)	ρ_ℓ Liquid density (kg/m^3)	ρ_v Vapor density (kg/m^3)	μ_ℓ Liquid viscosity (10^{-3} N s/m^2)	μ_v Vapor viscosity (10^{-7} N s/m^2)	k_ℓ Liquid thermal conductivity (W/m K)	k_v Vapor thermal conductivity[a] (W/m-K)	σ Liquid surface tension (10^{-3} N/m)	$c_{p,\ell}$ Liquid specific heat[b] (kJ/kg K)	$c_{p,v}$ Vapor specific heat[a] (kJ/kg K)
−40	0.01	660.0	860.0	0.03	0.800	68.0	0.200		31.0	2.04	1.109
−20	0.03	615.6	845.0	0.10	0.500	73.0	0.189	0.0082	27.6	2.07	1.160
0	0.10	564.0	812.0	0.26	0.395	78.0	0.183	0.0096	26.2	2.11	1.215
20	0.27	552.0	790.0	0.64	0.323	82.0	0.181	0.0110	23.7	2.16	1.271
40	0.60	536.0	768.0	1.05	0.269	86.0	0.175	0.0126	21.2	2.22	1.328
60	1.15	517.0	744.0	2.37	0.226	90.0	0.168	0.0143	18.6	2.29	1.386
80	2.15	495.0	719.0	4.30	0.192	95.0	0.160	0.0161	16.2	2.39	1.444
100	4.43	472.0	689.6	6.94	0.170	98.0	0.148	0.0178	13.4	2.49	1.502
120	6.70	425.1	660.3	11.02	0.148	99.0	0.135	0.0195	10.7	2.61	1.560
140	10.49	394.4	631.8	18.61	0.132	103.0	0.126	0.0215	8.1	2.77	1.616

[a]Interpolation from Rohsenow et al. (1998), [b]Interpolation from Vargaftik (1975)

Table B.27 Thermophysical properties at saturation for ammonia

Ammonia, NH_3 Molecular mass: 17.0, (T_{sat} = 239.9 K; T_m = 195.5 K; ASHRAE, 2009; Lemmon et al. 2016)

T Temp. (°C)	p_v Saturation pressure (10^6 Pa)	$h_{\ell v}$ Latent heat (kJ/kg)	ρ_ℓ Liquid density (kg/m^3)	ρ_v Vapor density (kg/m^3)	μ_ℓ Liquid viscosity (10^{-5} N s/m^2)	μ_v Vapor viscosity (10^{-7} N s/m^2)	k_ℓ Liquid thermal conductivity (W/m K)	k_v Vapor thermal conductivity (W/m K)	σ Liquid surface tension[a] (N/m)	$c_{p,\ell}$ Liquid specific heat (kJ/kg K)	$c_{p,v}$ Vapor specific heat[a] (kJ/kg K)
200	0.008646	1477	728.9	0.08899	50.728	69.5	0.803	0.0197	0.0604	4.227	2.076
210	0.017746	1451	717.5	0.1746	41.498	72.1	0.768	0.0199	0.0563	4.285	2.112
220	0.033811	1425	705.8	0.3190	34.668	74.8	0.733	0.0201	0.0523	4.342	2.160
230	0.060439	1398	693.7	0.5489	29.494	77.7	0.699	0.0205	0.0485	4.397	2.222
240	0.10226	1369	681.4	0.8972	25.485	80.6	0.665	0.0210	0.0447	4.449	2.298
250	0.16496	1339	668.9	1.404	22.308	83.6	0.632	0.0216	0.0410	4.498	2.392
260	0.25529	1307	656.1	2.115	19.734	86.6	0.600	0.0223	0.0374	4.548	2.503
270	0.38100	1273	642.9	3.086	17.606	89.6	0.569	0.0231	0.0340	4.599	2.634
280	0.55077	1237	629.2	4.380	15.812	92.7	0.539	0.0240	0.0306	4.656	2.788
290	0.77413	1198	615.0	6.071	14.274	95.8	0.509	0.0251	0.0274	4.722	2.967
300	1.0614	1159	600.2	8.247	12.933	98.9	0.480	0.0264	0.0242	4.800	3.177
310	1.4235	1113	584.6	11.01	11.749	102	0.452	0.0279	0.0212	4.897	3.423
320	1.8721	1066	568.2	14.51	10.691	106	0.425	0.0296	0.0183	5.018	3.718
330	2.4196	1014	550.9	18.89	9.73	109	0.398	0.0316	0.0155	5.176	4.078
340	3.0789	958	532.4	24.40	8.86	113	0.372	0.0339	0.0129	5.385	4.530
350	3.8641	895	512.3	31.34	8.04	118	0.345	0.0369	0.0104	5.671	5.125
360	4.7902	825	490.3	40.18	7.28	123	0.319	0.0408	0.0080	6.082	5.955
370	5.8740	745	465.5	51.65	6.55	131	0.293	0.0461	0.0058	6.715	7.214
380	7.1352	649	436.5	67.16	5.83	140	0.267	0.0546	0.0038	7.818	9.395
390	8.5977	529	400.2	89.85	5.09	155	0.240	0.0701	0.0020	10.31	14.19

Table B.28 Thermophysical properties at saturation for cesium

Cesium, Cs, Molecular mass: 132.9, (T_{sat} = 943 K; T_m = 201.6 K; Ivancvskii et al. 1982)

T Temp. (°C)	p_v Saturation pressure (10^5 Pa)	$h_{\ell v}$ Latent heat (kJ/kg)	ρ_ℓ Liquid density (10^3 kg/m^3)	ρ_v Vapor density (10^{-3} kg/m^3)	μ_ℓ Liquid viscosity (10^{-4} N s/m^2)	μ_v Vapor viscosity (10^{-5} N s/m^2)	k_ℓ Liquid thermal conductivity (W/m K)	k_v Vapor thermal conductivity[a] (W/m K)	σ Liquid surface tension (10^{-3} N/m)	$c_{p,\ell}$ Liquid specific heat[a] (kJ/kg K)	$c_{p,v}$ Vapor specific heat[a] (kJ/kg K)
500	0.0003	544.30	1.723	9.91	3.181	1.460	18.79		61.9	0.232	0.1982
600	0.0056	534.20	1.666	15.50	2.558	1.668	19.02	0.00530	57.1	0.224	0.2344
700	0.0437	523.30	1.609	105.20	2.163	1.893	18.79	0.00631	52.3	0.219	0.2645
800	0.2026	511.60	1.552	433.60	1.890	2.124	18.33	0.00724	47.5	0.217	0.2821
900	0.6580	499.50	1.495	1275.90	1.690	2.336	17.51	0.00807	42.7	0.222	0.2878
1000	1.6800	486.50	1.438	2990.40	1.536	2.567	16.47	0.00878	37.9	0.231	0.2850
1100	3.6000	472.60	1.377	5924.10	1.415	2.782	15.49	0.00942	33.1	0.239	0.2776
1200	6.7700	458.80	1.311	10,364.80	1.316	2.995	13.57	0.01000	28.3	0.248	0.2681
1300	11.5100	444.60	1.243	16,520.70	1.234	3.198	11.60	0.01060	23.5	0.256	0.2582
1400	18.0200	429.98	1.174	24,307.20	1.164	3.398	9.39	0.01110	18.0		
1500	26.7200	415.40	1.102	34,048.30	1.104	3.589	7.50	0.01150	14.0		

[a]Vargaftik (1975)

Table B.29 Thermophysical properties at saturation for Dowtherm®

Diphenyl mixture (Dowtherm[a]), Molecular mass: 166.0, (T_{sat} = 258 °C; T_m = 12 °C; Vargaftik 1975)

T Temp. (°C)	p_v Saturation pressure (10^5 Pa)	$h_{\ell v}$ Latent heat (kJ/kg)	ρ_ℓ Liquid density (kg/m^3)	ρ_v Vapor density (kg/m^3)	μ_ℓ Liquid viscosity (10^{-5} N s/m^2)	μ_v Vapor viscosity (10^{-5} N s/m^2)	k_ℓ Liquid thermal conductivity (W/m K)	k_v Vapor thermal conductivity (W/m K)	σ Liquid surface tension (10^{-3} N/m)	$c_{p,\ell}$ Liquid specific heat (kJ/kg K)	$c_{p,v}$ Vapor specific heat (kJ/kg K)
100	0.006	345	995	0.035	101.0	0.68	0.126		31.6	1.88	
150	0.051	329	953	0.24	60.3	0.77	0.119		26.5	2.14	
200	0.245	314	912	0.99	40.7	0.87	0.110		21.8	2.34	
250	0.843	291	871	3.20	29.7	0.97	0.104		17.3	2.60	
300	2.330	264	825	8.70	22.7	1.07	0.096		12.9	2.76	
350	5.200	235	772	20.0	18.2	1.17	0.090		8.9	2.89	
400	10.43	207	709	42.0	14.9	1.26	0.083		5.0	3.01	

[a]Dowtherm is an eutectic mixture of 73.5% phenulether and 26.5% diphenyl

Table B.30 Thermophysical properties at saturation for ethane

Ethane, C_2H_6, Molecular mass: 30.1, (T_{sat} = −88.6 °C, T_m = −183.3 °C; Ivanovskii *et al.*, 1982)

T Temp. (°C)	p_v Saturation pressure (10^5 Pa)	$h_{\ell v}$ Latent heat (kJ/kg)	ρ_ℓ Liquid density (kg/m^3)	ρ_v Vapor density (kg/m^3)	μ_ℓ Liquid viscosity (10^{-7} N s/m^2)	μ_v Vapor viscosity (10^{-7} N s/m^2)	k_ℓ Liquid thermal conductivity (W/m K)	k_v Vapor thermal conductivity[a] (W/m K)	σ Liquid surface tension (10^{-3} N/m)	$c_{p,\ell}$ Liquid specific heat (kJ/kg K)	$c_{p,v}$ Vapor specific heat (kJ/kg K)
−120	0.096	530	582	0.230	2580	49.0	0.149		21.23	2.82	1.297
−100	0.600	506	562	0.921	1800	55.0	0.137		17.93	2.94	1.349
−80	1.700	480	540	2.600	1360	61.0	0.125		14.60	3.05	1.401
−60	3.700	450	516	6.200	1100	67.0	0.113	0.0116	11.30	3.16	1.459
−40	7.200	414	488	12.700	900	73.0	0.100	0.0138	8.00	3.26	1.521
−20	14.000	368	454	25.500	760	79.0	0.088	0.0160	4.60	3.38	1.585
0	25.000	304	414	46.000	660	85.5	0.077	0.0185	1.20	3.48	1.660
20	38.000	200	360	85.000	600	91.0	0.066	0.0209	0.08		1.736

[a]Interpolation (Rohsenow et al. 1998)

Table B.31 Thermophysical properties at saturation for ethanol

Ethanol, C_2H_5CH, Molecular mass: 46.0, (T_{sat} = 78.3 °C; T_m = −114.5 °C; Ivanovskii et al. 1982)

T Temp. (°C)	p_v Saturation pressure (10^5 Pa)	$h_{\ell v}$ Latent heat (kJ/kg)	ρ_ℓ Liquid density (10^3 kg/m^3)	ρ_v Vapor density (kg/m^3)	μ_ℓ Liquid viscosity (10^{-3} N s/m^2)	μ_v Vapor viscosity (10^{-5} N s/m^2)	k_ℓ Liquid thermal conductivity (W/m K)	k_v Vapor thermal conductivity[a] (W/m K)	σ Liquid surface tension (10^{-3} N/m)	$c_{p,\ell}$ Liquid specific heat[b] (kJ/kg K)	$c_{p,v}$ Vapor specific heat[c] (kJ/kg K)
0	0.012	1048.4	0.901	0.036	1.7990	0.774	0.183	0.0117	24.4	2.27	1.34
20	0.058	1030.0	0.800	0.085	1.1980	0.835	0.179	0.0139	22.8	2.40	1.40
40	0.180	1011.9	0.789	0.316	0.8190	0.900	0.175	0.0160	21.0	2.57	1.48
60	0.472	988.9	0.770	0.748	0.5880	0.959	0.171	0.0179	19.2	2.78	1.54
80	1.036	960.0	0.757	1.430	0.4320	1.030	0.169	0.0199	17.3	3.03	1.61
100	2.250	927.0	0.730	3.410	0.3180	1.092	0.167	0.0219	15.5	3.30	1.68
120	4.290	885.5	0.710	6.010	0.2430	1.157	0.165	0.0238	13.4	3.61	1.75
140	7.530	834.0	0.680	10.670	0.1900	1.219	0.163	0.0256	11.2	3.96	
160	12.756	772.9	0.650	17.450	0.1500	1.293	0.161	0.0272	9.0		
180	19.600	698.8	0.610	27.650	0.1200	1.369	0.159	0.0288	6.7		
200	29.400	598.3	0.564	44.480	0.0950	1.464	0.157	0.0395	4.3		
220	42.800	468.5	0.510	74.350	0.0725	1.618	0.155	0.0321	2.2		
240	60.200	280.5	0.415	135.500	0.0488	1.948	0.153		0.1		

[a]Interpolation from Rohsenow et al. (1998), [b]Interpolation from Vargaftik (1975), [c]Reay et al. (2014)

Table B.32 Thermophysical properties at saturation for Freon®-113

Freon-113, $C_2F_3Cl_3$, Molecular mass: 187.4, (T_{sat} = 47.68 °C; T_m = −36.6 °C; Vargaftik 1975)

T Temp. (°C)	p_v Saturation pressure (10^5 Pa)	$h_{\ell v}$ Latent heat (kJ/kg)	ρ_ℓ Liquid density (10^3 kg/m^3)	ρ_v Vapor density (kg/m^3)	μ_ℓ Liquid viscosity (10^{-3} N s/m^2)	μ_v Vapor viscosity (10^{-7} N s/m^2)	k_ℓ Liquid thermal conductivity[a] (W/m K)	k_v Vapor thermal conductivity (W/m K)	σ Liquid surface tension (10^{-3} N/m)	$c_{p,\ell}$ Liquid specific heat[a] (kJ/kg K)	$c_{p,v}$ Vapor specific heat[a] (kJ/kg K)
−30	0.0283	156.88	1.687	0.2639	1.670	89.4	0.0889		25.3	0.855	0.587
−20	0.0905	161.48	1.643	0.7800	1.130	94.2	0.0867		22.8	0.882	0.597
0	0.1500	158.68	1.621	1.2510	0.948	96.7	0.0822		21.5	0.921	0.621
10	0.2387	155.83	1.598	1.9300	0.780	99.0	0.0799		20.6	0.937	0.627
30	0.5420	149.93	1.554	4.1500	0.590	104.0	0.0754		18.1	0.962	0.647
50	1.0943	143.82	1.508	8.0000	0.475	108.5	0.0709	0.00866	16.0	0.986	0.667
70	2.0120	137.46	1.455	14.3000	0.401	113.0	0.0664		13.9	1.004	0.689

[a]ASHRAE (2001)

Table B.33 Thermophysical properties at saturation for Freon®-123

Freon-123, $CHCl_2CF_3$, Molecular mass: 152.9, (T_{sat} = 27.8 °C; T_m = −107 °C; ASHRAE 2001)

T Temp. (°C)	p_v Saturation pressure (10^5 Pa)	$h_{\ell v}$ Latent heat (kJ/kg)	ρ_ℓ Liquid density (10^3 kg/m^3)	ρ_v Vapor density (kg/m^3)	μ_ℓ Liquid viscosity (10^{-3} N s/m^2)	μ_v Vapor viscosity (10^{-7} N s/m^2)	k_ℓ Liquid thermal conductivity (W/m K)	k_v Vapor thermal conductivity (W/m K)	σ Liquid surface tension (10^{-3} N/m)	$c_{p,\ell}$ Liquid specific heat (kJ/kg K)	$c_{p,v}$ Vapor specific heat (kJ/kg K)
−60	0.0081	204.20	1.665	0.070	1.383	75.0	0.1020	0.00435	25.78	0.932	0.553
−40	0.0358	196.63	1.620	0.283	0.986	83.1	0.0961	0.00549	23.19	0.948	0.585
−20	0.1200	189.11	1.574	0.880	0.735	90.9	0.0898	0.00661	20.66	0.968	0.617
0	0.3265	181.44	1.526	2.242	0.565	98.4	0.0837	0.00774	18.18	0.990	0.651
20	0.7561	173.44	1.477	4.905	0.443	105.6	0.0778	0.00889	15.77	1.014	0.686
40	1.5447	164.95	1.425	9.629	0.352	112.6	0.0724	0.01008	13.43	1.038	0.724
60	2.8589	155.73	1.370	17.331	0.284	119.4	0.0673	0.01134	11.16	1.066	0.767
80	4.8909	145.54	1.311	29.189	0.231	126.3	0.0626	0.01273	8.97	1.100	0.816

Table B.34 Thermophysical properties at saturation for Freon®-134a

Freon-134a, CF_3CH_2F, Molecular mass: 102.0, ($T_{sat} = -26.4$ °C; $T_m = -101$ °C; ASHRAE 2009)

T Temp. (°C)	p_v Saturation pressure (10^5 Pa)	$h_{\ell v}$ Latent heat (kJ/kg)	ρ_ℓ Liquid density (10^3 kg/m^3)	ρ_v Vapor density (kg/m^3)	μ_ℓ Liquid viscosity (10^{-3} N s/m^2)	μ_v Vapor viscosity (10^{-7} N s/m^2)	k_ℓ Liquid thermal conductivity (W/m K)	k_v Vapor thermal conductivity (W/m K)	σ Liquid surface tension (10^{-3} N/m)	$c_{p,\ell}$ Liquid specific heat (kJ/kg K)	$c_{p,v}$ Vapor specific heat (kJ/kg K)
−60	0.1591	237.95	1.474	0.9268	0.663	83.0	0.121	0.00656	20.80	1.223	0.692
−40	0.5121	225.86	1.418	2.769	0.472	91.2	0.111	0.00817	17.60	1.255	0.749
−20	1.3273	212.91	1.358	6.785	0.353	99.2	0.101	0.00982	14.51	1.293	0.816
0	2.9280	198.60	1.295	14.428	0.271	107.3	0.0920	0.01151	11.56	1.341	0.897
20	5.7171	182.28	1.225	27.778	0.211	115.81	0.0833	0.01333	8.76	1.405	1.001
40	10.166	163.02	1.147	50.075	0.163	125.5	0.0747	0.01544	6.13	1.498	1.145
60	16.818	139.13	1.053	81.413	0.124	137.9	0.0661	0.01831	3.72	1.660	1.387

Table B.35 Thermophysical properties at saturation for Freon®-21

Freon-21, $CHFCl_2$, Molecular mass: 102.9, (T_{sat} = 8.90 °C; T_m = −135 °C) (Vargaftik 1975)

T Temp. (°C)	p_v Saturation pressure (10^5 Pa)	$h_{\ell v}$ Latent heat (kJ/kg)	ρ_ℓ Liquid density (10^3 kg/m^3)	ρ_v Vapor density (kg/m^3)	μ_ℓ Liquid viscosity (10^{-3} N s/m^2)	μ_v Vapor viscosity (10^{-7} N s/m^2)	k_ℓ Liquid thermal conductivity (W/m K)	k_v Vapor thermal conductivity[a] (W/m K)	σ Liquid surface tension (10^{-3} N/m)	$c_{p,\ell}$ Liquid specific heat[a] (kJ/kg K)	$c_{p,v}$ Vapor specific heat[a] (kJ/kg K)
−60	0.0253	269	1.554	0.147	0.849	89	0.132		29.81		0.501
−40	0.0954	262	1.510	0.510	0.597	95	0.123		26.99		0.523
−20	0.2847	253	1.470	1.410	0.444	100	0.116		24.17	0.994	0.545
0	0.7085	243	1.420	3.310	0.345	106	0.109		21.35	1.015	0.566
20	1.5300	232	1.380	6.810	0.272	112	0.102		18.35	1.048	0.588
40	2.955	220	1.330	12.690	0.229	118	0.095	0.0094	15.71	1.093	0.606
60	5.216	206	1.280	21.930	0.200	124	0.087	0.0104	12.89	1.149	0.623
80	8.567	191	1.220	35.710	0.195	130	0.080	0.0113	10.07		0.641
100	13.283	174	1.160	55.860	0.180	136	0.072	0.0122	7.25		0.659
120	19.566	155	1.080	85.470	0.170	142	0.060	0.0132	4.43		0.677

[a]Interpolation from Rohsenow et al. (1998)

Table B.36 Thermophysical properties at saturation for Freon®-22

Freon-22, CHF_2Cl, Molecular mass: 86.5, (T_{sat} = −40.3 °C; T_m = −160 °C; Vargaftik 1975)											
T Temp. (°C)	p_v Saturation pressure (10^5 Pa)	$h_{\ell v}$ Latent heat (kJ/kg)	ρ_ℓ Liquid density (10^3 kg/m^3)	ρ_v Vapor density (kg/m^3)	μ_ℓ Liquid viscosity (10^{-4} N s/m^2)	μ_v Vapor viscosity (10^{-7} N s/m^2)	k_ℓ Liquid thermal conductivity (W/m K)	k_v Vapor thermal conductivity (W/m K)	σ Liquid surface tension[a] (10^{-3} N/m)	$c_{p,\ell}$ Liquid specific heat (kJ/kg K)	$c_{p,v}$ Vapor specific heat (kJ/kg K)
−100	0.0199	269.29	1.557	0.1196	6.00	80.0	0.1487	0.00446	28.1	1.075	0.497
−80	0.1034	257.43	1.514	0.561	5.00	87.5	0.1385	0.00525	24.8	1.083	0.528
−60	0.3752	245.42	1.465	1.865	4.14	95.0	0.1283	0.00612	21.5	1.091	0.564
−40	1.0540	232.92	1.412	4.885	3.49	101.7	0.1181	0.00831	18.5	1.105	0.611
−20	2.4560	219.40	1.351	10.821	3.02	110.4	0.1079	0.00929	15.0	1.130	0.654
0	4.9830	204.28	1.285	21.285	2.67	118.7	0.0977	0.01026	11.7	1.171	0.741
20	9.0970	185.89	1.214	38.550	2.40	126.8	0.0875	0.01123	8.7	1.232	0.854
40	15.3150	166.22	1.132	66.225	2.19	134.5	0.0772	0.01221	5.8	1.319	0.994
60	24.2360	139.94	1.030	111.65	2.00	142.1	0.0646	0.01318	3.3	1.526	1.243

[a]Ivanovskii et al. (1982)

Table B.37 Thermophysical properties at saturation for helium

Helium, He, Molecular mass: 4.0, (T_{sat} = −268 °C; T_m = −271 °C; Reay et al. 2014; Faghri 2016)

T Temp. (°C)	p_v Saturation pressure (10^5 Pa)	$h_{\ell v}$ Latent heat (kJ/kg)	ρ_ℓ Liquid density (kg/m^3)	ρ_v Vapor density (kg/m^3)	μ_ℓ Liquid viscosity (10^{-7} N s/m^2)	μ_v Vapor viscosity (10^{-8} N s/m^2)	k_ℓ Liquid thermal conductivity (W/m K)	k_v Vapor thermal conductivity[a] (W/m K)	σ Liquid surface tension (10^{-3} N/m)	$c_{p,\ell}$ Liquid specific heat[b] (kJ/kg K)	$c_{p,v}$ Vapor specific heat (kJ/kg K)
−271	0.06	22.8	148.3	26.0	390	20	0.0181	0.00393	0.26	5.18	2.045
−270	0.32	23.6	140.7	17.0	370	30	0.0224	0.00607	0.19	2.49	2.699
−269	1.00	20.9	128.0	10.0	290	60	0.0277	0.00803	0.09	3.99	4.619
−268	2.29	4.0	113.8	8.5	134	90	0.0350	0.00962	0.01	11.5	6.642

[a]Touloukian et al. (1970), [b]Vargaftik (1975)

Table B.38 Thermophysical properties at saturation for heptane

Heptane, C_7H_{16}, Molecular mass: 100.2, (T_{sat} = 98.43 °C; T_m = −90.59 °C; Vargaftik 1975)

T Temp. (°C)	p_v Saturation pressure (10^5 Pa)	$h_{\ell v}$ Latent heat (kJ/kg)	ρ_ℓ Liquid density (10^3 kg/m^3)	ρ_v Vapor density (10^3 kg/m^3)	μ_ℓ Liquid viscosity (10^{-3} N s/m^2)	μ_v Vapor viscosity (10^{-7} N s/m^2)	k_ℓ Liquid thermal conductivity (W/m K)	k_v Vapor thermal conductivity[a] (W/m K)	σ Liquid surface tension[a] (10^{-3} N/m)	$c_{p,\ell}$ Liquid specific heat (kJ/kg K)	$c_{p,v}$ Vapor specific heat[b] (kJ/kg K)
−20	0.003866	383.1	0.7172		0.6890		0.140	0.0084		2.10	0.83
0	0.0152	375.6	0.7005	0.000070	0.5260		0.134	0.0099		2.16	0.87
20	0.0472	366.0	0.6836	0.000200	0.4140		0.129	0.0115	20.86	2.23	0.92
40	0.1230	354.7	0.6665	0.000500	0.3380		0.123	0.0132	18.47	2.30	0.97
60	0.2800	342.6	0.6491	0.001100	0.2810		0.118	0.0151	16.39	2.39	1.02
80	0.5700	330.1	0.6311	0.002000	0.2390		0.113	0.0170	14.35	2.47	1.05
100	1.0606	316.7	0.6124	0.003597	0.1980	73.6		0.0189	12.47	2.57	1.09
120	1.8330	302.9	0.5926	0.006075	0.1672	78.2		0.0207	10.63	2.67	1.16
140	2.9790	287.4	0.5711	0.009785	0.1427	83.4		0.0228	8.87	2.78	
160	4.5990	269.5	0.5481	0.015110	0.1217	89.7		0.0251	7.19	2.89	

[a]Interpolation from Rohsenow et al. (1998), [b]Reay et al. (2014)

Table B.39 Thermophysical properties at saturation for lead

Lead, Pb, Molecular mass: 207.2, (T_{sat} = 1740 °C; T_m = 327.5 °C; Ivanovskii et al. 1982)

T Temp. (°C)	p_v Saturation pressure (10^2 Pa)	$h_{\ell v}$ Latent heat (kJ/kg)	ρ_ℓ Liquid density (10^3 kg/m^3)	ρ_v Vapor density (10^3 kg/m^3)	μ_ℓ Liquid viscosity (10^{-3} N s/m^2)	μ_v Vapor viscosity (10^{-7} N s/m^2)	k_ℓ Liquid thermal conductivity (W/m K)	k_v Vapor thermal conductivity (W/m K)	σ Liquid surface tension (10^{-3} N/m)	$c_{p,\ell}$ Liquid specific heat (kJ/kg K)	$c_{p,v}$ Vapor specific heat[a] (kJ/kg K)
1400	0.0986	920	9.27	0.147	0.9122	7.46			347.28		
1500	0.2108	920	9.14	0.296	0.8847	7.90			335.88		
1600	0.4200	920	9.01	0.559	0.8586	8.34			324.48		
1700	0.8010	920	8.89	1.011	0.8352	8.78			313.08		
1800	1.3620	920	8.76	1.635	0.8143	9.21			301.68		
1900	2.3100	920	8.63	2.648	0.7958	9.66			290.28		
2000	3.7410	920	8.51	4.106	0.7794	10.10			278.88		
2100	5.5500	920	8.37	5.817	0.7590	10.54			260.00		
2200	8.2000	920	8.25	8.256	0.7410	10.98			248.00		
2300	11.8500	920	8.12	11.480	0.7230	11.42			237.00		
2400	16.7500	920	7.99	15.600	0.7050	11.86			225.00		
2500	22.6000	920	7.86	20.280	0.6870	12.30			214.00		

[a]Reay et al. (2014)

Table B.40 Thermophysical properties at saturation for lithium

Lithium, Li, Molecular mass: 6.9, (T_{sat} = 1615 K; T_m = 453.7 K; Ivanovskii et al. 1982)

T Temp. (°C)	p_v Saturation pressure (10^2 Pa)	$h_{\ell v}$ Latent heat (kJ/kg)	ρ_ℓ Liquid density (kg/m^3)	ρ_v Vapor density (10^3 kg/m^3)	μ_ℓ Liquid viscosity (10^{-4} N s/m^2)	μ_v Vapor viscosity (10^{-8} N s/m^2)	k_ℓ Liquid thermal conductivity (W/m K)	k_v Vapor thermal conductivity (W/m K)	σ Liquid surface tension (10^{-3} N/m)	$c_{p,\ell}$ Liquid specific heat[a] (kJ/kg K)	$c_{p,v}$ Vapor specific heat[a] (kJ/kg K)
900	0.1256	21712	472.8	0.012	2.784	890.1	52.75		335.8	4.16	6.956
1000	0.9680	21400	462.6	0.085	2.472	975.2	55.10		321.8	4.16	8.171
1100	5.1200	21000	452.4	0.415	2.252	1055.0	57.42	0.120	307.8	4.15	9.114
1200	20.5000	20740	442.2	1.540	2.072	1128.0	59.62	0.138	293.8	4.14	9.723
1300	65.8600	20380	432.0	4.650	1.922	1213.0	61.94	0.156	279.8	4.16	10.019
1400	179.4000	20020	421.7	11.960	1.795	1289.0	64.00	0.172	266.0	4.19	10.049
1500	426.5000	19670	411.5	26.900	1.685	1368.0	66.50	0.183	252.0	4.20	9.891
1600	908.4000	19330	401.3	54.610	1.590	1442.0	68.50	0.192	238.0	4.23	9.611
1700	1769.3000	18990	391.1	101.500	1.506	1518.0	71.00	0.198	226.0	4.25	9.259
1800	3190.0000	18570	380.9	175.100	1.432	1587.0	73.00	0.202	212.0	4.27	8.871
1900	5397.0000	18370	370.0	283.900	1.380	1666.0	75.50	0.207	198.0	4.30	8.481
2000	8640.4000	18080	360.0	436.300	1.300	1746.0	77.00	0.209	182.0	4.32	8.098

[a]Vargaftik (1975)

Table B.41 Thermophysical properties at saturation for mercury

Mercury, Hg, Molecular mass: 200.6, (T_{sat} = 630.1 K; T_m = 234.3 K; Ivanovskii et al. 1982)

T Temp. (°C)	p_v Saturation pressure (10^5 Pa)	$h_{\ell v}$ Latent heat (kJ/kg)	ρ_ℓ Liquid density (kg/m^3)	ρ_v Vapor density (kg/m^3)	μ_ℓ Liquid viscosity (10^{-3} N s/m^2)	μ_v Vapor viscosity (10^{-7} N s/m^2)	k_ℓ Liquid thermal conductivity (W/m K)	k_v Vapor thermal conductivity[a] (W/m K)	σ Liquid surface tension (N/m)	$c_{p,\ell}$ Liquid specific heat[a] (kJ/kg K)	$c_{p,v}$ Vapor specific heat[a] (kJ/kg K)
100	0.0003745	303.317	13351.42	0.00242	1.241	360	9.475		0.4600	0.1371	1.04
200	0.02315	300.056	13111.97	0.11800	1.039	464	10.64		0.4360	0.1355	1.04
300	0.33015	296.824	12873.50	1.39100	0.926	562	11.69	0.0043	0.4050	0.1353	1.04
400	2.10240	293.314	12632.60	7.57200	0.853	662	12.60	0.0058	0.3770	0.1364	1.04
500	8.2220	289.116	12386.00	26.00000	0.804	762	13.39	0.0073	0.3290	0.1389	1.04
600	23.46000	283.769	12130.00	66.66000	0.767	862	14.04	0.0090	0.2989	0.1427	1.04
700	54.03000	276.845	11863.00	140.75000	0.739	961	14.58	0.0107	0.2687	0.1478	1.04

[a]Kakac et al. (1987)

Table B.42 Thermophysical properties at saturation for methanol

Methanol, CH_4O, Molecular mass: 32.0, (T_{sat} = 64.7 °C; T_m = −98 °C; Lemmon et al. 2016; Vargaftik 1975)

T Temp. (°C)	p_v Saturation pressure (10^5 Pa)	$h_{\ell v}$ Latent heat (kJ/kg)	ρ_ℓ Liquid density (10^3 kg/m^3)	ρ_v Vapor density (10^3 kg/m^3)	μ_ℓ Liquid viscosity (10^{-3} N s/m^2)	μ_v Vapor viscosity (10^{-7} N s/m^2)	k_ℓ Liquid thermal conductivity (W/m K)	k_v Vapor thermal conductivity[a] (W/m K)	σ Liquid surface tension (10^{-3} N/m)	$c_{p,\ell}$ Liquid specific heat (kJ/kg K)	$c_{p,v}$ Vapor specific heat (kJ/kg K)
0	0.0411	1210.0	0.8097	0.0000580	0.8170	88	0.205		24.5	2.40	2.99
20	0.103	1191.1	0.7909	0.0001751	0.5780	95	0.204		22.6	2.50	3.50
40	0.358	1163.9	0.7721	0.0004521	0.4460	101	0.203	0.00157	20.9	2.63	3.96
60	0.861	1130.4	0.7528	0.0010299	0.3470	108	0.202	0.00178	19.3	2.79	4.35
80	1.819	1084.4	0.7326	0.0021229	0.2710	115	0.200	0.00199	17.5	2.97	4.72
100	3.731	1030.0	0.7110	0.0040421	0.2140	123	0.198	0.00220	15.7	3.17	5.14
120	6.551	971.3	0.6873	0.0072359	0.1700	130	0.196	0.00241	13.6	3.40	5.69
140	10.810	904.3	0.6608	0.0123780	0.1360	136	0.194	0.00262	11.5	3.68	6.51
160	17.609	828.0	0.6304	0.0205290	0.1090	143		0.00283	9.3	4.02	7.61
180	16.869	741.1	0.5943	0.0331840	0.0883	150		0.00303	6.9	4.49	8.29
200	38.434	636.4	0.5492	0.0521240	0.0716	157		0.00324	4.5	5.27	8.06
220	56.728	473.1	0.4849	0.0881400	0.0583	166		0.00344	2.1	7.54	

Table B.43 Thermophysical properties at saturation for nitrogen

Nitrogen, N_2, Molecular mass: 28.0, (T_{sat} = −195.65 °C; T_m = 209.85 °C; Vargaftik, 1975)

T Temp. (°C)	p_v Saturation pressure (10^5 Pa)	$h_{\ell v}$ Latent heat (kJ/kg)	ρ_ℓ Liquid density (10^3 kg/m^3)	ρ_v Vapor density (10^3 kg/m^3)	μ_ℓ Liquid viscosity (10^{-5} N s/m^2)	μ_v Vapor viscosity (10^{-7} N s/m^2)	k_ℓ Liquid thermal conductivity (W/m K)	k_v Vapor thermal conductivity[a] (W/m K)	σ Liquid surface tension (10^{-3} N/m)	$c_{p,\ell}$ Liquid specific heat (kJ/kg K)	$c_{p,v}$ Vapor specific heat[a] (kJ/kg K)
70	0.3859	205.7	0.838	0.0019	2010	48.00	0.1420	0.0066	10.53	1.935	1.08
80	1.3690	194.5	0.790	0.0060	1390	55.20	0.1280	0.0077	8.27	1.964	1.14
90	3.6000	180.5	0.746	0.0150	1160	62.00	0.1120	0.0091	6.16	2.028	1.26
100	7.7750	162.2	0.691	0.0320	810	68.80	0.0955	0.0111	4.00	2.176	1.47
110	14.6700	137.0	0.626	0.0620	740	75.60	0.0802	0.0138	2.00	2.566	1.97
120	25.1500	95.7	0.528	0.1245	640	82.10	0.0628	0.0195	0.20		4.14

[a]ASHRAE (2001)

Table B.44 Thermophysical properties at saturation for potassium

Potassium, K, Molecular mass: 39.1, (T_{sat} = 1032.2 K; T_m = 336.4 K; Vargaftik 1975)

T Temp. (°C)	p_v Saturation pressure (10^5 Pa)	$h_{\ell v}$ Latent heat (kJ/kg)	ρ_ℓ Liquid density (kg/m^3)	ρ_v Vapor density (10^{-3} kg/m^3)	μ_ℓ Liquid viscosity (10^{-4} N s/m^2)	μ_v Vapor viscosity (10^{-7} N s/m^2)	k_ℓ Liquid thermal conductivity (W/m K)	k_v Vapor thermal conductivity[a] (W/m K)	σ Liquid surface tension (10^{-3} N/m)	$c_{p,\ell}$ Liquid specific heat (kJ/kg K)	$c_{p,v}$ Vapor specific heat (10^{-1} kJ/kg K)
600	0.0009258	2143	766.9	0.69	2.380		43.85		98.2	0.771	0.8194
700	0.01022	2108	743.3	6.68	1.981		40.72	0.0142	92.2	0.762	0.9646
800	0.06116	2068	719.6	36.44	1.707	134	37.58	0.0175	86.2	0.761	1.066
900	0.2441	2023	695.7	134.80	1.507	148	34.45	0.0205	80.2	0.769	1.116
1000	0.7322	1970	671.6	380.20	1.354	163	31.32	0.0228	74.2	0.792	1.121
1100	1.864	1924	647.3	871.90	1.233	178	28.19	0.0248	68.2	0.819	1.100
1200	3.913	1872	622.9	1703.00	1.135	196	25.05	0.0266	62.2	0.846	1.064
1300	7.304	1320	598.4	2969.10	1.053	212	22.00	0.0280	56.2	0.873	1.022
1400	12.44	1765	573.6	4768.70	0.984	228	19.00	0.0293	53.0	0.899	0.9796
1500	20.0	1711	548.8	7062.10	0.925	242	16.00	0.0303	47.0	0.924	

Table B.45 Thermophysical properties at saturation for rubidium

Rubidium, Rb, Molecular mass: 85.5, (T_{sat} = 959.2 K; T_m = 312.7 K; Vargaftik 1975)

T Temp. (°C)	p_v Saturation pressure (10^5 Pa)	$h_{\ell v}$ Latent heat (kJ/kg)	ρ_ℓ Liquid density (kg/m^3)	ρ_v Vapor density (kg/m^3)	μ_ℓ Liquid viscosity (10^{-4} N s/m^2)	μ_v Vapor viscosity (10^{-4} N s/m^2)	k_ℓ Liquid thermal conductivity (W/m K)	k_v Vapor thermal conductivity (W/m K)	σ Liquid surface tension (10^{-3} N/m)	$c_{p,\ell}$ Liquid specific heat (kJ/kg K)	$c_{p,v}$ Vapor specific heat (kJ/kg K)
500	0.0001733	889.6	1386	0.0003585	3.23		29.8		81.6	0.369	0.3353
600	0.003664	870.9	1340	0.006386	2.58	0.112	27.8	0.0073	75.7	0.362	0.4100
700	0.03174	849.7	1294	0.04819	2.18	0.135	25.9	0.0089	69.8	0.357	0.4679
800	0.1584	827.3	1248	0.2145	1.89	0.158	24.1	0.0103	63.9	0.353	0.4979
900	0.5476	804.6	1202	0.6726	1.69	0.183	22.2	0.0115	58.0	0.353	0.5035
1000	1.457	782.2	1156	1.658	1.53	0.208	20.3	0.0125	51.3	0.360	0.4937
1100	3.295	759.6	1110	3.437	1.40	0.244	18.5	0.0133	44.5	0.373	0.4762
1200	6.456	737.0	1064	6.274	1.30	0.268	16.7	0.0141	37.7	0.385	0.4558
1300	11.43	714.5	1018	10.36	1.21	0.289	15.0	0.0149	30.9	0.399	0.4354
1400	18.6	694.0	972	12.35	1.14	0.314	13.6	0.0156	26.0	0.408	0.4130
1500	28.5	674.0	926	22.22	1.08	0.336	12.0	0.0160	19.0	0.418	0.3900

Table B.46 Thermophysical properties at saturation for silver

Silver, Ag, Molecular mass: 107.9, (T_{sat} = 2212 °C; T_m = 960.5 °C; Ivanovskii et al. 1982)

T Temp. (°C)	p_v Saturation pressure (10^5 Pa)	$h_{\ell v}$ Latent heat (kJ/kg)	ρ_ℓ Liquid density (kg/m^3)	ρ_v Vapor density (kg/m^3)	μ_ℓ Liquid viscosity (10^{-3} N s/m^2)	μ_v Vapor viscosity (10^{-6} N s/m^2)	k_ℓ Liquid thermal conductivity[a] (W/m K)	k_v Vapor thermal conductivity (W/m K)	σ Liquid surface tension (10^{-3} N/m)	$c_{p,\ell}$ Liquid specific heat (kJ/kg K)	$c_{p,v}$ Vapor specific heat (kJ/kg K)
1500	0.01008	298	8782	0.0076	2.88	61.69	191.3		827.5		
1600	0.02420	298	8683	0.01698	2.47	64.69	192.7		810.1		
1700	0.05300	298	8585	0.03548	2.08	67.69	194.1		792.1		
1800	0.10800	298	8485	0.06823	1.75	70.69	195.5		775.3		
1900	0.20600	298	8385	0.12300	1.44	73.69	196.9		757.9		
2000	0.38300	298	8289	0.21880	1.17	76.69	198.3		740.5		
2100	0.63500	298	8190	0.35480	0.90	79.69	199.7		723.1		
2200	0.86000	298	8092	0.57540	0.67	82.69			705.7		
2300	1.36000	298	8000	0.87100	0.44	85.69			638.0		
2400	2.53000	298	7894	1.23000	0.24	88.69			680.0		
2500	3.84000	298	7796	1.82000	0.05	91.69			665.0		

[a]Brennan and Kroliczek (1979)

Table B.47 Thermophysical properties at saturation for sodium

Sodium, Na, Molecular mass: 23.0, (T_{sat} = 1151.2 K; T_m = 371.0 K; Ivanovskii et al. 1982; ANL 1995)

T Temp. (°C)	p_v Saturation pressure (10^2 Pa)	$h_{\ell v}$ Latent heat (kJ/kg)	ρ_ℓ Liquid density (kg/m^3)	ρ_v Vapor density (10^{-3} kg/m^3)	μ_ℓ Liquid viscosity (10^{-4} N s/m^2)	μ_v Vapor viscosity (10^{-8} N s/m^2)	k_ℓ Liquid thermal conductivity (W/m K)	k_v Vapor thermal conductivity[a] (W/m K)	σ Liquid surface tension (10^{-3} N/m)	$c_{p,\ell}$ Liquid specific heat[a] (kJ/kg K)	$c_{p,v}$ Vapor specific heat (10^{-1} kJ/kg K)
600	0.047	4429	873.2	0.022	3.276	1480	75.17		172.1	1.301	1.80
700	0.951	4341	849.4	0.396	2.690	1660	70.53	0.0277	162.1	1.277	2.28
800	8.760	4237	825.6	3.270	2.298	1827	65.88	0.0343	152.1	1.260	2.59
900	48.760	4131	801.8	16.500	2.018	2010	61.25	0.0406	142.1	1.252	2.72
1000	192.200	4026	778.0	59.980	1.809	2211	56.60	0.0455	132.1	1.252	2.70
1100	584.280	3925	754.2	168.100	1.645	2398	51.96	0.0492	122.1	1.261	2.62
1200	1465.400	3829	730.4	396.600	1.514	2577	47.00	0.0522	112.1	1.279	2.51
1300	3165.000	3742	706.6	804.500	1.407	2763	42.50	0.0547	102.1	1.305	2.43
1400	6097.400	3656	682.8	1459.200	1.317	2938	37.50	0.0570	92.1	1.340	2.39
1500	10716.600	3577	658.0	2424.800	1.240	3117	33.00	0.0592	82.0	1.384	2.36
1600	17495.900	3500	635.2	3750.900	1.176	3281	28.50		72.0	1.437	2.34
1700	26919.900	3425	611.4	5482.400	1.117	3449	24.00		62.0	1.500	2.41
1800	39250.000	3353	587.6	7627.700	1.067	3620	19.00		52.0	1.574	2.46

[a]Vargaftik (1975)

Table B.48 Thermophysical properties at saturation for water

Water, H_2O, Molecular mass: 18.0, (T_{sat} = 100 °C; T_m = 0.0 °C; Lemmon et al. 2016)

T Temp. (°C)	p_v Saturation pressure (10^5 Pa)	$h_{\ell v}$ Latent heat (kJ/kg)	ρ_ℓ Liquid density (kg/m^3)	ρ_v Vapor density (kg/m^3)	μ_ℓ Liquid viscosity (10^{-7} N s/m^2)	μ_v Vapor viscosity (10^{-7} N s/m^2)	k_ℓ Liquid thermal conductivity (W/m K)	k_v Vapor thermal conductivity (W/m K)	σ Liquid surface tension (10^{-3} N/m)	$c_{p,\ell}$ Liquid specific heat (kJ/kg K)	$c_{p,v}$ Vapor specific heat (kJ/kg K)
20	0.023368	2453.8	999.0	0.01729	10016	97.3	0.598	0.0182	72.88	4.184	1.906
40	0.073749	2406.5	993.05	0.05110	6530	103.1	0.631	0.0196	69.48	4.180	1.931
60	0.199190	2358.4	983.28	0.13020	4664	109.4	0.654	0.0212	66.07	4.185	1.965
80	0.473590	2308.9	971.82	0.29320	3543	115.9	0.670	0.0230	62.69	4.197	2.012
100	1.013250	2251.2	958.77	0.59740	2817	122.7	0.679	0.0251	58.91	4.216	2.080
120	1.985400	2202.9	943.39	1.12100	2321	129.6	0.683	0.0275	54.96	4.244	2.177
140	3.613600	2144.9	925.93	1.96560	1965	136.5	0.683	0.0301	50.79	4.283	2.311
160	6.180400	2082.2	907.44	3.25890	1702	143.4	0.680	0.0331	46.51	4.335	2.488
180	10.02700	2014.0	887.31	5.15970	1501	150.3	0.673	0.0364	42.19	4.405	2.713
200	15.55100	1939.0	865.05	7.86530	1343	157.2	0.663	0.0401	37.77	4.496	2.990

Table B.49 Binary diffusion coefficients at 1 atm[a] (Bergman and Lavine 2017)

	Substance *A*	Substance *B*	T (K)	D_{AB} (m^2/s)
Gases	NH_3	Air	298	0.28×10^{-4}
	H_2O	Air	298	0.26×10^{-4}
	CO_2	Air	298	0.16×10^{-4}
	H_2	Air	298	0.41×10^{-4}
	O_2	Air	298	0.21×10^{-4}
	Acetone	Air	273	0.11×10^{-4}
	Benzene	Air	298	0.88×10^{-5}
	Naphthalene	Air	300	0.62×10^{-5}
	Ar	N_2	293	0.19×10^{-4}
	H_2	O_2	273	0.70×10^{-4}
	H_2	N_2	273	0.68×10^{-4}
	H_2	CO_2	273	0.55×10^{-4}
	CO_2	N_2	293	0.16×10^{-4}
	CO_2	O_2	273	0.14×10^{-4}
	O_2	N_2	273	0.18×10^{-4}
Dilute solutions	Caffeine	H_2O	298	0.63×10^{-9}
	Ethanol	H_2O	298	0.12×10^{-8}
	Glucose	H_2O	298	0.69×10^{-9}
	Glycerol	H_2O	298	0.94×10^{-9}
	Acetone	H_2O	298	0.13×10^{-8}
	CO_2	H_2O	298	0.20×10^{-8}
	O_2	H_2O	298	0.24×10^{-8}
	H_2	H_2O	298	0.63×10^{-8}
	N_2	H_2O	298	0.26×10^{-8}
Solids	O_2	Rubber	298	0.21×10^{-9}
	N_2	Rubber	298	0.15×10^{-9}
	CO_2	Rubber	298	0.11×10^{-9}
	He	SiO_2	293	0.4×10^{-13}
	H_2	Fe	293	0.26×10^{-12}
	Cd	Cu	293	0.27×10^{-18}
	Al	Cu	293	0.13×10^{-33}

[a]Assuming ideal gas behavior, the pressure and temperature dependence of the diffusion coefficient for a binary mixture of gases may be estimated form the relation $D_{AB} \propto p^{-1}\, T^{3/2}$

Table B.50 Diffusion coefficients in air at 1 atm (1.013×10^5 Pa)[a] (Mills and Coimbra 2015)

T [K]	Binary diffusion coefficient ($m^2/s \times 10^4$)							
	O_2	CO_2	CO	C_7H_{16}	H_2	NO	SO_2	He
200	0.095	0.074	0.098	0.036	0.375	0.088	0.058	0.363
300	0.188	0.157	0.202	0.075	0.777	0.180	0.126	0.713
400	0.325	0.263	0.332	0.128	1.25	0.303	0.214	1.14
500	0.475	0.385	0.485	0.194	1.71	0.443	0.326	1.66
600	0.646	0.537	0.659	0.270	2.44	0.603	0.440	2.26
700	0.838	0.684	0.854	0.354	3.17	0.782	0.576	2.91
800	1.05	0.857	1.06	0.442	3.93	0.978	0.724	3.64
900	1.26	1.05	1.28	0.538	4.77	1.18	0.887	4.42
1000	1.52	1.24	1.54	0.641	5.69	1.41	1.06	5.26
1200	2.06	1.69	2.09	0.881	7.77	1.92	1.44	7.12
1400	2.66	2.17	2.70	1.13	9.90	2.45	1.87	9.20
1600	3.32	2.75	3.37	1.41	12.5	3.04	2.34	11.5
1800	4.03	3.28	4.10	1.72	15.2	3.70	2.85	13.9
2000	4.80	3.94	4.87	2.06	18.0	4.48	3.36	16.6

[a]Owing to the practical importance of water vapor-air mixtures, engineers have used convenient empirical formulas for $D_{H_2O\,air}$. A formula that has been widely used is

$$D_{H_2O,air} = 1.97 \times 10^{-5}\left(\frac{p_0}{p}\right)\left(\frac{T}{T_0}\right)^{1.685} \text{ m}^2/\text{s}; \quad 273 \text{ K} < T < 373 \text{ K}$$

where $p_0 = 1$ atm; $T_0 = 256$ K. The following formula has also found increasing use (Marrero and Mason 1972)

$$D_{H_2O\,air} = 1.87 \times 10^{-10}\frac{T^{2.072}}{p}; \quad 280 \text{ K} < T < 450 \text{ K}$$

$$= 2.75 \times 10^{-9}\frac{T^{1.632}}{p}; \quad 450 \text{ K} < T < 1070 \text{ K}$$

for p in atmospheres and T in Kelvins. Over the temperature range 290–330 K, the discrepancy between the two formulas is less than 2.5%. For small concentrations of water vapor in air, the older formula gives a constant value of $Sc_{H_2O\,air} = 0.61$ over the temperature range 273–373 K. On the other hand, the Marrero and Mason (1972) formula give values of $Sc_{H_2O\,air}$ that vary from 0.63 at 280 K to 0.57 at 373 K

Table B.51 Diffusion coefficients in solids, $D = D_0 \exp(-E_a/R_u T)$ (Mills and Coimbra 2015)

System	D_0 m²/s	E_a^a (kJ/kmol)
Oxygen-Pyrex glass	6.19×10^{-8}	4.69×10^4
Oxygen-fused silica glass	2.61×10^{-9}	3.77×10^4
Oxygen-titanium	5.0×10^{-3}	2.13×10^5
Oxygen-titanium alloy (Ti-6Al-4 V)	5.82×10^{-2}	2.59×10^5
Oxygen-zirconium	4.68×10^{-5}	7.06×10^5
Hydrogen-iron	7.60×10^{-8}	5.60×10^3
Hydrogen-α-titanium	1.80×10^{-6}	5.18×10^4
Hydrogen-β-titanium	1.95×10^{-7}	2.78×10^4
Hydrogen-zirconium	1.09×10^{-7}	4.81×10^4
Hydrogen-Zircaloy-4	1.27×10^{-5}	6.05×10^5
Deuterium-Pyrex glass	6.19×10^{-8}	4.69×10^4
Deuterium-fused silica glass	2.61×10^{-9}	3.77×10^4

(continued)

Table B.51 (continued)

System	D_0 m^2/s	E_a^a (kJ/kmol)
Helium-Pyrex glass	4.76×10^{-8}	2.72×10^4
Helium-fused silica glass	5.29×10^{-8}	2.55×10^4
Helium-borosilicate	1.94×10^{-8}	2.34×10^4
Neon-borosilicate	1.02×10^{-10}	3.77×10^4
Carbon-FCC iron	2.3×10^{-5}	1.378×10^5
Carbon-BCC iron	1.1×10^{-6}	8.75×10^4

[a]Activation energy

Table B.52 Schmidt number for vapors in dilute mixture in air at normal temperature, enthalpy of vaporization and boiling point at 1 atm[a] (Mills and Coimbra 2015)

Vapor	Chemical formula	Sc[b]	$h_{\ell v}$ J/kg $\times 10^{-6}$	Boiling point temperature K
Acetone	CH_3COCH_3	1.42	0.527	329
Ammonia	NH_3	0.61	1.370	240
Benzene	C_6H_6	1.79	0.395	354
Carbon dioxide	CO_2	1.00	0.398	194
Carbon monoxide	CO	0.77	0.217	81
Chlorine	Cl_2	1.42	0.288	238
Ethanol	CH_3CH_2OH	1.32	0.854	352
Helium	He	0.22		4.3
Heptane	C_7H_{16}	2.0	0.340	372
Hydrogen	H_2	0.20	0.454	20.3
Hydrogen sulfide	H_2S	0.94	0.548	213
Methanol	CH_3OH	0.98	1.100	338
Napthalene[c]	$C_{10}H_8$	2.35	0.567	491
Nitric oxide	NO	0.87	0.465	121
Octane	C_8H_{18}	2.66	0.303	399
Oxygen	O_2	0.83	0.214	90.6
Pentane	C_5H_{12}	1.49	0.357	309
Sulfur dioxide	SO_2	1.24	0.398	263
Water vapor	H_2O	0.61	2.257	373

[a]With the Clausius–Clapeyron relation, one may estimate vapor pressure as

$p_{sat} \simeq \exp\left\{-\frac{Mh_{\ell v}}{R_u}\left(\frac{1}{T}-\frac{1}{T_{BP}}\right)\right\}$atm, for $T \sim T_{BP}$

[b]The Schmidt number is defined as $Sc = \mu/\rho D = \nu/D$. Since the vapors are in small concentrations, values for μ, ρ and ν can be taken as pure air values

[c]Cho et al. (1992); $h_{\ell v} = 0.567 \times 10^6$ J/K is at 300 K

Table B.53 Schmidt numbers for dilute solution in water at 300 K[a] (Mills and Coimbra 2015)

Solute	Schmidt number, Sc	Molecular mass, M (kg/kmol)
Helium	120	4.003
Hydrogen	190	2.016
Nitrogen	280	28.02
Water	340	18.016
Nitric Oxide	350	30.01
Carbon monoxide	360	28.01
Oxygen	400	32.00
Ammonia	410	17.03
Carbon dioxide	420	44.01
Hydrogen sulfide	430	34.08
Ethylene	450	28.05
Methane	490	16.04
Nitrous oxide	490	44.02
Sulfur dioxide	520	64.06
Sodium chloride	540	58.45
Sodium hydroxide	490	40.00
Acetic acid	620	60.05
Acetone	630	58.08
Methanol	640	32.04
Ethanol	640	46.07
Chlorine	670	70.90
Benzene	720	78.11
Ethylene glycol	720	62.07
n-Propanol	730	60.09
i-Propanol	730	60.09
Propane	750	44.09
Aniline	800	93.13
Benzoic acid	830	122.12
Glycerol	1040	92.09
Sucrose	1670	342.3

[a]For other temperatures use $Sc/Sc_{300\,K} \simeq (\mu^2/\rho T)/(\mu^2/\rho T)_{300\,K}$, where μ and ρ are for water, and T is absolute temperature. For chemically similar solutes of different molecular weights use $Sc_2/Sc_1 \simeq (M_2/M_1)^{0.4}$. A table of $(\mu^2/\rho T)/(\mu^2/\rho T)_{300\,K}$ for water follows

T [K]	$(\mu^2/\rho T)/(\mu^2/\rho T)_{300\,K}$	T [K]	$(\mu^2/\rho T)/(\mu^2/\rho T)_{300\,K}$
290	1.66	340	0.221
300	1.00	350	0.167
310	0.623	360	0.123
320	0.429	370	0.097
330	0.296		

Spalding (1963)

Table B.54 Solubility and permeability of gases in solids (Mills and Coimbra 2015)

Gas	Solid	T (K)	S' [m^3 (STP)/m^3 atm] or S''[a]	Permeability[b] m^3(STP)/m^2s (atm/m)
H_2	Vulcanized rubber	300	$S' = 0.040$	0.34×10^{-10}
	Vulcanized neoprene	290	$S' = 0.051$	0.053×10^{-10}
	Silicone rubber	300		4.2×10^{-10}
	Natural rubber	300		0.37×10^{-10}
	Polyethylene	300		0.065×10^{-10}
	Polycarbonate	300		0.091×10^{-10}
	Fused silica	400	$S'' \cong 0.035$	
	Nickel	800	$S'' \cong 0.030$	
		360	$S'' \cong 0.202$	
		440	$S'' \cong 0.192$	
He	Silicone rubber	300		2.3×10^{-10}
	Natural rubber	300		0.24×10^{-10}
	Polycarbonate	300		0.11×10^{-10}
	Nylon 66	300		0.0076×10^{-10}
	Teflon	300		0.047×10^{-10}
	Fused silica	300	$S'' \cong 0.018$	
		800	$S'' \cong 0.026$	
	Pyrex glass	300	$S'' \cong 0.006$	
		800	$S'' \cong 0.024$	
	7740 glass (94% SiO_2 + B_2O_3 + P_2O_5 5% Na_2O + Li_2 + K_2O 1% other oxides)	470	$S' = 0.0084$	4.6×10^{-13}
		580	$S' = 0.0038$	1.6×10^{-12}
		720	$S' = 0.0046$	6.4×10^{-12}
	7056 glass (90% SiO_2 + B_2O_3 + P_2O_5 8% Na_2O + Li_2 + K_2O 1% PbO, 0.5% other oxides)	390	$S' = 0.0039$	1.2×10^{-14}
		680	$S' = 0.0059$	1.0×10^{-12}
O_2	Vulcanized rubber	300	$S' = 0.070$	0.15×10^{-10}
	Silicone rubber	300		3.8×10^{-10}
	Natural rubber	300		0.18×10^{-10}
	Polyethylene	300		4.2×10^{-12}
	Polycarbonate	300		0.011×10^{-10}
	Silicone-polycarbonate copolymer (57% silicone)	300		1.2×10^{-10}
	Ethyl cellulose	300		0.09×10^{-10}
N_2	Vulcanized rubber	300	$S' = 0.035$	0.054×10^{-10}
	Silicone rubber	300		1.9×10^{-10}
	Natural rubber	300		0.062×10^{-10}
	Silicone-polycarbonate copolymer (57% silicone)	300		0.53×10^{-10}
	Teflon®	300		0.019×10^{-10}

(continued)

Table B.54 (continued)

Gas	Solid	*T* (K)	S' [m^3 (STP)/m^3 atm] or S'^{a}	Permeability[b] m^3(STP)/m^{2}s (atm/m)
CO_2	Vulcanized rubber	300	$S' = 0.90$	1.0×10^{-10}
	Silicone rubber	300		21×10^{-10}
	Natural rubber	300		1.0×10^{-10}
	Silicone-polycarbonate copolymer (57% silicone)	300		7.4×10^{-10}
	Nylon 66	300		0.0013×10^{-10}
H_2O	Cellophane	310		$0.91–1.8 \times 10^{-10}$
Ne	Fused silica	300–1200	$S'' \cong 0.002$	
Ar	Fused silica	900–1200	$S'' \cong 0.01$	

[a]Solubility S' = Volume of solute gas (0 °C, 1 atm) dissolved in unit volume of solid when the gas is at 1 atm partial pressure. Solubility coefficient $S'' = c_{1,g}/c_2$
[b]Permeability $K = D_{AB}S'$
From various sources, including Geankoplis (1993), Doremus (1973) and Altemose (1961)

Table B.55 Henry's constant for selected gases in water at moderate pressure[a]

$H = p_{A,i}/x_{A,i}$ (bars)

T (K)	NH_3	Cl_2	H_2S	SO_2	CO_2	CH_4	O_2	H_2
273	21	265	260	165	710	22,880	25,500	58,000
280	23	365	335	210	960	27,800	30,500	61,500
290	26	480	450	315	1300	35,200	37,600	66,500
300	30	615	570	440	1730	42,800	45,700	71,600
310	–	755	700	600	2175	50,000	52,500	76,000
320	–	860	835	800	2650	56,300	56,800	78,600
323	–	890	870	850	2870	58,000	58,000	79,000

[a]Bergman and Lavine (2017) and Spalding (1963)

Table B.56 Solubility of selected gases and solids (Bergman and Lavine 2017)

Gas	Solid	T (K)	$S = c_{A,s}/p_{A,g}$ (k mol/m^3 bar)
O_2	Rubber	298	3.12×10^{-3}
N_2	Rubber	298	1.56×10^{-3}
CO_2	Rubber	298	40.15×10^{-3}
He	SiO_2	293	0.45×10^{-3}
H_2	Ni	358	9.01×10^{-3}

Table B.57 Solubility of inorganic compounds in water[a] (Mills and Coimbra 2015)

Solute	Formula	Solid Phase	T (K)											
			273.15	280	290	300	310	320	330	340	350	360	370	373.15
Aluminum sulfate	$Al_2(SO_4)_3$	$18H_2O$	31.2	32.8	35.5	39.1	44.3	50.3	57.0	63.9	70.8	78.3	84.6	89.0
Calcium bicarbonate	$Ca(HCO_3)_2$	–	16.15	16.30	16.53	16.75	16.98	17.20	17.43	17.65	17.88	18.10	18.33	18.40
Calcium chloride	$CaCl_2$	$6H_2O$	59.5	63.3	71.5	93.3	137.2	–	–	–	–	–	–	–
	$CaCl_2$	$2H_2O$	–	–	–	–	–	–	134.6	140.2	145.3	150.9	157.0	159.0
Calcium hydroxide	$Ca(OH)_2$	–	0.185	0.179	0.168	0.157	0.145	0.132	0.120	0.109	0.098	0.088	0.080	0.077
Potassium chloride	KCl	–	27.6	29.9	33.1	36.1	39.1	41.8	44.6	47.4	50.2	53.1	55.8	56.7
Potassium nitrate	KNO_3	–	13.3	18.5	28.2	41.3	58.2	78.7	102.3	129.2	159.2	191.6	232.1	246.0
Potassium sulfate	K_2SO_4	–	7.35	8.63	10.51	12.38	14.20	15.95	17.64	19.25	20.88	22.36	23.69	24.1
Sodium bicarbonate	$NaHCO_3$	–	6.9	7.76	9.14	10.63	12.20	13.90	15.79	–	–	–	–	–
Sodium carbonate	Na_2O_3	$10H_2O$	7	10.8	18.7	33.4	–	–	–	–	–	–	–	–
	Na_2CO_3	$1H_2O$	–	–	–	–	49.1	47.8	46.7	46.2	45.9	45.7	45.55	45.5
Sodium chloride	NaCl	–	35.7	35.8	35.9	36.2	36.5	36.9	37.2	37.6	38.2	38.8	39.5	39.8
Sodium nitrate	$NaNO_2$	–	73	78	85	93	101	111	121	132	144	159	175	180
Sodium sulfate	Na_2SO_4	$10H_2O$	5.0	7.7	16.1	34.1	–	–	–	–	–	–	–	–
	Na_2SO_4	$7H_2O$	19.5	26.7	39.6	–	–	–	–	–	–	–	–	–
	Na_2SO_4	–	–	–	–	–	49.6	47.4	45.7	43.7	44.0	43.3	42.7	42.5

[a]Solubility expressed in kilograms of anhydrous substance that is soluble in 100 kg water

Table B.58 Equilibrium compositions for the NH_3-water system (Mills and Coimbra 2015)

$p_{A,g}$ (atm)	$x_{A,\ell}$				
	290 K	300 K	310 K	320 K	330 K
0.02	0.030	0.019	0.012	0.008	0.006
0.04	0.056	0.036	0.024	0.016	0.012
0.06	0.078	0.052	0.035	0.024	0.017
0.08	0.096	0.064	0.046	0.032	0.023
0.1	0.11	0.079	0.056	0.040	0.029
0.2	0.18	0.14	0.099	0.057	0.052
0.4	0.26	0.21	0.16	0.12	0.092
0.6	0.31	0.26	0.20	0.16	0.13
0.8	0.35	0.29	0.23	0.19	0.15
1.0	–	0.32	0.27	0.22	0.17

Table B.59 Equilibrium compositions for the SO_2-water system[a] (Mills and Coimbra 2015)

$p_{A,g}$ (atm)	$x_{A,\ell} \times 10^3$			
	290 K	300 K	310 K	320 K
0.001	0.12	0.084	0.059	0.042
0.003	0.25	0.18	0.13	0.093
0.01	0.62	0.42	0.31	0.22
0.03	1.4	1.1	0.73	0.51
0.1	4.1	2.9	2.0	1.4
0.3	11.0	7.9	5.6	3.9
1.0	33.0	24.0	18.0	12.0

[a]Notice that Henry's law is invalid for the SO_2-water system, even at very dilute concentrations

Table B.60 Thermodynamic properties of water vapor-air mixtures at 1 atm (Mills and Coimbra 2015)

Temp. (°C)	Saturation mass fraction	Specific volume (m^3/kg)		Enthalpy[a, b] (KJ/kg)		
		Dry air	Saturated air	Liquid water	Dry air	Saturated air
10	0.007608	0.8018	0.8054	42.13	10.059	29.145
11	0.008136	0.8046	0.8086	46.32	11.065	31.481
12	0.008696	0.8075	0.8117	50.52	12.071	33.898
13	0.009289	0.8103	0.8148	54.71	13.077	36.401
14	0.009918	0.8131	0.8180	58.90	14.083	38.995
15	0.01058	0.8160	0.8212	63.08	15.089	41.684
16	0.01129	0.8188	0.8244	67.27	16.095	44.473
17	0.01204	0.8217	0.8276	71.45	17.101	47.367
18	0.01283	0.8245	0.8309	75.64	18.107	50.372
19	0.01366	0.8273	0.8341	79.82	19.113	53.493
20	0.01455	0.8302	0.8374	83.99	20.120	56.736
21	0.01548	0.8330	0.8408	88.17	21.128	60.107
22	0.01647	0.8359	0.8441	92.35	22.134	63.612
23	0.01751	0.8387	0.8475	96.53	23.140	67.259

(continued)

Table B.60 (continued)

Temp. (°C)	Saturation mass fraction	Specific volume (m^3/kg)		Enthalpy[a, b] (KJ/kg)		
		Dry air	Saturated air	Liquid water	Dry air	Saturated air
24	0.01861	0.8415	0.8510	100.71	24.147	71.054
25	0.01978	0.8444	0.8544	104.89	25.153	75.004
26	0.02100	0.8472	0.8579	109.07	26.159	79.116
27	0.02229	0.8500	0.8615	113.25	27.166	83.400
28	0.02366	0.8529	0.8650	117.43	28.172	87.862
29	0.02509	0.8557	0.8686	121.61	29.178	92.511
30	0.02660	0.8586	0.8723	125.79	30.185	97.357
31	0.02820	0.8614	0.8760	129.97	31.191	102.408
32	0.02987	0.8642	0.8798	134.15	32.198	107.674
33	0.03164	0.8671	0.8836	138.32	33.204	113.166
34	0.03350	0.8699	0.8874	142.50	34.211	118.893
35	0.03545	0.8728	0.8914	146.68	35.218	124.868
36	0.03751	0.8756	0.8953	150.86	36.224	131.100
37	0.03967	0.8784	0.8994	155.04	37.231	137.604
38	0.04194	0.8813	0.9035	159.22	38.238	144.389
39	0.04432	0.8841	0.9077	163.40	39.245	151.471
40	0.04683	0.8870	0.9119	167.58	40.252	158.862
41	0.04946	0.8898	0.9162	171.76	41.259	166.577
42	0.05222	0.8926	0.9206	175.94	42.266	174.630
43	0.05512	0.8955	0.9251	180.12	43.273	183.037
44	0.05817	0.8983	0.9297	184.29	44.280	191.815
45	0.06137	0.9012	0.9343	188.47	45.287	200.980
46	0.06472	0.9040	0.9391	192.65	46.294	210.550
47	0.06842	0.9068	0.9439	196.83	47.301	220.543
48	0.07193	0.9097	0.9489	201.01	48.308	230.980
49	0.07580	0.9125	0.9539	205.19	49.316	241.881

[a]The enthalpies of dry air and liquid water are set equal to zero at a datum temperature of 0 °C
[b]The enthalpy of an unsaturated water vapor-air mixture can be calculated as $h = h_{\text{dryair}} + (m_1/m_{1,\text{sat}})(h_{\text{sat}} - h_{\text{dryair}})$

References

Altemose, V. O. (1961). Helium diffusion through glass. *Journal of Applied Physics, 32*, 1309–1316.

American Society of Heating, Refrigerating and Air Conditioning Engineers (ASHRAE). (2009). *ASHRAE handbook of fundamentals*. New York, NY: ASHRAE.

Argonne National Laboratory (ANL). (1995). *Thermodynamic and transport properties of the sodium liquid and vapor*. ANL/RE-95/2, https://www.ne.anl.gov/eda/ANL-RE-95-2.pdf. Accessed on September 17, 2018.

Bejan, A. (2013). *Convection heat transfer* (4th ed.). New York, NY: Wiley.

Bennon, W. D., & Incropera, F. P. (1988). Developing laminar mixed convection with solidification in a vertical channel. *Journal of Heat Transfer, 110*, 410–415.

Bergman, T. L., & Lavine, A. S. (2017). *Fundamentals of heat and mass transfer* (8th ed.). Hoboken, NJ: Wiley.

Brennan, P. J., & Kroliczek, E. J. (1979). *Heat pipe design handbook* (Vol. 2). Towson, MD: NASA Goddard by B&K Engineering, Inc., Suite 825, One Investment Place.

Cengel, Y. A., Boles, M. A., & Kanoglu, M. (2019). *Thermodynamics—An engineering approach* (9th ed.). New York, NY: McGraw-Hill.

Cho, C., Irvine, T. F., Jr., & Karni, J. (1992). Measurement of the diffusion coefficient of naphthalene into air. *International Journal of Heat and Mass Transfer, 35*, 957–966.

Doremus, R. H. (1973). *Glass science*. New York: Wiley.
Faghri, A. (2016). *Heat pipe science and technology* (2nd ed.). Columbia, MO: Global Digital Press.
Geankoplis, C. J. (1993). *Transport processes and unit operations* (3rd ed.). Englewood Cliffs, NJ: Prentice-Hall.
Hale, D. V., Hoovers, M. J., & O'Nell, M. J. (1971). *Phase change materials handbook*. NASA-CR-61363.
Humphries, W. R., & Griggs, E. I. (1977). *A design handbook for phase change thermal control and energy storage devices*. NASA-TP-1074.
Iida, T., & Guthrie, R. I. L. (1988). *The physical properties of liquid metals*. Oxford, UK: Oxford University Press.
Ivanovskii, M.N., Sorokin, V.P., and Yagodkin, I.V., 1982, *The Physical Principles of Heat Pipes*, Clarendon Press, Oxford, UK.
Kakac, S., Shah, R. K., & Aung, W. (Eds.). (1987). *Handbook of single-phase convective heat transfer*. New York, NY: Wiley.
Lemmon, E. W., McLinden, M. O., & Friend, D. G. (2016). Thermophysical properties of fluid systems. In P. I. Linstrom & W. G. Mallard (Eds.), *NIST chemistry WebBook, NIST Standard Reference Database Number 69*. Gaithersburg, MD: National Institute of Standards and Technology, 20899. http://webbook.nist.gov.
Marrero, T. R., & Mason, E. A. (1972). Gaseous diffusion coefficients. *Journal of Physical and Chemical Reference Data, 1*, 3–118.
Mills, A. F., & Coimbra, C. F. M. (2015). *Basic heat and mass transfer* (3rd ed.). San Diego, CA: Temporal Publishing, LLC.
Reay, D. A., Kew, P. A., & McGlen, R. J. (2014). *Heat pipes—Theory, design, and applications* (6th ed.). New York: Elsevier.
Rohsenow, W. N., Hartnett, J. P., & Ganic, E. N. (Eds.). (1998). *Handbook of heat transfer fundamentals* (3rd ed.). New York, NY: McGraw-Hill.
Spalding, D. B. (1963). *Convective mass transfer*. New York, NY: McGraw-Hill.
Touloukian, Y. S., Liley, P. E., & Saxena, S. C., (Eds.). (1970). *Thermophysical properties of matter* (Vol. 3). New York, NY: Plenum.
Vargaftik, N. B. (1975). *Handbook of physical properties of liquids and gases*. New York, NY: Hemisphere.

Appendix C: Vectors and Tensors

C.1 Vectors

The term scalar refers to a single real number used to describe the magnitude of a quantity. Pressure, temperature, and internal energy are all scalar. A vector is defined as an entity that possesses both magnitude and direction or as a directed line segment subject to the parallelogram law of addition. In three-dimensional space, a vector may be specified by three vector components. A unit vector is a vector whose magnitude is unity. A vector in a three-dimensional Cartesian coordinate system can be expressed as

$$\mathbf{A} = \mathbf{i}A_x + \mathbf{j}A_y + \mathbf{k}A_z \tag{C.1}$$

where **i**, **j**, and **k** are unit vectors in the x-, y-, and z-directions. The vector components in the x-, y-, and z-directions are A_x, A_y, and A_z, respectively. A vector in a Cartesian coordinate system is shown in Fig. C.1. It can be seen that the magnitude of the vector is

$$|\mathbf{A}| = \sqrt{A_x^2 + A_y^2 + A_z^2} \tag{C.2}$$

and that its three components are the projections of the vector on the x, y, and z-axes.

A vector can also be represented in matrix form:

$$\mathbf{A} = \begin{bmatrix} A_x \\ A_y \\ A_z \end{bmatrix} \tag{C.3}$$

The dot product (also referred to as the scalar or inner product) of two vectors **A** and **B** is a scalar; it is obtained by summation of the product of each of their corresponding components, i.e.,

$$\mathbf{A} \cdot \mathbf{B} = A_x B_x + A_y B_y + A_z B_z \tag{C.4}$$

The cross product (or vector product) of two vectors **A** and **B** is a vector, i.e.,

$$\begin{aligned} \mathbf{A} \times \mathbf{B} &= \begin{vmatrix} \mathbf{i} & \mathbf{j} & \mathbf{k} \\ A_x & A_y & A_z \\ B_x & B_y & B_z \end{vmatrix} \\ &= \mathbf{i}(A_y B_z - A_z B_y) + \mathbf{j}(A_z B_x - A_x B_z) + \mathbf{k}(A_x B_y - A_y B_x) \end{aligned} \tag{C.5}$$

A. Faghri and Y. Zhang, *Fundamentals of Multiphase Heat Transfer and Flow*, https://doi.org/10.1007/978-3-030-22137-9

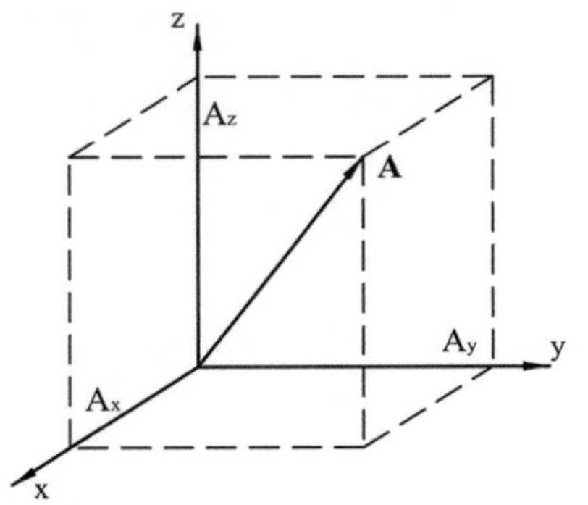

Figure C.1 Vector and its components

C.2 Operations with the ∇ Operator

C.2.1 Cartesian Coordinate System

An important vector for fluid mechanics and heat transfer is the ∇ operator (pronounced Del or nabla), which is defined as

$$\nabla = \mathbf{i}\frac{\partial}{\partial x} + \mathbf{j}\frac{\partial}{\partial y} + \mathbf{k}\frac{\partial}{\partial z} \tag{C.6}$$

in a three-dimensional Cartesian coordinate system. It can be applied to a scalar function, ϕ, to obtain its gradient,

$$\text{grad}\,\phi = \nabla\phi = \mathbf{i}\frac{\partial \phi}{\partial x} + \mathbf{j}\frac{\partial \phi}{\partial y} + \mathbf{k}\frac{\partial \phi}{\partial z} \tag{C.7}$$

which is a vector.

The ∇ operator can also be applied to a vector function such as velocity,

$$\mathbf{V} = \mathbf{i}u + \mathbf{j}v + \mathbf{k}w \tag{C.8}$$

to get its divergence:

$$\text{div}\mathbf{V} = \nabla \cdot \mathbf{V} = \frac{\partial u}{\partial x} + \frac{\partial v}{\partial y} + \frac{\partial w}{\partial z} \tag{C.9}$$

or its curl:

$$\text{curl}\mathbf{V} = \nabla \times \mathbf{V} = \mathbf{i}\left(\frac{\partial w}{\partial y} - \frac{\partial v}{\partial z}\right) + \mathbf{j}\left(\frac{\partial u}{\partial z} - \frac{\partial w}{\partial x}\right) + \mathbf{k}\left(\frac{\partial v}{\partial x} - \frac{\partial u}{\partial y}\right) \tag{C.10}$$

Another related operator that is very useful in fluid mechanics and heat transfer is Laplacian operator, defined as

$$\nabla^2 = \nabla \cdot \nabla = \frac{\partial^2}{\partial x^2} + \frac{\partial^2}{\partial y^2} + \frac{\partial^2}{\partial z^2} \tag{C.11}$$

Application of the Laplace operator to a scalar function ϕ results in

$$\nabla^2 \phi = \frac{\partial^2 \phi}{\partial x^2} + \frac{\partial^2 \phi}{\partial y^2} + \frac{\partial^2 \phi}{\partial z^2} \tag{C.12}$$

Forming the dot product of the velocity and the gradient of a scalar ϕ results in a scalar:

$$\mathbf{V} \cdot \nabla \phi = u \frac{\partial \phi}{\partial x} + v \frac{\partial \phi}{\partial y} + w \frac{\partial \phi}{\partial z} \tag{C.13}$$

which is used to describe the advection of the property ϕ.

The operation

$$\nabla \cdot (\alpha \nabla \phi) = \frac{\partial}{\partial x}\left(\alpha \frac{\partial \phi}{\partial x}\right) + \frac{\partial}{\partial y}\left(\alpha \frac{\partial \phi}{\partial y}\right) + \frac{\partial}{\partial z}\left(\alpha \frac{\partial \phi}{\partial z}\right) \tag{C.14}$$

results in a scalar that describe the diffusion of the property ϕ.

Application of the Laplace operator to a velocity vector, $\mathbf{V}$, results in a vector:

$$\nabla^2 \mathbf{V} = \mathbf{i} \nabla^2 u + \mathbf{j} \nabla^2 v + \mathbf{k} \nabla^2 w \tag{C.15}$$

C.2.2 Cylindrical Coordinate System

The cylindrical coordinate system (r, φ, z) shown in Fig. C.2 is related to Cartesian coordinates (x, y, z) by

$$x = r \cos \varphi \quad y = r \sin \varphi \quad z = z \tag{C.16}$$

The velocity vector, $\mathbf{V}$, in a cylindrical coordinate system has three components, i.e.,

$$\mathbf{V} = \mathbf{k}_r V_r + \mathbf{k}_\varphi V_\varphi + \mathbf{k}_z V_z \tag{C.17}$$

where $\mathbf{k}_r, \mathbf{k}_\varphi$ and $\mathbf{k}_z$ are the unit vectors in r-, φ-, and z-directions, respectively.

The operations involving the ∇ operator and either the general scalar function ϕ or the vector $\mathbf{V}$ in a cylindrical coordinate system is summarized below:

$$\text{grad}\, \phi = \nabla \phi = \mathbf{k}_r \frac{\partial \phi}{\partial r} + \mathbf{k}_\varphi \frac{1}{r} \frac{\partial \phi}{\partial \varphi} + \mathbf{k}_z \frac{\partial \phi}{\partial z} \tag{C.18}$$

$$\text{div}\mathbf{V} = \nabla \cdot \mathbf{V} = \frac{1}{r} \frac{\partial (r V_r)}{\partial r} + \frac{1}{r} \frac{\partial V_\varphi}{\partial \varphi} + \frac{\partial V_z}{\partial z} \tag{C.19}$$

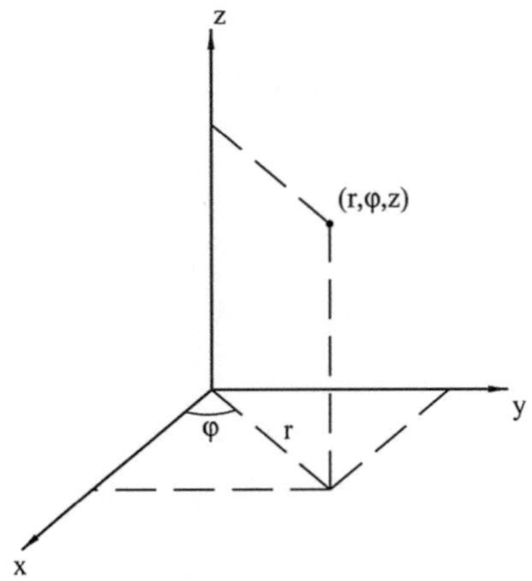

Figure C.2 Relationship between cylindrical and Cartesian coordinates

$$\nabla^2\phi = \frac{1}{r}\frac{\partial}{\partial r}\left(r\frac{\partial \phi}{\partial r}\right) + \frac{1}{r^2}\frac{\partial^2 \phi}{\partial \varphi^2} + \frac{\partial^2 \phi}{\partial z^2} \tag{C.20}$$

$$\mathbf{V}\cdot\nabla\phi = V_r\frac{\partial \phi}{\partial r} + \frac{V_\varphi}{r}\frac{\partial \phi}{\partial \varphi} + V_z\frac{\partial \phi}{\partial z} \tag{C.21}$$

$$\nabla\cdot\alpha\nabla\phi = \frac{1}{r}\frac{\partial}{\partial r}\left(r\alpha\frac{\partial \phi}{\partial r}\right) + \frac{1}{r}\frac{\partial}{\partial \varphi}\left(\frac{\alpha}{r}\frac{\partial \phi}{\partial \varphi}\right) + \frac{\partial}{\partial z}\left(\alpha\frac{\partial \phi}{\partial z}\right) \tag{C.22}$$

$$\begin{aligned}\nabla^2\mathbf{V} = {} & \mathbf{k}_r\left\{\frac{\partial}{\partial r}\left[\frac{1}{r}\frac{\partial}{\partial r}(rV_r)\right] + \frac{1}{r^2}\frac{\partial^2 V_r}{\partial \varphi^2} - \frac{2}{r^2}\frac{\partial V_\varphi}{\partial \varphi} + \frac{\partial^2 V_r}{\partial z^2}\right\} \\ & + \mathbf{k}_\varphi\left\{\frac{\partial}{\partial r}\left[\frac{1}{r}\frac{\partial}{\partial r}(rV_\varphi)\right] + \frac{1}{r^2}\frac{\partial^2 V_\varphi}{\partial \varphi^2} + \frac{2}{r^2}\frac{\partial V_r}{\partial \varphi} + \frac{\partial^2 V_\varphi}{\partial z^2}\right\} \\ & + \mathbf{k}_z\left\{\frac{1}{r}\frac{\partial}{\partial r}\left(r\frac{\partial V_z}{\partial r}\right) + \frac{1}{r^2}\frac{\partial^2 V_z}{\partial \varphi^2} + \frac{\partial^2 V_z}{\partial z^2}\right\}\end{aligned} \tag{C.23}$$

C.2.3 Spherical Coordinate System

The spherical coordinate system (r, θ, φ) shown in Fig. C.3 is related to the Cartesian coordinate system by

$$x = r\sin\theta\cos\varphi \quad y = r\sin\theta\sin\varphi \quad z = r\cos\theta \tag{C.24}$$

A velocity vector, $\mathbf{V}$, has three components, i.e.,

$$\mathbf{V} = \mathbf{k}_r V_r + \mathbf{k}_\theta V_\theta + \mathbf{k}_\varphi V_\varphi \tag{C.25}$$

where $\mathbf{k}_r, \mathbf{k}_\theta,$ and $\mathbf{k}_\varphi$ are the unit vectors in r, θ, and φ directions, respectively.

The operations involving the ∇ operator and either the general scalar function ϕ or the vector $\mathbf{V}$ in a spherical coordinate system are summarized below:

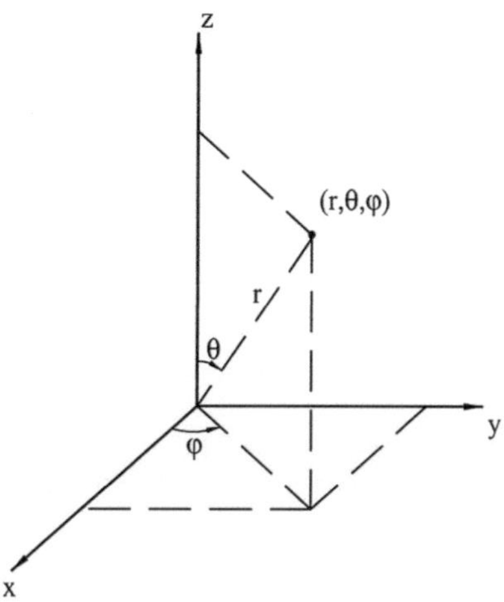

Figure C.3 Relationship between spherical and Cartesian coordinates

$$\operatorname{grad}\phi = \nabla\phi = \mathbf{k}_r\frac{\partial\phi}{\partial r} + \mathbf{k}_\theta\frac{1}{r}\frac{\partial\phi}{\partial\theta} + \mathbf{k}_\varphi\frac{1}{r\sin\theta}\frac{\partial\phi}{\partial\varphi} \tag{C.26}$$

$$\operatorname{div}\mathbf{V} = \nabla\cdot\mathbf{V} = \frac{1}{r^2}\frac{\partial(r^2V_r)}{\partial r} + \frac{1}{r\sin\theta}\frac{\partial(V_\theta\sin\theta)}{\partial\theta} + \frac{1}{r\sin\theta}\frac{\partial V_\varphi}{\partial\varphi} \tag{C.27}$$

$$\nabla^2\phi = \frac{1}{r^2}\frac{\partial}{\partial r}\left(r^2\frac{\partial\phi}{\partial r}\right) + \frac{1}{r^2\sin\theta}\frac{\partial}{\partial\theta}\left(\sin\theta\frac{\partial\phi}{\partial\theta}\right) + \frac{1}{r^2\sin^2\theta}\frac{\partial^2\phi}{\partial\varphi^2} \tag{C.28}$$

$$\mathbf{V}\cdot\nabla\phi = V_r\frac{\partial\phi}{\partial r} + \frac{V_\theta}{r}\frac{\partial\phi}{\partial\theta} + \frac{V_\varphi}{r\sin\theta}\frac{\partial\phi}{\partial\varphi} \tag{C.29}$$

$$\begin{aligned}\nabla\cdot(\alpha\nabla\phi) &= \frac{1}{r^2}\frac{\partial}{\partial r}\left(r^2\alpha\frac{\partial\phi}{\partial r}\right) + \frac{1}{r^2\sin\theta}\frac{\partial}{\partial\theta}\left[(\sin\theta)\alpha\frac{\partial\phi}{\partial\theta}\right]\\ &\quad + \frac{1}{r^2\sin^2\theta}\frac{\partial}{\partial\varphi}\left(\alpha\frac{\partial\phi}{\partial\varphi}\right)\end{aligned} \tag{C.30}$$

$$\begin{aligned}\nabla^2\mathbf{V} = \mathbf{k}_r&\left\{\nabla^2V_r - \frac{2V_r}{r^2} - \frac{2}{r^2}\frac{\partial V_\theta}{\partial\theta} - \frac{2V_\theta\cot\theta}{r^2} - \frac{2}{r^2\sin\theta}\frac{\partial V_\varphi}{\partial\varphi}\right\}\\ +\mathbf{k}_\theta&\left\{\nabla^2V_\theta + \frac{2}{r^2}\frac{\partial V_r}{\partial\theta} - \frac{V_\theta}{r^2\sin\theta} - \frac{2\cos\theta}{r^2\sin^2\theta}\frac{\partial V_\varphi}{\partial\varphi}\right\}\\ +\mathbf{k}_z&\left\{\nabla^2V_\phi - \frac{V_\varphi}{r^2\sin^2\theta} + \frac{2}{r^2\sin\theta}\frac{\partial V_r}{\partial\varphi} + \frac{2\cos\theta}{r^2\sin^2\theta}\frac{\partial V_\theta}{\partial\varphi}\right\}\end{aligned} \tag{C.31}$$

C.3 Tensors

A tensor of rank n in the Cartesian coordinate system has 3^n components. A scalar can be considered as a tensor of rank 0 because it has only one component. A vector is a tensor of rank 1 since it has three components. As was demonstrated in Sect. 1.3.2, the normal and shear stresses in a fluid can be described by a tensor of rank 2. The defining characteristics of a tensor are the manner in which its components transform under a rotation of the coordinate system where the components are defined. The transformation law for the components of tensor of a rank two is given as

$$\tau'_{ij} = \alpha_{ik}\alpha_{jl}\tau_{kl} \tag{C.32}$$

where the prime denotes the tensor components in the rotated coordinate, and α_{ik} and α_{jl} represent, respectively, the cosines of the angles between the ith rotated axis and the kth original axis, and between the jth rotated axis and the lth original axis. It should be noted that it is *not* the tensor itself that transforms under this change in the reference coordinate system but, rather, the coordinates that describe the tensor.

Application of the ∇ operator to each component of the velocity vector $\mathbf{V} = \mathbf{i}u + \mathbf{j}v + \mathbf{k}w$ also yields a tensor of rank two:

$$\nabla\mathbf{V} = \begin{bmatrix} \frac{\partial u}{\partial x} & \frac{\partial u}{\partial y} & \frac{\partial u}{\partial z} \\ \frac{\partial v}{\partial x} & \frac{\partial v}{\partial y} & \frac{\partial v}{\partial z} \\ \frac{\partial w}{\partial x} & \frac{\partial w}{\partial y} & \frac{\partial w}{\partial z} \end{bmatrix} \tag{C.33}$$

which is a dyadic product of two vectors ∇ and $\mathbf{V}$, and it can be used to determine the strain rate tensor.

The dot product of a vector and a tensor of rank 2 is a vector. For example, the dot product of the ∇ operator and a stress tensor is

$$\nabla\cdot\boldsymbol{\tau} = \boldsymbol{\tau}\cdot\nabla = \begin{bmatrix} \frac{\partial}{\partial x} \\ \frac{\partial}{\partial y} \\ \frac{\partial}{\partial z} \end{bmatrix} \begin{bmatrix} \tau_{xx} & \tau_{xy} & \tau_{xz} \\ \tau_{yx} & \tau_{yy} & \tau_{yz} \\ \tau_{zx} & \tau_{zy} & \tau_{zz} \end{bmatrix} = \begin{bmatrix} \frac{\partial \tau_{xx}}{\partial x} + \frac{\partial \tau_{xy}}{\partial y} + \frac{\partial \tau_{xz}}{\partial z} \\ \frac{\partial \tau_{yx}}{\partial x} + \frac{\partial \tau_{yy}}{\partial y} + \frac{\partial \tau_{yz}}{\partial z} \\ \frac{\partial \tau_{zx}}{\partial x} + \frac{\partial \tau_{zy}}{\partial y} + \frac{\partial \tau_{zz}}{\partial z} \end{bmatrix} \tag{C.34}$$

where $\nabla\cdot\boldsymbol{\tau} = \boldsymbol{\tau}\cdot\nabla$ is valid because the stress tensor is a symmetric tensor. For the case that the tensor is not symmetric, the order of vector and tensor cannot be switched in their dot product.

The contraction of two tensors of rank two **a** and **b** is obtained by summing the products of the corresponding components from both tensors:

$$\begin{aligned} \mathrm{a{:}b} = {} & a_{xx}b_{xx} + a_{xy}b_{xy} + a_{xz}b_{xz} + a_{yx}b_{yx} + a_{yy}b_{yy} + a_{yz}b_{yz} \\ & + a_{zx}b_{zx} + a_{zy}b_{zy} + a_{zz}b_{zz} \end{aligned} \tag{C.35}$$

which can also be written as

$$\mathrm{a{:}b} = a_{ij}b_{ij} \tag{C.36}$$

using the summation convention of tensors. According to the summation convention, the repetition of an index in a term denotes a summation with respect to that index over its range ($i, j = x, y, z$). The definitions and operations of the vectors and tensors reviewed here provide foundations for the governing equations for multiphase systems. Additional information about tensors and their associated operations can be found in a continuum mechanics textbook, such as Fung (1994).

Reference

Fung, Y. C. (1994). *First course in continuum mechanics* (3rd ed.). New York: Prentice Hall.

Appendix D: Convective Heat Transfer Correlations

See Table D.1.

Table D.1 Convective heat transfer correlations for various heat and mass transfer modes and geometries

Heat transfer mode	Geometry	Nusselt number	Comments and restrictions	Dimensionless numbers
Forced convection	Flow parallel to a flat plate 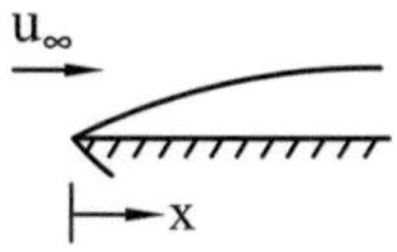	$Nu_x = 0.332 Re_x^{1/2} Pr^{1/3}$ $(Pr > 0.6)$ $Nu_x = 0.565 Re_x^{1/2} Pr^{1/2}$ $(Pr \le 0.05)$	Isothermal surface $Re_x < 5 \times 10^5$ (laminar)	$Nu_x = \frac{hx}{k}$ $Re_x = \frac{u_\infty x}{\nu}$
		$\overline{Nu} = 0.037(Re_L^{0.8} - 871)Pr^{0.33}$ $(0.6 \le Pr \le 60)$	$5 \times 10^5 < Re_L < 10^8$ (turbulent)	$\overline{Nu} = \frac{\bar{h}L}{k}$ $Re_L = \frac{u_\infty L}{\nu}$
	Flow in a pipe (conventional size) 	$\overline{Nu} = 3.66 + \frac{0.0668(D/L)RePr}{1+0.04[(D/L)RePr]^{2/3}}$	Isothermal surface $Re \le 2300$ Thermal entry region	$\overline{Nu} = \frac{\bar{h}D}{k}$ $Re = \frac{\bar{u}D}{\nu}$ $\bar{u}$ is mean velocity
		$Nu = 0.027 Re^{0.8} \times Pr^{0.33}(\mu/\mu_w)^{0.14}$ $(0.7 \le Pr \le 16{,}700)$	$L/D \ge 10$ $Re > 10{,}000$ (Fully developed turbulent) μ_w is viscosity evaluated at T_w	
	Flow in a pipe (miniature) 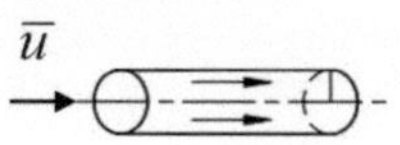	$\overline{Nu} = (1+F) \times \frac{(f/8)(Re - 1000)Pr}{1+12.7(f/8)^{0.5}(Pr^{2/3}-1)}$ $f = [1.82 \log Re) - 1.64]^{-2}$ $F = 7.6 \times 10^{-5} Re \times [1 - (D/D_0)^2]$	D_0 = 1.164 mm is reference diameter Correlation was obtained for water at D = 0.102, 0.76 and 1.09 mm	$\overline{Nu} = \frac{\bar{h}D}{k}$ $Re = \frac{\bar{u}D}{\nu}$
	Flow between parallel plates 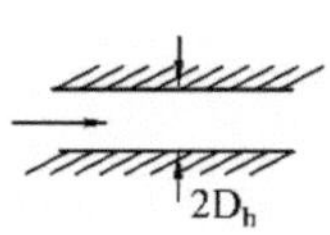	$\overline{Nu} = 7.54 + \frac{0.03(D_h/L)RePr}{1+0.016[(D_h/L)RePr]^{2/3}}$	Isothermal surface $Re \le 2800$ (laminar)	$Nu = \frac{\bar{h}D_h}{k}$ $Re = \frac{\bar{u}D_h}{\nu}$
		$\overline{Nu} = 0.023 Re^{0.8} Pr^{0.33}$ $(Pr > 0.5)$	$Re > 10{,}000$ (Turbulent)	

(continued)

A. Faghri and Y. Zhang, *Fundamentals of Multiphase Heat Transfer and Flow*, https://doi.org/10.1007/978-3-030-22137-9

Table D.1 (continued)

Heat transfer mode	Geometry	Nusselt number	Comments and restrictions	Dimensionless numbers
	Flow across a circular cylinder 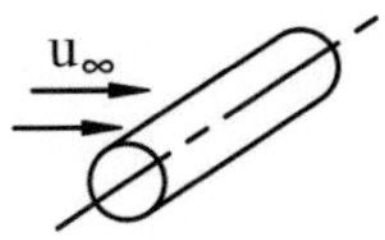	$\overline{Nu} = 0.3 + \frac{0.62Re^{1/2}Pr^{1/3}}{[1+(0.4/Pr/)^{2/3}]^{1/4}} \times \left[1+\left(\frac{Re}{282000}\right)^{5/8}\right]^{4/5}$	$RePr > 0.2$ (both laminar and turbulent)	$\overline{Nu} = \frac{\bar{h}D}{k}$ $Re = \frac{u_\infty D}{\nu}$
	Flow across a sphere 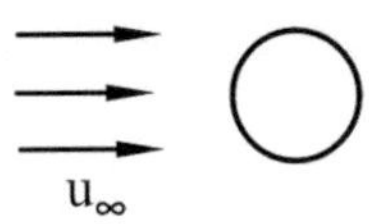	$\overline{Nu} = 2 + (0.4Re^{0.5} + 0.06Re^{2/3})Pr^{0.4}(\mu/\mu_w)^{\frac{1}{4}}$	$3.5 < Re < 76{,}000$ $0.71 \le Pr \le 380$ μ_w is viscosity evaluated at T_w	
	Flow through a packed bed of spheres 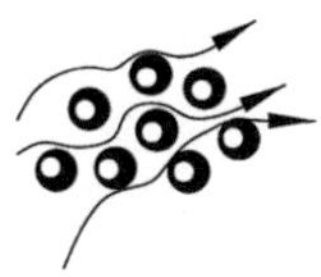	$\overline{Nu} = 1.625Re^{1/2}Pr^{1/3}$	$15 \le Re \le 120$ D—diameter of sphere A—bed cross-sectional area	$\overline{Nu} = \frac{\bar{h}D}{k}$ $Re = \frac{\dot{m}D}{A\mu}$
Free convection	On a vertical surface 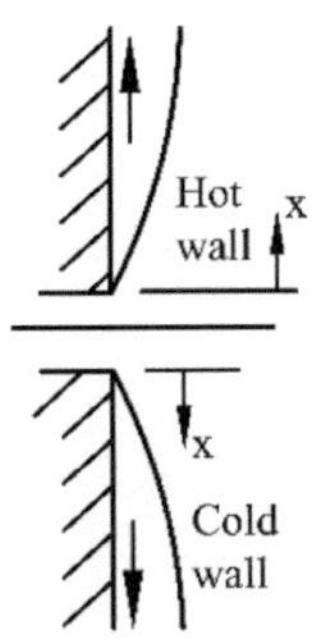	$\overline{Nu}^{1/2} = 0.825 + \frac{0.387\,Ra^{1/6}}{[1+(0.492/Pr)^{9/16}]^{8/27}}$	$\Delta T = \lvert T_w - T_\infty \rvert$ Applicable to both laminar and turbulent	$\overline{Nu} = \frac{\bar{h}L}{k}$ $Ra = \frac{g\beta\Delta TL^3}{\nu\alpha}$
	On a horizontal heated square facing up 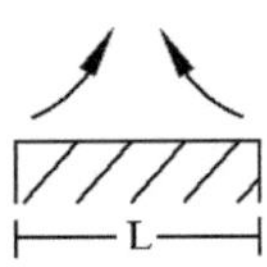	$\overline{Nu} = 0.54(Gr\,Pr)^{1/4}$	Isothermal surface $10^5 \le Gr \le 7 \times 10^7$ For rectangle, use shorter side of L	$\overline{Nu} = \frac{\bar{h}L}{k}$ $Gr = \frac{g\beta\Delta TL^3}{\nu^2}$

(continued)

Table D.1 (continued)

Heat transfer mode	Geometry	Nusselt number	Comments and restrictions	Dimensionless numbers
	On a horizontal heated square facing down 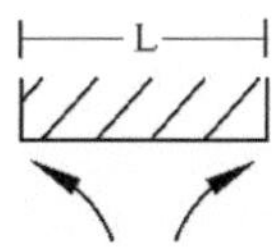	$\overline{Nu} = 0.27(Gr\,Pr)^{1/4}$	Isothermal surface $3 \times 10^5 \leq Gr \leq 3 \times 10^{10}$ For rectangle, use shorter side of L	$\overline{Nu} = \frac{\bar{h}L}{k}$ $Gr = \frac{g\beta\Delta T L^3}{\nu^2}$
	On a horizontal cylinder 	$\overline{Nu}^{1/2} = 0.60 + \frac{0.387\,Ra^{1/6}}{[1 + (0.559/Pr)^{9/16}]^{8/27}}$	$Ra < 10^{12}$	$\overline{Nu} = \frac{\bar{h}D}{k}$ $Ra = \frac{g\beta\Delta T D^3}{\nu\alpha}$
	On a sphere 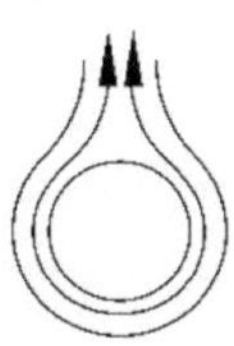	$\overline{Nu} = 2 + \frac{0.589\,Ra^{1/4}}{[1 + (0.469/Pr)^{9/16}]^{4/9}}$	$\Delta T = T_w - T_\infty$ $Ra < 10^{11}$ $Pr \geq 0.7$	$\overline{Nu} = \frac{\bar{h}D}{k}$ $Ra = \frac{g\beta\Delta T D^3}{\nu\alpha}$
Evaporation	Falling film evaporation 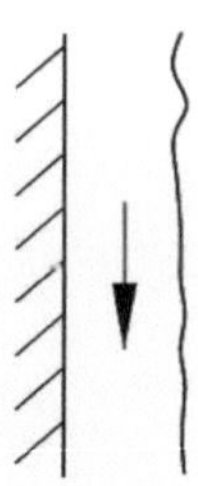	Laminar $Nu = 1.10 Re_\delta^{-1/3}$ $(Re_\delta \leq 30)$ Wavy laminar $Nu = 0.828\,Re_\delta^{-0.22}$ $(30 \leq Re_\delta \leq 1800)$ Turbulent $Nu = 0.0038\,Re_\delta^{0.4} Pr^{0.65}$ $(Re_\delta > 1800)$	Nu—local Nusselt number Γ—mass flow rate per unit width of the vertical surface	$Nu = \frac{h(\nu_\ell^2/g)^{\frac{1}{3}}}{k}$ $Re_\delta = \frac{4\Gamma}{\mu}$
Condensation	On a vertical surface 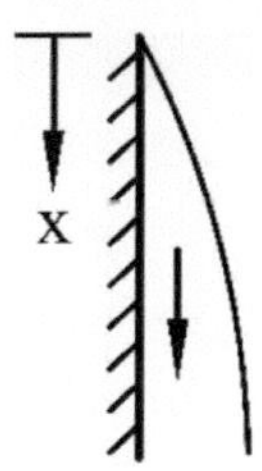	Laminar (Nusselt) $Nu = 1.10\,Re_\delta^{-1/3}$ $(Re_\delta \leq 30)$ Wavy laminar $Nu = \frac{Re_\delta}{Re_\delta^{1.22} - 5.22}$ $(30 \leq Re_\delta \leq 1800)$ Turbulent $Nu = 0.023\,Re_\delta^{0.25} Pr^{\ 0.5}$	Nu—local Nusselt number Γ—mass flow rate per unit width of the vertical surface	$Nu = \frac{h(\nu_\ell^2/g)^{\frac{1}{3}}}{k}$ $Re_\delta = \frac{4\Gamma}{\mu}$

(continued)

Table D.1 (continued)

Heat transfer mode	Geometry	Nusselt number	Comments and restrictions	Dimensionless numbers
	On tubes 	$\overline{Nu} = 0.729 \times \left[\frac{D^3 h_{\ell v} g(\rho_\ell - \rho_v)}{n k_\ell \nu_\ell \Delta T}\right]^{\frac{1}{4}}$	$\Delta T = T_{sat} - T_w$ n—number of tubes	$\overline{Nu} = \frac{\bar{h}D}{k_\ell}$
	In microscale channel ($D_h < 1.5$ mm) 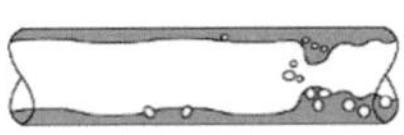	$Nu = We^{-Ja} Re\, Pr^Y$	$Y = 1.3$ for $Re \le 65$ $Y = (0.5D_h - 1)/(2D_h)$ for $Re > 65$	$We = \frac{\rho_\ell V^2 L}{\sigma}$ $Ja = \frac{c_{p\ell}(T_{sat} - T_w)}{h_{lv}}$ $Re = \frac{GD_h}{\mu_\ell}$ G—mass flux (kg/s m^2)
Boiling	Nucleate, saturated pool boiling 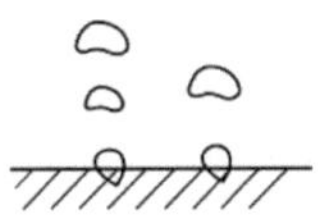	$\overline{Nu} = \frac{Ja_\ell^2}{C^3 Pr_\ell^m}$	m = 2 for water m = 4.1 for other fluids C = 0.013 water-copper or stainless steel C = 0.006 for water-nickel or brass	$\overline{Nu} = \frac{\bar{h}L_c}{k_\ell}$ $L_c = \sqrt{\frac{\sigma_\ell}{g(\rho_\ell - \rho_v)}}$ $Ja_\ell = \frac{c_{p,\ell}\Delta T}{h_{\ell v}}$ $\Delta T = T_w - T_{sat}$
	Film boiling on a horizontal plate 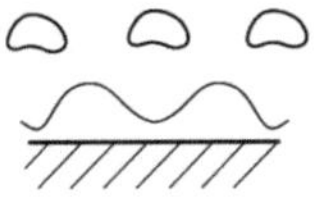	$\overline{Nu} = 0.425 \times \left[Gr Pr_v\left(\frac{1 + 0.4\, Ja_v}{Ja_v}\right)\right]^{\frac{1}{4}}$	Term in parentheses accounts for sensible heating effect in vapor film	$\overline{Nu} = \frac{\bar{h}L_c}{k_v}$ $L_c = \sqrt{\frac{\sigma_\ell}{g(\rho_\ell - \rho_v)}}$ $Gr = \frac{g[(\rho_\ell - \rho_v)/\rho_v]L_c^3}{\nu_v^2}$ $Ja_v = \frac{c_{p,v}\Delta T}{h_{\ell v}}$
	Film boiling on a horizontal cylinder 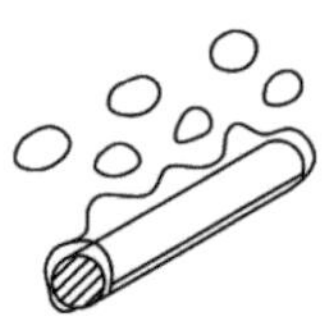	$\overline{Nu} = 0.62 \times \left[Gr\, Pr_v\left(\frac{1 + 0.4\, Ja_v}{Ja_v}\right)\right]^{\frac{1}{4}}$	$D \gg$ film thickness	$\overline{Nu} = \frac{\bar{h}D}{k_v}$ $Gr = \frac{g[(\rho_\ell - \rho_v)/\rho_v]D^3}{\nu_v^2}$ $Ja_v = \frac{c_{p,v}\Delta T}{h_{\ell v}}$
	Film boiling on a sphere 	$\overline{Nu} = 0.4 \times \left[Gr\, Pr_v\left(\frac{1 + 0.4\, Ja_v}{Ja_v}\right)\right]^{\frac{1}{3}}$	$D \gg$ film thickness	

(continued)

Table D.1 (continued)

Heat transfer mode	Geometry	Nusselt number	Comments and restrictions	Dimensionless numbers
	Boiling in microchannel (D = 1.39–1.69 mm) 	$Nu = 30\,Re^{0.857} \times Bo^{0.714}(1-x)^{-0.143}$	Correlation obtained by using Freon® 141 x is quality	$\overline{Nu} = \frac{\bar{h}D}{k_\ell}$ $Bo = \frac{q''}{h_{\ell v}G}$ G—mass flux (kg/s m^2)
Melting	Melting in a rectangular cavity 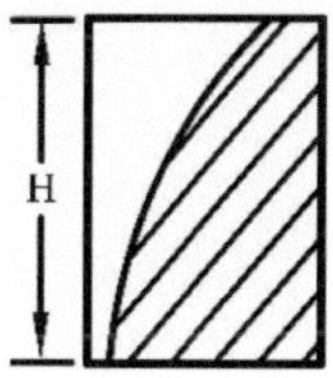	$Nu = (2\tau)^{-1/2} + [c_1 Ra^{1/4} - (2\tau)^{-1/2}] \times [1 + (c_2 Ra^{3/4}\tau^{3/2})^n]^{1/n}$ $c_1 = 0.35, c_2 = 0.175$	Nusselt number is function of time	$\overline{Nu} = \frac{\bar{h}H}{k}$ $Ra = \frac{g\beta\Delta T H^3}{\nu\alpha}$ $Fo = \frac{\alpha_\ell t}{H^2}$ $\tau = SteFo$
Solidification	Solidification around a horizontal tube 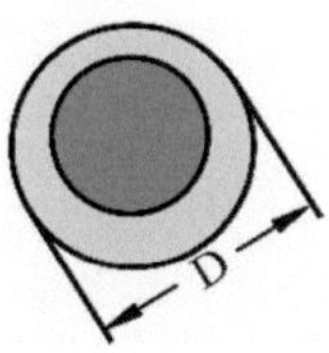	$\overline{Nu} = 0.52\,Ra^{1/4}$	D is transient equivalent outer diameter of the solid $Ra \le 10^9$	$\overline{Nu} = \frac{\bar{h}D}{k}$ $Ra = \frac{g\beta\Delta T D^3}{\nu\alpha}$
Sublimation	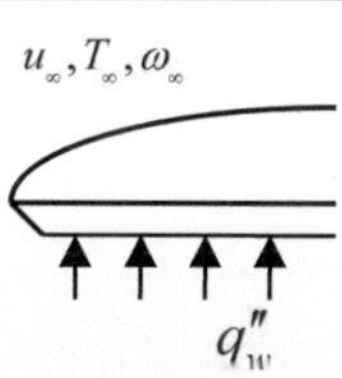	$Nu_x = 0.458\,Re_x^{1/2}Pr^{1/3}$ $Sh_x = 0.459\,Re_x^{1/2}Sc^{1/3}$	Uniform heat flux surface $Re_x < 5 \times 10^5$	$Nu_x = \frac{hx}{k}$ $Sh_x = \frac{h_m x}{D}$

Index

A. Faghri and Y. Zhang, *Fundamentals of Multiphase Heat Transfer and Flow*, https://doi.org/10.1007/978-3-030-22137-9

T

MIX
Papier aus verantwortungsvollen Quellen
Paper from responsible sources
FSC® C105338

If you have any concerns about our products,
you can contact us on
ProductSafety@springernature.com

In case Publisher is established outside the EU,
the EU authorized representative is:
Springer Nature Customer Service Center GmbH
Europaplatz 3, 69115 Heidelberg, Germany

Printed by Libri Plureos GmbH
in Hamburg, Germany